SOME COMMON CONVERSION FACTORS FOR GAS PHASE REACTIONS[a]

Concentrations

$1 \text{ mole L}^{-1} = 6.02 \times 10^{20} \text{ molecules cm}^{-3}$
$1 \text{ ppm} = 2.46 \times 10^{13} \text{ molecules cm}^{-3} = 40.9 \times (\text{MW})^{b}\ \mu\text{g m}^{-3}$
$1 \text{ ppb} = 2.46 \times 10^{10} \text{ molecules cm}^{-3} = 0.0409 \times (\text{MW})\ \mu\text{g m}^{-3}$
$1 \text{ ppt} = 2.46 \times 10^{7} \text{ molecules cm}^{-3} = (4.09 \times 10^{-5}) \times (\text{MW})\ \mu\text{g m}^{-3}$
$1 \text{ atm} = 760 \text{ Torr} = 4.09 \times 10^{-2} \text{ mole L}^{-1} = 2.46 \times 10^{19} \text{ molecules cm}^{-3}$

Second-Order Rate Constants

$\text{cm}^3 \text{ molecule}^{-1} \text{ s}^{-1} \times 6.02 \times 10^{20} = \text{L mole}^{-1} \text{ s}^{-1}$
$\text{ppm}^{-1} \text{ min}^{-1} \times 4.08 \times 10^{5} = \text{L mole}^{-1} \text{ s}^{-1}$
$\text{ppm}^{-1} \text{ min}^{-1} \times 6.77 \times 10^{-16} = \text{cm}^3 \text{ molecule}^{-1} \text{ s}^{-1}$
$\text{atm}^{-1} \text{ s}^{-1} \times 4.06 \times 10^{-20} = \text{cm}^3 \text{ molecule}^{-1} \text{ s}^{-1}$

Third-Order Rate Constants

$\text{cm}^6 \text{ molecule}^{-2} \text{ s}^{-1} \times 3.63 \times 10^{41} = \text{L}^2 \text{ mole}^{-2} \text{ s}^{-1}$
$\text{ppm}^{-2} \text{ min}^{-1} \times 9.97 \times 10^{12} = \text{L}^2 \text{ mole}^{-2} \text{ s}^{-1}$
$\text{ppm}^{-2} \text{ min}^{-1} \times 2.75 \times 10^{-29} = \text{cm}^6 \text{ molecule}^{-2} \text{ s}^{-1}$

[a]The concentrations ppm, ppb, and ppt are relative to air at 1 atm and 25°C, where 1 atm = 760 Torr total pressure.
[b]MW = molecular weight of species.

Atmospheric Chemistry

Atmospheric Chemistry:

FUNDAMENTALS AND EXPERIMENTAL TECHNIQUES

BARBARA J. FINLAYSON-PITTS
Professor of Chemistry
California State University, Fullerton

JAMES N. PITTS, JR.
Professor of Chemistry and Director
Statewide Air Pollution Research Center
University of California, Riverside

A WILEY-INTERSCIENCE PUBLICATION
JOHN WILEY & SONS
New York · Chichester · Brisbane · Toronto · Singapore

Library of Congress Cataloging-in-Publication Data

Finlayson-Pitts, B. J. (Barbara J.), 1948-
Atmospheric chemistry.

"A Wiley-Interscience publication."
Includes bibliographies and index.
1. Atmospheric chemistry. 2. Environmental chemistry. I. Pitts, James N. II. Title.

QC879.6.F56 1986 551.5′01′54 85-22743
ISBN 0-471-88227-5

Printed in the United States of America

10 9 8 7 6 5 4 3 2 1

To the memory of our "academic grandfather and great-grandfather," Philip A. Leighton, gentleman and scholar, who brought us and atmospheric chemistry together

and

To RD^2B, whose unflagging honesty, enthusiasm, and loyalty throughout the years have helped us maintain our perspective

"The play's the thing,
Wherein I'll catch the conscience of the king"

HAMLET
Act II, Scene II

Preface

Scientific and societal interest in the formation of photochemical oxidants, acid rain, and fogs and the persistence and fates of airborne toxic chemicals is manifest in industry, academia, and government, as well as the general public throughout the world. With this has come a growing scientific recognition that these atmospheric processes are closely interrelated; thus the complex chemical and physical reactions occurring in each case involve the photochemical and thermal reactions of a relatively small number of reactive atmospheric constituents. These drive the chemistry not only of these anthropogenic systems but also of biogenic and geogenic emissions in the natural troposphere. The study of these diverse reactions falls under the broad mantle of the rapidly emerging and maturing discipline of atmospheric chemistry, to which this book is devoted.

Thus, we have written it from the perspective that the chemistry of photochemical smog, acid deposition, and the fates of airborne toxic chemicals are closely related; they can be treated through a unified approach that encompasses conditions ranging from heavily polluted urban air parcels to "clean" atmospheres. Accordingly, this book is structured into the following nine parts:

- Part 1 provides an overview of interactions in the polluted troposphere. It includes a brief history of air pollution, a description of the earth's atmosphere, and a summary of the components (other than chemistry) of the "air pollution system" which are not treated in detail in this book.
- In Part 2 we discuss fundamental and applied aspects of those photochemical processes that drive the daytime chemistry of the lower atmosphere. These include quantitative aspects of the absorption of solar radiation and the resulting photolysis rates and products of the major light absorbing species.
- Experimental techniques employed in fundamental and applied studies of reactions in real and simulated atmospheres are considered in Part 3. Included is a brief treatment of the most commonly used monitoring techniques for gaseous criteria and non-criteria pollutants in ambient air.
- Part 4 is a survey of the present state of knowledge of the rates and mechanisms of the important homogeneous gas phase reactions in the troposphere; Chapter 7 treats hydrocarbons and oxygen-containing organics, while Chapter 8 deals with nitrogen-containing molecules.
- The chemical basis for the development of cost-effective control strategy

options for photochemical oxidants and gaseous airborne toxic chemicals is discussed in Part 5. This is relevant because accurate estimates of the lifetimes and fates of volatile toxic compounds are critical inputs into scientific risk assessment documents. For a given volatile toxic compound (and under specified meteorological conditions), these are generally controlled by the species involved with photochemical oxidant formation. As an example, in this section we illustrate how the fundamentals of atmospheric chemistry can be applied to estimate such lifetimes and to predict the fates of these volatile toxics.

- In Part 6 the chemistry associated with acid deposition is discussed; again the close relationship between the formation of photochemical oxidants and sulfuric and nitric acids in the troposphere is illustrated.
- Primary and secondary particles in the troposphere are discussed in Part 7. The sources, reactions, and sinks of combustion-generated, airborne polycyclic aromatic hydrocarbons (PAH) and their derivatives found in respirable soot particles from diesel exhaust, wood smoke, and coal fly ash are treated in a separate chapter. We also discuss the emerging significance of the reactions of gas phase PAH.
- Part 8 deals with the sources, atmospheric lifetimes, and chemical fates of key species in the natural troposphere.
- We conclude the book in Part 9 with a brief consideration of the impact of tropospheric chemical processes on the chemistry of the stratosphere. The latter area is not treated in depth; indeed, this subject could itself comprise an entire book.

We have written this book in the hope that it will prove useful, not only to the established researcher in the atmospheric sciences, but also to students at the senior and graduate levels in chemistry and related disciplines who wish to learn about the chemistry of the lower atmosphere and the experimental approaches used to study it. Therefore, where appropriate, we have included a discussion of the fundamentals behind the derivations of certain key equations and experimental techniques.

The problem of appropriately citing literature references in a field that is expanding as rapidly as atmospheric chemistry is complex. We have generally tried to cite the earliest literature references in a specific area, with additional references throughout as appropriate; our coverage extends through mid-1985. Generally, we have cited research published in peer reviewed journals, unless, to the best of our knowledge, the report is the first, or sole, reference to a significant scientific finding in a particular area. Our efforts do not constitute a comprehensive literature survey, and we apologize in advance to those of our colleagues whose publications may not have been cited. We trust they will understand our predicament.

BARBARA J. FINLAYSON-PITTS
JAMES N. PITTS, JR.

Fawnskin, California
October 1985

Acknowledgments

This book is not simply a product of our efforts. In particular, the professional abilities, personal attributes, enthusiasm, and overall "can do" approach of Ms. Ida Mae Minnich were absolutely essential to the preparation and completion of this manuscript. Her skills in word processing, editing, and overall manuscript preparation, as well as her sense of humor and tireless efforts, kept us going in difficult times.

A book such as this reflects the research and teaching experiences of both of us at our respective universities over a number of years. Without the support of our professional colleagues, including faculty, students, and university administrators, we could not have accomplished this task. We are particularly indebted to our colleagues in our Departments of Chemistry and in the University of California Statewide Air Pollution Research Center.

Additionally, we want to express our appreciation to our respective universities for granting us concurrent sabbatical leaves, and to Professors Robert Charlson and Alvin Kwiram of the University of Washington, who were generous hosts to us in the fall of 1982. Indeed, a course jointly sponsored by several departments, that one of us (BJF-P) gave during this period, formed the nucleus of our thinking for this book. We also extend our appreciation to the faculty, especially Professor B. S. Rabinovitch, and students who were professionally helpful and personally kind to us during our sabbatical stay at that campus.

Without the generous financial support and professional encouragement of a number of state and federal agencies, we would not have been able to conduct the research projects that provided us with the experience and impetus to write this book.

Specifically, we wish to express our appreciation to R. A. Carrigan and J. L. Moyers, Atmospheric Sciences Division, and K. Hancock of the Chemistry Division, of the National Science Foundation; to J. R. Holmes and J. K. Suder, Research Division, California Air Resources Board; to D. Ballantine and G. E. Stapleton of the U.S. Department of Energy; to J. J. Bufalini, M. C. Dodge, B. W. Gay, Jr., P. L. Hanst, and R. Papetti, of the U.S. Environmental Protection Agency; to H. H. Ramsey of the Research Corporation; to D. Stone of the U.S. Air Force; and to Dr. I. Sobolev of the Chemical Manufacturers Association.

We are especially grateful to those researchers in the field of atmospheric chemistry who were kind enough to review portions of the manuscript; their knowledgeable and detailed comments were helpful indeed. Reviewers of selected portions included R. Atkinson, W. P. L. Carter, G. J. Doyle, and A. M. Winer (UC Riverside); H. E. Hunziker (IBM); I. S. A. Isaksen (University of Oslo); L. R.

Martin (Aerospace Corporation); M. J. Molina (Jet Propulsion Laboratory); R. Nanes (California State University, Fullerton); R. F. Phalen (UC Irvine); and C. C. Wamser (Portland State University).

We also wish to thank those researchers who generously provided preprints, unpublished data, or interesting perspectives on various facets of atmospheric chemistry: B. N. Ames, (U. C. Berkeley); A. M. Bass and R. J. Cvetanovic (National Bureau of Standards); K. D. Bayes (UCLA); W. L. Belser, Jr., and E. R. Stephens (UC Riverside); A. Bjørseth (Norske Hydro); J. G. Calvert (National Center for Atmospheric Research); A. J. Diefenderfer and P. A. Wegner (California State University, Fullerton); T. E. Graedel and C. J. Weschler (Bell Labs); T. Greibrokk (University of Oslo); L. Grant and B. Jordan (U.S. Environmental Protection Agency); T. Hard and R. J. O'Brien (Portland State University); M. R. Hoffmann and J. H. Seinfeld (California Institute of Technology); O. Hov (Norwegian Institute for Air Research); T. Kleindienst (Northrup Services); A. C. Lloyd (Environmental Research & Technology, Inc.); G. I. Mackay (Unisearch Associates, Inc.); H. S. Rosenkranz (Case Western Reserve University); H. Niki and D. Schuetzle (Ford Motor Co.); D. Perner (Max Planck Institut für Chemie, Mainz, West Germany); U. Platt (Institut für Chemie 3, Atmosphärische Chemie, Jülich, West Germany); T. Nielson (RISØ National Laboratories, Denmark); T. T. Crocker, R. F. Phalen, F. S. Rowland, and J. L. Whittenberger (UC Irvine); A. R. Ravishankara (NOAA); S. P. Sander (Jet Propulsion Laboratory); and D. Singleton (National Research Council of Canada).

The assistance of the following individuals in completing the final manuscript was most helpful: C. Daly, J. Kriz, R. Ohta, T. Ramdahl, J. A. Sweetman, T. J. Wallington, A. M. Winer, and B. Zielinska. Without their hard and cheerful work in the final days, it would not have been completed within this time frame.

Additionally, it has been a pleasure to work with such skilled and thoughtful editors as Mr. James L. Smith and his colleagues at Wiley in the development of the concept and in the editing and production of this book.

Interactions with friends from widely differing backgrounds who share a common concern for our air environment provided us with a broader perspective of the multitude of problems associated with smog, acid rain, and volatile toxic organics which form the core of this book. They include Rev. and Mrs. Conrad H. Ciesel, Mr. and Mrs. Lee Lakes, and Dr. and Mrs. Fred Sattler of Big Bear Valley, and the family, managers, and special guests of the Arcularius Ranch, located on the upper Owens River near Mammoth Lakes, California, who are actively involved with the preservation of our environment.

We are indebted to our mentors, who sparked and developed our interest in fundamental photochemistry and atmospheric chemistry, especially Professor F. E. Blacet (UCLA), our "academic father and grandfather," and Professors R. E. March and I. Chapman (Trent University, Canada), who stimulated the initial curiosity in this field of one of the authors some 15 years ago.

Finally, we are grateful to our Canadian–American families for their patience, support, and belief that this book would indeed be completed some day—and it has!

B.J.F.-P.
J.N.P.

Contents

PART 2 TROPOSPHERIC PHOTOCHEMISTRY

PART 3 EXPERIMENTAL KINETIC, MECHANISTIC, AND SPECTROSCOPIC TECHNIQUES

PART 4 KINETICS AND MECHANISMS OF GAS PHASE REACTIONS IN REAL AND SIMULATED ATMOSPHERES

PART 5 PHOTOCHEMICAL AIR POLLUTION: MECHANISMS OF FORMATION AND CHEMICAL BASIS OF CONTROL STRATEGY OPTIONS FOR OXIDANT AND GASEOUS AIRBORNE TOXIC CHEMICALS

PART 7 PARTICULATE MATTER

PART 8 CHEMISTRY OF THE NATURAL TROPOSPHERE

PART 9 IMPACT OF TROPOSPHERIC CHEMICAL PROCESSES ON THE STRATOSPHERE

Atmospheric Chemistry

PART 1

Interactions in the Natural and Polluted Troposphere

1 An Overview of Our Natural and Polluted Air Environment

1-A. INTRODUCTION

Concern for our air environment is not a new phenomenon. Thus in the 12th century the philosopher, scientist, and jurist Moses Maimonides (1135–1204) wrote:

> Comparing the air of cities to the air of deserts and arid lands is like comparing waters that are befouled and turbid to waters that are fine and pure. In the city, because of the height of its buildings, the narrowness of its streets, and all that pours forth from its inhabitants and their superfluities. . . the air becomes stagnant, turbid, thick, misty, and foggy. . . . If there is no choice in this matter, for we have grown up in the cities and have become accustomed to them, you should. . . select from the cities one of open horizons. . . endeavor at least to dwell at the outskirts of the city. . . .

He then went on to state:

> If the air is altered ever so slightly, the state of the Psychic Spirit will be altered perceptibly. Therefore you find many men in whom you can notice defects in the actions of the psyche with the spoilage of the air, namely, that they develop dullness of understanding, failure of intelligence and defect of memory. . . . (from Goodhill, 1971)

1-A-1. "London" (Sulfurous) Smog

In the 13th century, coal began to replace wood for domestic heating and industrial uses in London; the impact of high-sulfur coal on air quality was dramatic. In his major 17th century treatise *Fumifugium* (see Lodge, 1969), the title page of which is shown in Fig. 1.1, John Evelyn wrote:

> It is this horrid Smoake which obscures our Church and makes our Palaces look old, which fouls our Cloth and corrupts the Waters, so as the very Rain, and refreshing

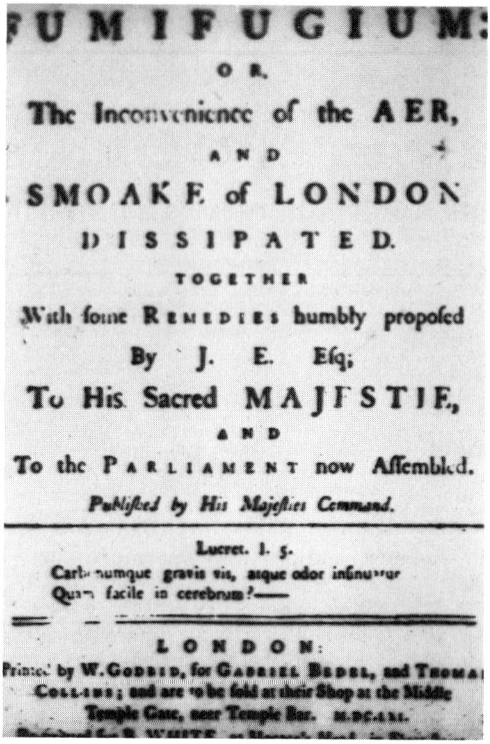
FUMIFUGIUM:
OR,
The Inconvenience of the AER,
AND
SMOAKE of LONDON
DISSIPATED.
TOGETHER
With ſome REMEDIES humbly propoſed
By J. E. Eſq;
To His Sacred MAJESTIE,
AND
To the PARLIAMENT now Aſſembled.
Publiſhed by His Majeſties Command.

Lucret. l. 5.
Carbonumque gravis vis, atque odor inſinuatur
Quam facile in cerebrum?——

LONDON:
Printed by W. GODBID, for GABRIEL BEDEL, and THOMAS COLLINS; and are to be ſold at their Shop at the Middle Temple Gate, neer Temple Bar. M.DC.LXI.

FIGURE 1.1. Front page of John Evelyn's classic work on air pollution in London, England, in the 17th century.

Dews which fall in the several Seasons, precipitate to impure vapour, which, with its black and tenacious quality, spots and contaminates whatever is exposed to it.

In addition to such effects on materials, rain, and dew, Evelyn noted:

But, without the use of Calculations it is evident to every one who looks on the yearly Bill of Mortality, that near half the children that are born and bred in London die under two years of age.[a] Some have attributed this amazing destruction to luxury and the abuse of Spirituous Liquors: These, no doubt, are powerful assistants; but the constant and unremitting Poison is communicated by the foul Air, which, as the Town still grows larger, has made regular and steady advances in its fatal influence.

There was renewed interest in air pollution and its health effects in the 20th century when a number of so-called killer smogs occurred. During these severe episodes, an excess number of deaths over that expected for the particular time of

[a]A child born in a Country Village has an even chance of living near 40 years. . . .

year and location was recorded, the most devastating being 4000 fatalities in London in 1952. The conditions during these episodes tended to be characterized by heavy fogs and by low inversion levels which concentrated the pollutants in a relatively small volume. In the 1952 London incident, for example, Wilkins (1954) describes the inversion heights as being so low that occasionally the tops of the stacks of a power station at Battersea, which were 337 feet high, could be seen above the fog/smog. In some locations, inversion heights as low as 150 feet occurred occasionally and visibility in the center of the affected area was below 22 yards.

The actual pollutants or combination of pollutants responsible for the excess deaths have not been identified, although in all cases there were increased levels of SO_2 and particulate matter in the presence of fog. In the 1952 London episode, concentrations of SO_2 as high as 1.3 ppm and total particles of 4.5 mg m^{-3} were recorded (Wilkins, 1954).

Subsequent to the 1952 London episode, Britain passed a Clean Air Act to reduce emissions. Although similar meteorological conditions as in December 1952 occurred in 1956 and 1962, the number of excess deaths that occurred declined dramatically.

Because the 1952 episode was the most dramatic recorded to date in terms of health effects, this type of air pollution, characterized by high SO_2 and particle concentrations in the presence of fog, has been since dubbed *London smog*. The term *smog*, in fact, derives from a combination of the words *smoke* and *fog*. It is also frequently referred to as *sulfurous smog*.

1-A-2. "Los Angeles" (Photochemical) Smog

Injury to certain vegetable crops grown in Los Angeles County was first observed in 1944. From the nature of the injury and the relative susceptibilities of various plants, it was established by the plant pathologists Middleton, Kendrick, and Schwalm (1950) at the University of California, Riverside, that established phytotoxic agent(s) such as sulfur dioxide and fluorine compounds were not responsible. Instead they described a new "gas-type injury," as well as another form of damage seen "only during periods when heavy air pollution is accompanied by fog particles—which presumably contain pollutants of an as yet undetermined nature."

This key paper was followed by that of Haagen-Smit, Darley, and co-workers (1951) who showed they could reproduce plant injury symptoms caused by ambient "Los Angeles" air pollution by exposing the crops to synthetic atmospheres containing mixtures of olefins and ozone, or to illuminated mixtures of NO_2 and olefins.

Clearly, one was dealing with a new type of oxidizing air pollution, very different than that responsible for the previous severe air pollution episodes associated with London smog. Because it was first recognized in the Los Angeles area, it is frequently dubbed *Los Angeles smog*. However, in the summertime virtually all major cities today suffer from this phenomenon including, for example, Mexico

City and Tokyo. Furthermore, the term smog is misleading since smoke and fog are not key components. The more appropriate term is *photochemical air pollution*, which we use, except when historical connotations are relevant.

In his classic series of papers, Haagen-Smit and his co-workers soon established that a major (but not sole) pollutant was ozone formed during the reaction of organics and oxides of nitrogen in air in the presence of sunlight (Haagen-Smit, 1952; Haagen-Smit and Fox, 1954, 1955, 1956; Haagen-Smit et al., 1951, 1953, 1959). The overall reaction, as written in the early 1950s, was

$$\text{NMOC} + \text{NO}_x + h\nu \rightarrow \text{O}_3 + \text{other pollutants} \tag{1}$$

where NMOC refers to non-methane organic compounds. NO_x represents nitric oxide (NO) plus nitrogen dioxide (NO_2); NO, however, generally forms the greatest fraction of directly emitted NO_x (see Section 1-D-1a). (Today NO_x is often used to denote the sum of all gaseous nitrogenous pollutants; these include not only NO and NO_2 but also such species as HNO_3, PAN, HONO etc, which are present in much smaller concentrations.)

Since hydrocarbons and NO_x are major constituents of the exhaust from uncontrolled motor vehicles and Los Angeles has year-round intense sunlight, appropriate meteorological and geographical characteristics, in retrospect it is clear why photochemical air pollution was first identified there.

After the broad outlines of photochemical smog were established in the early–mid-1950s, detailed studies were undertaken to elucidate the details of the meteorology, emissions, reactions, effects, and possible control strategies. Many of these were carried out under the auspices of the Air Pollution Foundation whose final report summarizes their excellent work in these areas (Faith, 1961). The individual reports issued by the Foundation provide fascinating reading; for example, Foundation-supported researchers examined the potential use of catalysts on automobiles and considered the use of alternate fuels such as gasohol as early as 1955!

Table 1.1 contrasts some of the general features of sulfurous and photochemical air pollution. Interestingly, in the 1950s and 1960s, air pollution researchers and control officials tended to view these as rather independent atmospheric phenomena. This view was reinforced by the common practice of regulatory agencies focusing on the control of the major criteria pollutants, that is, those for which ambient air quality standards were mandated [CO, SO_2, NO_2, O_3, non-methane hydrocarbons (NMHC), Pb, and total suspended particulates, TSP]. This resulted in the inadvertent neglect of the associated chemistry of a host of key, non-criteria pollutants such as gaseous nitric acid, nitrous acid (HONO), hydrogen peroxide, formaldehyde, peroxyacetyl nitrate (PAN), sulfate, nitrate, and organic aerosols.

About 1970 the situation changed dramatically. A major impetus was the suggestion that the gaseous hydroxyl radical played a critical role in driving the chemistry of the natural and polluted troposphere (Heicklen et al., 1969; Weinstock, 1969; Stedman et al., 1970; Levy, 1971). Subsequently, this was validated in laboratory experiments and in sophisticated smog chamber studies of simulated

TABLE 1.1. Comparison of General Characteristics of Sulfurous (London) and Photochemical (Los Angeles) Air Pollution

Characteristics	Sulfurous (London)	Photochemical (Los Angeles)
First rècognized	Centuries ago	Mid-1940s
Primary pollutants	SO_2, sooty particles	Organics, NO_x
Secondary pollutants	H_2SO_4, aerosols, sulfates, sulfonic acids, etc.	O_3, peroxyacetyl nitrate (PAN), HNO_3, aldehydes, particulate nitrates, sulfates, etc.
Temperature	Cool ($\leqslant$ 35°F)	Hot ($\gtrsim$75°F)
Relative humidity	High, usually foggy	Low, usually hot and dry
Type of inversion	Radiation (ground)	Subsidence (overhead)
Time air pollution peaks	Early morning	Noon–evening

polluted atmospheres containing NMOC–NO_x and NMOC–NO_x–SO_x mixtures in air. In addition, a range of trace nitrogenous and oxygenated species were identified and measured in ambient air. The chemistry involved in their formation and fates in the atmosphere is discussed in more detail in Section 1-D-3.

Concurrently, scientific and societal interest and concern began to go well beyond conventional air pollution problems per se. Thus began a growing recognition of the threats to our ecosystem of two other emerging problems of the polluted troposphere—acid rain and airborne toxic organic chemicals.

By the mid-1970s, it had become clear that these three major atmospheric phenomena, air pollution, acid rain, and airborne toxics, as well as the chemistry of the pristine troposphere were inextricably linked in one continuous cycle. In this complex chemical system, methane and a host of higher organics, oxides of nitrogen and sulfur, inorganics such as O_3 and CO, non-criteria pollutants such as H_2CO, HNO_3, H_2O_2, and HONO, and reactive intermediates such as the OH and hydroperoxyl (HO_2) radicals all play important roles.

We focus in this book on that portion of the earth's atmosphere in which we live, the troposphere. However, many of the reactions and reacting species discussed also play major roles in stratospheric chemistry. Thus, for example, with appropriate allowances for differences in pressure, temperature, and solar energy distribution, information on the kinetics and mechanisms of certain NO_x and HO_x reactions at the earth's surface can be utilized in describing such important phenomena as the formation and loss processes for stratospheric ozone. As we shall see, a prime example is the reaction

$$HO_2 + NO \rightarrow NO_2 + OH \tag{2}$$

which plays a critical role in the chemistry of both the troposphere and the stratosphere (Crutzen and Howard, 1978).

Relevant relationships between the chemistry of the troposphere and stratosphere are considered in Chapter 15. For example, we discuss the factors that determine the tropospheric lifetimes of such relatively inert pollutants as chlorofluorocarbons, and how this relates to their transport into the stratosphere where they can be photodissociated to produce Cl atoms by short-wavelength solar radiation as first suggested by Molina and Rowland in 1974. The projected impact on stratospheric ozone is discussed in that chapter, along with the coupling of ClO_x chemistry with NO_x, CH_4, and CO_2 (National Aeronautics and Space Administration, 1984; National Research Council, 1984).

1-B. EARTH'S ATMOSPHERE

Starting at the earth's surface, the atmosphere can be divided into several distinct regions based on the temperature–altitude profile. Figure 1.2 is a plot of average atmospheric temperature for mid-latitudes as a function of height above sea level. Also shown is the continual decrease of total pressure with altitude.

While the temperature initially falls with increasing altitude in the troposphere (ignoring localized radiation or subsidence inversions, Section 1-D-2), this is reversed approximately 15 km above the earth's surface at the tropopause. In the stratosphere, the temperature increases to an altitude of ~ 50 km. From 50 to ~ 85 km in the mesosphere, the temperature drops until ~ 85 km where the thermosphere begins. Here, once again the temperature increases with altitude.

The decreasing temperature with increasing altitude in the troposphere is due to the strong heating effect at the earth's surface from the absorption of visible and near ultraviolet solar radiation. Accompanying this, there is strong vertical mixing so that particles and gaseous air pollutants can move from the earth's surface to the top of the troposphere in a few days or less, depending on the meteorological conditions. Essentially all the water vapor, clouds, and precipitation in the earth's atmosphere are found in this region; as a result, removal of pollutants by precipitation scavenging can be an important process.

In the stratosphere, a series of photochemical reactions involving O_3 and molecular O_2 occur. Ozone strongly absorbs solar radiation in the region from ~ 210 to 290 nm, whereas O_2 absorbs at $\lesssim$ 200 nm (see Sections 3-C-1 and 3-C-2 in Chapter 3). The absorption of light, primarily by O_3, is a major factor causing the increase in temperature with altitude in the stratosphere. Excited O_2 and O_3 photodissociate, initiating a series of reactions in which O_3 is both formed and destroyed leading to a steady state concentration of O_3:

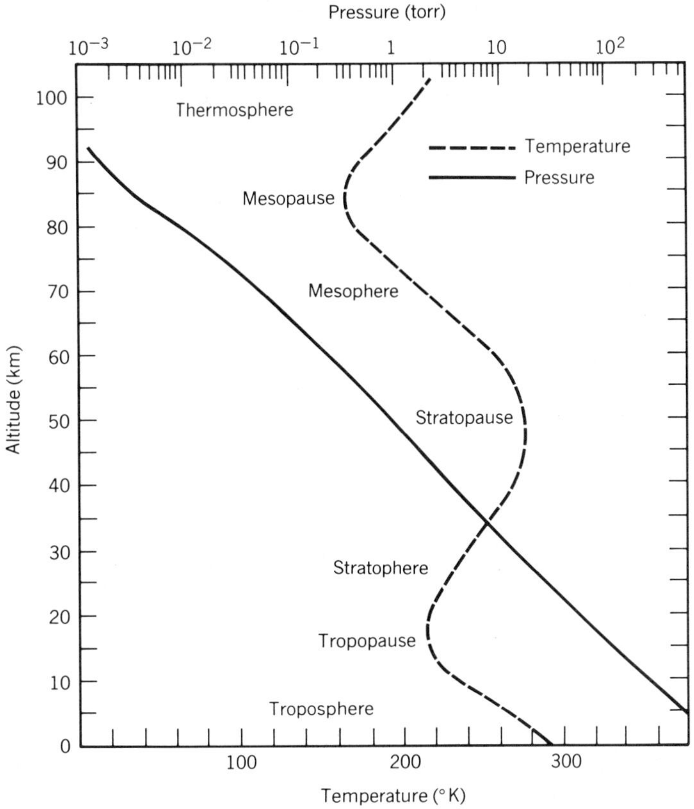

FIGURE 1.2. Variation of temperature with altitude as a basis for the divisions of the lower atmosphere into various regions. Also shown is the variation of total pressure (in Torr) with altitude (top-scale, base 10 logarithms) 1 standard atmosphere = 760 Torr.

$$O_2 + h\nu \rightarrow 2O \qquad (3)$$

$$O + O_2 \xrightarrow{M} O_3 \qquad (4)$$

$$O + O_3 \rightarrow 2O_2 \qquad (5)$$

$$O_3 + h\nu \rightarrow O_2 + O \qquad (6)$$

The steady state O_3 concentration in the stratosphere is essential for life on earth as we know it, because it strongly absorbs light of $\lambda \lesssim 290$ nm. As a result, sunlight reaching the troposphere, commonly referred to as actinic radiation, has wavelengths longer than ~290 nm. This short-wavelength cutoff sets limits on tropospheric photochemistry; thus only those molecules that absorb radiation > 290 nm can undergo photodissociation. If there was a net destruction of O_3 in the stratosphere by chlorofluorocarbons (''aerosols'') and oxides of nitrogen

(SSTs, fertilizers, etc.), as has been proposed, increased radiation of $\lambda < 290$ nm would reach the earth's surface, leading to concomitant increases in skin cancer in humans as well as other effects on plants, agriculture, and our climate.

Relatively little vertical mixing occurs in the stratosphere, and no precipitation scavenging occurs in this region. As a result, massive injections of particles, for example, from volcanic eruptions, often produce layers of particles in the stratosphere which persist for long periods of time, possibly a year or more.

In the mesosphere, from ~ 50 to ~ 85 km, the temperature again falls with altitude and vertical mixing within the region occurs. This temperature trend is due to the decrease in the O_3 concentration with altitude in this region; this in turn reduces the heat release through reactions (3)–(6).

At ~ 85 km the temperature starts to rise again because of increased absorption of solar radiation of wavelengths $\lesssim 200$ nm by O_2 and N_2 as well as by atomic species. This region is known as the thermosphere.

The portion of the atmosphere lying between about 20 and 110 km, which includes most of the stratosphere, the mesosphere, and the lower part of the thermosphere, is known as the chemosphere.

The transition zones between the various regions of the atmosphere are known as the tropopause, stratopause, and mesopause, respectively. Their locations, of course, are not fixed, but vary with latitude, season, and year. Thus Fig. 1.2 represents an average profile for mid-latitutdes.

Once a species is emitted into the troposphere, it can undergo three overall processes. The focus of this book is on the first—chemical reactions in the troposphere. The second fate, transport into the stratosphere, and the types of compounds for which this can be significant are discussed in Chapter 15. The third possible fate, deposition at the earth's surface, is considered briefly in Section 11-D in Chapter 11.

1-C. UNITS OF CONCENTRATION

A number of different units are used in expressing the concentrations of various species in the atmosphere; we shall review these here.

Parts Per Million, -Hundred Million, -Billion, and -Trillion. For gas phase species, the most commonly used units are parts per million (ppm), parts per hundred million (pphm), parts per billion (ppb), and parts per trillion (ppt). These units express the number of molecules of pollutant found in a million (10^6), a hundred million (10^8), a billion (an *American billion*—10^9), or a trillion (10^{12}) molecules of air, respectively. Specification of an American billion is necessary because a *British billion* is 10^{12}, not 10^9!

Alternatively, because numbers of molecules (or moles) are proportional to their volumes according to the ideal gas law ($PV = nRT$), these units may be thought of as the number of volumes of the pollutant found in 10^6, 10^8, 10^9, or 10^{12} volumes of air, respectively.

For example, a *background* concentration of O_3 may be 0.04 ppm; this is 4 pphm, 40 ppb, or 40,000 ppt. Thus in 10^8 molecules of air, only 4 are O_3; alternatively, in every 10^8 volumes of air, only 4 volumes are due to O_3. While it is most convenient to express the concentration of O_3 in ppm, pphm, or ppb, other important atmospheric species can be present in much smaller concentrations. For example, the hydroxyl free radical (OH), which, as we shall see, drives the daytime chemistry of both the clean and polluted troposphere, is believed to have typical concentrations of only $\lesssim 0.1$ ppt. Hence either ppt or an alternate unit discussed below (number per cm^3) is used.

Concentrations of pollutants in ambient air are normally sufficiently small that ppm is the largest unit in use. However, pollutant concentrations in stacks or exhaust trains prior to mixing and dilution with air are much higher, and percent (i.e., parts per hundred) is sometimes used in this case. For example, carbon monoxide concentrations in automobile exhaust are measured in percentages, reflecting the numbers of CO molecules (or volumes) per 100 molecules (or volumes) of exhaust.

Finally, it is important to remember that the number of molecules, or volumes, of a given gaseous species forms the basis of units in atmospheric chemistry; in water chemistry, mass rather than volume is used as the basis for expressing concentrations in ppm, and so on.

Number per Cubic Centimeter. A second type of concentration unit is generally used for species such as free radicals (e.g., OH) present at sub-ppt levels. It is the number of molecules, atoms, or free radicals present in a given volume of air, usually a cubic centimeter (cm^3). One can convert from units of ppm, pphm, ppb, or ppt to units of number cm^{-3} using the ideal gas law. Thus the number of molecules per cm^3 in air at one atmosphere pressure and 25°C (298°K) is given by

$$\frac{n}{V} = \frac{P}{RT} = \frac{1 \text{ atm}}{0.08206 \dfrac{\text{L atm}}{°\text{K mole}} \times 298°\text{K}} = 0.0409 \frac{\text{moles}}{\text{L}}$$

Converting to units of molecules cm^{-3}, one obtains

$$\frac{n}{V} = 0.0409 \frac{\text{moles}}{\text{L}} \times 10^{-3} \frac{\text{L}}{\text{cm}^3} \times 6.02 \times 10^{23} \frac{\text{molecules}}{\text{mole}}$$
$$= 2.46 \times 10^{19} \text{ molecules cm}^{-3}$$

From the definition of ppm as the number of pollutant molecules per 10^6 molecules of air, 1 ppm corresponds to $(2.46 \times 10^{19} \times 10^{-6}) = 2.46 \times 10^{13}$ molecules cm^{-3} at 25°C and one atmosphere total pressure. It follows that a concentration of the OH radical in polluted air of 0.1 ppt, is $2.46 \times 10^{19} \times 10^{-12} \times 0.1 = 2.46 \times 10^6$ molecules cm^{-3}. As we shall see, in clean atmospheres the OH concentration may be approximately an order of magnitude smaller than this, so

TABLE 1.2. Conversion Between Units of Concentration in ppm, pphm, ppb, ppt, and molecules cm^{-3}, Assuming One Atmosphere Pressure and 25°C[a]

Parts per	Unit	Molecules, Atoms, or Radicals per cm^3
10^6	1 ppm =	$2.46 \times 10^{13}\ cm^{-3}$
10^8	1 pphm =	$2.46 \times 10^{11}\ cm^{-3}$
10^9	1 ppb =	$2.46 \times 10^{10}\ cm^{-3}$
10^{12}	1 ppt =	$2.46 \times 10^{7}\ cm^{-3}$

[a] 1 ppm in units of mass = (40.9) × (MW) $\mu g\ m^{-3}$.

the use of fractional ppt units becomes inconvenient. Units of number per cm^3 are more commonly used in such cases.

Table 1.2 summarizes the relationship between these units at one atmosphere pressure and 25°C. Of course, corrections must be made if the temperature or pressure differs significantly.

Micrograms per Cubic Meter. A third unit of measurement for gaseous species is mass per unit volume, usually 10^{-6} g per cubic meter ($\mu g\ m^{-3}$). Since one atmosphere at 25°C contains 4.09×10^{-2} moles L^{-1}, 1 ppm must contain $(4.09 \times 10^{-2}) \times 10^{-6}$ or 4.09×10^{-8} moles L^{-1} or 4.09×10^{-5} moles m^{-3}. If the molecular weight of the pollutant is MW grams per mole, then 1 ppm in units of mass per m^3 is $(4.09 \times 10^{-5}) \times$ (MW) g m^{-3} or (40.9) × (MW) $\mu g\ m^{-3}$.

Returning to the example of a background O_3 of 0.04 ppm, this concentration in $\mu g\ m^{-3}$ is (40.9 × 48) × 0.04 (where 48 g $mole^{-1}$ is the molecular weight of O_3), or 79 $\mu g\ m^{-3}$. If the pollutant were 0.04 ppm SO_2 (MW = 64) rather than O_3, the concentration of SO_2 would be (40.9 × 64) × 0.04, or 105 $\mu g\ m^{-3}$.

The conversion between $\mu g\ m^{-3}$ and ppm, pphm, ppb, and ppt can be summarized as follows:

$$\begin{aligned} \mu g\ m^{-3} &= \text{ppm} \times 40.9(\text{MW}) \\ &= \text{pphm} \times 0.409(\text{MW}) \\ &= \text{ppb} \times 0.0409(\text{MW}) \\ &= \text{ppt} \times (4.09 \times 10^{-5})(\text{MW}) \end{aligned}$$

For atmospheric particulate matter, concentrations are expressed in mass per unit volume, commonly $\mu g\ m^{-3}$, or in the number of particles per unit volume, for example, per cm^3.

1-D. AIR POLLUTION SYSTEM

Since the foundations of tropospheric chemistry were established, to a large extent, by air pollution researchers, it is useful in this introductory chapter to view the general aspects of our subject within the conceptual framework of an *air pollution system*.

As shown in Fig. 1.3, the air pollution system starts with the various sources of anthropogenic and natural emissions. These are defined as *primary* pollutants since they are emitted directly into the air from their sources; they include, for example, SO_2, NO_x, CO, Pb, organics, and particulate matter. Once in the atmosphere, they are subjected to dispersion and transport, that is meteorology, and simultaneously to chemical and physical transformations into gaseous and particulate *secondary* pollutants, defined as those formed from reactions of the primary pollutants in air. The pollutants may be removed at the earth's surface via wet or

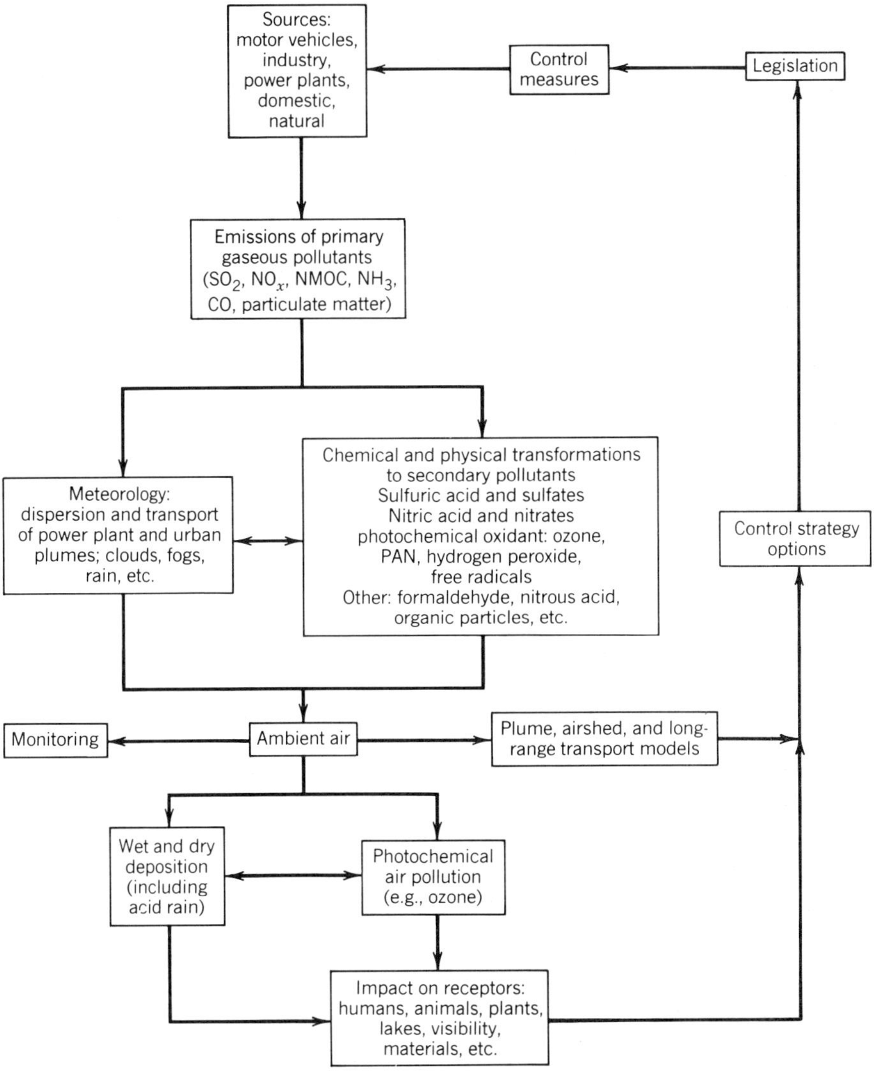

FIGURE 1.3. The air pollution system.

dry deposition and can impact a variety of receptors, for example, humans, animals, aquatic ecosystems, vegetation, and materials.

From a detailed knowledge of the emissions, topography, meteorology, and chemistry, one can develop mathematical models that predict the concentrations of primary and secondary pollutants as a function of time at various locations in a given airshed. These computer based models may describe the concentrations in a plume from a specific point source (plume models), in an air basin from a combination of diverse mobile and stationary sources (airshed models), or over a large geographical area downwind from a group of sources (long-range transport models). In order to validate these models, their predictions must be compared to the observed concentrations of the pollutants measured in appropriate ambient air monitoring programs; model inputs are adjusted to obtain acceptable agreement between the observed and predicted values.

These validated models can then be used, in combination with the documented impacts on receptors, to develop various control strategy options. Finally, through legislative and administrative action, control measures can be formulated and implemented which directly affect the starting point of our air pollution system, that is, the primary emissions and their sources.

We now consider briefly certain elements of the air pollution system from the overall perspective of their linkage to tropospheric chemistry. A detailed discussion of the element of the system, *Impact on Receptors*, is beyond the scope of this book and will not be treated.

1-D-1. Emissions

In describing a given air mass and the chemical reactions occurring therein, both natural and anthropogenic sources of primary emissions must be considered and their relative importance evaluated. Thus the impact on air quality of natural emissions can be an important issue because cost-effective control strategies for polluted air basins must take into account the relative strengths of emissions from all sources, not just those of anthropogenic origin. For example, the Edison Electric Institute has stated: "Today it is estimated that 93% of the global nitrogen emissions and 60% of the world's sulfur oxide emissions are naturally produced." However, in evaluating such statements, and their implications for control strategies, a number of factors must be taken into account. These include the chemical nature of the emissions, their temporal and spatial distributions, and the accuracy of published anthropogenic and natural emission inventories.

We now discuss the sources and strengths of several important classes of anthropogenic emissions. Natural sources are then briefly described and their source strengths compared to the estimated anthropogenic emissions. The units used are either teragrams (1 Tg = 1 teragram = 10^{12} grams) or metric tons, where 1 metric ton = 10^6 grams (i.e., 10^6 metric tons = 1 Tg).

1-D-1a. Anthropogenic Emissions

SO_2. The major source of SO_2 is the combustion of sulfur-containing fuels, so that its emissions can be calculated for a given source with some accuracy. Thus, knowing the rate of fuel consumption, the percent sulfur in the fuel, and the fact that 90% or more of the sulfur is emitted in the form of SO_2 (the remainder is primarily in the form of sulfates), the rate of SO_2 emission can be estimated.

Table 1.3 shows emission factors used in one estimate of global sulfur emissions. From these values and the amount of fuel consumed (or end product produced), the global SO_2 emissions in Table 1.4, expressed as teragrams of sulfur per year, were calculated. On a global scale, about 60% of the SO_2 was estimated to be from coal combustion and ~30% from petroleum refining and combustion.

The uncertainty involved in estimating anthropogenic emissions of even SO_2 is illustrated by the discrepancy between the estimates of Cullis and Hirschler (1980) and the more recent ones of Möller (1984a). Thus Möller estimates that the global anthropogenic production of SO_2 in 1976 was 72 Tg S yr^{-1}, compared to the Cullis and Hirschler estimate of 104 Tg S yr^{-1} (Table 1.4). A great deal of the discrepancy is due to different estimates of emission factors.

Total Suspended Particles. Air quality standards for particulate matter are now written in terms of total, suspended particulate (TSP), that is, non-size fractionated particles. However, a size-restricted standard may be mandated by the U.S. Environmental Protection Agency.

The rational for moving toward a size-related air quality standard based on health effects is evident from an examination of Fig. 1.4, a diagram of the human

TABLE 1.3. Emission Factors Used to Estimate Global Anthropogenic Emissions of Sulfur to the Atmosphere for the 1974–1976 Period

Source	Emission Factor (kg SO_2 per 10^3 kg either burned or produced)
Hard coal	48.2
Lignite	35.6
Coal coke	5.4
Petroleum refining	2.0
Gasoline	0.72
Kerosene	0.96
Jet fuel	0.96
Distillate fuel oils	4.47
Residual fuel oil	36.0
Petroleum coke	13.5
Copper smelting	2000
Copper refining	350
Lead	470
Zinc	200
Sulfuric acid	24
Pulp/paper	2
Sulfur	2

Source: Cullis and Hirschler, 1980.

TABLE 1.4. Estimated Global Anthropogenic Emissions of SO_2 in 1976 (Tg S yr^{-1})

Source	Total Global Emissions (Tg S yr^{-1})	Percentage of Total
Coal		
Hard coal	44.1	42.5
Lignite	16.6	16.0
Coal coke	1.3	1.2
Petroleum		
Refining	3.7	3.5
Gasoline	0.3	0.3
Kerosene	0.05	0.05
Jet fuel	0.05	0.05
Distillate fuel oil	1.9	1.8
Residual fuel oil	22.9	22.1
Petroleum coke	0.3	0.3
Non-ferrous Ores		
Copper	9.4	9.1
Lead	0.8	0.7
Zinc	0.5	0.5
Others		
H_2SO_4 production	1.3	1.3
Pulp/paper operations	0.3	0.3
Refuse incineration	0.3	0.3
Sulfur production	0.03	0.03
TOTALS	103.8	100%

Source: Cullis and Hirschler, 1980.

respiratory tract. Larger particles, if inhaled, are removed in the head or upper respiratory tract. The respiratory system from the nose through the tracheobronchial region is covered with a layer of mucus which is continuously moved upward by the motion of small hair-like projections called cilia; large particles deposit on the mucus, are moved up, and are ultimately swallowed. On the other hand, particles from fossil fuel combustion and gas-to-particle conversion are generally much smaller ($\lesssim 2.5$ μm diameter) and fall in the *respirable* size range. These particules can reach the alveolar region where gas exchange occurs and which is not coated with a protective mucus layer; here the clearance time for deposited particles is much greater than in the upper respiratory tract, and hence the potential for health effects is much greater (Phalen, 1984).

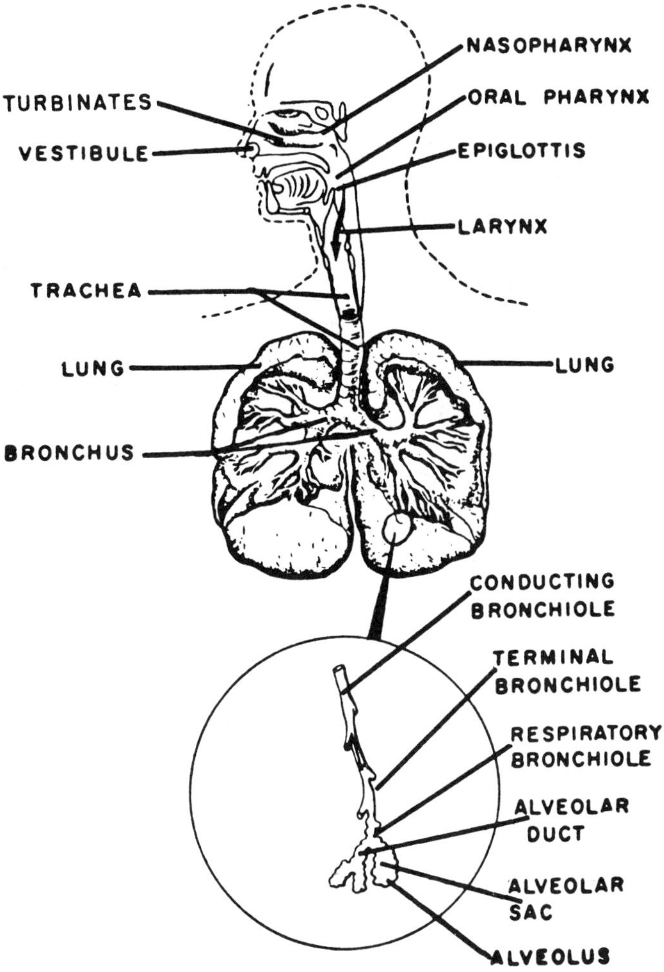

FIGURE 1.4. Schematic diagram of human respiratory tract (from Hinds, 1982).

Since small particles may contain more toxic species than the large particles (Section 12-D), they can contribute more to health effects than one might imply from their mass. For example, the volume of spherical particles of radius r increases with r^3, and, assuming equal densities, the mass of particles also increases with r^3. A particle of diameter 2 μm will thus weigh only 0.1% of a 20 μm diameter particle, that is, 1000 of the respirable small particles equals the mass of one of the large particles.

Interestingly, the fine particles with diameters $\lesssim 2.5$ μm are not only of greatest concern from the point of view of deposition in the respiratory tract, but they also are responsible for most of the light scattering, that is, visibility reduction (see Section 12-B-3). Thus an improvement in visibility in areas impacted by air pollution is likely to be accompanied by a reduction in total particle deposition in the alveolar region of the respiratory system as well.

Table 1.5 shows the sources of the emission of particulate matter, as well as of sulfur oxides in the United States in 1978. This will vary from country to country, of course, but for illustrative purposes it is probably reasonably representative of most developed nations.

CO. Carbon monoxide is produced by the incomplete combustion of fossil fuels, and in major urban areas of developed nations its major source is the exhaust from light duty motor vehicles (LDMV). For example, in the United States in 1977 ~75% of the CO emissions were estimated to be from motor vehicles, 8% from non-highway transportation such as railroads, aircraft, farm equipment, and so on, 9% from industrial processes, including petroleum refining, 7.5% from miscellaneous and solid waste combustion, and 0.6% from electric power generation and residential, commercial, and institutional fuel combustion (Informatics, 1979). However, this distribution is changing because of the effectiveness of CO removal by catalytic control devices on LDMV.

Non-methane Hydrocarbons (NMHC). It has been estimated that in 1974 transportation and associated petroleum refining, oil, and gas production, and marketing accounted for about ~50% of the NMHC released in the United States. Organic solvents account for another ~25% of the total emissions which were estimated to be ~ 3×10^7 metric tons (U.S. Environmental Protection Agency, 1978).

TABLE 1.5. Anthropogenic Emissions of Sulfur Oxides and Particles in the United States in 1978

Source Category	Sulfur Oxides (10^6 metric tons)	Particulates (10^6 metric tons)
Stationary fuel combustion	22.1	3.8
Industrial processes	4.1	6.2
Solid waste disposal	0.0	0.5
Transportation	0.8	1.3
Miscellaneous	0.0	0.7[a]
Industrial process fugitive emissions[b]	0.0	3.4
Non-industrial fugitive emissions[b]	0.0	110–370
TOTAL	27.0	~126–386

Source: U.S. Environmental Protection Agency, 1982a.

[a] Includes forest fires, agricultural burning, coal refuse burning, and structural fires.

[b] Fugitive particulate emissions include those from wind erosion, unpaved roads, and from vehicular traffic. Industrial process fugitive emissions result from materials handling, loading and unloading operations, etc. Non-industrial fugitive emissions are from unpaved and paved roads, construction activities, agricultural operations, surface mining, and fires.

NO_x. Nitric oxide is produced from the reaction of N_2 with O_2 in air during high-temperature combustion processes:

$$N_2 + O_2 \rightarrow 2NO \tag{7}$$

as well as from oxidation of nitrogen in the fuel. Smaller amounts of NO_2 are produced by the further oxidation of NO; trace amounts of other nitrogenous species such as HNO_3 are also formed.

The fraction of the total NO_x which is emitted as NO clearly depends on the conditions associated with the specific combustion process. It is often assumed to be $\gtrsim 90\%$; however, in one study of the JT3D and TF30 aircraft turbine engines at idle, only 13–28% of the NO_x was NO (U.S. Environmental Protection Agency, 1982b). Similarly, light and heavy duty diesel engines may emit more than 10% of their total NO_x as NO_2.

Table 1.6 gives the estimated annual global emissions of NO_x due to anthropogenic activities, while Table 1.7 shows the sources and anthropogenic NO_x emission levels for the United States in 1976; in the United States, fuel combustion in mobile and in stationary sources contribute about equally. A more recent estimate by Logan (1983) is in relatively good agreement with the estimate in Table 1.7, predicting $\sim 19 \times 10^6$ metric tons as NO_2 from anthropogenic activities in the United States.

Lead. Table 1.8 shows the estimated emissions of lead to the atmosphere on a global scale; clearly, gasoline combustion in motor vehicles is by far the greatest source.

Of the lead used in gasoline, only $\sim 25\%$ is retained in the automobile, the remainder being emitted in the form of particles. These particles are mainly ($> 90\%$ by mass) inorganic PbBrCl (the bromine and chlorine come from ethylene dibromide and ethylene dichloride in the fuel), with much smaller amounts being emitted in the form of organolead compounds. About one-third of the mass is in very small particles (mass median diameter < 0.5 μm) which remain suspended for long periods of time, whereas about one-half is in much larger particles (mass median

TABLE 1.6. Estimated Annual Global Anthropogenic Emissions of NO_x

Source Category	Emissions[a] (10^6 metric tons per year)
Fossil fuel combustion	69
Biomass burning[b]	39
TOTAL	108

Source: Logan, 1983.
[a] Expressed as NO_2.
[b] Includes both controlled (e.g. land clearance) and uncontrolled (e.g. forest wild fires) burning.

TABLE 1.7. Estimated NO_x Emissions in the United States in 1976[a]

Source Category	NO_x Emissions (10^6 metric tons, expressed as NO_2)	Percent of Total
Transportation		
Highway vehicles	7.8	33.9
Non-highway vehicles	2.3	10.0
Stationary fuel combustion		
Electric utilities	6.6	28.7
Industrial	4.5	19.6
Residential, commercial, and institutional	0.7	3.0
Industrial Processes		
Chemicals	0.3	1.3
Petroleum refining	0.3	1.3
Mineral products	0.1	0.4
Solid Waste Disposal	0.1	0.4
Miscellaneous		
Forest wildfires and managed burning	0.2	0.9
Coal refuse burning	0.1	0.4
TOTAL	23.0	100%

Source: U.S. Environmental Protection Agency, 1982b.
[a] Sources with emissions of < 50,000 metric tons per year are not included.

diameter > 5 μm) which fall out relatively close to the source (Ter Haar et al., 1972).

1-D-1b. Natural Emissions

The diverse nature of biogenic and geogenic sources of primary pollutants, as well as the wide variety of parameters that can alter their emission rates (e.g., temperature, sunlight), are responsible for a large uncertainty in estimates of their emissions. However, while these emissions are significant compared to anthropogenic emissions on a global scale and thus are very important in the "clean" troposphere, as discussed in Chapter 14, they generally do not play major roles in the chemistry of polluted urban areas. Here, we summarize the major sources of the natural emissions and their estimated source strengths.

Sulfur Compounds. Sulfur compounds are emitted to the atmosphere from a variety of sources, including volcanic eruptions, sea spray, and a host of biogenic

TABLE 1.8. Estimated Global Atmospheric Lead Emissions

Source Category	Emissions (metric tons per year)	Percentage of total
Oil and gasoline combustion	273,000	60.8
Coal combustion	14,000	3.1
Waste incineration	8,900	2.0
Wood combustion	4,500	1.0
Iron and steel production	50,000	11.1
Secondary non-ferrous metal production	770	0.2
Primary copper smelting	27,000	6.0
Mining non-ferrous metals	8,200	1.8
Primary lead smelting	31,000	6.9
Primary zinc smelting	16,000	3.5
Primary nickel smelting	2,500	0.5
Industrial applications	7,400	1.6
Miscellaneous	5,950	1.3
TOTAL	449,200	100%

Source: Nriagu, 1979.

processes. Most of the volcanic sulfur is emitted as SO_2, with smaller and highly variable amounts of hydrogen sulfide (H_2S) and dimethyl sulfide (CH_3SCH_3). Sea spray contains sulfate, some of which is carried over land masses. Biogenic processes emit reduced forms of sulfur, including H_2S and dimethyl sulfide, with lesser amounts of carbon disulfide (CS_2), dimethyl disulfide (CH_3SSCH_3), carbonyl sulfide (COS), and methyl mercaptan (CH_3SH). These reduced sulfur compounds are then oxidized in the atmosphere (see Section 14-C).

Table 1.9 compares one estimate of the natural emissions of sulfur compounds in the Northern and Southern Hemispheres with anthropogenic emissions, reported in terms of elemental sulfur in teragrams (1 Tg = 10^{12} g). On a global scale, human activities accounted for 41% of the total global emissions of sulfur in 1976. As expected, in the Northern Hemisphere where industrial activities are greatest, anthropogenic emissions accounted for 56% of the total; in the Southern Hemisphere, they were only 8%.

A more recent estimate of global biogenic sulfur emissions by Möller (1984b) is in the range 50–100 Tg S y^{-1}, in agreement with the estimate of 98 Tg S y^{-1} of Cullis and Hirschler (Table 1.9).

Particles. Major natural sources of particles include terrestial dust caused by winds, sea spray, biogenic emissions, volcanic eruptions, and wildfires. Table 1.10 contains estimates of the strengths of these natural sources globally and in the United States.

As with anthropogenic particulate emissions, particle size is important. Terrestial dust is generally in the large size range ($\gtrsim 2$ μm diameter) and is primarily

TABLE 1.9. Estimated Natural and Anthropogenic Global Emissions of Atmospheric Sulfur (Tg S yr^{-1})

Sources	1970 Northern and Southern Hemispheres	1976 Northern and Southern Hemispheres	1976 Northern Hemisphere[a]	1976 Southern Hemisphere[a]
NATURAL				
Volcanoes	5	5	3 (60)	2 (40)
Sea spray	44	44	19 (43)	25 (57)
Subtotal non-biogenic	49	49	22 (45)	27 (55)
Biogenic (land)	48	48	32 (67)	16 (33)
Biogenic (oceans)	50	50	22 (44)	28 (56)
Subtotal biogenic	98	98	54 (55)	44 (45)
TOTAL NATURAL	147	147	76 (52)	71 (48)
TOTAL ANTHROPOGENIC	86	104	98 (94)	6 (6)
TOTAL (natural + anthropogenic)	233	251	174 (69)	77 (31)
ANTHROPOGENIC (% of total)	37	41	56	8

Source: Cullis and Hirschler, 1980; 1 Tg = 10^{12} g.

[a] Figures in parentheses represent the percentages in each hemisphere.

TABLE 1.10. Estimated Strengths of Natural Sources of Particles Globally and in the United States

Source	Global Source Strength (10^6 metric tons)	Emissions in the United States[b] (10^6 metric tons)
Terrestrial dust	100–500[a]	57
Sea spray	900[c]	5.5
Biogenic processes	75–200[d]	20
Volcanic eruptions	Variable	Variable
Wildfires	1[b]	0.5–1.0
TOTAL	1076+[e] to 1600+[e]	83+[e]

[a] National Research Council, 1979.

[b] U.S. Environmental Protection Agency, 1982a.

[c] Robinson and Robbins, 1971.

[d] SMIC, 1971.

[e] Plus contribution from volcanic eruptions.

composed of crustal elements, including silicon, aluminum, iron, sodium, potassium, calcium, and magnesium.

Particles are generated at the surface of the ocean by the bursting of bubbles and some of these are carried inland. They, too, tend to be large, more than 50% having diameters $\gtrsim 3\ \mu m$. Their chemical composition includes the elements found in seawater (primarily chlorine, sodium, sulfate, magnesium, potassium, and calcium) and organic materials, perhaps including viruses, bacteria, and so on (see Section 12-D-1).

Biological emissions of particles may occur from plants and trees; additionally, volatile organics such as terpenes may react in the air to form small particles (see Section 14-B).

Volcanic eruptions are highly variable but, for example, for the St. Augustine (Alaska) eruption in 1976, particulate emissions over the period of 1 year were estimated to be $\sim 0.25 \times 10^6$ metric tons for particles in the size range 0.01–5 μm. When particle size range was extended from 0.01 to 66 μm, the emissions were $\sim 6 \times 10^6$ metric tons (Stith et al., 1978).

Wildfires also produce significant particulate matter, most of it in the respirable size range from ~ 0.1 to 1 μm. Elemental carbon and organics form the majority of these particles, with some minerals also being present.

CO. Table 1.11 gives a range of global CO emissions from both natural and anthropogenic sources. Major natural sources of CO include the oxidation of methane and other naturally occurring hydrocarbons in the troposphere (see Chapter 14), direct emissions from plants, and microbial activity in the ocean. While there are large uncertainties in these estimates, especially of the natural sources, on a

TABLE 1.11. Estimated Range of Emissions of CO Globally from Natural and Anthropogenic Sources

Source Category	Estimated Emissions (10^6 metric tons)
Natural	
CH_4 oxidation	60–5000
Oxidation of natural hydrocarbons	50–1300
Microbial activity in oceans	20–200
Emissions from plants	20–200
TOTAL	150–6700
Anthropogenic	
Fossil fuel combustion	250–1000
Forest fires	10–60
TOTAL	260–1060

Source: Informatics, 1979.

global basis, anthropogenic and natural emissions seem to be roughly of the same order of magnitude.

Non-methane Hydrocarbons. Biogenic processes release substantial quantities of reactive hydrocarbons such as isoprene and α-pinene, in addition to methane and other hydrocarbons (Altshuller, 1983). However, there is considerable controversy regarding the magnitude and chemical composition of such emissions (Section 14-B). In particular, emission inventories based on experimentally determined emission rates often do not agree with those derived from measurements of natural hydrocarbon concentrations in ambient air.

The total annual emissions of isoprene and some monoterpenes in the contiguous United States may approximate 65×10^6 metric tons per year, but this is possibly an underestimate of the biogenic source strengths. Perhaps the best that can be said today is that the total anthropogenic emissions of NMHC in the United States of $\sim 30 \times 10^6$ metric tons in 1974 is of the same order of magnitude as natural emissions. However, recall that these biogenic emissions are averaged over the entire United States, so that, for example, their contribution to oxidant formation in urban areas is generally small (Section 14-B-2).

NO_x. Major natural sources of NO_x and their emissions are shown in Table 1.12. The uncertainties in these estimates are illustrated by the range in Table 1.12. By comparison to the Stedman and Shetter estimate of *anthropogenic* NO_x sources of 66×10^6 metric tons per year (1983), and to the Logan estimate in Table 1.6 of 108×10^6 metric tons per year (both expressed as NO_2), natural sources are seen to be comparable in strength on a global scale to anthropogenic sources.

TABLE 1.12. Major Natural Sources of NO_x and Estimated Emissions[a]

Source	Emission[a] (10^6 metric tons per year) Stedman and Shetter (1983)	Logan (1983)
Lightning	10	26
Stratospheric injection	3	2
Ammonia oxidation	3	3–33
Biomass burning[b]	16	39
Soil emission	33	26
TOTAL	65	96–126

[a] The NO_x is expressed as NO_2.
[b] Includes both controlled (e.g. land clearance) and uncontrolled (e.g. forest wild fires) burning.

1-D-2. Meteorology

Clearly, the concentrations of pollutants in ambient air, and hence their impacts, are determined not only by the rates of emissions and chemical transformations, but to a very large extent by meteorology. For example, as we have seen, during the severe London-type smog episodes, meteorological conditions were such that the pollutants were effectively contained in relatively small volumes, leading to high pollutant concentrations.

In the lowest 10 km of the earth's atmosphere, the air temperature generally decreases with altitude at a rate of ~ 7°C per km; this is known as a positive lapse rate. In this "normal" troposphere, warm air close to the earth's surface, being less dense, rises and is replaced by cooler air from higher elevations. This results in mixing within the troposphere.

In some situations, however, the temperature of the air, at some height within the troposphere, may start to rise with increasing altitude before reversing itself again; that is, the lapse rate changes from positive to negative to positive (Fig. 1.5). This region, with a negative lapse rate, is known as an *inversion* layer. In effect, it acts as a "lid" on an air mass because the cooler air underneath it, being more dense, will not rise through it. In effect, pollutants trapped below the inversion layer are not mixed rapidly throughout the entire troposphere, but are confined to the much smaller volume beneath the inversion layer; this results in much higher ground-level concentrations.

The formation of thermal inversions is one of the most important meteorological factors contributing to air pollution problems in urban areas. The two classic types of air pollution, London or sulfurous and Los Angeles or photochemical, generally have different types of thermal inversions associated with them (Table 1.1).

The radiation (or ground) inversion characteristic of London smog is caused by the rapid cooling of the earth's surface, and the layer of air immediately above it,

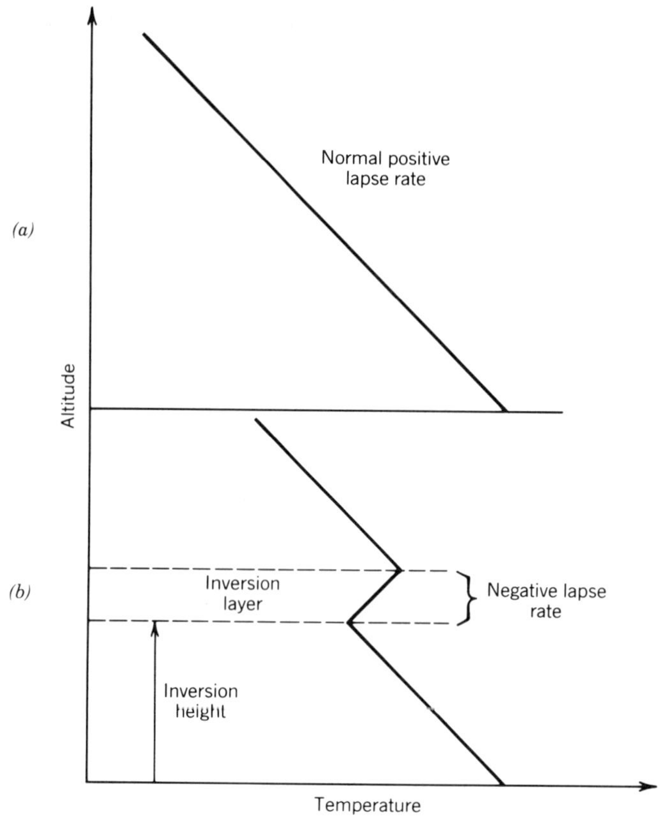

FIGURE 1.5. Variation of temperature with altitude within the troposphere: (*a*) normal lapse rate, (*b*) change in lapse rate from positive to negative, characteristic of a thermal inversion.

by the emission of infrared radiation immediately after sunset. On calm nights, this cooling may be sufficiently rapid that the layer of air adjacent to the surface becomes cooler than the air above it, that is, an inversion forms. This can persist until sufficient heating of the surface and the air above it occurs to ''break'' the inversion at dawn. With this type of inversion, the inversion height—the distance from the earth's surface to the point at which the lapse rate reverses—is often quite small. For example, as noted earlier, in the 1952 London smog episode, inversion heights as low as 150 ft were observed in some locations.

The overhead (or subsidence) inversion associated with photochemical air pollution is caused by the sinking motion of air masses as they pass over the continent. This leads to compression and heating of the air immediately below, resulting in a change in the lapse rate, that is, to the formation of an inversion layer. The inversion height is significantly higher than in the case of radiation inversions; for example, ~1500 ft would represent a relatively low subsidence inversion height.

Interestingly, the vertical distribution of photochemical oxidant may not be such that it falls off rapidly at the inversion layer. In fact, in a classic series of experiments, Edinger and co-workers (1972, 1973) showed that oxidant concentrations in the Los Angeles air basin could be *higher* within the inversion layer than at ground level. Thus, Fig. 1.6, for example, shows the temperature and oxidant profile for June 20, 1970, over Santa Monica, CA, a city adjacent to the Pacific Ocean. Several ''layers'' of oxidant exist within the inversion layer, reaching concentrations as high as 0.2 ppm, compared to a ground-level concentration of ~ 0.1 ppm.

The reason for this phenomenon is that the mountain slopes surrounding the Los Angeles air basin become heated by the sun; the layer of air in contact with the slopes is also heated and moves up the slopes. When the inversion layer is deep and strong, much of this rising polluted air does not get sufficiently warm to penetrate the inversion completely, so it moves out and away from the slopes, and remains within the inversion layer; multiple pollution peaks within the inversion layer result.

Because of such meteorological phenomena, pollutants can be trapped aloft and transported over large distances, undergoing essentially no deposition at the surfaces (Section 11-D-2); this has been documented, for example, by Blumenthal and co-workers (1984) in the midwest of the United States. Clearly, surface measurements are not adequate to document such high-level transport.

The role of meteorological transport in establishing the contributions of various

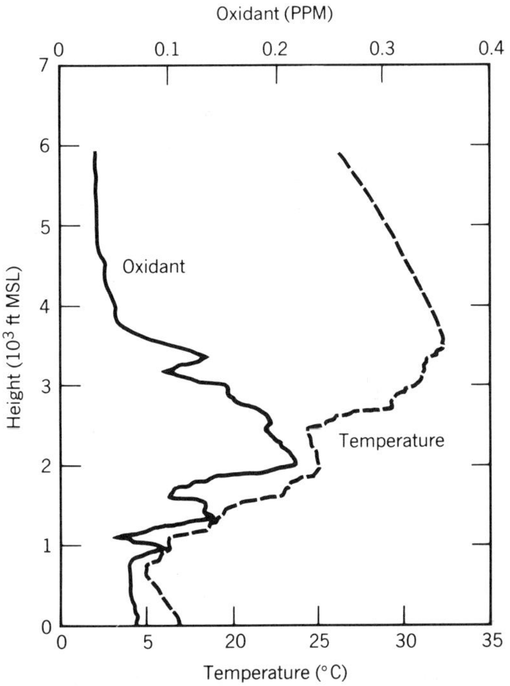

FIGURE 1.6. Temperature and oxidant profiles at 1:28 p.m. over Santa Monica, California, on June 20, 1970 (from Edinger, 1973).

sources to downwind ambient pollutant concentrations is clearly important not only for pollutants within a given air basin, but also for longer-range transport situations. For example, with the use of fluorescent particle tracers, air masses originating along the coast in the morning in southern California have been shown to travel long distances in the course of a day despite low wind speeds at ground level, probably because of transport by faster winds aloft (Metronics, 1973).

An important current problem is to find appropriate tracers for long-range transport (1000 km or more) studies of acid deposition. Such tracers must not have other natural or anthropogenic sources, must not react in the troposphere, and must be non-toxic. Furthermore, they must be capable of being detected and accurately measured at extremely small concentrations, because of the large dilution that occurs over these distances; the analytical problem is a difficult one. The "heavy" methanes $^{12}CD_4$ and $^{13}CD_4$ as well as perfluorocarbons look promising (Fowler and Barr, 1983) and are being used in such large-scale experiments.

For particles, some estimate of the contributions of various sources large distances upwind may be derived from their elemental composition if certain elements, or combinations of the elements, are characteristic of specific sources. For example, Rahn and Lowenthal (1984, 1985) describe a seven-element tracer system they postulate can be used to estimate contributions of regional sources ~ 1000 km upwind; however, a number of uncertainties can be associated with such an approach (Thurston and Laird, 1985) and hence it has not been widely used.

Long-range transport also plays a role in ambient O_3 concentrations. For example, Eliassen and co-workers (1982) have shown that some of the O_3 observed in Scandinavia must come from other European countries, including the USSR and Poland.

In addition to the direct dispersion and transport of pollutants, other meteorological factors play a role in tropospheric chemistry; these include temperature, sunlight intensity, and the presence of fogs. For example, fog water droplets may provide an aqueous medium for the liquid phase of SO_2 to sulfate conversion (Chapter 11). Additionally, the cycle of condensation of water vapor on aerosols at low temperatures, followed by evaporation during the day, is thought to be a major factor controlling the concentrations of pollutants within the droplets (Munger et al., 1983). Indeed, fogs with pH values < 2 have been observed in heavily populated coastal cities in southern California, and Hoffman and co-workers have also shown that fogs in the pH range of 2–3 are common throughout certain areas of the state (Chapter 11). Under certain meteorological conditions, such acid fogs may be important in other major polluted air basins throughout the world.

In summary, as we shall see throughout this book, meteorological parameters are extremely important, not only in determining the dispersion and transport of pollutants, but also in determining their chemistry.

1-D-3. Chemical Transformations

Having briefly considered emissions and meteorology, we turn to an overview of atmospheric chemical transformations. This section summarizes certain major features of air pollution chemistry. Hopefully, it will serve as useful background material before we set off on our detailed discussion of the fundamental and applied aspects of tropospheric chemistry.

As mentioned earlier, the primary pollutant NO_x (mainly NO) reacts in the presence of non-methane hydrocarbons (NMHC) and other organics, shown here as NMOC (non-methane organic compounds), and sunlight to form a host of secondary pollutants. Some of these are criteria pollutants for which air quality standards have been set, such as O_3; others are trace, non-criteria pollutants, for example, PAN, HNO_3, HCHO, and HCOOH. The overall reaction is

$$\text{NMOC} + \text{NO}_x + h\nu \rightarrow \text{O}_3\text{, PAN, etc.} \tag{1}$$

Certain reproducible features of time–concentration profiles for pollutants are observed in ambient air. Figure 1.7 shows such profiles for NO, NO_2, and total oxidant (mainly O_3) in Pasadena, CA, during a severe photochemical air pollution episode in July 1973. The reproducible features include the following: (1) In the early morning, the concentration of NO rises and reaches a maximum at a time that approximately coincides with the maximum emissions of NO, in this case, peak automobile traffic; (2) subsequently, NO_2 rises to a maximum; and (3) oxidant (e.g., O_3) levels, which are relatively low in the early morning, increase

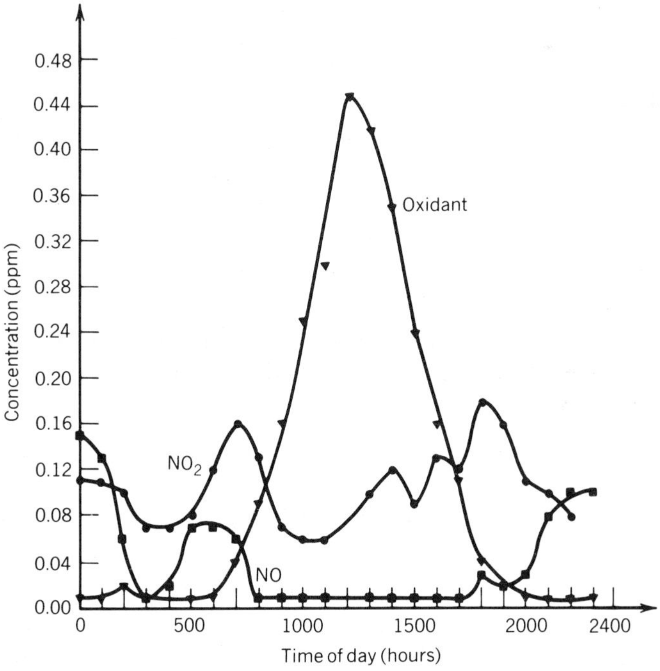

FIGURE 1.7. Diurnal variation of NO, NO_2, and total oxidant in Pasadena, California, on July 25, 1973 (from Finlayson-Pitts and Pitts, 1977).

significantly about noon when the NO concentration drops to a low value. The O_3 reaches a maximum after NO_2.

Downwind from urban centers, the profiles are shifted and O_3 may peak in the afternoon, or even after dark, depending on emissions and airshed transport phenomena. Thus, although O_3 is no longer formed after sunset, a dirty, urban air mass containing O_3 and other secondary pollutants formed during the day can be transported downwind to an otherwise "clean" rural site.

In the early 1950s, relatively soon after the new phenomenon of photochemical air pollution had been recognized, the chemistry responsible for many of these general features was established. Thus, as first suggested by Blacet in 1952, the photolysis of NO_2 in air was shown to form O_3:

$$NO_2 + h\nu\ (\lambda \lesssim 430\ \text{nm}) \rightarrow NO + O \tag{8}$$

$$O + O_2 \xrightarrow{M} O_3 \tag{9}$$

Reaction (9) still remains the sole known source of anthropogenically produced ozone.

NO was also found to react relatively rapidly with O_3, reforming NO_2:

$$NO + O_3 \rightarrow NO_2 + O_2 \tag{10}$$

Because of reaction (10), significant concentrations of O_3 and NO cannot co-exist, and the delay in the oxidant (O_3) peak until NO has fallen to low concentrations, shown in Fig. 1.7, is explained.

It should be noted that throughout this book, we use the terms *fast* or *slow* to reflect the magnitude of the reaction rate constants. Most reactions with which we deal are second-order or pseudo-second-order and hence have units of (concentration)$^{-1}$(time^{-1}). As discussed in detail in Section 4-A, a number of different units for rate constants are used in atmospheric chemistry. Some of these, and the conversion units between them, are summarized in Table 1.13.

TABLE 1.13. Some Units for Second-Order (or Pseudo-Second-Order) Rate Constants Commonly Used in Atmospheric Chemistry and Some Conversion Factors Between Them[a,b]

Unit	To Convert to	Multiply By
$cm^3\ molecule^{-1}\ s^{-1}$	$L\ mole^{-1}\ s^{-1}$	6.02×10^{20}
$L\ mole^{-1}\ s^{-1}$	$cm^3\ molecule^{-1}\ s^{-1}$	1.66×10^{-21}
$ppm^{-1}\ min^{-1}$	$cm^3\ molecule^{-1}\ s^{-1}$	6.77×10^{-16}
$ppb^{-1}\ min^{-1}$	$cm^3\ molecule^{-1}\ s^{-1}$	6.77×10^{-13}
$ppt^{-1}\ min^{-1}$	$cm^3\ molecule^{-1}\ s^{-1}$	6.77×10^{-10}
$atm^{-1}\ s^{-1}$	$cm^3\ molecule^{-1}\ s^{-1}$	4.06×10^{-20}

[a] Assuming 1 atm pressure and 298°K.

[b] L = dm^3, that is, $L\ mole^{-1}\ s^{-1} = dm^3\ mole^{-1}\ s^{-1}$.

As an example of the use of Table 1.13, the rate constant (k_{10}) at 25°C for the reaction of O_3 with NO at room temperature is 23.6 ppm^{-1} min^{-1}. In units of cm^3 $molecule^{-1}$ s^{-1}, this corresponds to $k_{10} = 23.6 \times 6.77 \times 10^{-16} = 1.6 \times 10^{-14}$ cm^3 $molecule^{-1}$ s^{-1}.

Throughout the book, we use units of cm^3 $molecule^{-1}$ s^{-1} for gas phase reactions and units of L $mole^{-1}$ s^{-1} for liquid phase reactions.

For our purposes, we define a fast second-order or pseudo-second-order gas phase reaction as one with a rate constant greater than $\sim 10^{-13}$ cm^3 $molecule^{-1}$ s^{-1}; note that the diffusion-controlled upper limit is $\sim 10^{-10}$ cm^3 $molecule^{-1}$ s^{-1} (see Section 4-A-7). A slow reaction is one with a rate constant $\lesssim 10^{-18}$ cm^3 $molecule^{-1}$ s^{-1}.

Three major questions on the overall chemistry of photochemical air pollution not readily answered in the early studies are the following:

1. How is NO oxidized to NO_2?
2. What is the role played by organics?
3. What reactions are responsible for the rapid loss of organics?

It was first suggested that NO was thermally oxidized by O_2:

$$2NO + O_2 \rightarrow 2NO_2 \qquad (11)$$

Indeed, in the laboratory, at Torr concentrations, the clear, colorless NO is virtually instantaneously oxidized in air to brown NO_2.

However, as discussed in detail in Section 4-A, the rate of this reaction is second order in NO, that is, the speed of oxidation increases as the square of the NO concentration. Thus when one lowers the NO from high concentrations to ambient ppb–ppm levels, the speed of the oxidation drops to the point where its rate is very small. For example, at 100 Torr NO ($\sim 1.3 \times 10^5$ ppm), approximately 85% of the NO is oxidized in ~ 15 s. However, at 0.1 ppm NO, approximately 226 days would be required to achieve the same net oxidation! As a result, the so-called thermal (i.e., non-photochemical) oxidation of NO in reaction (11) is generally too slow to be of importance in the atmosphere.

There is one exception to this, the case where high concentrations of NO (e.g., several thousand ppm) may be emitted from sources such as uncontrolled power plants. In the initial seconds as the plume enters the atmosphere before it has had a chance to become completely diluted with the surrounding air, the NO may be sufficiently concentrated that the oxidation (11) by O_2 is significant. For example, at 2000 ppm NO, 90% of the reactant would be oxidized to NO_2 within 30 min. *if* this high concentration were to be maintained for that long. Of course, the plume integrity usually is not maintained for this period of time; however, under some meterorological conditions the plume can be sufficiently stable that a significant fraction of the NO undergoes thermal oxidation by O_2, and NO_2 is directly formed at some distance from the stack.

In summary, it soon became evident that in ambient photochemical smog the thermal oxidation of NO could not explain the relatively rapid conversion of NO to NO_2.

With respect to the second and third problems, Fig. 1.8 shows the *observed* rate of propylene loss in a typical chamber run and the *calculated* loss rate of the organic based on the two reactions known to occur at the time of the experiment, circa 1970:

$$O_3 + C_3H_6 \rightarrow \text{products} \tag{12}$$

$$O + C_3H_6 \rightarrow \text{products} \tag{13}$$

Clearly, a large fraction of the hydrocarbon loss cannot be accounted for on the basis of reactions (12) and (13).

The answer to the problem of the reactions removing the hydrocarbon lay in a suggestion by Leighton, who, in his classic 1961 monograph *Photochemistry of Air Pollution*, speculated that free radicals such as R (alkyl), RO_2 (alkyl peroxy),

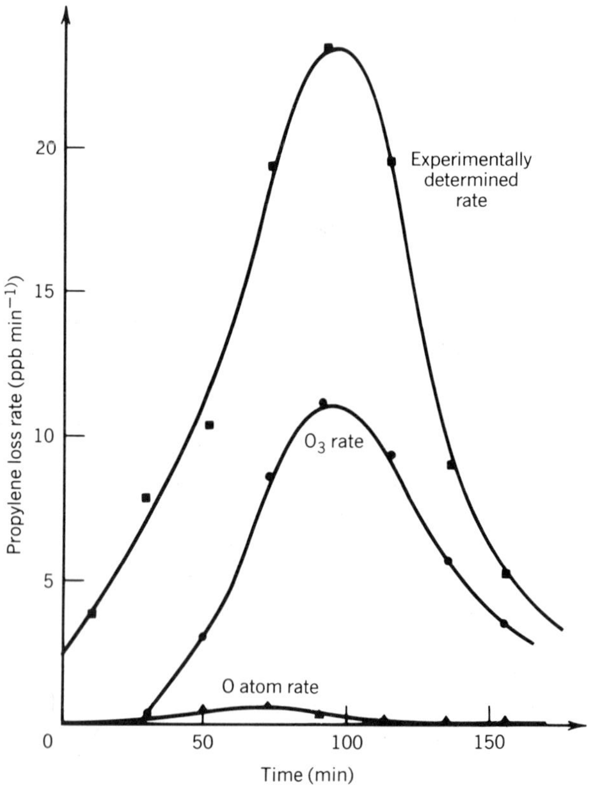

FIGURE 1.8. Experimentally observed rates of C_3H_6 loss and calculated loss rates due to its reaction with O_3 and O atoms. Conditions were 2.23 ppm C_3H_6, 0.97 ppm NO, and 0.05 ppm NO_2 at 50% relative humidity and 31.5 ± 2°C (from Niki et al., 1972).

RO (alkoxy), OH (hydroxyl), HO_2 (hydroperoxyl), and H might be formed from the organics, and furthermore, that these were involved in the NO oxidation. However, it was not until the mid- to late 1960s when OH reactions with hydrocarbons and CO were found experimentally to be fairly rapid (see Chapter 7) that this suggestion was developed further. Thus Heicklen and co-workers (1969), Weinstock, Niki, and co-workers (Weinstock, 1969; Stedman et al., 1970), and Levy (1971) proposed chain reactions of hydroxyl radicals in the atmosphere which would regenerate OH, such as reactions (14), (15), and (2), and which would convert NO to NO_2:

$$OH + CO \rightarrow H + CO_2 \tag{14}$$

$$H + O_2 \xrightarrow{M} HO_2 \tag{15}$$

$$HO_2 + NO \rightarrow OH + NO_2 \tag{2}$$

This reaction sequence has indeed proven important in relatively clean atmospheres containing small amounts of NO. Furthermore, in polluted urban atmospheres, organics were found to play a role similar to CO in carrying on chain reactions involving OH and HO_2 and the conversion of NO to NO_2. Thus, in Fig. 1.8, OH attack on the propene accounted for the discrepancy between its observed and calculated rates of disappearance.

We now recognize that the hydroxyl radical drives the daytime chemistry of both polluted and clean atmospheres; this has revolutionized our understanding of atmospheric chemistry. Thus OH initiates chain reactions by attack on organics or CO, and these chains are propagated through reactions such as those in Fig. 1.9, the oxidation of propane in air. In this cycle, propane is oxidized to propionaldehyde (C_2H_5CHO), two molecules of NO are converted to NO_2, and OH is regenerated. Of course, the propionaldehyde can itself then be attacked by OH and a similar cycle occurs leading to further NO oxidation (Chapter 7). (We show only abstraction of a primary hydrogen for simplicity, although these are abstracted more slowly than the secondary hydrogens; see Chapter 7-C-1.)

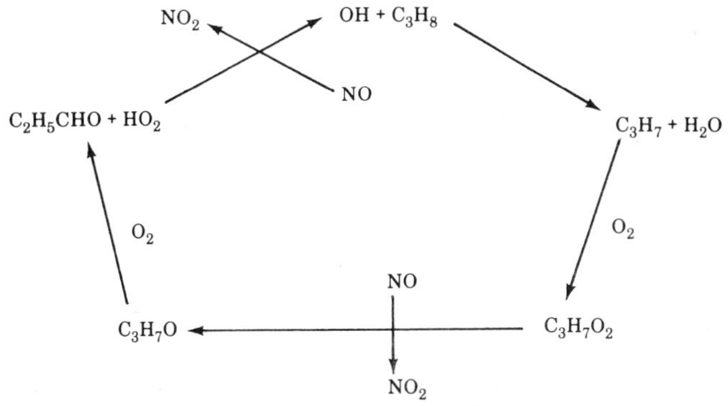

FIGURE 1.9. Oxidation of propane initiated by the hydroxyl radical.

The chain reactions are eventually terminated by such reactions as:

$$HO_2 + HO_2 \rightarrow H_2O_2 + O_2 \tag{16}$$

$$RO_2 + HO_2 \rightarrow ROOH + O_2 \tag{17}$$

$$OH + NO_2 \overset{M}{\rightarrow} HNO_3 \tag{18}$$

A major source of OH in both clean and polluted air is the photolysis of O_3 in sunlight to produce an electronically excited oxygen atom, $O(^1D)$:

$$O_3 + h\nu\ (\lambda \lesssim 320\ \text{nm}) \rightarrow O(^1D) + O_2 \tag{19}$$

followed by the reaction of the excited oxygen atom with water vapor, which is always present in the atmosphere:

$$O(^1D) + H_2O \rightarrow 2OH \tag{20}$$

In polluted areas other direct sources also form OH through photodissociation, including nitrous acid:

$$HONO + h\nu\ (\lambda < 400\ \text{nm}) \rightarrow OH + NO \tag{21}$$

and possibly hydrogen peroxide (H_2O_2)

$$H_2O_2 + h\nu\ (\lambda < 370\ \text{nm}) \rightarrow 2OH \tag{22}$$

An important source is the reaction of HO_2 with NO:

$$HO_2 + NO \rightarrow OH + NO_2 \tag{2}$$

Because HO_2 is intimately tied to OH through reaction (2) and cycles such as that in Fig. 1.9, when NO is present the sources and sinks of HO_2 are, in effect, sources or sinks of OH.

Sources of HO_2 include the reactions of O_2 with hydrogen atoms and formyl radicals, both of which are produced by formaldehyde photolysis:

$$HCHO + h\nu\ (\lambda < 370\ \text{nm}) \rightarrow H + HCO \tag{23}$$

$$H + O_2 \overset{M}{\rightarrow} HO_2 \tag{15}$$

$$HCO + O_2 \rightarrow HO_2 + CO \tag{24}$$

Another source is the abstraction of a hydrogen atom from alkoxy radicals:

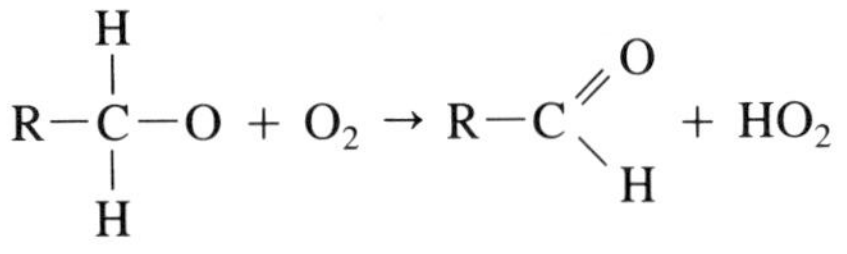

(25)

The relative importance of these sources of OH and HO_2 depends on the species present in the air mass, and hence on location and time of day. Figure 1.10, for example, shows the relative contributions, as a function of time of day, of three sources of OH/HO_2 in an urban air mass. In this case, nitrous acid is predicted to be the major OH source in the early morning hours, HCHO in mid-morning, and O_3 later in the day when its concentrations have built up significantly.

In summary, NO is now known to be converted to NO_2 during daylight hours in a reaction sequence initiated by HO attack on organics and involving HO_2 and RO_2 free radicals. These peroxy radicals are the species that actually convert NO to NO_2 at ambient concentrations where the thermal oxidation of NO by O_2 is negligible.

In the late 1970s, the importance of the nighttime chemistry of another trace reactive species emerged, that of the gaseous nitrate radical NO_3. Thus, although it had earlier been shown by Niki and co-workers that NO_3 reacted rapidly with simple olefins and acetaldehyde (Morris and Niki, 1974; Japar and Niki, 1975), it wasn't until the late 1970s that it was identified and measured by long-pathlength spectroscopic techniques first in clean continental air (Noxon et al., 1978, 1980) and then in polluted urban atmospheres (Platt et al., 1980).

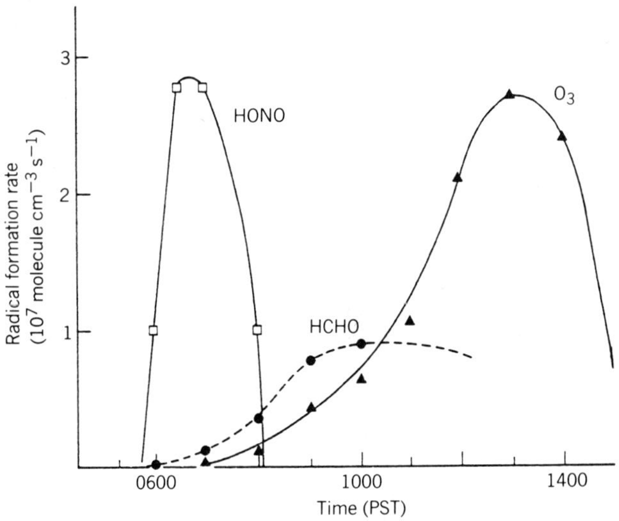

FIGURE 1.10 Predicted sources of OH in a polluted urban atmosphere as a function of the time of day (from Winer, 1985).

The NO_3 radical can be formed in ambient air under certain conditions; both O_3 and NO_2 must be present to form it via reaction (26):

$$O_3 + NO_2 \rightarrow NO_3 + O_3 \quad (26)$$

However, the concentrations of NO must be small because NO reacts quickly with NO_3 to reform NO_2 (see Section 8-A-2). Because NO_3 photolyzes rapidly (Sections 3-C-7 and 8-A-2), it exists at measurable concentrations only at night; levels ranging from a few ppt to over ~400 ppt have been observed spectroscopically at locations in Germany and the United States (Section 5-G).

Furthermore, a series of kinetic studies at the University of California Statewide Air Pollution Research Center (SAPRC) have shown that NO_3 reacts at significant rates with a variety of olefins, phenolic, and simple heterocyclic compounds, as well as with a variety of biogenic species including terpenes and dimethyl sulfide (see Chapters 7, 8, and 14). Indeed, in some cases, nighttime reaction with NO_3 is the major is the major tropospheric sink for these species. In addition, it is a source of nitric acid, through reaction (27)

$$NO_3 + RH \rightarrow HNO_3 + R \quad (27)$$

Reactions of NO_3 in water droplets may also be a source of HNO_3.

In the presence of NO_2, NO_3 forms an equilibrium producing dinitrogen pentoxide:

$$NO_3 + NO_2 \rightleftarrows N_2O_5 \quad (28)$$

In 1980 it was proposed that hydrolysis of N_2O_5 by several routes could be a significant nighttime source of nitric acid in polluted atmospheres (Platt et al., 1980); subsequent experimental and modeling studies have supported this idea. Additionally, N_2O_5 may be involved in recent studies of the impacts of mixtures of O_3 and NO_2 on the mutagenicity of wood smoke and lung damage to exposed animals. Since calculations show that ppb levels of N_2O_5 may exist under a variety of atmospheric conditions (Atkinson et al., 1985), its role in nighttime atmospheric chemistry may be significant.

With respect to the chemistry of SO_2, we now recognize that early distinctions between London and Los Angeles smogs were rather arbitrary. We deal with one atmospheric system containing not only SO_2 and combustion-generated soot as primary pollutants but also NO_x and NMOC. Thus during daylight hours the OH formed not only can convert NO_2 to HNO_3, reaction (18), but can also initiate the gas phase oxidation of SO_2 to sulfuric acid, reaction (29),

$$SO_2 + OH \xrightarrow{M} HOSO_2 \rightarrow \rightarrow H_2SO_4 \quad (29)$$

SO_2 can be oxidized not only in the gas phase via reaction (29), but also in liquid droplets to form sulfuric acid and sulfates, as well as other species, including sulfonic acids. However, again from the perspective of "one atmosphere," we note that one of the major oxidants of SO_2 in droplet chemistry may be H_2O_2, formed in part by the disproportionation of two HO_2 radicals, reaction (16). Details of these processes are briefly mentioned below in Section 1-E-1 and discussed in detail in Chapter 11.

1-D-4. Ambient Concentrations: Criteria and Non-criteria Pollutants and Air Quality Standards

Continuing through our air pollution system (Fig. 1.3), we indicate here the range of concentrations of the major criteria, as well as trace non-criteria, pollutants typically observed in clean to heavily polluted air. We also briefly discuss air quality standards.

A number of pollutants have documented effects on people, plants, or materials at concentrations, or approaching those, found in polluted air. In the United States, seven of them are known as *criteria pollutants* and national ambient air quality standards (NAAQS) have been set for them "to protect public health and welfare." Table 1.14 lists these compounds and their approximate ranges of concentrations observed from remote areas to severe air pollution episodes in urban areas. Figures 1.11 and 1.12 show the ranges of annual average concentrations of SO_2 and suspended particulate matter measured from 1976 to 1980 in various major cities around the world as part of the Global Environment Monitoring System supported by the World Health Organization, United Nations, and the World Mete-

TABLE 1.14. Criteria Pollutants and Typical Ranges of Concentrations Observed from Remote Areas to Severe Air Pollution Episodes in Urban Areas

Criteria Pollutant	Typical Range of Concentrations[a]
CO	$\leqslant$0.2–50 ppm
SO_2	$\leqslant$1 ppb–2 ppm[b]
O_3	0.01–0.5 ppm
NO_2	$\leqslant$1 ppb–0.5 ppm
NMHC	$\leqslant$65 ppbC–1.5 ppmC
Total suspended particulates (TSP)[b]	5–1500 $\mu g\ m^{-3}$
Lead[c]	0.0001–10 $\mu g\ m^{-3}$

[a] One hour average except where otherwise stated.

[b] 24 hour averages.

[c] Highest values generally recorded in vicinity of sources, for example, smelters.

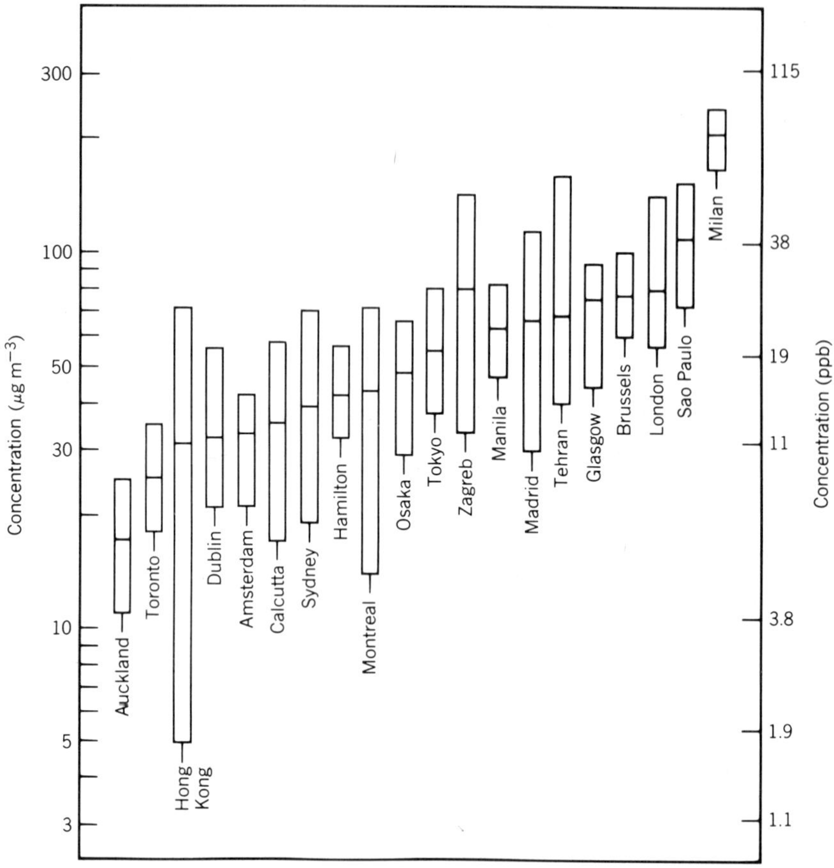

FIGURE 1.11. Ranges of annual average concentration of SO_2 in various cities around the world from 1976 to 1980. Composite average for each city is shown by horizontal line (from Bennett et al., 1985).

orological Organization; the concentrations span about an order of magnitude, depending on the city.

The U.S. national ambient air quality standards (NAAQS) and the *recommended values* or *limits* set by some other countries and the World Health Organization are given in Tables 1.15 and 1.16.

Ambient air quality standards have two components: a concentration and a time. For example, the U.S. NAAQS for O_3 is 0.12 ppm for 1 h. This means that to avoid deleterious effects, the O_3 concentration should not exceed 0.12 ppm for more than 1 h. An additional restriction is that the average number of days per year above the standard should be no more than one. Note, however, that the maximum pollutant levels and the associated restrictions on occurrence are subject to periodic review and modification (Padgett and Richmond, 1983); those cited in Table 1.15 were applicable in 1984.

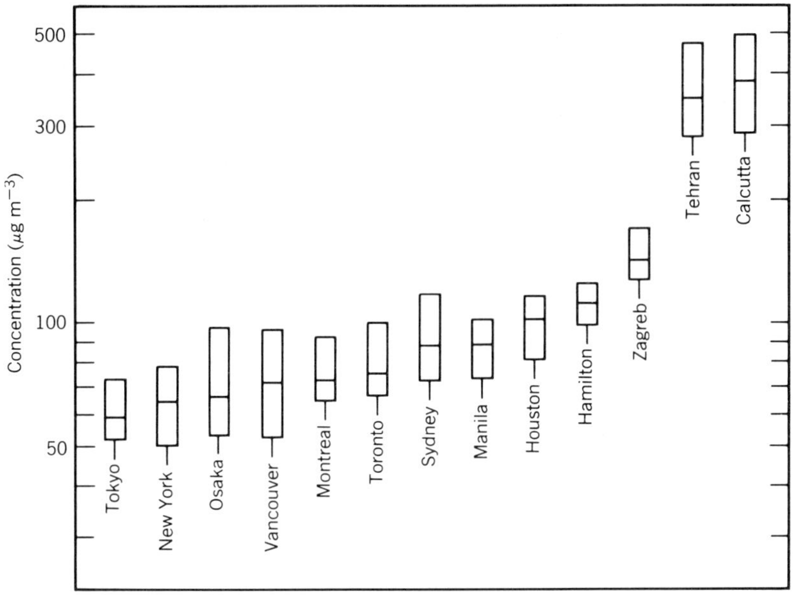

FIGURE 1.12. Ranges of annual average concentrations of suspended particulate matter in various cities around the world from 1976 to 1980. Composite average for each city is shown by horizontal line (from Bennett et al., 1985).

The United States sets two types of NAAQS, primary and secondary. They are based on information contained in air quality criteria documents (U.S. Environmental Protection Agency, 1978, 1979, 1982 a, b, 1984); these contain a wealth of information on all aspects of the criteria pollutants, as do the documents by the World Health Organization, and should be consulted for detailed information and references regarding pollutant sources, ambient levels, chemical transformations, effects, and so on.

Note that the definition of a U.S. *primary* ambient air quality standard is one designed "to protect the public health"—with "an adequate margin of safety." These standards are set to protect even the most susceptible groups in the population, including those with cardiac and respiratory disease, and newborn infants whose defense systems are not well developed.

Secondary NAAQS are set to protect "public welfare." This includes economic losses due to damage to plants and materials, aesthetic effects including visibility degradation, and "personal discomfort" (Jordon et al., 1983). For most pollutants, the primary and secondary NAAQS are the same. However, for particulates, they are not; interestingly, the *secondary* guidelines are more stringent than the *primary* standards.

It should be noted that the WHO standards in Table 1.16 are short-term goals; more stringent long-term goals for the various pollutants have also been set. These long-term goals, at which no ill effects would be anticipated, are concentrations falling between the short-term ones in Table 1.16 and the natural background of

TABLE 1.15. National Ambient Air Quality Standards (NAAQS) Set by the United States Federal Government[a]

	Primary		Secondary	
Pollutant	Concentration	Time	Concentration	Time
CO	9.0 ppm	8 h	9.0 ppm	8 h
	35.0 ppm	1 hr	35.0 ppm	1 h
SO_2	0.03 ppm	Annual arithmetic mean	—	—
	0.14 ppm	24 h	0.5 ppm	3 h
O_3	0.12 ppm	1 h	0.12 ppm	1 h
NO_2	0.05 ppm	Annual arithmetic mean	0.05 ppm	Annual arithmetic mean
NMHC[b]	0.24 ppm	Average from 6 to 9 a.m.	0.24 ppm	Average from 6 to 9 a.m.
Total suspended particles	75 $\mu g\ m^{-3}$	Annual geometric mean	60 $\mu g\ m^{-3}$	Annual geometric mean
	260 $\mu g\ m^{-3}$	24 h	150 $\mu g\ m^{-3}$	24 h
Lead	1.5 $\mu g\ m^{-3}$	Quarterly average	1.5 $\mu g\ m^{-3}$	Quarterly

[a] NAAQS in effect in 1984.

[b] NMHC = non-methane hydrocarbons; expressed as ppm carbon (ppmC).

the pollutant. For example, the long-term goals for CO are 9 and 35 ppm for 8 h and 1 h averages, respectively, the same as the U.S. NAAQS; these can be compared to the corresponding short-term goals of 30 and 100 ppm (Table 1.16).

In addition to the criteria pollutants, a wide variety of *trace* gaseous and particulate species are present in the polluted troposhere. Table 1.17 shows some of these gaseous non-criteria pollutants indentified in photochemical air pollution and gives typical concentrations under conditions ranging from those in remote areas to severely polluted urban air.

Although their peak concentrations are usually only in the ppb range, taken together they can form a substantial fraction of the concentration of the co-pollutant O_3, a criteria pollutant for which an air quality standard has been set and which, in southern California, is used to call *smog alerts*. For example, Fig. 1.13 shows the maximum concentrations of some of these secondary pollutants determined using FT–IR spectroscopy over a kilometer pathlength during a severe photochemical air pollution episode in October 1978 in Claremont, CA, a city approximately 30 miles east and generally downwind of central Los Angeles (Tuazon et al., 1981). Also shown is the observed peak ozone level, with the California Air Quality Standard (CAQS) for O_3 and the first, second, and third stage alert levels. These alert levels, or *episode criteria*, are set by the State of California to

TABLE 1.16. Selected Recommended Values or Limits[a] for Various Air Pollutants in Some Countries Around the World and by the World Health Organization

Pollutant	Country	Concentration	Time
CO	WHO[b,c]	25 ppm	24 h
		30 ppm	8 h
		100 ppm	1 h
	Canada	13 ppm	8 h
	Japan	20 ppm	8 h
	USSR	1.3 ppm	24 h
	W. Germany	26 ppm	0.5 h
SO_2	WHO	38–57 ppb[d]	24 h
		15–23 ppb[d]	Annual arithmetic mean
	Sweden	0.29 ppm	1 h
		0.12 ppm	24 h
	Netherlands	0.05 ppm	24 h
	USSR	0.02 ppm	24 h
	W. Germany	0.06 ppm	24 h
O_3	WHO	0.10 ppm[e]	1 h
	Japan	0.06 ppm	1 h
NO_2	Japan	0.04–0.06 ppm	24 h
	W. Germany	0.05 ppm	2–12 mo
	USSR	0.05 ppm	24 h
NMHC[f]	Canada	0.24 ppm	
Total suspended particulates	WHO	60–90 $\mu g\ m^{-3}$	Annual average
	Canada	120 $\mu g\ m^{-3}$	24 h
		70 $\mu g\ m^{-3}$	1 yr
	Sweden	260 $\mu g\ m^{-3}$	24 h
		75 $\mu g\ m^{-3}$	1 yr
	Japan	200 $\mu g\ m^{-3}$	1 h
	Israel	200 $\mu g\ m^{-3}$	24 h
	USSR	150 $\mu g\ m^{-3}$	24 h
Lead	USSR	0.7 $\mu g\ m^{-3}$	24 h

Source: Svenson et al., 1983; World Health Organization, 1972, 1978–1979.

[a] *Limits* are differentiated from *recommendations* in that limits are defined by legislation.

[b] WHO = World Health Organization guidelines.

[c] Based on a recommended limit in blood of 4% carboxyhemoglobin (COHb) or less. Since the percent COHb in blood depends on a number of factors (e.g., smoking, time of exposure, exercise), these are only estimates based on a typical exposure required to reach 4% COHb in the blood.

[d] Based on observations of community populations exposed simultaneously to mixtures of SO_2 and "smoke," that is, particulate matter.

[e] Oxidant as measured by KI method (see Chapter 5) and expressed as O_3.

[f] NMHC = non-methane hydrocarbons from 6:00 to 9:00 a.m.; expressed as ppm carbon (ppmC).

TABLE 1.17. Some Non-criteria Pollutants and Typical Concentration Ranges Observed from Remote Areas to Severe Air Pollution Episodes in Urban Areas

Non-criteria Pollutant	Typical Range of Concentrations
Nitrogenous Species	
Nitric oxide	0.02–2000 ppb
Peroxyacetyl nitrate	0.05–70 ppb
Nitric acid	<0.1–50 ppb
Nitrous acid	<0.03–8 ppb
Nitrate radical	<5–430 ppt
Dinitrogen pentoxide[a]	⩽15 ppb
Ammonia	<0.02–100 ppb
Oxygenated Organics	
Formaldehyde	<0.5–75 ppb
Formic acid	⩽20 ppb
Methanol	⩽40 ppb

[a] Calculated from ambient NO_3 and NO_2 concentrations.

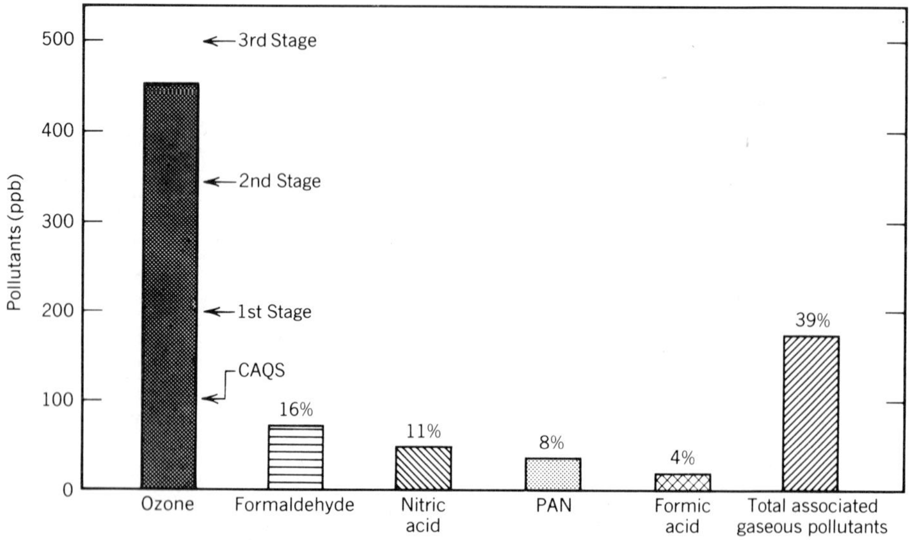

FIGURE 1.13. Maximum concentrations of some trace pollutants and the percentage of the peak ozone they form when summed. Also shown is the California Air Quality Standard for O_3 and the various alert levels (from Tuazon et al., 1981).

define the concentrations at which short-term exposures may affect susceptible portions of the population; various control actions are required when each episode level is reached. It is seen that the sum of the peak concentrations of formaldehyde, nitric acid, PAN, and formic acid totaled 39% of the peak ozone concentration.

1-D-5. Visibility

The loss of visibility which commonly accompanies high pollutant levels is perhaps the most obvious aspect of air pollution to the public; it is due to the scattering and absorption of light by pollutants. The total light extinction—the sum of scattering and absorption by gases and particles—and hence visibility reduction, depends on both the wavelength of the light and on the scattering angle, that is, the position of the sun, so that "haze" due to air pollution may appear to have different colors depending on the conditions.

Of the gaseous air pollutants, only NO_2 absorbs visible light to an significant extent and thus contributes to visibility reduction. It is an orange-brown gas that absorbs radiation strongly at $\lambda < 430$ nm; hence it acts as a filter for blue light. The brownish color of many polluted urban areas and the accompanying spectacular sunsets are at least partly due to the presence of NO_2. However, the contribution of NO_2 to the total extinction is generally small, for example $\lesssim 10\%$.

Scattering of light by gaseous air pollutants is negligible compared to the Rayleigh scattering by O_2 and N_2 because of the very low concentrations of pollutants compared to those of the major components of air.

Thus particulate matter suspended in air is generally responsible for the majority of light scattering and absorption and hence for visibility reduction associated with air pollution. Small particles with diameters of ~ 0.1–1 μm contribute the greatest amount to light scattering and hence visibility reduction is closely related to the fine-particle concentration (see Chapter 12). Thus the size range that has the potential for maximum health effects also contributes the greatest amount to aesthetic effects. It is believed that scattering by particles is responsible for something in the range of 50–95% of the total extinction due to air pollutants, depending on the conditions.

Elemental carbon is believed to be the only significant light-absorbing species in particles. In remote regions, absorption may only account for $\sim$5–10% of the total extinction, whereas in urban areas, its contribution is greater, up to $\sim$50%.

In terms of chemical composition, visibility reduction due to particulate matter is generally related most closely to scattering by sulfate and nitrate, and absorption and scattering by elemental carbon. Relative humidity (RH) is also an important factor, with a significant reduction in visibility occurring as the RH increases from 50 to 90%.

1-D-6. Models

Since about 1975 there has been increasing emphasis on the development of models that predict ambient pollutant concentrations at specific locations and times from a knowledge of the primary emissions, chemistry, meteorology, and topography.

Three general types of models have been developed: plume, airshed, and long-range transport (LRT) models.

As the name implies, plume models attempt to simulate the dispersion and chemical transformations of a plume downwind from a single point source, such as a power plant, or clustered group of point sources, such as an industrial complex. Airshed models, on the other hand, treat the effects of a diverse set of stationary and mobile sources scattered thoughout a geographical area with dimensions of the order of $\sim 10^2$ km; predictions of various pollutant concentrations both within the air basin and downwind are the goal. Long-range transport models are similar to airshed models except that the geographical area of interest is much greater, with dimensions of the order of 1000 km or more. LRT models are essential in assessing the acid deposition problem and the effects of various control strategies on reducing the impact (see Chapter 11).

For all these models, there are three basic components: emissions, including their spatial and temporal variation, meteorology, including such factors as the presence of clouds and fogs which can serve as a liquid reaction medium in the atmosphere, and chemistry, including reaction kinetics, mechanisms and products, and their temperature dependencies.

Because of the enormity of the computational task of trying to simulate accurately the details of each of these three major components over the large areas characteristic of LRT, until recently major simplifications were made in the chemistry. For example, individual oxidation steps of SO_2 were not considered, but rather, a simple first-order rate constant was assigned for the rate of conversion of SO_2 to H_2SO_4, the value being a "best guess" by the modeler. However, increasingly complex chemical submodels are now being incorporated into LRT models.

With airshed models, the conflict between incorporating explicit meteorology and detailed chemical kinetics and mechanisms is not as severe as for LRT models. As a result, airshed models have undergone substantial development over the last decade and are more advanced than LRT models. However, even in this case, accurate prediction of the *absolute* concentrations of all primary and secondary pollutants over a wide area is not yet achievable, especially for multi-day episodes which are associated with the generation of the highest pollutant levels. As a result, airshed models are best used at the present time to predict *relative* changes in pollutant concentrations expected from given changes in emissions. Nevertheless, today these models are important inputs into the development of cost-effective control strategies that will mitigate adverse effects of air pollutants on our health and environment.

The chemical kinetic submodels used for photochemical air pollution and for acid deposition are discussed in Chapters 9 and 11, respectively. Application of plume and airshed models to photochemical air pollution is discussed in Chapter 10.

1-E. SOME SPECIAL TOPICS IN ATMOSPHERIC CHEMISTRY

In this section, we cite several examples of major societal problems whose solutions require application of the fundamentals of atmospheric chemistry. Detailed discussions of these topics are found in Chapters 11, 13, and 10 which deal with acid rain and fogs, particulate mutagens and carcinogens, and airborne toxic organic chemicals, respectively.

1-E-1. Acid Rain and Fog

1-E-1a. History

In England and Sweden, the presence of sulfur compounds and acids in air and rain was recognized as early as the 18th century. Indeed, in 1692 Robert Boyle referred to "nitrous or salino-sulphureous spirits" in the air in his book *A General History of the Air* [see the excellent historical perspectives given by Brimblecombe (1978) and Cowling (1982)].

Remarkably, a treatise on acid rain was published in England in 1872 by Robert Angus Smith; 20 years earlier he had analyzed rain near Manchester and noted three types of areas as one moved from the city to the surrounding countryside:

> that with carbonate of ammonia in the fields at a distance, that with sulfate of ammonia in the suburbs and that with sulphuric acid or acid sulphate, in the town.

In his 1872 book, *Air and Rain: The Beginnings of a Chemical Climatology*, Smith coined the term *acid rain* and described many of the factors affecting this precipitation, such as coal combustion, and the amount and frequency of precipitation. He also suggested experimental protocols to be followed in sample collection and analysis and described acid rain damage to plants and materials. As Cowling (1982) points out, this remarkable book was essentially overlooked until it was critiqued by Gorham in 1981 for the U.S. National Academy of Sciences.

The more recent recognition of acid deposition in Europe, particularly the Scandinavian countries, and in the United States and Canada and the efforts to study its causes, effects, and controls are discussed in some detail by Cowling. One of the largest efforts was an 8 year program in Norway known as the SNSF Project: The Norwegian Interdisciplinary Research Programme, "Acid Precipitation—Effects on Forest and Fish" (Overrein et al., 1980). Its goals were to elucidate the effects of acid precipitation on fish and forests along with associated effects of air pollutants on soils, vegetation, and water.

Another important study, carried out by the European Organization for Economic Cooperation and Development (OECD, 1977), established the long-distance transport of pollutants and the impacts of European countries on their neighbors.

Individual researchers in North America studied the acid deposition problem beginning as early as 1923. In 1978 the Canadian and U.S. governments established a Bilateral Research Consultation Group on the "Long-Range Transport of Air Pollutants," and in 1980 signed a Memorandum of Intent to develop bilateral agreements on trans-boundary air pollution, including acid deposition. During this time, long-term precipitation monitoring programs began, as well as studies of the sources, chemistry, and effects; in the United States, many of these are being conducted under the umbrella of the National Acid Precipitation Assessment Program (NAPAP).

Since the early 1980s, scientists and the public alike have focused increasing attention on the expanding phenomenon of rapid forest decline, especially in such European countries as Germany and Switzerland. For example, the area of forest, both deciduous and conifer, affected in West Germany increased from ~8% in 1982 to ~34% in 1983 to ~50% in 1984 according to Schütt and Cowling (1985). In their comprehensive review, "*Waldsterben, A General Decline of Forests in Central Europe: Symptoms, Development and Possible Causes*," these authors note that:

> The stress factors inducing the Waldsterben syndrome are not known but it is widely assumed · · · that atmospheric deposition of toxic, nutrient, acidifying and/or growth-altering substances is involved.

It is important to note that the effects are attributed at least in part to *atmospheric* deposition, not only of acidic substances, but potentially of a variety of other species as well, such as ozone and possibly organics. While the term *acid deposition* has been used extensively in discussing probable causes of effects on forests and other plant ecosystems, it now seems more appropriate to use the term *atmospheric deposition,* since a variety of atmospheric pollutants may be involved, rather than some simple mixture of sulfuric and nitric acids. Similar considerations may apply to impacts of atmospheric pollutants on humans, animals and materials.

1-E-1b. Chemistry

Acid rain arises from the oxidation of SO_2 and NO_2 in the troposphere to form sulfuric and nitric acids, as well as other species, which are subsequently deposited at the earth's surface, either in precipitation (*wet deposition*) or in dry form (*dry deposition*). These oxidation and deposition processes can occur over relatively short distances from the primary pollutant sources or at distances of a thousand kilometers or more. Thus both short-range and long-range transport must be considered.

As we have seen, the gas phase oxidation of both SO_2 and NO_2 is initiated by reaction with hydroxyl radicals:

$$SO_2 + OH \xrightarrow{M} HOSO_2 \rightarrow\rightarrow H_2SO_4 \quad (29)$$

$$OH + NO_2 \xrightarrow{M} HNO_3 \quad (18)$$

In the case of SO_2, oxidation in the aqueous phase, present in the atmosphere in the form of aerosol particles, clouds, and fog, is also important. Thus SO_2 from the gas phase dissolves in these water droplets and may be oxidized within the droplet by such species as H_2O_2, O_3, O_2, and possibly free radicals such as OH. Oxidation of SO_2 on the surfaces of solids present either in the air or suspended in the water droplets is also possible.

In the case of nitric acid formation, we have already noted that several nighttime reactions of the NO_3 radical may be important, depending on meteorological conditions and the chemical composition of the polluted air mass. They include hydrogen abstraction,

$$NO_3 + RH \rightarrow HNO_3 + R \tag{27}$$

and the reaction of NO_3 with NO_2 to form N_2O_5,

$$NO_3 + NO_2 \rightleftarrows N_2O_5 \tag{28}$$

which then reacts with either liquid or gaseous water:

$$N_2O_5 + H_2O \rightarrow 2HNO_3 \tag{30}$$

Additionally, it has been suggested that NO_3 may also hydrolyze in water droplets to form HNO_3.

Acid fogs with pH values as low as 1.69 have been measured in coastal regions of southern California (Jacob and Hoffman, 1983). The chemistry of acid fog formation also involves SO_2 and NO_2 oxidation, in processes analogous to those summarized above for acid rain. In the case of fogs, however, evaporation of the fog droplets can result in very high concentrations of ions in a strongly acidic liquid phase. Such acid fogs, whether in London or Los Angeles, are a major health concern because the droplets are sufficiently small to be efficiently inhaled (Hoffman, 1984).

We describe the chemistry of acid rain and fogs in detail in Chapter 11.

1-E-2. Particulate Airborne Mutagens and Carcinogens

Among the great intuitive ideas in the history of epidemiology was the proposal in 1775 by Sir Percival Pott that the high rate of cancer of the scrotum incurred by London's chimney sweeps was due to the presence of a carcinogenic chemical in fireplace soot, combined with their infrequent bathing.

Today we recognize that particles from the combustion of fossil fuels contain polycyclic aromatic hydrocarbons (PAH), some of which, for example, benzo(a)pyrene, are animal, and possibly human, carcinogens.

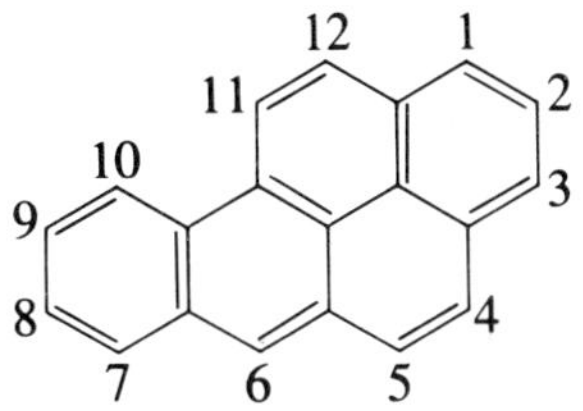

benzo(a)pyrene

It is also clear from short-term bioassays and animal studies of organic extracts of particulate polycyclic organic matter (POM) collected, for example, from auto exhaust, that other *as yet unidentified* animal carcinogens are also present in significant amounts. Similar considerations apply to extracts of fine particles in wood smoke and fly ash from coal-fired power plants, as well as to extracts of respirable particles collected from polluted ambient air in urban areas throughout the world. A major question is: What are the structures, sources, and sinks of these unknown carcinogens?

Interest in, and controversy over, the possible health effects of POM led Congress to include Section 122(a) in the 1977 amendments to the U.S. Clean Air Act. This required the Administrator of the EPA to:

> determine whether or not emissions of polycyclic organic matter (POM) into the ambient air will cause or contribute to air pollution which may reasonably be anticipated to endanger public health.

However, progress in isolating and identifying the chemical species responsible for the "excess" biological activity of airborne POM remained slow. This was due in large part to the chemical complexity of POM, and the associated expense in carrying out large numbers of animal tests for carcinogenicity on the samples collected during the chemical fractionation and analysis of the POM extracts.

In the mid-1970s, there was a major breakthrough; Ames and co-workers announced a rapid, and relatively inexpensive, *in vitro* microbiological test for chemical mutagens (Ames et al., 1973, 1975; McCann et al., 1975). Today, their *reverse* mutation system employing histidine-requiring (*His*$^-$) mutants of the bacterium *Salmonella typhimurium* has become universally recognized as a highly useful, though by no means exclusive, short-term test for the presence of mutagens in complex environmental mixtures.

For example, soon after the publication of the details of the *Ames test*, extracts of ambient particles collected in the Los Angeles basin, California; Okamura and Fukuoka, Japan; Berkeley, California; and Buffalo, New York were demonstrated to be strongly mutagenic (Ames, 1975; Tokiwa et al., 1976; Pitts et al., 1977; Tokiwa et al., 1977; Talcott and Wei, 1977). Furthermore, this activity was expressed *in the absence* (−S9), as well as presence (+S9), of metabolic activation by added mammalian enzymes (Pitts et al., 1977; Talcott and Wei, 1977).

The observation of *direct* mutagenicity (−S9) was important because carcinogenic PAH, such as BaP, known to be present in POM are *promutagens*; they are inactive per se in attacking the bacterial DNA and must be metabolized (+S9) to more reactive derivatives before their mutagenic activity can be expressed. Thus a second major question (related to the earlier one on excess carcinogens) became: What are the structures, sources, and sinks of these direct-acting chemical mutagens present in respirable ambient particles?

In Chapter 13, we discuss the use of combined chemical–microbiological procedures for following the environmental pathways of mutagenic species in airborne, combustion-generated POM. We show that extracts of primary combustion-generated fine particles, for example, diesel soot (Huisingh et al., 1978), fly ash

(Chrisp et al., 1978), and wood smoke (Lewtas, 1982; Ramdahl et al., 1982; Rudling et al., 1982) are strongly mutagenic without metabolic activation (−S9); furthermore, the activity resides in the submicron, respirable particles. Compounds responsible for this direct activity include a variety of mono- and dinitro-polycyclic aromatic hydrocarbons, as well as more polar hydroxy-NO_2–PAH derivatives and other oxidized PAH. The structures of many of these polar mutagens are unknown.

We also consider chemical transformations of PAH in simulated atmospheres containing such gaseous co-pollutants as NO_2 and HNO_3 (Pitts et al., 1978) and the photooxidation of PAH, mutagen formation during the sampling of POM (filter artifacts), and the increasing importance of gas phase reactions of volatile PAH.

1-E-3. Volatile Toxic Organic Compounds

Just as with particulate mutagens and carcinogens, data on the sources, atmospheric reactions, and sinks of toxic, volatile organic compounds (VOC) are critical inputs to *risk assessment* evaluations of their possible health effects. In formulating such risk assessments, atmospheric chemists and specialists in related areas must specify the emissions, ambient levels, and "persistence" of the toxic VOC to which the public is exposed. Biological scientists must identify and evaluate the nature and degree of their health and other ecological impacts (e.g., plant damage).

These assessments are important because they provide crucial scientific input to those public officials responsible for making difficult *risk management* decisions on the nature and degree of emission controls necessary to protect our health and welfare.

In 1970, the task of reliably estimating the lifetimes of a wide range of volatile VOC in remote or polluted atmospheres would have been difficult, if not impossible. The reactive intermediates and mechanisms of their degradation were not known in sufficient detail, and rate constants for many key reactions were unknown or highly uncertain.

Today, through major advances in our understanding of atmospheric chemistry, the situation has improved dramatically, especially with respect to homogeneous gas phase reactions. This is illustrated in Section 10-D where the principles developed throughout the text are applied to the prediction of the atmospheric lifetimes of such directly emitted toxic VOC as the carcinogens benzene and vinyl chloride, chlorinated olefinic solvents (e.g., trichloroethylene), and agricultural chemicals (e.g., ethylene dibromide, EDB).

We also discuss (Chapter 8) the facile formation, under certain conditions, of highly carcinogenic nitrosamines from precursors; these include volatile secondary amines plus nitrous acid and unsymmetrical dimethylhydrazine (UDMH) plus ozone. Interestingly, the latter reaction is now the basis for automatic monitoring of fuel leaks for the U.S. Titan missile.

1-E-4. Indoor Air Pollution

Until about 1975, relatively little attention was paid to the kinds and levels of gaseous and particulate pollutants that might be encountered in typical indoor situations, for example, during the winter in homes and offices. However, as heating by wood-burning stoves, kerosene heaters, natural gas cooking stoves, and so on increased in popularity, along with a trend toward "energy efficient" buildings with increased use of urea formaldehyde foam insulation and decreased ventilation rates, the situation changed.

Today it is recognized that indoor concentrations, not only of asbestos but also of the criteria pollutants CO, NO_2, and TSP and non-criteria pollutants such as formaldehyde, benzene, styrene, and nitrous acid, often exceed their outdoor levels in polluted urban atmospheres. Furthermore, in some parts of the United States with certain geological characteristics, some homes have high indoor levels of radon; radon is a daughter of uranium 238, a widespread component of the earth's crust, and being "inert" migrates from the soil to inside the dwelling. In some instances, the levels of this dangerous radioactive gas in homes exceed those permitted in uranium in the United States.

Given our long exposure times to these indoor pollutants, and their surprisingly high concentrations under certain conditions, the subdiscipline of *indoor atmospheric chemistry* clearly warrants detailed study.

Because the chemistry associated with indoor air pollution is the same as that in ambient air, we discuss it where relevant throughout the text.

1-F. CHEMICAL/PHOTOCHEMICAL PROCESSES IN THE NATURAL TROPOSPHERE

Understanding the chemistry of the natural, or unpolluted, troposphere is essential as a baseline from which the effects of human activities can be assessed. For example, concentrations of O_3 which exceed the earlier U.S. air quality standard of 0.08 ppm for 1 h occasionally have been observed in regions distant from pollution sources. There are three possible explanations: (1) long-range transport of pollutants from distant sources, (2) injection of stratospheric air masses containing O_3, and (3) formation of O_3 through natural cycles. Understanding chemical and photochemical cycles in the natural troposphere forms the basis for projections of O_3 concentrations expected from the third alternative. Such projections play a major role in setting air quality standards and in developing cost-effective control strategies.

The importance of the chemistry of the natural, or remote, troposphere is discussed in the report "*Global Tropospheric Chemistry; A Plan for Action*", produced by the United States National Academy of Sciences (1984). Subsequently in 1985 the National Science Foundation (NSF) and the National Aeronautics and Space Administration (NASA) brought together an internationally recognized group of atmospheric scientists who developed a "Global Tropospheric Chemistry Program", which proposed goals for future research and associated manpower and funding.

The overall features of the chemistry and photochemistry of the natural troposphere are now recognized to be analogous to those in polluted urban atmospheres. Thus OH drives the daytime chemistry and, for some organics, NO_3 may also play a nighttime role. The major differences are in the types of organics available for oxidation, the ambient concentrations of the species that participate in the chemistry, e.g. NO, and the relative importance of the termination versus propagation steps in the chain oxidations.

In the natural troposphere, the major organic present is methane. Today the *background* concentration of CH_4 is about 1.7 ppm and, as discussed in Section 14-B-1, is increasing. In addition to CH_4, other organics are emitted by plants and trees, including isoprene and monoterpenes (see Section 14-B); however, there is considerable controversy regarding their ambient concentrations in various geographical locations. In addition to organics, intermediates and products of their oxidations such as CO and HCHO are present. Finally, NO_x is also produced through natural processes in the troposphere such as lightning (Section 1-D-1b).

Like the oxidation scheme for propane in a polluted atmosphere (Fig. 1.9), the oxidation of CH_4 is initiated by attack by OH, generating CH_3 radicals. These are oxidized in air to form methylperoxy (CH_3O_2) radicals. However, an important difference between CH_4 oxidation in "clean" air and oxidation in polluted atmospheres is that in much of the remote troposphere, very low levels of NO may exist. In this case, the chain *propagating* step

$$CH_3O_2 + NO \rightarrow CH_3O + NO_2 \tag{31}$$

may be sufficiently slow that the chain *terminating* step

$$CH_3O_2 + HO_2 \rightarrow CH_3OOH + O_2 \tag{32}$$

becomes the major fate of the methylperoxy radical. This effectively terminates this particular cycle. For reaction (32) to compete with (31), the NO concentration would have to be ≤ 10 ppt. Whether or not this is the case over substantial regions of the troposphere is not known at present, although nitric oxide concentrations this low have been observed over the oceans.

In addition to naturally occurring organics and oxides of nitrogen, sulfur-containing species such as hydrogen sulfide and dimethyl sulfide are also produced in biological processes and released to the atmosphere. These are oxidized, again in reactions involving OH; under some conditions, attack by NO_3 radicals may also be a major nighttime sink for $(CH_3)_2S$. The mechanism, intermediates, and products of such reactions are not well established; however, these reduced sulfides appear to be transformed into CH_3SO_3H and SO_2, a portion of which forms sulfuric acid, contributing a natural component to acid deposition.

1-G. ATMOSPHERIC CHEMISTRY: AN EMERGING, INTERGRATED DISCIPLINE

In summary, the chemistry of the troposphere from pristine natural areas to heavily polluted urban areas is all part of one continuous cycle in which organics, oxides nitrogen and sulfur, and inorganics such as O_3 play major roles. The highly reactive hydroxyl free radical (OH) drives the daytime chemistry involving these species. We now recognize that at night the nitrate radical (NO_3) can play a major role in the chemistry of both remote and urban areas. Together, OH and NO_3 are central either directly or indirectly to such diverse phenomena as photochemical air pollution, acid rain and fogs, the fate of airborne toxic organics in the atmosphere, and the role of natural emissions in the chemistry of urban, rural, and pristine areas.

Thus, today we no longer speak separately of London smog, photochemical air pollution, acid deposition, airborne toxics, or the chemistry of the clean troposphere. They are all now recognized as component parts of the broader, emerging discipline of atmospheric chemistry. It is to this scientifically fascinating and societally relevant subject that we devote this book.

REFERENCES

Altshuller, A. P., "Review: Natural Volatile Organic Substances and Their Effect on Air Quality in the United States," *Atmos. Environ.*, **17,** 2131 (1983).

Ames, B. N.; private communication (1975).

Ames, B. N., W. E. Durston, E. Yamasaki, and F. D. Lee, "Carcinogens are Mutagens: A Simple Test System Combining Liver Homogenates for Activation and Bacteria for Detection," *Proc. Natl. Acad. Sci. USA*, **70,** 2281 (1973).

Ames, B. N., J. McCann, and E. Yamasaki, "Methods for Detecting Carcinogens and Mutagens with the *Salmonella*/Mammalian-Microsome Mutagenicity Test," *Mutat. Res.*, **31,** 347 (1975).

Atkinson, R., A. M. Winer, and J. N. Pitts, Jr., "Estimation of Nighttime N_2O_5 Concentrations from Ambient NO_2 and NO_3 Radical Concentrations and Its Role in Nighttime Chemistry," *Atmos. Environ.*, **19,** 000 (1985).

Bennett, B. G., J. G. Kretzschmar, G. G. Akland, and H. W. deKoning, "Urban Air Pollution Worldwide," *Environ. Sci. Technol.*, **19,** 298 (1985).

Blacet, F. E., "Photochemistry in the Lower Atmosphere," *Indust. Eng. Chem.*, **44,** 1339 (1952).

Blumenthal, D. L., J. A. McDonald, W. S. Keifer, J. B. Tommerdahl, M. L. Saeger, and J. H. White, "Three-Dimensional Pollutant Distribution and Mixing Layer Structure in the Northeast U.S., Summary of Sulfate Regional (SURE) Aircraft Measurements," *Atmos. Environ.*, **18,** 733 (1984).

Brimblecombe, P., "Interest in Air Pollution Among Early Fellows of the Royal Society," *Notes and Records of the Royal Society*, **32,** 123 (1978).

Chrisp, C. E., G. L. Fisher, and J. E. Lammert, "Mutagenicity of Filtrates from Respirable Coal Fly Ash," *Science*, **199,** 73 (1978).

Cowling, E. B., "Acid Precipitation in Historical Perspective," *Environ. Sci. Technol.*, **16,** 110A (1982).

Crutzen, P. J. and C. J. Howard, "The Effect of the HO_2 + NO Reaction Rate Constant on One-Dimensional Model Calculations of Stratospheric Ozone Perturbations," *Pure Appl. Geophys.*, **116,** 497 (1978).

Cullis, C. F. and M. M. Hirschler, "Atmospheric Sulphur: Natural and Man-Made Sources," *Atmos. Environ.*, **14,** 1263 (1980).

Edinger, J. G., "Vertical Distribution of Photochemical Smog in Los Angeles Basin," *Environ. Sci. Technol.*, **7,** 247 (1973).

Edinger, J. G., M. H. McCutchan, P. R. Miller, B. C. Ryan, M. J. Schroeder, and J. V. Behar,

"Penetration and Duration of Oxidant Air Pollution in the South Coast Air Basin of California," *J. Air Pollut. Control Assoc.*, **22,** 882 (1972).

Eliassen, A., O. Hov, I. S. A. Isaksen, J. Saltbones, and F. Stordal, "A Lagrangian Long-Range Transport Model with Atmospheric Boundary Layer Chemistry," *J. Appl. Meteorol.*, **21,** 1645 (1982).

Evelyn, J., *Fumifugium: or, The Inconveniencie of the Aer and Smoak of London Dissipated, Together with Some Remedies Humbly Proposed*, Bedel and Collins, London, 1661.

Faith, W. L., Final Report of the Air Pollution Foundation (Air Pollution Foundation, San Marino, California) May 1961.

Finlayson-Pitts, B. J. and J. N. Pitts, Jr., The Chemical Basis of Air Quality: Kinetics and Mechanism of Photochemical Air Pollution and Application to Control Strategies, in *Advances in Environmental Science and Technology*, Vol. 7, J. N. Pitts, Jr. and R. L. Metcalf, Eds., Wiley-Interscience, New York, 1977, pp. 75–162.

Fowler, M. M. and S. Barr, "A Long-Range Atmospheric Tracer Field Test," *Atmos. Environ.*, **17,** 1677 (1983).

Goodhill, V., "Maimonides—Modern Medical Relevance," XXVI Wherry Memorial Lecture, Transactions of the American Academy of Ophthalmology and Otolaryngology, May–June 1971, p. 463.

Haagen-Smit, A. J., "Chemistry and Physiology of Los Angeles Smog," *Indust. Eng. Chem.*, **44,** 1342 (1952).

Haagen-Smit, A. J. and M. M. Fox, "Photochemical Ozone Formation with Hydrocarbons and Automobile Exhaust," *J. Air Pollut. Control Assoc.*, **4,** 105 (1954).

Haagen-Smit, A. J. and M. M. Fox, "Automobile Exhaust and Ozone Formation," *SAE Trans.*, **63,** 575 (1955).

Haagen-Smit, A. J. and M. M. Fox, "Ozone Formation in Photochemical Oxidation of Organic Substances," *Indust. Eng. Chem.*, **48,** 1484 (1956).

Haagen-Smit, A. J., E. F. Darley, M. Zaitlin, H. Hull, and W. Noble, "Investigation on Injury to Plants from Air Pollution in the Los Angeles Area," *Plant Physiol.*, **27,** 18 (1951).

Haagen-Smit, A. J., C. E. Bradley, and M. M. Fox, "Ozone Formation in Photochemical Oxidation of Organic Substances," *Indust. Eng. Chem.*, **45,** 2086 (1953).

Haagen-Smit, A. J., M. F. Brunelle, and J. W. Haagen-Smit, "Ozone Cracking in the Los Angeles Area," *Rubber Chem Technol.*, **32,** 1134 (1959).

Heicklen, J., K. Westberg, and N. Cohen, Center for Air Environmental Studies, Report No. 115-69 (1969).

Hinds, W. C., *Aerosol Technology*, Wiley, New York, 1982.

Hoffman, M. R., "Response to Comment on 'Acid Fog,' " *Environ. Sci. Technol.*, **18,** 61 (1984).

Huisingh, J., R. Bradow, R. Jungers, L. Claxton, R. Zweidinger, S. Tejada, J. Bumgarner, F. Duffield, M. Waters, V. F. Simmon, C. Hare, C. Rodriguez, and L. Snow, Application of Bioassay to the Characterization of Diesel Particle Emissions, in *Application of Short-Term Bioassays in the Fractionation and Analysis of Complex Environmental Mixtures*, M. D. Waters, S. Nesnow, J. L. Huisingh, S. S. Sandhu, and L. Claxton, Eds., Plenum Press, New York, 1978, pp. 381–418.

Informatics, Inc., "Air Quality Criteria for Carbon Monoxide," Prepared for U.S. Environmental Protection Agency, EPA-600/8-79-022, October 1979.

Jacob, D. J. and M. R. Hoffman, "A Dynamic Model for the Production of H^+, NO_3^-, and SO_4^{2-} in Urban Fog," *J. Geophys. Res.*, **88,** 6611 (1983).

Japar, S. M. and H. Niki, "Gas-Phase Reactions of the Nitrate Radical with Olefins," *J. Phys. Chem.*, **79,** 1629 (1975).

Jordon, B. C., H. M. Richmond, and T. McCurdy, "The Use of Scientific Information in Setting Ambient Air Standards," *Environ. Health Perspect.*, **52,** 233 (1983).

Leighton, P. A., *Photochemistry of Air Pollution*, Academic Press, New York, 1961.

Levy, H., "Normal Atmosphere: Large Radical and Formaldehyde Concentrations Predicted," *Science*, **173,** 141 (1971).

Lewtas, J., Comparison of the Mutagenic and Potentially Carcinogenic Activity of Particle Bound Organics from Wood Stoves, Residential Oil Furnaces, and Other Combustion Sources, in *Residential Solid Fuels*, J. A. Cooper and D. Malek, Eds., Oregon Graduate Center, Beaverton, OR, 1982, pp. 606–619.

Lodge, J. P., Jr., Selections of *The Smoake of London, Two Prophecies*, Maxwell Reprint Co., Elmsford, NY, 1969.

Logan, J. A., "Nitrogen Oxides in the Troposphere: Global and Regional Budgets," *J. Geophys. Res.*, **88,** 10,785 (1983).

McCann, J., E. Choi, E. Yamasaki, and B. N. Ames, "Detection of Carcinogens as Mutagens in the *Salmonella*/Microsome Test: Assay of 300 Chemicals," *Proc. Natl. Acad. Sci. USA*, **72,** 5135 (1975).

Metronics Associates, Inc., "Field Studies of Air Pollution Transport in the South Coast Basin," Technical Report No. 186, May 15, 1973.

Middleton, J. T., J. B. Kendrick, Jr., and H. W. Schwalm, "Injury to Herbaceous Plants by Smog or Air Pollution," *Plants Dis. Rep.*, **34,** 245 (1950).

Molina, M. J. and F. S. Rowland, "Stratospheric Sink for Chlorofluoromethanes: Chlorine Atom-Catalysed Destruction of Ozone," *Nature*, **249,** 810 (1974).

Möller, D., "Estimation of the Global Man-Made Sulphur Emission," *Atmos. Environ.*, **18,** 19 (1984a).

Möller, D., "On the Global National Sulphur Emission," *Atmos. Environ.*, **18,** 29 (1984b).

Morris, E. D., Jr., and H. Niki, "Reaction of the Nitrate Radical with Acetaldehyde and Propylene," *J. Phys. Chem.*, **78,** 1337 (1974).

Munger, J. W., D. J. Jacob, J. M. Waldman, and M. R. Hoffman, "Fogwater Chemistry in an Urban Atmosphere," *J. Geophys. Res.*, **88,** 5109 (1983).

National Academy of Sciences, *Global Tropospheric Chemistry: A Plan for Action,* National Academy of Science Press, Washington, D.C., 1984.

National Aeronautics and Space Administration, *Present State of Knowledge of the Upper Atmosphere*, January 1984.

National Research Council, *Airborne Particles*, National Academy of Sciences, University Park Press Baltimore, MD, 1979.

National Research Council, *Causes and Effects of Changes in Stratospheric Ozone: Update, 1983*, National Academy Press, Washington, DC, 1984.

Niki, H., E. E. Daby, and B. Weinstock, "Mechanisms of Smog Reactions," *Adv. Chem. Ser.*, **113,** 16 (1972).

Noxon, J. F., R. B. Norton, and W. R. Henderson, "Observation of Atmospheric NO_3," *Geophys. Res. Lett.*, **5,** 675 (1978).

Noxon, J. F., R. B. Norton, and E. Marovich, "NO_3 in the Troposphere," *Geophys. Res. Lett.*, **7,** 125 (1980).

Nriagu, J. O., "Global Inventory of Natural and Anthropogenic Emissions of Trace Metals to the Atmosphere," *Nature*, **279,** 409 (1979).

Organization for Economic Cooperation and Development, "The OECD Programme in Long-Range Transport of Air Pollutants," OECD, Paris, 1977.

Overrein, L. N., H. M. Seip, and A. Tollan, "Acid Precipitation—Effects on Forest and Fish," Final Report of the SNSF Project, 1972–1980, RECLAMO, Oslo, Norway, 1980.

Padgett, J. and H. Richmond, "The Process of Establishing and Revising National Ambient Air Quality Standards," *J. Air Pollut. Control Assoc.*, **33,** 13 (1983).

Phalen, R. F., *Inhalation Studies: Foundations and Techniques*, CRC Press, Boca Raton, FL, 1984.

Pitts, J. N., Jr., D. Grosjean, T. M. Mischke, V. F. Simmon, and D. Poole, "Mutagenic Activity of Airborne Particulate Organic Pollutants," *Toxicol. Lett.*, **1,** 65 (1977).

Pitts, J. N., Jr., K. A. Van Cauwenberghe, D. Grosjean, J. P. Schmid, D. R. Fitz, W. L. Belser, Jr., G. B. Knudson, and P. M. Hynds, "Atmospheric Reactions of Polycyclic Aromatic Hydrocarbons: Facile Formation of Mutagenic Nitro Derivatives," *Science*, **202,** 515 (1978).

Platt, U., D. Perner, A. M. Winer, G. W. Harris, and J. N. Pitts, Jr., "Detection of NO_3 in the Polluted Troposphere by Differential Optical Absorption," *Geophys. Res. Lett.*, **7,** 89 (1980).

Rahn, K. and D. H. Lowenthal, "Elemental Tracers of Distant Regional Pollution Aerosols," *Science*, **223,** 132 (1984); **227,** 1407 (1985).

Ramdahl, T., I. Alfheim, S. Rustad, and T. Olsen, "Chemical and Biological Characterization of Emissions from Small Residential Stoves Burning Wood and Coal," *Chemosphere*, **11,** 601 (1982).

Robinson, E. and R. C. Robbins, "Emissions, Concentrations and Fate of Particulate Atmospheric Pollutants," Stanford Research Institute, Menlo Park, CA, March, 1971.

Robinson, E. and R. C. Robbins, Gaseous Atmospheric Pollution from Urban and Natural Sources, in *The Changing Global Environment*, S. F. Singer, Ed., Reidel, Boston, MA, 1975, pp. 111–123.

Rudling, L., B. Ahling, and G. Löfroth, "Chemical and Biological Characterization of Emissions from Combustion of Wood and Wood-Chips in Small Furnaces and Stoves," in *Residential Solid Fuels*, J. A. Cooper and D. Malek, Eds., Oregon Graduate Center, Beaverton, OR, 1982, pp. 34–53.

Schütt, P., and R. B. Cowling, "Waldsterben, A General Decline of Forests in Central Europe: Symptoms, Development and Possible Causes," *Plant Disease,* **69,** 548 (1985).

SMIC (Study of Man's Impact on Climate), *Inadvertent Climate Modification*, MIT Press, Cambridge, MA, 1971.

Smith, R. A., *Air and Rain: The Beginnings of a Chemical Climatology*, Longmans, Green, London, 1872.

Stedman, D. H., E. D. Morris, Jr., E. E. Daby, H. Niki, and B. Weinstock, "The Role of OH Radicals in Photochemical Smog Reactions," 160th National Meeting of the American Chemical Society, Chicago, IL, September 14–18, 1970.

Stedman, D. H. and R. E. Shetter, The Global Budget of Atmospheric Nitrogen Species, in *Advances in Environmental Science and Technology*, Vol. 12, S. E. Schwartz, Ed., Wiley-Interscience, New York, 1983, pp. 411–454.

Stith, J. L., P. V. Hobbs, and L. F. Radke, "Airborne Particle and Gas Measurements in the Emissions from Six Volcanoes," *J. Geophys. Res.*, **83,** 4009 (1978).

Svenson, G., B. Flyborg, B. Johansson, N. Häggström, S. E. Lorentzon, E. Marcusson, L. Friberg, J. Gawell, A. Hultkvist, L. Högberg, A. Kardell, E. Magnusson, T. Nordin, C. Striby, O. Åslander, B. Assarsson, A. Berggren, and G. Friberg, "Motor Vehicles and Cleaner Air," Report of the Swedich Government Committee on Automotive Air Pollution, Stockholm, 1983.

Talcott, R. and E. Wei, "Airborne Mutagens Bioassayed in *Salmonella Typhimurium*," *J. Natl. Cancer Inst.*, **58,** 449 (1977).

Ter Haar, G. L., D. L. Lenane, J. N. Hu, and M. Brandt, "Composition, Size, and Control of Automotive Exhaust Particulates," *J. Air Pollut. Control Assoc.*, **22,** 39 (1972).

Thurston, G. D. and N. M. Laird, "Tracing Aerosol Pollution," *Science*, **227,** 1406 (1985).

Tokiwa, H., H. Takeyoshi, K. Morita, K. Takahashi, N. Saruta and Y. Ohnishi, "Detection of Mutagenic Activity in Urban Air Pollutants," *Mutat. Res,* **38,** 351 (1976).

Tokiwa, H., K. Morita, H. Takeyoshi, K. Takahashi, and Y. Oshini, "Detection of Mutagenic Activity in Particulate Air Pollutants," *Mutat. Res.*, **48,** 237 (1977).

Tuazon, E. C., A. M. Winer, and J. N. Pitts. Jr., "Trace Pollutant Concentrations in a Multiday Smog Episode in the California South Coast Air Basin by Long Path Length Fourier Transform Infrared Spectroscopy," *Environ. Sci. Technol.*, **15,** 1232 (1981).

U.S. Environmental Protection Agency, "Air Quality Criteria for Particulate Matter and Sulfur Oxides,'' December 1982a; ''Air Quality Criteria for Oxides of Nitrogen,'' September 1982b; ''Air Quality Criteria for Ozone and Other Photochemical Oxidants,'' April 1978; ''Air Quality Criteria for Carbon Monoxide,'' October 1979; Addendum, ''Review of the NAAQS for Carbon Monoxide: 1983 Reassessment of Scientific and Technical Information,'' 1984; ''Revised Evaluation of Health Effects Associated with Carbon Monoxide Exposure,'' 1984; ''Air Quality Criteria for Lead,'' in review, 1984.

Weinstock, B., ''Carbon Monoxide: Residence Time in the Atmosphere,'' *Science*, **166,** 224 (1969).

Wilkins, E. T., ''Air Pollution and the London Fog of December 1952,'' *J. R. Sanitary Inst.*, **74,** 1 (1954).

Winer, A. M., Air Pollution Chemistry, in *Handbook of Air Pollution Analysis*, 2nd ed., R. M. Harrison and R. Perry, Eds., Chapman and Hall, London, 1985, Chap. 3.

World Health Organization, ''Air Quality Criteria and Guides for Urban Air Pollutants,'' Technical Report Series No. 506, World Health Organization, Geneva, 1972.

World Health Organization, ''Environmental Health Criteria. 3. Lead, 4. Oxides of Nitrogen, 7. Photochemical Oxidants, 8. Sulfur Oxides and Particulate Matter, 13. Carbon Monoxide,'' Geneva, 1978–1979.

PART 2

Tropospheric Photochemistry

2 Principles of Photochemistry and Spectroscopy Applicable to Tropospheric Chemistry

The chemistry of the troposphere is driven by solar radiation passing down through the stratospheric ozone layer. Consequently, the application of basic spectroscopic and photochemical principles and techniques to atmospheric processes, exemplified in Leighton's classic monograph, *Photochemistry of Air Pollution* (1961), has led to great advances in our understanding of tropospheric chemistry. Thus photochemistry and spectroscopy form the bases of the establishment of the rates and mechanisms of vapor phase photodissociative processes induced by absorption of solar radiation, which produce reactive atoms and free radicals such as OH. They are also essential to the development and application of long-pathlength, spectroscopic systems to unequivocally identify and measure highly labile, trace constituents in ambient air.

These and related topics are dealt with in detail in subsequent chapters. As background, we briefly review here some rather elementary, but fundamental, aspects of the spectroscopy and photochemistry of diatomic and polyatomic molecules relevant to our air environment. We focus on homogeneous, gas phase systems because they are of major importance and are currently best understood. However, we also emphasize that the application of spectroscopic and photochemical principles to the reactions of tropospheric species in condensed media, for example, within aqueous droplets or on the surfaces of synthetic or ambient fine particles, is a highly important subject; in the following chapters, we cite appropriate examples of current research in these complex, condensed phase systems. In particular, photochemistry in condensed media in the troposphere is discussed in Section 3-C-12.

Space considerations preclude a detailed discussion of theoretical and experimental aspects of photochemistry and spectroscopy. However, details of gas phase photochemistry can be found in Calvert and Pitts (1966), Wayne (1970), and Okabe (1978); organic photochemistry, especially in condensed media, is treated by Turro (1978), and inorganic photochemistry by Adamson and Fleischauer (1975). Spectroscopic matters are treated in Herzberg's classic series (1945, 1950, 1967). Experimental photochemical techniques are discussed by Calvert and Pitts (1966) and Rabek (1982).

2-A. ABSORPTION OF LIGHT AND THE FIRST LAW OF PHOTOCHEMISTRY

2-A-1 Basic Relationships

Light has both wave-like and particle-like properties. As a wave, it is a combination of oscillating electric and magnetic fields perpendicular to each other and to the direction of propagation (Fig. 2.1). The distance between consecutive peaks is the wavelength, λ, and the number of complete cycles passing a fixed point in 1 s is the frequency, ν. They are inversely proportional through the relationship

$$\lambda = \frac{c}{\nu} \tag{A}$$

where c is the speed of light in a vacuum, 2.9979×10^8 m s^{-1}.

Considered as a particle, the energy of a quantum of light E in ergs is

$$E = h\nu = \frac{hc}{\lambda} \tag{B}$$

where h is Planck's constant, 6.626×10^{-27} erg s per quantum (or 6.6262×10^{-34} J s per quantum), and the frequency ν is in s^{-1}. In the visible and ultraviolet regions of the spectrum, wavelength has been commonly expressed in angstrom units, 1 Å = 10^{-10} m; today the unit of a nanometer, 1 nm = 10^{-9} m, has also gained acceptance.

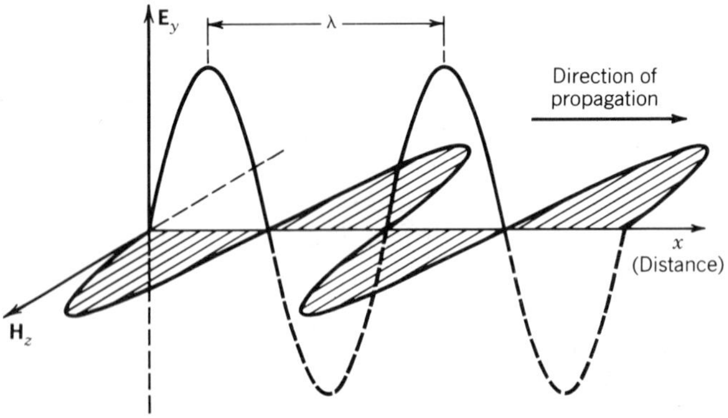

FIGURE 2.1. The instantaneous electric ($\mathbf{E}_y$) and magnetic ($\mathbf{H}_z$) field strength vectors of a plane-polarized light wave as a function of position along the axis of propagation (x) (from Calvert and Pitts, 1966).

In the infrared region both microns [1 micron (μ) = 1 micrometer (μm) = 10^{-6} m] and wavenumbers ω (in cm^{-1}) are employed; ω is the reciprocal of the wavelength λ expressed in centimeters and represents the number of waves per centimeter. It is directly related to energy through the Planck relationship

$$E = hc\omega \tag{C}$$

and today is generally the unit of choice in infrared spectroscopy.

Since chemists generally deal experimentally with moles of materials rather than molecules, a convenient unit is a mole of quanta, defined as 1 einstein. The energy of 1 einstein of light of wavelength λ in nm is

$$\begin{aligned} E &= (6.02 \times 10^{23})\, h\nu = 6.02 \times 10^{23} \frac{hc}{\lambda} \\ &= \frac{1.196 \times 10^5}{\lambda} \text{ kJ einstein}^{-1} \qquad \text{(D)} \\ &= \frac{2.859 \times 10^4}{\lambda} \text{ kcal einstein}^{-1} \end{aligned}$$

If the wavelength is in Å *units*, the values are

$$\begin{aligned} E &= \frac{1.196 \times 10^6}{\lambda} \text{ kJ einstein}^{-1} \\ &= \frac{2.859 \times 10^5}{\lambda} \text{ kcal einstein}^{-1} \end{aligned}$$

Another unit used in photochemistry to express the energy of a quantum of radiation is the electron volt; 1 eV = 96.49 kJ $mole^{-1}$ = 23.06 kcal $mole^{-1}$. Thus for λ *in nm*

$$\mathrm{E} = \frac{hc}{\lambda} = \frac{1.240 \times 10^3}{\lambda} \text{ eV} \tag{E}$$

and for λ *in* Å

$$E = \frac{1.240 \times 10^4}{\lambda}$$

To put these energies and wavelengths in perspective, Table 2.1 gives some typical wavelengths, frequencies, wavenumbers, and energies of various regions of the electromagnetic spectrum. The region of most direct interest in tropospheric photochemistry ranges from the visible at ~700 nm to the near ultraviolet at ~290 nm, the short-wavelength cutoff of the stratospheric ozone layer. The correspond-

TABLE 2.1. Typical Wavelengths, Frequencies, Wavenumbers, and Energies of Various Regions of the Electromagnetic Spectrum

Name	Typical Wavelength or Range of Wavelengths (nm)	Typical Range of Frequencies ν (s^{-1})	Typical Range of Wavenumbers ω (cm^{-1})	Typical Range of Energies (kJ $einstein^{-1}$)[a]
Radiowave	$\sim 10^8$–10^{13}	$\sim 3 \times 10^4$ –3×10^9	10^{-6}–0.1	$\sim 10^{-3}$–10^{-8}
Microwave	$\sim 10^7$–10^8	$\sim 3 \times 10^9$ –3×10^{10}	0.1–1	$\sim 10^{-2}$–10^{-3}
Far infrared	$\sim 10^5$–10^7	$\sim 3 \times 10^{10}$–3×10^{12}	1–100	$\sim 10^{-2}$–1
Near infrared	$\sim 10^3$–10^5	$\sim 3 \times 10^{12}$–3×10^{14}	10^2–10^4	$\sim$ 1–10^2
Visible				
Red	700	4.3×10^{14}	1.4×10^4	1.7×10^2
Orange	620	4.8×10^{14}	1.6×10^4	1.9×10^2
Yellow	580	5.2×10^{14}	1.7×10^4	2.1×10^2
Green	530	5.7×10^{14}	1.9×10^4	2.3×10^2
Blue	470	6.4×10^{14}	2.1×10^4	2.5×10^2
Violet	420	7.1×10^{14}	2.4×10^4	2.8×10^2
Near ultraviolet	400–200	$(7.5\text{–}15.0) \times 10^{14}$	$(2.5\text{–}5) \times 10^4$	$(3.0\text{–}6.0) \times 10^2$
Vacuum ultraviolet	$\sim$ 200–50	$(1.5\text{–}6.0) \times 10^{15}$	$(5\text{–}20) \times 10^4$	$\sim (6.0\text{–}24) \times 10^2$
X ray	$\sim$ 50–0.1	$\sim (0.6\text{–}300) \times 10^{16}$	$(0.2\text{–}100) \times 10^6$	$\sim 10^3$–10^6
γ ray	$\leqslant 0.1$	-3×10^{18}	$\geqslant 10^8$	$> 10^6$

[a] For kcal $einstein^{-1}$, divide by 4.184 (1 cal = 4.184 J).

ing energies [Eq. (D)], 170.9 and 412.4 kJ einstein^{-1} (or 40.8–98.6 kcal einstein^{-1}), are sufficient to break chemical bonds ranging from, for example, the weak O_2—O bond in ozone, ~100 kJ mole^{-1} (~25 kcal mole^{-1}), to the moderately strong C—H bond in formaldehyde, ~368 kJ mole^{-1} (~88 kcal mole^{-1}).

Other spectral regions are also important because the detection and quantification of small concentrations of labile molecular, free radical, and atomic species of tropospheric interest both in laboratory studies and in ambient air are based on a variety of spectroscopic techniques that cover a wide range of the electromagnetic spectrum. For example, the relevant region for FT–IR spectroscopy of stable molecules is generally from ~500 to 4000 cm^{-1} (20–2.5 μm), while the detection of atoms and free radicals by resonance fluorescence employs radiation down to 121.6 nm, the Lyman α line of the H atom.

Table 2.2 gives some relationships between commonly used energy units. Today the SI system of units is in general use, although much of the data in the literature is in the older units. Thus we use both types of units for energy, that is calories or kilocalories, and joules or kilojoules, where 1 cal = 4.184 J.

We generally express wavelengths in nanometers (nm), although angstrom units (1 Å = 1 × 10^{-10} m) are also frequently found in the literature.

2-A-2. Grotthus–Draper Law and the Beer–Lambert Relationship

The Grotthus–Draper law, or the first law of photochemistry as it is sometimes called, states that *only the light which is absorbed by a molecule can be effective in producing photochemical changes in the molecule*. Therefore, to assess the potential for photochemically induced changes in a photochemical system such as the daytime troposphere, it is essential to know the absorption spectra of the reactants. Indeed, to fully characterize the reaction mechanism, one must also know the absorption spectra of the intermediates and reaction products.

TABLE 2.2. Some Relationships Between Commonly Used Energy Units

(kJ mole^{-1})	× 0.2390	= kcal mole^{-1}
	× 0.0104	= eV
	× 83.59	= cm^{-1}
(kcal mole^{-1})	× 4.184	= kJ mole^{-1}
	× 0.04336	= eV
	× 349.8	= cm^{-1}
(cm^{-1})	× 1.196 × 10^{-2}	= kJ mole^{-1}
	× 2.859 × 10^{-3}	= kcal mole^{-1}
	× 1.240 × 10^{-4}	= eV
(eV)	× 96.49	= kJ mole^{-1}
	× 23.06	= kcal mole^{-1}
	× 8.066 × 10^{3}	= cm^{-1}

In addition to knowing the wavelengths of light absorbed by a molecule, it is important to know the strength of the absorption as a function of wavelength, λ. For absorbing solutes in solution, as customarily employed in spectrophotometric analyses, this is given by the well known Beer–Lambert law

$$\log\left(\frac{I_0}{I}\right) = \epsilon C \ell \tag{F}$$

where I_0 = intensity (e.g., quanta s^{-1} or energy per unit time) of monochromatic light of wavelength λ incident on the front of a column of a single absorbing species;
I = intensity of light transmitted through the column of material;
C = concentration of the absorbing species in moles L^{-1} (i.e., molar, M);
ℓ = pathlength of the absorbing column in cm.
ϵ = molar extinction coefficient in L $mole^{-1}$ cm^{-1}. It is a constant for a given species at the given wavelength λ. It is also referred to as the molar absorptivity (a), molar absorptivity index (a_M), or the absorption coefficient.
$\log(I_0/I)$ = absorbance (or molar absorbance), base 10.

Equation (F) can also be written in terms of the natural logarithm:

$$\ln\left(\frac{I_0}{I}\right) = \alpha C \ell \tag{G}$$

where $\alpha = 2.303\,\epsilon$. Unfortunately, in the literature the symbols ϵ and α often appear without specifying that they were derived from Eq. (F), that is, the base 10, or Eq. (G), that is, the base e. Compounding the confusion is a lack of consistency used for the units of concentration and pathlength. In this book, for liquid solutions we use the definitions given above.

The most common units used in gas phase tropospheric chemistry for concentration are the number of molecules cm^{-3}, symbolized by N, and for the pathlength, ℓ, units of cm. In addition, natural logarithms are most commonly used. The gas phase absorption coefficient, designated σ, is then in units of cm^2 $molecule^{-1}$; σ is also known as the absorption cross section. The Beer–Lambert law becomes

$$\frac{I}{I_0} = e^{-\sigma N \ell} \quad \text{or} \quad \ln\left(\frac{I_0}{I}\right) = \sigma N \ell \tag{H}$$

A second widely used set of units for gaseous species is their concentration in units of atmospheres (i.e., pressure, p), and the pathlength, ℓ, in cm. The absorption coefficient k is then in units of atm^{-1} cm^{-1}, and the Beer–Lambert law becomes

$$\frac{I}{I_0} = e^{-kp\ell} \quad \text{or} \quad \ln\left(\frac{I_0}{I}\right) = kp\ell \tag{I}$$

Since the pressure depends on temperature, a reference temperature, usually 273 or 298°K, must be specified. The logarithmic base 10 is also frequently encountered in the literature in combination with pressure units for concentration; thus one is also cautioned to verify the logarithmic base before using absorption coefficients.

Table 2.3 gives some conversion factors for converting absorption coefficients in one set of units to another set, and/or another logarithmic base number.

For a mixture of absorbing molecules where each molecule has an absorption coefficient ϵ_i and is present at concentration C_i ($i = 1, 2, 3, \ldots$), the Beer–Lambert law in decadic form becomes

$$\frac{I}{I_0} = 10^{-(\epsilon_1 C_1 + \epsilon_2 C_2 + \epsilon_3 C_3 + \cdots)\ell} \tag{J}$$

TABLE 2.3. Conversion Factors for Changing Absorption Coefficients from One Set of Units to Another

Both Units in Either Logarithmic Base e or Base 10	
(cm^2 molecule^{-1}) × 2.69 × 10^{19}	= (atm at 273°K)$^{-1}$(cm^{-1})
× 2.46 × 10^{19}	= (atm at 298°K)$^{-1}$(cm^{-1})
× 3.24 × 10^{16}	= (Torr at 298°K)$^{-1}$(cm^{-1})
× 6.02 × 10^{20}	= (L mole^{-1} cm^{-1})
(atm at 298 K)$^{-1}$(cm^{-1}) × 4.06 × 10^{-20}	= cm^2 molecule^{-1}
× 1.09	= (atm at 273°K)$^{-1}$(cm^{-1})
(L mole^{-1} cm^{-1}) × 4.46 × 10^{-2}	= (atm at 273°K)$^{-1}$(cm^{-1})
× 4.09 × 10^{-2}	= (atm at 298°K)$^{-1}$(cm^{-1})
× 5.38 × 10^{-5}	= (Torr at 298°K)$^{-1}$(cm^{-1})
× 1.66 × 10^{-21}	= (cm^2 molecule^{-1})
Change of Both Logarithmic Base and Units	
(cm^2 molecule^{-1}), base *e* × 1.17 × 10^{19}	= (atm at 273°K)$^{-1}$(cm^{-1}), base 10
× 1.07 × 10^{19}	= (atm at 298°K)$^{-1}$(cm^{-1}), base 10
× 1.41 × 10^{16}	= (Torr at 298°K)$^{-1}$(cm^{-1}), base 10
× 2.62 × 10^{20}	= L mole^{-1} cm^{-1}, base 10
(L mole^{-1} cm^{-1}), base 10 × 3.82 × 10^{-21}	= cm^2 molecule^{-1}, base *e*
× 0.103	= (atm at 273°K)$^{-1}$(cm^{-1}), base *e*
× 9.42 × 10^{-2}	= (atm at 298°K)$^{-1}$(cm^{-1}), base *e*
(atm at 273°K)$^{-1}$(cm^{-1}), base 10 × 8.57 × 10^{-20}	= cm^2 molecule^{-1}, base *e*
× 51.6	= L mole^{-1} cm^{-1}, base *e*
(Torr at 298°K)$^{-1}$(cm^{-1}), base 10 × 7.11 × 10^{-17}	= cm^2 molecule^{-1}, base *e*
×4.28 × 10^4	= L mole^{-1} cm^{-1}, base *e*
(atm at 298°K)$^{-1}$(cm^{-1}), base 10 × 9.35 × 10^{-20}	= cm^2 molecule^{-1}, base *e*
× 2.51	= (atm at 273°K)$^{-1}$(cm^{-1}), base *e*

For most tropospheric situations involving gaseous species, the Beer–Lambert law (or its approximation for conditions of weak absorption, Section 3-A-2) is an accurate method for treating light absorption; similar considerations apply to non-associated molecules in dilute solution. However, under laboratory conditions with relatively high concentrations of the absorbing species, deviations may arise from a variety of factors including: concentration-dependent association or dissociation reactions, deviations from the ideal gas law, saturation of the absorption bands with increasing concentration, pressure, and temperature effects on very narrow-line absorptions, and a situation in which a ‘‘monochromatic’’ analyzer beam actually has a bandwidth that is broad relative to very narrow lines of an absorbing species. Clearly, to be on the safe side, it is good practice to verify the linearity of $\ln(I_0/I)$ plots as a function of absorber concentration when experimentally determining absorption coefficients.

2-B. ENERGY LEVELS AND MOLECULAR ABSORPTION SPECTRA

Our simplified treatment of molecular spectra is based on the approximation that one can deal independently with the vibrational (V), rotational (R), and electronic (E) energy of a molecule, that is, $E_{\text{total}} = E_V + E_R + E_E$, and our discussion follows that general order.

2-B-1. Diatomic Molecules

2-B-1a. Vibrational Transitions

The bond between the two atoms in a diatomic molecule can be viewed as a vibrating spring in which, as the internuclear distance changes from the equilbrium value r_e, the atoms experience a force that tends to restore them to the equilibrium position. The *ideal*, or harmonic, oscillator is defined as one that obeys Hooke’s law, that is, the restoring force F on the atoms in a diatomic molecule is proportional to their displacement from the equilibrium position.

If one substitutes the potential energy of this harmonic oscillator into the Schrödinger wave equation, one finds that the vibrational energies are quantized, with the permissible eigenvalues (i.e., energies) E_v given by

$$E_v = h\nu_{\text{vib}}(v + \tfrac{1}{2}) \tag{K}$$

where ν_{vib} is a constant characteristic of the molecule and is related to the strength of the bond and the reduced mass of the molecule. The *vibrational quantum number* v can have the integral values 0, 1, 2, ... Thus the vibrational energy levels of this ideal oscillator are equally spaced.

However, as seen in Fig. 2.2, this idealized harmonic oscillator is satisfactory only for low vibrational energy levels (Fig. 2.2*a*). For real molecules, the potential energy rises sharply at small values of r when the atoms approach each other

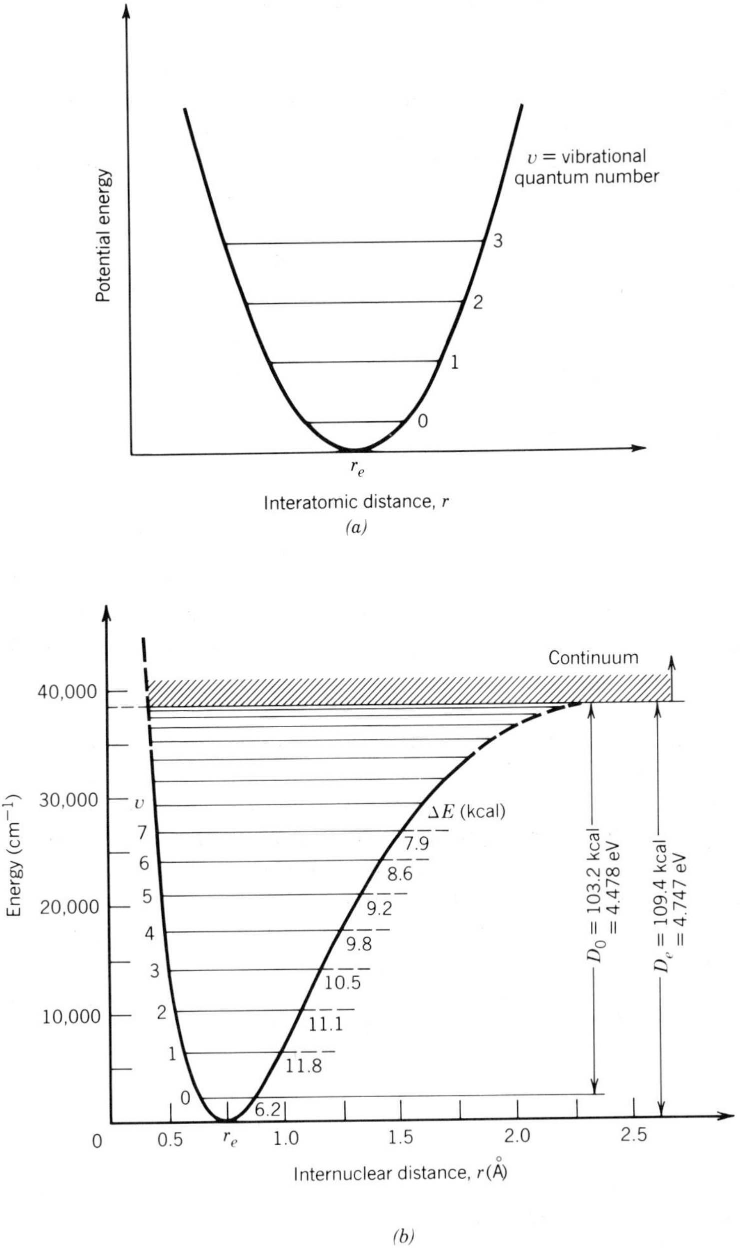

FIGURE 2.2. Potential energy of a vibrating diatomic molecule as a function of internuclear distance for (*a*) an ideal harmonic oscillator and (*b*) an anharmonic oscillator described by the Morse function, in this case the H_2 molecule (from Calvert and Pitts, 1966).

closely and experience significant charge repulsion. Furthermore, as the atoms move apart to large values of r, the bond stretches until it ultimately breaks and dissociation occurs. This is illustrated for the hydrogen molecule in Fig. 2.2*b*.

In order to obtain the allowed energy levels, E_v, for a real diatomic molecule, known as an *anharmonic* oscillator, one substitutes the potential energy function describing the curve in Fig. 2.2*b* into the Schrödinger equation; the allowed energy levels are

$$E_v = h\nu_{\text{vib}}(v + \tfrac{1}{2}) - h\nu_{\text{vib}}x_e\,(v + \tfrac{1}{2})^2 + h\nu_{\text{vib}}y_e(v + \tfrac{1}{2})^3 + \cdots \qquad \text{(L)}$$

Once again v is the vibrational quantum number with allowed values of 0, 1, 2, . . . , and x_e and y_e are anharmonicity constants characteristic of the molecule.

Equation (L) is often expressed in wavenumbers, ω_{vib}; the allowed energy states, $\bar{E}_v$, in units of wavenumbers (cm^{-1}) become

$$\bar{E}_v = \omega_e\,(v + \tfrac{1}{2}) - \omega_e x_e(v + \tfrac{1}{2})^2 + \omega_e y_e(v + \tfrac{1}{2})^3 + \cdots \qquad \text{(M)}$$

Note that throughout the book we use a bar over a parameter (e.g., $\bar{E}$) if it is expressed in units of wavenumbers. Values for ω_e, x_e, and y_e for a number of diatomic molecules are in Herzberg's classic *Molecular Spectra and Molecular Structure I. Spectra of Diatomic Molecules* (1950) and in Huber and Herzberg (1979).

An important consequence of using the potential energy for a real molecule in the Schrödinger equation is that the vibrational energy levels become more closely spaced with increasing quantum number v (Fig. 2.2*b* versus 2.2*a*).

When exposed to electromagnetic radiation of the appropriate energy, a molecule can interact with the radiation and absorb it, exciting the molecule into the next higher vibrational energy level. For the ideal harmonic oscillator, the selection rules are $\Delta v = \pm 1$; that is, the vibrational energy can only change by one quantum at a time. However, for anharmonic oscillators, weaker overtone transitions due to $\Delta v = \pm 2, \pm 3$, and so on may also be observed because of their non-ideal character. Because the vibrational energy level spacing is sufficiently large (typically of the order of 10^3 cm^{-1}) most molecules at room temperature are in their lowest vibrational energy level and light absorption normally occurs from $v = 0$.

For a purely vibrational transition, the selection rule for absorption of light requires that there be a changing dipole moment during the vibration. This oscillating dipole moment produces an electric field that can interact with the oscillating electric and magnetic fields of the electromagnetic radiation. Thus heteronuclear diatomic molecules such as NO, HCl, and CO absorb infrared radiation and undergo vibrational transitions, whereas homonuclear diatomic molecules such as O_2 and H_2, whose dipole moments remain constant during vibration, do not.

Since at room temperature most molecules are in $v'' = 0$, and $\Delta v = \pm 1$ is by far the strongest transition, the molecules go from $v'' = 0$ to $v' = 1$. (We follow the Herzberg convention that the quantum number of the upper state is designated by a prime, and the lower state by a double prime, e.g., in this example, $v'' = 0 \rightarrow v' = 1$). As a result, a single vibrational absorption band (with associated rotational structure) is normally observed in the infrared, the region corresponding to the energy level differences given by Eq. (M).

2-B-1b. *Rotational Transitions*

If a molecule has a permanent dipole moment, its rotation in space produces an oscillating electric field; this can interact with electromagnetic radiation, resulting in light absorption.

In the idealized case for rotation of a diatomic molecule, one assumes the molecule is analogous to a dumbbell with the atoms held at a fixed distance r from each other; that is, it is a *rigid rotor*. The simultaneous vibration of the molecule is ignored, as is the increase of internuclear distance at high rotational energies arising from the centrifugal force on the two atoms.

For this idealized case, the rotational energy levels (in cm^{-1}) are given by

$$\overline{E}_r = \overline{B}(J)(J + 1) \quad cm^{-1} \tag{N}$$

where $\overline{B}$, the rotational constant characteristic of the molecule, is given by

$$\overline{B} = \frac{h}{8\pi^2 Ic} \tag{O}$$

I is the moment of inertia of the molecule given by $I = \mu r^2$, where μ is the reduced mass defined by $\mu^{-1} = (M_A^{-1} + M_B^{-1})$ and M_A and M_B are the atomic masses, and r is the fixed, internuclear distance. J is the rotational quantum number; its allowed values are 0, 1, 2, ...

For a real rotating diatomic molecule, known as a non-rigid rotor, Eq. (N) becomes

$$\overline{E}_r = \overline{B}(J)(J + 1) - \overline{D}(J)^2(J + 1)^2 \tag{P}$$

The constant $\overline{D}$ is characteristic of the diatomic molecule and is much smaller than $\overline{B}$; generally $\overline{D} \leq 10^{-4}\,\overline{B}$. The second term in Eq. (P) generally becomes important at large values of J when centrifugal force increases the separation between atoms.

Because of the requirement of a permanent dipole moment, only heteronuclear molecules can absorb radiation and change their rotational energy. For the idealized case of a rigid rotor, the selection rule is $\Delta J = \pm 1$. For the energy levels given by Eq. (N), the energy level splitting between consecutive rotational energy levels is given by

$$\Delta\overline{E}_r = 2\overline{B}J' \tag{Q}$$

where J' is the quantum number of the *upper* rotational state involved in the transition. Thus the spacing between rotational energy levels increases with increasing rotational quantum number. Splittings are small compared to that between vibrational energy levels, typically of the order of 10 cm^{-1} in the lower levels; this corresponds to absorption in the microwave region. Indeed, these spacings are sufficiently small that the population of the rotational energy levels above $J = 0$ is significant at room temperature because the thermal energy available is sufficient to populate the higher rotational levels.

The Boltzmann expression can be used to calculate the relative populations of molecules in any rotational state J compared to the lowest rotational state $J = 0$ at temperature T (°K):

$$\frac{N_J}{N_0} = (2J + 1)\, e^{-E_J/kT} \tag{R}$$

In Eq. (R), k is the Boltzmann constant (1.381×10^{-23} J °K^{-1}) and E_J is the energy of the Jth rotational level given by Eq. (N) or Eq. (P).

The exponential energy factor in Eq. (R) gives decreasing populations with increasing J, but the degeneracy factor $(2J + 1)$ works in the opposite direction. As a result, rotational populations increase initially with increasing J, reach a peak, and subsequently decrease. For example, Fig. 2.3 shows the relative populations of the rotational energy levels N_J/N_0 for HCl at 300°K. In this case, the maximum population occurs around $J = 3$.

The combination of increasing spacing between energy levels as J increases and the significant population of molecules in higher rotational energy levels means that the absorption of microwave radiation results in a series of absorption lines, rather than a single line as seen in the pure vibrational infrared spectra of diatomic

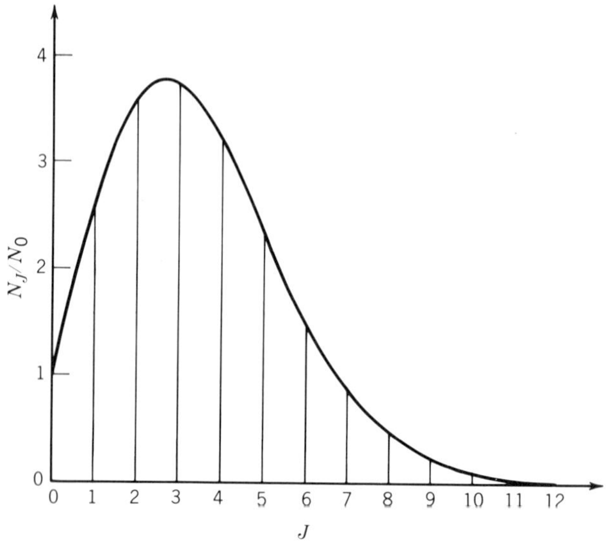

FIGURE 2.3. Relative populations, N_J/N_0, of rotational energy levels of HCl at 300°K (from Herzberg, 1950).

molecules. Figure 2.4 shows the pure rotational absorption spectrum of gaseous HCl. Since the energy level splitting is given by Eq. (Q) (for the ideal rigid rotor), one expects a series of lines separated by a spacing of $2\overline{B}$, as is seen in Fig. 2.4. From the spacing of these lines, the rotational constant $\overline{B}$, and hence the moment of inertia and the internuclear spacing, can be obtained.

Vibration–Rotation. Molecules, of course, vibrate and rotate simultaneously. The total change in energy is the sum of the changes in vibrational and rotational energy. The selection rules $\Delta v = \pm 1$ and $\Delta J = \pm 1$ apply for the ideal harmonic oscillator–rigid rotor. At room temperature, $v'' = 0 \rightarrow v' = 1$ is the only significant vibrational transition in absorption, but, as discussed above, a variety of rotational transitions can occur.

Figure 2.5 shows some of these possible transitions. Those with $\Delta J = +1$ are known as the R branch and occur at the high-energy side of the hypothetical transition $\Delta v = 1$, $\Delta J = 0$ (this is not allowed because of the selection rule, $\Delta J = \pm 1$). Those with $\Delta J = -1$ on the low-frequency side of the hypothetical transition form the P branch. Figure 2.6 shows the absorption spectrum of HCl at room temperature, with the rotational transitions responsible for each line. The relative intensities of the lines reflects the relative populations of the absorbing rotational levels; the peaks are doublets due to the separate absorptions of the two chlorine isotopes, that is, $H^{35}Cl$ and $H^{37}Cl$.

2-B-1c. *Electronic Transitions*

The electronic states of a diatomic molecule are described by several molecular quantum numbers, Λ, S, and Ω. Λ is the component of the total electronic orbital angular momentum L along the internuclear axis and can be determined from the electronic spectrum of the molecule (see Herzberg, 1950). Allowed values of Λ of 0, 1, 2, and 3 correspond to electronic states designated as Σ, Π, Δ, and Φ, respectively.

The spin quantum number S represents the net spin of the electrons. It has an integral value or zero for even numbers of electrons, and half-integral for odd numbers. The *multiplicity* of a molecular state is defined as $2S + 1$ and is written

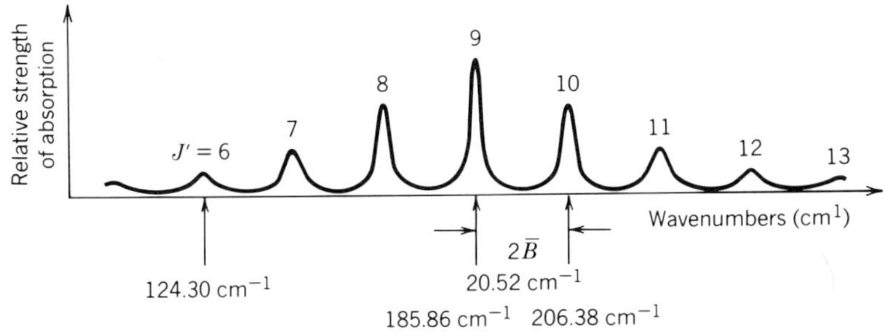

FIGURE 2.4. Portion of the pure rotational absorption spectrum of HCl. Rotational quantum numbers of final state, J', are given (from Herzberg, 1950).

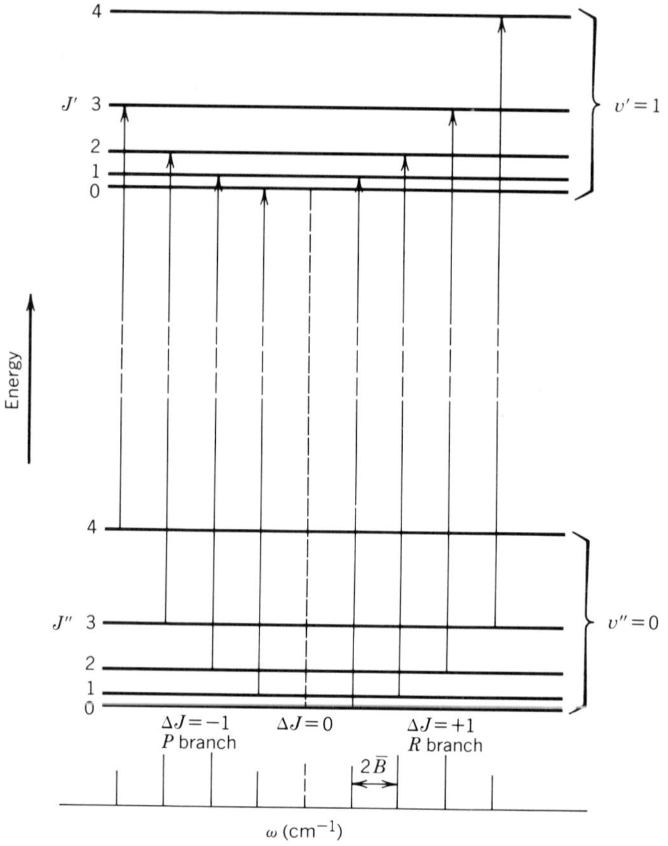

FIGURE 2.5. Schematic diagram of energy levels involved in HCl vibration–rotation transitions at room temperature (from Herzberg, 1950).

as a superscript to the left of the symbol corresponding to Λ. Values of S of 0, $\frac{1}{2}$, and 1, corresponding to multiplicities of 1, 2, and 3, are referred to as singlet, doublet, and triplet states, respectively.

For many molecules the quantum number Ω is defined and is given by the vector sum of Λ and Σ,

$$\Omega = |\Lambda + \Sigma|$$

where Σ is the vector component of S in the direction of the internuclear axis. Σ can have the values $+S, S - 1, \ldots, -S$, and can be positive, negative, or zero.

Two other symbols are used in designating electronic states according to their symmetry. For homonuclear diatomic molecules, states are designated "g" or "u" as a subscript to the right of the Λ symbol, depending on whether or not the wavefunction describing the molecular state changes sign when reflected through the center of symmetry of the molecule. If it does change sign, it is designated

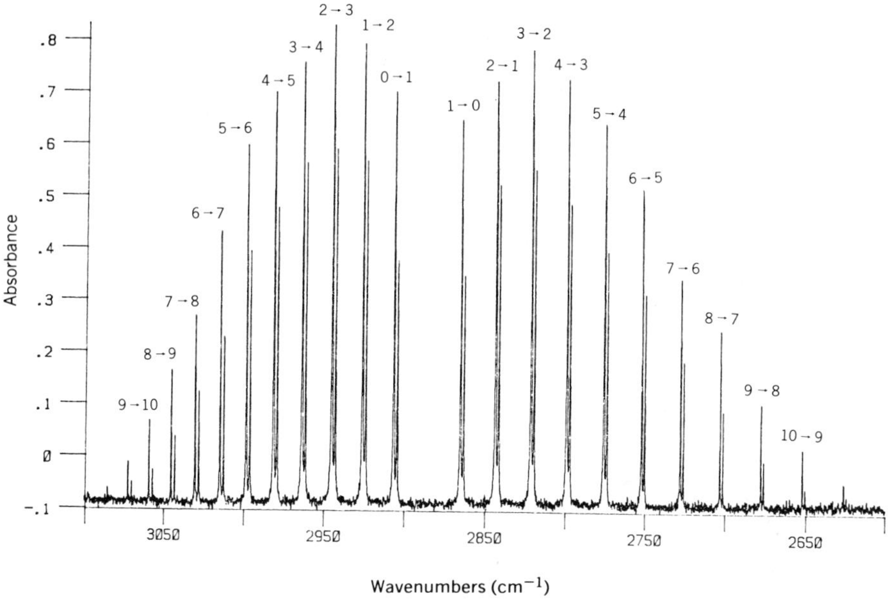

FIGURE 2.6. Vibration–rotation spectrum of 0.18 Torr HCl at room temperature. Resolution equals 0.25 cm^{-1}. Pathlength 19.2 m (from B. J. Finlayson-Pitts, unpublished data; spectrum taken by Mr. Stan Johnson).

"*u*" (for ungerade = uneven); if it does not, it is designated "*g*" (for gerade = even).

Finally, the symbols + and −, written as superscripts to the Λ symbol, refer to two types of sigma states, Σ^+ and Σ^-. If the wavefunction is unaltered by reflection through a plane passing through the two nuclei, the state is positive (+); if it changes sign, it is negative (−).

The selection rules for electronic transitions are not quite as clear-cut as in the case of vibration and rotation. In the case of molecules consisting of relatively light nuclei, which is the case for many molecules of tropospheric interest, the selection rules

$$\Delta\Lambda = 0, \pm 1 \quad \text{and} \quad \Delta S = 0$$

apply. Thus transitions between states of unlike multiplicity (e.g., singlet ↔ triplet) are "forbidden" (but in some cases may occur with a relatively small probability, most notably with the oxygen molecule which has a triplet ground state, Section 3-C-1). In terms of the symmetry of the wavefunctions, $u \leftrightarrow g$ transitions are allowed but $u \leftrightarrow u$ and $g \leftrightarrow g$ are forbidden. In addition, Σ^+ states cannot combine with Σ^- states, that is, $\Sigma^+ \leftrightarrow \Sigma^+$ and $\Sigma^- \leftrightarrow \Sigma^-$ transitions are allowed but $\Sigma^+ \leftrightarrow \Sigma^-$ are forbidden. Table 2.4 summarizes these selection rules for molecules with light nuclei.

Upon absorption of light of an appropriate wavelength, a diatomic molecule

TABLE 2.4 Allowed[a] Electronic Transitions of Diatomic Molecules Having Light Nuclei

Homonuclear Diatomic (Equal Nuclear Charge)	Heteronuclear Diatomic (Unequal Nuclear Charge)
$\Sigma_g^+ \leftrightarrow \Sigma_u^+$	$\Sigma^+ \leftrightarrow \Sigma^+$
$\Sigma_g^- \leftrightarrow \Sigma_u^-$	$\Sigma^- \leftrightarrow \Sigma^-$
$\Pi_g \leftrightarrow \Sigma_u^+, \Pi_u \leftrightarrow \Sigma_g^+$	$\Pi \leftrightarrow \Sigma^+$
$\Pi_g \leftrightarrow \Sigma_u^-, \Pi_u \leftrightarrow \Sigma_g^-$	$\Pi \leftrightarrow \Sigma^-$
$\Pi_g \leftrightarrow \Pi_u$	$\Pi \leftrightarrow \Pi$
$\Pi_g \leftrightarrow \Delta_u, \Pi_u \leftrightarrow \Delta_g$	$\Pi \leftrightarrow \Delta$
$\Delta_g \leftrightarrow \Delta_u$	$\Delta \leftrightarrow \Delta$

Source: Herzberg, 1950, p. 243.
[a]Presuming that the rule $\Delta S = 0$ is obeyed.

can undergo an *electronic* transition, along with simultaneous vibrational and rotational transitions. In this case, there is no restriction on Δv. That is, the selection rule $\Delta v = \pm 1$ valid for purely vibrational and vibrational–rotational transitions no longer applies; thus numerous vibrational transitions can occur. If the molecule is at room temperature, it will normally be in its lower state, $v'' = 0$; hence transitions corresponding to $v'' = 0$ to $v' = 0, 1, 2, 3, \ldots$ in the upper electronic state are usually observed.

The rotational selection rule is $\Delta J = 0, \pm 1$, except for the case of a transition involving $\Omega = 0$ for both the upper and lower states. Thus three sets of lines (known as the *P*, *Q*, and *R* branches) corresponding to $\Delta J = -1, 0,$ and $+1$, respectively, are observed for each band arising from a particular vibrational transition. Figure 2.7 illustrates these transitions schematically. However, if $\Omega = 0$ for both upper and lower states ($^1\Sigma \rightarrow {}^1\Sigma$ transition), the rotational selection rule is $\Delta J = \pm 1$, and the *Q* branch does not appear. For further details see Herzberg (1945, 1950, 1967).

As a general rule, the most probable, that is, most intense, vibrational transitions within a given electronic transition will be those in which the vibrational probabilities are maximum in both the initial and final states. An important restriction is that only *vertical* transitions are allowed. This is a consequence of the Franck–Condon principle which states that the time for an electronic transition to occur (typically 10^{-15} s) is so short relative to the time it takes for one vibration ($\sim 10^{-13}$ s) that the internuclear distance remains essentially constant during the electronic transition.

Figure 2.8 shows the Morse potential energy curves for two hypothetical electronic states of a diatomic molecule, the vibrational energy levels for each, and the shape of the vibrational wavefunctions (Ψ) within each vibrational energy level. At room temperature, most molecules will originate in $v'' = 0$; the vertical line is at the midpoint of $v'' = 0$ since the probability of finding the molecule at $r = r_e$ is a maximum here.

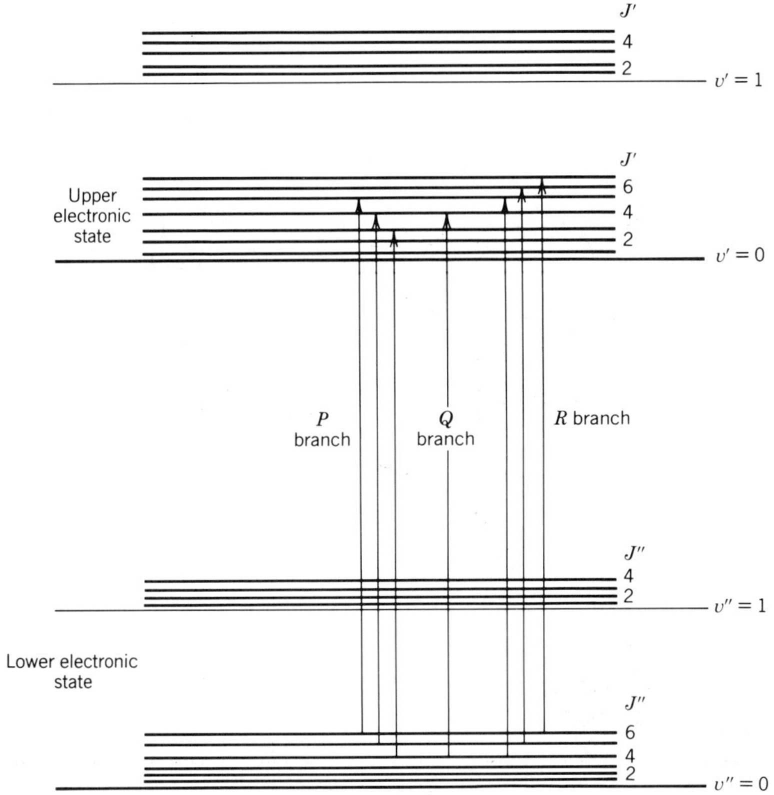

FIGURE 2.7. Schematic of some possible rotational and vibrational transitions involved during an electronic transition of a diatomic molecule from the ground electronic state.

The probability of a particular transition from $v'' = 0$ to an upper vibrational level v' is determined largely by the product of the wavefunctions for the two states, $\Psi_{v'}\Psi_{v''}$. A qualitative examination of the wavefunctions in the upper state, $\Psi_{v'}$, in Fig. 2.8 shows that $\Psi_{v'}$, and hence the product $\Psi_{v'}\Psi_{v''}$, is a maximum around $v' \simeq 4$; thus the vibrational transition corresponding to $v'' = 0 \rightarrow v' = 4$ is expected to be the most intense. On the other hand, the wavefunction at $v' = 0$ is very small; hence the $v'' = 0$ to $v' = 0$ transition should be weak. The right side of Fig. 2.8 shows the corresponding intensities expected for the various absorption lines in this electronic transition.

The overlap of vibrational wavefunctions in the upper and lower states clearly depends on the relative positions of the two states involved in the transition. For example, in the Fig. 2.9*a*, the equilibrium internuclear distances are about equal and the maximum intensity is expected in the $v'' = 0 \rightarrow v' = 0$ transition. This is indeed the case for the so-called atmospheric bands of O_2 near 762 nm and 1.27μ; they arise from the $^3\Sigma_g^-$ (ground state) $\rightarrow$ $^1\Sigma_g^+$ and the $^3\Sigma_g^- \rightarrow$ $^1\Delta_g$ transitions (see Section 3-C-1). Interestingly, although theoretically "forbidden" ac-

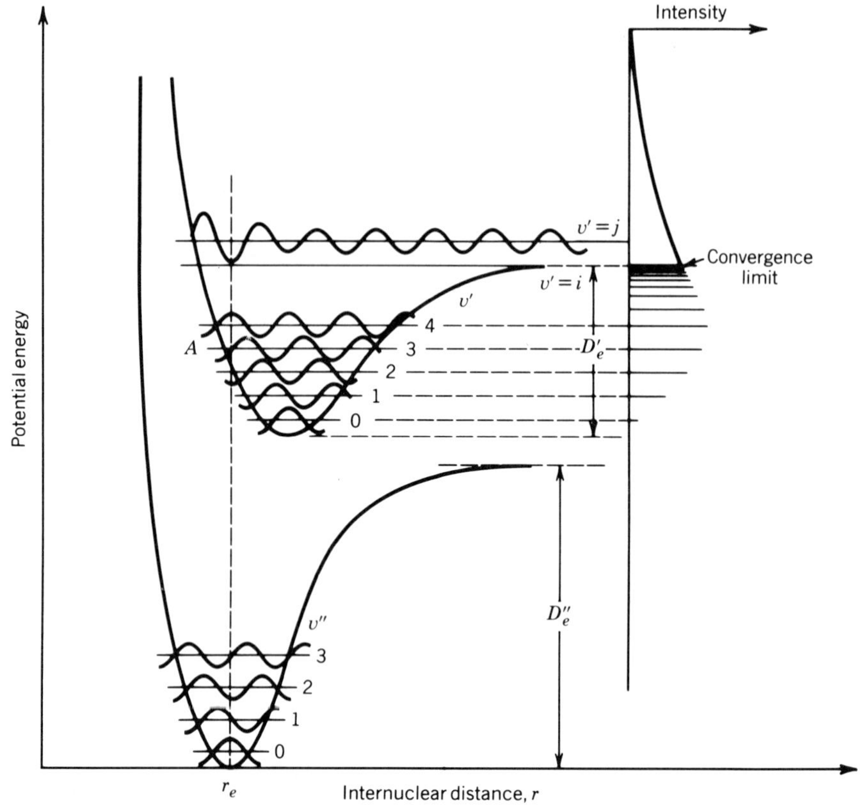

FIGURE 2.8. Potential energy curves for the ground state and an electronically excited state of a hypothetical diatomic molecule. Right-hand side shows relative intensities expected for absorption bands (from Calvert and Pitts, 1966).

cording to several of the selection rules (Table 2.4), they occur weakly due to collisional perturbations.

In Fig. 2.9*b*, the two curves are displaced from each other, but not as much as in Fig. 2.8; in this case the $v'' = 0$ to $v' = 1$ or 2 transition is expected to be the strongest. This is the case for the *fourth positive* band system of CO in the 120–160 nm (vacuum UV) region (Fig. 2.10). These arise by excitation from $v'' = 0$ of the ground, $X^1\Sigma^+$, state to $v' = 0, 1, 2, 3, \ldots.$ of the excited $A^1\Pi$ state. Transitions to final states corresponding to $v' = 1$ and 2 are the most intense.

The potential energy curves of excited electronic states need not have potential energy minima, such as those shown in Figs. 2.8 and 2.9. Thus Fig. 2.11 shows two hypothetical cases of such *repulsive states* where no minima are present. Dissociation occurs immediately following light absorption, giving rise to a spectrum with a structureless continuum. Transition *a* represents the case where dissociation of the molecule AB produces the atoms A and B in their ground states, and transition *b* the situation where dissociation produces one of the atoms in an electron-

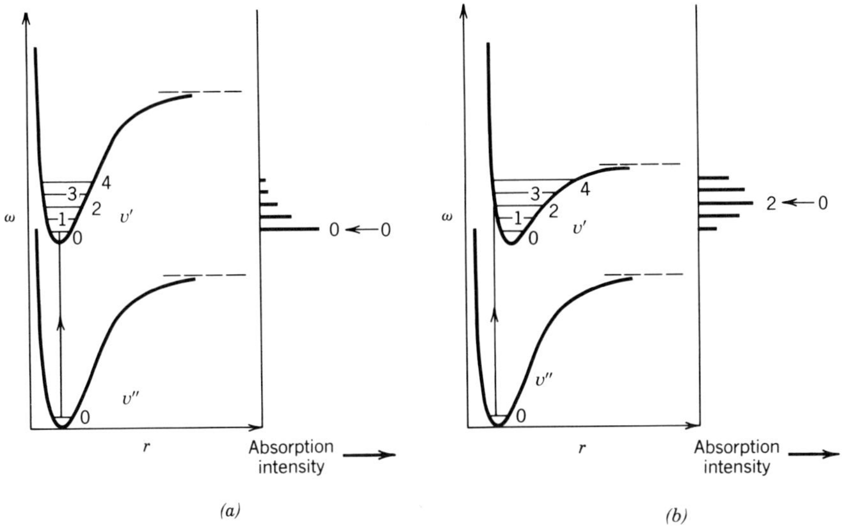

FIGURE 2.9. Typical potential energy curves with corresponding absorption lines expected for different separations of the equilibrium internuclear distances of the two electronic states involved (from Calvert and Pitts, 1966).

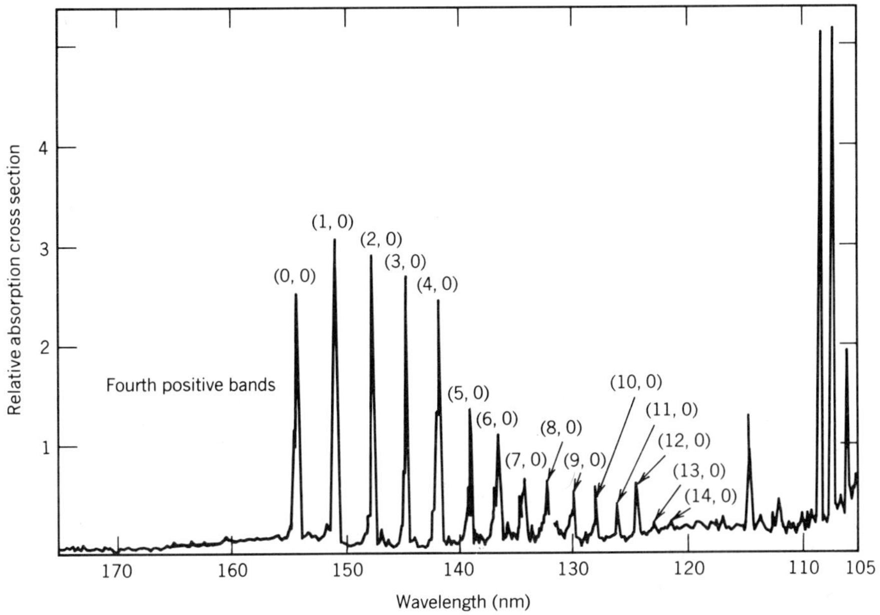

FIGURE 2.10. Relative absorption cross sections for CO at room temperature from ~105 to 170 nm. Numbers in parentheses for each line are the vibrational quantum numbers of the final vibrational state and the initial state, respectively; e.g., the (14, 0) line corresponds to a final vibrational level of $v' = 14$ and an initial level of $v'' = 0$ (from Myer and Samson, 1970).

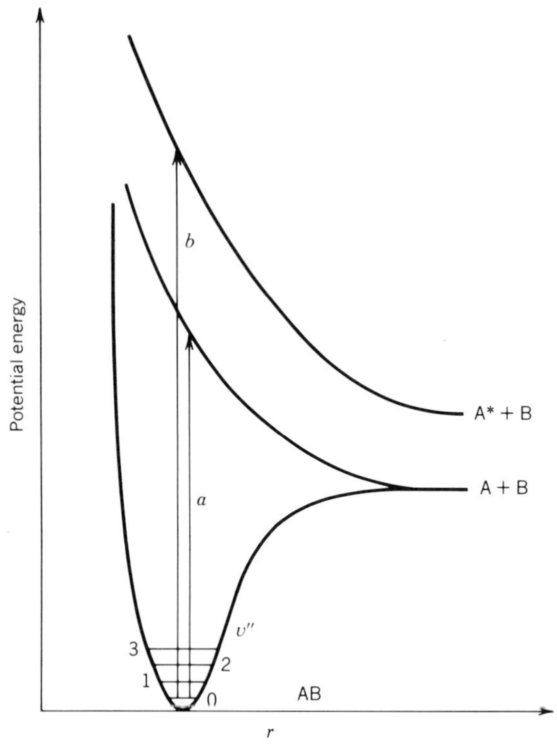

FIGURE 2.11. Potential energy curves for a hypothetical diatomic molecule showing electronic transitions to two repulsive excited states having no minima. A* is an electronically excited atom.

ically excited state, designated as A*. For example, the absorption of light by HI is continuous (Fig. 2.12). At $\lambda > 310$ nm, photodissociation produces the ground-state atoms $H(^2S) + I(^2P_{3/2})$; however, at $\lambda < 310$ nm, $H(^2S)$ and electronically excited $I(^2P_{1/2})$ are formed.

Some molecules may have a number of excited electronic states, some of which have potential minima as in Figs. 2.8 and 2.9, and some of which are wholly repulsive, as in Fig. 2.11. In this case, depending on the wavelength absorbed (i.e., the electronically excited state reached), the molecule may dissociate or undergo one of the photophysical processes described below.

A simplified, hypothetical example of this situation is shown in Fig. 2.13. If the molecule absorbs light corresponding to energies insufficient to produce vibrational energy levels above $v' = 2$ in the excited state E, a structured absorption (and emission) spectrum is observed. However, if the photon energy is greater than that required to produce A + B, that is, greater than the bond dissociation energy, then the molecule may be excited into either the repulsive state R, from which it immediately dissociates into ground-state atoms A + B, or alternatively into vibrational levels $v' > 2$ of excited state E. In the latter case, the excited molecule may undergo one of the photophysical processes discussed below (e.g.,

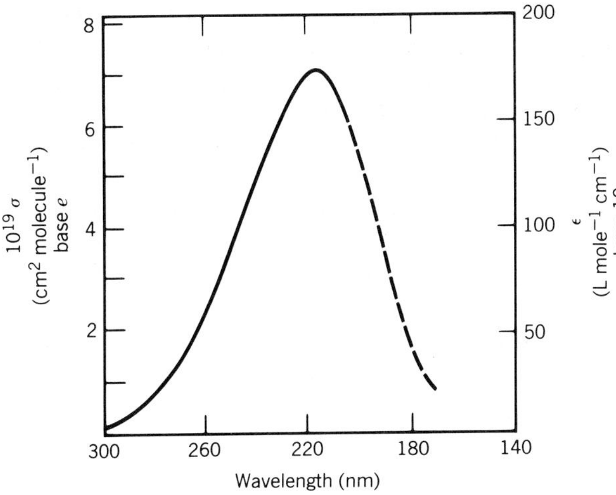

FIGURE 2.12. Absorption spectrum of HI (from Romand, 1949).

fluorescence, deactivation, etc.) or it may cross over from state E into the repulsive state R (point C in Fig. 2.13) and dissociate. This phenomenon is known as *predissociation.* In a case such as that in Fig. 2.13, the absorption spectrum would be expected to show well-defined rotational and vibrational structure up to a certain energy corresponding to the transition $v'' = 0 \rightarrow v' = 2$. For transitions to higher

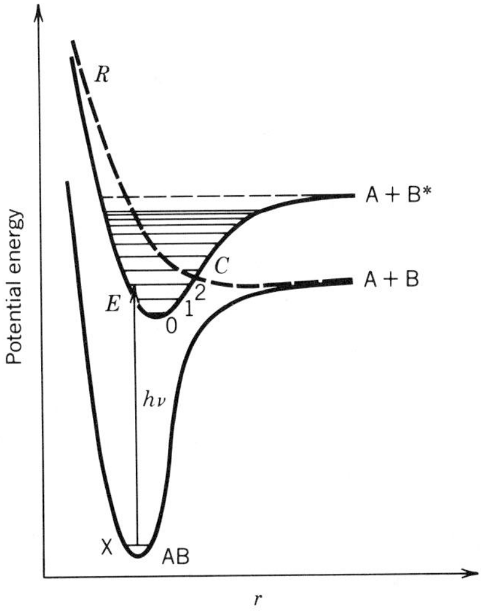

FIGURE 2.13. Potential energy curves for the ground state and two electronically excited states in a hypothetical diatomic molecule. Predissociation may occur when the molecule is excited into higher vibrational levels of the state E and crosses over to repulsive state R at the point C. (from Okabe, 1978).

vibrational levels, the rotational structure becomes blurred, and a *predissociation spectrum* is observed. I_2 is an example of a molecule with both a low-lying repulsive electronically excited state as well as bound excited states; excitation from the ground $^1\Sigma$ into the repulsive $^1\Pi$ state results in dissociation into the ground-state iodine atoms. On the other hand, excitation into the $B^3\Pi$ state below the dissociation limit gives electronically excited I_2 which returns to the ground state via fluorescence or crosses into the repulsive $^1\Pi$ state and predissociates (Okabe, 1978).

Finally, if the incident photon is sufficiently energetic and the appropriate selection rules are obeyed, the excited molecule may be produced in vibrational levels of state *E* that are sufficiently high that the molecule immediately dissociates into A + B*, where B* is an electronically excited state.

2-B-2. Polyatomic Molecules

The principles discussed above for diatomic molecules generally apply to polyatomic molecules, but their spectra are usually much more complex. For example, instead of considering rotation only about an axis perpendicular to the internuclear axis and passing through the center of mass, for non-linear molecules, one must think of rotation about three mutually perpendicular axes. Hence we have three rotational constants *A*, *B*, and *C* with respect to these three principal axes.

Furthermore, polyatomic molecules comprised of *n* atoms have 3*n*-6 vibrational degrees of freedom (or 3*n*-5 in the special case of a linear polyatomic molecule), instead of just one as in the case of a diatomic molecule. Some or all of these may absorb infrared radiation, leading to more than one infrared absorption band. In

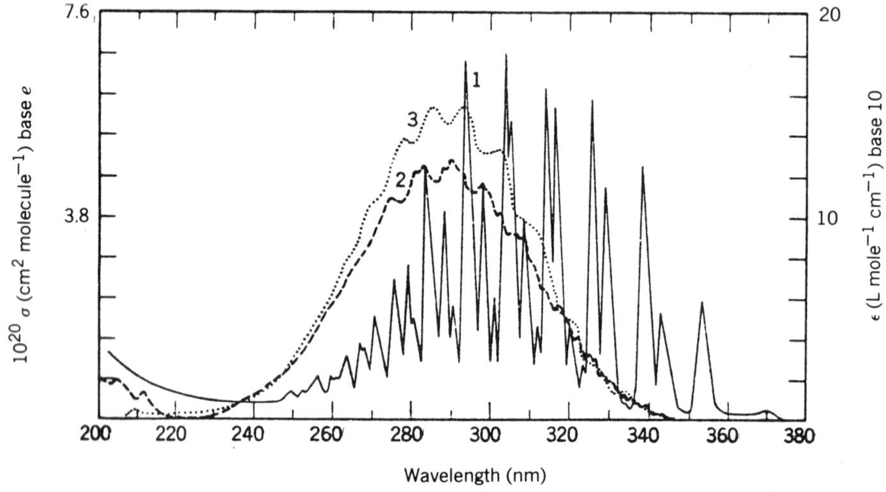

FIGURE 2.14. Absorption spectra for (1) HCHO at 75°C, (2) CH_3CHO at 25°C, and (3) C_2H_5CHO at 25°C. Note the loss of structure as the size of the molecule increases. (from Calvert and Pitts, 1966).

addition, overtone bands ($\Delta v > 1$) and combination bands (absorptions corresponding to the sum of two or more vibrations) are much more common.

An atmospherically relevant example of just how complex infrared spectra of polyatomic molecules can be is water vapor. As discussed in Section 5-C-2, a multitude of observed transitions occurs in the regions ~1300–2000 cm^{-1} and 3000–4000 cm^{-1}. This, combined with the relatively high, and often rapidly changing, concentrations, effectively rules out long-pathlength FT–IR spectroscopy of trace atmospheric species in ambient air in these wavelength regions. Similarly, absorption by CO_2 ''blanks out'' the region from ~2230 to ~2390 cm^{-1}.

In the electronic spectra of polyatomic molecules, the vibrational–rotational fine structure seen for gaseous diatomic molecules tends to disappear. The near ultraviolet absorption spectra of the simple aliphatic aldehydes illustrate the point. As seen in Fig. 2.14, the vibronic structure associated with this $n \rightarrow \pi^*$ transition, while pronounced with formaldehyde, becomes increasingly diffuse in its higher homologues.

2-C. POSSIBLE FATES OF AN ELECTRONICALLY EXCITED MOLECULE

2-C-1. Energy Level Diagrams and Primary Processes

Once a molecule is excited into an electronically excited state by absorption of a photon, it can undergo a number of different primary processes. *Photochemical* processes are those in which the excited species dissociates, isomerizes, rearranges, or reacts with another molecule. *Photophysical* processes include *radiative* transitions in which the excited molecule emits light in the form of fluorescence or phosphorescence and returns to the ground state, and intramolecular *non-radiative* transitions in which some or all of the energy of the absorbed photon is ultimately converted to heat.

These photophysical processes are often displayed in the form of the Jablonski-type energy level diagram shown in Fig. 2.15 (Calvert and Pitts, 1966). The common convention is that singlet states are labeled S_0, S_1, S_2, and so on, and the triplets are labeled T_1, T_2, T_3, and so on, in order of increasing energy. Vibrational and rotational states are shown as being approximately equally spaced for clarity of presentation. Radiative transitions, for example, fluorescence (F) and phosphorescence (P) are shown as solid lines, and non-radiative transitions as wavy lines. Vertical distances between the vibrational–rotational levels of the singlet ground state, S_0, and the two electronically excited states, the first excited singlet, S_1, and its triplet, T_1, correspond to their energy gaps.

Fluorescence is defined as the emission of light due to a transition between states of like multiplicity, that is, $S_1 \rightarrow S_0 + h\nu$. This is an allowed transition, and hence the lifetime of the upper state with respect to fluorescence is usually short, typically 10^{-6}–10^{-9} s. For example, the fluorescence lifetime of OH in the electronically excited $A^2\Sigma^+$ state is ~0.7 μs (McDermid and Laudenslager, 1982).

Phosphorescence is defined as the emission of light due to a transition between

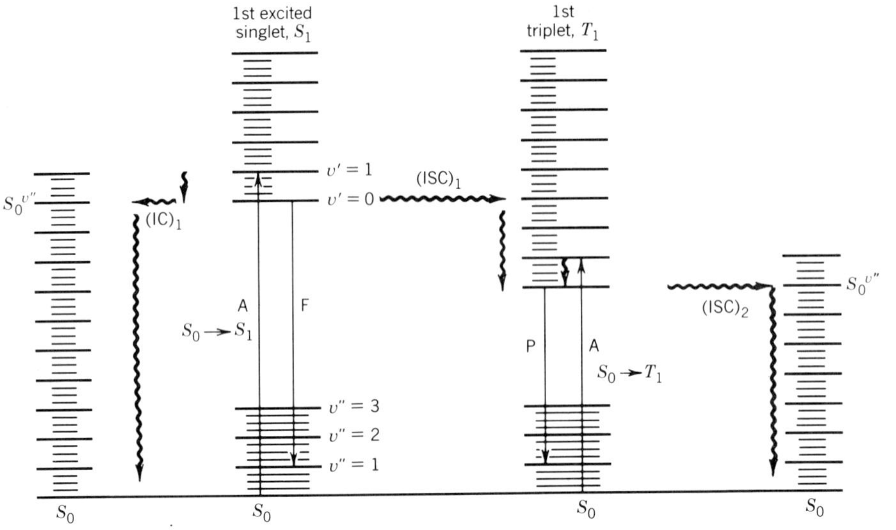

FIGURE 2.15 Jablonski diagram illustrating photophysical radiative and non-radiative transitions. S_0 = ground singlet state, S_1 = 1st excited singlet state, T_1 = 1st triplet state, A = absorption of light, F = fluorescence, P = phosphorescence, IC = internal conversion, ISC = intersystem crossing. Radiative transitions are shown by solid lines, non-radiative by wavy lines. Photochemical processes are not indicated.

states of different spin multiplicities. Because this is theoretically not an allowed transition for an ideal unperturbed molecule, phosphorescence lifetimes tend to be relatively long, typically 10^{-3}–10^{-2} s.

Intersystem crossing (ISC) is the intramolecular crossing from one state to another of different multiplicity without the emission of radiation. In Fig. 2.15 (ISC)$_1$ shows the transfer from the first excited singlet state S_1 to the first excited triplet state T_1. Since the process is horizontal, the total energy remains the same and the molecule initially is produced in upper vibrational and rotational levels of T_1, from which it is deactivated as shown by the vertical wavy line. Similarly, (ISC)$_2$ shows the intersystem crossing from T_1 to upper vibrational and rotational states of the ground state S_0 from which vibrational deactivation to $v'' = 0$ then occurs.

Internal conversion (IC) is the intramolecular crossing of an excited molecule from one state to another of the same multiplicity without the emission of radiation. As seen in Fig. 2.15, the horizontal wavy line (IC)$_1$ represents internal conversion from the lowest excited singlet state S_1 to high vibrational levels of the ground state S_0; this is generally followed by vibrational deactivation to $v'' = 0$.

Most important to atmospheric chemistry are primary *photochemical* processes. These, in contrast to the photophysical processes described above, produce new chemical species through photoreactions originating in electronically excited singlet or triplet states.

2-C-2. Primary and Overall Quantum Yields

The relative efficiencies of the various photophysical and photochemical primary processes are described in terms of quantum yields, ϕ.

In a molecule absorbing light, the primary quantum yield, ϕ, for the *i*th process, photophysical or photochemical, is given by Eq. (S):

$$\phi_i = \frac{\text{number of excited molecules proceeding by process } i}{\text{total number of photons absorbed}} \tag{S}$$

For chemical processes, two kinds of quantum yields are usually defined, a *primary* quantum yield, ϕ, and an *overall* quantum yield, Φ. For example, HCHO absorbs light from ~250 to ~370 nm (Fig. 2.14), decomposing either by a free radical process (1a) or by an intramolecular process producing stable molecules, (1b):

$$\begin{aligned} HCHO + h\nu &\xrightarrow[a]{} H + HCO \\ &\xrightarrow[b]{} H_2 + CO \end{aligned} \tag{1}$$

As discussed in Section 1-D-3, reaction (1a) is a significant source of HO_2 and hence ultimately OH in the troposphere because both H atoms and HCO radicals react with O_2 in air to form HO_2. The primary quantum yield of (1a) is

$$\phi_a = \frac{\text{number of H atoms (or HCO radicals) formed}}{\text{number of photons absorbed by HCHO}}$$

However, experimentally, one often measures the yields of stable products rather than of atoms and free radicals. The *overall* quantum yield of a stable product A, symbolized by Φ_A, is defined by

$$\Phi_A = \frac{\text{number of molecules of product A formed}}{\text{number of photons absorbed by reactant}} \tag{T}$$

For example, the overall quantum yields of H_2 and CO in the formaldehyde photolysis are defined by

$$\Phi_{H_2} = \frac{\text{number of molecules of } H_2 \text{ formed}}{\text{number of photons absorbed by HCHO}}$$

and

$$\Phi_{CO} = \frac{\text{number of molecules of CO formed}}{\text{number of photons absorbed by HCHO}}$$

It is important to recognize that CO and H_2 production in photolyses of pure H_2CO vapor arises not only directly from the primary molecular detachment (1b) but also from the subsequent *secondary* reactions of the reactive species H and HCO formed in (1a), reactions (2)–(4).

$$H + HCO \rightarrow H_2 + CO \tag{2}$$

$$2H \xrightarrow{M} H_2 \tag{3}$$

$$2HCO \underset{a}{\rightarrow} 2CO + H_2$$
$$\underset{b}{\rightarrow} CO + HCHO \tag{4}$$

The magnitude of *primary* and *overall* product quantum yields gives valuable indication of the relative importance of photophysical and photochemical processes and, in some cases, of the mechanism of a photochemical reaction. For example, small primary photochemical quantum yields ($\phi << 1$) indicate that photophysical processes must be important. On the other hand, overall product quantum yields greater than one ($\Phi > 1$) usually indicate that a chain reaction is occurring. For example, in the photolysis of mixtures of Cl_2 and H_2, Φ_{HCl} can be as high as 10^6. This is consistent with the following chain reaction:

$$Cl_2 + h\nu \rightarrow 2\ Cl \tag{5}$$

$$Cl + H_2 \rightarrow HCl + H \tag{6}$$

$$H + Cl_2 \rightarrow HCl + Cl \tag{7}$$

The chain consisting of reactions (6) and (7), initiated by reaction (5), the loss of Cl_2, will continue until one of the reactive species H or Cl is removed, for example, by reaction with another H or Cl, or at the walls of the reaction vessel. A value of Φ_{HCl} of $\sim 10^6$ suggests that reactions (6) and (7) occur on the order of a million times before such chain termination occurs.

Quantum yields for photophysical processes are defined in an analogous manner. For example, the primary quantum yield for fluorescence, ϕ_f, of an excited molecule is defined as

$$\phi_f = \frac{\text{number of excited molecules that fluoresce}}{\text{total number of photons absorbed}}$$

When all photophysical and photochemical processes are taken into account, the sum of their *primary* quantum yields must be unity, that is, $\Sigma\phi_i = 1.00$. This is one way of expressing the Stark–Einstein law or the *second law of photochemistry*, which states that the absorption of light by a molecule is a one-quantum process for low to moderate light intensities, for example, 10^{13}–10^{15} quanta s^{-1}. It is based on the fact that the probability of an electronically excited molecule absorbing a second quantum of light during its short lifetime of $\lesssim 10^{-8}$ s is very small under these conditions. Of course, with the very high quantum fluxes produced by lasers, multiphoton excitation is a common phenomenon of great importance in some areas of spectroscopy and photochemistry; in such systems the second law does not apply.

In short, under conditions of illumination relevant to the troposphere,

$$\Sigma(\phi_f + \phi_p + \phi_{deact} + \cdots + \phi_a + \phi_b + \cdots) = 1.00$$

where ϕ_f, ϕ_p, and ϕ_{deact} are the primary quantum yields for the photophysical processes of fluorescence, phosphorescence, and collisional deactivation, respectively, and ϕ_a, ϕ_b, and so on are the primary quantum yields for the various possible photochemically reactive decomposition paths of the excited molecule. Thus the sum of the quantum yields of primary photochemical processes determined experimentally cannot exceed 1 and may be substantially less if primary photophysical processes are important.

2-C-3. Intermolecular Non-radiative Processes: Collisional Deactivation and Energy Transfer

An electronically excited molecule in the gas phase at very low pressure suffers relatively few collisions per second with other molecules. Thus intramolecular primary photophysical and primary photochemical processes as discussed above predominate. However, at one atmosphere pressure, or in the liquid state, the excited molecule can undergo many collisions with ground-state molecules; this can lead to collisional deactivation of the excited species by several paths.

For example, an electronically excited molecule, A*, in the $S_1^{v'=i}$ state could undergo a series of collisions, be vibrationally (and rotationally) deactivated, and fall into the $S_1^{v'=0}$ state. The energy lost by A* is carried off as translational energy of the ground-state collision partner, B. From here A ($S_1^{v'=0}$) can undergo fluorescence, intersystem crossing, internal conversion, or photochemistry.

Alternatively, *energy transfer* from A* to the collision partner can occur in which the excitation energy appears as excess vibrational, rotational, and/or electronic energy of molecule B.

Collisional deactivation and energy transfer play important roles in tropospheric chemistry. For example, SO_2 absorbs light in the region between ~240 and 330 nm, forming the electronically excited states $^1A_2$ and $^1B_1$; a weaker absorption due to the spin forbidden transition to the $^3B_1$ state between 340 and 400 nm also occurs. The excited $^1A_2$ and $^1B_1$ states collide with O_2 and N_2 in air and are collisionally deactivated to the ground ($^1A_1$) state, to the relatively long-lived $^3B_1$ state, and possibly to the $^3A_2$ and $^3B_2$ states. The $^3B_1$ state is deactivated by O_2 (as well as by N_2 and H_2O) to the ground state, with part of this process occurring via triplet–triplet energy transfer. This produces two singlet, electronically excited states of O_2 which have many interesting spectroscopic, chemical, and biological properties (see Section 7-E-5):

$$SO_2(^3B_1) + O_2(^3\Sigma_g^-) \rightarrow SO_2(^1A_1) + O_2\ (^1\Sigma_g^+,\ ^1\Delta_g) \qquad (8)$$

Note that in the transfer of electronic energy between an excited atom or molecule and a second atom or molecule, the Wigner spin conservation rule generally applies. This states that the overall spin angular momentum of the system should not change during the energy transfer (see Herzberg for details).

Similarly, collisional deactivation is an important factor in trying to detect and measure various gaseous species in the troposphere using the technique of induced fluorescence. For example, as discussed in Section 5-B-2, induced fluorescence is one of the major techniques being developed in an attempt to determine the concentration of OH free radicals in the troposphere. The OH is excited by laser radiation to the $A^2\Sigma^+$ state, from which it fluoresces as it returns to the ground state. However, collisional deactivation of excited OH by O_2 and N_2 is significant at 1 atm pressure; this reduces the emitted light intensity and hence the detection sensitivity. One approach to increasing the sensitivity which is being developed is to lower the pressure of the air sample prior to excitation in order to minimize collisional deactivation (Hard et al., 1984).

2-D. TYPES OF PRIMARY PHOTOCHEMICAL PROCESSES

A variety of different types of primary photochemical processes can occur upon absorption of a light quantum, but by far the most important for gas phase, tropospheric chemistry is photodissociation into reactive fragments. In this section, we briefly illustrate several of these primary photoreactions, with emphasis on vapor phase photodissociation processes of relatively simple polyatomic molecules of atmospheric interest; for solution phase organic and inorganic photochemistry, Turro (1978) and Adamson and Fleischauer (1975) respectively should be consulted.

2-D-1. Photodissociation

The primary act of photodissociation can produce atoms and free radicals which, either directly or via secondary thermal reactions, are the source of free radicals such as OH, HO_2, and RO_2. As we have stated earlier, such species are central to the chemistry of both the clean and the polluted troposphere, as well as the stratosphere. These dissociation processes are discussed in detail in Chapter 3 and elsewhere throughout the book; we confine our brief discussion here to several important examples.

An example of a process producing fragments in their *ground electronic* states is the photodissociation of NO_2 in the actinic region at wavelengths below ~430 nm:

$$NO_2(\tilde{X}\,^2A_1) + h\nu\ (290 < \lambda < 430\ \text{nm}) \rightarrow NO(X^2\Pi) + O(^3P) \qquad (9)$$

Ground state NO and oxygen atoms are produced, and the process is spin-allowed. As discussed in Chapter 3, Section 3-C-3, the primary quantum yield is ~ 1 at 295 nm and decreases slightly to ~ 398 nm where it begins to fall off rapidly. Reaction (9) is quickly followed in one atmosphere of air by

$$O(^3P) + O_2 \xrightarrow{M} O_3 \tag{10}$$

This two-reaction sequence is the only known source of anthropogenically produced O_3 of tropospheric significance.

An example of photodissociation in the troposphere to produce *electronically excited* fragments is the spin-allowed dissociation of O_3 at wavelengths below ~ 320 nm:

$$O_3(^1B_2) + h\nu\ (\lambda < 320 \text{ nm}) \rightarrow O(^1D) + O_2\ (^1\Delta_g) \tag{11}$$

Both the oxygen atom and the molecular O_2 are electronically excited. In the natural and polluted troposphere, $O(^1D)$ atoms react rapidly with water vapor producing hydroxyl radicals (see Sections 3-D and 3-C-2b):

$$O(^1D) + H_2O \rightarrow 2OH \tag{12}$$

Examples of other molecules whose photodissociation to produce free radicals is important in the troposphere include HONO, HCHO, and higher aldehydes. We note here, however, that photodissociative processes following absorption of actinic UV light also include intramolecular rearrangements yielding stable smaller molecules, for example, primary processes (1b) for formaldehyde and (13b) and (13c) for *n*-butyraldehyde:

$$CH_3CH_2CH_2C\begin{matrix} /\!\!/^{O} \\ \diagdown_{H} \end{matrix} + h\nu\ (290 \lesssim \lambda \lesssim 340 \text{ nm}) \rightarrow C_3H_7 + CHO \tag{13a}$$

$$\rightarrow C_3H_8 + CO \tag{13b}$$

$$\rightarrow C_2H_4 + CH_3CHO \tag{13c}$$

While of interest in basic photochemistry, such gas phase dissociative processes as (13b) and (13c) are generally not significant in atmospheric chemistry per se.

2-D-2. Intramolecular Rearrangements

In some cases, the absorption of a light quantum causes an intramolecular rearrangement, as is the case for the photolysis of *o*-nitrobenzaldehyde in the vapor, solution, or solid phase:

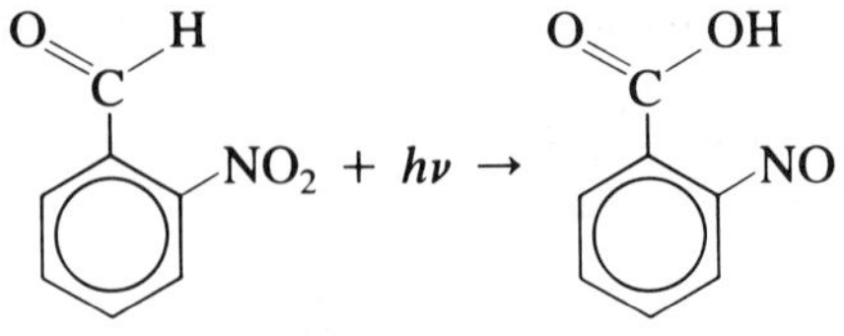

(14)

Indeed, this rearrangement has been used as an actinometer to measure light intensities in the actinic region (Pitts et al, 1965, 1968) since its quantum yield is virtually constant at ~0.5 over the wavelength region 300–410 nm. Used as a solution phase actinometer, the increase in the acidity of the solution has been followed with time as *o*-nitrosobenzoic acid is formed. Alternatively, when dispersed in thin polymeric films, the decrease with irradiation time of the infrared absorption peak due to NO_2 can be followed.

2-D-3. Photoisomerization

One of the photoinduced reaction paths available to gaseous *trans*-methyl propenyl ketone is isomerization to the *cis* form:

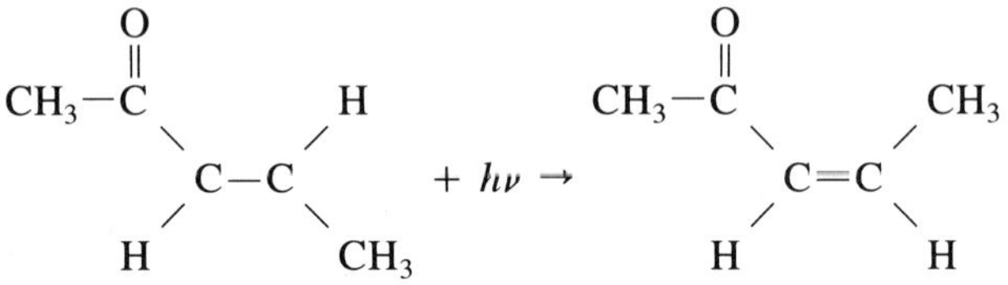

(15)

Similarly, gaseous alkyl and aryl azo compounds in the *trans* form photoisomerize readily to the *cis* isomer:

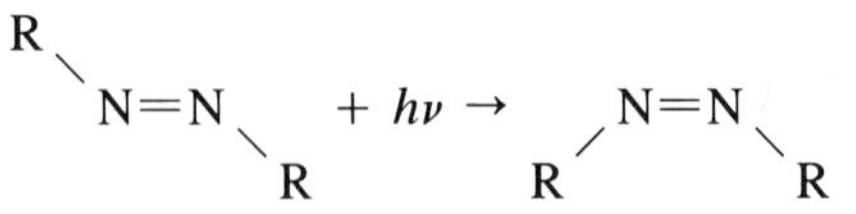

(16)

Gaseous *trans*-crotonaldehyde not only photoisomerizes to the *cis* isomer but also rearranges to give ethylketene and enol-crotonaldehyde:

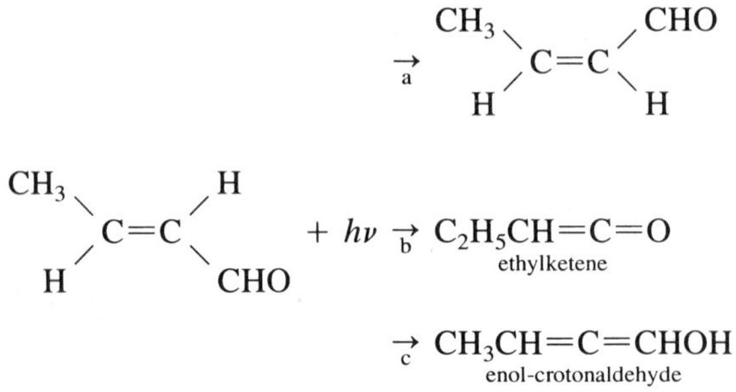

(17)

2-D-4. Photodimerization

Some molecules dimerize when exposed to light. Thus in the solid state, and in the absence of O_2, anthracene photodimerizes in the actinic UV

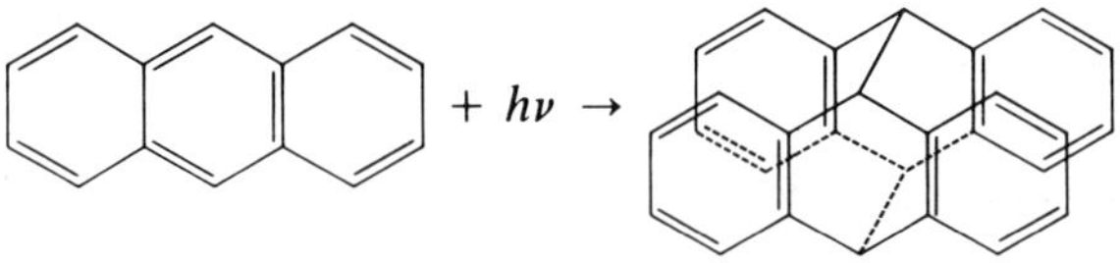

(18)

This occurs in competition with photooxidation when air is present, the latter leading to the endoperoxide. Such photooxidations of polycyclic aromatic hydrocarbons are discussed in Chapter 13.

2-D-5. Hydrogen Atom Abstraction

Carbonyl compounds may undergo an intramolecular hydrogen atom abstraction following excitation to an n, π^* state, especially if there is a hydrogen atom in the γ position. In this case, a six-membered ring can be formed in the transition state; this is energetically favorable over larger or smaller rings.

Under the appropriate conditions, some of which have environmental relevance, abstraction of a hydrogen atom by an electronically excited species may occur from a second molecule rather than intramolecularly. For example, nitrophenol has been observed over an agricultural field treated with parathion; this has been attributed to photolysis of the parathion to form nitrophenoxy radicals, followed by intermolecular hydrogen abstraction from a second, donor molecule (Crosby, 1983):

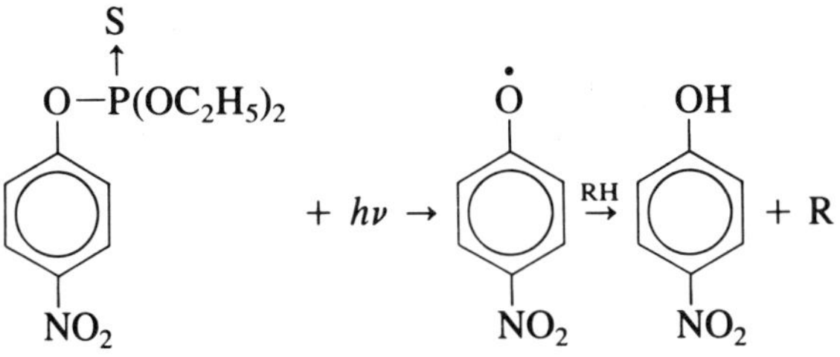

(19)

(As discussed in Section 10-D-8, nitrophenol may also be a product of OH attack on parathion).

A classic example of *intermolecular* hydrogen abstraction is the solution phase photoreduction of benzophenone in the presence of hydrogen-atom-donating solvents such as isopropyl alcohol:

$$\phi-\overset{\overset{\displaystyle O}{\|}}{C}-\phi + CH_3\overset{\overset{\displaystyle OH}{|}}{\underset{\underset{\displaystyle H}{|}}{C}}CH_3 + h\nu \rightarrow \phi-\overset{\overset{\displaystyle OH}{|}}{\underset{\cdot}{C}}-\phi + CH_3\overset{\overset{\displaystyle OH}{|}}{\underset{\cdot}{C}}CH_3 \quad (20)$$

Quantum yields for benzophenone removal can approach unity in solvents that are good hydrogen donors.

Although few data are available, hydrogen-atom abstractions may well be important in bimolecular photochemical processes of atmospheric interest occurring, for example, on surfaces or perhaps in aqueous droplets.

2-D-6. Photosensitized Reactions

An electronically excited molecule can transfer its energy to a second species which then undergoes a photochemical process even though it was not itself directly excited. The following sequence describes such a typical photosensitized process:

$$Sens(S_0) + h\nu \rightarrow Sens(S_1) \tag{21}$$

$$Sens(S_1) \xrightarrow[\text{crossing}]{\text{intersystem}} Sens(T_1) \tag{22}$$

$$Sens(T_1) + Acceptor(S_0) \xrightarrow[\text{transfer}]{\text{energy}} Sens(S_0) + Acceptor(T_1) \tag{23}$$

$$Acceptor(T_1) \longrightarrow Products \tag{24}$$

To function efficiently in this sequence, the sensitizer ideally will have high rates of intersystem crossing and of energy transfer to the acceptor. The latter requires that the triplet energy of the sensitizer be greater than that of the acceptor. In the optimum case, the sensitizer will not undergo photochemical processes which compete with formation of its triplet state and energy transfer.

Benzophenone, for example, has often been used in laboratory studies as a triplet photosensitizer. Thus when excited to its n, π^* singlet state, benzophenone efficiently undergoes intersystem crossing to its triplet state. This is of sufficiently high energy (286.6 kJ mole^{-1}) that it can transfer its electronic energy to a number of other organics (X) having triplet states lower in energy (Harris and Wamser, 1976):

$$\phi_2CO + h\nu \rightarrow \phi_2CO(S_1) \tag{25}$$

$$\phi_2CO(S_1) \xrightarrow{\text{ISC}} \phi_2CO(T_1) \tag{26}$$

$$\phi_2CO(T_1) + X(S_0) \rightarrow \phi_2CO(S_0) + X(T_1) \tag{27}$$

The acceptor molecule X in its excited triplet state (T_1) can then react or be deactivated. For example, *cis*-1,3-pentadiene isomerizes, giving a mixture of the *cis* and *trans* isomers:

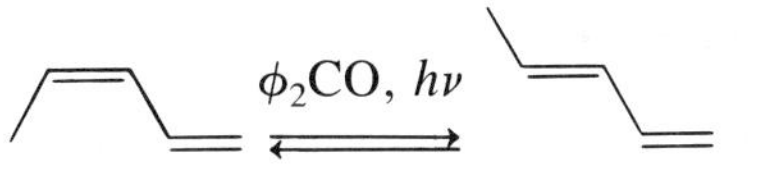

(28)

Similarly, in benzene solution, the pesticide aldrin rapidly converts to photo-aldrin in the presence of UV light and benzophenone:

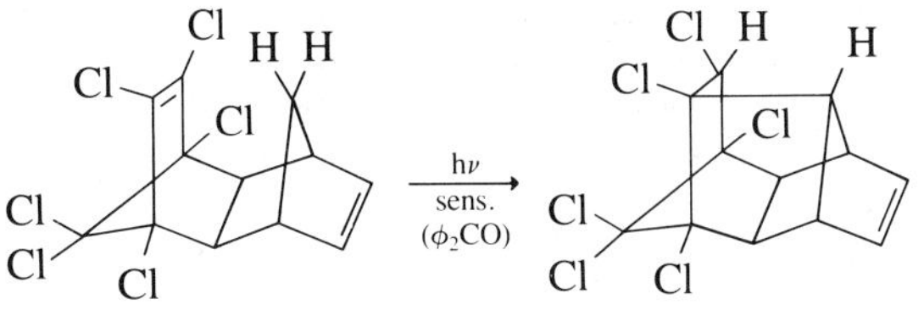

(29)

Aldrin Photo-aldrin

This does not occur in the absence of the sensitizer, benzophenone.

A classic example of a photosensitized reaction relevant to atmospheric chemistry is the formation of singlet molecular oxygen through the so-called Kautsky mechanism; this is analogous to reactions (21)–(24). However, in the case of $^1O_2$ production, the acceptor is ground-state oxygen, which is a triplet. Therefore, for the energy transfer step to be spin-allowed, the final state of the O_2 must be a singlet, assuming the sensitizer ground state is a singlet.

There are many sensitizers available in the atmosphere which can form $^1O_2$, including, as discussed earlier, gaseous SO_2. However, the sensitizer need not be gaseous but may exist in solutions or adsorbed on surfaces, for example, polycyclic aromatic hydrocarbons on combustion generated particles (Section 13-H-4). Soil surfaces exposed to O_2 also appear to form singlet oxygen, and the soil organics have been hypothesized to act as the sensitizer (Gohre and Miller, 1983).

Clearly, photosensitized reactions may play hitherto unrecognized but important roles in several aspects of atmospheric chemistry. More generally, the fields of surface photochemistry (including processes on the surfaces of plants and trees), and that of aqueous systems containing such species as dissolved SO_2, O_3, O_2, H_2O_2, NO_2, PAN and so on, both in the presence and absence of reactive organics, are becoming increasingly relevant and important to atmospheric chemistry (see Section 3-C-12); although very complex, they warrant major research efforts.

REFERENCES

Adamson, A. W. and P. D. Fleischauer, *Concepts of Inorganic Photochemistry*, Wiley, New York, 1975.

Calvert, J. G. and J. N. Pitts, Jr., *Photochemistry*, Wiley, New York, 1966.

Crosby, D. G., Atmospheric Reactions of Pesticides, in *Pesticide Chemistry: Human Welfare and the Environment*, J. Miyamoto and P. C. Kearney, Eds., Pergamon Press, New York, 1983, pp. 327–332.

Finlayson-Pitts, B. J., unpublished data.

Gohre, K. and G. C. Miller, "Singlet Oxygen Generation on Soil Surfaces," *J. Agric. Food Chem.*, **31,** 1104 (1983).

Hard, T. M., R. J. O'Brien, C. Y. Chan, and A. A. Mehrabzadeh, "Tropospheric Free Radical Determination by FAGE," *Environ. Sci. Technol.*, **18,** 768 (1984).

Harris, J. M. and C. C. Wamser, *Fundamentals of Organic Reaction Mechanisms*, Wiley, New York, 1976.

Herzberg, G., *Molecular Spectra and Molecular Structure. I. Spectra of Diatomic Molecules*, 1950; *II. Infrared and Raman Spectra of Polyatomic Molecules*, 1945; *III. Electronic Spectra and Electronic Structure of Polyatomic Molecules*, 1967, Van Nostrand, Princeton, NJ.

Huber, K. and G. Herzberg, *Molecular Spectra and Molecular Structure IV. Constants of Diatomic Molecules,* Van Nostrand, 1979.

Leighton, P. A., *Photochemistry of Air Pollution*, Academic Press, New York, 1961.

McDermid, I. S. and J. B. Laudenslager, "Radiative Lifetimes and Quenching Rate Coefficients for Directly Excited Rotational Levels of OH($A^2\Sigma^+$, $v' = 0$), *J. Chem. Phys.*, **76,** 1824 (1982).

Myer, J. A. and J. A. R. Samson, "Vacuum-Ultraviolet Absorption Cross Sections of CO, HCl, and ICN Between 1050 and 2100 Å," *J. Chem. Phys.*, **52,** 266 (1970).

Okabe, H., *Photochemistry of Small Molecules*, Wiley, New York, 1978.

Pitts, J. N., Jr., J. M. Vernon, and J. K. S. Wan, "A Rapid Actinometer for Photochemical Air Pollution Studies," *Int. J. Air Water Pollut.*, **9,** 595 (1965).

Pitts, J. N., Jr., G. W. Cowell, and D. R. Burley, "Film Actinometer for Measurement of Solar Ultraviolet Radiation Intensities in Urban Atmospheres," *Environ. Sci. Technol.*, **2,** 435 (1968).

Rabek, J. F., *Experimental Methods in Photochemistry and Photophysics*, Parts I and II, Wiley, New York, 1982.

Romand, J. "Absorption Ultraviolette Dans La Région de Schumann Étude de:ClH, BrH et IH Gazeux," *Ann. Phys. (Paris)*, **4,** 527 (1949).

Turro, N. J., *Modern Molecular Photochemistry*, Benjamin/Cummings, Menlo Park, CA, 1978.

Wayne, R. P., *Photochemistry*, Elsevier, New York, (1970).

3 Photochemistry in the Troposphere

3-A. SOLAR RADIATION

3-A-1. The Sun and Its Relationship to the Earth: Some Important Definitions for Atmospheric Chemistry

The sun can be considered a spherical light source of diameter 1.4×10^6 km located 1.5×10^8 km from the earth's surface. Incoming direct sunlight at the earth's surface is treated as a beam with an angle of collimation of $\sim 0.5°$ and thus is essentially parallel to $\pm 0.25°$.

The total intensity of sunlight outside the earth's atmosphere is characterized by the solar constant, defined as the total amount of light received per unit area normal to the direction of propagation of the light; the mean value is 1368 W m^{-2} (World Meteorological Organization, 1981).

Of more direct interest for atmospheric photochemistry is the solar flux per unit interval of wavelength. Table 3.1 gives one set of recently recommended values of the spectral irradiance outside the atmosphere as a function of wavelength (World Meteorological Organization, 1981). Figure 3.1 shows the solar flux as a function of wavelength outside the atmosphere and at sea level for a solar zenith angle of 0° (see below).

Outside the atmosphere, the solar flux approximates blackbody emission at 6000°K. However, light absorption or scattering by atmospheric constituents modifies the spectral distribution. The attenuation due to the presence of various naturally occurring atmospheric constituents is shown by the hatched areas in Figure 3.1.

Figure 3.2 shows the altitude corresponding to maximum light absorption by atomic and molecular oxygen and nitrogen, and by O_3, as a function of wavelength up to $\lambda = 300$ nm with the sun directly overhead. Table 3.2 summarizes the attenuation of solar radiation by the atmosphere in various wavelength regions. Because of the presence of these absorbing species in the upper atmosphere, only light of $\lambda \gtrsim 290$ nm is available for photochemical reactions in the troposphere. It is often expressed as the integrated radiation from all directions to a sphere and

TABLE 3.1. Solar Spectral Irradiance Outside the Atmosphere Integrated Over Intervals Varying from 2 to 5 nm Centered on the Given Wavelength

Wavelength Interval[a] (nm)	Solar Irradiance (photons cm^{-2} s^{-1})	Wavelength Interval[a] (nm)	Solar Irradiance (photons cm^{-2} s^{-1})
200.0–202.0	1.40×10^{12}	350	8.69
202.0–204.1	1.69	355	9.14
204.1–206.2	2.07	360	8.23
206.2–208.3	2.52	365	1.07×10^{15}
208.3–210.5	4.21	370	1.08×10^{15}
210.5–212.8	7.23	375	9.72×10^{14}
212.8–215.0	7.79	380	1.11×10^{15}
215.0–217.4	8.45	385	8.98×10^{14}
217.4–219.8	1.05×10^{13}	390	1.18×10^{15}
219.8–222.2	1.19	395	9.34×10^{14}
222.2–224.7	1.51	400	1.69×10^{15}
224.7–227.3	1.33	405	1.70
227.3–229.9	1.31	410	1.84
229.9–232.6	1.51	415	1.97
232.6–235.3	1.32	420	1.95
235.3–238.1	1.50	425	1.81
238.1–241.0	1.34	430	1.67
241.0–243.9	2.02	435	1.98
243.9–246.9	1.82	440	2.02
246.9–250.0	1.88	445	2.18
250.0–253.2	1.83	450	2.36
253.2–256.4	2.25	455	2.31
256.4–259.7	4.65	460	2.39
259.7–263.2	4.44	465	2.38
263.2–266.7	1.07×10^{14}	470	2.39
266.7–270.3	1.18	475	2.44
270.3–274.0	1.08	480	2.51
274.0–277.8	1.04	485	2.30
277.8–281.7	7.54×10^{13}	490	2.39
281.7–285.7	1.48×10^{14}	495	2.48
285.7–289.9	2.17	500	2.40
289.9–294.1	3.46	505	2.46
294.1–298.5	3.39	510	2.49
298.5–303.0	3.24	515	2.32
303.0–307.7	4.40	520	2.39
310	4.95×10^{14}	525	2.42
315	5.83	530	2.55
320	6.22	535	2.51
325	6.96	540	2.49
330	8.61	545	2.55
335	8.15	550	2.53
340	8.94	555	2.54
345	8.44	560	2.50

TABLE 3.1. (*Continued*)

Wavelength Interval[a] (nm)	Solar Irradiance (photons cm^{-2} s^{-1})	Wavelength Interval[a] (nm)	Solar Irradiance (photons cm^{-2} s^{-1})
565	2.57	710	2.51
570	2.58	715	2.48
575	2.67	720	2.45
580	2.67	725	2.48
585	2.70	730	2.45
590	2.62	735	2.44
595	2.69	740	2.39
600	2.63	745	2.40
605	2.68	750	2.41
610	2.66	755	2.40
615	2.59	760	2.38
620	2.69	765	2.34
625	2.61	770	2.32
630	2.62	775	2.30
635	2.62	780	2.33
640	2.63	785	2.34
645	2.60	790	2.29
650	2.55	795	2.29
655	2.48	800	2.27
660	2.57	805	2.27
665	2.61	810	2.20
670	2.61	815	2.22
675	2.62	820	2.18
680	2.62	825	2.20
685	2.57	830	2.14
690	2.52	835	2.14
695	2.60	840	2.13
700	2.58	845	2.09
705	2.52	850	2.05

Source: World Meteorological Organization, 1981.

[a] From 310 to 850 nm, the solar irradiance is for 5 nm intervals centered on the given wavelength. For example, the irradiance of 4.95×10^{14} photons cm^{-2} s^{-1} at 310 nm is the integrated irradiance from 307.5 to 312.5 nm.

is referred to as *actinic radiation*, although in the strictest sense, "actinic" means "capable of causing photochemical reactions."

The effect of light scattering and absorption by atmospheric constituents on the intensity and wavelength distribution of sunlight at the earth's surface depends on both the nature and concentration of the gases and particles and on the pathlength through which the light passes (Beer–Lambert law). The pathlength, that is, distance from the outer reaches of the atmosphere to an observer on the earth's sur-

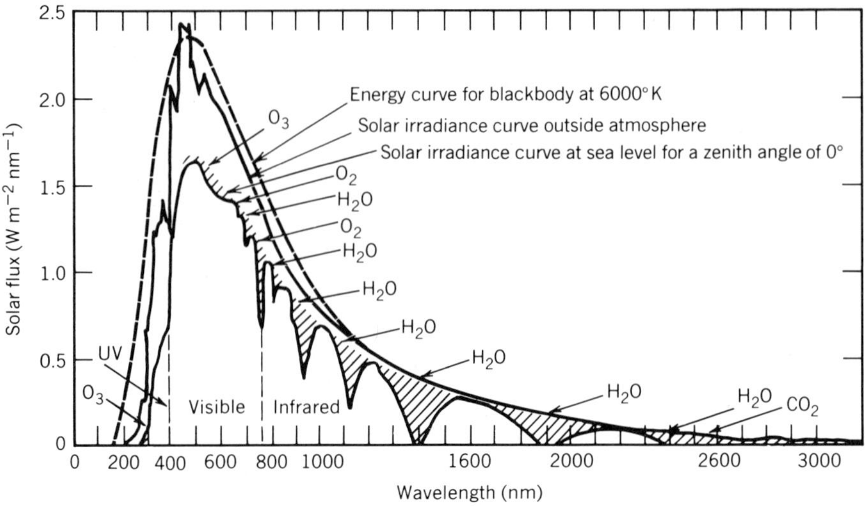

FIGURE 3.1. Solar flux outside the atmosphere and at sea level, respectively. The emission of a blackbody at 6000°K is also shown for comparison. The species responsible for light absorption in the various regions (O_3, H_2O, etc.) are also shown (from Howard et al., 1960).

face, is a function of the angle of the sun and hence time of day, latitude, and season. In addition, reflection of light from the earth's surface alters the light intensity at any given point in the atmosphere, as does the presence of clouds.

The angle of the sun relative to a fixed point on the surface of the earth is characterized by the solar zenith angle θ, defined, as shown in Fig. 3.3, as the angle between the direction of the sun and the vertical. Thus a zenith angle of zero

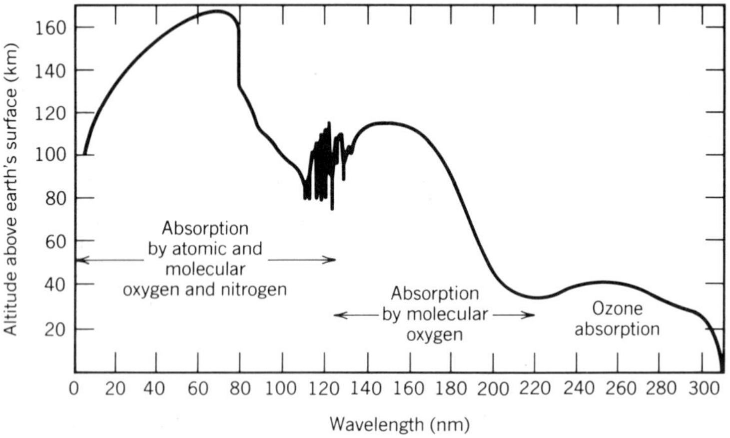

FIGURE 3.2. Approximate regions of maximum light absorption of solar radiation in the atmosphere by various atomic and molecular species as a function of altitude and wavelength with the sun overhead (from H. Friedman, in *Physics of the Upper Atmosphere*, J. A. Ratcliffe, Ed., Academic Press, New York, 1960).

corresponds to an overhead, noonday sun, and a zenith angle of $\sim 90°$ to sunrise and sunset. The greater the zenith angle, the longer is the pathlength through the atmosphere and hence the greater the reduction in solar intensity by absorption and scattering processes.

The pathlength L for direct solar radiation traveling through the earth's atmosphere, of height h, to a fixed point on the earth's surface can be estimated geometrically using Fig. 3.3. This flat earth approximation is accurate for zenith angles $\lesssim 60°$. One can approximate L using

$$\cos\theta \simeq \frac{h}{L} \tag{A}$$

or

$$L \simeq \frac{h}{\cos\theta} \simeq h \text{ secant } \theta \tag{B}$$

A common term used to express the pathlength traversed by solar radiation to reach the earth's surface is the air mass, m, defined as

$$m = \frac{\text{length of path of direct solar radiation through the atmosphere}}{\text{length of vertical path through the atmosphere}}$$

With reference to Fig. 3.3, for zenith angles less than 60°,

$$m = \frac{L}{h} = \text{secant } \theta \tag{C}$$

At larger angles, corrections for curvature of the atmosphere and refraction must be made to L and m.

Table 3.3 shows values of the air mass at various zenith angles θ, either estimated using m = secant θ or corrected for curvature of the atmosphere and for refraction; only for $\theta > 60°$ does this correction become significant.

3-A-2. Solar Spectral Distribution and Intensity in the Troposphere

3-A-2a. Derivation of the Actinic Flux $J(\lambda)$ at the Earth's Surface

When the radiation from the sun passes through the earth's atmosphere, it is modified both in intensity and in spectral distribution by absorption and scattering by gases as well as by particulate matter. In addition to these effects, the actual actinic flux to which a given volume of air is exposed is also affected by the zenith angle (i.e., time of day, latitude, and season), by the extent of surface reflections, and by the presence of clouds.

TABLE 3.2. Attenuation of Solar Radiation by the Atmosphere

Wavelength Regions (nm and μm)				
120–200 nm	200–290 nm	290–320 nm	320–350 nm	350–550 nm
Solar irradiation intensity approximates extra-atmospheric. Attenuation by scattering increases markedly toward shorter wavelengths.				
O_2 absorbs almost completely				
	190–210 nm Absorption by O_2 appreciable; absorption by O_3 appreciable	O_3 absorption not important		
	No radiation penetrates below ~11 km	O_3 absorption attenuates more than loss by scattering	O_3 absorption significantly attenuates radiation	Irradiation diminished mostly by scattering by permanent gases in atmosphere
			Highly variable dust, haze, H_2O, and smoke responsible for attenuation in regions 320–700 nm	
		Appreciable penetration through "clear" atmosphere to sea level	Penetration through "clear" atmosphere to sea level about 40%	Dust may rise to more than 4 km

Source: Howard et al., 1960.

To estimate the solar flux available for photochemistry in the troposphere then, one needs to know not only the flux outside the atmosphere, as given in Table 3.1, but also the extent of light absorption and scattering within the atmosphere. We discuss here the actinic flux $J(\lambda)$ at the earth's surface; the effects of elevation and of height above the surface are discussed in Sections 3-A-2d and 3-A-2e.

The reduction in solar intensity due to scattering and absorption can be estimated using a form of the Beer–Lambert law:

Wavelength Regions (nm and μm)					
550–900 nm	900 nm–2.5 μm	2.5–7 μm	7–20 μm	Altitude	Region of Atmosphere
Solar irradiation intensity approximates extra-atmospheric. Attenuation by scattering increases markedly toward shorter wavelengths.					
				Above 60 km	THERMO-SPHERE ↑ 85 km ↓
		Energy small	Energy very small	60–33 km	MESO-SPHERE ↓ 50 km ↑
H_2O responsible for major absorption; CO_2 absorbs slightly at 2 μm. Water vapor (or ice crystals) is found up to about 21 km			Strong O_3 absorption at 9.6 μm; strong CO_2 absorption at 12–17 μm	33–11 km	STRATO-SPHERE ↓ 11 km ↑
Energy transmitted with small loss down to 2 km Dust may rise to more than 4 km	Energy penetrates to sea level only through "windows" at ~1.2, 1.6 and 2.2 μ.	No significant penetration below 2 km except in "windows" at ~3.8 and 4.9 μm	Energy transmitted with moderate loss; many absorption bands due to atmospheric gases	2 km to sea level	TROPO-SPHERE

$$\frac{I}{I_0} = e^{-tm} \tag{D}$$

In Eq. (D), I_0 is the light intensity at a given wavelength incident at the top of the atmosphere and I is the intensity of the light transmitted to the earth's surface; m is the air mass as defined above.

Some confusion exists in the literature regarding the name and symbol used for the dimensionless term t in Eq. (D); for example, Leighton calls it an *attenuation coefficient*, symbolized σ, whereas Peterson and co-workers (Peterson, 1976; De-

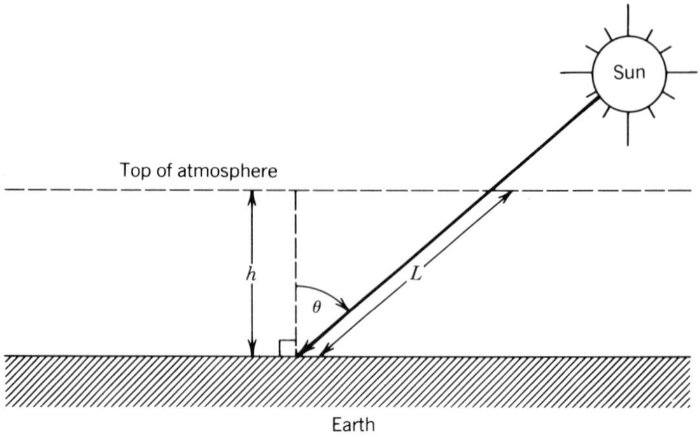

FIGURE 3.3. Definition of solar zenith angle θ at a point on the earth's surface.

merjian et al., 1980) refer to it as the *optical thickness*. We prefer *attenuation coefficient* but with the symbol t; the latter avoids confusing it with the symbol σ used for the absorption cross section of gas molecules. Thus throughout this book we will refer to the term t as the attenuation coefficient, where the natural logarithmic base e is understood to apply, as in Eq. (D).

For the sun directly overhead (i.e., zenith cycle $\theta = 0$) the air mass is unity ($m = 1.0$); the attenuation coefficient then reflects the minimum possible attenuation by the atmosphere. As θ increases until the sun is on the horizon (i.e., sunset or sunrise), m also increases (Table 3.3); thus the attenuation of the sunlight increases due to the increased pathlength in the atmosphere through which the light must travel to reach the earth's surface.

The attenuation coefficient, t, represents a combination of light scattering and absorption by gases and particles and is actually a sum of four terms,

$$t = t_{sg} + t_{ag} + t_{sp} + t_{ap} \tag{E}$$

where sg = light *s*cattering by *g*ases, ag = light *a*bsorption by *g*ases, sp = light *s*cattering by *p*articles, and ap = light *a*bsorption by *p*articles. Let us consider gases and particles separately.

Gases scatter light by molecular, or Rayleigh, scattering. The intensity, $I(\theta)$, of light of wavelength λ scattered at an angle θ to the direction of incident light of intensity I_0 at a distance R from the scattering species from polarized incident light is given by the following:

$$I(\theta) = \frac{I_0 \pi^4 d^6}{8R^2\lambda^4} \frac{(a^2 - 1)^2}{(a^2 + 2)^2} (1 + \cos^2 \theta) \tag{F}$$

a is the index of refraction of the scattering molecule and d is its diameter.

TABLE 3.3. Values of the Air Mass m at the Earth's Surface for Various Zenith Angles: (a) Calculated from m = secant θ and (b) Corrected for Atmospheric Curvature and for Refraction

Zenith Angle (θ)	Air Mass (m)	
	m = secant θ	Corrected
0°	1.00	1.00
10°	1.02	1.02
20°	1.06	1.06
30°	1.15	1.15
40°	1.31	1.31
50°	1.56	1.56
60°	2.00	2.00
70°	2.92	2.90
78°	4.81	4.72
86°	14.3	12.4

Source: Demerjian et al., 1980.

Making the simplifying assumptions of a homogeneous atmosphere of fixed height of 7.996×10^5 cm, and of uniform temperature and pressure throughout, this equation for Rayleigh scattering can be simplified for application to the atmosphere; as discussed in detail by Leighton (1961), this becomes

$$t_{sg} = \frac{1.044 \times 10^5 (n_{0\lambda} - 1)^2}{\lambda^4} \tag{G}$$

where $n_{0\lambda}$ is the index of refraction of air at wavelength λ and the pressure and temperature of interest. The dependence of Rayleigh scattering on λ^{-4} is evident in Fig. 3.4 which shows the attenuation coefficient for Rayleigh scattering as a function of wavelength from 290 to 700 nm; shorter wavelengths (i.e., in the blue–ultraviolet region) are scattered much more strongly than the longer wavelengths.

In the atmosphere light absorption in the ultraviolet region is predominantly due to O_3 and this is predominantly in the stratosphere (Figs. 3.1 and 3.2). Since the absorption coefficients σ of O_3 are reasonably well established a variant of the Beer–Lambert law can be applied to determined how much of the incident light is absorbed by O_3:

$$\frac{I}{I_0} = e^{-\sigma N_{O_3} m}$$

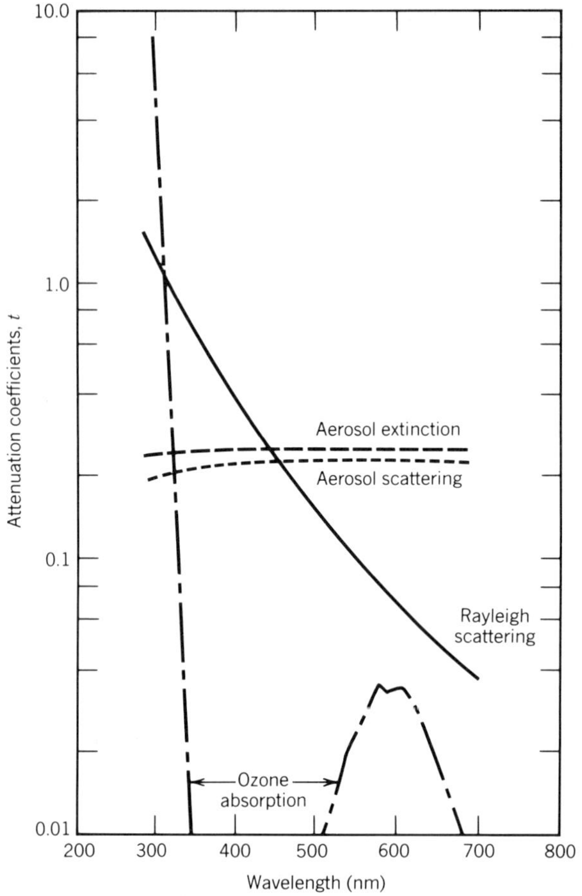

FIGURE 3.4. Attenuation coefficients (t) for light scattering (Rayleigh scattering) and absorption (ozone absorption) by gases and for scattering (aerosol scattering) and scattering plus absorption (aerosol extinction) by particulates for an overhead sun ($\theta = 0°$, $m = 1.0$) (from Peterson, 1976, and Demerjian et al., 1980).

One needs to know, in addition to σ, the O_3 concentration as a function of altitude (z), that is, $N_{O_3} = \int_{z=0}^{\infty} O_3(z)\,dz$. Figure 3.5 shows a typical distribution of O_3 with altitude used by Peterson and co-workers to estimate the attenuation coefficient for absorption of light by O_3 which they equate to t_{ag}.

From such distribution data, the total ozone in a vertical atmospheric column extending from the earth's surface to the top of the atmosphere can be calculated; this is equivalent to N_{O_3}. Using the published absorption coefficients (σ) as a function of wavelength, one can then apply the Beer–Lambert law to calculate the intensity of light transmitted through such a vertical column to the earth's surface. This applies to the intensity transmitted for a solar zenith cycle of $\theta = 0°$, that is, $m = 1.0$:

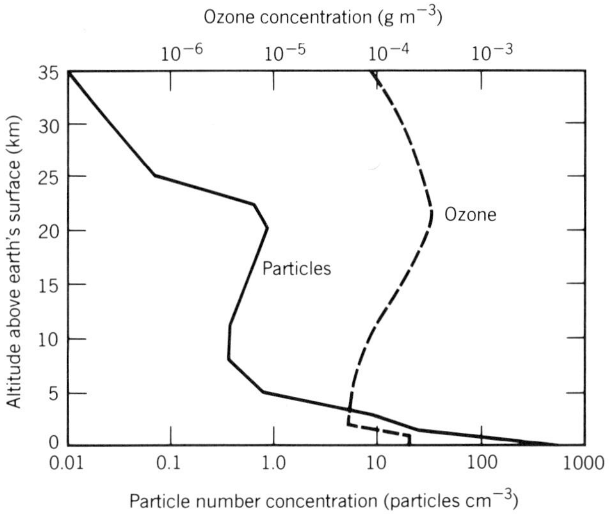

FIGURE 3.5. Variation with height (km) of particle number concentration (particles cm^{-3}) and ozone concentration (g m^{-3}) used as input for the actinic flux concentrations (from Peterson, 1976, and Demerjian et al., 1980).

$$\frac{I}{I_0} = e^{-\sigma N_{O_3} m} = e^{-tm} = e^{-t(1.0)}$$

Thus the attenuation coefficient for light absorption by gases is essentially that due to O_3, $\sigma N_{O_3} m = t_{ag} m$. The resulting attenuation coefficients for O_3 are shown in Fig. 3.4; clearly, for an overhead sun, O_3 is responsible for most of the attenuation of light directly from the sun of $\lambda \lesssim 310$ nm reaching the earth's surface.

This region of the spectrum around 300 nm is a crucial one for tropospheric photochemistry in both clean and polluted atmospheres. As we have indicated in Chapter 2, it is here that species such as ozone and aldehydes photolyze to produce atoms and free radicals critical to the chemistry of the troposphere. For example, environmental chamber studies have confirmed that increasing the solar intensity in the 290–330 nm region increases rates of formation of O_3 and PAN in irradiated HC–NO_x mixtures in air (Winer et al., 1979) (see Section 6-E-3).

Scattering and absorption of light by particulate matter (Section 12-B-3) is much more complex and will not be treated in detail here. Clearly, the size distribution and chemical composition, as well as the concentration of the particles, are very important in determining the extent of light scattering and absorption. Since these parameters will vary significantly geographically, seasonally, and diurnally, accurately estimating their impact on light intensities at a particular location at the earth's surface is difficult. Simplifications for the attenuation coefficient for scattering by particles such as

$$t_{sp} = \frac{b}{\lambda^n} \qquad \text{(H)}$$

are often made, where b depends on the concentration of particles and n on their size; for example, n decreases from ~ 4 to 0 as the particle size increases (Leighton, 1961).

One estimate of the attenuation coefficients for light scattering by particles, t_{sp}, is given in Fig. 3.4. Also shown are these researchers' estimates of total scattering plus absorption due to particulate matter, known as the *aerosol extinction*:

$$t_{ap} + t_{sp} = \text{aerosol extinction} \quad \text{(I)}$$

In this case the radii of the particles were assumed to fall between 0.01 and 2.0 μm; the peak in the number versus size distribution was at 0.07 μm. The vertical distribution of the concentrations of the particles was that shown in Fig. 3.5.

Given estimated values for the attenuation coefficients for scattering and absorption of light by gases and particles (i.e., t_{sg}, t_{ag}, t_{sp}, t_{ap}), one can calculate from Eq. (D) the fraction of the direct solar intensity incident on the top of the atmosphere which is transmitted to the earth's surface at any given wavelength. However, when one considers the *actual* light intensity that reaches a given volume of gas in the troposphere, one must take into account not only this direct sunlight but also two other sources of indirect light: (1) light, either direct from the sun or reflected from the earth's surface, which is scattered to the volume by gases or particles, and (2) light which is reflected from the earth's surface. These are illustrated in Fig. 3.6.

Estimating the intensity of the scattered light, or so-called *sky radiation*, at a given point in the atmosphere is difficult because of the substantial uncertainties and variability involved in the factors that contribute to light scattering; for example, the size distribution, concentration, and composition of particles which to a large extent cause this scattering are highly variable from location to location and from time to time and are not always well known for a particular point.

The amount of light reflected from the earth's surface to a volume of air clearly depends on the type of surface, as well as the wavelength of light; thus snow is highly reflecting, whereas black lava rock reflects very little of the incident radiation. The term used to describe the extent of this reflection is the *surface albedo*, which is the fraction of light incident on the surface which is reflected. Reflection can be specular (e.g., a water surface at large zenith angles) or diffuse (e.g., white rocks or buildings). Table 3.4 gives some reported values of surface albedos for different types of surfaces.

One can thus estimate the total light intensity incident on a given volume of air in the troposphere due to direct solar radiation, scattering, and reflection. The light absorbed in that volume can then be calculated using the Beer–Lambert law, if the concentrations and absorption coefficients of all absorbing species are known.

Consider the case shown in Fig. 3.7 where light of wavelength λ and total intensity I_i' [the sum of direct (I_d'), scattered (I_s'), and reflected (I_r') light in units of photons cm^{-2} s^{-1} (i.e., $I_i' = I_d' + I_s' + I_r'$)] is incident on a box of air 1 cm^2 in area and of height ℓ. For simplicity in the following derivation, the light is treated as if it is vertical to the flat face of the box; however, its components include the *total* direct, scattered, and reflected light incident on the box from all

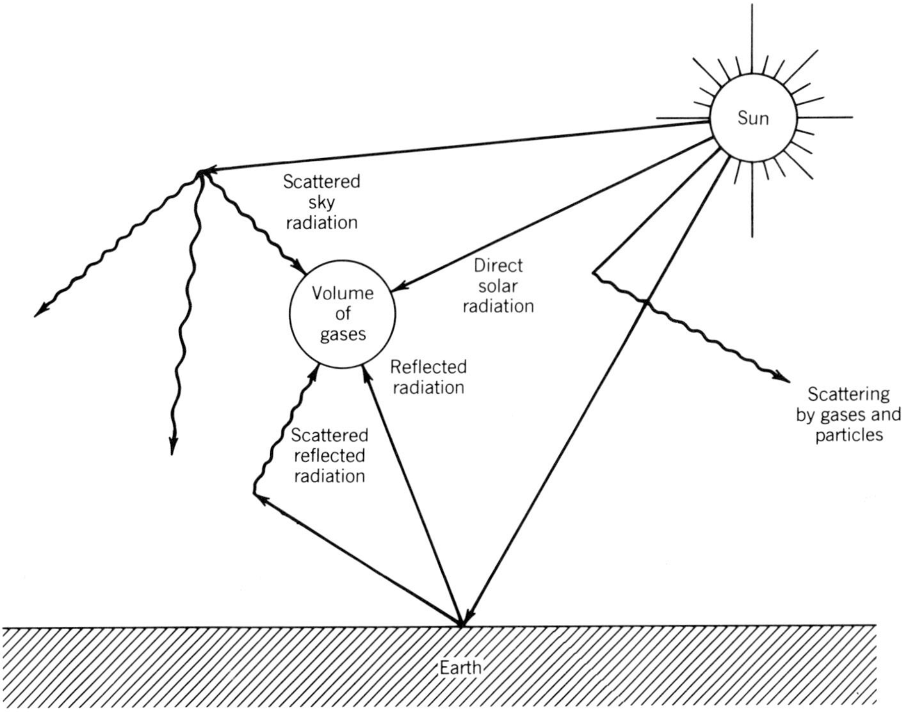

FIGURE 3.6. Different sources of radiation striking a volume of gas in the atmosphere. These include direct radiation from the sun, radiation scattered by gases and particles, and radiation reflected from the earth's surface.

TABLE 3.4. Some Typical Albedos for Various Types of Surfaces

	Albedo		
Type of Surface	Visible	Actinic UV	Reference
Snow	0.46–0.85	0.46–0.85[a]	Leighton, 1961
	—	0.93 ± 0.05	Dickerson et al., 1982
Ocean	0.03–0.46	—	Leighton, 1961
	0.24–0.45	—	Dickerson, 1982
Forests	0.05–0.18	—	Leighton, 1961
	0.06–0.73	—	Dickerson et al., 1982
Fields and meadows	0.15–0.30	—	Leighton, 1961
Bare ground	0.1–0.2	—	Leighton, 1961
Stone	—	0.22–0.25[a]	Leighton, 1961
Dune sand	0.17	0.1–0.25[a]	Leighton, 1961
Clouds	0.70–0.95	—	Dickerson et al., 1982
White cement	—	0.17[b]	Harvey et al., 1977
White plywood	—	0.07[b]	Harvey et al., 1977
Black cloth	—	0.02[b]	Harvey et al., 1977

[a] Below 360 nm.

[b] 300–385 nm; corrected data as reported by Dickerson et al. (1982).

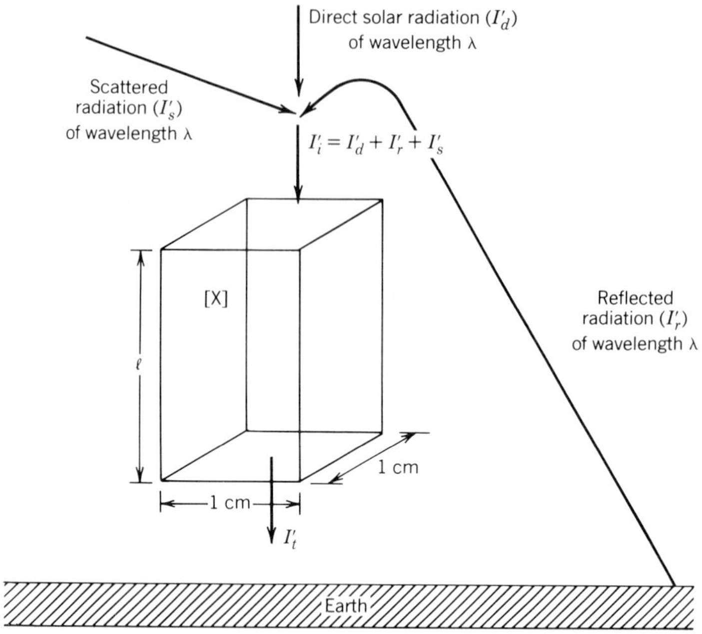

FIGURE 3.7. Schematic of absorption of light by a molecule X in a box in the troposphere. The light consists of direct scattered and reflected radiation, which for simplicity is treated as striking the 1 cm^2 face of the box from a vertical direction.

directions. Thus the incident intensity I'_i is really equivalent to a spherically integrated flux, that is, the flux incident from all directions which would be measured by an actinometer at a given location.

This box of air contains a light absorbing species X whose concentration is [X] (number cm^{-3}) and whose absorption cross section is σ (cm^2 $molecule^{-1}$). If the light transmitted through the volume is I'_t (photons cm^{-2} s^{-1}) as shown in Fig. 3.7, then the light absorbed by X, I'_a (photons cm^{-2} s^{-1}), must be given by

$$I'_a = I'_i - I'_t = I'_i \left[1 - \frac{I'_t}{I'_i} \right] \tag{J}$$

From the Beer–Lambert law [Eq. (H) of Chapter 2],

$$\frac{I}{I_0} = \frac{I'_t}{I'_i} = e^{-\sigma[\mathrm{X}]\ell}$$

Thus

$$I'_a = I'_i \left[1 - e^{-\sigma[\mathrm{X}]\ell}\right] \tag{K}$$

For weak absorption, which is applicable under atmospheric conditions, the approximation

$$(1 - e^{-a}) \simeq a$$

can be used. Hence Eq. (K) becomes

$$I'_a \left(\frac{\text{photons}}{\text{cm}^2\ \text{s}}\right) \simeq I'_i\,(\sigma[\text{X}]\ \ell) \tag{L}$$

It is important to note that the use of the weak absorption approximation to derive Eq. (L) results in the absorption cross section σ in Eq. (L) being that to the base e. In Section 3-B below, we show how photolysis rate constants (k_p) can be derived based on Eq. (N) which is derived from Eq. (L); in calculating such photolysis rate constants, σ must be in the base e, *not* base 10.

Referring to Fig. 3.7, the light absorbed *per unit volume* per second, I''_a, is given by

$$I''_a \left(\frac{\text{photons}}{\text{cm}^3\ \text{s}}\right) = \frac{I'_i\,(\sigma[\text{X}]\ \ell)}{\ell} = I'_i\,\sigma[\text{X}] \tag{M}$$

The most common form of Eq. (M) is

$$I_a(\lambda) = \sigma(\lambda)\ J(\lambda)\ [\text{X}] \tag{N}$$

$$\text{Units:}\quad \frac{\text{photons}}{\text{cm}^3\ \text{s}} = \frac{\text{cm}^2}{\text{molecule}} \times \frac{\text{photons}}{\text{cm}^2\ \text{s}} \times \frac{\text{molecules}}{\text{cm}^3}$$

where the dependence on wavelength has been explicitly included. The actinic irradiance $J(\lambda)$ (in units of photons cm^{-2} s^{-1}), usually used in place of I'_i, is the total light intensity from direct, scattered, and reflected light which is incident on a horizontal surface of unit area. It is basic to all calculations of atmospheric photolysis rates.

It is important to note that while $J(\lambda)$ is defined in units of photons cm^{-2} s^{-1}, it is in fact a *spherically* integrated flux. This can be seen by considering that, in reality, quanta may impinge on a given point in space from any direction in a diffuse radiation field. The light intensity striking this point can then be treated in terms of quanta per second per cm^2 per steradian per unit wavelength or wavelength interval; let us designate this j_λ. Thus the integral

$$\int_0^{4\pi\ \text{steradians}} j_\lambda\ d\Omega$$

represents the effective intensity (in units of quanta s^{-1} cm^{-2} per unit wavelength) on a target that presents the same area to the radiation regardless of the direction of incidence of the light. If the target is a molecule with absorption cross section σ (Section 2-A-2) in units of cm^2 $molecule^{-1}$, the molecule absorbs quanta at a rate given by

$$\sigma \int_0^{4\pi \text{ steradians}} j_\lambda \, d\Omega \quad \text{(units of quanta per second per unit wavelength)}$$

The rate of light absorption in photons per cm^3 per second per unit wavelength or wavelength interval at a concentration of the absorber, [X], is then

$$I_a(\lambda) = \sigma[X] \int_0^{4\pi \text{ steradians}} j_\lambda \, d\Omega$$

This is the same as Eq. (N) where

$$J(\lambda) = \int_0^{4\pi \text{ steradians}} j_\lambda \, d\Omega$$

Thus $J(\lambda)$ is indeed a spherically integrated flux.

Because the actinic irradiance $J(\lambda)$ is the sum of direct, scattered, and reflected light, its value depends on a number of parameters. Perhaps the most important of these (after the solar constants, Table 3.1) is the zenith angle, θ. Thus increasing the zenith angle increases the total pathlength through the atmosphere traversed by the light; this increases both absorption and scattering and hence decreases $J(\lambda)$.

Any parameter that affects the extent of light scattering or absorption throughout the atmosphere also alters $J(\lambda)$; thus the properties of the particles present and the total column O_3 are important in determining $J(\lambda)$. Factors that alter the surface albedo or the incoming solar intensity (e.g., clouds) also affect $J(\lambda)$. Finally, the height above the earth's surface of the air parcel being considered will affect $J(\lambda)$ because of the effects of light absorption, scattering, and reflection by the surface.

3-A-2b. Estimates of the Actinic Flux, $J(\lambda)$

In his classic monograph in 1961, Leighton calculated the actinic flux as a function of wavelength and zenith angle; these values have been used for years to calculate the rates of photolysis of species in the troposphere. More recently, Peterson (1976) recalculated these solar fluxes from 290 to 700 nm using a radiative transfer model developed by Dave (1972). Demerjian et al. (1980) then applied them to the photolysis of some important atmospheric species. In this model, molecular scattering, absorption due to O_3, H_2O, O_2, and CO_2, as well as scattering and absorption by particles are taken into account.

We concentrate here on the actinic flux estimates at the earth's surface by Peterson, Demerjian and co-workers because they are most often used in tropospheric calculations today. However, other recent flux estimates take various approaches to incorporating light scattering, as well as reflection at the earth's surface (e.g., Isaksen et al., 1977; Nicolet et al., 1982). Somewhat different calculated photolysis rates may result from different assumptions regarding scattering and reflection.

In the "average" case for which the Peterson model calculations were carried out, absorption and scattering as the light traveled from the top of the atmosphere to the earth's surface was assumed to be due to O_3 (UV light absorption), air molecules (scattering), and particles (scattering and absorption). The vertical profile assumed for O_3 and particles in the atmosphere was shown in Fig. 3.5. The particles were assumed to be spherical, homogeneous, and partly light absorbing, although, as seen in Fig. 3.4, ~90% of the total light attenuation due to particles was assumed to be due to scattering and only ~10% to absorption.

Scattering of light by gases (i.e., Rayleigh scattering) was treated as discussed above [and treated in detail by Leighton (1961)]. The surface albedo was taken as either 0, 0.1, or a set of wavelength-dependent "best estimate" values which varied from 0.05 in the 290–400 nm region to 0.15 in the 660–700 nm region.

The assumed concentrations and vertical distributions of O_3 and particles in the atmosphere were judged by Peterson and co-workers to be characteristic of general conditions over the continental United States, "with some emphasis on typical urban concentrations." However, the effects of varying the input parameters to more closely correspond to either cleaner or more polluted air were also investigated, as discussed below.

In the model, the region from the earth's surface to the top of the atmosphere is divided into 40 layers and the spectrum into 48 intervals. The solar constant data of DeLuisi (1975) were used, although corrections to other estimates of the solar constants can be made using a linear adjustment. From 290 to 420 nm, where the absorption due to O_3 and the Rayleigh scattering due to air molecules changes rapidly, the spectral intervals are 5 nm wide, whereas from 420 to 580 nm they are 10 nm wide, and from 580 to 700 nm, 20 nm wide. It should be noted that, in some cases, it may be desirable to use a more narrow spectral resolution than 5 nm in the 290–320 nm region where the strong O_3 absorption is falling off rapidly (Section 3-C-2). Photolysis in this region produces electronically excited oxygen atoms, $O(^1D)$, which, through their reaction with water, are a major source of OH radicals in the troposphere (Section 3-D-1). Hence model calculations may be improved by using a smaller spectral resolution in this region (Isaksen, 1985).

Table 3.5 shows the calculated actinic flux at the earth's surface as a function of zenith angle averaged over the wavelength intervals described above, assuming their best estimate for the surface albedo. These data are plotted for six wavelength intervals as a function of zenith angle in Fig. 3.8. The initially small change in actinic flux with zenith angle as it increases from 0° to ~50° at a given wavelength followed by the rapid drop of intensity from 50° to 90° (Fig. 3.8) is due to the fact that the air mass m changes only gradually to ~50° but then increases much more rapidly to $\theta = 90°$ [see Table 3.3 and Eq. (D)]. At the shorter wavelengths at a fixed zenith angle, the rapid increase in actinic flux with wavelength is primarily due to the strongly decreasing O_3 absorption in this region (Section 3-C-2a).

TABLE 3.5. Calculated Actinic Flux $J(\lambda)$ at the Earth's Surface as a Function of Wavelength and Zenith Angle Within Specified Wavelength Intervals for Best Estimate Surface Albedos

Wavelength (nm)	Exp[a]	Actinic Flux (photons cm^{-2} s^{-1}) 0°[b]	10°	20°	30°	40°	50°	60°	70°	78°	86°
290–295	14	0.001	0.001	—	—	—	—	—	—	—	—
295–300	14	0.041	0.038	0.030	0.019	0.009	0.003	—	—	—	—
300–305	14	0.398	0.381	0.331	0.255	0.167	0.084	0.027	0.004	0.001	—
305–310	14	1.41	1.37	1.25	1.05	0.800	0.513	0.244	0.064	0.011	0.002
310–315	14	3.14	3.10	2.91	2.58	2.13	1.56	0.922	0.357	0.090	0.009
315–320	14	4.35	4.31	4.10	3.74	3.21	2.52	1.67	0.793	0.264	0.030
320–325	14	5.48	5.41	5.19	4.80	4.23	3.43	2.43	1.29	0.502	0.073
325–330	14	7.89	7.79	7.51	7.01	6.27	5.21	3.83	2.17	0.928	0.167
330–335	14	8.35	8.25	7.98	7.50	6.76	5.72	4.30	2.54	1.15	0.241
335–340	14	8.24	8.16	7.91	7.46	6.78	5.79	4.43	2.69	1.25	0.282
340–345	14	8.89	8.80	8.54	8.09	7.38	6.36	4.93	3.04	1.44	0.333
345–350	14	8.87	8.79	8.54	8.11	7.43	6.44	5.04	3.15	1.51	0.352
350–355	14	10.05	9.96	9.70	9.22	8.48	7.39	5.83	3.69	1.77	0.414
355–360	14	9.26	9.18	8.94	8.52	7.86	6.88	5.47	3.50	1.69	0.391
360–365	14	10.25	10.16	9.91	9.46	8.76	7.71	6.17	3.99	1.94	0.444
365–370	15	1.26	1.25	1.22	1.17	1.08	0.958	0.772	0.505	0.247	0.055
370–375	15	1.14	1.13	1.10	1.06	0.983	0.873	0.708	0.467	0.230	0.051
375–380	15	1.27	1.26	1.23	1.18	1.10	0.983	0.802	0.535	0.265	0.058
380–385	15	1.05	1.04	1.02	0.980	0.917	0.820	0.673	0.453	0.226	0.049
385–390	15	1.15	1.15	1.12	1.08	1.01	0.909	0.750	0.510	0.257	0.054
390–395	15	1.19	1.18	1.16	1.11	1.05	0.943	0.783	0.537	0.273	0.057
395–400	15	1.44	1.43	1.40	1.35	1.28	1.15	0.962	0.666	0.341	0.070
400–405	15	1.73	1.72	1.69	1.63	1.53	1.39	1.16	0.809	0.418	0.085

405–410	15	1.94	1.93	1.90	1.83	1.73	1.57	1.32	0.926	0.482	0.097
410–415	15	2.05	2.04	2.00	1.93	1.83	1.66	1.41	0.993	0.522	0.104
415–420	15	2.08	2.07	2.03	1.96	1.86	1.70	1.44	1.03	0.543	0.107
420–430	15	4.08	4.06	3.99	3.87	3.67	3.36	2.87	2.07	1.11	0.216
430–440	15	4.20	4.18	4.11	3.99	3.80	3.49	3.01	2.19	1.20	0.229
440–450	15	4.87	4.85	4.77	4.64	4.43	4.09	3.54	2.61	1.45	0.272
450–460	15	5.55	5.51	5.43	5.27	5.03	4.64	4.02	2.99	1.67	0.312
460–470	15	5.68	5.65	5.57	5.42	5.17	4.79	4.17	3.12	1.77	0.325
470–480	15	5.82	5.79	5.70	5.55	5.31	4.91	4.32	3.26	1.87	0.341
480–490	15	5.78	5.75	5.67	5.53	5.29	4.93	4.33	3.29	1.90	0.339
490–500	15	5.79	5.76	5.68	5.54	5.31	4.96	4.37	3.34	1.95	0.344
500–510	15	5.99	5.96	5.87	5.71	5.47	5.09	4.47	3.41	1.99	0.340
510–520	15	5.88	5.86	5.77	5.62	5.38	5.02	4.43	3.40	2.00	0.340
520–530	15	5.98	5.95	5.87	5.72	5.48	5.11	4.52	3.47	2.04	0.336
530–540	15	5.98	5.95	5.87	5.72	5.48	5.12	4.52	3.48	2.05	0.326
540–550	15	5.88	5.85	5.77	5.62	5.40	5.04	4.46	3.44	2.03	0.317
550–560	15	5.94	5.91	5.83	5.68	5.44	5.08	4.49	3.46	2.04	0.312
560–570	15	5.99	5.96	5.88	5.73	5.49	5.13	4.54	3.50	2.06	0.306
570–580	15	6.12	6.09	6.00	5.85	5.61	5.24	4.63	3.57	2.10	0.301
580–600	16	1.25	1.24	1.22	1.19	1.14	1.07	0.951	0.737	0.439	0.064
600–620	16	1.26	1.26	1.24	1.21	1.16	1.08	0.963	0.748	0.448	0.065
620–640	16	1.27	1.26	1.24	1.17	1.10	1.10	0.980	0.771	0.473	0.074
640–660	16	1.30	1.30	1.28	1.25	1.20	1.13	1.01	0.803	0.502	0.086
660–680	16	1.33	1.33	1.31	1.28	1.23	1.16	1.04	0.828	0.527	0.096
680–700	16	1.33	1.32	1.30	1.27	1.23	1.16	1.04	0.839	0.541	0.104

Source: Peterson, 1976, and Demerjian et al., 1980.

[a]This column (Exp) lists the power of 10 by which all entries should be multiplied. For example, at $\theta = 0°$, $J(\lambda) = 1.41 \times 10^{14}$ photons cm^{-2} s^{-1} from 305 to 310 nm.

[b]Zenith angles.

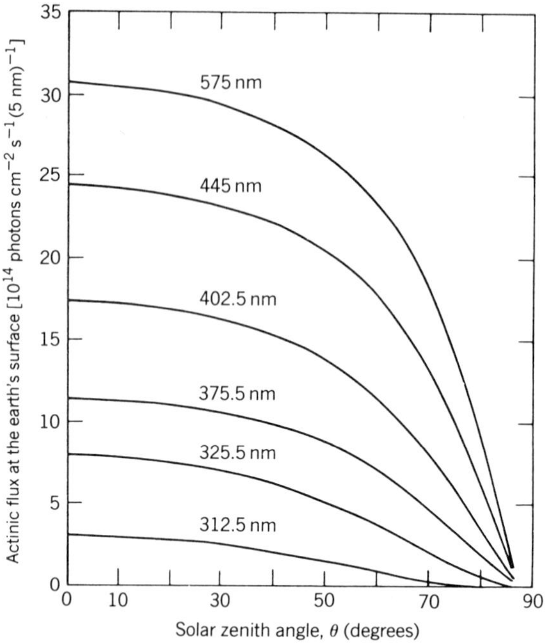

FIGURE 3.8. Calculated actinic flux within 5 nm wavelength intervals, centered on the indicated wavelengths, at the earth's surface using best estimate albedos as function of solar zenith angle (from Peterson, 1976, and Demerjian et al., 1980).

3-A-2c. Effects of Latitude, Season, and Time of Day on J(λ)

To estimate photolysis rates for a given geographical location, one must take into account the latitude and season, as well as the time of day.

The data in Table 3.5 are representative for the average earth–sun distance characteristic of early April and October. The change in the earth–sun distance with season causes a small change ($\lesssim 3\%$) in the solar flux. Correction factors for this seasonal variation are given in Table 3.6.

Table 3.7 gives the solar zenith angles at latitudes of 20°N, 30°N, 40°N, and 50°N as a function of month and true solar time. True solar time, also known as apparent solar time or apparent local solar time, is defined as the time scale referenced to the sun crossing the meridian at noon. For example, at a latitude of 50°N at the beginning of January, 2 h before the sun crosses the meridian corresponds to a true solar time of 10 a.m.; from Table 3.7, the solar zenith angle at this time is 77.7°. To obtain the actinic flux at this time at any wavelength, one takes the fluxes in Table 3.5 listed under 78°; thus the flux in the 400–405 nm wavelength interval at 10 a.m. at 50°N latitude is 0.418×10^{15} photons cm^{-2} s^{-1}.

Afternoon values of θ are not given in Table 3.7 as the data are symmetric about noon. Thus at a time of 2 p.m. at 50°N latitude, the flux would be the same as calculated above for 10 a.m.

Figure 3.9 shows the solar angle θ as a function of true solar time for several

TABLE 3.6. Correction Factors for Extraterrestrial Solar Flux Values Depending on Earth–Sun Distance at Various Times of the Year

Date	Correction Factor	Date	Correction Factor
January 1	1.033	July 1	0.966
January 15	1.032	July 15	0.967
February 1	1.029	August 1	0.970
February 15	1.024	August 15	0.974
March 1	1.018	September 1	0.982
March 15	1.011	September 15	0.989
April 1	1.001	October 1	0.998
April 15	0.993	October 15	1.006
May 1	0.984	November 1	1.015
May 15	0.978	November 15	1.022
June 1	0.971	December 1	1.027
June 15	0.968	December 15	1.031

Source: Demerjian et al., 1980.

latitudes and different times of the year. As expected, only for the lower latitudes at the summer solstice does the solar zenith angle approach 0° at noon. For a latitude of 50°N even at the summer solstice, θ is 27°.

Figure 3.10 shows the diurnal variation of the solar zenith angle as a function of season for Los Angeles, which is located at a latitude of 34.1°N. Clearly, the peak solar zenith angle varies dramatically with season.

The differences in light intensity, and in its diurnal variation at different latitudes and seasons, are expected to alter the atmospheric chemistry at various geographical locations because photochemistry is the major source of the free radicals such as OH which drive the chemistry. The results of computer modeling studies support this expectation.

For example, Bottenheim and co-workers (1977) carried out simulation studies of photochemical air pollution for temperatures and solar zenith angles characteristic of winter, spring, and summer at 60°N latitude, corresponding to northern Alberta. They found that in their model the change in solar zenith angle with season (and hence changes in the rate of formation of free radicals) was primarily responsible for the predicted decreased rates of O_3 and NO_2 production in the winter, as well as much lower calculated peak O_3 concentrations.

The effect of latitude has also been examined using computer models. Thus Nieboer et al. (1976) calculated the O_3 and PAN dosages [dosage = integrated (concentration × time)] predicted for latitudes of 34°N, 52°N, and 65°N, corresponding to Los Angeles, Rotterdam, and Nome or Fairbanks, Alaska. Calculations were carried out for both the summer solstice (June 21) and the equinox (March 21/September 23). Interestingly, at the solstice, while the diurnal variations and peak light intensities differed for the three locations, the volumetric light

TABLE 3.7. Tabulation of Solar Zenith Angles as a Function of True Solar Time and Month

	0400	0430	0500	0530	0600	0630	0700	0730	0800	0830	0900	0930	1000	1030	1100	1130	1200
Latitude 20°N																	
January 1							84.9	78.7	72.7	66.1	61.5	56.5	52.1	48.3	45.5	43.6	43.0
February 1						88.9	82.5	75.8	69.6	63.3	57.7	52.2	47.4	43.1	40.0	37.8	37.2
March 1						85.7	78.8	72.0	65.2	58.6	52.3	46.2	40.5	35.5	31.4	28.6	27.7
April 1					88.5	81.5	74.4	67.4	60.3	53.4	46.5	39.7	33.2	26.9	21.3	17.2	15.5
May 1					85.0	78.2	71.2	64.3	57.2	50.2	43.2	36.1	29.1	26.1	15.2	8.8	5.0
June 1				89.2	82.7	76.0	69.3	62.5	55.7	48.8	41.9	35.0	28.1	21.1	14.2	7.3	2.0
July 1				88.8	82.3	75.7	69.1	62.3	55.5	48.7	41.8	35.0	28.1	21.2	14.3	7.7	3.1
August 1					83.8	77.1	70.2	63.3	56.4	49.4	42.4	35.4	28.3	21.3	14.3	7.3	1.9
September 1					87.2	80.2	73.2	66.1	59.1	52.1	45.1	38.1	31.3	24.7	18.6	13.7	11.6
October 1						84.1	77.1	70.2	63.3	56.5	49.9	43.5	37.5	32.0	27.4	24.3	23.1
November 1						87.8	81.3	74.5	68.3	61.8	56.0	50.2	45.3	40.7	37.4	35.1	34.4
December 1							84.3	78.0	71.8	66.1	60.5	55.6	50.9	47.2	44.2	42.4	41.8
Latitude 30°N																	
January 1							89.4	83.7	78.3	73.2	68.4	64.1	60.4	57.3	55.0	53.5	53.0
February 1							86.2	80.3	74.6	69.1	64.1	59.4	55.3	51.9	49.3	47.7	47.2
March 1						87.5	81.1	74.9	68.8	62.9	57.4	52.2	47.5	43.5	40.3	38.4	37.7
April 1					87.2	81.4	74.9	68.5	62.1	55.8	49.6	43.8	38.2	33.3	29.3	26.5	25.5
May 1				88.9	82.7	76.3	69.9	63.4	56.9	50.4	44.0	37.6	31.4	25.6	20.4	16.5	15.0
June 1				85.3	79.2	73.1	66.8	60.4	54.0	47.5	41.0	34.5	28.1	21.7	15.7	10.5	8.0
July 1				84.7	78.7	72.5	66.3	60.0	53.5	47.1	40.6	34.1	27.7	21.2	15.1	9.6	6.9
August 1				87.2	81.0	74.7	68.3	61.9	55.4	48.9	42.4	36.0	29.7	23.6	18.1	13.7	11.9
September 1					85.9	79.4	72.9	66.4	60.0	53.6	47.3	41.2	35.5	30.2	25.8	22.8	21.6
October 1						85.1	78.7	72.4	66.2	60.1	54.4	48.8	43.8	39.5	36.1	33.9	33.1
November 1							84.6	78.4	72.8	67.1	62.0	57.0	53.0	49.3	46.7	44.9	44.4
December 1							88.7	82.8	77.3	72.2	67.3	63.0	59.2	56.0	53.7	52.2	51.8

Latitude 40°N																	
January 1								89.0	84.2	79.8	75.7	72.1	69.0	66.4	64.6	63.4	63.0
February 1								84.8	79.8	75.2	70.7	67.0	63.5	60.9	58.8	57.6	57.2
March 1						89.1	83.7	78.1	72.8	67.8	63.1	58.8	55.1	51.9	49.6	48.1	47.7
April 1					87.1	81.4	75.6	70.0	64.4	59.0	53.8	49.0	44.6	40.9	38.0	36.2	35.5
May 1				85.9	80.5	74.7	68.9	63.2	57.5	51.8	46.3	41.0	36.1	31.7	28.2	25.8	25.0
June 1			86.8	81.5	76.1	70.5	64.9	59.2	53.4	47.7	42.0	36.5	31.2	26.2	22.1	19.1	18.0
July 1			86.0	80.8	75.4	69.9	64.3	58.6	52.8	47.1	41.4	35.8	30.4	25.4	21.1	18.0	16.9
August 1			89.3	83.9	78.4	72.8	67.1	61.4	55.6	49.9	44.4	39.0	33.8	29.2	25.4	22.8	21.9
September 1					84.6	78.9	73.2	67.5	61.8	56.3	51.0	46.0	41.4	37.5	34.4	32.3	31.6
October 1						86.3	80.6	75.1	69.7	64.5	59.6	55.1	51.2	47.8	45.3	43.6	43.1
November 1							88.1	82.6	77.8	72.8	68.6	64.4	61.1	58.2	56.1	54.7	54.4
December 1								88.1	83.3	78.8	74.7	71.0	67.8	65.3	63.3	62.2	61.8
Latitude 50°N																	
January 1										86.5	83.2	80.2	77.7	75.7	74.2	73.3	73.0
February 1								89.5	85.3	81.5	78.0	74.9	72.2	70.0	68.5	67.5	67.2
March 1							86.3	81.8	77.4	73.3	69.6	66.2	63.2	60.9	59.1	58.0	57.7
April 1					86.6	81.8	76.9	72.2	67.6	63.2	59.1	55.4	52.0	49.3	47.2	45.9	45.5
May 1			87.8	83.2	78.5	73.7	68.9	64.1	59.4	54.7	50.3	46.2	42.5	39.4	37.0	35.5	35.0
June 1		86.7	82.4	78.0	73.3	68.6	63.8	59.0	54.2	49.5	44.9	40.6	36.6	33.1	30.4	28.6	28.0
July 1	89.7	85.7	81.5	77.1	72.5	67.8	63.0	58.2	53.4	48.7	44.1	39.7	35.6	32.1	29.3	27.5	26.9
August 1		89.7	85.4	80.8	76.2	71.4	66.6	61.8	57.0	52.4	47.9	43.7	39.9	36.6	34.1	32.5	31.9
September 1				88.3	83.6	78.8	74.0	69.2	64.6	60.1	55.9	52.0	48.5	45.7	43.5	42.1	41.6
October 1						87.6	82.8	78.2	73.8	69.6	65.7	62.1	59.1	56.5	54.7	53.5	53.1
November 1								87.1	83.0	79.0	75.5	71.5[a]	69.5	67.3	65.7	64.7	64.4
December 1									89.2	85.5	82.1	79.1	76.5	74.5	73.0	72.1	71.8

Source: Peterson, 1976, and Demerjian et al., 1980.
[a]Estimated value based on interpolation.

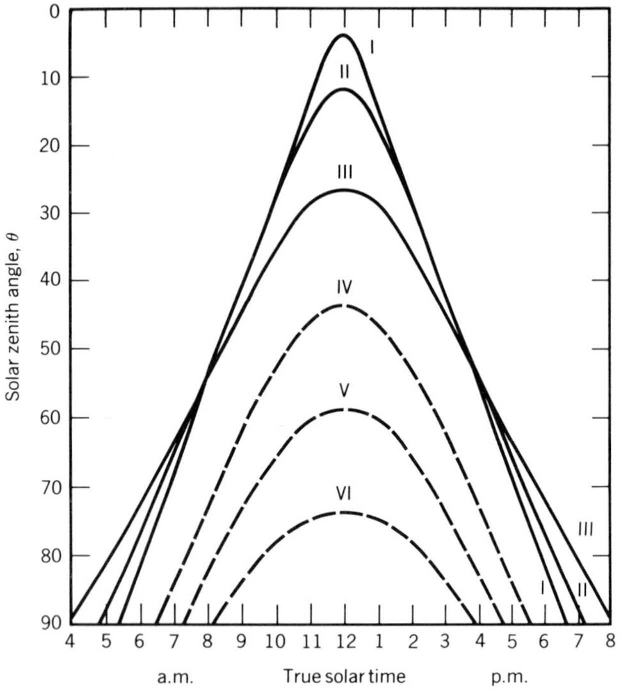

FIGURE 3.9. Effect of latitude on solar zenith angle. On the scale of true solar time, also called apparent solar time and apparent local solar time, the sun crosses the meridian at noon. The latitudes and seasons represented are: I, 20°N latitude, summer solstice; II, 35°N latitude, summer solstice; III, 50°N latitude, summer solstice; IV, 20°N latitude, winter solstice; V, 35°N latitude, winter solstice; and VI, 50°N latitude, winter solstice (from Leighton, 1961).

intensities integrated over the entire day were relatively constant. Furthermore, the calculated O_3 dosages only varied from 72.6 to 62.6 ppm min on June 21 in going from 34°N to 65°N latitude. This indicated that these dosages depended more on the integrated light intensity than on its diurnal variation. The PAN dosage was somewhat more sensitive, varying from 2.75 to 1.92 ppm min from 34°N to 65°N.

Under conditions characteristic of the equinox, however, the integrated light intensities fell off as one moved north, and the predicted O_3 and PAN dosages showed a relatively linear variation with the integrated intensities. Thus at these times of year, the overall chemistry, as well as peak pollutant levels, may depend significantly on the latitude.

3-A-2d. Effect of Surface Elevation on $J(\lambda)$

The variation in the actinic flux with *surface* elevation is important because some of the world's major cities are located substantially above sea level. For example, Mexico City and Denver, Colorado, are at elevations of 2.2 and 1.6 km, respectively.

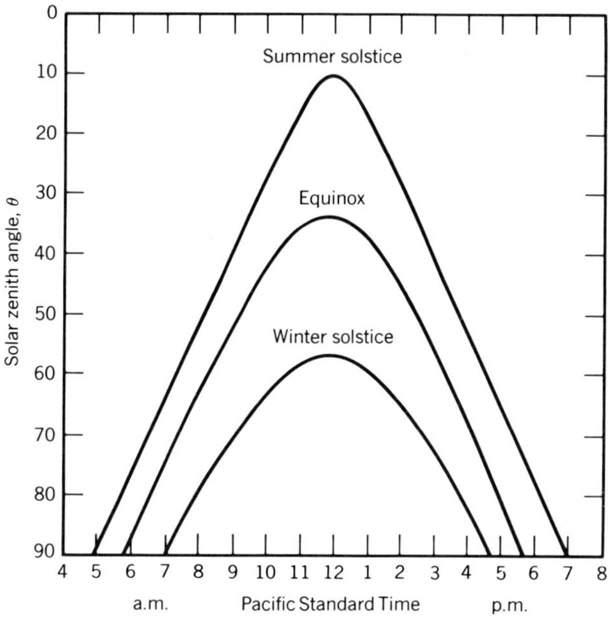

FIGURE 3.10. Relation between solar zenith angle and time of day at Los Angeles, California (from Leighton, 1961).

Table 3.8 shows the calculated percentage increase in the actinic flux at the earth's surface for an elevation of 1.5 km and atmospheric pressure of 0.84 atm (corresponding approximately to Denver), as a function of zenith angle for four wavelength intervals. In this calculation, it was assumed that the vertical O_3 and particle concentrations were the same, but that the Rayleigh scattering was reduced due to the lowered gas concentrations. The increase in actinic flux in the UV is relatively small ($\lesssim 5\%$) for zenith angles less than ~45°; at larger zenith angles, the change is less than 13%. For the longer wavelengths, it is small at all zenith angles.

3-A-2e. *Effect of Height Above Earth's Surface on J(λ)*

While Leighton's original calculations were carried out for sea level only, Peterson and co-workers have examined the variation in the actinic flux as a function of distance *above the earth's surface* using their multi-layer model. Figures 3.11, 3.12, and 3.13 show the *relative* changes in the actinic flux as a function of altitude from 0 to 25 km at zenith angles of 20° (curve *a*), 50° (curve *b*), and 78° (curve *c*) at wavelengths of 332.5 nm (Fig. 3.11), 412.5 nm (Fig. 3.12), and 575 nm (Fig. 3.13).

The calculated actinic flux typically increases significantly in the first few kil-

TABLE 3.8. Percentage Increase of Calculated Actinic Flux at a Surface Elevation of 1.5 km Using Best Estimate Albedos as a Function of Solar Zenith Angle and Selected Wavelengths[a] (Relative to Sea Level)

	Actinic Flux Increase (%)									
Wavelength	0°[b]	10°	20°	30°	40°	50°	60°	70°	78°	86°
340–345	2.1	2.3	2.6	3.2	4.2	5.7	8.1	11.4	12.4	7.5
400–405	0.9	0.9	1.1	1.5	2.1	3.0	4.6	7.6	10.9	6.7
540–550	0.2	0.2	0.2	0.4	0.5	0.9	1.4	2.4	4.3	4.7
680–700	0.02	0.02	0.05	0.1	0.2	0.3	0.5	1.0	1.7	2.8

Source: Peterson, 1976, and Demerjian et al., 1980.

[a] Relative to the values of Table 3.5 when surface elevation is increased to 1500 m.

[b] Zenith angles.

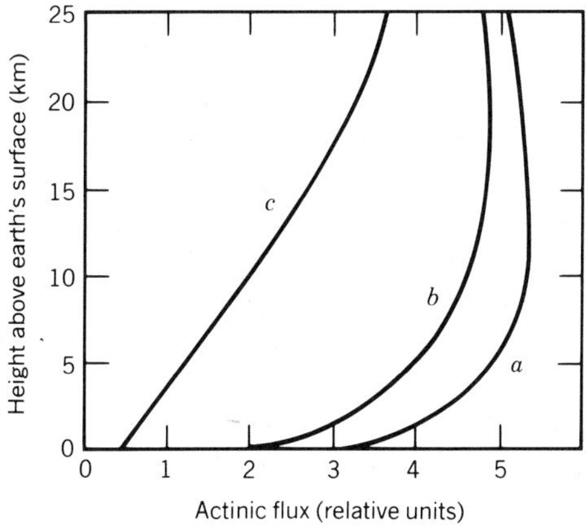

FIGURE 3.11. Calculated relative actinic flux using best estimate albedos as function of height above the earth's surface at 332.5 nm wavelength for solar zenith angle θ of (*a*) 20°, (*b*) 50°, and (*c*) 78° (from Peterson, 1976, and Demerjian et al., 1980).

ometers. This is partly due to backscattering of light by particulate matter and to light absorption by tropospheric O_3 close to the surface. The effect of O_3 can be seen by comparing the total fluxes at 332.5 nm (Fig. 3.11) where O_3 absorbs to those at 412.5 nm (Fig. 3.12) where it does not.

Peterson and co-workers have examined the percentage increase in total actinic

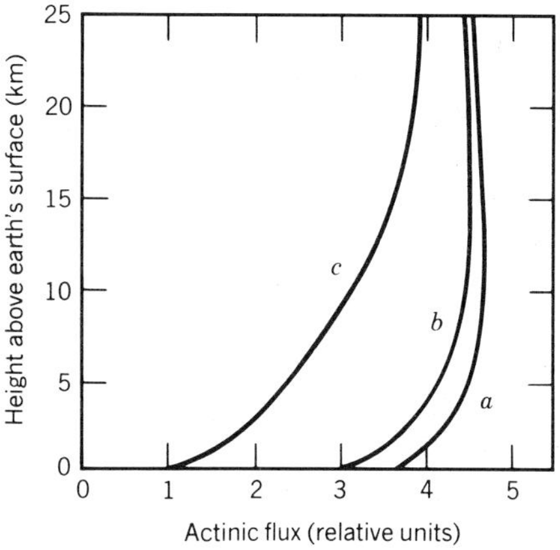

FIGURE 3.12. Calculated relative actinic flux using best estimate albedos as function of height above the earth's surface at 412.5 nm wavelength for solar zenith angle θ of (*a*) 20°, (*b*) 50°, and (*c*) 78° (from Peterson, 1976, and Demerjian et al., 1980).

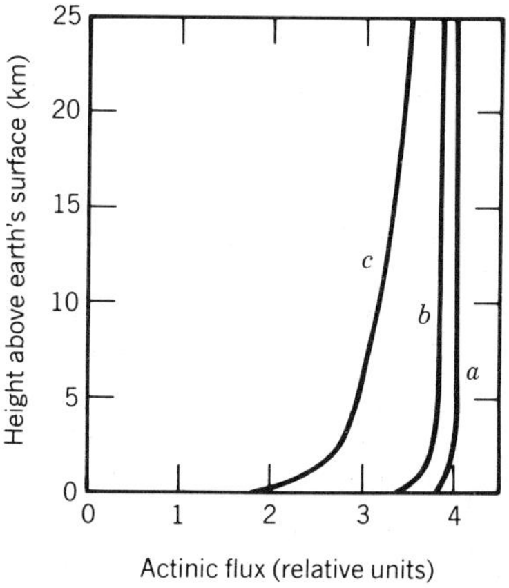

FIGURE 3.13. Calculated relative actinic flux using best estimate albedos as function of height above the earth's surface at 575 nm wavelength for solar zenith angle θ of (*a*) 20°, (*b*) 50°, and (*c*) 78° (from Peterson, 1976, and Demerjian et al., 1980).

flux going from the surface to ~ 1 km; they estimate that at short wavelengths ($\lambda \leq 310$ nm), the increase is $\geq 37.5\%$ for all zenith angles. This increase in flux with altitude at short wavelengths could be particularly significant in photochemical smog formation. Thus pollutants trapped in an inversion layer aloft may be exposed to higher actinic fluxes than at ground level and photolyze more rapidly, hastening the formation of various secondary pollutants. The increased actinic flux with altitude close to the earth's surface is the basis for their suggestion that the presence of increased O_3 in, or close to, the inversion layer may be at least partially the result of the height dependence of $J(\lambda)$.

3-A-2f. Sensitivity of Calculated Actinic Fluxes to Input Values for Surface Albedo and Particle and Ozone Concentrations

As discussed earlier, the net actinic flux incident on a volume of air is sensitive to a number of parameters, including surface reflection (i.e., albedo) and the concentrations of particulate matter and ozone which scatter and/or absorb light. Peterson and co-workers have examined the effects of changing these parameters on the total actinic flux, as a function of wavelength and zenith angle. Tables 3.9–3.12 show the results of these calculations.

In Table 3.9 the calculated actinic fluxes are given for a surface albedo of zero. By comparison to the data in Table 3.5, somewhat lower actinic fluxes result if the surface reflectivity is taken to be zero.

Table 3.10 shows the percentage increase in the calculated actinic flux for the lowest level of the model (i.e., closest to the earth's surface) if the surface albedo is changed from zero (Table 3.9) to 10%. For albedos $\lesssim 20\%$, the fluxes vary in an approximately linear fashion with the albedo so that the flux for albedos other than those given by Peterson and co-workers can be estimated. It is seen from Table 3.10 that the largest increases in the flux occur for the smallest zenith angles.

Table 3.11 shows the percentage change in the actinic flux compared to the data in Table 3.5 for two cases: (1) a particle concentration of zero, corresponding to a very clean atmosphere, and (2) a particle concentration double that shown in Fig. 3.5 (and used in calculating the data in Table 3.5). This would correspond to heavily polluted conditions over large cities.

As seen in Table 3.11, the actinic flux in general, would increase if the particulate concentration was lowered to zero and decrease if it were doubled. The surprising exception is at long wavelengths where, for example, at a zenith angle of 0°, the actinic flux is predicted to *decrease* on lowering the particulate concentration to zero but *increase* on doubling it. This is due to the fact that, near the surface, multiple scattering by the particles causes some light to pass through a given region several times, effectively increasing the pathlength; hence lowered particle concentrations result in lowered multiple scattering, and vice versa.

Since O_3 absorbs light primarily in the near ultraviolet, a change in its concentration will have the greatest effect in this wavelength region. Table 3.12 shows the percentage decrease in actinic flux from 300 to 305 nm and from 320 to 325 nm compared to those in Table 3.5, if the vertical O_3 profiles shown in Fig. 3.5 were increased uniformly by 5%. Clearly, the UV flux is quite sensitive to changes

TABLE 3.9. Calculated Actinic Flux at the Earth's Surface as a Function of Wavelength and Zenith Angle Within Specified Wavelength Intervals for Surface Albedo of Zero

Wavelength (nm)	Exp[a]	Actinic Flux (photons cm^{-2} s^{-1}) 0°[b]	10°	20°	30°	40°	50°	60°	70°	78°	86°
290–295	14	0.001	0.001	—	—	—	—	—	—	—	—
295–300	14	0.037	0.035	0.027	0.017	0.008	0.003	—	—	—	—
300–305	14	0.362	0.347	0.302	0.234	0.154	0.078	0.025	0.004	0.001	—
305–310	14	1.28	1.24	1.14	0.965	0.737	0.476	0.227	0.060	0.010	0.001
310–315	14	2.85	2.81	2.65	2.36	1.96	1.44	0.858	0.332	0.083	0.008
315–320	14	3.94	3.91	3.73	3.42	2.96	2.33	1.56	0.739	0.245	0.027
320–325	14	4.96	4.90	4.71	4.38	3.88	3.18	2.26	1.20	0.466	0.067
325–330	14	7.14	7.06	6.82	6.40	5.76	4.83	3.57	2.02	0.864	0.155
330–335	14	7.54	7.47	7.25	6.84	6.22	5.30	4.01	2.38	1.07	0.221
335–340	14	7.45	7.38	7.18	6.81	6.23	5.37	4.14	2.52	1.17	0.262
340–345	14	8.03	7.97	7.76	7.38	6.79	5.90	4.60	2.85	1.34	0.310
345–350	14	8.02	7.95	7.75	7.40	6.84	5.98	4.71	2.97	1.41	0.327
350–355	14	9.08	9.01	8.80	8.42	7.80	6.86	5.45	3.47	1.66	0.385
355–360	14	8.36	8.30	8.12	7.78	7.23	6.39	5.12	3.30	1.59	0.364
360–365	14	9.25	9.19	9.00	8.64	8.06	7.16	5.78	3.77	1.83	0.413
365–370	15	1.14	1.13	1.11	1.07	0.997	0.890	0.724	0.477	0.233	0.052
370–375	15	1.03	1.02	1.00	0.965	0.906	0.811	0.664	0.422	0.217	0.048
375–380	15	1.15	1.14	1.12	1.07	1.02	0.914	0.753	0.506	0.251	0.054
380–385	15	0.948	0.942	0.926	0.895	0.845	0.763	0.633	0.429	0.214	0.045
385–390	15	1.04	1.04	1.02	0.987	0.933	0.846	0.705	0.484	0.244	0.051
390–395	15	1.07	1.07	1.05	1.02	0.965	0.878	0.737	0.510	0.259	0.053
395–400	15	1.30	1.29	1.28	1.24	1.18	1.07	0.905	0.632	0.324	0.066
400–405	15	1.54	1.53	1.51	1.46	1.39	1.28	1.08	0.761	0.394	0.079
405–410	15	1.72	1.71	1.69	1.64	1.57	1.44	1.23	0.872	0.455	0.090
410–415	15	1.82	1.81	1.78	1.74	1.66	1.53	1.31	0.937	0.493	0.097
415–420	15	1.84	1.83	1.81	1.77	1.69	1.56	1.34	0.968	0.514	0.100

TABLE 3.9. **(*Continued*)**

Wavelength (nm)	Exp[a]	Actinic Flux (photons cm^{-2} s^{-1}) 0°[b]	10°	20°	30°	40°	50°	60°	70°	78°	86°
420–430	15	3.62	3.60	3.56	3.48	3.34	3.09	2.68	1.95	1.05	0.201
430–440	15	3.72	3.71	3.67	3.59	3.45	3.22	2.80	2.07	1.13	0.214
440–450	15	4.32	4.30	4.26	4.18	4.03	3.77	3.31	2.47	1.38	0.255
450–460	15	4.73	4.72	4.67	4.59	4.43	4.16	3.68	2.78	1.57	0.287
460–470	15	4.85	4.84	4.80	4.71	4.56	4.30	3.82	2.91	1.66	0.299
470–480	15	4.96	4.95	4.91	4.83	4.69	4.43	3.96	3.04	1.76	0.314
480–490	15	4.93	4.93	4.88	4.81	4.68	4.43	3.97	3.07	1.79	0.341
490–500	15	4.94	4.93	4.89	4.82	4.69	4.46	4.01	3.12	1.84	0.319
500–510	15	4.93	4.92	4.88	4.82	4.69	4.46	4.02	3.14	1.85	0.310
510–520	15	4.85	4.83	4.80	4.74	4.62	4.41	3.98	3.13	1.86	0.309
520–530	15	4.93	4.92	4.88	4.82	4.71	4.49	4.06	3.20	1.91	0.524
530–540	15	4.92	4.92	4.89	4.83	4.71	4.49	4.07	3.21	1.92	0.299
540–550	15	4.84	4.83	4.80	4.75	4.64	4.43	4.02	3.17	1.90	0.291
550–560	15	4.81	4.80	4.78	4.72	4.61	4.41	4.00	3.17	1.90	0.285
560–570	15	4.86	4.85	4.82	4.76	4.66	4.45	4.04	3.20	1.92	0.279
570–580	15	4.96	4.95	4.92	4.87	4.76	4.55	4.13	3.27	1.96	0.276
580–600	16	1.01	1.01	1.00	0.992	0.971	0.931	0.848	0.676	0.409	0.059
600–620	16	1.01	1.00	1.00	0.990	0.970	0.931	0.852	0.682	0.417	0.060
620–640	16	1.01	1.01	1.00	0.995	0.978	0.943	0.867	0.703	0.439	0.068
640–660	16	1.01	1.01	1.01	1.00	0.986	0.954	0.883	0.724	0.463	0.078
660–680	16	1.01	1.01	1.01	1.00	0.990	0.960	0.893	0.740	0.482	0.087
680–700	16	1.01	1.01	1.01	1.00	0.988	0.961	0.897	0.749	0.496	0.094

Source: Peterson, 1976, and Demerjian et al., 1980.

[a]This column (Exp) lists the power of 10 by which all entries should be multiplied. For example, at $\theta = 0°$, $J(\lambda) = 1.28 \times 10^{14}$ photons cm^{-2} s^{-1} from 305 to 310 nm

[b]Zenith angles.

TABLE 3.10. Percentage Increase of Calculated Actinic Flux at the Earth's Surface as a Function of Solar Zenith Angle and Selected Wavelengths Relative to the Values in Table 3.9 when Surface Albedo is Increased from 0.0 to 10%

Wavelength (nm)	Actinic Flux Increase (%)									
	0°[a]	10°	20°	30°	40°	50°	60°	70°	78°	86°
290–295	19.0	18.8	—	—	—	—	—	—	—	—
295–300	19.6	19.3	18.8	17.3	15.2	15.0	—	—	—	—
300–305	20.0	19.8	19.1	18.1	16.8	15.6	14.4	15.6	17.2	—
305–310	20.5	20.3	19.7	18.6	17.3	16.0	14.9	15.0	17.0	17.3
315–320	21.2	21.0	20.3	19.2	17.8	16.3	15.0	14.7	15.8	16.5
330–335	21.6	21.4	20.6	19.4	17.9	16.2	14.7	13.9	14.9	16.0
350–355	21.7	21.5	20.6	19.3	17.7	15.8	13.9	12.6	13.3	15.2
370–375	21.7	21.5	20.6	19.2	17.4	15.4	13.3	11.6	11.9	14.3
390–395	21.7	21.4	20.5	19.1	17.2	15.0	12.7	10.8	10.7	13.4
410–415	21.7	21.4	20.5	19.0	17.0	14.8	12.4	10.2	9.7	12.7
440–450	21.6	21.3	20.3	18.8	16.8	14.5	11.9	9.5	8.5	11.6
480–490	21.5	21.2	20.2	18.7	16.6	14.1	11.5	8.9	7.6	10.3
520–530	21.4	21.1	20.1	18.5	16.4	13.9	11.2	8.5	7.0	9.3
560–570	21.3	21.0	20.0	18.4	16.3	13.8	11.0	8.3	6.7	9.2
620–640	21.2	20.9	19.9	18.2	16.1	13.6	10.8	8.1	6.3	7.6
680–700	21.1	20.8	19.8	18.2	16.0	13.5	10.7	7.9	6.1	6.9

Source: Peterson, 1976, and Demerjian et al., 1980.
[a]Zenith angles.

TABLE 3.11. Percentage Change of Calculated Actinic Flux at the Earth's Surface Using Best Estimate Albedos as a Function of Solar Zenith Angle and Selected Wavelengths Relative to the Values in Table 3.5 when Model Aerosol Concentrations Are Either Zero or Doubled

Wavelength (nm)	Actinic Flux Change (%)									
	0°[a]	10°	20°	30°	40°	50°	60°	70°	78°	86°
340–345										
No Aerosol	+8.2	+8.4	+8.8	+9.5	+10.7	+12.7	+16.1	+22.3	+26.5	+17.6
Double	−6.1	−6.3	−6.6	−7.3	−8.3	−10.1	−12.8	−16.1	−16.4	−12.5
400–405										
No Aerosol	+5.8	+6.0	+6.4	+7.1	+8.3	+10.7	+15.3	+26.2	+46.8	+35.7
Double	−4.0	−4.1	−4.5	−5.3	−6.6	−8.8	−12.6	−19.4	−24.9	−15.9
540–550										
No Aerosol	+0.9	+1.0	+1.2	+1.8	+2.9	+5.1	+10.4	+25.4	+67.1	+261.
Double	−0.8	−0.9	−1.4	−2.2	−3.7	−6.4	−11.6	−21.4	−33.6	−27.4
680–700										
No Aerosol	−2.4	−2.4	−2.2	−1.8	−0.9	+1.1	+6.2	+21.7	+67.0	+447
Double	+0.7	+0.6	+0.2	−0.6	−2.1	−4.9	−10.4	−20.8	−34.9	−35.5

Source: Peterson, 1976, and Demerjian et al., 1980.
[a]Zenith angles.

TABLE 3.12. Percentage Decrease of Calculated Actinic Flux at the Earth's Surface Using Best Estimate Albedos as a Function of Solar Zenith Angle and Selected Wavelengths Relative to the Values in Table 3.5 when Model Ozone Concentrations Are Increased by 5%

Wavelength (nm)	Actinic Flux Decrease (%)									
	0°[a]	10°	20°	30°	40°	50°	60°	70°	78°	86°
300–305	9.9	10.1	10.5	11.3	12.6	14.6	17.7	21.3	18.3	16.3
320–325	1.7	1.7	1.8	1.9	2.2	2.5	3.1	4.3	6.3	10.5

Source: Peterson, 1976, and Demerjian et al., 1980.
[a] Zenith angles.

in the O_3 concentration. This is particularly important since the total column abundance of O_3 can change by 10% or more within a season or latitude belt.

3-A-2g. *Effects of Clouds on J(λ)*

All the calculated actinic fluxes discussed above refer to a cloudless sky. At present, there is no universally accepted, accurate method for taking into account the effect of clouds. Both simple cloud transmission models and the more complex radiative transfer models have advantages and disadvantages. In addition, neither has been shown to give good fits to experimentally measured atmospheric data over a wide range of conditions and cloud types.

Peterson and co-workers used the cloud transmission method of Atwater and Brown (1974) in which the clear sky fluxes are multiplied by a factor C, where

$$C = \sum_{i=1}^{n} [1 - c_i(1 - T)] \tag{O}$$

In Eq. (O), n is the number of cloud layers present, c_i is the amount of cloud in each layer, and T is the transmission of solar radiation through the cloud. The value of T depends on the type of cloud; the estimated values of T in Table 3.13 are based on measurements of total solar flux under clear and cloudy conditions. However, as discussed by Peterson and co-workers, any such approximations for clouds are likely inaccurate.

The large uncertainty in the changes in the actinic flux due to cloud cover makes it impossible to accurately estimate rates of photolysis under cloudy conditions. This is unfortunate as the *relative* rates of photolysis of NO_2 (which leads to O_3

TABLE 3.13. Transmission (T) of Solar Radiation Through Various Cloud Types as a Function of Optical Air Mass (M)

Cloud Type	Equation[a]
Fog	$T = 0.1626 + 0.0054M$
Stratus	$T = 0.2684 - 0.0101M$
Stratocumulus	$T = 0.3658 - 0.0149M$
Cumulus	$T = 0.3658 - 0.0149M$
Cumulonimbus	$T = 0.2363 + 0.0145M$
Altostratus	$T = 0.4130 - 0.0014M$
Altocumulus	$T = 0.5456 - 0.0236M$
Cirrus	$T = 0.8717 - 0.0179M$
Cirrostratus	$T = 0.9055 - 0.0638M$

Source: Atwater and Brown, 1974.

[a] M = surface pressure/(1013 cos θ), where θ = solar zenith angle.

formation) and of O_3 itself are predicted to be sensitive to the presence of clouds; thus clouds may possibly alter the relative rates of O_3 formation and destruction, i.e. the O_3 concentration, in the unpolluted troposphere (Thompson, 1984).

3-A-2h. Comparison of Calculated Actinic Fluxes to Experimentally Measured Fluxes

As described in Chapter 6, a common method of measuring the solar intensity in environmental chambers, as well as in ambient air, involves measuring the rate of photolysis of NO_2. Demerjian and co-workers have compared the data of Sickles et al. (1977) for the rate of photolysis of NO_2 as a function of time during the day at Research Triangle Park, North Carolina (35.8°N, 78.6°W) to the rates of NO_2 photolysis expected using the actinic flux data from their model calculations (Table 3.5). Figure 3.14 shows the results of this comparison for a clear day. The agreement between the calculated and experimental rate constants, k_p, is quite good, in general. The overprediction for the rate of NO_2 photolysis around noon by as much as 15% has been attributed by Demerjian et al. to experimental errors in the measured photolysis rates.

Other workers have also found good agreement between the experimentally observed and theoretically predicted rates of NO_2 photolysis on clear sky days. For example, Fig. 3.15 shows the measured NO_2 photolysis rate constants at a Colorado site as a function of solar zenith angle on three separate days. The solid

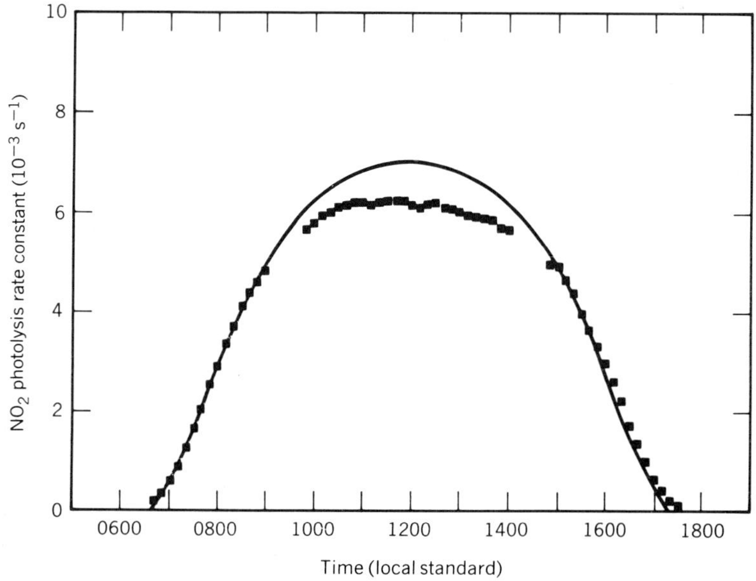

FIGURE 3.14. Comparison of the experimental (■) and clear-sky theoretical (—) diurnal variation of the photolytic rate constant k_p (units of 10^{-3} s^{-1}) for the photolysis of NO_2 near Raleigh, North Carolina (35.8°N, 78.6°W), on a clear day, October 22, 1975 (from Demerjian et al., 1980).

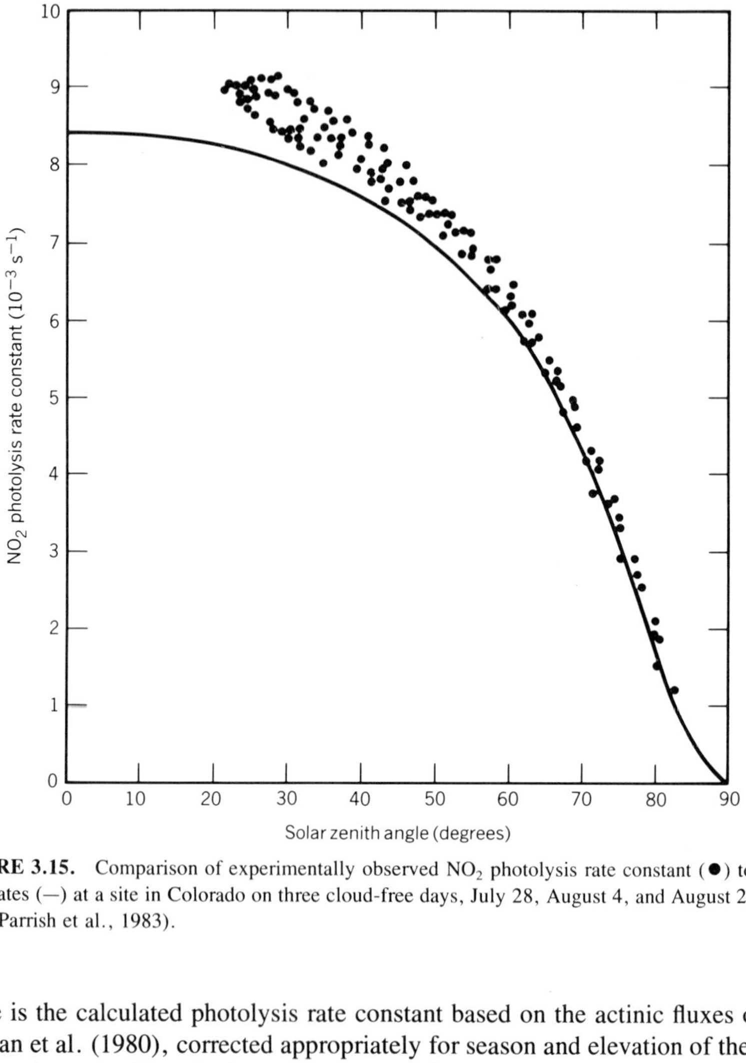

FIGURE 3.15. Comparison of experimentally observed NO_2 photolysis rate constant (●) to calculated rates (—) at a site in Colorado on three cloud-free days, July 28, August 4, and August 21, 1980 (from Parrish et al., 1983).

curve is the calculated photolysis rate constant based on the actinic fluxes of Demerjian et al. (1980), corrected appropriately for season and elevation of the monitoring site as well as for best estimates of the atmospheric aerosol and surface albedo. The agreement is excellent, except at smaller zenith angles where the calculated rates are a few percent smaller than those observed.

However, as might be expected from Section 3-A-2g, accurately predicting photolysis rates when clouds are present is much more difficult, and the results, of necessity, are highly uncertain. Rather than attempting to predict from first principles the change due to clouds, an alternate approach is frequently taken in which the calculated actinic fluxes and photolysis rates are scaled using the intensity of UV radiation measured with a photometer. Figure 3.16 compares the observed NO_2 photolysis rate constant with the calculated rates that have been scaled using the decrease in UV intensity on an overcast day with some rain.

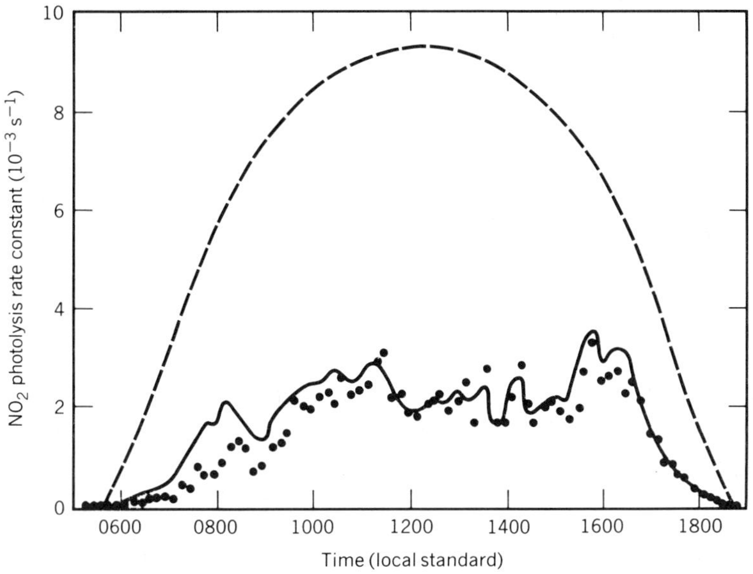

FIGURE 3.16. Comparison of the experimental (●), theoretical (---), and UV-scaled theoretical (—) diurnal variation of the photolytic rate constant for the photolysis of NO_2 near Raleigh, North Carolina (35.8°N, 78.6°W), on an overcast day with some rain, April 28, 1975 (from Demerjian et al., 1980).

3-B. CALCULATING RATES OF PHOTOLYSIS

3-B-1. Calculation Procedure

In Section 3-A-2a we developed an expression, Eq. (N), for $I_a(\lambda)$, the number of photons cm^{-3} s^{-1} absorbed by a species X:

$$I_a(\lambda) = \sigma(\lambda)\ J(\lambda)\ [X]\ \frac{\text{photons}}{\text{cm}^3\ \text{s}} \tag{N}$$

In Eq. (N), $\sigma(\lambda)$ is the absorption cross section of X at wavelength λ (see Section 2-A-2) in units of cm^2 $molecule^{-1}$, $J(\lambda)$ is the actinic flux in units of photons cm^{-2} s^{-1} given in Table 3.5 (with appropriate corrections for latitude, surface albedo, etc.), and [X] is the concentration of the absorbing species in molecules cm^{-3}.

To calculate rates of photolysis of X from Eq. (N), one needs to take into account only the fraction of light absorbed that leads to photochemical changes (as opposed to photophysical processes such as fluorescence in which the absorbing molecule is not chemically altered); thus, as discussed in Chapter 2, upon absorbing light, the electronically excited molecule X* may either form products

$$X + h\nu \rightarrow X^* \rightarrow \text{products} \tag{1a}$$

or return to the ground state by photophysical processes such as fluorescence or collisional deactivation:

$$X + h\nu \rightarrow X^* \rightarrow X + h\nu' \qquad \text{(1b)}$$
$$X^* \xrightarrow{M} X$$

The rate of photolysis of X at one wavelength λ, in units of molecules cm^{-3} s^{-1}, is given by

$$\text{rate of photolysis of X at wavelength } \lambda = \sigma(\lambda)\ \phi(\lambda)\ J(\lambda)\ [X] \qquad \text{(P)}$$

where $\phi(\lambda)$ is the total primary quantum yield at the wavelength λ for the *loss* of X due to its conversion into products by one or more primary photochemical processes (reaction 1a). For example, in the case of formaldehyde (Section 3-B-2), there are two primary photochemical processes, one producing H + HCO, and the other H_2 + CO. Thus $\phi(\lambda) = \phi_{H+HCO} + \phi_{H_2+CO}$. However, in the troposphere, the total rate of photolysis must include all wavelengths ≳290 nm. This is given by Eq. (Q):

$$\text{Total rate of photolysis of X} = \int_{\lambda=290\,\text{nm}}^{\lambda_i} \sigma(\lambda)\ \phi(\lambda)\ J(\lambda)\ d\lambda\ [X] \qquad \text{(Q)}$$
$$(\text{molecules cm}^{-3}\ \text{s}^{-1})$$

λ_i is the longest wavelength at which the light absorption occurs and/or ϕ is nonzero.

In practice, rather than integrating Eq. (Q) in a continuous fashion, the sum of the product $[\sigma(\lambda)\ \phi(\lambda)\ J(\lambda)]$ over discrete wavelength intervals $\Delta\lambda$ is used. Choice of the width of these intervals is usually based on the intervals for which $J(\lambda)$ is reported. Thus the data in Table 3.5 are reported in 5 nm intervals from 290 to 420 nm, 10 nm intervals from 420 to 580 nm, and 20 nm intervals from 580 to 700 nm; we use these in our calculations. Since absorption coefficients (σ) and primary quantum yields (ϕ) are normally not measured using identical intervals, representative averages of these parameters must often be calculated from the literature.

In summary, the total rate of photolysis of a species X in the atmosphere, in units of molecules cm^{-3} s^{-1}, can be calculated using Eq. (R):

$$\text{Total rate of photolysis of X} = \sum_{\lambda=290\,\text{nm}}^{\lambda_i} \sigma(\lambda)\ \phi(\lambda)\ J(\lambda)\ [X] \qquad \text{(R)}$$
$$(\text{molecules cm}^{-3}\ \text{s}^{-1})$$

where

$\sigma(\lambda)$ = absorption cross section base *e* of X in cm^2 $molecule^{-1}$ averaged over a wavelength interval $\Delta\lambda$, centered at λ;

$\phi(\lambda)$ = primary quantum yield for the loss of X averaged over the wavelength interval $\Delta\lambda$, centered at λ;

$J(\lambda)$ = actinic flux in photons cm^{-2} s^{-1} summed over the wavelength interval $\Delta\lambda$ centered at λ, at a solar zenith angle θ (Table 3.5), corrected for season (Table 3.6) and latitude (Table 3.7). If desired, corrections for altitude, surface albedo, and so on can be included.

This total rate of photolysis is also equal to an effective first-order rate constant, k_p (s^{-1}), times the concentration of X:

$$\text{Total rate of photolysis of X} = k_p[\text{X}] \quad (\text{molecules cm}^{-3}\ \text{s}^{-1}) \qquad \text{(S)}$$

Combining Eqs. (R) and (S), one obtains

$$k_p\ (\text{s}^{-1}) = \sum_{\lambda = 290\,\text{nm}}^{\lambda_i} \sigma(\lambda)\ \phi(\lambda)\ J(\lambda) \qquad \text{(T)}$$

The integral form of Eq. (T) is, of course,

$$k_p\ (\text{s}^{-1}) = \int_{\lambda = 290\,\text{nm}}^{\lambda_i} \sigma(\lambda)\ \phi(\lambda)\ J(\lambda)\ d\lambda \qquad \text{(U)}$$

The first-order rate constant for the photolysis of X, k_p, is sometimes referred to as the photolytic rate constant. We again emphasize that the absorption cross sections $\sigma(\lambda)$ used in Eq's (T) and (U) must be to the base *e* because of the weak absorption approximation used in deriving these equations (see above).

Experimental protocols have been developed both in the United States and Europe for determining the absorption cross sections, σ, in units of cm^2 $molecule^{-1}$ of organics that photolyze in the troposphere [Pitts et al., 1981; European Chemical Industry Ecology and Toxicology Center (ECETOC), 1983], as well as for measuring the photolytic rate constants (k_p) for substances that absorb light of $\lambda \geq 290$ nm (ECETOC, 1983; Carter et al., 1984). The absorption cross sections are determined from 285 to 825 nm at various gas phase concentrations in 1 atm of ultra-pure air using a conventional UV–visible spectrophotometer (Pitts et al., 1981).

An alternate, perhaps experimentally more convenient method, is to obtain the absorption spectrum of the compound dissolved in an inert organic solvent. Of course, this introduces greater uncertainty into the results (ECETOC, 1983); the effects of solvent on absorption spectra are discussed in Chapter 13. The measured absorbances are averaged over 10 nm wavelength intervals, and, using the known concentrations, these are converted to average absorption cross sections over 10 nm intervals.

There are several procedures for determining photolysis reaction rate constants. One experimental approach involves following the simultaneous rates of disappearance of the species of interest and of a reference organic in an irradiated NO–organic–air mixture; the rate constants for OH attack on the unknown and reference must be known to correct the observed disappearance rates for the OH reaction (Carter et al., 1984). A limitation of this method is that the minimum value of k_p which can be accurately measured using this technique is about 1×10^{-5} s^{-1}, which is relatively rapid; thus while nitrite, α-dicarbonyl, and nitrosamine photolysis rates could be measured, that of simple aldehydes and ketones could not.

An alternate approach is to use the experimentally determined absorption cross sections, $\sigma(\lambda)$, in combination with actinic flux estimates, $J(\lambda)$, to calculate k_p from Eq. (T), using literature values for $\phi(\lambda)$. This procedure may be quite accurate for studies carried out at one atmosphere in air, but subject to significant errors for quantum yields determined at reduced pressures in the absence of air. If no primary quantum yield data are available for the compound in question, one can *assume* a quantum yield of 1.0. However, since $\phi \leq 1.0$ for primary photochemical processes, this will provide only an *upper limit* to the photolysis rate constant. The rationale behind one's choice of the value for ϕ should be stated since it can range from unity down to zero, even though the molecule absorbs strongly.

3-B-2. Example: HCHO Photolysis

As an example, let us calculate the rate of photolysis of formaldehyde in the troposphere at sea level, at a latitude of 40°N. As we have already noted, HCHO plays an important role in both clean and polluted atmospheres because one of its primary photodissociative paths, (2a), produces hydrogen atoms and formyl radicals:

$$\begin{aligned} HCHO + h\nu &\xrightarrow{a} H + HCO \\ &\xrightarrow{b} H_2 + CO \end{aligned} \qquad (2)$$

In air both H and HCO rapidly react with O_2 to form HO_2 which, in the presence of NO, forms hydroxyl radicals.

We calculate the noontime photolysis rates for January 1 and July 1, and in the latter case also at sunrise and sunset. We assume cloudless conditions and that the "best estimates" of the surface albedo of Peterson and co-workers apply, as do their estimated O_3 and particle concentrations.

The absorption spectrum, absorption cross sections, and quantum yields for HCHO photolysis are discussed in detail in Section 3-C-9. It absorbs actinic UV out to approximately 370 nm, with well-defined vibronic bands. Table 3.14 summarizes the currently recommended absorption cross sections and primary quantum yields for reactions (2a) and (2b) from 290 to 360 nm in 10 nm intervals centered on the given wavelengths. Thus from 285 to 290 nm and from 290 to 295 nm, $\sigma = 2.51 \times 10^{-20}$ cm^2 $molecule^{-1}$, $\phi_{2a} = 0.71$ and $\phi_{2b} = 0.26$. While the cross sections and quantum yields are for 300°K (27°C), which will not be typical of all the conditions at noon on January 1 and July 1, as well as at sunrise/sunset on July 1 at 40°N latitude, they are not extremely temperature sensitive over the range normally encountered in the atmosphere (Section 3-C-9); hence their use in

TABLE 3.14. Absorption Cross Sections (σ) and Primary Quantum Yields (ϕ) for Reactions (2a) and (2b) for HCHO at 300°K and One Atmosphere Pressure

λ (nm)	Absorption Cross Section[a] σ (10^{-20} cm^{-2} molecule^{-1})	Quantum Yields ϕ_{2a} (H + HCO)	ϕ_{2b} (H_2 + CO)
290	2.51	0.71	0.26
300	2.62	0.78	0.22
310	2.45	0.77	0.23
320	1.85	0.62	0.38
330[b]	1.76	0.31	0.80
340	1.18	~0 (<0.1)	0.69
350	0.42	0	0.40
360	0.06	0	0.12

Source: DeMore et al., 1983; values are averages at 290°K for 10 nm intervals centered on the indicated wavelengths.

[a] $(\ln (I_o/I))/[X]\ell$, in units of cm^2 molecule^{-1}, [X] = concentration in molecules cm^{-3} and ℓ is the pathlength in cm.

[b] $(\phi_{2a} + \phi_{2b}) = 1.11$ at 330 nm; this is not possible for primary quantum yields whose sum must be ≤ 1.0 (Section 2-C-2). $\phi_{2b} = 0.61$, as recommended by Baulch et al. (1982), would be more consistent.

situations where the temperature is lower or higher should not introduce large errors in the calculated photolysis rates in this case.

To apply the actinic flux data in Table 3.5, we must first correct the solar zenith angles corresponding to noon and sunrise/sunset at a latitude of 40°N and at the two seasons to be treated. Let us first consider the actinic flux at noon. From Table 3.7, at a true solar time of 1200 and at a latitude of 40°N, the solar zenith angle is 63.0° on January 1 and 16.9° on July 1. To obtain the actinic flux on January 1 corresponding to $\theta = 63°$, we must interpolate between the $\theta = 60°$ and 70° data in Table 3.5. For example, in the wavelength interval from 305 to 310 nm the actinic flux will lie between 0.244×10^{14} ($\theta = 60°$) and 0.064×10^{14} photons cm^{-2} s^{-1} ($\theta = 70°$); we estimate J (305–310 nm) $\simeq 0.190 \times 10^{14}$ photons cm^{-2} s^{-1} for $\theta = 63°$. Similarly, the actinic flux at $\theta = 63°$ can be estimated for the other wavelength intervals.

However, an additional correction for season must be included. Thus the mean earth–sun distance changes with season, which changes the solar flux incident on the top of the atmosphere. These correction factors are given in Table 3.6. For January 1, the correction factor is 1.033 so that the actinic flux from 305 to 310 nm at $\theta = 63°$ on January 1 is $1.033 \times 0.190 \times 10^{14} = 0.196 \times 10^{14}$ photons cm^{-2} s^{-1}. The actinic fluxes for noon on January 1 at 40° latitude thus calculated are shown in the second column of Table 3.15 for the wavelength region 290–365 nm.

To obtain estimates of the solar flux at noon on July 1, one must estimate, using Table 3.5, the flux for $\theta = 16.9°$ by interpolating between $\theta = 10°$ and $\theta = 20°$; we estimate $J = 1.29 \times 10^{14}$ photons cm^{-2} s^{-1} from 305 to 310 nm. The correction factor for July 1 for the mean sun–earth distance is 0.966 from Table 3.6, so that the final estimate of the flux is $0.966 \times 1.29 \times 10^{14} = 1.25 \times 10^{14}$ photons cm^{-2} s^{-1}. This and the estimates for the remaining wavelength intervals are shown in the third column of Table 3.15.

The much smaller fluxes at sunrise and sunset similarly can be obtained using Tables 3.5 and 3.7. Thus from Table 3.7 on January 1 at a latitude of 40°, the sun is essentially on the horizon at 0730 and 1630, where $\theta = 89°$. On July 1, sunrise occurs at approximately 0436. Approximating sunrise by $\theta = 86°$, corresponding to the last entry in Table 3.7, one obtains the data in the fourth column of Table 3.15. Taking the actinic fluxes from Table 3.5 for $\theta = 86°$ and multiplying by the correction factor of 0.966 from Table 3.6 to take into account the change in the mean earth–sun distance, one obtains the fluxes shown in column 4 of Table 3.15; these apply to sunrise/sunset on July 1 at a latitude of 40°N.

As seen in Fig 3.8, the flux at a given wavelength changes much more rapidly at large zenith angles than at small ones. This introduces greater uncertainty into extrapolating between the larger values of θ characteristic of sunrise/sunset, in

TABLE 3.15. Estimated Actinic Fluxes[a] $J(\lambda)$ for a Latitude of 40°N

	Estimated Actinic Flux (10^{14} photons cm^{-2}s^{-1}) at 40°N		
$\Delta\lambda$ (nm)	Noon January 1	Noon July 1	Near Sunrise/Sunset ($\theta = 86°$) July 1
290–295	0.0	0.0	0.0
295–300	0.0	0.031	0.0
300–305	0.021	0.335	0.0
305–310	0.196	1.25	0.002
310–315	0.777	2.87	0.009
315–320	1.45	4.02	0.029
320–325	2.16	5.08	0.071
325–330	3.44	7.34	0.161
330–335	3.90	7.79	0.233
335–340	4.04	7.72	0.272
340–345	4.51	8.33	0.321
345–350	4.62	8.33	0.340
350–355	5.36	9.45	0.400
355–360	5.04	8.71	0.378
360–365	5.70	9.65	0.429

[a]Calculated as described in text using data from Tables 3.5, 3.6, and 3.7.

TABLE 3.16. Calculated Photolytic Rate Constants for the Two Paths in HCHO Photolysis at 40° N Latitude at Noon January 1 and July 1, and at Sunrise/Sunset on July 1

	Noon, January 1		Noon, July 1		Sunrise/Sunset July 1	
$\Delta\lambda$ (nm)	(2a) H+ HCO	(2b) H_2 + CO	(2a) H+ HCO	(2b) H_2 + CO	(2a) H + HCO	(2b) H_2 + CO
290–295	0	0	0	0	0	0
295–300	0	0	6.3(−8)	1.8(−8)	0	0
300–305	4.3(−8)[a]	1.2(−8)	6.8(−7)	1.9(−7)	0	0
305–310	3.7(−7)	1.1(−7)	2.3(−6)	7.0(−7)	3.8(−9)	1.1(−9)
310–315	1.5(−6)	4.4(−7)	5.4(−6)	1.6(−6)	1.7(−8)	5.1(−9)
315–320	1.7(−6)	1.0(−6)	4.6(−6)	2.8(−6)	3.3(−8)	2.0(−8)
320–325	2.5(−6)	1.5(−6)	5.8(−6)	3.6(−6)	8.1(−8)	5.0(−8)
325–330	1.9(−6)	4.8(−6)	4.0(−6)	1.0(−5)	8.8(−8)	2.3(−7)
330–335	2.1(−6)	5.5(−6)	4.3(−6)	1.1(−5)	1.3(−7)	3.3(−7)
335–340	~0	3.3(−6)	0	6.3(−6)	0	2.2(−7)
340–345	0	3.7(−6)	0	6.8(−6)	0	2.6(−7)
345–350	0	7.8(−7)	0	1.4(−6)	0	5.7(−8)
350–355	0	9.0(−7)	0	1.6(−6)	0	6.7(−8)
355–360	0	3.6(−8)	0	6.3(−8)	0	2.7(−9)
360–365	0	4.1(−8)	0	6.9(−8)	0	3.1(−9)
TOTAL[a] = k_p (s^{-1}):	10.1(−6)	22.1(−6)	27.1(−6)	46.1(−6)	3.5(−7)	12.5(−7)
Overall photolytic lifetime (τ):	8.6 h		3.8 h		174 h	

[a] 10.1(−6) ≡ 10.1×10^{-6}, and so on.

Table 3.5. Additionally, as Peterson and co-workers point out, the fluxes at greater zenith angles in Table 3.5 are themselves subject to greater errors because of the large air masses. The small fluxes at large θ are due mainly to multiple scattered radiation; thus small percentage changes in the atmospheric gases and particles responsible for light scattering can lead to larger percentage errors in the calculated fluxes.

Once the actinic fluxes, $J(\lambda)$, have been calculated as in Table 3.15, and the absorption cross sections $\sigma(\lambda)$ and quantum yields ϕ_{2a} and ϕ_{2b} tabulated as in Table 3.14, the photolytic rate constant k_p can be calculated from their product:

$$k_p\ (\mathrm{s}^{-1}) = \sum_{\lambda = 290\,\mathrm{nm}}^{\lambda = 365\,\mathrm{nm}} \sigma(\lambda)\ \phi(\lambda)\ J(\lambda) \qquad \text{(T)}$$

Table 3.16 shows the product $\sigma(\lambda)\ \phi(\lambda)\ J(\lambda)$ calculated for each of the two possible paths, (2a) and (2b), for formaldehyde photolysis at noon January 1 and July 1, or at sunrise/sunset on July 1, at a latitude of 40°N. The photolytic rate constant k_p is the sum of these, as shown by the totals given at the bottom of each column. Thus at noon on January 1, HCHO photolyzes to H + HCO with a first-order rate constant of $k_p^{2a} = 10.1 \times 10^{-6}\ \mathrm{s}^{-1}$, and to H_2 + CO with $k_p^{2b} = 22.1 \times 10^{-6}\ \mathrm{s}^{-1}$. Overall, the *total* photolysis occurs with a rate constant $k_p = k_p^{2a} + k_p^{2b} = (10.1 + 22.1) \times 10^{-6}\ \mathrm{s}^{-1} = 32.2 \times 10^{-6}\ \mathrm{s}^{-1}$.

As discussed in Section 4-A-6, this rate can also be expressed in terms of a *natural lifetime* (τ) with respect to photolysis; the lifetime is defined as the time required for the HCHO concentration to fall to $1/e$ of its initial value. The last row of Table 3.16 gives the lifetimes of HCHO with respect to photolysis via both channels (2a) and (2b) at 40°N latitude at the three chosen times. As expected, photolysis is much more rapid on July 1 than on January 1, and much faster at noon than at sunrise or sunset.

In summary, if the absorption cross sections and primary quantum yields are known for a light-absorbing molecule as a function of wavelength, the actinic fluxes given in Table 3.5 (with appropriate correction factors for season, latitude, etc.) can be applied to calculate the first-order rate constant for photolysis under the desired set of conditions. As noted above, such calculations are extremely important in atmospheric chemistry because photolysis is the major source of key species such as OH and HO_2.

3-C. IMPORTANT LIGHT-ABSORBING SPECIES IN THE CLEAN AND POLLUTED TROPOSPHERE: ABSORPTION SPECTRA, PRIMARY PHOTOCHEMICAL PROCESSES, AND QUANTUM YIELDS

3-C-1. Molecular Oxygen

3-C-1a. Absorption Spectra

Molecular oxygen absorbs light strongly in the ultraviolet at wavelengths below ~200 nm (Fig. 3.2).

Although this is not important in the troposphere itself, the photodissociation of O_2 at $\lambda \lesssim 220$ nm is important in the stratosphere (Chapter 15), and hence we include a brief discussion of its absorption spectra and photochemistry here. The potential energy curves for the ground state and for the first four electronically excited states of O_2 are shown in Fig. 3.17. The ground state, $X^3\Sigma_g^-$, is unusual in that it is a triplet; as a result, only transitions to upper triplet states are spin-allowed. The transition from the ground state to the $A^3\Sigma_u^+$ state is also theoretically forbidden because it involves a $(+) \rightarrow (-)$ transition (see Section 2-B-1c); however, this $X^3\Sigma_g^- \rightarrow A^3\Sigma_u^+$ transition does occur weakly, resulting in weak absorption bands ($\sigma \lesssim 10^{-23}$ cm^2 molecule^{-1}), known as the Herzberg continuum, at wavelengths between ~200 and 300 nm. At present, there is considerable uncertainty in the absolute values of the cross sections in this region (DeMore et al., 1985).

The $X^3\Sigma_g^- \rightarrow B^3\Sigma_u^-$ transition is allowed and results in an absorption in the 130–200 nm region (Figs. 3.18 and 3.19) known as the Schumann–Runge system. As shown in Fig. 3.19, a banded structure is seen from about 175 to 200 nm; this corresponds to transitions from $v'' = 0$ and $v'' = 1$ (i.e., hot bands) of the ground ($X^3\Sigma_g^-$) state to different vibrational levels of the upper state. A recent high-resolution study has established the wavenumbers and rotational and vibrational assignments of the bands in the 175–205 nm region (Yoshino et al., 1984). The upper $^3\Sigma_u^-$ state is crossed by the repulsive $^3\Pi_u$ state (Fig. 3.17) at $\sim v' = 4$,

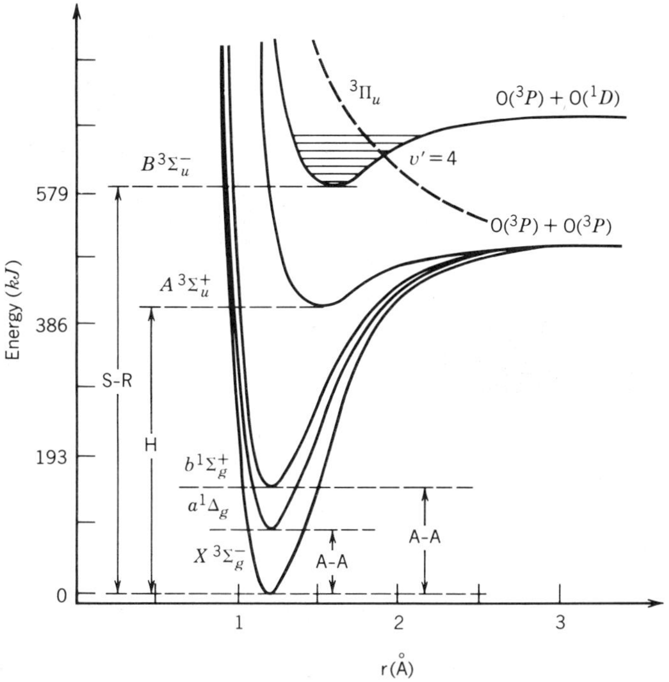

FIGURE 3.17. Potential energy curves for ground and first four excited states of O_2. S–R = Schumann–Runge system, H = Herzberg continuum, A–A = atmospheric bands (from Gaydon, 1968).

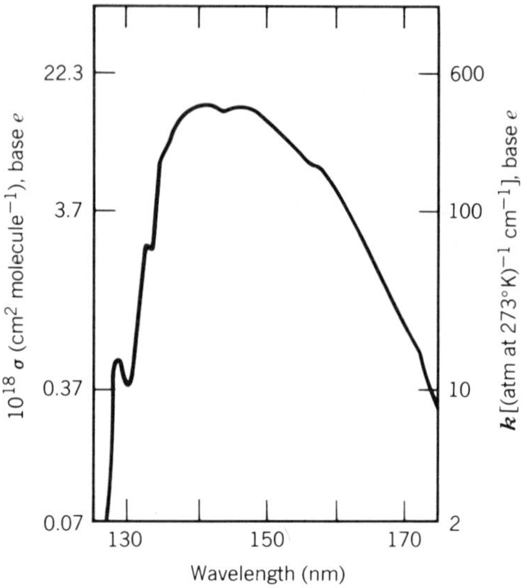

FIGURE 3.18. Absorption coefficients for O_2 in the Schumann–Runge continuum. Left axis is the absorption cross section, σ (cm^2 $molecule^{-1}$), base e; right axis is k [(atm at 273°K)$^{-1}$ cm^{-1}], base e (from Inn, 1955–56).

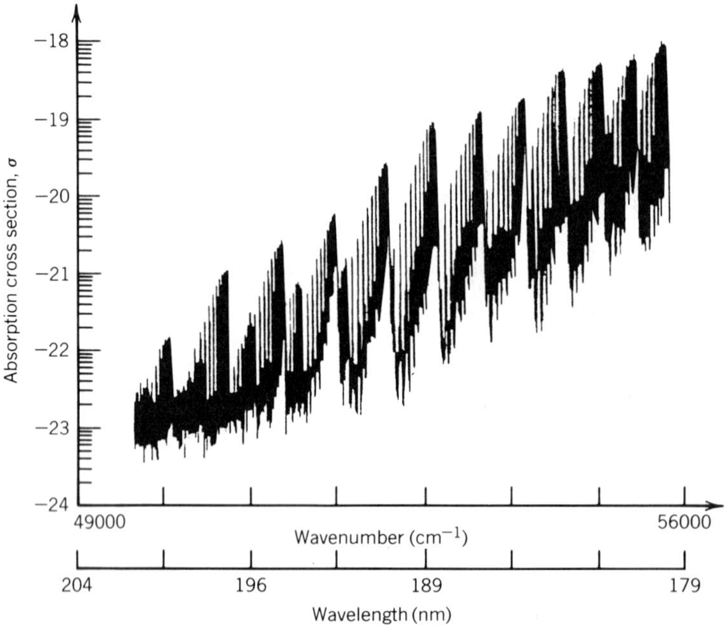

FIGURE 3.19. Semi-logarithmic plot (base 10) of measured absorption coefficients in terms of the absorption cross section, σ (cm^2 $molecule^{-1}$), base e, for O_2 at 300°K in the 179.3-201.5 nm region. The structure seen for $\sigma \gtrsim 10^{-22}$ cm^2 $molecule^{-1}$ is real; at smaller cross sections, some noise is present (from Yoshino et al., 1983).

providing a mechanism for the production of two ground-state $O(^3P)$ atoms from the $B^3\Sigma_u^-$ state.

The absorption spectrum becomes continuous at ~175 nm, with a strong absorption down to ~130 nm. This continuum is believed to be due to dissociation of the $B^3\Sigma_u^-$ state to $O(^3P) + O(^1D)$. Below 130 nm, a banded absorption again appears, as seen in Fig. 3.20; at wavelengths below 133.2 nm, there is sufficient energy to produce $O(^1S)$ atoms.

There is a minimum in the absorption at 121.6 nm, coincident with the Lyman α line of the hydrogen atom (Fig. 3.20). This is fortunate as O_2 can thus be used as a filter for Lyman α radiation. For example, if one wishes to detect H atoms by resonance fluorescence or resonance absorption at 121.6 nm (Section 4-B-2c), radiation at other wavelengths which might interfere (e.g., from oxygen atoms at 130.5 nm) can be selectively blocked using a stream of O_2 in front of the detector.

In addition to the absorptions in the ultraviolet, O_2 also has very weak absorptions in the red (762 nm) and infrared (1.27 and 1.07 μm), known as the *atmospheric oxygen bands*. These produce O_2 in the singlet excited states $b^1\Sigma_g^+$ and $a^1\Delta_g$, respectively. The two bands in the infrared correspond to transitions to two different vibrational levels, $v' = 0$ (1.27 μm) and $v' = 1$ (1.07 μm) within the $a^1\Delta_g$ state.

The absorption cross sections for O_2 in various spectral regions are discussed in detail in the 1983 NASA document (De More et al., 1983).

3-C-1b. *Photochemistry*

Table 3.17 summarizes the threshold wavelengths for production of ground-state oxygen atoms, $O(^3P)$, as well as electronically excited $O(^1D)$ and $O(^1S)$ atoms.

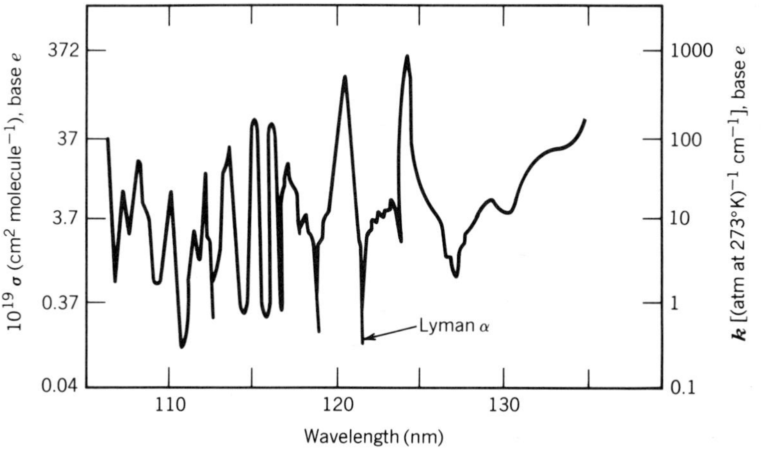

FIGURE 3.20. Absorption coefficients of O_2 in the 105–130 nm region. Left axis is σ (cm^2 molecule^{-1}), base e; Right axis is k [(atm at 273°K)$^{-1}$ cm^{-1}], base e. The O_2 absorption line corresponding to the Lyman-α line of the H atom is shown by the arrow (from Inn, 1955–56).

TABLE 3.17. Threshold Wavelengths for the Production of Ground-State or Electronically Excited Oxygen Atoms from O_2 Photolysis

Electronic State of Oxygen Atoms[a]	Threshold Wavelength (nm)
$O(^3P) + O(^3P)$	242.4
$O(^3P) + O(^1D)$	175.0
$O(^3P) + O(^1S)$	133.2

Source: Okabe, 1978.
[a] $O(^3P)$ is the ground-state species.

Dissociation of O_2 in the 175–242 nm region to produce atoms is particularly important in the stratosphere because it is the only significant source of O_3 (see Chapter 15).

Although light absorption to excite O_2 into the $a^1\Delta_g$ and $b^1\Sigma_g^+$ states (collectively referred to as *singlet oxygen*) is very weak, there are other sources of singlet oxygen, including energy transfer (e.g., from electronically excited NO_2^*), O_3 photolysis, and exothermic chemical reactions. When $O_2(a^1\Delta_g)$ and $O_2(b^1\Sigma_g^+)$ are formed in such processes, they do not readily undergo radiative transitions to the ground state because, as in absorption, the processes are "forbidden." Thus the radiative lifetime of the $b^1\Sigma_g^+$ state, which lies 37.5 kcal above the ground state, is ~12 s (Wallace and Hunten, 1968), whereas that for the $a^1\Delta_g$ state, 22.5 kcal above the ground state, is ~65 min (Badger et al., 1965). As a result, when these electronic states are formed in the atmosphere they are primarily collisionally deactivated to ground-state O_2 (Section 7-E-5).

Under low-pressure conditions in the laboratory or in the upper atmosphere, where collisional deactivation is slow, weak radiative transitions from these two excited singlet states to the ground state are observed. The (0,0) emission bands of the $b^1\Sigma_g^+ \rightarrow X^3\Sigma_g^-$ and the $a^1\Delta_g \rightarrow X^3\Sigma_g^-$ transitions occur at 761.9 and 1269 nm, respectively. The 761.9 nm band due to $b^1\Sigma_g^+$ is often observed in systems containing the $a^1\Delta_g$ state because of the energy pooling reaction:

$$O_2(a^1\Delta_g) + O_2(a^1\Delta_g) \rightarrow O_2(X^3\Sigma_g^-) + O_2(b^1\Sigma_g^+) \quad (3)$$

Possible sources of $^1O_2$ in the atmosphere and its fates are discussed in detail in Section 7-E-5.

3-C-2. Ozone

3-C-2a. Absorption Spectra

The absorption spectrum of O_3 from 200 to 300 nm, known as the Hartley bands, is shown in Fig. 3.21. As we saw earlier, this strong light absorption by O_3 in the

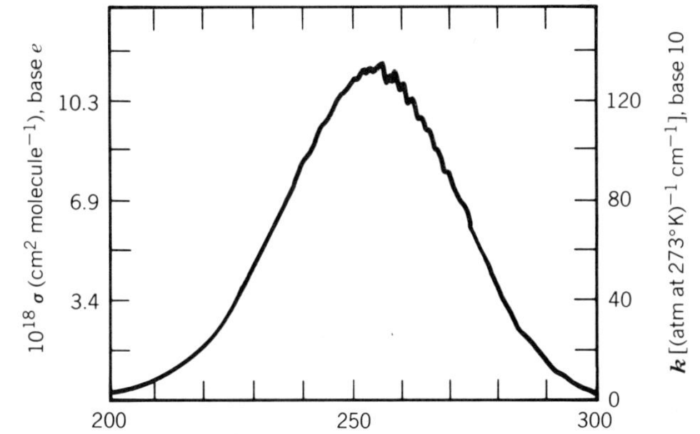

FIGURE 3.21. Absorption spectrum of O_3 known as the Hartley bands. Left axis is σ (cm^2 molecule^{-1}), base e, right axis is k [(atm at 273)$^{-1}$ cm^{-1}], base 10 (from Griggs, 1968).

stratosphere controls the short-wavelength limit of light that reaches the troposphere. Absorption in the region 300–360 nm, the Huggins bands, is shown in Fig. 3.22, while 3.23 shows the absorption in the 440–850 nm region, known as the Chappius bands.

Table 3.18 shows the absorption cross sections (base e) in 1 nm intervals from 253 to 360 nm as a function of temperature in the range 298–206°K (Davenport,

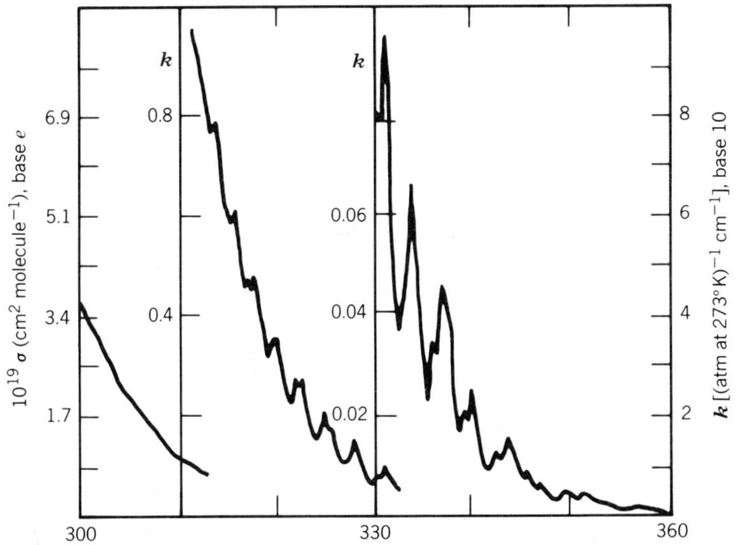

FIGURE 3.22. Absorption spectrum of O_3 known as the Huggins bands. Left axis is σ (cm^2 molecule^{-1}), base e; right axis is k [(atm at 273)$^{-1}$ cm^{-1}], base 10. Note that the units of the expanded spectra from 310–330 nm and 330–360 nm are k[(atm at 273)$^{-1}$ cm^{-1}], base 10 (from Griggs, 1968).

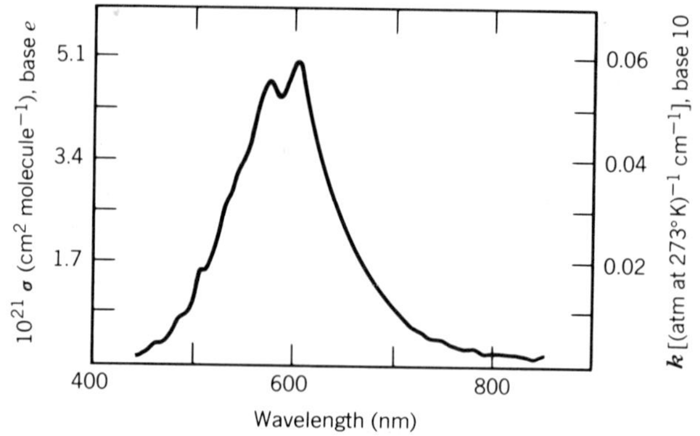

FIGURE 3.23. Absorption spectrum of O_3 known as the Chappius bands. Left axis is σ (cm^2 molecule^{-1}), base e; right axis is k [(atm at 273)$^{-1}$ cm^{-1}], base 10 (from Griggs, 1968).

1982). These values agree reasonably well with earlier published absorption cross sections, as well as with a more recent set determined in 0.05 nm increments from 203 to 298°K (Bass, 1984). In order to use these absorption cross sections to calculate photolysis rate constants for O_3 using Eq. (T), the reader will have to average the data over the appropriate wavelength intervals as used in the flux data of Table 3.5.

3-C-2b. *Photochemistry*

The photolysis of O_3 produces molecular oxygen and atomic oxygen, either or both of which may be in excited states, depending on the excitation energy. For example, at $\lambda < 320$ nm, the primary photochemical process is

$$O_3 + h\nu\ (\lambda \lesssim 320 \text{ nm}) \rightarrow O_2(^1\Delta_g) + O(^1D) \tag{4}$$

Table 3.19 shows the wavelength threshold below which each combination of products may be formed.

The most important aspect of O_3 photochemistry for the troposphere is the yield and wavelength dependence of $O(^1D)$ production in reaction (4) since it is a source of hydroxyl free radicals via its reaction with water:

$$O(^1D) + H_2O \rightarrow 2\ OH \tag{5}$$

$$k_5^{298°K} = 2.2 \times 10^{-10} \text{ cm}^3 \text{ molecule}^{-1} \text{ s}^{-1}$$

This very fast reaction occurs in competition with the deactivation of $O(^1D)$ by air:

$$O(^1D) + M(M = \text{air}) \rightarrow O(^3P) + M \tag{6}$$

$$k_6^{298°K} = 2.9 \times 10^{-11} \text{ cm}^3 \text{ molecule}^{-1} \text{ s}^{-1}$$

TABLE 3.18. **Absolute Absorption Cross Sections σ (cm^2 $molecule^{-1}$), Base *e*, for Ozone at 253–330 nm from 206 to 298°K**

Wavelength (nm)	298°K	271°K	225°K	206°K
253	1.21×10^{-17}	1.10×10^{-17}	1.08×10^{-17}	1.09×10^{-17}
254	1.41×10^{-17}	1.37×10^{-17}	1.33×10^{-17}	1.31×10^{-17}
255	1.15×10^{-17}	1.06×10^{-17}	1.09×10^{-17}	1.08×10^{-17}
256	1.32×10^{-17}	1.34×10^{-17}	1.31×10^{-17}	1.29×10^{-17}
257	9.71×10^{-18}	9.90×10^{-18}	9.61×10^{-18}	9.60×10^{-18}
258	1.11×10^{-17}	1.11×10^{-17}	1.08×10^{-17}	1.02×10^{-17}
259	1.13×10^{-17}	1.10×10^{-17}	1.09×10^{-17}	1.07×10^{-17}
260	1.14×10^{-17}	1.11×10^{-17}	1.08×10^{-17}	1.04×10^{-17}
261	1.08×10^{-17}	1.05×10^{-17}	1.01×10^{-17}	1.00×10^{-17}
262	1.05×10^{-17}	1.06×10^{-17}	1.00×10^{-17}	9.68×10^{-18}
263	1.03×10^{-17}	9.98×10^{-18}	9.38×10^{-18}	9.31×10^{-18}
264	1.03×10^{-17}	9.81×10^{-18}	9.42×10^{-18}	9.26×10^{-18}
265	9.67×10^{-18}	9.29×10^{-18}	9.11×10^{-18}	9.10×10^{-18}
266	9.10×10^{-18}	9.07×10^{-18}	9.05×10^{-18}	9.06×10^{-18}
267	8.7×10^{-18}	8.54×10^{-18}	8.49×10^{-18}	8.39×10^{-18}
268	7.57×10^{-18}	8.18×10^{-18}	8.07×10^{-18}	8.02×10^{-18}
269	8.24×10^{-18}	7.85×10^{-18}	7.81×10^{-18}	7.79×10^{-18}
270	7.81×10^{-18}	7.76×10^{-18}	7.61×10^{-18}	7.58×10^{-18}
271	7.82×10^{-18}	7.26×10^{-18}	7.20×10^{-18}	7.21×10^{-18}
272	6.95×10^{-18}	6.91×10^{-18}	6.90×10^{-18}	6.83×10^{-18}
273	6.24×10^{-18}	6.30×10^{-18}	6.21×10^{-18}	6.12×10^{-18}
274	5.82×10^{-18}	5.85×10^{-18}	5.51×10^{-18}	5.43×10^{-18}
275	5.51×10^{-18}	5.31×10^{-18}	5.20×10^{-18}	5.13×10^{-18}
276	5.16×10^{-18}	5.17×10^{-18}	5.13×10^{-18}	5.01×10^{-18}
277	5.03×10^{-18}	4.84×10^{-18}	4.80×10^{-18}	4.71×10^{-18}
278	4.58×10^{-18}	4.42×10^{-18}	4.31×10^{-18}	4.30×10^{-18}
279	4.04×10^{-18}	4.08×10^{-18}	4.01×10^{-18}	3.99×10^{-18}
280	3.85×10^{-18}	3.81×10^{-18}	3.81×10^{-18}	3.78×10^{-18}
281	3.45×10^{-18}	3.46×10^{-18}	3.39×10^{-18}	3.40×10^{-18}
282	3.40×10^{-18}	3.20×10^{-18}	3.21×10^{-18}	3.16×10^{-18}
283	3.04×10^{-18}	3.00×10^{-18}	3.01×10^{-18}	2.94×10^{-18}
284	2.66×10^{-18}	2.61×10^{-18}	2.55×10^{-18}	2.56×10^{-18}
285	2.42×10^{-18}	2.34×10^{-18}	2.31×10^{-18}	2.30×10^{-18}
286	2.07×10^{-18}	2.19×10^{-18}	2.02×10^{-18}	1.98×10^{-18}
287	2.00×10^{-18}	1.94×10^{-18}	1.92×10^{-18}	1.90×10^{-18}
288	1.70×10^{-18}	1.51×10^{-18}	1.48×10^{-18}	1.46×10^{-18}
289	1.54×10^{-18}	1.50×10^{-18}	1.41×10^{-18}	1.40×10^{-18}
290	1.35×10^{-18}	1.31×10^{-18}	1.21×10^{-18}	1.20×10^{-18}
291	1.21×10^{-18}	1.19×10^{-18}	1.00×10^{-18}	9.98×10^{-19}
292	1.10×10^{-18}	1.05×10^{-18}	1.01×10^{-18}	9.99×10^{-19}
293	8.39×10^{-19}	8.96×10^{-19}	8.45×10^{-19}	8.20×10^{-19}
294	8.31×10^{-19}	8.30×10^{-19}	8.29×10^{-19}	8.21×10^{-19}
295	7.88×10^{-19}	7.79×10^{-19}	7.71×10^{-19}	7.70×10^{-19}
296	6.47×10^{-19}	6.40×10^{-19}	6.40×10^{-19}	6.38×10^{-19}

TABLE 3.18. (*Continued*)

Wavelength (nm)	298°K	271°K	225°K	206°K
297	5.77×10^{-19}	5.70×10^{-19}	5.64×10^{-19}	5.60×10^{-19}
298	4.45×10^{-19}	4.18×10^{-19}	4.07×10^{-19}	4.04×10^{-19}
299	4.48×10^{-19}	4.15×10^{-19}	4.10×10^{-19}	4.03×10^{-19}
300	4.09×10^{-19}	3.96×10^{-19}	3.90×10^{-19}	3.85×10^{-19}
301	3.49×10^{-19}	3.31×10^{-19}	3.21×10^{-19}	3.20×10^{-19}
302	3.13×10^{-19}	2.96×10^{-19}	2.92×10^{-19}	2.91×10^{-19}
303	2.68×10^{-19}	2.60×10^{-19}	2.51×10^{-19}	2.48×10^{-19}
304	2.15×10^{-19}	2.28×10^{-19}	2.17×10^{-19}	2.08×10^{-19}
305	2.11×10^{-19}	1.99×10^{-19}	1.95×10^{-19}	1.91×10^{-19}
306	1.83×10^{-19}	1.79×10^{-19}	1.75×10^{-19}	1.73×10^{-19}
307	1.61×10^{-19}	1.55×10^{-19}	1.53×10^{-19}	1.49×10^{-19}
308	1.48×10^{-19}	1.51×10^{-19}	1.49×10^{-19}	1.51×10^{-19}
309	1.21×10^{-19}	1.23×10^{-19}	1.19×10^{-19}	1.12×10^{-19}
310	9.92×10^{-20}	9.81×10^{-20}	9.42×10^{-20}	9.89×10^{-20}
311	9.52×10^{-20}	9.48×10^{-20}	9.30×10^{-20}	9.10×10^{-20}
312	8.07×10^{-20}	7.41×10^{-20}	7.25×10^{-20}	6.92×10^{-20}
313	6.93×10^{-20}	6.73×10^{-20}	6.61×10^{-20}	6.54×10^{-20}
314	6.35×10^{-20}	5.14×10^{-20}	4.50×10^{-20}	3.96×10^{-20}
315	5.24×10^{-20}	5.01×10^{-20}	4.62×10^{-20}	4.54×10^{-20}
316	5.16×10^{-20}	4.98×10^{-20}	4.22×10^{-20}	4.19×10^{-20}
317	4.12×10^{-20}	3.83×10^{-20}	3.36×10^{-20}	2.98×10^{-20}
318	3.74×10^{-20}	3.41×10^{-20}	3.21×10^{-20}	2.79×10^{-20}
319	2.80×10^{-20}	2.51×10^{-20}	2.32×10^{-20}	2.11×10^{-20}
320	2.59×10^{-20}	2.49×10^{-20}	2.43×10^{-20}	2.40×10^{-20}
321	2.43×10^{-20}	2.26×10^{-20}	9.91×10^{-21}	8.96×10^{-21}
322	2.65×10^{-20}	2.13×10^{-20}	2.02×10^{-20}	1.98×10^{-20}
323	2.54×10^{-20}	1.92×10^{-20}	1.36×10^{-20}	1.38×10^{-20}
324	2.25×10^{-20}	1.80×10^{-20}	1.53×10^{-20}	1.38×10^{-20}
325	1.84×10^{-20}	1.61×10^{-20}	1.38×10^{-20}	1.04×10^{-20}
326	1.16×10^{-20}	1.11×10^{-20}	9.82×10^{-21}	9.63×10^{-21}
327	9.27×10^{-21}	7.94×10^{-21}	6.03×10^{-21}	5.93×10^{-21}
328	1.31×10^{-20}	1.29×10^{-20}	1.29×10^{-20}	1.30×10^{-20}
329	9.38×10^{-21}	6.86×10^{-21}	5.61×10^{-21}	4.64×10^{-21}
330	6.68×10^{-21}	6.01×10^{-21}	5.03×10^{-21}	4.10×10^{-21}
331	7.29×10^{-21}	6.94×10^{-21}	5.80×10^{-21}	5.39×10^{-21}
332	6.97×10^{-21}	4.84×10^{-21}	3.12×10^{-21}	2.50×10^{-21}
333	4.74×10^{-21}	3.63×10^{-21}	2.71×10^{-21}	1.77×10^{-21}
334	5.66×10^{-21}	5.62×10^{-21}	5.61×10^{-21}	5.60×10^{-21}
335	2.53×10^{-21}	2.21×10^{-21}	1.95×10^{-21}	1.60×10^{-21}
336	2.60×10^{-21}	2.11×10^{-21}	1.01×10^{-21}	7.33×10^{-22}
337	3.81×10^{-21}	3.61×10^{-21}	1.82×10^{-21}	1.78×10^{-21}
338	2.57×10^{-21}	2.47×10^{-21}	1.92×10^{-21}	1.33×10^{-21}
339	1.32×10^{-21}	9.84×10^{-22}	6.12×10^{-22}	4.99×10^{-22}
340	2.04×10^{-21}	1.13×10^{-21}	7.26×10^{-22}	6.81×10^{-22}
341	1.01×10^{-21}	8.64×10^{-22}	7.41×10^{-22}	7.30×10^{-22}

TABLE 3.18. (*Continued*)

Wavelength (nm)	298°K	271°K	225°K	206°K
342	7.65×10^{-22}	6.21×10^{-22}	3.19×10^{-22}	2.25×10^{-22}
343	9.64×10^{-22}	7.01×10^{-22}	4.29×10^{-22}	1.43×10^{-22}
344	1.24×10^{-21}	1.02×10^{-21}	9.9×10^{-23}	9.0×10^{-23}
345	6.26×10^{-22}	5.74×10^{-22}	4.25×10^{-22}	3.01×10^{-22}
346	5.77×10^{-22}	1.21×10^{-22}	8.4×10^{-23}	5.9×10^{-23}
347	5.22×10^{-22}	5.01×10^{-22}	4.56×10^{-22}	4.49×10^{-22}
348	3.62×10^{-22}	3.64×10^{-22}	3.20×10^{-22}	2.73×10^{-22}
349	3.06×10^{-22}	1.09×10^{-22}	7.8×10^{-23}	5.2×10^{-23}
350	3.80×10^{-22}	9.21×10^{-23}	4.3×10^{-23}	2.0×10^{-23}
351	3.14×10^{-22}	9.48×10^{-23}	6.1×10^{-23}	5.6×10^{-23}
352	2.47×10^{-22}	2.36×10^{-22}	2.13×10^{-22}	2.17×10^{-22}
353	2.44×10^{-22}	1.02×10^{-22}	6.9×10^{-23}	5.2×10^{-23}
354	1.34×10^{-22}	6.73×10^{-23}	1.1×10^{-23}	1.00×10^{-23}
355	1.00×10^{-22}	3.16×10^{-23}	6×10^{-24}	5.0×10^{-24}
356	7.6×10^{-23}	5.26×10^{-23}	2.6×10^{-23}	1.0×10^{-23}
357	1.35×10^{-22}	9.58×10^{-23}	3.2×10^{-23}	9×10^{-24}
358	1.03×10^{-22}	5.34×10^{-23}	6×10^{-24}	$< 10^{-24}$
359	5.4×10^{-23}	2.1×10^{-24}	$< 10^{-24}$	$< 10^{-24}$

Source: Davenport, 1982.

In 1 atm of air at 50% relative humidity and 298°K, approximately 10% of the $O(^1D)$ produced via reaction (4) reacts with water to form hydroxyl radicals.

One can calculate this ratio using the kinetic principles outlined in Section 4-A. Thus the rate of reaction with H_2O is given by

$$-\frac{d[O(^1D)]}{dt} = k_5[H_2O][O(^1D)]$$

TABLE 3.19. Wavelength Threshold (nm) Below Which Indicated Reactions are Energetically Possible in the Photolysis of O_3

Electronic State of Oxygen Atom	Electronic State of Molecular O_2				
	$^3\Sigma_g^-$	$^1\Delta_g$	$^1\Sigma_g^+$	$^3\Sigma_u^+$	$^3\Sigma_u^-$
^3P	1180	611	463	230	173
^1D	411	310	266	168	136
^1S	237	200	180	129	109

Source: Okabe, 1978.

and that for deactivation by air by

$$-\frac{d[O(^1D)]}{dt} = k_6[M][O(^1D)]$$

The amount of $O(^1D)$ removed by reaction with H_2O compared to that deactivated by air is given by

$$\frac{k_5[H_2O][O(^1D)]}{k_6[M][O(^1D)]} = \frac{k_5[H_2O]}{k_6[M]}$$

Using the values of k_5 and k_6 above, $[H_2O] = 3.9 \times 10^{17}$ molecules cm^{-3} at 25°C for 50% relative humidity and $[M] = 2.46 \times 10^{19}$ molecules cm^{-3} (1 atm), one obtains a ratio of 0.12.

The reaction of $O(^1D)$ with H_2O proceeds primarily to produce 2 OH as shown by reaction (5). However, a small percentage of the $O(^1D)$ (4.9 ± 3.2%) is deactivated by H_2O vapor (Wine and Ravishankara, 1982):

$$O(^1D) + H_2O \rightarrow O(^3P) + H_2O \qquad (7)$$

An even smaller percentage (~1%) of the $O(^1D)$–H_2O interaction leads to the products $H_2 + O_2$ (Zellner and Wagner, 1980).

While the production of $O(^1D)$ by O_3 photolysis is clearly very important, there has been considerable controversy concerning the absolute primary quantum yield of $O(^1D)$ production, as well as its wavelength dependence. Thus until the late 1970s, it was believed that the quantum yield for $O(^1D)$ production at wavelengths less than 300 nm was, within experimental error, unity. Furthermore, most studies in the tropospherically important region 290–320 nm determined the yields *relative* to those at shorter wavelengths where the quantum yield for $O(^1D)$ production was *assumed* to be 1.0.

However, in 1978, Fairchild, Stone, and Lawrence used photofragment spectroscopy to study the photolysis of O_3 in the 270–310 nm region and reported that the primary quantum yield for $O(^3P)$ atoms was 0.1 at 274 nm. This significant production of ground-state oxygen atoms at these shorter wavelengths meant that much of the data based on *relative* values for the $O(^1D)$ yields had to be corrected to lower values.

Since that time, a number of studies have confirmed that $\phi[O(^1D)]$ is indeed less than unity at shorter wavelengths; Table 3.20 gives selected absolute values of the quantum yields of $O(^1D)$ measured at $\lambda < 300$ nm. From these data, it appears that as much as 10% of the O_3 photolysis at certain wavelengths below 300 nm proceeds via reaction (8):

$$O_3 + h\nu\ (\lambda \lesssim 300\text{ nm}) \rightarrow O_2(^3\Sigma_g^-) + O(^3P) \qquad (8)$$

Results of some recent measurements of the $O(^1D)$ yields as a function of wave-

TABLE 3.20. Selected Literature Values of the Absolute Quantum Yields of $O(^1D)$ from the Photolysis of O_3 in the Hartley Bands

Photolysis Wavelength (nm)	$\phi\ [O(^1D)]$	Reference
290	0.95 ± 0.02	Davenport, 1982
274	~0.9	Fairchild et al., 1978
270	0.92 ± 0.03	Davenport, 1982
230–280	1.00 ± 0.05	Kajimoto and Cvetanovic, 1979
230–280	1.0	Fairchild and Lee, 1978
266	~0.9	Sparks et al., 1980
266	0.88	Brock and Watson, 1980a
254	0.92 ± 0.04	Cobos et al., 1983
248	0.85 ± 0.02	Amimoto et al., 1980
248	0.91 ± 0.03	Wine and Ravishankara, 1982
248	0.94 ± 0.01	Greenblatt and Wiesenfeld, 1983

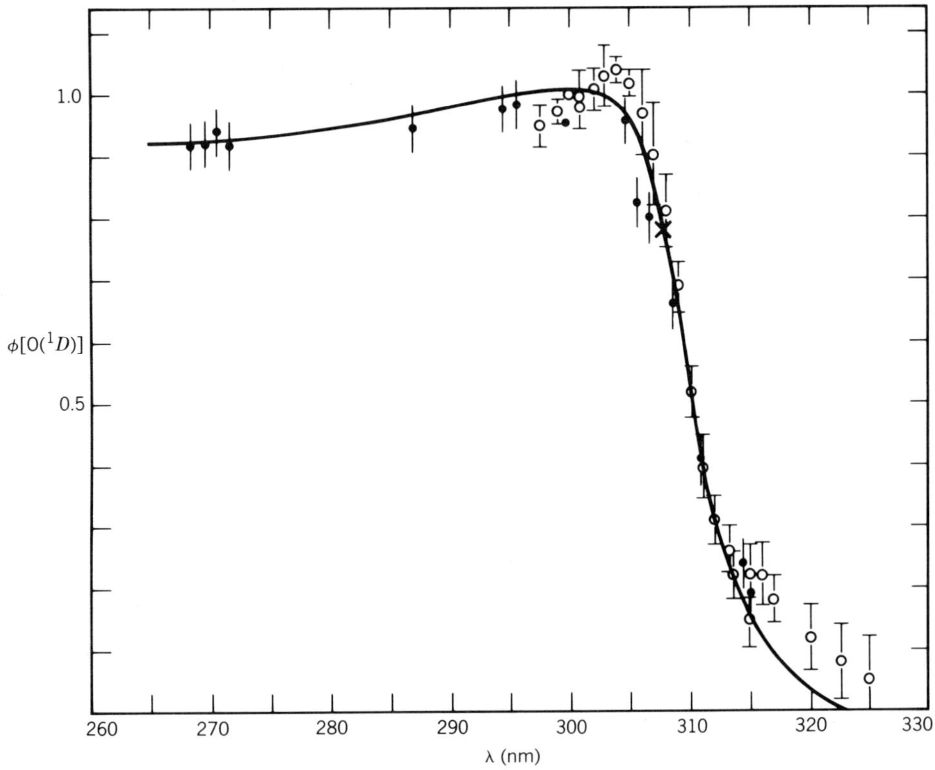

FIGURE 3.24. Quantum yield for $O(^1D)$ formation from the photolysis of O_3 in the Hartley band. ●, Davenport (1982); ○, Brock and Watson (1980b); ×, Greenblatt and Wiesenfeld (1983).

length from ~270 to 325 nm are shown in Fig. 3.24. It appears that φ is unity, within experimental error, at 300 nm, but drops off slightly to lower values on the short-wavelength side. On the long-wavelength side, ϕ drops off rapidly.

It is not clear whether the tail of the $O(^1D)$ production above 315 nm observed by some workers (e.g., see Brock and Watson, 1980b) is real. If so, it may arise from a small contribution from the spin-forbidden reaction (9):

$$O_3 + h\nu\ (\lambda < 410\ \text{nm}) \rightarrow O(^1D) + O_2(^3\Sigma_g^-) \tag{9}$$

Table 3.21 gives the mathematical expression for the $O(^1D)$ quantum yields (ϕ) as a function of temperature and pressure in the falloff region around 300 nm, recommended in a recent NASA evaluation (DeMore et al., 1983, 1985). The form of this expression was originally developed by Moortgat and Kudszus (1978) but has been corrected for the more recent discovery of significant (~10%) $O(^3P)$ production at shorter wavelengths (Table 3.20).

Some measurements of the rate of O_3 photolysis in the troposphere have been carried out and compared to the rates calculated using Eq. (U) above (e.g., see Dickerson et al., 1982). Considering the uncertainties in σ, ϕ, and J, as well as experimental errors in measuring photolysis rates, the agreement between the two is reasonably good.

In the Huggins bands at $\lambda \gtrsim 320$ nm, the products appear to be $O_2(^1\Delta_g)$ and ground-state oxygen atoms (Okabe, 1978), indicating that the spin-forbidden process (10) is occurring:

$$O_3 + h\nu \rightarrow O_2(^1\Sigma_g^+ \text{ or } ^1\Delta_g) + O(^3P) \tag{10}$$

TABLE 3.21. Mathematical Expression for $O(^1D)$ Quantum Yields, ϕ, in the Photolysis of O_3, Recommended by NASA Evaluations, 1983 and 1985.

$$\phi(\lambda, T) = A(\tau)\arctan[B(\tau)(\lambda - \lambda_0(\tau))] + C(\tau)$$

Where $\tau = T - 230$ is a temperature function with T given in Kelvin, λ is expressed in nm, and arctan in radians.

The coefficients $A(\tau)$, $B(\tau)$, $\lambda_0(\tau)$, and $C(\tau)$ are expressed as interpolation polynomials of the third order:

$$A(\tau) = 0.332 + 2.565 \times 10^{-4}\,\tau + 1.152 \times 10^{-5}\,\tau^2 + 2.313 \times 10^{-8}\,\tau^3$$
$$B(\tau) = -0.575 + 5.59 \times 10^{-3}\,\tau - 1.439 \times 10^{-5}\,\tau^2 - 3.27 \times 10^{-8}\,\tau^3$$
$$\lambda_0(\tau) = 308.20 + 4.4871 \times 10^{-2}\,\tau + 6.9380 \times 10^{-5}\,\tau^2 - 2.5452 \times 10^{-6}\,\tau^3$$
$$C(\tau) = 0.466 + 8.883 \times 10^{-4}\,\tau - 3.546 \times 10^{-5}\,\tau^2 + 3.519 \times 10^{-7}\,\tau^3$$

In the limits where $\phi(\lambda, T) > 0.9$, the quantum yield is set $\phi = 0.9$. Similarly, if $\phi(\lambda, T)$ calculated from this expression is < 0, the quantum yield is set to zero.

Source: DeMore et al., 1983, 1985; based on Moortgat and Kudszus, 1978.

In the Chappius region (440–850 nm), the products appear to be the ground-state species (Okabe, 1978):

$$O_3 + h\nu \rightarrow O(^3P) + O_2(X^3\Sigma_g^-) \tag{11}$$

In both these regions with $\lambda \gtrsim 320$ nm, the absorption coefficient is one to two orders of magnitude less than that at 300 nm. In addition, they produce $O(^3P)$ rather than $O(^1D)$. Under conditions typical of urban and suburban areas where sufficient NO_2 is present, the production of $O(^3P)$ from O_3 photolysis will be much less than that from NO_2; in addition, since the $O(^3P)$ reacts with O_2 to reform O_3, this leads to no net loss of O_3. As a result, these photolysis processes at $\lambda > 320$ nm are not significant in tropospheric chemistry.

3-C-3. Nitrogen Dioxide

3-C-3a. Absorption Spectra

The absorption spectrum of NO_2 from 180 to 410 nm is shown in Fig. 3.25; Table 3.22 gives the absorption cross sections, σ, at selected wavelengths at 298°K. Although we show the cross sections at 5 nm intervals and at 298°K, Bass and coworkers (1976), whose data are cited in Table 3.22, report high-resolution data (every 0.125 nm) at two temperatures, 235 and 298°K. To calculate photolysis rate constants for NO_2 averaged over certain wavelength intervals using Eq. (T), the reader should consult the original reference and average the data over the appropriate interval.

A major experimental difficulty in determining the absorption spectrum and photochemistry of NO_2 is that it is in equilibrium with the dimeric N_2O_4 and under many laboratory conditions with NO_2 in the Torr pressure range, significant amounts of N_2O_4 can be present (*vide infra*).

$$2\ NO_2 \rightleftarrows N_2O_4 \tag{12}$$

As seen in Fig. 3.26, N_2O_4 also absorbs light strongly in the near ultraviolet, so that corrections must be made for its presence.

The equilibrium constant for (12) at 298°K is 6.84 atm^{-1} (Stull and Prophet, 1971). From this equilibrium constant, the total pressure of an NO_2–N_2O_4 mixture, and the stoichiometry of (12), the equilibrium partial pressures of NO_2 and N_2O_4 at various total pressures can be calculated. Some of these are shown in Table 3.23 for total pressures of an NO_2–N_2O_4 mixture from 5 Torr to 4 μTorr. In accord with Le Chatelier's principle, the fraction of the total due to N_2O_4 decreases rapidly as the total pressure falls reaching 3.5×10^{-5} and 3.5×10^{-8} at NO_2 concentrations equivalent to 5 ppm and 5 ppb, respectively in 1 atm of air.

Because of the presence of N_2O_4, one must exercise caution in extrapolating results from NO_2 studies in the Torr pressure region to ambient ppb levels. In some laboratory situations, the presence of N_2O_4 is not a problem because the

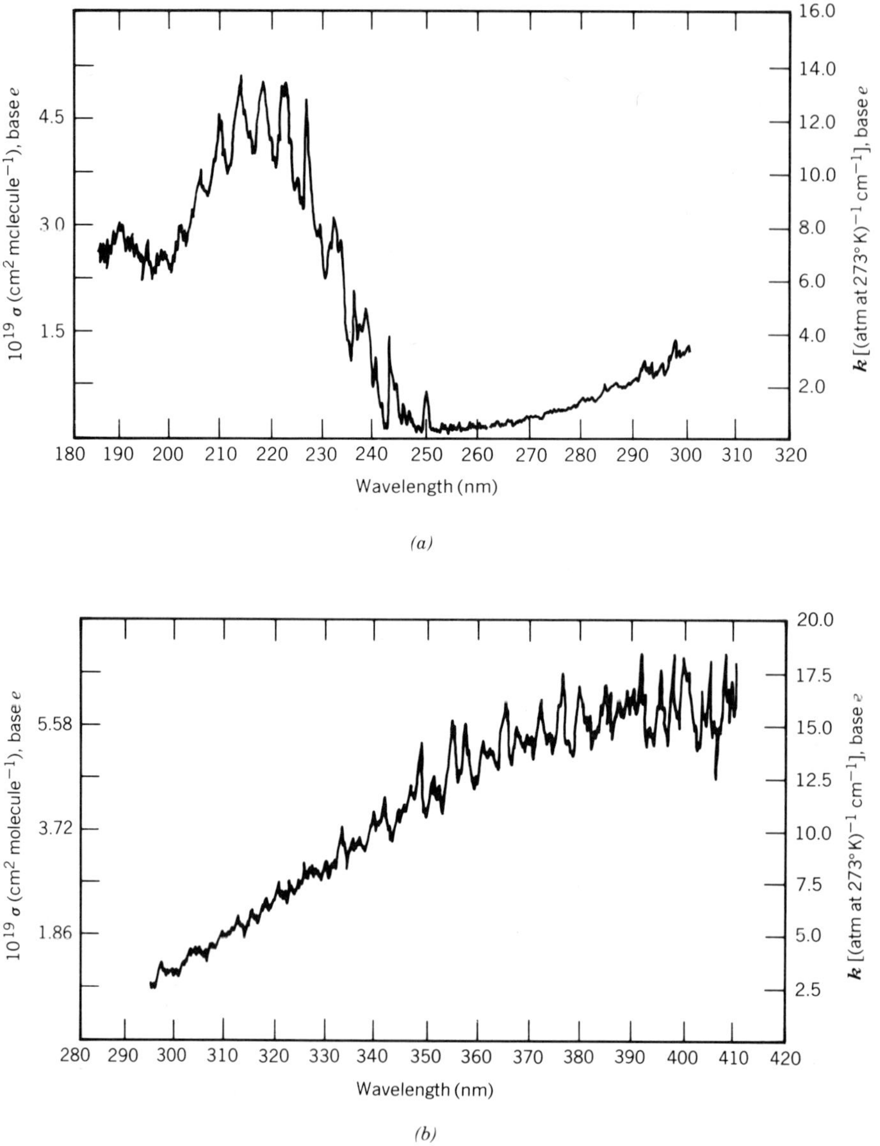

FIGURE 3.25. Absorption spectrum of NO_2 from (*a*) 180–300 nm and (*b*) 295–410 nm. Units of the absorption coefficient are, on the left axis, σ (cm^2 molecule^{-1}), base *e*, and, on the right axis, *k* [(atm at 273)$^{-1}$ cm^{-1}], base 10 (from Bass et al., 1976).

TABLE 3.22. Absorption Cross Sections, Base *e*, for NO_2 at 298°K

λ (nm)	10^{20} σ (cm^2 $molecule^{-1}$)	λ (nm)	10^{20} σ (cm^2 $molecule^{-1}$)
185	26.0	300	11.7
190	29.3	305	16.6
195	24.2	310	17.6
200	25.0	315	22.5
205	37.5	320	25.4
210	38.5	325	27.9
215	40.2	330	29.9
220	39.6	335	34.5
225	32.4	340	38.8
230	24.3	345	40.7
235	14.8	350	41.0
240	6.70	355	51.3
245	4.35	360	45.1
250	2.83	365	57.8
255	1.45	370	54.2
260	1.90	375	53.5
265	2.05	380	59.9
270	3.13	385	59.4
275	4.02	390	60.0
280	5.54	395	58.9
285	6.99	400	67.6
290	8.18	405	63.2
295	9.67	410	57.7

Source: De More et al., 1983; data from Bass et al., 1976.

equilibrium shifts rapidly; in fast flow discharge systems, for example (see Section 4-B-2), when the NO_2–N_2O_4 mixture is added and the NO_2 reacts, for example, with H atoms, the N_2O_4 rapidly dissociates to form more NO_2. Thus, essentially all the NO_2–N_2O_4 reacts on the millisecond time scales characteristic of these systems, and the initial presence of N_2O_4 does not affect the system.

3-C-3b. Photochemistry

Nitrogen dioxide exposed to tropospheric radiation photodissociates at $\lambda \lesssim 420$ nm to give nitric oxide and an oxygen atom:

$$NO_2 + h\nu \rightarrow NO + O \tag{13}$$

Table 3.24 gives the calculated wavelengths below which it is energetically possible to produce the fragments in each of the electronic states shown if there is no contribution from internal energy of the molecule.

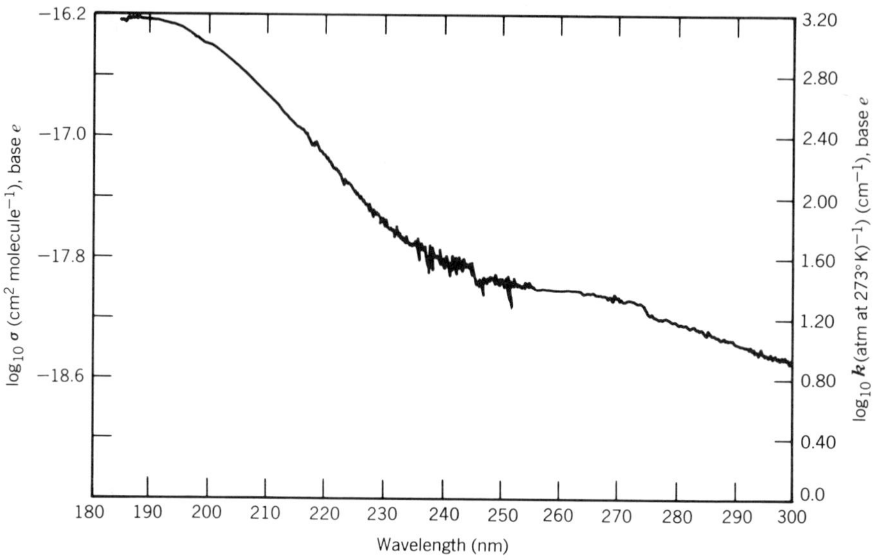

(a)

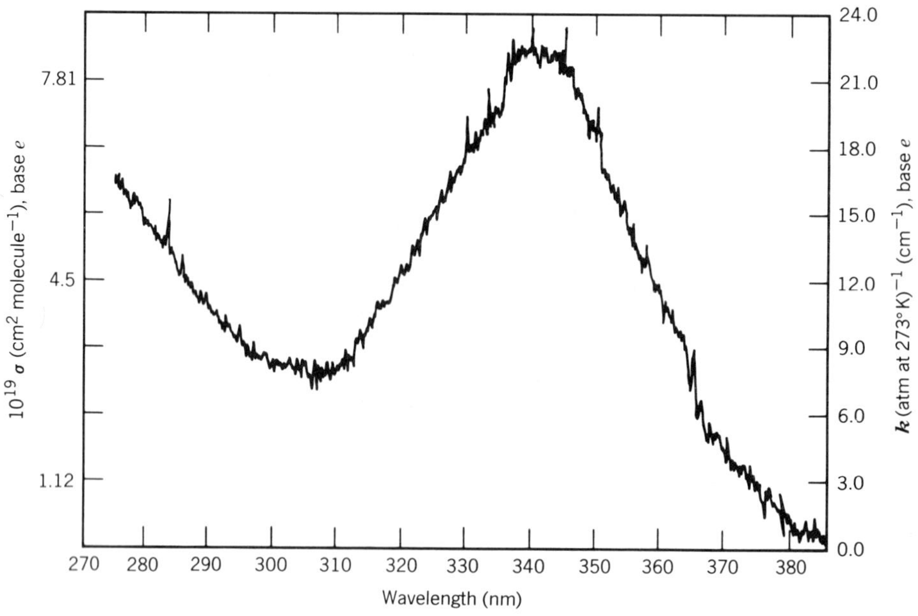

(b)

FIGURE 3.26. Absorption spectrum of N_2O_4 from (*a*) 180–300 nm and (b) 270–380 nm. (*a*) $\log_{10}$ (absorption coefficient) in units of σ (cm^2 $molecule^{-1}$), base *e*, on the left axis and *k* [(atm at $273)^{-1}$ cm^{-1}], base *e*, on the right axis. (*b*) Absorption coefficient in units of σ (cm^2 $molecule^{-1}$), base *e*, on the left axis, and *k* [(atm at 273°K$)^{-1}$ cm^{-1}], base *e*, on the right axis (from Bass et al., 1976).

TABLE 3.23. Partial Pressures of NO_2 and N_2O_4 in NO_2–N_2O_4 Mixtures at Equilibrium at 298°K and at Various Total Pressures

Total Pressure (Torr)	$P^{eq}_{NO_2}$ (Torr)	$P^{eq}_{N_2O_4}$ (Torr)	Fraction of N_2O_4 at Equilibrium
5.00	4.79	0.21	4.2×10^{-2}
4×10^{-3} (5 ppm)	4×10^{-3}	1.4×10^{-7} (0.19 ppb)	3.5×10^{-5}
4×10^{-6} (5 ppb)	4×10^{-6}	1.4×10^{-13} (1.8×10^{-4} ppt)	3.5×10^{-8}

The quantum yield for oxygen atom production in (13) has been studied extensively because of its role as the only significant anthropogenic source of O_3 in the troposphere via (13) followed by (14):

$$O + O_2 \xrightarrow{M} O_3 \tag{14}$$

As seen in Table 3.24, the threshold wavelength for production of ground-state NO and O atoms is 397.8 nm. Figure 3.27 shows selected values of the experimentally determined primary quantum yields for reaction (13) as a function of wavelength in the region 295 to 445 nm. Also shown are expressions for the quantum yield dependence on wavelength given by Jones and Bayes (1973a) and recommended by De More et al. (1983, 1985):

$$\phi(295 \leq \lambda \leq 365 \text{ nm}) = 1.0 - 0.0008\ (\lambda - 275)$$

as well as that recommended by Atkinson and Lloyd (1984):

$$\phi(295 \leq \lambda \leq 375 \text{ nm}) = 1.0 - 0.0025\ (\lambda - 295)$$

TABLE 3.24. Calculated Wavelengths (nm) for NO_2 Photolysis Below Which the Fragments Shown Can be Produced[a]

	Oxygen Atoms		
NO	^3P	^1D	^1S
$X^2\Pi$	397.8	243.9	169.7
$A^2\Sigma^+$	144.2	117.4	97

Source: Okabe, 1978.

[a]Assuming no contribution from internal energy of the molecule.

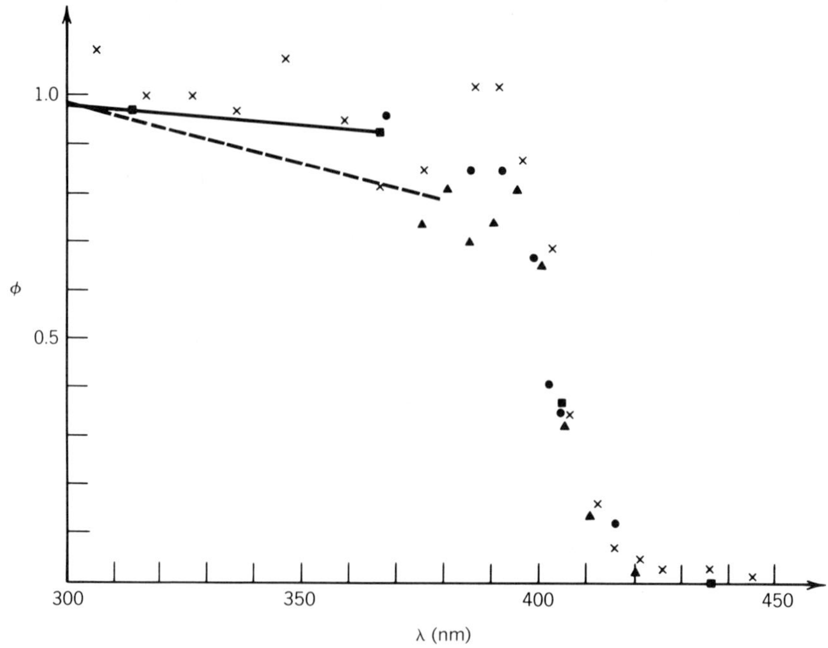

FIGURE 3.27. Selected primary quantum yields as a function of wavelength for the photolysis of NO_2. Solid line is expression recommended by De More et al. (1983, 1985), dashed line by Atkinson and Lloyd (1984). Data from: ■, Pitts et al. (1964); ×, Jones and Bayes (1973a); ●, Gaedtke and Troe (1975); ▲, Harker et al. (1977).

Table 3.25 shows the primary quantum yields from 375 to 420 nm for NO_2 photolysis as a function of wavelength recommended in the most recent NASA evaluation (De More et al., 1985).

The quantum yield declines slightly from approximately 1 at 300 nm to ~0.8 at the theoretical threshold for dissociation, 300.6 kJ mole^{-1} (71.8 kcal mole^{-1}), corresponding to 397.8 nm. This has been attributed to the formation of a nondissociative excited state of NO_2. In ambient air, electronically excited NO_2, which does not dissociate to form $O(^3P)$, is collisionally deactivated. When O_2 is the collision partner, energy transfer may occur a fraction of the time to form $O_2(a^1\Delta_g)$:

$$NO_2^* + O_2 \rightarrow NO_2 + O_2(a^1\Delta_g) \tag{15}$$

Figure 3.28 shows the efficiency of reaction (15), that is, the percentage of molecules of $O_2(a^1\Delta_g)$ formed per photon absorbed by NO_2, as a function of wavelength in the 330–590 nm region (Jones and Bayes, 1973b). Up to 4.5% of NO_2^*–O_2 collisions form electronically excited O_2; in an independent study, efficiencies of 7.5 ± 2.5% were reported for exciting light with wavelengths between 405 and 700 nm (Frankiewicz and Berry, 1973). As discussed in Section 7-E-5, this may be a significant source of $O_2(^1\Delta_g)$ formed in polluted urban atmospheres.

While ϕ drops rapidly at the expected threshold wavelength, there is still significant dissociation above the thermodynamically calculated 397.8 nm cutoff (Fig.

TABLE 3.25. Primary Quantum Yields, ϕ, for NO_2 Photolysis into NO + $O(^3P)$

λ (nm)	ϕ	λ (nm)	ϕ	λ (nm)	ϕ
375	0.77	389	0.78	400	0.68
376	0.78	390	0.80	401	0.65
377	0.92	391	0.88	402	0.62
378	0.82	392	0.84	403	0.57
379	0.87	393	0.90	404	0.42
380	0.90	394	0.90	405	0.32
381	0.81	394.5	0.86	406	0.33
382	0.70	395	0.84	407	0.25
383	0.68	395.5	0.81	408	0.20
384	0.70	396	0.83	409	0.19
385	0.77	396.5	0.88	410	0.15
386	0.84	397	0.82	411	0.10
387	0.75	398	0.77	415	0.067
388	0.81	399	0.78	420	0.023

Source: De More et al., 1985; data based on product ($\phi\sigma$) of Harker et al., (1977) and values of σ of Bass et al. (1976).

3.27). The non-zero $O(^3P)$ yield between 397.8 and 420 nm may be due to a contribution from internal rotational energy (Pitts et al., 1964). Figure 3.29 shows the experimental primary quantum yield and those calculated (solid line) assuming there is no centrifugal barrier to dissociation and that all molecules whose total energy, comprised of photon energy + vibrational + rotational energy, equals or exceeds that needed for dissociation do indeed dissociate to NO + $O(^3P)$ (Gaedtke

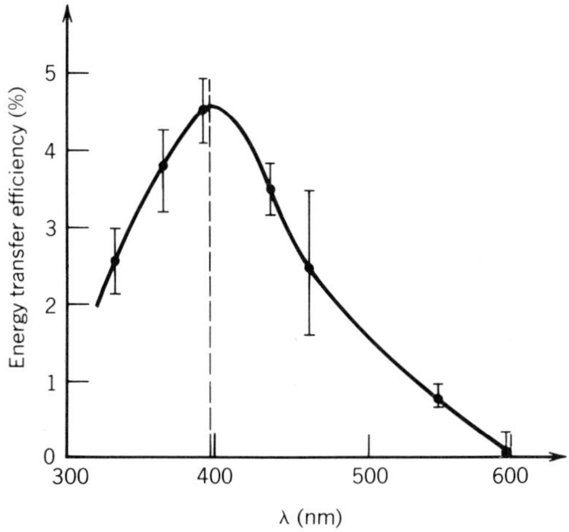

FIGURE 3.28. The energy transfer efficiency for production of $O_2(a^1\Delta_g)$ from electronically excited NO_2, expressed as a percentage, plotted against the wavelength of the radiation used to excite NO_2. The error limits are ± 2 standard deviations for the observed counts. The dashed line marks the accepted dissociation limit for NO_2 at 398 nm (from Jones and Bayes, 1973b).

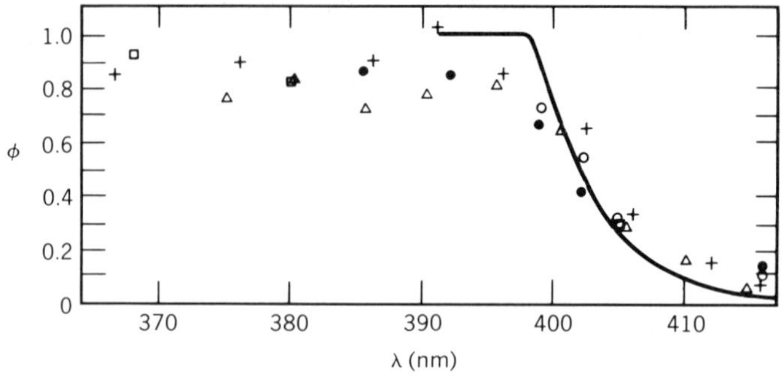

FIGURE 3.29. Primary quantum yields for $O(^3P)$ formation from NO_2 photolysis at 296°K. ●, Measured by Gaedtke and Troe (1975); +, measured by Jones and Bayes (1973a); □, measured by Pitts et al. (1964); △, measured by Harker et al. (1977); solid line calculated by Gaedtke and Troe (1975) (adapted from Gaedtke and Troe, 1975).

and Troe, 1975). It is seen that a reasonably good fit to the data is obtained, supporting the hypothesis that internal energy in the absorbing NO_2 contributes to $O(^3P)$ formation beyond the thermodynamically calculated threshold.

Beyond 430 nm, very small yields of NO and O_2 have been observed. These have been attributed to reaction of an excited molecule of NO_2 (NO_2*) with a ground-state NO_2 molecule, giving 2 NO + O_2 (Okabe, 1978), or possibly NO_3 + NO (Jones and Bayes, 1973a). This process will not occur in the troposphere where the NO_2 concentrations are too small for such a bimolecular process to occur before NO_2* is deactivated by collisions with air.

A number of measurements of the photodissociation rate of NO_2 in the troposphere have been carried out (e.g., see Madronich et al., 1983, and Parrish et al., 1983, and the references therein). As seen in Figs. 3.13–3.15, and discussed in Section 3-A-2h, reasonably good agreement has been obtained between the experimental measurements and calculated photolysis rates under clear sky conditions.

3-C-4. Sulfur Dioxide

3-C-4a. Absorption Spectra

As seen in Fig. 3.30, SO_2 absorbs strongly in the 240–330 nm region, with a much weaker absorption from ~340 to 400 nm. The strong absorption peaking at ~290 nm forms two singlet excited states of SO_2,

$$SO_2(X^1A_1) + h\nu\ (240 < \lambda < 330\ \text{nm}) \rightarrow SO_2(^1A_2,\ ^1B_1) \qquad (16)$$

while the weak absorption involves a spin-forbidden transition to a triplet state:

$$SO_2(X^1A_1) + h\nu\ (340 < \lambda < 400\ \text{nm}) \rightarrow SO_2(^3B_1) \qquad (17)$$

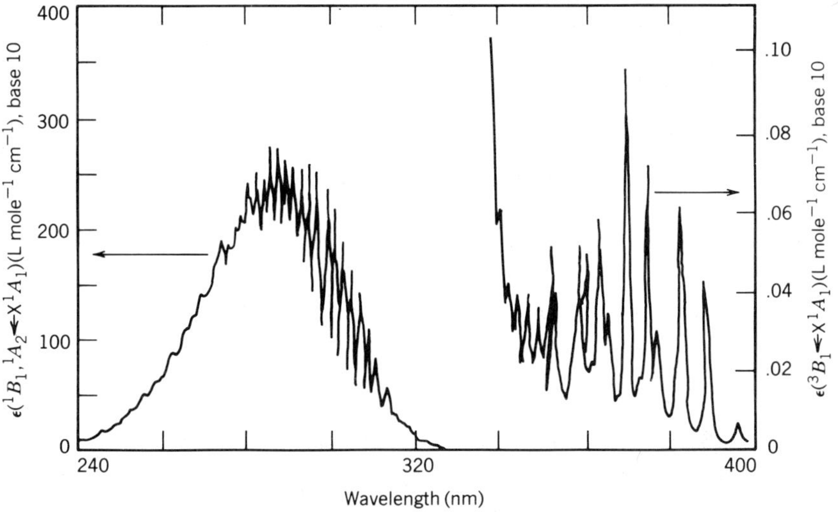

FIGURE 3.30. Absorption spectrum of SO_2 at 25°C., ε given in units of L mole^{-1} cm^{-1}, base 10 (from Calvert and Stockwell, 1984).

3-C-4b. *Photochemistry*

SO_2 can dissociate to SO + O only at wavelengths below 218 nm:

$$SO_2(X^1A_1) + h\nu\ (\lambda < 218\ \text{nm}) \rightarrow SO(^3\Sigma_u^-) + O(^3P) \qquad (18)$$

Thus reaction (18) cannot occur in the troposphere where only wavelengths of 290 nm and above are present.

The singlet excited states of SO_2 formed in (16) are collisionally deactivated in 1 atm air to the $^3B_1$ state, possibly the $^3A_2$ and $^3B_2$ states, and to the ground state. The $^3B_1$ state is then deactivated to the ground state by collisions with N_2, O_2, and H_2O. At 25°C and 50% relative humidity, these gases account for 45.7, 41.7, and 12.2% of the $SO_2(^3B_1)$ deactivation, respectively (Calvert and Stockwell, 1984). In the interaction with O_2, as we have seen, $SO_2(^3B_1)$ can transfer energy to form singlet molecular oxygen:

$$SO_2(^3B_1) + O_2 \rightarrow SO_2 + O_2(^1\Sigma_g^+, ^1\Delta_g) \qquad (19)$$

The tropospheric relevance of these spectroscopic observations is that, in air, the gas phase oxidation of SO_2 to sulfuric acid is *not* initiated by direct photochemical processes resulting from the absorption of solar radiation. Rather, as discussed in Chapter 11, it is initiated by attack by hydroxyl radicals, which themselves are formed as a result of primary photochemical processes.

The photophysics and photochemistry of SO_2 are reviewed in detail by Heicklen and co-workers (1980).

3-C-5. Nitrous Acid and Organic Nitrites

The absorption spectrum of gaseous HONO from 310 to 390 nm, determined in three separate studies, is shown in Fig. 3.31.

The most common method of generating known concentrations of HONO is to let mixtures of NO, NO_2, and water vapor come to equilibrium:

$$NO + NO_2 + H_2O \rightleftarrows 2\ HONO \tag{20}$$

$$K_{eq}\ (\text{atm}^{-1}) = \exp\left(-15.56 + \frac{4.73 \times 10^3}{T}\right)$$

(Stockwell and Calvert, 1978)

Since NO_2 absorbs light in the same region as HONO (see Section 3-C-3 above), corrections must be made for its contribution to the absorbance in order to extract that due to HONO. The substantial discrepancies seen in Fig. 3.31 have thus been attributed by Stockwell and Calvert (1978) to errors in accurately correcting the spectrum for absorption by NO_2. A further complication is that reaction (20) comes to equilibrium fairly slowly (the rate may be dependent on surface catalysis) so that one must be careful to ensure that equilibrium is, in fact, established or the HONO concentrations calculated using the equilibrium constant for reaction (20) will be in error.

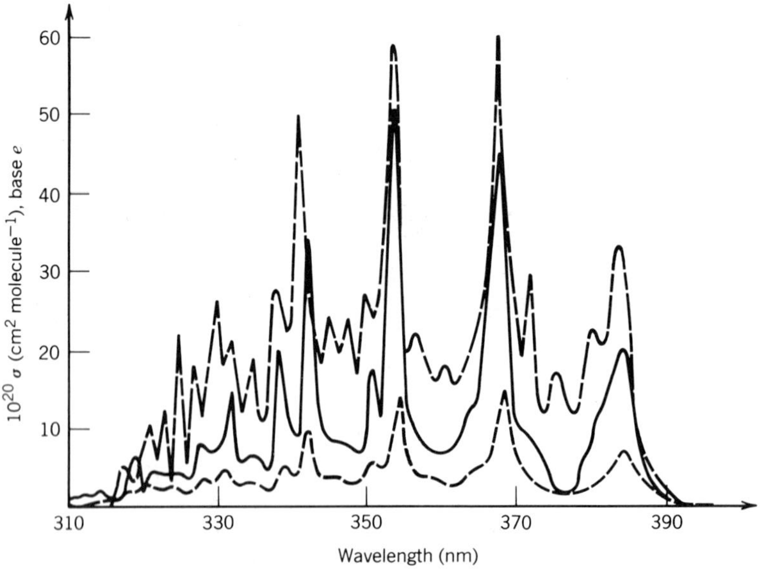

FIGURE 3.31. The absorption cross section of $HONO_{(g)}$ σ (cm^2 molecule^{-1}), base e, as a function of wavelength at 25°C. Data from Stockwell and Calvert (1978) are shown as a solid curve; those of Cox and Derwent (1976/77) and Johnston and Graham (1974) are given by the lower and upper broken curves, respectively (from Stockwell and Calvert, 1978).

TABLE 3.26 Nitrous Acid Absorption Cross Sections

Wavelength (nm)	$10^{20}\ \sigma$ (cm^2 molecule^{-1}, base e)	Wavelength (nm)	$10^{20}\ \sigma$ (cm^2 molecule^{-1}, base e)	Wavelength (nm)	$10^{20}\ \sigma$ (cm^2 molecule^{-1}, base e)
310	0.0	342	33.5	371	9.46
311	0.0	343	20.1	372	8.85
312	0.2	344	10.2	373	7.44
313	0.42	345	8.54	374	4.77
314	0.46	346	8.32	375	2.7
315	0.42	347	8.20	376	1.9
316	0.3	348	7.49	377	1.5
317	0.46	349	7.13	378	1.9
318	3.6	350	6.83	379	5.8
319	6.10	351	17.4	380	7.78
320	2.1	352	11.4	381	11.4
321	4.27	353	37.1	382	14.0
322	4.01	354	49.6	383	17.2
323	3.93	355	24.6	384	19.9
324	4.01	356	11.9	385	19.0
325	4.04	357	9.35	386	11.9
326	3.13	358	7.78	387	5.65
327	4.12	359	7.29	388	3.2
328	7.55	360	6.83	389	1.9
329	6.64	361	6.90	390	1.2
330	7.29	362	7.32	391	0.5
331	8.70	363	9.00	392	0.0
332	13.8	364	12.1	393	0.0
333	5.91	365	13.3	394	0.0
334	5.91	366	21.3	395	0.0
335	6.45	367	35.2	396	0.0
336	5.91	368	45.0		
337	4.58	369	29.3		
338	19.1	370	11.9		
339	16.3				
340	10.5				
341	8.70				

Source: Stockwell and Calvert, 1978.

The cross sections of Stockwell and Calvert for gaseous HONO, given in Table 3.26, are recommended in the NASA evaluation (DeMore et al., 1983, 1985).

As discussed in more detail later (Section 3-D), the photodissociation of nitrous acid is a major source of OH radicals in polluted urban atmospheres:

$$HONO + h\nu\ (\lambda < 400\ \text{nm}) \rightarrow OH + NO \qquad (21)$$

The primary quantum yield of reaction (21) has been shown by Cox and Derwent (1976/77) to be essentially unity.

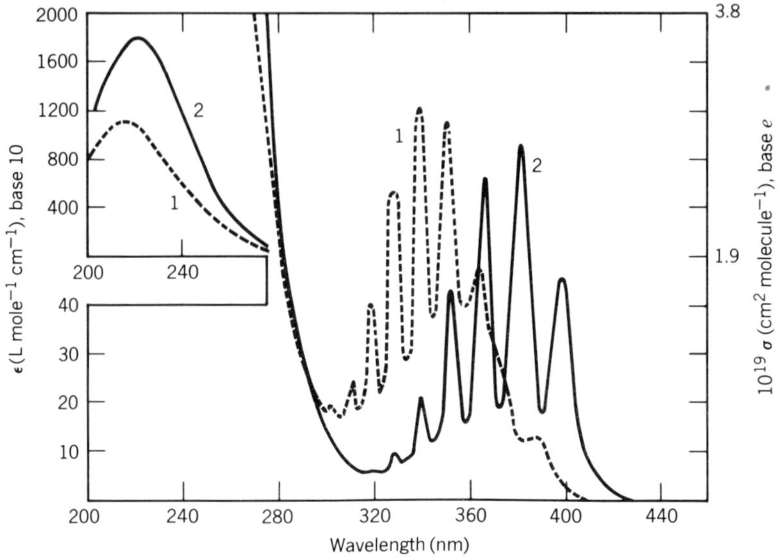

FIGURE 3.32. Absorption spectra of (1) methyl nitrite [$CH_3ONO_{(g)}$], 25°C; (2) *tert*-butyl nitrite [$(CH_3)_3CONO_{(g)}$], 25°C. Left axis gives ϵ in units of L mole^{-1} cm^{-1}, base 10. Right axis is σ (cm^2 molecule^{-1}), base *e* (from Calvert and Pitts, 1966).

As seen in Fig. 3.32, gaseous organic nitrites also absorb strongly in the actinic UV, showing distinct vibronic bands. Following absorption of light, these nitrites photodissociate into alkoxy radicals and NO, in a manner analogous to HONO:

$$\text{RONO} \rightarrow \text{RO} + \text{NO} \qquad (22)$$

While the primary quantum yields of dissociation are usually taken to be unity (Atkinson and Lloyd, 1984), they may be less than this for wavelengths above 300 nm (Morabito and Heicklen, 1985); more work needs to be done to clarify the quantum yields in the actinic UV (see Section 7-D-3).

Thus, as discussed in Section 7-D-3d, alkyl nitrites are short-lived in the troposphere and their formation in ambient air by the reaction of alkoxy radicals with NO [i.e., the reverse of reaction (22)] is believed to serve only as a temporary nighttime storage until sunrise, when photolysis occurs.

3-C-6 Nitric Acid and Other Organic Nitrates, Including PAN

Gaseous HNO_3 absorbs light weakly in the shorter-wavelength region of the actinic UV (Fig. 3.33). The absorption cross sections determined recently by Molina and Molina (1981) in the region 290–330 nm at 25°C are given in Table 3.27; these are the values recommended in the NASA evaluation (DeMore et al., 1983, 1985).

Johnston and co-workers (1974) established that the HNO_3 photolysis produces OH and NO_2,

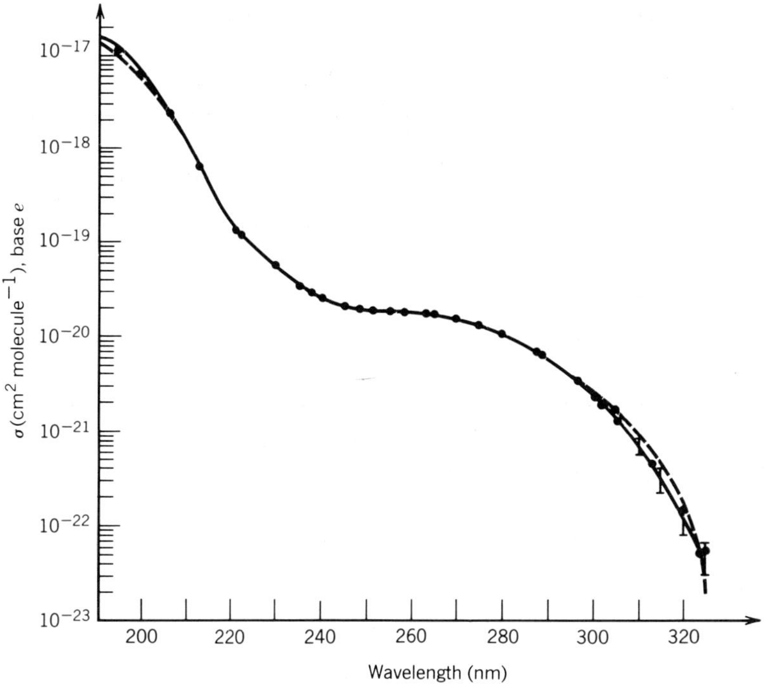

FIGURE 3.33. Semi-logarithmic plot of the absorption spectrum of HNO_3 vapor with σ in units of cm^2 $molecule^{-1}$, base *e*: —, from Molina and Molina (1981); - - -, from Johnston and Graham (1973); ●, from Biaume (1973) (from Molina and Molina, 1981).

TABLE 3.27. Absorption Cross Sections of HNO_3 in the Actinic Region

λ (nm)	10^{20} σ (cm^2 $molecule^{-1}$, base *e*)
285	0.848
290	0.607
295	0.409
300	0.241
305	0.146
310	0.071
315	0.032
320	0.012
325	0.005
330	0.002

Source: Molina and Molina, 1981.

$$HNO_3 + h\nu\ (200 < \lambda < 320\text{ nm}) \xrightarrow[a]{} OH + NO_2 \quad (23)$$

with a quantum yield of approximately unity. This has been confirmed by Margitan and Watson (1982) who find very small quantum yields for oxygen atom (ϕ = 0.03) and hydrogen atom ($\phi < 0.002$) production at λ = 266 nm; these atoms would be expected as products of the alternate photolysis routes (23b) and (23c):

$$\begin{aligned} HNO_3 + h\nu &\xrightarrow[b]{} O + HNO_2 \\ &\xrightarrow[c]{} H + NO_3 \end{aligned} \quad (23)$$

Organic nitrates have an absorption band that tails into the actinic UV (i.e., $\lambda \geq 290$ nm). Figure 3.34 shows the absorption spectra of some organic nitrates in the region 200–330 nm; the absorption coefficients at $\lambda > 290$ nm are fairly small and drop off rapidly with increasing wavelength.

Three primary processes proposed for organic nitrates are

$$\begin{aligned} RCH_2ONO_2 + h\nu &\xrightarrow[a]{} RCH_2O + NO_2 \\ &\xrightarrow[b]{} RC(=O)H + HONO \\ &\xrightarrow[c]{} RCH_2ONO + O \end{aligned} \quad (24)$$

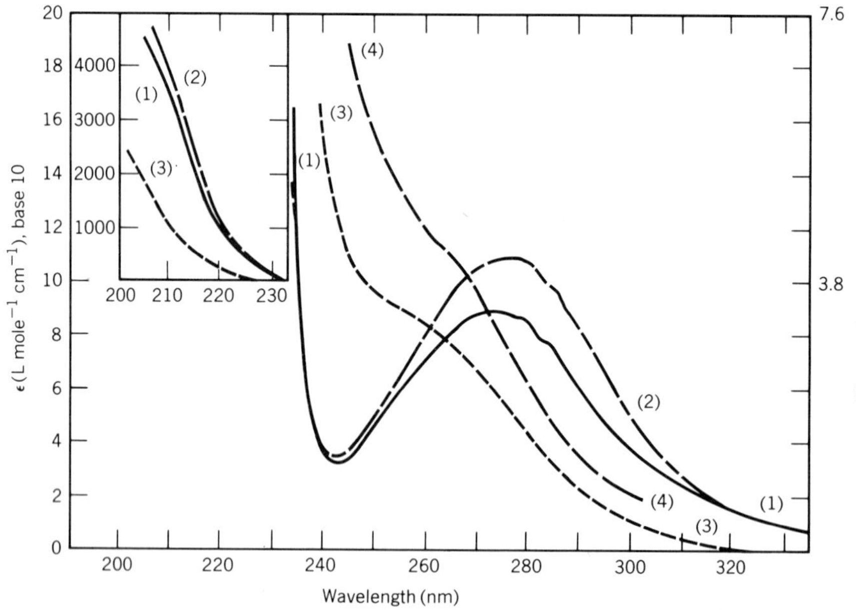

FIGURE 3.34. Absorption spectra at 25°C of (1) nitromethane [$CH_3NO_{2(g)}$], (2) nitroethane [$C_2H_5NO_{2(g)}$], (3) methyl nitrate [$CH_3ONO_{2(g)}$], (4) ethyl nitrate [$C_2H_5ONO_{2(g)}$], Left axis is ϵ in units of L mole^{-1} cm^{-1}, base 10; right axis is σ (cm^2 molecule^{-1}), base e (from Calvert and Pitts, 1966).

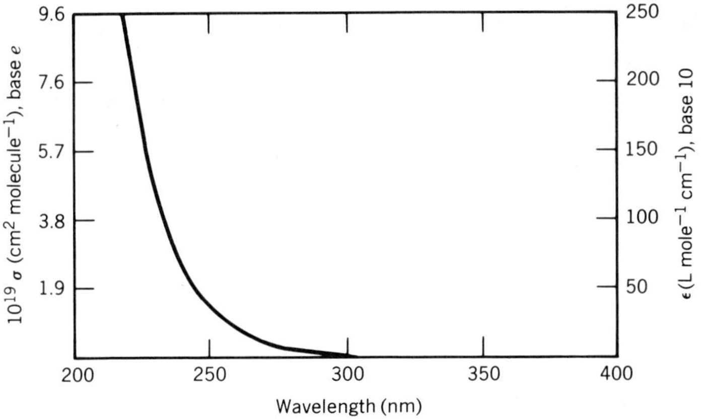

FIGURE 3.35. Ultraviolet spectrum of peroxyacetyl nitrate vapor diluted with 1 atm air. Left axis is σ (cm^2 $molecule^{-1}$), base e; right axis is ϵ in units of L $mole^{-1}$ cm^{-1}, base 10 (from Stephens, 1969).

It is believed that (24a), the formation of an alkoxy radical and NO_2, is the most important decomposition route (Calvert and Pitts, 1966).

Perhaps the most important organic nitrate in tropospheric chemistry is peroxyacetyl nitrate (PAN); the UV spectrum of PAN in air is shown in Fig. 3.35. Table 3.28 shows the absorption coefficients determined recently for gaseous PAN in the region 200–300 nm at 25°C. Also shown are the cross sections for peroxypropionyl nitrate (PPN) which is also produced in ambient air, but at lesser concentrations than PAN (see Chapter 8). Cross sections at $\lambda > 300$ nm were not reported because light absorption was negligible in this region at the sample pressures (2.5–25 Torr for PAN and 1.1–12.3 Torr for PPN) and pathlength (10.1 cm) used.

These absorption coefficients are sufficiently small that photolysis of PAN and PPN is expected to be negligible compared to the thermal decomposition (see Section 4-A-7c and Chapter 8). For example, Senum and co-workers (1984) estimate that PAN lifetimes with respect to photolysis are 55–193 days (assuming constant sunlight) for zenith angles of 0°–60°.

3-C-7. Other Inorganic Nitrogen-Containing Compounds: HO_2NO_2, NO_3, N_2O_5, N_2O_3, N_2O and ClNO

3-C-7a. HO_2NO_2

As discussed in Chapter 8, peroxynitric acid can serve as a reservoir for NO_2 at low temperatures:

$$HO_2 + NO_2 \rightleftarrows HO_2NO_2 \tag{25}$$

It's vapor phase absorption spectrum is shown in Fig. 3.36. Table 3.29 contains the NASA recommended absorption cross section of Molina and Molina (1981);

TABLE 3.28. UV Absorption Cross Sections for Gaseous PAN and PPN

λ (nm)	$10^{20}\sigma$ (cm^2 molecule^{-1}, base e)[a] PAN	PPN
200	317 ± 23	269[b]
205	237 ± 22	211 ± 4
210	165 ± 14	155 ± 6
215	115 ± 9	101 ± 6
220	77 ± 6	69 ± 5
225	55 ± 4	47.9 ± 3.4
230	39.9 ± 3.1	34.7 ± 1.4
235	29.0 ± 2.1	24.2 ± 1.8
240	20.9 ± 1.6	17.3 ± 1.3
245	15.0 ± 1.2	12.4 ± 0.8
250	10.9 ± 0.9	8.9 ± 0.6
255	7.9 ± 0.6	6.5 ± 0.4
260	5.7 ± 0.4	4.6 ± 0.4
265	4.04 ± 0.30	3.24 ± 0.33
270	2.79 ± 0.17	2.29 ± 0.22
275	1.82 ± 0.12	1.51 ± 0.15
280	1.14 ± 0.08	0.99 ± 0.15
285	0.716 ± 0.023	0.60 ± 0.11
290	0.414 ± 0.025	0.33 ± 0.04
295	0.221 ± 0.016	0.16 ± 0.04
300	0.105 ± 0.023	0.097 ± 0.009

Source: Senum et al., 1984.

[a] $\sigma = \ln(I_0/I)/C\ell$, where C = concentration in molecules cm^{-3} and ℓ is the pathlength in cm. PAN = peroxyacetyl nitrate, PPN = peroxypropionyl nitrate.

[b] Based on a single measurement.

they suggest that the photolytic lifetime of HO_2NO_2 in the troposphere will be of the order of 1 day. The quantum yields and products of the photodissociation have not yet been established.

3-C-7b. NO_3

The absorption spectrum of the gaseous nitrate radical NO_3 from ~570 to 670 nm is shown in Fig. 3.37. The absorption cross sections of NO_3 are important in that they are used in measuring NO_3 in the troposphere by long-pathlength UV/visible absorption spectrometry (Section 5-D) and in quantifying the yields of NO_3 from various reactions in laboratory studies.

Table 3.30 shows the values of the absorption cross section at the 662 nm peak reported by four groups, as well as the integrated absorption across the 662 nm band. The integrated absorption intensities are in good agreement, as are the peak

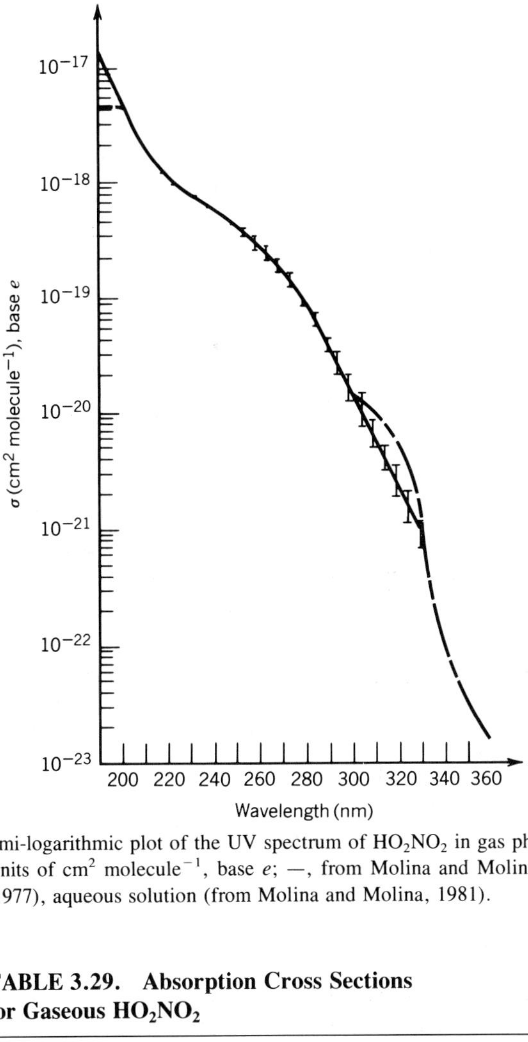

FIGURE 3.36. Semi-logarithmic plot of the UV spectrum of HO_2NO_2 in gas phase and in aqueous solution with σ in units of cm^2 $molecule^{-1}$, base e; —, from Molina and Molina (1981); — - — -, from Jesson et al. (1977), aqueous solution (from Molina and Molina, 1981).

TABLE 3.29. Absorption Cross Sections for Gaseous HO_2NO_2

Wavelength (nm)	10^{20} σ (cm^2 $molecule^{-1}$, base e)
290	4.0
295	2.6
300	1.6
305	1.1
310	0.7
315	0.4
320	0.3
325	0.2
330	0.1

Source: Molina and Molina, 1981.

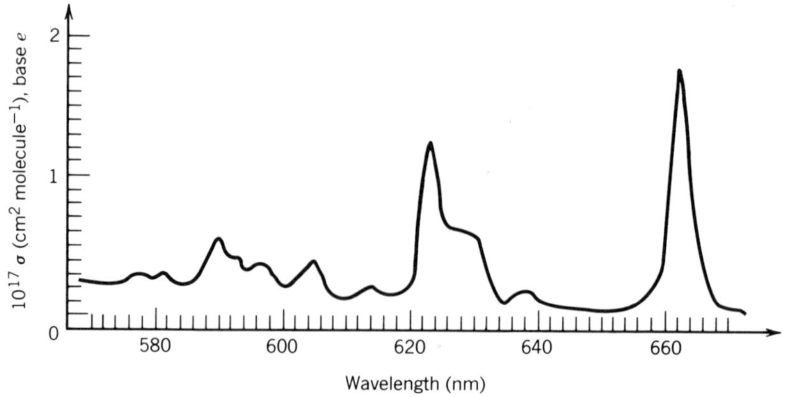

FIGURE 3.37. Absorption spectrum of NO_3 from 510 to 670 nm. σ in units of cm^2 $molecule^{-1}$, base e (from Ravishankara and Wine, 1983).

absorption cross sections except for that of Mitchell et al. (1980). However, the agreement may be somewhat fortuitous, since many of these studies required calculation of the NO_3 concentration in very complex systems.

Thus Sander (1985) has measured an absorption cross section at the 662 nm peak of 2.28×10^{-17} cm^2 $molecule^{-1}$ at 298°K and shown that the intensity of this peak increases by ~18% at 230°K compared to that at 298°K; the intensity of the 679 nm and 637 nm bands decreased in intensity in going from 298° to 230°K. Thus the cross sections in certain regions of the spectrum may be strongly temperature dependent, and the presently recommended cross sections (Table 3.31) may be too small by ~25%.

There are two possible sets of photodissociation products:

$$NO_3 + h\nu \underset{a}{\rightarrow} NO + O_2 \qquad (26)$$
$$\underset{b}{\rightarrow} NO_2 + O$$

TABLE 3.30. Selected Peak Absorption Cross Sections and Integrated Absorptions of the NO_3 Band at 662 nm Reported by Several Groups

10^{17} σ_{max} at 662 nm (cm^2 $molecule^{-1}$) base e	Integrated Absorption of 662 nm Band[a] (cm)	Reference
1.86	1.99×10^{-15}	Graham and Johnston, 1978
1.21	2.06×10^{-15}	Mitchell et al., 1980
1.90	2.02×10^{-15}	Marinelli et al., 1982
1.78	1.88×10^{-15}	Ravishankara and Wine, 1983
1.63	n.r.[b]	Cox et al., 1984
1.85	1.82×10^{-15}	Burrows et al., 1985

[a] Integrated from 670.7 to 654.0 nm.
[b] n.r. = not reported

TABLE 3.31. Absorption Cross Sections (cm^2 molecule^{-1}, base *e*) Averaged Over Each Nanometer for the Gaseous Nitrogen Trioxide (NO_3) Free Radical at 298°K

λ (nm)	$10^{20}\sigma$ (cm^2)	λ (nm)	$10^{20}\sigma$ (cm^2)	λ (nm)	$10^{20}\sigma$ (cm^2)
571	226	605	365	639	157
572	224	606	291	640	111
573	220	607	194	641	92
574	221	608	143	642	85
575	240	609	125	643	83
576	270	610	116	644	84
577	288	611	139	645	80
578	286	612	166	646	65
579	263	613	203	647	65
580	277	614	213	648	55
581	305	615	180	649	46
582	259	616	157	650	46
583	231	617	153	651	46
584	213	618	162	652	47
585	203	619	185	653	48
586	263	620	236	654	65
587	319	621	342	655	83
588	407	622	795	656	122
589	504	623	1238	657	162
590	490	624	998	658	185
591	434	625	698	659	278
592	397	626	628	660	522
593	397	627	628	661	1063
594	323	628	619	662	1756
595	351	629	601	663	1618
596	368	630	555	664	1017
597	351	631	425	665	615
598	305	632	342	666	397
599	250	633	157	667	185
600	222	634	110	668	125
601	222	635	102	669	92
602	251	636	139	670	70
603	296	637	162		
604	360	638	171		

Source: Ravishankara and Wine, 1983.

Magnotta and Johnston (1980) measured the quantum yields for O(^3P) and NO formation, respectively, as a function of wavelength from 470 to 685 nm; the results of their studies are shown in Fig. 3.38. The O(^3P) yield is reasonably constant from 470 to 580 nm and then drops to zero at ~620 nm. The quantum yield of NO, on the other hand, rises from zero at ~580 nm to a peak at ~590

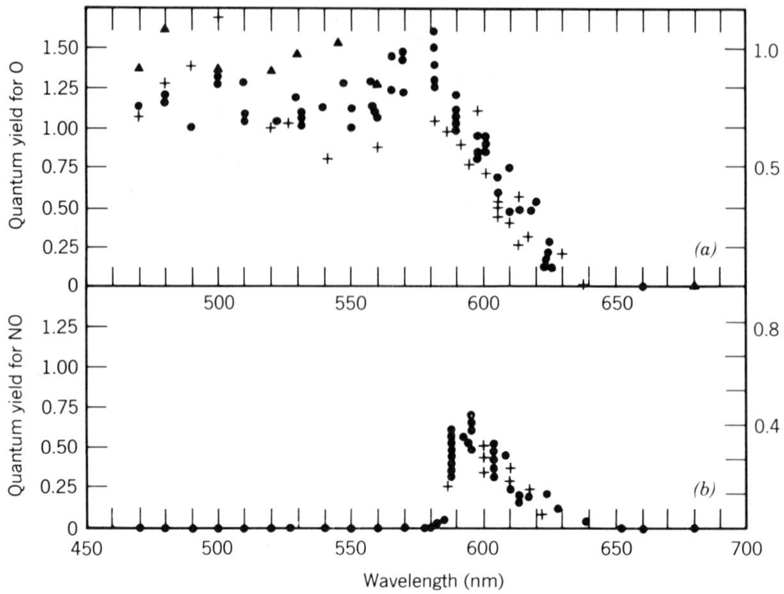

FIGURE 3.38. Primary quantum yields for (*a*) $O(^3P)$ and (*b*) NO production from the photolysis of the gaseous NO_3 radical as a function of wavelength (from Magnotta and Johnston, 1980). The different symbols represent data taken at different fluences of the photolyzing laser. The two scales shown on the left and right sides respectively show the two different sets of quantum yields which can be derived, depending on the laser beam fluence.

nm and then falls again to zero at ~620 nm. Production of $O(^3P)$ exceeds that of NO at all wavelengths. These results are consistent with the fluorescence spectrum of NO_3 which occurs with a large (approaching unity) fluorescence quantum yield for excitation wavelengths $\gtrsim$ 625 nm; the fluorescence is not observed, however, when excited at ~600 nm (Nelson et al., 1983; Ishiwata et al., 1983). This is consistent with the requirement that the sum of all primary quantum yields, including photochemical and photophysical processes, must be unity (Section 2-C-2).

The possible existence of *primary* quantum yields greater than 1.0 shown by the data in Fig. 3.38 is clearly impossible (see Section 2-C-2) and, as discussed by Magnotta and Johnston, implies the possible presence of systematic errors. Thus further work in this important area is needed. In addition to the need to better define the relative importance of the two possible channels, the electronic state of the O_2 needs to be defined; based solely on energetic considerations, at wavelengths below 1100 and 680 nm, O_2 could be produced in the $(a^1\Delta_g)$ state or the $(b^1\Sigma_g^+)$ state, respectively.

In the absence of such data, DeMore et al. (1983, 1985) recommend the Magnotta and Johnston (1980) estimates that, for an overhead sun and for the wavelength region $470 < \lambda < 650$ nm, the photolytic rate constants, k_{26a} for ($NO + O_2$) and k_{26b} for ($NO_2 + O$) production, be taken as $0.022 \pm 0.007\ s^{-1}$ and $0.18 \pm 0.06\ s^{-1}$, respectively.

3-C-7c. N_2O_5

N_2O_5 is expected to exist in ambient air when both NO_2 and NO_3 are present because of the equilibrium (27):

$$NO_3 + NO_2 \rightleftarrows N_2O_5 \tag{27}$$

Figure 3.39 shows the absorption spectrum for N_2O_5 at 25°C and Table 3.32 lists the absorption cross sections at this temperature obtained from the temperature-dependent expression recommended by Yao et al., (1982) for the region 290–380 nm and temperatures (T, °K) 241–300°K:

$$\ln \sigma = 0.432537 + \frac{(4728.48\text{-}17.1269\ \lambda)}{T}$$

In this expression, σ is in units of 10^{-19} cm^2 molecule^{-1} and λ is the absorbing wavelength in nm; the final result for σ is significant to two figures only.

NO_3 has been measured as a photolysis product of N_2O_5 with a primary quantum yield of approximately unity at wavelengths from 249 to 350 nm (Swanson et al., 1984; Burrows et al., 1984; Barker et al., 1985). Thus reaction (28a), which

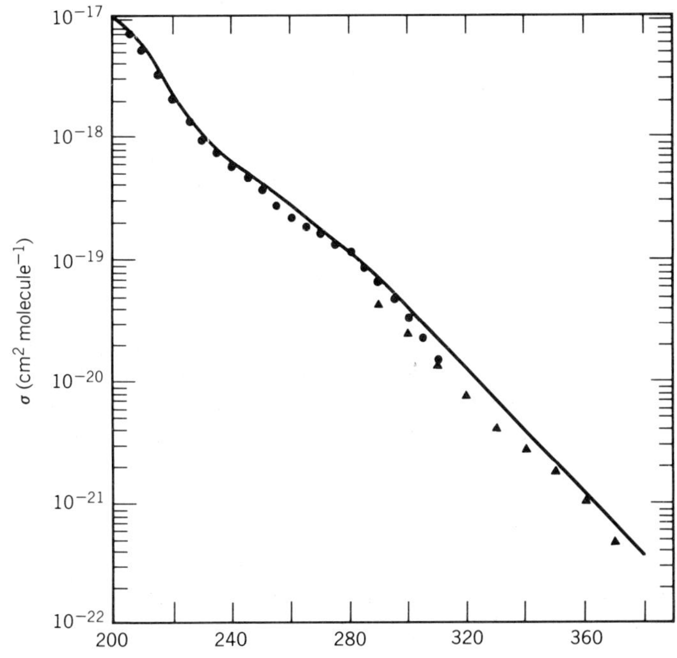

FIGURE 3.39. Semi-logarithmic plot of the absorption coefficient, σ, of N_2O_5 in cm^2 molecule^{-1}, base *e*. —, from Yao et al. (1982); ●, from Johnston and Graham (1974); ▲, from Jones and Wulf (1937) (from Yao et al., 1982).

TABLE 3.32. Absorption Cross Sections for Gaseous N_2O_5 Calculated for 298°K

Wavelength (nm)	$10^{20}\sigma$ (cm^2 molecule^{-1}, base e)
290	6.9
295	5.2
300	3.9
305	2.9
310	2.2
315	1.6
320	1.2
325	0.93
330	0.70
335	0.52
340	0.39
345	0.29
350	0.22
355	0.17
360	0.12
365	0.09
370	0.07
375	0.05
380	0.04

Source: Yao et al., 1982.

is energetically possible at $\lambda < 1270$ nm, appears to be important at all wavelengths:

$$N_2O_5 + h\nu \rightarrow NO_3 + NO_2 \tag{28a}$$

The yields of $O(^3P)$ have been measured in unpublished work as 0.15, 0.22, 0.35, and 0.7 at 291, 287, 266, and 248 nm, respectively (Margitan, 1984; Ravishankara, 1985); recently an upper limit to the $O(^3P)$ quantum yield of 0.1 at 290 nm has been reported (Barker et al., 1985). These data suggest that reaction (28b),

$$N_2O_5 + h\nu \rightarrow 2NO_2 + O \tag{28b}$$

which is theoretically possible at $\lambda < 393$ nm is not important in the troposphere but becomes more important at shorter wavelengths. There is no direct evidence for a third possible channel in the photolysis,

$$N_2O_5 + h\nu \rightarrow NO_2 + NO + O_2$$

which is energetically possible at $\lambda < 1070$ nm, or for a fourth channel,

$$N_2O_5 + h\nu \rightarrow NO + O + NO_3$$

which is possible at $\lambda < 300$ nm.

The likelihood of N_2O_5 being present in polluted ambient air and possible implications of this are discussed in Section 8-A-2b.

3-C-7d. N_2O_3

The absorption spectrum for gaseous N_2O_3 is shown in Fig. 3.40, while Table 3.33 lists the corresponding room temperature absorption cross sections; N_2O_3 was formed from the reaction of NO with NO_2:

$$NO + NO_2 \rightleftarrows N_2O_3 \tag{29}$$

$$K_{29}^{298°K} = 0.53 \text{ atm}^{-1} \quad \text{(Stull and Prophet, 1971)}$$

The possible involvement of N_2O_3 in studies relevant to atmospheric chemistry is discussed in Section 8-A-2c.

3-C-7e. N_2O

The "apparent" absorption spectrum of nitrous oxide is shown in Fig. 3.41. The cross sections are so close to those expected for Rayleigh scattering for this mol-

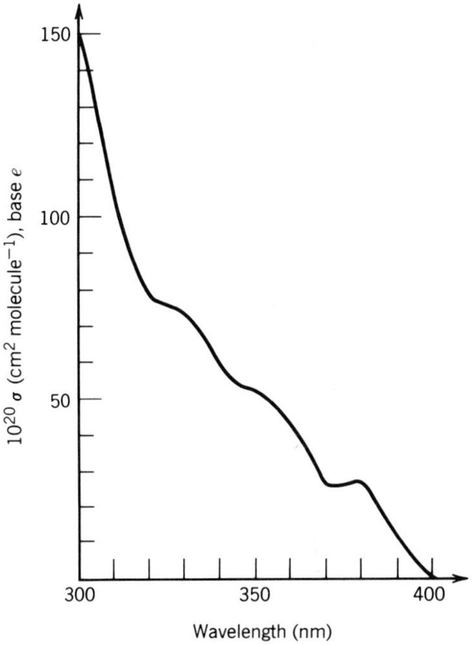

FIGURE 3.40. The absorption cross section σ (cm^2 molecule^{-1}), base e, for gaseous N_2O_3 as a function of wavelength (from Stockwell and Calvert, 1978).

TABLE 3.33. Dinitrogen Trioxide Absorption Cross Sections

Wavelength (nm)	$10^{20}\sigma$ (cm^2 molecule^{-1}, base e)	Wavelength (nm)	$10^{20}\sigma$ (cm^2 molecule^{-1}, base e)	Wavelength (nm)	$10^{20}\sigma$ (cm^2 molecule^{-1}, base e)
300	150	334	68.4	368	29
301	146	335	66.8	369	27
302	141	336	65.7	370	26
303	136	337	63.8	371	26
304	132	338	62.3	372	26
305	128	339	60.3	373	26
306	123	340	58.1	374	26
307	118	341	57.3	375	26
308	113	342	56.1	376	26
309	109	343	55.4	377	27
310	106	344	54.2	378	27
311	101	345	53.5	379	27
312	97.4	346	53.1	380	27
313	94.0	347	52.7	381	25
314	90.9	348	52.3	382	24
315	88.6	349	51.9	383	23

316	85.6	350	51.6	384	21
317	83.3	351	50.4	385	20
318	81.7	352	49.7	386	18
319	79.8	353	48.9	387	17
320	78.3	354	48.1	388	15
321	77.5	355	47.7	389	13
322	76.8	356	47.0	390	12
323	76.4	357	45.8	391	10
324	76.0	358	44.7	392	8.8
325	75.6	359	43.9	393	7.3
326	75.2	360	43.2	394	5.7
327	74.9	361	41.2	395	4.6
328	74.5	362	39.7	396	3.8
329	74.1	363	38.2	397	3
330	73.7	364	36	398	2
331	71.8	365	35	399	1
332	70.7	366	33	400	0
333	69.9	367	31		

Source: Stockwell and Calvert, 1978.

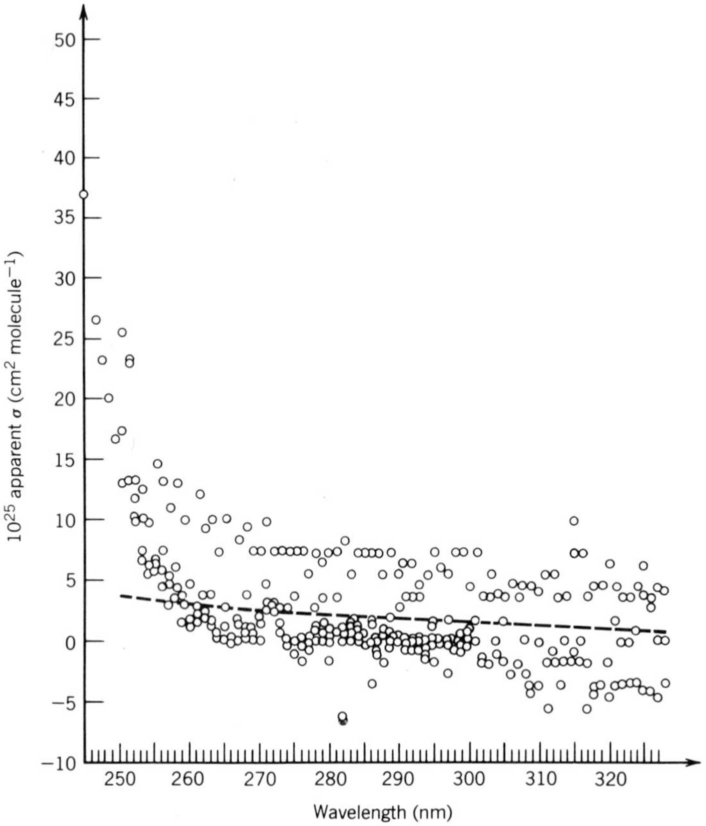

FIGURE 3.41. *Apparent* absorption cross sections for N_2O from 245 to 325 nm. Dashed line is an estimate for the Rayleigh scattering cross section for N_2O. *Apparent* is used because cross sections are so close to Rayleigh scattering that they must contain a significant scattering component (from Johnston and Selwyn, 1975).

ecule that it probably does not absorb in the actinic UV. Thus N_2O does not contribute to tropospheric photochemistry by direct photolysis; that it might do so by heterogeneous photosensitized processes is an intriguing idea (Rebbert and Ausloos, 1978).

3-C-7f ClNO

As discussed in Section 12-D-1, under laboratory conditions, ppm concentrations of NO_2(g) react with NaCl(s) to form nitrosyl chloride, ClNO; this may also occur in marine urban environments (Finlayson-Pitts, 1983). The absorption spectrum of ClNO in the actinic UV is shown in Fig. 3.42 and the absorption cross sections are given in Table 3.34. These cross sections are sufficiently large that photolysis in sunlight is quite rapid, producing Cl + NO with a quantum yield of unity (Calvert and Pitts, 1966):

$$ClNO + h\nu \rightarrow Cl + NO$$

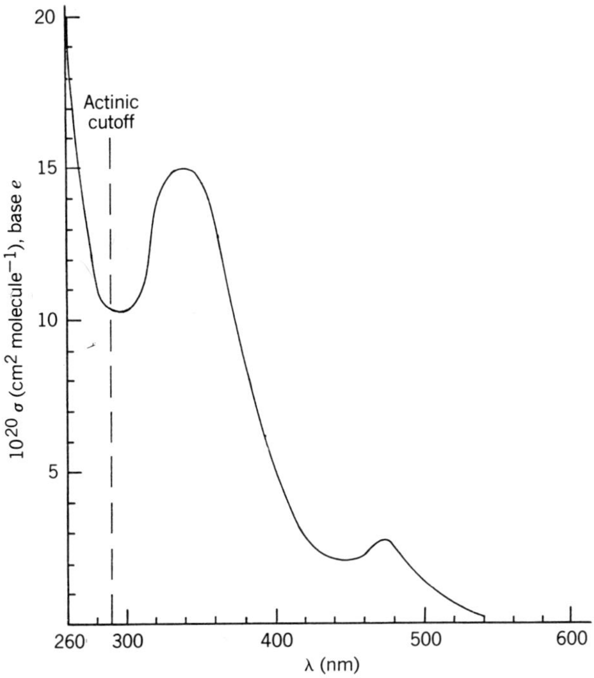

FIGURE 3.42. Absorption spectrum of ClNO vapor with σ in cm^2 $molecule^{-1}$, base e. Data from Ballash and Armstrong (1974).

3-C-8. Hydrogen Peroxide, Organic Peroxides, and Hydroperoxides

Hydrogen peroxide absorbs weakly in the actinic UV, as do the organic peroxides (Figs. 3.43 and 3.44). Table 3.35 gives the absorption cross sections from 290 to 350 nm recommended in the NASA evaluation (DeMore, 1983, 1985) for gaseous H_2O_2 and methyl hydroperoxide.

TABLE 3.34. Absorption Cross Sections, σ, for Gaseous ClNO

λ (nm)	10^{20} σ (cm^2 $molecule^{-1}$, base e)
280	10.3
300	9.5
320	12.1
340	13.7
360	12.2
380	8.32
400	5.14

Source: DeMore et al., 1983; based on Ballash and Armstrong, 1974, and Illies and Takacs, 1976/77.

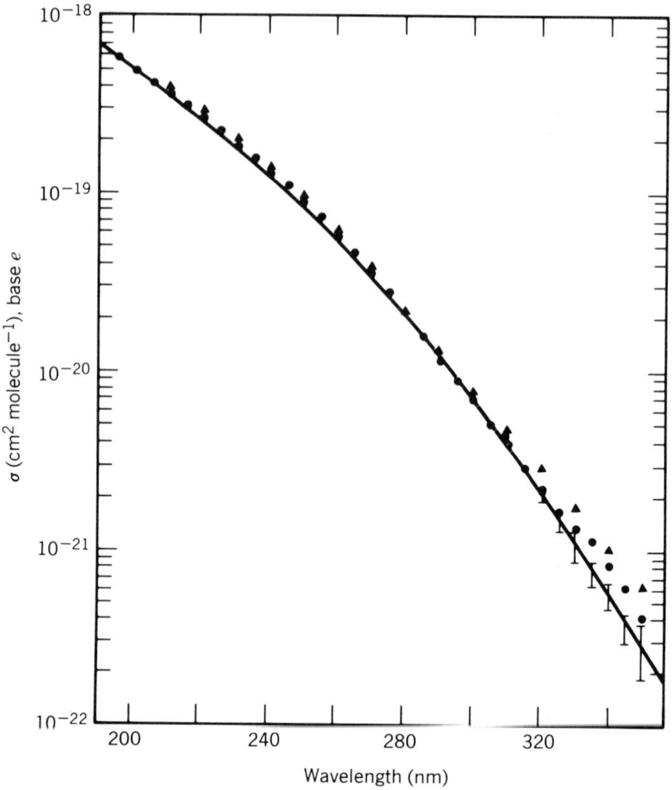

FIGURE 3.43. Semi-logarithmic plot of the UV spectrum of H_2O_2 vapor with σ in cm^2 $molecule^{-1}$, base e: —, from Molina and Molina (1981); ●, from Lin et al. (1978), ▲, from Molina et al. (1977) (from Molina and Molina, 1981).

The photolysis of H_2O_2 gives hydroxyl free radicals

$$H_2O_2 + h\nu\ (\lambda \leq 360 \text{ nm}) \rightarrow 2\text{ OH} \tag{30}$$

with a quantum yield of unity (Atkinson and Lloyd, 1984). H_2O_2 is also believed to play an important role in suspended droplets in the atmosphere; its photochemistry in solution is discussed in Section 3-C-12.

Organic peroxides photodissociate like H_2O_2, that is, by rupture of the weak peroxidic O—O bond:

$$R_aOOR_b + h\nu \rightarrow R_aO + R_bO \tag{31}$$

However, the alkoxy radicals produced may posses sufficient excess energy to undergo further fragmentation (Calvert and Pitts, 1966). Thus a typical RO—OR bond strength is ~40 kcal $mole^{-1}$, whereas light of wavelength 300 nm has an energy of 95 kcal $mole^{-1}$; the excess energy initially resides as internal energy in

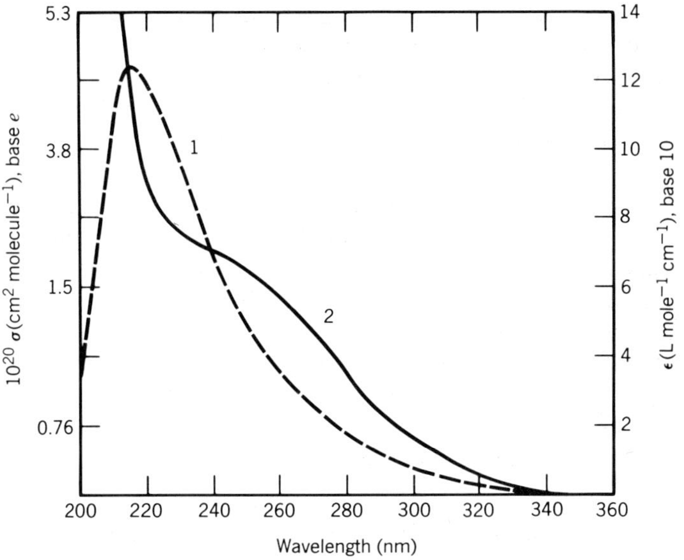

FIGURE 3.44. Absorption spectra of (1) dimethyl peroxide [$CH_3OOCH_{3(g)}$] from Takezaki et al. (1956); (2) di-*tert*-butyl peroxide [$(CH_3)_3COOC(CH_3)_{3(g)}$], all at 25°C. Left axis is σ (cm^2 $molecule^{-1}$), base *e*; right axis is ϵ (L $mole^{-1}$ cm^{-1}, base 10 (from Calvert and Pitts, 1966).

TABLE 3.35. Absorption Cross Sections for Gaseous H_2O_2 and CH_3OOH

Wavelength (nm)	10^{20} σ (cm^2 $molecule^{-1}$, base *e*)	
	H_2O_2[a]	CH_3OOH[b]
290	1.13	0.90
295	0.87	
300	0.66	0.58
305	0.49	
310	0.37	0.34
315	0.28	
320	0.20	0.19
325	0.15	
330	0.12	0.11
335	0.09	
340	0.07	0.06
345	0.05	
350	0.03	0.04

[a] *Source:* DeMore et al., 1983, 1985; from the data of Lin et al., 1978, and Molina and Molina, 1981.

[b] *Source:* Molina and Arguello, 1979.

the fragments and this may lead to their dissociation. However, whether this can occur under atmospheric conditions where deactivation is rapid is not known. Solution phase photolysis of hydroperoxides is discussed in Section 3-C-12.

3-C-9. Formaldehyde and the Higher Aldehydes

The absorption spectra of formaldehyde, acetaldehyde, propionaldehyde, *n*-butyraldehyde, and isobutyraldehyde are shown in Figs. 3.45 and 3.46. The sharply banded structure for HCHO is in sharp contrast to the continuous spectra for the higher aldehydes. It is this banded structure for HCHO which allows the application of differential optical absorption spectrometry to the detection and measurement of HCHO in the troposphere (Section 5-D). It is also noteworthy that the absorption of HCHO extends out to ~370 nm, whereas the higher aldehydes only absorb to ~345 nm.

As we have seen, formaldehyde has two photodissociation paths:

$$\begin{aligned} \mathrm{HCHO} + h\nu\ (\lambda < 370\ \mathrm{nm}) &\xrightarrow[a]{} \mathrm{H} + \mathrm{HCO} \\ &\xrightarrow[b]{} \mathrm{H_2} + \mathrm{CO} \end{aligned} \tag{2}$$

Path (2a) is particularly important in tropospheric chemistry since in air it provides a source of HO_2 radicals and hence ultimately OH as discussed in Section 3-D-2.

Because of the importance of HCHO as a source of OH in the troposphere, a

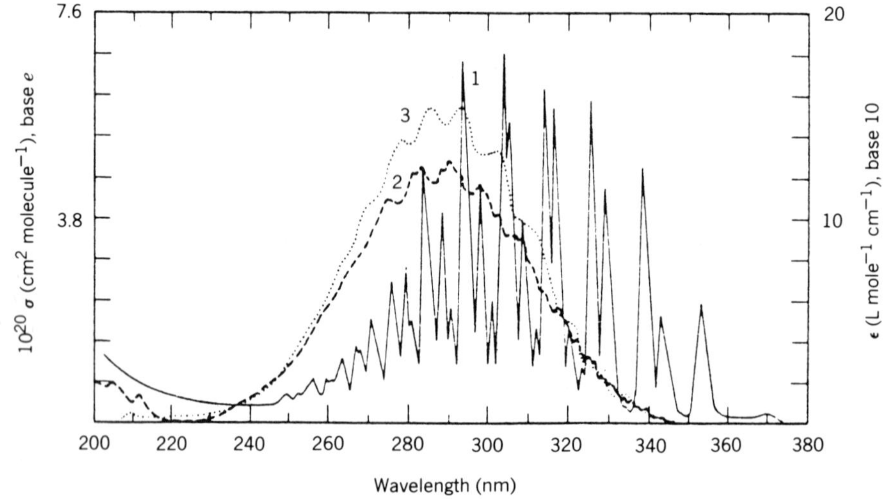

FIGURE 3.45. Vapor phase absorption spectra for (1) formaldehyde [$CH_2O_{(g)}$], ~75°C; (2) acetaldehyde [$CH_2CHO_{(g)}$], 25°C; (3) propionaldehyde [$C_2H_5CHO_{(g)}$], 25°C. Left axis is σ (cm^2 molecule^{-1}), base *e*; right axis is ε (L mole^{-1} cm^{-1}), base 10 (from Calvert and Pitts, 1966).

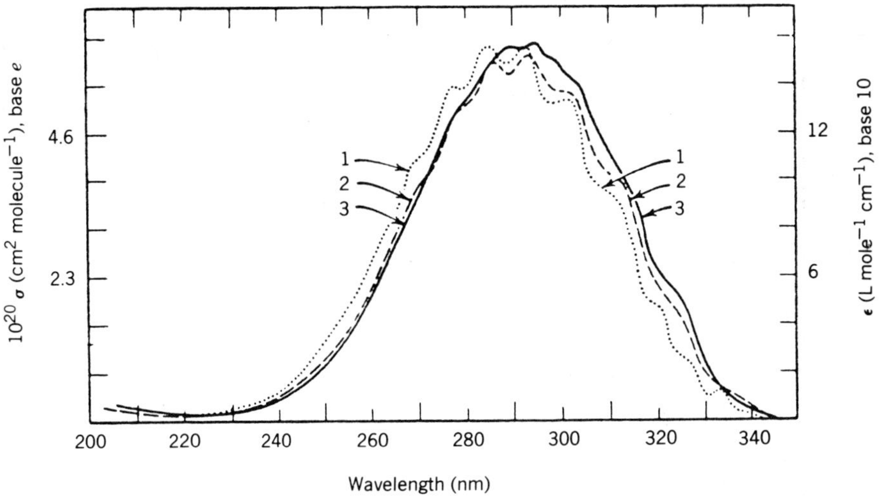

FIGURE 3.46. Vapor phase absorption spectra for (1) propionaldehyde [$C_2H_5CHO_{(g)}$], (2) *n*-butyraldehyde [*n*-$C_3H_7CHO_{(g)}$], (3) isobutyraldehyde [iso-$C_3H_7CHO_{(g)}$], all at 25°C. Left axis is σ (cm^2 molecule^{-1}), base *e*; right axis is ϵ (L mole^{-1} cm^{-1}), base 10 (from Calvert and Pitts, 1966).

great deal of work has been carried out on the absorption cross sections and the primary quantum yields. The data through 1979 are reviewed by Calvert (1980). The most recent measurements of the absorption cross sections by Bass et al. (1980) and by Moortgat et al. (1980, 1983) differ by ~30% in the actinic UV. The NASA evaluation recommends using the mean of the two sets of data.

These recommended absorption cross sections and the quantum yields for (2a) and (2b) were given in Table 3.14, where HCHO was used as an example of how to calculate photolysis rates. Because ϕ_{2b} depends on pressure at $\lambda \geq 330$ nm, the values given in Table 3.14 are for 1 atm. For reduced pressures the NASA evaluation recommends using the expression

$$\phi_{2b} = \frac{1 - \exp(112.8 - 0.347\lambda)}{1 + (p/760)(\lambda - 329)/(364 - \lambda)}$$

where λ is in nm and the pressure p is in Torr.

Figure 3.47 shows ϕ_{2a} and ϕ_{2b} as functions of wavelength from 250 to 360 nm; the sum of the primary quantum yields ($\phi_{2a} + \phi_{2b}$) is approximately unity up to ~330 nm; it then falls off at longer wavelengths.

For further details of the chemical dynamics, electronic states, and product energy distributions involved in HCHO photolysis, the reader is referred to the reviews by Moore and Weisshaar (1983) and Lee and Lewis (1980).

Acetaldehyde has four theoretically possible sets of photodecomposition products:

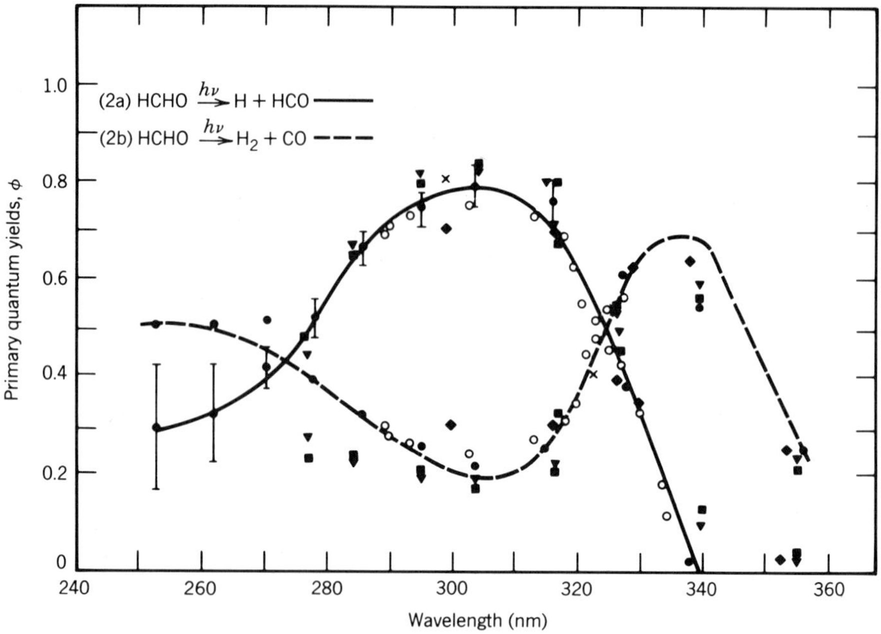

FIGURE 3.47. Wavelength dependence of HCHO photolysis quantum yields. ●, Moortgat et al. (1980); ×, Tang et al. (1979); ○, Horowitz and Calvert (1978); ■, ▼ Moortgat and Warneck (1979); ♦, Clark et al. (1978). Data of Moortgat and coworkers in 1 atm air (from Baulch et al., 1982).

$$
\begin{aligned}
CH_3CHO + h\nu &\xrightarrow[a]{} CH_3 + HCO \\
&\xrightarrow[b]{} CH_4 + CO \\
&\xrightarrow[c]{} H + CH_3CO \\
&\xrightarrow[d]{} H_2 + CH_2CO
\end{aligned}
\qquad (32)
$$

While a number of laboratory studies of the gas phase photochemistry of the simple aliphatic aldehydes have been carried out over the past five decades, including pioneering studies in the laboratory of F. E. Blacet (Calvert and Pitts, 1966; Lee and Lewis, 1980), relatively few have been under conditions either typical of the atmosphere or which can be readily extrapolated to atmospheric conditions. This is particularly important since deactivation of excited states can be important at higher pressures. Furthermore, it has been suggested that a triplet state of CH_3CHO formed on photolysis in the actinic UV may form a complex with O_2 (Weaver et al., 1976/77).

In the one study of acetaldehyde photochemistry carried out in 1 atm of air (Meyrahn et al., 1981), H_2 was not found in significant amounts, ruling out (32d). The two major paths were found to be (32a) and (32b), with the primary quantum yields varying with wavelength as shown in Fig. 3.48; also shown in Fig. 3.48

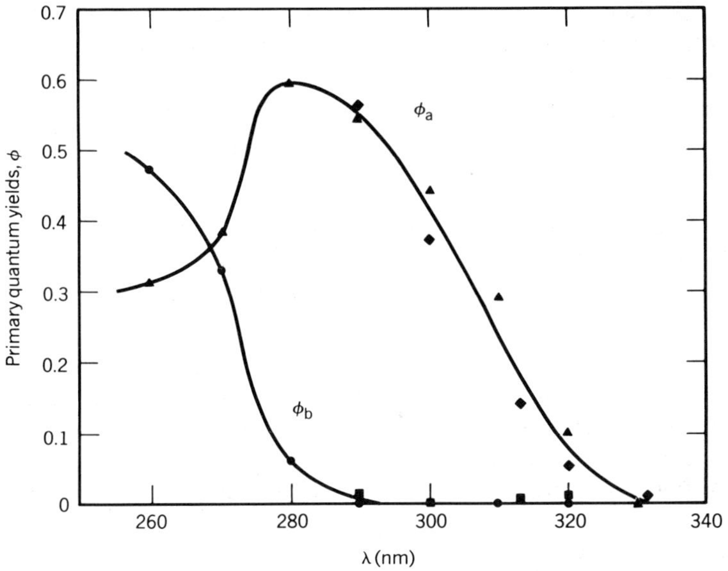

FIGURE 3.48. Primary quantum yields ϕ_a and ϕ_b for photodissociation of CH_3CHO to CH_3 + HCO (ϕ_a) and CH_4 + CO (ϕ_b). ♦, Horowitz and Calvert (1982); ▲, ●, Meyrahn et al. (1981). Line is set recommended by Atkinson and Lloyd (1984) (from Atkinson and Lloyd, 1984).

are the quantum yield estimates for (32a) and (32b) of Horowitz and Calvert (1982), obtained by extrapolating the results of their laboratory study to 1 atm of air. These two studies are used as the basis of the recommendations of Atkinson and Lloyd (1984) for the quantum yields for (32a), (32b), and (32c) given in Table 3.36.

The majority of studies of aldehyde photolysis relevant to atmospheric conditions involve formaldehyde or acetaldehyde, since HCHO and, to a lesser extent, CH_3CHO are the aldehydes typically found in the highest concentrations in am-

TABLE 3.36. Primary Quantum Yields in CH_3CHO Photolysis in One Atmosphere of Air at 298°K[a]

λ (nm)	ϕ_{32a} (CH_3 + HCO)	ϕ_{32b} (CH_4 + CO)	ϕ_{32c} (H + CH_3CO)
280	0.59	0.06	n.r.[b]
290	0.55	0.01	0.026
300	0.415	0.00	0.009
310	0.235	0.00	0.002[c]
320	0.08	0.00	0.00
330	0.00	0.00	0.00

[a] Recommended by Atkinson and Lloyd, 1984.

[b] n.r. = not reported.

[c] At λ = 313 nm instead of 310 nm.

bient air (Section 5-G). Additionally, the radical formation rate from the photolysis of acetaldehyde is much less (approximately an order of magnitude) than from formaldehyde (Horowitz and Calvert, 1982); this is expected to be true of the higher aldehydes as well which are present in even smaller concentrations. For example, for propionaldehyde photolysis in air, Shepson and Heicklen (1982a,b) report that for the two routes (33a) and (33b)

$$CH_3CH_2CHO + h\nu \xrightarrow{a} C_2H_5 + CHO$$
$$\xrightarrow{b} C_2H_6 + CO \qquad (33)$$

$\phi_{33a} \simeq 0.20 \pm 0.08$ from 254 to 334 nm and $\phi_{33b} \simeq 0$ for $\lambda \geq 280$ nm.

Gaseous α, β-unsaturated aldehydes such as acrolein appear to be quite stable towards photooxidation under laboratory conditions. Their photochemistry in simulated atmospheres warrants further study since certain of them are combustion products from diesel engines and non-catalyst equipped motor vehicles.

3-C-10. Ketones

Figure 3.49 shows the vapor phase absorption spectra of four simple ketones; they absorb light in the actinic UV out to ~330 nm.

Relatively few studies of simple ketone photolysis have been carried out under atmospheric conditions, although the vapor phase photooxidations of simple ke-

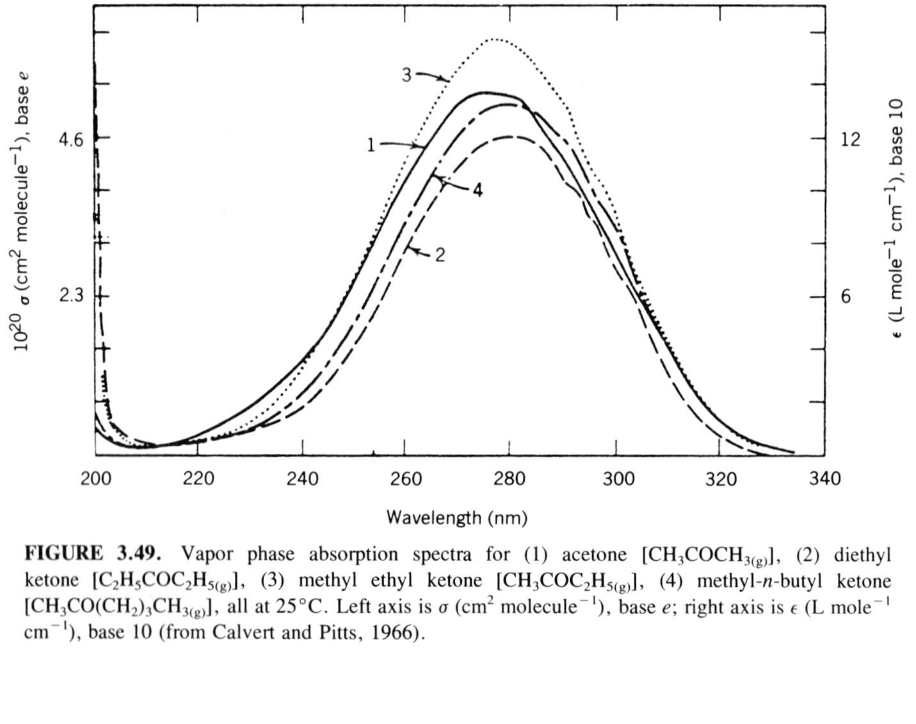

FIGURE 3.49. Vapor phase absorption spectra for (1) acetone [$CH_3COCH_{3(g)}$], (2) diethyl ketone [$C_2H_5COC_2H_{5(g)}$], (3) methyl ethyl ketone [$CH_3COC_2H_{5(g)}$], (4) methyl-*n*-butyl ketone [$CH_3CO(CH_2)_3CH_{3(g)}$], all at 25°C. Left axis is σ (cm^2 $molecule^{-1}$), base *e*; right axis is ϵ (L $mole^{-1}$ cm^{-1}), base 10 (from Calvert and Pitts, 1966).

tones in the presence of O_2 have been investigated in detail in classic studies in the laboratory of W. A. Noyes, Jr. and coworkers (Calvert and Pitts, 1966; Lee and Lewis, 1980). Upper limits to the rates of photolysis of three ketones have been calculated by Cox et al. (1981) using the solar flux data of Peterson (1976), the absorption spectra shown in Fig. 3.49, and taking the quantum yield for the loss of the carbonyl compound to be unity; these are summarized in Table 3.37 for a solar zenith angle of 30°. If, in fact, the quantum yields for photodissociation at 1 atm in air were unity, these photolysis rates would apply and photodissociation would predominate over reaction with OH as a sink for the removal of acetone and methyl ethyl ketone in the atmosphere (Cox et al., 1981).

The primary modes of photodecomposition of simple ketones under laboratory conditions have been well established, in part through detailed studies by W. A. Noyes, Jr. and his co-workers (Calvert and Pitts, 1966; Lee and Lewis, 1980). Thus acetone undergoes an α-cleavage reaction to form methyl and acetyl radicals (the Norrish type I split):

$$CH_3\overset{\overset{\displaystyle O}{\|}}{C}CH_3 + h\nu \rightarrow CH_3 + CH_3CO \qquad (34)$$

For larger ketones such as methyl ethyl ketone, there are two possible bond cleavages, producing different sets of free radicals:

$$CH_3\overset{\overset{\displaystyle O}{\|}}{C}C_2H_5 + h\nu \xrightarrow{a} CH_3CO + C_2H_5 \qquad (35)$$
$$\xrightarrow{b} CH_3 + C_2H_5CO$$

Rupture of the bond forming the most stable radical usually predominates. For example, under laboratory conditions at 313 nm with I_2(g) present as a free radical

TABLE 3.37. Upper Limits[a] to Calculated Rates of Photolysis for Simple Ketones

Ketone	$10^6\,k_p$ (s^{-1})
CH_3COCH_3	12.4
$CH_3COC_2H_5$	12.5
$CH_3COCH_2CH(CH_3)_2$	12.5[b]

Source: Cox et al., 1981.

[a] Upper limits assuming a photodissociation quantum yield of unity.

[b] Assumed by Cox et al. (1981) to be identical to the rate of methyl ethyl ketone photolysis.

trap, $\phi_{35a}/\phi_{35b} \sim 40$ (Pitts and Blacet, 1950). However care should be taken if one extrapolates these values for ϕ_a and ϕ_b to the photolysis of gaseous methyl ethyl ketone in air.

For longer-chain ketones containing a γ hydrogen, photodissociation into a lower ketone and an olefin (Norrish type II split) analogous to that shown for n-butyraldehyde [Section 2-D-1, Eq. (13c)], can occur, as well as an intramolecular reduction of the carbonyl group to form a cyclobutyl alcohol. However, such reactions are not generally important in the chemistry of the troposphere, other than being sinks for the relatively low levels of such ketones found there.

The photolysis of diethyl ketone vapor at temperatures above 125° and pressures less than 50 Torr has been used as an actinometer for the 250–320 nm region, with the quantum yield of CO being near unity and that of the C_2H_5 radicals ~ 2 (Calvert and Pitts, 1966):

$$C_2H_2COC_2H_5 + h\nu \rightarrow CO + 2C_2H_5 \tag{36}$$

The gas phase photooxidation of diethyl ketone carried out at 120°C under laboratory conditions in the presence of oxygen proceeds efficiently and yields a variety of products including CH_3CHO, CO_2, C_2H_5OH, and CO, as well as much smaller amounts of oxygen-containing organics such as perpropionic acid and propionic acid (Kallend and Pitts, 1969). Certainly many of these might also be found in atmospheric photooxidations, but the ambient levels of diethyl ketone are generally too low for such processes to be important, except perhaps in certain occupational settings.

3-C-11. Dicarbonyl Compounds

Rates and mechanisms of photolysis of both saturated and unsaturated dicarbonyls are of interest because they are formed in the air photooxidation of aromatic hydrocarbons in the presence of NO_x; for example, such α-dicarbonyls as glyoxal, methyl glyoxal, and biacetyl have been observed, as have some unsaturated species such as 3-hexene-2,5-dione (see Section 7-G). As seen in Fig. 3.50, these dicarbonyls absorb in the actinic region. A number of laboratory studies of their photochemistry have been carried out (Lee and Lewis, 1980); however, relatively little is known about the photochemistry under solar actinic radiation and in 1 atm of air.

The photolyses of glyoxal $(CHO)_2$, methyl glyoxal (CH_3COCHO), and biacetyl $(CH_3CO)_2$ in air have been studied in an experimental chamber and the rates found to be relatively fast (Plum et al., 1983). Thus for the spectral distribution and intensity used, which approximated that at the earth's surface, the photolysis rate constants (k_p) varied from 0.8% of that of NO_2 for glyoxal to 3.6% of that of NO_2 for biacetyl; this is sufficiently fast to make photolysis predominate over reaction with OH or O_3 in the removal of these dicarbonyls from the atmosphere. Comparison of the experimentally determined photolysis rates to those calculated using Eq. (T) above with the assumption of unit quantum yield over the entire absorption

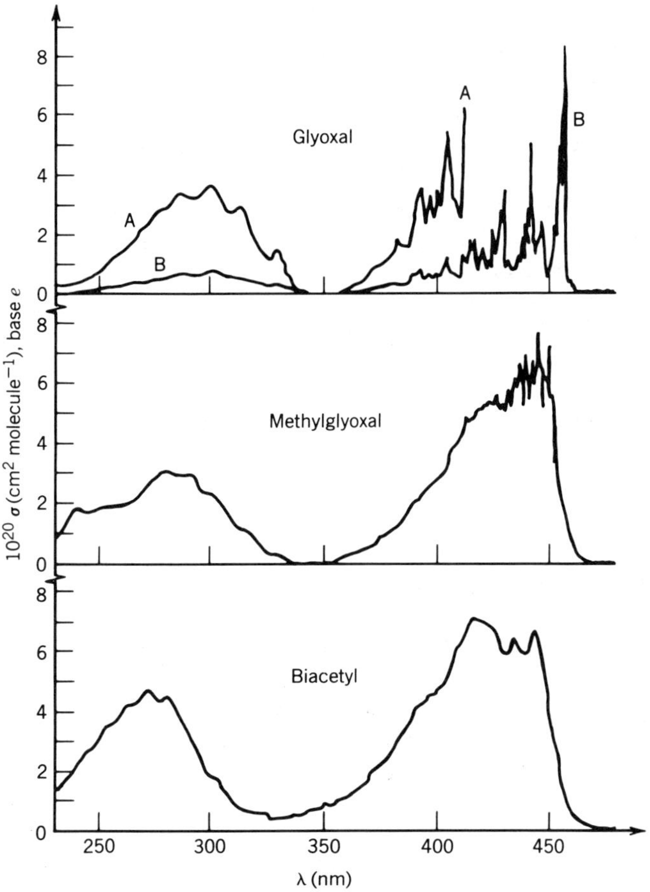

FIGURE 3.50. Absorption spectra of gaseous glyoxal, methylglyoxal, and biacetyl at room temperature. Absorption cross sections σ are given in units of cm^2 molecule^{-1}, base e. Curve B = A × 0.2 (from Plum et al., 1983).

showed that the average quantum yields for photodissociation for this spectral distribution were 0.029 ± 0.018 for glyoxal, 0.11 ± 0.03 for methyl glyoxal, and 0.16 ± 0.02 for biacetyl, with essentially all the photochemical activity occurring in the 340–470 nm absorption band (Fig. 3.50).

The yield of HCHO in the glyoxal photolysis suggested that ~13% proceeded via (37b), with the rest likely proceeding via (37a) (Osamura et al., 1981):

$$
\begin{aligned}
(CHO)_2 + h\nu &\underset{a}{\rightarrow} 2CO + H_2 \\
&\underset{b}{\rightarrow} HCHO + CO
\end{aligned}
\tag{37}
$$

In the methyl glyoxal and biacetyl photolysis, at least part of the dissociation produces acetyl radicals (CH_3CO), as indicated by the formation of PAN (Plum et al., 1983). However, the extent of radical formation is subject to some controversy (Killus and Whitten, 1983; Besemer and Nieboer, 1983).

Both *cis*- and *trans*-3-hexene-2,5-dione photolyze in air to give an equilibrium mixture of the two with a *trans*/*cis* ratio of 0.55 ± 0.05; this photoisomerization accounted for ≳80% of the observed loss rate of each isomer (Tuazon et al., 1985).

Dicarboxylic acids may be formed in the atmosphere from oxidation of the corresponding dicarbonyl compounds (eg see Section 12-D-2); however, little is known about their photochemistry in the actinic UV, especially under atmospheric conditions. Laboratory studies of the photolysis of pure oxalic acid vapor at 115°C and from 257 to 313 nm have been carried out and two primary processes identified (Yamamoto and Back, 1985):

$$(COOH)_2 + h\nu(\lambda \leq 313 \text{ nm}) \xrightarrow[a]{} CO_2 + HCOOH$$

$$\xrightarrow[b]{} CO_2 + CO + H_2O$$

The ratio of the quantum yields was found to be $\phi_a/\phi_b = 2.6$.

3-C-12. Photochemical Reactions in Condensed Phases

The solid phase photooxidations in air of certain polycyclic aromatic hydrocarbons (PAH) exposed to actinic radiation are an important loss process. Some of these apparently proceed through processes involving $O_2(^1\Delta_g)$, but, in general, their mechanisms of reaction on complex surfaces (e.g., combustion-generated aerosols) are not well understood; this is discussed in Chapter 13.

Similarly, photochemical reactions of both organics and inorganics in mixtures commonly found in polluted droplets involved in acidic rain and fog are very important in tropospheric chemistry. However, relatively few quantitative data are currently available on the solution phase photochemistry of complex aqueous systems containing such inorganic species as O_3, NO_2, PAN, HONO, H_2O_2, metals, and so on, in the absence and presence of reactive organic compounds; indeed, as discussed in Section 11-A-2a, there is even some disagreement concerning which reactions might be important.

Photochemistry in atmospheric droplets, as in the gas phase, is determined by the intensity and spectral distribution of the actinic radiation, by the photochemical quantum yields, and by the absorption coefficients. While the incoming solar radiation has the intensity and spectral distribution discussed in Section 3-A-2, reflection and scattering occur at the surface of the drop. It is believed, however, that these will not reduce the radiation within the droplet by more than 10–20%. As a result, the actinic flux in these solutions is often taken to be ~85% of the clear sky values given earlier.

Photolysis rate constants $k_p(s^{-1})$ for solution phase photochemical processes are thus calculated as described for gases in Section 3-B-1, using Eq. (T), and the actinic fluxes of Table 3.5 appropriately corrected for season, latitude, clouds, and so on. However, to take into account the reduction in light intensity due to reflection and scattering at the air–water interface, the actinic fluxes are multiplied by a factor of ~0.8–0.9 (Graedel and Goldberg, 1983).

The photochemical quantum yields for photolysis in solution are usually significantly lower (i.e., $\phi \lesssim 0.1$) than those in the gas phase because of the solvent cage effect (see Section 4-D). When a species such as HONO photodissociates in solution, the two fragments cannot rapidly move apart as they do in the gas phase. Thus they are likely to recombine in the solvent cage, leading to a low effective quantum yield. In the case of HONO photolysis from 300 to 400 nm in aqueous solution, for example, $\phi = 0.095 \pm 0.005$ for photolysis and escape of the fragments from the solvent cage (Rettich, 1978).

Because of this solvent cage effect, the term *effective quantum yield* is frequently used with regard to solution phase photochemistry, where "effective" is used to indicate that it includes both photolysis and escape of the fragments from the solvent cage.

Tables 3.38 to 3.45 give the absorption coefficients recommended by Graedel

TABLE 3.38. Absorption Coefficients of Ozone in Aqueous Solution

λ (nm)	$10^{21}\ \sigma$ (cm^2 molecule^{-1}) Base 10	Base *e*	ϵ (L mole^{-1} cm^{-1}) Base 10
290	1500	3.45×10^3	905
300	500	1.15×10^3	301
305	270	6.2×10^2	163
310	140	3.2×10^2	84
313	100	2.3×10^2	60
320	42	97	25
330	13	30	7.8
340	3.3	7.6	2.0
480	1.2	2.8	0.72
500	2.8	6.4	1.7
520	3.8	8.8	2.3
540	5.5	13	3.3
546	6.0	14	3.6
560	7.1	16	4.3
580	8.0	18	4.8
590	8.5	20	5.1
600	7.6	17.5	4.6
625	5.5	13	3.3
650	4.2	9.7	2.5
675	2.7	6.2	1.6

Source: Recommended by Graedel and Weschler, 1981.

TABLE 3.39. Absorption Coefficients of Hydrogen Peroxide in Aqueous Solution

λ (nm)	10^{23} σ (cm^2 molecule^{-1}) Base 10	Base *e*	ε (L mole^{-1} cm^{-1}) Base 10
290	760	1.75×10^3	4.6
300	380	875	2.3
310	180	415	1.1
320	88	203	0.53
330	41	94	0.25
340	19	44	0.11
350	8.3	19	0.11
360	3.7	8.5	2.2×10^{-2}
370	1.6	3.7	9.6×10^{-3}
380	0.80	1.8	4.8×10^{-3}
390	0.66	1.5	4.0×10^{-3}

Source: Recommended by Graedel and Weschler, 1981; based on the data for H_2O_2 of Taylor and Cross, 1949, and Phibbs and Giguère, 1951.

and Weschler (1981) for O_3, H_2O_2, NO_2, HONO, HNO_3, NO_2^-, NO_3^-, and the nitrate radical (NO_3) in aqueous solution; their article should be consulted for the original references. It is important to note that Graedel and Weschler give the absorption coefficients as cross sections (σ), *but in the logarithmic base* 10, rather than base *e* as is done for the gas phase cross sections (see Section 2-A-2). We

TABLE 3.40. Absorption Coefficients of Nitrogen Dioxide in Aqueous Solution

λ (nm)	10^{19} σ (cm^2 molecule^{-1}) Base 10	Base *e*	ε (L mole^{-1} cm^{-1}) Base 10
290	8.18	18.8	492
300	6.94	16.0	418
320	5.74	13.2	346
330	5.93	13.7	357
340	5.96	13.7	359
360	5.70	13.1	343
370	6.41	14.8	386
380	7.54	17.4	454
400	7.50	17.3	452
420	7.13	16.4	429

Source: Recommended by Graedel and Weschler, 1981.

TABLE 3.41. Absorption Coefficients of Nitrous Acid in Aqueous Solution

λ (nm)	$10^{20}\sigma$ (cm^2 $molecule^{-1}$) Base 10	Base *e*	ε (L $mole^{-1}$ cm^{-1}) Base 10
305–310	1.50	3.45	9.03
310–315	4.26	9.81	25.6
315–320	4.16	9.58	25.0
320–325	4.26	9.81	25.6
325–330	4.62	10.6	27.8
330–335	5.41	12.5	32.6
335–340	6.29	14.5	37.9
340–345	7.44	17.1	44.8
345–350	9.18	21.1	55.3
350–355	9.01	20.8	54.2
355–360	12.3	28.3	74.0
360–365	9.73	22.4	58.6
365–370	10.2	23.5	61.4
370–375	11.8	27.2	71.0
375–380	6.29	14.5	37.9
380–385	5.63	13.0	33.9
385–390	5.76	13.3	34.7
390–395	2.41	5.55	14.5
395–400	0.63	1.45	3.8
400–405	0.21	0.48	1.3
405–410	0.10	0.23	0.60

Source: Recommended by Graedel and Weschler, 1981.

show in Tables 3.38 to 3.45 the absorption cross sections in base 10 as reported by Graedel and Weschler, as well as ε (L $mole^{-1}$ cm^{-1}) and σ in base *e*.

It is essential to note that when calculating the photolysis rate constant k_p using Eq's (T) or (U) and the actinic flux data of Table 3.5, that the absorption cross sections to be used are those to the base *e*; if those to the base 10 (i.e. σ, base 10) are used directly, the photolysis rate constants so derived will be too small by a factor of 2.303.

HO_2^- and O_2^- also absorb into the actinic region. The molar extinction coefficients (base 10) reported for HO_2^- are ~45 and 12 L $mole^{-1}$ cm^{-1} at 290 and 313.5 nm, respectively (Baxendale and Wilson, 1957); in water, the products are $OH + OH^- + O_2$ (Treinen, 1970; Graedel et al., 1985a). For O_2^-, the absorption coefficients fall continuously from $\sim 6 \times 10^2$ at 290 nm to ~75 L $mole^{-1}$ cm^{-1} (base 10) at 320 nm (Behar et al., 1970).

Some of the absorption coefficients are similar to those measured for the gas phase species, and their photochemical reactions might be expected to play a significant role in the chemistry of atmospheric droplets.

TABLE 3.42. Absorption Coefficient of HNO_3 in Aqueous Solution[a]

λ (nm)	$10^{20}\sigma$ (cm^2 $molecule^{-1}$) Base 10	Base e	ε (L $mole^{-1}$ cm^{-1}) Base 10
285	0.368	0.848	2.22
290	0.264	0.607	1.59
295	0.178	0.409	1.06
300	0.105	0.241	0.63
305	0.063	0.146	0.38
310	0.031	0.071	0.19
315	0.014	0.032	0.084
320	0.005	0.012	0.031
325	0.002	0.005	0.01
330	0.0009	0.002	0.005

[a]In the absence of other data, Graedel and Weschler (1981) suggest assuming these are the same as the gas phase cross sections measured by Molina and Molina (1981).

TABLE 3.43. Absorption Coefficients of Nitrite Ion in Aqueous Solution

λ (nm)	$10^{21}\sigma$ (cm^2 $molecule^{-1}$) Base 10	Base e	ε (L $mole^{-1}$ cm^{-1}) Base 10
290	15	35	9.0
300	15	35	9.0
310	16	37	9.6
320	17	39	10
330	23	53	14
340	29	67	17
350	36	83	22
360	37	85	22
370	27	62	16
380	15	35	9.0
390	8.6	20	5.2
400	1.8	4.1	1.1
410	0.17	0.40	0.10

Source: Recommended by Graedel and Weschler, 1981; based on data from Strickler and Kasha, 1963.

TABLE 3.44. Absorption Coefficients of Nitrate Ion in Aqueous Solution

λ (nm)	$10^{20}\sigma$ (cm^2 $molecule^{-1}$) Base 10	Base *e*	ϵ (L $mole^{-1}$ cm^{-1}) Base 10
290	0.95	2.2	5.7
295	1.1	2.5	6.6
300	1.2	2.8	7.2
305	1.2	2.8	7.2
310	1.0	2.3	6.0
315	0.81	1.9	4.9
320	0.55	1.3	3.3
325	0.28	0.65	1.7
330	0.15	0.35	0.90

Source: Recommended by Graedel and Weschler, 1981; based on the data of Meyerstein and Treinin, 1961, and Rotlevi and Treinin, 1965.

It is noteworthy that the ionized forms of SO_2 in aqueous solution, that is, HSO_3^- and SO_3^{2-} (see Section 11-A-2), do not absorb light in the actinic region (Golding, 1960; Hayon et al., 1972). Thus, just as in the gas phase, the direct photochemistry of SO_2 does not play a role in solution phase chemistry in the troposphere.

Table 3.46 summarizes in reactions (38)–(46) what is known about the photochemistry of these species in aqueous solutions. As in the gas phase, O_3 photolysis at $\lambda < 320$ nm produces $O(^1D)$ which reacts rapidly with the water solvent to produce H_2O_2:

$$O(^1D) + H_2O(aq) \rightarrow H_2O_2 \tag{45}$$

The presence of electronically excited singlet molecular oxygen, $O_2(^1\Delta_g)$, at concentrations of $\sim 10^{-11}$–10^{-14} moles L^{-1} at the surface of both fresh and coastal waters has been suggested (Zafiriou et al., 1984) and may arise from reaction (38). While it is expected to be rapidly deactivated to the ground state (with a lifetime of the order of μs) by water (Merkel and Kearns, 1972; Rodgers and Snowden, 1982), its role in the chemistry of atmospheric droplets is not known; it may have interesting biological implications.

The photolysis of H_2O_2 gives 2OH, as in the gas phase; the quantum yield in Table 3.46 of ~ 0.50 is an estimate of the true primary quantum yield since chain reactions seemed to be occurring, resulting in higher overall quantum yields in the laboratory studies.

Nitrous acid photolyzes to give OH + NO; however, as mentioned above, solvent cage effects cause significant recombination of the fragments, leading to

TABLE 3.45. Absorption Coefficients of Nitrogen Trioxide in Aqueous Solution

λ (nm)	$10^{18}\sigma$ (cm^2 molecule^{-1}) Base 10	Base e	ϵ (L mole^{-1} cm^{-1}) Base 10
535–540	1.03	2.37	620
540–545	1.33	3.06	801
545–550	1.69	3.89	1.02×10^3
550–555	2.12	4.88	1.28×10^3
555–560	2.65	6.10	1.60×10^3
560–565	3.01	6.93	1.81×10^3
565–570	3.14	7.23	1.89×10^3
570–575	3.28	7.55	1.98×10^3
575–580	3.42	7.88	2.06×10^3
580–585	3.71	8.54	2.23×10^3
585–590	4.03	9.28	2.43×10^3
590–595	4.54	10.5	2.73×10^3
595–600	4.73	10.9	2.85×10^3
600–605	4.54	10.5	2.73×10^3
605–610	4.19	9.65	2.52×10^3
610–615	4.03	9.28	2.43×10^3
615–620	4.03	9.28	2.43×10^3
620–625	4.37	10.1	2.63×10^3
625–630	5.11	11.8	3.08×10^3
630–635	5.96	13.7	3.59×10^3
635–640	5.74	13.2	3.46×10^3
640–645	4.37	10.1	2.63×10^3
645–650	3.14	7.23	1.89×10^3
650–655	2.03	4.68	1.22×10^3
655–660	1.53	3.52	0.92×10^3
660–665	1.77	4.08	1.07×10^3
665–670	3.01	6.93	1.81×10^3
670–675	3.71	8.54	2.23×10^3
675–680	2.43	5.60	1.46×10^3

Source: Recommended by Graedel and Weschler, 1981; based on the data of Martin et al., 1963.

an overall efficiency for their production and escape from the solvent cage of ~10% (Rettich, 1978).

Relatively little is known about nitric acid photolysis in aqueous solutions, and indeed it may not be entirely relevant to the atmosphere because it is essentially completely ionized in aqueous solution. Graedel and Weschler (1981) suggest that, in the absence of other data, the gas phase absorption cross sections be used and that, by analogy to HONO, the effective quantum yield (including escape from the solvent cage) be taken as ~0.1.

The nitrite ion (NO_2^-) has an absorption spectrum peaking at ~350 nm and

TABLE 3.46. Summary of Photochemistry of Some Species in Aqueous Solutions

Photochemical Dissociation	Quantum Yield[a]	Reference[b]
(38) $O_3 + h\nu\ (\lambda \le 320\text{ nm}) \rightarrow O(^1D) + O_2(^1\Delta_g)$	$\Phi(-O_3)^c = 0.23$ at 310 nm	1
$\overset{H_2O}{\rightarrow} H_2O_2$	$= 0.002 - 0.005$ at 600 nm	
(39) $H_2O_2 + h\nu\ (\lambda \le 380\text{ nm}) \rightarrow 2OH$	$\phi \simeq 0.50$	2–4
(40) $HONO + h\nu\ (\lambda \le 390\text{ nm}) \rightarrow OH + NO$	$\phi \simeq 0.095$	5
(41) $HNO_3 + h\nu\ (\lambda \le 320\text{ nm}) \rightarrow OH + NO_2$	$\phi \simeq 0.1$	6
(42) $NO_2^- + h\nu\ (\lambda \le 410\text{ nm}) \underset{a}{\rightarrow} (NO_2^-)^* \overset{H_2O}{\rightarrow} NO + OH + OH^-$	$\phi \simeq 0.05$	7
$\underset{b}{\overset{?}{\rightarrow}} NO + O^-$		
(43) $NO_3^- + h\nu\ (\lambda \le 350\text{ nm}) \underset{a}{\rightarrow} NO_2^- + O$	$\phi_a + \phi_b \simeq 0.04$	8–12
$\underset{b}{\rightarrow} NO_2 + O^-$		
(44) $NO_3 + h\nu\ (\lambda \le 690\text{ nm}) \underset{a}{\rightarrow} NO_2 + O$	$\phi_a \simeq 0.09$	6,13
$\underset{b}{\rightarrow} NO + O_2$	$\phi_b \simeq 0.01$	
(46) $HO_2^- + h\nu(\lambda < 390\text{ nm}) \overset{H_2O}{\rightarrow} OH + OH^- + O_2$		14,15

[a]These are *effective quantum yields*, that is, those for photolysis and escape of the species from the solvent cage.

[b]References are as follows: (1) Taube, 1957; (2) Hunt and Taube, 1952; Taube, 1957; (3) Baxendale and Wilson (1957); (4) Phibbs and Giguère (1951); (5) Rettich, 1978; (6) Graedel and Weschler, 1981; (7) Zafiriou and True, 1979a; (8) Zafiriou and True, 1979b; (9) Bayliss and Bucat, 1975; (10) Daniels et al., 1968; (11) Barat et al., 1969; (12) Shuali et al., 1969; (13) Hayon and Saito, 1965; (14) Treinen, 1970; (15) Graedel et al, 1985a.

[c]Quantum yield for O_3 disappearance.

an effective quantum yield of ~0.05. Although it is generally accepted that the photolysis product in aqueous solutions are NO + OH + OH^-, the primary products (NO + O^-) cannot be ruled out (Zafiriou, 1983). The photolyses of nitrite and to a lesser extent nitrate (see below) have been hypothesized to play a significant role in the chemistry of natural waters (Zafiriou et al., 1984) and may be important in atmospheric aqueous systems as well.

The nitrate ion (NO_3^-) also absorbs significantly in the actinic region; the two possible sets of photolysis products are shown in Table 3.46. There is some controversy concerning the relative importance of paths (43a) and (43b) in the photolysis (see refs. 8–12 at bottom of Table 3.46), although Graedel and Weschler suggest that path (43a) to produce NO_2^- + O is likely most important. The reaction of the oxygen atom with O_2 present in the solvent cage may lead to the formation of significant amounts of O_3 in atmospheric droplets (Graedel and Goldberg, 1983).

The effective quantum yields for photolysis of the nitrate radical (NO_3) in aqueous solution are not known; Graedel and Weschler (1981) suggest $\phi_a \simeq 0.09$ and $\phi_b \simeq 0.01$.

Not included in Table 3.46 is the photolysis of NO_2 which might be expected to produce $O(^3P)$ and NO_2 at $\lambda < 430$ nm as in the gas phase. However, it is not clear that an overall product quantum yield significantly greater than zero results, although in some studies O_3 formation, presumably from the reaction of $O(^3P)$ with O_2 present in the solvent cage, has been observed (Graedel and Weschler, 1981). The photolysis products may recombine in the solvent cage, leading to no net chemical change.

In solution, as in the gas phase, photolysis competes with other reactions. Possible competing chemical reactions are discussed in detail in Section 11-A-2. However, we note here that Graedel and Weschler (1981) estimate that, under conditions typical of aqueous solutions in the atmosphere, photolysis is expected to be a significant or predominant loss process for H_2O_2, HO_2^-, HONO, and NO_2^-. For O_3, photolysis is likely a minor process compared to its reactions with OH^- and HSO_3^- (see Section 11-A-2).

By analogy to gas phase systems, aldehydes and ketones in aqueous solutions should also undergo photochemical reactions, although few quantitative data on their absorption coefficients and quantum yields exist. The situation is complicated by the rapid formation in solution of diols, which do not absorb actinic radiation:

$$R_1R_2CO + H_2O \rightleftarrows R_1-\underset{\underset{\displaystyle OH}{|}}{\overset{\overset{\displaystyle R_2}{|}}{C}}-OH \qquad (47)$$

For example, HCHO is essentially entirely in the form of methylene glycol, $CH_2(OH)_2$. As discussed in Section 11-A-2, the equilibrium for larger aldehydes lies further to the left; thus for the C_2–C_6 straight chain aldehydes, approximately half the aqueous aldehyde is in the form of the diol.

In addition to the reaction with water to form diols, a further complication is the rapid deactivation of the electronically excited carbonyl compounds by the water solvent, and the possible reaction of the excited species with water itself.

For example, the simple carbonyl compounds CH_3CHO, C_2H_5CHO, and CH_3COCH_3 have been observed under certain laboratory conditions in the condensed phase to undergo hydrogen abstraction rather than α cleavage (see Sections 3-C-9 and 3-C-10) at $\lambda \geq 290$ nm (Henne and Fischer, 1975; Blank et al., 1975). Very little work has been reported in aqueous solutions typical of atmospheric droplets, however. Graedel and Weschler (1981) suggest that such reactions will lead ultimately to carboxylic acids of the same carbon number, rather than to reactive free radicals as occurs in the gas phase.

Organic hydroperoxides (ROOH), if formed in solution, will photodissociate, cleaving at the weak peroxidic O—O bond as in the gas phase. However, the absorption coefficients and quantum yields under conditions relevant to the atmosphere are not established (Graedel and Goldberg, 1983).

An area which has not been explored in detail but which clearly warrants further study is the role of the photochemistry of transition metal complexes in the chemistry of atmospheric droplets. Thus Graedel, Weschler and Mandlich have suggested that the absorption of light by dissolved iron(III) aqueous hydroxy complexes which absorb in the actinic UV out to ~400 nm (Weschler et al., 1985), eg.

$$[Fe^{III}(OH^-)(H_2O)_5]^{2+} \xrightarrow{h\nu,\ H_2O} [Fe^{II}(H_2O)_6]^{2+} + OH \qquad (48)$$

may be the predominant source of OH free radicals in these droplets during the day (Weschler et al, 1985; Graedel et al, 1985a,b). These photoreactions involve a charge-transfer from the ligand to the metal,

$$[Fe^{III}(OH^-)(H_2O)_5]^{2+} \xrightarrow{h\nu} [Fe^{II}(OH)(H_2O)_5]^{2+}$$

followed by dissociation into a reduced metal complex and a free radical (Weschler et al, 1985):

$$[Fe^{II}(OH)(H_2O)_5]^{2+} \xrightarrow{H_2O} [Fe^{II}(H_2O)_6]^{2+} + OH$$

The quantum yields for production of OH from such complexes are not known, but Weschler et al propose a lower limit of 0.02. They also suggest that aquated Fe^{III}-S(IV) complexes may undergo photolysis during the day as well (see Section 11-A-2).

Photochemical reactions may occur not only in the liquid phase in the atmosphere but at the surfaces of solids as well. This relatively unexplored area may involve reactions at the surfaces of particles suspended either in air or in the aqueous solutions found in many aerosols, fogs, and clouds (see Chapters 11 and 12). Thus the composition of atmospheric particulate matter, depending on the particular location and time, can include some metal oxides such as α-Fe_2O_3 (hematite) and

MnO_2 (see Section 12-D) which absorb radiation in the actinic region. This suggests that semiconductor type photochemistry may occur in the atmosphere; for example, colloidal solutions of α-Fe_2O_3 (hematite) photolyzed at 347 nm have been shown to oxidize I^- very efficiently, producing I_2^- with a quantum yield of ≥ 0.9 (Moser and Grätzel, 1982).

Surface complexes of species adsorbed on particulate matter also seem likely to undergo photochemical reactions in the atmosphere. For example, as discussed in Chapters 11 and 12, S(IV) is known to adsorb on solids such as elemental carbon and Fe_2O_3. The absorption spectra of species adsorbed on surfaces often undergo significant spectral shifts compared to their dissolved form, and this may bring some absorptions into the actinic region. For example, the aqueous forms of S(IV) all absorb light only at $\lambda < 290$ nm (Hayon et al., 1972; Golding, 1960); however, the aqueous Fe(III)–S(IV) complex has an absorption peak at 367 nm, and the surface complex likely does as well (Faust and Hoffmann, 1985). The possible role of such complexes in the oxidation of S(IV) in the atmosphere is discussed in Section 11-A-3.

3-D. MAJOR PHOTOLYTIC SOURCES OF OH AND HO_2 RADICALS IN THE GAS PHASE

Although the tropospheric sources of OH and HO_2 in the gas phase are discussed at various points throughout the book, we summarize them in this section to provide an overview of these important photochemical processes. As we saw in Chapter 1, OH and HO_2 are interconverted in the atmosphere through a series of reactions involving hydrocarbons and oxides of nitrogen. Thus sources of HO_2 are, in effect, sources of OH, under most tropospheric conditions where NO is present in sufficient concentrations.

Possible sources of OH and HO_2 in aqueous atmospheric droplets are discussed in Section 11-A-2; at the present time there is considerable uncertainty concerning the sources of these and other free radicals in such solutions.

3-D-1. Sources of OH

A major source of OH is the photolysis of O_3 to form $O(^1D)$ followed by its reaction with water (Section 3-C-2b):

$$O_3 + h\nu\ (\lambda < 320\text{ nm}) \rightarrow O_2(^1\Delta_g) + O(^1D) \tag{4}$$

$$O(^1D) + H_2O \rightarrow 2OH \tag{5}$$

Photolysis of HONO and H_2O_2 produces OH directly:

$$HONO + h\nu\ (\lambda < 400\text{ nm}) \rightarrow OH + NO \tag{21}$$

$$H_2O_2 + h\nu\ (\lambda < 360\text{ nm}) \rightarrow 2OH \tag{30}$$

Possible sources of HONO are discussed in detail in Section 8-A-3; these include the reactions of NO_2 with H_2O, of OH with NO, of NO + NO_2 + H_2O, possibly a contribution from a minor channel in the $HO_2 + NO_2$ reaction, and, finally, direct emissions, for example, from automobiles.

H_2O_2 is formed from the reaction of two hydroperoxyl free radicals:

$$HO_2 + HO_2 \rightarrow H_2O_2 + O_2 \qquad (49)$$

The rate constant k_{49} has two components, a pressure-independent, bimolecular component and a pressure-dependent, termolecular component. In addition to the pressure dependence, the rate constant has been shown to increase in the presence of water vapor, suggesting that the mechanism of the reaction in the presence of water might involve the formation of a radical–water complex:

$$HO_2 + H_2O \rightleftarrows (HO_2 \cdot H_2O) \qquad (50)$$

$$HO_2 + (HO_2 \cdot H_2O) \rightarrow\rightarrow H_2O_2 + O_2 + H_2O \qquad (51)$$

$$2(HO_2 \cdot H_2O) \rightarrow H_2O_2 + O_2 + 2H_2O \qquad (52)$$

The effects of pressure and water vapor are reviewed in detail by Atkinson and Lloyd (1984) and by DeMore et al. (1983, 1985). In an extensive study of the effects of pressure, temperature, and water vapor concentration, Kircher and Sander (1984) studied reaction (49) using flash photolysis/UV absorption; they recommend the following expression for the rate constant for the $HO_2 + HO_2$ reaction under atmospheric conditions which explicitly takes into account the total pressure and the water vapor concentration:

$$k_{49} = \left[2.2 \times 10^{-13} \exp\left(\frac{620}{T}\right) + 1.9 \times 10^{-33} [M] \exp\left(\frac{980}{T}\right)\right] \times \left[1 + 1.4 \times 10^{-21} [H_2O] \exp\left(\frac{2200}{T}\right)\right] \text{ cm}^3 \text{ molecule}^{-1} \text{ s}^{-1}$$

At 1 atm, 298°K, and 50% relative humidity, $k_{49} = 5.6 \times 10^{-12}$ cm^3 molecule^{-1} s^{-1}. Possible reaction mechanisms which explain the observed water effect, and the effects of pressure and temperature, are discussed by Mozurkewich and Benson (1985) and references therein.

Finally, conversion of HO_2 to OH via reaction (53) is the key to the HO_2–OH interconversion:

$$HO_2 + NO \rightarrow OH + NO_2 \qquad (53)$$

3-D-2. Sources of HO_2

Formaldehyde photolysis is a major source of HO_2 during the daylight hours via the reactions of H and HCO with O_2:

$$HCHO + h\nu \rightarrow H + HCO \tag{2a}$$

$$H + O_2 \xrightarrow{M} HO_2 \tag{54}$$

$$HCO + O_2 \rightarrow HO_2 + CO \tag{55}$$

Indeed, any process that produces HCO or H is a source of HO_2 in the troposphere. As we saw in Section 3-C-9, this includes higher aldehydes such as CH_3CHO, although their absorption cutoff (~ 345 nm) combined with their lower atmospheric concentration compared to HCHO (see Section 5-G) generally makes them less important sources of HO_2 than HCHO:

$$RCHO + h\nu \rightarrow R + CHO \tag{56}$$

$$HCO + O_2 \rightarrow HO_2 + CO \tag{55}$$

Alkoxy radicals also serve as a source of HO_2 through their reaction with O_2, as discussed in Section 7-D-3a:

$$RCH_2O + O_2 \rightarrow RCHO + HO_2 \tag{57}$$

These alkoxy radicals are formed in the chain oxidation of hydrocarbons in the atmosphere initiated primarily by OH; thus reaction (57) serves as one link in the interconversion of OH and HO_2.

There is an increasing recognition that significant nighttime NO_x reactions occur, and many of these generate free radicals which lead to HO_2 as well (Chapter 8). Thus, although it may seem somewhat surprising at first glance, the free radical concentration does not drop to zero at night. Reactions producing free radicals involve the thermal decomposition of peroxyacetyl nitrate (PAN) and reactions of the nitrate (NO_3) radical formed in the reaction of O_3 with NO_2.

As discussed in detail in Chapter 8, the peroxyacetyl radical produced from PAN decomposition can undergo a series of reactions in the presence of NO, ultimately yielding HO_2:

$$CH_3\overset{\overset{\displaystyle O}{\|}}{C}OONO_2 \rightleftarrows CH_3\overset{\overset{\displaystyle O}{\|}}{C}OO + NO_2 \tag{58}$$

$$CH_3\overset{\overset{\displaystyle O}{\|}}{C}OO + NO \rightarrow CH_3\overset{\overset{\displaystyle O}{\|}}{C}O + NO_2 \tag{59}$$

$$CH_3\overset{\overset{\displaystyle O}{\|}}{C}O \rightarrow CH_3 + CO_2 \tag{60}$$

$$CH_3 + O_2 \rightarrow CH_3O_2 \tag{61}$$

$$CH_3O_2 + NO \rightarrow CH_3O + NO_2 \tag{62}$$

$$CH_3O + O_2 \rightarrow HCHO + HO_2 \tag{63}$$

The relative importance of this route for free radical generation depends on the temperature, since the rate of decomposition (58) increases rapidly with temperature (see Sections 4-A-7c and 8-B-1).

Similarly, the strongly temperature-dependent decomposition of peroxynitric acid, HO_2NO_2,

$$HO_2NO_2 \rightleftarrows HO_2 + NO_2 \tag{64}$$

produces HO_2.

Finally, hydrogen abstractions by the nitrate radical, NO_3, at night (see Section 8-A-2a) also lead to HO_2 via the formation of formyl radicals and alkoxy radicals:

$$NO_3 + HCHO \rightarrow HNO_3 + HCO \tag{65}$$

$$NO_3 + RH \rightarrow HNO_3 + R \tag{66}$$

$$R + O_2 \rightarrow RO_2 \tag{67}$$

$$RO_2 + NO \rightarrow RO + NO_2 \tag{68}$$

The formyl and alkoxy radicals produced in reactions (65) and (68), respectively, then react with O_2, producing HO_2 via reactions (55) and (57).

The relative importance of the various sources of OH and HO_2 will clearly depend on the actual conditions, in particular, on whether it is day or night, as well as on the ambient concentrations of hydrocarbons, NO, NO_2, and O_3. For example, Fig. 1.10 (Chapter 1) shows the calculated rates of radical formation from three sources, HONO, HCHO, and O_3 photolyses, respectively, in a typical polluted urban air mass as a function of time of day during daylight hours. HONO contributes the largest amount at sunrise, whereas HCHO and then O_3 become important at later times as their concentrations increase. In a pristine area, however, where NO_x and hydrocarbon concentrations are low, O_3 photolysis may serve as the major daytime OH source.

REFERENCES

Amimoto, S. T., A. P. Force, J. R. Wiensenfeld, and R. H. Young, "Direct Observation of $O(^3P_J)$ in the Photolysis of O_3 at 248 nm," *J. Chem. Phys.*, **73,** 1244 (1980).

Atkinson, R. and A. C. Lloyd, "Evaluation of Kinetic and Mechanistic Data for Modeling of Photochemical Smog," *J. Phys. Chem. Ref. Data*, **13,** 315 (1984).

Atwater, M. A. and P. S. Brown, "Numerical Computations of the Latitudinal Variation of Solar Radiation for an Atmosphere of Varying Opacity," *J. Appl. Meteorol.*, **13,** 289 (1974).

Badger, R. M., A. C. Wright, and R. F. Whitlock, "Absolute Intensities of the Discrete and Continuous Absorption Bands of Oxygen Gas at 1.26 and 1.065 μ and the Radiative Lifetime of the $^1\Delta_g$ State of Oxygen," *J. Chem. Phys.*, **43,** 4345 (1965).

Ballash, N. M. and D. A. Armstrong, "On the Ultraviolet and Visible Absorption Spectrum of ClNO," *Spectrochim. Acta A*, **30,** 941 (1974).

Barat, F., B. Hickel, and J. Sutton, "Flash Photolysis of Aqueous Solutions of Azide and Nitrate Ions," *Chem. Commun.*, 125 (1969).

Barker, J. R., L. Brouwer, R. Patrick, M. J. Rossi, P. L. Trevor, and D. M. Golden, *Int. J. Chem. Kinet.*, **17,** 991 (1985).

Bass, A. M., personal communication, 1984.

Bass, A. M., A. E. Ledford, Jr., and A. H. Laufer, "Extinction Coefficients of NO_2 and N_2O_4," *J. Res. Natl. Bur. Stand. Sect. A*, **80,** 143 (1976).

Bass, A. M., L. C. Glasgow, C. Miller, J. P. Jesson, and D. L. Filkin, "Temperature Dependent Absorption Cross Sections for Formaldehyde (CH_2O): The Effect of Formaldehyde on Stratospheric Chlorine Chemistry," *Planet. Space Sci.*, **28,** 675 (1980).

Baulch, D. L., R. A. Cox, P. J. Crutzen, R. F. Hampson, Jr., J. A. Kerr, J. Troe, and R. T. Watson, "Evaluated Kinetic and Photochemical Data for Atmospheric Chemistry: Supplement I. CODATA Task Group on Chemical Kinetics," *J. Phys. Chem. Ref. Data*, **11,** 327 (1982).

Baxendale, J. H. and J. A. Wilson, "The Photolysis of Hydrogen Peroxide at High Light Intensities," *Trans. Faraday Soc.*, **53,** 344 (1957).

Bayliss, N. S. and R. B. Bucat, "The Photolysis of Aqueous Nitrate Solutions," *Aust. J. Chem.*, **28,** 1865 (1975).

Behar, D., G. Czapski, J. Rabani, L. M. Dorfman, and H. A. Schwarz, "The Acid Dissociation Constant and Decay Kinetics of the Perhydroxyl Radical," *J. Phys. Chem.*, **74,** 3209 (1970).

Besemer, A. C. and H. Nieboer, Authors' Reply to Comments on "Formation of Chemical Compounds from Irradiated Mixtures of Aromatic Hydrocarbons and Nitrogen Oxides," *Atmos. Environ.*, **17,** 1598 (1983).

Biaume, F., "Nitric Acid Vapor Absorption Cross-Section Spectrum and Its Photodissociation in the Stratosphere," *J. Photochem.*, **2,** 139 (1973).

Blank, B., A. Henne, G. P. Laroff, and H. Fischer, "Enol Intermediates in Photoreduction and Type I Cleavage Reactions of Aliphatic Aldehydes and Ketones," *Pure Appl. Chem.*, **41,** 475 (1975).

Bottenheim, J. W., S. E. Braslavsky, and O. P. Strausz, "Modeling Study of Seasonal Effect On Air Pollution at 60°N Latitude," *Environ. Sci. Technol.*, **11,** 801 (1977).

Brock, J. C. and R. T. Watson, "Ozone Photolysis: Determination of the $O(^3P)$ Quantum Yield at 266 nm," *Chem. Phys. Lett.*, **71,** 371 (1980a).

Brock, J. C. and R. T. Watson, "Laser Flash Photolysis of Ozone: $O(^1D)$ Quantum Yields in the Fall-Off Region 297–325 nm," *Chem. Phys.*, **46,** 477 (1980b).

Burrows, J. P., G. S. Tyndall, and G. K. Moortgat, "Photolysis Products of $ClONO_2$ and N_2O_5, Equilibrium Constant of N_2O_5," Abstract TH-16, XVI Informal Conference on Photochemistry, Harvard University, August 20–24, 1984.

Burrows, J. P., G. S. Tyndall, and G. K. Moortgat, "Absorption Spectrum of NO_3 and Kinetics of the Reactions of NO_3 with NO_2, Cl, and Several Stable Atmospheric Species at 298 K," *J. Phys. Chem.*, **89,** 4848 (1985).

Calvert, J. G., "The Homogeneous Chemistry of Formaldehyde Generation and Destruction within the Atmosphere," in *Proceedings of the NATO Advanced Study Institute on Atmospheric Ozone*, U.S. Department of Transportation, FAA Office of the Environment and Energy, High Altitude Program, Report No. FAA-EE-80-20, Washington, DC, May 1980, pp. 153–190.

Calvert, J. G. and J. N. Pitts, Jr., *Photochemistry*, Wiley, New York, 1966.

Calvert, J. G. and W. R. Stockwell, "Mechanism and Rates of the Gas-Phase Oxidations of Sulphur Dioxide and Nitrogen Oxides in the Atmosphere," Chapter 1 in *SO_2, NO, and NO_2 Oxidation Mechanisms*, J. G. Calvert, Ed., Vol. 3 in *Acid Precipitation Series*, J. I. Teasley, Series Ed., Butterworth, Storeham, MA, 1984.

Carter, W. P. L., R. Atkinson, A. M. Winer, and J. N. Pitts, Jr., "Experimental Protocol for Determining Photolysis Reaction Rate Constants," EPA-600/3-83-100, January 1984.

Clark, J. H., C. B. Moore, and N. S. Nogar, "The Photochemistry of Formaldehyde: Absolute Quantum Yields, Radical Reactions, and NO Reactions," *J. Chem. Phys.*, **68,** 1264 (1978).

Cobos, C., E. Castellano, and H. J. Schumaker, "The Kinetics and the Mechanism of Ozone Photolysis at 253.7 nm," *J. Photochem.*, **21,** 291 (1983).

Cox, R. A. and R. G. Derwent, "The Ultra-Violet Absorption Spectrum of Gaseous Nitrous Acid," *J. Photochem.*, **6,** 23 (1976/77).

Cox, R. A. and K. Patrick, ''Kinetics of the Reaction $HO_2 + NO_2 (+M) = HO_2NO_2$ Using Molecular Modulation Spectrometry,'' *Int. J. Chem. Kinet.*, **11,** 635 (1979).

Cox, R. A., K. F. Patrick, and S. A. Chant, ''Mechanism of Atmospheric Photooxidation of Organic Compounds. Reactions of Alkoxy Radicals in Oxidation of *n*-Butane and Simple Ketones,'' *Environ. Sci. Technol.*, **15,** 587 (1981).

Cox, R. A., R. A. Barton, E. Ljungstrom, and D. W. Stocker, ''The Reactions of Cl and ClO with the NO_3 Radical,'' *Chem. Phys. Lett.*, **108,** 228 (1984).

Daniels, M., R. V. Meyers, and E. V. Belardo, ''Photochemistry of the Aqueous Nitrate System. I. Excitation in the 300-mμ Band,'' *J. Phys. Chem.*, **72,** 389 (1968).

Dave, J. V., *Development of Programs for Computing Characteristics of Ultraviolet Radiation*, Final Report under Contract NAS 5-21680, NASA Report CR-139134, National Aeronautics and Space Administration, Goddard Space Flight Center, Greenbelt, Maryland, NTIS No. N75-10746/6SL (1972).

Davenport, J. E., ''Parameters for Ozone Photolysis as a Function of Temperature at 280–330 nm,'' FAA Report No. FAA-EE-80-44R, May 1982.

DeLuisi, J. J., ''Measurements of the Extraterrestrial Solar Radiant Flux from 2981 to 4000 Å and Its Transmission Through the Earth's Atmosphere as It Is Affected by Dust and Ozone,'' *J. Geophys. Res.*, **80,** 345 (1975).

Demerjian, K. L., K. L. Schere, and J. T. Peterson, ''Theoretical Estimates of Actinic (Spherically Integrated) Flux and Photolytic Rate Constants of Atmospheric Species in the Lower Troposphere,'' *Adv. Environ. Sci. Technol.*, **10,** 369 (1980).

DeMore, W. B., M. J. Molina, R. T. Watson, D. M. Golden, R. F. Hampson, M. J. Kurylo, C. J. Howard, and A. R. Ravishankara, ''Chemical Kinetics and Photochemical Data for Use in Stratospheric Modeling, Evaluation No. 6,'' JPL Publication No. 83-62, September 15, 1983.

DeMore, W. B., J. J. Margitan, M. J. Molina, R. T. Watson, D. M. Golden, R. F. Hampson, M. J. Kurylo C. J. Howard, and A. R. Ravishankara, ''Chemical Kinetics and Photochemical Data for Use in Stratospheric Modeling, Evaluation No. 7,'' JPL Publication No. 85-37, 1985.

Dickerson, R. R., D. H. Stedman, and A. C. Delany, ''Direct Measurements of Ozone and Nitrogen Dioxide Photolysis Rates in the Troposphere,'' *J. Geophys. Res.*, **87,** 4933 (1982).

European Chemical Industry Ecology and Toxicology Center (ECETOC), ''Experimental Assessment of the Phototransformation of Chemicals in the Atmosphere,'' Technical Report No. 7, Brussels, Belgium, September 30, 1983.

Fairchild, P. W. and E. K. C. Lee, ''Relative Quantum Yields of $O(^1D)$ in Ozone Photolysis in the Region Between 250 and 300 nm,'' *Chem. Phys. Lett.*, **60,** 36 (1978).

Fairchild, C. E., E. J. Stone, and G. M. Lawrence, ''Photofragment Spectroscopy of Ozone in the UV Region 270–310 nm and at 600 nm,'' *J. Chem. Phys.*, **69,** 3632 (1978).

Faust, B. C. and M. R. Hoffmann, ''Photo-Induced Reductive Dissolution of Hematite (α-Fe_2O_3) by S(IV) Oxyanions,'' *Environ. Sci. Technol.*, submitted for publication (1985).

Finlayson-Pitts, B. J., ''The Reaction of NO_2 with NaCl and Atmospheric Implications of NOCl Formation,'' *Nature*, **306,** 676 (1983).

Frankiewicz, T. C. and R. S. Berry, ''Production of Metastable Singlet O_2 Photosensitized by NO_2,'' *J. Chem. Phys.*, **58,** 1787 (1973).

Friedman, H., in *Physics of the Upper Atmosphere*, J. A. Ratcliffe, Ed., Academic Press, New York, 1960.

Gaedtke, H. and J. Troe, ''Primary Processes in the Photolysis of NO_2,'' *Ber. Bunsenges. Phys. Chem.*, **79,** 184 (1975).

Gaydon, A. G., *Dissociation Energies and the Spectra of Diatomic Molecules*, 3rd ed., Chapman and Hall, London, 1968.

Golding, R. M., ''Ultraviolet Absorption Studies of the Bisulphite–Pyrosulphite Equilibrium,'' *J. Chem. Soc.*, 3711 (1960).

Graedel, T. E. and K. I. Goldberg, ''Kinetic Studies of Raindrop Chemistry. I. Inorganic and Organic Processes,'' *J. Geophys. Res.*, **88,** 10,865 (1983).

Graedel, T. E. and C. J. Weschler, "Chemistry within Aqueous Atmospheric Aerosols and Raindrops," *Rev. Geophys. Space Phys.*, **19,** 505 (1981).

Graedel, T. E., M. L. Mandich and C. J. Weschler, "Kinetic Model Studies of Atmospheric Droplet Chemistry 2. Homogeneous Transition Metal Chemistry in Raindrops," *J. Geophys. Res., 90* in press (1985a).

Graedel, T. E., C. J. Weschler and M. L. Mandich, "The Influence of Transition Metal Complexes on Atmospheric Droplet Acidity," *Nature,* **317,** 240 (1985b).

Graham, R. A. and H. S. Johnston, "The Photochemistry of NO_3 and the Kinetics of the N_2O_5–O_3 System," *J. Phys. Chem.*, **82,** 254 (1978).

Graham, R. A., A. M. Winer, and J. N. Pitts, Jr., "Ultraviolet and Infrared Absorption Cross Sections of Gas Phase HO_2NO_2," *Geophys. Res. Lett.*, **5,** 909 (1978).

Greenblatt, G. D. and J. R. Wiesenfeld, "Time-Resolved Resonance Fluorescence Studies of $O(^1D_2)$ Yields in the Photodissociation of O_3 at 248 and 308 nm," *J. Chem. Phys.*, **78,** 4924 (1983).

Griggs, M., "Absorption Coefficients of Ozone in the Ultraviolet and Visible Regions," *J. Chem. Phys.*, **49,** 857 (1968).

Harker, A. B., W. Ho, and J. J. Ratto, "Photodissiciation Quantum Yield of NO_2 in the Region 375 to 420 nm," *Chem. Phys. Lett.*, **50,** 394 (1977).

Harvey, R. B., D. H. Stedman, and W. Chameides, "Determination of the Absolute Rate of Solar Photolysis of NO_2," *J. Air Pollut. Control Assoc.*, **27,** 663 (1977).

Hayon, E. and E. Saito, "ESR Study of NO_3 and NO_2 Produced on UV Irradiation of HNO_3 Solutions at 77°K," *J. Chem. Phys.*, **43,** 4314 (1965).

Hayon, E., E. Treinin, and J. Wilf, "Electronic Spectra, Photochemistry and Autooxidation Mechanism of the Sulfite–Bisulfite–Pyrosulfite Systems. The SO_2^-, SO_3^-, SO_4^-, and SO_5^- Radicals," *J. Am. Chem. Soc.*, **94,** 47 (1972).

Heicklen, J., N. Kelly, and K. Partymiller, "The Photophysics and Photochemistry of SO_2," *Rev. Chem. Int.*, **3,** 315 (1980).

Henne, A. and H. Fischer, "Low Temperature Photochemistry of the Acetone/2-Propanol System," *Helv. Chim. Acta*, **58,** 1598 (1975).

Horowitz, A. and J. G. Calvert, "Wavelength Dependence of the Quantum Efficiencies of the Primary Processes in Formaldehyde Photolysis at 25°C," *Int. J. Chem. Kinet.*, **10,** 805 (1978).

Horowitz, A. and J. G. Calvert, "Wavelength Dependence of the Primary Processes in Acetaldehyde Photolysis," *J. Phys. Chem.*, **86,** 3105 (1982).

Howard, J. N., J. I. F. King, and P. R. Gast, "Thermal Radiation," *Handbook of Geophysics*, Macmillan, New York, 1960, Chap. 16.

Hunt, J. P. and H. Taube, "The Photochemical Decomposition of Hydrogen Peroxide: Quantum Yields, Tracer, and Fractionation Effects," *J. Am. Chem. Soc.*, **74,** 5999 (1952).

Illies, A. J. and G. A. Takacs, "Gas Phase Ultra-Violet Photoabsorption Cross-Sections for Nitrosyl Chloride and Nitryl Chloride," *J. Photochem.*, **6,** 35 (1976/77).

Inn, E. C. Y., "Vacuum Ultraviolet Spectroscopy," *Spectrochim. Acta*, **7,** 65 (1955–56).

Isaksen, I. S. A., personal communication, 1985.

Isaksen, I. S. A., K. H. Midtbø, J. Sunde, and P. J. Crutzen, "A Simplified Method to Include Molecular Scattering and Reflection in Calculations of Photon Fluxes and Photodissociation Rates," *Geophys. Norveg.*, **31,** 11 (1977).

Ishiwata, T., I. Fujiwara, Y. Naruge, K. Obi, and I. Tanaka, "Study of NO_3 by Laser Induced Fluorescence," *J. Phys. Chem.*, **87,** 1349 (1983).

Jesson, J. P., L. C. Glasgow, D. L. Filkin, and C. Miller, "The Stratospheric Abundance of Peroxynitric Acid," *Geophys. Res. Lett.*, **4,** 513 (1977).

Johnston, H. S. and R. A. Graham, "Gas-Phase Ultraviolet Absorption Spectrum of Nitric Acid Vapor," *J. Phys. Chem.*, **77,** 62 (1973).

Johnston, H. S. and R. A. Graham, "Photochemistry of NO_x and HNO_x Compounds," *Can. J. Chem.*, **52,** 1415 (1974).

Johnston, H. S. and G. S. Selwyn, "New Cross Sections for the Absorption of Near Ultraviolet Radiation by Nitrous Oxide," *Geophys. Res. Lett.*, **2,** 549 (1975).

Johnston, H. S., S.-G. Chang, and G. Whitten, "Photolysis of Nitric Acid Vapor," *J. Phys. Chem.*, **78,** 1 (1974).

Jones, E. L. and O. R. Wulf, "The Absorption Coefficient of Nitrogen Pentoxide in the Ultraviolet and the Visible Absorption Spectrum of NO_3," *J. Chem. Phys.*, **5,** 873 (1937).

Jones, I. T. N. and K. D. Bayes, "Photolysis of Nitrogen Dioxide *J. Chem. Phys.*, **59,** 4836 (1973a).

Jones, I. T. N. and K. D. Bayes, "Formation of $O_2(a^1\Delta_g)$ by Electronic Energy Transfer in Mixtures of NO_2 and O_2," *J. Chem. Phys.*, **59,** 3119 (1973b).

Kajimoto, O. and R. J. Cvetanovic, "Absolute Quantum Yield of $0(^1D_2)$ in the Photolysis of Ozone in the Hartley Band," *Int. J. Chem. Kinet.*, **11,** 605 (1979).

Kallend, A. S. and J. N. Pitts, Jr., "Vapor Phase Photooxidation of Diethyl Ketone," *J. Am. Chem. Soc.*, **91,** 1269 (1969).

Killus, J. P. and G. Z. Whitten, "Formation of Chemical Compounds from Irradiated Mixtures of Aromatic Hydrocarbons and Nitrogen Oxides," *Atmos. Environ.*, **8,** 1597 (1983).

Kircher, C. C. and S. P. Sander, "Kinetics and Mechanism of HO_2 and DO_2 Disproportionations," *J. Phys. Chem.*, **88,** 2082 (1984).

Lee, E. K. C. and R. S. Lewis, "Photochemistry of Simple Aldehydes and Ketones in the Gas Phase," *Adv. Photochem.*, **12,** 1 (1980).

Leighton, P. A., *Photochemistry of Air Pollution*, Academic Press, New York, 1961.

Lin, C. L., N. K. Rohatgi, and W. B. DeMore, "Ultraviolet Absorption Cross Sections of Hydrogen Peroxide," *Geophys. Res. Lett.*, **5,** 113 (1978).

Madronich, S., D. R. Hastie, B. A. Ridley, and H. I. Schiff, "Measurement of the Photodissociation Coefficient of NO_2 in the Atmosphere: I. Method and Surface Measurements," *J. Atmos. Chem.*, **1,** 3 (1983).

Magnotta, F. and H. S. Johnston, "Photodissociation Quantum Yields for the NO_3 Free Radical," *Geophys. Res. Lett.*, **7,** 769 (1980).

Margitan, J. J., "N_2O_5 Photolysis: Quantum Yields for O-Atom Production," Abstract TH-14, XVI Informal Conference on Photochemistry, Harvard University, August 20–24, 1984.

Margitan, J. J. and R. T. Watson, "Kinetics of the Reaction of Hydroxyl Radicals with Nitric Acid," *J. Phys. Chem.*, **86,** 3819 (1982).

Marinelli, W. J., D. M. Swanson, and H. S. Johnston, "Absorption Cross Sections, and Line Shape for the NO_3(0—0) Band, *J. Chem. Phys.*, **76,** 2864 (1982).

Martin, T. W., A. Henshall, and R. C. Gross, "Spectroscopic and Chemical Evidence for the NO_3 Free Radical in Solution at Room Temperature," *J. Amer. Chem. Soc.*, **85,** 113 (1963).

Merkel, P. B. and D. R. Kearns, "Remarkable Solvent Effects on the Lifetime of $^1\Delta_g$," *J. Am. Chem. Soc.*, **94,** 1029 (1972).

Meyerstein, D. and A. Treinin, "Absorption Spectra of NO_3^- in Solution," *Trans. Faraday Soc.*, **57,** 2104 (1961).

Meyrahn, H., G. K. Moortgat, and P. Warneck, "The Photolysis of Acetaldehyde Under Atmospheric Conditions," in *Atmospheric Trace Constituents*, F. Herbert, Ed., Proceedings of the 5th Two-Annual Colloquium of the Sonderforschungsbereich 73 of the Universities Frankfurt and Mainz and the Max-Planck-Institute, Mainz, Germany, July 1, 1981.

Mitchell, D. N., R. P. Wayne, P. J. Allen, R. P. Harrison, and R. J. Twin, "Kinetics and Photochemisty of NO_3, Part 1. Absolute Absorption Cross-Section," *J. Chem. Soc. Faraday Trans. 2*, **76,** 785 (1980).

Molina, M. J. and G. Arguello, "Ultraviolet Absorption Spectrum of Methylhydroperoxide Vapor," *Geophys. Res. Lett.*, **6,** 953 (1979).

Molina, L. T. and M. J. Molina, "UV Absorption Cross Sections of HO_2NO_2 Vapor," *J. Photochem.*, **15,** 97 (1981).

Molina, L. T., S. D. Schinke, and M. J. Molina, "Ultraviolet Absorption Spectrum of Hydrogen Peroxide Vapor, *Geophys. Res. Lett.*, **4,** 580 (1977).

Moore, C. B. and J. C. Weisshaar, "Formaldehyde Photochemistry," *Ann. Rev. Phys. Chem.*, **34,** 525 (1983).

Moortgat, G. K. and E. Kudszus, "Mathematical Expression for the $O(^1D)$ Quantum Yields from O_3 Photolysis as a Function of Temperature (230–320 K) and Wavelength (295–320 nm), *Geophys. Res. Lett.*, **5,** 191 (1978).

Moortgat, G. K. and P. Warneck, "CO and H_2 Quantum Yields in the Photodecomposition of Formaldehyde in Air," *J. Chem. Phys.*, **70,** 3639 (1979).

Moortgat, G. K., W. Klippel, K. H. Möbus, W. Seiler, and P. Warneck, "Laboratory Measurements of Photolytic Parameters for Formaldehyde," FAA-EE-80-47, U.S. Department of Transportation, Office of Environment and Energy, Washington, DC, November 1980.

Moortgat, G. K., W. Seiler, and P. Warneck, "Photodissociation of HCHO in Air: CO and H_2 Quantum Yields at 220° and 330°K,"*J. Chem. Phys.*, **78,** 1185 (1983).

Morabito, P. and J. Heicklen, "Primary Photochemical Processes in the Photolysis of Alkyl Nitrites at 366 nm and 23°C in the Presence of ^{15}NO," *Int. J. Chem. Kinet.*, **17,** 535 (1985).

Morel, O, R. Simonaitis, and J. Heicklen, "Ultraviolet Absorption Spectra of HO_2NO_2, $CCl_3O_2NO_2$, $CCl_2FO_2NO_2$, and $CH_3O_2NO_2$," *Chem. Phys. Lett.*, **73,** 38 (1980).

Moser, J. and M. Grätzel, "Photochemistry with Colloidal Semiconductors. Laser Studies of Halide Oxidation in Colloidal Dispersions of TiO_2 and α-Fe_2O_3," *Helv. Chim. Acta*, **65,** 1436 (1982).

Mozurkewich, M. and S. W. Benson, "Self-Reaction of HO_2 and DO_2: Negative Temperature Dependence and Pressure Effects," *Int. J. Chem. Kinet.*, **17,** 787 (1985).

Nelson, H. H., L. Pasternack, and J. R. McDonald, "Laser-Induced Excitation and Emission Spectra of NO_3," *J. Phys. Chem.*, **87,** 1286 (1983).

Nicolet, M., R. R. Meier, and D. E. Anderson, Jr., "Radiation Field in the Troposphere and Stratosphere. II. Numerical Analysis," *Planet. Space Sci.*, **30,** 935 (1982).

Nieboer, H., W. P. L. Carter, A. C. Lloyd, and J. N. Pitts, Jr., "The Effect of Latitude on the Potential for Formation of Photochemical Smog," *Atmos. Environ.*, **10,** 731 (1976).

Okabe, H., *Photochemistry of Small Molecules*, Wiley, New York 1978.

Osamura, Y., H. F. Schaefer III, M. Dupuis, and W. A. Lester, Jr., "A Unimolecular Reaction ABC $\rightarrow$ A + B + C Involving Three Product Molecules and a Single Transition State. Photodissociation of Glyoxal: HCOHCO $\rightarrow$ H_2 + CO + CO," *J. Phys. Chem.*, **75,** 5828 (1981).

Parrish, D. D., P. C. Murphy, D. L. Albritton, and F. C. Fehsenfeld, "The Measurement of the Photodissociation Rate of NO_2 in the Atmosphere," *Atmos. Environ.*, **17,** 1365 (1983).

Peterson, J. T., "Calculated Actinic Fluxes (290–700 nm) for Air Pollution Photochemistry Applications," U.S. Environmental Protection Agency Report No. EPA-600/4-76-025, June 1976.

Phibbs, M. K. and P. A. Giguère, "Hydrogen Peroxide and Its Analogues. III. Absorption Spectrum of Hydrogen and Deuterium Peroxides in the Near Ultraviolet,"*Can. J. Chem.*, **29,** 490 (1951).

Pitts, J. N., Jr. and F. E. Blacet, "Methyl Ethyl Ketone Photochemical Processes," *J. Am. Chem. Soc.*, **72,** 2810 (1950).

Pitts, J. N., Jr., J. H. Sharp, and S. I. Chan, "Effects of Wavelength and Temperature on Primary Processes in the Photolysis of Nitrogen Dioxide and a Spectroscopic–Photochemical Determination of the Dissociation Energy," *J. Phys. Chem.*, **42,** 3655 (1964).

Pitts, J. N., Jr., A. M. Winer, D. R. Fitz, A. K. Knudsen, and R. Atkinson, "Experimental Protocol for Determining Absorption Cross Sections of Organic Compounds," EPA-600/3-81-051, December 1981.

Plum, C. N., E. Sanhueza, R. Atkinson, W. P. L. Carter, and J. N. Pitts, Jr., "OH Radical Rate Constants and Photolysis Rates of α-Dicarbonyls," *Environ. Sci. Technol.*, **17,** 479 (1983).

Ravishankara, A. R., private communication, 1985; cited in DeMore et al, 1985.

Ravishankara, A. R. and P. H. Wine, "Absorption Cross Sections for NO_3 Between 565 and 673 nm," *Chem. Phys. Lett.*, **101,** 73 (1983).

Rebbert, R. E. and P. Ausloos, "Decomposition of N_2O Over Particulate Matter," *Geophys S. Res. Lett.*, **5,** 76 (1978)

Rettich, T. R., "Some Photochemical Reactions of Aqueous Nitrous Acid," Ph.D. Dissertation, Case Western Reserve University (*Diss. Int. Abstr. B*, **38,** 5968), Cleveland, Ohio, 1978.

Rodgers, M. A. J. and P. T. Snowden, "Lifetime of $O_2(^1\Delta_g)$ in Liquid Water as Determined by Time-Resolved Infrared Luminescence Measurements," *J. Am. Chem. Soc.*, **104,** 5541 (1982).

Rotlevi, E. and A. Treinin, "The 300-mμ Band of NO_3^-," *J. Phys. Chem.*, **69,** 2645 (1965).

Sander, S. P., *J. Phys. Chem.*, submitted for publication (1985).

Senum, G. I., Y.-N. Lee, and J. S. Gaffney, "Ultraviolet Absorption Spectrum of Peroxyacetyl Nitrate and Peroxypropionyl Nitrate," *J. Phys. Chem.*, **88,** 1269 (1984).

Shepson, P. B. and J. Heicklen, "The Photo-oxidation of Propionaldehyde," *J. Photochem.*, **18,** 169 (1982a); "The Wavelength and Pressure Dependence of the Photolysis of Propionaldehyde in Air," *J. Photochem.*, **19,** 215 (1982b).

Shuali, U., M. Ottolenghi, J. Rabani, and Z. Yelin, "On the Photochemistry of Aqueous Nitrate Solution Excited in the 195-nm Band," *J. Phys. Chem.*, **73,** 3445 (1969).

Sickles, J. E., L. A. Ripperton, W. C. Eaton, and R. S. Wright, "Diurnal Sunlight Intensity Determined by Nitrogen Dioxide Photolysis. A Field Study," Research Triangle Institute, Research Triangle Park, North Carolina, Report No. RTI-1163/F3, July 1977.

Sparks, R. K., L. R. Carlson, K. Shobatake, M. L. Kowalczyk, and Y. T. Lee, "Ozone Photolysis: A Determination of the Electronic and Vibrational State Distributions of Primary Products," *J. Phys. Chem.*, **72,** 1401 (1980).

Stephens, E. R., "The Formation, Reactions, and Properties of Peroxyacyl Nitrates in Photochemical Air Pollution," *Adv. Environ. Sci. Technol.*, **1,** 119 (1969).

Stockwell, W. R. and J. G. Calvert, "The Near Ultraviolet Absorption Spectrum of Gaseous HONO and N_2O_3," *J. Photochem.*, **8,** 193 (1978).

Stockwell, W. R. and J. G. Calvert, "The Mechanism of NO_3 and HONO Formation in the Nighttime Chemistry of the Urban Atmosphere," *J. Geophys Res.*, **88,** 6673 (1983).

Strickler, S. J. and M. Kasha, "Solvent Effects on the Electronic Absorption Spectrum of Nitrite Ion," *J. Am. Chem. Soc.*, **85,** 2899 (1963).

Stull, D. R. and H. Prophet, Eds., *JANAF Thermochemical Tables*, 2nd ed., NSRDS-NBS37, June 1971.

Swanson, D., B. Kan, and H. S. Johnston, "NO_3 Quantum Yields from N_2O_5 Photolysis *J. Phys. Chem.*, **88,** 3115 (1984).

Takezaki, Y., T. Miyazaki, and N. Nakakara, "Photolysis of Dimethyl Peroxide," *J. Chem. Phys.*, *25*, 536 (1956).

Tang, K. Y., P. W. Fairchild, and E. K. C. Lee, "Laser-Induced Photodecomposition of Formaldehyde ($\tilde{A}^1 A_2$) from Its Single Vibronic Levels. Determination of the Quantum Yield of H Atom by HNO* ($\tilde{A}^1 A''$) Chemiluminescence," *J. Phys. Chem.*, **83,** 569 (1979).

Taube, H., "Photochemical Reactions of Ozone in Solution," *Trans. Faraday Soc.*, **53,** 656 (1957).

Taylor, R. C. and P. C. Cross, "Light Absorption of Aqueous Hydrogen Peroxide Solutions in the Near Ultraviolet Region," *J. Am. Chem. Soc.*, **71,** 2266 (1949).

Thompson, A. M., "The Effect of Clouds on Photolysis Rates and Ozone Formation in the Unpolluted Troposphere," *J. Geophy. Res.*, **89,** 1341 (1984).

Treinin, A., "The Photochemistry of Oxyanions," *Israel J. Chem.*, **8,** 103 (1970).

Tuazon, E. C., R. Atkinson, and W. P. L. Carter, "Atmospheric Chemistry of *cis*- and *trans*-3-Hexene-2, 5-Dione," *Environ. Sci. Technol.*, **19,** 265 (1985).

Turro, N. J., *Modern Molecular Photochemistry*, Benjamin Cummings Publishing, Menlo Park, CA, 1978.

Wallace, L. and D. M. Hunten, ''Dayglow of the Oxygen *A* Band,'' *J. Geophys. Res. Space Phys.*, **73,** 4813 (1968).

Weaver, J., J. Meagher, and J. Heicklen, ''Photo-oxidation of CH_3CHO Vapor at 3130 Å,'' *J. Photochem.*, **6,** 111 (1976/77).

Weschler, C. J., M. L. Mandich and T. E. Graedel, ''Speciation, Photosensitivity, and Reactions of Transition Metal Ions in Atmospheric Droplets,'' *J. Geophys. Res.*, **90,** in press (1985).

Wine, P. H. and A. R. Ravishankara, ''O_3 Photolysis at 248 nm and $O(^1D_2)$ Quenching by H_2O, CH_4, H_2, and N_2O: $O(^3P_J)$ Yields,'' *J. Chem. Phys.*, **69,** 365 (1982).

Winer, A. M., G. M. Breuer, W. P. L. Carter, K. R. Darnall, and J. N. Pitts, Jr., ''Effects of Ultraviolet Spectral Distribution on the Photochemistry of Simulated Polluted Atmospheres,'' *Atmos. Environ.*, **13,** 989 (1979).

World Meteorological Organization, ''The Stratosphere, 1981, Theory and Measurements,'' WMO Global Ozone Research and Monitoring Project, Report No. 11, May 1981.

Yamamoto, S. and R. A. Back, ''The Gas-Phase Photochemistry of Oxalic Acid,'' *J. Phys. Chem.*, **89,** 622 (1985).

Yao, F., I. Wilson, and H. Johnston, ''Temperature-Dependent Ultraviolet Absorption Spectrum for Dinitrogen Pentoxide,'' *J. Phys. Chem.*, **86,** 3611 (1982).

Yoshino, K., D. E. Freeman, J. R. Esmond, and W. H. Parkinson, ''High Resolution Absorption Cross Section Measurements and Band Oscillator Strengths of the (1,0)–(12,0) Schumann–Runge Bands of O_2,'' *Planet. Space Sci.*, **31,** 339 (1983).

Yoshino, K., D. E. Freeman, and W. H. Parkinson, ''Atlas of the Schumann–Runge Absorption Bands of O_2 in the Wavelength Region 175–205 nm,'' *J. Phys. Chem. Ref. Data.*, **13,** 207 (1984).

Zafiriou, O. C., ''Natural Water Photochemistry, in *Chemical Oceanography*, Vol. 8, Academic Press, London, 1983, pp. 339–379.

Zafiriou, O. C. and M. B. True, ''Nitrite Photolysis in Seawater by Sunlight, *Mar. Chem.*, **8,** 9 (1979a).

Zafiriou, O. C. and M. B. True, ''Nitrate Photolysis in Seawater by Sunlight,'' *Mar. Chem.*, **8,** 33 (1979b).

Zafiriou, O. C., J. Joussot-Dubien, R. G. Zepp, and R. G. Zika, ''Photochemistry of Natural Waters,'' *Environ. Sci. Technol.*, **18,** 358A (1984).

Zellner, R. and G. Wagner, Paper presented at The Sixth Symposium on Gas Kinetics, University of Southhampton, July 14–17, 1980.

PART 3

Experimental Kinetic, Mechanistic, and Spectroscopic Techniques

4 Fundamentals of Kinetics Applied to Atmospheric Reactions

4-A. FUNDAMENTAL PRINCIPLES OF GAS PHASE KINETICS

We briefly review here the basic principles of gas phase kinetics needed to assess the importance of various reactions in the troposphere. In Sections 4-D and 4-E, these principles are extended to reactions in solution and on the surfaces of solids.

4-A-1. Elementary and Overall Reactions

Elementary reactions are defined as those that cannot be broken down into two or more simpler reactions. Generally, they consist of one or two reactant species and are referred to as unimolecular and bimolecular processes, respectively. However, there are a number of important gas phase processes in which three different species participate; these are termolecular reactions. In the troposphere they usually involve N_2 and/or O_2 as one of the three participants; the role of the third molecule is generally to act as an "inert gas" which stabilizes the energy-rich product from a highly exothermic bimolecular reaction by siphoning off the excess energy and thus preventing dissociation back into the reactants. In such cases, rather than being specific as to the colliding third body, the symbol "M" is used.

Examples of these three classes of gas phase reactions are

Unimolecular. The thermal decomposition of PAN:

$$CH_3\overset{\overset{\displaystyle O}{\|}}{C}{-}OONO_2 \rightarrow CH_3\overset{\overset{\displaystyle O}{\|}}{C}{-}OO + NO_2 \tag{1}$$

Bimolecular. Formation of the gaseous nitrate radical:

$$O_3 + NO_2 \rightarrow NO_3 + O_2 \tag{2}$$

Termolecular. The formation of ozone by the reaction of a ground-state oxygen atom, $O(^3P)$, with O_2:

$$O(^3P) + O_2 + M \rightarrow O_3 + M \tag{3}$$

Because "M" does not enter into the reaction chemically, such reactions are often written with the "M" above the arrow:

$$O(^3P) + O_2 \xrightarrow{M} O_3$$

Each of these types of elementary processes will be treated in some detail in subsequent sections of this chapter.

While two-body collisions are common in the gas phase, three-body collisions are much less probable and four-body collisions can essentially be ignored because of their low probability. Thus the majority of the reactions we deal with in the atmosphere are bimolecular, with a lesser number being termolecular.

An *overall* reaction includes two or more elementary reactions; indeed there is no limit to the number of reactants or elementary reactions comprising an overall reaction. Thus if a single reaction step *as written* has four or more reactants, it cannot be an elementary process, and it must occur via two or more consecutive steps. If a reaction step contains two to three reactants, it may or may not be an elementary reaction.

An example of an overall reaction is the oxidation in air of methane by OH radicals, in the presence of nitric oxide. The major stable products, formaldehyde and NO_2, are formed in a sequence of five elementary reactions:

$$CH_4 + OH \rightarrow CH_3 + H_2O \tag{4}$$

$$CH_3 + O_2 \xrightarrow{M} CH_3O_2 \tag{5}$$

$$CH_3O_2 + NO \rightarrow CH_3O + NO_2 \tag{6}$$

$$CH_3O + O_2 \rightarrow CH_2O + HO_2 \tag{7}$$

$$HO_2 + NO \rightarrow NO_2 + OH \tag{8}$$

In fact, this is an example of an overall chain reaction, with the *initiating* step being reaction (4) and the *chain carrying* steps which regenerate the OH radical being reactions (5)–(8).

However, this reaction sequence does not go on recycling OH radicals indefinitely. Thus OH can react with NO_2 to give nitric acid and HO_2 can self-recombine to form hydrogen peroxide and O_2

$$OH + NO_2 \xrightarrow{M} HONO_2 \tag{9}$$

$$2\ HO_2 \xrightarrow{M} H_2O_2 + O_2 \tag{10}$$

These are elementary *chain termination* processes because the OH and HO_2 radicals form molecular products that are relatively stable in the troposphere, that is, $HONO_2$ and H_2O_2.

A *mechanism* of a reaction is the sequence of elementary reactions leading from reactants to products and includes labile intermediates which, in atmospheric oxidations, are generally free radicals. Thus the reaction sequence shown above, reactions (4)–(8), (9), and (10), is a *mechanism* for the oxidation of CH_4 (and other alkanes) in the troposphere.

4-A-2. Reaction Rates

The rate of a reaction is defined as the change in the concentration of a reactant or product with time. For simple reactions occurring with unit stoichiometry, the rate expressed in terms of reactant disappearance is the same as the rate in terms of product formation. For example, for reaction (11), the reaction of ozone with nitric oxide,

$$NO + O_3 \rightarrow NO_2 + O_2 \tag{11}$$

the rate is defined as

$$\text{rate} = -\frac{d[O_3]}{dt} = -\frac{d[NO]}{dt} = +\frac{d[O_2]}{dt} = +\frac{d[NO_2]}{dt} \tag{A}$$

For reactions of the more general form

$$aA + bB \rightarrow cC + dD \tag{12}$$

where the stoichiometric coefficients *a*, *b*, *c*, and *d* are not all unity, the rate in terms of disappearance of A may not be equal to the rate in terms of disappearance of B, or the appearance of C or D. To take such differences in stoichiometry into account, the rate of the generalized reaction (12) is defined *by convention* as

$$\text{rate} = -\frac{1}{a}\frac{d[A]}{dt} = -\frac{1}{b}\frac{d[B]}{dt} = +\frac{1}{c}\frac{d[C]}{dt} = +\frac{1}{d}\frac{d[D]}{dt} \tag{B}$$

For example, in the thermal oxidation of NO by oxygen,

$$2NO + O_2 \rightarrow 2NO_2 \tag{13}$$

two molecules of NO disappear for each molecule of O_2 reacted, and the rate of loss of NO is twice that of O_2:

$$\text{rate} = -\frac{1}{2}\frac{d[NO]}{dt} = -\frac{d[O_2]}{dt} = +\frac{1}{2}\frac{d[NO_2]}{dt}$$

While this convention is now widely used, it was not in some earlier kinetic studies. Thus one must be careful to note exactly how the rate is defined so that the reported rate constants are interpreted and applied correctly.

4-A-3. Rate Laws, Reaction Order, and the Rate Constant

In systems of atmospheric interest, the rate law or rate expression for a reaction, either elementary or overall, is the equation expressing the dependence of the rate on the concentrations of reactants. In a few reactions (mainly those in solution), products may also appear in the rate law (*vide infra*).

The importance of distinguishing between elementary and overall reactions comes in formulating such rate laws. For *elementary reactions only*, the rate law may be written directly from the stoichiometric equation. Thus for the general *elementary* gas phase reaction

$$aA + bB \rightarrow cC + dD \tag{12}$$

$$\text{rate} = k[A]^a[B]^b$$

where $a + b \leq 3$ by definition of an elementary reaction. For example, the rate expression for the elementary reaction (11) is given by

$$NO + O_3 \rightarrow NO_2 + O_2 \tag{11}$$

$$\text{rate} = k_{11}[O_3][NO]$$

The *rate constant*, k, is simply the constant of proportionality in the expression relating the rate of a reaction to the concentrations of reactants and/or products, each expressed with the appropriate exponent. The *order* of a reaction is defined as the sum of the exponents in the rate law. Thus reaction (11) is $(1 + 1) =$ second order. The order with respect to each species appearing in the rate law is the exponent of the concentration of that species; thus reaction (11) is first order in both O_3 and NO.

The basis of predicting rate laws for elementary reactions lies in the fact that they must occur during a single collision (although the *probability* of reaction during any one collision is equal to or less than unity). Thus doubling the concentration of O_3, say in reaction (11), will double the number of collisions per second of O_3 with NO. If the probability of reaction per collision remains constant, then the number of O_3 and NO molecules reacting, and O_2 and NO_2 formed per unit time (i.e., the rate), must double.

The thermal oxidation of NO by molecular oxygen, reaction (13), is another example where the stoichiometry and the molecularity of the reaction are directly related, and the rate law is

$$\text{rate} = k_{13}^{III}[NO]^2[O_2]$$

Thus the rate is proportional to the first power of the oxygen concentration and the square of the nitric oxide concentration and the reaction order is 1 + 2 = 3. However, in the troposphere, the O_2 concentration is always so large relative to NO that it is effectively a constant and thus can be incorporated into the rate constant k_{13}^{III}. The rate law is now written

$$\text{rate} = k_{13}^{bi}[\text{NO}]^2$$

and the reaction is referred to as *pseudo*-second-order. We adopt the convention of writing a third-order rate constant as k^{III} and a pseudo-second-order rate constant as k^{bi}, as illustrated above.

For the general *overall* reaction (12), the rate law has the form

$$\text{rate} = k[\text{A}]^m[\text{B}]^n[\text{C}]^p[\text{D}]^q$$

where, depending on the mechanism of the reaction, m, n, p, and q may be zero, integers, or fractions. As noted earlier, in most gas phase atmospheric reactions, the exponents of the product concentration (i.e., p and q) are zero and the rate laws involve only the reactant species. It is important to stress here that in contrast to elementary reactions, in *overall* reactions the exponents in the rate laws (e.g., m, n, p, q) *do not* necessarily bear a relationship to the stoichiometric coefficients of the reaction (e.g., a, b, c, or d).

The rate law and the reaction order can often be used to show that a reaction cannot be an elementary reaction since, in the latter case, the exponents must be integers and the overall reaction order must be ≤ 3. However, it should be noted that these kinetic parameters cannot be used to confirm that a particular reaction *is* elementary; they can only indicate that the kinetic data do not rule out the possibility that the reaction is elementary.

For example, the thermal decomposition of O_3 to O_2 is represented by the equation:

$$2\text{O}_3 \rightarrow 3\text{O}_2 \tag{14}$$

This reaction could be an elementary process since only two reactant molecules are involved. If it were an elementary reaction, the rate law would be given by

$$\text{rate} = -\frac{1}{2}\frac{d[\text{O}_3]}{dt} = +\frac{1}{3}\frac{d[\text{O}_2]}{dt} = k[\text{O}_3]^2$$

Experimentally, however, it is found that the rate law depends on the reaction conditions. For example, if sufficient O_2 is present initially, the rate is given by

$$\text{rate} = k[\text{O}_3]^2[\text{O}_2]^{-1}$$

The dependence of the rate law on the reaction conditions and the appearance of the concentration of the product O_2 in the rate expression show that reaction (14) cannot be an elementary reaction.

The units of the rate constant k depend on the reaction order. From the rate law for the general overall reaction (12), k is given by

$$k = \frac{\text{rate}}{[A]^m [B]^n [C]^p [D]^q}$$

Since the units of the rate are concentration $\times$ time^{-1}, the units of k must be concentrations$^{[1-(m+n+p+q)]}$ $\times$ time^{-1}. For the usual gas phase process, p and q are zero so that

$$k = \frac{\text{rate}}{[A]^m [B]^n}$$

and the units of k are concentration$^{[1-(m+n)]}$ $\times$ time^{-1}.

In gas phase reactions, concentrations are usually expressed in molecules cm^{-3} and time in seconds, the convention we employ in this book. Thus the units of k are

First order	s^{-1}
Second order	cm^3 molecule^{-1} s^{-1}
Third order	cm^6 molecule^{-2} s^{-1}

As we have seen (Section 1-C), concentrations of gaseous pollutants are often expressed in terms of parts per million (ppm) by volume, and time is expressed in minutes. Use of these concentration units must be reflected in the units used for the rate constants as well; for example, second-order rate constants are in units of ppm^{-1} min^{-1}.

To convert from a rate constant given in cm^3 molecule^{-1} s^{-1} to one in ppm^{-1} min^{-1}, both the units of concentration and time must be changed. Using the conversion factors given in Table 4.1, 1 ppm = 2.46×10^{13} molecules cm^{-3} at 760 Torr total pressure and 25°C (298°K), one can convert k = 1 cm^3 molecule^{-1} s^{-1} into ppm^{-1} min^{-1} in the following manner:

$$1 \text{ cm}^3 \text{ molecule}^{-1} \text{ s}^{-1} = \frac{1 \text{ cm}^3}{\text{molecule s}} \times 2.46 \times 10^{13} \frac{\text{molecules}}{\text{cm}^3 \text{ ppm}} \times \frac{60 \text{ s}}{\text{min}}$$

$$= 1.48 \times 10^{15} \text{ ppm}^{-1} \text{ min}^{-1}$$

Therefore, to convert second-order rate constants in units of cm^3 molecule^{-1} s^{-1} into units of ppm^{-1} min^{-1}, one must multiply by 1.48×10^{15}. Conversely, to convert k in ppm^{-1} min^{-1} to k in cm^3 molecule^{-1} s^{-1}, one divides by 1.48×10^{15} or multiplies by $1/(1.48 \times 10^{15}) = 6.77 \times 10^{-16}$ (Table 4.1). Similar consider-

TABLE 4.1. Some Common Conversion Factors for Gas Phase Reactions

Concentrations[a]

1 mole L^{-1} = 6.02×10^{20} molecules cm^{-3}
1 ppm = 2.46×10^{13} molecules cm^{-3}
1 ppb = 2.46×10^{10} molecules cm^{-3}
1 ppt = 2.46×10^{7} molecules cm^{-3}
1 atm = 760 Torr = 4.09×10^{-2} mole L^{-1} = 2.46×10^{19} molecules cm^{-3}

Second-Order Rate Constants

cm^3 molecule^{-1} s^{-1} $\times 6.02 \times 10^{20}$ = L mole^{-1} s^{-1}
ppm^{-1} min^{-1} $\times 4.08 \times 10^{5}$ = L mole^{-1} s^{-1}
ppm^{-1} min^{-1} $\times 6.77 \times 10^{-16}$ = cm^3 molecule^{-1} s^{-1}
atm^{-1} s^{-1} $\times 4.06 \times 10^{-20}$ = cm^3 molecule^{-1} s^{-1}

Third-Order Rate Constants

cm^6 molecule^{-2} s^{-1} $\times 3.63 \times 10^{41}$ = L^2 mole^{-2} s^{-1}
ppm^{-2} min^{-1} $\times 9.97 \times 10^{12}$ = L^2 mole^{-2} s^{-1}
ppm^{-2} min^{-1} $\times 2.75 \times 10^{-29}$ = cm^6 molecule^{-2} s^{-1}

[a] The concentrations ppm, ppb, and ppt are relative to air at 1 atm and 25°C, where 1 atm = 760 Torr total pressure.

ations apply to the conversion of third-order rate constants from units of ppm^{-2} min^{-1} to cm^6 molecule^{-2} s^{-1}.

Occasionally, gas concentrations are given in units of moles L^{-1} or in units of pressure such as Torr, atmospheres, or pascals; these can be converted to the more conventional units in tropospheric chemistry using the ideal gas law. Table 4.1 gives some common conversion factors for gas phase concentrations and rate constants at 1 atm pressure (760 Torr total pressure) and 25°C.

For solution phase reactions, we use concentrations units of moles L^{-1}, giving corresponding rate constant units of L mole^{-1} s^{-1} (second order) and L^2 mole^{-2} s^{-1} (third order).

4-A-4. Termolecular Reactions and Pressure Dependence of Rates

Termolecular elementary reactions whose rates depend on the total pressure are important in the atmosphere. Examples include the formation of O_3,

$$O(^3P) + O_2 + M \rightarrow O_3 + M \quad (3)$$

and the oxidation of SO_2 and NO_2 to sulfuric and nitric acids via gas phase OH reactions:

$$OH + SO_2 + M \rightarrow HOSO_2 + M \quad (15)$$

$$OH + NO_2 \xrightarrow{M} HONO_2 \quad (9)$$

Let us take the formation of O_3 as an example. The exothermic bond formation between $O(^3P)$ and O_2 releases energy which must be removed in order to form a stable O_3 molecule; if the energy remains as internal energy, the O_3 will quickly fly apart to reform $O + O_2$. The third molecule, M, is any molecule that stabilizes the O_3 by colliding with it and removing some of its excess internal energy.

From the role which M plays in termolecular reactions as well as from the rate law,

$$\text{rate} = \frac{-d[O]}{dt} = \frac{-d[O_2]}{dt} = k_3[O][O_2][M]$$

one might expect the rate to increase with the concentration or pressure of the third body M. However, there clearly must be some limit since the rate cannot increase to infinity but only to some upper limit determined by how fast the two reactive species can combine chemically. As a result, one might intuitively expect the rates of reactions such as (3), (9), and (15) to increase initially as the pressure of M is increased from zero, and then to plateau off at some limiting value at high pressures.

In the atmosphere, N_2 and O_2 are the major species acting as the third bodies (M), and at the earth's surface the pressure is ~ 1 atm. One might therefore assume that rate constants of interest for tropospheric reactions need only be measured in 1 atm of air, and that establishing the pressure dependence over a range of pressures is not necessary. However, in order to model the chemistry of the troposphere accurately, the reaction rates should be computed as a function of altitude, with the changing pressure and temperature taken explicitly into account. In addition, there are both experimental and practical problems with restricting kinetic studies to 1 atm in air:

- Some kinetic techniques can only be used at relatively low total pressures ($\sim$0.5–100 Torr). Application of the results to the troposphere requires a knowledge of the pressure dependence of the reaction, with extrapolation of the experimental results to higher pressures.
- In some experimental techniques, air cannot be present because O_2 reacts with some of the reactants or intermediates or photolyzes to oxygen atoms which then react further, leading to interferences for which the data cannot be reliably corrected.
- Many of the termolecular reactions of interest for tropospheric chemistry also occur in the stratosphere and mesosphere; assessing the importance of such reactions throughout the atmosphere thus requires a knowledge of their pressure dependencies.

One thus needs to establish the reaction rates over a range of pressures, ideally up to 1 atm, and in the presence of air.

Let us take the reaction (15) of OH with SO_2 as an example of a termolecular reaction of atmospheric interest and examine how its pressure dependence is established. It is common in kinetic studies to follow the decay of one reactant in an excess of the second reactant. In the case of reaction (15),

$$OH + SO_2 + M \rightarrow HOSO_2 + M \tag{15}$$

the decay of OH is followed in the presence of excess SO_2 and the third body M, where M is an inert bath gas such as He, Ar, or N_2. Since it is an elementary reaction, the rate law for reaction (15) can be written for low pressures:

$$\frac{-d[OH]}{dt} = k_{15}^{III}[OH][SO_2][M]$$

If [M] is constant, k_{15}^{III} and [M] can be combined to form an effective bimolecular rate constant, k_{15}^{bi}:

$$\frac{-d[OH]}{dt} = k_{15}^{III}[OH][SO_2][M]$$

$$= k_{15}^{bi}[OH][SO_2]$$

Since SO_2 is in great excess, its concentration does not change significantly even when all the OH has reacted and hence it remains approximately constant throughout the reaction at its initial value, $[SO_2]_0$. Rearranging the rate law and integrating from time $t = 0$ when the initial concentration of OH is $[OH]_0$ to time t when the OH concentration is [OH], one obtains

$$\ln \frac{[OH]}{[OH]_0} = -k_{15}^{bi}[SO_2]_0\, t$$

Since the initial concentration of OH, $[OH]_0$, is a constant, a plot of ln [OH] against reaction time t should be a straight line with slope or decay rate given by

$$\text{decay rate} = -k_{15}^{bi}[SO_2]_0$$

A plot of these decay rates against $[SO_2]_0$ should thus be linear, with the slopes increasing with pressure since k_{15}^{bi} depends on [M].

Figure 4.1 shows such a plot of the absolute values of the observed OH decay rates against $[SO_2]_0$ at total pressures of Ar from 50 to 402 Torr. As expected, the decay rates are linear with $[SO_2]_0$ and increase with the pressure of M.

To obtain the termolecular rate constant k_{15}^{III}, the effective bimolecular rate constant $k_{15}^{bi} = k_{15}[M]$ is plotted in Fig. 4.2 as a function of [M] (i.e., of total pressure). As expected from the earlier discussion, k_{15}^{bi} increases with [M] at low pressures but approaches a plateau at higher pressures.

Termolecular reactions can be treated, as a first approximation, as if they consist of several elementary steps, for example, for reaction (15),

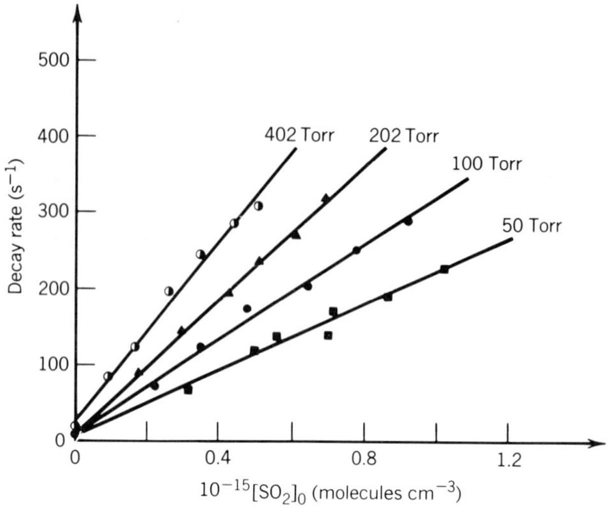

FIGURE 4.1. Plots of the OH decay rates against the initial SO_2 concentration at total pressures of Ar from 50 to 402 Torr (from Atkinson, et al., 1976).

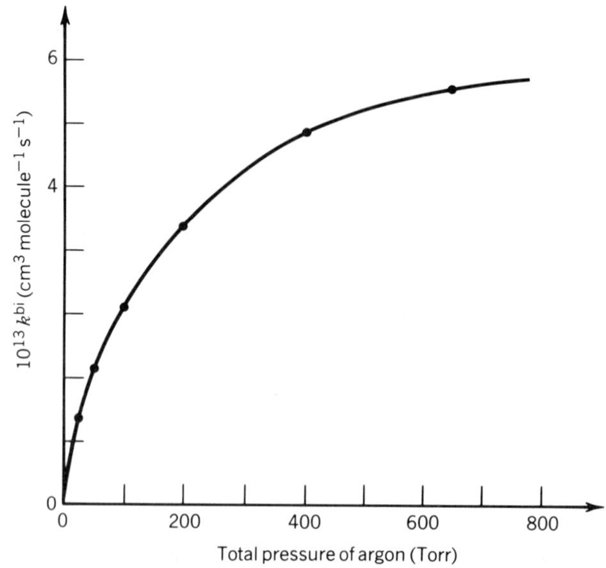

FIGURE 4.2. Plot of k_{15}^{bi} against total pressure for M = Ar for the reaction of OH with SO_2 (from Atkinson et al., 1976).

$$OH + SO_2 \underset{k_b}{\overset{k_a}{\rightleftarrows}} HOSO_2^* \tag{16,−16}$$

$$HOSO_2^* + M \xrightarrow{k_c} HOSO_2 + M \tag{17}$$

$HOSO_2^*$ is the OH–SO_2 adduct which contains the excess internal energy from bond formation in (16), and $HOSO_2$ is the stabilized adduct resulting when some of this internal energy is removed by a collision with M.

If the system is treated as if the concentration of the energized adduct ($HOSO_2^*$) remains constant with time, then its rates of formation and loss are equal. These rates can be written from Eq. (16, −16) and (17) since these are assumed to be elementary reactions. Thus

$$\frac{d[HOSO_2^*]}{dt} = 0 = k_a[OH][SO_2] - k_b[HOSO_2^*] - k_c[HOSO_2^*][M]$$

This is an example of the *steady state approximation*, widely employed in gas phase kinetics and mechanistic studies.

Rearranging, an expression for $[HOSO_2^*]$ is obtained in terms of the reactants OH and SO_2:

$$[HOSO_2^*] = \frac{k_a[OH][SO_2]}{k_b + k_c[M]}$$

The rate of the reaction in terms of product formation is given by

$$\begin{aligned}\frac{d[HOSO_2]}{dt} &= k_c[M][HOSO_2^*] \\ &= k_c[M]\frac{k_a[OH][SO_2]}{k_b + k_c[M]} \\ &= \frac{k_a k_c[M]}{k_b + k_c[M]}[OH][SO_2]\end{aligned}$$

In this form, k_{15}^{bi} can be rationalized by the combination of rate constants and [M] given in brackets. Alternatively, $1/k_{15}^{bi}$ is given by

$$\frac{1}{k_{15}^{bi}} = \frac{k_b + k_c[M]}{k_a k_c[M]} = \frac{k_b}{k_a k_c[M]} + \frac{1}{k_a}$$

At "infinite" pressure where $1/[M] = 0$, the rate constant should have its high-pressure limiting value. It is seen that this high-pressure limit value, k_{15}^{∞}, is equal to k_a. One would also qualitatively expect $k_{15}^{\infty} = k_a$ from the reaction scheme consisting of (16, −16) and (17); thus in the limit of infinite pressure, all the energized adducts formed in (16) will be stabilized in (17) and none will have a chance to decompose back to reactants via (−16). In this case, the rate constant will just be that for formation of $HOSO_2{}^*$, that is, k_a.

This approximate treatment of termolecular reactions can also be used to examine how the third-order low-pressure rate constant k^{III} relates to the rate constants k_a, k_b, and k_c for the elementary reactions assumed to be involved. As [M] approaches zero, k_{15}^{bi} approaches $k_a k_c[M]/k_b$, so that k_{15}^{III} is given by

$$k_{15}^{III} = \frac{k_a k_c}{k_b}$$

Many addition reactions such as the OH–SO_2 reaction are in the falloff region between second and third order in the range of total pressures encountered from the troposphere through the stratosphere. Troe and co-workers (Troe, 1977a,b; 1979; 1981; 1983a,b; Gilbert, et al., 1983) have carried out extensive theoretical studies of addition reactions and their reverse unimolecular decompositions as a function of pressure. In this work they have developed expressions for the rate constants in the falloff region; these are now most commonly used to derive the limiting low- and high-pressure rate constants from experimental data, as well as to report the temperature and pressure dependence of termolecular reactions in compilations of kinetic data (Section 4-F).

Equation (C) gives the most commonly used form of the rate constant expression of Troe and co-workers

$$k = \frac{k_0[M]}{1 + k_0[M]/k_\infty} F_C^{[1 + (N^{-1}\log k_0[M]/k_\infty)^2]^{-1}} \qquad \text{(C)}$$

In Eq. (C), k_0 (or k^{III} as used earlier) is the low-pressure limiting rate constant and k_∞ is the high-pressure limiting rate constant. F_C is known as the broadening factor of the falloff curve; while its actual value depends on the particular reaction and can be calculated theoretically, Troe (1979) suggests that for reactions under atmospheric conditions, the value of F_C will be ~0.7–0.9, independent of temperature. However, values as low as 0.4 are often observed. The recent NASA evaluations of stratospheric reactions (DeMore et al., 1983, 1985) takes $F_C = 0.6$. The factor N in Eq. (C) is given by

$$N \simeq 0.75 - 1.27 \log F_C \qquad \text{(D)}$$

and is approximated by $N \simeq 1$ for reactions in the atmosphere, as suggested by Troe (1979). Note that the logarithms used in Eq.'s (C) and (D) are to the *base 10.*

The temperature dependence of k arises primarily in the temperature dependence of k_0 and k_∞, which are expressed in terms of their values at 300°K, k_0^{300} and k_∞^{300}:

$$k_0(T) = k_0^{300}\left(\frac{T}{300}\right)^{-n} \tag{E}$$

$$k_\infty(T) = k_\infty^{300}\left(\frac{T}{300}\right)^{-m} \tag{F}$$

F_C may also be temperature dependent; for example, in the reaction

$$NO_3 + NO_2 + M \rightarrow N_2O_5 + M$$

Kircher et al. (1984) use for F_c the temperature-dependent expression

$$F_C = 0.565 - 0.000697\,T \tag{G}$$

This will also make N temperature dependent [Eq. (D)].

Thus there are a number of parameters—k_0^{300}, k_∞^{300}, n, m, and F_C (which depends on N)—which must be known in order to calculate k throughout the falloff region. Experimentally, these are determined by obtaining a best fit to the observed values of k using a calculated or assumed value of F_C.

Recent kinetic evaluations (e.g., DeMore et al., 1985) list values of k_0^{300}, k_∞^{300}, n, and m with a assumed value of $F_C = 0.6$ (and $N = 1$), so that rate constants for the termolecular reactions can be calculated as a function of pressure and temperature. For example, for the reaction (15) of OH with SO_2 discussed earlier, $k_0^{300} = (3.0 \pm 1.5) \times 10^{-31}$ cm^6 molecule^{-2} s^{-1}, $n = 3.4 \pm 1.5$, $k_\infty^{300} = (2.0 \pm 1.5) \times 10^{-12}$ cm^{-3} molecule^{-1} s^{-1}, and $m = 0 \pm 1.0$ (DeMore et al. 1983). At 300°K and 760 Torr pressure, [M] $= 2.46 \times 10^{19}$ molecules cm^{-3} and k_0[M] $= 7.4 \times 10^{-12}$ cm^{-3} molecule^{-1} s^{-1}. Thus the value of the rate constant under these conditions can be calculated as

$$k = \left[\frac{7.4 \times 10^{-12}}{1 + (7.4 \times 10^{-12})/(2.0 \times 10^{-12})}\right] 0.6^{[1 + (\log[(7.4 \times 10^{-12})/(2.0 \times 10^{-12})])^2]^{-1}}$$

$$= [1.6 \times 10^{-12}]\, 0.6^{0.76} = 1.1 \times 10^{-12} \text{ cm}^3 \text{ molecule}^{-1} \text{ s}^{-1}$$

However, under conditions typical of the lower stratosphere, (~20 km), the temperature and pressure are much lower. Let us calculate k_{15} for conditions where the temperature is −219°K and the total pressure is ~39 Torr; thus [M] = 1.7 × 10^{18} molecules cm^{-3}. The low- and high-pressure limiting rate constants at this temperature are given by Eqs. (E) and (F):

$$k_0^{219} = 3.0 \times 10^{-31}\left(\frac{219}{300}\right)^{-3.4} = 8.7 \times 10^{-31} \text{ cm}^6 \text{ molecule}^{-2} \text{ s}^{-1}$$

$$k_\infty^{219} = 2.0 \times 10^{-12}\left(\frac{219}{300}\right)^{-0} = 2.0 \times 10^{-12} \text{ cm}^3 \text{ molecule}^{-1} \text{ s}^{-1}$$

From Eq. (C), the rate constant at 219°K and 39 Torr pressure is

$$k = \left[\frac{1.5 \times 10^{-12}}{1 + (1.5 \times 10^{-12})/(2.0 \times 10^{-12})}\right] 0.6^{[1+(\log[(1.5\times10^{-12})/(2.0\times10^{-12})])^2]^{-1}}$$

$$= [8.6 \times 10^{-13}]\, 0.6^{0.98} = 5.2 \times 10^{-13}\ \text{cm}^3\ \text{molecule}^{-1}\ \text{s}^{-1}$$

This is about a factor of two slower than at 300°K and 760 torr pressure.

In summary, rate constants for addition reactions in the atmosphere can be estimated as a function of temperature and pressure if values are available for the low- and high-pressure limiting rate constants as a function of temperature, that is, if k_0^{300}, k_∞^{300}, n, m, and F_C are known.

4-A-5. Aren't Most Bimolecular Reactions of Atmospheric Interest Simple Concerted Processes?

At first glance, it might appear that the vast majority of the bimolecular reactions with which one deals in the troposphere are simple concerted reactions, that is, during the collision of the reactants there is a reorganization of the atoms, leading directly to the formation of the products. However, it has become increasingly apparent in recent years that some important reactions which *appeared* to be concerted exhibit characteristics such as pressure dependencies which are not consistent with a direct concerted process.

A classic case is the reaction of OH with CO:

$$OH + CO \rightarrow H + CO_2 \tag{18}$$

This reaction appears to be an elementary bimolecular reaction involving a simple transfer of an oxygen atom from OH to CO. In accord with the definition of an elementary reaction, one can imagine it occurring during one collision of an OH radical with a CO molecule.

A number of studies of the kinetics of this reaction were carried out in the 1960s and the early 1970s, and the room temperature rate constants, measured at total pressures up to ~200 Torr in inert gases such as He, Ar, and N_2, were generally in good agreement (Table 4.2). In fact, this reaction was often used (and still is) to test whether a newly constructed kinetic apparatus was functioning properly.

In the mid-1970s, however, an apparent pressure effect on the rate constants was reported by Cvetanovic and co-workers (Overend et al., 1974). Subsequently, two groups of investigators examined the reaction at higher pressures and in the presence of diluents such as air, H_2, and SF_6 (Cox et al., 1976; Sie et al., 1976) and confirmed an increase in the rate constant as the total pressure increased; this is unexpected for a simple concerted process. This led to the suggestion that the reaction being observed experimentally was, in fact, not a direct concerted reaction, but rather a combination of several steps in which a bound adduct is formed as an intermediate:

TABLE 4.2. Selected Values Reported Before 1975 of the Rate Constant at 298°K for the OH + CO Reaction

$k \times 10^{13}$ (cm^3 $molecule^{-1}$ s^{-1})	Experimental Conditions	Reference
0.85 ± 0.33	1.8 Torr in Ar	Herron, 1966
1.9 ± 0.1	1 Torr in He or Ar	Dixon-Lewis et al., 1966
1.5 ± 0.2	100–200 Torr in Ar	Greiner, 1967
1.73 ± 0.02	1.8 Torr in Ar	Wilson and O'Donovan, 1967
1.66 ± 0.5	0.2–1 Torr in Ar	Mulcahy and Smith, 1971
1.35 ± 0.20	20 Torr in He	Stuhl and Niki, 1972
1.45 ± 0.22	10–20 Torr in N_2O/H_2 or H_2O	Smith and Zellner, 1973
1.3 ± 0.1	1–3 Torr in He	Westenberg and de Haas, 1973
1.59 ± 0.16	20–100 Torr in He or 20 Torr N_2	Davis et al., 1974
1.56 ± 0.2	0.4–5 Torr in He, Ar, or N_2	Howard and Evenson, 1974

$$\dot{\mathrm{O}}\mathrm{H} + \mathrm{CO} \underset{-a}{\overset{a}{\rightleftarrows}} \mathrm{HOCO}^* \xrightarrow{b} \mathrm{H} + \mathrm{CO_2}$$
$$\mathrm{HOCO}^* \xrightarrow[+\mathrm{M}]{c} \mathrm{HOCO} \qquad (19)$$

In (19) HOCO is the radical adduct of OH + CO, and HOCO* is the adduct containing excess internal energy resulting from the energy released by bond formation between OH and CO. As described earlier, M is any molecule or atom that collides with the HOCO*, removing some of its excess energy; in practice, it is usually an inert bath gas such as He or Ar which is present in great excess over the reactants. Such a mechanism had been suggested more than a decade earlier by Ung and Back (1964).

Reactions such as (19), which proceed with the formation of a bound adduct between the reactants, are known as indirect or non-concerted reactions. The adduct is "stable" in the sense that it corresponds to a well on the potential energy surface connecting the reactants and products; as such it has a finite lifetime and should be capable of being detected using appropriate techniques. Because of the complex nature of the mechanism, such reactions can exhibit a relatively complex temperature dependence. In addition, if the rate of collisional stabilization of the excited adduct is comparable to its rate of decomposition, a pressure dependence may result, as in the OH + CO reaction. The distinction between bimolecular and termolecular reactions blurs in such cases.

If the excited radical adduct HOCO* is assumed to be in a steady state (i.e., its rate of formation and loss are equal) then using the elementary reactions of (19), the steady state concentration of HOCO* is given by (H):

$$[\mathrm{HOCO}^*] = \frac{k_a[\mathrm{OH}][\mathrm{CO}]}{k_{-a} + k_b + k_c[\mathrm{M}]} \tag{H}$$

In most kinetic experiments, rates and rate constants are derived by following the loss of one reactant in the presence of a great excess of the other reactants (see Section 4-B). For example, in experiments to derive k_{18}, OH is followed in an excess of CO. The rate of loss of OH predicted on the basis of the mechanism (19) is

$$\frac{-d[\mathrm{OH}]}{dt} = k_a[\mathrm{OH}][\mathrm{CO}] - k_{-a}[\mathrm{HOCO}^*] \tag{I}$$

Using (H) this becomes

$$\frac{-d[\mathrm{OH}]}{dt} = k_a[\mathrm{OH}][\mathrm{CO}] \frac{k_b + k_c[\mathrm{M}]}{k_{-a} + k_b + k_c[\mathrm{M}]} \tag{J}$$

Comparing (J) to the expected form (K) of the rate law for a simple bimolecular reaction,

$$\frac{-d[\mathrm{OH}]}{dt} = k_{18}[\mathrm{OH}][\mathrm{CO}] \tag{K}$$

it is seen that the measured rate constant k_{18} for the *overall* reaction actually consists of a combination of rate constants for the *elementary* reactions in the mechanism (19):

$$k_{18} = k_a \frac{k_b + k_c[\mathrm{M}]}{k_{-a} + k_b + k_c[\mathrm{M}]} \tag{L}$$

In the limits of either zero or infinite pressure, k_{18} becomes independent of the pressure M. In the intermediate regime, k_{18} should show a complex dependence on pressure, as has been observed. Benson and coworkers have shown that the formation of an intermediate complex as in reaction (19) is consistent with the observed pressure and temperature dependence (Mozurkewich et al., 1984).

While it had been suggested that k_{18} might depend not only on the total pressure but also on the presence of O_2 (Biermann et al., 1978), recent experiments by Hofzumahaus and Stuhl (1984), Paraskevopoulos and Irwin (1984), and DeMore (1984) do not show such a dependence of the rate constant on O_2; any effects observed are likely due to secondary reactions of species such as HO_2.

In any event, the OH + CO rate constant at 1 atm in air is now thought to be significantly greater than the low-pressure values in Table 4.2; the most recent studies of Hofzumahaus and Stuhl (1984), Paraskevopoulos and Irwin (1984), Niki et al. (1984), and DeMore (1984) suggest that it is $\sim(2.2\text{–}2.5) \times 10^{-13}$ cm^3 molecule^{-1} s^{-1} at 25°C and 1 atm air.

4-A-6. Half-Lives and Lifetimes

A rate constant is a quantitative measure of how fast reactions proceed and therefore is an indicator of how long a given set of reactants will survive in the atmosphere under a particular set of reactant concentrations. However, the rate constant *per se* is not a parameter which by itself is readily related to the average length of time a species will survive in the atmosphere before reacting. A more intuitively meaningful parameter is the *half-life* ($t_{1/2}$) or the *natural lifetime* (τ) of a pollutant with respect to reaction with a labile species such as OH or NO_3 radicals.

The half-life ($t_{1/2}$) is defined as the time required for the concentration of a reactant to fall to one-half of its initial value, whereas the lifetime is defined as the time it takes for the reactant concentration to fall to $1/e$ of its initial value (e is the base of natural logarithms, 2.718). Both $t_{1/2}$ and τ are directly related to the rate constant and to the concentrations of any other reactants involved in the reactions. These relationships are given in general form in Table 4.3 for first-, second-, and third-order reactions.

These expressions can be readily derived from the rate laws. For a first-order reaction of a pollutant species A, the rate law for the reaction

$$\mathrm{A} \xrightarrow{k_1} \text{products}$$

is given by

$$\frac{-d[\mathrm{A}]}{dt} = k_1[\mathrm{A}]$$

Rearranging, this becomes

$$\frac{-d[\mathrm{A}]}{[\mathrm{A}]} = k_1\,dt$$

TABLE 4.3. Relationships Between the Rate Constant, Half-Lives and Lifetimes for First-, Second-, and Third-Order Reactions

Reaction Order	Reaction	Half-Life of A	Lifetime of A
First	$\mathrm{A} \xrightarrow{k_1} \text{products}$ (1)	$t^{\mathrm{A}}_{1/2} = \dfrac{0.693}{k_1}$	$\tau^{\mathrm{A}} = \dfrac{1}{k_1}$
Second	$\mathrm{A} + \mathrm{B} \xrightarrow{k_2} \text{products}$ (2)	$t^{\mathrm{A}}_{1/2} = \dfrac{0.693}{k_2[\mathrm{B}]}$	$\tau^{\mathrm{A}} = \dfrac{1}{k_2[\mathrm{B}]}$
Third	$\mathrm{A} + \mathrm{B} + \mathrm{C} \xrightarrow{k_3} \text{products}$ (3)	$t^{\mathrm{A}}_{1/2} = \dfrac{0.693}{k_3[\mathrm{B}][\mathrm{C}]}$	$\tau^{\mathrm{A}} = \dfrac{1}{k_3[\mathrm{B}][\mathrm{C}]}$

Integrating from time $t = 0$ when the initial concentration of A is $[A]_0$ to time t when the concentration is [A], one obtains

$$\ln \frac{[A]}{[A]_0} = -k_1 t$$

After one half-life (i.e., at $t = t_{1/2}$) by definition $[A] = 0.5[A]_0$. Substituting into the integrated rate expression, one obtains

$$t_{1/2} = -\frac{\ln 0.5}{k_1} = \frac{0.693}{k_1}$$

For second- and third-order reactions, *if one assumes the concentrations of the reactants other than A are constant with time*, the derivation is the same except that k is replaced by $k[B]$ (second order) or $k[B][C]$ (third order).

In most practical situations, however, the concentration of at least one of the other reactants is not constant, but changes with time due to reactions, fresh injections of pollutants, and so on. As a result, using half-lives (or lifetimes) of a pollutant with respect to second- or third-order reactions is an approximation that involves *assumed* constant concentrations of the other reactants. These half-lives for bimolecular and termolecular reactions are thus directly affected by the concentrations of the other reactant.

Derivation of the relationship between the rate constant k and the lifetime τ follows that for $t_{1/2}$, except that, from the definition of τ, at $t = \tau$, $[A] = [A]_0/e$.

Let us use these to calculate the natural lifetimes of some species in three atmospherically relevant examples. We assume 25°C, 1 atm in air, and average OH concentrations of 5×10^5 and 5×10^6 radicals cm^{-3}, typical of a relatively clean atmosphere or one polluted with photochemical smog, respectively.

EXAMPLE 1: FIRST-ORDER REACTION—THERMAL DECOMPOSITION OF PAN

$$\underset{\text{PAN}}{CH_3\overset{\overset{O}{\|}}{C}-OONO_2} \xrightarrow{k_1} CH_3\overset{\overset{O}{\|}}{C}-OO + NO_2 \qquad k_1 = 3.6 \times 10^{-4}\ s^{-1} \qquad (1)$$

$$\tau^{PAN} = \frac{1}{k_1} = \frac{1}{3.6 \times 10^{-4}\ s^{-1}} = 2.78 \times 10^3\ s = \boxed{46\ min}$$

EXAMPLE 2: SECOND-ORDER REACTION—OXIDATION OF METHANE BY OH

$$CH_4 + OH \xrightarrow{k_2} CH_3 + HOH \qquad k_2 = 8.4 \times 10^{-15}\ cm^3\ molecule\ s^{-1} \qquad (2)$$

For $[OH] = 5 \times 10^5\ \text{cm}^{-3}$

$$\tau_{OH}^{CH_4} = \frac{1}{k_2[OH]} = \frac{1}{(8.4 \times 10^{-15})(5 \times 10^5)} = 2.38 \times 10^8\ \text{s} = \boxed{7.5\ \text{yr}}$$

For $[OH] = 5 \times 10^6\ \text{cm}^{-3}$

$$\tau_{OH}^{CH_4} = 0.75\ \text{yr} = \boxed{276\ \text{days}}$$

EXAMPLE 3: THIRD-ORDER REACTION—FORMATION OF O_3

$$O(^3P) + O_2 \underset{M}{\overset{k_3^{bi}}{\rightarrow}} O_3 \qquad k_3^{III} = 6.0 \times 10^{-34}\ \text{cm}^6\ \text{molecule}^{-2}\ \text{s}^{-1}$$

At 1 atm pressure, this reaction is still in the low pressure limit (DeMore et al., 1985). Thus the effective second-order rate constant, k_3^{bi}, where 1 atm of air as the third body has already been taken into account, is $1.5 \times 10^{-14}\ \text{cm}^3\ \text{molecule}^{-1}\ \text{s}^{-1}$. Thus for $[O_2] = 5.2 \times 10^{18}\ \text{cm}^{-3}$,

$$\tau_{O_2}^{O(^3P)} = \frac{1}{k_3^{bi}[O_2]} = \frac{1}{(1.5 \times 10^{-14})(5.2 \times 10^{18})}$$
$$= 1.3 \times 10^{-5}\ \text{s} = \boxed{13\ \mu\text{s}}$$

Two points should be made about such calculations of the tropospheric lifetime. First, they are valid only for the specified reaction; if there are other competing loss processes such as photolysis, the actual overall lifetime will be shortened accordingly. On the other hand, for a species such as CH_4 which does not photolyze or react significantly with other atmospheric species such as O_3 or NO_3, $\tau_{OH}^{CH_4}$ is indeed close to the overall lifetime of CH_4.

Second, in bi- and termolecular reactions, $t_{1/2}$ and τ depend on the concentration of other reactants; this is particularly important when interpreting atmospheric lifetime. For example, as discussed earlier, reaction with the OH radical is a major fate of most organics during daylight in both the clean and polluted troposphere. However, the actual concentrations of OH at various geographical locations and under a variety of conditions are quite uncertain at the present time; in addition, its concentration varies diurnally since it is produced primarily by photochemical processes (Section 3-D). Finally, the concentration of OH varies with altitude as well so that the lifetime will depend on where in the troposphere the reaction occurs.

Thus when a lifetime of an organic in the atmosphere is cited with respect to OH attack, one should examine carefully the concentration of OH that was *assumed* in arriving at that lifetime; the substantial uncertainties in these estimated lifetimes which arise from the uncertainties in the estimated atmospheric OH concentrations should be clearly recognized.

4-A-7. Temperature Dependence of Rate Constants

Many rate constants show an exponential dependence on temperature which, over a relatively narrow temperature range, can be empirically fit by the *Arrhenius equation*

$$k = Ae^{-E_a/RT} \tag{M}$$

where R is the gas constant and the temperature T is in degrees Kelvin (°K = °C + 273.15). A, the preexponential factor, and E_a, the activation energy, are parameters characteristic of the particular reaction.

To a first approximation over the relatively small temperature range encountered in the troposphere, A is found to be independent of temperature for many reactions, so that a plot of ln k versus $1/T$ gives a straight line of slope $-E_a/R$ and intercept equal to ln A. However, the Arrhenius expression for the temperature dependence of the rate constant is empirically based. As the temperature range over which experiments could be carried out was extended, nonlinear *Arrhenius plots* of ln k against $1/T$ were observed for some reactions. This is not unexpected when the predictions of the two major kinetic theories in common use today, collision theory and transition state theory, are considered. A brief summary of the essential elements of these is found below, as we refer to them periodically throughout the text.

For many reactions, the temperature dependence of A is small (e.g., varies with $T^{1/2}$) compared to the exponential term so that Eq. (M) is a good approximation, at least over a limited temperature range. For some reactions encountered in tropospheric chemistry, however, this is not the case. For example, for reactions in which the activation energy is small or zero, the temperature dependence of A can become significant. As a result, the Arrhenius expression (M) is not appropriate to describe the temperature dependence, and the form

$$k = BT^n e^{-E_a/RT} \tag{N}$$

is frequently used, where B is a temperature-independent constant characteristic of the reaction and n is a number adjusted to provide a best fit to the data.

As the variety of reactions studied and the temperature ranges covered are extended, an increasing number of reactions are being found to show *non-Arrhenius* behavior. Indeed, the comment has been made. "we may soon have to come to grips with the converse question: why do most Arrhenius graphs look so straight?" (Gardiner, 1977).

While most reactions with which we deal increase in rate as the temperature increases, there are several notable exceptions. The first is the case of termolecular reactions, which generally slow down as the temperature increases. This can be rationalized qualitatively on the basis that the lifetime of the excited bimolecular complex formed by two of the reactants with respect to decomposition back to reactants decreases as the temperature increases, so that the probability of the excited complex being stabilized by a collision with a third body falls with temperature.

An alternate explanation can be seen by treating termolecular reactions as the sum of bimolecular reactions, as was illustrated in Section 4-A-4 for the OH + SO_2 + M reaction. Recall that the third-order low-pressure rate constant k^{III} can be expressed as the product of the three rate constants k_a, k_b, and k_c for the three individual reaction steps (16), (−16), and (17):

$$k_{15}^{III} = \frac{k_a k_c}{k_b}$$

Expressing each of the component rate constants in the Arrhenius form, k_{15}^{III} becomes

$$k_{15}^{III} = A_{15} e^{-E_{15}/RT} = \left(\frac{A_a A_c}{A_b}\right) e^{-(E_a + E_c - E_b)/RT}$$

Thus the activation energy for the reaction, E_{15}, is a combination of the activation energies for the individual steps, $(E_a + E_c - E_b)$. If $E_b > (E_a + E_c)$, that is, if the activation energy for decomposition of the energized adduct $HOSO_2^*$ back to reactants is greater than the sum of those for its formation (a) and deactivation to the stabilized adduct $HOSO_2$ (c), then the effective activation energy E_{15} for the termolecular reaction becomes negative and the rate constant decreases as the temperature increases.

Benson and coworkers (Mozurkewich and Benson, 1984; Mozurkewich et al., 1984; Lamb et al., 1984) have examined some elementary gas phase reactions in which negative activation energies and/or curved Arrhenius plots have been been observed experimentally. They have shown that activation energies as negative as −1.2 to −1.8 kcal mol^{-1} near 300°K as well as curved Arrhenius plots can be explained by the formation of an intermediate complex; they have also shown that the Arrhenius plots for the reactions of OH with CO, HNO_3 and HNO_4 are consistent with such intermediate complex formation.

Another type of reaction important in atmospheric chemistry which shows this somewhat unusual inverse relationship between temperature and rate is that between alkenes and OH and, in some cases, $O(^3P)$, where negative activation energies also occur. Possible explanations for this are discussed in detail in Section 7-E-1; they include the formation of an intermediate complex in a manner analogous to the OH + SO_2 + M reaction, or a significant temperature dependence of the preexponential factor combined with an approximately zero activation energy.

We now turn to the predictions of collision theory and transition state theory for the expected form of the rate constant and its temperature dependence.

4-A-7a. *Collision Theory*

Collision theory is based on the concept that molecules behave like hard spheres; during a collision of two species, a reaction may occur. To estimate a rate constant for a bimolecular reaction between reactants A and B based on this theory one needs first to calculate the number of collisions occurring in a unit volume per second (Z_{AB}) when the two species, A and B, having radii r_A and r_B, are present in concentrations N_A and N_B, respectively. From gas kinetic theory this can be shown to be given by Eq. (O):

$$Z_{AB} = (r_A + r_B)^2 \left(\frac{8\pi kT}{\mu}\right)^{1/2} N_A N_B \tag{O}$$

μ is the reduced mass of A and B [$\mu = m_A m_B/(m_A + m_B)$ where m is the mass of the species], k is the Boltzmann constant (1.38×10^{-23} J°K^{-1}), and T is the temperature in °K. In deriving Eq. (O), it is assumed that A and B are hard spheres and only collide when their centers come within a distance $\leq (r_A + r_B)$ of each other.

In considering reactions between colliding molecules, two additional factors must be taken into account. First, different collisions will occur with different amounts of energy depending on the speed of the molecules as they collide. Most reactions are expected to have an energy barrier which must be surmounted in order for a reaction to occur. This energy barrier arises from the net effect of simultaneous bond breaking and formation; at the transition state in the reaction, the energy released from bond formation is generally less than that required for bond breaking. The difference, which is the energy barrier, must be supplied in other ways if reaction is to take place.

Second, even if the reactants collide with enough energy to surmount the energy barrier, they may not react if they are not in the proper orientation with respect to each other. The importance of this so called *steric factor* can be illustrated using the reaction of ground-state oxygen atoms with the hydroxyl free radical. If the oxygen atom collides with the oxygen atom end of the OH, the orientation is correct for an overall reaction to O_2 + H:

$$O + O{-}H \rightarrow [O{-}{-}{-}O{-}{-}{-}H] \rightarrow O_2 + H \tag{20a}$$

However, if it collides with the hydrogen atom end, no net chemical change will result, although, in fact, an exchange reaction (i.e., exhange of the oxygen atoms) may occur:

$$O + H{-}O \rightarrow [O{-}{-}{-}H{-}{-}{-}O] \rightarrow O{-}H + O \quad \text{(no net chemical change)} \tag{20b}$$

To take into account the energy requirement, one can modify the result in Eq. (O) by calculating only the number of collisions between A and B which have a certain minimum energy, E_0. In the simplest approach, it is assumed that no reaction occurs if the energy of the colliding pair is less than E_0, and reaction occurs 100% of the time for energies $\geq E_0$. Alternatively, it can be assumed that for energies $\geq E_0$, the probability of reaction increases as the relative collision energy increases.

To take into account this dependence on energy, the concept of A and B being hard spheres with radii r_A and r_A can be modified. Let $\sigma_{AB} = \pi(r_A + r_B)^2$ be the *collisional* cross section for hard-sphere collisions between A and B and σ_R the *reaction* cross section for reaction between A and B. At energies $< E_0$, $\sigma_R = O$, that is, no reaction occurs. For energies $> E_0$, σ_R could be taken as a constant (e.g., equal to σ_{AB}), which would correspond to assuming all collisions having energies above the threshold energy lead to reaction. The alternate approach of an increase in reaction probability with increasing energy above the threshold corresponds to assuming an expression for σ_R which is a function of total energy E, for example,

$$\begin{aligned} \sigma_R &= \sigma_{AB}\left(1 - \frac{E_0}{E}\right) && \text{for } E \geq E_0 \\ \sigma_R &= 0 && \text{for } E < E_0 \end{aligned} \tag{P}$$

For the form of σ_R shown in Eq. (P), integrating over all total energies from 0 to ∞, the rate of reactive collisions becomes

$$Z_R = \sigma_{AB}\left(\frac{8kT}{\pi\mu}\right)^{1/2} e^{-E_0/kT} N_A N_B \tag{Q}$$

The rate constant k can thus be identified with Eq. (R):

$$k = \sigma_{AB}\left(\frac{8kT}{\pi\mu}\right)^{1/2} e^{-E_0/kT} \tag{R}$$

Equation (R), however, does not take into account the need for proper orientation of the colliding molecules in order for a reaction to occur. This is commonly done by including an extra factor, P, (the steric factor) the probability that the colliding molecules will have the correct orientation. This leads to Eq. (S):

$$k = P\sigma_{AB}\left(\frac{8kT}{\pi\mu}\right)^{1/2} e^{-E_0/kT} \tag{S}$$

By comparison of (S) and (M), it can be seen that the preexponential factor A in the Arrhenius equation can be identified with $P\sigma_{AB}(8kT/\pi\mu)^{1/2}$ and the activation energy, E_a, with the threshold energy E_0. It is important to note that collision theory predicts that the preexponential factor should indeed be dependent on temperature ($T^{1/2}$). The reason that so many reactions appear to follow the Arrhenius equation with A being temperature independent is that the temperature dependence contained in the exponential term normally swamps the smaller $T^{1/2}$ dependence, if the activation energy is significant. However, for reactions where E_a approaches zero, the temperature dependence of the preexponential factor can be significant.

Collision theory is used mainly as a reference for the efficiency of reactions. Thus at a temperature of 25°C (298°K), the rate constant for a reaction between two molecules each having a radius of 2 Å and a molecular weight of 50 g mole^{-1} would, according to Eq. (S), be 2.5×10^{-10} cm^3 molecule^{-1} s^{-1} for $P = 1$ and $E_0 = 0$. That is, when there are no steric or energy barriers to the reaction, the rate constant should be of the order of 10^{-10} cm^3 molecule^{-1} s^{-1}. A reaction with a rate constant of 10^{-15} cm^3 molecule^{-1} s^{-1} is therefore one that goes in approximately every 10^5 collisions.

4-A-7b. Transition State Theory

Transition state theory is more commonly applied today than collision theory. It is especially useful in examining reactions in solution (Section 4-D) and avoids the problem of introducing arbitrary factors such as the steric factor P [Eq. (S)] to take into account steric requirements.

Transition state theory treats a reacting system thermodynamically. Let us again take a bimolecular reaction between A and B. Transition state theory assumes that as A and B collide and start to react, they form a species called the activated complex, which corresponds to the A–B adduct at the peak of the energy hill lying between reactants and products. This activated complex is thus in a "transition state" and can either fall back to reactants or go on to form products. The activated complex is normally indicated with a double dagger symbol, AB‡. The reaction can thus be given as

$$\mathrm{A} + \mathrm{B} \underset{k_r}{\overset{k_f}{\rightleftarrows}} \mathrm{AB}^\ddagger \xrightarrow{k_p} \text{products}$$

Assuming the activated complex is in equilibrium with reactants, one can define the equilibrium constant $K^\ddagger$:

$$K^\ddagger = \frac{k_f}{k_r} = \frac{[\mathrm{AB}^\ddagger]}{[\mathrm{A}][\mathrm{B}]} \tag{T}$$

$K^\ddagger$ is related to the standard free energy change in going from reactants to the transition state, $\Delta G^{\circ\ddagger}$, by the usual thermodynamic relationship:

$$K^\ddagger = e^{-\Delta G^{\circ\ddagger}/RT} = (e^{-\Delta H^{\circ\ddagger}/RT})\,(e^{\Delta S^{\circ\ddagger}/R}) \tag{U}$$

$\Delta H^{\circ\ddagger}$ and $\Delta S^{\circ\ddagger}$ are the standard enthalpy and entropy changes in going from reactants to the transition state.

The reaction rate is determined by the rate at which AB‡ forms products

$$\text{rate of product formation} = \nu[\mathrm{AB}^\ddagger]$$

where ν is the frequency with which AB‡ breaks up into products. Substituting for [AB‡] from Eq. (T) and for $K^\ddagger$ from (U) one obtains

$$\text{rate} = \nu K^{\ddagger}[\text{A}][\text{B}]$$
$$= \nu(e^{\Delta S^{\circ\ddagger}/R})\,(e^{-\Delta H^{\circ\ddagger}/RT})\,[\text{A}][\text{B}] \qquad \text{(V)}$$

The enthalpy change $\Delta H^{\circ\ddagger}$ is of course related to the energy change in going from reactants to the transition state, that is, to the activation energy. The frequency ν of breakup of the activated complex into products is often approximated by $\nu \simeq kT/h$ where k and h are the Boltzmann and Planck constants, respectively. Comparison of Eq. (V) to the rate equation

$$\text{rate} = k[\text{A}][\text{B}]$$

shows that

$$k = \frac{kT}{h}\,(e^{\Delta S^{\circ\ddagger}/R})\,(e^{-\Delta H^{\circ\ddagger}/RT}) \qquad \text{(W)}$$

Again the preexponential factor is seen to be temperature dependent, but for large activation energies, the exponential term dominates the temperature dependence of the rate constant.

The preexponential factor involves the entropy change in going from reactants to the transition state; the more highly ordered and tightly bound is the transition state, the more negatives $\Delta S^{\circ\ddagger}$ will be and the lower the preexponential factor will be. Transition state theory thus automatically takes into account the effect of steric factors on rate constants, in contrast to collision theory.

An alternate form of the rate constant predicted by transition state theory using a statistical mechanical approach for the equilibrium constant $K^{\ddagger}$ is Eq. (X):

$$k = \left(\frac{kT}{h}\right)\left(\frac{q_{\text{AB}^{\ddagger}}}{q_{\text{A}}q_{\text{B}}}\right)(e^{-E^{\ddagger}/RT}) \qquad \text{(X)}$$

Here the q's are partition functions for the reactants (q_{A} and q_{B}) and the transition state ($q_{\text{AB}^{\ddagger}}$), and $E^{\ddagger}$ is the energy difference between the reactants and the transition state.

The partition functions include contributions from translational, rotational, vibrational, and electronic degrees of freedom,

$$q = q_{\text{trans}}q_{\text{rot}}q_{\text{vib}}q_{\text{elect}}$$

Those for the reactants can be evaluated using conventional techniques discussed in physical chemistry texts; estimating the partition functions for the transition state requires making assumptions concerning the nature of the transition state.

Transition state theory can be used to test reaction dynamics on a molecular scale. Thus one can hypothesize a spatial configuration of the atoms in the transition state and from this calculate $\Delta S^{\circ\ddagger}$; the predicted rate constant can then be compared to that observed. If the agreement is not acceptable, the molecular configuration of the transition state can be adjusted until such agreement is obtained. Assuming this molecular configuration approximates the actual form of the intermediate in the reaction, we can learn something about the chemical dynamics of the reaction.

An example of the application of transition state theory to atmospheric reactions is the reaction of OH with CO. As discussed earlier, this reaction is now believed to proceed by the formation of a radical adduct HOCO, which can decompose back to reactants or go on to form the products $H + CO_2$. For complex reactions such as this, transition state theory can be applied to the individual reaction steps, that is, to the steps shown in reaction (19). Figure 4.3 shows schematically the potential energy surface proposed for this reaction; the adduct HOCO, corresponding to a well on the potential energy surface, can either decompose back to reactants via the transition state shown as $HOCO_a^{\ddagger}$, or form products via transition state $HOCO_b^{\ddagger}$. Possible structures for these two transition states are discussed in detail by Mozurkewich et al. (1984).

A second example of the application of transition state theory to reactions of atmospheric interest is the work of Kaufman and co-workers (Jeong and Kaufman, 1982) on the reactions of OH with methane and a series of nine halogen substituted methanes. The agreement, an average, between the experimentally determined A factors (A^{exptl}) and those predicted theoretically (A^{th}) using a 150° C—H—O angle was relatively good, the ratio $A^{\text{th}}/A^{\text{exptl}}$ being 1.14 ± 0.71; of course, in individual cases the discrepancies were frequently much larger.

4-A-7c. *Atmospheric Example of Importance of Temperature Dependence: PAN Decomposition*

As we have seen in Chapter 1, peroxyacetyl nitrate (PAN) is a severe plant phytotoxicant formed in irradiated NMOC–NO_x mixtures from the reaction of peroxyacetyl radicals with NO_2:

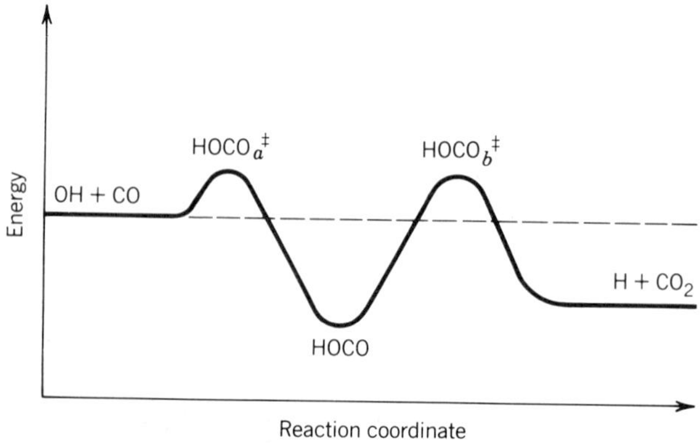

FIGURE 4.3. Typical energy diagram proposed for reaction of OH with CO (from Mozurkewich et al., 1984).

$$\mathrm{CH_3\overset{\overset{\displaystyle O}{\|}}{C}OO + NO_2 \rightarrow \underset{\text{PAN}}{CH_3\overset{\overset{\displaystyle O}{\|}}{C}OONO_2}} \tag{21}$$

PAN is thermally unstable, however, and decomposes at higher temperatures to reform peroxyacetyl radicals and NO_2, that is, the reverse of reaction (21):

$$\mathrm{CH_3\overset{\overset{\displaystyle O}{\|}}{C}OONO_2 \overset{\Delta}{\rightarrow} CH_3\overset{\overset{\displaystyle O}{\|}}{C}OO + NO_2} \tag{-21}$$

The rate constant for PAN decomposition (k_{-21}) is strongly temperature dependent. This is important in the atmosphere because it suggests that PAN may act as a reservoir for NO_2. Thus, as discussed in Section 8-B-1, at low temperatures, much of the total NO_x may be tied up in PAN; as the atmosphere warms, either due to diurnal temperature variations or to the air mass being transported into warmer regions, the rate of PAN decomposition increases. This releases NO_2 which can then form other secondary pollutants such as O_3 and HNO_3.

The rate constant k_{-21} can be fit to the Arrhenius form

$$k_{-21} = 1.95 \times 10^{16}\, e^{-112.6 \text{ kJ mole}^{-1}/RT} \text{ s}^{-1}$$

over the temperature range normally encountered in the troposphere. At room temperature (25°C = 77°F) k_{-21} is 3.6×10^{-4} s^{-1} giving a natural lifetime with respect to decomposition of 46 min. However, at 0°C (32°F), k_{-21} is 5.6×10^{-6} s^{-1}, corresponding to a lifetime of ~50 h. At 35°C (95°F), the rate constant k_{-21} is larger than that at 0°C by almost three orders of magnitude (1.6×10^{-3} s^{-1}), and the lifetime for PAN is correspondingly shorter, only ~11 min! Clearly, the temperature dependence of rate constants can be extremely important in determining the lifetimes and fates of certain species in the atmosphere, as well as their contribution to secondary pollutant formation.

4-B. LABORATORY TECHNIQUES FOR DETERMINING ABSOLUTE RATE CONSTANTS FOR GAS PHASE REACTIONS

In this section we discuss the major experimental methods currently in use to determine *absolute* rate constants for gas phase reactions relevant to atmospheric chemistry. These include fast flow systems (FFS), flash photolysis (FP), molecular modulation and pulse radiolysis techniques, and static reaction systems. The determination of *relative* rate constants is discussed in Section 4-C.

In general, we use simple bimolecular reactions of the type

$$\mathrm{A + B \overset{k_{22}}{\rightarrow} products} \tag{22}$$

TABLE 4.4. Some Reactants of Interest in Tropospheric Chemistry[a]

"A" Highly to Moderately Reactive Atoms, Free Radicals, and Molecules	"B" More Stable Reactants
H	Alkanes
O	Alkenes
OH	Alkynes
HO_2	Aromatics
NO_3	Organics containing a heteroatom (S,N,O)
O_3	NO
$O_2\ (^1\Delta_g)$	NO_2
	SO_2
	"Trace" species

[a] The current status of reactions between these species, A and B, is discussed in detail in Chapters 7, 8, and 11.

as illustrations. However, the techniques can be modified to study termolecular reactions, as discussed earlier, as well as unimolecular reactions. Table 4.4 shows some classes of relatively stable molecules (B) whose reactions with highly to moderately reactive atoms, free radicals, and molecules listed under "A" are of interest in tropospheric chemistry.

In order to study the reaction kinetics of a relatively reactive species A with a second reactant B, one normally follows the loss of small amounts of A in the presence of a great excess of B. This requires then that one be able first to generate A, and second to monitor its concentration as a function of time. Ideally, to fully elucidate the reaction mechanism, one would also monitor the concentrations of intermediates and products. As we shall see, in practice for many reactions this proves to be much more difficult than to simply determine the rate constant itself.

4-B-1. Kinetic Analysis

The rate law for a simple bimolecular reaction such as (22) is given by

$$\frac{-d[\mathrm{A}]}{dt} = k_{22}[\mathrm{A}][\mathrm{B}] \qquad \text{(Y)}$$

If a small concentration of A is generated in a great excess of B, then even if (22) is allowed to go to completion, the concentration of B will remain essentially constant at its initial concentration $[\mathrm{B}]_0$. Integrating (Y) and treating $[\mathrm{B}]_0$ as constant, one obtains

$$\ln \frac{[A]}{[A]_0} = -(k_{22}[B]_0)t \tag{Z}$$

That is, A decays exponentially with time with a rate given by $-k_{22}[B]_0$, that is, as if it were a first-order reaction. Thus under these so-called *pseudo-first-order* conditions, a plot of ln [A] against time for a given value of $[B]_0$ should be linear with a slope equal to $(-k_{22}[B]_0)$. These plots are carried out for a series of concentrations of $[B]_0$ and the values of the corresponding decays determined. Finally, the absolute rate constant of interest, k_{22}, is the slope of a plot of these decay rates against the corresponding values of $[B]_0$. Some examples are discussed below.

As we have seen earlier, even third-order reactions can be reduced to pseudo-first-order reactions by keeping the concentrations of all species except A constant, and in great excess compared to A. This technique of using pseudo-first-order conditions is by far the most common technique for generating rate constants. Not only does it require monitoring only one species, A, as a function of time, but even absolute concentrations of A need not be measured. Because the ratio $[A]/[A]_0$ appears in Eq. (Z) the measurement of any parameter that is *proportional* to the concentration of A will suffice in determining k_{12}, since the proportionality constant between the parameter and [A] cancels out in Eq. (Z). For example, if A absorbs light in a convenient spectral region and Beer's law is obeyed (Section 2-A-2), then the absorbance (Abs) of a given concentration of A, N (number cm^{-3}), is given by

$$\ln \frac{I_0}{I} = \text{Abs} = \sigma N \ell \tag{AA}$$

where I_0 and I are the intensities of the incident and transmitted light respectively, ℓ is the optical pathlength, and σ is the absorption cross section of A (to the base e).

Substituting into Eq. (Z) for $[A] = N = \text{Abs}/\sigma\ell$, one obtains

$$\ln \frac{(\text{Abs}/\sigma\ell)}{(\text{Abs}/\sigma\ell)_0} = \ln \frac{(\text{Abs})}{(\text{Abs})_0} = -k_{22}[B]_0 t \tag{BB}$$

where (Abs) and $(\text{Abs})_0$ are the absorbance of the light by A at times t and $t = 0$, respectively. For example, O_3 has a strong absorption at 254 nm (Section 3-C-2) which can be used to monitor its concentration.

This ability to monitor a parameter that is *proportional* to concentration, rather than the absolute concentration itself, affords a substantial experimental advantage in most kinetic studies, since determining absolute concentrations of atoms and free radicals is difficult in most cases.

This pseudo-first-order kinetic analysis is generally applied regardless of the *experimental* system used. The fundamentals of the most common experimental systems are discussed next.

4-B-2. Fast Flow Systems

4-B-2a. *Basis of Technique*

Fast flow systems (FFS) consist of a flow tube typically 2–5 cm in diameter in which the reactants A and B are mixed in the presence of a large amount of an inert "bath gas" such as He or Ar. As the mixture travels down the flow tube at relatively high linear flow speeds (typically 1000 cm s^{-1}), A and B react. The decay of A along the length of the flow tube is followed and Eq. (Z) applied to obtain the rate constant of interest.

The term *fast flow* comes from the high flow speeds. In most of these systems, discharges are used to generate A or another species that is a precursor to A; hence the term *fast flow discharge system* (FFDS) is also commonly used. Since fast flow discharge systems have been applied in many kinetic and mechanistic studies relevant to tropospheric chemistry (e.g., see Kaufman, 1981, 1984), we concentrate on them. However, all fast flow systems rely on the same experimental and theoretical principles.

Experimentally, two different approaches can be used. In the first (Fig. 4.4*a*), A enters the flow tube at the upstream end and is mixed with B. The decay of A is followed using a detector which moves along the length of the flow tube. In the second approach (Fig. 4.4*b*), one of the reactants enters at the upstream end of the flow tube and the second reactant is added through a movable inlet; in this case the detector is fixed at the downstream end of the flow tube and the reaction time is varied by moving the mixing point for A and B (i.e., the movable inlet) relative to the fixed detector. In a modification of this second approach, a series of fixed inlets along the length of the flow tube can be used for adding the second reactant at various distances upstream of the detector (Fig. 4.4*c*).

Under conditions where the *plug flow* assumption is valid, that is, concentration gradients are negligible so that the linear flow velocity of the carrier gas is the same as that of the reactants, the time (t) for A and B to travel a distance d along the flow tube is given by

$$t = \frac{d}{v} \qquad \text{(CC)}$$

Here v is the linear flow speed which can be calculated from the cross sectional area of the flow tube (A), the total pressure (P) in the flow tube, the temperature (T), and the molar flow rates (dn/dt) of the reactants and the diluent gas:

$$v = \frac{RT}{AP}\frac{dn}{dt} \qquad \text{(DD)}$$

At typical linear flow speeds of 1000 cm s^{-1}, 1 cm along the tube corresponds to 1 ms reaction time. Thus a flow tube of length 1 m can be used to study reactions at reaction times up to 100 ms.

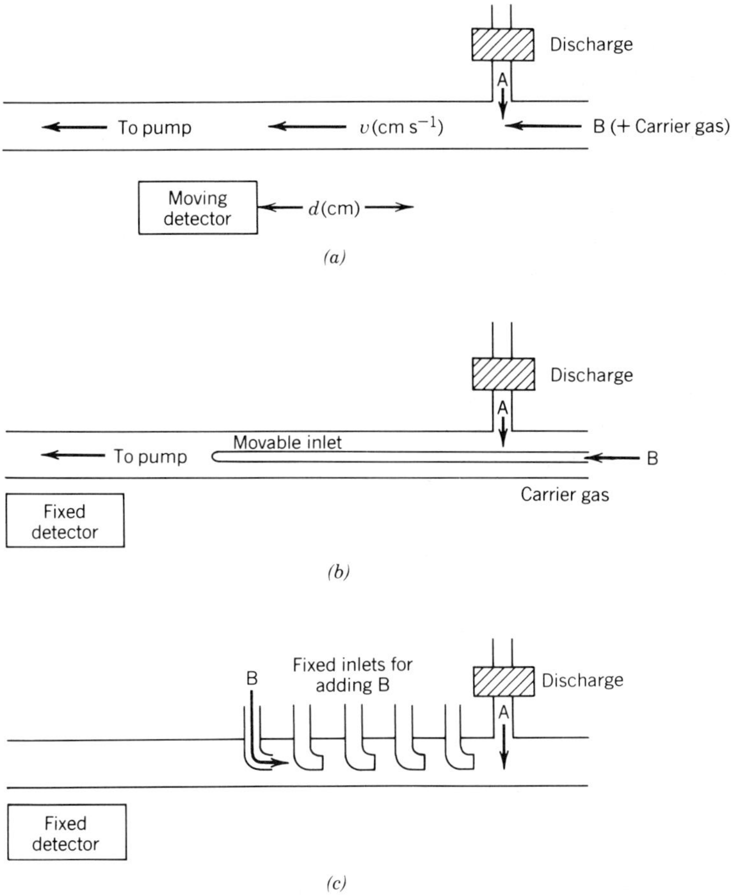

FIGURE 4.4. Schematic diagrams of typical fast flow discharge systems: (*a*) Fixed mixing point for A and B, movable detector; (*b*) moving mixing point for A and B, fixed detector; and (*c*) fixed series of mixing points and fixed detector. v is the linear flow speed of the gases along the flow tube and d is the distance from the detector to the point of mixing of the gases.

The total pressure in FFDS is typically 0.5–10 Torr. The upper end of the pressure range is determined by the need to have relatively rapid diffusion across the flow tube so that the concentration profile of the reactants across the tube is relatively flat, and hence Eq. (CC) is valid. In addition, maintaining the discharge used to generate A is difficult above a few Torr total pressure. Recently, however, flow tube studies have been extended to much higher pressures, up to ~100 Torr (Keyser, 1984) with appropriate corrections for axial and radial diffusion.

The lower end of the pressure range is determined by the need to maintain viscous flow and to avoid significant axial concentration gradients. The latter may arise because of the lower concentrations of reactants at the downstream end of the flow tube; these can cause the true flow speed of the reactants to be greater than the calculated linear flow speed, due to their axial diffusion. Techniques for estimating errors due to such factors and for correcting measured rate constants for them are discussed in detail by Mulcahy (1973), Brown (1978) and Lambert et al. (1985).

A major factor in many FFDS studies is diffusion of the reactive species accompanied by their loss at the walls of the flow tube. Unfortunately, OH radicals are particularly sensitive to removal by wall reactions. While the mechanism and products of these wall reactions are unknown, it has been established that the rate of loss at the walls can be minimized by using various flow tube wall coatings or treatments. These include substances such as halocarbon waxes, which simply cover the entire surface so the incoming reactive species are only exposed to relatively unreactive carbon–halogen bonds, or treatment with boric or phosphoric acids. While such treatments have been shown to lower the rates of removal at the walls, *why* they do so is unknown.

Fortunately, the kinetics of the wall loss, measured from the decay of the reactive species in the absence of added reactant, are generally observed to be first order, so that corrections for these processes can be made. When these wall losses are significant, the integrated form of the rate expression (Z) becomes

$$\ln \frac{[A]}{[A]_0} = -(k_{22}[B]_0 + k_w)t \qquad \text{(EE)}$$

where k_w is the observed loss of A at the walls of the flow tube in the absence of B. The rate constant k_{22} can then be extracted from the slopes of plots of the pseudo-first-order rates of decay, $R = (k_{22}[B]_0 + k_w)$, against $[B]_0$.

An example is shown in Figs. 4.5 and 4.6 for the for the reaction (18) of OH with CO. Figure 4.5 shows the decay of OH resonance fluorescence emission intensity (proportional to the OH concentration) as a function of reaction time in a fast flow discharge system (see below) at ~1 Torr total pressure as the concentration of CO is increased from 0 to 5.25×10^{14} molecules cm^{-3}. As expected from Eq. (EE), the absolute value of the slope of the lines increases as $[CO]_0$ increases. Figure 4.6 shows the plot of the absolute values of the slopes (S_1) in Fig. 4.5 against $[CO]_0$. The slope of this plot gives the rate constant k_{18} under these conditions, $k_{18} = 1.43 \times 10^{-13}$ cm^3 molecule^{-1} s^{-1}, in good agreement with low-pressure values of Table 4.2. The non-zero decay of OH when the CO concentration is zero is due to loss of OH at the walls of the flow tube. From the decay of OH in the absence of CO (Fig. 4.6) and the intercept in Fig. 4.6 at $[CO]_0 = 0$, k_w is found to be 14.1 s^{-1}, typical for such flow tubes whose walls have been treated to minimize the wall loss.

These wall reactions can be a problem in FFDS studies. To avoid unrecognized interferences in the data associated with these heterogeneous reactions, as well as other secondary reactions, it is generally recommended that flow tube studies of a particular reaction be carried out using as many different wall coatings as possible. In addition, the use of different carrier gases and flow tubes of different diameters is recommended. The nature of the carrier gas affects the rates of diffusion of the

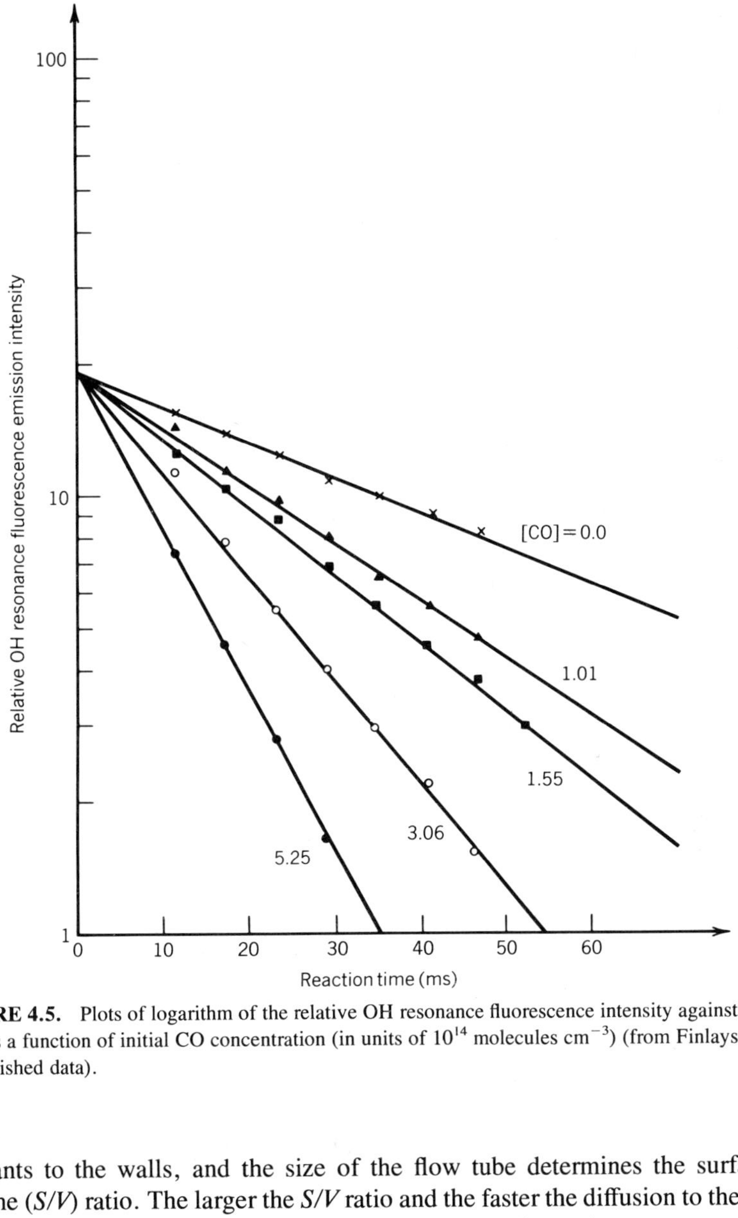

FIGURE 4.5. Plots of logarithm of the relative OH resonance fluorescence intensity against reaction time as a function of initial CO concentration (in units of 10^{14} molecules cm^{-3}) (from Finlayson-Pitts, unpublished data).

reactants to the walls, and the size of the flow tube determines the surface-to-volume (S/V) ratio. The larger the S/V ratio and the faster the diffusion to the walls, the greater the interference one might expect from wall reactions.

4-B-2b. *Methods of Generation of Atoms and Free Radicals*

As discussed earlier, discharges are usually used in FFS to generate the highly reactive species A, either directly or indirectly. The discharges are usually electrodeless (e.g., microwave discharges).

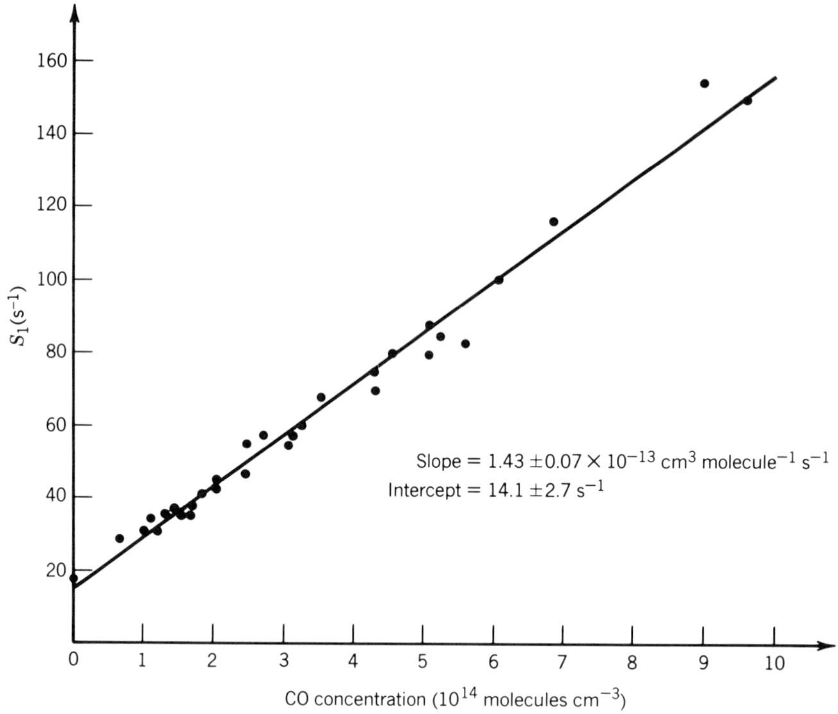

FIGURE 4.6. Absolute values of the slopes (S_1) of the lines in Fig. 4.5 as a function of initial CO concentration. Slope gives the rate constant (1.43×10^{-13} cm^3 molecule^{-1} s^{-1}), and the intercept gives the wall loss of OH (from Finlayson-Pitts, unpublished data).

For atoms such as hydrogen, oxygen, and nitrogen, discharging the corresponding stable diatomic molecule (i.e., H_2, O_2, or N_2) produces a mixture of the atomic species and the undissociated parent molecule. To produce the small concentrations of atoms usually desired, a dilute mixture of the parent molecule in an inert gas such as helium is discharged. Some caution must be exercised, however, because excited states of the parent molecule are also frequently produced and these may cause unrecognized interferences. For example, discharging O_2 produces not only oxygen atoms but also electronically excited O_2 (e.g., the $^1\Delta_g$ and $^1\Sigma_g^+$ states), and discharging H_2 produces vibrationally excited H_2 as well as hydrogen atoms.

In addition, any impurities in the parent molecule gas will also be discharged. For example, even if ultra-high-purity H_2 is used and is passed through cold traps at liquid nitrogen temperature prior to entering the discharge, small concentrations of oxygen impurities remain. Thus when the H_2 is discharged, small concentrations of oxygen atoms are produced in addition to hydrogen atoms; in some systems this can cause problems.

Some species of interest, for example, OH, cannot be cleanly generated by a discharge of a single precursor parent molecule. In this case, titration reactions can be used to generate the reactant. For example, for OH and HO_2, the following titrations are frequently employed:

$$H + NO_2 \rightarrow OH + NO \tag{23}$$

$$H + O_2 + M \rightarrow HO_2 + M \tag{24}$$

Other reactions used to generate OH and HO_2 include, for example,

$$F + H_2O \rightarrow HF + OH$$

$$F + H_2O_2 \rightarrow HF + HO_2$$

Similarly, NO_3 can be generated from the reaction of F atoms with HNO_3:

$$F + HNO_3 \rightarrow HF + NO_3$$

The atoms are generated from a discharge of the parent diatomic, e.g., H from H_2 and F from F_2. Ground-state oxygen atoms, $O(^3P)$, can be generated by titrating nitrogen atoms with NO:

$$N + NO \rightarrow N_2 + O(^3P) \tag{25}$$

As discussed in more detail below, reaction (25) is commonly used to generate known concentrations of ground-state oxygen atoms since, in excess atomic nitrogen, the concentration of $O(^3P)$ is just equal to the initial NO concentration. For example, reaction (25) was used to generate known concentrations of $O(^3P)$ to calibrate a resonance fluorescence lamp (see below). This lamp was then used to quantitatively determine the yields of $O(^3P)$ in the H + O_3 system (Finlayson-Pitts et al., 1981), which is important in the mesosphere.

Some reactive species can be generated in fast flow systems using more specialized and specific techniques. For example, if H_2 is passed over a hot tungsten wire, some of it dissociates to produce H atoms. This technique has the advantage that impurities such as O_2 and N_2 are not dissociated simutaneously, as in a discharge, and small concentrations of hydrogen atoms can be reproducibly generated. On the other hand, it cannot be used to produce large concentrations of atoms which are sometimes desired. This technique is also useful for halogen atoms. Howard and Evenson (1977), for example, in their studies of the HO_2 + NO reaction (see below) used a hot tungsten wire to generate hydrogen atoms from H_2 and then generated HO_2 via reaction (24).

A recent technique used to generate free radicals in flow systems involves the multi-photon-induced decomposition (MPD) of relatively large parent molecules. In this technique, an infrared laser (commonly a CO_2 laser) is used to pump the parent molecule to such high vibrational–rotational levels that it dissociates into free radicals. For example, Gutman and co-workers (Yamada et al., 1981) have used the MPD of $C_6F_5OCH_3$ as a source of methyl radicals to study the reaction of CH_3 with NO_2:

$$C_6H_5OCH_3 + nh\nu \rightarrow C_6H_5O + CH_3 \quad (26)$$

Similarly, MPD of allyl bromide,

$$C_3H_5Br + nh\nu \rightarrow C_3H_5 + Br \quad (27)$$

was used to generate allyl radicals (C_3H_5) to study their reactions (Slagle et al., 1981).

4-B-2c. Methods of Detection of Atoms, Free Radicals, and Stable Molecules

A number of different methods of monitoring the loss of the reactive species have been applied in FFDS. Methods commonly used include (1) optical absorption, (2) fluorescence, (3) electron spin resonance, (4) laser magnetic resonance, (5) chemiluminescence, (6) chemical titration, and (7) mass spectrometry. The basis of each technique and some typical applications are described below.

Optical Absorption. If the reactive species absorbs light in a convenient spectral region, the fraction of the incident light which is absorbed may be used to monitor the species A by application of Beer's law. As discussed earlier, only the absorbance need be measured since under the appropriate conditions this is proportional to [A]. Many reactive species of atmospheric interest have optical absorptions in convenient spectral regions; OH, for example, absorbs light at ~306 nm to generate its first electronically excited state, and $O(^3P)$ absorbs at 130.5 nm.

However, in flow tubes the absorbing pathlength is limited by the diameter of the flow tube, typically 2–5 cm, although this can be extended somewhat by multiple pass optics. In addition, the absorption coefficients, especially for molecules, are often low. Finally, in measuring absorbance at small concentrations, one is typically measuring the difference between two large numbers, that is, between the incident light intensity I_0 and the light intensity I after passing through the absorber; accurately measuring such small differences is difficult. As a result, the sensitivity of this technique is usually relatively low. Thus it is less commonly used in kinetic studies if more sensitive methods are available (see below).

The lamps used to produce the radiation can be either continuous sources or line sources; they are discussed in detail by Clyne and Nip (1979).

Fluorescence. Many atoms and molecules of atmospheric interest absorb light in the visible and ultraviolet regions to produce electronically excited species (see Chapters 2 and 3). Some of these electronically excited states are very short-lived and rapidly emit fluorescence as they return to the ground state. In theory then, one can excite the molecules using a beam of radiation of the appropriate wavelength which is focussed in one direction. Since the excited molecules emit the light in all directions, one can monitor the emission in a direction perpendicular to the exciting radiation, thereby monitoring only the re-emitted light and not the exciting radiation itself. In practice, it is impossible to avoid detecting some scattered light from the exciting source, but this can be subtracted from the total signal to obtain that due only to the excited species of interest.

Figure 4.7 shows the potential energy curves for the two lowest electronic states of the important atmospheric species, the hydroxyl radical, which can be monitored using this technique of induced fluorescence (IF). Thus, from its ground state, OH can be excited into the three lowest vibrational energy levels ($v' = 0$, 1, 2) of the first electronically excited state ($A^2\Sigma^+$). [At higher levels the molecule predissociates into $O(^3P) + H(^2S)$.] The excited molecule then emits radiation and returns to the ground state. The intensity of this emission is proportional to the concentration of the excited state and thus to that of the ground state from which it was formed. (Molecules formed in the $v' = 1$, 2 levels of the $A^2\Sigma^+$ state may relax to the lower vibrational levels before fluorescing.) This technique was used to monitor OH in the OH + CO study shown in Figs. 4.5 and 4.6.

Table 4.5 shows some of the species of atmospheric interest which have been measured using IF, and the wavelengths used.

Induced fluorescence is intrinsically much more sensitive than absorption spectrometry because low fluorescence intensities are easily measured if the scattered radiation from the exciting lamp is minimized. This is experimentally much easier than measuring the very small fraction of the exciting radiation *absorbed* by the atoms or molecules.

IF can be made even more sensitive by the use of lasers as the exciting source. As discussed in Chapter 5, one of the first techniques for measuring OH in ambient air and in environmental chambers involved IF where the OH is excited using a

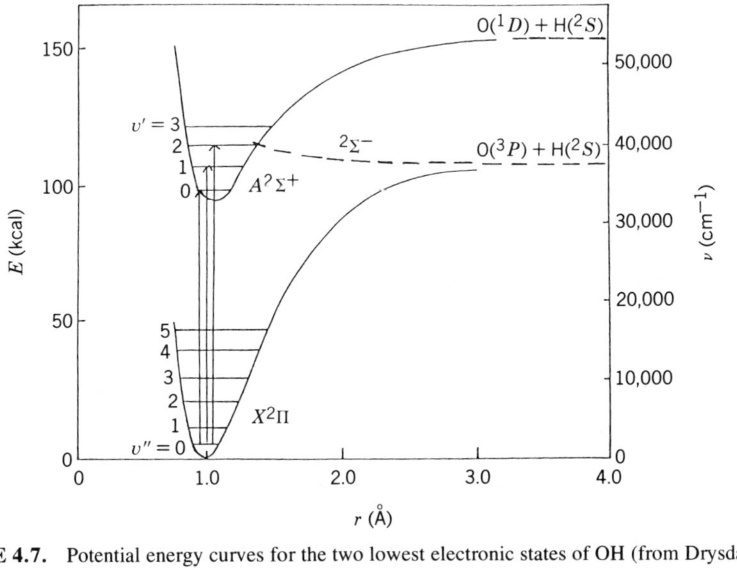

FIGURE 4.7. Potential energy curves for the two lowest electronic states of OH (from Drysdale and Lloyd, 1970).

TABLE 4.5. Some Atoms and Molecules of Atmospheric Interest Measured by Induced Fluorescence and the Wavelengths Used

Atom or Molecule	Wavelength (nm)
OH	306.4
O (^3P)	130.5
H (^2S)	121.6
Cl	139.0
CH_3O	303.9
C_2H_5O	322.8
NO_3	662.9

frequency-doubled dye laser. This *laser-induced fluorescence* (LIF) technique excites a greater proportion of the ground-state molecules due to the high incident radiation intensity; the emitted fluorescence intensity is correspondingly greater. As discussed in Chapter 5, this can present problems because the exciting beam may also induce some undesired interferences due to light absorption by other species present. For example, O_3 also absorbs in the same region as OH (see Section 3-C-2); at wavelengths less than ~320 nm electronically excited oxygen atoms, $O(^1D)$, are formed:

$$O_3 + h\nu\ (\lambda < 320\text{ nm}) \rightarrow O_2 + O(^1D) \qquad (28)$$

These excited oxygen atoms react with water vapor, producing OH:

$$O(^1D) + H_2O \rightarrow 2\ OH \qquad (29)$$

Thus the use of intense exciting radiation can actually generate the species one is trying to measure—a sort of atmospheric uncertainty principle!

Simple alkoxy radicals have been observed to undergo LIF and this has been used to study their reaction kinetics. For example, the LIF of CH_3O and C_2H_5O at 303.9 and 322.8 nm, respectively, has been used to monitor the decay of these species in flash photolysis–resonance fluorescence (FP–RF) studies (see below).

In most laboratory studies, non-laser sources of exciting radiation are used for IF because they are relatively inexpensive, easily constructed, versatile, and provide sufficient sensitivity for most kinetic studies. A microwave discharge of the parent molecule as a dilute mixture in an inert gas is typically used as the exciting lamp. For example, for H atom detection, a discharge containing H_2 produces the desired Lyman α radiation at 121.6 nm whereas a discharge in O_2 produces O atom radiation at ~130.5 nm. For OH, a discharge of a water-saturated inert gas is used. The experimenter must often be careful to use appropriately dilute mixtures of the parent molecule. Otherwise, such high atom concentrations may be produced in the lamp that *self-reversal* occurs, that is, the desired radiation emitted by excited atoms in the lamp is absorbed by ground-state atoms in the lamp before the light exits the lamp.

Frequently, the small concentrations of impurities present in inert gases such as He produce sufficient radiation to be used as the lamp for IF and yet are small enough to avoid the problem of self-reversal. Figure 4.8, for example, is a vacuum ultraviolet spectrum, taken at low resolution, of the emission from discharged He at ~1 Torr pressure which had been passed through a glass-wool-packed trap immersed in liquid nitrogen (77°K). Clearly, enough impurities passed through the trap to give emission lines due to O, H, and N. The carbon atom lines are probably due to a small amount of pump oil that back-diffused. A typical configuration for an IF lamp is shown in Fig. 4.9.

Using such lamps, which are easily and inexpensively constructed, one can obtain relatively high sensitivities. For OH, $O(^3P)$, and $H(^2S)$, for example, typical sensitivities obtained are $\lesssim 10^{10}$, 10^9, and 10^{10} cm^{-3}, respectively.

Electron Spin Resonance. The technique of electron spin resonance (ESR) or electron paramagnetic resonance (EPR) has been in use for decades to detect atoms and molecular species having an unpaired electron. When a magnetic field is applied to the paramagnetic species (i.e., one having an unpaired electron), the energy levels corresponding to the two possible spin orientations of the electron with respect to the magnetic field are split. The difference in energy between the two spin states (ΔE) is proportional to the strength of the applied magnetic field, H, so that the splitting increases as H increases. A transition can be induced between these two states by the absorption of electromagnetic radiation (EMR) which, for typical magnetic field strengths, lies in the microwave region of the spectrum.

While in theory a constant magnetic field strength could be applied and the frequency of the EMR varied until an absorption was observed, in practice, the

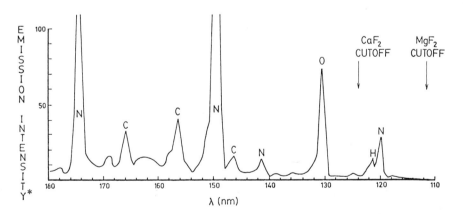

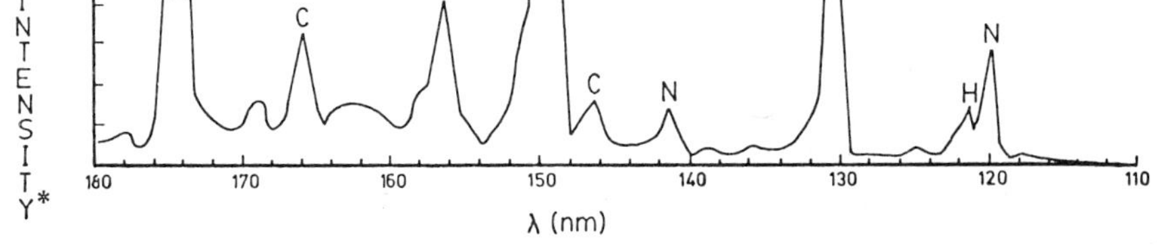

*(Uncorrected for Spectral Response)

FIGURE 4.8. Vacuum-ultraviolet spectral emission (uncorrected for spectral response) from discharged He recorded at low resolution. Atoms responsible for the emissions are identified as are the transmission limits of two materials commonly used for windows in this spectral region (from Finlayson-Pitts, unpublished data).

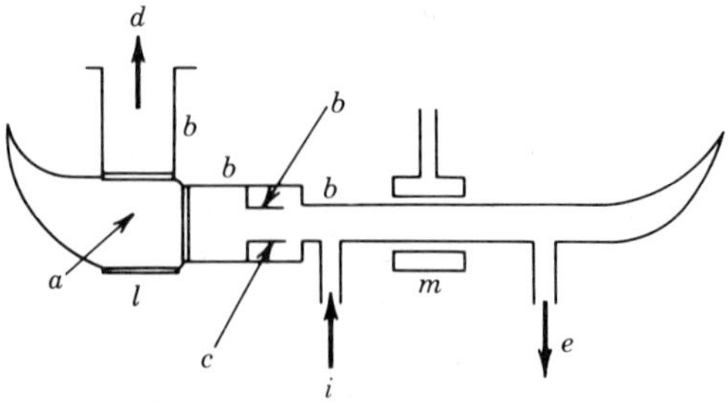

FIGURE 4.9. Typical lamp used in resonance fluorescence studies in FFDS. *a*, axis of flow tube; *b*, areas internally blackened; *c*, collimating tube; *d*, to monochromator and detection system; *e*, lamp exhaust; *i*, lamp inlet; *m*, microwave cavity, *l*, LiF window (from Clyne and Nip, 1979).

EMR is held fixed (typically at ~9–10 GHz) while the magnetic field strength (typically 3–4 kG) is varied.

The advantages of EPR include its versatility, since it can be used to detect any paramagnetic species; thus a number of different reactants, intermediates, and products can often be monitored in the same experiment. In addition, absolute concentrations of these species can be obtained by comparison of the transition intensities to those of known standards such as NO and O_2. For example, in one of the earliest studies of the OH + CO reaction, Dixon-Lewis and co-workers (1966) monitored the decay of OH by ESR in a great excess of CO; concentrations of OH and the H atoms used to generate it were determined by comparing their signal intensities to that from pure NO added in separate studies.

The major disadvantages are that it is less sensitive than other techniques (e.g., IF) and is limited to paramagnetic species. The sensitivity for $O(^3P)$, for example, is $\sim 10^{12}$ cm^{-3} compared to $\sim 10^9$ cm^{-3} for IF. Using lasers, concentrations of OH down to $\sim 10^6$ cm^{-3} can be detected by LIF (laser-induced fluorescence), whereas the detection limit for OH by EPR is $\sim 10^{10}$ cm^{-3}.

Laser Magnetic Resonance. A variation on the principle of EPR is used in laser magnetic resonance (LMR), which was developed in the late 1960s by Evenson and his colleagues. A recent review is given by Davies (1981). Indeed, the technique is sufficiently similar in principle to EPR that LMR was first known by the name *laser electron paramagnetic resonance spectroscopy*. Like EPR, LMR is based on the Zeeman effect, that is, on the shifts in energy levels induced by the presence of a magnetic field.

Figure 4.10 illustrates the principle of LMR for a molecule having two rotational levels with quantum numbers $J'' = \frac{3}{2}$ and $J' = \frac{5}{2}$, respectively. These states, which are non-degenerate in the absence of an applied magnetic field, split into their sublevels (shown under m_J) when a magnetic field is applied. As in the case of EPR, the magnitude of the splitting increases with the magnetic field strength.

If one were to try to detect the molecule by its absorption of electromagnetic radiation in going from the lower state $J'' = \frac{3}{2}$ to the upper state $J' = \frac{5}{2}$ in the absence of a magnetic field, one would need a source of radiation whose frequency ν was exactly

$$\nu = \frac{\Delta E}{h} = \frac{E'_J - E''_J}{h}$$

Suppose this frequency was not available, but there was a frequency, corresponding to energy E_L, close (within ~0.2–2 cm^{-1}) to ν. As long as the Zeeman shifts of the two levels are different, one could then "tune" the molecular energy levels by increasing the applied magnetic field strength until the transition matched the laser frequency ν and energy E_L.

In Fig. 4.10, for example, as the magnetic field strength is increased, two allowed transitions come into resonance with the applied frequency of the radiation; hence absorption can occur. The magnetic field strengths at which these absorptions occur are determined by the energy levels and hence by the specific molecule. This spectroscopic technique can thus be used to identify paramagnetic species, as well as to monitor their concentration using the intensity of the absorption.

The rotational energy levels involved in the transition may belong to the same or different vibrational energy levels. LMR may also be used detect and measure paramagnetic atoms; in this case the transitions involved are those between spin–orbit states (e.g., the $^3P_2$ and $^3P_1$ states for ground-state oxygen atoms).

The biggest advantage of LMR compared to EPR is its sensitivity. The sensi-

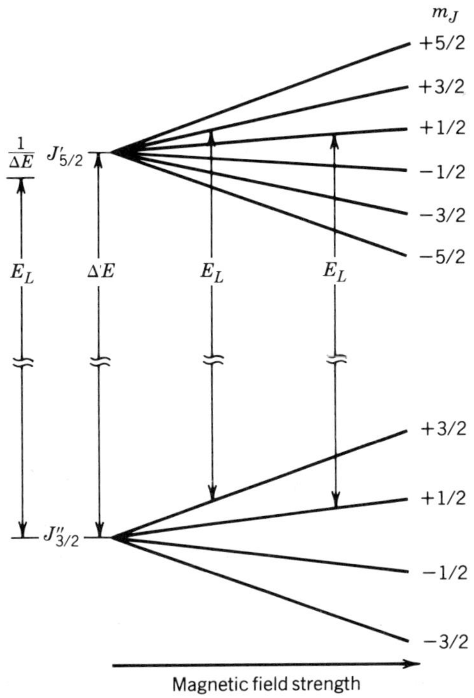

FIGURE 4.10. Rotational energy levels and sublevels which are split in the presence of a magnetic field. Typical LMR transitions are indicated (from Howard and Evenson, 1974).

tivity increases roughly in proportion to either ν or ν^2. Since the absorbed frequencies in LMR are in the infrared ($\gtrsim 30\ cm^{-1}$), whereas in EPR they are in the microwave region ($\sim 0.3\ cm^{-1}$), a gain in LMR sensitivity of a factor of $\sim 10^2$–10^4 is obtained with LMR. For example, the detection limit for OH is $\sim 10^{10}\ cm^{-3}$ by EPR but $\sim 10^8\ cm^{-3}$ by LMR. The high LMR sensitivity for HO_2 ($\sim 10^8\ cm^{-3}$) has proven to be especially valuable (Radford et al., 1974) in studying the kinetics of its reactions since there were no other techniques which were both highly sensitive *and* specific for HO_2 when its importance in atmospheric chemistry became recognized.

An additional advantage of LMR is that saturation effects (i.e., large populations in the upper state compared to the lower state) are less severe because of the larger energy splitting between the levels involved in LMR compared to EPR.

For example, LMR was used by Howard and Evenson (1977) to study the reaction of HO_2 with NO

$$HO_2 + NO \rightarrow OH + NO_2 \tag{30}$$

a key reaction in both the troposphere and stratosphere. The rate constant for the reaction was found in these studies to be much greater ($8.1 \times 10^{-12}\ cm^3\ molecule^{-1}\ s^{-1}$ at 296°K) than had been reported earlier ($\leq 10^{-12}\ cm^3\ molecule^{-1}\ s^{-1}$). This revision in the rate constant had a significant impact on the predictions of computer models of both the troposphere (Carter et al., 1979) and stratosphere (Crutzen and Howard, 1978). Indeed, the change in this one rate constant changed the predicted loss of O_3 in the stratosphere due to NO_x emissions (e.g., from supersonic transports, SSTs) to such an extent that at low-altitudes of NO_x injection, under some conditions, an *increase* in total O_3 rather than a loss was even projected!

The accuracy of the work of Howard and Evenson, which has since been confirmed in a number of laboratories, was possible because of the specificity and sensitivity of LMR for HO_2.

The limitations of LMR arise from the fact that, with conventional magnetic field strengths, the splitting of the levels is relatively small so that the laser frequency must closely match ($\lesssim 2\ cm^{-1}$) the energy level gap in the absence of the magnetic field. This limits the number of species which can be detected using LMR. A second limitation is that the sensitivity for atoms is much poorer than that obtained using other techniques (e.g., IF) so that it is not especially useful for atomic species.

Chemiluminescence. A number of chemical reactions generate products not in their lowest energy states, but rather in upper levels. That is, some of the exothermicity of the reaction is channeled internally into electronic (E), vibrational (V), or rotational (R) energy of one or more of the products, rather than being released as heat. The excited product molecules may emit this energy as light, known as *chemiluminescence* because of the chemical source of the energy.

The earliest kinetic and mechanistic studies in FFDS relied on chemiluminescence to detect and measure a number of species. Oxygen atoms were detected most commonly through their reaction with NO, which produces electronically excited NO_2:

$$\begin{aligned} O(^3P) + NO + M \rightarrow\ & NO_2^* + M \\ & \downarrow \\ & NO_2 + h\nu \end{aligned} \tag{31}$$

The NO_2^* emission extends throughout the visible into the infrared, with the peak intensity occurring at ~630 nm. As a result, it can easily be monitored using conventional photomultipliers sensitive in the visible region of the spectrum. Since the intensity of the NO_2^* emission is proportional to [O] and [NO] at pressures $\gtrsim$ 1 Torr, as long as the concentration of NO is constant, oxygen atoms can be monitored quantitatively by following the emission intensity. The sensitivity of this detection technique for $O(^3P)$ is surprisingly high, $\sim 10^{10}$ cm^{-3}.

As discussed earlier, oxygen atoms are frequently generated by the titration of nitrogen atoms with NO:

$$N(^4S) + NO \rightarrow N_2 + O(^3P) \tag{25}$$

If one's eyes are dark-adapted, a discharge of N_2 produces a long-lived yellow afterglow due to emissions from electronically excited states of N_2 produced on recombination of nitrogen atoms. When small amounts of NO are added, reaction (25) occurs and produces oxygen atoms. This diminishes the intensity of the so-called *active nitrogen* emissions and produces blue chemiluminescence from electronically excited NO* formed by reaction (32):

$$O + N + M \rightarrow NO^* + M \tag{32}$$

The color of the gases thus changes from bright yellow to a yellowish-blue. When NO is added to the endpoint (i.e., when [N] = [NO]), only N_2 and O are present and no visible emissions result. As more NO is added, the excess NO reacts with O via reaction (31), producing the yellowish-green NO_2^* emission. Thus one can follow the course of this titration quite easily with the naked eye; indeed, for the uninitiated this sequence provides quite a spectacular demonstration of FFDS and titration reactions!

Hydrogen atoms can also be measured using a chemiluminescent reaction with NO:

$$\begin{aligned} H + NO + M \rightarrow\ & HNO^* + M \\ & \downarrow \\ & HNO + h\nu \end{aligned} \tag{33}$$

The HNO* emission falls in the red to infrared regions and thus can also be measured quantitatively using conventional photomultipliers. For example, Clyne and Stedman (1966) used the HNO emission intensity to study the H + NOCl reaction:

$$H + NOCl \rightarrow HCl + NO$$

In excess H, that is, when [H] > [NOCl], the excess H reacted with the product NO to produce the HNO* emission. When [H] = [NOCl], the emission intensity was zero, and the concentration of hydrogen atoms could be obtained from the measured NOCl flow.

While these chemiluminescent reactions were historically very important in FFDS studies, they are not used extensively today in such laboratory studies. (They are, however, used as the basis of several methods for monitoring pollutants in ambient air; see Chapter 5-B-1.) Thus, as we have seen, more sensitive and specific techniques have generally supplanted chemiluminescence. In the latter case, most of the light emissions are broad (i.e., cover a relatively large spectral region) and thus are subject to interferences from other emissions. In addition, these techniques generally require adding a reactive species such as NO in order to generate the chemiluminescence, and these may react further with other species present in the system to cause interferences.

Titration. Some species for which no suitable specific and sensitive detection technique is available can be monitored by reacting them to form a second species for which such a technique is available. An example of atmospheric importance is the detection of HO_2 by titration with NO to form OH:

$$HO_2 + NO \rightarrow OH + NO_2 \qquad (30)$$

The OH is then detected by one of the techniques (e.g., IF) discussed above. For example, Kaufman and co-workers (Kaufman, 1981) have used this titration to monitor HO_2 in studying its reactions using fast flow techniques. While this avoids very expensive instrumentation as is needed for LMR, for example, some care must be exercised in interpretation of the data to be sure that any increased OH observed is truly due to reaction (30). For example, many systems in which HO_2 is present also contain OH initially. If some of the OH is in excited states (e.g., upper vibrational levels), the addition of NO can lead to increased ground-state OH via physical deactivation of the excited species. The increase in ground-state OH measured upon addition of NO can be misinterpreted as due to reaction (30), especially if the presence of the vibrationally excited OH is not recognized.

Thus while titration reactions may provide a means of monitoring certain species that cannot otherwise be followed, great care must be taken to understand the detailed chemistry of the system before the application of such techniques.

Mass Spectrometry. Conventional mass spectrometers can be coupled to FFDS and used to monitor a variety of species, particularly the more stable molecules. Detecting and monitoring atoms and free radicals is much more difficult because of the problems of sampling, as well as of fragmentation of parent molecules during electron impact excitation.

Sampling problems have been reduced through the use of multi-stage systems in which the gases are first sampled from the flow tube into a vacuum chamber operating at pressures of $\sim 10^{-4}$–10^{-5} Torr. This reduces the collisions and hence further reactions of the sampled species to negligible levels. The central portion of the beam entering this first chamber is then sampled into a second vacuum chamber in which the mass spectrometer resides, operated at $\lesssim 10^{-6}$ Torr.

Figure 4.11 is a schematic diagram of a typical fast flow discharge system which is interfaced to an electron impact mass spectrometer with a two-stage sampling system. In this particular case, the beam of molecules entering the second vacuum chamber housing the quadrupole MS unit is chopped by a tuning fork and phase sensitive detection of the ions is used to enhance the signal-to-noise ratio. The flow system is equipped with ports for IF detection of atoms and free radicals and has an envelope around the flow tube to circulate heated or cooled liquids in order to determine temperature dependencies of kinetics and mechanisms.

The problem of detecting atoms and free radicals due to interferences from fragments produced upon electron impact of a stable molecule can be reduced by lowering the ionization energy. Unfortunately, this also lowers the sensitivity.

One technique that minimizes the problems of interference from fragmentation in detecting free radicals is that of photoionization mass spectrometry (PIMS). This technique uses photons instead of electron impact to cause ionization. The wavelength and hence energy of the ionizing photon can be chosen so that it is just sufficient to ionize the free radical of interest, but not large enough to ionize or fragment larger species in the system. This technique, whose kinetic application was pioneered by Bayes and co-workers (Jones and Bayes, 1971) and subsequently by Gutman and co-workers (Kanofsky and Gutman, 1972), allows one to detect small concentrations of free radicals in the presence of large concentrations of the parent compound.

Figure 4.12 shows the energies of the photons produced by some common PIMS lamps and the ionization potentials of some stable organic molecules, free radicals, and inorganics. For example, if one wanted to detect the formyl radical , HCO, produced by a reaction of formaldehyde (HCHO), one could use the Kr lines at 10.03 and 10.64 eV; from Fig. 4.12 it is seen that these energies are sufficient to ionize HCO but not HCHO. Detecting small amounts of the formyl radical in the presence of large amounts of the parent HCHO is extremely difficult using conventional MS since the stable parent molecule fragments under electron impact to give formyl radicals. For example, this technique was used to establish conclusively for the first time that O_3 reactions with olefins generate free radicals (Atkinson et al., 1973) such as HCO, C_2H_5, HO_2, and CH_3O_2 in the case of *cis*-2-butene (see Section 7-E-2). This could not have been done with conventional MS since stable products of the O_3–C_4H_8 reaction could fragment in the MS to give these radicals.

While PIMS clearly presents significant advantages for identifying and monitoring free radicals, it suffers the disadvantages that the ionizing efficiency and

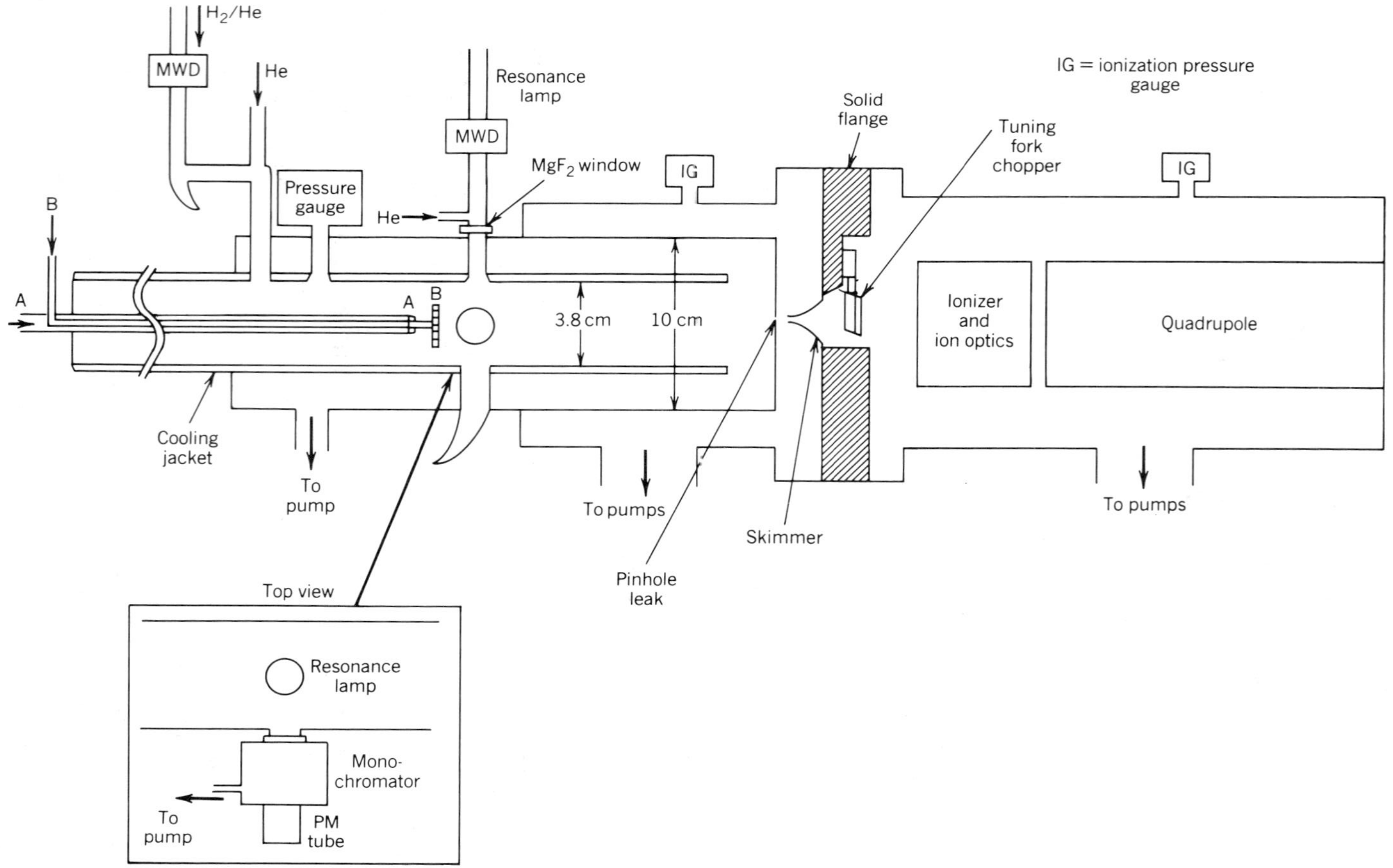

FIGURE 4.11. Schematic diagram of a typical fast flow discharge system interfaced to a mass spectrometer and equipped with resonance fluorescence detection (from Finlayson-Pitts, unpublished data).

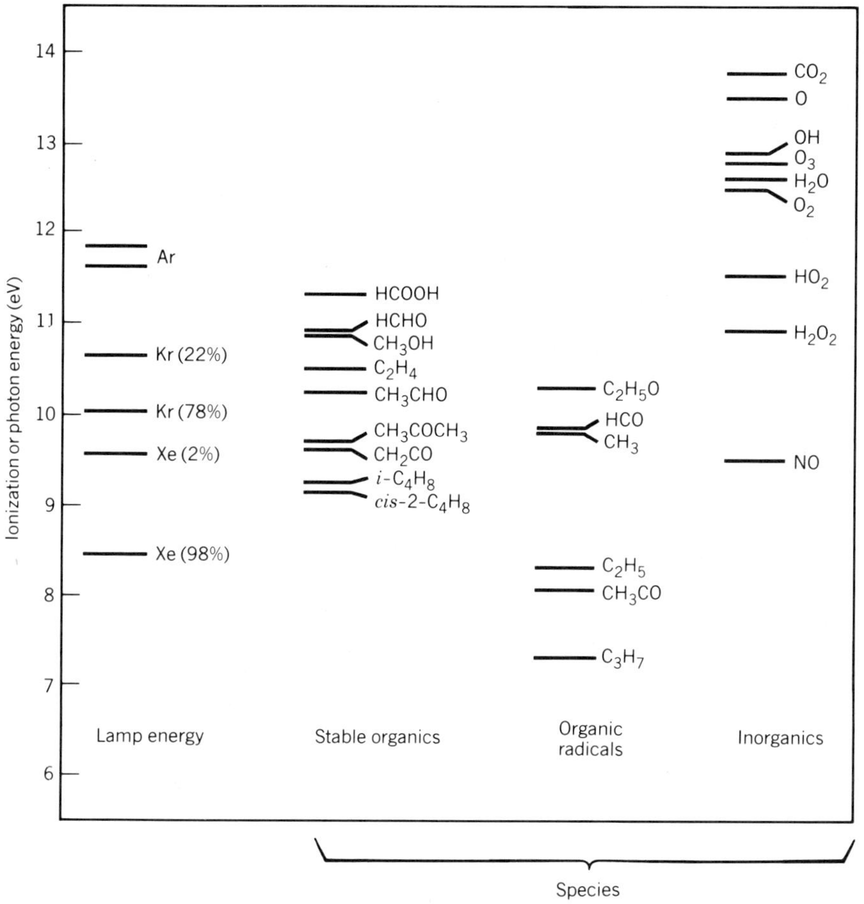

FIGURE 4.12. Energies of photons emitted by some commonly used lamps in photoionization mass spectrometry and ionization potentials of some species of interest (from Atkinson et al., 1973).

hence sensitivity is not as great as with electron impact, and calibrating to obtain absolute radical concentrations is difficult for many species of interest. In addition, although fragmentation is minimized with PIMS, it still occurs to a small extent in some cases, making data interpretation difficult.

4-B-3. Flash Photolysis Systems

4-B-3a. Basis of Technique

As the name implies, this technique relies on flash photolysis to generate the reactive species A. In the most common configuration, resonance or induced fluorescence is used to monitor the decay of A, hence the name *flash photolysis–resonance fluorescence* (FP–RF).

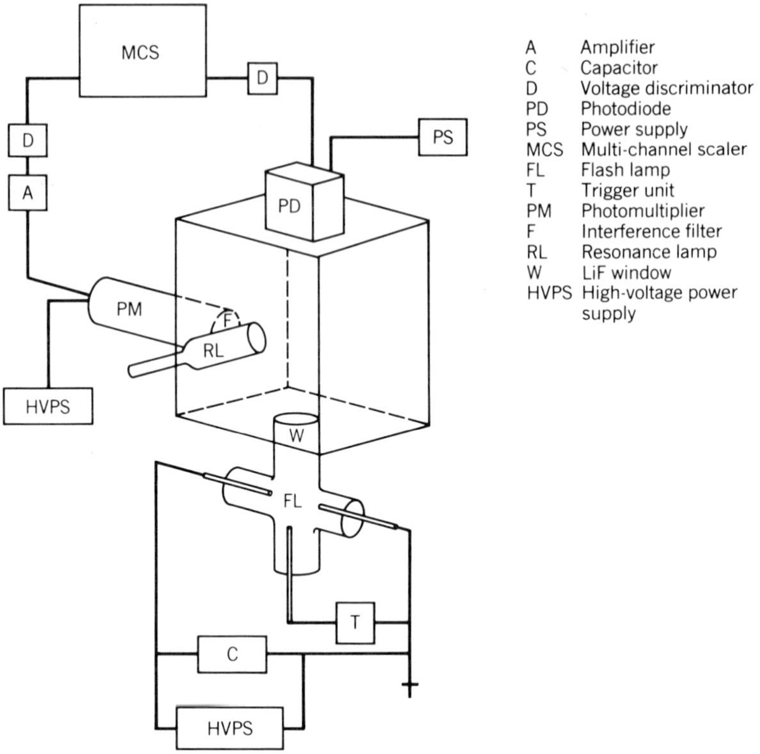

FIGURE 4.13. Schematic diagram of a flash photolysis–resonance fluorescence system (from Atkinson et al., 1979).

Figure 4.13 is a schematic diagram of a typical FP–RF apparatus. The flash lamp is triggered and the pulse of light emitted dissociates a parent molecule, producing the reactive species, A, of interest. After a preset time following the photolytic flash, the time decay of the reactive species is monitored using IF excited by a resonance lamp of the type described above for FFDS. As seen in Fig. 4.13, the IF signal is monitored in a direction perpendicular to both the flash lamp and the resonance lamp.

As in fast flow systems, the concentration of the stable reactant B is determined from its flow rate; thus, using the ideal gas law, the partial pressure of B divided by the total pressure is equal to the flow rate of B divided by the total flow rate (i.e., the sum of all flow rates). Since B is present in concentrations in great excess compared to A, care must be taken to avoid impurities that may react with A or photolyze to produce reactive species that do. A restriction on the nature of B is that it must not photolyze significantly itself. This means that reactions of such important pollutants as NO_2 and O_3 which dissociate to product highly reactive oxygen atoms are often difficult to study using FP–RF.

In theory, FP–RF systems can be used in a static mode, that is, the reactants can be admitted to the cell in a buffer gas and the cell sealed. The mixture is then repeatedly flashed; the decay of the reactive species can be stored in a multi-channel analyzer, for example, and signal averaged to improve the signal-to-noise ratio. In practice, however, static operation leads to the buildup of reaction products or of photolysis products in the photolysis cell, and some of these can photolyze and produce interfering secondary reactions. As a result, a slow flow of gas through the cell is usually maintained to avoid such an accumulation of products.

The limitations on the total pressure in the FP–RF cell are far less severe than those for FFDS. Thus the lower end of the pressure range that can be used is determined by the need to minimize diffusion of the reactants out of the viewing zone. The upper end is determined primarily by the need to minimize both the absorption of the flash lamp radiation by the carrier gas and the quenching of the excited species being monitored by IF. In practice, pressures of ~5 Torr up to several atmospheres are used.

The kinetic analysis is again typically pseudo-first-order with the "stable" reactant molecule B in great excess over the reactive species A generated by flash photolysis. The total pseudo-first-order rate of decay, R, obtained by plotting ln [A] against reaction time, is a combination of decay due to reaction and a second, usually much smaller, term, which is the rate of decay of A observed in the absence of added reactant (i.e., in carrier gas alone). The latter is due to diffusion out of the viewing zone and to reaction with impurities in the carrier gas. Thus for the reaction

$$A + B \xrightarrow{k_{22}} \text{products} \tag{22}$$

the integrated form of the rate law corresponding to Eq. (Z) becomes

$$\ln \frac{[A]}{[A]_0} = -(k_{22}[B]_0 + k^{B=0})\, t \tag{FF}$$

where $k^{B=0}$ is the first-order rate of decay of A in the absence of B.

The rate constant of interest, k_{22}, can be extracted by plotting the slopes, R, of plots of ln [A] against reaction time as a function of the initial concentration of B ($[B]_0$):

$$R = k_{22}[B]_0 + k^{B=0} \tag{GG}$$

k_{22} is then just the slope of such plots.

4-B-3b. *Methods of Generation of Atoms and Free Radicals*

Typical precursors to OH radicals in FP–RF systems include H_2O and mixtures of H_2 with O_3, N_2O, or NO_2; in the latter cases, photolysis of the O_3, N_2O, or NO_2 is used to produce electronically excited O (^1D) atoms which then react with H_2 to generate OH. Similarly, the photolysis of O_3 in the presence of H_2O has been used. Other common photolytic sources of OH include HNO_3 and H_2O_2. Ground-state oxygen atoms can be generated by the vacuum UV (VUV) photolysis of O_2;

H atoms are frequently generated using the VUV photolysis of hydrocarbons such as C_3H_8.

Laser dissociation of larger molecules has also been used to generate more complex radicals. For example, Gutman and co-workers (1982) used photolysis of the simple alkyl nitrites CH_3ONO and C_2H_5ONO at 266 nm to produce the alkoxy radicals CH_3O and C_2H_5O. The concentrations of these alkoxy radicals were then followed using their laser-induced fluorescence signals at 303.9 and 322.8 nm, respectively.

4-B-3c. *Methods of Detection of Atoms and Free Radicals*

In the case of FP–RF, the detection technique is that of IF, described above in regard to FFDS (Section 4-B-2c). Resonance absorption (RA) has also been applied to detect species such as OH. While inherently less sensitive, it offers some

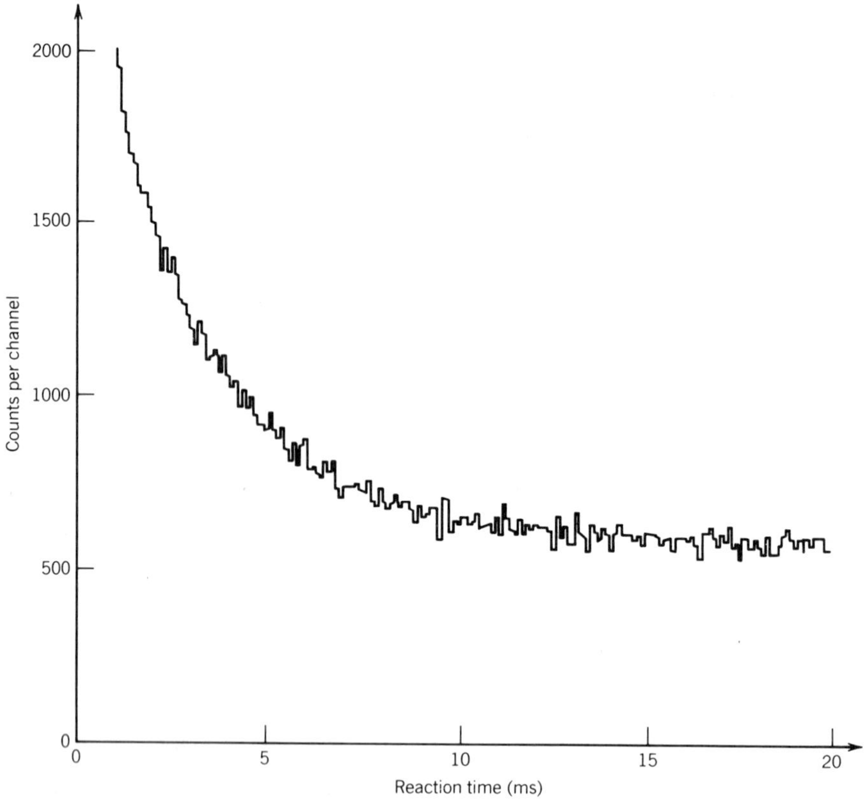

FIGURE 4.14. Results of typical FP–RF experiment on OH radical kinetics; time dependence of the OH radical resonance fluorescence signal intensity accumulated from 1600 flashes of an isobutene (0.000484 Torr)–H_2O (0.013 Torr)–argon (50.3 Torr) mixture at 423.9°K with a multi-channel scaler channel width of 100 μs and a flash energy of 50 J per flash (from Atkinson and Pitts, 1975).

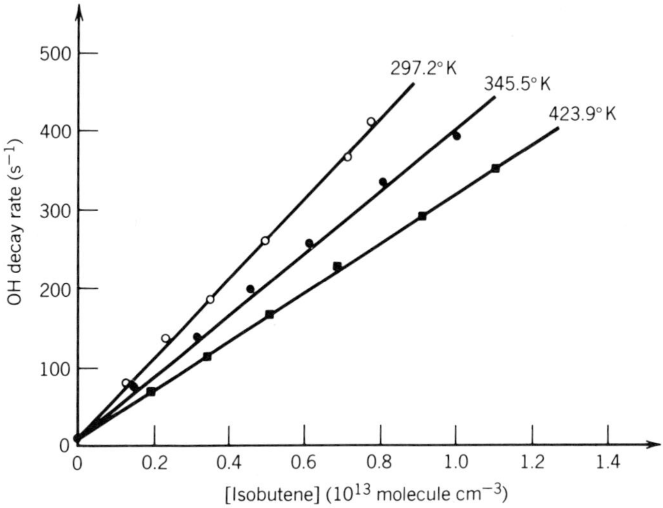

FIGURE 4.15. Plots of the OH radical decay rate monitored in experiments at several temperatures, one of which is shown in Fig. 4.14, against the isobutene concentration at 297.2, 345.5, and 423.9°K. Total pressure ~50 Torr with argon diluent. Note the negative temperature dependence in the reaction (from Atkinson and Pitts, 1975).

advantages in certain applications. For example, at high total pressures, quenching of electronically excited OH monitored by IF can become significant; this can be avoided by using RA since the ground electronic state is directly monitored.

An example of the use of FP–RF to study the kinetics of an atmospherically relevant reaction is found in Figs. 4.14 and 4.15. OH was generated by the pulsed vacuum ultraviolet photolysis of H_2O and monitored by IF at 308 ± 1 nm using the apparatus shown in Fig. 4.13 (Atkinson and Pitts, 1975). Figure 4.14 shows the decay of resonance fluorescence counts from OH in the presence of excess isobutene. This was generated by signal averaging the OH decay from 1600 separate flashes. The decay of OH is seen to be exponential as expected from Eq. (FF). When these decay rates are plotted as a function of the isobutene concentration (Fig. 4.15), one obtains a straight line whose slope, according to Eq. (GG), is the OH–isobutene rate constant and whose intercept is the first-order decay of OH in the absence of isobutene. Note that the slope of the line in Fig. 4.15, i.e. the rate constant, decreases as the temperature increases. This is an example of a negative temperature dependence of a rate constant, as discussed in more detail in Section 4-A-7.

4-B-4. Comparison of Fast Flow Discharge System and Flash Photolysis–Resonance Fluorescence Techniques

Table 4.6 summarizes the parameters used in FFDS and FP–RF. As described earlier, the pressure range for FFDS is more limited than that for FP–RF because

TABLE 4.6. Comparison of Fast Flow Discharge System (FFDS) and Flash Photolysis–Resonance Fluorescence (FP–RF) Techniques

Parameter	FFDS	FP–RF
Pressure range	0.5 to ~100 Torr	5 Torr to ⩾1 atm
Temperature range	200 to ⩾ 1000°K	100 to ⩾1000°K
Rate constant range[a]	10^{-10}–10^{-16}	10^{-10}–10^{-18}
Detectors	Many	Must be fast
Reactants	Many	More limited
Wall reactions	Yes	Negligible

Source: Adapted from Howard, 1979.
[a] Units of cm^3 molecule^{-1} s^{-1}.

of the need to maintain certain flow dynamics in the tube. However, FFDS allows pressures down to ~0.5 Torr to be studied, whereas ~5 Torr is the lower limit of FP–RF.

The available temperature range is approximately the same for both techniques. FP–RF, however, allows a more extended range of rate constants to be measured, from $\sim 10^{-10}$ to as low as $\sim 10^{-18}$ cm^3 molecule^{-1} s^{-1}. This ability to measure rate constants of relatively slow reactions arises because higher total pressures, and hence higher reactant concentrations, can be used to produce measurable rates of decay. In FFDS, the reactant concentrations and hence measured decay rates are limited by the lower total pressures.

An additional advantage of FP–RF is that the larger reaction cell dimensions and the higher total pressures used usually result in wall reactions being negligible. In FFDS, on the other hand, removal of the reactive species A at the walls of flow tubes, as well as other secondary reactions which can occur there, present perhaps the greatest uncertainty in data interpretation in these studies.

On the other hand, FFDS is much more versatile in terms of the detectors as well as the kinds of reactants which can be used. Thus FP–RF requires a detector that responds relatively rapidly (on a μs time scale) because the decay of the reactive species must be followed in real time. However, FFDS experiments are steady state types of experiments; the detector need not be fast since the reaction time is determined by the position of the mixing point of A and B relative to the detector.

In addition, flash photolysis systems are limited with respect to detectors, whereas, as we have seen, a wide variety of detection systems can be used with FFDS. This is especially an advantage in investigating reaction mechanisms. Unless a product can be followed by IF or absorption spectrometry, FP–RF systems give little mechanistic information, whereas the use of techniques such as LMR, MS, and so on, combined with FFDS, allows the identification of some intermediates and products.

Finally, the types of reactants which can be studied by FP–RF are limited to those that can be generated using flash photolysis. As we have seen in the case of fast flow systems, a variety of techniques for generating reactants is available. Thus fast flow systems can be used relatively easily to study radical–radical reactions such as $OH + HO_2$, whereas FP–RF is most readily used to study reactions between a stable molecule and a reactive species.

Furthermore, the stable species present in excess or other added gases must not photolyze. For example, Paraskevopoulos and Irwin (1984) used flash photolysis–resonance absorption to study the reaction of OH with CO and the effects of added gases, including O_2, on the rate constant. However, they were restricted to the addition of only small pressures (0.1–0.2 Torr) of O_2 because it photolyzes at the wavelength used to produce OH. This forms O atoms and these can react with the OH, thus causing an interference with the measurement of OH + CO reaction.

4-B-5. Other Techniques for Atoms and Free Radicals

While the vast majority of absolute rate constants for atom and free radical reactions with species in the natural and polluted troposphere have been generated using FFDS and FP–RF, other methods have also been applied. These include molecular modulation and pulse radiolysis.

4-B-5a. Molecular Modulation

This technique has been applied primarily to the determination of the kinetics of ground-state oxygen atom reactions. The production of the reactive species, $O(^3P)$, is sinusoidally modulated by modulating the photolyzing light which leads to the formation of $O(^3P)$.

The $O(^3P)$ is produced by the mercury photosensitized decomposition of N_2O, with 253.7 nm resonance radiation being used to excite the ground-state Hg (6^1S_0) to the metastable (6^3P_1) state (symbolized by Hg*):

$$Hg(6^1S_0) + h\nu\ (253.7\ \text{nm}) \rightarrow Hg(6^3P_1) \tag{34}$$

$$Hg^* + N_2O \rightarrow Hg + N_2 + O(^3P) \tag{35}$$

The concentration of $O(^3P)$ atoms is followed by adding small concentrations of NO and monitoring the chemiluminescence from reaction (31):

$$\begin{array}{c} O + NO + M \xrightarrow{k_{31}} NO_2^* + M \\ \downarrow \\ NO_2 + h\nu \end{array} \tag{31}$$

The $O(^3P)$ concentration (i.e., the NO_2^* chemiluminescent emission intensity) is phase shifted from the exciting 253.7 nm radiation by an amount that depends on the frequency of modulation, f, and on the rate of loss of $O(^3P)$ from the system. In the case where $O(^3P)$ is removed by reaction (31) as well as by reaction (36) with a reactant B,

$$O(^3P) + B \xrightarrow{k_{36}} \text{products} \tag{36}$$

the phase shift ϕ is given by the relationship

$$\tan \phi = 2\pi f(k_{31}[NO][M] + k_{36}[B]) \tag{HH}$$

Thus k_{36} can be determined by monitoring the phase shift ϕ as a function of the concentration of B, which, when present in great excess, remains constant.

This technique has also been applied in one study to OH kinetics in systems where B is a hydrocarbon such as propane. Thus reactions (34)–(36) produce OH radicals if the organic B is such that $O(^3P)$ can abstract a hydrogen atom from it. The OH so produced then reacts with the parent organic B, and the rate constant for this reaction can be determined by following the concentration of OH radicals (e.g., by absorption spectrometry).

The molecular modulation technique has been applied recently to more complex systems. For example, the rate constant of the reaction

$$OH + HO_2 \rightarrow H_2O + O_2 \tag{37}$$

was determined by Burrows, Cox, and Derwent (1981) using the modulated photolysis at 253.7 nm of an O_3–H_2O–O_2–N_2 (or He) mixture. In this complex system, the reactions are

$$O_3 + h\nu\ (253.7\ \text{nm}) \rightarrow O(^1D) + O_2 \tag{28}$$

$$O(^1D) + M \rightarrow O(^3P) + M \tag{38}$$

$$O(^1D) + H_2O \rightarrow 2OH \tag{29}$$

$$OH + O_3 \rightarrow HO_2 + O_2 \tag{39}$$

$$HO_2 + O_3 \rightarrow OH + 2O_2 \tag{40}$$

$$HO_2 + HO_2 \rightarrow H_2O_2 + O_2 \tag{41}$$

$$OH + HO_2 \rightarrow H_2O + O_2 \tag{37}$$

The OH and HO_2 radicals were followed by absorption spectrometry at 308.2 and 210 nm, respectively. (The 308.2 nm absorption line is one of the rotational lines in the $A^2\Sigma^+ - X^2\Pi$ system of OH shown in Fig. 4.7.) In this system, which is relatively complex compared to the $O(^3P)$ atom system described earlier, computer simulation was required to extract the value of k_{37} from the data. k_{37} was found to be 6.2×10^{-11} cm^3 molecule^{-1} s^{-1} independent of temperature over the range 288–348°K.

Thus molecular modulation is a useful technique, especially for species such as $O(^3P)$ which can be produced relatively easily by a short series of modulated photolysis reactions. However, it has also been successfully applied to the reactions of other species such as OH, HO_2, and RO_2 (e.g., see Parkes, 1977; Cox and Tyndall, 1980; Burrows et al., 1981; Anastasi et al., 1983).

4-B-5b. *Pulse Radiolysis*

In a few studies, reactive species have been produced by a pulse of high-energy electrons impacting on stable vapor phase precursors. For example, OH has been produced by the pulse radiolysis of water vapor, and its kinetics studied using optical absorption to follow its decay in a great excess of reactant. A variety of radicals such as NH, CN, and NH_2 can be generated and studied using pulse radiolysis. This technique is especially convenient for high temperature work (Jonah et al., 1984) and for application to atmospheric chemistry since it can be used at total pressures of one atmosphere and more. However, this technique has not found widespread use because of the expense and complexity of the equipment and for some radicals, the difficulty of finding suitable precursors such that the other fragments produced in the radiolysis do not cause interfering secondary reactions.

4-B-6. Static Techniques for Determining O_3 Rate Constants

Of the reactive species A in Table 4.4 which are of most interest for atmospheric chemistry, all are free radicals except O_3 and $O_2(^1\Delta_g)$. As a result of the high reactivity of free radicals and the high physical quenching rates of electronically excited O_2, all except O_3 must be generated and monitored *in situ* using somewhat specialized techniques such as those described above.

Ozone, however, is a relatively stable molecule that can be generated outside a kinetic apparatus and stored for later use. In addition, its reactions are generally much slower than those of atoms and free radicals such as OH. As a result, the experimental techniques used for determining absolute rate constants for O_3 reactions are generally static in nature and do not require *in situ* generation of the O_3.

In these static techniques, O_3 is first prepared outside the reaction cell. Mixtures of O_3 in O_2 or in air are easily prepared using electrical discharges (e.g., as in commercial ozonizers) or vacuum ultraviolet photolysis of O_2; in both cases, oxygen atoms are formed which react with the excess O_2 present:

$$O_2 \xrightarrow[\text{or } h\nu]{\text{discharge}} 2O \tag{42}$$

$$O + O_2 \xrightarrow{M} O_3 \tag{43}$$

The mixture of O_3 in the precursor gas can sometimes be used as prepared. Alternatively, the O_3 may be purified by trapping it at low temperatures from the discharged O_3–O_2 or O_3–air mixture and pumping off the O_2 and/or N_2. A preferable procedure involves trapping the O_3 on cooled silica gel (e.g., in a dry ice–acetone bath) and pumping off the diluent gases. This second technique involving adsorbed O_3 is relatively safe compared to trapping O_3 as a liquid, as the latter is highly explosive. However, *great care also should be taken in handling* adsorbed O_3. All applicable safety precautions should be followed *rigorously*!

Ozone decomposes on surfaces, the rate depending on the nature of the particular surface and whether it has been previously ''conditioned'' by exposure to O_3. While this heterogeneous decomposition is much slower than the wall loss of OH, the homogeneous gas phase reactions of O_3 of interest are also slower than those of OH. Thus when a reactant (X) is introduced into the cell in great excess over O_3, both reaction (44)

$$O_3 + X \rightarrow \text{products} \tag{44}$$

and wall loss (45)

$$O_3 \xrightarrow{k_w} \text{wall loss} \tag{45}$$

may occur. If $[X]_0 >> [O_3]_0$, then the concentration of X remains approximately constant at $[X]_0$, and the rate law is

$$\frac{-d[O_3]}{dt} = k_{44}[O_3][X] + k_w[O_3]$$

or in the integrated form

$$\ln \frac{[O_3]}{[O_3]_0} = -(k_w + k_{44}[X])\, t$$

Determining the rate constant k_{44} thus requires measuring O_3 concentrations (or a proportional parameter) as a function of reaction time. Monitoring O_3 accurately and inexpensively in situ is reasonably straightforward because it absorbs strongly at 253.7 nm, a stable, inexpensive, and readily available mercury line. Alternatively, samples of the reaction mixture can be withdrawn and the O_3 analyzed using a commercial O_3 monitoring instrument.

A requirement of the Toxic Substances Control Act was that experimental protocols be developed for determining the kinetics of reactions of toxic chemicals with O_3 and OH radicals, as well as their photolysis rates, so that there are standard procedures available for assessing various environmental fates and lifetimes of these chemicals. The kinetic approach outlined above for O_3 is now used as a standard protocol by the U.S. Environmental Protection Agency for determining O_3 rate constants (Pitts et al., 1981). Similar protocols have been developed in Europe by the European Chemical Industry, Ecology, and Toxicology Center (Turner, 1983). Figure 4.16 shows the results of one experiment in which this protocol for O_3 reactions was tested for the reaction with *o*-cresol. An FEP Teflon bag was subdivided into two chambers using metal rods. O_3 was then introduced into one side of the bag at concentrations that give ~1 ppm in the entire bag when

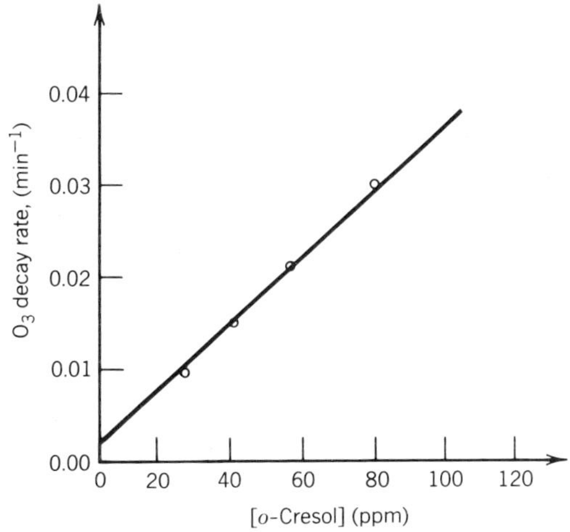

FIGURE 4.16. Plots of ozone decay rate against *o*-cresol concentration using U.S. EPA protocol for determining O_3 rate constants (from Pitts et al., 1981).

mixed. The organic in excess concentration was injected into the other side, both the organic and O_3 being in ultra-high-purity air. After removing the metal barriers and mixing, O_3 is monitored as a function of time using a commercial chemiluminescence ozone analyzer. In this particular case, a rate constant for the O_3–*o*-cresol reaction of 2.55×10^{-19} cm^3 $molecule^{-1}$ s^{-1} was obtained, in good agreement with the literature.

Such experiments can also be carried out with O_3 in great excess; however, a technique must be available for following the concentration of the reactant X with time, and corrections may have to be made for changing ozone concentrations due to wall losses during the experiment. In addition, interferences from secondary reactions are more likely under these conditions.

4-C. LABORATORY TECHNIQUES FOR DETERMINING RELATIVE RATE CONSTANTS FOR GAS PHASE REACTIONS

4-C-1. Basis of Kinetic Analysis

Many of the rate constants for gas phase reactions of atmospheric interest reported in the literature were actually determined not as *absolute* values, but rather as a *ratio of rate constants*. Thus if the absolute value for one of the rate constants has been determined independently, the second one can then be calculated from the experimentally determined ratio.

In the simplest case, determining relative rate constants for a reactive species A is based on a competition between two reactions:

$$A + X_1 \xrightarrow{k_1} P_1 \tag{46}$$

$$A + X_2 \xrightarrow{k_2} P_2 \tag{47}$$

X_1 and X_2 are the reactants that compete for A, and P_1 and P_2 are the respective products of these reactions. By monitoring the concentrations of the reactants X_1 and X_2, the products P_1 and P_2, or the change in one of these as the second reactant is added with time, the rate constant ratio k_1/k_2 can be obtained.

For example, if X_1 and X_2 are monitored, the relevant rate laws are as follows:

$$\frac{-d[X_1]}{dt} = k_1[A][X_1]$$

$$\frac{-d[X_2]}{dt} = k_2[A][X_2]$$

Rearranging one obtains

$$\frac{-d \ln [X_1]}{dt} = k_1[A] \tag{II}$$

$$\frac{-d \ln [X_2]}{dt} = k_2[A] \tag{JJ}$$

Combining (II) and (JJ) to eliminate [A] yields Eq. (KK):

$$[A] = -\frac{1}{k_1}\frac{d \ln[X_1]}{dt} = -\frac{1}{k_2}\frac{d \ln[X_2]}{dt} \tag{KK}$$

Integrating from time $t = 0$ when the initial concentrations are $[X_1]_0$ and $[X_2]_0$, respectively, to time t when the concentrations are $[X_1]$ and $[X_2]$, gives Eq. (LL):

$$\ln \frac{[X_1]_0}{[X_1]} = \frac{k_1}{k_2} \ln \frac{[X_2]_0}{[X_2]} \tag{LL}$$

Thus the concentrations of X_1 and X_2 as a function of reaction time, plotted as given by Eq. (LL) (i.e., $\ln [X_1]_0/[X_1]$ versus $\ln [X_2]_0/[X_2]$) can be used to derive the rate constant ratio k_1/k_2. If an absolute value is known for one of the two rate constants from independent studies, then an absolute value for the second one can be obtained.

For example, the rate constants for the reaction of OH with a series of alkanes have been determined using such a relative rate technique (Atkinson et al., 1982). The photolysis of methyl nitrite (CH_3ONO) in air containing NO was used as the source of OH (see below) and the loss of the added alkanes was followed with time using gas chromatography. Figure 4.17 shows the results of an experiment on a mixture of propane and *n*-butane. As expected from Eq. (LL), the plot is linear and the slope gives the ratio of rate constants for reaction of OH with these

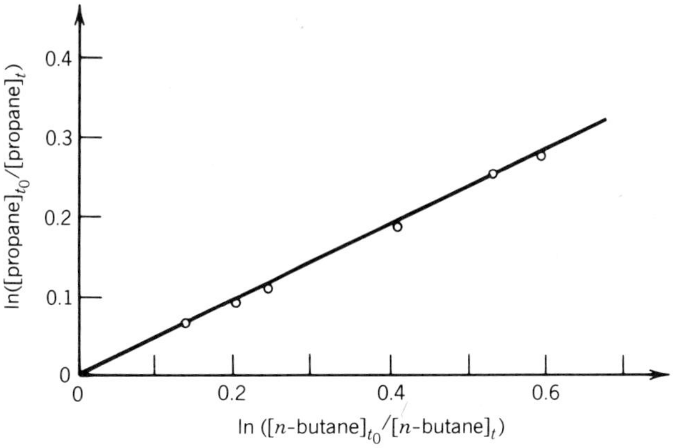

FIGURE 4.17. Plot of ln ([propane]$_{t0}$/[propane]$_t$) against ln ([n-butane]$_{t0}$/[n-butane]$_t$) for reactions with OH in a relative rate experiment with n-butane as the reference organic (from Atkinson et al., 1982).

two alkanes. In this particular case k_1/k_2 was found to be $0.473 \pm 0.016(2\sigma)$, in excellent agreement with the literature (Atkinson et al., 1979). The absolute rate constant for the propane reaction was then obtained assuming $k_{OH\ +\ n\text{-}C_4H_{10}} = 2.58 \times 10^{-12}$ cm^3 molecule^{-1} s^{-1}.

Examples of similar kinetic analyses used if the products P_1 and P_2 are followed, or if the decrease in the rate of production of one of the products (e.g., P_1) is followed as the second reactant X_2 is added, are given in Sections 4-C-2 and 4-C-3. However, care must be exercised since products may undergo secondary reactions.

Relative rate techniques have the advantage that such relative measurements can often be made with greater precision than absolute rate constant measurements because only relative, not absolute, concentrations of X_1 and X_2 need be measured. (Increased *precision,* however, does not *necessarily* imply increased accuracy!)

In addition, the species A, which is frequently a highly reactive free radical such as OH which is difficult to measure, need not be monitored in such experiments; only X_1 and X_2 which are usually stable and easily measured molecules, such as hydrocarbons, are followed.

Finally, as seen from the examples in Sections 4-C-2 and 4-C-3, relative rate experiments can often be carried out under conditions directly relevant to the troposphere (i.e., in 1 atm of air).

The accuracy of the results, however, depends critically on knowing enough of the mechanistic details of the reaction system to be sure that the kinetic analysis applied is valid; this is not always straightforward in complex systems. Furthermore, obtaining an accurate rate constant from the rate constant ratio (k_1/k_2) requires accurate knowledge of the other rate constant.

This is not as trivial a problem as it might seem. For example, many relative rate constants have been determined using as a reference the absolute value of the rate constant for the OH + CO reaction. As discussed in Section 4-A-5, until the mid- to late 1970s $k_{OH + CO}$ was taken as being 1.4×10^{-13} cm^3 molecule^{-1} s^{-1}, independent of total pressure and O_2 concentrations; however, the value of $k_{OH + CO}$ at 760 Torr total pressure in air is now thought to be ~50–100% larger. Thus the rate constants derived from relative techniques and based on the earlier value of $k_{OH + CO}$ may not be correct as reported. In such cases, the experimental conditions in the relative rate studies must be examined, and the relative rate constants *corrected* based on a "best guess" for the appropriate value of the reference OH + CO rate constant under these conditions.

4-C-2. Typical Laboratory Studies Relevant to Atmospheric Chemistry

The reaction of OH with CO

$$OH + CO \rightarrow CO_2 + H \tag{18}$$

was cited in Section 4-A-5 as an example of a reaction that initially appeared to be a simple bimolecular reaction, but that is now believed to involve the intermediate formation of a complex. Some of the first experiments suggesting that (18) was not an elementary reaction were relative rate studies by Sie, Simonaitis, and Heicklen (1976). They photolyzed N_2O to produce electronically excited oxygen atoms, $O(^1D)$. In the presence of a great excess of H_2, $O^1(D)$ reacts to produce OH:

$$N_2O + h\nu\ (213.9\ \text{nm}) \rightarrow N_2 + O(^1D) \tag{48}$$

$$O(^1D) + H_2 \rightarrow OH + H \tag{49}$$

The yield of N_2 is a measure of the amount of N_2O photolyzed and hence of $O(^1D)$ and OH produced.

In the presence of CO, reaction (18) competes with reaction (50) which occurs when CO is absent:

$$OH + CO \rightarrow CO_2 + H \tag{18}$$

$$OH + H_2 \rightarrow H_2O + H \tag{50}$$

The yield of the product CO_2 at different ratios of the CO and H_2 concentrations should be a measure of the rate of (18) compared to that of (50).

The CO_2 yield can be quantitatively related to the relative rate constant ratio k_{50}/k_{18} in the following manner. If OH reacts only with CO or H_2, the fraction (f) of the total OH produced which ultimately produces CO_2 is simply the fraction that reacts with CO, that is,

$$f = \frac{k_{18}[CO]}{k_{18}[CO] + k_{50}[H_2]} \tag{MM}$$

One molecule of N_2 is produced in reaction (48) for each atom of $O(^1D)$, and according to the above reaction scheme, each $O(^1D)$ produces one OH radical. Thus the yields of OH (Y_{OH}) and N_2 (Y_{N_2}) are equal. The yield of CO_2 (Y_{CO_2}) should be given by

$$Y_{CO_2} = fY_{OH} = fY_{N_2}$$

Substituting from Eq. (MM) and rearranging, one obtains the following expression for the relative yields of N_2 and CO_2 as a function of the H_2 and CO reactant concentrations:

$$\frac{Y_{N_2}}{Y_{CO_2}} = 1 + \frac{k_{50}[H_2]}{k_{18}[CO]} \tag{NN}$$

Thus the ratio of rate constants k_{50}/k_{18} can be obtained by measuring the yields of N_2 and CO_2 as functions of the concentrations of H_2 and CO, both of which are present in sufficiently great excess that they remain approximately constant at their initial concentrations.

In practice, under realistic experimental conditions, not all the $O(^1D)$ formed in the N_2O photolysis reacts with H_2. For example, even in a great excess of H_2, some $O(^1D)$ reacts with the N_2O. However, corrections to the expected yield of N_2 can be made for such secondary reactions of $O(^1D)$ since the rate constants for these $O(^1D)$ reactions are reasonably well established.

Figure 4.18 shows some typical plots of the relative yields of N_2 and CO_2 [corrected for $O(^1D)$ reactions other than with CO and H_2] as a function of $[H_2]$/[CO] at different total pressures at 298°K. As expected from Eq. (NN), the plots are linear with an intercept of 1.0. From the slopes of these plots, k_{50}/k_{18} can be obtained. It is seen that the slope (i.e., the rate constant ratio) decreases as the total pressure increases. Since k_{50} is known to be independent of total pressure, k_{18} must be increasing with total pressure. As discussed in Section 4-A-5, this was one of the first indications that reaction (18) was indeed pressure dependent.

Other complex systems have also been used to generate relative rate constants with considerable success. Cox and co-workers, for example, pioneered the use of nitrous acid photolysis in the presence of various reactant gases to generate relative rate constants for OH reactions (Cox, 1975). In these early studies, the concentrations of NO and NO_2 (which are present initially in small concentrations as impurities) were followed as a function of time as the HONO photolyzed:

$$HONO + h\nu \rightarrow OH + NO \tag{51}$$

The change in the rates of formation of NO and NO_2 upon the addition of various concentrations of a reactant gas can be related to the ratio of rate constants for the reaction of OH with the reactant gas, compared to that for reaction with $NO + NO_2 + HONO$. Alternatively, as used in more recent studies, reaction (51) can be used to generate OH and relative rate constants for two or more organics

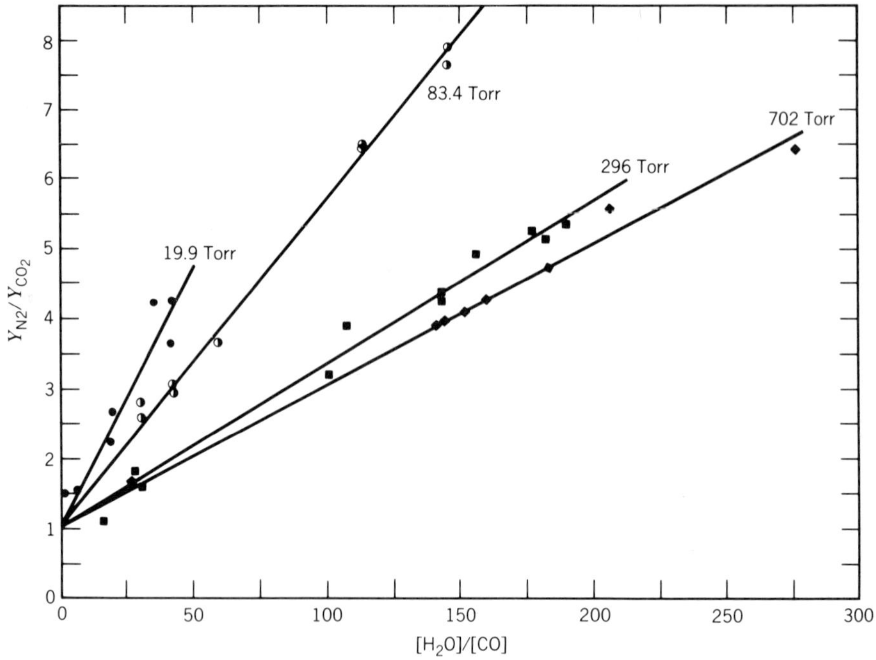

FIGURE 4.18. Relative yields of N_2 and CO_2 [corrected for loss of $O(^1D)$ other than by reaction with H_2] as a function of the reactant ratio $[H_2]/[CO]$ at 298°K at total pressures from 19.9 to 702 Torr (from Sie et al., 1976).

obtained by following the decay of the organics in a manner similar to the methyl nitrite technique discussed above. Support for this technique comes from the generally good agreement of their results with those derived using very different techniques (e.g., see Cox et al., 1980).

While the accuracy of relative rate studies depends on the validity of the assumed reaction mechanism and resulting kinetic analysis, the range of conditions (e.g., total pressure, nature of diluent gases) that can be covered is generally much greater than is possible in absolute rate constant determinations. In addition, the monitoring techniques required to follow species such as N_2, CO_2, NO, and NO_2 are usually straightforward and relatively inexpensive since they involve equipment such as gas chromatographs which are readily available in most laboratories. Finally, the concentrations of reactants and products in such laboratory studies are usually sufficient to be easily followed.

4-C-3. Environmental Chamber Studies

In typical laboratory relative rate studies, large concentrations of reactants (relative to atmospheric conditions) are generally used, and the kinetic schemes involved are often complex. To avoid possible errors in applying the results of such studies to tropospheric conditions, it is desirable whenever possible to confirm the rate constants under conditions as close as possible to those found in ambient air. Environmental chambers (see Chapter 6) were first used for determining relative rate constants for OH reactions under typical atmospheric conditions.

The use of environmental chambers for OH kinetic studies was based on the recognition that in irradiated NMOC–NO_x mixtures attack by OH is the predominant mode of removal for most organics, at least at short irradiation times when the O_3 levels are low. Figure 4.19, for example, shows plots of calculations of the fraction of the total rate of attack on propylene due to OH, O_3, HO_2, $O(^3P)$, and $O_2(^1\Delta_g)$ during irradiation of a NO_x–propylene–air mixture at 25°C in an environmental chamber. Clearly, OH attack is by far the predominant mode of removal of propylene in the first hour when O_3 levels are low. However, after 6 h irradiation, O_3 levels build up sufficiently that O_3 attack also becomes a major loss process for propylene. For species such as alkanes, which react negligibly slowly with O_3, attack by OH predominates at much longer times.

Given this information, we know that if two hydrocarbon species X_1 and X_2 are irradiated in a chamber in a mixture with NO_x and air, then only attack by OH need be considered at short irradiation times:

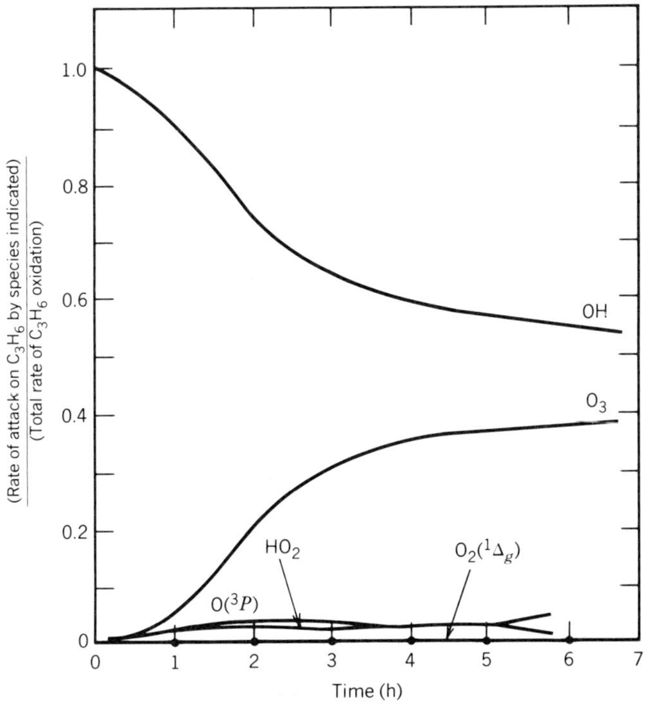

FIGURE 4.19. Calculated fraction of total rate of attack on propylene by OH, O_3, HO_2, $O(^3P)$ and $O_2(^1\Delta_g)$ during irradiation of NO_x–propylene–air mixture at 25°C. Initial concentrations are 0.437 ppm NO, 0.05 ppm NO_2, and 0.53 ppm C_3H_6 (from Doyle et al., 1975).

$$OH + X_1 \xrightarrow{k_{52}} \text{products} \quad (52)$$

$$OH + X_2 \xrightarrow{k_{53}} \text{products} \quad (53)$$

The rate of disappearance of X_1 is thus given by

$$\frac{-d[X_1]}{dt} = k_{52}[OH][X_1]$$

or, rearranging,

$$\frac{-d\ \ln[X_1]}{dt} = k_{52}[OH]$$

A similar expression applies for X_2. The slopes of plots of $\ln[X_1]$ and $\ln[X_2]$ against reaction time should thus be proportional to k_{52} and k_{53}, respectively, and the rate constant ratio k_{52}/k_{53} can be derived from the ratio of the slopes of such plots.

An alternate kinetic analyses for such systems is based on Eq. (II)–(LL) where A = OH. In the approach, $\ln[X_1]_0/[X_1]$ is plotted against $\ln[X_2]_0/[X_2]$ to derive the relative rate constants.

While such relative rate studies using environmental chambers have provided a variety of accurate and useful OH rate constants under typical tropospheric conditions (Atkinson et al., 1979), there are some disadvantages to their use. Thus the rate constant of interest must be relatively large ($\gtrsim 3 \times 10^{-13}$ cm^3 molecule^{-1} s^{-1}) so that measurable rates of decay of the reactant result; the concentrations of OH generated in such environmental chambers are relatively small ($\lesssim 5 \times 10^6$ cm^{-3}) and hence the rate of removal of the reactant by OH attack is too small to measure compared to loss via dilution unless $k \gtrsim 3 \times 10^{-13}$ cm^3 molecule^{-1} s^{-1}. Dilution can be minimized, however, through the use of flexible Teflon bags which collapse as samples are withdrawn (Darnell et al., 1978). Additionally, the method is based on the assumption that only OH attack is important in removing the reactant of interest. Therefore, this technique cannot be applied to compounds that photolyze at a significant rate.

Because of these disadvantages, this method is no longer extensively used to obtain rate constants under conditions typical of the troposphere. Instead, the photolysis of HONO (discussed above) or of organic nitrites (see following) have supplanted the irradiated NMOC–NO_x–air method.

Thus, more recently, the photolysis of methyl nitrite in the presence of NO and air has been applied by Atkinson and co-workers (Atkinson et al., 1981) to OH relative rate studies:

$$CH_3ONO + h\nu \rightarrow CH_3O + NO \quad (54)$$

$$CH_3O + O_2 \rightarrow HCHO + HO_2 \quad (55)$$

$$HO_2 + NO \rightarrow OH + NO_2 \quad (30)$$

The higher OH concentrations generated by such sources extends the rate constant range to $\lesssim 3 \times 10^{-13}$ cm^3 $molecule^{-1}$ s^{-1} (Atkinson et al., 1981).

This technique forms the basis of the experimental protocol recommended by the U.S. Environmental Protection Agency for determining OH rate constants for reaction with organics (Atkinson et al., 1981; Pitts et al., 1982).

Determination of OH relative rate constants for compounds that photolyze significantly in actinic radiation requires a nonphotolytic source of OH. Three such OH sources are H_2O_2–NO_2–CO mixtures (Campbell et al., 1975, 1979; Audley et al., 1982), the thermal decomposition of HO_2NO_2 in the presence of NO (Barnes et al., 1982), and O_3–hydrazine reactions (Tuazon et al., 1983). However, in these cases, the reactant must not react with O_3, HO_2, or H_2O_2, and care must be taken in interpreting the data since these systems have the potential of being rather complex. Indeed, the rate constants derived have not always agreed well with literature values. Thus, until the general features of the mechanisms involved in the production of OH in these systems have been fully elucidated, the simultaneous production of other highly reactive species, and hence possible interfering secondary reactions, cannot be firmly ruled out.

4-D. REACTIONS IN SOLUTION

4-D-1. Interactions of Gaseous Air Pollutants with Atmospheric Aqueous Solutions

Because of the gaseous nature of many of the important primary and secondary pollutants, the past emphasis in kinetic studies of atmospheric reactions has been on gas phase systems. Increasingly, however, it is being recognized that reactions which occur in the liquid phase and on the surface of solids may play important roles in such problems as acid rain and fogs (Chapter 11) and in the growth and properties of aerosols (Chapter 12). We therefore briefly discuss reaction kinetics in solution in this section and on solid surfaces in Section 4-E. The basic principles of kinetics discussed in Sections 4-A-1, 4-A-2, and 4-A-3, of course, still apply.

The aqueous phase which serves as a reaction medium in the atmosphere is present in the form of clouds, fogs, and rain. In addition, particulate matter consisting either of an aqueous solution containing pollutants or a film of water surrounding an insoluble core (see Chapter 12) may provide the liquid phase for reactions; thus at typical relative humidities, ~30–50% of the aerosol mass is due to water (Graedel and Weschler, 1981). However, many of the species which are believed to react in such atmospheric solutions, for example, SO_2, O_3, H_2O_2, and NO_x, are emitted or formed in the gas phase. Before reactions can occur in solution, then, several steps must first take place:

1. Diffusion of the gas to the surface of the droplet.
2. Transport of the gas across the air–water interface and establishment of a gas–liquid equilibrium.

3. Establishment of ionized and/or hydrolyzed species and the equilibria between them.
4. Diffusion of the solvated species into the bulk phase of the droplet.

Diffusion of gases to the surface of atmospheric droplets is fast relative to diffusion in the aqueous phase, i.e., step (1) is fast relative to step (4). Thus diffusion coefficients for gases at 1 atm pressure are ~ 0.1–$1\ \text{cm}^2\ \text{s}^{-1}$, whereas in liquids they are $\sim 10^{-5}\ \text{cm}^2\ \text{s}^{-1}$ for small molecules. As discussed in detail by Schwartz and Freiberg (1981), gas phase diffusion, in most cases, will not be the slowest (i.e., rate-determining) step.

Gases dissolve in aqueous solution to various extents, depending on the nature of the gas. An equilibrium between the gas and liquid phases can be described by Henry's law:

$$[\text{X}] = H_x P_x \qquad \text{(OO)}$$

where [X] is the equilibrium concentration of X in solution (in M = moles L^{-1}), P_x is the gas phase equilibrium pressure (in atm), and H_x is the Henry's law constant (in $M\ \text{atm}^{-1}$). Values of H for some common gaseous air pollutants dissolving in aqueous solutions at 25°C are found in Chapter 11. They range from $\sim 10^{-3}$ $M\ \text{atm}^{-1}$ for relatively insoluble gases such as O_2 to $\sim 10^5\ M\ \text{atm}^{-1}$ for highly soluble gases such as H_2O_2 and HNO_3.

Henry's law can be applied to predict solution concentrations only if certain conditions are met. Thus it assumes that there are no irreversible chemical reactions which are so fast that the equilibrium cannot be established. It also assumes that the surface of the droplet is an unimpeded air–water interface; it has been suggested that some atmospheric aerosols have an organic surface film (Chang and Hill, 1980; Graedel and Weschler, 1981; Gill et al., 1983) which could alter the establishment of the equilibrium anticipated by Henry's law.

Ionized and/or hydrolyzed species may be formed from the dissolved gas in some cases. An important example is SO_2 which dissolves to set up equilibria involving HSO_3^- and SO_3^{2-} in a manner similar to CO_2.

As mentioned earlier, diffusion within the bulk aqueous phase is much slower than gas phase diffusion; it can be rate-limiting under conditions of high reactant concentrations where the rate of the chemical reaction is high. This appears to have been a problem in experimental studies of some aqueous phase reactions relevant to the atmosphere where either bulk solutions or large droplets and reactant concentrations higher than atmospheric were used (Freiberg and Schwartz, 1981). However, except under relatively extreme conditions, this is not expected to be the case under atmospheric conditions.

One major difference between reactions in the gas phase and in solution is the presence of solvent molecules in the latter case. In the liquid phase, molecules are in close contact, with the space between molecules being $\sim 10\%$ of the distance between their centers. Reactants thus have a number of nearest neighbors, in the range of ~4–12, with which they can collide. The reactants can then be thought of as existing in a solvent "cage," in which they undergo many collisions before breaking out of that particular environment. If two participants required for a chemical reaction diffuse into such a solvent cage, they will then be held together for a period of time and undergo a number of collisions with each other; such a series of collisons is known as an *encounter.* Because of this *cage effect,* highly reactive species such as atoms and free radicals which are formed in the cage, for example, by photolysis, have a much higher efficiency of recombination than if they were in the gas phase.

Compared to the gas phase, then, reactants take longer to diffuse together, but once they find themselves as nearest neighbors, they undergo a series of collisions rather than separating after one collision; this difference is illustrated in Fig. 4.20. As a result, for neutral non-polar reactants (as opposed to ions, see below), the rate constants in solution are expected to be approximately equal to those in the gas phase.

4-D-2. Diffusion-Controlled Reactions of Uncharged Non-polar Species in Solution

Let us first consider a very fast reaction between uncharged non-polar reactants in solution. In this case, the rate is controlled by the number of encounters. Once A and B diffuse into the same solvent cage they will react; hence the rate of these diffusion-controlled reactions is determined by how fast A and B diffuse together in solution.

Fick's first law describes the rate of diffusion of a species A in solution across an area E in the direction of the x axis, for example. The rate of diffusion, $J = dn/dt$ (in molecules s^{-1}), is given, according to Fick's first law, by

$$J = \frac{dn}{dt} = DE\frac{\delta[N_A]}{\delta x}$$

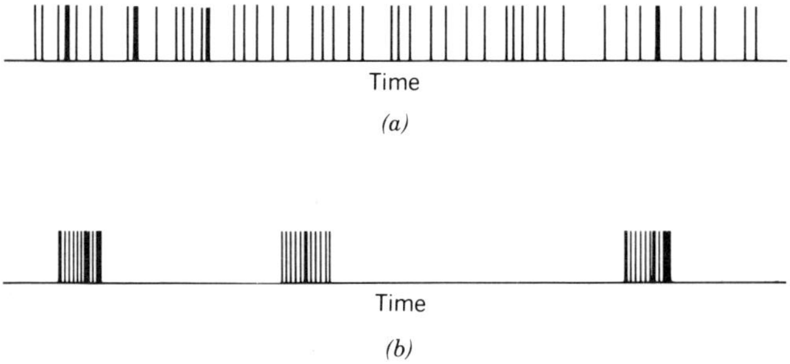

FIGURE 4.20. Patterns of A-B collisions expected (*a*) in the gas phase and (*b*) in solution (from Adamson, 1973).

where dn is the amount of A crossing the area E (cm^2) in time dt, D is the diffusion coefficient (in units of $cm^2\ s^{-1}$) and $\delta[N_A]/\delta x$ is the gradient (in units of molecules cm^{-4}) of the concentration of A in the "x" direction.

Starting with Fick's first law, one can calculate for a solution of two reactants A and B, the frequency of A–B encounters, which is in effect the reaction rate constant for diffusion-controlled reactions. This is given by the following, in units of L $mole^{-1}\ s^{-1}$:

$$k = 4\pi r_{AB} D_{AB}(6.02 \times 10^{20}) \qquad \text{(PP)}$$

r_{AB} is the distance between the centers of the molecules when they react and $D_{AB} = D_A + D_B$, where D_A and D_B are the diffusion coefficients of A and B, respectively. For typical values of $r_{AB} = 4$ Å and $D_A = D_B = 2 \times 10^{-5}\ cm^2\ s^{-1}$, a rate constant of $\sim 10^{10}$ L $mole^{-1}\ s^{-1}$ is obtained. In solution then, diffusion-controlled bimolecular reactions between uncharged species occur with rate constants $\sim 10^{10}$ L $mole^{-1}\ s^{-1}$. For reactions with significant activation energies and/or steric requirements, the rate constants are correspondingly lower.

In this case of uncharged, non-polar reactions, there is little interaction between the reactants and the solvent. As a result, the solvent does not play an important role in the kinetics but acts mainly as a diluent. The rate constants for such reactions therefore tend to be similar to those for the same reactions occurring in the gas phase. Thus as we saw in Section 4-A-7, diffusion-controlled reactions in the gas phase have rate constants of $\sim 10^{-10}\ cm^3$ $molecule^{-1}\ s^{-1}$, which in units of L $mole^{-1}\ s^{-1}$ corresponds to $\sim 6 \times 10^{10}$ L $mole^{-1}\ s^{-1}$, about equal to that for diffusion-controlled reactions in solution.

4-D-3. Reactions of Charged Species in Solution

If the reactants are ionic with opposite charges, the rate constant can be greater than 10^{10} L $mole^{-1}\ s^{-1}$ due to the favorable attractive forces. For example, the rate constant for the reaction of H^+ with OH^- in aqueous solutions at 25°C is $\sim 10^{11}$ L $mole^{-1}\ s^{-1}$. On the other hand, the electrostatic repulsion between ions of like sign can significantly slow their reaction. Similarly, if the reactants are polar molecules, electrostatic forces between them and the solvent may come into play.

For ions and polar molecules, the nature of the solvent is an important factor in solution phase reactions. Following the derivation of Laidler and Meiser (1982), we first consider the reaction between two ions A and B with charges $Z_A e$ and $Z_B e$, respectively, where e is unit electronic charge and Z_A and Z_B are the number of unit charges on the ions, i.e., are whole positive or negative numbers. The electrostatic force (F) between these two ions separated by a distance r in a vacuum is given by Coulombs law,

$$F = \frac{Z_A Z_B e^2}{4\pi\epsilon_0 r^2} \qquad \text{(QQ)}$$

where $\epsilon_0 = 8.85 \times 10^{-12}\ C^2\ J^{-1}\ m^{-1}$ is the *permittivity* of a vacuum. However, if the ions are immersed in a solvent, having a dielectric constant ϵ, the electrostatic force between them is modified by the properties of the solvent. Equation (QQ) thus becomes

$$F = \frac{Z_A Z_B e^2}{4\pi\epsilon_0 \epsilon r^2} \tag{RR}$$

The higher the solvent dielectric constant ϵ, the more the electrostatic force between the ions is reduced. From this expression for the force between two ions, one can calculate the work done to bring the two ions from infinite distance to the distance necessary to react, $r = d_{AB}$; this is equal to the change in free energy due to the electrostatic forces as the ions approach each other, ΔG_{es}. The total free energy change in bringing the ions together is the sum of this electrostatic term and a non-electrostatic one, ΔG_0:

$$\begin{aligned} \Delta G_{TOT} &= \Delta G_0 + \Delta G_{es} \\ &= \Delta G_0 + \frac{(6.02 \times 10^{23}) Z_A Z_B e^2}{4\pi\epsilon_0 \epsilon d_{AB}} \end{aligned} \tag{SS}$$

(Avogadro's number is included in the electrostatic term to convert to units of per mole rather than per molecule.)

As seen in Section 4-A-7b, the transition state form of the rate constant is given by

$$k = \frac{kT}{h} e^{-\Delta G^{\circ\ddagger}/RT}$$

The free energy of activation $\Delta G^{\circ\ddagger}$ for bringing two ions to the necessary distance d_{AB} in order to react is given by Eq. (SS). Thus the natural logarithm of the rate constant becomes

$$\ln k = \ln \frac{kT}{h} - \frac{\Delta G_0}{RT} - \frac{(6.02 \times 10^{23})\, Z_A Z_B e^2}{4\pi\epsilon_0 \epsilon d_{AB} RT}$$

$$\ln k = \ln k_0 - \frac{Z_A Z_B e^2}{4\pi\epsilon_0 \epsilon d_{AB} kT} \tag{TT}$$

where

$$\ln k_0 = \ln \frac{kT}{h} - \frac{\Delta G_0}{RT}$$

and the Boltzmann constant $\mathbf{k}$ has been substituted in the electrostatic term for $(R/6.02 \times 10^{23}) = \mathbf{k}$. The term k_0 is the rate constant for the ion reactions in a medium where $\epsilon = \infty$, that is, when the electrostatic forces have become zero.

Equation (TT) predicts that the rate constant for the reaction between two ions in solution will depend on the dielectric constant ϵ, and hence the nature, of the solvent. A plot of ln k against $1/\epsilon$ should be a straight line with slope

$$-\frac{Z_A Z_B e^2}{4\pi\epsilon_0 d_{AB} \mathbf{k} T}$$

From the slope, d_{AB} can be obtained. Experimentally, it is found that Eq. (TT) is indeed followed in many cases.

In the atmosphere, the solutions available for reaction are aqueous. The dielectric constant for water at 25°C is $\epsilon = 78.3$.

A second important factor for reactions in solution between ions or polar molecules is the ionic strength (I) of the solution. I is defined as

$$I = \tfrac{1}{2} \Sigma C_i Z_i^2$$

where C_i is the molar concentration of the ith ion and Z_i is its charge. For a 1 M solution of $NaNO_3$, for example, $I = 1\ M$, whereas for a 1 M solution of Na_2SO_4, $I = 3\ M$. ($C_{NA}^{+} = 2$, $Z_{NA^+} = +1$, $C_{SO_4{}^{2-}} = 1$, $Z_{SO_4{}^{2-}} = -2$).

In solution thermodynamics, the concentration (C) of ions is replaced by their activity, a, where

$$a = C\gamma$$

and γ is the activity coefficient that takes into account non-ideal behavior due to ion–solvent and ion–ion interactions. The Debye–Hückel limiting law predicts the relationship between the ionic strength of a solution and γ for an ion of charge Z in dilute solutions:

$$\log \gamma = -BZ^2\sqrt{I} \qquad \text{(UU)}$$

B is a constant which depends on the properties of the solution, for example, on its dielectric constant, and on the temperature. For water at 25°C, $B \simeq 0.51$ $L^{1/2}$ mole$^{-1/2}$. The Debye–Hückel limiting law applies only for solutions of low ionic strength, for example, below ~0.01 M for 1:1 electrolytes, such as $NaNO_3$, and below ~0.001 for electrolytes of higher charge.

The influence of ionic strength on solution rate constants can be anticipated by again resorting to transition state theory. For the reaction

$$A + B \rightleftharpoons X^{\ddagger} \xrightarrow{k_p} \text{products}$$

the rate of the reaction is given by $k_p[X^\ddagger]$, where $[X^\ddagger]$ is the concentration of the activated complex in the transition state. The concentration of the activated complexes can be obtained from the equilibrium assumed between the reactants and $X^\ddagger$:

$$K^\ddagger = \frac{a_{X^\ddagger}}{a_A a_B} = \frac{\gamma_{X^\ddagger}[X^\ddagger]}{\gamma_A[A]\gamma_B[B]}$$

Thus

$$\text{rate} = k_p K^\ddagger \frac{\gamma_A \gamma_B}{\gamma_{X^\ddagger}} [A][B]$$
$$= k\,[A][B]$$

where the reaction rate constant is given by

$$k = k_p K^\ddagger \frac{\gamma_A \gamma_B}{\gamma_{X^\ddagger}}$$

Thus

$$\log k = \log (k_p K^\ddagger) + \log \frac{\gamma_A \gamma_B}{\gamma_{X^\ddagger}}$$

Using the Debye–Hückel limiting law for the relationship between the activity coefficients γ and the ionic strength of the solution, one finds

$$\log k = \log k_0 + \log \gamma_A + \log \gamma_B - \log \gamma_{X^\ddagger}$$
$$= \log k_0 + 2BZ_A Z_B \sqrt{I}$$

Using $B = 0.51\ \text{L}^{1/2}\ \text{mole}^{-1/2}$ for aqueous solutions at 25°C, this becomes

$$\log k = \log k_0 + 1.02\ Z_A Z_B \sqrt{I} \qquad \text{(VV)}$$

Thus a plot of $\log k$ against $\sqrt{I}$ should give straight lines of slope $1.02\ Z_A Z_B$ and intercepts of $\log k_0$. The constant k_0 is seen to be the rate constant in a solution of zero ionic strength, that is, at infinite dilution. Equation (VV) also predicts that reactions between ions of the same sign should speed up as the ionic strength increases, whereas reactions between oppositely charged ions should slow down with increasing ionic strength. For reactions between an ion and an uncharged molecule ($Z_B = 0$), ionic strength should not alter the rate constant. These relationships have been confirmed for solutions that are sufficiently dilute so that the

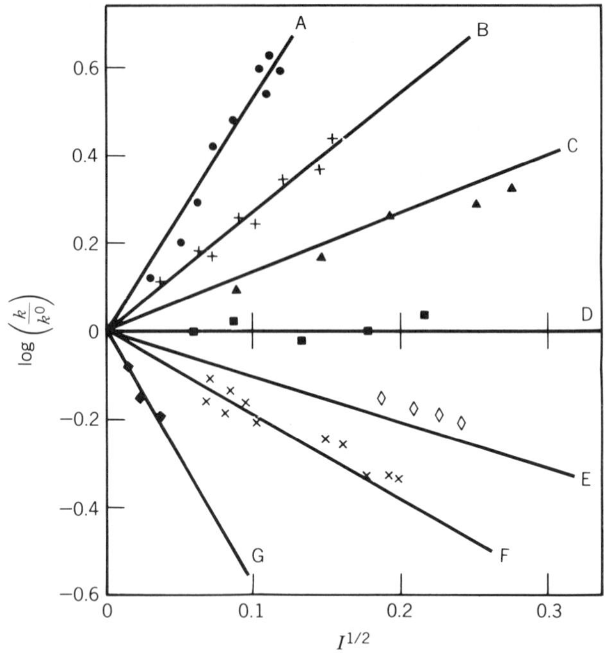

FIGURE 4.21. Variation of rate constant with ionic strength (I) of the solution for reactants having different charges (from Benson, 1960). Reactions: A: $[Co(NH_3)_5Br]^{2+} + Hg^{2+} + H_2O \rightarrow [Co(NH_3)_5(H_2O)]^{3+} + (HgBr)^+$, B: $S_2O_8^{2-} + I^- \rightarrow$ A: $\rightarrow I_3^- + 2SO_4^{2-}$ (not balanced), C: $[O_2N\text{—}N\text{—}COOEt]^- + OH^- \rightarrow N_2O + CO_3^{2-} + EtOH$, D: cane sugar + $OH^- \rightarrow$ invert sugar (hydrolysis reaction), E: $H_2O_2 + H^+ + Br^- \rightarrow H_2O + \frac{1}{2} Br_2$ (not balanced), F: $[Co(NH_3)_5Br]^{2+} + OH^- \rightarrow [Co(NH_3)_5(OH)]^{2+} + Br^-$, G: $Fe^{2+} + Co(C_2O_4)_3^{3-} \rightarrow Fe^{3+} + [Co(C_2O_4)_3]^{4-}$.

Debye–Hückel law is applicable (Fig. 4.21). As might be expected, deviations are observed at higher ionic strengths [e.g., see the text by Benson (1960) for a more detailed discussion].

An example of the application of the effect of changing ionic strength to test a reaction mechanism is that of Maahs (1983); he showed that the increase in log k with ionic strength in the O_3–S(IV) reaction is consistent with the reaction of two univalent negative species (i.e., OH^- and HSO_3^-), an important result that suggests an ionic rather than the previously assumed free radical mechanism (see Section 11-A-2d).

This effect of ionic strength on solution rate constants is extremely important in studying reactions relevant to atmospheric chemistry. Thus care must be taken to study the effects of ionic strength over a range that approximates those found in the atmosphere. Aerosols in polluted urban areas can be highly concentrated solutions with ionic strengths in the range of 8–19 M (Stelson and Seinfeld, 1981); reactions in these solutions will not follow the "ideal" relationships discussed above. On the other hand, cloud water and rainwater in clean areas contain much lower solute concentrations; for example, from the ionic composition of precipitation samples in the maritime area of Cape Grim, Australia (Ayers, 1982), the ionic strength can be calculated to be $\sim 10^{-3}$ M. An example of the application of solution thermodynamics to atmospheric problems is found in the article by Stelson et al. (1984), which treats the thermodynamics of aqueous solutions containing nitrate, sulfate, and ammonium ions.

4-D-4. Experimental Techniques Used for Studying Solution Reactions

In principle, the kinetic techniques for studying aqueous phase reactions are analogous to those used in the gas phase. Thus pseudo-first-order conditions are commonly used to simplify the data analysis, and the decrease in the reactant present in the smaller concentration is followed with time.

For reactions occurring on the millisecond time scale or longer, flow methods are often used. Liquids can be mixed in $\sim 10^{-3}$ s and allowed to flow down a tube; measurement of the reactant concentration along the length of the tube can then be translated into concentrations as a function of time if the flow rate is known.

A second method is the stopped-flow technique in which the reactant solutions are rapidly mixed and the flow then suddenly stopped in a reaction cell. The decrease in concentration of one of the reactants is then followed with time using standard analytical techniques such as spectrophotometry.

The stopped-flow method has been used in many studies of the S(IV) oxidation by other atmospheric pollutants. For example, Martin and co-workers (Martin et al., 1981) studied the reaction of S(IV) in aqueous solutions with some oxides of nitrogen. Solutions of the bisulfite ion (SO_3^{2-}) were mixed with solutions containing an excess of the second reactant. The decay of S(IV) was followed using the absorbance of the aquated form of SO_2 ($SO_2 \cdot aq$) (see Section 11-A-2) at 280 nm. Because $SO_2 \cdot aq$ is in rapid equilibrium with the other forms of S(IV), such as HSO_3^-, the decay of the absorbance at 280 nm can be used to follow the loss of S(IV).

Figure 4.22 shows the raw data in one run for the reaction of S(IV) with HNO_3; in this case the absorbance (i.e., optical density = OD) at 280 nm did not change significantly with time up to ~ 1000 s, from which an upper limit to the rate constant of 0.01 L mole^{-1} s^{-1} could be obtained.

For much faster reactions, flow methods cannot be used because the time of mixing (~ 1 ms) becomes limiting. In this case relaxation methods can be employed. This involves perturbing a system in equilibrium by rapidly changing a parameter, such as the temperature, and following the response of the system to the shock as it returns to equilibrium; alternatively the parameter can be varied sinusoidally and the lag in response of the chemistry followed.

As discussed at the beginning of this section, the reactions of gases with solutions involves a number of steps which include mass transport as well as chemical reaction. Studies of such gas–liquid reactions relevant to atmospheric chemistry have been carried out using reaction cells such as that in Fig. 4.23. The gas is introduced into the solution in the form of finely divided bubbles by passing it through a fritted disk. The course of the reaction is followed by monitoring changes

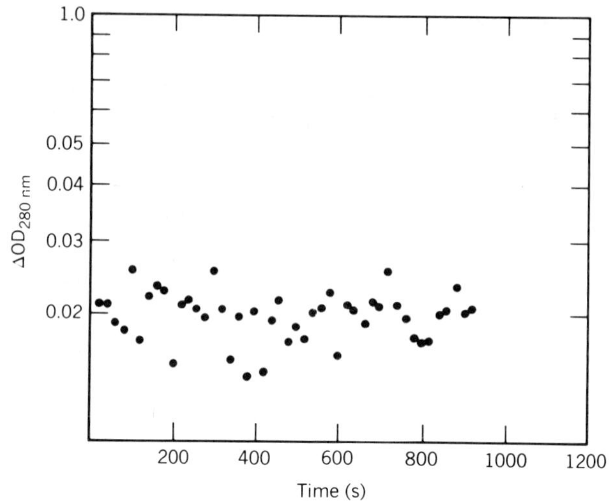

FIGURE 4.22. Change in optical density at 280 nm ($\Delta OD_{280\ nm}$) versus reaction time for the reaction of 2.5×10^{-3} M SO_3^{2-} with 5×10^{-3} M HNO_3 in a stopped flow system (from Martin et al., 1981).

in the solution concentrations either using conventional chemical analytical techniques or parameters such as the electrical conductivity which can be related to the solution concentrations. Data analysis is somewhat complex in such systems because mass transport as well as the chemical kinetics must be taken into account; this is discussed in detail by Lee and Schwartz (1981).

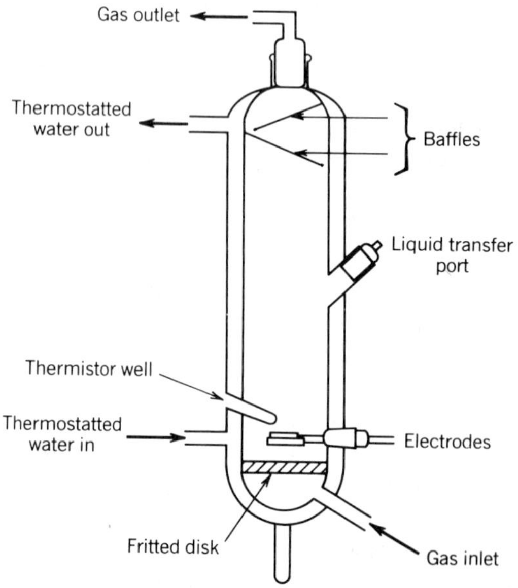

FIGURE 4.23. Gas-liquid reaction cell used in NO_2–H_2O studies (from Lee and Schwartz, 1981).

Finally, some gas–liquid reactions have been studied using cloud chambers. In these experiments condensation nuclei are generated and put into the chamber to serve as sites for the formation and growth of cloud droplets initiated by a drop in the chamber pressure. When carried out in the presence of gases, these cloud droplets provide the aqueous phase for reactions under conditions which are closer to atmospheric than using bulk solutions. Such cloud chamber experiments, when carried out at concentrations approximating those in the atmosphere, minimize the problems associated with mass transport. Rates of chemical reaction in the droplets can be derived by chemical analysis of the cloud water collected after the experiment.

Figure 4.24 is a schematic diagram of one such cloud chamber apparatus used to study reactions of atmospheric interest (Gertler et al., 1984).

For example, the reaction

$$2NO_{2(g)} + H_2O_{(l)} \rightarrow 2H^+ + NO_3^- + NO_2^- \quad (56)$$

has been postulated over the years to be responsible for such observations as the oxidation of NO_2 in clouds and the formation of nitrous acid in the atmosphere. This reaction has been studied by Lee and Schwartz (1981) using the apparatus shown in Fig. 4.23 and by Gertler and co-workers (1984) using the cloud chamber shown in Fig. 4.24. The results of the two studies, shown in Fig. 4.25, are in agreement within a factor of 5, which Gertler and co-workers state may be acceptable for such studies considering that different experimental methods and temperatures are used; for example, the the temperature was held fixed at 22°C in the experiments of Lee and Schwartz, but fell from 23°C to 10°C during the cloud chamber experiments.

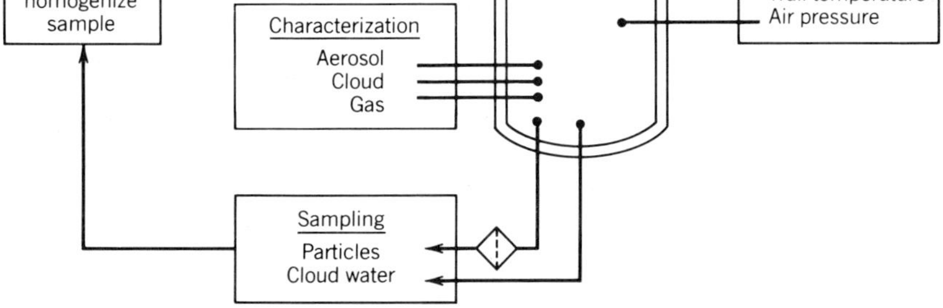

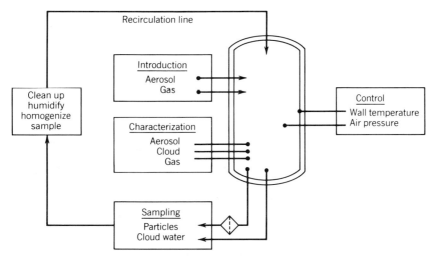

FIGURE 4.24. Schematic diagram of cloud chamber apparatus used in atmospheric chemistry studies of reactions in droplets (from Gertler et al., 1984).

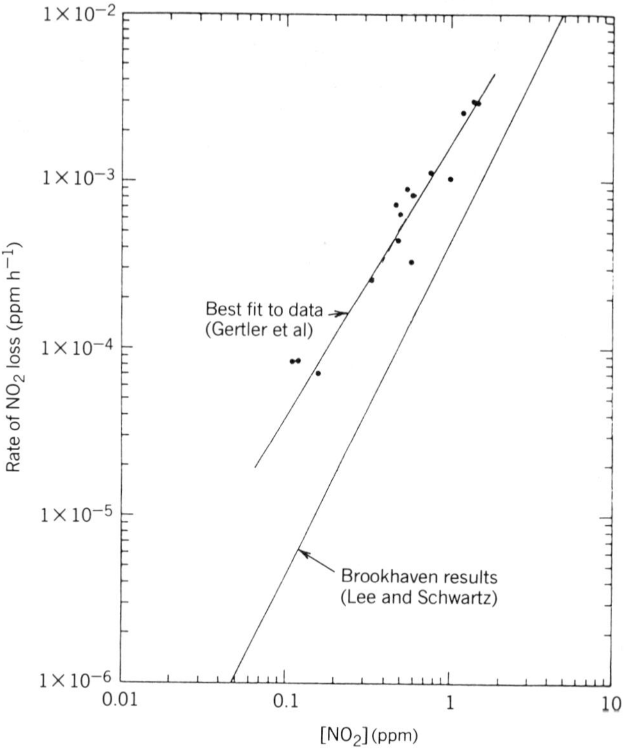

FIGURE 4.25. Comparison of the results of two studies of the rate of the reaction $2NO_{2(g)} + H_2O_{(l)} \rightarrow 2H^+ + NO_2^- + NO_3^-$. Upper line through data points from cloud chamber experiments; lower line from bulk phase studies of Lee and Schwartz (1981) (from Gertler et al., 1984).

Finally, pulse radiolysis, discussed in Secion 4-B-5b, and flash photolysis are also frequently used to study solution phase reactions. These techniques will undoubtedly grow in application to solution phase atmospheric chemistry in the future, as our need to understand these condensed phase reactions is growing rapidly (see Section 11-A-2a).

4-E. REACTIONS ON SOLID SURFACES

Some reactions of atmospheric interest can occur on the surfaces of solids. For example, SO_2 is known to be oxidized to sulfate on surfaces such as graphite, soot, fly ash, MgO, V_2O_5, and Fe_2O_3 (Section 11-A-3), and particle-bound polycyclic aromatic hydrocarbons (PAH) are known to photooxidize and react with gaseous pollutants such as (NO_2 + HNO_3) and O_3 (Chapter 13).

4-E-1. Gas–Solid Reactions

Relatively little is known at present about the detailed molecular mechanisms involved in reactions of gases on surfaces in the atmosphere. In some cases the reactions leave the interface chemically altered at the end of the reaction. For example, SO_2 is initially rapidly oxidized to sulfate on the surface of freshly formed soot particles, but the subsequent formation of sulfate by this surface reaction is much slower; this has been attributed to sulfate and other species covering the surface, rendering it less active for the oxidation reaction (Novakov, 1984). Similarly, the reaction of NO_2 with dry sea salt aerosols (NaCl mainly) leaves a layer of $NaNO_3$ on the surface (Chung et al., 1978). As a result, as the reaction proceeds and the surface becomes increasingly covered with product, the reaction tends to slow down.

On the other hand, there are some surface processes that may leave the interface unaltered chemically after the reaction has occurred. A relevant example is laboratory studies of the loss of atoms and free radicals on surfaces. The hydroxyl radical has been found to be particularly sensitive to loss at surfaces. Thus in flow tube studies (see Section 4-B-2), the tube must be coated with a relatively inert material such as a halocarbon wax or phosphoric acid to inhibit the loss of OH from the gas phase even in the absence of other reactants. If the walls of the flow tube are not treated, the rate of loss of OH at the surface is so large that it is experimentally impossible to obtain accurate and reliable rate constants for its reactions in the gas phase. Although no definitive studies of the mechanism of the surface reactions of OH under such conditions have been reported, the fact that such untreated surfaces do not seem to become conditioned with exposure to OH suggests that the surface may not be altered chemically.

Adsorption of species on solids is usually classed as either physical adsorption (also known as physisorption or van der Waals adsorption) or chemisorption. Physical adsorption is relatively weak, the heat of adsorption generally being $\lesssim 5$ kcal mole^{-1}. The forces involved in the adsorption are thought to be similar to van der Waals forces between gas molecules.

Chemisorption, on the other hand, involves much stronger binding forces, the heat released being of the order of that in chemical reactions, that is, typically 10–1000 kcal mole^{-1}. This type of process is usually slower than physical adsorption and is often irreversible. A classic example is the sorption of H_2 on metals where it is believed the H_2 dissociates to form H atoms which become attached to the surface in the form of metal hydrides.

Adsorption of a gas at a surface must be the first step in reactions occurring on solids. This can be described in its simplest form by Eq. (57):

$$\mathrm{G} + \underset{/////}{-\mathrm{S}-} \underset{k_{-57}}{\overset{k_{57}}{\rightleftharpoons}} \underset{/////}{-\overset{\overset{\mathrm{G}}{|}}{\mathrm{S}}-} \tag{57}$$

where G is the gas phase molecule and S is the surface. This type of mechanism leads to what is known as the Langmuir isotherm:

$$\theta = \frac{K[\mathrm{G}]}{1 + K[\mathrm{G}]} \tag{WW}$$

This adsorption isotherm gives the relationship between the gas phase concentration, [G] and θ, the fraction of the surface covered by adsorbed molecules at equilibrium; K is a constant for a given temperature.

The Langmuir isotherm can be developed by considering the rates of adsorption of G onto the surface and its rate of desorption back into the gas phase. Thus the rate of adsorption is expected to be proportional to the concentration of G in the gas phase (which by gas kinetic theory is proportional to the number of collisions of G with the surface) and to the fraction of the surface available for adsorption. If θ is the fraction of the surface already covered by G, then the fraction still available for reaction is $1 - \theta$. Thus the rate of adsorption should be given by

$$\text{rate of adsorption} = k_{57}[\mathrm{G}][1 - \theta]$$

Similarly, the rate of desorption should be proportional to the concentration of surface adsorbed species:

$$\text{rate of desorption} = k_{-57}\theta$$

Taking $k_{57}/k_{-57} = K$, where K is the equilibrium constant for reaction (57), and rearranging, one arrives at the Langmuir isotherm:

$$\theta = \frac{K[\mathrm{G}]}{1 + K[\mathrm{G}]} \tag{WW}$$

The Langmuir isotherm describes the adsorption of gases on solids in many systems. However, a number do not follow this predicted behavior, and other isotherms have been postulated for such cases. An example is the Freundlich isotherm:

$$\theta = k[\mathrm{G}]^n \tag{XX}$$

Advantages and disadvantages of these and other isotherms are discussed in detail elsewhere (e.g., see Benson, 1960; Adamson, 1982). One of the advantages of the Langmuir isotherm is the logical kinetic basis from which it can be derived. Hence we shall use it in further considering reactions at surfaces.

The Langmuir isotherm can be applied to predicting rates of unimolecular and bimolecular reactions on surfaces. If a molecule G is rapidly adsorbed on the surface, and then undergoes a slow chemical reaction, the latter will be the rate-determining step. For a unimolecular reaction then, the concentration of the reactant surface species will be proportional to the fraction θ of the surface which is covered with G, and the rate of reaction is given by

$$\text{rate of unimolecular reaction} = k\theta$$

where k is the reaction rate constant. Substituting for θ from the Langmuir adsorption isotherm, one obtains

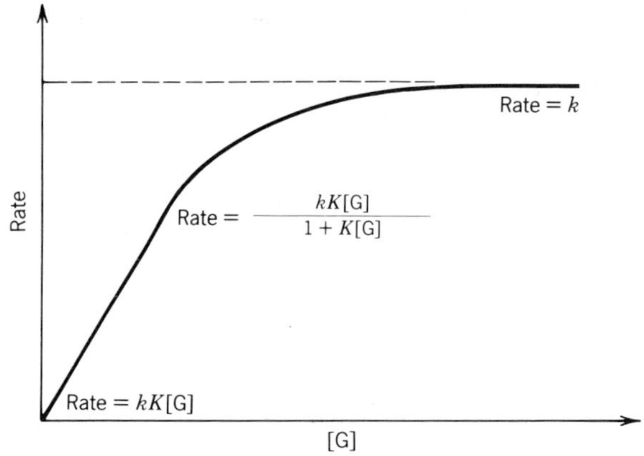

FIGURE 4.26. Dependence of the rate of a unimolecular surface reaction of an adsorbed species G on its gas phase concentration [G].

$$\text{rate} = \frac{kK[G]}{1 + K[G]} \tag{YY}$$

Figure 4.26 is a plot of the rate against the gas phase concentration [G]. As expected from the form of (YY), at low concentrations of G, $1 \gg K[G]$ and the rate is first order in G. In the other extreme at very high concentrations of G, $K[G] \gg 1$ and the rate is zero order in G.

Bimolecular reactions on surfaces in the atmosphere have also been suggested. There are two possible mechanisms for such reactions. In one type, the two gaseous species G and A first become adsorbed on the surface. When they become adsorbed on adjacent surface sites, they can react, and the products formed can then desorb into the gas phase. This is known as the Langmuir–Hinshelwood mechanism. A second mechanism, known as the Langmuir–Rideal mechanism, involves only one of the two reactants becoming adsorbed on the surface prior to reaction. The second molecule in the gas phase then collides with the surface-adsorbed species and reacts. The two mechanisms can be represented as follows:

Langmuir–Hinshelwood

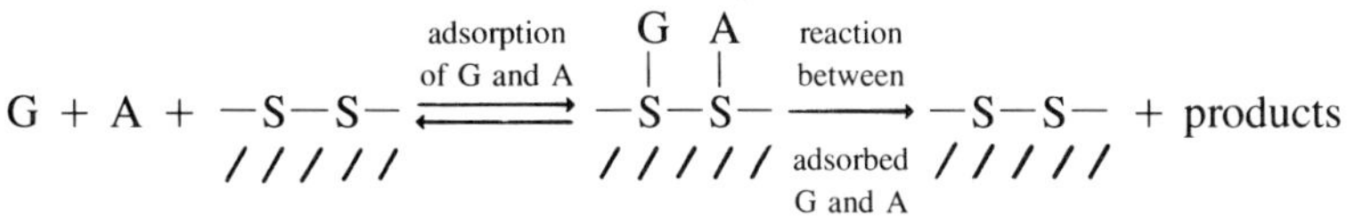

(58)

Langmuir–Rideal

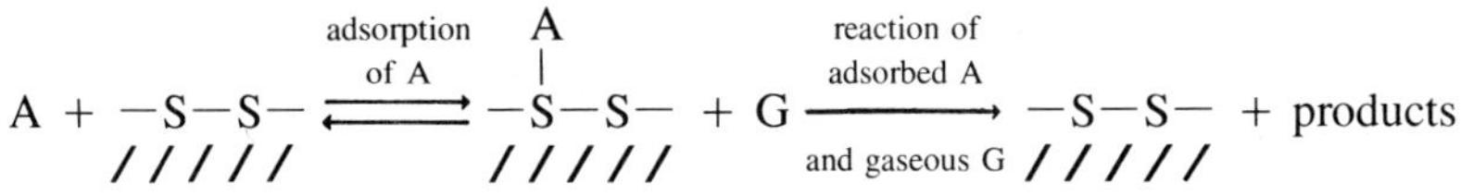

(59)

For the Langmuir–Hinshelwood mechanism, the rate of the bimolecular reaction should be proportional to the fractions of the surface covered by the two adsorbed reactants, θ_G and θ_A, respectively. From the Langmuir isotherm, it can be shown (Laidler and Meiser, 1982) that these fractions are given by

$$\theta_G = \frac{K_G[G]}{1 + K_G[G] + K_A[A]} \tag{ZZ}$$

and

$$\theta_A = \frac{K_A[A]}{1 + K_G[G] + K_A[A]} \tag{AAA}$$

The rate of a bimolecular reaction between two adsorbed species G and A, is thus, according to the Langmuir–Hinshelwood mechanism, given by

$$\begin{aligned} \text{rate} &= k\theta_G\theta_A \\ &= \frac{kK_GK_A[G][A]}{(1 + K_G[G] + K_A[A])^2} \end{aligned} \tag{BBB}$$

If the concentration of A is held constant and [G] is varied, the rate first increases with [G] and then decreases, as shown in Fig. 4.27. The decrease arises because very high concentrations of G result in large amounts of adsorbed G and correspondingly less of the second reactant A.

The Langmuir–Rideal mechanism involves reaction between an adsorbed species A and a molecule from the gas phase, G. The rate of the reaction should thus be given by

$$\text{rate} = k[G]\,\theta_A$$

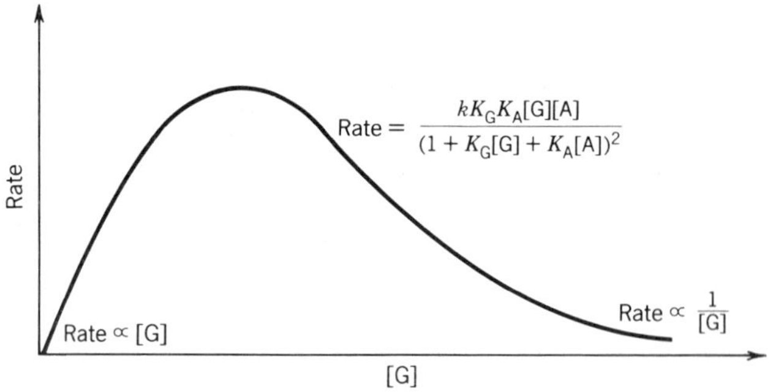

FIGURE 4.27. Variation of rate of a bimolecular surface reaction between two adsorbed species G and A according to the Langmuir–Hinshelwood mechanism, when [A] is held constant and [G] is varied.

Since G may be adsorbed on the surface in addition to A (but is assumed not to react when adsorbed), the expression for θ_A is as given earlier:

$$\theta_A = \frac{K_A[A]}{1 + K_G[G] + K_A[A]} \tag{CCC}$$

The rate is thus

$$\text{rate} = \frac{kK_A[A][G]}{1 + K_G[G] + K_A[A]} \tag{DDD}$$

This is plotted against [G] in Fig. 4.28. At low [G], the rate is proportional to [G], but at high concentrations the rate levels off. This is because the number of collisions of G with the surface is proportional to [G], so that at high concentrations the number of collisions is sufficiently great that it no longer determines the rate; instead, the availability of surface-adsorbed A becomes controlling.

The different shapes of the curves in Figs. 4.27 and 4.28 provide a means of differentiating between the two mechanisms. Thus in the Langmuir–Hinshelwood case, the rate falls at high [G], whereas in the Langmuir–Rideal case, it plateaus off to a constant value.

4-E-2. Liquid–Solid Reactions

The considerations discussed above for gas–solid interactions also apply to the interactions of species in the liquid phase with surfaces. However, in this case the presence of ions in solution and the resulting electrostatic forces when they approach a surface must also be considered. For example, if the surface of a solid suspended in solution is negatively charged, the positively charged ions will tend

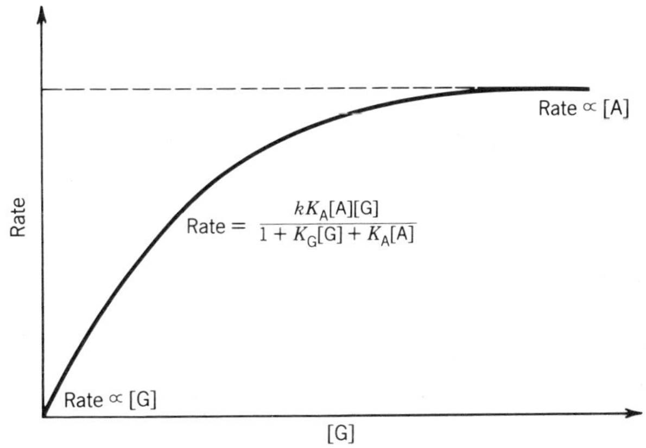

FIGURE 4.28. Variation of rate of a bimolecular surface reaction between an adsorbed species A and a molecule G from the gas phase according to the Langmuir–Rideal mechanism when [A] is held constant and [G] is varied.

to be attracted to it, forming an electric "double layer," that is, a layer of negative charges on the solid with a layer of positive charges from the solution lined up along the surface.

Such electrostatic interactions may play an important role in reactions on the surfaces of solids that are suspended in atmospheric droplets. For example, Novakov and co-workers have found that the rate of S(IV) oxidation on carbon surfaces suspended in aqueous solutions depends on the pH; in one study the rate of oxidation to S(IV) was observed to fall to zero at pH values above 7.6 but to be independent of pH at pH < 7.6 (Brodzinsky et al., 1980). They postulated that this was due to an alteration of the reactive site on the carbon by the change in pH. This is reasonable if at pH > 7.6 the surface became negatively charged; since S(IV) exists primarily in solution in the form HSO_3^- and SO_3^{2-}, electrostatic repulsion between the reactive ions and the surface would then greatly decrease the rate of the surface reaction.

The ionic nature of surfaces is determined not only by the particular environment, but also by its prior history. Thus the surface of black carbon particles may be either acidic or basic. For example, as discussed by Chang and Novakov (1983), exposure of carbon to O_2 at temperatures of 200–400°C gives an acidic surface, probably because of the formation of carboxyl and hydroxyl groups on the surface. On the other hand, heating carbon under vacuum or in CO_2 gives a basic surface on exposure to O_2 at room temperature; the nature of the bases has not been elucidated. Some idea of the chemical complexity of even relatively simple surfaces such as carbon is seen in Fig. 4.29, which gives some of the functional groups that have been postulated to exist on carbon surfaces (Chang et al., 1982).

4-E-3. Experimental Techniques Used in Studying Reactions on Surfaces

4-E-3a. Gas–Solid Reactions

Many elegant techniques such as Auger electron spectroscopy (AES), low-energy electron diffraction spectroscopy (LEEDS), and X-ray electron spectroscopy (XPS) (also known as electron spectroscopy for chemical analysis—ESCA) are currently in use to study the interactions of gases with and on surfaces. While a great deal of progress in understanding the detailed chemical reaction mechanisms of single molecules on surfaces has been made using them, they are unfortunately not directly applicable to conditions relevant to the atmosphere. Thus, as we have seen in a number of cases, data applicable to atmospheric reactions must either be gathered under those conditions or be capable of being reliably extrapolated to such conditions. Techniques such as AES and LEEDS require ultra-high vacuums, very "clean" systems, and generally involve only a few very-high-purity reactants, conditions that are far removed from those found in the atmosphere. As a result, these have not been widely applied to following the course of atmospheric reactions on surfaces in real time.

Such spectroscopic techniques, however, have been used to examine the surface of solids *after* exposure to various pollutants. For example, soot particles have been exposed to pollutants such as NH_3 or SO_2 in air and the surfaces then ex-

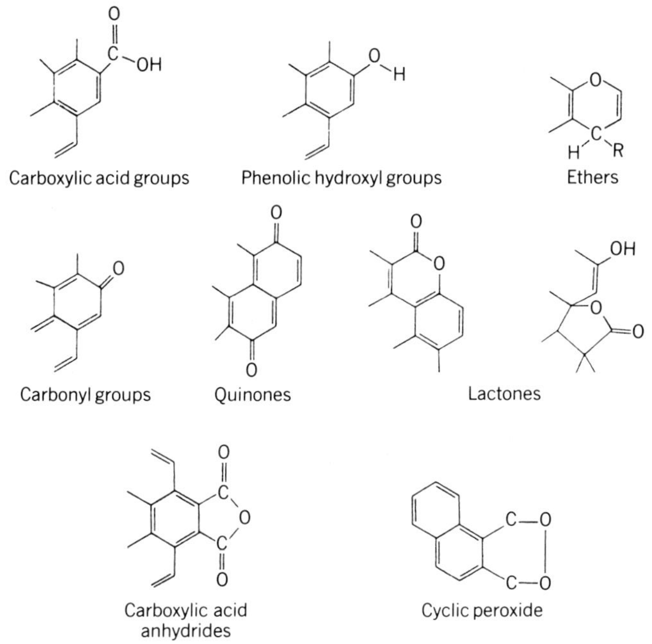

FIGURE 4.29. Some oxygenated functional groups proposed to exist on the surface of carbon particles (from Chang et al., 1982).

amined by electron spectroscopy for chemical analysis (ESCA). The same technique has been used to examine ambient particles. The formation on such surfaces of sulfate, nitrate, NH_4^+, and reduced nitrogen species has been clearly identified in this manner (Chang and Novakov, 1983). Figure 4.30, for example, shows the ESCA spectrum of graphite exposed to SO_2 in air; peaks due to sulfate and sulfide can be clearly seen.

Other spectroscopic techniques such as infrared (e.g., see Goodsel et al., 1972; Lin and Lunsford, 1975) and EPR (Lin and Lunsford, 1975) have also been applied to examining the surfaces of solids. Because IR can be applied under atmospheric conditions, it has the potential for following the changes in surfaces during the course of a reaction. Figure 4.31, for example, shows the formation of $NaNO_3$ on the surface of NaCl and NOCl in the gas phase upon exposure to NO_2. While high NO_2 concentrations (Torr) were used in these particular studies, the same reaction has been shown to occur at ppm NO_2 concentrations (Finlayson-Pitts, 1983). The relatively recent development of reflectance techniques such as diffuse reflectance combined with Fourier transform infrared spectrometry (commonly known as DRIFTS) will undoubtedly assist in following chemical changes on solid surfaces during a reaction under conditions more realistically approaching atmospheric (Griffiths et al., 1982).

A less direct means of studying gas–solid reactions involves passing the gaseous air pollutant over the solid and measuring the rate of loss of the gaseous species. The most common way of expressing the rate of the loss is the fraction, ϕ, of

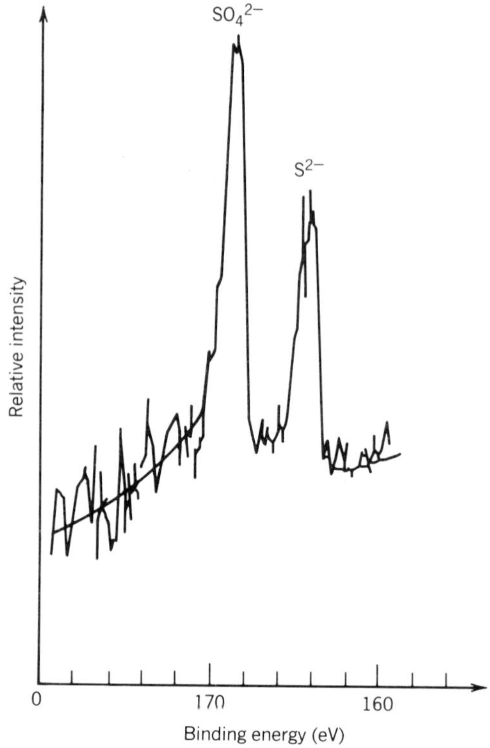

FIGURE 4.30. ESCA spectrum of graphite particles exposed to SO_2 in air (from Novakov et al., 1974).

collisions of the gas with the surface which results in removal of the gaseous species. The loss of the gas can be measured along the length of a flow tube, for example (see Section 4-B-2), and ϕ can be calculated from these data by using a knowledge of the flow dynamics.

The coefficient ϕ is very commonly reported in the literature for the loss of species from the gas phase at surfaces. It can be directly applied to atmospheric conditions using conventional gas kinetic molecular theory. Thus the rate of collisions between a gas molecule of molecular weight M, present at a concentration [G] in the gas phase, and a particle having surface area A is given by

$$\text{rate of collisions} = \left(\frac{RT}{2\pi M}\right)^{1/2} A[\text{G}]$$

However, not all collisions lead to reaction or incorporation into the particle; many collisions simply result in reflection off the surface. The rate of removal of the gaseous species by collisions with the particle is thus given by

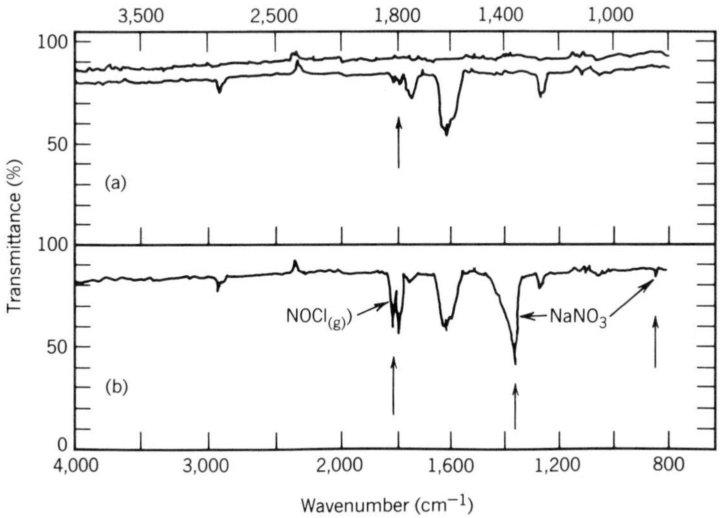

FIGURE 4.31. Formation of $NaNO_3$ on the surface of NaCl and of NOCl in the gas phase due to the reaction of NO_2(g) with NaCl, followed by infrared spectroscopy. Bands at ~1360 and 840 cm^{-1} are due to $NaNO_3$; that at ~1800 cm^{-1} is due to NOCl in gas phase (a) at a reaction time < 10 min, (b) after 96 min, (from Finlayson-Pitts, 1983).

$$\text{rate of removal of gas} = \left(\frac{RT}{2\pi M}\right)^{1/2} A[\mathrm{G}]\,\phi \qquad \text{(EEE)}$$

ϕ is variously referred to as a *surface recombination coefficient*, *accommodation coefficient*, or *sticking coefficient*.

An atmospherically relevant example of the determination of ϕ is a study by Judeikis and co-workers (1978). They measured the loss of SO_2 at a number of solid surfaces of atmospheric interest using a tubular flow reactor of radius r. The tube was coated with the solid of interest and the loss of SO_2 followed using mass spectrometry. They applied a model to the data which is based on equating the diffusive flux of SO_2 to the wall, given by $-D\ \delta c/\delta r$ (where D is the diffusion coefficient and c the gas phase SO_2 concentration), to the rate of removal of SO_2 at the surface:

$$-D\frac{\delta c}{\delta r} = \phi k_r c \qquad \text{(FFF)}$$

k_r is the molecular speed of SO_2 in the radial direction, so that $(k_r c)$ is the rate of gas–solid collisions. Figure 4.32 shows some typical data for the loss of SO_2 along the length of the flow tube when it was coated with fly ash.

Another technique that measures the loss of gases on contact with solids or with

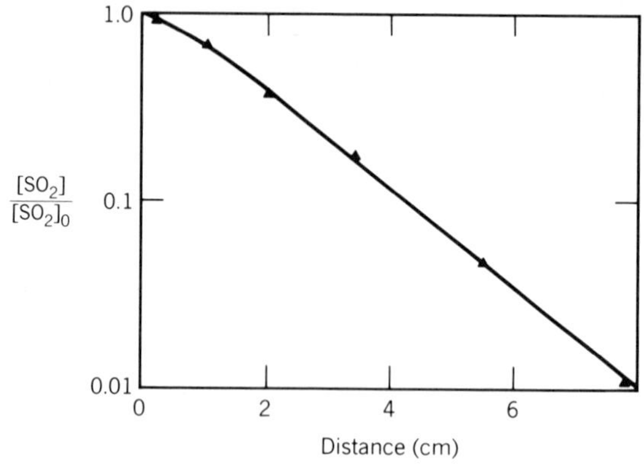

FIGURE 4.32. Typical data for the loss of SO_2 along the length of a tubular reactor coated with Mohave fly ash. The experimental data are shown by the triangles, and the solid curve was calculated for $\phi = 4.4 \times 10^{-4}$. Total pressure was 55 Torr with an O_2 pressure of 6 Torr and SO_2 of 9 mTorr. Relative humidity was 0% (from Judeikis et al., 1978).

liquid surfaces such as H_2SO_4 is the Knudsen cell reactor. Figure 4.33 is a diagram of a two-chamber reactor. The lower chamber contains the surface, while the upper chamber contains the gas inlets and the aperture through which the gas mixture containing products and unreacted materials is sampled into a mass spectrometer.

Figure 4.34 shows the results from a typical experiment in which a carbon surface was exposed to SO_2 (Baldwin, 1982). The residence time of SO_2 in the reactor could be varied by changing the diameter of the escape aperture into the mass spectrometer; the loss of SO_2 as a function of time, that is, the kinetics of the reaction, could thus be obtained.

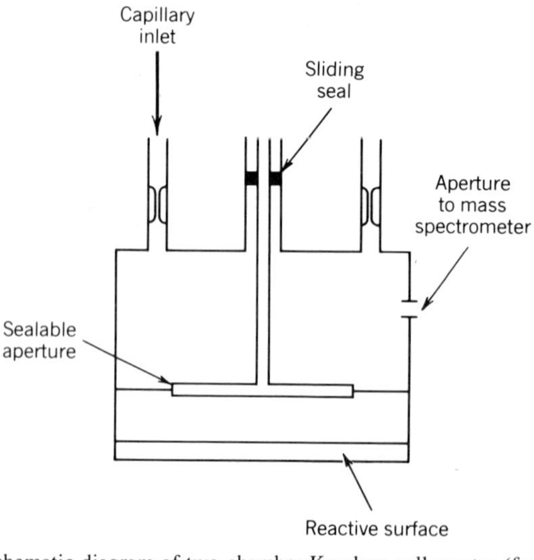

FIGURE 4.33. Schematic diagram of two-chamber Knudsen cell reactor (from Baldwin, 1982).

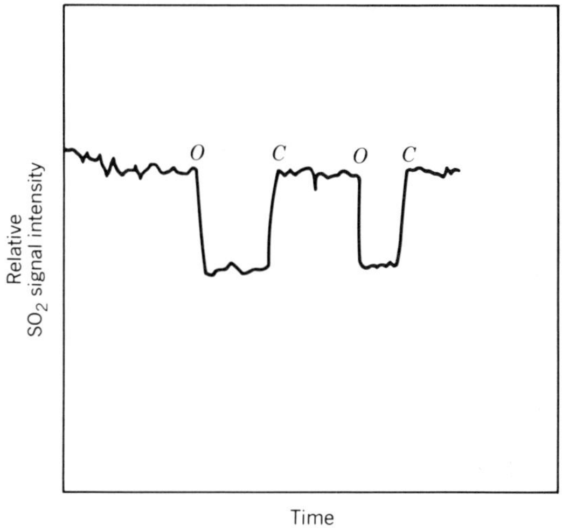

FIGURE 4.34. Typical data for SO_2 loss on carbon surfaces obtained using Knudsen cell reactor of Fig. 4.35. O = aperture open to mass spectrometer, C = aperture closed (from Baldwin, 1982).

A third technique used to study gas–solid reactions involves following the increase in weight of the solid as it is exposed to a gas. The increase in weight after the exposure has stopped and the volatile species have been desorbed (e.g., by passing a flow of inert gas over the solid) is taken as a measure of products of the gas–solid reaction left on the surface (e.g., see Cofer et al., 1981).

Techniques in which the loss of a gas upon exposure to a solid is measured, or net gain in weight of the solid followed, yield data on the kinetics of the loss but not the products or mechanisms involved. However, such kinetic studies are frequently extended by studying the chemical characteristics of the surface before and after the exposure. Spectroscopic studies of the surfaces, as described earlier, have been used. In addition, wet chemical techniques have been applied in which the surface species are extracted and measured using methods such as wet chemical analyses or ion chromatography.

4-E-3b. Liquid–Solid Reactions

Relatively few studies of the atmospheric reactions of dissolved species in aqueous solutions with solids have been carried out. Techniques that have been used involve suspending the solid particles in bulk solutions and analyzing aliquots of the solution at various times for reactants and/or products to obtain kinetics and/or mechanisms (e.g., see Brodzinsky et al., 1980). Alternatively, the particles have been suspended in air, humidified, and then cooled to cause water to condense on the particles to form liquid droplets in which the solid is suspended. Samples of these droplets are then periodically withdrawn from the chamber and collected on filters for conventional chemical analysis of reactants and products. For example, Fig. 4.35 is a schematic of such a "fog chamber" used in studying S(IV) reactions on carbon surfaces in aqueous solutions.

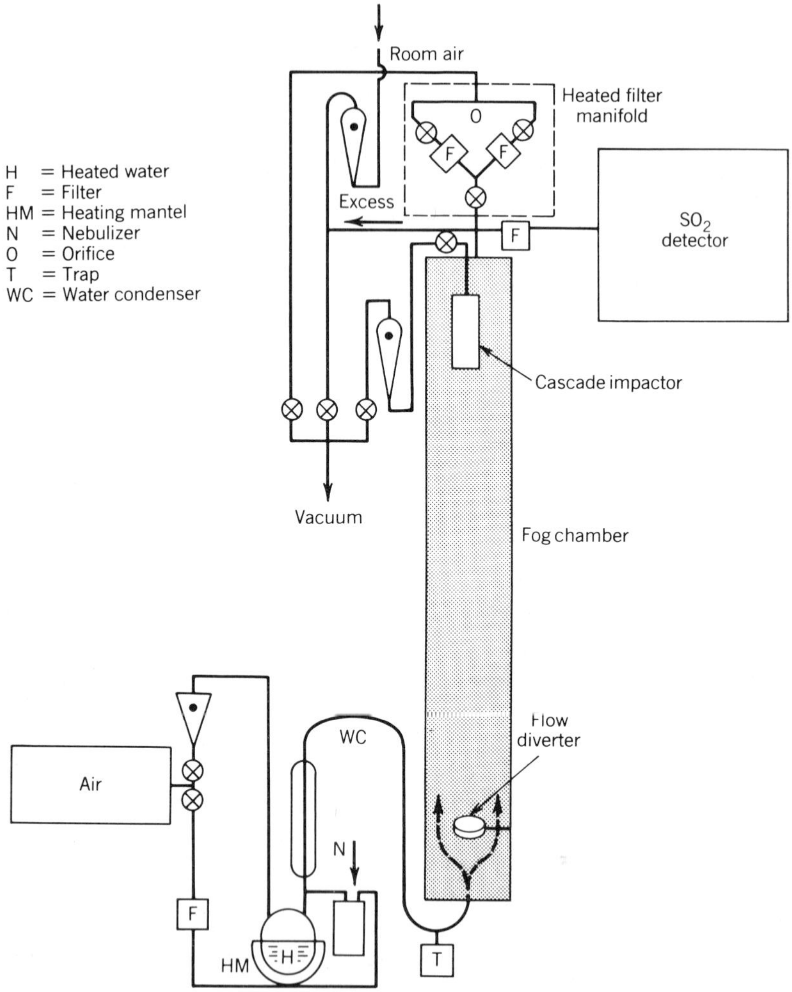

FIGURE 4.35. Schematic of fog chamber apparatus used to study surface reactions in aqueous solutions (from Benner et al., 1982).

Reactions on surfaces, whether suspended in air or in solution, are sensitive to a variety of factors, including the nature of the surface, its surface area, the temperature, and the relative humidity.

The surface area of the solid is a parameter that is routinely determined in all such studies. Because many surfaces are irregular and highly porous, their surface area is much greater than what might be expected on the basis of geometric considerations. For example, solids such as "porous glass," silica, and activated carbon typically have surface areas of the order of 10^2 m^2 per gram of material.

In addition to the need to determine surface areas in lab studies, measurements of the surface areas of atmospheric particulate matter can be used along with the total particulate mass to calculate the surface area mean diameter of the particles, if their density is known. Thus surface area determinations are important in characterizing size distributions of atmospheric particulate matter (see Section 12-B-1).

The most common method for determining these surface areas is the so-called BET method, named after Brunauer, Emmett, and Teller who first described it (Brunauer, Emmett, and Teller, 1938). The procedure can be found in most physical chemistry laboratory textbooks (e.g., see Shoemaker and Garland, 1967).

In brief, the BET method is based on an extension of Langmuir's isotherm, with a number of assumptions; these include that all adsorption sites on the solid are identical, once adsorbed the molecules can't move, layers of molecules can be adsorbed on top of the first layer, and so on. Based on this model, the so-called BET isotherm can be developed:

$$\frac{X}{V(1 - X)} = \frac{1}{V_m C} + \frac{(C - 1)\,X}{V_m C} \qquad \text{(GGG)}$$

Here $X = P/P_0$, where P is the pressure of the gas and P_0 is the vapor pressure of the corresponding liquid at that temperature, V is the volume of gas adsorbed on the surface of the solid at that temperature and pressure (expressed as the volume in cm^3 at 1 atm pressure and 273°K), V_m is the adsorbed volume that covers the entire surface with one monolayer, and C is a constant that depends on temperature.

If one can determine V_m, then the number of moles of gas required to form a monolayer on the surface of the solid can be calculated from the ideal gas law. Knowing the area of one gas molecule, this can be used to give the surface area of the solid, and knowing the weight of the sample, this gives the surface area in m^2 per g of solid.

In practice, a gas such as N_2, whose molecular area is known and which does not react with the surface, is introduced into a vacuum chamber containing a measured weight of the solid. The solid has been previously degassed by heating to remove adsorbed species such as H_2O. A measured pressure of N_2 is introduced into the sample chamber which is in a bath held at a known temperature; usually a liquid nitrogen bath is used. Knowing the temperature, P_0 can be obtained from the literature. The volume of gas adsorbed, V, at that pressure, P, can be measured gravimetrically or volumetrically. A plot of $X/V(1 - X)$ against X has a slope of $(C - 1)/V_m C$ and an intercept of $1/V_m C$. Thus values for V_m and C can be calculated from the slope and intercept, and the surface area calculated from V_m.

4-F. COMPILATIONS OF KINETIC DATA FOR ATMOSPHERIC REACTIONS

Starting in the early 1970s, there was a veritable explosion of kinetic studies relevant to atmospheric reactions. As a result, approximately two decades of data in atmospheric chemistry and kinetics now confront the non-specialist who now wishes to estimate atmospheric lifetimes of various molecules. Indeed, the voluminous literature and often conflicting results may be sufficient to cause even the most determined to blanch.

Fortunately, in recent years a number of excellent reviews of the available kinetic data on gas phase reactions have been compiled. These compilations generally summarize the values of the room temperature rate constants as well as the temperature and pressure dependencies reported in the literature. Where justified by the data, recommendations are made concerning which values are considered most reliable. These reviews are an invaluable source of kinetic data for atmospheric reactions and are highly recommended. The most recent gas phase kinetic compilations are those by Hampson and Garvin (1978), Hampson (1980), Baulch et al. (1980, 1982, 1984), and DeMore et al. (1983, 1985). The excellent reviews by Atkinson and Lloyd (1984), Atkinson and Carter (1984), and Atkinson (1985) are recommended for both kinetic and mechanistic data, especially with regard to gas phase organic reactions of atmospheric interest.

The recommendations from these reviews, as well as much of the data considered in deriving them, are included in the appropriate sections of this book.

As discussed earlier, protocols approved by the U.S. Environmental Protection Agency and by the European Chemical Industry Ecology and Toxicology Centre for determining the rate constants for O_3 and OH reactions, as well as photolysis rates for photochemically active compounds, have been developed to assist chemical manufacturers and others in assessing environmental lifetimes and fates of various chemicals.

It is only recently that substantial attention has been devoted to reactions of atmospheric interest in solutions. As a result, a body of accepted kinetic and mechanistic data comparable to that for gas phase reactions has not been developed. However, there are a number of reviews of solution phase chemistry relevant to the atmosphere; for an overview the reader is referred to those by Graedel and Weschler (1981), Graedel and Goldberg (1983) and Graedel et al. (1985). The aqueous phase chemistry of SO_2 is reviewed in the volumes edited by Schwartz (1983), Calvert (1984) and Durham (1984); the individual articles in these volumes are referred to in the appropriate sections of this book. The liquid phase oxidations of NO_x are also reviewed in these volumes and in articles by Schwartz and White (1981, 1983). Reviews of the chemistry of OH, HO_2, and aliphatic carbon-centered radicals in solution are also available (Farahataziz and Ross, 1977; Ross and Neta, 1979, 1982), as is a review of photochemical and photophysical processes in solution (Brummer et al., 1980). The current understanding of both photochemical and thermal reactions in atmospheric droplets is discussed in Sections 3-C-12 and 11-A-2, respectively.

REFERENCES

Adamson, A. W., *A Textbook of Physical Chemistry*, Academic Press, New York, 1973.

Adamson, A. W., *Physical Chemistry of Surfaces*, 4th ed., Wiley, New York, 1982.

Anastasi, C., D. J. Waddington, and A. Woolley, "Reactions of Oxygenated Radicals in the Gas Phase. Part 10.—Self-Reactions of Ethylperoxy Radicals," *J. Chem. Soc. Faraday Trans 1*, **79,** 505 (1983).

Atkinson, R., "Kinetics and Mechanisms of the Gas Phase Reactions of the Hydroxyl Radical with Organic Compounds Under Atmospheric Conditions," *Chem. Rev.*, in press (1985).

Atkinson, R. and W. P. L. Carter, "Kinetics and Mechanisms of the Gas Phase Reactions of Ozone with Organic Compounds Under Atmospheric Conditions," *Chem. Rev.*, **84,** 437 (1984).

Atkinson, R. and A. C. Lloyd, "Evaluation of Kinetic and Mechanistic Data for Modeling of Photochemical Smog," *J. Phys. Chem. Ref. Data*, **13,** 315 (1984).

Atkinson, R. A. and J. N. Pitts, Jr., "Rate Constants for the Reaction of OH Radicals with Propylene and the Butenes Over the Temperature Range 297–425°K," *J. Chem. Phys.*, **63,** 3591 (1975).

Atkinson, R., B. J. Finlayson, and J. N. Pitts, Jr., "Photoionization Mass Spectrometer Studies of Gas Phase Ozone–Olefin Reactions," *J. Am. Chem. Soc.*, **95,** 7592 (1973).

Atkinson, R., R. Perry, and J. N. Pitts, Jr., "Rate Constants for the Reactions of the OH Radical with NO_2 ($M = Ar$ and N_2) and SO_2 ($M = Ar$), *J. Chem. Phys.*, **65,** 306 (1976).

Atkinson, R., K. R. Darnall, A. C. Lloyd, A. M. Winer, and J. N. Pitts, Jr., "Kinetics and Mechanisms of the Reaction of the Hydroxyl Radical with Organic Compounds in the Gas Phase," *Adv. Photochem.*, **11,** 375 (1979).

Atkinson, R., W. P. L. Carter, A. M. Winer, and J. N. Pitts, Jr., "An Experimental Protocol for the Determination of OH Radical Rate Constants with Organics Using Methyl Nitrite Photolysis as an OH Radical Source," *J. Air Pollut. Control Assoc.*, **31,** 1090 (1981).

Atkinson, R., S. M. Aschmann, W. P. L. Carter, A. M. Winer, and J. N. Pitts, Jr., "Kinetics of the Reactions of OH Radicals with *n*-Alkanes at 299 ± 2°K," *Int. J. Chem. Kinet.*, **14,** 781 (1982).

Audley, G. J., D. L. Baulch, I. M. Campbell, D. J. Waters, and G. Watling, "Gas-Phase Reactions of Hydroxyl Radicals with Alkyl Nitrite Vapours in $H_2O_2 + NO_2 + CO$ Mixtures," *J. Chem. Soc., Faraday Trans. 1*, **78,** 611 (1982).

Ayers, G. P., The Chemical Composition of Precipitation: A Southern Hemisphere Perspective, in *Atmospheric Chemistry*, E. D. Goldberg, Ed., Springer-Verlag, New York, 1982, pp. 41–56.

Baldwin, A. C., "Heterogeneous Reactions of Sulfur Dioxide with Carbonaceous Particles," *Int. J. Chem. Kinet.*, **14,** 269 (1982).

Barnes, I, V. Bastian, K. H. Becker, E. H. Fink, and F. Zabel, "Reactivity Studies of Organic Substances Towards Hydroxyl Radicals Under Atmospheric Conditions," *Atmos. Environ.*, **16,** 545 (1982).

Baulch, D. L., R. A. Cox, R. F. Hampson, Jr., J. A. Kerr, J. Troe, and R. T. Watson, "Evaluated Kinetic and Photochemical Data for Atmospheric Chemistry," *J. Phys. Chem. Ref. Data*, **9,** 295 (1980).

Baulch, D. L., R. A. Cox, P. J. Crutzen, R. F. Hampson, Jr., J. A. Kerr, J. Troe, and R. T. Watson, "Evaluated Kinetic and Photochemical Data for Atmospheric Chemistry: Supplement I." CODATA Task Group on Chemical Kinetics, *J. Phys. Chem. Ref. Data*, **11,** 327 (1982).

Baulch, D. L., R. A. Cox, R. F. Hampson, Jr., J. A. Kerr, J. Troe, and R. T. Watson, "Evaluated Kinetic and Photochemical Data for Atmospheric Chemistry: Supplement II." CODATA Task Group on Gas Phase Chemical Kinetics, *J. Phys. Chem. Ref. Data.*, **13,** 1259 (1984).

Benner, W. H., R. Brodzinsky, and T. Novakov, "Oxidation of SO_2 in Droplets Which Contain Soot Particles," *Atmos. Environ.*, **16,** 1333 (1982).

Benson, S. W., *The Foundations of Chemical Kinetics*, McGraw-Hill, New York, 1960.

Biermann, H. W., C. Zetzsch, and F. Stuhl, "On the Pressure Dependence of the Reaction of OH with CO," *Ber. Bunsenges. Phys. Chem.*, **82,** 633 (1978).

Brodzinsky, R., S.-G. Chang, S. S. Markowitz, and T. Novakov, "Kinetics and Mechanism for the Catalytic Oxidation of Sulfur Dioxide on Carbon in Aqueous Suspensions," *J. Phys. Chem.*, **84,** 3354 (1980).

Brown, R. L., "Tubular Flow Reactors with First-Order Kinetics," *J. Res. Natl. Bur. Stand.*, **83,** 1 (1978).

Brummer, J. G., W. P. Helman, and A. B. Ross, "A Catalog of Data Compilations on Photochemical and Photophysical Processes in Solution," NBS Spec. Publ. (1980).

Brunauer, S., P. H. Emmett, and E. Teller, "Adsorption of Gases in Multimolecular Layers," *J. Am. Chem. Soc.*, **60,** 309 (1938).

Burrows, J. P., R. A. Cox, and R. G. Derwent, "Modulated Photolysis of the Ozone–Water Vapor System: Kinetics of the Reaction of OH with HO_2," *J. Photochem.*, **16,** 147 (1981).

Calvert, J. G., Ed., *SO_2, NO and NO_2 Oxidation Mechanisms: Atmospheric Considerations*, Acid Precipitation Series, J. I. Treasley, Series Ed., Butterworth, Boston, 1984.

Campbell, I. M., B. J. Handy, and R. M. Kirby, "Gas Phase Chain Reaction of H_2O_2 + NO_2 + CO," *J. Chem. Soc. Faraday Trans. 1*, **71,** 867 (1975).

Campbell, I. M. and P. E. Parkinson, "Mechanism and Kinetics of the Chain Reaction in H_2O_2 + NO_2 + CO Systems," *J. Chem. Soc. Faraday Trans. 1*, **75,** 2048 (1979).

Carter W. P. L., A. C. Lloyd, J. L. Sprung, and J. N. Pitts, Jr., "Computer Modeling of Smog Chamber Data: Progress in Validation of a Detailed Mechanism for the Photooxidation of Propene and *n*-Butane in Photochemical Smog," *Int. J. Chem. Kinet.*, **11,** 45 (1979).

Chang, D. P. Y. and R. C. Hill, "Retardation of Aqueous Droplet Evaporation by Air Pollutants," *Atmos. Environ.*, **14,** 803 (1980).

Chang, S.-G. and T. Novakov, "Role of Carbon Particles in Atmospheric Chemistry," *Adv. Environ. Sci. Technol.*, **12,** 191 (1983).

Chang, S.-G., R. Brodzinsky, L. A. Gundel, and T. Novakov, Chemical and Catalytic Properties of Elemental Carbon, in *Particulate Carbon: Atmospheric Life Cycle*, G. T. Wolff and R. L. Klimisch, Eds., Plenum Press, New York, 1982, pp. 159–181.

Chung, T. T., J. Dash, and R. J. O'Brien, "In Situ TEM Studies of NaCl Gas Reactions," *9th Int. Cong. Electron Microsc.*, **1,** 440 (1978).

Clyne, M. A. A. and D. H. Stedman, "Reactions of Atomic Hydrogen with Hydrogen Chloride and Nitrosyl Chloride," *Trans. Faraday Soc.*, **62,** 2164 (1966).

Clyne, M. A. A. and W. S. Nip, Generation and Measurement of Atom and Radical Concentrations in Flow Systems, in *Reactive Intermediates in the Gas Phase: Generation and Monitoring*, D. W. Setser, Ed., Academic Press, New York, 1979, pp. 1–58.

Cofer, W. R. III, D. R. Schryer, and R. S. Rogowski, "The Oxidation of SO_2 on Carbon Particles in the Presence of O_3, NO_2 and N_2O," *Atmos. Environ.*, **15,** 1281 (1981).

Cox, R. A., "The Photolysis of Gaseous Nitrous Acid—A Technique for Obtaining Kinetic Data on Atmospheric Photooxidation Reactions," *Int. J. Chem. Kinet. Symp.*, **1,** 379 (1975).

Cox, R. A. and G. S. Tyndall, "Rate Constants for the Reactions of CH_3O_2 with HO_2, NO and NO_2 Using Molecular Modulation Spectrometry," *J. Chem. Soc. Faraday Trans. 2*, **76,** 153 (1980).

Cox, R. A., R. G. Derwent, and P. M. Holt, "Relative Rate Constants for the Reactions of OH Radicals wih H_2, CH_4, CO, NO and HONO at Atmospheric Pressure and 296 K," *J. Chem. Soc. Faraday Trans. 1*, **72,** 2031 (1976).

Cox, R. A., R. G. Derwent, and M. R. Williams, "Atmospheric Photooxidation Reactions. Rates, Reactivity, and Mechanisms for Reaction of Organic Compounds with Hydroxyl Radicals," *Environ. Sci. Technol.*, **14,** 57 (1980).

Crutzen, P. J. and C. J. Howard, "The Effect of the HO_2 + NO Reaction Rate Constant on One-Dimensional Model Calculations of Stratospheric Ozone Perturbations," *Pure Appl. Geophys.*, **116,** 497 (1978).

Darnall, K. R., R. Atkinson, and J. N. Pitts, Jr., "Rate Constants for the Reaction of the OH Radical with Selected Alkanes at 300°K, *J. Phys. Chem.*, **82,** 1581 (1978).

Davies, P. B., "Laser Magnetic Resonance Spectroscopy," *J. Phys. Chem.*, **85,** 2599 (1981).

Davis, D. D., S. Fischer, and R. Schiff, "Flash Photolysis–Resonance Fluorescence Kinetics Study:

Temperature Dependence of the Reactions OH + CO → CO_2 + H and OH + CH_4 → H_2O + CH_3," *J. Chem. Phys.*, **61,** 2213 (1974).

DeMore, W. B., "Rate Constant for the OH + CO Reaction: Pressure Dependence and the Effect of Oxygen," *Int. J. Chem. Kinet.*, **16,** 1187 (1984).

DeMore, W. B., M. J. Molina, R. T. Watson, D. M. Golden, R. F. Hampson, M. J. Kurylo, C. J. Howard, and A. R. Ravishankara, "Chemical Kinetics and Photochemical Data for Use in Stratospheric Modeling Evaluation No. 6," JPL Publ. No. 83-62, National Aeronautics and Space Administration, September 15, 1983.

DeMore, W. B., J. J. Margitan, M. J. Molina, R. T. Watson, D. M. Golden, R. F. Hampson, M. J. Kurylo, C. J. Howard, and A. R. Ravishankara, "Chemical Kinetics and Photochemical Data for Use in Stratospheric Modeling, Evaluation No. 7," JPL Publication No. 85-37, July 1, 1985.

Dixon-Lewis, G., W. E. Wilson, and A. A. Westenberg, "Studies of Hydroxyl Radical Kinetics by Quantitative ESR," *J. Chem. Phys.*, **44,** 2877 (1966).

Doyle, G. J., A. C. Lloyd, K. R. Darnall, A. M. Winer, and J. N. Pitts, Jr., "Gas Phase Kinetic Study of Relative Rates of Reaction of Selected Aromatic Compounds with Hydroxyl Radicals in an Environmental Chamber," *Environ. Sci. Technol.*, **9,** 237 (1975).

Drysdale, D. D. and A. C. Lloyd, "Gas Phase Reactions of the Hydroxyl Radical," *Oxid. Combust. Rev.*, **4,** 157 (1970).

Durham, J. L. Ed., *Chemistry of Particles, Fogs and Rain*, Acid Precipitation Series, J. I. Teasley, Series Ed., Butterworth, Boston, 1984.

Durham, J. L., J. H. Overton, Jr., and V. P. Aneja, "Influences of Gaseous Nitric Acid on Sulfur Production and Acidity in Rain," *Atmos. Environ.*, **15,** 1059 (1981).

Farahataziz and A. B. Ross, "Selected Specific Rates of Reactions of Transients from Water in Aqueous Solution. III. Hydroxyl Radical and Perhydroxyl Radical and Their Radical Ions," NSRDS-NBS59 (1977), PB263-198 (NTIS).

Finlayson-Pitts, B. J., "Reaction of NO_2 with NaC1 and Atmospheric Implications of NOC1 Formation," *Nature*, **306,** 676 (1983).

Finlayson-Pitts, B. J., T. E. Kleindienst, M. J. Ezell, and D. W. Toohey, "The Production of $O(^3P)$ and Ground State OH in the Reaction of Hydrogen Atoms with Ozone," *J. Chem. Phys.*, **74,** 4533 (1981).

Freiberg, J. E. and S. E. Schwartz, "Oxidation of SO_2 in Aqueous Droplets: Mass-Transport Limitation in Laboratory Studies and the Ambient Atmosphere," *Atmos. Environ.*, **15,** 1145 (1981).

Gardiner, W. C., Jr., "Temperature Dependence of Bimolecular Gas Reaction Rates," *Acc. Chem. Res.*, **10,** 326 (1977).

Gertler, A. W., D. F. Miller, D. Lamb, and U. Katz, Studies of Sulfur Dioxide and Nitrogen Dioxide Reactions in Haze and Cloud, in *Chemistry of Particles, Fogs, and Rain*, J. L. Durham, Ed., Acid Precipitation Series, Vol. 2, J. I. Teasley, Series Ed., Butterworth, Boston, MA, 1984, pp. 131–160.

Gilbert, R. G., K. Luther, and J. Troe, "Theory of Thermal Unimolecular Reactions in the Fall-Off Range. II. Weak Collision Rate Constants," *Ber. Bunsenges. Phys. Chem.*, **87,** 169 (1983).

Gill, P. S., T. E. Graedel, and C. J. Weschler, "Organic Films on Atmospheric Aerosol Particles, Fog Droplets, Cloud Droplets, Raindrops, and Snowflakes," *Rev. Geophys. Space Phys.*, **21,** 903 (1983).

Golden, D. M., "Experimental and Theoretical Examples of the Value and Limitations of Transition State Theory," *J. Phys. Chem.*, **83,** 108 (1979).

Goodsel, A. J., M. J. D. Low, and N. Takezawa, "Reactions of Gaseous Pollutants with Solids. II. Infrared Study of Sorption of SO_2 on MgO," *Environ. Sci. Technol.*, **6,** 268 (1972).

Graedel, T. E. and C. J. Weschler, "Chemistry within Aqueous Atmospheric Aerosols and Raindrops," *Rev. Geophys. Space Phys.*, **19,** 505 (1981).

Graedel, T. E. and K. I. Goldberg, "Kinetic Studies of Raindrop Chemistry. I. Inorganic and Organic Processes," *J. Geophys. Res.*, **88,** 10,865 (1983).

Graedel, T. E., M. L. Mandlich and C. J. Weschler, "Kinetic Model Studies of Atmospheric Droplet Chemistry. 2. Homogeneous Transition Metal Chemistry in Raindrops," *J. Geophys. Res.*, in press (1985).

Greiner, N. R., "Hydroxyl-Radical Kinetics by Kinetic Spectroscopy. I. Reactions with H_2, CO, and CH_4 at 300°K," *J. Chem. Phys.*, **46,** 2795 (1967).

Griffiths, P. R., S. A. Yeboah, I. M. Hamadeh, P. J. Duff, W-J. Yang, and K. W. Van Emery, "Analytical Applications of Diffuse Reflectance Infrared Fourier Transform Spectroscopy," *Recent Adv. Anal. Spectrosc.*, K. Fuwa, Ed., Pergamon, New York, 1982, pp. 299–309.

Gutman, D., N. Sanders, and J. E. Butler, "Kinetics of the Reactions of Methoxy and Ethoxy Radicals with Oxygen," *J. Chem. Phys.*, **86,** 66 (1982).

Hampson, R. F., Jr. and D. Garvin, "Reaction Rate and Photochemical Data for Atmospheric Chemistry—1977," NBS Special Publ. 513 (1978).

Hampson, R. F., Ed., "Chemical Kinetic and Photochemical Data Sheets for Atmospheric Reactions," FAA Report No. FAA-EE-80-17, April 1980.

Herron, J. T., "Mass-Spectrometric Study of the Rate of the Reaction CO + OH," *J. Chem. Phys.*, **45,** 1854 (1966).

Hofzumahaus, A. and F. Stuhl, "Rate Constant of the Reaction OH + CO in the Presence of N_2 and O_2," *Ber. Bunsenges. Phys. Chem.*, **88,** 557 (1984).

Howard, C. J., "Kinetic Measurements Using Flow Tubes," *J. Chem. Phys.*, **83,** 3 (1979).

Howard, C. J. and K. M. Evenson, "Laser Magnetic Resonance Study of the Gas Phase Reactions of OH with CO, NO, and NO_2, *J. Chem. Phys.*, **61,** 1943 (1974).

Howard, C. J. and K. M. Evenson, "Kinetics of the Reactions of HO_2 with NO," *Geophys. Res. Lett.*, **4,** 437 (1977).

Jeong, K.-M. and F. Kaufman, "Kinetics of the Reaction of Hydroxyl Radical with Methane and with Nine C1- and F-Substituted Methanes. 2. Calculation of Rate Parameters as a Test of Transition State Theory," *J. Phys. Chem.*, **86,** 1816 (1982).

Jonah, C. D., W. A. Mulac and P. Zeglinski, "Rate Constants for the Reaction of OH + CO, OD + CO, and OH + Methane as a Function of Temperature," *J. Phys. Chem.*, **88,** 4100 (1984).

Jones, I. T. N. and K. D. Bayes, "Energy Transfer from Electronically Excited NO_2," *Chem. Phys. Lett.*, **11,** 163 (1971).

Judeikis, H. S., T. B. Stewart, and A. G. Wren, "Laboratory Studies of Heterogeneous Reactions of SO_2," *Atmos. Environ.*, **12,** 1633 (1978).

Kanofsky, J. R. and D. Gutman, "Direct Observation of the Products Produced by the O-Atom Reactions with Ethylene and Propylene Studied in High Intensity Molecular Beams," *Chem. Phys. Lett.*, **15,** 236 (1972).

Kaufman, F., "Laser Studies of Atmospheric Reactions," *J. Photochem.*, **17,** 397 (1981).

Kaufman, F., "Kinetics of Elementary Radical Reactions in the Gas Phase," *J. Phys. Chem.*, **88,** 4909 (1984).

Keyser, L. F., "High-Pressure Flow Kinetics. A Study of the OH + HC1 Reaction from 2 to 100 Torr," *J. Phys. Chem.*, **88,** 4750 (1984).

Kircher, C. C., J. J. Margitan, and S. P. Sander, "Pressure and Temperature Dependence of the Reaction $NO_2 + NO_3 + M \rightarrow N_2O_5 + M$," *J. Phys. Chem.*, **88,** 4370 (1984).

Laidler, K. J. and J. H. Meiser, *Physical Chemistry*, Benjamin/Cummings, Menlo Park, CA, 1982.

Lamb, J. J., M. Mozurkewich, M., and S. W. Benson, "Negative Activation Energies and Curved Arrhenius Plots. 3. OH + HNO_3 and OH + HNO_4," *J. Phys. Chem.*, **88,** 6441 (1984).

Lambert, M., C. M. Sadowski and T. Carrington, "Uses of the Transit Time Distribution in Kinetic Flow Systems," *Int. J. Chem. Kinet.*, **7,** 685 (1985).

Lee, Y.-N. and S. E. Schwartz, "Reaction Kinetics of Nitrogen Dioxide with Liquid Water at Low Partial Pressure," *J. Phys. Chem.*, **85,** 840 (1981).

Lin, M. J. and J. H. Lunsford, "Photooxidation of Sulfur Dioxide on the Surface of Magnesium Oxide," *J. Phys. Chem.*, **79,** 892 (1975).

Maahs, H. G., "Kinetics and Mechanism of the Oxidation of S(IV) by Ozone in Aqueous Solution with Particular Reference to SO_2 Conversion in Nonurban Tropospheric Clouds," *J. Geophys. Res.*, **88,** 10,721 (1983).

Martin, L. R., Kinetic Studies of Sulfite Oxidation in Aqueous Solution, in *SO_2 NO and NO_2 Oxidation Mechanisms: Atmospheric Considerations*, J. G. Calvert, Ed., Acid Precipitation Series, Vol. 3, J. I. Teasley, Series Ed., Butterworth, Boston, MA, 1984.

Martin, L. R., D. E. Damschen, and H. S. Judeikis, "The Reactions of Nitrogen Oxides with SO_2 in Aqueous Aerosols," *Atmos. Environ.*, **15,** 191 (1981).

Mozurkewich, M. and S. W. Benson, "Negative Activation Energies and Curved Arrhenius Plots. 1. Theory of Reactions over Potential Wells," *J. Chem. Phys.*, **88,** 6429 (1984).

Mozurkewich, M., J. J. Lamb, and S. W. Benson, "Negative Activation Energies and Curved Arrhenius Plots. 2. OH + CO," *J. Phys. Chem.*, **88,** 6435 (1984).

Mulcahy, M. F. R., *Gas Kinetics*, Halsted Press, New York, 1973.

Mulcahy, M. F. R. and R. H. Smith, "Reactions of OH Radicals in the H–NO_2 and H–NO_2–CO Systems," *J. Chem. Phys.*, **54,** 5215 (1971).

Niki, H., P. D. Maker, C. M. Savage, and L. P. Breitenbach, "Fourier Transform Infrared Spectroscopic Study of the Kinetics for the HO Radical Reaction of $^{13}C^{16}O$ and $^{12}C^{18}O$," *J. Phys. Chem.*, **88,** 2116 (1984).

Novakov, T., "The Role of Soot and Primary Oxidants in Atmospheric Chemistry," *Sci. Tot. Environ.*, **36,** 1 (1984).

Novakov, T., S. G. Chang, and A. B. Harker, "Sulfates as Pollution Particulates: Catalytic Formation on Carbon (soot) Particles," *Science*, **186,** 259 (1974).

Overend, R., G. Paraskevopoulos, and R. J. Cvetanovic, Abstracts, 11th Informal Conference on-Photochemistry, Vanderbilt University, Nashville, Tennessee, 1974, p. 248.

Paraskevopoulos, G. and R. S. Irwin, "The Pressure Dependence of the Rate Constant of the Reaction of OH Radicals with CO," *J. Chem. Phys.*, **80,** 259 (1984).

Parkes, D. A., "The Oxidation of Methyl Radicals at Room Temperature," *Int. J. Chem. Kinet.*, **9,** 451 (1977).

Pitts, J. N., Jr., A. M. Winer, D. R. Fitz, S. M. Aschmann, and R. Atkinson, "Experimental Protocol for Determining Ozone Reaction Rate Constants," U.S. Environmental Protection Agency, Report No. EPA-600/S3-81-024, May 1981.

Pitts, J. N., Jr., A. M. Winer, S. M. Aschmann, W. P. L. Carter, and R. Atkinson, "Experimental Protocol for Determining Hydroxyl Radical Reaction Rate Constants," U.S. Environmental Protection Agency, Report No. EPA-600/S3-82-038, October 1982.

Radford, H. E., K. M. Evenson, and C. J. Howard, "HO_2 Detected by Laser Magnetic Resonance," *J. Chem. Phys.*, **60,** 3178 (1974).

Ross, A. B. and P. Neta, "Rate Constants for Reactions of Aliphatic Carbon-Centered Radicals in Aqueous Solution," NSRDS-NBS70 October (1982).

Ross, A. B. and P. Neta, "Rate Constants for Reactions of Inorganic Radicals in Aqueous Solution," NSRDS-NBS 65, June, 1979.

Schwartz, S. E., Ed., *Trace Atmospheric Constituents*, Vol. 12, *Adv. Environ. Sci. Tech.*, Wiley, New York, 1983.

Schwartz, S. E. and J. E. Freiberg, "Mass-Transport Limitation to the Rate of Reaction of Gases in Liquid Droplets: Application to Oxidation of SO_2 in Aqueous Solutions," *Atmos. Environ.*, **15,** 1129 (1981).

Schwartz, S. E. and W. H. White, "Solubility Equilibria of the Nitrogen Oxides and Oxyacids in Dilute Aqueous Solution," *Adv. Environ. Sci. Eng.* **4,** 1 (1981).

Schwartz, S. E. and W. H. White, "Kinetics of Reactive Dissolution of Nitrogen Oxides into Aqueous Solution," *Adv. Env. Sci. Technol.*, **12,** 1 (1983).

Shoemaker, D. P. and C. W. Garland, *Experiments in Physical Chemistry*, 2nd ed, McGraw-Hill, New York, 1967.

Sie, B. K. T., R. Simonaitis, and J. Heicklen, "The Reaction of OH with CO," *Int. J. Chem. Kinet.*, **8,** 85 (1976).

Slagle, I. R., F. Yamada, and D. Gutman, "Kinetics of Free Radicals Produced by Infrared Multiphoton-Induced Decompositions. 1. Reactions of Allyl Radicals with Nitrogen Dioxide and Bromine," *J. Am. Chem. Soc.*, **103,** 149 (1981).

Smith, I. W. M. and R. Zellner, "Rate Measurements of Reactions of OH by Resonance Absorption. Part 2.—Reactions of OH with CO, C_2H_4 and C_2H_2," *J. Chem. Soc. Faraday Trans. 2*, **69,** 1617 (1973).

Stelson, A. W. and J. H. Seinfeld, "Chemical Mass Accounting of Uban Aerosol," *Environ. Sci. Technol.*, **15,** 671 (1981).

Stelson, A. W., M. E. Bassett, and J. H. Seinfeld, Thermodynamic Equilibrium Properties of Aqueous Solutions of Nitrate, Sulfate and Ammonium, in *Chemistry of Particles, Fogs, and Rain*, J. L. Durham, Ed., Acid Precipitation Series, Vol. 2, J. I. Teasley, Series Ed., Butterworth, Boston, MA, 1984, pp. 1–52.

Stuhl, F. and H. Niki, "Pulsed Vacuum-UV Photochemical Study of Reactions of OH with H_2, D_2, and CO Using a Resonance-Fluorescent Detection Method," *J. Chem. Phys.*, **57,** 3671 (1972).

Troe, J., "Theory of Thermal Unimolecular Reactions at Low Pressures. I. Solutions of the Master Equation," *J. Chem. Phys.*, **66,** 4745 (1977a) "II. Strong Collision Rate Constants. Applications." *ibid.*, **66,** 4758 (1977b).

Troe, J., "Predictive Possibilities of Unimolecular Rate Theory," *J. Phys. Chem.*, **83,** 114 (1979).

Troe, J., "Theory of Thermal Unimolecular Reactions at High Pressures," *J. Chem. Phys.*, **75,** 226 (1981).

Troe, J., "Theory of Thermal Unimolecular Reactions in the Fall-Off Range. Strong Collision Rate Constants," *Ber. Bunsenges. Phys. Chem.*, **87,** 161 (1983a).

Troe, J., "Specific Rate Constants $k(E,J)$ for Unimolecular Bond Fissions," *J. Chem. Phys.*, **79,** 6017 (1983b).

Tuazon, E. C.,, W. P. L. Carter, R. Atkinson, and J. N. Pitts, Jr., "The Gas Phase Reaction of Hydrazine and Ozone: A Nonphotolytic Source of OH Radicals for Measurement of Relative OH Radical Rate Constants," *Int. J. Chem. Kinet.*, **15,** 619 (1983).

Turner, L., Ed., "Experimental Assessment of the Phototransformation of Chemicals in the Atmosphere," Technical Report No. 7, European Chemical Industry, Ecology, and Toxicology Center, Brussels, Belgium, September 30,1983.

Ung, A. Y.-M. and R. A. Back, "The Photolysis of Water Vapor and Reactions of Hydroxyl Radicals," *Can J. Chem.*, **42,** 753 (1964).

Westenberg, A. A. and N. deHaas, "Rates of CO + OH and H_2 + OH Over an Extended Temperature Range," *J. Chem. Phys.*, **58,** 4061 (1973).

Wilson, W. E., Jr. and J. T. O'Donovan, "Mass-Spectrometric Study of the Reaction Rate of OH with Itself and with CO," *J. Chem. Phys.*, **47,** 5455 (1967).

Yamada, F., I. R. Slagle, and D. Gutman, "Kinetics of the Reaction of Methyl Radicals with Nitrogen Dioxide," *Chem. Phys. Lett.*, **83,** 409 (1981).

5 Monitoring Techniques for Gaseous Criteria and Non-criteria Pollutants

5.A. HISTORICAL PERSPECTIVE: NEED FOR SENSITIVE, SPECIFIC, AND ACCURATE MONITORING TECHNIQUES

5-A-1. Introduction

Chemical techniques were applied to detecting airborne pollutants long before the relatively recent surge of interest in the chemistry of the atmosphere. For example, the presence of nitric acid in air was inferred in the 1800s by its reaction with calcium carbonate. However, with the recognition of the severity of many of the so-called London smog episodes and the identification in the 1940s of photochemical smog came the need to monitor a whole host of primary and secondary pollutants. In this chapter, we deal with some common techniques for detecting and measuring gas phase pollutants at a given point or over relatively short pathlengths; for discussions of remote sensing techniques to measure the column abundance of various species (i.e., the integrated concentration in a column from the surface of the earth through the atmosphere), the reader is referred to Hanst (1971), Hinkley et al. (1976a), and Killinger and Mooradian (1983).

Our emphasis in this chapter is on the bases of the monitoring techniques, their advantages and disadvantages, and the gaseous pollutants that have been successfully monitored in ambient air using these methods. We do not attempt to include all pollutants and all methods reported in the literature, nor do we discuss source monitoring. We restrict the discussion to neutral species in the troposphere; it is noteworthy, however, that the detection of several gaseous ionic species, NH_4^+ and NO_3^-, has been reported (Perkins and Eisele, 1984).

Thus we concentrate in this chapter on the major criteria and non-criteria pollutants whose chemistry forms the focus of this book and on some of the methods to measure their concentrations which either have been successfully used in ambient air studies or are promising and in the field-testing stage. For detailed treatment of important associated aspects of monitoring such as calibration techniques, the reader is referred to the U.S. Environmental Protection Agency Air Quality Criteria Documents and to literature reviews (e.g., Nelson, 1980).

Detecting various species, that is, establishing that they are indeed present in the atmosphere, is an important first step in understanding the chemistry. However, measuring them, that is, quantitatively determining their concentrations, is essential for the elucidation of the relative importance of various reactions in the atmosphere, and hence for formulating cost-effective control strategies. Detection and measurement do not always go hand-in-hand; for example, formic acid was detected in the Los Angeles air basin in the 1950s, but its accurate measurement awaited the development and application of spectroscopic techniques.

5-A-2. Oxidant Monitoring: Historical Example

One of the first obvious manifestations of photochemical air pollution was the effect on rubber products that rapidly hardened and cracked. This was soon used as a means of monitoring the major recognized secondary pollutants, photochemical oxidants, which caused this effect (Bradley and Haagen-Smit, 1951). Thus, Haagen-Smit and co-workers (Haagen-Smit et al., 1952) in the 1950s monitored oxidants in ambient air by measuring the time for cracking to appear in a piece of rubber placed in ambient air; this time was inversely proportional to the O_3 concentration. By comparison to the effects of laboratory-produced O_3, an estimate of the oxidant concentration in ambient air could be made.

Subsequently, several wet chemical techniques for monitoring photochemical oxidants were developed. The one that gained the most widespread use was the potassium iodide (KI) method, which is still operative today in some programs. The basis of this method is the oxidation of the colorless iodide ion in solution to form brown I_2:

$$2H^+ + 2I^- + O_3 \rightarrow I_2 + O_2 + H_2O \qquad (1)$$

The formation of I_2 can be monitored colorimetrically or by other techniques such as coulometry.

While O_3 efficiently oxidizes I^- in solution, some other air pollutants also contribute to the $I^- \rightarrow I_2$ conversion. Thus both NO_2 and PAN will react in a manner similar to O_3, although with lesser efficiencies; for example, a 1 ppm concentration of NO_2 gives a response of ~5–10% (i.e., equivalent to 0.05–0.10 ppm O_3).

Bubbling ambient air through an iodide solution gives a reading of *total oxidant* which includes a weighted combination of various pollutants such as O_3, NO_2, and PAN. In a typical Los Angeles atmosphere, this is usually predominantly O_3; under certain conditions, however, O_3 may make only a minor contribution. For example, one of the authors has experienced "pegging the needle" on such an instrument due primarily to NO_2 while flying in the plume from a fossil-fuel-fired power plant.

Just as a variety of air pollutants give a positive response, some interfere negatively. For example, SO_2 gives a 100% negative response and must therefore be removed from the gas stream prior to analysis.

In summary, this wet chemical technique is not *specific* for one pollutant. In fact, the term ''oxidant'' is defined by the technique itself; that is, oxidant is any species giving a positive response in the KI method. This technique of measuring and reporting total oxidants was used almost exclusively until the mid-1970s. In addition, the Federal Air Quality Standard was written in terms of total oxidant (0.08 ppm oxidant for 1 h) rather than O_3 specifically.

In the early to mid-1970s, however data were obtained suggesting that this technique was not as accurate and reliable as had been previously accepted. Thus, simultaneous monitoring of oxidants in the Los Angeles air basin was carried out during a smog episode by a state agency, the California Air Resources Board (CARB), and by a local agency, the Los Angeles County Air Pollution Control District (LAAPCD). The two sets of oxidant measurements were found to differ by ~30%, despite use of what was thought to be the same monitoring technique. The discrepancy was ultimately traced to a difference in calibration procedure. CARB used a neutral buffered KI solution (2%) and absorption spectrophotometry to determine the endpoint, whereas the LAAPCD used an unbuffered solution and established the endpoint visually.

A subsequent detailed comparison of the readings on laboratory-generated O_3 samples using the KI technique with those obtained by infrared (IR) and ultraviolet (UV) spectroscopy (which are specific for O_3) was carried out (DeMore and Patapoff, 1976; Pitts et al., 1976a). Significant discrepancies between the KI method and the spectroscopic methods were found, with the magnitude of the discrepancies depending in part on relative humidity. While the reasons for the discrepancy remain obscure, it is clear that the wet chemical method is not as accurate or reliable as desired, even for pure O_3 under laboratory conditions. These problems, combined with the lack of specificity, raised serious questions about its application to the much more complex mixtures found in ambient air.

The implications of the ~30% difference in measured concentrations using the two calibration procedures were even more far-reaching than simply illustrating the inaccuracies of wet chemical methods. For example, a distorted view of the geographical oxidant distribution developed based on these data. Thus, cities in the eastern most part of the air basin (Fig. 5.1)—Riverside and San Bernardino, which used the CARB technique—had historically been thought to suffer the highest oxidant concentrations. In fact, when the ambient air data were corrected to a common scale based on a spectroscopic technique, cities considerably further west which used the LAAPCD method were found to have the most severe oxidant problem.

Figure 5.2, for example, shows the oxidant dosages (i.e., concentration × exposure time) equal to or greater than the first-stage alert level of 0.20 ppm for six air monitoring stations in the Los Angeles air basin in 1973. Part (*a*) shows the uncorrected data as originally reported, and part (*b*) shows the corrected data referenced to the spectroscopic technique. It is seen that the correction causes a significant shift in the geographical oxidant trend.

This shift in the oxidant distribution not only has socioeconomic implications, but impacts control strategy options as well. As described in Chapters 9 and 10,

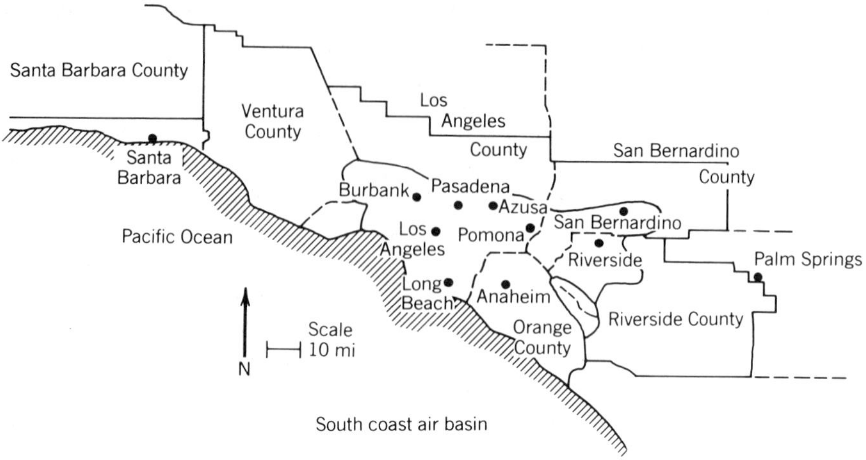

FIGURE 5.1. Map of Los Angeles air basin (white region) with locations of nine of the air monitoring stations shown (From Pitts et al., 1976a).

one means of developing such options is the use of air quality models. A crucial step in the use of these models, which mathematically describe the emissions, meteorology, and chemistry in an air basin, is validation against existing ambient air data. That is, the model is used to predict pollutant concentrations for various days in the past for which monitoring data exist; adjustments are made in the model to optimize agreement between the observed and predicted pollutant concentrations. The impact on air quality of various control strategy options is then predicted using the model. Clearly, the impacts of the resulting control strategy options cannot be more reliably predicted than the accuracy of the model itself, and this depends critically on the ambient air data against which it is validated.

Today, the UV method is most commonly used to monitor O_3 in ambient air, and is accepted as an ''equivalent method'' by the EPA (see below). It should be noted, however, that even with this spectroscopic technique, care must be taken to avoid interferences under some conditions; for example, compounds such as some aromatic hydrocarbons absorb light at 254 nm, and small particles scatter light (see Chapter 12). As a result, if they are present in significant concentrations, they may interfere in the ozone measurement (Grosjean and Harrison, 1985a).

The recognition of the problems with the wet chemical KI technique and the simultaneous development of physical techniques for monitoring the major oxidant O_3 specifically led to a change in the Federal Air Quality Standard from oxidant to O_3; simultaneously the standard was relaxed to higher concentrations, 0.12 ppm O_3 for 1 h.

The problems with oxidant measurement are not unique; similar situations have arisen with other pollutants such as NO_2 (Hauser and Shy, 1972). This serves to illustrate the great need for accurate, sensitive, and specific methods of measurement for air pollutants.

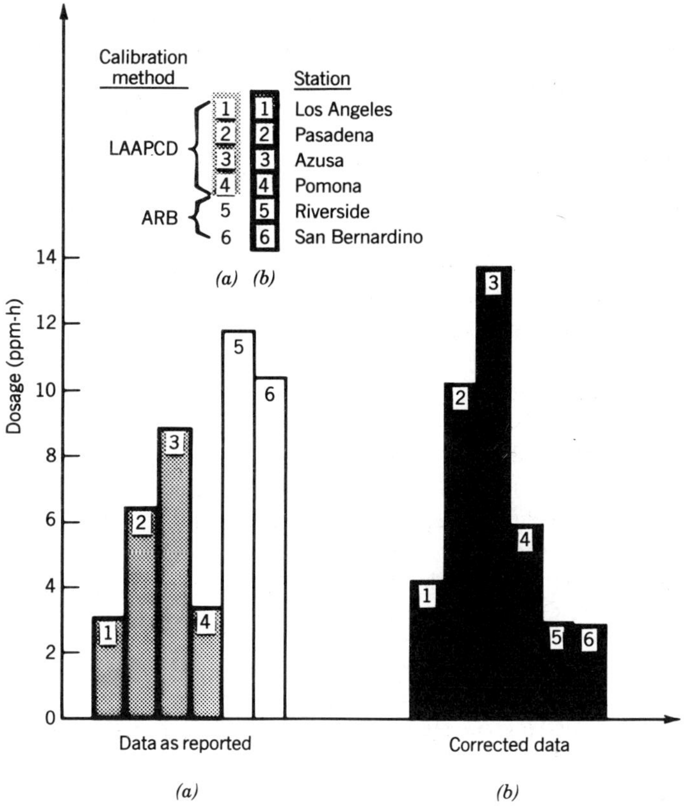

FIGURE 5.2. Oxidant dosage ≥ 0.20 ppm at six monitoring stations in the Los Angeles area in 1973: (*a*) Data as reported; (*b*) corrected data referenced to spectroscopic technique (from Pitts et al., 1976b).

Only physical techniques satisfy these requirements, and, as a result, such techniques have undergone substantial development over the last few years. Even then, as we shall see, there remain a number of important pollutants for which no sufficiently sensitive physical monitoring technique yet exists.

Ambient air data taken by wet chemical methods prior to the mid-1970's should be used with caution, both with regard to the absolute values and when comparing them to data taken using other calibration or monitoring techniques. This must be done to ensure that all data are placed on a common basis before using them in studies of long-term trends and/or comparisons of data between air basins.

In the remainder of this chapter, we discuss the major techniques currently in use to monitor gas phase air pollutants, including typical results from both laboratory and ambient air studies. Because many of these are spectroscopic in nature, we first briefly review the principles on which these monitoring techniques are based. No attempt is made to review every technique that has ever been applied to the monitoring of all possible air pollutants; rather, emphasis is on the most commonly used techniques for criteria pollutants as well as on promising methods used to monitor trace, non-criteria species such as OH, HONO, and NO_3, which play key roles in atmospheric chemistry.

5-B. SPECTROSCOPIC MONITORING TECHNIQUES: EMISSION SPECTROMETRY

A number of the techniques discussed in this and the following sections are also used in kinetic studies and have already been described in Chapter 4; the reader will be referred to the relevant sections.

5-B-1. Chemiluminescence

Chemiluminescence, the production of light from the energy released in a chemical reaction (see Section 4-B-2c), is used to monitor a variety of pollutants, including O_3 and NO. In this technique, air containing the pollutant to be measured is drawn into the monitoring instrument where it is mixed with a species with which it reacts to produce light. The emission intensity is proportional to the pollutant concentration. In the case of O_3, the two reactions used are those with ethylene,

$$O_3 + C_2H_4 \rightarrow HCHO^* + \text{other products} \quad (2)$$

$$HCHO^* \rightarrow HCHO + h\nu \quad (300 \lesssim \lambda \lesssim 550 \text{ nm}) \quad (3)$$

or with nitric oxide,

$$O_3 + NO \rightarrow NO_2^* + O_2 \quad (4)$$

$$NO_2^* \rightarrow NO_2 + h\nu \quad (590 \lesssim \lambda \lesssim 3000 \text{ nm}) \quad (5)$$

The light-emitting species are electronically excited formaldehyde and nitrogen dioxide, respectively.

Not only ethylene but all simple olefins emit light upon reaction with O_3 at low pressures (Fig. 5.3). The emitting species have been identified as electronically excited formaldehyde and, in some cases, dicarbonyl compounds, as well as vibrationally excited OH in its ground electronic state (Kummer et al., 1971; Finlayson et al., 1972, 1974). At pressures approaching atmospheric in commercial instruments based on the O_3–C_2H_4 reactions, all emissions except that due to HCHO, are effectively quenched.

NO is also monitored using reactions (4) and (5). In this case, the ratio of reactants is reversed compared to the O_3 monitoring case (i.e., to monitor NO, excess O_3 is used).

Several other important gaseous nitrogen-containing species can be monitored by converting them first to NO. Thus NO_2 can be converted to NO thermally (e.g., with molybdenum at 200°C) or chemically (using $FeSO_4$). Photolysis (UV pho-

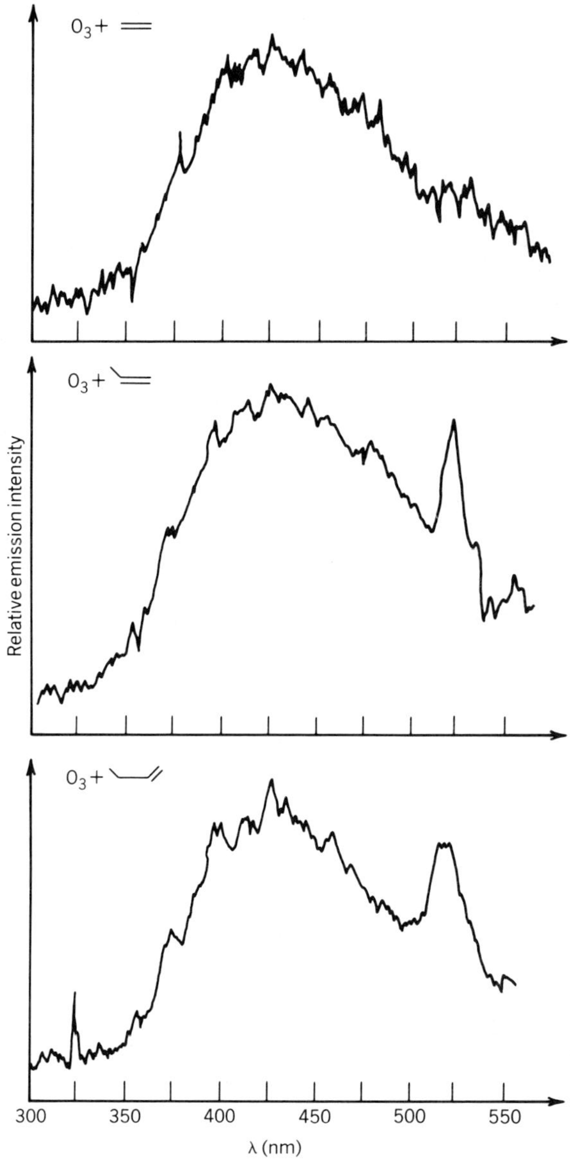

FIGURE 5.3. Chemiluminescence observed from the reaction of O_3 with some simple alkenes at low (Torr) pressures (from Finlayson et al., 1972).

todissociation) has been less commonly used; this is only about 50% efficient, whereas the first two techniques are ~100% efficient.

In commercial chemiluminescent NO_x analyzers, the instrument cycles automatically between the NO (i.e., direct injection) mode and the NO_x (i.e., passing the airstream over a converter) mode. The difference between the two is total NO_x

minus NO, which is often mainly NO_2. However, it has been shown that in the NO_x mode, other gaseous nitrogenous species, such as PAN, organic nitrates, NH_3, HNO_3, N_2O_5, $ClNO_x$, and HONO, are measured since these are also reduced to NO (Winer et al., 1974; Matthews et al., 1977; Sanhueza et al., 1984; Grosjean and Harrison, 1985a). Indeed HNO_3 has been measured by using a nylon filter to remove the HNO_3; the HNO_3 concentration is then determined by the difference in the NO_x–NO concentration with and without the nylon filter in the incoming airstream. However, Sanhueza et al. (1984) have shown recently that HONO is also removed quantitatively by these nylon filters and hence may act as an interference in HNO_3 measurements under some conditions. Since HONO rapidly photolyzes during the day, this is most likely to present a problem at night when significant HONO concentrations may accumulate.

Chemiluminescence techniques are much more specific than wet chemical methods because of the relatively few reactions that give off light. In addition, filters or monochromators can be used to isolate only the wavelengths of interest in order to discriminate against interfering emissions. However, as we have seen, when catalysts are used for NO_x reduction, some selectivity is lost because of the number of compounds that can be reduced to NO and hence give positive responses.

The factors affecting the response of chemiluminescent analyzers and optimization of their sensitivity are discussed in detail by O'Brien and co-workers (Mehrabzadeh et al., 1983; O'Brien et al., 1983).

5-B-2. Fluorescence

The basis of the fluorescence technique is described in detail in Section 4-B. Briefly, the pollutant is excited to a higher electronic state using an appropriate light source, and the intensity of the emitted radiation as it returns to the ground state is monitored. Only species that have bound upper electronic states with sufficiently short radiative lifetimes can be measured using this technique; otherwise, other processes (e.g., dissociation, physical deactivation to the ground state by air) compete significantly with radiation, and the emitted intensity becomes too weak to monitor at the low concentrations typical of ambient air.

This technique is the basis of some commercial SO_2 monitoring instruments; the SO_2 is excited using a zinc (213.8 nm) or cadmium (228.8 nm) lamp, and the resulting fluorescence in the ~220–400 nm region is monitored (Okabe et al., 1973).

A second important application of fluorescence is the detection and measurement of the key free radical in the atmosphere, OH. As discussed earlier with respect to Fig. 4.7, the ground state species is excited into the first electronically excited $A^2\Sigma^+$ state, and the light emitted as the molecule returns to the ground state monitored. While the basis of the technique is the same as that used to monitor OH in laboratory kinetic studies, the experimental details differ because of the need for much higher sensitivity in ambient air.

OH concentrations in ambient air are believed to be typically in the range of

$\sim 10^5$–10^7 radicals cm^{-3} during the daylight hours, whereas $\sim 10^9$–10^{12} radicals cm^{-3} is typical in kinetic studies. In addition, deactivation of the electronically excited OH to the ground state is frequently not as severe in kinetic studies due to the use of low pressures and/or choice of carrier gases to minimize quenching of excited OH (Chan et al., 1983, 1984; Wang, 1984; Davis et al., 1984).

In the detection of OH in ambient air, an increase in sensitivity is obtained by using an intense laser beam as the exciting light source, giving rise to the term *laser-induced fluorescence* (LIF). One type of experimental configuration used in a typical LIF apparatus is shown in Fig. 5.4. To minimize the problem of stray light, the OH is excited using light of wavelength 282 nm into the first vibrational level ($v' = 1$) rather than into $v' = 0$ of the $A^2\Sigma^+$ state. At atmospheric pressure, vibrational relaxation within the $A^2\Sigma^+$ state is very rapid so that significant concentrations of $v' = 0$ are quickly generated by deactivation of $v' = 1$. The $v' = 0$ state then undergoes fluorescence at 309 nm to the ground electronic state. By

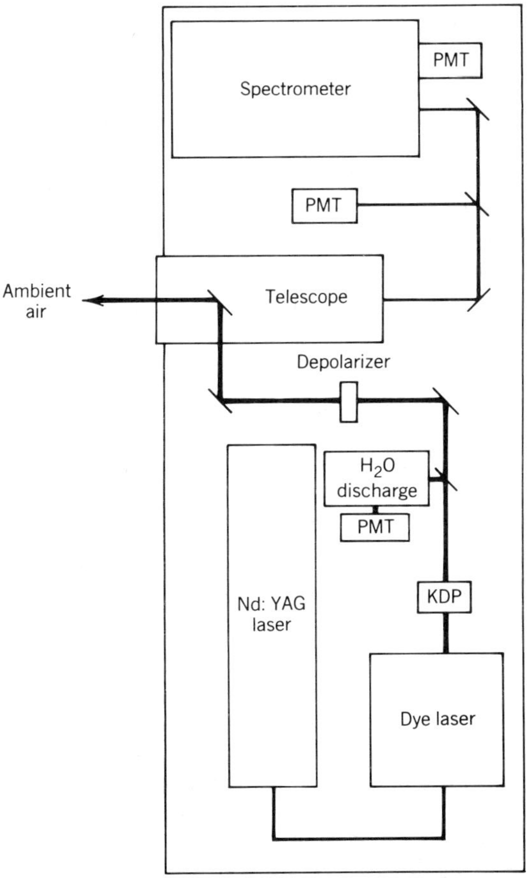

FIGURE 5.4. Example of one experimental configuration for the laser-induced fluorescence detection of OH in ambient air (from Wang, 1983).

exciting at one wavelength and monitoring at another, interference from scattered laser light is minimized.

While the use of lasers increases the fraction of OH which is excited and hence the intensity of the emitted fluorescence, it also generates a number of measurement problems in itself (Ortgies et al., 1980, 1981). First, the exciting radiation is absorbed by other molecules which are present in ambient air. The most important of these is O_3 which is always present at "background" concentrations of ~0.01–0.05 ppm. When it photolyzes, it produces an electronically excited oxygen atom, $O(^1D)$ (see Section 3-C-2), which rapidly reacts with water vapor to generate OH:

$$O_3 + h\nu\ (\lambda < 320\ \text{nm}) \rightarrow O(^1D) + O_2 \tag{6}$$

$$O(^1D) + H_2O \rightarrow 2OH \tag{7}$$

Thus the very act of trying to measure OH generates substantial concentrations of this radical, which are then detected. While O_3 is the major recognized co-pollutant that causes this effect, others may also contribute.

Second, the use of lasers generates interfering emissions via Rayleigh, Mie, and Raman scattering by O_2, N_2, and H_2O. Fluorescence from unidentified species in air has also been observed (Hard et al., 1979).

Finally, the laser intensities that can be used, and hence the sensitivity of LIF, are limited by saturation effects at high laser intensities. This effect causes the absorbed light intensity and the observed fluorescence intensity to become non-linear with the intensity of the exciting radiation. The rate of light absorption is linear with laser intensity only if the concentration of the absorbing species is constant; at high laser intensities, a significant fraction of the ground-state OH becomes excited into the upper electronic state. The concentration of the absorbing molecules is thus changed, resulting in non-linear light absorption.

Of these problems, the secondary generation of OH from O_3 photolysis is the most severe. Several methods have been proposed or employed to minimize this interference. The most promising of these, developed by O'Brien, Hard, and co-workers, appears to be the use of reduced pressures by expansion of the air sample (Hard et al., 1980, 1984). While this lowers the concentration of OH and hence the LIF signal, the photolytic secondary generation of OH is reduced to a greater extent. In addition, the air fluorescence and light scattering drop off more rapidly with total pressure than the OH fluorescence, resulting in a net gain in the OH signal.

Figure 5.5 shows the calculated pressure dependence of the OH fluorescence signal compared to that of the interfering signals from air fluorescence, scattering, and secondary generation of OH from photolysis of O_3 or hydroxyl-containing species (shown as ROH) such as HONO. Lowering the sample pressure to sub-Torr pressures should lower these potential interferences to $\lesssim 10\%$ of the total observed signal when OH is present.

Another method of minimizing secondary production of OH by O_3 photolysis is the use of shorter laser pulses. Typical laser pulse lengths used in LIF have been

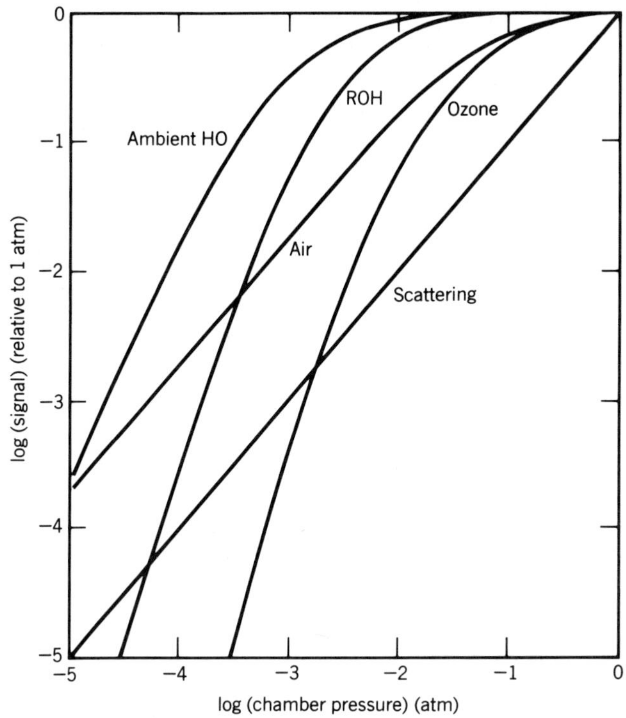

FIGURE 5.5. Pressure dependence of pulsed OH fluorescence and of the principal interferences and backgrounds in the detection, following expansion from atmospheric pressure (from Hard et al., 1984).

~7 nanoseconds (ns). However, reaction (7) forms OH within ~1 ns under atmospheric conditions (Ortgies et al., 1980, 1981), so that generation and detection of secondary OH can occur within the time span of one laser pulse. Making the laser pulse length much shorter than the time to generate OH via reactions (6) and (7) would minimize this problem.

An additional problem in the measurement of OH in the atmosphere (but which is not associated with the use of lasers) is that of accurately calibrating the instrument. While comparison of the observed signal intensity to that of the N_2 Raman signal at 302.5 nm or to a signal from OH generated independently in a calibration chambers have both been used, there is considerable disagreement as to which, if either, is accurate (Davis et al., 1981a, b, 1982; Wang and Davis, 1982). Finally, converting the fluorescence signal to absolute OH concentrations is complicated by the need in fluorescence measurements to know the fluorescence quantum yield for OH excited into $v' = 1$ of the excited state, a subject of some debate (Chan et al., 1983, 1984; Wang, 1984; Davis et al., 1984).

Even with the use of lasers, the sensitivity of LIF detection for OH borders on those concentrations typical of ambient air in many regions. While OH levels may approach 10^7 radicals cm^{-3} in heavily polluted regions at midday in the summer, more typical levels are one to two orders of magnitude lower. The current sensitivity of the LIF technique for detecting OH from all sources (including the artifactual formation from O_3) is ~(0.5–3) × 10^6 radicals cm^{-3} (Hübler et al., 1984 and references therein; O'Brien, 1985), although future developments will hopefully lower this detection limit. Some of the measurement methods and tropospheric concentrations of OH are reviewed by Hewitt and Harrison (1985).

5-B-3. Flame Photometry

A third light-emission technique used for detecting and measuring some gaseous pollutants, particularly sulfur compounds, is that of flame photometry. In this method, the substance is introduced into a flame where it is burned. In the flame, electronically excited species are formed which emit light in the visible and ultraviolet. The emission wavelength is characteristic of the species, which provides selectivity, and the intensity is proportional to the excited species and hence to the pollutant concentration. The basic apparatus required consists of a flame, a filter or monochromator to isolate various wavelengths, a photomultiplier to measure the emission intensity, and a readout device.

Flame photometric detection is usually used after a prior separation has been achieved, such as by gas chromatograph (GC). For example, organic sulfur compounds can be separated by GC and then detected at the column exit using flame photometry. In a hydrogen-rich flame, the organic sulfur compounds produce electronically excited S_2 which emits at 394 nm.

5-C. SPECTROSCOPIC MONITORING TECHNIQUES: INFRARED ABSORPTION SPECTROMETRY

5-C-1. Beer–Lambert Absorption Law and Multiple Pass Cells

The basis of light absorption techniques for monitoring air pollutants is the Beer–Lambert law discussed in Section 2-A-2:

$$\frac{I}{I_0} = e^{-\sigma[N]\ell} \qquad \text{(A)}$$

I_0 and I are the incident and transmitted light intensities, respectively, $[N]$ is the concentration of the absorbing species in number cm^{-3}, ℓ is the pathlength (in cm), and σ is the absorption cross section per molecule (in cm^2 $molecule^{-1}$).

Air pollutants are present in such small concentrations that the fraction of light absorbed is generally low at the pathlengths of conventional laboratory spectrometers. For example, an extremely large O_3 concentration in ambient air would be approximately 0.5 ppm; in a 10 cm gas cell, this would give rise to $I/I_0 \simeq 0.99998$, corresponding to an absorbance of 2.1×10^{-5} (base e), at the strong infrared absorption at 9.48 μm. Experimentally, it is not possible to measure this small absorption accurately. While the experimenter has no control over the concentration of the air pollutant to be measured, nor over its absorption cross section, the pathlength over which the measurement is made can be increased in order to produce measurable light absorptions.

There are two ways of increasing the pathlength ℓ. One is to simply increase the distance between the light source and the detector. The second is to use a set of reflecting mirrors on either end of a cell of fixed length and bounce the light beam back and forth many times, giving an effective absorbing pathlength for the pollutant which is some multiple of the cell length. Since their first application to atmospheric chemistry studies in the mid-1950s by Stephens, Hanst, Doerr, and Scott, these multiple pass cells have found great applicability in studies of both ambient air and laboratory systems (Stephens et al., 1956a,b; Scott et al., 1957; Stephens, 1958).

Figure 5.6 is a schematic diagram of a three-mirror multiple pass cell known as a *White cell* after the individual who first put forth the basic design (White, 1942). The light is first focused on the entrance to the cell. The beam diverges and falls on spherical mirror M_1 which reflects the image and refocuses it on mirror M_2. The diverging beam from M_2 is reflected to spherical mirror M_3 which, like M_1, reflects and refocuses the image at the opposite end of the cell. If the mirrors are adjusted so that this image is at the exit aperture of the cell, the light beam leaves and strikes a detector. A total of four passes of the cell has therefore been made and the effective pathlength for absorption is $\ell = 4a$, where a is the length of the cell. However, the mirrors may be adjusted so that the reflected image from M_3 falls on mirror M_2 and is again reflected to M_1 at a small angle to the original input light beam, leading to another set of four passes along the length of the cell.

The advantage of such a White cell is that the source is reimaged on M_2 after each double traversal of the cell. This keeps the energy that enters the cell within the mirror system so that energy losses occur only through light absorption by the mirrors and, of course, by the gases in the cell. The loss of light energy through absorption by the mirrors imposes a major limitation on the number of passes which can be used, in practice. The fraction of the energy lost after n reflections from a mirror whose reflectivity is R is given by $1 - R^n$. Thus, if a mirror reflects 98% of the incident light and absorbs 2%, only 36% of the incident intensity will remain after 50 reflections from the mirror. After 100 reflections, only 13% of the incident intensity is left. While the pathlength and hence absorbance have increased, the energy loss may be so severe that such a large number of reflections becomes impractical.

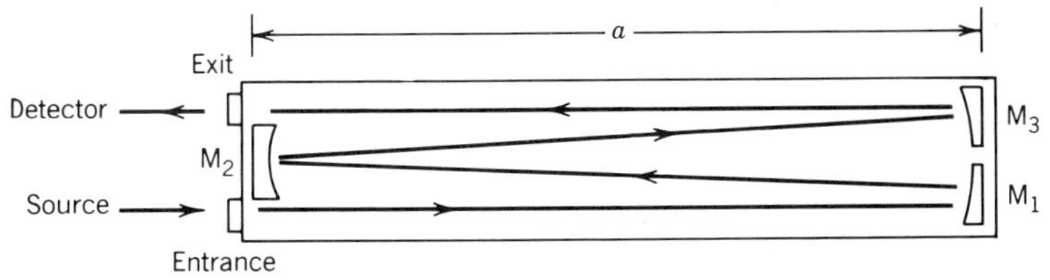

FIGURE 5.6. Schematic diagram of a multiple pass *White cell* used to increase the pathlength in absorption spectrometry studies.

The number of reflections is also limited by the size of the image striking the entrance and the size of the mirror M_2. The images which are refocused from M_1 and M_3 onto M_2 are "stacked" beside each other. The width of M_2 therefore determines how many of these images can be accommodated (i.e., how many reflections are possible). A practical problem arising when the images are too closely spaced (i.e., at long pathlength) is one of adjustment; temperature changes, for example, can cause very small changes in the mirror adjustments which result in moving the exit beam away from the exit aperture.

More complex multiple reflection systems which give a much greater number of traversals have been developed recently. Figure 5.7 is a schematic diagram of an eight-mirror system used in conjunction with Fourier transform infrared spectroscopy (FTIR) studies of ambient air. This design, described in detail by Hanst (1971), consists of a set of four collecting mirrors and four nesting mirrors spaced 22.5 m apart. The optics follow that of the White cell, that is, the image is focused on the nesting mirrors at one end of the cell (the *in-focus* end) and the divergent reflected beam strikes the collecting mirrors at the opposite (*out of focus*) end before being refocused back onto the nesting mirrors.

In Fig. 5.7 the path of the light beam is indicated by the numbered dots. Thus the light beam enters at "IN" at the left end of nesting mirror N4 which is shorter than the three other nesting mirrors in order to provide entrance and exit apertures. The image focused at the entrance forms a divergent beam which strikes collecting mirror C1 at the point 1 on C1. This mirror reflects and focuses the light to form an image at the point 2 on nesting mirror N1. The divergent beam from N1 then strikes collecting mirror C2 (marked "3" on C2) from which it is refocused to point 4 on nesting mirror N2. The sequence of reflections can be followed using the numbered dots in Fig. 5.7. Ultimately, the beam is refocused onto the exit

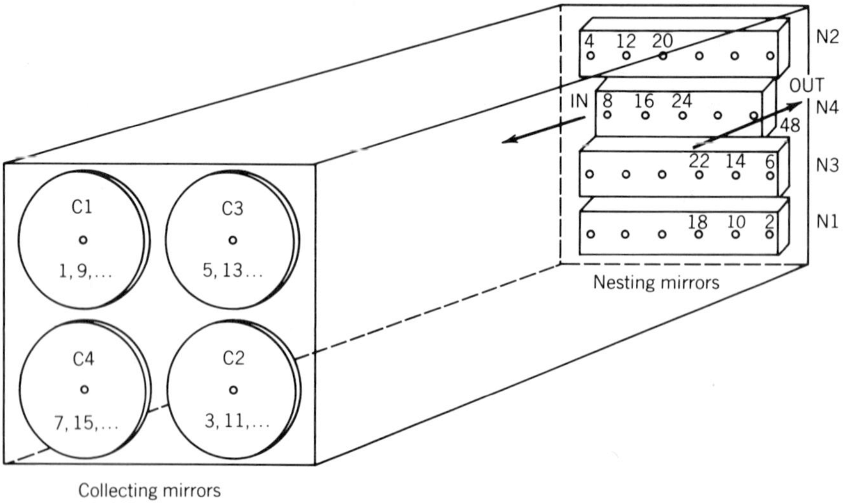

FIGURE 5.7. Eight-mirror multiple reflection system. Numbered dots indicate sequence of light reflections (from Tuazon et al., 1980).

aperture marked "OUT" at the right side of nesting mirror N4 and strikes a detector. The number of internal reflections and hence the pathlength can be controlled (in multiples of eight passes) through individual adjustment of the mirrors. The advantages and disadvantages of such multiple mirror cells are discussed in detail by Hanst (1971).

5-C-2. Fourier Transform Infrared Spectrometry

5-C-2a. Basis of Technique

Long-pathlength infrared spectrometry was first applied to monitoring the reactants, intermediates, and products of hydrocarbon–NO_x–air systems in the mid-1950s by Stephens, Scott, Hanst, and Doerr (Stephens et al., 1956a,b; Scott et al., 1957; Stephens, 1958). Indeed, it was by this technique that PAN, known as *compound X*, was identified first in laboratory systems (Stephens et al., 1956a,b) and then in ambient air (Scott et al., 1957). These traditional dispersive infrared instruments provided the first spectroscopic confirmation that O_3 was the major oxidizing secondary pollutant formed in these systems and allowed upper limits to be placed on the concentrations of other secondary pollutants in ambient air such as formic acid (Scott et al., 1957).

However, the application of IR to ambient air studies was limited by several factors. First, water vapor, carbon dioxide, and methane are present in significant concentrations in ambient air, and they absorb strongly in many regions of the IR spectrum. As a result, the spectral regions that can easily be used to search for pollutants are limited to 760–1300 cm^{-1}, 2000–2230 cm^{-1}, and 2390–3000 cm^{-1}. Second, the sensitivity was insufficient to detect pollutants at the low, sub-pphm levels at which many important species exist in ambient air. Finally, spectral analysis, especially for very small peaks, was difficult since subtraction of background spectra, or individual components, had to be carried out manually.

The development of Fourier transform infrared spectrometry (FTIR) in the early 1970s provided a quantum leap in our infrared analytical capabilities for monitoring trace pollutants in ambient air. This technique offers a number of advantages over conventional infrared dispersive systems, including sensitivity, speed, and improved data processing. The basis of this method and the resulting improvements in pollutant detection are discussed here; typical results from both simulated and real atmospheres are discussed in the following section. For a detailed treatment of FTIR, the reader is referred to the books and articles by Griffiths (1975, 1983a,b; Griffiths and de Haseth, 1986). Applications to atmospheric measurements are discussed by Maker et al. (1979) and Tuazon et al. (1980).

The configuration of one type of commercially available FTIR spectrometer is shown in Fig. 5.8. The basis of the instrument is a Michelson interferometer consisting of a fixed mirror, a moving mirror, and a beam splitter. The light beam is shown in Fig. 5.8 approaching from the right and striking the beam splitter at point O. Half of the beam is reflected at 90° to the fixed mirror, while the remainder passes through to the moving mirror. The light beams are reflected off mirrors A

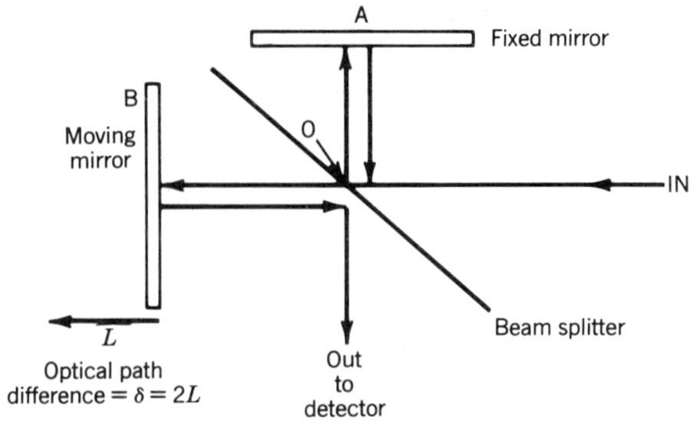

FIGURE 5.8. Schematic diagram of Michelson interferometer.

and B, respectively, back to the beam splitter where they recombine and are directed out to the detector.

Let us first take the case where the input light is a monochromatic beam of wavelength λ and the distances from the center of the beam splitter to the two mirrors are equal (i.e., OA = OB). The light beams which are split at O travel the same distance in going to A or B and back. When they recombine at O, the two light waves will have traveled the same number of wavelengths and will constructively interfere. Now let the moving mirror B be displaced away from the beam splitter by a distance (L) equal to one-half a wavelength ($L = \lambda/2$). The beam reflected from B now travels an extra distance of $\lambda/2$ before striking B and $\lambda/2$ in returning to O. When it recombines with the portion of the beam reflected off A, it has traveled an extra distance equal to one wavelength. At this optical path difference $\delta = 2L = \lambda$, then, constructive interference will again result at O and the detector will show a signal equal in strength to that before B was displaced.

However, if B is moved a distance $L = \lambda/4$, the optical path difference will be $\delta = 2L = \lambda/2$. When the beams recombine at O, they will be out of phase by 180° and hence destructively interfere, giving rise to no net signal. At mirror displacements between $\lambda/2$ and $\lambda/4$, the combining waves will be out of phase by an amount between 0° and 180° and destructive interference will occur to some extent less than the maximum. The signal at the detector will therefore be between zero and the maximum at L = O, $\lambda/2$, and so on. Figure 5.9*a* shows this signal as a function of the optical path difference δ for this case of monochromatic, or single frequency, radiation of wavelength λ as the moving mirror moves back and forth about the position where OA = OB.

For monochromatic radiation of a different wavelength, the same pattern of signal intensity as a function of optical displacement will result, except that the peaks and troughs will occur at different values of δ. Figure 5.9*b*, for example, is a typical case for radiation of wavelength $\lambda_2 < \lambda_1$. Only when the mirrors are equidistant from point O does a maximum signal result for *both* wavelengths λ_1 and λ_2.

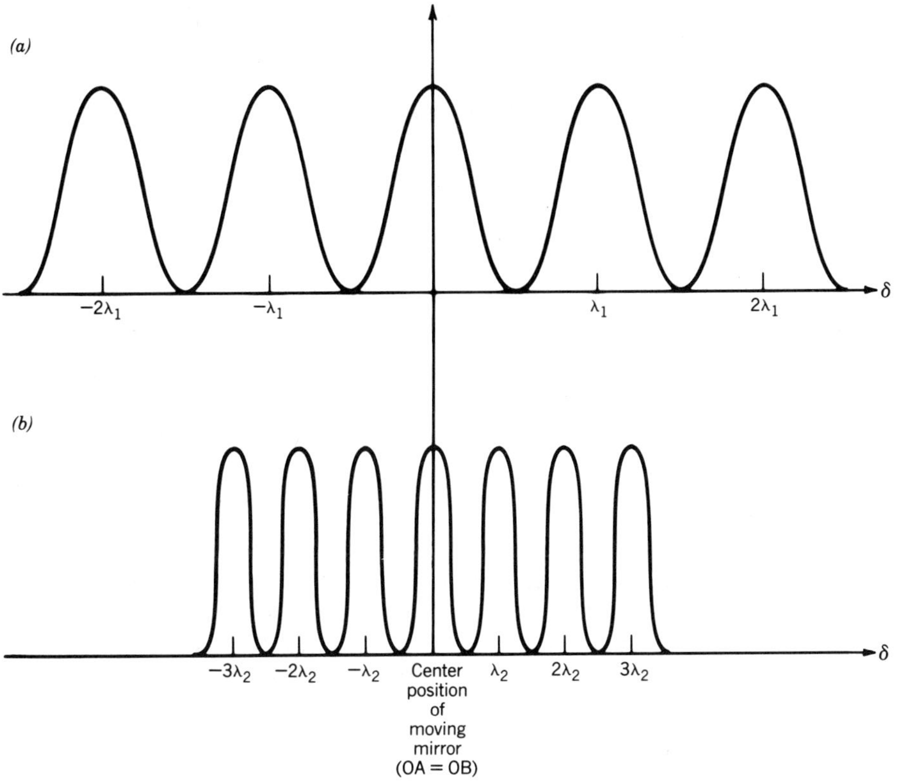

FIGURE 5.9. Signal expected from Michelson interferometer for monochromatic radiation as a function of optical path difference, $\delta = 2L$, where L is the displacement of the moving mirror. (*a*) wavelength λ_1, (*b*) wavelength λ_2, where $\lambda_2 < \lambda_1$.

With a typical infrared light source such as a Nernst glower, a larger number of wavelengths are emitted. While each wavelength alone would result in a signal as a function of the position of mirror B similar to those shown in Fig. 5.9, it is only when OA = OB that all will constructively interfere and produce a maximum signal. As a result, the analog to Fig. 5.9 for a real infrared light source is the interferogram shown in Fig. 5.10.

This interferogram in an empty cell is characteristic of the source and the beam splitter. The conversion of the signal intensity as a function of optical path difference (the interferogram) into signal intensity as a function of wavelength (the infrared spectrum) requires applying a mathematical formalism known as a Fourier transform, hence the name FTIR.

If a sample that selectively absorbs certain of the wavelengths is now added, the interferogram, which is the combination of all the wavelengths reaching the detector, and hence the infrared spectrum, will change. The difference between the two (i.e., with and without an absorbing sample) is characteristic of the infrared-absorbing species in the sample.

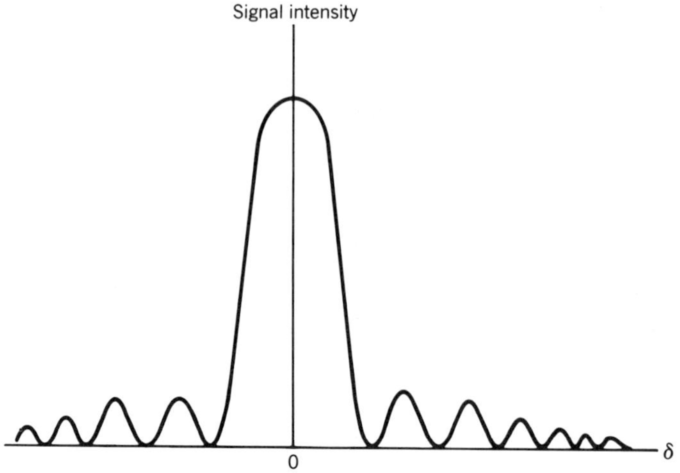

FIGURE 5.10. Interferogram for typical infrared light source.

The characteristics of commercially available FTIR instruments are discussed in detail by Griffiths (1983a).

There are a number of advantages of FTIR over conventional dispersive systems which make them especially useful in atmospheric studies:

- ***Fellgett's Advantage*** (multiplex advantage). FT systems sample all wavelengths at once whereas in dispersive systems, slits and gratings restrict the wavelength intervals reaching the detector at any one time. The simultaneous sampling of all wavelengths in FTIR results in much faster data collection to get the same signal-to-noise (S/N) ratio. Alternatively, an increased S/N ratio can be obtained for the same total scan time.
- ***Jacquinot's Advantage*** (throughput advantage). On a dispersive instrument only a small portion of the source energy falls on the detector at a given time because the grating and slit act to isolate a limited wavelength interval. In an FT instrument, all wavelengths strike the detector simultaneously. The energy throughput is therefore much greater, which makes FT especially attractive for analyzing weak light sources, for example, in astronomy, or for highly absorbing samples.
- ***Connes' Advantage.*** Very accurate frequency determinations are a third advantage of FTIR, known as Connes' advantage. A He–Ne laser is typically used as a reference for the position of the moving mirror. Since it is a monochromatic light source, the interferogram is a sine wave similar to those in Fig. 5.9, and the optical path difference δ can be measured accurately by counting the number of wavelengths. This also allows data collection to be carried out reproducibly at equal distance intervals.
- ***Resolution.*** The resolution of FT instruments is constant over the entire spectrum and is high compared to dispersive instruments since it theoreti-

cally depends only on the maximum displacement (L_{max}) of the moving mirror, that is, $\Delta\nu = 1/2L_{max}$. For example, resolution of $\frac{1}{16}$ cm^{-1} is relatively routine in commercially available research grade FT instruments.

- ***Computer Data Processing Capabilities.*** An essential component of FTIR is a computer to carry out the Fourier transform of the interferogram to produce the traditional infrared spectrum. This requirement means that substantial data processing capabilities are available for manipulation (e.g., subtraction, addition) of spectral data. This is especially useful in atmospheric studies as the *changes* in the spectrum with time are of greatest interest. In addition, this allows one to subtract out from the total spectrum the absorptions due to known pollutants, leaving the peaks due to unidentified pollutants relatively unobscured. This greatly assists in identifying such species.

In summary, the development of FTIR with its significant advantages over dispersive systems has provided substantial impetus for the increased application of infrared absorption spectrometry to atmospheric systems. These studies, some of which are described below, have provided data on the concentrations of a number of trace, non-criteria pollutants formed in ambient air as well as in laboratory systems (see Chapter 7, for example).

5-C-2b. Applications to Atmospheric Studies

Figure 5.11 is a schematic diagram of an FTIR system used to monitor pollutants in ambient air. The multiple pass reflection system is that shown in Fig. 5.7. The collecting and nesting mirrors are separated by a distance of 22.5; multiple reflections within this cell lead to total absorbing pathlengths of up to 2 km. However, at this number of multiple reflections, some partial overlapping of the images can occur at the exit aperture; thus a 1.08 km pathlength is more commonly used. Two different infrared detectors are used, a photovoltaic InSb detector for the 2000–3900 cm^{-1} range or a photoconductive HgCdTe detector for the 600–2000 cm^{-1} region.

Figure 5.12 shows some reference spectra of three species of atmospheric interest, NH_3, HCHO, and HNO_3, obtained using the apparatus in Fig. 5.11. Figure 5.13 shows the spectra of ambient air samples taken approximately an hour apart. While most of the absorption peaks are due to atmospheric H_2O, the spectrum taken at later times (Fig. 5.13*b*) shows some additional weak absorption bands not present in the earlier spectrum (Fig. 5.13*a*). These become more obvious when the two spectra are ratioed, that is, the spectrum in Fig. 5.13*b* is divided by that in 5.13*a*; the results are shown in Fig. 5.13*c*. By comparison to the spectra in Fig. 5.12, the doublet due to NH_3 at 931 and 967 cm^{-1} can be clearly identified. Application of the Beer–Lambert law to these peaks indicated that the NH_3 concentration in the ambient air sample was 32 ppb.

The computer data processing capabilities of these instruments makes ratioing and/or subtraction of spectra routine. However, it is not possible in practice to

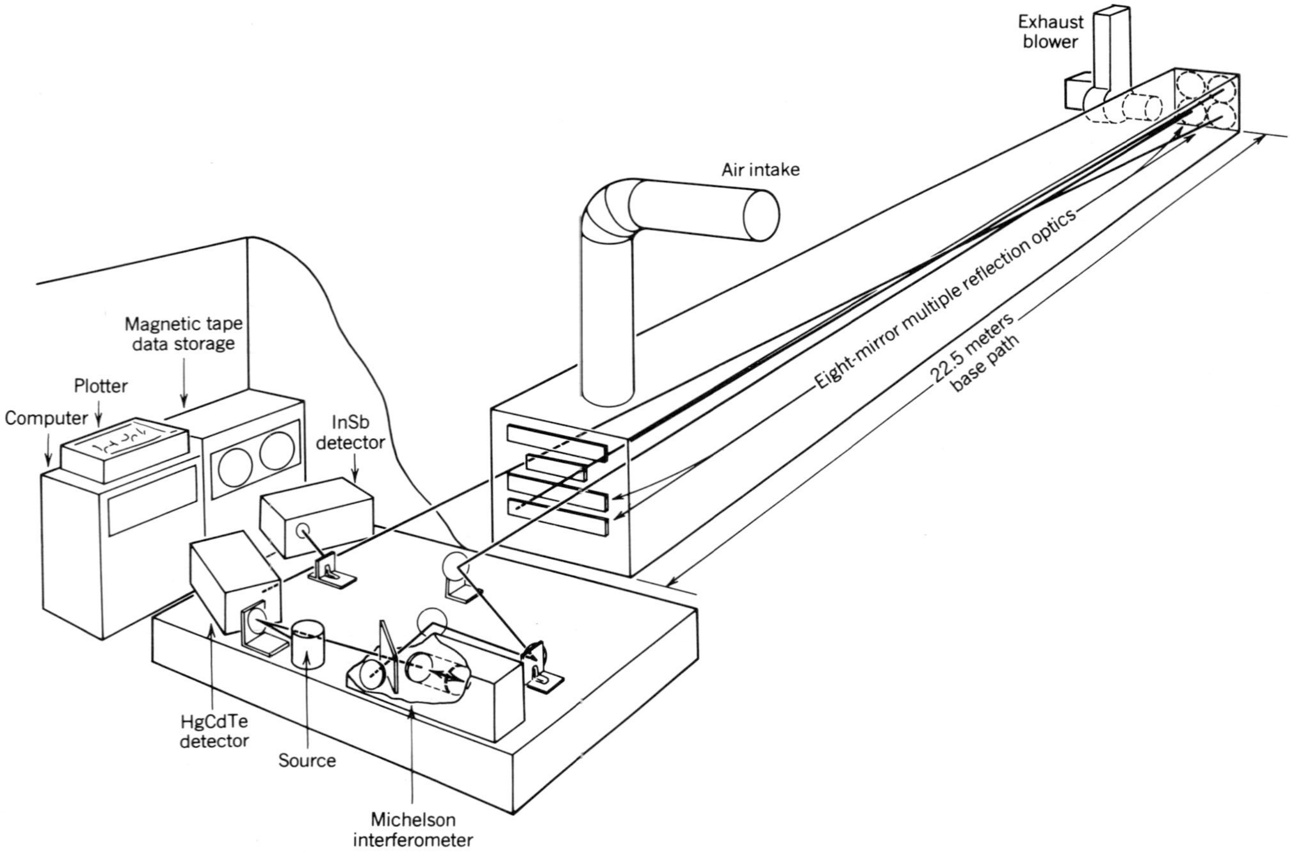

FIGURE 5.11. Schematic diagram of an FTIR system used in ambient air studies (from Tuazon et al., 1978).

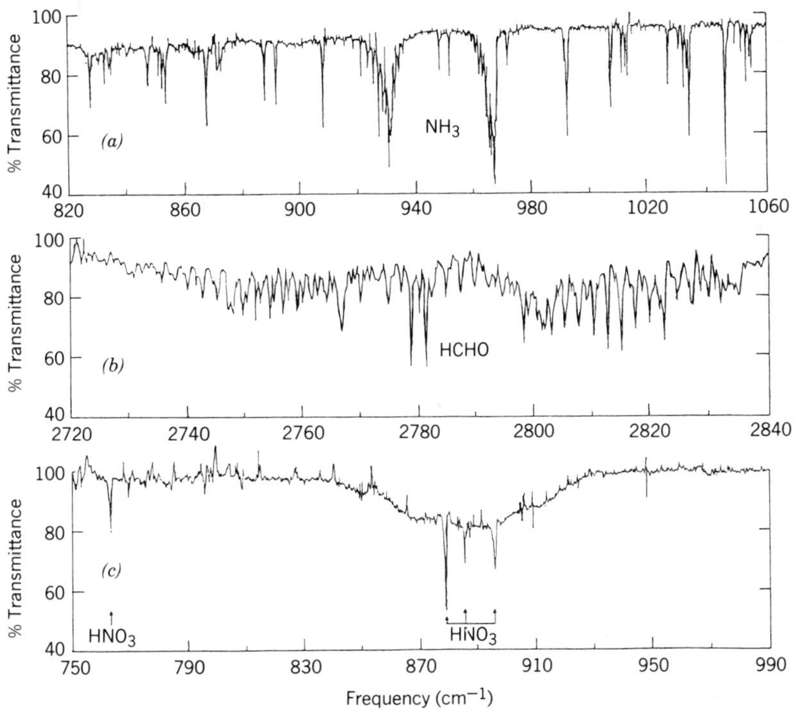

FIGURE 5.12. Reference FTIR spectra of three species of atmospheric interest taken using the apparatus of Fig. 5.11. Pathlength was 1.08 km and spectral resolution was 0.5 cm^{-1}. (*a*) ~200 ppb NH_3, (*b*) ~270 ppb HCHO, (*c*) ~150 ppb HNO_3 (from Tuazon et al., 1978).

cancel out all the H_2O lines completely because of changing water concentrations in the atmosphere. In Fig. 5.13*c*, for example, absorptions due to H_2O, such as that marked with an ×, can still be observed.

Figures 5.14 and 5.15 show ambient air spectra and the results of ratioing in a fashion similar to that in Fig. 5.13. In these cases, the background used for obtaining the ratio of the ambient air spectra was a synthetic mixture of water in N_2, chosen to match the relative humidity of the ambient air as closely as possible. The peaks due to 8 ppb HCHO and 16 ppb HNO_3 can be identified by comparison to the reference spectra in Fig. 5.12.

Figure 5.16 gives the portion of the infrared spectrum from 780 to 1260 cm^{-1} of ambient air taken in Los Angeles in 1980. The peaks due to PAN, HNO_3, C_2H_4, O_3, CH_3OH, HCOOH, and alkyl nitrates (RNO_3) are shown on the spectrum. The remaining peaks are primarily due to atmospheric H_2O and CO_2 which were not completely canceled when the ratio was obtained. From these studies, the pollutant concentration–time profiles of Figs. 5.17 and 5.18 were obtained. Clearly, using FTIR alone, a wide variety of both criteria and non-criteria pollutants can be relatively easily monitored with sufficient sensitivity for many ambient air studies. This versatility is one of the significant advantages of FTIR (Pitts et al., 1977).

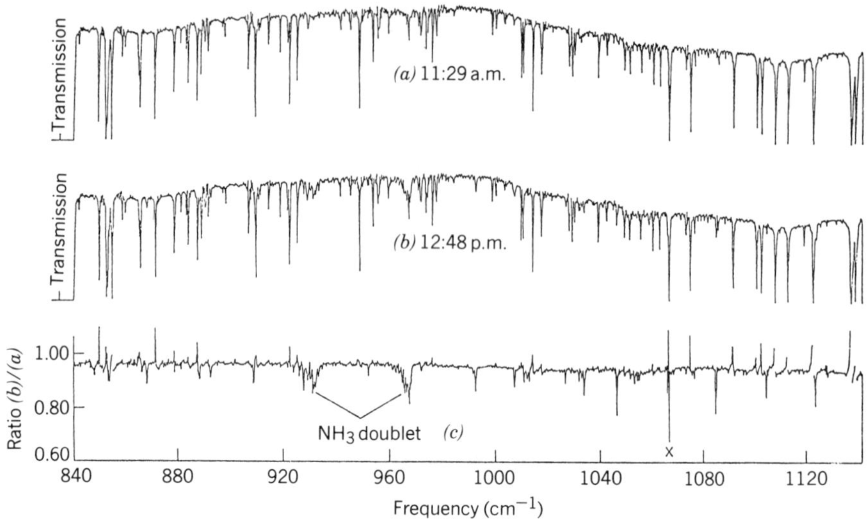

FIGURE 5.13. Spectra of ambient air in Riverside, California, on October 1, 1976 at (*a*) 11:29 a.m. and (*b*) 12:48 p.m. The ratio of the spectra, (*b*)/(*a*), is shown in part (*c*) and shows the presence of 32 ppb NH_3. The spectral resolution was 0.5 cm^{-1} and the pathlength was 1.08 km (from Tuazon et al., 1978).

The major disadvantage is the sensitivity limits for some important species, including H_2O_2, whose concentrations are lower than the current FTIR detection limit. Additionally, the spectrometer and associated optical system are relatively expensive and require trained operators.

5-C-3. Tunable Diode Laser Spectrometry

5-C-3a. Basis of Technique

A second technique based on infrared absorption spectrometry currently undergoing substantial development is tunable diode laser spectrometry. As in the case of FTIR, this technique relies on measuring the absorbance at specific wavelengths due to the absorption of IR radiation by various pollutants. However, rather than using a continuous wavelength light source and scanning the entire infrared spectrum, tunable diode laser spectroscopy (TDLS) employs a laser light source of very narrow linewidth which is tunable over a smaller (e.g., 300 cm^{-1}) wavelength range.

The major theoretical advantage of TDLS over FTIR is increased sensitivity. The major disadvantage is that scanning the entire IR spectrum quickly is not possible. Thus TDLS is more useful for following specific pollutants known to be present than for searching for previously unidentified species.

Different types of lasers used in monitoring are discussed in detail elsewhere (e.g., see Hinkley and co-workers, 1976a,b). Commonly used tunable diode lasers

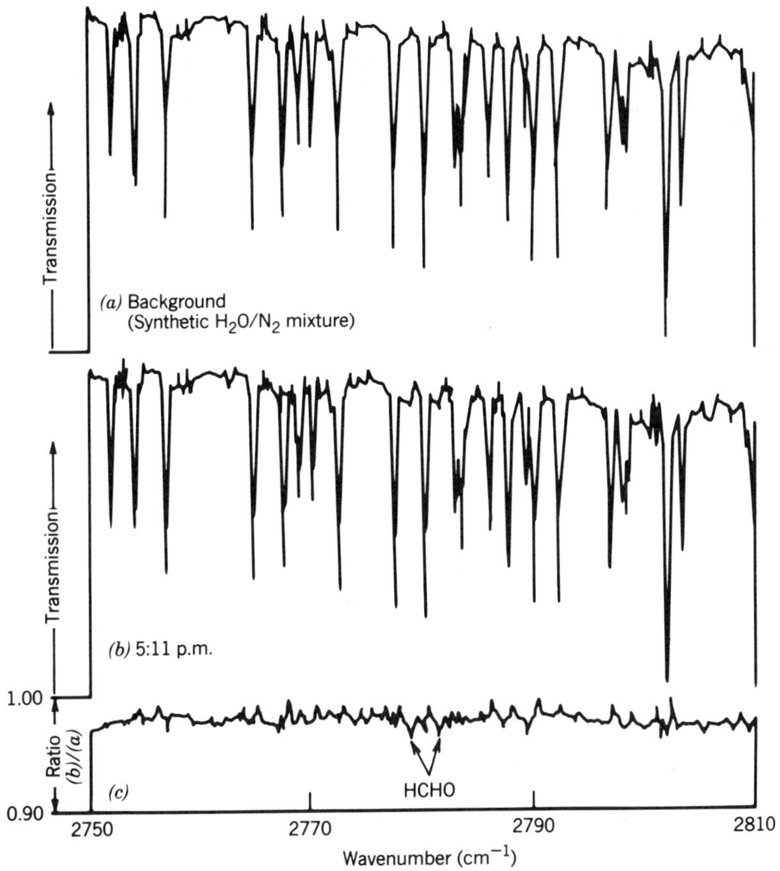

FIGURE 5.14. Spectra of ambient air in Riverside, California, on October 6, 1976: (*a*) Synthetic H_2O/N_2 mixture used for background spectrum, (*b*) ambient air at 5:11 p.m., (*c*) ratio of (*b*)/(*a*) showing peaks due to 8 ppb HCHO. Spectral resolution was 0.5 cm^{-1} and the pathlength was 1.08 km (from Tuazon et al., 1978).

are made of lead-salt compounds such as $PbS_{1-x}Se_x$, $Pb_{1-x}Sn_xTe$, $Pb_{1-x}Ge_xTe$, $Pb_{1-x}Sn_xSe$, and $Pb_{1-x}Cd_xS$. A p-n junction is formed in the crystal which is then cut into pieces typically 1 mm in length and 0.3 mm × 0.2 mm in the other two dimensions. Electrical contacts are attached, and the diode is mounted onto a support such as copper which serves as a temperature controller during operation.

When an electrical current is applied, the diode emits light spontaneously in a laser action at a wavelength corresponding to the energy band gap in the semiconductor. This gap depends on the chemical composition of the laser and hence different wavelengths can be produced by altering the diode composition. Tuning of the emitted wavelength can be accomplished through variation of one of three possible parameters: applied magnetic field strength, diode temperature, and hy-

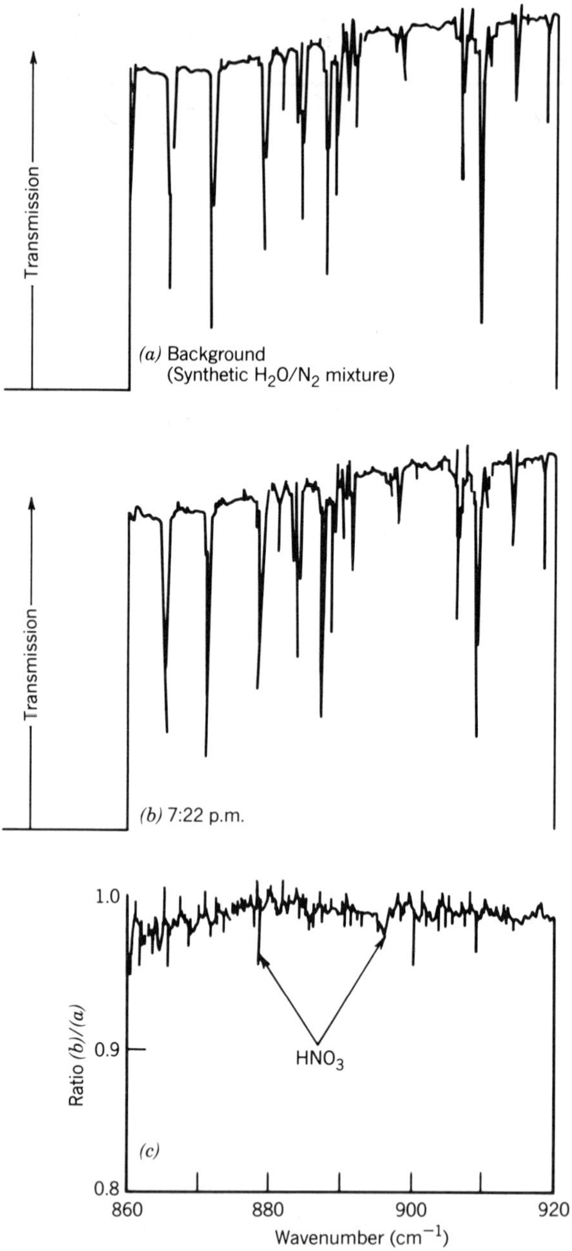

FIGURE 5.15. Spectra of ambient air in Riverside, California, on October 5, 1976: (*a*) Synthetic H_2O/N_2 mixture used for background spectrum, (*b*) ambient air at 7:22 p.m., (*c*) ratio of (*b*)/(*a*) showing peaks due to 16 ppb HNO_3. Spectral resolution was 0.5 cm^{-1} and pathlength was 1.08 km (from Tuazon et al., 1978).

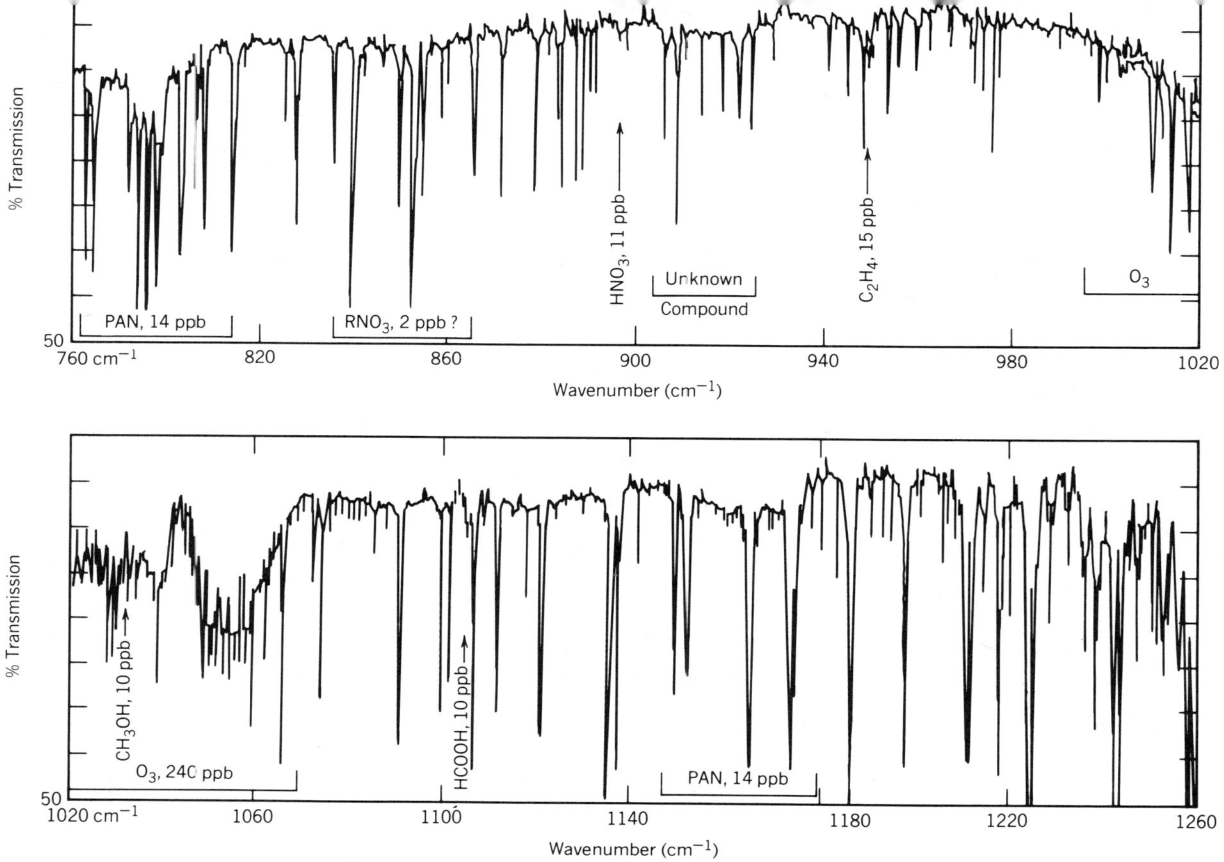

FIGURE 5.16. Ambient air FTIR spectrum from 780 to 1260 cm^{-1} in Los Angeles on June 26, 1980 at 2:45 p.m. at a pathlength of 1.26 km and spectral resolution of 0.25 cm^{-1}. The spectrum has been ratioed to that of reference air made up of 1 atm of N_2 with CO_2, CH_4, and N_2O equivalent to the amounts in clean air and H_2O equal to ~30% relative humidity (from Hanst et al., 1982).

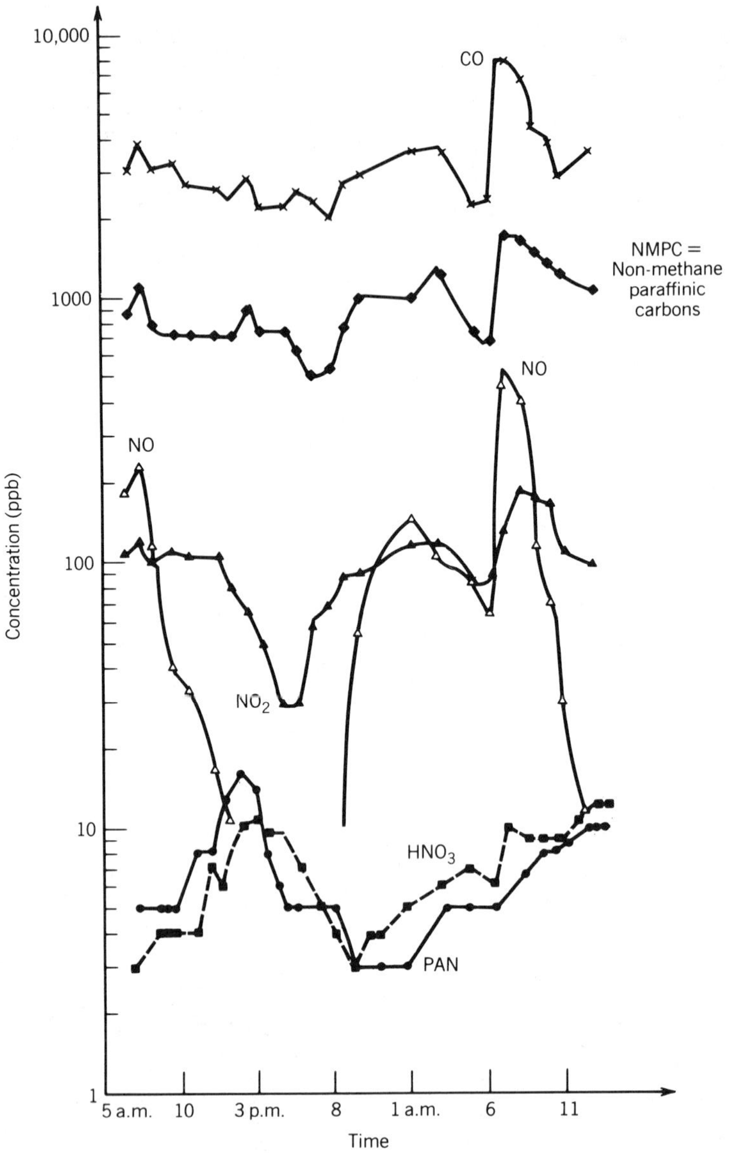

FIGURE 5.17. Concentration–time profiles for some pollutants monitored using FTIR (see Fig. 5.16) at Los Angeles, California, on June 26 and 27, 1980. NMPC = non-methane paraffinic carbons monitored using the 2975 cm^{-1} absorption peak due to $—CH_2—$ and $—CH_3$ groups (from Hanst et al., 1982).

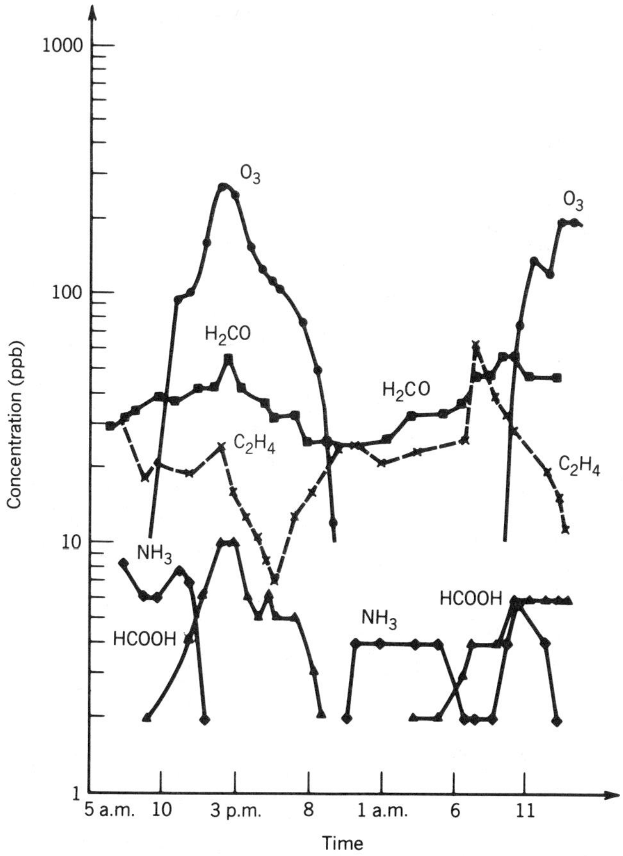

FIGURE 5.18. Concentration–time profile for some pollutants monitored using FTIR (see Fig. 5.16) at Los Angeles, California, on June 26 and 27, 1980 (from Hanst et al., 1982).

drostatic pressure. Temperature, which can be controlled by changing the current through the diode, is most commonly used. Figure 5.19, for example, shows the output of laser frequency as a function of temperature from a quaternary $(PbSe)_{1-x}(SnTe)_x$ semiconductor diode laser; also shown are absorption bands for some common air pollutants. Temperature variation can permit tuning of the laser over several hundred wavenumbers, sufficient to measure more than one pollutant with a single laser.

Several different modulation techniques can be used to increase the signal-to-noise ratio. For example, the laser beam can be mechanically chopped and detected using phase-sensitive detection with a lock-in amplifier. A more commonly used method for accurately measuring small absorbances is to modulate the frequency output of the laser by modulating the current and thus the temperature of the diode (Reid et al., 1978a). Absorbances down to $\sim 10^{-5}$ can be measured, correspond-

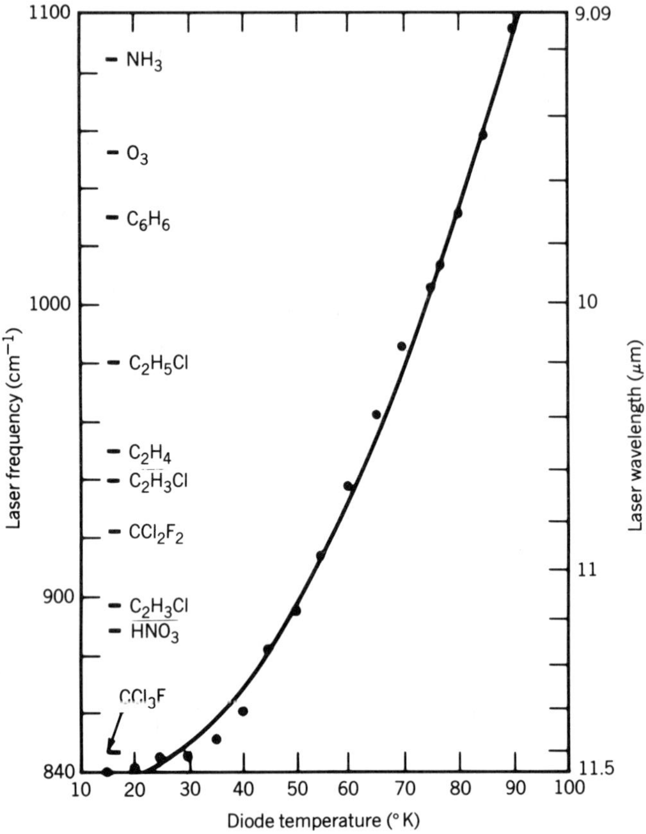

FIGURE 5.19. Variation of laser frequency with temeprature for one type of tunable diode laser, with some absorption bands due to some common air pollutants (from Hinkley et al., 1976c).

ing to ppb to sub-ppb concentrations at pathlength of ~40 m (e.g., using White cell optics) for many pollutants of atmospheric interest.

The linewidth of the output of tunable diode lasers is theoretically quite narrow but, in practice, is frequently limited by instrumental characteristics (e.g., vibration). Typical linewidths in practice are in the range 0.6–500 MHz, corresponding to 2×10^{-5}–1.7×10^{-2} cm^{-1} (Reid et al., 1982). This can be compared to typical pressure-broadened halfwidths of infrared absorption bands of species of atmospheric interest which are of the order of 0.05 cm^{-1} at atmospheric pressure; at low pressures (e.g., $\leq$ 1 Torr) where the linewidth is limited by Doppler broadening, typical halfwidths are 0.0005–0.005 cm^{-1}. Thus the TDL output is usually sufficiently narrow to scan rotational absorption lines even at low pressures where Doppler broadening is the limiting factor on line shape.

This narrow laser linewidth allows one to measure weak absorptions between the ambient H_2O and CO_2 lines. Thus one can measure accurately small absorbances due to specific rotational lines in a vibration–rotation spectrum with high selectivity. However, for many molecules of interest, the presence of such rotational fine structure requires lowering the total pressure of the sample to the low Torr pressure region to avoid pressure broadening of the absorption lines. (For larger molecules, the absorption spectrum appears as a continuum even at these lowered pressures.) Alternatively, low pressures can be used to identify the various absorbing species present and to choose a monitoring wavelength free of interferences. This wavelength can then be used to measure the pressure-broadened absorbance of the pollutant at atmospheric pressure (Hinkley et al., 1976c).

Figure 5.20, for example, shows the rotational lines of NH_3 in the 1085 cm^{-1} region at low pressure and at two higher pressures obtained using a TDL; also shown is one of the CO_2 laser lines which could be used for atmospheric monitoring of NH_3 (Hinkley et al., 1976c).

Table 5.1 shows estimated detection limits for some pollutants of interest using TDLS and FTIR. TDLS is expected to be more sensitive than FTIR and hence is potentially promising. To date it has not been as widely applied in experimental atmospheric studies. Part of the reason for this is the relatively small wavelength range covered by one TDL, and hence the limitation on the range of species which can be detected simultaneously. An additional problem has been that of calibration which must be carried out for each species; because the incident laser power is not measured, Beer's law cannot be applied directly. Calibration for ''sticky'' species such as HNO_3 which are difficult to generate accurately and reproducibly in known concentrations is thus a significant problem using TDLS (Schiff et al., 1983).

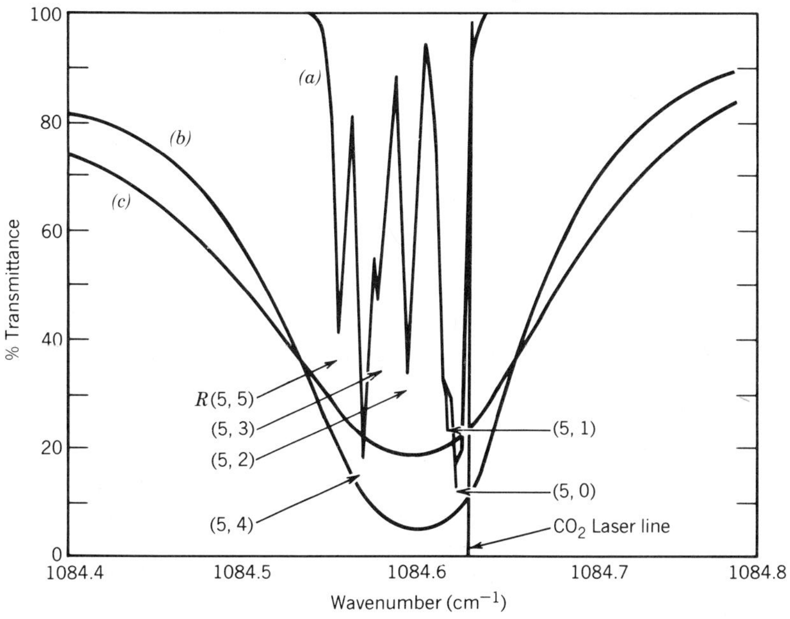

FIGURE 5.20. Laser spectroscopy of NH_3 using a diode laser in a closed cycle refrigerator operating at ~90°K. Trace (*a*) represents 0.1 Torr NH_3, (*b*) 0.46 Torr NH_3 with 360 Torr air added, and (*c*) 0.46 Torr NH_3 with 750 Torr (1 atm) air added. The CO_2 laser line is *R*(30) at 1084.635 cm^{-1}. Cell length is 30 cm (from Hinkley et al., 1976c).

TABLE 5.1. Estimated Detection Limits for Some Pollutants Using Tunable Diode Laser Spectroscopy (TDLS) and Fourier Transform Infrared Spectrometry (FTIR)

	Sensitivity (ppb)	
Pollutant	TDLS	FTIR
SO_2	3–12[a], 1[f]	[h]
O_3	0.5[b]	10[e]
NH_3	0.5–0.75[a], 0.1[f]	3[e], 1[g]
PAN	~0.3[b]	3[e]
N_2O	0.1–20[a]	[h]
CO	0.25[a]	[h]
NO	0.5[a,i]	[h]
NO_2	0.5[a,c,i], 0.1[f]	[h]
HNO_3	0.35[a,c,f]; ≤0.5[d,i]	6[e], 3[g]
HONO	10–20[f]	10[e]
HCHO	0.3[f]	6[e]
H_2O_2	0.6[f]	[h]

[a] Cassidy and Reid, 1982a; calculated for pathlength of 200 m.

[b] Reid et al., 1978a; calculated for pathlength of 200 m.

[c] May be lower limit due to severe attenuation of laser beam by the atmosphere at the monitoring wavelength.

[d] Schiff et al., 1983.

[e] Tuazon et al., 1980; 1 km pathlength, 0.5 cm^{-1} resolution.

[f] Mackay and Schiff, 1985; pathlength 35 m.

[g] Tuazon, E., 1985; at 1.15 km pathlength and 0.125 cm^{-1} resolution.

[h] Not reported.

[i] Walega et al., 1984.

However, as discussed below, it has been successful in detecting and measuring HNO_3 in ambient air in recent studies by Schiff and coworkers (Schiff et al., 1983; Walega et al., 1984; Anlauf et al., 1985); in preliminary studies, they have also successfully utilized this technique to detect and measure H_2O_2 in the air of Toronto, Canada (Mackay and Schiff, 1985). This is an important advance because H_2O_2 is believed to play an important role in the chemistry of acid deposition (see Chapter 11-A-2), yet no sufficiently sensitive method has been available to measure its concentrations in the gas phase.

5-C-3b. Applications to Atmospheric Studies

Hinkley and co-workers first used TDLS to monitor CO in ambient air (Ku et al., 1975). Since then, this technique has been applied to other species including O_3,

SO_2, N_2O, NH_3, CH_4, HCl, NO, NO_2 and HNO_3. (See, for example, Reid et al., 1978a,b, 1980; El-Sherbiny et al., 1979; Connell et al., 1980; Pokrowsky and Herrmann, 1981; Cassidy and Reid, 1982a,b; Schiff et al., 1983; Walega et al., 1984.) Figure 5.21, for example, shows the absorption signal as a function of diode current, that is, laser emission wavelength, for ambient air in Hamilton, Ontario, Canada (Reid et al., 1978a). The laser used was $Pb_{1-x}Sn_xSe$ diode with an output from 1050 to 1150 cm^{-1}, chosen to span the ν_1 absorption band of SO_2. The absorption bands due to SO_2 (40 ppb) can be seen clearly in Fig. 5.21; also shown are bands due to N_2O in the air as well as an unidentified contaminant in the White cell used in the measurement.

Figure 5.22 shows the signals from NO, NO_2, and HNO_3 in laboratory air in Toronto, Ontario, Canada. These laboratory studies were carried out using frequency modulation of the laser output by applying a 1–8 kHz sine wave to the laser current; with this mode of operation, a single absorption peak appears with a shape similar to a derivative curve (e.g., as in electron spin resonance, ESR). The detector output is followed at twice the modulation frequency to reduce the zero offset inherent in this technique and to maximize the peak signal.

To the left of each ambient air signal is the signal from bottled air containing the minimum detectable concentration of each species, that is, 0.26 ppb NO, 0.35 ppb NO_2, and 0.35 ppb HNO_3.

Subsequently, Walega et al. (1984) reported the application of TDLS to the measurement of NO, NO_2, and HNO_3 in ambient air in Los Angeles. For NO the results were in good agreement with the concentrations determined using a chemiluminescence analyzer. For NO_2, the TDLS gave lower concentrations than the chemiluminescence instrument, and the difference between the two increased

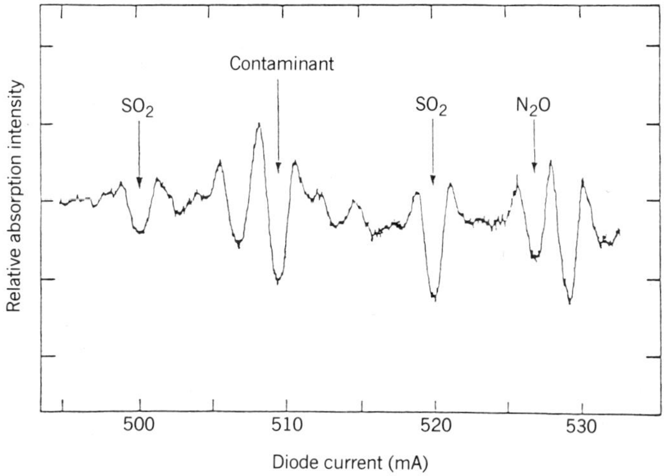

FIGURE 5.21. TDL absorption spectrum of ambient air in Hamilton, Ontario, Canada. A White cell of pathlength 300 m was used, and the air pressure was reduced to 10 Torr (from Reid et al., 1978a).

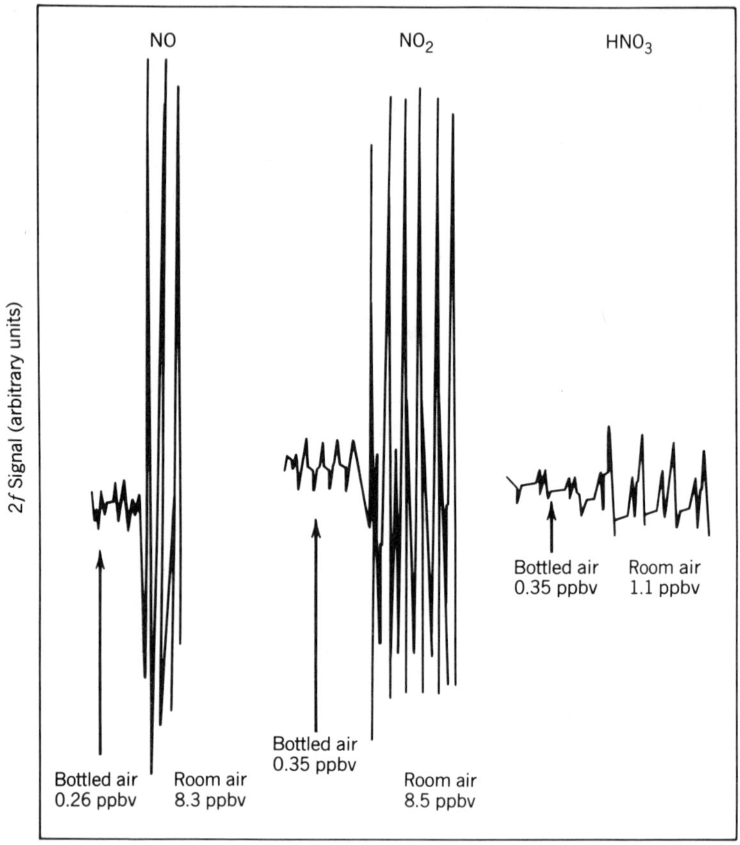

FIGURE 5.22. TDL frequency modulated (2*f*) signals for NO, NO_2, and HNO_3 in laboratory air in Toronto, Ontario, Canada. Minimum detectable concentrations are shown as bottled air to left of each trace (from Schiff et al., 1983).

throughout the day; the higher results from the chemiluminescence instrument were attributed to contributions from other nitrogenous species such as PAN, which also gave positive responses in this instrument (see Section 5-B-1). For HNO_3, good agreement was obtained on ambient air samples in the morning, but in the afternoon; significantly lower concentrations were obtained using TDLS compared to the chemiluminescence method. While the reasons for this are not clear, Walega et al., (1984) suggested that the subtraction of two large signals to obtain HNO_3 by the chemiluminescence method may make the chemiluminescence measurements inaccurate.

The application of TDLS to the measurement of H_2O_2, HCHO, and NH_3 in air is also underway (NASA, 1984a). As shown in Table 5.1, detection limits for these three species of 0.6 ppb, 0.3 ppb, and 0.1 ppb have recently been attained (Mackay and Schiff, 1985).

In summary, TDLS appears especially promising for the measurement of certain key species which cannot be monitored with sufficient sensitivity and/or selectivity in other ways. However, its potential remains to be fully explored.

5-D. SPECTROSCOPIC MONITORING TECHNIQUES: ULTRAVIOLET/VISIBLE ABSORPTION SPECTROMETRY

5-D-1. Introduction

While infrared absorption spectrometry has proven highly useful in atmospheric studies, it cannot measure all pollutants of interest with the desired specificity and sensitivity. Thus absorption spectrometry in the UV/visible regions of the spectrum is now being used increasingly as well.

UV/visible absorption spectrometry does not measure as wide a range of pollutants as IR spectrometry because many molecules do not possess the fine structure in the UV/visible region which is needed for specificity. Thus molecular absorption bands in the UV and visible regions tend to be rather broad because simultaneous electronic, vibrational, and rotational transitions are involved (Section 2-B). Fortunately, however, certain species which play key roles in the chemistry of both clean and polluted atmospheres do have sufficiently structured electronic absorption spectra that UV/visible spectrometry can be used for their detection and measurement.

Absorption coefficients in the UV/visible are generally much greater than those in the infrared, leading to intrinsically higher sensitivities for those pollutants which can be measured by this technique. This is particularly important for species such as NO_3 and HONO where other reliable, accurate, and specific techniques with the required high sensitivity for a wide range of ambient conditions from pristine areas to heavily polluted ones are not readily available.

In principle, one could measure the transmitted intensity at an absorption peak (I) and the intensity without light absorption (I_0), and apply Beer's law to obtain the concentration of the pollutant. However, the wide range of both gases and particulates in ambient air gives rise to a complex mixture of light scattering and absorption when a beam is propagated through ambient air. It is therefore difficult to actually measure I_0. As a result, a modification known as differential optical absorption is used in which I_0 is replaced by I_0', the light intensity in the absence of a structured absorption band.

5-D-2. Differential Optical Absorption Spectrometry

5-D-2a. Basis of Technique

Differential optical absorption spectrometry (DOAS) is based on measuring the difference between the absorbance at some wavelength where the species of interest has a distinct absorption peak, and another wavelength on either side of the

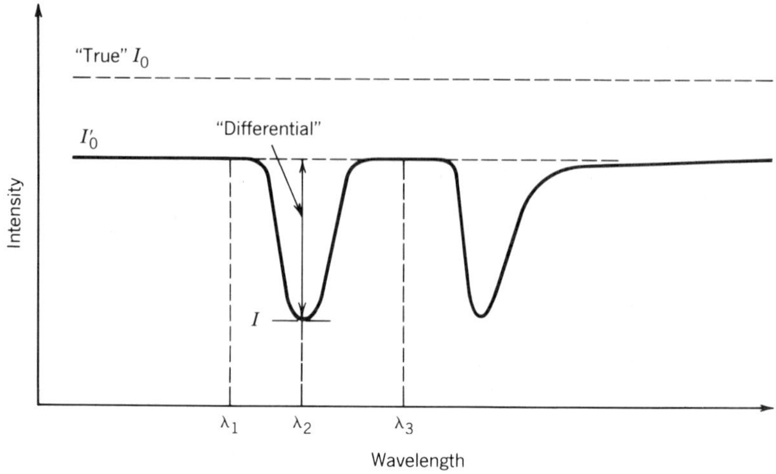

FIGURE 5.23. Basis of differential optical absorption spectrometry (DOAS) (from Platt and Perner, 1982).

peak. Figure 5.23 illustrates this technique for a molecule that has a structured absorption peaking at wavelength λ_2. On either side of the peak, that is, at λ_1 and λ_3, the absorption spectrum is not structured; however, some light absorption is occurring at these wavelengths, so that the true light intensity without any absorption (I_0), shown by the dotted line in Fig. 5.23, cannot be measured by scanning the spectrum. The concentration of the pollutant is thus determined not from log (I_0/I), but rather from log (I'_0/I), where I'_0 is the measured intensity at λ_1 and/or λ_3. In practice, a reference spectrum of the species being measured is fit to the observed spectrum so that all the absorption bands due to that molecule, including their relative strengths and shapes, are used in the identification and measurement.

Figure 5.24 is a schematic diagram of a DOAS spectrometer. The white light from a continuous light source is formed into a parallel beam using a focusing mirror. High-pressure Xe lamps or incandescent quartz-iodine lamps are typically used. The parallel beam traverses the parcel of air to be monitored before striking a collection mirror which focuses the light onto the entrance slit of a monochromator. The grating of the monochromator disperses the light so that the spectrum spreads out across the exit slit, that is, different wavelengths are projected continuously across the exit slit.

In conventional spectroscopy, the exit slit of the monochromator is stationary and wavelength scanning is achieved by slowly rotating the grating; however, this is not suitable for ambient air studies where atmospheric turbulences with frequencies of $\lesssim$ 10 Hz make it desirable that spectra be scanned at rates $\gtrsim$ 100 Hz.

Platt, Perner, and co-workers have recently developed and applied to atmospheric studies a technique known as differential optical absorption spectroscopy, DOAS, in which such a rapid scanning rate is readily achievable. In this technique, the conventional exit slit has been replaced by a mask which allows a 6–40 nm segment of the dispersed spectrum to fall on a rotating wheel, with the central

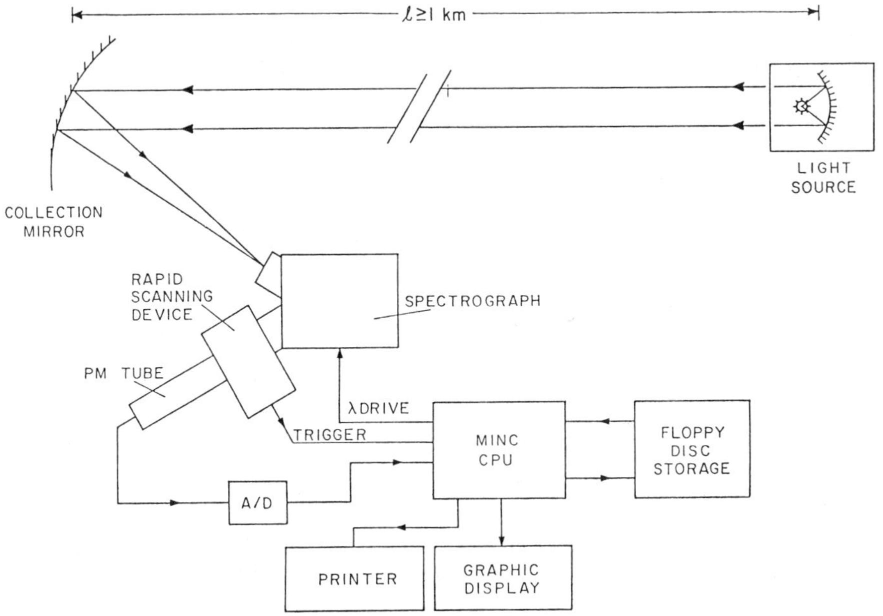

FIGURE 5.24. Schematic diagram of DOAS spectrometer.

wavelength set by the monochromator wavelength setting. The rotating wheel contains a number of narrow slits (typically 50) around its perimeter. As seen in Fig. 5.25, as the wheel rotates, the slits "scan" the portion of the spectum dispersed across the monochromator exit slit. The slits in the rotating wheel are sufficiently well spaced that only one rotating slit is in the aperture at one time, and also sufficiently narrow that only the light from a small portion of the dispersed spectrum passes through the rotating slit to the detector.

The signal, detected using a photomultiplier, is measured at several hundred different locations of the rotating slit across the exit aperture (i.e., at several hundred different wavelength intervals), and these signals are stored in different channels of a computer for subsequent data analysis. The light barrier on the edge of the mask shown in Fig. 5.25 serves to trigger the computer so that as a rotating slit enters the mask aperture, data accumulation is started. As each rotating slit crosses the exit plane of the monochromator and performs one scan, the signals are added to the appropriate channels in the computer, resulting in many scans being superimposed; this signal averaging increases the signal-to-noise ratio.

For example, if a spectral region 40 nm wide fills the exit aperture of the monochromator, and 200 channels are used to store the measured spectral intensity as the rotating slit crosses the monochromator exit, each channel will correspond to a wavelength interval of approximately $40/200 = 0.2$ nm. For a wheel with 50 slits rotating at a speed of 2 Hz, a total of 100 complete scans will be carried out in 1 s. Accumulation of signals for 10 min, say, results in the averaging of 60,000 scans, which decreases the random noise as well as signal fluctuations due to tem-

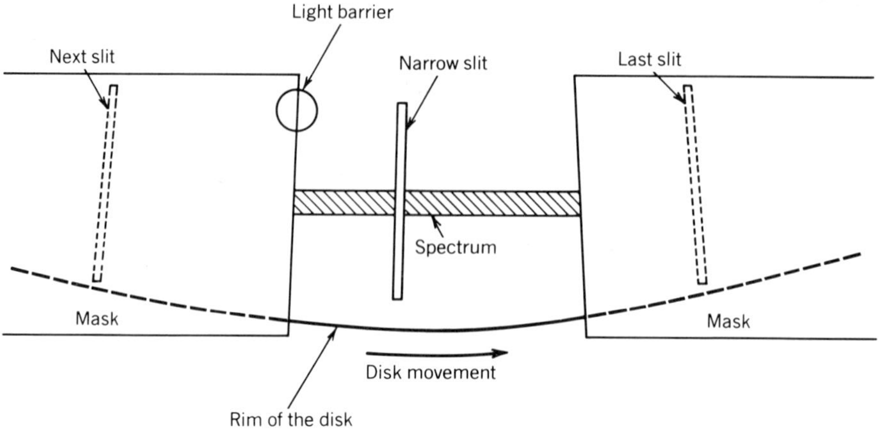

FIGURE 5.25. Schematic diagram of rotating wheel device of rapid scanning mechanism used in atmospheric DOAS studies (from Platt and Perner, 1982).

poral variations in the light source. Indeed, Platt and Perner (1982) point out that even temporarily blocking the light beam (e.g., by a vehicle driving through it) does not cause noticeable effects on the spectrum obtained. In addition, the short time required for one scan (1/100 s = 10 ms in the example given above) minimizes temporal changes in atmospheric refraction due to turbulence since these changes typically occur on a time scale of $\lesssim 0.1$ s.

The minimum optical density (base 10) detectable using the DOAS system de-

TABLE 5.2. Wavelength Range, Differential Absorption Coefficients, and Detection Limits for Some Species of Atmospheric Interest Using DOAS

Substance	Wavelength Range (nm)	Differential Absorption Coefficient[a] (cm^2 molecule^{-1})[e]	at λ (nm)	Detection Limit for 10 km Path (ppt)
SO_2	200–230, 290–310	5.7×10^{-19}	300	85
NO	215, 226	2.3×10^{-18}	226	400[b]
NO_2	330–500	1.0×10^{-19}	363	500
NO_3	623, 662	1.8×10^{-17}	662	2.5
HONO	330–380	4.2×10^{-19}	354	100
O_3	220–330	4.5×10^{-21}	~328	1×10^4
HCHO	250–360	7.8×10^{-20}	340	600
OH	308	2×10^{-16}[c]	308	0.25, 0.03[d]

Source: Platt and Perner, 1982, adjusted to minimum detectable optical densities of 5×10^{-4} (Biermann, 1985).

[a] 0.3 nm spectral resolution.

[b] 1 km light path, minimum detectable O.D. = 10^{-3}.

[c] 0.003 nm spectral resolution.

[d] Hübler et al., 1982, 1984

[e] Base *e*

scribed above is typically (5–10) $\times$ 10^{-4} for $\lambda \geq 300$ nm and a signal averaging time of several minutes. Table 5.2 gives some differential absorption coefficients as well as detection limits using DOAS for some species of atmospheric interest at a pathlength of 10 km; many of these species can be detected in the ppt concentration range. More importantly, key species in the troposphere, such as NO_3 and HONO, which are difficult, if not impossible, to measure accurately using other techniques at the low concentrations found in the atmosphere, can be measured with specificity and sensitivity using DOAS.

Because ambient air typically contains a variety of light-absorbing species such as those in Table 5.2, the contributions of various pollutants to the observed absorption spectrum must be obtained by a process of sequential subtractions. Figure 5.26 illustrates how this is accomplished. Part (*a*) is the spectrum of ambient air observed near Jülich, West Germany, in the wavelength region from 324 to 336 nm. Part (*b*) is a reference absorption spectrum of NO_2; a comparison to the ob-

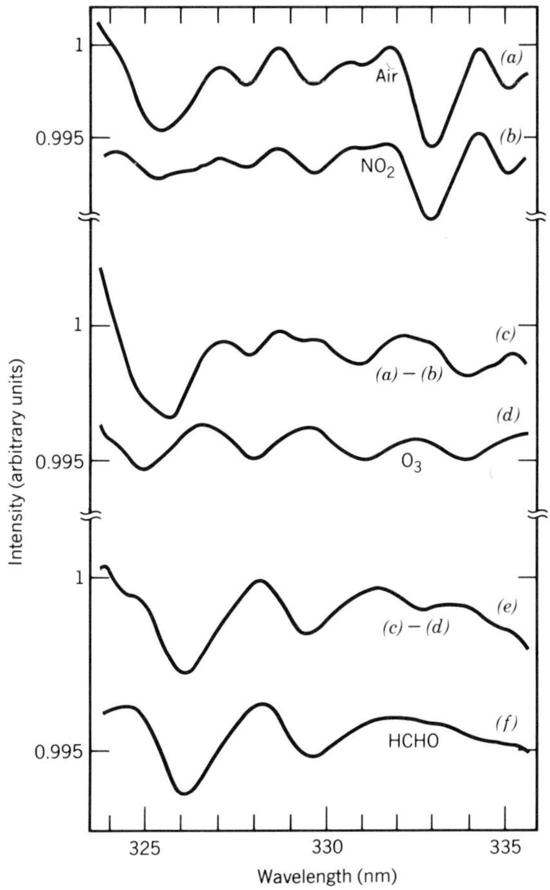

FIGURE 5.26. (*a*) Spectrum of air obtained near Jülich (September 18, 1978, 16:20) with a 5 km light path using DOAS; (*b*) nitrogen dioxide comparison spectrum; (*c*) spectrum (*b*) subtracted from spectrum (*a*); (*d*) ozone comparison spectrum; (*e*) spectrum (*d*) subtracted from spectrum (*c*), (*f*) formaldehyde comparison spectrum (from Platt et al., 1979).

served ambient air spectrum shows that some of the bands (e.g., ~329, ~333, and ~335 nm) are due to NO_2. The spectrum in part (*c*) is that remaining after subtraction of the NO_2 reference from the total observed spectrum. The NO_2 reference spectrum is multiplied by an appropriate factor prior to the subtraction to adjust for its concentration in air relative to the reference; the factor is chosen so that the major NO_2 absorption bands, which are free of other interfering absorptions, are completely removed.

Part (*d*) is the absorption spectrum of a reference sample of O_3; comparison to the remaining air spectrum after the contribution due to NO_2 has been subtracted out (part *c*) shows that some of the remaining bands may be attributed to O_3. Subtracting these bands out leaves the spectrum shown in part (*e*). By comparison to a spectrum of an authentic sample of HCHO (part *f*), the remaining strong absorption bands are identified as due to HCHO.

The concentration of NO_2, O_3, and HCHO were determined using a modified form of the Beer–Lambert law

$$C = \frac{1}{\epsilon' L} \ln \frac{I_0'}{I}$$

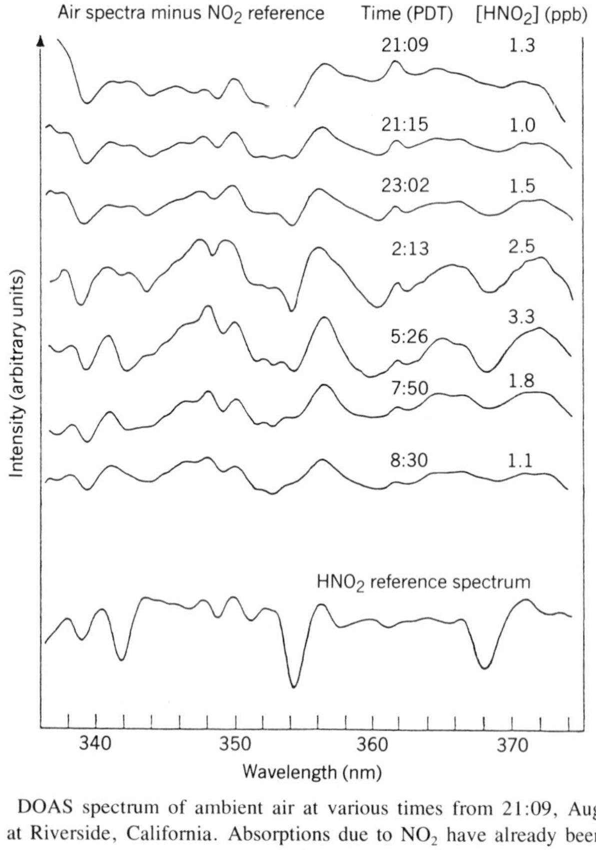

FIGURE 5.27. DOAS spectrum of ambient air at various times from 21:09, August 4, to 8:30, August 5, 1979 at Riverside, California. Absorptions due to NO_2 have already been subtracted out (from Platt et al., 1980a).

where I_0' is the observed intensity to either side of the absorption peak and ϵ' is the differential absorption coefficient of the species of interest at its absorption peak. Because the absorption peak is being measured relative to the surrounding structureless absorption (Fig. 5.23), ϵ' is generally less than the total absorption coefficient ϵ at that wavelength.

5-D-2b. *Applications to Atmospheric Studies*

In the first application of DOAS to ambient air studies (Platt et al., 1979), HCHO, O_3, and NO_2 were monitored simultaneously. Since then, this technique has been applied to detecting and measuring a number on non-criteria pollutants in various locations in the Northern Hemisphere.

Figure 5.27 shows the results of the first detection of HONO in ambient air in Riverside, California, using DOAS. (The absorptions due to NO_2 have already been subtracted out of these ambient air spectra.) By comparison to the HONO reference spectrum shown at the bottom, the peaks due to HONO at ~354.1 and ~368.1 nm can be clearly identified. As seen in Fig. 5.27, and in Fig. 5.28 which shows HONO data obtained using DOAS in the Los Angeles area, the concentration of HONO rises throughout the night. However, at dawn, the concentration falls sharply as expected due to its rapid photolysis (see Section 3-C-5).

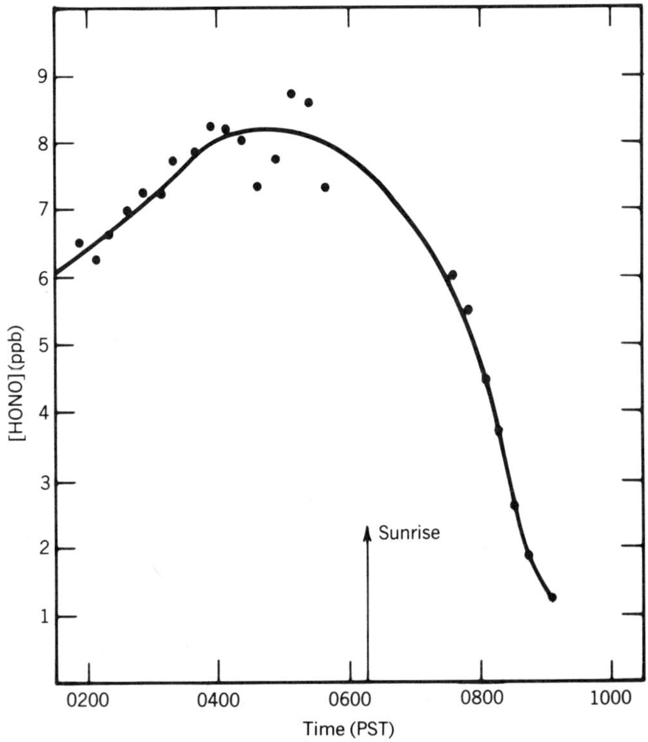

FIGURE 5.28. Observed concentration–time profile for HONO at a central Los Angeles location near the intersection of two heavily traveled freeways (from Harris et al., 1982).

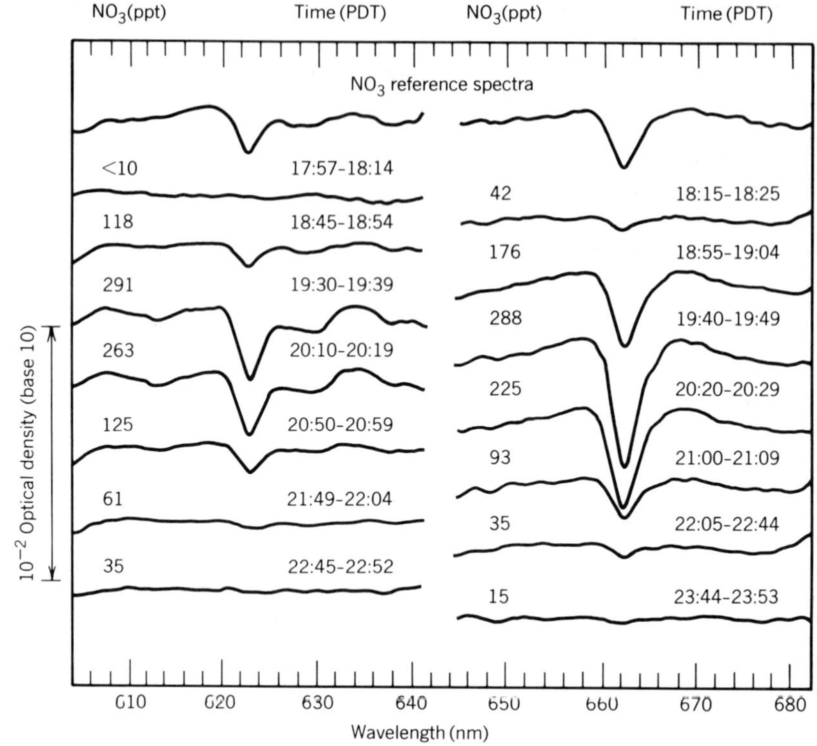

FIGURE 5.29. Development of absorption bands due to NO_3 in ambient air in Riverside, California, on the evening of September 12, 1979 (from Platt et al., 1980b).

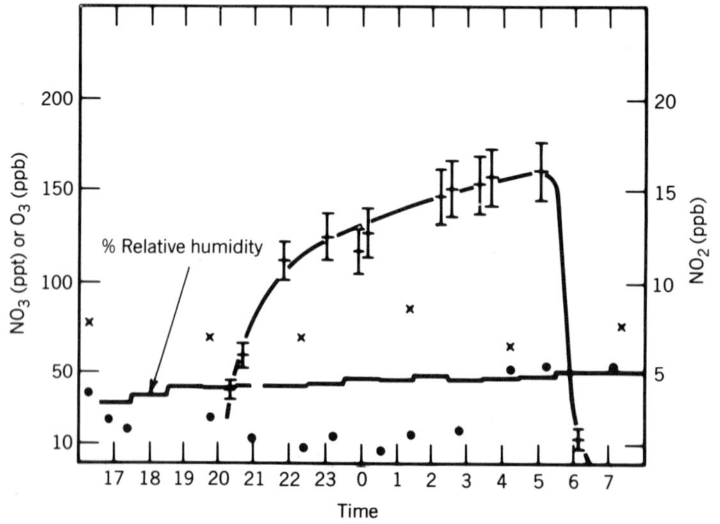

FIGURE 5.30. Typical nighttime profile of NO_3 (±), O_3 (×), and NO_2 (●) at Deuselbach, West Germany, measured using DOAS on April 15/16, 1980; RH read on left scale (from Platt et al., 1981).

Identification of the nitrate radical (NO_3) in the atmosphere was first reported by Noxon and co-workers (1978, 1980) who measured the column abundance (i.e., integrated concentration in a column extending through the upper atmosphere) using the moon as the light source. Although they interpreted their original data in terms of NO_3 predominantly in the stratosphere, tropospheric NO_3 was subsequently identified and measured using DOAS (Platt et al., 1980b).

Figure 5.29 shows the development of the characteristic NO_3 absorption peaks at 623 and 662 nm (see Section 3-C-7b) during the night. Typical concentration–time profiles for NO_3 in Deuselbach, West Germany, are shown in Figs. 5.30 and 5.31. The concentration of NO_3 rises after sunset and then falls rapidly due to photolysis (Section 3-C-7b). However, the shape of the diurnal profile during the night can vary substantially (Figs. 5.30 and 5.31).

Application of the DOAS technique to ambient OH measurements is discussed in Section 5-F.

Table 5.3 summarizes some peak concentrations of trace species measured in selected studies using DOAS at various locations in the Northern Hemisphere. As expected, the peak concentrations generally increase as one moves from a maritime air mass to a rural continental air mass to a polluted urban area.

While most of the DOAS studies reported to date have used single pass systems such as that shown in Fig. 5.24, more recently a multiple-pass White-cell type of optical system has been developed. A 37-layer dielectric coating on the mirrors gives a reflectivity of >98% in the important region around 350 nm and in the 600–700 nm region, allowing for multiple reflections within a 25 m base path to obtain a total pathlength of up to approximately 2 km (Biermann, 1985). This is a significant development because it allows measurement of pollutants in a single air mass, rather than integrated over a long distance (Winer, 1986). The detection limits obtained with this system for NO_2, HONO and NO_3 are 2 ppb, 0.5 ppb, and 10 ppt respectively, in its most recently used configuration of a total pathlength of 800 m and minimum optical density detection of 1.5×10^{-4} (Biermann, 1985).

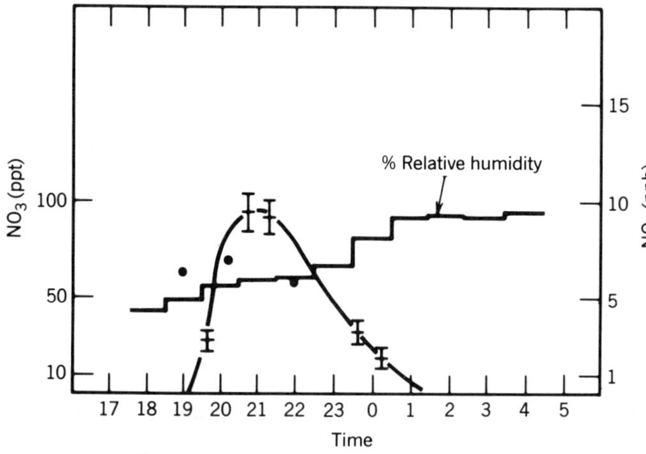

FIGURE 5.31. Typical nighttime profile of NO_3 (Ī) and NO_2 (●) at Deuselbach, West Germany, measured using DOAS on April 17/18, 1980; RH read on left scale (from Platt et al., 1981).

TABLE 5.3. Peak Concentrations of Trace Species Measured Using DOAS at Various Locations in the Northern Hemisphere

Measurement Location	Air Type	Monitoring Dates	Species	Observed Concentration Range (ppb)
Loop Head, Ireland[a]	Maritime	Apr. 1979	HCHO	≤ 0.3
			HONO	≤ 0.03
			O_3	23–50
			NO_2	≤ 0.1–2.6
			SO_2	0.4–5.3
			NO_3	≤ 0.014
Jülich, West Germany[a]	Rural continental	Feb.–Mar. 1979	HONO	≤ 0.05–0.8
			O_3	2–32
			NO_2	5–58
			SO_2	11–59
			NO_3	*b*
		June–Sept. 1980	OH[c]	$(<3\text{–}7.4) \times 10^{-5}$
Whitewater, California[e]	Rural continental	Oct.–Dec. 1981	NO_3	<0.002–0.11
Death Valley, California[e]	Remote continental	Apr.–May 1982	NO_3	0.009–0.029
Deuselbach, West Germany	Urban	Apr. 1980	NO_3	<0.006–0.28
Los Angeles area	Polluted urban	July–Aug. 1980[d]	HONO	0.4–8.0
			NO_2	22–120
		Aug–Sept. 1979[d]	NO_3	<0.006–0.36
		Sept. 1984[f]	NO_3	0.071–0.421

[a] Platt and Perner, 1980.

[b] Always below detection limit which is given as 0.0005 ppb under optimum measurement conditions; it is not clear if this limit was feasible during these studies.

[c] Hübler et al., 1982.

[d] Harris et al., 1982.

[e] Platt et al., 1984.

[f] Atkinson et al., 1985.

In summary, the development of the rapid scanning device in conjunction with absorption spectrometry has made DOAS a very powerful technique for monitoring certain key species which play important roles in tropospheric chemistry. The application of this as well as other spectroscopic techniques such as FTIR and TDLS in the measurement of non-criteria pollutants in a variety of locations and under a variety of conditions around the world has already greatly increased our understanding of the chemistry of both clean and polluted atmospheres.

5-E. MONITORING CRITERIA GASEOUS POLLUTANTS

As discussed in Chapter 1, the five criteria gaseous pollutants, that is, those for which federal air quality standards are set, are CO, NO_2, O_3, SO_2, and non-methane hydrocarbons (NMHC). Table 5.4 summarizes some of the techniques used to monitor these.

Those techniques designated as a "Reference Method" by the United States Environmental Protection Agency are indicated in the table. Thus, for each of the criteria pollutants, the EPA designates one measurement method to be used by EPA and by state and local agencies to measure the ambient concentration of that pollutant. The Agency also specifies approved "Equivalent Methods" which have been shown to have a "consistent relationship to the reference method" and meet EPA's performance specifications; these are also acceptable methods of monitoring the criteria pollutants. The World Health Organization (WHO) also has published recommended sampling and measurement methods for SO_2, CO, NO_x, oxidants, and suspended particulate matter. The reader is encouraged to consult the U.S. EPA Air Quality Criteria Documents and the WHO summary (1976) for futher details on monitoring and calibration procedures.

Table 5.4 includes both physical/spectroscopic techniques as well as wet chemical techniques. At the present time, only for SO_2 is a wet chemical technique designated as the EPA reference method. However, many of the instruments based on physical/spectroscopic techniques are expensive and not easily available in all parts of the world. We therefore also include in Table 5.4 reference to some of the wet chemical methods which are commonly used to monitor these criteria gaseous pollutants. Comments on the techniques for CO and NMHC which have not been discussed earlier follow.

CO. Non-dispersive infrared spectrometry (NDIR) is the U.S. EPA reference method for measuring carbon monoxide in ambient air. Infrared light from a conventional source passes through two parallel cells. The reference cell contains a gas which does not absorb the radiation, while ambient air is continuously pulled through the sample cell. CO in the sample cell absorbs the IR at ~ 4.6 μm, resulting in less light reaching the detector for the sample cell than that reaching the detector for the reference cell. The difference in signals from the two cells is a measure of the CO present.

Selectivity for CO is obtained by using a detector consisting of a dual compartment cell, each side containing CO. The light passing through the reference cell strikes one compartment, while that from the sample cell strikes the second compartment. The two sides of the detector are separated by a thin diaphragm whose movement can be monitored electronically. Light absorbed by the CO in the detector is converted to heat; when more light reaches one side than the other and is converted to heat, the pressure rises and the diaphragm distorts, resulting in a signal. Thus, when CO is present in the sample cell, less radiation strikes that

TABLE 5.4. Some Techniques Used to Monitor Criteria Gaseous Pollutants

Pollutant	Technique	Detection Limit	Reference
CO	Non-dispersive infrared spectrometry^{a}	1 ppm	Informatics, 1979
	Gas chromatography	0.02 ppm	Informatics, 1979
	TDLS^{b}	250 ppt	Cassidy and Reid, 1982a
	Electrochemical sensors	1 ppm	Informatics, 1979
	Mercury replacement	0.02 ppm	Informatics, 1979
	Colorimetric analysis	$\geqslant$1 ppm	Informatics, 1979
NMHC^{e}	Gas chromatography with flame ionization detection^{a}	0.1 ppmC	Sexton et al., 1982
	Photoionization	>0.1 ppmC^{f}	Sexton et al., 1982
NO_2	Chemiluminescence (O_3)^{a}	10 ppt	NASA, 1983^{c}
	TDLS^{b}	100 ppt	Mackay and Schiff, 1985
	DOAS (10 km pathlength)	500 ppt	Platt and Perner, 1982
	Griess–Saltzman colorimetric	10 ppb	U.S. EPA, 1982a
	Sodium arsenite	5 ppb	U.S. EPA, 1982a
	TG–ANSA	8 ppb	U.S. EPA, 1982a
O_3	Chemiluminescence^{a}	2 ppb	U.S. EPA, 1978
	UV absorption	1 ppb	U.S. EPA, 1978
	TDLS^{b}	500 ppt	Cassidy and Reid, 1982a
	Neutral buffered KI^{d}	3 ppb	U.S. EPA, 1978
SO_2	West Gaeke (pararosaniline)^{a}	10 ppb	U.S. EPA, 1982b
	Fluorescence	5 ppb	U.S. EPA, 1982b
	GC–flame photometry	2–10 ppb	U.S. EPA, 1982b
	TDLS^{b}	1 ppb	Mackay and Schiff, 1985
	Second derivative spectrophotometer	10 ppb	U.S. EPA, 1982b
	DOAS (10 km pathlength)	85 ppt	Platt and Perner, 1982
	Coulometric	2–50 ppb	U.S. EPA, 1982b
	Conductometric	5–40 ppb	U.S. EPA, 1982b
	Colorimetric	2–10 ppb	U.S. EPA, 1982b
	H_2O_2 titrimetric	10 ppb	U.S. EPA, 1982b

[a] U.S. Environmental Protection Agency Reference Method.

[b] TDLS = Tunable diode laser spectrometry. See Section 5-C-3.

[c] NASA = U.S. National Aeronautics and Space Administration.

[d] Measures total oxidant; see discussion in Section 5-A.

[e] NMHC = Non-methane hydrocarbons.

[f] Based only on olefins and aromatics to which this detector responds; as discussed in text, response to smalle alkanes is zero.

side of the detector. The reference side therefore heats up more, and the resulting distortion of the diaphragm is a measure of the amount of CO present in the sample cell.

Variants of this technique such as gas filter correlation spectroscopy (Hanst, 1971) are also in use, although these are not currently accepted as an equivalent CO monitoring technique by the EPA.

Gas chromatography (GC) with flame ionization detection (FID) has also been used extensively to monitor CO. CO is first reduced to methane so that the sensitive FID (which does not respond to CO) can be used for its detection. Because, as discussed below, GC–FID is also used for monitoring methane and non-methane hydrocarbons (NMHC), the capability of monitoring CO as well as CH_4 and NMHC is usually incorporated into the same instrument. Thus injection of an ambient air sample into an FID gives total hydrocarbons, including methane. A portion of the same sample can be passed through an absorbant column to remove all NMHC and the CH_4 and CO eluting from that passed through a GC column to separate them from each other. The CH_4 can then be measured separately by FID, and the CO reduced catalytically to CH_4 before being injected into the FID for measurement.

An intercomparison study of the measurement of CO by different techniques, a TDLS system and three grab sample–gas chromatography methods, was carried out in 1983. The relative agreement between the techniques under ambient conditions where the concentration of CO was in the 200 ppbv range was found to be approximately 14% (Hoell et al., 1984).

NMHC. The very low reactivity of methane is such that it is classed as the only non-reactive hydrocarbon (see discussion of reactivity scales in Section 10-C). As a result, measuring the concentrations of all hydrocarbons except methane, that is, non-methane hydrocarbons (NMHC), as well as of other reactive organics such as aldehydes is important.

Individual hydrocarbons can be separated and measured using gas chromatography with flame ionization detection, which is very sensitive for hydrocarbons. Mass spectrometry interfaced to GC can be used to confirm the identification of individual compounds suggested by their retention times on the GC column. The results of some typical studies of the distribution of NMHC in ambient air are discussed below.

For routine monitoring purposes, however, such analyses are not feasible due to the time required for each analysis and the need for highly trained personnel. In addition, the air quality standards are currently set in terms of total NMHC rather than individual hydrocarbons, making separation of the components unnecessary from a compliance viewpoint. As a result, several automated methods of measuring total NMHC without separation of the individual components are now in use.

Conventional NMHC analyzers use FID to measure the total hydrocarbon concentration in the air sample. The CH_4 is separated and measured independently and the NMHC obtained by difference. Separation of the CH_4 is accomplished in one of two ways: (1) gas chromatography to separate CH_4 from all other hydrocarbons, or (2) all organics except CH_4 are oxidized (presumably to $CO_2 + H_2O$ which give no response in FID), leaving the CH_4 to be measured separately.

Several types of so-called non-conventional NMHC analyzers are also available (Sexton et al., 1982). In one type, the air is passed through a chromatographic column that retains all NMHC, letting CH_4 pass through. The adsorbed NMHC are then backflushed into the FID for their measurement. In some commercial instruments, the backflushed NMHC are first oxidized to CO_2 and then reduced to CH_4 prior to measurement by FID; the theoretical advantage of these extra steps to convert all NMHC to CH_4 is that the instrument response per C atom is then constant, regardless of the original hydrocarbon composition (see below). However, in one evaluation of this type of instrument (Sexton et al., 1982), some problems were encountered with interference from CO_2 in the original sample, which elutes from the column close to the C_2 peaks.

A second type of non-conventional analyzer is based on the principle of photoionization. Hydrocarbons are ionized by exposure to ultraviolet light with 10.2 eV energy, and the ions are detected and measured using conventional techniques. The advantage of this type of approach is that H_2, needed for FID and which is explosive in certain mixtures with O_2, is not required and an extensive valving and GC column system is not needed. However, although it is particularly sensitive to aromatics and other unsaturated compounds, it does not respond to C_2–C_4 (and possibly higher) alkanes, so that the total NMHC reported for ambient air samples using this method are generally significantly lower than those from conventional analyzers.

TABLE 5.5. Average NMHC Concentrations Measured in Some Urban and Rural Areas

City	Average [NMHC] (ppmC)	Reference
	Urban[a]	
Los Angeles, California	1.93	Mayrsohn et al., 1975
Houston, Texas	1.36	Sexton and Westberg, 1981, 1984
Philadelphia, Pennsylvania	0.58	Sexton and Westberg, 1981, 1984
Denver, Colorado	1.40	Ferman et al., 1981
Sydney, Australia	0.61	Nelson et al., 1983
Delft, Netherlands	0.39	Bos et al., 1978
	Rural	
Belfast, Maine	0.013	Sexton and Westberg, 1981, 1984
Janesville, Wisconsin	0.014	Sexton and Westberg, 1981, 1984
Robinson, Illinois	0.065	Sexton and Westberg, 1981, 1984

[a] 6–9 a.m. average concentrations measured because air quality standards are generally set for this period; see Section 1-D-4.

(ppbC)

Aliphatic Hydrocarbon	Denver Colorado[a] (Urban)	Los Angeles, California[b] (Urban)	Boston, Massachusetts[c] (Urban)	Sydney, Australia[d] (Urban)	Ever-glades, Florida[e] (Rural)	Jones State Forest, Texas[f] (Rural)
	Location and Type of Air Mass					
Ethane	7	91	8	15	2.3	27
Propane	104	88	9	18	0.4	18
Isobutane	58	45	12	19	0.3	6
n-Butane	153	102	29	30	0.6	10
Isopentane	97	140	35	45	0.9	6
n-Pentane	74	78	16	25	0.4	4
Cyclopentane	n.r.[g]	n.r.	n.r.	4	1.1	0.6
2-Methylpentane	95	102	12	16	0	1
3-Methylpentane	51	61	8	10	0.7	0.9
n-Hexane	85	117	9	13	0.2	2
Other alkanes	n.r.	573	n.r.	88	6.7	11
Ethene	n.r.	69	n.r.	25	0.7	4
Propene	n.r.	25	4	22	0.6	1
C-4 Olefins	n.r.	n.r.	≤ 4	14	17.2	2
C-5 Olefins	n.r.	n.r.	n.r.	18	0	0.6
Acetylene	48	64	9	20	1.2	3

[a]Ferman et al., 1981.

[b]Mayrsohn et al., 1975; average of three daily sampling periods (0200–0500 hours, 0600–0900 hours; 1100–1400 hours) in July 1974.

[c]Sexton and Westberg, 1981, 1984.

[d]Nelson and Quigley, 1982; Nelson et al., 1983.

[e]Lonneman et al., 1978.

[f]Seila, 1979.

[g]n.r. = not reported.

TABLE 5.7. Typical Ambient Concentrations of Individual Aromatic Hydrocarbons in Various Locations (ppbC)

	Location and Type of Air Mass				
Aromatic	Denver, Colorado[a] (Urban)	Los Angeles, California[b] (Urban)	Boston, Massachusetts[c] (Urban)	Oslo, Norway[d] (Urban)	Sydney, Australia[e] (Urban)
Benzene	44	60	8	8–212	16
Toluene	135	248	28	11–397	62
o-Xylene	34	80	5	2–65	12
(*m* + *p*)-Xylene	100	173	11	4–230	31
Ethylbenzene	n.r.	n.r.	4	n.r.	10
Other aromatics	n.r.	515	5	n.r.	33

[a] Ferman et al., 1981.
[b] Mayrsohn et al., 1975
[c] Sexton and Westberg, 1981, 1984.
[d] Wathne, 1983.
[e] Nelson and Quigley, 1982; Nelson et al., 1983.

Interpretation of the data from NMHC analyzers is somewhat uncertain due to the varying responses of different organic compounds in FID. Thus, the response per carbon atom for a given hydrocarbon depends on the structure of the organic. Isopentane and methane, for example, have been observed to give per carbon atom responses 50 and 30% larger, respectively, than propane; on the other hand, toluene and *m*-xylene gave responses lower than propane by approximately 20 and 40%, respectively (Sexton et al., 1982).

While we have referred to hydrocarbon measurement, FIDs respond to a variety of organics, not only hydrocarbons, albeit with different responses per C atom. Thus they are also frequently referred to as non-methane organic carbon (NMOC) analyzers.

Instrument operating parameters such as the ratio of the flows of hydrogen to that of the sample, the air flow rate, and the oxygen content of the sample air may also alter the response of FIDs to a given hydrocarbon. Fluctuation in the first two parameters have been shown to cause relatively minor changes in FID response; while the oxygen content of the sample can have a significant effect, especially if it falls below 15% (Sexton et al., 1982), this is not a significant drawback in typical monitoring applications since the O_2 content of tropospheric ambient air remains constant.

The results of NMHC measurements are normally reported as ppmC or ppbC, defined as follows:

$$\text{ppmC} = \text{ppm of the organic} \times \text{number of carbon atoms per organic molecule}$$

TABLE 5.7. (*Continued*)

Location and Type of Air Mass					
London, U.K.[f] (Urban)	Berlin, W. Germany[g] (Urban)	Johannesburg, South Africa[h] (Urban)	Jones State Forest, Texas[i] (Rural)	Silwood, U.K.[f] (Rural)	Everglades, Florida,[a] (Rural)
53	91	20	n.r.	9	n.r.
95	183	72	5	9	2.5
10	n.r.	13	5[j]	1	2.8
22	145	36	—	4	4.5
7	n.r.	15	1	1	2.0
n.r.	n.r.	n.r.	13	n.r.	16.7

Clark et al., 1984.
Seifert and Ullrich, 1978.
Louw et al., 1977.
Seila, 1979.
Total xylenes.
Lonneman et al., 1978.

For example, 2 ppm of propane (C_3H_8) would be reported as 6 ppmC:

$$\text{ppmC} = 2 \text{ ppm } C_3H_8 \times 3 \text{ C atoms} = 6 \text{ ppmC}$$

Calibration of the NMHC analyses is carried out using a known concentration of a reference hydrocarbon such as methane or propane in air.

There are several techniques for sampling NMHC prior to analysis. If the concentrations are sufficiently high, for example, those found in polluted urban areas, direct sampling without preconcentration can be used. Care must be taken to avoid losses due to absorption on the walls of the sampling container, or due to reactions occurring after sampling. Contamination from the container itself (e.g., solvents and plasticizers from bags made of polymers) can also be a problem (e.g., see Seila et al., 1976; van Ham and Lems, 1978).

At lower concentrations, some form of concentration prior to analysis may be needed. Cryogenic techniques as well as the use of solid sorbents (e.g., Tenax, XAD-2 resin, activated charcoal) from which the organic can be either thermally desorbed or extracted with a solvent have been used (Pellizzari, 1978; Hughes et al., 1980; Schlitt et al., 1980; Lamb et al., 1980; Keith, 1984). Cryogenic techniques are generally used for the C_2–C_8 hydrocarbons, while adsorption onto solid sorbents is used for the C_6–C_{15} gaseous organics.

Sampling and analytical techniques for organics, including those containing heteroatoms such as N, S, O, and halogens, are reviewed by Singh (1980), Lamb et al. (1980), and in the volumes edited by Natusch and Hopke, (1983), and Keith (1984).

Table 5.5 shows typical NMHC concentrations in a number of urban and rural areas. The urban concentrations are one to two orders of magnitude greater than the non-urban concentrations.

Tables 5.6 and 5.7 show the concentrations of some individual aliphatic and aromatic hydrocarbons measured in various urban centers as well as in some rural areas. The C_1–C_5 alkanes, as well as ethene and propene, are present in significant concentrations. For the aromatic hydrocarbons, toluene and the xylenes are the major species present.

The actual concentrations of hydrocarbons which are observed and their relative amounts are sensitive to the sampling location. Such factors as whether the sampling site is close to sources or whether it is downwind of the sources, and hence

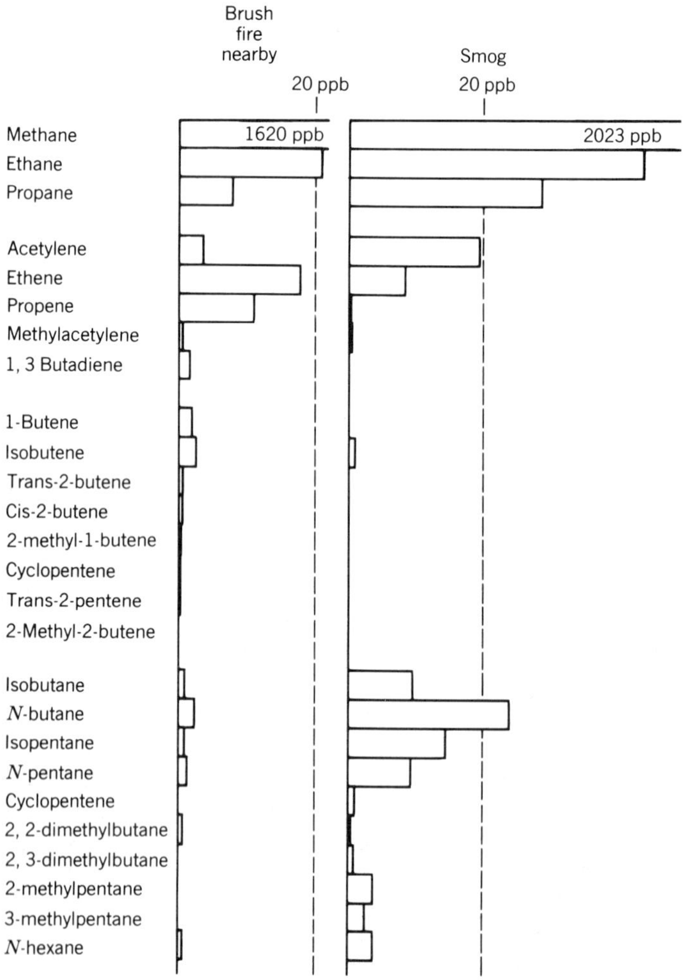

FIGURE 5.32. Distribution of aliphatic hydrocarbons observed when there was a nearby brush fire and in typical smoggy air. Note methane is not to scale (from Stephens and Burleson, 1969).

has ''aged air'' in which the most reactive hydrocarbons have been partially consumed, are important in determining the composition of ambient organics.

Figures 5.32 and 5.33, for example, show the observed distribution of some aliphatics in ambient air in Riverside, California, approximately 60 mi east of Los Angeles. Figure 5.32 shows the concentrations for two situations: first when the smoke from a nearby brush fire was present, and second on a typical smoggy day with no known nearby strong hydrocarbon sources. Figure 5.33 shows analogous data for three different sampling locations: (*a*) near an industrial area in Los Angeles County, (*b*) in a parking lot, and (*c*) at a busy intersection where both heavy traffic and three gasoline stations are located. Acetylene, which is used as a tracer

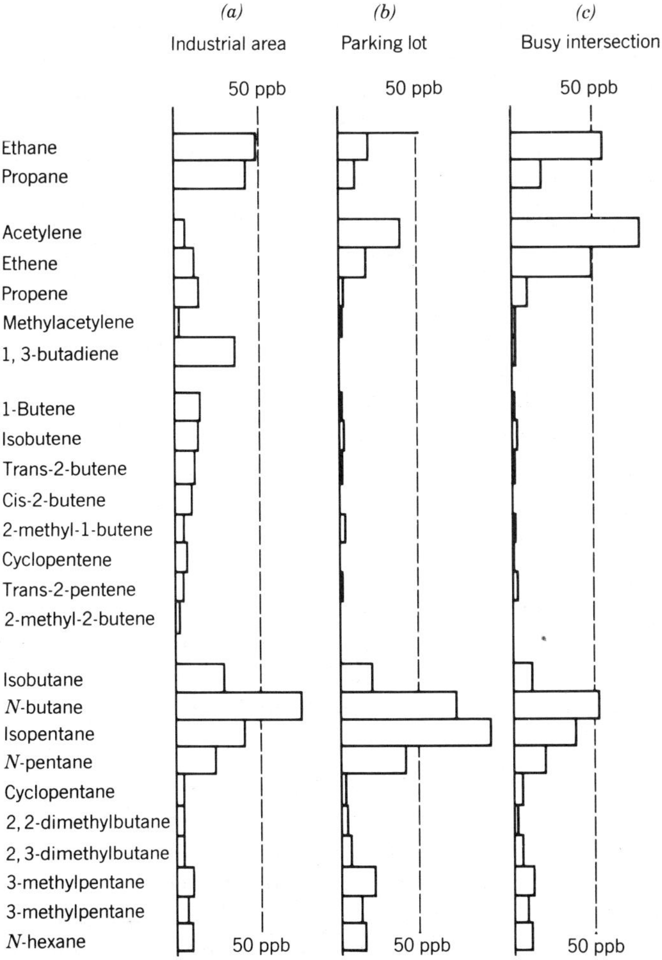

FIGURE 5.33. Concentrations of aliphatic hydrocarbons near three types of sources: (*a*) an industrial area, (*b*) in a parking lot, and (*c*) at a busy intersection. (from Stephens and Burleson, 1969).

for automobile exhaust, is present at high concentrations at the busy intersection as might have been expected. On the other hand, significant concentrations of 1,3-butadiene were present in the industrial area. Thus the mixture of hydrocarbons representative of a given location depends very much on the distribution of the surrounding sources.

The bases of many of the techniques used for NO_2, O_3, and SO_2 measurements have already been treated here or in Chapter 4 and will not be discussed further here.

5-F. MONITORING NON-CRITERIA GASEOUS POLLUTANTS

As we have seen in Chapter 1 and discussed in more detail in the following chapters, a wide variety of secondary pollutants are formed in ambient air simultaneously with the criteria pollutants. Although many of them are formed in relatively small (e.g., ppt) concentrations, they may play a key role in tropospheric chemistry, as is the case for NO_3, for example. In addition, they may, either alone or in concert with other pollutants, have deleterious effects on human health, plants, animals, or materials. PAN, for example, is a well known plant phytotoxicant in ppb concentrations and causes eye irritation in the ppm–ppb concentration range.

Table 5.8 summarizes some of the techniques used to monitor these pollutants. This list is by no means an exhaustive account of all methods that have been used or are under development, but rather, it is indicative of some of the common monitoring techniques either in use or under development and undergoing field testing. An intensive discussion of the details of each technique, including potential interferences, accuracy, time resolution, and calibration techniques, is beyond the scope of this book; the reader is encouraged to consult the literature for these and other relevant details.

Specific comments on certain of these pollutants follows.

NO. An intercomparison study of a laser-induced fluorescence (LIF) and two chemiluminescence methods for NO was carried out in 1983 simultaneously with the CO measurements discussed earlier. Over a range of concentrations from ~10 to 180 ppt, the relative agreement between the techniques was found to be ~17% (Hoell et al., 1984). This provides some indication of the agreement to be expected when different methods are used to measure even the relatively simple pollutant NO.

PAN. PAN is a particularly important secondary pollutant because of its plant phytotoxicity and potential for eye irritation. It is routinely monitored using gas chromatography (GC) with electron capture detection (ECD), a technique that is particularly sensitive for compounds containing oxygen, nitrogen, and halogens. In this technique, pioneered for application to PAN by Stephens and co-workers (Stephens et al., 1960, 1961; Darley et al., 1963), a 9 in. GC column made of $\frac{1}{8}$ in. ID Teflon tubing and packed with 5% Carbowax E400 on a solid support of 100–120 mesh Chromosorb-W that had been previously treated with hexamethyldisilazane (HMDS) was used; other GC columns have also been used with success. At an oven temperature of 25°C and N_2 carrier gas flow rate of ~40 mL min^{-1}, a retention time of the order of a minute is obtained for PAN (Smith et al., 1972). The sample of ambient air is introduced into the GC column using a gas sampling valve which may be automated to sample and inject air periodically (e.g., every 15 min). With a typical sample volume of 2 mL air, a sensitivity of ~1 ppb for PAN can be obtained. The detector response is linear up to concentrations of ~50 ppb (Smith et al., 1972).

Calibration of the GC column requires generating and injecting known concentrations of PAN into the GC column. There are several methods for generating samples of PAN in air (Stephens et al., 1965; Stephens, 1969):

- Photolysis of ethyl nitrite in oxygen. Although the mechanism of formation of PAN is not known with certainty, it likely involves the generation of the acetyl radical (CH_3CO) from the ethoxy (C_2H_5O) radical produced on photolysis. PAN can then be formed by the following reactions:

$$CH_3\overset{\overset{\displaystyle O}{\|}}{C} + O_2 \rightarrow CH_3\overset{\overset{\displaystyle O}{\|}}{C}OO \tag{8}$$

$$CH_3\overset{\overset{\displaystyle O}{\|}}{C}OO + NO_2 \rightleftarrows CH_3\overset{\overset{\displaystyle O}{\|}}{C}OONO_2 \tag{9}$$

- Photolysis of NO_x with organic compounds in air or oxygen. Biacetyl ($CH_3COCOCH_3$) or 2-butene at a concentration of a few hundred ppm is commonly used. Biacetyl photolyzes to produce the precursor acetyl radicals. Presumably 2-butene also forms CH_3CO, although the sequence of reactions is clearly more complex.
- Reaction of acetaldehyde (CH_3CHO) with dinitrogen pentoxide (N_2O_5) or a combination of NO_2 and O_3 in the dark. The mechanism likely involves generation of the nitrate radical (NO_3) followed by its reaction with CH_3CHO to form $CH_3\overset{\overset{O}{\|}}{C}$ (Section 7-H-3) which reacts as discussed above to form PAN.

All three techniques generate PAN as a mixture in air with other products or unreacted starting materials. There are two methods of using such mixtures for calibration of the GC column for PAN. In the first, high concentrations ($\lesssim 1000$ ppm) of PAN are generated by one of the above methods in a 10 cm gas cell, and the known infrared absorption coefficients for PAN (Stephens, 1964, 1969; Gaffney et al., 1984) are applied using Beer's law to determine its concentration. This mixture, containing a known PAN concentration large enough to be measured in a 10 cm cell by IR, is then diluted sequentially into the low-ppb range using measured flows of the mixture and diluent pure air (Stephens and Price, 1973).

TABLE 5.8. **Summary of Techniques Used to Monitor Non-Criteria Gaseous Pollutants**[a]

Pollutant	Technique	Detection Limit	Reference
NO	Chemiluminescence	5 ppt	NASA, 1983[d]
	Single photon laser-induced fluorescence	4–240 ppt[e] (0.1 km pathlength)	NASA, 1983[d]
	Lidar laser-induced fluorescence	10 ppt	NASA, 1983[d]
	TDLS	500 ppt	Cassidy and Reid, 1982a
	DOAS	400 ppt	Platt and Perner, 1982
PAN	GC with electron capture detection	10 ppt	Stephens, 1969
	FTIR (1 km pathlength)	3 ppb	Tuazon et al., 1980
	TDLS	0.3 ppb	Cassidy and Reid, 1982a
NH_3	TDLS	0.1 ppb	Mackay and Schiff, 1985
	Tungstic acid denuder	0.1 ppb	Braman et al., 1982; NASA, 1983
	Oxalic acid denuder	10 ppt	Ferm, 1979
	Fluorescence derivatization	0.1 ppb	Abbas and Tanner, 1981
	Photoacoustic detection after adsorption on a solid	0.5 ppb	McClenney and Bennett, 1980
	FTIR (1 km pathlength)	1 ppb	Tuazon, 1985
	Chemiluminescence with thermal converter	1 ppb	Baumgardner et al., 1979
HNO_3	Chemiluminescence	200 ppt	NASA, 1983
	Tungstic acid	70 ppt	Braman et al., 1982; NASA, 1983
	Denuder tubes (Na_2CO_3, MgO, etc.)	80 ppt	Shaw et al., 1982; Forrest et al., 1982
	Nylon filter	2 ppt	Joseph and Spicer, 1978
	TDLS	350 ppt	Mackay and Schiff, 1985
	FTIR	3 ppb	Tuazon, 1985
NO_3	DOAS	3 ppt	Platt and Perner, 1982

	FTIR (1 km pathlength)	10 ppb	Tuazon et al., 1980
Organic nitrosamines	Thermal energy analyzer–gas chromatography	8 ppt	Fine et al., 1977; Rounbehler et al., 1980
HCHO	DOAS (10 km pathlength)	600 ppt	Platt and Perner, 1982
	FTIR (1 km pathlength)	6 ppb	Tuazon et al., 1980
	TDLS	300 ppt	Mackay and Schiff, 1985
	Sampling onto solid sorbent followed by thermal desorption into GC–MS	0.3 ppb	Yokouchi et al., 1979
	Derivatization (DNPH)[b] followed by HPLC[f]	1.5 ppb	Kuwata et al., 1979; Fung & Grosjean, 1981; Grosjean & Fung, 1982
	Wet chemistry spectrophotometric methods (e.g., chromotropic acid, pararosaniline)	40 ppb	National Research Council, 1981
H_2O_2	TDLS	600 ppt	Mackay and Schiff, 1985
OH	Laser-induced fluorescence	0.02 ppt	O'Brien, 1985
	UV absorption (DOAS)	0.03 ppt	Hübler et al., 1982
	Radiocarbon technique	1.6×10^{-3} ppt	Campbell et al., 1979
	Spin trapping/ESR and GC–MS	0.02 ppt	Watanabe et al., 1982
HO_2	Conversion to OH by reaction with NO and measurement of OH	0.02 ppt	O'Brien, 1985
RO_2	Chemical amplification technique by reaction with NO	n.d.[c]	Cantrell and Stedman, 1982

[a] As discussed in the text, the techniques listed here are either currently in use or promising methods in the field-testing stage.

[b] DNPH = 2,4-dinitrophenylhydrazine.

[c] n.d. = not determined.

[d] NASA = National Aeronautics and Space Administration.

[e] Range for averaging times of 1 s to 1 h.

[f] HPLC = High-performance liquid chromatography.

Alternatively, the PAN may be separated and purified from the mixture using a preparative scale GC column described in detail by Stephens and co-workers (Stephens et al., 1965). The purified PAN eluting from the GC column can be trapped in a cold-trap and stored as a liquid. However, since liquid PAN is *extremely explosive* (Stephens et al., 1969), it is preferable to store it as a gas diluted with an inert carrier such as He or N_2. Typically, PAN is stored at a temperature of 15°C in cylinders pressurized to 100 psi with N_2 and with PAN concentrations of 500–1000 ppm (Smith et al., 1972). *Great care* should also be taken when handling these cylinders containing mixtures of PAN in nitrogen.

PAN has also been synthesized in solution for use in calibrating the GC column (e.g., see Kravetz et al., 1980; Nielsen et al., 1982; Holdren and Spicer, 1984; Gaffney et al., 1984). For example, nitration of peracetic acid followed by extraction into an organic solvent such as *n*-heptane or *n*-tridecane and purification with high-performance liquid chromatography has been used to generate solutions of PAN. The solution concentration can be determined by hydrolysis of PAN and measurement of the product nitrite, nitrate and acetate using conventional techniques, e.g., ion chromatography or infrared spectroscopic analysis using published absorption coefficients (Stephens, 1964; Holdren and Spicer, 1984; Grosjean and Harrison, 1985b). Once the solution concentration of PAN is known, measured volumes of the solution can be evaporated into known volumes of pure air to yield the calibration standards.

A calibration procedure based on the thermal decomposition of PAN in the presence of NO (see Section 8-B-1) has also been suggested by Lonneman and coworkers (Lonneman and Bufalini, 1982; Lonneman et al., 1982). The stoichiometry $\Delta[NO]/\Delta[PAN]$ in the presence of benzaldehyde was shown to be 4.7 ± 0.2, so that the change in NO monitored using a standard chemiluminescence instrument should be relatable to the PAN concentration. This procedure is based on a complex chemical system, however, and should be used with some caution.

Figure 5.34 shows a typical gas chromatogram of (*a*) ambient air in which PAN is present and (*b*) ambient air that has been passed through activated carbon to remove oxidants, including PAN. In the unfiltered ambient air sample, a smaller peak elutes after PAN; this is the next larger homologue in the peroxyacyl nitrate series, peroxypropionyl nitrate (PPN).

NH_3. Ammonia can be measured using the standard chemiluminescence technique for NO_x. For example, the gas stream can be passed through a thermal converter which converts the NH_3 to NO; this gives a combined $NO_x + NH_3$ measurement. The gas stream is then passed through an acidic scrubber, which removes NH_3, and into the chemiluminescence instrument to obtain NO_x alone. The difference then gives the NH_3 concentration (Baumgardner et al., 1979).

A technique known as the diffusion denuder technique (Stevens et al., 1978; Durham et al., 1978) has also been applied to the determination of ambient concentrations of ammonia as well as of such other gases as nitric acid and SO_2. The basis of this technique involves flowing ambient air through a hollow tube which is coated with a substance that will absorb the gaseous species of interest. The flow conditions are such that gases diffuse relatively rapidly to the coated walls, whereas particulate matter passes through the tube. After sampling a suitable vol-

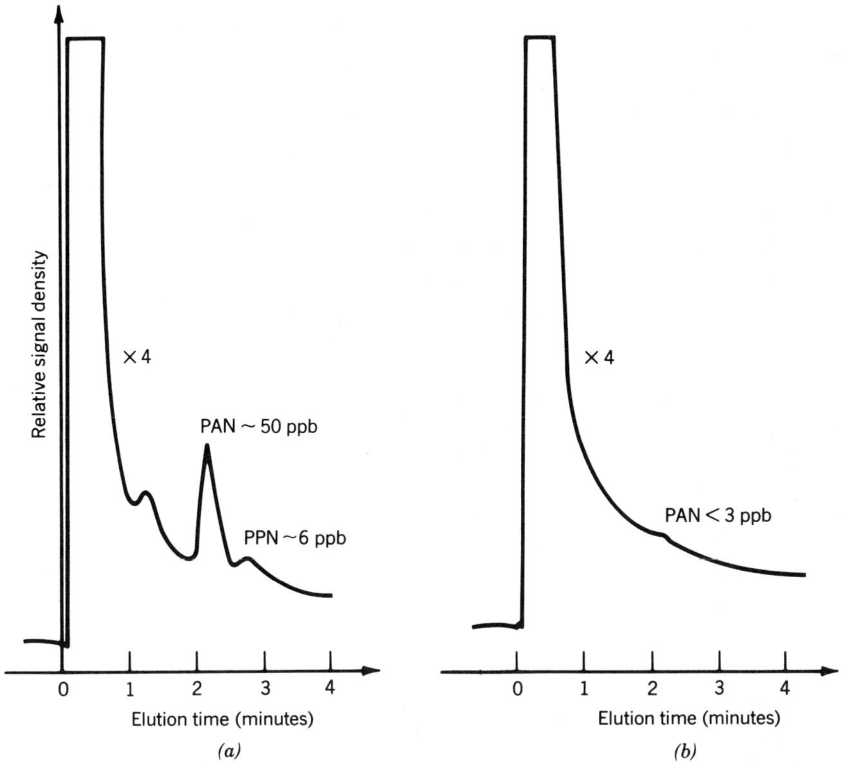

FIGURE 5.34. Typical gas chromatogram used to measure PAN in ambient air. (*a*) Unfiltered ambient air, (*b*) ambient air filtered using activated carbon to remove oxidants including PAN (from Darley et al., 1963).

ume of air, the gas is released from the denuder tube for analysis. Alternatively, two parallel sampling trains, one containing a denuder and the other without it, can be used and a difference measurement used to calculate the gas phase concentration.

Coatings used for NH_3 measurements take advantage of the fact that NH_3 is the major common, basic gas in the troposphere. Thus oxalic acid, for example, has been used to coat diffusion denuder tubes (Ferm, 1979), trapping the ammonia in the form of the ammonium ion which can be eluted and analyzed using conventional techniques such as ion chromatography or specific ion electrodes.

One diffusion denuder coating used to measure both NH_3 and HNO_3 simultaneously is tungstic acid. Braman and co-workers (1982) suggest that the absorption of ammonia is also due to a reversible acid–base reaction:

$$NH_3 + H_2WO_4 \rightleftarrows NH_4HWO_4 \qquad (10)$$

The reactions involved in HNO_3 absorption are not clear but appear to be irreversible; however, on heating the denuder tube, the HNO_3 produces NO_2 in the gas phase which can then be measured.

To monitor NH_3 and HNO_3 simultaneously the airstream is passed through the coated tube. Both HNO_3 and NH_3 diffuse to, and absorb on, the coated walls and can be subsequently analyzed separately using a two-step process. First, the collection tube is heated, releasing NH_3 and NO_2 (from the HNO_3). These gases pass through a hollow transfer tube which re-adsorbs the NH_3, thus separating it from the NO_2 which was originally produced from the HNO_3. The airstream then passes over a heated catalyst that converts the NO_2 to NO, and the total NO_x (equal to the initial HNO_3 concentration) is measured using a conventional NO_x chemiluminescence analyzer. The second stage of the analysis involves thermally desorbing NH_3 from the transfer tube and catalytically oxidizing it to NO, which is detected using the chemiluminescence technique.

While good sensitivity for both NH_3 and HNO_3 is claimed (see Table 5.8), interference from some organic amines may be a problem under some conditions.

HNO_3. In addition to the spectroscopic techniques, HNO_3 has been determined using two other methods—diffusion denuder techniques and filter techniques. In the case of the diffusion denuder method, tungstic acid coatings, as discussed above, have been used, as have others such as MgO and Na_2CO_3.

It is noteworthy that HNO_3 has been shown to exist as a hydrate, $HNO_3 \cdot XH_2O_{(g)}$, where the value of X depends on the relative humidity (Eatough et al., 1985); this lowers the diffusion coefficient of HNO_3 in air and was shown to decrease the collection efficiency of a tungstic acid denuder for $HNO_{3(g)}$.

Filter methods for gaseous nitric acid determination generally involve passing the airstream first through a prefilter to remove particulate matter, including particulate nitrates such as ammonium nitrate. The air then flows through a filter which is impregnated with a compound that will react with the HNO_3 to form an involatile salt; for example, NaCl and tetra-*n*-butylammonium hydroxide have been used. Alternatively, the filter material itself may absorb HNO_3. This is the case for nylon filters, for example. Gaseous nitric acid then can be determined by elution of the filters and analysis for nitrate using conventional analytical techniques. A second approach is to measure total NO_x (e.g., by the chemiluminescence technique with O_3) with and without the filter in place; the difference is assumed to be gaseous nitric acid.

Filter techniques can be subject to both positive and negative errors. Thus some gaseous HNO_3 may be removed by the prefilter or by the particulate matter collected on it, leading to a negative error. On the other hand, positive errors can result if particulate nitrate is displaced as HNO_3 from the prefilter, for example, by the reaction of strong acids such as H_2SO_4; NH_4NO_3 collected on the prefilter can revolatilize to form $NH_3 + HNO_3$, which is then collected on the second filter (e.g., Appel et al., 1980, 1981; Cadle, 1985).

In addition to these sampling problems, it is now becoming evident that some of these filters absorb not only nitric acid, but other nitrogenous acids as well. For example, nitrous acid (HONO) has been shown to be adsorbed quantitatively by nylon filters (Sanhueza et al., 1984); hence measurement of HNO_3 by difference using such filters during the night when significant concentrations of HONO may be present in reality gives total HNO_3 + HONO (and probably any other gaseous acidic nitrogenous species as well).

For a discussion of the result of intercomparison studies carried out on nitric acid in ambient air, the reader is referred to Spicer et al. (1982), Anlauf et al., (1985) and Mulawa and Cadle (1985). The results of an intercomparison study carried out in Claremont, Ca. in September, 1985, sponsored by the California Air Resources Board, will be published in the future.

HCHO. Because of its importance as a free radical source in urban atmospheres, formaldehyde has been monitored for a number of years using wet chemical techniques (see the National Research Council Report of 1981 for details of these methods). More recently, as shown in Table 5.8, spectroscopic techniques (DOAS and FTIR) have been applied. Another relatively new technique included in Table 5.8 which looks promising is derivatization using 2,4-dinitrophenylhydrazine followed by high-performance liquid chromatography. This has the advantage that various aldehydes and ketones larger than HCHO can be separated and identified as well. For example, Fig. 5.35 shows the distribution of some simple oxygenated organics observed recently in the Los Angeles air basin (Grosjean, 1982). Formaldehyde is typically present in the largest concentration, followed by acetaldehyde. The larger oxygenates are present at lower concentrations.

H_2O_2. Gaseous hydrogen peroxide is thought to be a potentially very important oxidant, for example, for SO_2 in the aqueous phase (see Section 11-A-2). While some measurements of its concentration in the gas phase using wet chemical techniques have been reported in the literature, it appears that these were subject to interferences from unknown species; this again illustrates the importance of having specific and sensitive spectroscopic techniques. Several spectroscopic methods, including TDLS (see above) and photofragmentation/differential fluorescence (NASA, 1984b) for H_2O_2 look very promising. Thus as discussed earlier, H_2O_2 has been detected and measured both in laboratory studies and in ambient air using TDLS, with an estimated detection limit of 0.6 ppb at a pathlength of 35 m (Mackay and Schiff, 1985).

Note that the above comments apply to the measurement of H_2O_2 in *the gas phase*. Methods of measurement of H_2O_2 in the liquid phase, e.g. clouds, fogs and rainwater, which is very important in acid deposition, are discussed in Chapter 11.

OH. In addition to laser-induced fluorescence discussed in Section 5-B-2, three other techniques have been used to measure OH concentrations. The most extensive work has been done on UV absorption and a radiocarbon technique. One report of trapping of OH by reaction with a spin label followed by electron spin resonance and GC–MS detection and measurement of the OH–organic adduct has also been made. Various methods of OH detection and measurement are discussed in detail in the NASA document (1984b).

UV Absorption

This method was first applied at the earth's surface by Perner and co-workers (Perner et al., 1976) to detect and measure OH near Jülich, West Germany. A frequency-doubled dye laser is tuned to the rotational transition of OH at 307.9951

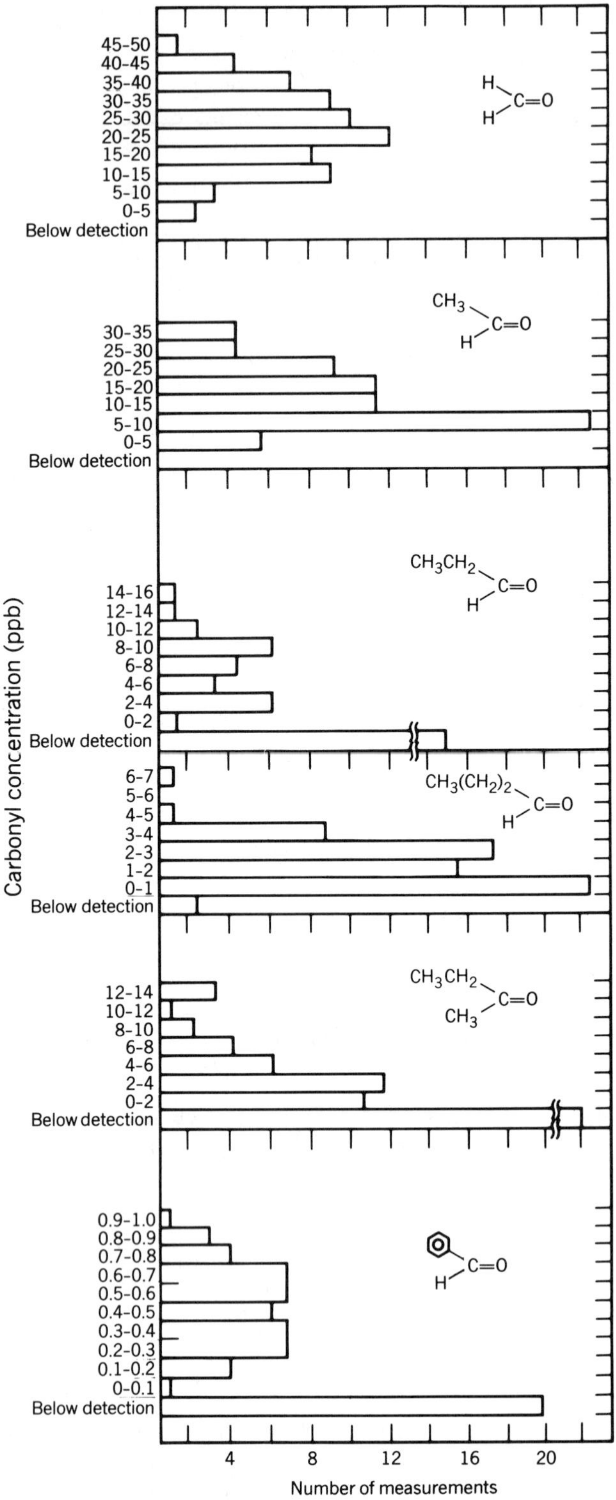

FIGURE 5.35. Frequency distributions of ambient levels of some oxygenated hydrocarbons observed at Claremont, California, in 1980 (from Grosjean, 1982).

nm, corresponding to a transition from the $X^2\Pi$ ground state to the first electronically excited $A^2\Sigma^+$ state. From the known oscillator strengths for that transition and the fraction of the total OH in the absorbing rotational level, the total average OH concentration over the 7.8 km light path was determined.

This technique has since been adapted to measure OH by DOAS (see Section 5-D-2) using their rapid scanning system (Hübler et al., 1982, 1984). Since both SO_2 and HCHO have absorption bands in the same region, the absorptions due to these species must be subtracted from the air spectrum first. Indeed, in applying this technique to measure OH in Jülich, the SO_2 absorption bands were usually found to be sufficiently large to obscure those due to OH. A detection limit of $\sim 0.6 \times 10^6$ cm^{-3} OH was obtained in these studies.

A different experimental configuration to measure OH by long-path absorption has been reported by Ortgies and Comes (1984); the limit of detection was reported as 3×10^6 cm^{-3}, which is close to the peak OH levels reported in other studies (Hübler et al., 1984).

Radiocarbon Technique

This method, developed by Campbell and Sheppard (Campbell et al., 1979), is based on the oxidation by OH of isotopically labeled ^{14}CO introduced into the air sample. Assuming that all the CO is oxidized to CO_2 by reaction with OH, the rate of formation of CO_2 is given by the following expression:

$$\frac{d[CO_2]}{dt} = k[CO][OH]$$

Integrating this expression and applying it to the ^{14}CO artificially introduced into the system, one obtains

$$\frac{[^{14}CO_2]_t}{[^{14}CO]_0} = \frac{[^{14}CO_2]_0}{[^{14}CO]_0} + k[OH]t$$

where k is the rate constant for the OH + CO reaction under the experimental conditions (see discussion of this reaction in Section 4-A-5), $[^{14}CO]_0$ and $[^{14}CO_2]_0$ (present as an impurity in the ^{14}CO) are the initial concentrations of CO and CO_2, and $[^{14}CO_2]_t$ is the $^{14}CO_2$ concentration at time t. Knowing the initial concentrations of ^{14}CO and $^{14}CO_2$ as well as the value of k, one can obtain [OH] by following the formulation of $^{14}CO_2$ with time.

While this technique avoids the problems associated with the use of lasers, for example, secondary production of OH via photolysis of other species such as O_3, there are a number of experimental difficulties. The most important is that the composition and chemistry of the air sample must be perturbed only minimally by the sampling train and addition of ^{14}CO. Wall losses of OH and HO_2 in the sampling apparatus must therefore be negligible and the amount of ^{14}CO added must be small compared to the naturally occurring CO in the sample. Additionally, since the $^{14}CO_2$ is measured by its radioactive emission intensity using either a gas-proportional counting tube or a liquid scintillation counter, it is important to separate the $^{14}CO_2$ formed from any unreacted ^{14}CO which would give an interfering signal. This separation must be accomplished with negligible loss of the $^{14}CO_2$, or at least reproducible sampling efficiencies under a wide range of experimental conditions, for example, relative humidity. Other sources of ^{14}C must be minimized as well.

Spin Trapping

Recently, Watanabe and co-workers (1982) measured OH in the upper troposphere (6–10 km region) by reacting OH with a stable substrate to form an adduct that can be detected by ESR and GC–MS. The substrate, α-4-pyridyl-*N-tert*-butylnitrone α-1-oxide (4-POBN), was adsorbed on filter paper and the air containing OH passed through the paper. It reacts to form an OH adduct:

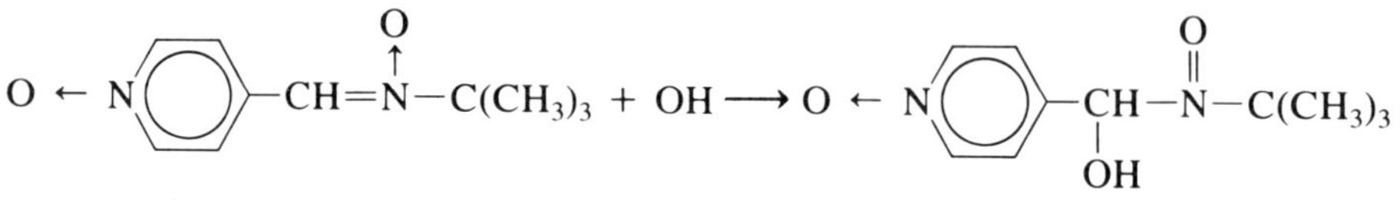

(11)

After exposure to air, the organic on the filter was extracted into benzene, and the OH-4-POBN adduct detected by ESR. Further identification and measurement were carried out by forming a trimethylsilyl derivative of the adduct and measuring it using GC–MS.

While this technique looks interesting, a number of potential problems remain to be addressed in detail. These include possible interferences by reaction of other species with 4-POBN, as well as knowledge of the trapping efficiency for OH and the loss of organic from the filter during sampling, including how ambient conditions (e.g., relative humidity) affect these parameters. The solvent extraction efficiency for the OH-4-POBN adduct must also be determined.

OH measurements by the radiocarbon method and two LIF methods were included in the 1983 intercomparison study of NO and CO discussed earlier (Hoell et al., 1984). Unfortunately, instrumentation problems precluded carrying out intercomparisons as all three instruments were not operational simultaneously during the sampling periods. This is indicative of the difficulty and sophistication involved in making such measurements, which, however, are critical to our understanding of the chemistry of the natural and polluted troposphere.

HO_2 and RO_2. HO_2 can be measured by conversion to OH via reaction (12)

$$HO_2 + NO \rightarrow OH + NO \quad (12)$$

followed by measurement of OH using LIF. Thus O'Brien and coworkers (Hard et al., 1984) have reported measuring HO_2 concentrations up to 2×10^7 cm^{-3} by this method in downtown Portland, Oregon. The sensitivity of the method is the same as that for detection of OH, and is currently approximately 5×10^5 radicals cm^{-3} (O'Brien, 1985).

An indirect technique for application in the troposphere known as *chemical amplification* by reaction with NO is in the development stage (Cantrell and Stedman, 1982). Ambient air is mixed with NO and CO, and the NO_2 formed is detected by its chemiluminescent surface reaction with an aqueous luminol solution. The formation of NO_2 is due to the reaction of HO_2 (or RO_2, see below) with NO followed by the regeneration of HO_2 from OH using CO:

$$OH + CO \xrightarrow{O_2,\,M} CO_2 + HO_2 \tag{13}$$

When chain termination of OH and HO_2 in the system is dominated by the reaction of OH with NO,

$$OH + NO \xrightarrow{M} HONO \tag{14}$$

the change in the NO_2 concentration with time can be shown to be given by the following expression:

$$\Delta[NO_2] = \frac{k_{13}[CO]\,[HO_2]_0}{k_{14}[NO]}\,[1 - \exp(-\gamma)]$$

where $\gamma = k_{12}k_{14}[NO]^2\, t/k_{13}[CO]$. Knowing the rate constants k_{12}, k_{13}, and k_{14} under the experimental conditions as well as the concentrations of CO and NO, the initial HO_2 concentration can be determined by following the change in the NO_2 concentration with time. When the HO_2 concentrations are sufficiently large that the additional termination step

$$HO_2 + HO_2 \rightarrow H_2O_2 + O_2 \tag{15}$$

must also be taken into account, a more complex expression for the change in NO_2 with time must be used.

Since RO_2 radicals also react with NO to form NO_2 (Section 7-D-2),

$$RO_2 + NO \rightarrow RO + NO_2 \tag{16}$$

RO_2 is also measured by this technique. However, the regeneration of HO_2 from RO requires a sequence of reactions. For the larger ($\geq C_4$) alkoxy radicals where a variety of potential fates are possible in air in addition to reaction with O_2 (including spontaneous decomposition or isomerization), the system becomes quite complex (see Section 7-D-3). Thus it appears that this method may give concentrations that are approximately the sum of $HO_2 + RO_2$ (NASA, 1984b).

It is interesting to note that free radicals such as HO_2 were detected in the stratosphere before they were successfully measured in the troposphere (Mihelcic et al., 1978a,b; Anderson et al., 1981).

TABLE 5.9. Typical Peak Concentrations of Gas Phase Criteria Pollutants Observed in the Troposphere Over the Continents[a]

Pollutant	Type of Atmosphere: Remote	Rural	Moderately Polluted	Heavily Polluted	U.S. Primary Federal Air Quality Standard
CO	≤0.2 ppm[f]	0.2–1 ppm[h]	~1–10 ppm[g]	10–50 ppm	9.0 ppm for 8 h 35.0 ppm for 1 h
NO_2	≤1 ppb[b,f,j]	1–20 ppb[c,d]	0.02–0.2 ppm[g]	0.2–0.5 ppm	0.05 ppm annual average
O_3	≤0.05 ppm[b,f]	0.02–0.08 ppm	0.1–0.2 ppm	0.2–0.5 ppm	0.12 ppm for 1 h
SO_2	≤1 ppb[e]	~1–30 ppb[d]	0.03–0.2 ppm	0.2–2 ppm	0.14 ppm for 24 h 0.03 ppm annual average
NMHC	≤65 ppbC[f]	100–500 ppbC[h]	300–1500 ppbC[i]	≥1.5 ppmC	0.24 ppmC average from 6–9 a.m.

[a] As 1 h averages.
[b] Kelly et al., 1980.
[c] Spicer et al., 1982; Pratt et al., 1983.
[d] Martin and Barber, 1981.
[e] Ludwick et al., 1980; Maroulis et al., 1980.
[f] Kelly et al., 1982; Hoell et al., 1984.
[g] Ferman et al., 1981.
[h] Seila, 1979.
[i] Sexton et al., 1982.
[j] Johnston and McKenzie, 1984.

TABLE 5.10. Typical Concentrations of Some Gas Phase Trace Species and Non-Criteria Pollutants Reported in the Troposphere Over the Continents

Pollutant	Type of Atmosphere: Remote	Rural	Moderately Polluted	Heavily Polluted
NO	$\leqslant$50 ppt[a,v]	~0.05–20 ppb[g,i,v]	0.02–1 ppm[j]	~1–2 ppm
PAN	$\leqslant$50 ppt[q]	2 ppb[g]	2–20[g] ppb	20–70 ppb[e]
NH_3	15 ppt[n]	1–10 ppb[c,o,r]	1–10 ppb[c,o,r]	10–100 ppb[f]
HNO_3	$\leq$0.03–0.1 ppb[a]	~0.1–4 ppb[b,r]	1–10 ppb[p,r]	10–50 ppb[e]
NO_3	$\leqslant$5 ppt[l]	5–10 ppt	10–100 ppt[k,l]	100–430 ppt[m]
HONO	<30 ppt[o]	0.03–0.8 ppb[o]	0.8–2 ppb	2–8 ppb[h,n]
HCHO	$\leqslant$0.5–2 ppb[d]	2–10[o] ppb	10–20 ppb[d,u]	20–75 ppb[e]
OH	$(4\text{–}40) \times 10^{-3}$ ppt[t]	0.01–0.10 ppt[t]	0.05–0.4 ppt[s,t]	$\geqslant$0.4 ppt
(Midday average peak for sunny conditions)	$(1\text{–}10) \times 10^5 \frac{\text{molec}}{\text{cm}^3}$	$(2.5\text{–}25) \times 10^5 \frac{\text{molec}}{\text{cm}^3}$	$(1\text{–}10) \times 10^6 \frac{\text{molec}}{\text{cm}^3}$	$\geqslant 1 \times 10^7 \frac{\text{molec}}{\text{cm}^3}$

[a] Kelly et al., 1980; Huebert and Lazrus, 1980; Bollinger et al., 1984.
[b] Shaw et al., 1982; Huebert and Lazrus, 1980.
[c] Levine et al., 1980.
[d] National Research Council, 1981.
[e] Tuazon et al., 1981.
[f] Doyle et al., 1979.
[g] Temple and Taylor, 1983; Spicer, 1982.
[h] Pitts et al., 1983.
[i] Martin and Barber, 1981; Pratt et al., 1983.
[j] Ferman et al., 1981.
[k] Platt et al., 1981.
[l] Platt et al., 1982.
[m] Platt et al., 1980b; Atkinson et al., 1985.
[n] Gras, 1983.
[o] Harward et al., 1982.
[p] Spicer, 1977.
[q] Singh and Salas, 1983.
[r] Cadle et al., 1982.
[s] Hard et al., 1984.
[t] Hübler et al., 1984, and references therein.
[u] Schjoldager, 1984.
[v] Logan, 1983.

5-G. TYPICAL CONCENTRATIONS OF KEY SPECIES IN THE REMOTE AND POLLUTED TROPOSPHERE

Let us briefly examine the range of pollutant concentrations which have been reported over land in the troposphere using these techniques.

Table 5.9 gives typical concentration ranges for the gaseous criteria pollutants, as well as the primary ambient air quality standards for comparison, for four types of ambient conditions: (1) remote, (2) rural, (3) moderately polluted, and (4) heavily polluted. Measurements over the oceans have not been considered here because it is suspected that for some pollutants, at least, the sources, sinks, and ambient concentrations may be quite different than those over land (e.g, see Kelly et al., 1980). However, some of these maritime concentrations are discussed with regard to the chemistry of the natural troposphere in Chapter 14. The typical ranges of concentrations given are intended as general guidelines only, since the difference between "moderately" and "heavily" polluted, for example, cannot be rigorously defined; indeed, the definitions of such air masses may vary depending on the point of view of the person defining them! In addition, for some of these pollutants, as well as for the non-criteria pollutants (see below), there are very few measurements reported in remote, relatively pristine areas. As a result, some of the concentrations given for these regions represent "best guesses" on the part of the authors.

Table 5.10 gives typical concentrations of some gaseous trace species and non-criteria pollutants in the same types of atmospheres. The same caveats as discussed for Table 5.9, of course, apply here as well. The important non-criteria pollutant H_2O_2 is not included in Table 5.10 because there are no existing, accurate, and reliable gas phase measurements available.

REFERENCES

Abbas, R. and R. L. Tanner, "Continuous Determination of Gaseous Ammonia in the Ambient Atmosphere Using Fluorescence Derivatization," *Atmos. Environ.*, **15,** 277 (1981).

Anderson, J. G., H. J. Grassl, R. E. Shetter, and J. J. Margitan, "HO_2 in the Stratosphere: Three *In Situ* Observations," *Geophys. Res. Lett.*, **8,** 289 (1981).

Anlauf, K. G., P. Fellin, H. A. Wiebe, H. I. Schiff, G. I. Mackay, R. S. Braman and R. Gilbert, "A Comparison of Three Methods for Measurement of Atmospheric Nitric Acid and Aerosol Nitrate and Ammonium," *Atmos. Environ.*, **19,** 325 (1985).

Altshuller, A. P., W. A. Lonneman, F. D. Sutterfield, and S. L. Kopczynski, "Hydrocarbon Composition of the Atmosphere of the Los Angeles Basin—1967," *Environ. Sci. Technol.*, **5,** 1009 (1971).

Appel, B. R., S. M. Wall, Y. Tokiwa, and M. Haik, "Simultaneous Nitric Acid, Particulate Nitrate and Acidity Measurements in Ambient Air," *Atmos. Environ.*, **14,** 549 (1980).

Appel, B. R., Y. Tokiwa, and M. Haik, "Sampling of Nitrates in Ambient Air," *Atmos. Environ.*, **15,** 283 (1981).

Atkinson, R., A. M. Winer, and J. N. Pitts, Jr., "Estimation of Nighttime N_2O_5 Concentrations from Ambient NO_2 and NO_3 Radical Concentrations and The Role of N_2O_5 in Nighttime Chemistry," *Atmos. Environ.*, **19,** 000 (1985).

Baumgardner R. E., W. A. McClenny, and R. K. Stevens, "Optimized Chemiluminescence System for Measuring Atmospheric Ammonia," EPA Report No. EPA-600/2-79-028, February 1979.

Biermann, H. W., personal communication (1985).

Bollinger, M. J., C. J. Hahn, D. D. Parrish, P. C. Murphy, D. L. Albritton, and F. C. Fehsenfeld, "NO_x Measurements in Clean Continental Air and Analysis of the Contributing Meterology," *J. Geophys. Res.*, **89,** 9623 (1984).

Bos, R., E. Goudena, R. Guicherit, A. Hoogeveen, and A. F. de Vreede, Atmospheric Precursors and Oxidants Concentrations in the Netherlands, in *Photochemical Smog Formation in the Netherlands*, R. Guicherit, Ed., TNO's, Gravenhage, Netherlands, 1978, pp. 20–59.

Bradley, C. E. and A. J. Haagen-Smit, "The Application of Rubber in the Quantitative Determination of Ozone," *Rubber Chem. & Technol.*, **24,** 750 (1951).

Braman, R. S., T. J. Shelley, and W. A. McClenny, "Tungstic Acid for Preconcentration and Determination of Gaseous and Particulate Ammonia and Nitric Acid in Ambient Air," *Anal. Chem.*, **54,** 358 (1982).

Cadle, S. H., "Seasonal Variations in Nitric Acid, Nitrate, Strong Aerosol Acidity, and Ammonia in an Urban Area," *Atmos. Environ.*, **19,** 181 (1985).

Cadle, S. H., R. J. Countess, and N. A. Kelly, "Nitric Acid and Ammonia in Urban and Rural Locations," *Atmos. Environ.*, **16,** 2501 (1982).

Campbell, M. J., J. C. Sheppard, and B. F. Au, "Measurement of Hydroxyl Concentration in Boundary Layer Air by Monitoring CO Oxidation," *Geophys. Res. Lett.*, **6,** 175 (1979).

Cantrell, C. A. and D. H. Stedman, "A Possible Technique for the Measurement of Atmospheric Peroxy Radicals," *Geophys. Res. Lett.*, **9,** 846 (1982).

Cassidy, D. T. and J. Reid, "Atmospheric Pressure Monitoring of Trace Gases Using Tunable Diode Lasers," *Appl. Opt.*, **21,** 1185 (1982a).

Cassidy, D. T. and J. Reid, "High-Sensitivity Detection of Trace Gases Using Sweep Integration and Tunable Diode Lasers," *Appl. Opt.*, **21,** 2527 (1982b).

Chan, C. Y., R. J. O'Brien, T. M. Hard, and T. B. Cook, "Laser-Excited Fluorescence of the Hydroxyl Radical: Relaxation Coefficients at Atmospheric Pressure," *J. Phys. Chem.*, **87,** 4966 (1983); *J. Phys. Chem.*, **88,** 2924 (1984) and references therein.

Clark, A. I., A. E. McIntyre, R. Perry, and J. N. Lester, "Monitoring and Assessment of Ambient Atmospheric Concentrations of Aromatic and Halogenated Hydrocarbons at Urban, Rural and Motorway Locations," *Environ. Pollut.* (*Ser. B*), **7,** 141 (1984).

Connell, P. S., R. A. Perry, and C. J. Howard, "Tunable Diode Laser Measurement of Nitrous Oxide in Air," *Geophys. Res. Lett.*, **7,** 1093 (1980).

Darley, E. F., K. A. Kettner, and E. R. Stephens, "Analysis of Peroxyacyl Nitrates by Gas Chromatography with Electron Capture Detection," *Anal. Chem.*, **35,** 589 (1963).

Davis, D. D., M. O. Rodgers, S. D. Fischer, and K. Asai, "An Experimental Assessment of the O_3/H_2O Interference Problem in the Detection of Natural Levels of OH Via Laser Induced Fluorescence," *Geophys. Res. Lett.*, **8,** 69 (1981a).

Davis, D. D., M. O. Rodgers, S. D. Fischer, and W. S. Heaps, "A Theoretical Assessment of the O_3/H_2O Interference Problem in the Detection of Natural Levels of OH Via Laser Induced Fluorescence," *Geophys. Res. Lett.*, **8,** 73 (1981b).

Davis, D. D., S. D. Fischer, and M. O. Rodgers, "Reply to Wang and Davis," *Geophys. Res. Lett.*, **9,** 101 (1982).

Davis, D. D., J. D. Bradshaw, and M. O. Rodgers, "Comments on 'Laser Excited Fluorescence of the Hydroxyl Radical: Relaxation Coefficients at Atmospheric Pressure'," *J. Phys. Chem.*, **88,** 2923 (1984).

DeMore, W. B. and M. Patapoff, "Comparison of Ozone Determinations by Ultraviolet Photometry and Gas-Phase Titration," *Environ. Sci. Technol.*, **10,** 897 (1976).

Doyle, G. J., E. C. Tuazon, R. A. Graham, T. M. Mischke, A. M. Winer, and J. N. Pitts, Jr., "Simultaneous Concentrations of Ammonia and Nitric Acid in a Polluted Atmosphere and Their Equilibrium Relationship to Particulate Ammonium Nitrate," *Environ. Sci. Technol.*, **13,** 1416 (1979).

Durham, J. L., W. E. Wilson, and E. B. Bailey, "Application of an SO_2-Denuder for Continuous Measurement of Sulfur in Submicrometric Aerosols," *Atmos. Environ.*, **12,** 883 (1978).

Eatough, D. J., V. F. White, L. D. Hansen, N. L. Eatough, and E. C. Ellis, "Hydration of Nitric Acid and Its Collection in the Atmosphere by Diffusion Denuders," *Anal. Chem.*, **57,** 743 (1985).

El-Sherbiny, M., E. A. Ballik, J. Shewchun, B. K. Garside, and J. Reid, "High Sensitivity Point Monitoring of Ozone, and High Resolution Spectroscopy of the ν_3 Band of Ozone Using a Tunable Semiconductor Diode Laser," *Appl. Opt.*, **18,** 1198 (1979).

Ferm, M., "Method for Determination of Atmospheric Ammonia," *Atmos. Environ.*, **13,** 1385 (1979).

Ferman, M. A., G. T. Wolff, and N. A. Kelly, "An Assessment of the Gaseous Pollutants and Meterological Conditions Associated with Denver's Brown Cloud," *J. Environ. Sci. Health,* **A16,** 315 (1981).

Fine, D. H., D. P. Rounbehler, E. Sawicki, and K. Krost, "Determination of Dimethylnitrosamine in Air and Water by Thermal Energy Analysis: Validation of Analytical Procedures," *Environ. Sci. Technol.*, **11,** 577 (1977).

Finlayson, B. J., J. N. Pitts, Jr., and H. Akimoto, "Production of Vibrationally Excited OH in Chemiluminescent Ozone–Olefin Reactions," *Chem. Phys. Lett.*, **12,** 495 (1972).

Finlayson, B. J., J. N. Pitts, Jr., and R. Atkinson, "Low Pressure Gas-Phase Ozone–Olefin Reactions: Chemiluminescence, Kinetics, and Mechanisms," *J. Am. Chem. Soc.*, **96,** 5356 (1974).

Forrest, J., D. J. Spandau, R. L. Tanner, and L. Newman, "Determination of Atmospheric Nitrate and Nitric Acid Employing a Diffusion Denuder with Filter Pack," *Atmos. Environ.*, **16,** 1473 (1982).

Fung, K. and D. Grosjean, "Determination of Nanogram Amounts of Carbonyls as 2, 4-Dinitrophenylhydrazones by High-Performance Liquid Chromatography," *Anal. Chem.*, **53,** 168 (1981).

Gaffney, J. S., R. Fager, and G. I. Senum, "An Improved Procedure for High Purity Gaseous Peroxyacyl Nitrate Production: Use of Heavy Lipid Solvents," *Atmos. Environ.*, **18,** 215 (1984).

Gras, J. L., "Ammonia and Ammonium Concentrations in the Antarctic Atmosphere," *Atmos. Environ.*, **17,** 815 (1983).

Griffiths, P. R., *Chemical Infrared Fourier Transform Spectroscopy*, Wiley, New York, 1975.

Griffiths, P. R., "Recent Commercial Instrumental Developments in Fourier Transform Infrared Spectrometry," *Adv. Infrared Raman Spectrosc.*, **10,** 277 (1983a).

Griffiths, P. R., "Fourier Transform Infrared Spectrometry," *Science*, **222,** 297 (1983b).

Griffiths, P. R. and J. A. de Haseth, *Fourier Transform Infrared Spectrometry,* Wiley, New York, 1986.

Grosjean, D., "Formaldehyde and Other Carbonyls in Los Angeles Ambient Air," *Environ. Sci. Technol.*, **16,** 254 (1982).

Grosjean, D. and K. Fung, "Collection Efficiencies of Cartridges and Microimpingers for Sampling of Aldehydes in Air as 2, 4-Dinitrophenylhydrazones," *Anal. Chem.*, **54,** 1221 (1982).

Grosjean, D. and J. Harrison, "Response of Chemiluminescence NO Analyzers and Ultraviolet Ozone Analyzers to Organic Air Pollutants," *Environ. Sci. Technol.*, **19,** 862 (1985a).

Grosjean, D. and J. Harrison, "Peroxyacetyl Nitrate: Comparison of Alkaline Hydrolysis and Chemiluminescence Methods," *Environ. Sci. Technol.*, **19,** 749 (1985b).

Haagen-Smit, A. J., C. E. Bradley, and M. M. Fox, "Formation of Ozone in Los Angeles Smog," *Proc. Second Natl. Air Pollut. Symp.*, 1952, pp. 54–56.

Hanst, P. L., "Spectroscopic Methods for Air Pollution Measurement," *Adv. Environ. Sci. Technol.*, **2,** 91 (1971).

Hanst, P. L., N. W. Wong, and J. Bragin, "A Long-Path Infra-Red Study of Los Angeles Smog," *Atmos. Environ.*, **16,** 969 (1982).

Hard, T. M., R. J. O'Brien, T. B. Cook, and G. A. Tsongas, "Interference Suppression in HO Fluorescence Detection," *Appl. Opt.*, **18,** 3216 (1979).

Hard, T. M., R. J. O'Brien, and T. B. Cook, "Pressure Dependence of Fluorescent and Photolytic Interferences in OH Detection by Laser-Excited Fluorescence," *J. Appl. Phys.*, **51,** 3459 (1980).

Hard, T. M., R. J. O'Brien, C. Y. Chan, and A. A. Mehrabzadeh, "Tropospheric Free Radical Determination by FAGE," *Environ. Sci. Technol.*, **18,** 768 (1984).

Harris, G. W., W. P. L. Carter, A. M. Winer, J. N. Pitts, Jr., U. Platt, and D. Perner, "Observations of Nitrous Acid in the Los Angeles Atmosphere and Implications for Predictions of Ozone-Precursor Relationships," *Environ. Sci. Technol.*, **16,** 414 (1982).

Harward, C. N., W. A. McClenny, J. M. Hoell, J. A. Williams, and B. S. Williams, "Ambient Ammonia Measurements in Coastal Southeastern Virginia," *Atmos. Environ.*, **16,** 2497 (1982).

Hauser, T. R. and C. M. Shy, "Position Paper: NO_x Measurement," *Environ. Sci. Technol.*, **6,** 890 (1972).

Hewitt, C. N. and R. M. Harrison, "Tropospheric Concentrations of the Hydroxyl Radical-A Review," *Atmos. Environ.*, **19,** 545 (1985).

Hinkley, E. D., Ed., *Laser Monitoring of the Atmosphere*, Springer-Verlag, New York, 1976.

Hinkley, E. D., R. T. Ku, and P. L. Kelley, "Techniques for Detection of Molecular Pollutants by Absorption of Laser Radiation," in *Laser Monitoring of the Atmosphere*, E. D. Hinkley, Ed., Springer-Verlag, New York, 1976a, pp. 228–290.

Hinkley, E. D., K. W. Nill, and F. A. Blum, "Tunable Lasers in the Infrared," *Laser Focus*, 47–51 (April 1976b).

Hinkley, E. D., R. T. Ku, K. W. Nill, and J. F. Butler, "Long-Path Monitoring: Advanced Instrumentation with a Tunable Diode Laser," *Appl. Opt.*, **15,** 1653 (1976c).

Hoell, J. M., G. L. Gregory, M. A. Carroll, M. McFarland, B. A. Ridley, D. D. Davis, J. Bradshaw, M. O. Rodgers, A. L. Torres, G. W. Sachse, G. F. Hill, E. P. Condon, R. A. Rasmussen, M. C. Campbell, J. C. Farmer, J. C. Sheppard, C. C. Wang, and L. I. Davis, "An Intercomparison of Carbon Monoxide, Nitric Oxide, and Hydroxyl Measurement Techniques: Overview of Results, " *J. Geophys. Res.*, **89,** 11,819 (1984).

Holdren, M. W. and C. W. Spicer, "Field Compatible Calibration Procedure for Peroxyacetyl Nitrate," *Environ. Sci. Technol.*, **18,** 113 (1984).

Hübler, G., D. Perner, U. Platt, A. Toennissen, and D. H. Ehhalt, "Ground Level OH Radical Concentration: New Measurements by Optical Absorption," Proceedings of the 2nd Symposium on the Composition of the Non-urban Troposphere, Williamsburg, Virginia, May 25–28, 1982, pp. 315–318.

Hübler, G., D. Perner, U. Platt, A. Toennissen, and D. H. Ehhalt, "Ground Level OH Radical Concentration: New Measurements by Optical Absorption," *J. Geophys. Res.*, **89,** 1309 (1984).

Huebert, B. J. and A. L. Lazrus, "Tropospheric Gas-Phase and Particulate Nitrate Measurements," *J. Geophys. Res.*, **85C,** 7322 (1980).

Hughes, T. J., E. Pellizzari, L. Little, C. Sparacino, and A. Kobler, "Ambient Air Pollutants: Collection, Chemical Characterization, and Mutagenicity Testing," *Mutat. Res.*, **76,** 51 (1980).

Informatics, Inc., *Air Quality Criteria for Carbon Monoxide*, U.S. Environmental Protection Agency Report EPA-600/8-79-0922, October 1979.

Johnston, P. V. and R. L. McKenzie, "Long-Path Absorption Measurements of Tropospheric NO_2 in Rural New Zealand," *Geophys. Res. Lett.*, **11,** 69 (1984).

Joseph, D. W. and C. W. Spicer, "Chemiluminescence Method for Atmospheric Monitoring of Nitric Acid and Nitrogen Oxides," *Anal. Chem.*, **50,** 1400 (1978).

Keith, L. H., *Identification and Analysis of Organic Pollutants in Air*, Butterworth, Boston, 1984.

Kelly, N. A., G. T. Wolff, and M. A. Ferman, ''Background Pollutant Measurements in Air Masses Affecting the Eastern Half of the United States. I. Air Masses Arriving from the Northwest,'' *Atmos. Environ.*, **16,** 1077 (1982).

Kelly, T. J., D. H. Stedman, J. A. Ritter, and R. B. Harvey, ''Measurements of Oxides of Nitrogen and Nitric Acid in Clean Air,'' *J. Geophys. Res.*, **85C,** 7417 (1980).

Killinger, D. K. and A. Mooradian, Eds., *Optical and Laser Remote Sensing*, Springer-Verlag, New York, 1983.

Kravetz, T. M., S. W. Martin, and G. D. Mendenhall, ''Synthesis of Peroxyacetyl and Peroxyaroyl Nitrates. Complexation of Peroxacetyl Nitrate with Benzene,'' *Environ. Sci. Technol.*, **14,** 1262 (1980).

Ku, R. T., E. D. Hinkley, and J. O. Sample, ''Long-Path Monitoring of Atmospheric Carbon Monoxide with a Tunable Diode Laser System,'' *Appl. Opt.*, **14,** 854 (1975).

Kummer, W. A., J. N. Pitts, Jr., and R. P. Steer, ''Chemiluminescent Reactions of Ozone with Olefins and Sulfides,'' *Environ. Sci. Technol.*, **5,** 1045 (1971).

Kuwata, K., M. Uebori, and Y. Yamasaki, ''Determination of Aliphatic and Aromatic Aldehydes in Polluted Air as Their 2, 4-Dinitrophenyldrazones by High Performance Liquid Chromatography,'' *J. Chromatogr. Sci.*, **17,** 264 (1979).

Lamb, S. I., C. Petrowski, I. R. Kaplan, and B. R. T. Simoneit, ''Organic Compounds in Urban Atmospheres; A Review of Distribution, Collection, and Analysis,'' *J. Air Pollut. Control Assoc.*, **30,** 1098 (1980).

Levine, J. S., T. R. Augustsson, and J. M. Hoell, ''The Vertical Distribution of Tropospheric Ammonia,'' *Geophys. Res. Lett.*, **7,** 317 (1980).

Logan, J. A., ''Nitrogen Oxides in the Troposphere: Global and Regional Budgets,'' *J. Geophys. Res.*, **88,** 10,785 (1983).

Lonneman, W. A. and J. J. Bufalini, ''A Convenient Method for Preparation of Pure Standards of Peroxyacetyl Nitrate for Atmospheric Analysis,'' *Atmos. Environ.*, **16,** 2755 (1982).

Lonneman, W. A., T. A. Bellar, and A. P. Altshuller, ''Aromatic Hydrocarbons in the Atmosphere of the Los Angeles Basin,'' *Environ. Sci. Technol.*, **2,** 1017 (1968).

Lonneman, W. A., R. L. Seila, and J. J. Bufalini, ''Ambient Air Hydrocarbon Concentrations in Florida,'' *Environ. Sci. Technol.*, **12,** 459 (1978).

Lonneman, W. A., J. J. Bufalini, and G. R. Namie, ''Calibration Procedure for PAN Based on Its Thermal Decomposition in the Presence of Nitric Oxide,'' *Environ. Sci. Technol.*, **16,** 655 (1982).

Louw, C. W., J. F. Richards, and P. K. Faure, ''The Determination of Volatile Organic Compounds in City Air by Gas Chromatography Combined with Standard Addition, Selective Subtraction, Infrared Spectrometry, and Mass Spectrometry,'' *Atmos. Environ.*, **11,** 703 (1977).

Ludwick, J. D., D. B. Weber, K. B. Olsen, and S. R. Garcia, ''Air Quality Measurements in the Coal Fired Power Plant Environment of Colstrip, Montana,'' *Atmos. Environ.*, **14,** 523 (1980).

Mackay, G. I. and H. I. Schiff, personal communication (1985)

Maker, P. D., H. Niki, C. M. Savage, and L. P. Breitenbach, Fourier Transform Infrared Analysis of Trace Gases in the Atmosphere, in *Monitoring Toxic Substances*, D. Schuetzle, Ed., American Chemical Society Symposium Series 94, American Chemical Society, Washington, DC, 1979, pp. 161–175.

Maroulis, P. J., A. L. Torres, A. B. Goldberg, and A. R. Bandy, ''Atmospheric SO_2 Measurements on Project Gametag,'' *J. Geophys. Res.*, **85C,** 7345 (1980).

Martin, A. and F. R. Barber, ''Sulphur Dioxide, Oxides of Nitrogen and Ozone Measured Continuously for Two Years at a Rural Site,'' *Atmos. Environ.*, **15,** 567 (1981).

Matthews, R. D., R. F. Sawyer, and R. W. Schefer, ''Interferences in Chemiluminescent Measurement of NO and NO_2 Emissions from Combustion Systems,'' *Environ. Sci. Technol.*, **11,** 1092 (1977).

Mayrsohn, H., M. Kuramoto, J. H. Crabtree, R. D. Sothern, and S. H. Mano, ''Atmospheric Hydrocarbon Concentrations, June–September, 1974,'' Report to State of California Air Resources Board, April 1975.

McClenny, W. A. and C. A. Bennett, Jr., ''Integrative Technique for Detection of Atmospheric Ammonia,'' *Atmos. Environ.*, **14,** 641 (1980).

Mehrabzadeh, A. A., R. J. O'Brien, and T. M. Hard, ''Optimization of Response of Chemiluminescence Analyzers,'' *Anal. Chem.*, **55,** 1660 (1983).

Mihelcic, D., D. H. Ehhalt, J. Klomfaβ, G. F. Kulessa, U. Schmidt, and M. Trainer, ''Measurements of Free Radicals in the Atmosphere by Matrix Isolation and Electron Paramagnetic Resonance,'' *Ber. Bunsenges. Phys. Chem.*, **82,** 16 (1978a).

Mihelcic, D., D. H. Ehhalt, G. F. Kulessa, J. Klomfaβ, M. Trainer, U. Schmidt, and H. Röhrs, ''Measurements of Free Radicals in the Atmosphere by Matrix Isolation and Electron Paramagnetic Resonance,'' *Pure Appl. Geophys.*, **116,** 530 (1978b).

Mulawa, P. A. and S. H. Cadle, ''A Comparison of Nitric Acid and Particulate Nitrate Measurements by the Penetration and Denuder Difference Methods,'' *Atmos. Environ.*, **19,** 1317 (1985).

National Aeronautics and Space Administration, ''Assessment of Techniques for Measuring Tropospheric N_xO_y,'' NASA Conference Publ. 2292 (1983).

National Aeronautics and Space Administration, ''Research Needs in Heterogeneous Tropospheric Chemistry,'' NASA Conference Publ. 2320 (1984a).

National Aeronautics and Space Administration, ''Assessment of Techniques for Measuring Tropospheric H_xO_y,'' NASA Conference Publ. 2332 (1984b).

National Research Council, *Formaldehyde and Other Aldehydes*, National Academy Press, Washington, DC, 1981.

Natusch, D. F. and P. K. Hopke, *Analytical Aspects of Environmental Chemistry,* Wiley, New York, 1983.

Nelson, G. O., Calibration Techniques, in *Sampling and Analysis of Toxic Organics in the Atmosphere,* ASTM STP 721, American Society for Testing and Materials, Philadelphia, 1980, pp. 184–192.

Nelson, P. F. and S. M. Quigley, ''Non-Methane Hydrocarbons in the Atmosphere of Sydney, Australia,'' *Environ. Sci. Technol.*, **16,** 650 (1982).

Nelson, P. F., S. M. Quigley, and M. Y. Smith, Atmospheric Hydrocarbons in Sydney: Ambient Concentrations and Relative Source Strengths, in Proceedings of the Conference *The Urban Atmosphere—Sydney, A Case Study,* CSIRO, Melbourne, Australia, 1983, pp. 329–349.

Nielsen, T., A. M. Hansen, and E. L. Thomsen, ''A Convenient Method for Preperation of Pure Standards of Peroxyacetyl Nitrate for Atmospheric Analysis,'' *Atmos. Environ.*, **16,** 2447 (1982).

Noxon, J. F., R. B. Norton, and W. R. Henderson, ''Observation of Atmospheric NO_3,'' *Geophys. Res. Lett.*, **5,** 675 (1978).

Noxon, J. F., R. B. Norton, and E. Marovich, ''NO_3 in the Troposphere,'' *Geophys. Res. Lett.*, **7,** 125 (1980).

O'Brien, R. J., personal communication (1985).

O'Brien, R. J., T. M. Hard, and A. A. Mehrabzadeh, ''Comment on 'Laboratory Evaluation of An Airborne Ozone Instrument that Compensates for Altitude/Sensitivity Effects','' *Environ. Sci. Technol.*, **17,** 560 (1983).

Okabe, H., P. L. Splitstone, and J. J. Ball, ''Ambient and Source SO_2 Detector Based on a Fluorescence Method,'' *J. Air Pollut. Control Assoc.*, **23,** 514 (1973).

Ortgies, G. and F. J. Comes, ''A Laser Optical Method for the Determination of Tropospheric OH Concentrations,'' *Appl. Phys.*, **33,** 103 (1984).

Ortgies, G., K.-H. Gericke, and F. J. Comes, ''Is UV Laser Induced Fluorescence a Method to Monitor Tropospheric OH?,'' *Geophys. Res. Lett.*, **7,** 905 (1980); ''Optical Measurements of Tropospheric Hydroxyl with Lasers,'' *Z. Naturforsch.*, **36a,** 177 (1981).

Pellizzari, E. D., State-of-the-Art Analytical Techniques for Ambient Vapor Phase Organics and Volatile Organics in Aqueous Samples from Energy-Related Activities, in *Application of Short-Term Bioassays in the Fractionation and Analysis of Complex Environmental Mixtures*, M. D. Waters, S. Nesnow, J. L. Huisingh, S. S. Sandhu, and L. Claxton, Eds., EPA Report No. EPA-600/9-78-027, September 1978.

Perkins, M. D. and F. L. Eisele, "First Mass Spectrometric Measurements of Atmospheric Ions at Ground Level," *J. Geophys. Res.*, **89,** 9649 (1984).

Perner, D., D. H. Ehhalt, H. W. Pätz, U. Platt, E. P. Röth, and A. Volz, "OH-Radicals in the Lower Troposphere," *Geophys. Res. Lett.*, **3,** 466 (1976).

Pitts, J. N., Jr., J. M. McAfee, W. D. Long, and A. M. Winer, "Long-Path Infrared Spectroscopic Investigation of Ambient Concentrations of the 2% Neutral Buffered Potassium Iodide Method for Determination of Ozone," *Environ. Sci. Technol.*, **10,** 787 (1976a).

Pitts, J. N., Jr., J. L. Sprung, M. Poe, M. C. Carpelan, and A. C. Lloyd, "Corrected South Coast Air Basin Oxidant Data: Some Conclusions and Implications," *Environ. Sci. Technol.*, **10,** 794 (1976b).

Pitts, J.. N. Jr., B. J. Finlayson-Pitts and A. M. Winer, "Optical Systems Unravel Smog Chemistry," *Environ. Sci. Technol.*, **11,** 568 (1977).

Pitts, J. N., Jr., A. M. Winer, G. W. Harris, W. P. L. Carter, and E. C. Tuazon, "Trace Nitrogenous Species in Urban Atmospheres," *Environ. Health Perspect.*, **52,** 153 (1983).

Platt, U. and D. Perner, "Direct Measurements of Atmospheric CH_2O, HNO_2, O_3, NO_2, and SO_2 by Differential Optical Absorption in the Near UV," *J. Geophys. Res.*, **85C,** 7453 (1980).

Platt, U. and D. Perner, "Measurements of Atmospheric Trace Gases by Long Path Differential UV/Visible Absorption Spectoscopy," Paper Presented at the Workshop on Optical and Laser Remote Sensing, Monterey, California, February 9–11, 1982.

Platt, U., D. Perner, and H. W. Patz, "Simultaneous Measurement of Atmospheric CH_2O, O_3, and NO_2 by Differential Optical Absorption," *J. Geophys. Res.*, **84C,** 6329 (1979).

Platt, U., D. Perner, G. W. Harris, A. M. Winer and J. N. Pitts, Jr., "Observations of Nitrous Acid in an Urban Atmosphere by Differential Optical Absorption," *Nature*, **285,** 312 (1980a).

Platt, U., D. Perner, A. M. Winer, G. W. Harris, and J. N. Pitts, Jr., "Detection of NO_3 in the Polluted Troposphere by Differential Optical Absorption," *Geophys. Res. Lett.*, **7,** 89 (1980b).

Platt, U., D. Perner, J. Schröder, C. Kessler, and A. Toennisen, "The Diurnal Variation of NO_3," *J. Geophys. Res.*, **86,** 11,965 (1981).

Platt, U., D. Perner, and C. Kessler, "The Importance of NO_3 for the Atmospheric NO_x Cycle from Experimental Observations," Proceedings of the 2nd Symposium on the Non-urban Troposphere, Williamsburg, Virginia, May 25–28, 1982, pp. 21–24.

Platt, U., A. M. Winer, H. W. Biermann, R. Atkinson, and J. N. Pitts, Jr., "Measurement of Nitrate Radical Concentrations in Continental Air," *Environ. Sci. Technol.*, **18,** 365 (1984).

Pokrowsky, P. and W. Herrmann, "Sensitive Detection of Hydrogen Chloride by Derivative Spectroscopy with a Diode Laser," SPIE *J*, **286,** 33 (1981).

Pratt, G. C., R. C. Hendrickson, B. I. Chevone, D. A. Christopherson, M. V. O'Brien, and S. V. Krupa, "Ozone and Oxides of Nitrogen in the Rural Upper-Midwestern U.S.A.," *Atmos. Environ.*, **17,** 2013 (1983).

Reid, J., J. Shewchun, B. K. Garside, and E. A. Ballik, "High Sensitivity Pollution Detection Employing Tunable Diode Lasers," *Appl. Opt.*, **17,** 300 (1978a).

Reid, J., B. K. Garside, J. Shewchun, M. El-Sherbiny, and E. A. Ballik, "High Sensitivity Point Monitoring of Atmospheric Gases Employing Tunable Diode Lasers," *Appl. Opt.*, **17,** 1806 (1978b).

Reid, J., M. El-Sherbiny, B. K. Garside, and E. A. Ballik, "Sensitivity Limits of a Tunable Diode Laser Spectrometer, with Application to the Detection of NO_2 at the 100-ppt Level," *Appl. Opt.*, **19,** 3349 (1980).

Reid, J., D. T. Cassidy, and R. T. Menzies, "Linewidth Measurements of Tunable Diode Lasers Using Heterodyne and Etalon Techniques," *Appl. Opt.*, **21,** 3961 (1982).

Rounbehler, D. P., J. W. Reisch, and D. H. Fine, Nitrosamine Air Sampling Using a New Artifact-Resistant Solid Sorbent System, in *Sampling and Analysis of Toxic Organics in the Atmosphere*, ASTM STP 721, American Society for Testing and Materials, Philadelphia, 1980, pp. 80–91.

Sanhueza, E., C. N. Plum, and J. N. Pitts, Jr., "Positive Interference of Nitrous Acid in the Determination of Gaseous HNO_3 by the NO_x Chemiluminescence–Nylon Cartridge Method: Applications to Measurements of ppb Levels of HONO in Air,"*Atmos. Environ.*, **18,** 1029 (1984).

Schiff, H. I., D. R. Hastie, G. I. Mackay, T. Iguchi, and B. A. Ridley, "Tunable Diode Laser Systems for Measuring Trace Gases in Tropospheric Air," *Environ. Sci. Technol.*, **17,** 352A (1983).

Schjoldager, J., "Aldehyder I Byluft," Presented at the First Nordic Conference on Organic Pollutants, Geiranger, Norway, September 10–13, 1984.

Schlitt, H., H. Knoeppel, B. Versino, A. Peil, H. Schauenburg, and H. Vissers, Organics in Air: Sampling and Identification, in *Sampling and Analysis of Toxic Organics in the Atmosphere*, ASTM STP 721, American Society for Testing and Materials, Philadelphia, 1980, pp. 22–35.

Scott, W. E., E. R. Stephens, P. L. Hanst, and R. C. Doerr, "Further Developments in the Chemistry of the Atmosphere," Paper presented at the 22nd Midyear Meeting of the American Petroleum Institute's Division of Refining, Philadelphia, PA, May 14, 1957.

Seifert, V. B. and D. Ullrich, "Konzentration Anorganischer und Organischer Luftschadstoffe an Einer Straβenkreuzung in Berlin," *Staub Reinhalt. Luft*, **38,** 359 (1978).

Seila, R. L., "Non-urban Hydrocarbon Concentrations in Ambient Air North of Houston, Texas," EPA Report No. EPA-600-3-79-010, February 1979.

Seila, R. L., W. A. Lonneman, and S. A. Meeks, "Evaluation of Polyvinyl Fluoride as a Container Material for Air Pollution Samples," *J. Environ. Sci. Health Environ. Sci. Eng.*, **A11,** 121 (1976).

Sexton, K. and H. Westberg, "Nonmethane Hydrocarbon Composition of Urban and Rural Atmospheres," Paper presented at the 74th Air Pollution Control Association Meeting, Philadelphia, PA, 1981, Paper No. 81-47.3; *Atmos. Environ.*, **18,** 1125 (1984).

Sexton, F. W., R. M. Michie, Jr., F. F. McElroy, and V. L. Thompson, "A Comparative Evaluation of Seven Automated Ambient Non-Methane Organic Compound Analyzers," EPA Report No. EPA-600/4-02-046, June 1982.

Shaw, R. W. Jr., R. K. Stevens, J. Bowermaster, J. W. Tesch and E. Tew, "Measurements of Atmospheric Nitrate and Nitric Acid: The Denuder Difference Experiment," *Atmos. Environ.*, **16,** 845 (1982).

Singh, H. B. "Guidance for the Collection and Use of Ambient Hydrocarbon Species. Data in Development of Ozone Control Strategies," EPA Report No. EPA-450/4-80-008, April 1980.

Singh, H. B. and L. J. Salas, "Peroxyacetyl Nitrate in the Free Troposphere," *Nature,* **302,** 326 (1983).

Smith, R. G., R. J. Bryan, M. Feldstein, B. Levadie, F. A. Miller, and E. R. Stephens, Tentative Method of Analysis for Peroxyacetyl Nitrate (PAN) in the Atmosphere (Gas Chromatographic Method), in *Methods of Air Sampling and Analysis*, American Public Health Association, Washington, DC, 1972, pp. 319–323.

Spicer, C. W., "Photochemical Atmospheric Pollutants Derived from Nitrogen Oxides," *Atmos. Environ.*, **11,** 1089 (1977).

Spicer, C. W., J. E. Howes, Jr., T. A. Bishop, L. H. Arnold, and R. K. Stevens, "Nitric Acid Measurement Methods: An Intercomparison," *Atmos. Environ.*, **16,** 1487 (1982).

Stephens, E. R., "Long-Path Infrared Spectroscopy for Air Pollution Research," *Soc. Appl. Spectrosc.*, **12,** 80 (1958).

Stephens, E. R., "Absorptivities for Infrared Determination of Peroxyacyl Nitrates," *Anal. Chem.*, **36,** 928 (1964).

Stephens, E. R., "The Formation, Reactions, and Properties of Peroxyacyl Nitrates (PAN's) in Photochemical Air Pollution," *Adv. Environ. Sci. Technol.*, **1,** 119 (1969).

Stephens, E. R. and F. R. Burleson, "Distribution of Light Hydrocarbons in Ambient Air," *J. Air Pollut. Control Assoc.*, **19,** 929 (1969).

Stephens, E. R. and M. A. Price, "Analysis of an Important Air Pollutant: Peroxyacetyl Nitrate," *J. Chem. Ed.*, **50,** 351 (1973).

Stephens, E. R., P. L. Hanst, R. C. Doerr, and W. E. Scott, "Reactions of Nitrogen Dioxide and Organic Compounds in Air," *Ind. Eng. Chem.*, **48,** 1498 (1956a).

Stephens, E. R., W. E. Scott, P. L. Hanst, and R. C. Doerr, "Recent Developments in the Study of the Organic Chemistry of the Atmosphere," *J. Air. Pollut. Control Assoc.*, **6,** 159 (1956b).

Stephens, E. R., E. F. Darley, O. C. Taylor, and W. E. Scott, "Photochemical Reaction Products in Air Pollution," *Proc. Am. Petrol. Inst.*, **40,** III, 325 (1960); *Int. J. Air Water Pollut.*, **4,** 79 (1961).

Stephens, E. R., F. R. Burleson, and E. A. Cardiff, "The Production of Pure Peroxyacyl Nitrates," *J. Air Pollut. Control Assoc.*, **15,** 87 (1965).

Stephens, E. R., F. R. Burleson, and K. M. Holtzclaw, "A Damaging Explosion of Peroxyacetyl Nitrate," *J. Air Pollut. Control Assoc.*, **19,** 261 (1969).

Stevens, R. K., T. G. Dzubay, G. Russwurm, and D. Rickel, "Sampling and Analysis of Atmospheric Sulfates and Related Species," *Atmos. Environ.*, **12,** 55 (1978).

Temple, P. J. and O. C. Taylor, "World-Wide Ambient Measurements of Peroxyacetyl Nitrate (PAN) and Implications for Plant Injury," *Atmos. Environ.*, **17,** 1583 (1983).

Tuazon, E. C., personal communication (1985).

Tuazon, E. C., R. A. Graham, A. M. Winer, R. R. Easton, J. N. Pitts, Jr., and P. L. Hanst, "A Kilometer Pathlength Fourier-Transform Infrared System for the Study of Trace Pollutants in Ambient and Synthetic Atmospheres," *Atmos. Environ.*, **12,** 865 (1978).

Tuazon, E. C., A. M. Winer, R. A. Graham, and J. N. Pitts, Jr., "Atmospheric Measurements of Trace Pollutants by Kilometer-Pathlength FT–IR Spectroscopy," *Adv. Environ. Sci. Technol.*, **10,** 259 (1980).

Tuazon, E. C., A. M. Winer, and J. N. Pitts, Jr., "Trace Pollutant Concentrations in a Multiday Smog Episode in the California South Coast Air Basin by Long Path Length Fourier Transform Infrared Spectroscopy," *Environ. Sci. Technol.*, **15,** 1232 (1981).

U.S. Environmental Protection Agency, "Air Quality Criteria for Ozone and Other Photochemical Oxidants," EPA-600/8-78-004, April 1978; revised document expected in 1986.

U.S. Environmental Protection Agency, "Air Quality Criteria for Oxides of Nitrogen," EPA-600/8-82-026, September 1982a.

U.S. Environmental Protection Agency, "Air Quality Criteria for Particulate Matter and Sulfur Oxides," EPA-600/8-82-029, 1982b.

van Ham, J. and Th. Lems, Objections to the Use of Polyvinyl Fluoride in Smog Chamber Experiments, in *Photochemical Smog Formation in the Netherlands*, R. Guicherit, Ed., TNO's, Gravenhage, Netherlands, 1978, pp. 172–177.

Walega, J. G., D. H. Stedman, R. E. Shetter, G. I. Mackay, T. Iguchi, and H. I, Schiff, "Comparison of a Chemiluminescent and a Tunable Diode Laser Absorption Technique for the Measurement of Nitrogen Oxide, Nitrogen Dioxide, and Nitric Acid," *Environ. Sci. Technol.*, **18,** 823 (1984).

Wang, C. C., "Comments on 'Laser-Excited Fluorescence of the Hydroxyl Radical: Relaxation Coefficients at Atmospheric Pressure,'" *J. Phys. Chem.*, **88,** 2924 (1984).

Wang, C. C. Remote Sensing of OH in the Atmosphere Using the Technique of Laser-Induced Fluorescence, in *Optical and Laser Remote Sensing*, D. K. Killinger and A. Mooradian, Eds., Springer-Verlag, New York, 1983, pp. 205–212.

Wang, C. C. and L. I. Davis, Jr., "Comments on 'Theoretical and Experimental Assessment of the

O_3/H_2O Interference Problem...In the Detection of OH...' by Davis et al.,'' *Geophys. Res. Lett.*, **9,** 98 (1982).

Wang, C. C., L. I. Davis, Jr., P. M. Selzer, and R. Munoz, ''Improved Airborne Measurements of OH in the Atmosphere Using the Technique of Laser-Induced Fluorescence,'' *J. Geophys. Res.*, **86,** 1181 (1981).

Watanabe, T., M. Yoshida, S. Fujiwara, K. Abe, A. Onoe, M. Hirota, and S. Igarashi, ''Spin Trapping of Hydroxyl Radical in the Troposphere for Determination by Electron Spin Resonance and Gas Chromatography/Mass Spectrometry,'' *Anal. Chem.*, **54,** 2470 (1982).

Wathne, B. M., ''Measurements of Benzene, Toluene, and Xylenes in Urban Air,'' *Atmos. Environ.*, **17,** 1713 (1983).

White, J. U., ''Long Optical Paths of Large Aperture,'' *J. Opt. Soc. Am.*, **32,** 285 (1942).

Winer, A. M., ''Spectroscopic Observations of Previously Undetected Atmospheric Species: Implications for Air Pollution, Atmospheric Chemistry and Acid Deposition,'' *J. Clean Air Soc. Australia and New Zealand,* in press (1986).

Winer, A.M., J. W. Peters, J. P. Smith, and J. N. Pitts, Jr., ''Response of Commercial Chemiluminescent NO–NO_2 Analyzers to Other Nitrogen-Containing Compounds,'' *Environ. Sci. Technol.*, **8,** 1118 (1974).

World Health Organization, *Selected Methods of Measuring Air Pollutants*, WHO Offset Publication No. 24, World Health Organization, Geneva, 1976.

Yokouchi, Y., T. Fujii, Y. Ambe, and K. Fuwa, ''Gas Chromatographic–Mass Spectrometric Analysis of Formaldehyde in Ambient Air Using a Sampling Tube,'' *J. Chromatogr.*, **180,** 133 (1979).

6 Environmental Chambers

6-A OBJECTIVES OF SMOG CHAMBER STUDIES OF SIMULATED ATMOSPHERES

Predicting quantitatively the relationship between emissions and air quality is an ultimate goal of atmospheric scientists. Perhaps the most direct experimental means of examining this relationship between emissions and air quality is to simulate atmospheric conditions using large chambers. Measured concentrations of the primary pollutants are injected into these environmental (or *smog*) chambers, as they are called. These are then irradiated with sunlight or lamps used to mimic the sun, and the concentrations of the primary pollutants as well as the resulting secondary pollutants are measured. The primary pollutant concentrations as well as temperature, relative humidity, and so on can be systematically varied in order to establish the relationship between emissions and air quality, free from the complexities of continuously injected pollutant emissions and meteorology, both of which complicate the interpretation of ambient air data.

The results from such chamber studies are frequently used to test the chemical submodels of various airshed models for photochemical air pollution in order to provide a scientific basis for control strategies. While the interpretation of the results of smog chamber studies and their extrapolation to atmospheric conditions also have some limitations, such studies do provide a relatively inexpensive means of at least initially examining the emissions–air quality relationship under controlled conditions.

In addition, smog chambers are useful as large reaction vessels to generate kinetic and mechanistic data on individual reactions believed to be important in the atmosphere. They have also been used extensively as exposure chambers to study the effects of air pollution on people, animals, and plants.

In this chapter, the general characteristics of chambers, typical results, and the advantages and limitations of chamber studies are discussed.

6-B. DESIGN CRITERIA AND TYPES OF ENVIRONMENTAL CHAMBERS

Environmental chambers have been widely used in atmospheric chemistry studies since the first discovery of photochemical air pollution. Design criteria for these chambers are based on reproducing as faithfully as possible conditions in ambient air, excluding meteorology and the uncontrolled addition of pollutants. Although the general aims of all chamber studies are similar (i.e., to simulate reactions in ambient air under controlled conditions), the chamber designs and capabilities used to meet this goal vary widely. Thus chambers can differ in any or all of the following characteristics: (1) size and shape, (2) surface materials to which the pollutants are exposed, (3) range of pressures and temperatures which can be attained, (4) methods of preparation of reactants, including "clean" air, (5) conditions, (i.e., static or flow) under which experiments can be carried out, (6) analytical capabilities, and (7) spectral characteristics of the light source.

In reviewing the types and uses of the wide range of chambers employed in atmospheric chemistry, one major caveat should be kept in mind. The use of chambers by necessity involves the presence of surfaces in the form of the chamber walls, and this is the single largest uncertainty in using them as a surrogate for ambient air studies. Contributing to this uncertainty are possible unknown heterogeneous reactions occurring on both fresh (i.e. "clean") and conditioned chamber surfaces. Additionally, the outgassing of uncharacterized reactive vapors either deposited there during previous experiments or released from the plastic films used to make the chambers (e.g. non-polymerized organics) can have pronounced effects in certain reaction systems, especially kinetic studies of low reactivity organics (Joshi et al., 1982). Another, less severe, problem is reproducing the actinic radiation to which pollutants are exposed in ambient atmospheres. The design of environmental chamber facilities attempts insofar as possible to minimize these variations from "real" air masses.

A general discussion of the nature and importance of these chamber characteristics, including "wall effects" follows. For detailed descriptions of various types of smog chamber facilities and their operation, one should consult the original literature, including for example, indoor studies utilizing: a 440 L borosilicate glass cylinder (Joshi et al., 1982); a 17.3 m^3 chamber lined with 5-mil Teflon (FEP) film (Spicer, 1983); and two similar 6 m^3 evacuable and thermostatted chambers (Darnall et al., 1976 and Winer et al., 1980; Akimoto et al., 1979). Fluorocarbon-film bags used as chambers range in size from 15–40 L (Lonneman et al., 1981) to 450 −2000 L bags (Kelly et al., 1985) to large outdoor chambers with capacities of 25 m^3 (Jeffries et al., 1976, and Kamens et al., 1985), 40 m^3 (Fitz et al., 1981) and 65 m^3 (Leone et al., 1985).

6-B-1. Glass Reactors

Many early studies were carried out in borosilicate glass reactors similar to those used in typical laboratory studies of gas phase reactions. These were usually relatively small, a few liters up to approximately 100 L. While they were convenient, inexpensive, and readily available, there are some problems associated with their use. For example, Pyrex glass absorbs light at wavelengths $\leq$ 350 nm (the cutoff depends on the thickness of the glass; see below), which we have seen is a critical region for atmospheric photochemistry. Thus species such as O_3 and HCHO which produce the free radicals OH and HO_2 are exposed in such vessels to less actinic ultraviolet radiation than is present under atmospheric conditions, and hence undergo less photolysis (see Section 6-E).

In addition, such small vessels have high surface-to-volume (S/V) ratios which may increase the relative contributions of reactions which occur on the surface. One such heterogeneous reaction is the decomposition on surfaces of O_3 to form O_2; obviously the faster this decomposition, the lower the concentrations of O_3 that will be observed during the chamber run.

A second important reaction which may occur on surfaces in smog chambers (see Section 8-A-3a) is one generating HONO:

$$2NO_2 + H_2O \xrightarrow{\text{surface}} HONO + (HNO_3?)$$

As discussed in Section 8-A-3a, this reaction has been shown to be too slow in aqueous solution to be significant in the atmosphere; while it may occur faster on surfaces and has been proposed as a source of HONO in smog chambers (Sakamaki et al., 1983; Pitts et al., 1984; Leone et al., 1985), this remains speculative at present. However, since HONO is a major OH source in the early stages of irradiation in smog chambers, it is important to understand the mechanism of its formation and to quantify its rate of production under various experimental conditions. Thus, if this reaction only occurs at a significant rate on the surfaces typically encountered in chambers, the effects it has on the overall reactions should be removed when extrapolating the results of chamber studies to ambient air. In the case of HONO generation, studies carried out in a mobile laboratory where the surfaces were very different than in environmental chambers also showed HONO formation (Pitts et al., 1985) suggesting that this reaction may not be unique to smog chamber surfaces.

The belief generally has been that the smaller the S/V ratio (i.e., the larger the smog chamber), the less important such surface reactions will be, and hence the more representative of the ambient atmosphere the results. While there is doubtless some justification for this approach, it must also be kept in mind that there are a variety of surfaces present in real atmospheres as well. These include not only the surfaces of the earth, buildings, and so on but also the surfaces of particulate matter suspended in air (Chapter 12). If the heterogeneous formation of HONO occurs not only on chamber surfaces but also on those found in urban atmospheres as well, then it is important to include it in extrapolating the chamber results to ambient air. In this case, the effects on the kinetics due to the different types and available amounts of surfaces in air compared to chambers must, of course, be taken into account.

6-B-2. Collapsible Bags

As a result of these problems, larger smog chambers with surfaces thought to be relatively inert are more generally in use today. Thus conditioned FEP Teflon films for example, have been shown to have relatively low rates of surface destruction of a variety of reactive species including O_3. In addition, they typically transmit

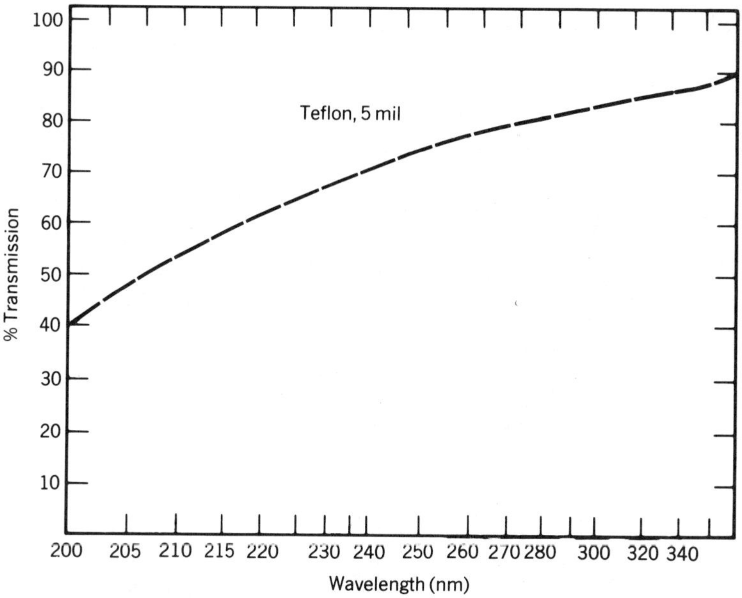

FIGURE 6.1. Typical transmission spectrum in the UV region of a Teflon film used to construct environmental chambers (from Dimitriades, 1967).

solar radiation in the 290–800 nm region (Fig. 6.1) and have low rates of hydrocarbon offgassing.

Bag type smog chambers are easily constructed using flexible thin films of this material. In addition to the low rates of destruction of reactive species such as O_3, and their transparency to actinic UV, they have the advantage that the size of the bag can be easily varied. An additional advantage of bag chambers is that they may be easily divided into two sections simply by putting a heavy divider (e.g., bar) across the middle of the bag. One can then use one side of the bag as a "control" and the other to study the effects of varying one parameter such as the injection of additional pollutants.

Figure 6.2 shows a schematic diagram of such a chamber. Ports are included for the introduction of the primary pollutants and for sampling for product analysis. Because such bags do not have a rigid shape, they are operated at atmospheric pressure. The pressure inside the chamber may be maintained during a run by introducing clean air at the same rate as sampling removes air from the chamber, thus diluting the mixture. Alternatively, these soft bags can be allowed to collapse as air is removed for analysis; this maintains the pressure at 1 atm but results in an increasing S/V ratio during a run. Such a 65 m^3 outdoor smog chamber fabricated from 2-mil thick Teflon film has recently been employed in an experimental and modeling study of the photooxidation of toluene-NO_x-air mixtures (Leone et al., 1985).

A potential problem with the use of these collapsible bags is contamination by outgassing of organics from the bag material itself; for example, the release of

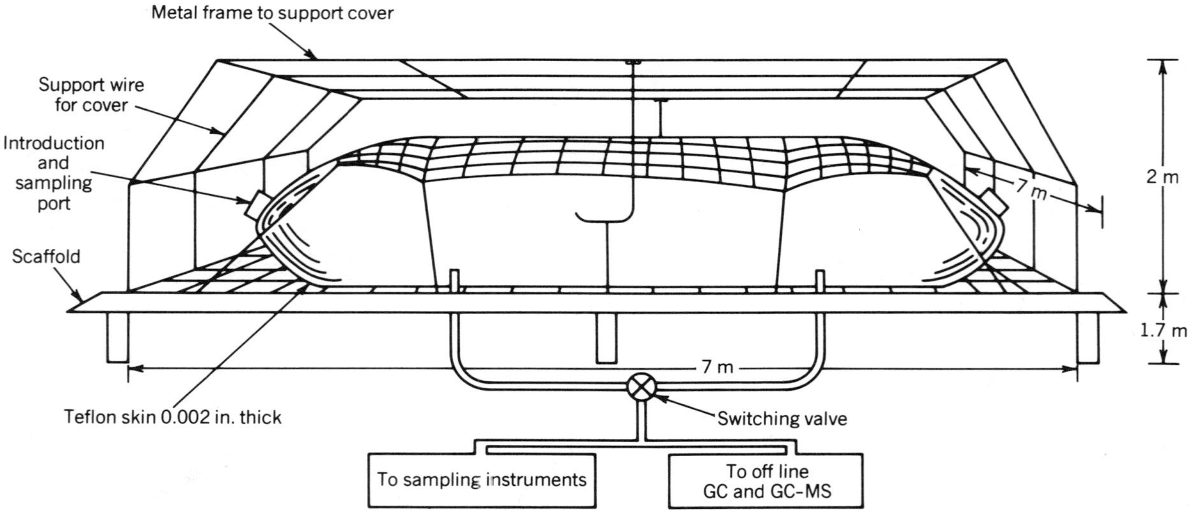

FIGURE 6.2. Schematic diagram of typical outdoor 40 m^3 collapsible bag smog chamber (from Fitz et al., 1981).

significant amounts of low molecular weight fluorocarbons from FEP Teflon has been reported (Lonneman et al., 1981). While this problem does not seem to occur with all bag samples (Kelly et al., 1985), investigators should clearly exercise caution with regard to possible contamination from this source.

Another possible problem is that the surfaces of fresh, nonconditioned Teflon films may be quite electrostatic. Transport to, and adsorption on, such surfaces might deplete the concentrations of particles in such "new" chambers.

In any case, experience suggests that they first be conditioned by filling them with air containing O_3 and leaving them in the dark for several hours, and/or filling them with "clean" air containing added NO_x and irradiating them for a period of time (Kelly, 1982). Such chamber "conditioning" and characterization studies are an essential initial step in detailed chamber studies of the kinetics and mechanisms of atmospheric reactions.

6-B-3. Evacuable Chambers

Ideally, one would like to be able to vary the pressure and temperature during smog chamber runs in order to simulate various geographical locations, seasons, and meteorology, and to establish the pressure and temperature dependencies of reactions. Varying the pressure and temperature also allows one to simulate the upper atmosphere (e.g., to study stratospheric and mesospheric chemistry).

Temperature control has an additional advantage with respect to the problem of chamber contamination. Thus, after a smog chamber has been used, some hydrocarbons and nitrogen compounds may remain adsorbed on the chamber walls. These may desorb in subsequent runs and, in some cases (e.g., HCHO), act as free radical sources to accelerate the photooxidation processes. The ability to "bake out" smog chambers while pumping to low pressures is therefore useful in reducing chamber contamination effects.

While glass reactors can be easily designed to include pressure and temperature control, they suffer from other limitations discussed earlier. In addition, the use of very large glass evacuable chambers at low pressures presents a safety problem. On the other hand, pressure and temperature are not easily controlled using collapsible bags.

As a result, some evacuable smog chambers have been designed and constructed in order to control both temperature and pressure. The one shown schematically in Fig. 6.3 is constructed with an aluminum alloy and the walls are coated with FEP Teflon. The end windows through which the mixture is irradiated are ultraviolet grade quartz to allow transmission of the actinic UV. The radiation cutoff and spectral distribution in the 290–350 nm region can be varied using filters between the irradiation source and the chamber windows. The pumping system used to evacuate the chamber is hydrocarbon free; back-diffusion of organics from the use of conventional pump fluids can produce NMHC concentrations that by themselves exceed the NMHC air quality standard! In the particular case shown in Fig. 6.3, the pumping system consists of a liquid ring roughing pump (the fluid is water) with an air injection pump, cryosorption roughing pumps, a diffusion

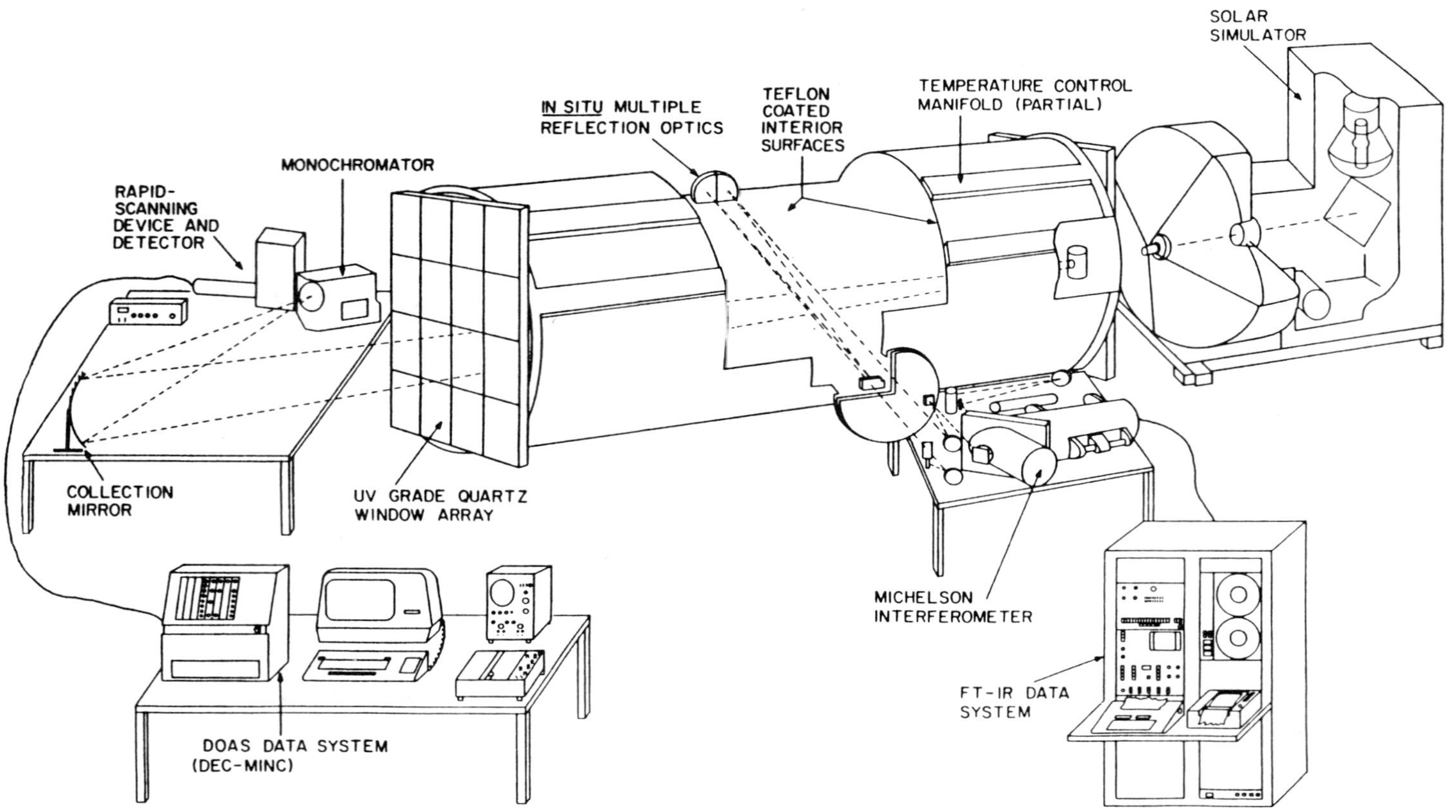

FIGURE 6.3. Schematic diagram of evacuable chamber at the Statewide Air Pollution Research Center, University of California, Riverside.

pump, and a mechanical pump. (Both the diffusion and mechanical pumps use unreactive perfluorinated oils rather than hydrocarbons as the working fluid).

Ports are included in the chamber for air and pollutant injection, for sampling (e.g., for GC or GC–MS analysis), and for in situ analysis using optical absorption spectroscopy (Darnall et al., 1976; Winer et al., 1980).

This type of chamber satisfies most design criteria in that both pressure and temperature can be varied, the intensity and spectrum of the irradiation can be altered, and the surface can be coated with a relatively inert material to minimize heterogeneous reactions and pollutant adsorption and offgassing. In addition, ports for both in situ spectroscopic product analysis and for sampling can be easily included. The disadvantages are that they are relatively expensive, varying the S/V ratio through changing the dimensions of the chamber is not practical, and changing the nature of the chamber surface is difficult and time consuming. Two such chambers are described by Akimoto et al. (1979) and Winer et al. (1980).

6-B-4. Preparation of Reactants, Including "Clean Air"

Different laboratories often use different sources and/or methods of preparation of the reactants NO_x, NMOC, and "clean air." Frequently, the desired compounds can be purchased commercially. However, care must be taken when using commercially supplied materials to be sure that trace impurities, which can alter the experiment, are not present or are removed by purification before use. For example, gaseous HNO_3 at concentrations up to several percent are commonly present in commercially produced NO_2 which is stored in a gas cylinder. Similarly, small concentrations of alkenes are frequently present in commercial cylinders of the alkanes.

By far the largest component in smog chamber studies is air. It is especially imperative therefore that the air used to dilute the NO_x and NMOC is "clean." Ambient air and commercially supplied air generally contain sufficient organic and NO_x impurities that extensive purification is needed to reduce these contaminants to low-ppb levels or below. One such purification system is described by Doyle, et al. (1977); others are described in the papers on chamber facilities cited earlier.

Regardless of the source of the "clean" air used in smog chamber studies, it should be regularly tested and characterized to ensure that it does not contain trace impurities that could affect the overall experiment, especially when systems of relatively low reactivity are involved.

6-B-5. Mode of Chamber Operation: Static Versus Dynamic

Most chamber experiments are conducted under static conditions i.e., the reactants and air are injected into the chamber; the chamber is sealed except for sample withdrawal, and the mixture is irradiated. The advantages of this technique are that it is simple, allows for good control of experimental parameters, and permits long reaction times and hence the formation of minor products can be more readily studied.

In the dynamic mode of operation, the pollutants in air flow continuously through the chamber at a constant flow rate. This has the advantage that large amounts of the irradiated mixture can be generated, which are sometimes needed, for example, in exposure studies. The net contribution of surface reactions may in some cases be less than in the static mode because of the shorter residence time for the pollutants in the chamber. However, counterbalancing these advantages is the problem of obtaining sufficient residence time to produce measurable concentrations of secondary pollutants. In addition, large quantities of the purified reactants and clean air must be continuously and reliably produced. These problems are sufficiently great that the static mode of operation is most frequently used today, including a protocol used to simulate multi-day smog episodes. In this procedure, a fresh charge of reactants is introduced into the chamber each morning and allowed to mix with products of the previous day's irradiation, as well as those which may have formed overnight. The mixture is then irradiated to correspond to a second or third day of the simulated multi-day episode.

6-C. LIGHT SOURCES: SPECTRAL DISTRIBUTIONS AND INTENSITIES OF SOLAR SIMULATORS

6-C-1. The Sun

At first glance it might appear that the sun would be the ideal irradiation source for environmental chambers, and, indeed, it has been used successfully in outdoor chamber facilities (e.g., see Jeffries et al., 1976; Leone et al., 1985). However, there are a number of practical problems. First, it requires either that the chamber be built outdoors or that the building housing the chamber have a suitable opening to admit the sunlight. Under these conditions, independent temperature control of the chamber becomes difficult. Second, the intensity of the sunlight can be altered by passing clouds in a manner that is difficult to measure and describe accurately (see Section 3-A-2g). Third, experiments are limited to days of appropriate meteorology (e.g., rainy days are generally excluded). As an alternative, three types of lamps have been used to mimic irradiation from the sun. These are black fluorescent lamps, sun lamps, and xenon lamps.

6-C-2. Black Lamps

A black lamp is a low-pressure mercury lamp whose envelope is covered with a phosphor such as strontium fluoroborate or barium disilicate. The type of phosphor determines the spectral distribution of the lamp output (Forbes et al., 1976). Figure 6.4 shows a typical spectral distribution for $\lambda \leq 500$ nm from a black lamp as well as the solar spectrum at three zenith angles, where all curves have been normalized to the same NO_2 photolysis rate constant k_1. (See below for a discussion of k_1, which is a measure of the total light intensity.) Superimposed on the broad emission continuum from the phosphor are the low-pressure mercury lines at 313, 334, 365–366, 405–408, and 436 nm.

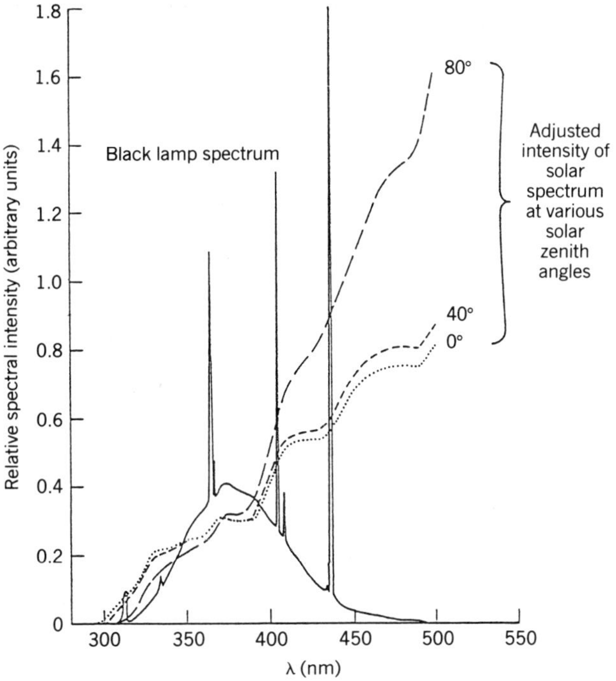

FIGURE 6.4. Relative spectral distributions for a typical black lamp and the solar spectrum, at zenith angles of 0°, 40° and 80° normalized to the same NO_2 photolysis rate constant (from Carter et al., 1984).

While such lamps provide good light intensity in the ~340–400 nm region (and at the 313 nm mercury line) where important atmospheric photochemistry occurs, their spectral distribution is very different than that of the sun. Specifically, much of the intensity resides in the sharp mercury lines, and the output is poor in the critically important actinic UV region from 290 to 340 nm. In addition, the intensity falls off at $\lambda \geq 375$ nm, whereas the intensity of solar radiation is increasing significantly in this region. Such differences may significantly alter the photochemistry of important species such as O_3, NO_3 and HCHO, even if the lamp output is normalized to give the same NO_2 photolysis rate constant (k_1) as the sun.

6-C-3. Sun Lamps

A sun lamp is similar in construction to a black lamp, except that a different type of phosphor is used and the lamp envelope transmits radiation in the UV. Figure 6.5 shows a typical spectral distribution from a commercial sun lamp. The wavelength corresponding to maximum power is shifted to lower wavelengths (~310 nm), compared to black lamps, and there is significant intensity in the photochemically important actinic UV, down to ~270 nm. The mercury lines can again be seen superimposed on the phosphor fluorescence.

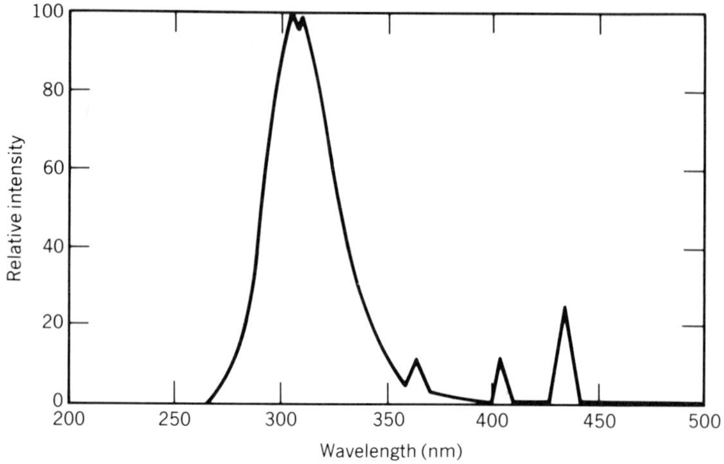

FIGURE 6.5. Typical spectral distribution from a sun lamp (Reprinted with permission of North American Philips Lighting Corporation).

6-C-4. Xenon Lamps

High-pressure xenon lamps provide the most faithful artificial reproduction of the solar energy distribution at the earth's surface in the wavelength region 290–700 nm. Figure 6.6 compares the output of an unfiltered xenon lamp to the zero air mass solar spectral irradiance. Unlike black lamps, xenon lamps have substantial

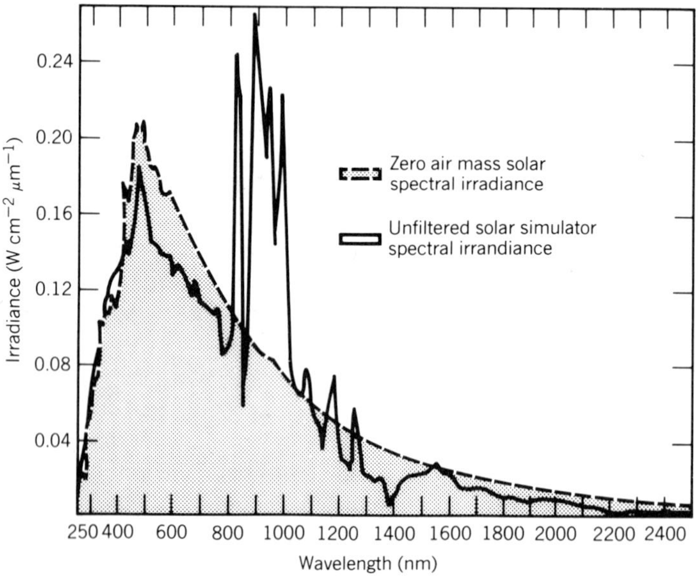

FIGURE 6.6. Spectral irradiation of filtered and unfiltered solar simulator compared to zero air mass solar spectral irradiance (from Winer et al., 1979).

intensity in the critical region around 300 nm; the region ≤290 nm can be filtered to match the solar energy distribution at the earth's surface using Pyrex of varying thickness. In contrast to black lamps, the xenon lamp has substantial intensity at wavelengths above 400 nm.

The Xe lamp also has a series of peaks in the 800–1000 nm region which do not appear in sunlight. This relatively low-energy radiation does not cause significant photochemistry in the troposphere. However, if desired, the intensity of these peaks can be decreased with the use of appropriate filters.

6-C-5. Measurement of Light Intensity

In chamber studies, the spectral distribution of the irradiation source must be measured periodically (e.g., using a calibrated monochromator–photomultiplier combination) because the lamp and the windows in the chamber "age" (i.e., change with time). In addition to the spectral energy distribution of the lamp, the total absolute light intensity also must be measured. In particular, the intensity of the region below 430 nm where the most important photochemistry (e.g., of NO_2, O_3, HCHO) occurs is of greatest interest. Both of these calibrations are tedious and must be carried out with care. However, they are sufficiently critical to data interpretation that they are carried out frequently. For example, measurement of the total absolute light intensity is typically carried out after every four or five runs, and in some cases where knowledge of the light intensity is essential, after every run.

The photolysis rate for NO_2 might be expected to be a good (although non-specific) indicator of the intensity in the region ≲430 nm since it absorbs strongly (see Section 3-C-3) and is also one of the major photochemically active species in NMOC–NO_x systems. Thus a standard procedure in smog chamber studies is to measure the rate of photolysis of NO_2 (k_1) as a relative measure of the total light intensity:

$$NO_2 + h\nu \xrightarrow{k_1} O(^3P) + NO \tag{1}$$

Determining k_1 is not as simple as measuring the loss of NO_2, however, since secondary reactions of the O and NO produced in (1) lead to non-exponential decays of NO_2. Thus plots of ln $[NO_2]$ against irradiation time are observed to be curved. In early smog chamber studies, a parameter known as k_d was reported as a measure of the light intensity, where k_d was defined by

$$k_d = \left(\frac{-d \ln [NO_2]}{dt}\right)_{\lim t \to 0}$$

k_d was thus obtained experimentally by extrapolating the NO_2 concentration–time profile back to the beginning of the irradiation, $t = 0$.

Since k_1 is the fundamental parameter of interest, however, there has been emphasis on measuring and reporting k_1 rather than k_d in smog chamber studies. A procedure for determining k_1 from measured rates of photolysis of NO_2 is described in detail by Holmes and co-workers (1973). In this procedure, NO_2 is photolyzed in the smog chamber. When O_2 is absent (i.e., in 1 atm of N_2), the reactions of interest are (1) and (2)–(7):

$$O + NO_2 \xrightarrow{k_2} NO + O_2 \tag{2}$$

$$O + NO_2 + M \xrightarrow{k_3} NO_3 + M \tag{3}$$

$$O + NO + M \xrightarrow{k_4} NO_2 + M \tag{4}$$

$$NO_3 + NO \rightarrow 2\ NO_2 \tag{5}$$

$$NO_3 + NO_2 \overset{M}{\rightleftarrows} N_2O_5 \tag{6, −6}$$

$$NO_3 + NO_2 \rightarrow NO + NO_2 + O_2 \tag{7}$$

Using appropriate steady state assumptions for O, NO_3, and N_2O_5, the kinetic expressions for these reasons can be solved to yield the following equation for k_1:

$$k_1 = \frac{1}{2t}\left[(1 + R_1 - R_2)\ln\frac{[NO_2]_0}{[NO_2]} + R_2\left(\frac{[NO_2]_0}{[NO_2]} - 1\right)\right]$$

where

$$R_1 = \frac{k_3[M]}{k_2} \quad \text{and} \quad R_2 = \frac{k_4[M]}{k_2}$$

Thus, knowing the total pressure of N_2 and the rate constants k_2, k_3, and k_4, one can determine k_1 from the initial concentration of NO_2 and its loss with time.

The photolysis of NO_2 can also be carried out in the presence of O_2. However, in this case additional reactions (e.g., $O + O_2 + M \rightarrow O_3 + M$) must be considered and the kinetic expression for k_1 is more complex.

Of course, k_1 is not an absolute light intensity measurement per se, but merely an indication of the intensity in the wavelength region of greatest interest for atmospheric chemistry. It has the advantage of being simple, convenient, and inexpensive, since only monitoring instruments for NO_2 are needed and these are generally standard components of the analytical apparatus. The disadvantage is that a number of photochemically active species (e.g., O_3, HCHO) have absorption coefficients and wavelength dependencies different from NO_2 and, for these, k_1 will not necessarily be a good measurement of the rates of their photochemical reactions, depending on the spectral distribution of the light source.

6-D. ANALYTICAL METHODS FOR MAJOR AND TRACE SPECIES

The analytical methods used in smog chamber studies are generally the same as those used in ambient air studies. Thus ports are included for withdrawing samples for GC or GC–MS analysis (e.g., organics) and/or wet chemical analysis (e.g., formaldehyde). Provisions are also generally made for continuous sampling for measurement of NO, NO_2, O_3, and so on.

Included in some chambers (e.g., the evacuable chamber of Fig. 6.3) are ports for in situ spectroscopic measurements using techniques such as FTIR and DOAS (see Chapter 5). Because of the relatively small pathlength across such chambers, multiple pass optical systems must be used. Examples of the use of spectroscopic detection in chamber studies are discussed below and throughout the book.

6-E. TIME–CONCENTRATION PROFILES FOR TYPICAL NMOC–NO_x–AIR IRRADIATIONS IN SMOG CHAMBERS

6-E-1. Reactants and Products in Typical Chamber Runs

Figure 6.7 shows some typical results of an irradiation of a propene–NO_x mixture in the evacuable chamber of Fig. 6.3. This shows the loss of the reactants and the formation of the most commonly monitored secondary pollutants O_3, PAN, and the oxygenates HCHO and CH_3CHO.

With the addition of in situ spectroscopic techniques, critical data on the formation of such species as HONO, HNO_3, and NO_3 which are essential to understanding the chemistry of these systems can now be obtained. Figure 6.8, for

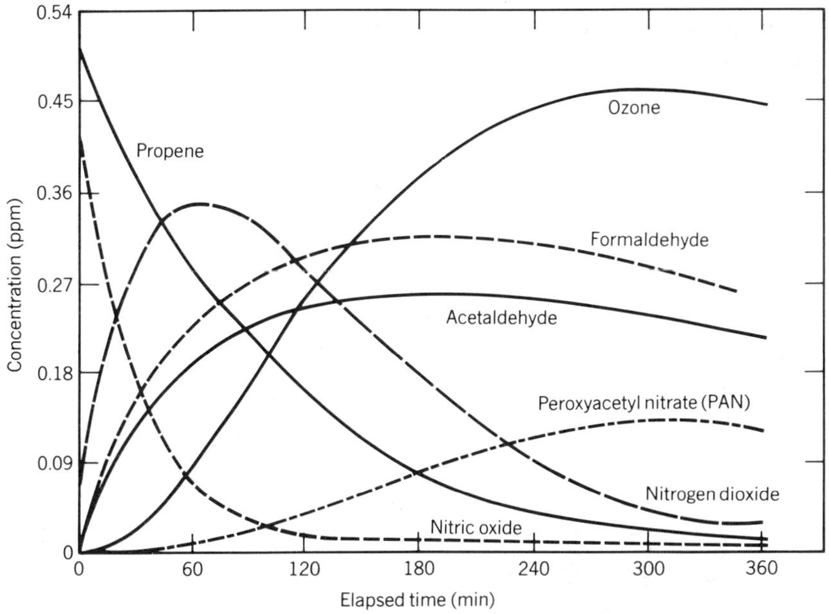

FIGURE 6.7. Typical primary and secondary pollutant profiles in a propene–NO_x irradiation in a smog chamber (from Pitts et al., 1975).

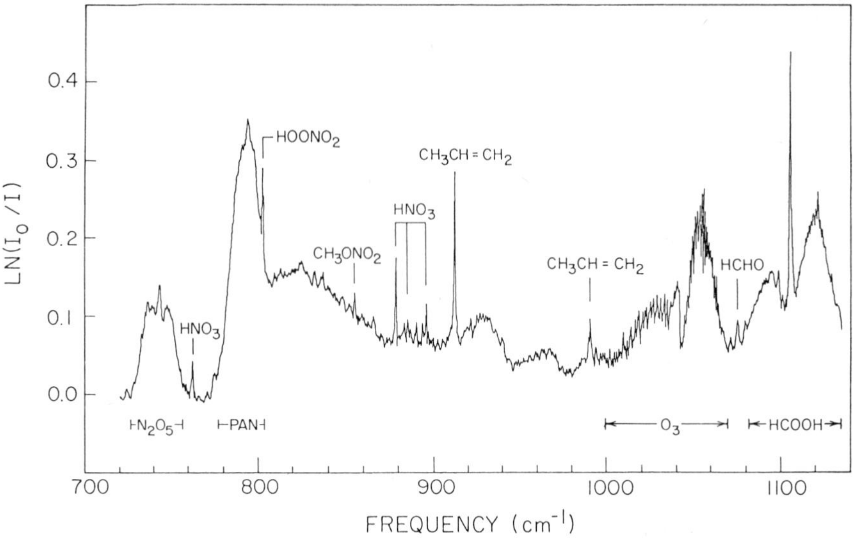

FIGURE 6.8. Infrared spectrum in the 700–1100 cm^{-1} region of a hydrocarbon–NO_x mixture irradiated for 139 min. Initial conditions were 10 ppm propene, 1 ppm *n*-butane, 1 ppm neopentane, and 5.4 ppm NO_x at 48°F. Pathlength, 85 m; resolution, 0.125 cm^{-1} (from Tuazon, unpublished data and Pitts et al., 1977).

example, shows one portion of an FTIR spectrum obtained in a chamber run for a propene–NO_x mixture carried out at relatively high reactant concentrations and 48°F. A variety of species which would be difficult (or impossible) to monitor with other techniques, such as HNO_3 and peroxynitric acid (HO_2NO_2), are easily identified and measured by FTIR. Clearly, such data can be used to examine the

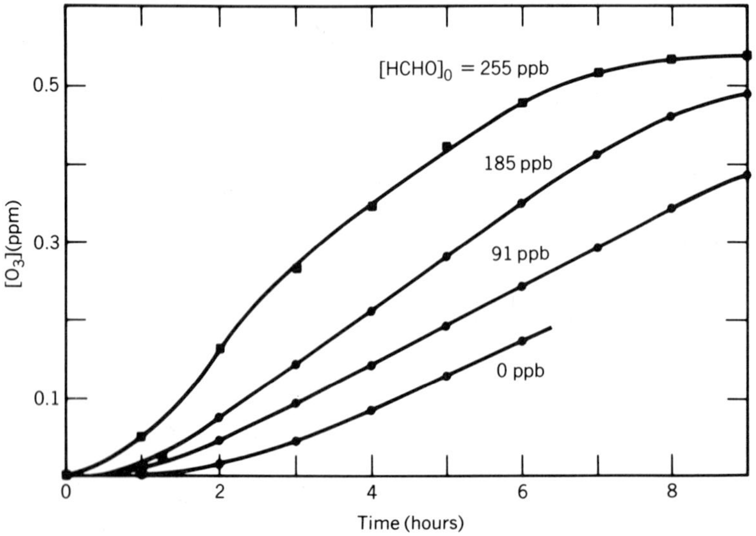

FIGURE 6.9. Effect of added HCHO on ozone formation in irradiations of a hydrocarbon–NO_x mixture. Average NMHC was 2450 ppbc; average NO_x was 0.33 ppm (from Pitts et al., 1976).

relationship between primary emissions and the formation of a host of secondary pollutants. For example, runs can be carried out at varying initial concentrations of hydrocarbon and NO_x, and the effects on the formation of secondary pollutants such as O_3 studied (see Chapter 10). The reactivity of various hydrocarbons can be examined by studying them singly or in combination. In addition, such parameters as temperature, relative humidity and total pressure, presence of co-pollutants, and spectral distribution of the light source can be systematically varied.

6-E-2. Effects of Added HCHO

One example of the use of chambers to study the effects of addition of co-pollutants is seen in Fig. 6.9. As discussed in Section 3-D-2, one source of HO_2 free radicals in ambient air is the photolysis of formaldehyde:

$$HCHO + h\nu \rightarrow H + HCO$$
$$H + O_2 \xrightarrow{M} HO_2$$
$$HCO + O_2 \rightarrow HO_2 + CO$$

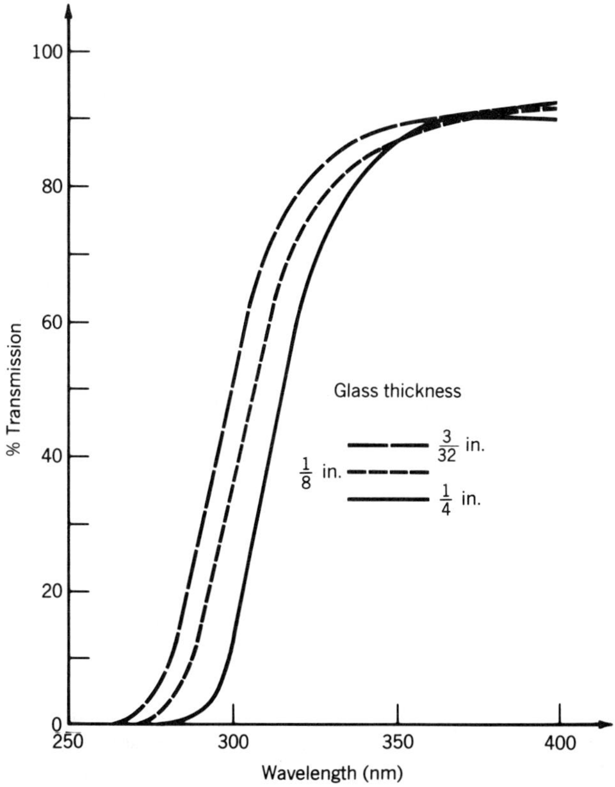

FIGURE 6.10. Transmission spectra of three Pyrex filters of different thicknesses (from Winer et al., 1979).

One might anticipate that the addition of HCHO prior to irradiation would increase the rate of conversion of the reactants to secondary pollutants such as O_3 by providing an immediate source of the free radicals needed for the chain oxidations. Indeed, as seen in Fig. 6.9, this is precisely what is observed.

6-E-3. Effects of Actinic UV Irradiation

Chambers also allow such parameters as the spectral distribution of the light source to be varied systematically. While this is impossible in ambient air studies, it is an important variable in order to simulate atmospheric chemistry at various altitudes. Since Pyrex glass absorbs radiation with $\lambda \lesssim 350$ nm (Calvert and Pitts, 1966), different thicknesses of Pyrex can be used to provide different amounts of filtering in this portion of the actinic UV. Figure 6.10, for example, shows the percent transmission of three different thicknesses of glass ($\frac{3}{32}$ in., $\frac{1}{8}$ in., and $\frac{1}{4}$ in.) in the 250–400 nm region; as expected, the thicker the glass, the less transmitting it is below 350 nm. Since much of the important photochemistry (e.g., O_3, HCHO) occurs in this region, one might expect the rates of free radical formation and hence the overall rates of product formation to be less for the most filtered case.

Figure 6.11 shows the formation of O_3 in a chamber as a function of time for a propene–*n*-butane–NO_x mixture irradiated using a xenon lamp (see below) with

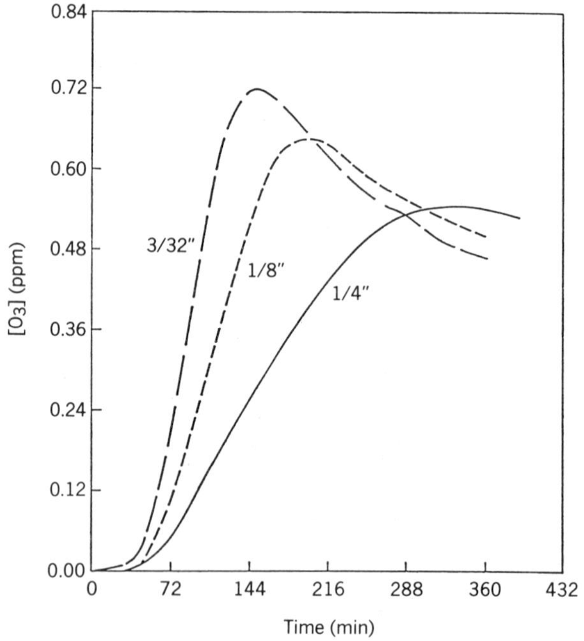

FIGURE 6.11. Ozone–time profiles from the irradiation of a hydrocarbon-NO_x mixture (initial concentrations were 0.4 ppm propene, 2.0 ppm *n*-butane, 0.4 ppm NO and 0.1 ppm NO_2) in a smog chamber where the light source has been filtered with Pyrex of three different thickenesses, $\frac{3}{32}$ in., $\frac{1}{8}$ in., and $\frac{1}{4}$ in., respectively (from Winer et al., 1979).

three glass filters of different thicknesses. As expected, O_3 is formed slower and the peak concentration is lower for the most highly filtered light source.

In summary, chamber studies are a highly valuable experimental technique for studying atmospheric chemistry and the effects of varying parameters under controlled conditions.

6-F. ADVANTAGES AND LIMITATIONS OF CHAMBER STUDIES

Clearly, smog chamber studies are very useful tools in examining the chemical relationships between emissions and air quality and for carrying out related (e.g., exposure) studies. Use of these chambers has permitted the systematic variation of individual parameters under controlled conditions, unlike ambient air studies where the continuous injection of pollutants and the effects of meteorology are difficult to measure and to quantitatively incorporate in the data analysis. Chamber studies have also provided the basis for the validation of computer kinetic models (Chapter 9). Finally, they have provided important kinetic and mechanistic information on some of the individual reactions occurring during photochemical smog formation.

However, as is true of any experimental system, there are limitations to chamber studies. One major problem has proven to be chamber contamination (e.g., Bufalini et al., 1977). For example, when clean air is irradiated in a smog chamber that has been in use, many of the characteristics of photochemical air pollution are observed, for example, nitrogen-containing compounds, and sometimes organics are observed, and O_3 is formed. This is attributed to the adsorption of nitrogen compounds and organics on the walls of the smog chamber during a run, followed by desorption into the gas phase at a later time.

Indeed, offgassing of nitrogenous inorganics has been measured from the Teflon wall of the evacuable chamber shown in Fig. 6.3 (Carter, 1981). After a typical series of runs, the chamber was filled with pure air at ~5% relative humidity and air samples then removed for analysis. Offgassing of NO and a second nitrogen-containing species, probably HNO_3, were observed; only in one experiment was a trace of HONO observed. The rate of offgassing increased with temperature but fell substantially after an evacuated bakeout. Interestingly, in this particular case, no release of organics was observed, although it has been observed in other chambers.

Clearly, such adsorption–desorption processes on the surfaces of chambers potentially can have substantial effects on the observed levels of O_3 and other trace pollutants and on their rates of formation. While such effects can be minimized using bakeout while pumping if the chamber is evacuable, relatively few smog chambers have such capabilities at present. Even for evacuable chambers, contamination from adsorption on, and desorption from, the walls occurs. How to correct the results for this and reliably extrapolate the data to "real" atmospheres remains a problem.

A second major source of concern in recent years has been the accumulation of indirect evidence pointing to the occurrence of chemical reactions on the walls of the chamber followed by desorption of some of the products into the gas phase. These reactions manifest themselves by changing the observed gas phase concentrations of one or more species in a manner that cannot be explained solely on the basis of known homogeneous gas phase reactions.

For example, Fig. 6.12 shows the concentrations of OH as a function of irradiation time when an NMOC–NO_x mixture was irradiated in the evacuable smog chamber shown in Fig. 6.3 (Carter et al., 1981, 1982). The rates of decay of propane and propene were used to estimate the concentration of OH shown by the horizontal bars in Fig. 6.12, as a function of reaction time.

In this system the concentrations of propane, propene, and formaldehyde were kept sufficiently small that the predominant reactions were those of the NO_x species. By keeping the NO concentrations high, O_3 concentrations could be kept low because reaction (8) is fast:

$$NO + O_3 \rightarrow NO_2 + O_2 \qquad (8)$$

Under these conditions, reactions (1)–(8) above and (9)–(13) below are the major reactions to be considered:

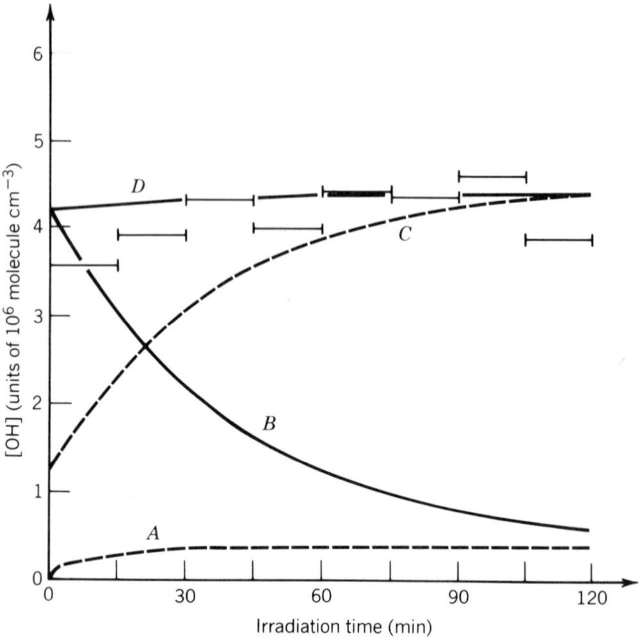

FIGURE 6.12. Concentrations of OH as a function of irradiation time during irradiation of a NMOC–NO_x mixture in the evacuable smog chamber of Fig. 6.3. The initial conditions were as follows: $[NO]_0$ = 0.499 ppm, $[NO_2]_0$ = 0.115 ppm, $[C_3H_8]_0$ = 0.013 ppm, $[C_3H_6]_0$ = 0.010 ppm, $[HCHO]_0$ = 0.02 ppm, T = 303°K, RH = 50%, k_1 (NO_2 photolysis) = 0.49 min^{-1}. Bars are experimental data; see text for description of curves *A–D* (from Carter et al., 1982).

$$O + O_2 + M \rightarrow O_3 + M \qquad (9)$$

$$2\ NO + O_2 \rightarrow 2\ NO_2 \qquad (10)$$

$$OH + NO + M \rightarrow HONO + M \qquad (11)$$

$$HONO + h\nu \rightarrow OH + NO \qquad (12)$$

$$OH + NO_2 + M \rightarrow HNO_3 + M \qquad (13)$$

The rate constants for all these reactions are reasonably well established, so that the concentrations of OH expected can be predicted with some degree of reliability. (This would not be true if the concentration of organics was high enough to perturb the NO_x chemistry because of the many uncertainties in the mechanisms of the organic reactions.)

The dashed line *A* in Fig. 6.12 shows the OH concentrations predicted using only the homogeneous gas phase chemistry of reactions (1)–(13). Clearly, much larger concentrations of OH are observed than can be rationalized on the basis of gas phase chemistry alone.

As discussed in Chapter 3, one source of OH is the photolysis of HONO, which as discussed earlier, is thought to be formed at least partially on surfaces by reaction (14),

$$2\ NO_2 + H_2O \xrightarrow{\text{surfaces}} HONO + (HNO_3?) \qquad (14)$$

If one assumes that 10 ppb of HONO were present initially, the OH radical profile would follow curve *B*. This is in agreement with the initial OH concentrations, but not with the larger concentrations at later times. The shape of curve *B* is expected since any initial HONO present will rapidly photolyze, producing a "pulse" of OH; however, after the HONO has photolyzed, lower concentrations of OH follow.

The shape of the observed OH profile suggests that some unrecognized source of OH must be present. Curve *C* is the predicted OH profile if it is assumed that a constant OH source exists during the run which produces OH at a rate of 0.245 ppb min^{-1}. In contrast to curve *B*, this leads to agreement with the observed OH levels and curve shape at long reaction times, but not at short times. (The lower OH concentrations predicted at short reaction times are due to the net reaction of OH with NO while HONO is building up to its equilibrium concentration).

Combining curves *B* and *C* (i.e., using both the assumption of an initial HONO concentration as well as a constant radical source) one obtains curve *D*, which matches the observations quite well.

While a constant flux of OH was assumed in generating curves *C* and *D* of Fig. 6.12, this is obviously only a simplified construct taken to represent some species that forms OH in subsequent reactions. The flux of this unknown species appears to depend on the particular chamber and chamber history and is roughly proportional to the light intensity. It also increases significantly with temperature, relative humidity, and NO_2 concentrations, but is independent of total pressure and the NO concentration (Carter et al., 1982).

At present, the identity and mechanism of formation of the precursor to OH are unknown and subject to some controversy (Killus and Whitten, 1981; Besemer and Nieboer, 1985; Carter et al., 1985; Leone et al., 1985), although HONO appears to be the most likely candidate. The dependence of its flux on the particular chamber characteristics suggests the involvement of unrecognized heterogenous reactions on the chamber surfaces. However, without knowing the source of the increased OH flux, extrapolation of the concentration–time profiles of both the primary and secondary pollutants observed in such smog chamber studies to real atmospheres is highly uncertain.

For example, the reactions leading to the unknown precursor to OH may occur only in smog chambers. Extrapolation to ambient air would thus require subtracting out this radical source. On the other hand, the same reactions may occur in ambient air where surfaces are available in the form of particulate matter, buildings, the earth, and so on; if this is true, then the rates would be expected to depend on the nature and types of surfaces available and may thus differ quantitatively from the smog chamber observations.

While chamber contamination and the presence of unknown surface reactions are probably the most important problems in interpreting smog chamber data and extrapolating to atmospheric conditions, other minor problems exist as well. These include the need to measure carefully and frequently a number of chamber-specific parameters such as the decay rate of O_3 on the chamber walls and the initial formation of HONO. Such chamber-specific parameters raise the question again of how best to modify these parameters to describe ambient air.

In summary, despite these complications smog chambers have proven extremely useful in studying the chemistry of photochemical air pollution under controlled conditions in which emissions and meteorology are not complicating factors. While there are some uncertainties and limitations in quantitatively extrapolating the results to ambient air, we may find in the future that what appear to be chamber-specific complications may in fact apply in ambient air as well.

REFERENCES

Akimoto, H., M. Hoshino, G. Inoue, F. Sakamaki, N. Washida and M. Okuda, "Design and Characterization of the Evacuable and Bakable Photochemical Smog Chamber," *Environ. Sci. Technol.*, **13,** 471 (1979).

Besemer, A. C. and H. Nieboer, "The Wall as a Source of Hydroxyl Radicals in Smog Chambers," *Atmos. Environ.*, **19,** 507 (1985).

Bufalini, J. J., T. A. Walter, and M. M. Bufalini, "Contamination Effects on Ozone Formation in Smog Chambers," *Environ. Sci. Technol.*, **11,** 1181 (1977).

Calvert, J. G. and J. N. Pitts, Jr., *Photochemistry,* Wiley, New York, 1966.

Carter, W. P. L., unpublished results, 1981.

Carter, W. P. L., R. Atkinson, A. M. Winer, and J. N. Pitts, Jr., "Evidence for Chamber-Dependent Radical Sources: Impact on Kinetic Computer Models for Air Pollution," *Int. J. Chem. Kinet.*, **13,** 735 (1981).

Carter, W. P. L., R. Atkinson, A. M. Winer, and J. N. Pitts, Jr., "Experimental Investigation of Chamber-Dependent Radical Sources," *Int. J. Chem. Kinet.*, **14,** 1071 (1982).

Carter, W. P. L., R. Atkinson, A. M. Winer, and J. N. Pitts, Jr., "Experimental Protocol for Determining Photolysis Reaction Rate Constants," Report to the U.S. Environmental Protection Agency, Contract No. R80666-01, 1984.

Carter, W. P. L., R. Atkinson, A. M. Winer and J. N. Pitts Jr., "Comments on 'The Wall as a Source of Hydroxyl Radicals in Smog Chambers'," *Atmos. Environ.*, **19,** 000 (1985).

Darnall, K. R., W. P. L. Carter, A. C. Lloyd, A. M. Winer and J. N. Pitts Jr., "Importance of RO_2 + NO in Alkyl Nitrate Formation from $C_4 - C_6$ Alkane Photooxidation Under Simulated Atmospheric Conditions," *J. Phys. Chem.*, **80,** 1948 (1976).

Dimitriades, B., "Methodology in Air Pollution Studies Using Irradiation Chambers," *J. Air Pollut. Control. Assoc.*, **17,** 460 (1967).

Doyle, G. J., P. J. Bekowies, A. M. Winer, and J. N. Pitts, Jr., "Charcoal-Adsorption Air Purification System for Chamber Studies Investigating Atmospheric Photochemistry," *Environ. Sci. Technol.*, **11,** 45 (1977).

Fitz, D. R., M. C. Dodd and A. M. Winer, "Photooxidation of α-Pinene at Near-Ambient Concentrations Under Simulated Atmospheric Conditions," Paper No. 81-27.3, 74th Annual Meeting of the Air Pollution Control Association, Philadelphia, Pa., June 21–26, 1981.

Forbes, P. D., R. E. Davies, L. C. D'Aloisio, and C. Cole, "Emission Spectrum Differences in Fluorescent Blacklight Lamps," *Photochem. Photobiol.*, **24,** 613 (1976).

Holmes, J. R., R. J. O'Brien, J. H. Crabtree, T. A. Hecht, and J. H. Seinfeld, "Measurement of Ultraviolet Radiation Intensity in Photochemical Smog Studies," *Environ. Sci. Technol.*, **7,** 519 (1973), and references therein.

Jeffries, H. E., D. L. Fox, and R. Kamens, "Photochemical Conversion of NO to NO_2 by Hydrocarbons in an Outdoor Chamber," *J. Air Pollut. Control Assoc.*, **26,** 480 (1976).

Joshi, S. B., M. C. Dodge and J. J. Bufalini, "Reactivities of Selected Organic Compounds and Contamination Effects," *Atmos. Environ.*, **16,** 1301 (1982).

Kamens, R., D. Bell, A. Dietrich, J. Perry, R. Goodman, L. Claxton, and S. Tejada, "Mutagenic Transformations of Dilute Wood Smoke Systems in the Presence of Ozone and Nitrogen Dioxide. Analysis of Selected High-Pressure Liquid Chromatography Fractions from Wood Smoke Particle Extracts," *Environ. Sci. Technol.*, **19,** 63 (1985).

Kelly, N. A., "Characterization of Fluorocarbon-Film Bags as Smog Chambers," *Environ. Sci. Technol.*, **16,** 763 (1982).

Kelly, N. A., K. L. Olson, and C. A. Wong, "Tests for Fluorocarbon and Other Organic Vapor Release by Fluorocarbon Film Bags," *Environ. Sci. Technol.*, **19,** 361 (1985).

Killus, J. P. and G. Z. Whitten, "Comments on 'A Smog Chamber and Modeling Study of the Gas Phase NO_x–Air Photooxidation of Toluene and the Cresols'," *Int. J. Chem. Kinet.*, **13,** 1101 (1981).

Leone, J. A., R. C. Flagan, D. Grosjean and J. H. Seinfeld, "An Outdoor Smog Chamber and Modeling Study of Toluene-NO_x Photooxidation," *Int. J. Chem. Kinet.*, **17,** 177 (1985).

Lonneman, W. A., J. J. Bufalini, R. L. Kuntz, and S. A. Meeks, "Contamination from Fluorocarbon Films," *Environ. Sci. Technol.*, **15,** 99 (1981).

Pitts, J. N., Jr., A. C. Lloyd, and J. L. Sprung, "Ecology, Energy and Economics," *Chem. Br.* **11,** 247 (1975).

Pitts, J. N., Jr., A. M. Winer, K. R. Darnall, G. J. Doyle, and J. M. McAfee, "Chemical Consequences of Air Quality Standards and of Control Implementation Programs: Role of Hydrocarbons, Oxides of Nitrogen and Aged Smog in the Production of Photochemical Oxidant," Final Report to the California Air Resources Board, Contract No. 4-214, May 1976.

Pitts. J. N., Jr., B. J. Finlayson-Pitts, and A. M. Winer, "Optical Systems Unravel Smog Chemistry," *Environ. Sci. Technol.*, **11,** 568 (1977).

Pitts, J. N., Jr., E. Sanhueza, R. Atkinson, W. P. L. Carter, A. M. Winer, G. W. Harris, and C. N.

Plum, ''An Investigation of the Dark Formation of Nitrous Acid in Environmental Chambers,'' *Int. J. Chem. Kinet.*, **16,** 919 (1984).

Pitts, J. N., Jr., T. J. Wallington, H. W. Biermann, and A. M. Winer, ''Identification and Measurement of Nitrous Acid in an Indoor Environment,'' *Atmos. Environ.*, **19,** 763 (1985).

Sakamaki, F., S. Hatakeyama, and H. Akimoto, ''Formation of Nitrous Acid and Nitric Oxide in the Heterogeneous Dark Reaction of Nitrogen Dioxide and Water Vapor in a Smog Chamber,'' *Int. J. Chem. Kinet.*, **15,** 1013 (1983).

Spicer, C. W., ''Smog Chamber Studies of NO_x Transformation Rate and Nitrate/Precursor Relationships,'' *Environ. Sci. Technol.*, **17,** 112 (1983).

Winer, A. M., G. M. Breuer, W. P. L. Carter, K. R. Darnall, and J. N. Pitts, Jr., ''Effects of Ultraviolet Spectral Distribution on the Photochemistry of Simulated Polluted Atmospheres,'' *Atmos. Environ.*, **13,** 989 (1979).

Winer, A. M., R. A. Graham, G. J. Doyle, P. J. Bekowies, J. M. McAfee, and J. N. Pitts, Jr., ''An Evacuable Environmental Chamber and Solar Simulator Facility for the Study of Atmospheric Photochemistry,'' *Adv. Environ. Sci. Technol.*, **10,** 461 (1980).

PART 4

Kinetics and Mechanisms of Gas Phase Reactions in Real and Simulated Atmospheres

7 Rates and Mechanisms of Gas Phase Reactions of Hydrocarbons and Oxygen-Containing Organics with Labile Species Present in Irradiated Organic–NO_x–Air Mixtures

7-A. INTRODUCTION

In previous chapters we described the experimental techniques employed to measure various parameters, for example, relative and absolute rate constants as well as intermediates, products, and mechanisms, which are important in homogeneous gas phase atmospheric processes. We now present the results from a wide range of kinetic and mechanistic studies employing these experimental methods and show how they are utilized in determining the lifetimes and fates of important classes of biogenic and anthropogenic organics released into the natural and polluted troposphere. Thus in this chapter we treat the reactions of highly reactive trace atmospheric constituents such as OH radicals with hydrocarbons and oxygen-containing organics, which we shall refer to as oxygenates. Reactions of some halogen-containing species are also included; the long-lived halogenated compounds of importance to stratospheric chemistry are discussed in more detail in Chapter 15. In Chapter 8 we consider nitrogen-containing organics, and the concurrent reactions of the inorganic nitrogenous pollutants present in these mixtures. Organics containing sulfur atoms are dealt with in Chapter 14.

In citing rate constant and associated data relevant to these reactions, we generally use the latest recommended values from several sources, including the reviews of Atkinson and Lloyd (1984), Baulch et al. (1984), Atkinson and Carter (1984), and Atkinson (1985), and the tropospheric and stratospheric atmospheric kinetic data reviews cited in Section 4-F of Chapter 4. When these are not available, we cite those values which, in our judgment, seem most reasonable. Unless stated otherwise, concentrations continue to be in units of molecules cm^{-3} and time in seconds.

7-B CALCULATED TROPOSPHERIC LIFETIMES OF REPRESENTATIVE ORGANICS

As we have seen, the troposphere contains, in addition to O_3, a variety of trace, but highly reactive, oxidizing species; one or more of these may participate in the daytime photooxidation of natural or anthropogenically emitted organics. Labile species include the hydroxyl (OH), hydroperoxyl (HO_2), and alkylperoxy (RO_2) radicals, ground-state [$O(^3P)$] and electronically excited [$O(^1D)$] oxygen atoms, and electronically excited singlet molecular oxygen [$O_2(^1\Delta_g)$]. Recently, the gaseous nitrate radical (NO_3) has been detected in urban and rural airsheds and, as seen in Sections 7-E-4 and 7-H-3, its nighttime reactions now appear to be a further atmospheric sink for certain classes of organics.

Adding to the complexity of the atmospheric system is the fact that there are hundreds, if not thousands, of organics present in the natural and polluted troposphere, including alkanes, alkenes, alkynes, and aromatics, and their derivatives containing oxygen, nitrogen, sulfur, and halogen atoms. Furthermore, depending on the physical state of the organic, for example, gaseous or adsorbed on the surface of an aerosol particle, the reactions may be homogeneous and/or heterogeneous in nature.

Clearly, given this vast array of possible reactions, the determination of those specific processes that may have significant impact on the physical, chemical, and biological properties of a given air parcel—remote or polluted—is a formidable task. However, particularly since around 1970 and the discovery of the key role of the OH radical in atmospheric photooxidations, our knowledge of the kinetics and mechanisms of the elementary reactions of the labile oxidizing species with organics and inorganics has increased dramatically. For example, we now have a reasonably extensive data bank of accurate absolute rate constants for gas phase reactions of O_3, OH, and ground-state $O(^3P)$ and electronically excited $O(^1D)$ atoms; rate constants for the reactions of NO_3 are becoming increasingly available. Furthermore, for OH reactions, we have an emerging understanding of structure–reactivity relationships for various classes of organics. Such information now permits one to estimate with some confidence the relative contributions of each of the oxidizing species to the net rates of removal of most organics and inorganics, such as SO_2 or NO_2, from the atmosphere by homogeneous gas phase processes. This, of course, narrows down significantly the number of different types of reactions that should be considered when evaluating the chemistry of a specific air mass, whether it is the plume of a fossil-fuel-fired power plant, a parcel of ambient air, or the interface between the two.

Let us illustrate this point by calculating the tropospheric lifetimes (τ) of some "model" organics of anthropogenic origin. Recall from Section 4-A-6 that for a second-order reaction between an organic, A, and an oxidizing species, Ox,

$$A + Ox \xrightarrow{k_{ox}} \text{products}$$

$\tau_A = 1/k_{ox}[Ox]$. Thus to evaluate τ, we need not only the value of the absolute rate constant, k_{ox}, but estimates of the concentrations of the specific oxidant involved. For the purposes of this discussion, we have chosen the maximum values for $[O_3]$, [OH], $[HO_2]$, and $[O(^3P)]$ *calculated* from a computer simulation of an irradiated propylene–n-butane–NO_x mixture. The concentration assumed for NO_3 radicals, 1.1×10^{10} cm^{-3} (430 ppt), corresponds to one of the highest *experimentally* observed to date in an urban atmosphere (Atkinson et al., 1985a). These are very high concentrations, representative of the peaks which in most cases will not exist for substantial periods of time and hence are not representative of averages even in heavily polluted areas; in addition, these calculations ignore the fact that each oxidizing species may be important at different times of the day, under different conditions, or at different stages of the overall reaction. We use them merely to illustrate that some reactions in the matrix of organics and oxidizing species in the atmosphere are sufficiently slow that they can be omitted from further consideration. As discussed in Section 5-G, however, OH concentrations up to 1×10^7 cm^{-3} in ambient air have been reported (although the uncertainties in the reported values are very large due to potential interferences), and peak concentrations of O_3 of 200 ppb are common in polluted urban atmospheres. Of course if one took more typical "average" values for the concentrations of these oxidants, for example, 1×10^6 cm^{-3} for OH and 1×10^9 cm^{-3} for NO_3, the lifetimes of the organics would be an order of magnitude longer.

While our choices of the concentrations of the various oxidants are somewhat arbitrary, it is clear from the data in Table 7.1 that, during daylight hours in ambient air, OH radicals are a prime oxidizing species for all the organics. Interestingly, one measurement of k_{HO_2} suggests that the HO_2 radical competes with OH in the oxidation of formaldehyde.

The olefin, *trans*-2-butene, is attacked rapidly by O_3 and OH and NO_3 radicals with lifetimes of 17, 26, and 4 min, respectively. However, it should be kept in mind that O_3 attack can occur 24 hours a day, while that of OH radicals is a daytime photooxidation phenomenon because photolysis is needed to produce OH. Reactions of the NO_3 radical are only important at night because it readily photodissociates in sunlight and daytime steady state concentrations are very low (see Section 3-C-7b).

The organics and their lifetimes cited in Table 7.1 are, of course, only examples. As we shall see, they should not be taken as representative of *all* compounds in their classes nor of all atmospherically important processes. Thus we have not considered here the direct photolysis of those organics such as HCHO which absorb actinic UV radiation; this can be a major atmospheric loss process. Further-

TABLE 7.1. Calculated Minimum Tropospheric Lifetimes[a] (τ) of Some Hydrocarbons and Oxygenates with Respect to Oxidation by Calculated Peak Concentrations[c] of OH, O_3, HO_2, $O(^3P)$, and Experimentally Measured Peak Concentrations of NO_3 Radicals

	Oxidizing Species[b,c]				
Concentration (molecules cm^{-3})	O_3 5×10^{12} (200 ppb)	OH 1×10^{7} (0.4 ppt)	HO_2 2×10^{9} (81 ppt)	$O(^3P)$ 8×10^{4} (0.003 ppt)	NO_3[m] 1.1×10^{10} (430 ppt)
n-Butane					
τ	$\gtrsim 6.5 \times 10^2$ yr	11 h	$\sim 10^3$ yr	18 yr	29 days
(k)	($\lesssim 9.8 \times 10^{-24}$)[e]	(2.5×10^{-12})[l]	(1.6×10^{-20})[h]	(2.2×10^{-14})[d]	(3.6×10^{-17})[i]
trans-2-Butene					
τ	17 min	26 min	$\geqslant 4$ yr	6.3 days	4 min
(k)	(2.0×10^{-16})[e]	(6.5×10^{-11})[l]	($\leqslant 4 \times 10^{-18}$)[g]	(2.3×10^{-11})[d]	(3.8×10^{-13})[f]
Acetylene					
τ	$\gtrsim 2 \times 10^2$ days	1.5 days	n.d.[n]	2.5 yr	n.d.[n]
(k)	($\lesssim 10^{-20}$)[e]	(7.8×10^{-13})[l]		(1.6×10^{-13})[j]	
Toluene					
τ	$\gtrsim 2 \times 10^2$ days	4.5 h	n.d.[n]	5.6 yr	33 days
(k)	($\lesssim 10^{-20}$)[e]	(6.2×10^{-12})[l]		(7.1×10^{-14})[d]	(3.2×10^{-17})[k]

Formaldehyde					
τ	$\gtrsim 3.2 \times 10^3$ yr	3.1 h	1.9 h	2.5 yr	1.8 days
(k)	$(\leqslant 2 \times 10^{-24})^e$	$(9.0 \times 10^{-12})^l$	$(7.5 \times 10^{-14})^o$	$(1.6 \times 10^{-13})^d$	$(5.8 \times 10^{-16})^k$

[a]The natural lifetime, $\tau = 1/k[\mathrm{Ox}]$, where k is the rate constant for the reaction of the organic with the oxidizing species Ox, where concentration is [Ox] in molecules or radicals cm^{-3}. τ = the time for the concentration of the organic to drop to $1/e = 1/2.72$ of its initial concentration.

[b]The numbers in parentheses below the lifetimes are the rate constants, k, at 298°K, in units of cm^3 $molecule^{-1}$ s^{-1}.

[c]These are the *peak* concentrations predicted for a simulated atmosphere, *calculated* using a propene–*n*-butane model based on the mechanism of Atkinson and Lloyd (1984), with an initial reactive hydrocarbon concentration of 0.60 ppm and an initial NO_x of 0.060 ppm. The NO_3 concentration was taken to be 1.1×10^{10} cm^{-3} (430 ppt), the maximum nighttime level *experimentally* observed to date in ambient air studies (Atkinson et al., 1985a). If one takes typical average levels of 1×10^6 OH and 1.0×10^9 NO_3 radicals cm^{-3}, the corresponding "average" lifetimes are, of course, an order of magnitude longer.

[d]Recommended by Atkinson and Lloyd, 1984.

[e]From Atkinson and Carter, 1984.

[f]From Ravishankara and Mauldin, 1985.

[g]From Graham et al., 1979.

[h]From Lloyd, 1974.

[i]From Atkinson et al., 1984e, adjusted as described in Section 7-C-2 for $K_{eq} = 3.35 \times 10^{-11}$ cm^3 $molecule^{-1}$ $= [N_2O_5]/[NO_3][NO_2]$.

[j]From Herron and Huie, 1973.

[k]From Atkinson et al., 1984a, adjusted as described in Section 7-C-2 for $K_{eq} = 3.35 \times 10^{-11}$ cm^3 $molecule^{-1}$ $= [N_2O_5]/[NO_3][NO_2]$.

[l]Recommended by Atkinson, 1985.

[m]NO_3 reactions are significant only at night since NO_3 photodecomposes readily during daylight hours.

[n]n.d. = not determined.

[o]Veyret et al., 1982.

TABLE 7.2. Estimated Tropospheric Lifetimes, τ, for Phenol and Two Naturally Occurring Organics, α-Pinene and Dimethyl Sulfide, in Heavily Polluted and Remote Atmospheres[a]

	Oxidizing Species and Concentrations (molecules cm^{-3})					
	O_3		OH		NO_3	
Organic	5×10^{12} (200 ppb)	1×10^{12} (40 ppb)	1×10^{7} (0.4 ppt)	5×10^{5} (0.02 ppt)	1.1×10^{10} (430 ppt)	2.5×10^{8} (10 ppt)
Phenol						
τ	>4.6 days	>23 days	1 h	20 h	24 s	18 min
(k)	($\leqslant 5 \times 10^{-19}$)[f]		(2.8×10^{-11})[g]		(3.8×10^{-12})[c,i]	
α-Pinene						
τ	33 min	2.8 h	31 min	10 h	15 s	11 min
(k)	($\sim 1 \times 10^{-16}$)[b]		(5.3×10^{-11})[h]		(6.1×10^{-12})[d,i]	
Dimethyl sulfide CH_3SCH_3						
τ	>3 days	>15 days	4.4 h	3.7 days	1.6 min	1.1 h
(k)	($< 8 \times 10^{-19}$)[b]		(6.3×10^{-12})[h]		(9.7×10^{-13})[e,i]	

[a] For comparison, O_3, OH, and NO_3 levels chosen for a ''polluted'' atmosphere were those used in Table 7.1; those for the natural troposphere are approximations (see Section 5-G).

[b] See review of Atkinson and Carter, 1984, and references therein.

[c] Atkinson et al., 1984c.

[d] Atkinson et al., 1984b.

[e] Atkinson et al., 1984d.

[f] Not reported; approximate upper-limit estimate given here is based on data for the cresols (see Atkinson and Carter, 1984).

[g] Zetzsch, 1982; Rinke and Zetzsch, 1984.

[h] Recommended by Atkinson, 1985.

[i] Rate constants adjusted as described in Section 7-C-2 for $K_{eq} = 3.35 \times 10^{-11}$ cm^3 $molecule^{-1}$ $= [N_2O_5]/[NO_3][NO_2]$.

more, structural effects in a homologous series can significantly alter the relative rates of attack by the oxidizing species (e.g, $k_{OH} = 2.5 \times 10^{-12}$ for n-C_4H_{10} versus 0.27×10^{-12} cm^3 molecule^{-1} s^{-1} for ethane).

We have not included in Table 7.1 several classes of anthropogenic and biogenic organics that are important in the troposphere. For example, as seen in Table 7.2, the NO_3 radical not only reacts rapidly with phenol, but it is sufficiently reactive to be a major nighttime loss process for such biogenic organics as α-pinene and dimethyl sulfide [$(CH_3)_2S$].

Finally, the calculations reflect only the contribution of a specific oxidizing species to the *overall* rate of removal of the organic; they do not take into account the potential chemical or biological importance of the products of the various reactions. For example, singlet molecular oxygen is not included in Table 7.1 since it is not believed to be important in the overall gas phase removal of organics in ambient air. Thus, as seen in Section 7-E-5, $O_2(^1\Delta_g)$ is deactivated rapidly by air to ground-state oxygen, and its reactions with alkenes are relatively slow. However, the mechanisms of its reactions with alkenes are included in this chapter because of its historical role in our understanding of the chemistry of the troposphere and because of the nature and potential biological significance of such products as allylic hydroperoxides which are formed in reactions such as (1) with 2,3-dimethyl-2-butene:

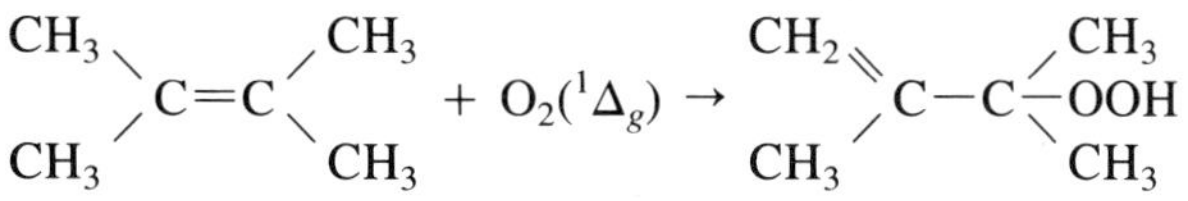

(1)

7-C. REACTIONS OF ALKANES

7-C-1. Hydroxyl Radical, OH: Kinetics and Mechanism

The initial attack of OH radicals on alkanes is a hydrogen-atom abstraction reaction forming alkyl radicals (R) and water:

$$RH + OH \rightarrow R + H_2O \tag{2}$$

Table 7.3 gives recommended values of these rate constants for selected alkanes and haloalkanes, as well as their temperature dependencies; the review of Atkinson (1985) should be consulted for the original references on which the recommendations are based, for the temperature range over which they are applicable, and for the uncertainties associated with them.

The C—H bond strength depends on whether the hydrogen is primary, secondary, or tertiary; these bond dissociation energies are approximately 98, 94, and 92 kcal mole^{-1}, respectively. Since a C—H bond is being broken in reaction (2), one would expect the reaction to be fastest for the weaker, tertiary C—H bonds and slowest for the stronger, primary C—H bonds.

This, in fact, is the case. Greiner (1970), for example, in his early kinetic studies of OH reactions (which ultimately led to the recognition of the importance of OH in the atmosphere) proposed that the rate data for a series of alkanes could be expressed as the sum of rate constants for primary, secondary, and tertiary C—H bonds, respectively:

$$k = N_p k_p + N_s k_s + N_t k_t \tag{A}$$

In this expression, k is the overall rate constant for the OH–alkane reaction, k_p, k_s, and k_t are the rate constants per primary, secondary, and tertiary C—H bond, respectively, and N_p, N_s, and N_t are the corresponding numbers of each kind of bond in the molecule. Using Eq. (A), the room temperature kinetic data for the simple alkanes were subsequently fit with $k_p = 6.5 \times 10^{-14}$, $k_s = 5.8 \times 10^{-13}$, and $k_t = 2.1 \times 10^{-12}$ cm^3 molecule^{-1} s^{-1} (Darnell et al., 1978), the trend to increasing values reflecting the trend in decreasing C—H bond strengths. On this simplied basis (which, as discussed below, has been further refined), the calculated value for k for n-butane is $[6 \times (6.5 \times 10^{-14}) + 4 \times (5.8 \times 10^{-13}) + 0 \times (2.1 \times 10^{-12})] = 2.71 \times 10^{-12}$ cm^3 molecule^{-1} s^{-1}, in good agreement with the experimental value of 2.55×10^{-12} shown in Table 7.3. The rate constant for methane cannot be predicted from Eq. (A) since it does not contain a primary carbon (which is defined as being attached to one other carbon).

As expected from a consideration of the detailed thermochemical kinetics (e.g., see Benson, 1976), in addition to the numbers of each type of C—H bond, the detailed structure of the molecule has an influence on the rate constants. For example, cycloalkanes deviate from Eq. (A) if the ring strain is $\gtrsim$4–5 kcal mole^{-1} (Atkinson et al., 1983a; Jolly et al., 1985). In addition, the experimental kinetic data show that even in simple alkanes a secondary C—H bond reacts ~40% faster if it is bonded to two other —CH_2— groups rather than to one —CH_2— and one —CH_3 (Atkinson et al., 1982a, 1983a). This is illustrated in Fig. 7.1 which shows the observed overall OH–alkane rate constants for a series of straight chain alkanes up to decane. The dashed line represents the predicted rate constants using Eq. (A) and the solid line those predicted if the reactivities of the two types of secondary C—H are taken into account.

The capability to predict rate constants for reactions of the OH radical, and indeed of other important atmospheric species such as O_3 and NO_3, is very important. Thus in recent years it has been increasingly recognized that a wide variety of toxic volatile organics are emitted into the troposphere (Section 10-D), yet little is known of their persistence and fates. Because the major possible chemical fates are reaction with OH, O_3, and NO_3, as well as photolysis, the lifetimes of these volatile organics can be estimated, if the rate constants and absorption cross sections are known or can be estimated. Having reliable methods of estimating reaction rate constants for compounds for which no experimental data exist is therefore an important aspect of risk assessment for volatile toxic chemicals.

In the case of OH–alkane reactions, a number of methods of estimating rate

TABLE 7.3. Selected Rate Constants for the Reaction of OH with Some Alkanes and Haloalkanes at 298°K and Their Temperature Dependencies[a]

Alkane	k (298°K)[e] (cm^3 molecule^{-1} s^{-1})	Temperature Dependence
CH_4	0.0084×10^{-12}[e]	$6.95 \times 10^{-18}\ T^2\ e^{-1280/T}$
C_2H_6.	0.274	$1.37 \times 10^{-17}\ T^2\ e^{-444/T}$
C_3H_8	1.18	$1.27 \times 10^{-17}\ T^2 e^{14/T}$
C_4H_{10}	2.55	$1.55 \times 10^{-11}\ e^{-540/T}$
$CH(CH_3)_3$	2.37	$9.58 \times 10^{-18}\ T^2\ e^{305/T}$
2-Methylbutane	3.9	n.a.[c]
2,3-Dimethylbutane	6.2	6.2×10^{-12}[b]
2,2,3-Trimethylbutane	4.1	$1.12 \times 10^{-11}\ e^{-300/T}$
2,2,3,3-Tetramethylbutane	1.06	$1.87 \times 10^{-17}\ T^2\ e^{-133/T}$
n-Pentane	4.1	n.a.
2-Methylpentane	5.5	n.a.
3-Methylpentane	5.6	n.a.
Cyclopentane	5.2	n.a.
2,2,4-Trimethylpentane	3.66	$1.62 \times 10^{-11}\ e^{-443/T}$
n-Hexane	5.58	n.a.
Cyclohexane	7.38	$2.73 \times 10^{-11}\ e^{-390/T}$
n-Octane	8.72	$3.12 \times 10^{-11}\ e^{-380/T}$
CCl_3F	$<5 \times 10^{-16}$[d]	n.a.
CCl_2F_2	$<6 \times 10^{-16}$[d]	n.a.
$CHClF_2$	4.68×10^{-15}	$1.51 \times 10^{-18}\ T^2\ e^{-1000/T}$
CH_3CCl_3	1.19×10^{-14}	$5.92 \times 10^{-18}\ T^2\ e^{-1129/T}$
CH_2Cl_2	1.42×10^{-13}	$8.54 \times 10^{-18}\ T^2\ e^{-500/T}$
CCl_4	$<1 \times 10^{-15}$	n.a.
CH_3Cl	4.36×10^{-14}	$3.5 \times 10^{-18}\ T^2\ e^{-585/T}$
CH_3Br	3.85×10^{-14}	$1.17 \times 10^{-18}\ T^2\ e^{-296/T}$

[a] All rate constants and temperature dependencies recommended by Atkinson (1985); this reference should be consulted for the original references on which the recommendations are based, temperature ranges over which temperature expressions are valid, and the uncertainties in these rate constants.

[b] Independent of temperature in the range 300–500°K.

[c] n.a. = not available.

[d] At 478–480°K; see Chang and Kaufman, 1977.

[e] All rate constants for the alkanes are in units of 10^{-12} cm^3 molecule^{-1} s^{-1}.

constants have been proposed. Most are based on the correlation between the rate constant and the C—H bond dissociation energy, or some parameter related to it such as the C—H bond stretching frequency (e.g., see Greiner, 1970; Darnall et al., 1978; Gaffney and Levine, 1979; Heicklen, 1981; Cohen, 1982; Jolly et al., 1985). Approaches based on extrapolation of experimental data as well as more theoretical approaches (e.g., see Cohen, 1982; Jeong et al., 1984) based, for example, on transition state theory, have been used. These are briefly reviewed by

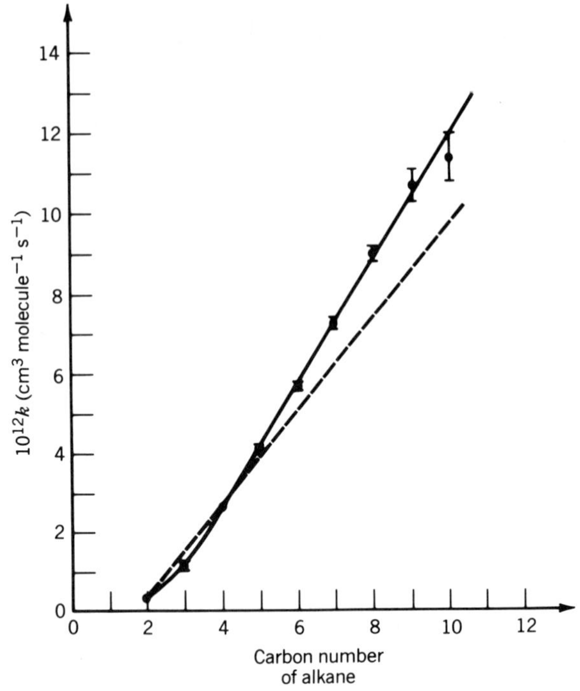

FIGURE 7.1. Overall room temperature rate constant for the reaction of OH with a series of straight chain alkanes: ●, experimental points; – – –, predicted values using (A); —, prediction corrected for molecular structure (from Atkinson et al., 1982a).

Atkinson (1985); their temperature dependencies have been reviewed by Walker (1985). The application of such techniques to the estimation of rate constants for some volatile toxic organics is discussed in Section 10-D.

Included in Table 7.3 are rate data for a number of haloalkanes. The species CF_2Cl_2 and CCl_3F react so slowly with OH radicals that only an upper limit can be placed on the rate constant even at 480°K. In these cases, of course, there are no hydrogen atoms available for abstraction, and halogen atom abstraction is thermodynamically unfavorable. If one fluorine atom is replaced by a hydrogen atom, however, the rate constant increases to a measurable (albeit, low) value. It is the lack of reactivity of OH (or other oxidizing species) with fully chlorinated and fluorinated alkanes that is responsible for their long lifetimes, and hence ubiquity, in the troposphere. Thus they survive sufficiently long to be slowly transported through the tropopause to the stratosphere where they are postulated to play a major role in the destruction of O_3 (see Chapter 15).

7-C-2. Nitrate Radical, NO_3: Kinetics and Mechanism

The pioneering research of Niki and co-workers on the kinetics of the gas phase reactions of the nitrate radical, and their implications to tropospheric chemistry, was carried out in the early to mid-1970s. Since then little additional work in this area was reported until the comprehensive studies of Atkinson et al. (1984a–e) on the kinetics of the reactions of NO_3 radicals in air with a variety of classes of organics, including alkanes, alkenes, aromatics, simple aldehydes, and biogenic species such as isoprene, monoterpenes, and dimethylsulfide.

Until 1985, the rate constants reported for NO_3 reactions were obtained by indirect methods that required knowledge of the absolute values of rate and/or equilibrium constants of other reactions. In particular, all the reported values depend in a linear manner on the equilibrium constant (K_{eq}) for the NO_3–NO_2–N_2O_5 equilibrium:

$$NO_3 + NO_2 \overset{M}{\rightleftarrows} N_2O_5 \tag{80}$$

$$K_{eq} = \frac{[N_2O_5]}{[NO_3][NO_2]}$$

In one type of experiment, the NO_3 rate constants were obtained by following the decay of N_2O_5 in the presence of the reactant. Interpretation of the data requires taking into account the equilibrium (80) as well as the following reactions:

$$N_2O_5 \rightarrow \text{loss} \tag{81}$$

$$NO_3 + \text{reactant} \rightarrow \text{products} \tag{82}$$

The reader is referred to the papers by Japar and Niki (1975) and Atkinson et al. (1984a,b) for the details of the kinetic analysis.

A second technique used for measuring the kinetics of NO_3 reactions is based on measuring the relative rates of decay of two or more species under conditions where NO_3 is the sole reactant with the organics (Atkinson et al., 1984a,b). The kinetic analysis for this relative rate technique is completely analogous to that outlined for OH reactions in Section 4-C. The relative rate constants thus generated can be converted to absolute values if one of the NO_3 reaction rate constants is known. However, until 1985, the "known" rate constant was again based on an assumed value of K_{eq} for reaction (80).

As discussed in Section 8-A-2, values measured or calculated for the equilibrium constant for reaction (80) at 298°K differ by almost a factor of 2, 1.87×10^{-11} cm^3 molecule^{-1} (Malko and Troe, 1982), 2.2×10^{-11} (Graham and Johnston, 1978), 3.44×10^{-11} (Tuazon et al., 1984b), and 3.26×10^{-11} cm^3 molecule^{-1} (Kircher et al., 1984); two very recent values of Perner et al. (1985) and Burrows et al. (1985) are 1.8×10^{-11} (average of all 6 values from their chamber studies), and 2.5×10^{-11} cm^3 molecule^{-1}, respectively. Thus the rate constant for NO_3 reactions derived from one study can also vary by this factor, depending on the value chosen for K_{eq}.

Because of this, we report in this chapter several possible values of the rate constants for NO_3 reactions, corresponding to different values of K_{eq}. However, until further work is carried out, we recommend the NO_3 rate constants based on a value of $K_{eq} = 3.35 \times 10^{-11}$ cm^3 molecule^{-1}, which is the average of the values reported by Tuazon et al. (1984b), 3.44×10^{-11} cm^3 molecule^{-1}, and by Kircher et al. (1984), 3.26×10^{-11} cm^3 molecule^{-1}, at 298°K. The reason for this is that *absolute* rate constants for the reaction of NO_3 with two alkenes have been reported in 1985 for the first time by Ravishankara and Mauldin (1985); they studied the decay of NO_3, generated from the reaction of fluorine atoms with HNO_3, and in some experiments by N_2O_5 thermolysis, in the presence of an excess of *trans*-2-butene or isobutene in a fast flow system (Chapter 4). In this case, the kinetics of NO_3 decay do not depend on K_{eq}. Their rate constants are in excellent agreement with the indirect values derived using $K_{eq} = 3.35 \times 10^{-11}$ cm^3 molecule^{-1}. Ravishankara and Mauldin suggest that the reported relative rate constants may thus be converted to absolute values using this value of K_{eq}. Whether this value of K_{eq} is correct remains uncertain, since the reported values vary by about -30% to $+40\%$ about their mean (see Chapter 8-A-2). If 3.35×10^{-11} cm^3 molecule^{-1} for K_{eq} is established in the future not to be correct, the agreement of rate constants from relative rate studies derived from this value with the absolute rate constants of Ravishankara and Mauldin may be due to a fortuitous cancellation of unrecognized errors in the relative rate studies. We therefore recommend, until further absolute data are generated, that the NO_3 rate constants, which have been measured indirectly but interpreted using this value of K_{eq} of 3.35×10^{-11} cm^3 molecule^{-1}, be used. However, should this value of K_{eq} be found in the future to be incorrect, the relative rate data can be readily corrected since they are linearly dependent on K_{eq}. For example, if the value of Burrows et al. (1985), $K_{eq} = 2.5 \times 10^{-11}$ cm^3 molecule^{-1}, is found to be correct, the rate constants based on $K_{eq} = 3.35 \times 10^{-11}$ cm^3 molecule^{-1} should be multiplied by the factor (2.5/3.35) = 0.75 (subject, of course, to the caveat about possible cancelling errors in the relative rate experiments, *vide supra*).

As seen in Tables 7.1 and 7.2, regardless of which value of K_{eq} is used, the rate constants for some of the NO_3 radical reactions are sufficiently fast that they can be important nighttime loss processes for many organics, as well as sources of HNO_3 in acid deposition. In this context we first discuss the reaction of NO_3 with alkanes and compare its reactivity with that of the OH radicals.

The rate constants for the reaction of NO_3 with alkanes fall in the range from $\sim 10^{-17}$ for the small alkanes to $\sim 10^{-16}$ cm^3 molecule^{-1} s^{-1} for the branched 2,3-dimethylbutane (Table 7.4). These can now be used in conjunction with ambient OH and NO_3 radical concentration data to obtain the relative importance of NO_3 compared to OH in removing alkanes from the troposphere (this is an extension of our *lifetime* calculations in Section 7-B).

As we have seen, the daytime concentrations of OH and nighttime concentrations of NO_3 radicals vary over wide ranges (Tables 7.1 and 7.2), but for discussion one can assume the ratio $[NO_3]/[OH]$ is $\sim 10^3$. Recalling that the relative atmospheric lifetimes of an organic to attack by OH or NO_3 radicals, τ_{OH}/τ_{NO_3}, is given by the ratio $k_{NO_3}[NO_3]/k_{OH}[OH]$, we see that for the branched alkane 2,3-dimethylbutane, $[k_{NO_3}/k_{OH}] \times 10^3 \cong 0.04$. Thus attack by NO_3 is responsible for only a small portion of the total removal of the organic. For the simple, straight

TABLE 7.4. Values of the Rate Constants[a] for the Reaction of NO_3 with Some Alkanes at 298°K Derived Using Different Values of the Equilibrium Constant for $NO_2 + NO_3 \overset{M}{\rightleftarrows} N_2O_5$

	$10^{17}\, k$ (cm^3 molecule^{-1} s^{-1})		
		K_{eq}[b] =	
Alkane	1.87×10^{-11}[c]	2.20×10^{-11}[d]	3.35×10^{-11}[e,f]
n-Butane	2.0	2.4	3.6
n-Pentane	2.4	2.8	4.3
n-Hexane	3.2	3.8	5.7
n-Heptane	4.1	4.8	7.3
n-Octane	5.5	6.5	9.8
n-Nonane	7.2	8.5	13
Isobutane	2.9	3.4	5.2
2,3-Dimethylbutane	12	14	22
Cyclohexane	4.0	4.7	7.2

[a] From Atkinson et al., 1984e.

[b] $K_{eq} = [N_2O_5]/[NO_2][NO_3]$ in units of cm^3 molecule^{-1}.

[c] Malko and Troe, 1982.

[d] Revised from Graham and Johnston (1978) using value of DeMore et al. (1983) for $k(O_3 + NO_2) = 3.2 \times 10^{-17}$ cm^3 molecule^{-1} s^{-1}.

[e] Average values of Tuazon et al. (1984), 3.44×10^{-11} and of Kircher et al. (1984), 3.26×10^{-11} cm^3 molecule^{-1}.

[f] Recommended values; see discussion in text.

chain alkanes such as *n*-butane which react more slowly with NO_3, their removal can be attributed almost entirely to reaction with OH.

While NO_3 is a relatively minor contributor compared to OH to the overall oxidation of alkanes, the reaction is interesting because the mechanism must involve a hydrogen atom abstraction which forms gaseous nitric acid at night.

$$RH + NO_3 \rightarrow R + HNO_3 \tag{3}$$

In Section 11-B-1b we estimate the potential contribution of the homogeneous gas phase NO_3–alkane reaction (3) versus that of the daytime process (4)

$$NO_2 + OH \xrightarrow{M} HNO_3 \tag{4}$$

to the formation of gaseous HNO_3 in the troposphere.

7-D. REACTIONS OF ALKYL (R), ALKYLPEROXY (RO_2), AND ALKOXY (RO) RADICALS IN AMBIENT AIR

As we have seen, the reactions of alkanes in the troposphere with OH and, to a lesser extent, NO_3 radicals produce alkyl radicals. In this section we discuss the fates of this class of radicals in ambient air and follow the subsequent radical chain reactions through to the formation of stable products.

7-D-1. Alkyl Radicals

The only significant atmospheric fate of alkyl radicals is the reaction with O_2 to form alkylperoxy radicals (RO_2):

$$R + O_2 \xrightarrow{M} RO_2 \tag{5}$$

These are three-body reactions, in which the collision partner M acts to remove the excess energy from bond formation (Section 4-A-4). As might be expected for such combination reactions, the high-pressure limit where the reaction becomes effectively second order is reached at lower total pressures as the size of R increases, that is, the radical has more degrees of freedom to absorb the excess energy from bond formation. Thus the high-pressure limits for the methyl and ethyl radical reactions are ~700 and ~100 Torr, respectively, whereas they are only a few Torr for the reactions of propyl and butyl radicals (Ruiz and Bayes, 1984; Cobos et al., 1985; Borrell et al., 1985). Since the high-pressure limits for all the $R + O_2$ reactions are less than 760 Torr, in tropospheric reaction mechanisms at the earth's surface they are treated as second-order processes, or pseudo-first-order processes in R (since O_2 is also constant).

Table 7.5 gives the effective second-order rate constants k_5 at 1 atm pressure for some simple alkyl radicals at room temperature. They are relatively rapid, increasing from $\sim 10^{-12}$ to $\sim 10^{-11}$ cm^3 molecule^{-1} s^{-1} as the size and nature

TABLE 7.5. Selected Rate Constants (k_5) for the Reaction of Alkyl Radicals with O_2 at Room Temperature and One Atmosphere Pressure of Air

R	$10^{12}\,k_5$ (cm^3 molecule^{-1} s^{-1})	Reference
CH_3	1.1[a]	Recommended by DeMore et al., 1985
C_2H_5	4.8[b]	Recommended by Baulch et al., 1984
1-C_3H_7	5.5 ± 0.9	Ruiz and Bayes, 1984
2-C_3H_7	14.1 ± 2.4	Ruiz and Bayes, 1984
1-C_4H_9	7.5 ± 1.4	Lenhardt et al., 1980
2-C_4H_9	16.6 ± 2.2	Lenhardt et al., 1980
t-C_4H_9	23.4 ± 3.9	Lenhardt et al., 1980
3-Hydroxy-2-butyl	28 ± 18	Lenhardt et al., 1980

[a] Calculated using Eq. (C), Chapter 4, and recommended values of DeMore et al. (1985) of $k_0 = 4.5 \times 10^{-31} (T/300)^{-2}$ cm^6 molecule^{-2} s^{-1}, $k_\infty = 1.8 \times 10^{-12}$ cm^3 molecule^{-1} s^{-1}, and $F_c = 0.6$.

[b] Calculated using Eq. (C), Chapter 4, and recommended values of Baulch et al. (1984) of $k_0 = 2.0 \times 10^{-28} (T/300)^{-3.8}$ cm^6 molecule^{-2} s^{-1}, $k_\infty = 5 \times 10^{-12}$ cm^3 molecule^{-1} s^{-1}, and $F_c = 0.7$.

(i.e., primary, secondary, or tertiary) of the radical changes. Clearly, because of these fast reactions with O_2, the lifetimes of alkyl radicals in the troposphere are short and reaction with O_2 is their sole fate.

Interestingly, the rate constant for the hydroxy-substituted 2-butyl radical is, within the relatively large error limits given, not significantly different from the unsubstituted 2-butyl radical. This is important in that such hydroxy-substituted alkyl radicals are formed in OH–alkene reactions (Section 7-E-1b), but relatively little is known about their fate in the atmosphere. The similarity in rate constants for reaction with O_2 suggests (Lenhardt et al., 1980) that to a first approximation, the OH–alkene radical adducts may react at rates similar to the corresponding alkyl radicals for which more data are available. Product studies of irradiated HONO–NO–alkene mixtures, where the reaction of OH with the alkene initiates the reaction sequence (Niki et al., 1978a) and of O_3–alkene reactions where OH is believed to be an intermediate (Martinez et al., 1980), also strongly suggest that hydroxy-substituted alkyl radicals react in the same way as unsubstituted alkyl radicals.

It has been suggested that alternate bimolecular reaction channels might exist, for example, for CH_3

$$CH_3 + O_2 \rightarrow HCHO + OH$$

However, the experimental evidence available at present indicates that this channel is negligible ($k < 5 \times 10^{-17}$ cm^3 molecule^{-1} s^{-1} at 298°K) under atmospheric conditions (e.g., see Plumb and Ryan, 1981, 1982; Selzer and Bayes, 1983; and Baulch et al., 1984). The alternate channel in the $C_2H_5 + O_2$ reaction

$$C_2H_5 + O_2 \rightarrow C_2H_4 + HO_2$$

is also significantly slower than the addition reaction, with a rate constant of 2×10^{-13} cm^3 molecule^{-1} s^{-1} at 298°K (Plumb and Ryan, 1981), but a small but significant fraction ($\sim$0.05) of the reaction proceeds via this channel. Slagle et al. (1984) have studied the kinetics and C_2H_4 product yields in the $C_2H_5 + O_2$ reaction as a function of temperature from 294 to 1002°K and show that the yield of C_2H_4 increases with temperature. They suggest that the reaction to form C_2H_4 proceeds through the adduct, that is,

$$C_2H_5 + O_2 \rightarrow C_2H_5O_2 \rightarrow C_2H_4 + HO_2$$

rather than as a parallel path to adduct formation.

7-D-2. Alkylperoxy Radicals

7-D-2a. Reaction with NO and NO_2

In all but the remote troposphere, the alkylperoxy radicals formed from the alkyl radicals react with NO. The only significant path for the smaller radicals ($\lesssim C_4$) is oxidation of NO to NO_2 and formation of alkoxy radicals:

$$RO_2 + NO \rightarrow RO + NO_2 \quad (6)$$

Only in the last few years have techniques become available for measuring the rate constants of these reactions. Hence a wide range of alkyl radicals have not been studied, and there are considerable discrepancies in the rate constants reported in the literature. However, the sparse data available indicate that the rate constants for (6) are not very sensitive to the nature of R. As a result, until further data become available, a value of 7.6×10^{-12} cm^3 $molecule^{-1}$ s^{-1} is recommended (see Atkinson and Lloyd, 1984) for all alkylperoxy radicals.

While essentially all of the reaction of CH_3O_2 with NO is known to proceed as in (6) above, for larger alkyl radicals a second reaction channel becomes significant (Darnall et al., 1976). This is the addition reaction to form stable alkyl nitrates:

$$RO_2 + NO \xrightarrow{M} RONO_2 \quad (7)$$

A possible mechanism is addition followed by isomerization:

$$RO_2 + NO \xrightarrow{M} (ROONO)^* \rightarrow (RONO_2)^* \rightarrow RO + NO_2$$
$$(RONO_2)^* \xrightarrow{M} RONO_2$$

Studies of a variety of alkanes have established that the addition channel can become quite significant ($\geq 30\%$) for certain alkyl radicals depending on their structure and size (Atkinson et al., 1982b).

Table 7.6 shows, for some simple alkyl radicals, the fraction of the overall reaction of RO_2 with NO which results in alkyl nitrate formation at atmospheric pressure, that is, the ratio $k_7/(k_6 + k_7)$. The increasing importance of addition as the R group increases in size reflects the longer lifetimes of the excited intermediates, which allows collisional deactivation to the stable nitrate to compete with its decomposition. As expected, the observed alkyl nitrates from straight chain alkane reactions correspond generally to abstraction of secondary hydrogen atoms rather than of the stronger primary H atoms. Somewhat surprisingly, the rate constant ratio $(k_7/(k_6 + k_7))$ is smaller for tertiary alkylperoxy radicals; the reasons for this are unclear (Atkinson et al., 1984f).

Alkylperoxy radicals also react rapidly with NO_2, but, as discussed for PAN (Section 4-A-7c), the nitrates formed thermally decompose back to the reactants:

$$RO_2 + NO_2 \rightleftarrows RO_2NO_2$$

Thus no overall reaction occurs, and this process acts primarily as a low-temperature storage for NO_2.

7-D-2b. *Bimolecular Self-Reaction of RO_2*

At the very low NO concentrations believed to be characteristic of remote areas, the reaction of RO_2 with NO is sufficiently slow that certain reactions of alkyl-

TABLE 7.6. Fraction of Reaction of RO_2 + NO Producing Alkyl Nitrates for Individual Alkylperoxy Radicals at Room Temperature and 735–740 Torr Total Pressure

	Primary		Secondary		Tertiary	
Alkane	RO_2	$k_2/(k_6 + k_7)$	RO_2	$k_7/(k_6 + k_7)$	RO_2	$k_7/(k_6 + k_7)$
Ethane	Ethyl	$\leqslant 0.014$				
Propane	1-Propyl	0.020	2-Propyl	0.042		
n-Butane	1-Butyl	$\leqslant 0.041$	2-Butyl	0.090		
n-Pentane			2-Pentyl	0.129, 0.134		
			3-Pentyl	0.131, 0.146		
Neopentane	Neopentyl	0.051				
2-Methylbutane			2-Methyl-3-butyl	0.141	2-Methyl-2-butyl	0.047
n-Hexane			2-Hexyl	0.209		
			3-Hexyl	0.230		
Cyclohexane			Cyclohexyl	0.160		
2-Methylpentane			2-Methyl-3-pentyl +2-Methyl-4-pentyl	0.190	2-Methyl-2-pentyl	0.031
3-Methylpentane			3-Methyl-2-pentyl	0.178		
n-Heptane			2-Heptyl	0.301, 0.291		
			3-Heptyl	0.323, 0.325		
			4-Heptyl	0.301, 0.285		
n-Octane			2-Octyl	0.323		
			3-Octyl	0.348		
			4-Octyl	0.329		

Source: Carter and Atkinson, 1985.

peroxy radicals with HO_2 could very well be important. As we shall see, the self-reaction of RO_2 is unlikely to be important, even in the clean troposphere. (RO_2 will be produced in these areas by the reactions of CH_4 and other naturally produced organics; see Chapter 14.) Studies of the kinetics of these reactions were delayed for a number of years by the lack of sensitive and specific techniques for directly monitoring alkylperoxy radicals. However, in 1973 Parkes and co-workers characterized the ultraviolet absorption spectrum of CH_3O_2 and $C_2H_5O_2$, as shown in Fig. 7.2. In addition, absorption spectra in the near infrared have been observed (Hunziker and Wendt, 1976). This allowed the application of spectroscopic techniques to follow the concentration–time profile of RO_2 and hence determine the kinetics of its reactions. Since then, the absorption spectra for other simple alkylperoxy radicals have also been recorded and applied to studies of their reaction kinetics (e.g., see Adachi and Basco, 1982; Anastasi et al., 1983). As might be expected, the absorption spectra are all similar, suggesting that the electronic transition is relatively insensitive to the nature of the alkyl groups attached (Kirsch et al., 1978).

While most of the kinetic work reported to date on RO_2 reactions has relied on absorption spectrometry to monitor the RO_2 concentration, some work has been carried out using mass spectrometric detection of RO_2 (Plumb et al., 1979). Indirect techniques such as monitoring the formation of products rather than loss of the reactant RO_2 have also been applied (e.g., see Ravishankara et al., 1981).

The kinetics of the self-reaction of RO_2, that is,

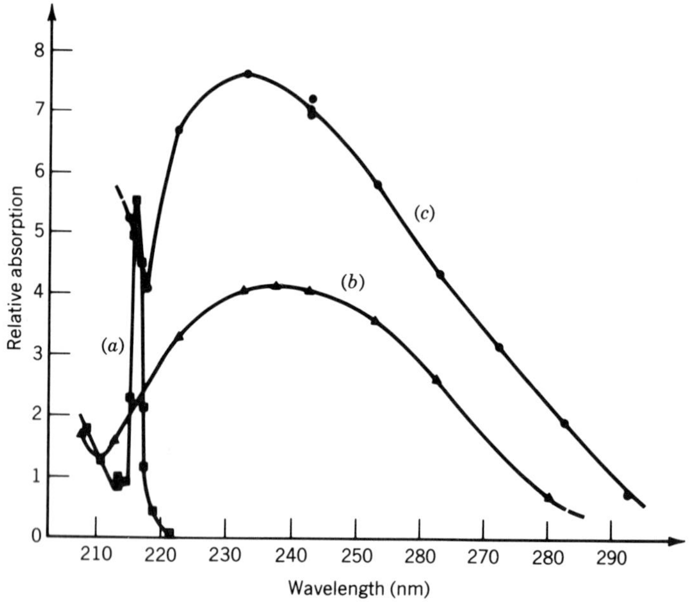

FIGURE 7.2. Ultraviolet absorption spectrum of (*a*) CH_3, (*b*) CH_3O_2, and (*c*) $C_2H_5O_2$ (from Parkes et al., 1973).

$$RO_2 + RO_2 \rightarrow \text{products} \tag{8}$$

have been established for the simple alkylperoxy radicals (R = CH_3, C_2H_5, C_3H_7, i-C_3H_7, t-C_4H_9) and selected values of the room temperature rate constants are given in Table 7.7; they are in the range $\sim(10^{-13}$–$10^{-17})$ cm^3 $molecule^{-1}$ s^{-1}.

It is interesting that, in contrast to most reactions of alkyl-substituted species, the rate constants decrease as R goes from primary to secondary to tertiary. There is no accepted theoretical explanation for this trend at present, although it has been suggested that it is due to steric effects that increase the activation energies for decomposition of the tetroxides which are thought to be formed on dimerization of the alkylperoxy radicals (Bartlett and Guaraldi, 1967).

Concentrations of NO, RO_2, and HO_2 in remote areas are not well established. However, estimates of 30 ppt NO (7.4×10^8 cm^{-3}) (see Section 5-G), 2.5×10^8 cm^{-3} RO_2, and 7×10^8 cm^{-3} HO_2 (Bottenheim and Strausz, 1980) seem reasonable when comparing relative loss rates of alkylperoxy radicals by reaction with NO or with RO_2 and HO_2. Taking a rate constant k_8 for self-reaction of 3×10^{-13} cm^3 $molecule^{-1}$ s^{-1}, which is typical for R = CH_3 and C_3H_7 (Table 7.7), then $k_8[RO_2]/(k_6 + k_7)[NO] = 0.01$.

For the reaction of RO_2 with HO_2 using a rate constant of 3×10^{-12} (Table 7.7), this ratio is ~ 0.4. Thus the reaction of HO_2 with RO_2 is much more important than the self-reaction of RO_2 radicals. However, in either case, they contribute

TABLE 7.7 Some Selected Room Temperature Rate Constants for the Bimolecular Reaction of RO_2 Radica s[a]

RO_2	k (cm^3 $molecule^{-1}$ s^{-1})	Reference
CH_3O_2	3.1×10^{-13}	Recommended by Atkinson and Lloyd, 1984
	3.7×10^{-13}	Recommended by Baulch et al., 1984
$C_2H_5O_2$	1.0×10^{-13}	Adachi et al., 1979
	5.8×10^{-14}	Anastasi et al., 1983
	1×10^{-13}	Baulch et al., 1984
n-$C_3H_7O_2$	3.3×10^{-13}	Adachi and Basco, 1982
$(CH_3)_2CHO_2$	1.3×10^{-15}	Kirsch et al., 1978
	1.3×10^{-15}	Adachi and Basco, 1982
$(CH_3)_3CO_2$	2.6×10^{-17}	Recommended by Atkinson and Lloyd, 1984
$HO_2 + RO_2$[b]	$\sim 3 \times 10^{-12}$	Recommended by Atkinson and Lloyd, 1984

[a] Rate constant k defined by $-d[RO_2]/dt = 2k[RO_2]^2$.

[b] RO_2 = Typical alkylperoxy radical.

to the removal of RO_2 from the troposphere only in remote areas, where the NO concentration is small.

As discussed in Chapter 3-D-1, the self-reaction of HO_2 is quite fast, occurring with a rate constant of 5.6×10^{-12} cm^3 molecule^{-1} s^{-1} at one atmosphere pressure, 298°K and 50% relative humidity. The reaction of HO_2 with NO is also fast, with a rate constant of 8.3×10^{-12} cm^3 molecule^{-1} s^{-1} at 298°K (see Section 8-A-1); thus, at the radical concentrations cited above, the removal of HO_2 by the self-reaction with HO_2 becomes comparable to that by reaction with NO at an NO concentration of 20 ppt.

The mechanism and products of these reactions of alkylperoxy radicals have been studied in greatest detail for CH_3O_2 radicals where the possible reaction paths are as follows:

$$\begin{aligned} CH_3O_2 + CH_3O_2 &\xrightarrow[a]{} 2CH_3O + O_2 \\ &\xrightarrow[b]{} CH_3OH + HCHO + O_2 \\ &\xrightarrow[c]{} CH_3OOCH_3 + O_2 \\ &\xrightarrow[d]{} CH_3O_2H + H_2COO \end{aligned} \tag{9}$$

The most recent studies (Parkes, 1975; Kan et al., 1980; Niki et al., 1981a) are in good agreement and indicate that only paths (a), (b), and (c) are important. From a composite of these three studies, the percentages of the total reaction proceeding via (a), (b), and (c) are ~35%, 57%, and 8%, respectively (Atkinson and Lloyd, 1984; Baulch et al., 1984). Thus the majority of the reaction proceeds via formation of the alcohol and aldehyde, with a lesser, but still significant portion, producing methoxy radicals. Path (d), proposed by Nangia and Benson (1980) to produce a Criegee biradical similar to that from ozone–olefin reactions (see Section 7-E-2), appears to be negligible.

For the large alkyl radicals, the two reaction paths producing alkoxy radicals (a) or aldehyde plus alcohol (b), respectively, also appear to be the major reaction paths (e.g., see Niki et al., 1982b; Anastasi et al., 1983; and reviews by Atkinson and Lloyd, 1984, and Baulch et al., 1984). For example, in the self-reaction of ethylperoxy radicals, the ratio of the path producing $2C_2H_5O + O_2$ to that forming $CH_3CHO + C_2H_5OH + O_2$ is estimated to be 1.75 ± 0.05 at 302°K; diethyl peroxide was produced at an upper limit 5% of the other major products (Anastasi et al., 1983). The recommendation of Baulch et al. (1984) is that 52% produces $2C_2H_5O + O_2$ and $\geq 39\%$ forms $(CH_3CHO + C_2H_5OH + O_2)$.

The reaction of HO_2 with CH_3O_2 forms a product that absorbs strongly in the UV (210–280 nm). This has been tentatively identified as methylhydroperoxide, suggesting that at least part of this reaction produces $CH_3OOH + O_2$ (Cox and Tyndall, 1979, 1980). Alkyl hydroperoxides have also been observed using Fourier transform infrared spectroscopy as products of the chlorine atom initiated oxidation of ethane, propane, and butane in air in the absence of NO_x. These have been attributed to the RO_2–HO_2 reaction (Hanst and Gay, 1983).

To summarize, in much of the troposphere where NO is present at sufficient concentrations, alkyl radicals add O_2 to form alkylperoxy radicals (RO_2); these oxidize NO to NO_2 producing alkoxy radicals (RO), or add NO to form stable nitrates. The free radical chain carrier at this point is thus the alkoxy radical. Under very clean conditions where the NO concentration is low, reaction of the alkylperoxy radicals with HO_2 (and of HO_2 with itself) becomes important.

7-D-3. Alkoxy Radicals

Alkoxy radicals have a more diverse chemistry than either of their precursors, R and RO_2. This includes (1) reaction with O_2, (2) isomerization, (3) decomposition, (4) reaction with NO, and (5) reaction with NO_2.

7-D-3a. Reaction with O_2

Although relatively few rate data are available, alkoxy radicals react with O_2 if an abstractable hydrogen atom is available on the neighboring carbon, for example,

$$CH_3O + O_2 \rightarrow HCHO + HO_2 \qquad (10)$$

$$CH_3-\underset{\displaystyle CH_3}{\underset{|}{\overset{\displaystyle H}{\overset{|}{C}}}}-O + O_2 \rightarrow CH_3-\underset{\displaystyle CH_3}{\underset{|}{C}}=O + HO_2 \qquad (11)$$

Direct determinations of k_{10} and the corresponding rate constant for C_2H_5O have been reported using flash photolysis to generate the alkoxy radicals and resonance fluorescence to monitor their concentrations (Gutman et al., 1982). The results for CH_3O, $\sim 1.3 \times 10^{-15}$ cm^3 molecule^{-1} s^{-1} at 298°K, are in relatively good agreement with rate constants reported earlier obtained using indirect techniques, and with a more recent value of 1.9×10^{-15} cm^3 molecule^{-1} s^{-1}, obtained using laser photolysis followed by laser-induced fluorescence detection of CH_3O (Lorenz et al., 1985); the rate constant for C_2H_5O, 6.7×10^{-15} cm^3 molecule^{-1} s^{-1}, is the first direct value for this radical. In 1 atm air, then, the lifetime of CH_3O with respect to the O_2 reaction is only ~ 0.1 ms.

These two rate constants were then used as a basis for calculating the corresponding rate constants for the reaction of larger alkoxy radicals. The results are given in Table 7.8.

In extrapolating their results for the methoxy (CH_3O) and ethoxy (C_2H_5O) radicals to the larger alkoxy radicals given in Table 7.8, Gutman et al. (1982), using thermochemical kinetic techniques, estimated that activation energies for the reactions of secondary alkoxy radicals such as s-C_4H_9O ($CH_3-\overset{\displaystyle O}{\overset{|}{C}}HC_2H_5$) were close to zero. In the case of the s-butoxy radical, this gives a room temperature rate constant of 5×10^{-14} cm^3 molecule^{-1} s^{-1} (Table 7.8).

TABLE 7.8. Arrhenius Parameters and Room Temperature Rate Constants Measured or Calculated for the Reaction of RO with O_2[a]

Alkoxy Radical	10^{14} A (cm^3 molecule^{-1} s^{-1})	E (kcal mole^{-1})	10^{15} k at 298°K (cm^3 molecule^{-1} s^{-1})
CH_3O	10^b(8)	2.6^b (2.6)	1.3^b (1.1)
C_2H_5O	7^b(5)	1.3^b (1.5)	8^b (4.3)
C_3H_7O	7(5)	1.3 (1.5)	7^c (4.3)
i-C_3H_7O	$3(2)^b$	$0.0 (0.39)^b$	30^c $(7.8)^b$
C_4H_9O	7(5)	1.4 (1.7)	7^c (3.0)
i-C_4H_9O	7(5)	1.1 (1.3)	10^c (6.0)
s-C_4H_9O	3(3)	−0.2 (0.2)	50^c (18)

[a] Gutman et al., 1982 and in parentheses, Balla et al., 1985.

[b] Direct measurement.

[c] Calculated.

Balla et al. (1985) studied the reactions of the isopropoxy radical with O_2, NO and NO_2 using pulsed laser photolysis of isopropyl nitrite and laser-induced fluorescence to follow the decay of the alkoxy radical; at 298°K they find a rate constant for reaction with O_2 of 7.8×10^{-15} cm^3 molecule^{-1} s^{-1}. They also estimated the rate constants and Arrhenius parameters for the reactions of a series of small alkoxy radicals with O_2, and these are also shown in Table 7.8. The rate constants measured or estimated in the two studies differ by up to a factor of approximately three, except for the methoxy radical where the agreement is better.

In indirect studies of the reactions of C_2H_5O, n-C_3H_7O, and i-C_4H_9O with O_2, Zabarnick and Heicklen (1985a,b,c) report rate constants for the reactions with O_2 of 1.4, 1.5, and 1.9×10^{-14} cm^3 molecule^{-1} s^{-1} respectively, based on an assumed rate constant for the reactions of these radicals with NO of 4.4×10^{-11} cm^3 molecule^{-1} s^{-1}. They also report in these systems the quantum yields for photolysis at 366 nm of the corresponding alkyl nitrites used as precursors to the alkoxy radicals, as 0.29 (C_2H_5ONO), 0.38 (n-C_3H_7ONO) and 0.24 (i-C_4H_9ONO) respectively, approximately the same as those found in companion studies (Morabito and Heicklen, 1985). These are clearly much less than the unit quantum yields often assumed for organic nitrite photolysis (see Section 3-C-5). However, these quantum yields need to be confirmed in independent studies, since the system from which they were drawn is complex.

Carter and Atkinson (1985) suggest that for the abstraction of hydrogen atoms from primary alkoxy radicals, $k = 3.7 \times 10^{-14}\, e\,(-483/T)$ cm^3 molecule^{-1} s^{-1} $= 7.3 \times 10^{-15}$ cm^3 molecule^{-1} s^{-1} at 298°K. For abstraction from secondary alkoxy radicals, they recommend $k = 1.8 \times 10^{-14}\, e\,(-196/T) = 9.3 \times 10^{-15}$ cm^3 molecule^{-1} s^{-1} at 298 K.

While the reaction of alkoxy radicals with O_2 has been generally accepted to proceed by hydrogen atom abstraction as shown in reactions (10) and (11), Lorenz et al. (1985) have suggested that the possibility of other, more complex pathways, e.g. to produce $HCO + H_2O_2$ in the case of CH_3O, should be investigated.

Clearly, much more experimental work clearly needs to be done in this difficult but important area.

7-D-3b. *Isomerization*

A second route available to larger alkoxy radicals is isomerization. This intramolecular process can become fast when a 1,4 or 1,5 hydrogen atom shift is possible, since relatively stable five- or six-membered transition states are possible:

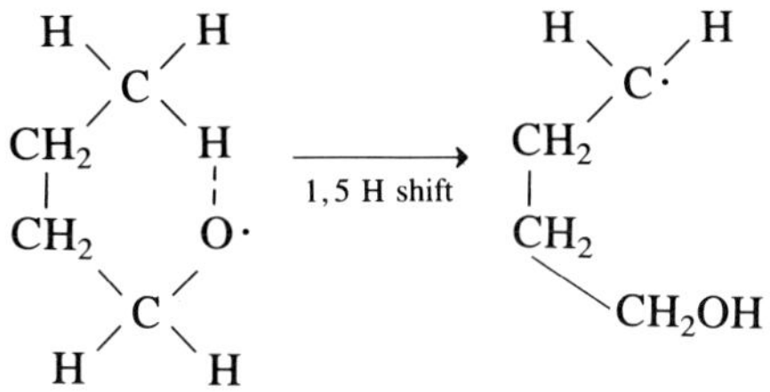

(12)

The products of such isomerizations are β- and α-substituted alkyl radicals; these will react with O_2 to form alkylperoxy radicals as discussed earlier.

Isomerizations that involve a 1,4 hydrogen atom shift, such as that of the 2-butoxy radical,

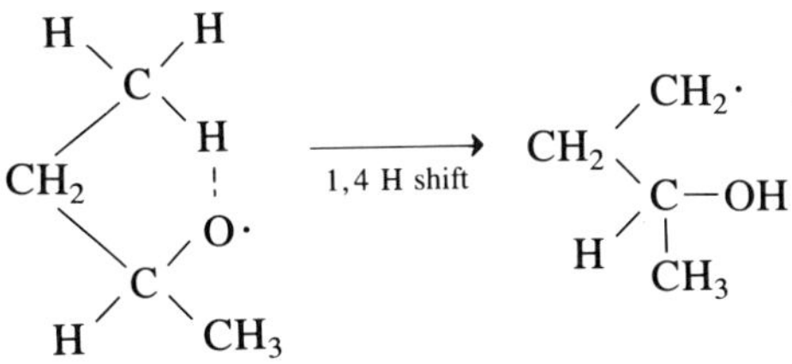

(13)

are slower due to the less favorable five-member ring transition state. This can be seen from the data in Table 7.9 which shows three estimates of the rates of isomerization of various types of alkoxy radicals. The isomerizations involving 1,4 hydrogen shifts are estimated to be at least an order of magnitude slower than those involving 1,5 shifts.

7-D-3c. *Decomposition*

Both experimental and theoretical studies indicate that larger alkoxy radicals can undergo unimolecular decomposition to form carbonyl compounds and free radicals, for example,

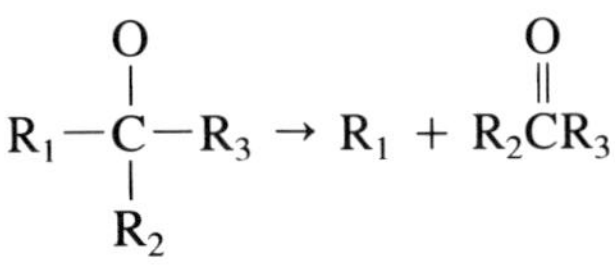

(14)

TABLE 7.9. Estimated Rate Constants for Isomerization of Alkoxy Radicals of Various Types at Room Temperature

	Rate Constant (s^{-1}) per Abstractable H Atom[a]		
	Baldwin et al. (1977)	Carter et al. (1976)	Carter and Atkinson (1985)
1,4 H Shift			
RCH_2—H	39	1.5×10^2	7
RCH(OH)—H	3.0×10^2	—	70
R_1R_2CH—H	7.3×10^3	2.2×10^3	1.7×10^3
$R_1R_2R_3C$—H	7.3×10^3	3.9×10^3	1.7×10^3
$RC(OH)_2$—H	7.3×10^3	—	—
1,5 H Shift			
RCH_2—H	1.8×10^5	5.9×10^4	4.3×10^4
RCH(OH)—H	1.4×10^6	—	3.1×10^5
R_1R_2CH—H	3.4×10^7	8.8×10^5	7.5×10^6
$R_1R_2R_3C$—H	3.4×10^7	1.6×10^7	7.6×10^6
$RC(OH)_2$—H	3.4×10^7	—	—

Source: Adapted from Atkinson and Lloyd, 1984, and Carter and Atkinson, 1985.
[a] Note that these are per abstractable H atom; thus for abstraction of H from a CH_3 group, these should be multiplied by 3, and for abstraction from a $-CH_2-$ group, they should be multiplied by 2.

where R_1 = alkyl group and R_2, R_3 = H or an alkyl group. Bond cleavage generally produces the most stable alkyl radical (e.g. see Drew et al., 1985).

Table 7.10 gives estimated decomposition rates for some simple alkoxy radicals at 298°K in 1 atm of air. It is seen they can be quite substantial for the larger secondary and tertiary alkoxy radicals. However, it is also clear that there is considerable discrepancy in the calculated rates of decomposition. Experimental work directed toward elucidating the rates of alkoxy radical decomposition is warranted.

7-D-3d. *Reaction with NO*

A fourth reaction channel available to alkoxy radicals in typical urban atmospheres is addition to NO to form alkyl nitrites, reactions (15a) and (16a), or abstraction by NO to form carbonyl compounds, reactions (15b) and (16b):

$$CH_3O + NO \xrightarrow{a} CH_3ONO \qquad \xrightarrow{b} HCHO + HNO \tag{15}$$

$$R_1{-}\underset{R_2}{\overset{H}{C}}{-}O + NO \xrightarrow{a} R_1{-}\underset{R_2}{\overset{H}{C}}{-}ONO \qquad \xrightarrow{b} R_1R_2CO + HNO \tag{16}$$

Rate constants for these reactions are highly uncertain but are estimated to be in the region of $\sim 3 \times 10^{-11}$ cm^3 $molecule^{-1}$ s^{-1}, with $\sim 20\%$ of the reaction proceeding via (16b) for C_2 and larger radicals (Atkinson and Lloyd, 1984). For example, Morabito and Heicklen (1985) have shown that at 175°C approximately 20–30% of the reaction of simple alkoxy radicals with NO proceeds by abstraction rather than addition. For CH_3O, the fraction proceeding via (15b) is less than 20% and could be as small as 1% or less at atmospheric pressure.

Recently the first direct rate measurement for the reaction of an alkoxy radical with NO has been reported; in this study, Balla et al. (1985) report that for the isopropoxy radical, the preexponential factor for the overall reaction, i.e. for reactions (16a) + (16b), is 1.22×10^{-11} cm^3 $molecule^{-1}$ s^{-1} and the activation energy is 620 cal mol^{-1}. The rate constant at 298°K is thus 3.5×10^{-11} cm^3 $molecule^{-1}$ s^{-1}.

Uncertainties in the kinetics leading to alkyl nitrite formation will not cause great uncertainties in our modeling of atmospheric chemistry, however, because in daylight nitrites rapidly photolyze back to the original reactants (Section 3-C-5):

$$\text{RONO} + h\nu\ (\lambda \lesssim 440\ \text{nm}) \rightarrow \text{RO} + \text{NO} \tag{17}$$

Thus the reaction of alkoxy radicals with NO serves primarily as a temporary storage of these radicals during the night. At dawn, photolysis regenerates RO + NO, a step that contributes to the *aged smog* phenomenon, that is, the contribution of pollutants from one day to the enhancement of photochemical air pollution on the next day.

TABLE 7.10. Estimated Rates of Decomposition of Alkoxy Radicals

	$k(s^{-1})$ (1 atm Air, 298°K)		
Alkoxy Radicals	Baldwin et al. (1977)[a]	Batt (1979a, b, 1980)	Carter and Atkinson (1985)
$CH_3{-}CH_2O$	2×10^{-5}	<0.14	0.16
$CH_3CH_2CH_2{-}CH_2O$	0.3	—	26
$CH_3CH_2{-}\overset{\overset{O}{\vert}}{C}HCH_3$	3×10^3	5×10^3	4.8×10^3
$HOCH_2CH_2CH_2{-}CH_2O$	0.2	—	—
$CH_3{-}\underset{\underset{CH_3}{\vert}}{\overset{\overset{CH_3}{\vert}}{C}}O$	2×10^3	$\leqslant 1 \times 10^3$	5.3×10^2

Source: Adapted from Atkinson and Lloyd, 1984, and Carter and Atkinson, 1985.
[a] As recalculated by Atkinson and Lloyd (1984).

7-D-3e. *Reaction with NO_2*

The reaction of alkoxy radicals with NO_2 may proceed in a manner analogous to the NO reaction, that is, via addition to form alkyl nitrates (path a) or abstraction to form carbonyl compounds and nitrous acid (path b):

$$R_1-\overset{\displaystyle H}{\underset{\displaystyle R_2}{\overset{|}{\underset{|}{C}}}}-O + NO_2 \xrightarrow[a]{} R_1-\overset{\displaystyle H}{\underset{\displaystyle R_2}{\overset{|}{\underset{|}{C}}}}-ONO_2 \tag{18}$$

$$\xrightarrow[b]{} HONO + R_1R_2CO$$

The overall rate constants again are uncertain but appear to be ~50% of those for the reaction with NO i.e., $\sim 1.5 \times 10^{-11}$ cm^3 molecule^{-1} s^{-1} for the reaction of all RO radicals with NO_2 (Atkinson and Lloyd, 1984). While the contribution of the abstraction reaction (18b) appears to be negligible at room temperature, more work is needed both on the relative importance of paths (a) and (b) as well as on the overall rate constants. For example, Balla et al. (1985) report using their direct measurement technique, that in contrast to previous studies, the reaction of the isopropoxy radical with NO_2 is faster than that with NO at pressures of less than 50 Torr. McCaulley et al. (1985) report a rate constant for the reaction of CH_3O with NO_2 of 1.5×10^{-12} cm^3 molecule^{-1} s^{-1} at 298°K; since they state that this is at or near the low pressure limit, it will certainly be significantly greater at one atmosphere (see Chapter 4-A-4).

7-D-3f. *Relative Importance of Various Fates of Alkoxy Radicals in the Troposphere*

The relative importance of each of the reaction paths available to RO under typical tropospheric conditions is summarized in Table 7.11. Reaction with NO is not included, since like NO_2, it is fairly slow and may act as a temporary nighttime storage for alkoxy radicals rather than as a permanent sink.

It is seen that for the simplest alkoxy radicals, reaction with O_2 to form aldehydes and HO_2 is overwhelmingly the fate. For the larger radicals, however, isomerization or decomposition may become significant, the relative importance of these paths depending critically on the structure of the radical. Even with NO_2 at 0.1 ppm, corresponding to a moderately polluted atmosphere, the addition reaction to form alkyl nitrates is insignificant. The three major atmospheric fates of alkoxy radicals as a class, then, are reaction with O_2, isomerization, and decomposition. In all three cases, free radicals are produced which then carry on the chain reactions.

While the reactions discussed above are presumed to be initiated by alkane oxidations in the troposphere, analogous reactions occur regardless of the source of these types of radicals, although the relative rates of the various reactions will clearly depend on the structure of the radicals. As we shall see in the following

TABLE 7.11. Relative Importance of Various Fates of Alkoxy Radicals Under Typical Polluted Tropospheric Conditions

	Reaction Rate (s^{-1})			
Alkoxy Radical	Reaction with O_2	Decomposition	Isomerization	Reaction with NO_2[a]
CH_3CH_2O	4.1×10^4[b]	<0.16[c]	Negligible	38[c]
$CH_3CH_2CH_2CH_2O$	3.0×10^4[c]	26	1.3×10^5 (1.3×10^5)[d]	38[c]
$CH_3CH_2CH(O)CH_3$	3.8×10^4[c]	4.8×10^3[c] $(\sim 1.3 \times 10^5)$[d]	21[c]	38[c]

Source: Adapted from Carter and Atkinson, 1985.
[a]Assuming $[NO_2] = 2.5 \times 10^{12}$ cm^{-3} (0.1 ppm) and equal rate constants of 1.5×10^{-11} cm^3 molecule^{-1} s^{-1} for all the alkoxy radicals (see text).
[b]Experimental value (Gutman et al., 1982).
[c]Estimated values; see text.
[d]Experimental value, relative to reaction with O_2 (Carter et al., 1979b; Niki et al., 1981d; Cox et al., 1981).

sections, the oxidation of all types of organics leads to similar types of radicals. Their fates in the troposphere will not be treated explicitly unless reactions other than those discussed above may be important.

7-E REACTIONS OF ALKENES

7-E-1. Hydroxyl Radical, OH

7-E-1a. Kinetics

The reactions of OH radicals with alkenes are very fast, with the rate constants for the larger alkenes having values approximately equal to the diffusion-controlled limits (i.e., reaction occurs on every collision). Table 7.12 gives selected rate constants at room temperature, as well as their temperature dependencies as recommended by Atkinson (1985) for some alkenes and haloalkenes.

One striking feature of the data in Table 7.12 is the *negative* activation energies (i.e., the rates of these reactions *decrease* as the temperature increases). Recall that the activation energy is usually thought of as the energy barrier which must be surmounted as two reactants collide in order for a reaction to occur (see Section 4-A-7). While it is easy to see that the existence of a zero energy barrier (i.e., $E_a = 0$) is possible, leading to reaction on every collision, a negative activation energy for a bimolecular reaction is not easily rationalized in the physical terms of energy barriers.

TABLE 7.12. Selected OH–Alkene Rate Constants at 298°K and Their Temperature Dependencies[a]

Alkenes	$10^{12}\,k$ (298°K) (cm^3 $molecule^{-1}$ s^{-1})	Temperature Dependence
Ethene	8.54	$2.15 \times 10^{-12}\, e^{411/T}$
Propene	26.3	$4.85 \times 10^{-12}\, e^{504/T}$
1-Butene	31.4	$6.53 \times 10^{-12}\, e^{468/T}$
cis-2-Butene	56.1	$1.09 \times 10^{-11}\, e^{488/T}$
trans-2-Butene	63.7	$1.01 \times 10^{-11}\, e^{549/T}$
2-Methylpropene	51.4	$9.51 \times 10^{-12}\, e^{503/T}$
3-Methyl-1-butene	31.8	$5.32 \times 10^{-12}\, e^{533/T}$
2-Methyl-2-butene	86.9	$1.92 \times 10^{-11}\, e^{450/T}$
Cyclohexene	67.4	n.a.[b]
2,3-Dimethyl-2-butene	110	n.a.[b]
Propadiene	9.79	$5.86 \times 10^{-12}\, e^{153/T}$
1,3-Butadiene	66.8	$1.39 \times 10^{-11}\, e^{468/T}$
β-Pinene	78.2	$2.36 \times 10^{-11}\, e^{357/T}$
α-Pinene	53.2	$1.20 \times 10^{-11}\, e^{444/T}$
Isoprene (2-methyl-1,3-butadiene)	101	$2.55 \times 10^{-11}\, e^{409/T}$
C_2HCl_3	2.36	$5.63 \times 10^{-13}\, e^{427/T}$
C_2Cl_4	0.17	$9.64 \times 10^{-12}\, e^{-1209/T}$

[a] Recommended by Atkinson (1985) which should be consulted for the original references on which the recommendations are based, temperature ranges over which the temperature expressions are valid, and the uncertainties in these rate constants.

[b] n.a. = not available.

Several explanations have been proposed to explain negative activation energies in these and other reactions. In the first, the reaction is postulated to occur via a weakly bound complex (Singleton and Cvetanovic, 1976). The individual elementary reactions then become similar to those discussed in Sections 4-A-4 and 4-A-5 for the OH + CO and OH + SO_2 reactions:

$$OH + \;{>}C{=}C{<} \;\rightleftarrows\; \left[{>}C{=}C{<}\right]\cdot OH \qquad (19,\text{-}19)$$

$$\left[{>}C{=}C{<}\right]\cdot OH \rightarrow {>}\overset{HO}{\overset{|}{C}}{-}\dot{C}{<} \qquad (20)$$

$\left[{>}C{=}C{<}\right]\cdot OH$ is the weakly bound complex postulated as an intermediate between the separated reactants and the covalently bound free radical adduct ${>}\overset{HO}{\overset{|}{C}}{-}\dot{C}{<}$. If k_{19}, k_{-19}, and k_{20} are the rate constants for the individual reactions, then following the same steady state analysis used in Sections 4-A-4 and 4-A-5, one obtains the following expression for the rate constant (k) for the overall reaction:

$$k = \frac{k_{19}k_{20}}{k_{-19} + k_{20}}$$

If $k_{-19} >> k_{20}$, then

$$k = \frac{k_{19}k_{20}}{k_{-19}} = \left(\frac{A_{19}A_{20}}{A_{-19}}\right) e^{-(E_{19}+E_{20}-E_{-19})/RT}$$

and the activation energy for the OH–alkene reaction is the combination ($E_{19} + E_{20} - E_{-19}$) of activation energies for the individual steps. If $E_{-19} > (E_{19} + E_{20})$, a negative activation energy results. E_{19}, the activation energy for complex formation, is likely close to zero, and hence E_{-19} must be greater than E_{20}; if $k_{-19} >> k_{20}$, then $A_{-19} >> A_{20}$.

The second possible explanation for thc observed negative activation energies lies in the use of the Arrhenius expression. As described in Section 4-A-7, the Arrhenius expression for the temperature dependence of rate constants is an empirical expression found to describe most kinetic data over limited temperature ranges. However, it is not surprising that for some reactions it does not fit the data with the expected positive activation energy and temperature-independent preexponential A factor. As a result, data are sometimes fit to the form $k = BT^n e^{-E_a/RT}$. Indeed, most of the temperature dependence data for OH–alkene reactions can be fit to this alternate form with $n \simeq -1.5$.

As discussed in Chapter 4, kinetic theories predict that the A factor does depend on temperature, and this may also explain the OH–alkene kinetics. Thus one can apply transition state theory (see the kinetics texts cited in Chapter 4, for example) to the OH–alkene reaction (Atkinson et al., 1977):

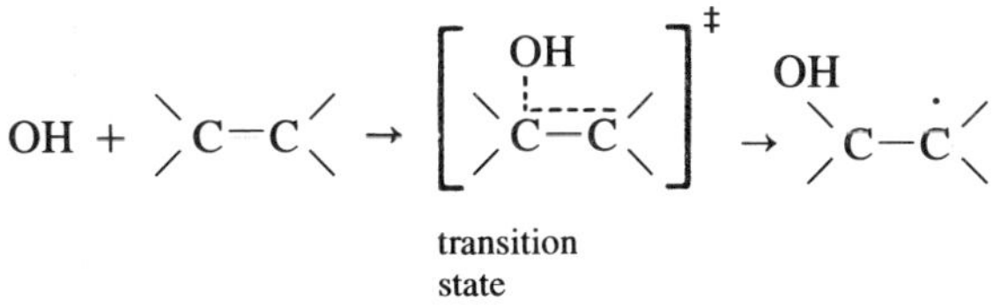

(21)

This theory predicts that the A factor is given by

$$A = \tau \frac{\mathbf{k}T}{h} \frac{Q^{\ddagger}}{Q_{C=C}Q_{OH}}$$

where τ is the transmission coefficient, $Q^{\ddagger}$, $Q_{C=C}$, and Q_{OH} are the partition functions for the transition state, alkene, and OH, respectively, $\mathbf{k}$ and h are the Boltzmann and Planck constants, and T is the temperature. Q_{OH} is expected to vary with

$T^{5/2}$. If the partition functions for the transition state and the alkene have the same temperature dependence, then A will vary with $T^{-3/2}$, as will k if the true activation energy is zero or nearly so. This is consistent with the OH–alkene data where $n \simeq -1.5$ when the rate constants are fit to the form $k = BT^n e^{-E_a/RT}$.

Finally, Zellner and co-workers suggest that these negative activation energies can be rationalized on the basis of collision theory (Section 4-A-7a) as well; thus a $T^{-1.5}$ temperature dependence is predicted if the reaction cross section is assumed to increase rapidly at the threshold energy, reach a sharp maximum, and then decrease as the energy increases (Zellner and Lorenz, 1984).

7-E-1b. Mechanisms

As might be expected for an electron-deficient free radical, OH adds to the double bond to form a radical adduct:

$$OH + \gt C{=}C\lt \rightarrow \gt \underset{}{\overset{HO}{\overset{|}{C}}}{-}\dot{C}\lt \qquad (21)$$

This adduct was first directly observed by Niki and co-workers (Morris et al., 1971) in the ethene and propene reactions using mass spectrometry (MS). Figure 7.3 shows the mass spectra observed when various isotopically substituted propenes were reacted with OH or OD radicals. In all cases, peaks corresponding to the adducts were observed. (Both OH and OD adducts are seen in the OD reaction because some impurity OH was also present.) The adducts have also been observed

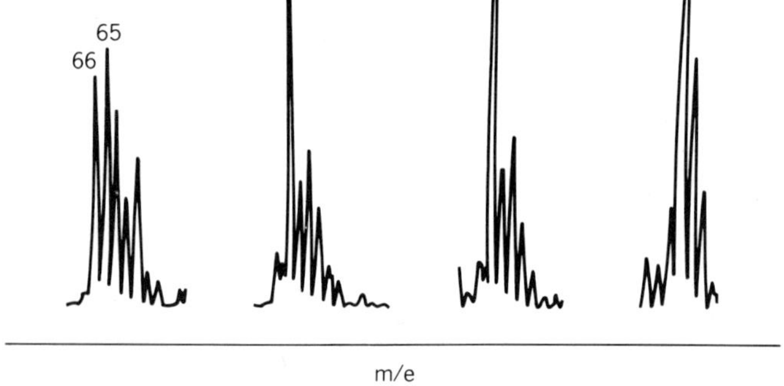

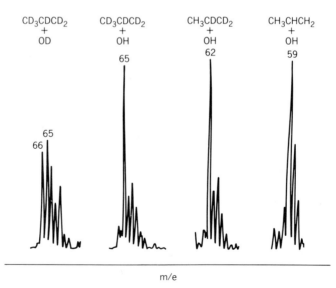

FIGURE 7.3. Mass spectra of intermediates in the reaction of propene with the hydroxyl radical (from Morris et al., 1971).

using MS by Hoyermann and Sievert (1979, 1983) in the reactions of OH with propene and the various isomers of butene.

The bond dissociation energy for the allylic C—H is only ~80 kcal $mole^{-1}$ compared to ~90–100 kcal $mole^{-1}$ for most other C—H bonds. Thus for long chain alkenes containing weak secondary and tertiary allylic C—H bonds, hydrogen atom abstraction in competition with addition has been suggested:

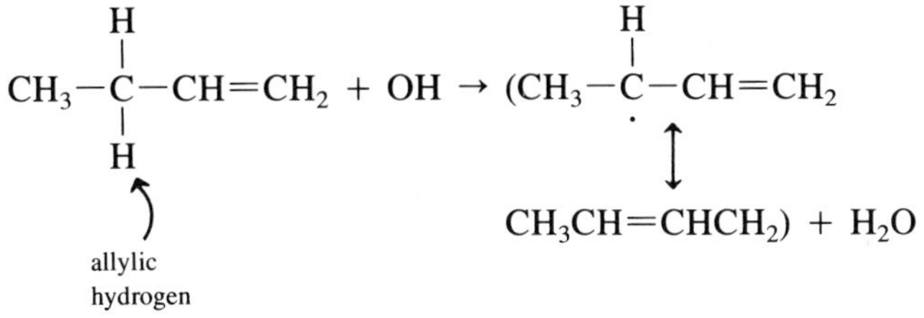

(22)

At room temperature, the results of a low-pressure study suggested that as much as 20% of the reaction with 1-butene proceeds via abstraction (Biermann et al., 1982); however, more recent work suggests that abstraction is negligible, ≲10% of the reaction (Hoyermann and Sievert, 1983; Atkinson et al., 1985b). Thus hydrogen abstraction can essentially be ignored at ambient tropospheric pressures and temperatures.

In the early literature on OH–alkene reactions there was some confusion regarding the mechanisms because the relative importance of addition and abstraction depends very much on the experimental conditions. The addition of OH to the alkene is exothermic by approximately 32 kcal $mole^{-1}$ so that the adduct contains excess energy when initially formed. This energized adduct may decompose back to reactants, be stabilized by collision with another molecule, or, for the larger alkenes, fragment into smaller species. At low pressures (<< 1 Torr) characteristic of some experimental techniques, the decomposition back to reactants can predominate for the smaller alkenes, so that the addition reaction is not observed. Competing pathways such as abstraction are then relatively more important.

However, as the total pressure is increased, stabilization of the excited adduct increases. For example, Niki and co-workers observed, during the experiments shown in Fig. 7.3, that the OH–olefin adduct peaks increased as the total pressure increased from 1 to 4 Torr. This has since been confirmed by a number of researchers. Thus at atmospheric pressure, decomposition of the adduct back to the reactants is now believed to be insignificant.

This example once again illustrates the importance of studying reaction mechanisms under atmospheric conditions, or, if this is not possible, under a sufficiently widely varying set of experimental conditions that extrapolation to the atmosphere can be carried out reliably. The term *atmospheric conditions* means not only in 1 atm of air, but also with reactant and co-pollutant concentrations typical of those found in the troposphere. However, as seen in Chapter 4, many of the techniques used to study the kinetics and mechanisms of reactions of atmospheric interest cannot be applied under realistic atmospheric conditions. Therefore great care must be taken in extrapolating the results of such studies to the troposphere. In more than one case, considerable confusion has arisen in the field of atmospheric chemistry because of invalid extrapolations of experimental data taken under conditions far from atmospheric (e.g., low pressures in the absence of air).

For asymmetrical alkenes such as propene, addition of OH may occur at either end of the double bond, giving rise to different radical adducts:

$$OH + CH_3CH=CH_2 \xrightarrow{a} CH_3CH-CH_2OH$$

$$\xrightarrow{b} CH_3\overset{OH}{\underset{|}{CH}}-CH_2 \quad (23)$$

From basic organic chemical principles (analogous to the Markovnikov rule for electrophilic additions), one might expect the formation of the more stable secondary radical in (a) to dominate, and this appears to be the case. For example, Cvetanovic (1976) has estimated from his studies of the stable products ultimately formed in the OH–propene reaction that ~65% of reaction (23) proceeds via (a). This is also consistent with the product yields observed in C_3H_6–NO_x–air irradiations (e.g. see Shepson et al., 1985).

In summary, the initial steps in OH–alkene reactions are now agreed to be almost exclusively addition to the double bond.

The fate of the OH–olefin adduct under atmospheric conditions has been the subject of a number of studies. Since it effectively is an alkyl radical, as discussed in Section 7-D, it is expected to add O_2 to form an alkylperoxy radical which, under most conditions, will oxidize NO to NO_2 and form a hydroxy-substituted alkoxy radical.

For ethene, for example, the following reactions are anticipated:

$$OH + C_2H_4 \rightarrow H-\underset{H}{\overset{OH}{\underset{|}{C}}}-\underset{H}{\overset{|}{C}}-H \xrightarrow{O_2} H-\underset{H}{\overset{HO}{\underset{|}{C}}}-\underset{H}{\overset{O}{\overset{\diagup}{C}}}-H$$

$$H-\underset{H}{\overset{HO}{\underset{|}{C}}}-\underset{H}{\overset{O}{\overset{\diagup}{\underset{|}{C}}}}-H + NO \rightarrow NO_2 + H-\underset{H}{\overset{HO}{\underset{|}{C}}}-\underset{H}{\overset{O}{\underset{|}{C}}}-H \quad (24)$$
$$(I)$$

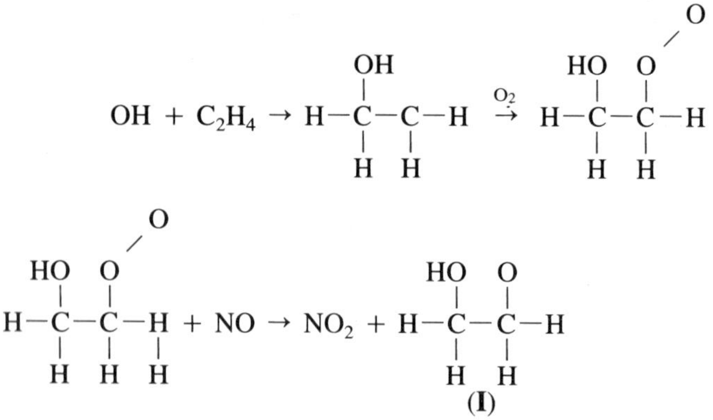

The alkoxy radical (**I**) has available the reaction paths discussed in Section 7-D-3, that is, reaction with O_2, decomposition, and reaction with NO_2. Isomerization can be ignored since 1,4 and 1,5 hydrogen shifts are not possible with this simple radical. As seen earlier, reaction with NO_2 and NO under typical atmospheric conditions is also negligible. However, decomposition and reaction with O_2 remain as possibilities:

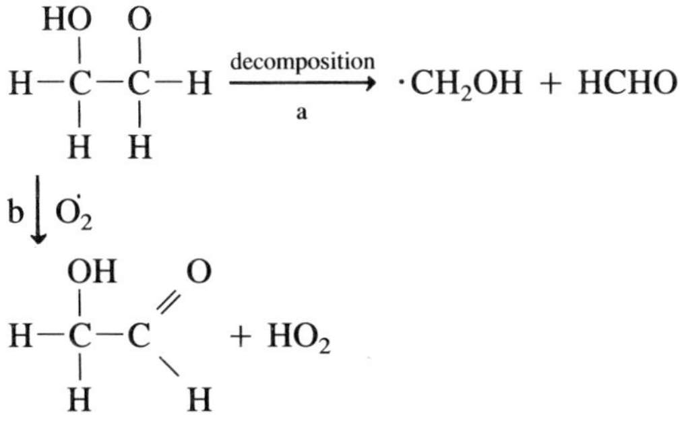

glycolaldehyde

(25)

Niki and co-workers (1978a) using FTIR observed that the reaction of OH with C_2H_4 in 700 Torr of air produced two HCHO molecules about 80% of the time. This is consistent with reaction (24) followed by (25), if the decomposition (25a) accounts for approximately 80% of the loss of the alkoxy radical and if the CH_2OH radical formed reacts with O_2 to produce HCHO:

$$CH_2OH + O_2 \rightarrow HCHO + HO_2 \tag{26}$$

By following the relative yields of glycolaldehyde and formaldehyde as a function of the pressure of O_2, Niki et al. (1981c) established that $k_{25a}/k_{25b} \cong (1.8 \pm 0.4) \times 10^{19}$ molecules cm^{-3}; thus $k_{25a}/k_{25b}\,[O_2] \simeq 4$ at 700 Torr in air. This is consistent with the yields of HCHO observed earlier.

Evidence that the reaction of the CH_2OH radical with O_2 proceeds via (26) rather than by addition of O_2 to form an alkylperoxy radical ($OOCH_2OH$) was also obtained by Niki and co-workers (1978a) using the chlorine-atom-initiated oxidation of CH_3OH in the presence of isotopically labeled $^{18}O_2$. None of the ^{18}O was found to be incorporated into the HCHO, supporting reaction (26). Similar conclusions have been reached by studying the products of the chlorine-atom-initiated photooxidation of methanol in air using FTIR (Whitbeck, 1983).

Laser magnetic resonance was used to measure the rate constant for reaction (26) by following the production of HO_2; the reaction was found to be quite fast, $k_{26} \simeq 2 \times 10^{-12}$ cm^3 $molecule^{-1}$ s^{-1} at room temperature (Radford, 1980). This has been confirmed recently by Wang et al. (1984) who find $k_{26} = (1.4 \pm 0.4) \times 10^{-12}$ cm^3 $molecule^{-1}$ s^{-1}. The lifetime of the CH_2OH radical is thus only ~0.1 μs in 1 atm of air.

Recently a rate constant of 9.5×10^{-12} cm^3 $molecule^{-1}$ s^{-1} has been reported for this reaction of CH_2OH with O_2, significantly greater than the values determined earlier (Grotheer et al., 1985); more work on this reaction thus seems warranted.

Studies of the secondary and tertiary radicals CH_3CHOH and $(CH_3)_2COH$ indicate that they too react with O_2 predominantly by abstraction rather than by addition (e.g., Niki et al., 1978a; Carter et al., 1979a; Washida, 1981; Ohta et al., 1982):

$$R_1R_2COH + O_2 \rightarrow R_1\overset{\overset{\displaystyle O}{\|}}{C}R_2 + HO_2 \tag{27}$$

In summary, it appears that in the ethene–OH photooxidation in air, the two-carbon alkoxy radical produced decomposes most of the time via reaction (25a); however, reaction with O_2 to form a two-carbon multi-functional compound, reaction (25b), may be sufficiently fast to account for a small (~20%) amount of the reaction.

The larger olefins are expected to undergo a similar set of reactions with some minor modifications. For example, the reaction of NO with the alkylperoxy radicals will form alkyl nitrates a small fraction of the time (see Section 7-D-2a).

Figure 7.4, for example, shows the mechanism for the OH–*trans*-2-butene–NO_x photooxidation recommended by Atkinson and Lloyd (1984). This is based to a great extent on the FTIR work of Niki and co-workers (1978a) who studied the photolysis of olefin–NO_x–HONO mixtures in air. Figure 7.5 shows a portion of the infrared spectrum from the photolysis of 21.3 ppm *trans*-2-butene, 18.7 ppm HONO, 49.2 ppm NO, and 5.7 ppm NO_2 in 700 Torr of air; (*a*) is the spectrum before irradiation, (*b*) the spectrum after 1.5 min of irradiation, and (*c*) is the difference between (*a*) and (*b*). Thus "negative" peaks in (*c*) correspond to loss of reactant, and "positive" peaks to the formation of products.

Figure 7.6 shows how the difference spectrum in Fig. 7.5 can be attributed to a combination of the loss of 6.1 ppm of the reactant *trans*-2-butene and the formation of 12.8 ppm CH_3CHO and 12.6 ppm NO_2. Thus the sum of (*b*), (*c*), and (*d*) in Fig. 7.6 results in a spectrum that matches the difference spectrum (*c*) in Fig. 7.5, showing that approximately two molecules each of CH_3CHO and NO_2 are formed for each molecule of *trans*-2-butene lost. Although the alkyl nitrate was not detected in these studies, the small yield (8%) expected is consistent with the stoichiometry observed for CH_3CHO formation when experimental error is taken into account. The high yield of CH_3CHO shows that the alkoxy radical $CH_3\underset{}{\overset{\overset{\displaystyle OH}{|}}{C}}H-\overset{\overset{\displaystyle O}{|}}{C}HCH_3$ decomposes rapidly compared to reaction with O_2, as in the case of ethene. (Note that this may not be the case for the simple, unsubstituted alkoxy radical $CH_3CH_2\overset{\overset{\displaystyle O}{|}}{C}HCH_3$; see Table 7.11.)

Figure 7.7 shows the reaction scheme postulated for the 1-butene photooxidation by Atkinson and Lloyd (1984). Abstraction of an allylic hydrogen was included since it was considered to be important at that time; as seen in Fig. 7.7,

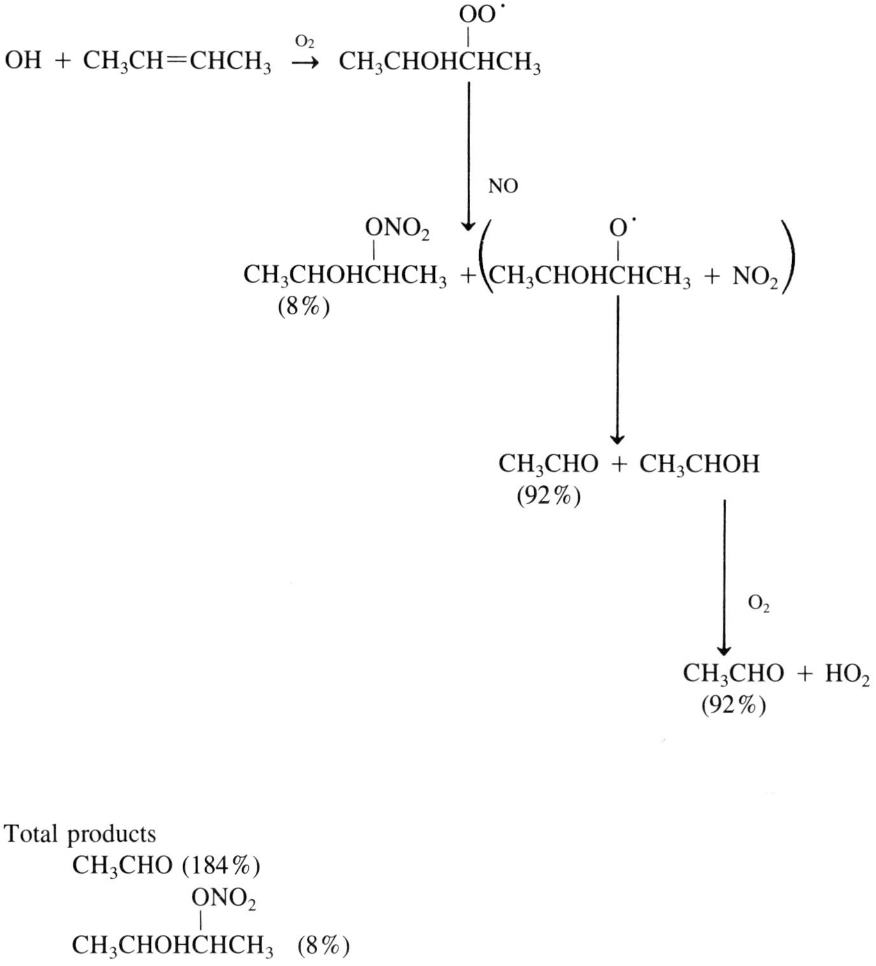

FIGURE 7.4. Recommended pathways for the OH-*trans*-2-butene-NO_x photooxidation in air (from Atkinson and Lloyd,. 1984).

the subsequent reactions of the resulting radical are expected to lead to the formation of methyl vinyl ketone, $CH_3COCH{=}CH_2$, and much smaller amounts of nitrate. However, since this scheme was postulated, methyl vinyl ketone has been shown not to be a significant product of the 1-butene oxidation, from which an upper limit of ~1% for allylic hydrogen abstraction can be derived (Atkinson et al., 1985b).

As seen above, the OH-substituted alkoxy radicals formed from the OH reactions with ethene and *trans*-2-butene seem to undergo unimolecular decomposition faster than they react with O_2. However, in the case of 1-butene, one of the alkoxy radicals formed, $CH_3CH_2CHOHCH_2O$, is sufficiently large that isomerization by a 1,5 H shift could also occur:

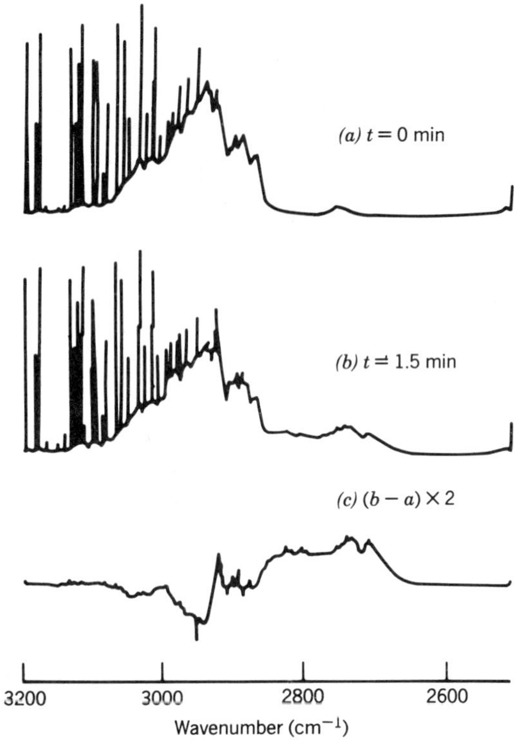

FIGURE 7.5. Fourier transform infrared spectrum of the photolysis of a mixture of 21.3 ppm *trans*-2-butene, 18.7 ppm HONO, 49.2 ppm NO, and 5.7 ppm NO_2: (*a*) Before irradiation, (*b*) after 1.5 min irradiation, (*c*) difference spectrum (*b*)–(*a*) (from Niki et al., 1978a).

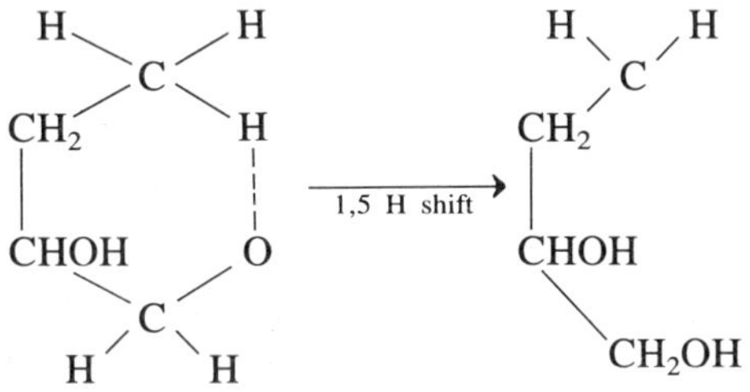

However, this isomerization must be slow relative to decomposition; thus the yield of propionaldehyde in the OH–1-butene reaction was measured as 94 ± 12%, which is only consistent with decomposition being the major fate of this radical (Atkinson et al., 1985b).

Relatively few data are available on the mechanisms of reaction of OH with alkenes containing a halogen atom attached to the double bonded carbon. While addition of OH to the double bond is still possible, a second reaction pathway, elimination of the halogen atom (X), must also be considered:

$$CH_2{=}CHX + OH \rightleftarrows [CH_2CHXOH]^* \rightarrow CH_2{=}CHOH + X$$

While the elimination of Cl or Br atoms from the adduct is thermochemically favored, Atkinson (1985) points out that the pressure dependence of the rate constants suggests that this is a relatively minor pathway. Thus addition of OH to the double bond seems to be the major route. Possible subsequent reactions of the adduct in the troposphere are discussed in Section 10-D.

7-E-2. Ozone, O_3

7-E-2a. Kinetics

Table 7.13 gives the room temperature rate constants and the Arrhenius parameters for the reaction of O_3 with selected alkenes. The rate constants increase significantly with increasing alkyl substitution, consistent with an electrophilic attack of O_3 on the double bond. Although these rate constants are substantially smaller than those for attack by OH, the concentrations of O_3 in the natural and polluted trop-

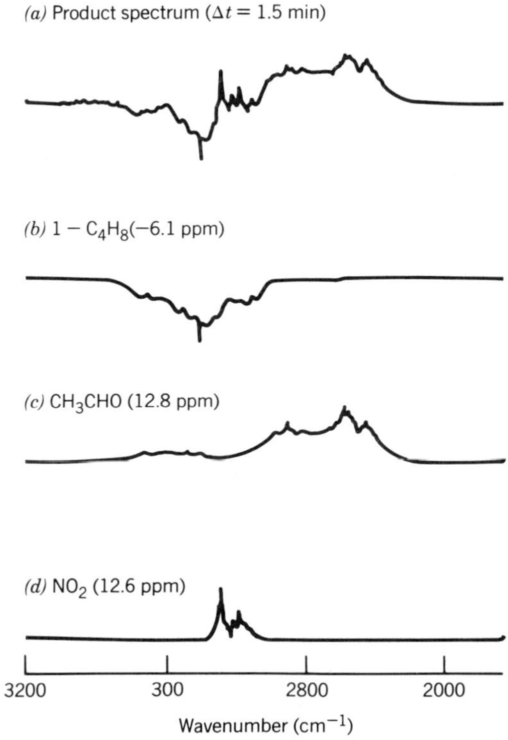

FIGURE 7.6. Analysis of difference spectrum shown in Fig. 7.5*c* (*a*) Same as Figure 7.5*c*. (*b*)–(*d*) spectra of reactant loss (*b*) and product formation (*c* and *d*) which account for the difference spectrum (*a*) (from Niki et al., 1978a).

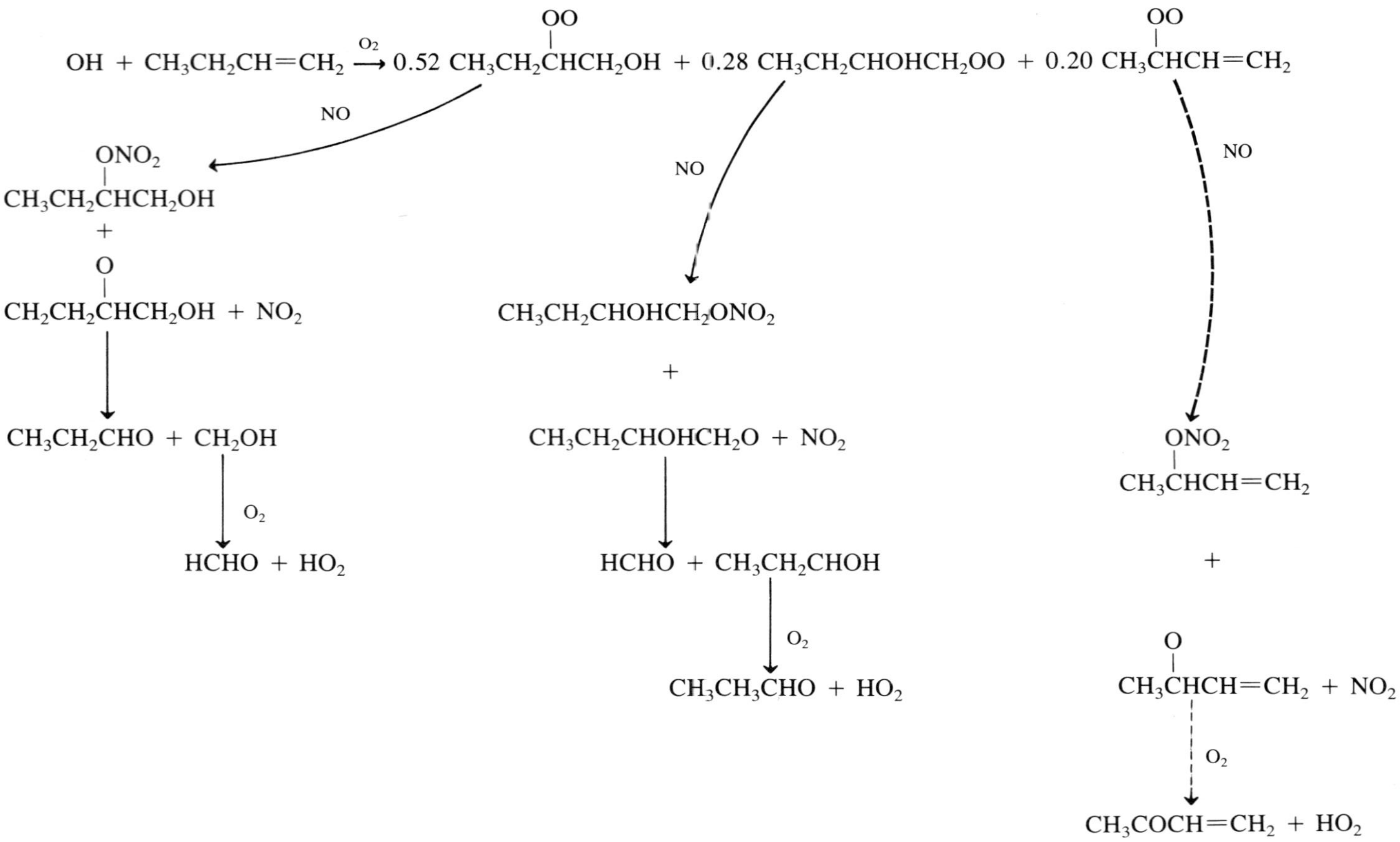

FIGURE 7.7. Reaction scheme postulated for oxidation of 1-butene by OH radicals in an irradiated NO_x-air mixture. This was from Atkinson and Lloyd, (1984); however, note that since publication of their article, the abstraction of an allylic hydrogen has been shown to account for $\leq 1\%$ of the OH-1-butene reaction, so that the path shown by the dashed arrows is negligible.

osphere are much greater than OH radicals so that reaction of alkenes with O_3 can be very important. Thus for a typical *background* concentration of 0.03–0.04 ppm O_3, $(7–10) \times 10^{11}$ cm^{-3}, the lifetimes of the alkenes in Table 7.13 range from 7 to 9 days for ethene to only 14 to 21 min for 2,3-dimethyl-2-butene.

For a detailed review of the kinetics of reaction of O_3 with these and other alkenes (including cycloalkenes and haloalkenes), the reader is referred to the review by Atkinson and Carter (1984).

7-E-2b. Mechanisms

The gas phase reaction of O_3 with alkenes has been known for years to produce aldehydes, ketones, and acids, as well as inorganics such as CO, CO_2, and H_2O. For a thorough review of the literature through 1960, the reader is referred to Leighton's monograph (1961).

By analogy with the well studied reaction in the liquid phase, the gas phase mechanism is believed to involve an initial electrophilic addition of O_3 to the double bond to form a primary ozonide, or molozonide, (**I**):

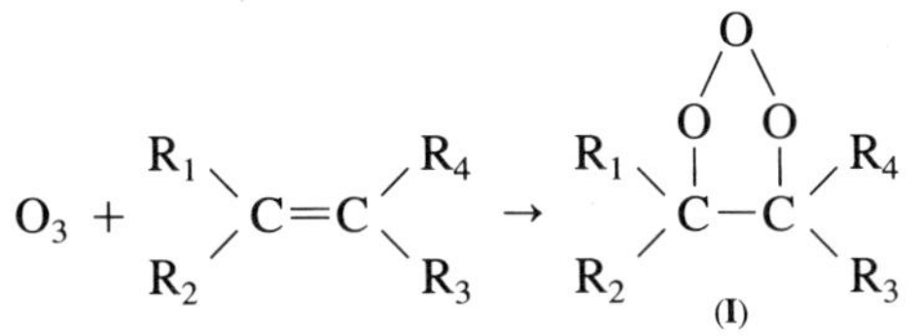

(28)

(**I**)
primary ozonide
or
molozonide

TABLE 7.13. Rate Constants at 25°C and Arrhenius Parameters for the Reaction of O_3 with Some Alkenes

Alkene	$10^{18}\,k$ (cm^3 $molecule^{-1}$ s^{-1}) at 298°K	$10^{15}\,A$ (cm^3 $molecule^{-1}$ s^{-1})	E_a (cal $mole^{-1}$)
Ethene	1.75	12.0	5232 ± 117
Propene	11.3	13.2	4182 ± 648
1-Butene	11.0	3.46	3403 ± 325
2-Methylpropene	12.1	3.55	3364 ± 182
cis-2-Butene	130	3.52	1953 ± 229
trans-2-Butene	200	9.08	2258 ± 435
2-Methyl-2-butene	423	6.17	1586 ± 389
2,3-Dimethyl-2-butene	1.16×10^3	3.71	690 ± 703
1-Hexene	11.7	n.r.[a]	n.r.
1,3-Butadiene	8.1	88	5500
Isoprene (2-methyl-1,3-butadiene)	14.3	12.3	4000

Source: Recommended by Atkinson and Carter, 1984.
[a] n.r. = not reported.

However, to date there is no direct experimental evidence for the existence of such primary ozonides in the gas phase.

Subsequent cleavage of a peroxide O—O bond and the C—C bond in **I** leads to a carbonyl compound and a *Criegee intermediate*, named after the investigator who originally proposed this mechanism (e.g., see Criegee, 1975):

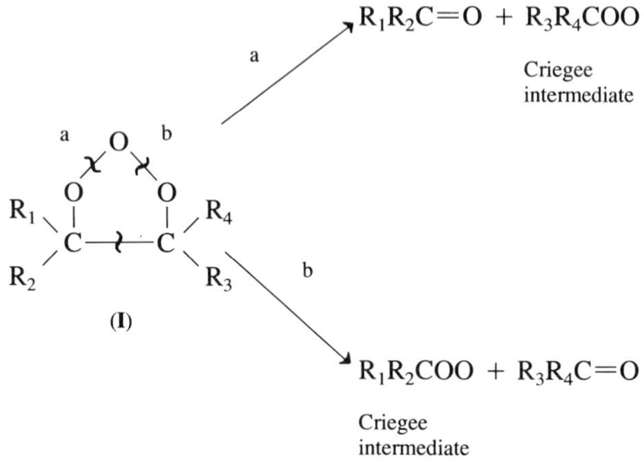

(29)

The structure of the Criegee intermediate has been the subject of recent studies because of the relevance to its subsequent reactions and fates in the atmosphere. Although traditionally it has been known as a *Criegee zwitterion* in liquid phase reactions and usually written $R_1R_2\overset{+}{C}{-}O{-}O^-$ or $R_1R_2C{=}\overset{+}{O}{-}O^-$, such an ionic structure is clearly unlikely in the gas phase where it is not surrounded by polar solvent molecules which can stabilize the charge separation.

In the gas phase, the species often has been written $R_2R_2\dot{C}O\dot{O}$ and called a *Criegee biradical.* However, four possible non-zwitterionic structures have been proposed based on theoretical studies. For the Criegee intermediate from the ethene reaction, these are (Ha et al., 1974; Wadt and Goddard, 1975):

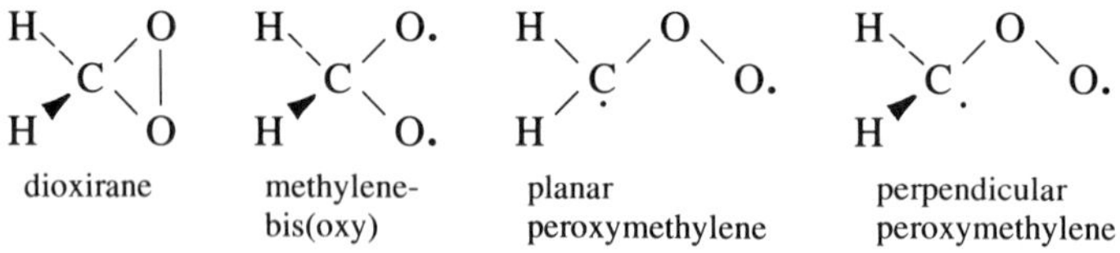

The calculated heats of formation increase continuously from dioxirane to the perpendicular form of peroxymethylene, that is, the stability decreases in this direction. The theoretically most stable form, dioxirane, has been observed both by microwave and mass spectroscopy from the reaction of ethene with O_3 at low temperatures (−196°K) (Lovas and Suenram, 1977; Martinez et al., 1977). However, as seen in Fig. 7.8, all four structures are energetically accessible in the O_3–ethene reaction and, for the sake of convenience, we use the notation R_1R_2COO without specifying the precise structure(s).

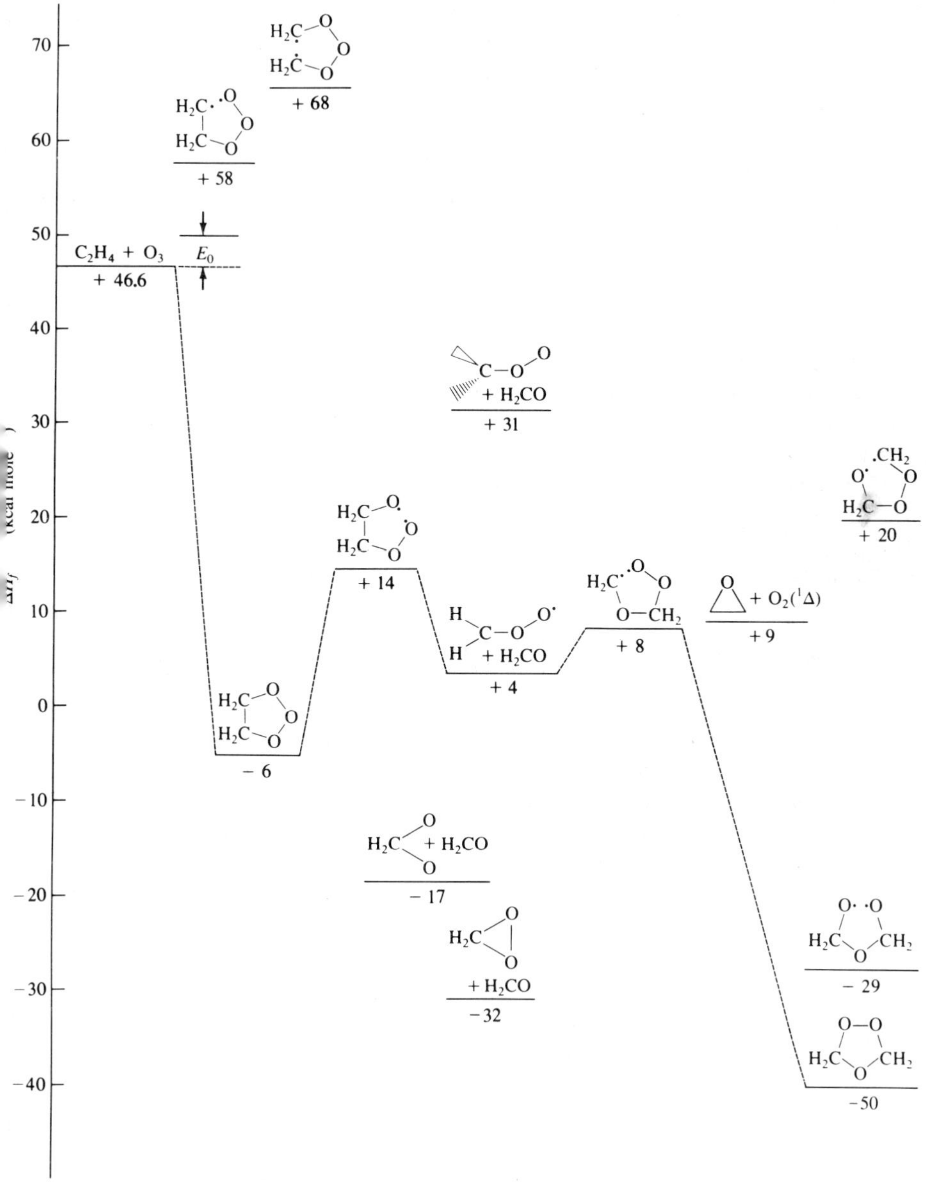

FIGURE 7.8. Calculated thermochemistry for the O_3–ethene reaction. Heats of formation in kcal mole^{-1} are shown below each set of molecules. The Criegee mechanism is shown by the dashed line (from Harding and Goddard, 1978).

In the solution phase, the solvent acts to remove any excess energy in the fragments formed in reaction (29) and cage effects are operative. Thus the fragments can recombine to form secondary ozonides (**II**). These are observed as stable reaction products:

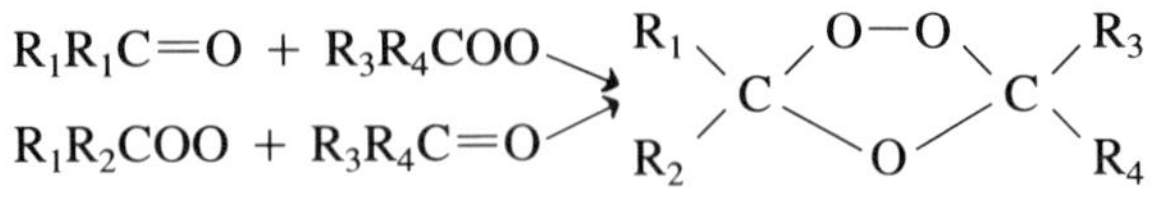

$$R_1R_1C=O + R_3R_4COO \atop R_1R_2COO + R_3R_4C=O \rightleftharpoons {R_1 \atop R_2}\!\!>\!\!C{O-O \atop O}\!\!C\!\!<{R_3 \atop R_4} \tag{30}$$

secondary ozonide
(**II**)

In the gas phase, however, stabilization of energetic species is much less efficient and cage effects are not present to keep the fragments in close proximity. Thus the two species formed in (29) fly apart rapidly when they are formed and because the Criegee intermediate will likely contain some excess energy from the reaction exothermicity, it may subsequently decompose.

In early studies of gas phase O_3–olefin reactions, only stable products could be observed. In 1958, however, Saltzman used mass spectrometry to observe free radicals from the reaction of O_3 with 1-hexene. While this was the first observation of free radical production in these reactions, no quantitative assessment of their yields could be made. More importantly, their formation within the mass spectrometer itself by fragmentation of larger species could not be ruled out with certainty. However, the peaks were observed to persist down to low ionizing voltages, suggesting that they were indeed products of the O_3–olefin reaction rather than fragmentation peaks. This observation was important because it raised the question as to whether these reactions contribute directly to the free radical chains in the atmosphere rather than forming stable products (which, however, may later form free radicals, e.g., via photolysis).

It was not until the early 1970s that direct confirming evidence for free radical production in O_3–olefin reactions became available. The first support was the observation of chemiluminescence from vibrationally excited OH radicals formed as products in a variety of O_3–olefin reactions in the Torr pressure range (Finlayson et al., 1972, 1974). As seen in Fig. 7.9, the emission spectrum was identical to the Meinel bands from the reaction of hydrogen atoms with O_3,

$$H + O_3 \rightarrow OH_{v \leq 9} + O_2 \tag{31}$$

strongly suggesting that hydrogen atoms were produced in O_3–olefin reactions. Simultaneously, chemiluminescent emissions due to excited carbonyls and dicarbonyls were observed, some of which also suggested free radical reactions (Kummer et al., 1971; Finlayson et al., 1974).

The second piece of evidence for free radical production was the direct observation of free radicals in these reactions using photoionization mass spectrometry (Atkinson et al., 1973); as discussed in Section 4-B-2c judicious choice of the wavelength (i.e., energy) of the ionizing radiation minimizes the fragmentation of

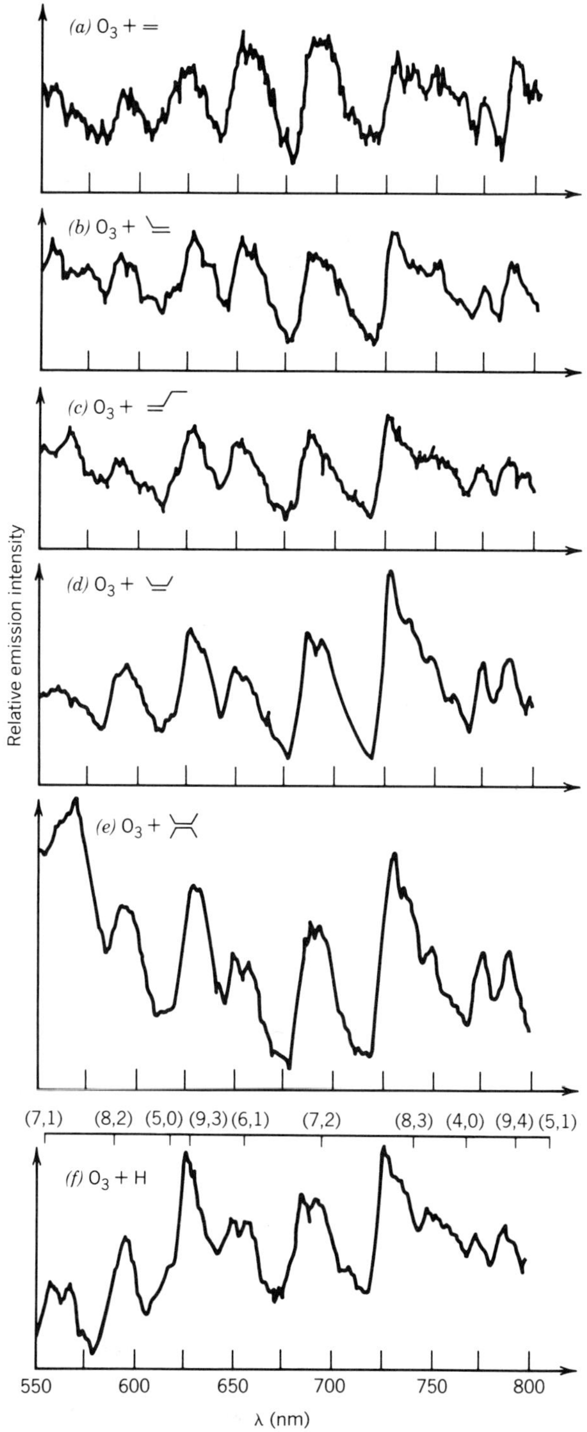

FIGURE 7.9. Chemiluminescent emission in the range 550–800 nm from O_3–olefin reactions and from the H + O_3 reaction (from Finlayson et al., 1972).

larger molecules so that small concentrations of free radicals can be detected relatively free from interferences in the presence of large concentrations of stable molecules. Thus free radicals such as HCO, C_2H_5, HO_2, and CH_3O_2 were observed for the first time from the O_3–*cis*-2-butene reaction at total pressures of ~2 Torr. While some fragmentation of larger species may not be definitively ruled out, it could not account for all the observed free radicals.

These data strongly suggested that energetic intermediates produced in the initial O_3–olefin reaction decompose to some extent to produce free radicals, at least at low pressures. It is now believed that an energetic Criegee intermediate is responsible for this free radical production. Decomposition is generally proposed to proceed via the initial isomerization to an excited molecule of a carboxylic acid. For example, for the Criegee intermediate from the O_3–ethene reaction, the following is proposed:

$O_3 + H_2C{=}CH_2 \rightarrow$

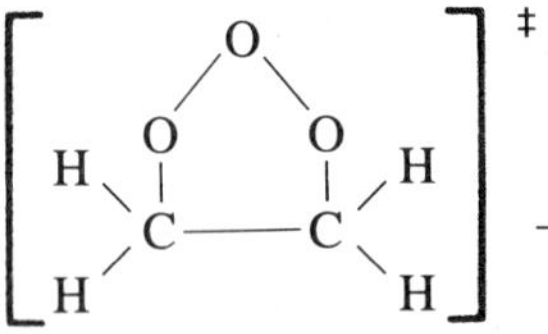

$\rightarrow HCHO + H_2COO^*$ (32)

$$H_2COO^* + M \rightarrow H_2COO \qquad (33)$$

$$H_2COO^* \rightarrow (H\overset{\overset{\displaystyle O}{\|}}{C}OH)^* \xrightarrow[a]{} CO + H_2O$$

$$\xrightarrow[b]{} CO_2 + H_2$$

$$\xrightarrow[c]{} CO_2 + 2H \qquad (34)$$

$$\xrightarrow[d]{M} H\overset{\overset{\displaystyle O}{\|}}{C}OH$$

In this scheme, H_2COO^* is the Criegee intermediate containing excess internal energy, and H_2COO is the form stabilized through collisions; it is possible that the initial O_3–alkene reaction produces biradicals in two different states, one of which can be stabilized while the other decomposes. At 700 Torr in (O_2 + N_2) or air, approximately 35–40% of the excited CH_2OO biradicals appear to be stabilized (Su et al., 1980; Kan et al., 1981; Niki et al., 1981b; Hatakeyama et al., 1984); the remainder decompose or isomerize via (34a–d). For the biradicals formed from the reaction of alkenes larger than ethene, the yield of stabilized biradicals is lower (Hatakeyama et al., 1984) (see below).

The fraction of reaction (34) proceeding via each of the paths (a)–(d) has been estimated from the observed stable product yields in two separate studies. The two estimates (Table 7.14) are in surprisingly good agreement considering that the study of Herron and Huie (1977) was carried out at ~ 8 Torr total pressure with

TABLE 7.14. Estimated Percentages of the Various Decomposition Paths for the Excited Criegee Intermediate in the O_3–C_2H_4 Reaction at Room Temperature

Products (Reaction)	Percentage of Overall Decomposition Pathway: Herron and Huie (1977)	Percentage of Overall Decomposition Pathway: Su, Calvert, and Shaw (1980)
$CO + H_2O$ (34a)	67%	58 ± 10%
$CO_2 + H_2$ (34b)	18	→ 35 ± 6 (34b + 34c)
$CO_2 + 2H$ (34c)	9	
$HC(=O)OH$ (34d)	6	7 ± 1

O_2 as the carrier gas, whereas that of Calvert and co-workers (Su et al., 1980) was carried out in 700 Torr of an O_2–N_2 mixture. Approximately 10% of the excited Criegee intermediate is proposed to decompose to free radicals (hydrogen atoms) by reaction (34c), the remainder giving stable products.

The production of free radicals in O_3–alkene reactions explains to a great extent why conflicting results have been reported over the years in the literature for the kinetics, stoichiometry, and products of these reactions. These radicals, unless scavenged, can react further with the reactants to such an extent that secondary reactions become quite significant. For example, Japar and co-workers (Japar et al., 1976) showed that the measured bimolecular rate constants for the reactions of propene and *trans*-2-butene were sensitive to added O_2; this is shown in Fig. 7.10 for propene. This phenomenon can be explained on the basis that free radical intermediates produced in the initial O_3–alkene reaction are scavenged by O_2 before they can attack the reactants.

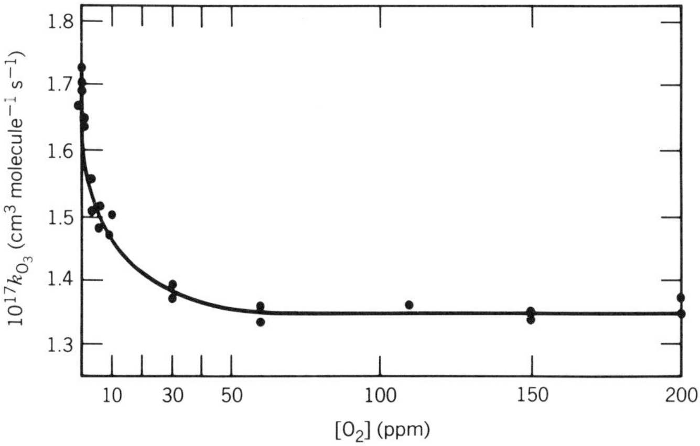

FIGURE 7.10. Effect of O_2 on the measured bimolecular rate constant for the O_3–C_3H_6 reaction at 299°C and 760 Torr total pressure in He (from Japar et al., 1976).

Because of the potential importance of secondary attack by free radicals on the reactants, these reactions are especially sensitive to the concentration regime studied; the higher the initial concentrations, the more likely secondary reactions are to be important. This is another case then where studying the reactions under conditions as close as possible to atmospheric (i.e., ppm–ppb reactant concentrations in 1 atm of air) is critical.

While the O_3–ethene reaction has been most thoroughly studied in recent years, data on the reactions of the larger olefins suggest that analogous reactions occur, that is, reaction to form a *thermalized* form of the Criegee intermediate, as well as an energized form that decomposes. Hatakeyama and co-workers (1984) measured the yields of sulfuric acid aerosol produced in the reactions of O_3 with 14 olefins in the presence of SO_2 at 1 atm in air. Because the thermalized Criegee biradical is thought to be the species oxidizing SO_2 (see below), the yields of H_2SO_4 were hypothesized to correspond to the yields of stabilized biradicals produced in the O_3–alkene reaction. These yields were 0.39, 0.25, 0.19, and 0.17 for C_2H_4, C_3H_6, *trans*-2-C_4H_8, and 2-methylpropene, respectively; for the cyclic alkenes cyclopentene, cyclohexene, and cycloheptene, the yields are much lower, 0.052, 0.032, and 0.029, respectively. In the case of *trans*-2-butene, the yield of H_2SO_4 (and hence of the thermalized biradicals) increased with total pressure from 10 Torr with the yield leveling off at ~ 600 Torr. This is consistent with competition between stabilization and decomposition of an excited Criegee biradical.

For the symmetrical alkenes, only one Criegee biradical is produced and the yields of H_2SO_4 thus reflect the stabilization of this intermediate. For *trans*-2-butene, for example, the yield of H_2SO_4 and hence stabilized CH_3CHOO biradicals is 0.185; this is in agreement with the yield of propylene ozonide of 0.18 from the *cis*-2-butene–O_3–HCHO system where the stabilized biradicals react with HCHO to form the ozonide (Niki et al., 1977). However, it is lower than the yield of 0.43 estimated by Atkinson and Lloyd (1984) from the data of Cox and Penkett (1972).

For the asymmetrical alkenes, the yield of H_2SO_4 represents a combination of the two different thermalized biradicals formed. The fraction of the reaction giving each one is thus unknown. For propene, for example, the following mechanism has been recommended (Atkinson and Carter, 1984):

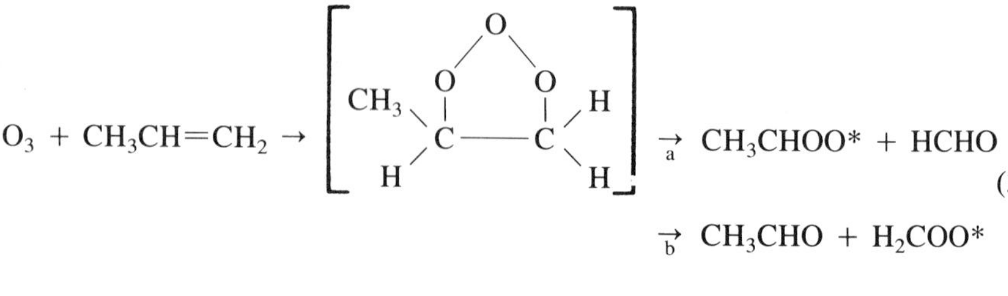

$$O_3 + CH_3CH=CH_2 \rightarrow \left[\begin{array}{c} \text{CH}_3 \diagdown \quad \overset{O}{\underset{|}{\diagup \diagdown}} \quad \diagup H \\ \diagdown C \!-\!-\! C \diagdown \\ H H \end{array} \right] \begin{array}{c} \xrightarrow{a} CH_3CHOO^* + HCHO \\ \\ \xrightarrow{b} CH_3CHO + H_2COO^* \end{array} \quad (35)$$

where $k_a \simeq k_b$ (Herron and Huie, 1978), followed by the reactions of the Criegee intermediates:

$$\mathrm{CH_3CHOO^* + M \xrightarrow[40\%]{} CH_3CHOO} \tag{36}$$

$$\begin{aligned} \mathrm{CH_3CHOO^*} &\xrightarrow[12\%]{a} \mathrm{CH_4 + CO_2} \\ &\xrightarrow[19\%]{b} \mathrm{CH_3 + CO + OH} \\ &\xrightarrow[24\%]{c} \mathrm{CH_3 + CO_2 + H} \\ &\xrightarrow[5\%]{d} \mathrm{HCO + CH_3O} \end{aligned} \tag{37}$$

$$\mathrm{H_2COO^* + M \longrightarrow H_2COO + M} \tag{38}$$

$$\mathrm{H_2COO^* \xrightarrow[\substack{a \\ \text{etc.}}]{} CO + H_2O} \tag{39}$$

The structures of the Criegee intermediates involved are highly uncertain, as are the fractions of the reaction proceeding by each of the alternate paths. Again, we have written the Criegee intermediate in the form R_1R_2COO, although it may exist in any or all of the four different electronic structures discussed earlier.

Indeed, there is some evidence that different forms of the Criegee intermediate are responsible for different observed stable products. For example, in the ethene reaction, formic anhydride, $\mathrm{H\overset{\overset{O}{\|}}{C}-O-\overset{\overset{O}{\|}}{C}H}$, has been observed as a stable product (Kühne et al., 1976; Su et al., 1980; Kan et al., 1981; Niki et al., 1981b). It appears that it is produced by the decomposition (perhaps partly on the surface of the reaction cell) of a compound tentatively identified as $HOCH_2—O—CHO$ (**III**). Formation of (**III**) is suppressed by the addition of species such as CO, CH_3CHO, and SO_2 which are thought to react with Criegee intermediate H_2COO. However, formation of (**III**) is enhanced by the addition of HCHO, leading to the hypothesis that it is produced by the reaction of some isomeric form of the thermalized Criegee intermediate with HCHO.

A simplified reaction scheme proposed by Niki and co-workers (1983) to account for their observations is

$$\mathrm{O_3 + C_2H_4 \rightarrow H_2COO^* + HCHO} \tag{40}$$

$$\mathrm{H_2COO^* \rightarrow \text{decomposition products}} \tag{41}$$

$$\begin{aligned} \mathrm{H_2COO^* + M} &\overset{a}{\rightarrow} \mathrm{H_2COO + M} \\ &\overset{b}{\rightarrow} \mathrm{Y + M} \end{aligned} \tag{42}$$

$$\mathrm{Y + HCHO \rightarrow \underset{\textbf{(III)}}{HOCH_2-O-CHO}} \tag{43}$$

$$\mathrm{\underset{\textbf{(III)}}{HOCH_2-O-CHO} \rightarrow \rightarrow H\overset{\overset{O}{\|}}{C}-O-\overset{\overset{O}{\|}}{C}H} \tag{44}$$

In this scheme Y is an isomeric form of the Criegee intermediate, for example, OCH_2O, as suggested by Kan et al. (1981).

This discussion applies to Criegee intermediates formed from unsubstituted alkenes. There is some recent evidence that substitution of chlorine atoms on the double bond results in a chlorinated Criegee intermediate which decomposes in a manner different from that described above (Niki et al., 1982a,b, 1983, 1984a; Zang et al., 1983). For example, the product distribution in the O_3 + *cis*-CHCl=CHCl reaction in air was found to be very similar to that from the chlorine-atom-initiated photooxidation of this organic. Furthermore, the addition of chlorine atom scavengers markedly reduced the observed rate of loss of the reactants (Niki et al., 1982a, 1983). This could be explained on the basis of a decomposition of a Criegee intermediate in a manner analogous to reaction (34c):

$$\mathrm{CHClOO^* \rightarrow Cl + H + CO_2} \tag{45}$$

or possibly

$$\mathrm{CHClOO^* \rightarrow Cl + OH + CO} \tag{46}$$

However, the regeneration of O_3 was also observed and attributed to the following:

$$\mathrm{CHClOO^* \rightarrow O(^3\mathit{P}) + CHClO} \tag{47}$$

$$\mathrm{O(^3\mathit{P}) + O_2 \xrightarrow{M} O_3} \tag{48}$$

Approximately 20% of the biradical CHClOO was found to decompose via reaction (47) (Niki et al., 1984a).

There is no evidence for a decomposition comparable to (47) when the Criegee intermediate contains only hydrogen atoms. Thus extrapolation of our present knowledge to alkenes containing heteroatoms should be undertaken with some care, recognizing present uncertainties.

Relatively few studies have been carried out directed to the mechanism of reaction of cycloalkenes with ozone. As discussed in Section 12-D-2d, these reactions produce some condensed phase products in laboratory systems and hence may contribute to the organic component of urban aerosols (Grosjean and Friedlander, 1980; Hatakeyama et al., 1985); the reaction of cyclohexene with O_3, for example, produces a variety of C_5 and C_6 multi-functional oxygenated compounds in aerosols (Tables 12.25 and 12.26, Chapter 12). The yield of aerosol, however, may depend on the reactant concentrations since it is well known that O_3–alkene reactions carried out at high concentrations give significant yields of products in the condensed phase. For example, Hatakeyama et al. (1985) report an increase in the yield of aerosol organic carbon in the cyclohexene-ozone reaction as the cyclohexene concentration was increased from 17 to 245 ppm, with the ratio of the initial reactant concentrations held constant. Extrapolation of the yields to low concentrations gave the fraction of initial cyclohexene converted to aerosol organic carbon as $13 \pm 3\%$ at ppm reactant concentrations.

Many of the aerosol products can be rationalized on the basis of the formation and decomposition or stabilization of an excited Criegee intermediate. The low yield of H_2SO_4 when the reaction is carried out in the presence of SO_2 discussed above (Hatakeyama et al., 1984) and the production of significant yields of HCOOH, CO, CO_2, and C_5 products suggest that decomposition of the excited Criegee intermediate is relatively important (Niki et al., 1983). However, further work is needed to clarify the formation mechanisms of the observed products, including the importance of secondary reactions in these systems.

Collisionally stabilized Criegee intermediates can react with a number of different species in the atmosphere. In the presence of an aldehyde, for example, they form secondary ozonides, as in solution. In fact, secondary ozonides as small as **(II)** have been observed (Niki et al., 1977):

$CH_3CHOO + HCHO \rightarrow$ 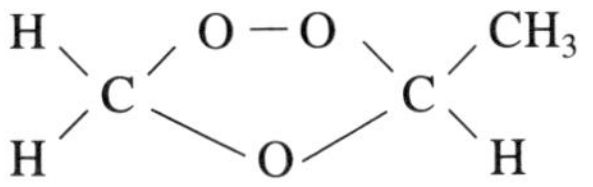 (49)

(II)

secondary propylene ozonide

Criegee intermediates also appear to react with SO_2, although the mechanism and products are not well established. For example, Cox and Penkett (1971, 1972) established that SO_2 was rapidly oxidized to form an aerosol (presumably at least partly H_2SO_4) when both O_3 and olefins were present, whereas oxidation was slow in the presence of either reactant alone. Thus an intermediate or product of the O_3–olefin reaction must be the oxidant, and the most likely species is the Criegee intermediate. This has been confirmed by Hatakeyama et al. (1984) who measured yields of H_2SO_4 in a series of O_3–alkene–SO_2 reactions; as discussed earlier, the H_2SO_4 yields ranged from 39% of the O_3 consumed for C_2H_4 to 3% for cycloheptene.

Further support comes from experiments in which added SO_2 quenched secondary ozonide formation in an O_3–alkene–aldehyde mixture to the same extent that SO_2 was consumed [i.e., one less ozonide molecule was observed for each SO_2 lost (Niki et al., 1980a)]. This is consistent with SO_2 intercepting the Criegee intermediate. Finally, as discussed earlier, SO_2 suppresses the formation of formic anhydride for which some form of the Criegee intermediate is thought to be responsible.

The products of the reaction, however, are unknown; SO_3 and a carbonyl compound, or possibly an organosulfur addition product, have been proposed.

$$H_2COO + SO_2 \xrightarrow{a} HCHO + SO_3 \xrightarrow{H_2O} \rightarrow \rightarrow H_2SO_4 \qquad (50)$$

$$\xrightarrow{b} (H_2COOSO_2) \rightarrow \rightarrow \text{organosulfur aerosol}$$

While only H_2SO_4 was detected in the experiments of Hatakeyama et al. (1984), an infrared absorption due to C=O has been observed by FTIR to be associated with an H_2SO_4 aerosol (Niki et al., 1983); this was proposed to be due to an organosulfur compound formed in (50b) or to the incorporation of stable carbonyl products into a previously formed sulfuric acid aerosol.

An alternate mechanism for production of H_2SO_4 has been proposed by Martinez and Herron (1981). In this mechanism, SO_2 reacts with the Criegee intermediate (e.g., RCHOO) to form an ozonide-like intermediate:

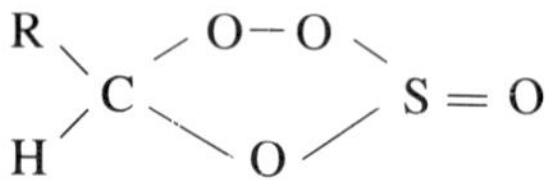

This intermediate is then proposed to react directly with water vapor to produce RCHO + H_2SO_4.

There is experimental evidence that Criegee intermediates also rearrange to acids in the presence of water vapor.

$$CH_2OO + H_2O \to \to H\overset{\overset{O}{\|}}{C}OH + H_2O \qquad (51)$$

This interaction with H_2O, suggested by Calvert and co-workers (1978), could explain the inhibition of SO_2 oxidation by water vapor in the O_3–olefin system observed earlier by Cox and Penkett (1972).

Hatakeyama et al. (1981) generated H_2COO independent of O_3–alkene reactions by photolyzing ketene in the presence of O_2:

$$CH_2CO + h\nu \to CH_2 + CO \qquad (52)$$

$$CH_2 + O_2 \xrightarrow{M} CH_2O_2 \qquad (53)$$

The addition of water vapor to the system increased the yield of formic acid. The use of ^{18}O isotopically labeled water showed that two mechanisms are likely operative, one an *exchange* reaction (54)

$$CH_2OO + H_2{}^{18}O \to [CH_2OO \cdot H_2{}^{18}O] \to HC^{18}OOH, HCO^{18}OH + H_2O \qquad (54)$$

and the other a deactivation of an excited formic acid formed by isomerization of excited CH_2OO:

$$CH_2OO^* \to HCOOH^* \qquad (55)$$

$$HCOOH^* + H_2O \to HCOOH + H_2O \qquad (56)$$

$$HCOOH^* + M \rightarrow HCOOH + M \tag{57}$$

In the latter case, ^{18}O is not incorporated into the formic acid.

Given these reactions with aldehydes, H_2O, and SO_2, it is highly likely that the Criegee intermediates also react with NO and NO_2. Although no absolute rate data are available for these reactions, oxidation of these nitrogen oxides seems reasonable:

$$R_1R_2COO + NO \rightarrow R_1\underset{\underset{\displaystyle O}{\|}}{C}R_2 + NO_2 \tag{58}$$

$$R_1R_2COO + NO_2 \rightarrow R_1\underset{\underset{\displaystyle O}{\|}}{C}R_2 + NO_3 \tag{59}$$

Finally, based on the suppression of $HOCH_2—O—CHO$ formation by CO (Su et al., 1980), the reaction of the Criegee intermediate with this pollutant should also be considered.

While the Criegee intermediate *may* thus react with NO, NO_2, SO_2, H_2O, aldehydes, and CO in the atmosphere, no absolute rate data exist for any of these reactions. Some relative rate data have been derived from studying the effects on product formation of adding molecules which compete for the Criegee intermediate. From these relative rate data and kinetic data for analogous reactions, several estimates of the absolute values of these rate constants have been made, two of which are shown in Table 7.15.

Recently, however, Lee and co-workers (Suto et al., 1985; Manzanares et al., 1985) have obtained the relative rates of reaction of the biradical CH_2OO with SO_2, H_2O, and NO_2 by following the suppression in aerosol formation in the $SO_2—O_3—C_2H_4$ system upon addition of NO_2. They obtain $k(CH_2OO + SO_2)/k(CH_2OO + NO_2) = 74 \pm 20$ and $k(CH_2OO + SO_2)/k(CH_2OO + H_2O) = (4.4 \pm 2) \times 10^3$, significantly different than the previous estimates (Table 7.15).

TABLE 7.15. Estimated Rate Constants for Reactions of the Criegee Intermediate with Species of Atmospheric Interest at 25°C

	k (cm^3 molecule^{-1} s^{-1})	
Reactant	Herron et al. (1982)	Atkinson and Lloyd (1984)
NO	—	7×10^{-12}
NO_2	1×10^{-17}–7×10^{-14}	7×10^{-13}
SO_2	3×10^{-15}–2×10^{-11}	7×10^{-14}
H_2O	2×10^{-19}–1×10^{-15}	4×10^{-18}
CO	—	1×10^{-16}
RCHO	2×10^{-16}–8×10^{-13}	2×10^{-14}
O_2	$<8 \times 10^{-22}$	—

TABLE 7.16. Estimated Relative Loss Rates for Criegee Intermediates Under Typical Smog Chamber Conditions and Ambient Air in the Los Angeles Air Basin

Reactant	Rate Constant Relative to SO_2	Smog Chamber		Ambient Air	
		Concentration (ppm)	Relative Rate[a]	Concentration (ppm)	Relative Rate[a]
NO	~100	0.002	0.2	0.01	1.0
NO_2	~0.01[b]	0.09	9×10^{-4}	0.1	10^{-3}
SO_2	1	—	—	0.05	0.05
H_2O	2.3×10^{-4}[b]	2×10^4	4.6	1×10^4	2.3
CO	2×10^{-3}	1.0	2×10^{-3}	10.0	0.02
RCHO	0.25	3.0	0.75	0.02	0.005
O_2	$<1 \times 10^{-8}$	2×10^5	$<2 \times 10^{-3}$	2×10^5	$<2 \times 10^{-3}$

Source: Adapted from Atkinson and Lloyd, 1984.

[a]Relative rate = relative rate constant × concentration (ppm).

[b]Based on relative rate data of Suto et al. (1985) and Manzanares et al. (1985).

Table 7.16 shows the relative rates of the various loss processes for the Criegee intermediate estimated under conditions typical of smog chambers and of ambient air in the Los Angeles air basin. Under the chamber conditions, reactions with NO, H_2O, and RCHO are all predicted to be important. In ambient air, NO and H_2O appear to be the most likely reactants.

Interestingly, as seen in Table 7.16, the reaction of the Criegee intermediate with O_2 is estimated to be too slow to be of importance in the atmosphere. This reaction (60)

$$R_1R_2COO + O_2 \rightarrow R_1\underset{\underset{\displaystyle O}{\|}}{C}R_2 + O_3 \tag{60}$$

had been postulated some years ago as a potential source of O_3 in polluted urban atmospheres. However, there is no evidence, either kinetic or mechanistic, to support this. As noted in Chapter 1, the only known source of O_3 is the reaction of $O(^3P)$ with O_2, and $O(^3P)$ is not produced in O_3–alkene reactions except in special cases, for example, for chlorine containing alkenes (see above). The small formation of O_3 observed in CH_3CO_3 self-reaction (Addison et al., 1980) will not be important at the low radical concentrations in the atmosphere.

While the Criegee mechanism is currently accepted as describing reasonably well the main features of the reaction of O_3 with olefins in the gas phase, other reactions contribute to a minor extent. For example, Atkinson et al. (1973), Finlayson et al. (1974), Günthard and co-workers (Kühne et al., 1976, 1978, 1980), and Martinez et al. (1981) have detected compounds in which the carbon atoms of the original C=C remain connected as minor gas phase products. Thus, 2,3-epoxybutane and butanone were detected as products of the O_3–*cis*-2-butene reaction.

Theoretical thermochemical kinetic studies of these reactions (O'Neal and Blumstein, 1973) also indicated that alternate reaction paths to the Criegee split might exist, although their importance now appears to be less than the original calculations suggested. For example, breaking of a peroxide bond in the molozonide followed by internal hydrogen atom abstractions and rearrangements were proposed by O'Neal and Blumstein.

Figure 7.11 shows one suggested overall reaction scheme for the ozone–ethene reaction (Harding and Goddard, 1978) which incorporates not only the Criegee route (path B) and one suggested by O'Neal and Blumstein (path D), but also two additional paths (A and C). The combination of these was proposed to explain the observed formation of oxygenated species in which the carbon atoms that were originally joined by the double bond remain connected (e.g., epoxide and aldehydes or ketones), as well as the observation of HCHO chemiluminescence. However, it should be noted that the products and chemiluminescence observed in O_3–olefin reactions which cannot be easily rationalized by the Criegee route are generally minor contributors to the overall reaction. Thus, confirmation of the mechanisms responsible for these minor routes has not been possible to date. Indeed, given their minor role, further work in these areas is mainly of mechanistic, rather than atmospheric, interest.

In summary, the initial formation of an unstable primary ozonide whose major fate is decomposition into a carbonyl compound and a Criegee intermediate is consistent with the experimentally observed major stable products in O_3–alkene reactions under atmospheric conditions. However, some of the Criegee intermediates contain sufficient excess energy from the reaction exothermicity that they spontaneously decompose even in 1 atm of air. The remaining *thermalized* species are expected to react in the atmosphere with pollutants such as NO, NO_2, SO_2, carbonyl compounds, and H_2O vapor, although the rate constants and mechanisms of these reactions are presently uncertain. It seems likely that several, if not all, of the four electronic structures postulated theoretically for the Criegee intermediate are formed in these reactions; different products may arise from the subsequent reactions of each of these structures, and, because of the different thermodynamics involved in each case, some pressure dependencies of the products may be observed. In addition, minor products may be formed from non-Criegee reactions of the primary ozonide.

A number of questions remain unanswered. For example, for asymmetrical alkenes, the fraction of the primary ozonide which follows each of the two possible Criegee splits has not been established, except for the reaction of vinyl chloride where 76% of the reaction was estimated to give CH_2OO + CHClO and 24% to give CHClOO + HCHO (Zhang et al., 1983). The relative importance of decomposition versus stabilization of the excited Criegee intermediate is not known for a variety of olefins over a wide range of temperatures and pressures, nor are the available pathways for decomposition of the intermediate quantitatively established. Virtually no absolute rate data or mechanistic information exist on the re-

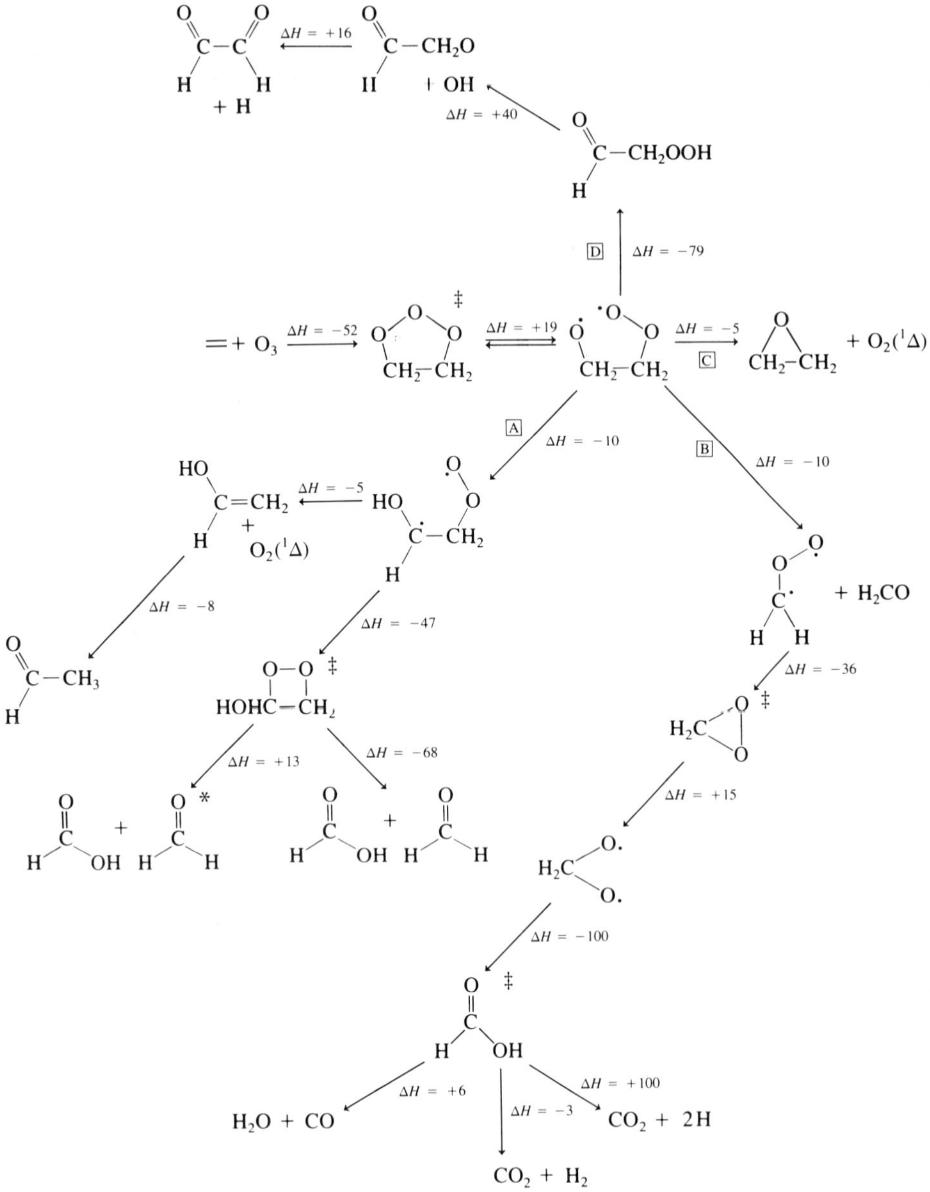

FIGURE 7.11. Suggested mechanisms for the gas phase ozone–ethene reaction. Species formed with a large excess of vibrational energy are marked with ‡ and those with excess electronic energy by * (from Harding and Goddard, 1978).

actions of the thermalized Criegee intermediates with important atmospheric species. Finally, the relative importance and mechanisms of alternate minor pathways to the Criegee route need to be more fully explored.

7-E-3. Ground-State Oxygen Atoms, $O(^3P)$

7-E-3a. *Kinetics*

Table 7.17 gives selected rate constants and the Arrhenius parameters for some $O(^3P)$–olefin reactions. As in the OH–alkene system, some of the $O(^3P)$ reactions with the larger, more highly substituted olefins show negative activation energies.

There are four possible explanations for these negative activation energies discussed in detail by Cvetanovic and Singleton (1984):

1. A simple bimolecular reaction with an activation energy that is approximately zero so that the temperature dependence of the preexponential factor becomes important.
2. The formation of a transient $O(^3P)$–alkene complex that can decompose back to the reactants or form products. However, in the latter case, the complex is likely an electron donor–acceptor complex with the electron donor being the alkene π bond and the electron acceptor being $O(^3P)$ (e.g., see Singleton and Cvetanovic, 1976). This differs from the OH–alkene case where the intermediate is a stable radical adduct.
3. The presence of two or more different reaction channels whose activation energies are significantly different; two such paths might be addition to the double bond and atom or free radical displacement (see Section 7-E-3b).

TABLE 7.17. Selected Rate Constants at 25°C and Arrhenius Parameters for $O(^3P)$–Alkene Reactions

Alkene	$10^{12}k$ (cm^3 $molecule^{-1}$ s^{-1})[a]	$10^{11}A$ (cm^3 $molecule^{-1}$ s^{-1})[b]	E_a (cal $mole^{-1}$)[b]
Ethene	0.76 ± 0.03	1.2	1670
Propene	4.05 ± 0.20	1.2	620
1-Butene	4.24 ± 0.12	1.3	660
trans-2-Butene	22.3 ± 1.3	2.3	−20
2,3-Dimethyl-2-butene	76.9 ± 1.2	2.0	−760

Source: Cvetanovic and Singleton, 1984.

[a]Mean values from studies using photochemical techniques (competitive method, phase shift, flash photolysis–chemiluminescence and flash photolysis–resonance fluorescence) at 25°C.

[b]Mean values from competitive techniques, phase shift method, and flash photolysis–chemiluminescence. Results from flash photolysis–resonance fluorescence studies not averaged in, because as discussed by Cvetanovic and Singleton (1984) they give consistently lower results than the other three methods.

4. Crossing of potential energy surfaces. If the $O(^3P)$–olefin initial interaction has a shallow minimum and the rate determining step involves crossing in a repulsive region to an attractive potential energy surface, then intersection of the surfaces at energies lower than that of the separated reactants would result in negative activation energies.

As pointed out by Cvetanovic and Singleton (1984), a combination of these four possibilities may also be occurring.

In air, most $O(^3P)$ react with O_2 to form O_3. However the $O(^3P)$-alkene reactions are discussed here because of their general interest.

7-E-3b. *Mechanisms*

The pioneering work on the mechanisms and products of $O(^3P)$–alkene reactions was carried out by Cvetanovic and co-workers and the reader is referred to his comprehensive 1963 article for a review of the field up to that time.

Oxygen atoms were generated using several techniques. The most common was the mercury photosensitized decomposition of N_2O:

$$Hg(6^1S_0) + h\nu(253.7\text{ nm}) \rightarrow Hg(6^3P_1)$$

$$Hg(6^3P_1) + N_2O \rightarrow Hg(6^1S_0) + N_2 + O(^3P)$$

The N_2 which is co-produced with the $O(^3P)$ serves as a convenient measure of the number of oxygen atoms produced. Because of its inert nature, problems with secondary reactions of the N_2O are minimal except at elevated temperatures. Other techniques were also used as a check; for example, the photolysis of NO_2 (Section 3-C-3) served as an $O(^3P)$ source in some of the studies. Using the analysis of stable products as a function of total pressure, temperature, and the presence of added free radical traps, Cvetanovic postulated the following reaction scheme:

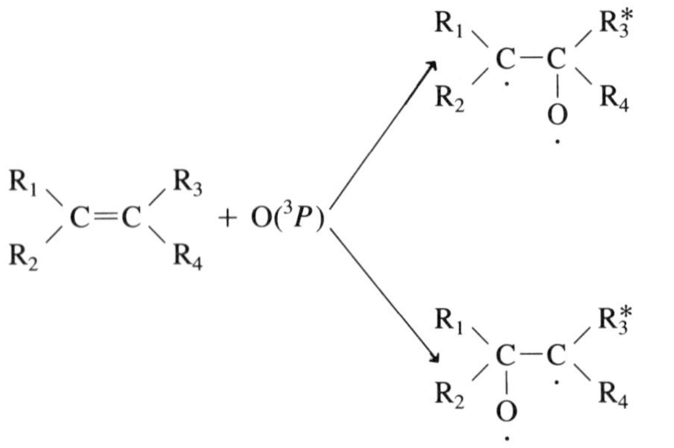

(61)

This addition step is followed by

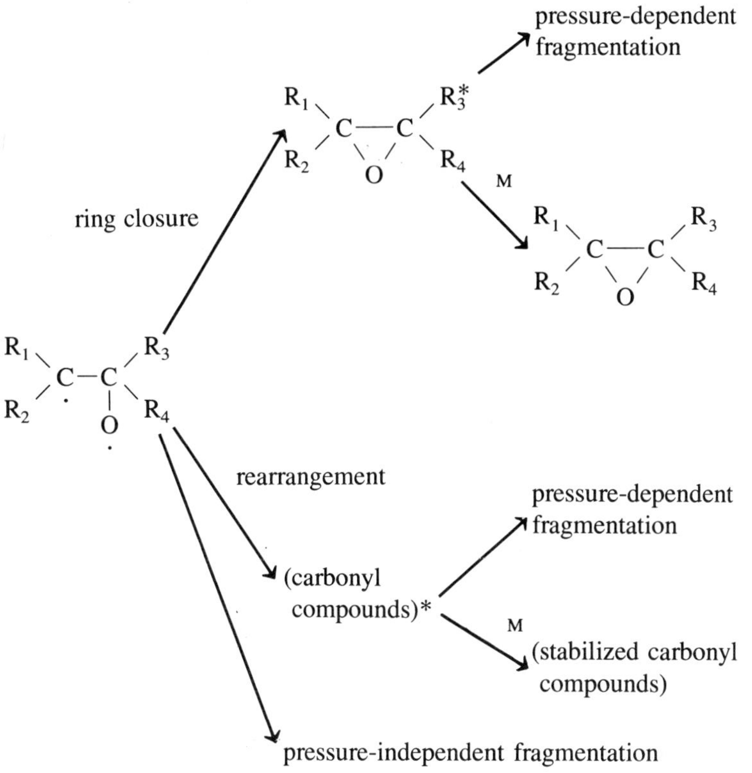

(62)
(63)
(64)
(65)
(66)

The proposed reaction mechanism consists of an initial electrophilic addition of $O(^3P)$ to the electron-rich π bond to form a triplet biradical. The triplet nature of the initial intermediate follows from the spin conservation rule, since the reactants are a triplet [the $O(^3P)$] and a singlet (the alkene), respectively. The biradical formed contains excess energy because the addition is exothermic overall; the exact amount of excess energy it contains, however, is not well defined at present (see below).

The excited biradical is then hypothesized to follow one of three possible paths: (1) ring closure to an excited epoxide, (2) rearrangement to an excited carbonyl compound, or (3) decomposition into free radicals, the so-called *pressure-independent fragmentation* route. The first two routes were postulated to result either in the stabilized epoxide or carbonyl compound, respectively, or alternatively in free radicals from the decomposition of these molecules. Formation of ground-state epoxide or carbonyl compounds requires a spin conversion since the biradical is a triplet and these addition products are singlets; the mechanism by which this is hypothesized to occur is discussed below. The relative importance of the paths is a function of the size and structure of the alkene, the total pressure, and the temperature.

Figure 7.12, for example, shows the effect of total pressure on the yields of products from the $O(^3P)$–propene reaction. This can be understood in the frame-

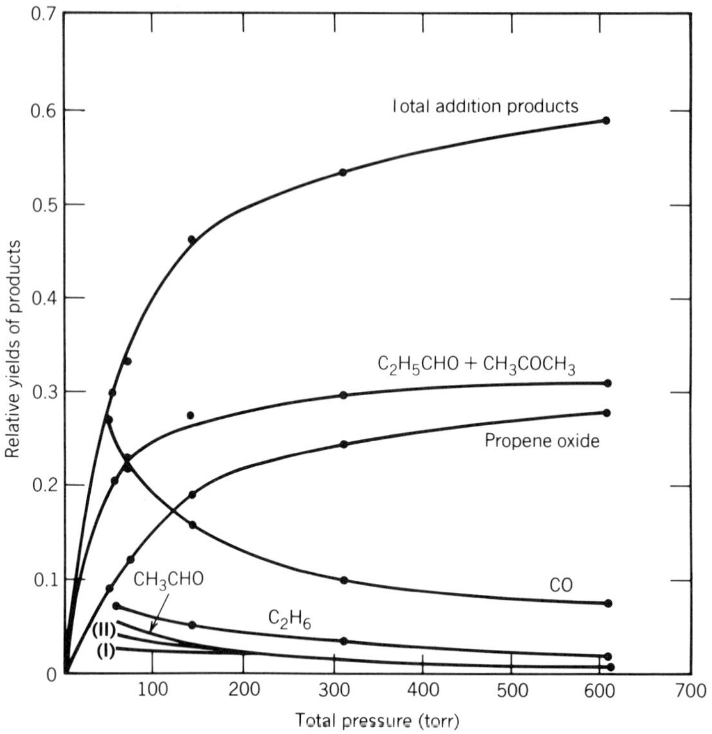

FIGURE 7.12. Effect of total pressure on the product yields in the $O(^3P)$–propene reaction (from Cvetanovic, 1963).

work of the above overall reaction scheme. Thus $O(^3P)$ can add to either end of the double bond:

$$O(^3P) + CH_3CH{=}CH_2 \xrightarrow[a]{} (CH_3\dot{C}H-CH_2\dot{O})^* \quad \textbf{(I)} \qquad (67)$$

$$\xrightarrow[b]{} (CH_3CH(\dot{O})-\dot{C}H_2)^* \quad \textbf{(II)}$$

Both (**I**) and (**II**) can ring close to the excited epoxide, which can either be stabilized by collisions with a third body to form the epoxide or undergo fragmentation to produce free radicals:

$$\textbf{(I)} \text{ or } \textbf{(II)} \xrightarrow{M} CH_3\overset{O}{\overbrace{CH-CH_2}} \qquad (68)$$

$$(\mathbf{I}) \text{ or } (\mathbf{II}) \rightarrow \text{free radicals} \tag{69}$$

The free radicals will then react further to produce stable products. For example, acetaldehyde and ethane in Fig. 7.12 were postulated to be formed by free radical reactions such as

$$2CH_3 \xrightarrow{M} C_2H_6 \tag{70}$$

and

$$CH_3 + CHO \xrightarrow{M} CH_3CHO \tag{71}$$

Since the fragmentation to free radicals occurs in competition with stabilization to the epoxide, one would expect increasing epoxide yields and decreasing C_2H_6 and CH_3CHO yields as the total pressure, which acts to stabilize the excited intermediate, increases. This is, in fact, observed (Fig. 7.12).

A major portion of the overall mechanism involves internal rearrangement of the biradical to excited carbonyl compounds by migration of a hydrogen atom or alkyl group from the carbon atom to which the oxygen atom is attached. The migration of hydrogen atoms was postulated always to be internal, whereas alkyl group migration was thought to be at least partly external, that is, the alkyl group leaves the intermediate completely and hence is capable of being trapped as an independent free radical. For (**I**) and (**II**), these rearrangements can be written as follows:

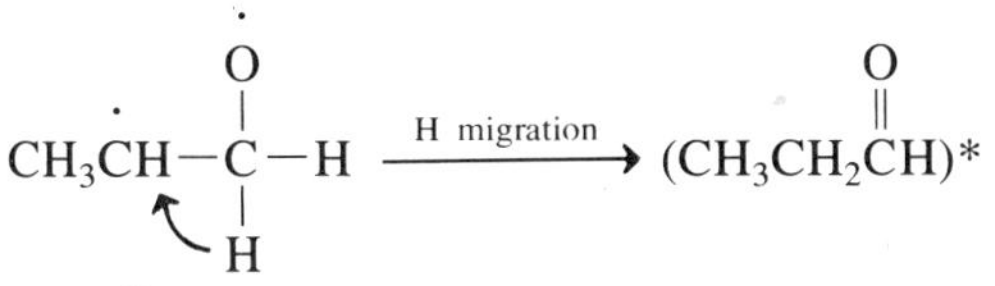

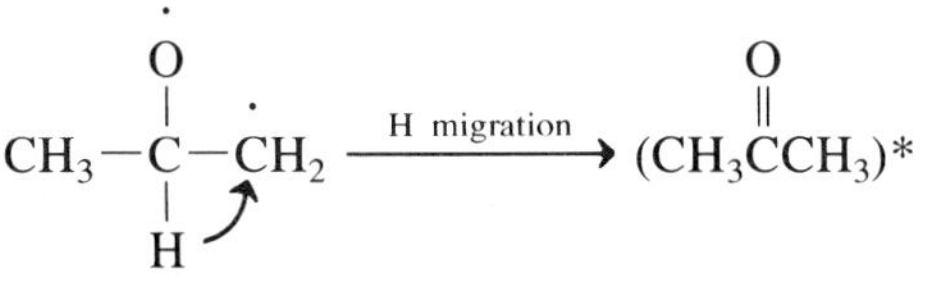

$$CH_3\overset{\cdot}{C}H-\underset{\underset{H}{|}}{\overset{\overset{\cdot}{O}}{\underset{|}{C}}}-H \xrightarrow{\text{H migration}} (CH_3CH_2\overset{\overset{O}{\|}}{C}H)^* \tag{72}$$

(I)

$$CH_3-\underset{\underset{H}{|}}{\overset{\overset{\cdot}{O}}{\underset{|}{C}}}-\overset{\cdot}{C}H_2 \xrightarrow{\text{H migration}} (CH_3\overset{\overset{O}{\|}}{C}CH_3)^* \tag{73}$$

(II)

$$CH_3-\underset{\underset{H}{|}}{\overset{\overset{\cdot}{O}}{\underset{|}{C}}}-\overset{\cdot}{C}H_2 \rightarrow (CH_3 + CH_2CHO) \rightarrow (H\overset{\overset{O}{\|}}{C}-CH_2CH_3)^* \tag{74}$$

(II)

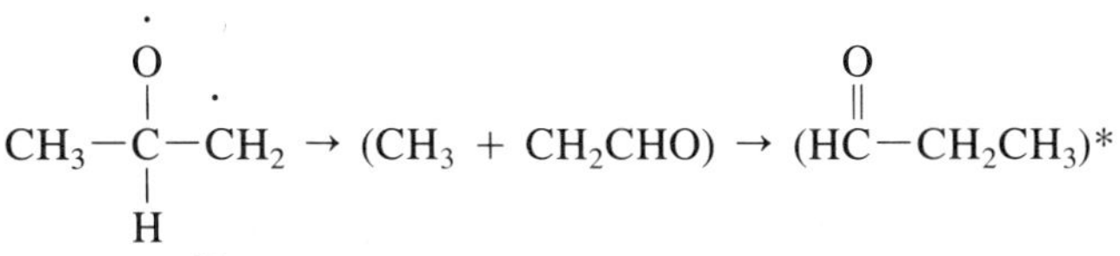

Stabilization of the excited carbonyls thus leads to propionaldehyde and acetone, whose yields should increase with the concentration of stabilizing gases (i.e, with the total pressure) as is observed in Fig. 7.12. The excited carbonyls which are not stabilized will undergo a *pressure-dependent fragmentation*, again to form free radicals such as C_2H_5 and HCO (from C_2H_5—CHO*) and CH_3, which react further to form stable products.

Finally, a portion of the overall reaction of the biradical was postulated to consist of fragmentation to free radicals in a manner not dependent on total pressure. That is, even when the excited epoxides and carbonyl compounds are essentially totally deactivated by collisions to the stable compounds, some free radical production persists. In Fig. 7.12, for example, small but finite yields of C_2H_6 and CH_3CHO are observed at higher presssures where the yields of the epoxide and carbonyl compounds appear to have leveled off.

From such a mechanism, one would expect that the addition products from the larger alkenes would be more readily stabilized since there are more degrees of freedom to absorb the excess energy. This is indeed found to be the case, with ethylene continuing to show considerable fragmentation at 1 atm total pressure, whereas the butenes and higher alkenes are stabilized at pressures below 100 Torr.

While this mechanism is consistent with a wide body of experimental data on the stable products of $O(^3P)$–alkene reactions, the subsequent development and application of sensitive and specific detection techniques for free radical internediates (see Chapter 4) have led recently to some further elucidation of the details of this proposed mechanism. Thus while the free radicals expected from the reaction scheme of Cvetanovic were detected in mass spectrometric studies (Kanofsky et al., 1973; Blumenberg et al., 1977), researchers recently have reported high yields of vinoxy radicals CH_2CHO, from the reaction of $O(^3P)$ with a number of simple olefins. [As Cvetanovic and Singleton (1984) point out, the name *vinoxy* is misleading since calculations by Dupuis et al. (1982a) show that the structure is close to $CH_2{-}C(=O)H$ (with the C=O drawn as a double bond), that is, formyl methylene, rather than $CH_2{=}C(-O)H$ (with the C–O drawn as a single bond). However, for consistency with the literature, we shall use the term vinoxy radical]. This radical was first observed by Lee and co-workers (Buss et al., 1981) and has now been detected by three different methods: UV absorption spectrometry (Hunziker et al., 1981), laser-induced fluorescence (Kleinermanns and Luntz, 1981; Inoue and Akimoto, 1981), and cross-beam mass spectrometry (Buss et al., 1981; Clemo et al., 1982). It has been observed in the reactions of $O(^3P)$ with ethene, propene, and 1-butene. In addition, high yields of hydrogen atoms, the second product expected to accompany CH_2CHO, have been observed (Sridharan and Kaufman, 1983) in the reaction of $O(^3P)$ with ethene.

Figure 7.13 shows the yields of CH_2CHO as a function of total pressure for ethene, propene, and 1-butene; the yields are significant ($\gtrsim 17\%$) and, at least in the case of propene, appear to depend on pressure.

These results are consistent with the reaction scheme based on stable product

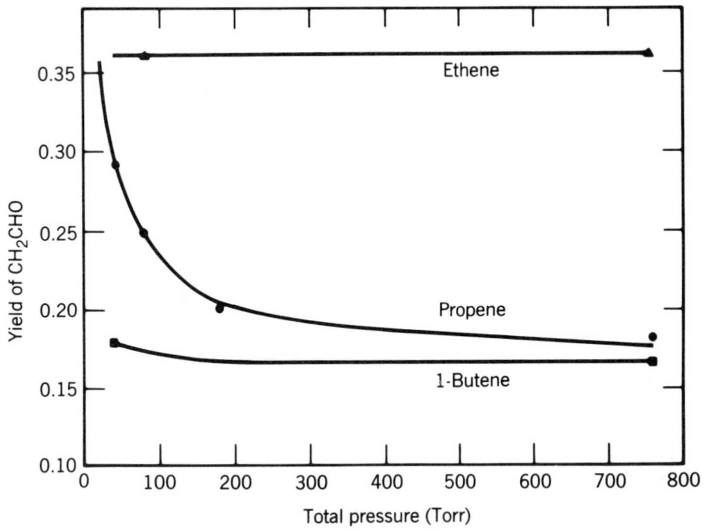

FIGURE 7.13. Yields of the vinoxy radical (CH_2CHO) as a function of total pressure for the reactions of $O(^3P)$ with ethene, propene, and 1-butene (from Hunziker et al., 1981).

analysis proposed by Cvetanovic if the CH_2CHO radical is associated partly with the pressure-independent fragmentation route, and if this route accounts for 15–20% of the reaction for these three alkenes in the 40–760 Torr pressure range. The remainder of the CH_2CHO observed at low pressures is then associated with a pressure-dependent fragmentation which, for propene and possibly 1-butene, is quenched as the total pressure increases. This pressure-independent fragmentation path can be thought of as a displacement reaction (75), as opposed to the addition reaction leading to stable products:

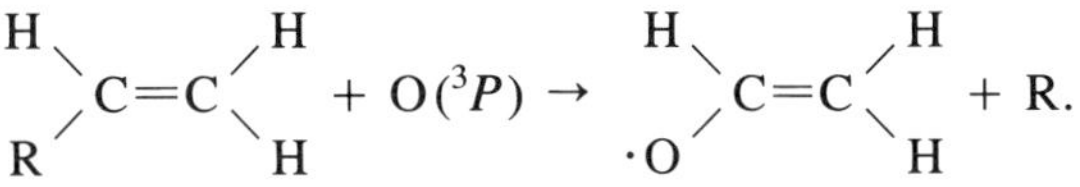

$$\begin{array}{c} H \\ R \end{array} C=C \begin{array}{c} H \\ H \end{array} + O(^3P) \rightarrow \begin{array}{c} H \\ \cdot O \end{array} C=C \begin{array}{c} H \\ H \end{array} + R. \qquad (75)$$

A further refinement in the Cvetanovic mechanism concerns the chemical dynamics involved in the reaction. Based on the *ab initio* calculations of Dupuis and co-workers (Dupuis et al., 1982a,b), Hunziker et al. (1981) and Kleinermanns and Luntz (1981) proposed that diradicals in two different electronic states may be formed in $O(^3P)$–alkene reactions. For example, the energetics for the $O(^3P)$–ethene reaction proposed by Hunziker and co-workers is shown in Fig. 7.14. The two triplet biradicals correspond to the (π,σ) and (π,π) electronic states, respectively. These states are expected to dissociate adiabatically to a hydrogen atom and the vinoxy radical, but the electronic state of the vinoxy radical formed will depend on that of the biradical precursor. Thus the (π,π) biradical will form CH_2CHO in its ground state (designated as X in Fig. 7.14), while the (π,σ) bir-

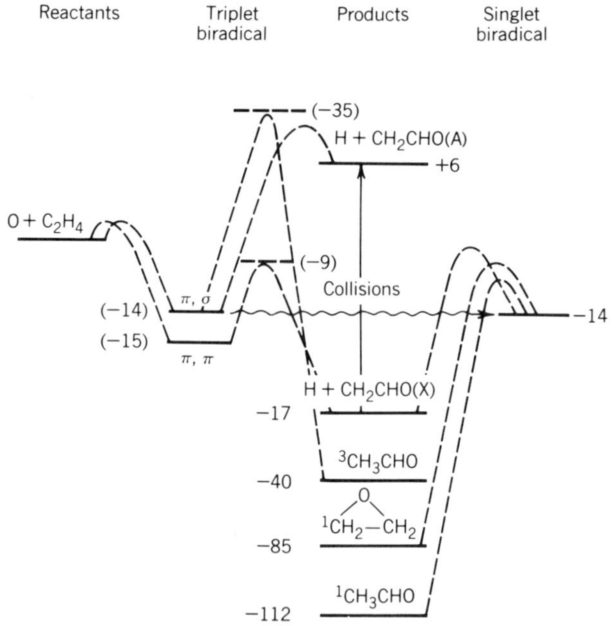

FIGURE 7.14. Proposed mechanism for the reaction of $O(^3P)$ with ethene where energies (kcal mole^{-1}) are given in parentheses relative to the reactants. Energy levels are not shown to scale (from Hunziker et al., 1981).

adical correlates with electronically excited CH_2CHO (designated A in Fig. 7.14) which lies 6 kcal mole^{-1} above the original reactants.

The (π,σ) state also correlates with a triplet state of acetaldehyde (shown as $^3CH_3CHO$ in Fig. 7.14), a reaction that would be exothermic overall. However, as seen in Fig. 7.14, the activation energy for this reaction is large (Dupuis et al., 1982a,b) and thus it is unlikely to occur to a significant extent at room temperature.

For energetic reasons, then, the triplet (π,σ) biradical is not expected to form CH_2CHO or CH_3CHO. Its possible fates are therefore redissociation to the reactants or formation of a singlet-state biradical. The latter would involve a spin conversion and hence requires collisions with other molecules as shown in Fig. 7.14 by the wavy line. [For larger olefins, the triplet biradical may be in the large molecule limit where collisions are not required for intersystem crossing (Hunziker, 1984)]. Once in the singlet state, three allowed exothermic reactions become possible, to form (1) H + CH_2CHO in the X ground state, (2) the epoxide, or (3) ground-state acetaldehyde ($^1CH_3CHO$ in Fig. 7.14).

The (π, π) biradical is shown by the calculations of Dupuis and co-workers to have available an allowed, exothermic path to form H + CH_2CHO(X). This would explain why the formation of the vinoxy radical is observed to be the major process at the low pressures ($\sim 10^{-7}$ Torr) used in molecular beam experiments, whereas at higher pressures (1 Torr to 1 atm) the addition products (i.e., epoxides and carbonyl compounds) or their decomposition products are formed. At the very low pressures, the (π, σ) biradical would redissociate to reactants, leading to no net reaction, and only the reaction of the (π, π) biradical would be observed. At higher pressures characteristics of the troposphere, the (π, σ) biradical would be converted via collision-induced processes to the singlet biradical, the precursor to the *addition* products.

In the experiments of Hunziker and co-workers, for example, the lowest total pressure studied was 40 Torr. At this pressure, the CH_2CHO observed is proposed to be formed both in a collision-free (π, π) biradical reaction and in a collision-induced (π, σ) biradical crossing from a triplet to a singlet state. The singlet state then produces CH_2CHO, as well as the addition products which contain sufficient exothermicity to fragment to other free radicals, primarily CH_3 + HCO. The formyl radical was in fact observed in the ethene reaction, its yield being ~0.55 in the 80–760 Torr pressure range. The observed yields of CH_2CHO and HCO thus accounted for more than 90% of the overall $O(^3P)$–C_2H_4 reaction.

Figure 7.14 is consistent with the pressure dependence of the CH_2CHO yields seen in Fig. 7.13 if, over the pressure range studied by Hunziker and co-workers (40 Torr to 1 atm), both the pressure-independent path (i.e., the π, π biradical) and the pressure-dependent path (i.e., the π, σ biradical) are operative and the formation of CH_2CHO from the singlet biradical is pressure dependent. A pressure dependence of CH_2CHO formation from the singlet biradical could occur if the fraction reacting by each of the routes shown in Fig. 7.14 is dependent on pressure—higher pressures leading to less CH_2CHO + H formation.

No pressure dependence of the CH_2CHO yield from ethene is then observed in Fig. 7.13 because ethene and the singlet biradical produced from it are sufficiently small that they decompose too rapidly for collisional processes to be important. For propene, increased pressure could increase the fraction of the singlet biradical which forms the addition products, leading to a decrease in CH_2CHO yield with pressure. For 1-butene, even at 40 Torr the fraction of the singlet biradical forming addition products could be at its maximum, leading to negligible pressure effects over the pressure range studied by Hunziker and co-workers.

The new insights into the mechanism of $O(^3P)$–alkene reactions provided by application of these spectroscopic techniques illustrate that there remains a substantial amount to learn about these reactions, especially under atmospheric conditions. In particular, the nature and amounts of free radicals produced in 1 atm of air compared to the addition products need to be investigated. As stressed by Cvetanovic and Singleton (1984), "the relative importance of different reaction channels . . . may be strongly influenced by the energy content of the reacting species and thus by experimental conditions, such as temperature, pressure, phase, and the nature of the chemical species present in the reaction system." Again we see the importance of studying reactions under conditions as close to atmospheric as possible or extrapolating the results of studies under laboratory conditions with great care.

The fates of the free radicals formed in the atmosphere are those discussed in

Section 7-D. For example, the free radicals CH_3 and HCO formed in the $O(^3P)$-C_2H_4 reaction react with O_2:

$$CH_3 + O_2 \xrightarrow{M} CH_3O_2 \qquad (76)$$

$$HCO + O_2 \xrightarrow{M} HO_2 + CO \qquad (77)$$

The CH_3O_2 and HO_2 then oxidize NO to NO_2 to form alkoxy radicals, and so on.

The production of vinoxy radicals has only recently been recognized and hence little is known about its atmospheric fate. The reaction of CH_2CHO with O_2 shows kinetic behavior that suggests the existence of two reaction paths—one dependent on pressure and one independent of pressure. The observation of OH and a mass spectrometer signal at $m/e = 30$, attributed to HCHO, has led Gutman and Nelson (1983) to postulate the following as one possible set of reactions:

$$CH_2CHO + O_2 \leftrightarrows \left[\begin{array}{c} H \\ \diagdown \\ H \diagup \end{array} \!\! C\!\!-\!\!\overset{\overset{\displaystyle O\cdot}{\overset{\displaystyle |}{O}}}{\underset{\underset{\displaystyle O}{\displaystyle \|}}{C}}\!\!-\!\!H \right]^* \rightarrow \begin{array}{c} H \\ \diagdown \\ H \diagup \end{array} \!\! C\!\!-\!\!C \!\!\begin{array}{c} \diagup O \cdots \\ \diagdown \\ O \end{array} \!\!\! H$$

$$\downarrow [M] \qquad\qquad\qquad\qquad\qquad \searrow \text{fast}$$

$$\qquad\qquad\qquad\qquad\qquad\qquad\qquad HCHO + CO + OH \qquad (78)$$

$$\begin{array}{c} H \\ \diagdown \\ H \diagup \end{array} \!\! C\!\!-\!\!\overset{\overset{\displaystyle O\cdot}{\overset{\displaystyle |}{O}}}{\underset{\underset{\displaystyle O}{\displaystyle \|}}{C}}\!\!-\!\!H$$

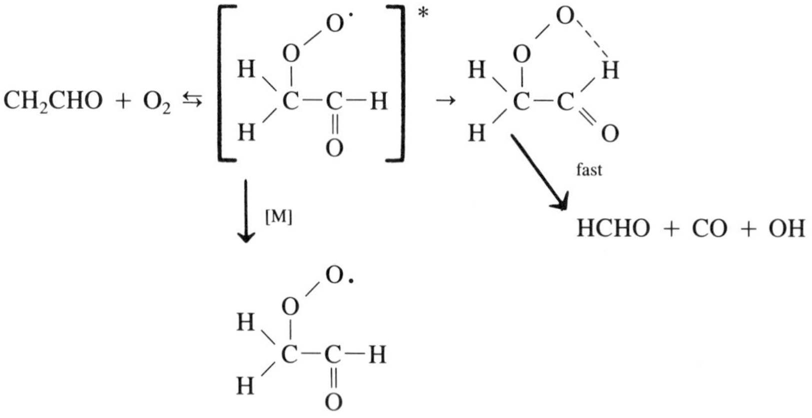

(III)

Unfortunately, the fate of vinoxy radicals in air remains somewhat speculative at present. Thus Gutman and Nelson point out that while both OH and HCHO were observed as products of the vinoxy radical–O_2 reaction, the OH appeared at a slower rate than the vinoxy radical was consumed, and the HCHO could not be followed with sufficient time resolution to conclude that it was an initial product. Therefore, secondary reactions rather than (78) may be responsible for the OH and HCHO observed. Thus, while the production of OH has been observed by other workers as well (Schmidt et al., 1984, and Lorenz et al., 1985), alternate reaction paths leading to OH and $(CHO)_2$ have been proposed (see Section 7-F-1b below). However, it seems clear from the observed pressure dependence of the rate constant at 298°K, that at least a portion of the reaction proceeds via addition of the O_2 to the radical as shown in reaction (78); the high-pressure limiting value, which is reached at about 150 Torr, has been reported as 2.6×10^{-13} cm^3 molecule^{-1} s^{-1} in He as the carrier gas (Lorenz et al., 1985).

(III) is an alkylperoxy radical and, if formed, is expected to react in a manner analogous to the reactions discussed earlier, for example,

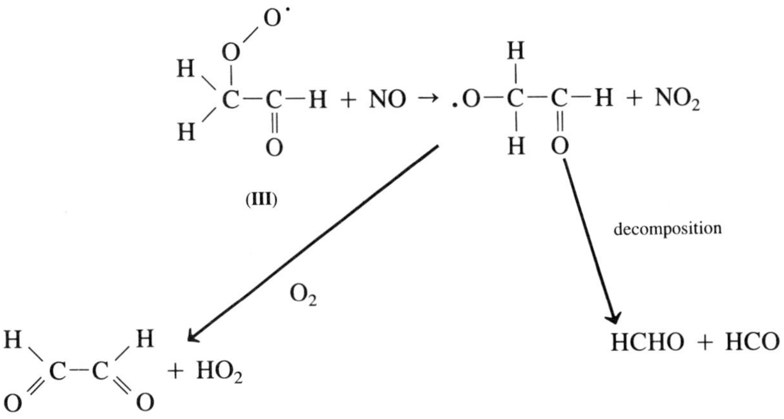

(79)

7-E-4. Nitrate Radical, NO_3

7-E-4a. Kinetics

The nitrate radical (NO_3) was first shown to react rapidly with alkenes in the classic work of Niki and co-workers (Morris and Niki, 1974; Japar and Niki, 1975). They showed that the room temperature rate constants were two to four orders of magnitude greater than those for the corresponding O_3 reactions and that the trend with the degree of alkyl substitution at the double bond strongly suggested an attack by the electron-deficient free radical NO_3 on the double bond.

Table 7.18 gives the values of room temperature rate constants for NO_3–alkene reactions, assuming the three different values of K_{eq} for the reaction

$$NO_2 + NO_3 \overset{M}{\rightleftarrows} N_2O_5 \tag{80}$$

Those of Niki and co-workers have also been adjusted to conform to these new values of K_{eq}. Also shown are the absolute rate constants for the reactions of NO_3 with 2-methylpropene and *trans*-2-butene determined by Ravishankara and Mauldin (1985). As discussed in Section 7-C-2, these are the first absolute rate constants reported for NO_3 reactions; the preferred values of the remaining rate constants are based on $K_{eq} = 3.35 \times 10^{-11}$ cm^3 molecule^{-1} because of the excellent agreement with the absolute rate constants for the 2-methylpropene and *trans*-2-butene reactions.

It is seen from Table 7.18 that the rate constants obtained in the two most complete studies to date generally agree within a factor of 2 (except for ethene); this is in contrast to OH–alkene reactions, for example, where agreement to $\pm 25\%$ is common. The poorer agreement for the NO_3–alkene reactions likely arises from the complex nature of the system used to generate the NO_3 in which NO_2 and N_2O_5 were present simultaneously. Furthermore, wall losses and/or reactions that are not well understood frequently plague NO_x studies. However, the data in Table 7.18 certainly show that these reactions are relatively fast and illustrate the dramatic increase in the rate constant with increasing substitution on the double bond.

TABLE 7.18. Rate Constants for the Reaction of NO_3 with Some Alkenes at 298°K Derived Using Different Values of the Equilibrium Constant[a] for $NO_2 + NO_3 \overset{M}{\rightleftarrows} N_2O_5$

Alkene	Exponential Factor[b]	Relative Rate Constant k (cm^3 $molecule^{-1}$ s^{-1}), K_{eq} = 1.87×10^{-11}[c]	2.20×10^{-11}[d]	3.35×10^{-11}[e,f]	Absolute Rate Constant	Reference
Ethene	10^{-16}	0.61	0.72	1.1	n.d.[l]	*g*
		11	13	20	n.d.	*h*
Propene	10^{-15}	4.2	4.9	7.5	n.d.	*g*
		6.2	7.3	11	n.d.	*h*
1-Butene	10^{-15}	5.4	6.4	9.7	n.d.	*g*
		9.1	11	16	n.d.	*h*
2-Methylpropene	10^{-13}	1.7	2.0	3.0	n.d.	*g*
		1.3	1.5	2.3	n.d.	*h*
		n.a.[k]			3.3	*i*
cis-2-Butene	10^{-13}	1.9	2.2	3.4	n.d.	*g*
		2.1	2.5	3.8	n.d.	*h*
trans-2-Butene	10^{-13}	2.1	2.5	3.8	n.d.	*g*
		1.6	1.9	2.9	n.d.	*h*
					3.78	*i*
2-Methyl-2-butene	10^{-12}	5.5	6.5	9.9	n.d.	*j*
		6.4	7.5	11	n.d.	*h*
2,3-Dimethyl-2-butene	10^{-11}	3.4	4.0	6.1	n.d.	*j*
		4.3	5.1	7.7	n.d.	*h*

Cyclohexene	10^{-13}	2.9	3.4	5.1	n.d.	*j*
2-Methyl-1,3-butadiene (isoprene)	10^{-13}	3.2	3.8	5.8	n.d.	*j*
α-Pinene	10^{-12}	3.4	4.0	6.1	n.d.	*j*
β-Pinene	10^{-12}	1.4	1.7	2.5	n.d.	*j*
d-Limonene	10^{-12}	7.7	9.1	14	n.d.	*j*
1,3-Butadiene	10^{-14}	5.3	6.3	9.6	n.d.	*j*
1,3-Cyclohexadiene	10^{-12}	7.2	8.5	13	n.d.	*j*
1,4-Cyclohexadiene	10^{-13}	2.9	3.4	5.2	n.d.	*j*

[a] Most rate constants derived from experimental data depend on value of equilibrium constant assumed for $K_{eq} = [N_2O_5]/[NO_2][NO_3]$. For example, the rate constants derived for ethene are 6.1, 7.2, or 11 × 10^{-17} cm^3 $molecule^{-1}$ s^{-1}, respectively, depending on the value taken for K_{eq}.

[b] The numbers given under each value of K_{eq} should be multiplied by this factor. For example, the rate constants derived for propene using the three different values of K_{eq} are 4.2×10^{-15}, 4.9×10^{-15}, and 7.5×10^{-15} cm^3 $molecule^{-1}$ s^{-1}, respectively.

[c] Malko and Troe, 1982.

[d] Graham and Johnston (1978), using $k(O_3 + NO_2) = 3.2 \times 10^{-17}$ cm^3 $molecule^{-1}$ s^{-1} at 298°K recommended by DeMore et al. (1983).

[e] Average of values reported by Tuazon et al. (1984b), 3.44×10^{-11}, and by Kircher et al. (1984), 3.26×10^{-11} cm^3 $molecule^{-1}$, at 298°K.

[f] Preferred values based on absolute rate constants reported by Ravishankara and Mauldin (1985). See Section 7-C-2.

[g] Atkinson et al., 1984a.

[h] Japar and Niki, 1975.

[i] Absolute values reported by Ravishankara and Mauldin (1985); do not depend on value assumed for K_{eq}.

[j] Atkinson et al., 1984b.

[k] n.a. = not applicable.

[l] n.d. = not determined.

For example, at an NO_3 concentration of 100 ppt (2.5×10^9 cm^{-3}), which was seen in polluted urban atmospheres at night, the atmospheric lifetimes range from ~42 days for ethene to only 7 s for 2,3-dimethyl-2-butene.

In relatively clean atmospheres, the nighttime NO_3 concentration seems more likely to be ~10 ppt (2.5×10^8 cm^{-3}). The *natural hydrocarbons* isoprene and α-pinene will have lifetimes of only 2 h and 11 min, respectively, with respect to reaction with NO_3 under these conditions. This can be compared to the corresponding lifetimes with respect to reaction with OH at a concentration of 1×10^6 cm^{-3} during daylight hours; these lifetimes are ~2.8 h for isoprene and 5.2 h for α-pinene. Thus, for these naturally emitted alkenes, reaction with NO_3 may be a major loss process (Winer et al., 1984; Atkinson et al., 1984d) (see also Section 14-B-2).

To date, no data on the temperature dependence of these NO_3-alkene reactions have been reported.

7-E-4b. Mechanisms

The increasing trend in rate constants with increasing substitution on the double bond (Table 7.18) parallels the trend for $O(^3P)$ and OH reactions with alkenes, suggesting that the initial addition step involves a substantial amount of charge transfer from the alkene to the electron-deficient nitrate during the rate-determining transition state.

In Fourier transform infrared spectrometry (FTIR) studies of the reaction of propene with NO_x systems by Akimoto and co-workers (Akimoto et al., 1978a; Hoshino et al., 1978a; Akimoto et al., 1980; Bandow et al., 1980), 1,2-propanediol dinitrate (PDDN)

$$CH_3-CH-CH_2$$
$$\quad\quad | \quad\quad |$$
$$\quad O_2NO \quad ONO_2$$

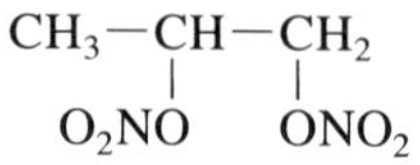

PDDN

was identified as a reaction product and attributed to the reaction of NO_3 with propene. The same infrared bands have been observed by Morris and Niki (1974) and by Atkinson et al. (1984a).

Figure 7.15a, for example, shows the infrared spectrum in the region 780–1880 cm^{-1} of the products of the reaction of C_3H_6 with NO_3 in an $N_2 + O_2$ mixture after 7 min. By comparison to an authenic spectrum of PDDN (Fig. 7.15b), the dinitrate can clearly be seen to be one of the products. The spectrum in Fig. 7.15c is that remaining after the PDDN spectrum has been substracted from that of the products (i.e., from Fig. 7.15a). Many of the remaining peaks are attributed to an intermediate, nitroxyperoxypropyl nitrate (NPPN):

$$CH_3-CH-CH_2 \quad\quad CH_3-CH-CH_2$$
$$\quad | \quad\quad | \quad\text{and/or}\quad \quad | \quad\quad |$$
$$O_2NOO \quad ONO_2 \quad\quad\quad O_2NO \quad OONO_2$$

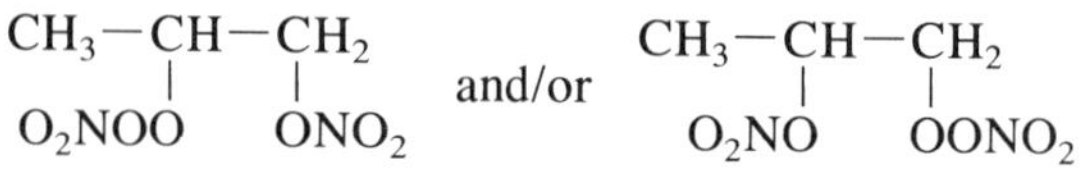

NPPN

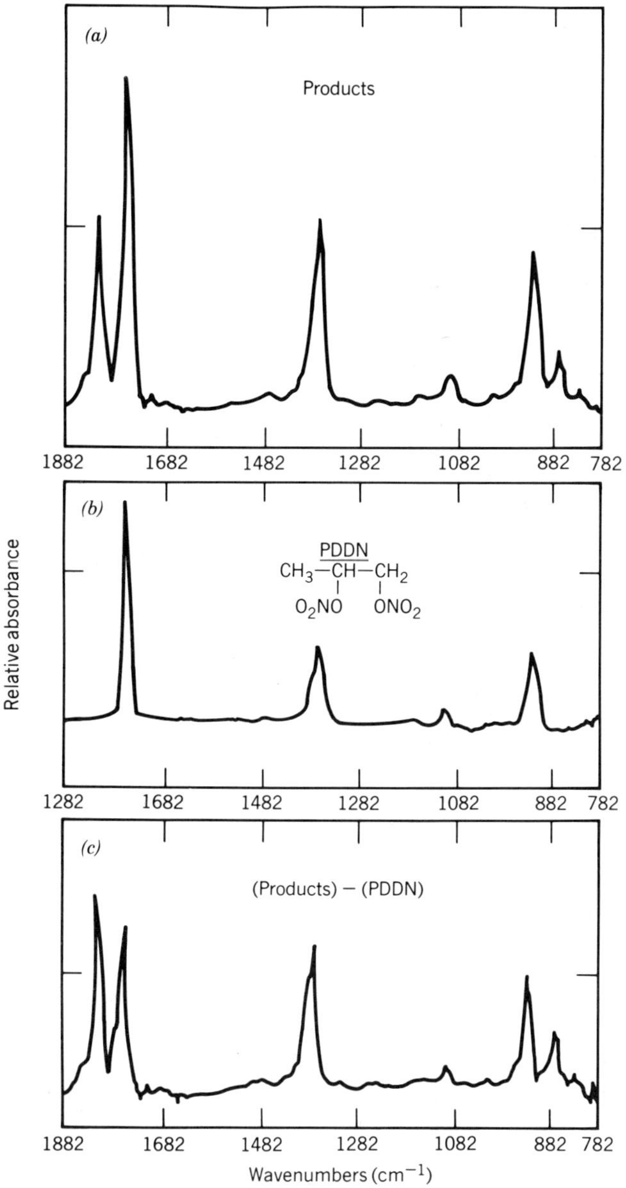

FIGURE 7.15. (*a*) FTIR spectrum of products from the reaction of 35.8 mTorr C_3H_6 with 22.1 mTorr N_2O_5 in the presence of 650 Torr N_2 and 100 Torr O_2 after 7 min at 26°C. (*b*) Spectrum of authentic sample of 1,2-propanediol dinitrate (PDDN). (*c*) Product spectrum remaining when PDDN is subtracted from (*a*) (from Bandow et al., 1980).

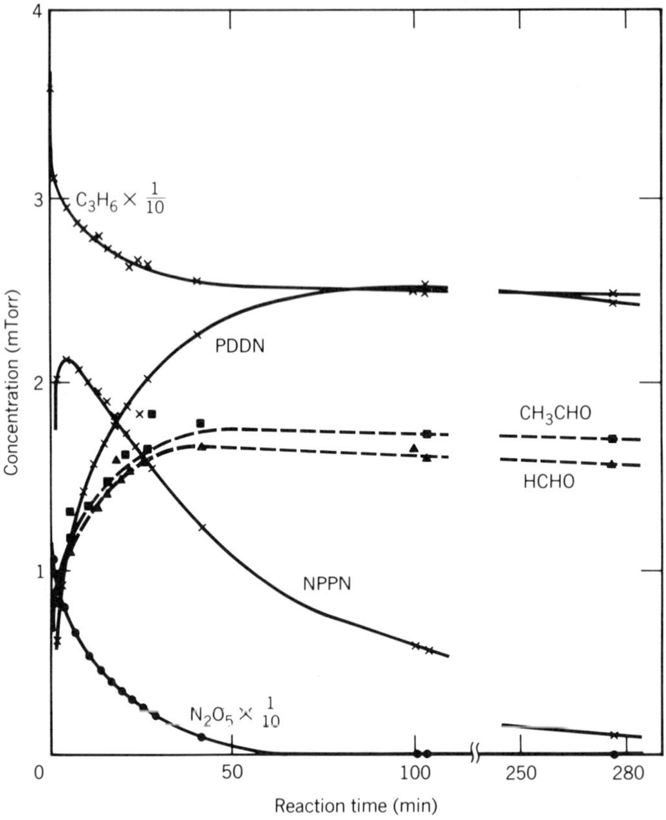

FIGURE 7.16. Concentrations of reactants and products as a function of reaction time in the C_3H_6 (35.8 mTorr)–N_2O_5 (22.1 mTorr)–N_2 (650 Torr)–O_2 (100 Torr) system at 26°C (from Bandow et al., 1980).

The kinetic behavior of NPPN strongly suggested that it was a precursor to the final product PDDN. Thus Fig. 7.16 shows the concentration–time profile of reactants and products in the C_3H_6–N_2O_5–N_2–O_2 system. The concentration of NPPN peaks early in the reaction and then falls, while the PDDN concentration rises. Based on this, Akimoto and co-workers proposed the following mechanism:

$$CH_3CH{=}CH_2 + NO_3 \xrightarrow{a} \underset{\textbf{(IV)}}{CH_3CH{-}CH_2(ONO_2)}$$

$$\xrightarrow{b} \underset{\textbf{(V)}}{CH_3CH(O_2NO){-}CH_2} \qquad (83)$$

As is characteristic of free radical additions to a double bond, formation of the more stable secondary radical (**IV**) is expected to predominate over the formation of the primary radical (**V**). Subsequently, reaction (84), the intramolecular decomposition of (**IV**), was proposed to account for the formation of propylene oxide:

$$\underset{\textbf{(IV)}}{CH_3CH{-}\underset{\displaystyle ONO_2}{CH_2}} \rightarrow NO_2 + \underset{\text{propylene oxide}}{CH_3CH{-}CH_2\ (\text{–O–})} \quad (84)$$

Radical (**IV**) is an alkyl radical and hence in air is also expected to add O_2 (see Section 7-D):

$$\underset{\textbf{(IV)}}{CH_3CH{-}CH_2(ONO_2)} + O_2 \rightarrow \underset{\textbf{(VI)}}{CH_3{-}CH(OO){-}CH_2(ONO_2)} \quad (85)$$

Akimoto and co-workers proposed that under their laboratory reaction conditions, the alkylperoxy type radical formed in (85) can add NO_2, forming NPPN:

$$\underset{\textbf{(VI)}}{CH_3{-}CH(OO){-}CH_2(ONO_2)} + NO_2 \rightleftarrows \underset{\text{NPPN}}{CH_3{-}CH(O_2NOO){-}CH_2(ONO_2)} \quad (86)$$

Like other peroxyalkyl nitrates, NPPN is expected to thermally decompose back to NO_2 and the alkylperoxy radical (see Section 4-A-7c); hence (86) is written as an equilibrium.

The major fate of alkylperoxy radicals (**VI**) in the experimental system of Akimoto and co-workers is expected to be self-reaction since the NO concentration is low (see Section 7-D-2). Although there are three possible reaction paths [e.g., see Eq. (9) for methylperoxy radicals], one produces an alkoxy radical (**VII** which can then react with NO_2 to give the observed stable product PDDN, or alternatively, split off NO_2 to form the major aldehydes observed, CH_3CHO and HCHO:

$$2\textbf{(VI)} \rightarrow \underset{\textbf{(VII)}}{2CH_3{-}CH(O){-}CH_2(ONO_2)} + O_2 \quad (87)$$

$$\textbf{(VII)} \xrightarrow{+NO_2} \text{PDDN}$$

$$\textbf{(VII)} \rightarrow CH_3CHO + HCHO + NO_2$$

In ambient air, the alkoxy radical (**VII**) would also be produced from (**VI**), but in this case most likely by reaction with NO:

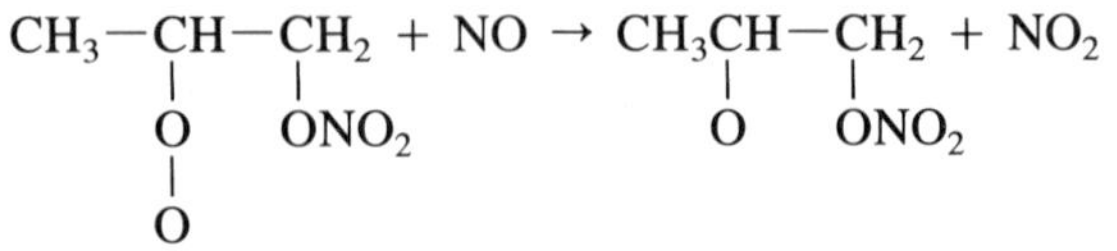

(**VI**) (**VII**) (88)

Akimoto and co-workers proposed that the dinitrate PDDN could then be formed from the addition of NO_2 to the alkoxy radical (**VII**). However, as we have seen earlier (Table 7.11), the addition of NO_2 to simple alkoxy radicals under atmospheric conditions is slow compared to the other available reaction paths, that is, isomerization, decomposition, and reaction with O_2. It is somewhat surprising, therefore, that (**VII**) would react significantly with NO_2; this casts some doubt on the proposed mechanism of production of PDDN.

More recently, however, much lower yields of PDDN have been observed in studies of the NO_3–C_3H_6 reaction carried out at lower reactant concentrations (Shepson et al., 1985). Simultaneously, the formation of α-(nitrooxy)acetone ($CH_3COCH_2ONO_2$) was observed. They suggest that the alkoxy radical (**VII**) reacts via hydrogen abstraction with O_2 to form the nitrated acetone [reaction (89a)], rather than with NO_2 [reaction (89b)] to form PDDN:

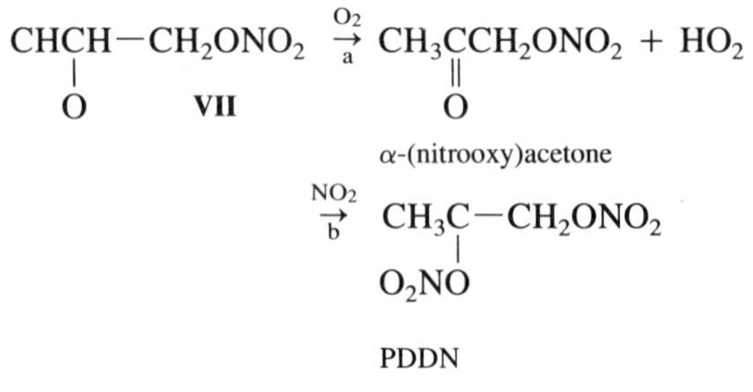

(89)

Such a mechanism is more consistent with our expectations for alkoxy radical reactions (Section 7-D). The higher yields of PDDN observed by Akimoto and co-workers are likely due to the much higher (by factor of $\sim 10^2$) concentrations of NO_2 used in their studies.

While 2-hydroxypropyl nitrate and hydroxypropyl nitrate were observed in the NO_3–C_3H_6 reaction in addition to PDDN and α-(nitrooxy)acetone, these products were also formed in C_3H_6–NO_x–air photooxidations, perhaps due to the OH–C_3H_6 reaction (Shepson et al., 1985).

Similar reactions were postulated for the primary radical (**V**) formed on the initial addition of NO_3 to the central carbon atom. Addition of O_2 and NO_2 would lead to the isomeric structure of NPPN, $CH_3CH(ONO_2)CH_2(OONO_2)$. The alkoxy radical (**VIII**) would be formed from the subsequent reactions of (**V**)

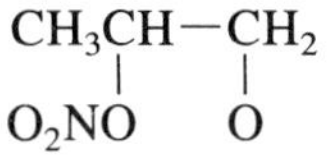

(VIII)

and was proposed by Akimoto and co-workers to react with O_2 to form 1-formylethyl nitrate, which was also tentatively identified by GC–MS and FTIR:

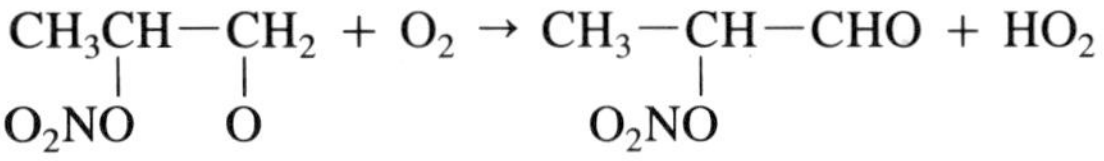

1-formylethyl nitrate

Such a reaction is in accord with expectations for alkoxy radicals in ambient air (Section 7-D-3).

The formation of nitrated organics in these laboratory studies and the possibility that they are also formed in ambient air are of interest from the point of view of health effects. For example, Akimoto et al. (1978a) state that dinitrates have been reported to cause "headaches, nasal congestion, dizziness, eye irritation," all of which have been reported during severe photochemical smog episodes. In addition, Shepson et al. (1985) indicate that α-(nitrooxy)acetone is a mutagen in the Ames assay (see Section 13-E).

Clearly, more detailed studies of a wide variety of NO_3–alkene reaction mechanisms and products and the mechanism of dinitrate formation are warranted on grounds of both chemical and health effects. Additionally, measurement of nitrates in ambient air during severe smog episodes would be useful.

7-E-5. Singlet Molecular Oxygen, $O_2(^1\Delta_g)$

7-E-5a. *Kinetics*

Figure 7.17 shows the potential energy curves for the lowest electronic states of molecular oxygen. The lowest excited states are the $a^1\Delta_g$ state and the $b^1\Sigma_g^+$ states; these are ~22.5 and 37.5 kcal, respectively, above the ground state. Because these are singlet states and the ground state of O_2 is a triplet state, radiative transitions to the ground state are slow and the natural radiative lifetimes (i.e., in the absence of collisions) are relatively long, ~12 s for the $^1\Sigma_g^+$ state and 65 min for the $a^1\Delta_g$ state (see Kearns et al., 1971 and Section 3-C-1). However, the $^1\Sigma_g^+$ state is rapidly deactivated by atmospheric gases (Baulch et al., 1984) probably mainly to the $a^1\Delta_g$ state, so that only the $a^1\Delta_g$ state need be considered for tropospheric chemistry.

While the production of electronically excited molecular oxygen in the atmosphere and the potential for reaction with other species were suggested by Leighton (1961), Bayes (1964) was the first to suggest that $O_2(^1\Delta_g)$ might be important in hydrocarbon oxidation and hence play a role in photochemical air pollution. At that time, the reactions of hydrocarbons and the nature of the reactive species responsible for the conversion of NO to NO_2 in photochemical smog were un-

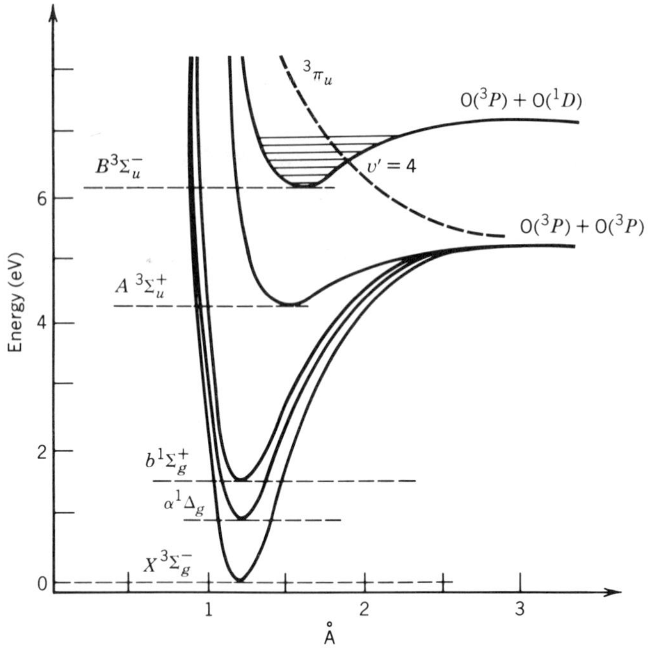

FIGURE 7.17. Potential energy curves for ground and first four excited states of O_2 (from Gaydon, 1968).

known. Bayes pointed out that if collisional deactivation of $O_2(^1\Delta_g)$ by air was sufficiently slow, it might react with alkenes at a rate comparable to the $O(^3P)$–alkene reactions. Recall from Chapter 1 that the only significant reactants consuming organics known at that time were $O(^3P)$ and O_3; the importance of OH had yet to be recognized. Bayes' suggestions thus led to a flurry of research activity into the kinetics and mechanisms of gas phase $O_2(^1\Delta_g)$ reactions.

There are a number of potential mechanisms of production of $O_2(^1\Delta_g)$ [and $O_2(^1\Sigma_g^+)$ which may be deactivated to $O_2(^1\Delta_g)$] in the troposphere as discussed initially by Pitts et al. (1969, 1971) and Calvert (1973). These include the following:

1. *Direct absorption of solar radiation.* This is inefficient since it involves a spin conversion (see Section 2-B), although collisional line broadening as suggested by Bayes (1964) increases the absorption coefficients.

2. *Energy transfer from electronically excited molecules.* For example, an organic molecule in a triplet state (T_1) may be formed via the direct absorption of sunlight to a singlet state (S_1) followed by intersystem crossing to the triplet state (see Section 2-C):

$$S_0 + h\nu \rightarrow S_1 \tag{90}$$

$$S_1 \xrightarrow[\text{crossing}]{\text{intersystem}} T_1 \tag{91}$$

$$T_1 + {}^3O_2 \rightarrow S_0 + {}^1O_2 \tag{92}$$

This is expected to be fast since net spin can be conserved. For example, such energy transfer to ground state O_2 to form singlet oxygen has been observed from benzaldehyde, benzene, and naphthalene in the gas phase (Kummler and Bortner, 1969; Steer et al., 1969a; Coomber and Pitts, 1970).

Energy transfer from electronically excited inorganics has also been demonstrated. The most important reaction of this type shown to date is transfer from excited NO_2 (Jones and Bayes, 1971, 1973; Frankiewicz and Berry, 1973):

$$NO_2^* + O_2(X^3\Sigma_g^-) \rightarrow NO_2 + O_2(a^1\Delta_g)$$

The efficiency of this process depends on the wavelength of light used to excite the NO_2 and may be as high as 7.5% per collision with excited NO_2.

Energy transfer to ground state oxygen may also occur from solid surfaces such as polycyclic aromatic hydrocarbons (PAH) adsorbed on combustion-generated particles, which are irradiated in the presence of O_2, as suggested by Khan et al. (1967) and Pitts et al. (1969). This was confirmed in experimental studies by Eisenberg et al. (1984) for the case of adsorbed PAH, and by Gohre and Miller (1983, 1985) for soils, and solids such as silica gel, alumina oxide and magnesium oxide.

3. *Direct photolysis of ozone.* As discussed in Section 3-C-2, this produces $O_2(^1\Delta_g)$ at wavelengths ≤ 320 nm.

4. *Exothermic chemical reactions producing O_2.* Some of the reaction exothermicity may be channeled into O_2 in the form of electronic energy. A number of such reactions are known in solution. For example, the adduct formed between O_3 and triphenyl phosphite in solution decomposes to form singlet oxygen (Murray and Kaplan, 1969):

$$(C_6H_5O)_3P + O_3 \xrightarrow{-70°C} (C_6H_5O)_3P\begin{smallmatrix}O-O\\|\ \ \ |\\O\end{smallmatrix} \tag{93}$$

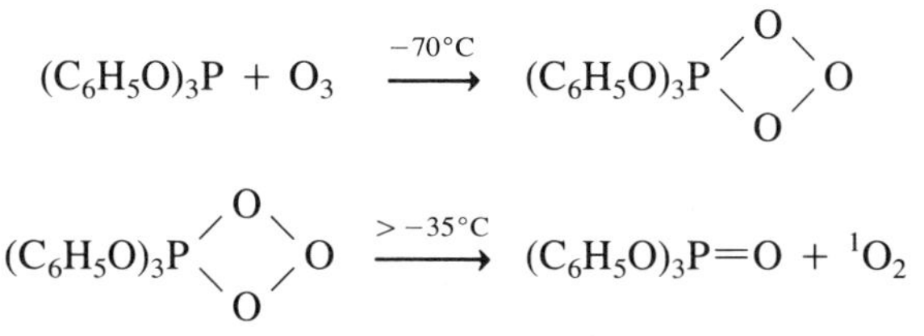

$$(C_6H_5O)_3P\begin{smallmatrix}O-O\\|\ \ \ |\\O\end{smallmatrix} \xrightarrow{>-35°C} (C_6H_5O)_3P=O + {}^1O_2 \tag{94}$$

A second example of the production of $^1O_2$ in the solution phase is the reaction of PAN with base (Steer et al., 1969b):

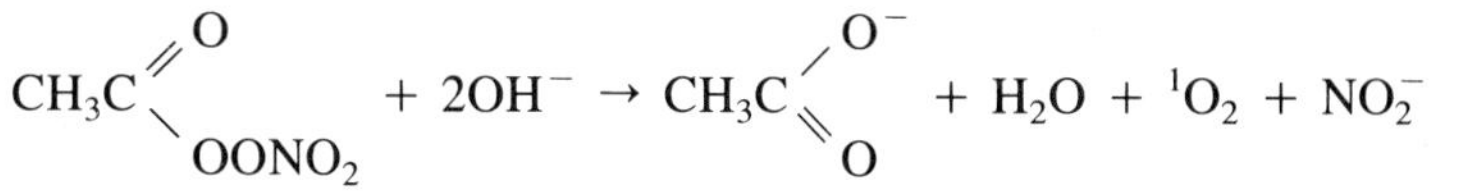

$$CH_3C\begin{smallmatrix}O\\\|\\OONO_2\end{smallmatrix} + 2OH^- \rightarrow CH_3C\begin{smallmatrix}O^-\\|\\O\end{smallmatrix} + H_2O + {}^1O_2 + NO_2^- \tag{95}$$

However, at present there appear to be no significant gas phase exothermic chemical reactions which produce substantial concentrations of $O_2(^1\Delta_g)$ in the troposphere.

While it is difficult to make realistic estimates for some of these processes, modeling studies that incorporate $O_2(^1\Delta_g)$ formation by direct absorption of sunlight, O_3 photolysis, energy transfer from excited NO_2, $O(^1D)$ energy transfer to O_2

$$O(^1D) + O_2(X^3\Sigma_g^-) \rightarrow O(^3P) + O_2(^1\Sigma_g^+) \tag{96}$$

and the potential reactions of O_3 with $O(^1D)$ and $O(^3P)$

$$O(^1D) + O_3 \rightarrow 2\ O_2(^1\Sigma_g^+) \tag{97}$$

$$O(^3P) + O_3 \rightarrow 2\ O_2(^1\Sigma_g^+), (^3\Sigma_g^-) \tag{98}$$

suggest that peak $O_2(^1\Delta_g)$ concentrations under moderately polluted conditions will be of the order of 10^8 cm^{-3} (Calvert, 1973; Demerjian et al., 1974).

Rate constants for $O_2(^1\Delta_g)$ reactions and/or deactivation with air and some alkenes are given in Table 7.19. The rate constants for the alkenes are relatively small, except for the most highly substituted ones. Even here, deactivation by air is sufficiently fast that most of the $O_2(^1\Delta_g)$ will be deactivated to the ground state before it can react.

However, even the small fraction of the $O_2(^1\Delta_g)$ which does survive contributes negligibly to the overall removal of alkenes. For example, the rate constant for the reaction of 2,3-dimethyl-2-butene with OH at room temperature is 1.1×10^{-10} cm^3 molecule^{-1} s^{-1} (Atkinson, 1985). Taking peak OH and $O_2(^1\Delta_g)$ concentrations in ambient air to be $\sim 1 \times 10^6$ cm^{-3} and 1×10^8 cm^{-3}, respectively, the relative rates of removal of the alkene by OH and $O_2(^1\Delta_g)$ are

TABLE 7.19. Room Temperature Rate Constants for Reaction and/or Deactivation of O_2 ($^1\Delta_g$) by Air and Some Alkenes

Molecule	k (cm^3 molecule^{-1} s^{-1})	Reference
O_2	1.7×10^{-18}	Baulch et al., 1984
N_2	$\leq 1.4 \times 10^{-19}$	Baulch et al., 1984
H_2O	5×10^{-18}	Baulch et al., 1984
Dry synthetic air	$(3.0 \pm 0.1) \times 10^{-19}$	Penzhorn et al., 1974
Ambient air	$(3.4 \pm 0.4) \times 10^{-19}$	Penzhorn et al., 1974
Ethene	1.8×10^{-18}	Ackerman et al., 1970
Propene	$\leq 2 \times 10^{-17}$	Herron and Huie, 1970
	2.2×10^{-18}	Ackerman et al., 1970
2-Butene	$< 2 \times 10^{-17}$	Herron and Huie, 1970
1-Butene	2.3×10^{-18}	Ackerman et al., 1970
1-Pentene	3.2×10^{-18}	Ackerman et al., 1970
trans-3-Methyl-2-pentene	$\sim 2 \times 10^{-17}$	Huie and Herron, 1973
2-Methyl-2-butene	5.5×10^{-17}	Huie and Herron, 1973
2,3-Dimethyl-2-butene	1.3×10^{-15}	Huie and Herron, 1973
	8.1×10^{-16}	Ackerman et al., 1972
2,5-Dimethylfuran	2.5×10^{-14}	Huie and Herron, 1973

$$\frac{k[\mathrm{OH}]}{k[\mathrm{O_2}(^1\Delta_g)]} \simeq \frac{1 \times 10^6 \times 1.1 \times 10^{-10}}{1 \times 10^8 \times 1.3 \times 10^{-15}} \simeq 10^3$$

In summary then, $O_2(^1\Delta_g)$ is now believed not to contribute to a significant extent to the overall gas phase oxidation of hydrocarbons in the troposphere.

7-E-5b. *Mechanisms*

There are three general types of reactions which $O_2(^1\Delta_g)$ undergoes with alkenes. These are (1) the *ene* reaction with olefins containing allylic hydrogen atoms, (2) 1,4 cycloaddition to dienes and heterocycles, and (3) 1,2 cycloaddition to certain electron-rich alkenes.

The *ene* reaction involves a shift of the C=C double bond with the formation of an allylic hydroperoxide. The first example of this in the gas phase was the classic work of Winer and Bayes (1966) who showed that 2,3-dimethyl-2-butene reacted with singlet oxygen in a manner analogous to the established solution phase reaction (Foote and Wexler, 1964):

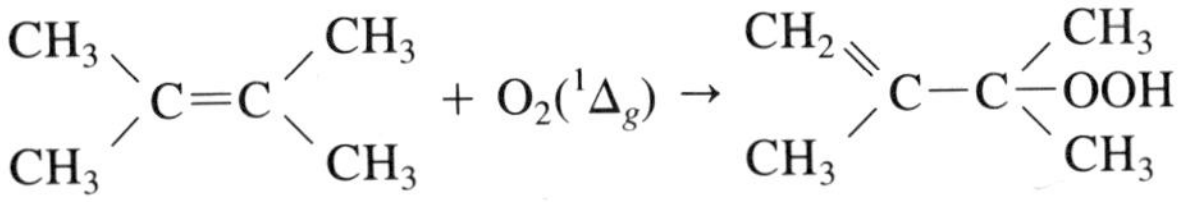

(1)

The 1,4 cycloaddition of $O_2(^1\Delta_g)$ to dienes and heterocycles is analogous to the Diels–Alder reaction. Interestingly, the reaction with 2,5-dimethylfuran is sufficiently rapid (Table 7.19) that it is frequently used to titrate $O_2(^1\Delta_g)$ in the gas phase:

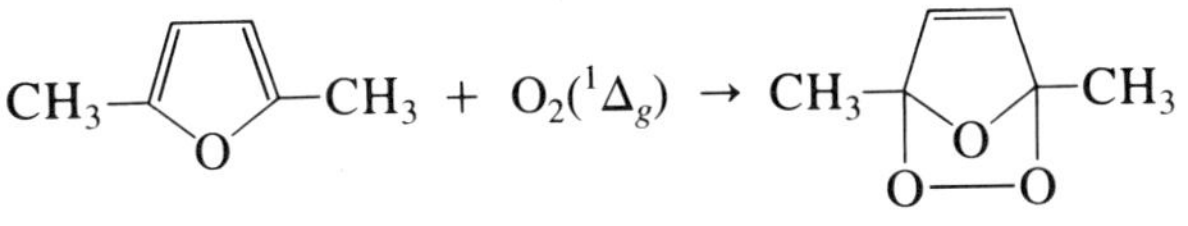

(99)

(I)

There is some infrared spectroscopic data (Gleason et al., 1970) supporting the formation of **(I)** in the gas phase at room temperature, and it has been identified in solution at low temperatures. Because it is unstable at room temperature, however, it is difficult to isolate and thus a secondary product formed from its reaction with methanol is most commonly identified at room temperature:

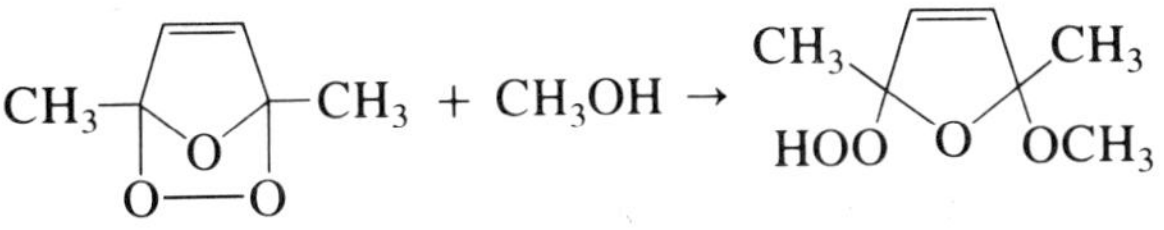

(100)

The third type of reaction common to singlet oxygen reactions is the 1,2 cycloaddition to alkenes which are electron-rich (e.g., vinyl ethers and enamines) or to simple alkenes which cannot undergo the ene reaction (i.e., which do not have

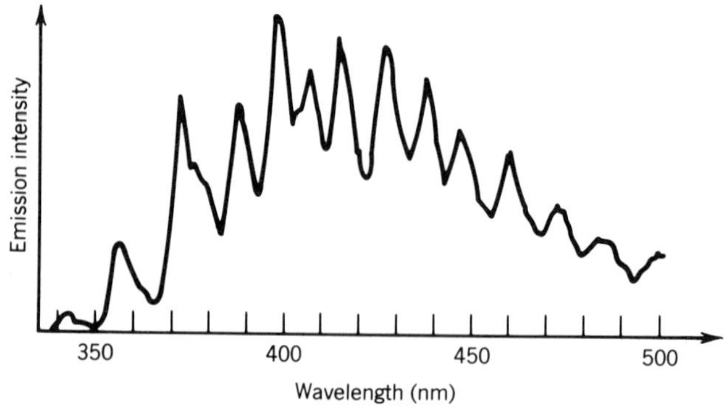

FIGURE 7.18. Chemiluminescence from electronically excited formaldehyde produced in the reaction of $O_2(^1\Delta_g)$ with ethene (from Bogan et al., 1976).

allylic hydrogen atoms). These reactions form dioxetanes which contain the excess reaction exothermicity and hence decompose to carbonyl compounds.

Chemiluminescence is frequently observed from these reactions since the excess energy can be channeled into electronic energy of the products. Thus formaldehyde chemiluminescence has been observed from the gas phase reaction of $O_2(^1\Delta_g)$ with ethene as well as with methyl-, ethyl-, and *n*-butyl-vinyl ethers (Bogan et al., 1975, 1976). For example, Fig. 7.18 shows the emission spectrum observed from the $O_2(^1\Delta_g)$–C_2H_4 reaction (Bogan et al., 1976); the bands are clearly assignable to electronically excited $HCHO(A^1A_2 \rightarrow X^1A_2)$. The emission is attributed to the intermediate formation of dioxetane followed by its decomposition to produce an excited formaldehyde:

$H_2C{=}CH_2 + O_2(^1\Delta_g) \rightarrow$ 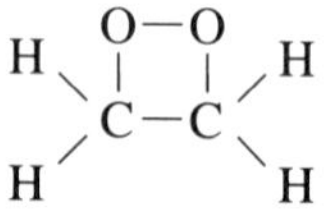 $\rightarrow HCHO^* + HCHO$ (101)

$\downarrow$

$HCHO + h\nu$ (102)

In summary, $O_2(^1\Delta_g)$ does not participate directly to a significant extent in the overall oxidation of hydrocarbons in photochemical air pollution. However, because of the nature of the products of its reaction, the role of $O_2(^1\Delta_g)$ in photosensitized oxidations, particularly in biological systems, remains of considerable interest. The reader is referred to the reviews by Foote (1976) and the volumes edited by Schaap (1976) and Rånby and Rabek (1978) for further details.

7-F. REACTIONS OF ALKYNES

7-F-1. Hydroxyl Radical, OH

7-F-1a. Kinetics

As expected for the relatively inert C≡C triple bond, reaction with OH is slow compared to the alkene reactions. Table 7.20 gives the rate data for three smallest

TABLE 7.20. Kinetic Data for the Reaction of OH with Acetylene, Propyne, and 1-Butyne at One Atmosphere Pressure

Alkyne	$10^{12}\,k$ (298°K) (cm^3 molecule^{-1} s^{-1})	Temperature Dependence	Reference
C_2H_2	0.78	$1.7 \times 10^{-12}\,e^{-233/T}$	*a*
$CH_3C{\equiv}CH$	0.95	n.a.	*b*
	6.06	n.a.	*c*
$CH_3CH_2C{\equiv}CH$	8.04	n.a.	*c*

[a] Recommended by Atkinson (1985).

[b] Bradley et al. (1973); carried out at a total pressure of 2.5 Torr in He, hence reaction is likely in fall-off kinetic region (see discussion in text).

[c] Atkinson and Aschmann (1984a); relative rate based on $k(OH + \text{cyclohexane}) = 7.38 \times 10^{-12}$ cm^3 molecule^{-1} s^{-1}.

alkynes. Acetylene (C_2H_2) is the only alkyne that has been studied by a number of groups and by several different methods. At an OH concentration of 1×10^6 cm^{-3}, assuming a rate constant of 7.3×10^{-13} cm^3 molecule^{-1} s^{-1}, the lifetime of acetylene would be ~16 days, too slow to be of much consequence in photochemical air pollution. In fact, acetylene is used as a tracer of automobile exhaust from non-catalyst equipped vehicles because of its relatively inert nature.

7-F-1b. Mechanisms

Relatively little direct evidence is available concerning the mechanism of the reaction of OH with alkynes. An observed pressure dependence of the rate constant for the OH–C_2H_2 reaction (Perry et al., 1977c; Michael et al., 1980; Perry and Williamson, 1982; Schmidt et al., 1984) suggests that addition to form an OH–alkyne adduct is the first step, similar to that in OH–alkene reactions (see Section 7-E-1); thus the excess energy in the OH–alkyne adduct must be removed by collision with a third body or the adduct will decompose back to reactants.

Figure 7.19, for example, shows the experimentally determined effective bimolecular rate constant for the OH–C_2H_2 reaction as a function of pressure of argon diluent for five temperatures from 228 to 413°K. At room temperature, which is of greatest interest for the troposphere, the rate constant rises to a plateau at pressures above approximately 150 Torr. Addition to the triple bond of acetylene is reasonable as the C—H bond is very strong (~132 kcal mole^{-1}), making abstraction of a hydrogen atom difficult.

Recent kinetic studies (Schmidt et al., 1985; Wahner and Zetsch, 1985) confirm the pressure dependence of the rate constant; they report values for the low-pressure limiting rate constant k_0 (see Section 4-A-4) of 2.5×10^{-30} and 5.0×10^{-30} cm^6 molecule^{-2} s^{-1}, respectively, and a high-pressure limiting rate constant k_∞ of 9×10^{-13} cm^3 molecule^{-1} s^{-1}. The small bimolecular component was reported by Schmidt and coworkers to be 0.5×10^{-13} cm^3 molecule^{-1} s^{-1}, about an order of magnitude lower than that reported by Michael et al. (1980).

The fate of the OH–C_2H_2 adduct in air is not known. Schmidt et al. (1984, 1985) have observed vinoxy radicals (CH_2CHO) in the OH–C_2H_2 reaction, and

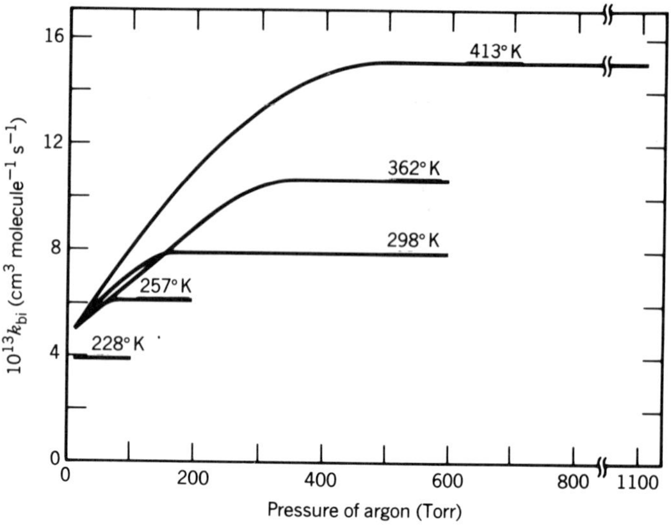

FIGURE 7.19. Effective bimolecular rate constants k_{bi} for the OH–C_2H_2 reaction as a function of argon pressure at temperatures from 228 to 413°K (from Michael et al., 1980).

their decay rate increased when O_2 was added; simultaneously, glyoxal $(CHO)_2$ was formed and some OH was regenerated. They suggested that the mechanism might proceed at least partly by reactions (103) and (104):

$$C_2H_2 + OH \rightleftarrows (HC{=}CHOH)^* \xrightarrow{M} \rightarrow \cdot CH_2C\begin{matrix} \diagup O \\ \diagdown H \end{matrix} \qquad (103)$$

$$CH_2C\begin{matrix} \diagup O \\ \diagdown H \end{matrix} + O_2 \rightarrow \rightarrow (CHO)_2 + OH \qquad (104)$$

As discussed in Section 7-E-3b, the vinoxy radical has also been proposed by Gutman and Nelson (1983) to react with O_2 to form HO, plus HCHO + CO. Clearly, much work remains to be done on the mechanism of this reaction.

The similarity in the rate constants for propyne and 1-butyne suggests that addition of OH to the triple bond predominates for alkynes larger than acetylene; thus if abstraction of a hydrogen atom were occurring to a significant extent, one might expect by analogy to the OH–alkane reactions (Section 7-C-1) that the rate constant would increase as the number and nature (i.e., primary, secondary, or tertiary) of abstractable C—H bonds increased.

To illustrate this point, Eq. (A) given in Section 7-C-1 for estimating OH–alkane rate constants can be applied as a first approximation. Ignoring the strong ≡C—H bonds, one would predict that abstraction of a methyl group hydrogen in propyne would occur with a rate constant of $\sim 3(6.5 \times 10^{-14})$, or $\sim 2 \times 10^{-13}$ cm^3 $molecule^{-1}$ s^{-1}. The addition of two secondary C—H bonds in 1-butyne would be expected to increase this by $2(5.8 \times 10^{-13})$ cm^3 $molecule^{-1}$ s^{-1}, or to approximately 1.4×10^{-12} cm^3 $molecule^{-1}$ s^{-1}. However, rather than an increase of a factor of ~5–10 in going from propyne to 1-butyne, the rate constant only increases by approximately 30% (Table 7.18). This suggests that addition to the triple bond is occurring rather than hydrogen abstraction.

A second piece of evidence supporting the addition route is that the rate constant for the reaction of OH with 1-butyne is faster than that for *n*-butane (Table 7.3) despite the fact it has fewer abstractable C—H hydrogens and the C—H bond is stronger. However, OH–alkyne reactions are clearly an area that needs more attention in the future.

7-G. REACTIONS OF AROMATICS

7-G-1. Hydroxyl Radical, OH

Aromatic hydrocarbons comprise a significant fraction of the organics in automobile exhaust (Mayrsohn et al., 1977; Nelson and Quigley, 1984) and in ambient air (Lonneman et al., 1968; Altshuller et al., 1971), with toluene usually being the aromatic present in the highest concentration (see Section 5-E and Table 5.7). Because, with the exception of benzene, these aromatics are moderately reactive in photochemical smog formation (Section 10-C-2), their chemistry in the atmosphere is of great interest. As discussed in Section 7-B, OH radial attack is the only significant loss process for simple aromatic hydrocarbons. We thus review here the kinetics and mechanisms of reaction of OH with aromatic hydrocarbons.

7-G-1a. Kinetics

Table 7.21 gives the room temperature rate constants for the reaction of OH radicals with some common aromatic hydrocarbons and cresols. The reactions are fast, except for the simplest aromatic, benzene which is still quite rapid.

The OH–aromatic reactions provide a good illustration of how kinetic data can provide insight into reaction mechanisms. Early studies of the OH reaction with benzene and toluene indicated that the rate constants exhibited a pressure dependence below 100 Torr total pressure of helium as a carrier gas (Davis et al., 1975), although not above ~50 Torr in argon (Hansen et al., 1975; Tully et al., 1981). This suggested that addition of OH radicals to the ring contributed significantly to the overall reaction.

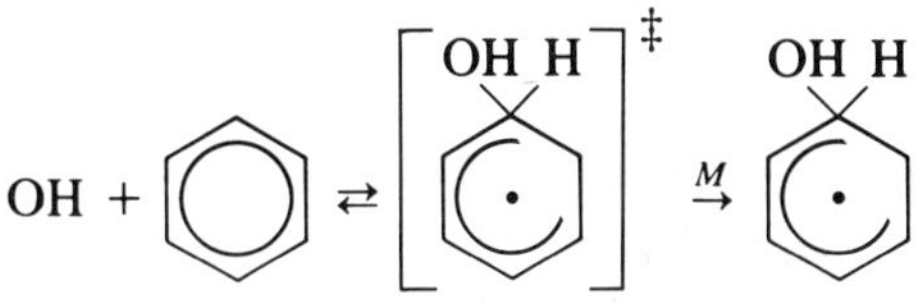

(105)

TABLE 7.21. Rate Constants at 298°K and One Atmosphere Pressure and Expressions for Their Temperature Dependence in the Reactions of OH with Some Aromatic Hydrocarbons[a]

Aromatic Hydrocarbon	$10^{12}\ k$ (298°K) (cm^3 $molecule^{-1}$ s^{-1})	Temperature Dependence
Benzene[b]	1.3	$7.57 \times 10^{-12}\ e^{-529/T}$
Toluene[b]	6.2	$2.10 \times 10^{-12}\ e^{322/T}$
o-Xylene	14.7	*c*
m-Xylene[b]	24.5	$1.66 \times 10^{-11}\ e^{116/T}$
p-Xylene[b]	15.2	*c*
Ethylbenzene	7.5	n.a.[d]
n-Propylbenzene	5.7	n.a.
Isopropylbenzene	~6.6	n.a.
o-Cresol	40	n.a.
m-Cresol	57	n.a.
p-Cresol	44	n.a.

[a] Recommended by Atkinson (1985), which should be consulted for original references on which these are based and for uncertainties. Note expressions for temperature dependencies only apply at lower temperatures (⩽325°K) where addition of OH to the ring predominates (see text).

[b] Temperature dependence applies only to $T \leqslant 325$°K; see text.

[c] Approximately independent of temperature from 298 to 320°K.

[d] n.a. = not available.

Subsequent temperature studies of these rate constants provided clear support for the addition pathway as well as for abstraction. For the higher aromatics where abstractable hydrogens are available on the alkyl side chains, abstraction occurs from the substituent alkyl groups since these C—H bonds are much weaker, for example ~85 kcal $mole^{-1}$ compared to ~110 kcal mole for C—H bonds in benzene:

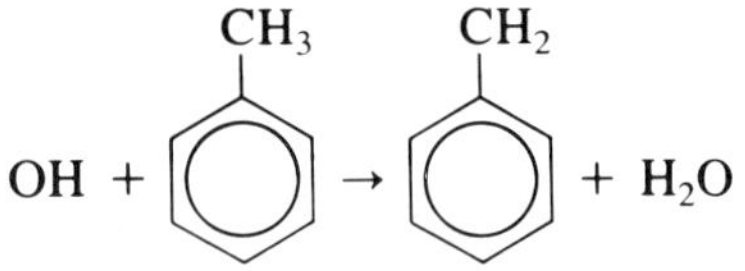

(106)

In these temperature studies, rate constants were obtained as a function of temperature in the usual manner by following the decay of OH in the presence of a great excess of the organic (see Chapter 4). Under these conditions, the decay of OH is normally exponential, that is, plots of log [OH] against reaction time are normally linear. As seen in Fig. 7.20 for the OH–toluene reaction, such plots are indeed linear at temperature below ~325°K and above ~380°K. However, between these two temperatures, non-exponential decays are obtained. When the exponential data for temperatures <325°K and >380°K are plotted in the Arrhenius form, the discontinuous curves shown in Figs. 7.21 and 7.22 are obtained.

Both the non-exponential OH decays in the intermediate temperature region (~325–380°K) as well as the unusual Arrhenius plots can be rationalized on the

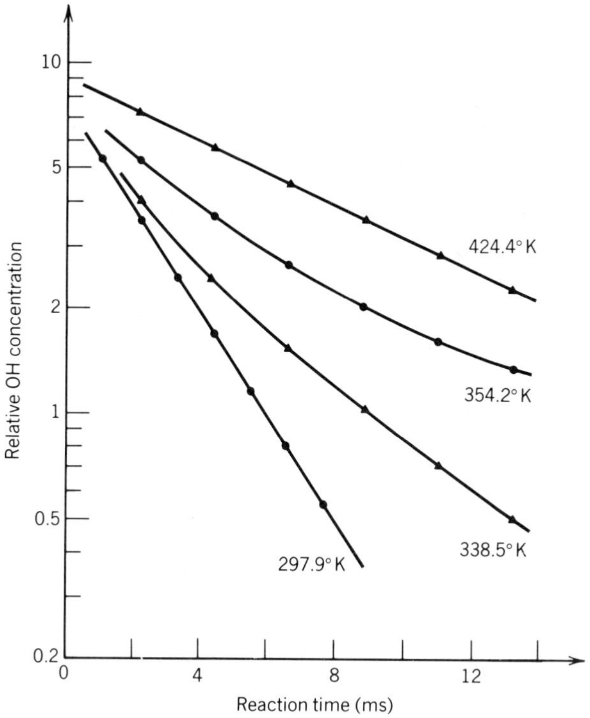

FIGURE 7.20. Semi-log plots of OH decays as a function of reaction time in the presence of a great excess of toluene ($\sim 5 \times 10^{13}$ cm^{-3}) at temperatures from 297.9 to 424.4°K and in ~100 Torr argon (from Perry et al., 1977a).

basis of two reaction paths, addition and abstraction, occurring simultaneously. At room temperature, both routes occur and hence the rate constant is the sum of those for abstraction and addition. For toluene, for example, which is a major component of auto exhaust (Mayrsohn et al., 1977; Nelson and Quigley, 1978) as well as of the aromatic hydrocarbons found in ambient air (see Section 5-G), the rate constant in Table 7.21 (6.2×10^{-12} cm^3 $molecule^{-1}$ s^{-1}) is the sum of abstraction, reaction (106), and addition, reaction (107):

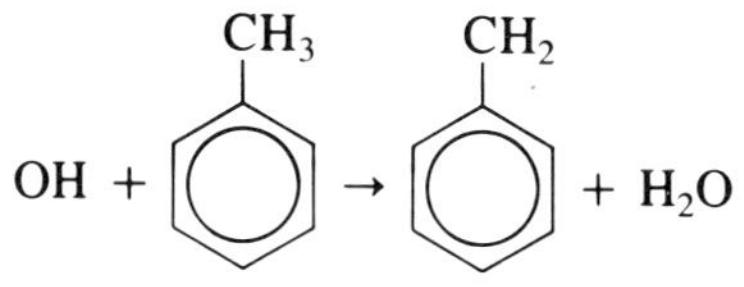

(106)

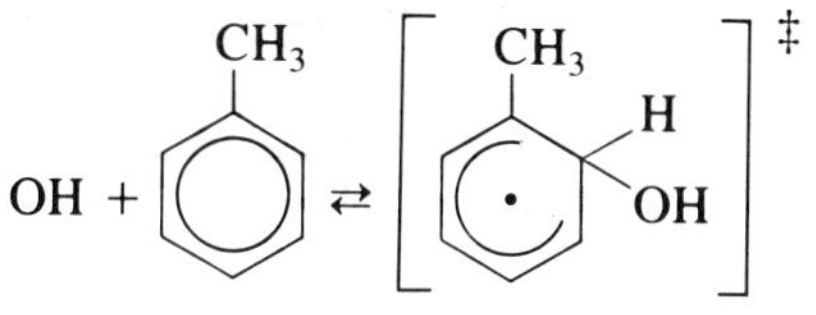

(107, −107)

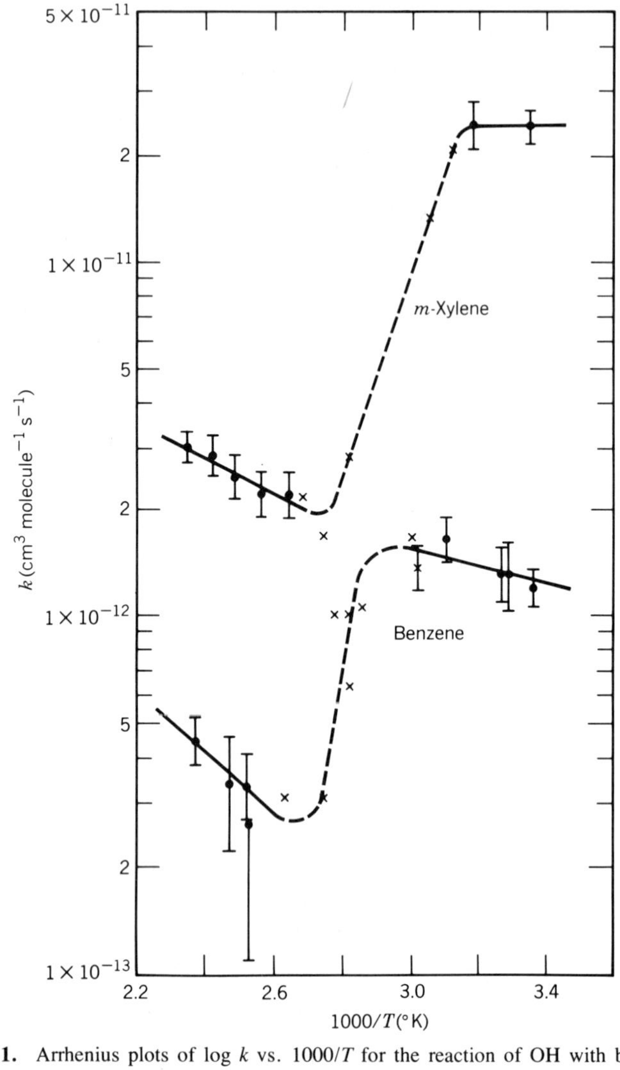

FIGURE 7.21. Arrhenius plots of log k vs. $1000/T$ for the reaction of OH with benzene and m-xylene: ●, Exponential OH decays observed, ×, non-exponential OH decays observed (from Perry et al., 1977a).

If not collisionally stabilized, the OH–aromatic adduct is postulated to decompose back to the reactants, reaction (−107). This is reasonable in view of the excess energy, ~ 18 kcal mole^{-1}, the adduct contains from the exothermicity of the forward addition reaction. As the temperature is increased, the rate of decomposition of the adduct (i.e., k_{-107}) increases, leading to non-exponential OH decays (Fig. 7.20) as well as to a net decrease in the overall rate of removal of OH and hence in the measured overall rate constant. At temperatures above 380°K,

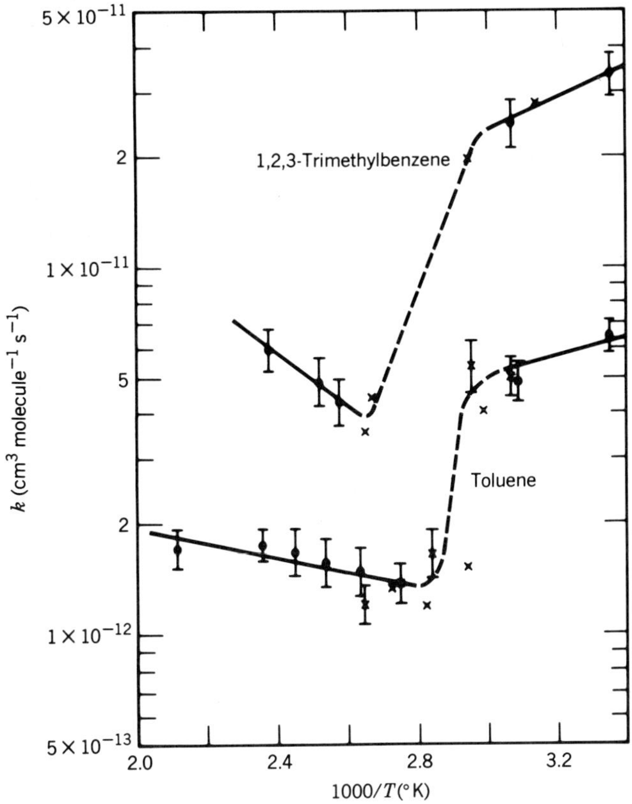

FIGURE 7.22. Arrhenius plots of log k vs. $1000/T$ for the reaction of OH with toluene and 1,2,3-trimethylbenzene: ●, Exponential OH decays observed, ×, non-exponential OH decays observed (from Perry et al., 1977a).

the decomposition of the adduct is so fast compared to the time scale of these experiments (~ 1–30 ms) that only the abstraction reaction is observed. As expected for such an abstraction reaction, as the temperature is increased further, the rate constant increases (Figs. 7.21 and 7.22).

There are two pieces of evidence which show that, at room temperature, addition predominates over abstraction. The first is that the reaction of OH with perdeuteroluene ($CD_3C_6D_5$) at room temperature is, within experimental error, identical to that for toluene, whereas at 423°K, it is a factor of approximately 2 lower (Perry et al., 1977a; Lorenz and Zellner, 1983). This is consistent with addition at room temperature since no C—H or C—D bond is broken and hence no significant isotope effect is expected. On the other hand, at the higher temperature where abstraction predominates, breaking of a C—D bond is expected to have a higher activation energy and hence be slower, as is observed.

The second piece of evidence supporting the predominance of addition at room

TABLE 7.22. Selected Values of the Fraction of the Overall Reaction of OH with Some Aromatics Which Proceeds by Hydrogen Atom Abstraction at Room Temperature

Compound	Fraction Proceeding by Abstraction	Reference
Benzene	0.05	Perry et al., 1977a
	0.01	Tully et al., 1981
	0.01	Lorenz and Zellner, 1983
Toluene	0.16	Perry et al., 1977a
	~0.25–0.35	Hoshino et al., 1978b
	0.15	Kenley et al., 1978
	0.04	Tully et al., 1981
	0.08	Atkinson et al., 1983b
	0.13	Gery et al., 1985
o-Xylene	0.20	Perry et al., 1977a
	0.08	Takagi et al., 1980
m-Xylene	0.04	Perry et al., 1977a
	0.02	Nicovich et al., 1981
p-Xylene	0.07	Perry et al., 1977a
	0.04	Nicovich et al., 1981
1,2,3-Trimethylbenzene	0.035	Perry et al., 1977a
1,3,5-Trimethylbenzene	0.02	Perry et al., 1977a

temperature arises from extrapolating the abstraction data above 400°K to ~300°K. The ratio of this calculated rate constant for abstraction to the total observed rate constant gives the fraction of the reaction proceeding by abstraction at room temperature. Table 7.22 gives some values for this fraction for several aromatics reported in the literature. Most studies show abstraction contributing less than 20% to the room temperature rate constant; indeed, 10% may be a more reasonable upper limit.

7-G-1b. Mechanisms

From the kinetic work discussed above, addition of OH to the aromatic ring is anticipated to predominate in atmospheric reactions of aromatics, since temperatures above 325°K are not encountered in the troposphere; abstraction of a hydrogen atom should account for less than ~10–20% of the overall reaction. Confirming evidence for this is found in the case of benzene, where the OH adduct has been observed in the gas phase using UV absorption at 308 nm (Fritz et al., 1985).

Taking toluene as an example then, the major products of its atmospheric reactions are expected to be those resulting from the subsequent reactions of the OH–aromatic adduct.

While OH may theoretically add to the ortho, meta, or para positions in toluene or even to the carbon atom which bears the methyl group, attack is observed experimentally (see below) to occur primarily at the ortho position:

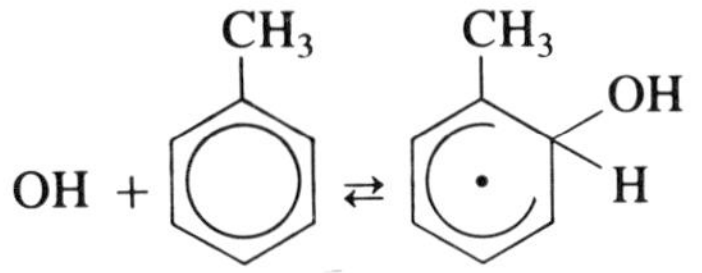

(108)

The reasons for this very large predominance of attack at the ortho position compared to the para position are not clear; some possible mechanisms to explain this are addressed by Kenley et al. (1981). Reaction with O_2 is expected to be sufficiently fast that, at atmospheric pressure, the decomposition of the adduct back to reactants is negligible. A reasonable fate of this adduct in 1 atm of air would be reaction with O_2 to form *o*-cresol:

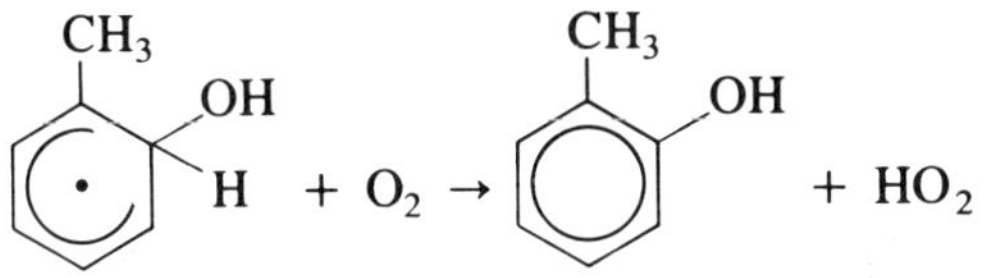

(109)

It should be noted, however, that reaction (109) is likely not, in fact, a simple hydrogen atom abstraction by O_2. In solution, for example, isotope labeling studies have shown that some of the cresol product contains an oxygen atom from the O_2 (Narita and Tezuka, 1982).

Minor amounts of *m*-cresol and *p*-cresol are expected from OH attack at the meta and para positions, respectively. This has been confirmed experimentally by Kenley et al. (1981) who measured the product ratios of the isomeric cresols. They established that the relative rates of attack at the ortho, meta, and para positions were 0.806 ± 0.022, 0.051 ± 0.009, and 0.143 ± 0.019, respectively, in agreement with the work of Hoshino et al. (1978b) who found that approximately 80% of the total cresol formed was *o*-cresol.

The minor portion of the reaction which occurs by abstraction from the methyl group

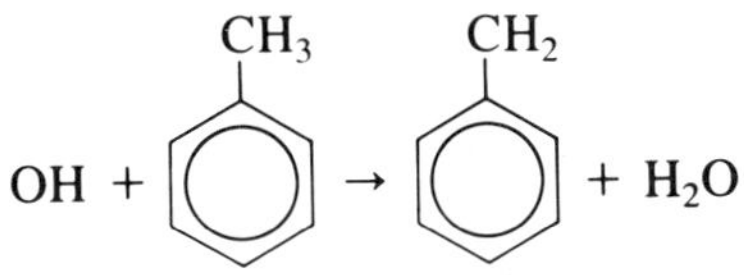

(106)

produces a benzyl radical whose atmospheric reactions are analogous to those of alkyl radicals (Section 7-D). The stable end products anticipated are benzaldehyde and much smaller amounts of benzylnitrate:

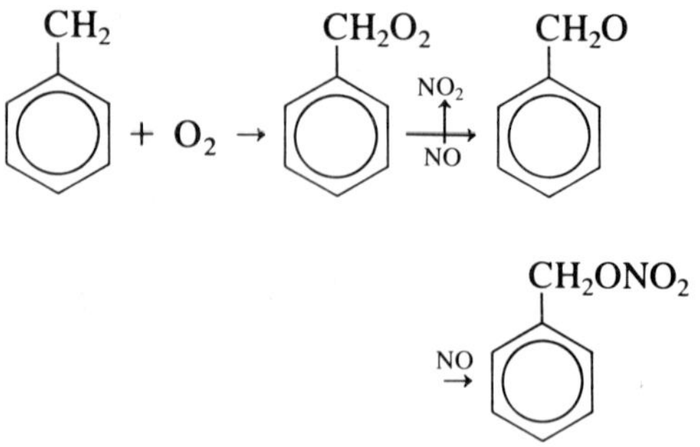

(110)

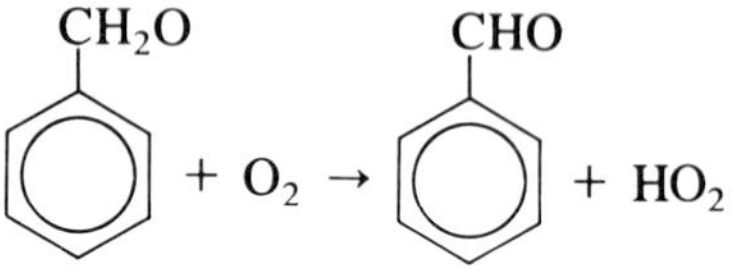

(111)

On the basis of the above mechanism, one might thus anticipate *o*-cresol as a major product of the atmospheric oxidation of toluene, and the meta and para cresols, as well as benzaldehyde, as minor products.

Table 7.23 shows the result of some studies of the yields of *o*-cresol and ben-

TABLE 7.23. Some Reported Yields of the Expected Products *o*-Cresol and Benzaldehyde from the Photooxidation of Toluene at Room Temperature

Product	Yield (%)	Reference
o-Cresol	12[a]	Spicer and Jones, 1977
	51[b]	Hoshino et al., 1978b
	68[c]	Kenley et al., 1978, 1981
	1–2[d]	Besemer, 1982
	5	O'Brien et al., 1979a,b
	~21	Atkinson et al., 1980
	13.1 ± 7.2	Atkinson et al., 1983b
Benzaldehyde	~50	Spicer and Jones, 1977
	30[b]	Hoshino et al., 1978b
	15 ± 2	Kenley et al., 1978, 1981
	3–7	Besemer, 1982
	2.5	O'Brien et al., 1979a,b
	12	Atkinson et al., 1980
	7.3 ± 2.2	Atkinson et al., 1983b
	10.4 ± 2.9	Gery et al., 1985

[a] Based on 80.6% of total cresol yield reported being *o*-cresol as found by Kenley et al. (1981).

[b] Based on relative product yields and extrapolated to atmospheric conditions.

[c] Extrapolated to atmospheric conditions assuming observed products account for 100% of toluene consumed; maximum possible reacted toluene, which is not accounted for, was reported as 22%.

[d] Combined yields of *o*- and *p*-cresol.

zaldehyde from the photooxidation of toluene at room temperature. There is considerable disagreement on the absolute yields of both products. However, most of the reported benzaldehyde yields are less than ~20%, in agreement with the kinetic data on the fraction of the overall OH-toluene reaction which proceeds by abstraction (Table 7.22).

The reported yields of *o*-cresol seem to fall into two groups: $\lesssim$20% and $\gtrsim$50%. Possible explanations for these discrepancies include experimental problems in sampling and analysis or differences in reaction conditions. For example, Kenley and co-workers used a flow system at 6–12 Torr total pressure whereas the other studies were carried out at 1 atm. However, pressure differences alone cannot account for the discrepancies since no change in the yields of *o*-cresol and benzaldehyde were seen as a function of pressure over the 6–12 Torr range in these studies, nor over the 62–740 Torr range in the work by Atkinson and co-workers (1983b).

The data in Table 7.23 show that the ratio of the yield of *o*-cresol to benzaldehyde measured in each of these studies is usually $\lesssim$2; however, if abstraction of a methyl hydrogen accounts for 10–20% of the overall reaction, leading to benzaldehyde, and addition to the ring to form *o*-cresol accounts for most of the rest, then the ratio should be approximately 5–10. The much lower ratios observed for *o*-cresol to benzaldehyde suggest that there are other reactions occurring in addition to the two reaction sequences (106) + (110) + (111) and (108) + (109). Supporting this are the data in Table 7.23 obtained under conditions closest to those in the troposphere, where the yields of the cresols and benzaldehyde account for less than half of the reacted toluene.

In short, there must be other fates of the OH–toluene adduct formed in reaction (108) than formation of cresols. Similar conclusions have been reached in the photooxidations of this and other aromatics where a variety of *ring cleavage* products have also been observed (e.g., see Schwartz, 1974; Nojima et al., 1974; Takagi et al., 1980, 1982; Darnall et al., 1979; Atkinson et al., 1983b; Tuazon et al., 1984a; O'Brien et al., 1984; Dumdei and O'Brien, 1984; Shepson et al., 1984).

To give the reader some appreciation for the multitude and complexity of products observed in aromatic oxidations, the higher molecular weight products observed in toluene–NO_x photooxidations, in addition to the cresols and benzaldehyde, are summarized in Table 7.24; smaller compounds such as CO, CO_2, HCHO, HCOOH, and PAN have also been observed. In the NO_x–air photooxidation, OH is believed to be the only significant oxidant for toluene (Table 7.1), so the products arise from the OH–toluene reaction followed by secondary reactions of the initial products. Clearly, ring opening reactions occur to form a variety of oxygenated products, some of which are unsaturated.

However, one word of caution is appropriate in that, as in all laboratory studies, conditions differ from those in the atmosphere and this may alter the products either qualitatively or quantitatively. For example, reactions can potentially occur on the walls of the reaction vessel; these will be more severe the smaller the vessel, that is, the larger the surface-to-volume ratio. Another potential problem is the use of long photolysis times which increases the extent of secondary reactions; while this

TABLE 7.24. Some of the High Molecular Weight Products Identified in the Toluene–NO_x Photooxidation

Compound	Structure	Reference
Methyglyoxal	CH_3COCHO	*a–e,o*
Glyoxal	$(CHO)_2$	*a–d,o*
Acetic acid	CH_3COOH	*c*
2-Oxo-3-butene	$CH_3COCH{=}CH_2$	*c,n*
1,4-Butenedial	$CHOCH{=}CHCHO$	*b,c,n*
3-Hydroxy-3-butene-2-one	$CH_3COC(OH){=}CH_2$	*c*
3,5-Hexadiene-2-one	$CH_3COCH{=}CH{-}CH{=}CH_2$	*c,e*
4-Oxo-2-pentenal	$CH_3COCH{=}CHCHO$	*b,c,e,n*
2-Hydroxy-1,4-butenedial	$CHOC(OH){=}CHCHO$	*c,e*
Phenol	C_6H_5OH	*c,l*
2-Hydroxy-3-oxobutanal	$CH_3COCH(OH)CHO$	*c*
3-Hydroxy-3,5-Hexadiene-2-one	$CH_3COC(OH){=}CH{-}CH{=}CH_2$	*c*
3-Hydroxy-4-oxo-2-pentenal	$CH_3COC(OH){=}CHCHO$	*c*
Benzoic acid	C_6H_6COOH	*c*
Hydroxybenzaldehyde	$C_6H_4(CHO)OH$	*c*
Dihydroxytoluene	$C_6H_3(CH_3)(OH)_2$	*c*
6-Oxo-2,4-heptadienal	$CH_3COCH{=}CH{-}CH{=}CHCHO$	*c,e*
4,5-Dioxo-2-hexenal	$CH_3COCOCH{=}CHCHO$	*c*
Nitrotoluene	$C_6H_4(CH_3)NO_2$	*b,c,f–j,m,o*
1-Pentene-3,4-dione	$CH_3COCOCH{=}CH_2$	*n*
2-Oxo-3-butenal	$OHCCOCH{=}CH_2$	*n*
Furan		*n*
2-Methylfuran	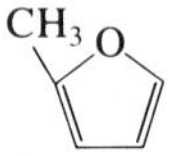	*n*
Furfural	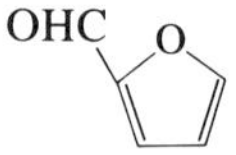	*n*
Nitrophenol	$C_6H_4(OH)NO_2$	*c,j,o*
6-Oxo-5-hydroxy-2,4-heptadienal	$CH_3COC(OH){=}CH{-}CH{=}CHCHO$	*c*
Nitrobenzaldehyde	$C_6H_4(CHO)NO_2$	*c*
Nitrocresol	$C_6H_3(CH_3)(OH)NO_2$	*b,c,f–k,o*
Benzene dicarboxaldehyde	$C_6H_4(CHO)_2$	*l*
Tolualdehyde	$C_6H_4(CH_3)CHO$	*l*
Benzyl nitrate	$C_6H_5CH_2ONO_2$	*b,c,g–i,o*
Benzyl alcohol	$C_6H_5CH_2OH$	*l,m*
Dinitrotoluene	$C_6H_3(CH_3)(NO_2)_2$	*c*

TABLE 7.24. (*Continued*)

Compound	Structure	Reference
2-Methyl-1,4-benzoquinone	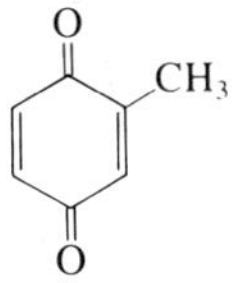	*m,o*

Source: Adapted from Dumdei and O'Brien, 1984.
[a]Nojima et al., 1974.
[b]Besemer, 1982.
[c]Dumdei and O'Brien, 1984.
[d]Tuazon et al., 1984a.
[e]O'Brien et al., 1984.
[f]Akimoto et al., 1978b.
[g]Hoshino et al., 1978b.
[h]Ishikawa et al., 1978.
[i]O'Brien et al., 1979b.
[j]Atkinson et al., 1980.
[k]Schwartz et al., 1974.
[l]Spicer and Jones, 1977.
[m]Kenley et al., 1981.
[n]Shepson et al., 1984.
[o]Gery et al., 1985.

is frequently necessary in order to form sufficient products for analysis, it obviously complicates the task of mechanistic interpretation.

While a variety of products in the toluene–NO_x photooxidation have been qualitatively identified, in most studies the carbon balance is poor. For example, in one study, the formation of *o*-cresol, benzaldehyde, and two of the major ring cleavage products, methylglyoxal and glyoxal, together accounted for only about half (50 $\pm$ 15%) of the reacted toluene (Tuazon et al., 1984a; Atkinson et al., 1983b).

To account for these ring cleavage products, it appears that O_2 does not simply abstract hydrogen atoms from the OH–aromatic adduct, but also must add to it:

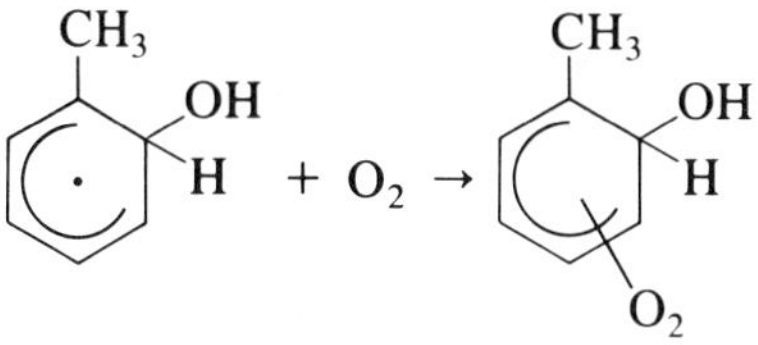

(112)

Addition could occur at the 1, 3, or 5 positions (relative to OH) to yield the following peroxy radicals:

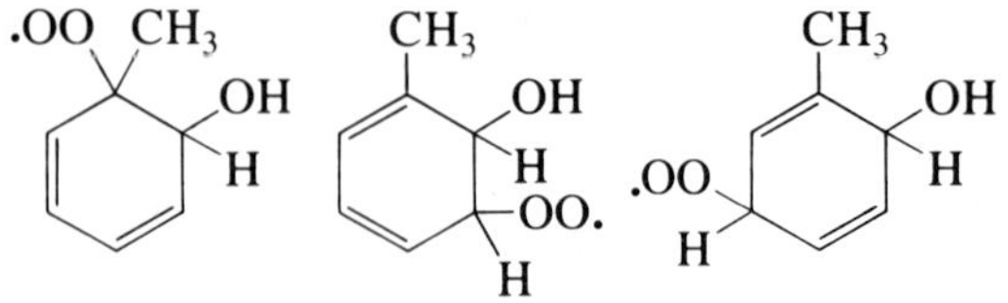

However, the relative amounts of each radical formed are unknown, as are their subsequent atmospheric fates. For example, estimates of the ratio of the rate constants for hydrogen atom abstraction from the OH-aromatic adduct, reaction (109), to that for addition, reaction (112), vary from 0.2 (Leone and Seinfeld, 1984; Leone et al., 1985) to ≥0.5 (Gery et al., 1985).

There have been several attempts to simulate the concentration–time profiles of the major reactants and products in the toluene–NO_x–air photooxidations using computer kinetic models (Atkinson et al., 1980; Killus and Whitten, 1982; Leone and Seinfeld, 1984; Leone et al., 1985). By adjustment of the mechanism and/or rate constants for various steps, the best match to the experimental observations is found; from such interplay between experiments and hypothetical reaction schemes, support is derived for various mechanisms. However, as might be expected for such a complex system, obtaining good calculated fits to the observed major reactants and products does not guarantee that the mechanism is correct and will predict the formation and time dependencies of the minor products such as those in Table 7.24. For example, the formation of methylglyoxal can be predicted from the peroxy radical adduct through reaction sequences such as the following (Atkinson and Lloyd, 1984); while at present these are highly speculative, they illustrate the type of ring opening reactions that may occur:

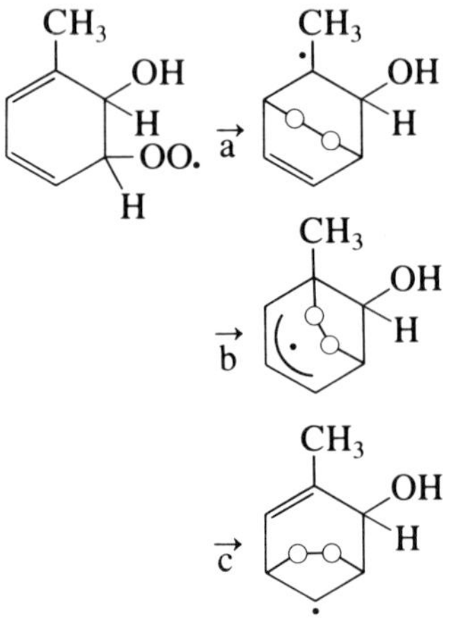

(113)

Following only the first radical, a reaction sequence involving a number of bond scissions in the β position relative to the radical center was proposed; these result in the formation of methylglyoxal and an unsaturated 1, 4-dicarbonyl:

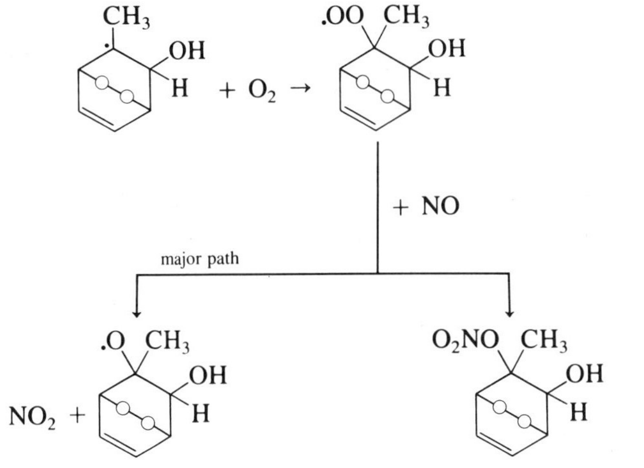

(114)

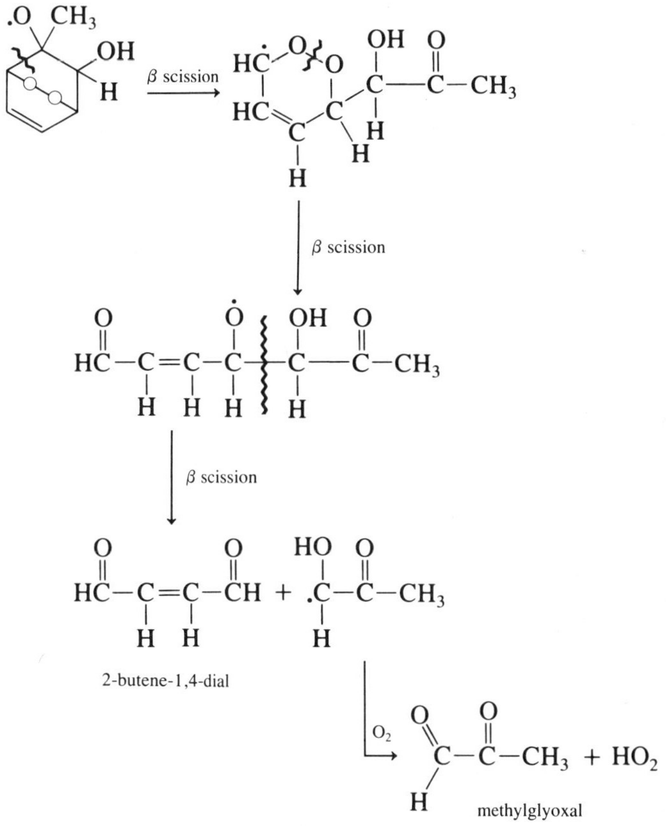

(115)

Other products such as those in Table 7.24 may arise from competing reactions of the peroxybicyclic radicals formed in reaction (113). For example, Gery et al. (1985) observed ~5% yields of methyl-*p*-benzoquinone, which had been observed earlier in small yields by Kenley et al. (1981), and attributed it to the elimination of water from one of these biradicals, followed by hydrogen abstraction by O_2. Leone et al. (1985) suggest that a unimolecular decomposition of the OH-toluene-O_2 adduct may be responsible for C_7 unsaturated dicarbonyls such as 6-oxo-2,4-heptadienal which has been observed by O'Brien and coworkers (Table 7.24).

Similar reactions of the other two cyclized radicals are also proposed, with products of a similar nature predicted.

The α-dicarbonyl compounds glyoxal, $(CHO)_2$, and methylglyoxal, appear to be characteristic products of the reaction of OH with simple aromatics; thus these two compounds have been observed from the photooxidations of toluene, *m*-and *p*-xylene, 1,2,3-trimethylbenzene, and 1,2,4-trimethylbenzene, and biacetyl has been observed from *o*-xylene, and the 1,2,3- and 1,2,4-trimethylbenzenes (Tuazon et al., 1986). For example, the yields in the toluene reaction, defined as the number of product molecules formed, corrected for reaction with OH, and photolysis, per molecule of toluene reacted, have been reported as being in the range of 7.5 to 14.6% for methyglyoxal and 8.0–11.1% for glyoxal (Tuazon et al., 1984a; Shepson et al., 1984. Gery et al., 1985). While the formation of glyoxal by OH attack at the *meta* or *para* positions of the ring can be rationalized through a reaction sequence similar to the complex one outlined above, the high yields relative to those of methylglyoxal are not expected given the predominance of OH attack at the *ortho* position (see above). Thus the source of glyoxal remains uncertain, as do the yields, nature, and mechanisms of formation of the remaining products. In addition, the experimentally determined α-dicarbonyl yields are much smaller than expected from some models, suggesting that as much as 45% of the toluene reaction pathways are at present unaccounted for (Leone et al., 1985).

Thus all that can be said with certainty at the present time is that ring opening reaction(s) must occur to explain the products observed in smog chamber studies. Such reactions have also been invoked as a possible explanation for the observation of the oxalate ion, $(COO)_2^{2-}$, in aerosols and precipitation (Norton et al., 1983); thus oxalic acid might be expected from the oxidation of glyoxal, $(CHO)_2$, which has been observed as a product of aromatic hydrocarbon oxidations. Clearly, the aromatic photooxidations are complex and will require a great deal of further study before the atmospheric reaction pathways are fully elucidated.

In summary, as seen in Table 7.1 and discussed in Section 7-B, only OH reactions are sufficiently fast to contribute significantly to the removal of aromatic hydrocarbons from the troposphere. These reactions are expected to lead to oxygenated aromatics as well as to oxygenated ring fragmentation products. The atmospheric fates of these products can be predicted using the same kinetic and mechanistic approach applied to various types of organics in this chapter.

7-H. REACTIONS OF OXYGENATES

7-H-1. Hydroxyl Radical, OH

7-H-1a. Kinetics

Table 7.25 gives selected room temperature rate constants for the reaction of OH with some oxygenated organics.

Relatively few data, particularly on the temperature dependence, exist for most of these reactions. As a result, for many of the compounds in Table 7.25, Atkinson (1985) does not give a recommendation regarding the rate constant; in these cases we cite the result of one, or in some cases two, of the published values to provide the reader with some idea of the order of magnitude of the rate constant.

The reactions range from fast to moderately fast, i.e. within one to three orders of magnitude of diffusion controlled. As a result, the atmospheric lifetimes of some of these organics with respect to reaction with OH are relatively short, of the order of 28 h (for an OH concentration of 1×10^6 cm^{-3} and $k \approx 10^{-11}$ cm^3 molecule^{-1} s^{-1}).

7-H-1b. Mechanisms

The reactions of OH with oxygenated organics which do not contain C=C double bonds proceed by hydrogen atom abstraction analogous to the alkane reactions. As discussed by Atkinson (1985) for aldehydes, abstraction of the weak aldehydic hydrogen (C—H bond strength is ~87 kcal mole^{-1}) occurs (Kerr and Sheppard, 1981):

$$OH + RCHO \rightarrow RCO + H_2O \tag{116}$$

In the case of HCHO, for example, only hydrogen abstraction occurs; a hydrogen displacement reaction to give HC(O)OH + H has been ruled out (Morrison and Heicklen, 1980; Niki et al., 1984b).

While the overall reaction appears to result in the abstraction of an aldehydic hydrogen, Semmes et al. (1985) suggest, based on their kinetic studies of the reactions of OH with a series of aldehydes, that the reaction may in fact proceed via a long-lived complex rather than by a concerted bimolecular reaction. Thus the activation energies determined for the series of aldehydes from acetaldehyde to pivaldehyde are negative (Semmes et al., 1985; Michael et al., 1985), consistent with a three-body recombination (see Section 4-A-4). While the negative temperature dependence can also be explained on the basis of a temperature-dependent preexponential factor (see Section 7-E-1a), Semmes et al. find the assumptions necessary to explain the results by collision or transition state theory applied to a simple concerted bimolecular reaction unsatisfactory. Clearly more work needs to be done on these reactions.

The subsequent reactions of the free radical produced are analogous to those described in Section 7-D; for example, for acetaldehyde, reaction (116) is followed by (117)–(119):

$$CH_3\overset{\overset{\displaystyle O}{\|}}{C} + O_2 \rightarrow CH_3\overset{\overset{\displaystyle O}{\|}}{C}OO \tag{117}$$

$$CH_3\overset{\overset{\displaystyle O}{\|}}{C}OO + NO \rightarrow CH_3\overset{\overset{\displaystyle O}{\|}}{C}O + NO_2$$

$$\downarrow$$

$$CH_3 + CO_2 \tag{118}$$

TABLE 7.25. Selected Room Temperature Rate Constants for the Reaction of OH with Some Oxygenated Organics

Compound	k^a (cm^3 molecule^{-1} s^{-1})	Reference
Aromatic Oxygenates		
Phenol	2.8×10^{-11}	*l*
o-Cresol	4.0×10^{-11}	*a*
m-Cresol	5.7×10^{-11}	*a*
p-Cresol	5.7×10^{-11}	*a*
Methoxybenzene (anisole)	2.0×10^{-11}	*m*
Aldehydes		
HCHO	9.0×10^{-12}	*a*
CH_3CHO	1.6×10^{-11}	*a*
C_2H_5CHO	2.0×10^{-11}	*a*
n-C_3H_7CHO	2.1×10^{-11}	*o*
i-C_3H_7CHO	2.4×10^{-11}	*o*
n-C_4H_9CHO	2.7×10^{-11}	*o*
i-C_4H_9CHO	2.6×10^{-11}	*o*
$(CH_3)_3CCHO$	3.1×10^{-11}	*o*
C_6H_5CHO	1.3×10^{-11}	*a*
$(CHO)_2$	1.2×10^{-11}	*c*
Ketones		
Acetone	2.3×10^{-13}	*a*
Butanone	1.0×10^{-12}	*a*
4-Methyl-2-pentanone	1.4×10^{-11}	*a*
2,6-Dimethyl-4-heptanone	2.7×10^{-11}	*a*
Biacetyl	2.4×10^{-13}	*b*
CH_3COCHO	1.7×10^{-11}	*c*
	0.7×10^{-11}	*h*
α,β-Unsaturated Carbonyls		
Acrolein	2.0×10^{-11}	*a*
Methacrolein	3.1×10^{-11}	*a*
Crotonaldehyde	3.6×10^{-11}	*a*
Methylvinyl ketone	1.9×10^{-11}	*a*
Ethers		
CH_3OCH_3	3.0×10^{-12}	*a,e*
$C_2H_5OC_2H_5$	9.2×10^{-12}	*f*
Tetrahydrofuran	1.5×10^{-11}	*a*
Furan	4.1×10^{-11}	*a*

TABLE 7.25. (*Continued*)

Compound	k^a (cm^3 $molecule^{-1}$ s^{-1})	Reference
Alcohols		
CH_3OH	9×10^{-13}	*a*
C_2H_5OH	2.9×10^{-12}	*a*
C_3H_7OH	5.3×10^{-12}	*d*
$(CH_3)_2CHOH$	5.5×10^{-12}	*d*
Esters		
Methyl acetate	1.8×10^{-13}	*g*
Ethyl acetate	1.9×10^{-12}	*g*
Methyl propionate	2.8×10^{-13}	*g*
Epoxides		
Ethene oxide	8.1×10^{-14}	*i*
	5.3×10^{-14}	*j*
Propene oxide	1.2×10^{-12}	*k*
	0.5×10^{-12}	*j*
1,2-Epoxybutane	2.1×10^{-12}	*k*

[a]Recommended by Atkinson, 1985.
[b]Darnall et al., 1979.
[c]Plum et al., 1983.
[d]Overend and Paraskevopoulos, 1978.
[e]Perry et al., 1977d.
[f]Lloyd et al., 1976.
[g]Campbell and Parkinson, 1978.
[h]Kleindienst et al., 1982.
[i]Lorenz and Zellner, 1984.
[j]Zetzsch, private communication to Atkinson, 1985.
[k]Winer et al., 1978.
[l]Zetzsch, 1982.
[m]Perry et al., 1977b.
[o]Semmes et al., 1985.

$$\underset{}{CH_3\overset{\overset{\displaystyle O}{\|}}{C}OO} + NO_2 \rightleftarrows \underset{\text{PAN}}{CH_3\overset{\overset{\displaystyle O}{\|}}{C}OONO_2} \tag{119}$$

For formaldehyde, formation of HO_2 and thus eventually regeneration of OH occurs:

$$HCO + O_2 \rightarrow HO_2 + CO \tag{120}$$

$$HO_2 + NO \rightarrow OH + NO_2 \tag{121}$$

etc.

Kinetic and product data for the reaction of OH with ketones are consistent with hydrogen atom abstraction from C—H bonds (Atkinson, 1985).

α, β-Unsaturated carbonyl compounds have available two possible reaction paths, addition to the double bond or hydrogen atom abstraction. For α, β-unsaturated ketones, addition to the double bond is also expected to predominate. However, for the unsaturated aldehydes where a weak aldehydic hydrogen is present, abstraction is expected to occur. For methacrolein, which is an important intermediate in isoprene oxidation (see Chapter 14), for example, Atkinson (1985) suggests that ~55–70% of the overall reaction occurs by hydrogen abstraction, reaction (122a), and the rest via addition:

$$OH + \underset{\text{methacrolein}}{CH_2{=}\overset{\overset{\displaystyle CH_3}{|}}{C}CHO} \xrightarrow{a} CH_2{=}\overset{\overset{\displaystyle CH_3}{|}}{C}CO + H_2O$$

$$\xrightarrow{b} HOCH_2{-}\overset{\overset{\displaystyle CH_3}{|}}{C}CHO \tag{122}$$

In the case of alcohols, it is expected that abstraction from the weaker (~94 kcal mole^{-1}) C—H bonds (McMillen and Golden, 1982), for example,

$$CH_3OH + OH \rightarrow CH_2OH + H_2O \tag{123}$$

will predominate over that from the stronger O—H bond (~104 kcal mole^{-1}, McMillen and Golden, 1982):

$$CH_3OH + OH \rightarrow CH_3O + H_2O \tag{124}$$

However, there is considerable scatter in the values reported for the ratio $k_{124}/(k_{123} + k_{124})$ at room temperature. It has been reported as 0.11 ± 0.03 (Hägele et al., 1983) and 0.17–0.25 (Meier et al., 1984, 1985), respectively. Meier et al. (1984) suggest their data indicate that all hydrogens are abstracted with equal probability rather than showing a preference for the weaker C—H bonds. In the case of ethanol, abstraction of a secondary hydrogen from the $-CH_2-$ group appears to occur about 75% of the time (Meier et al., 1985), as might be expected from the methanol reaction.

As discussed earlier, the α-hydroxy radical CH_2OH formed in (123) reacts with O_2 via overall abstraction to form HO_2 and HCHO. Since CH_3O also reacts with O_2 to form HCHO, the position of hydrogen abstraction in the case of methanol is not important mechanistically.

For a discussion of the mechanisms of reaction of OH with other oxygenated organics, the reader is referred to the Atkinson review (1985) and references therein.

7-H-2. Reaction of HCHO with HO_2

As seen in Table 7.1, HO_2 reacts sufficiently rapidly with HCHO that the reaction is competitive with the attack by OH under certain circumstances. This is somewhat surprising since HO_2 reactions with organics are generally relatively slow at room temperature (Kaufman and Sherwell, 1983). The reaction proceeds via the initial addition of HO_2 to the double bond to form an alkoxy radical which can rapidly isomerize to a peroxy radical (Su et al., 1979a,b; Niki et al., 1980b,c):

$$HO_2 + HCHO \rightleftarrows (HOOCH_2O) \rightleftarrows OOCH_2OH \qquad (125)$$

This peroxy radical is reasonably stable as shown by the formation of the peroxynitrate $HOCH_2OONO_2$ (Niki et al., 1980b; Barnes et al., 1985). What is particularly interesting is that the same peroxy radical should be formed by the addition of O_2 to the radical CH_2OH. However, as discussed in Section 7-E-1b, there is a great deal of evidence that the $O_2 + CH_2OH$ reaction gives HO_2 + HCHO. Atkinson (1985) suggests that in the $O_2 + CH_2OH$ reaction the adduct containing the ~33 kcal mole^{-1} excess energy from the addition reaction decomposes rapidly compared to collisional deactivation:

$$O_2 + CH_2OH \rightarrow [OOCH_2OH]^* \rightarrow HO_2 + HCHO$$
$$\xrightarrow{M} OOCH_2OH$$

This peroxy radical may react with NO (Section 7-D-2) and then O_2, ultimately forming formic acid (Atkinson and Lloyd, 1984):

$$OOCH_2OH + NO \rightarrow OCH_2OH + NO_2 \qquad (126)$$

$$OCH_2OH + O_2 \rightarrow HCOOH + HO_2 \qquad (127)$$

This may contribute, for example, to the formation of formic acid which Dawson et al. (1980) suggest occurs via a photochemical route. Alternatively, at low NO concentrations, the peroxy radical will likely react with HO_2 or RO_2 as discussed in Section 7-D-2b. Based on kinetic simulations of photolyzed $HCHO-O_2-NO$ mixtures, Veyret et al. (1982) derived values of the rate constants $k_{125} = (7.5 \pm 3.5) \times 10^{-14}$ cm^3 molecule^{-1} s^{-1} and $k_{127} = (3.5 \pm 1.6) \times 10^{-14}$ cm^3 molecule^{-1} s^{-1}. However, the rate constants for these reactions remain highly uncertain; for example, that reported for reaction (125), i.e. k_{125}, varies from 1×10^{-14} (Su et al., 1979b) to 11×10^{-14} cm^3 molecule^{-1} s^{-1} (Barnes et al., 1985), and estimates of the rate of decomposition of the adduct, i.e. k_{-125}, vary from 1.5 s^{-1} (Su et al., 1979b) to 20 s^{-1} (Barnes et al., 1985) to 30 s^{-1} (Veyret et al., 1982).

While hydrogen abstraction by HO_2 from higher aldehydes is known to be slow at ambient temperatures (Kaufman and Sherwell, 1983), it is not known whether addition reactions of HO_2 also occur with a variety of larger aldehydes, analogous to reaction (125) for HCHO. However, recent studies by Niki et al. (1985) suggest that in fact an analogous reaction of HO_2 with $(CHO)_2$ also occurs.

7-H-3. Nitrate Radical, NO_3: Kinetics and Mechanisms

Niki and co-workers (Morris and Niki, 1974) were the first to report a rate constant for an NO_3–aldehyde reaction. Only recently has further work been reported on these reactions; the limited kinetic data available are given in Table 7.26. As before, most of the rate constants for oxygenates are derived assuming three different values of the equilibrium constant for reaction (80) ($NO_2 + NO_3 \rightleftarrows N_2O_5$). However, as discussed in Section 7-C-2, those based on $K_{eq} = 3.35 \times 10^{-11}$ cm^3 molecule^{-1} are preferred because of the agreement with very recent absolute rate measurements of the NO_3 reactions with *trans*-butene and 2-methylpropene by Ravishankara and Mauldin (1985). We focus on the H_2CO-NO_3 reaction.

In the work of the Calvert group (Cantrell et al., 1985), nine experiments were carried out in which reactants and products were followed with time using FTIR (Section 5-C-2) and DOAS (Section 5-D-2). The results were analyzed using a computer kinetic model consisting of 36 elementary reactions. Three approaches to the data analysis were used; these were based on (1) the rate of formation of CO and/or HCHO loss, (2) the rates of N_2O_5 and NO_3 loss, and (3) estimated best visual fits to the experimental product data. Application of these three approaches to the nine experiments yielded 27 separate estimates of the rate constant for the reaction of NO_3 with HCHO; these were averaged to give the reported value of 6.3×10^{-16} cm^3 molecule^{-1} s^{-1}. However, of these 27 estimates, 19 depended on the value chosen for the $NO_2 + NO_3 \rightleftarrows N_2O_5$ equilibrium, which they took as 2.3×10^{-11} cm^3 molecule^{-1}. The remaining eight experimental values cited for the rate constant did not depend on the value of K_{eq}. We have therefore reported separately in Table 7.26 the averages of the 19 experiments adjusted to the values of K_{eq} shown there, and the average of the eight experiments which do not depend on K_{eq}. Their rate constant estimate which is independent of the equilibrium constant, 5.6×10^{-16} cm^3 molecule^{-1} s^{-1}, is in excellent agreement with that of Atkinson et al. (1984a), 5.8×10^{-16} cm^3 molecule^{-1} s^{-1}, which is based on our preferred value of $K_{eq} = 3.35 \times 10^{-11}$ cm^3 molecule^{-1} (see Section 7-C-2).

While no product studies of the NO_3–aldehyde reactions have been reported,

TABLE 7.26. Values of the Rate Constants for the Reactions of NO_3 with Some Oxygenated Compounds at 298°K Derived Using Different Values of the Equilibrium Constant for $NO_2 + NO_3 \overset{M}{\rightleftarrows} N_2O_5$

		k (cm^3 $molecule^{-1}$ s^{-1})				
		K_{eq} =				
Alkane	Exponential Factor[a]	1.87×10^{-11}[b]	2.20×10^{-11}[c]	3.35×10^{-11}[d,e]	Rate Constant Not Depending on K_{eq}	Reference
Formaldehyde	10^{-16}	3.2	3.8	5.8		*f*
	10^{-16}	5.3	6.2	9.4	5.6	*g*
Acetaldehyde	10^{-15}	1.3	1.6	2.4		*f*
	10^{-15}	1.4	1.7	2.5		*h*
Benzaldehyde	10^{-15}	1.1	1.3	2.0		*i*
Phenol	10^{-12}	2.1	2.5	3.8		*i*
o-Cresol	10^{-11}	1.2	1.4	2.2		*i*
m-Cresol	10^{-11}	0.92	1.1	1.7		*i*
p-Cresol	10^{-11}	1.3	1.5	2.3		*i*
Methoxybenzene (anisole)	10^{-17}	5.0	5.9	9.0		*i*

[a] These are the factors by which the numbers given for the rate constant must be multiplied; for example, the rate constants for formaldehyde are 3.2×10^{-16}, 3.8×10^{-16}, and 5.8×10^{-16} cm^3 $molecule^{-1}$ s^{-1}, respectively, depending on the value chosen for K_{eq}.

[b] Malko and Troe, 1982.

[c] Graham and Johnston (1978) using value for $k(O_3 + NO_2) = 3.2 \times 10^{-17}$ cm^3 $molecule^{-1}$ s^{-1} at 298°K recommended by DeMore et al. (1983).

[d] Average of values reported by Tuazon et al. (1984b), 3.44×10^{-11}, and Kircher et al. (1984), 3.26×10^{-11} cm^3 $molecule^{-1}$.

[e] Preferred values; see text.

[f] Atkinson et al., 1984a.

[g] Cantrell et al., 1985.

[h] Morris and Niki, 1974.

[i] Atkinson et al., 1984c.

the reaction most likely involves hydrogen abstraction from the aldehydic C—H bond in a manner analogous to OH–aldehyde reactions:

$$NO_3 + RCHO \rightarrow RCO + HNO_3 \tag{128}$$

Note that the reactions of NO_3 with phenol and the cresols are much faster than the reactions with unsubstituted aromatic hydrocarbons (Carter et al., 1981). For example, the NO_3–toluene reaction is approximately 10^6 times slower than the reaction with *o*-cresol at room temperature. This dramatic increase in the rate constant for reaction with NO_3 with substitution of a hydroxyl group on the aromatic ring suggests that the mechanism involves abstraction of a hydrogen atom from the phenolic group:

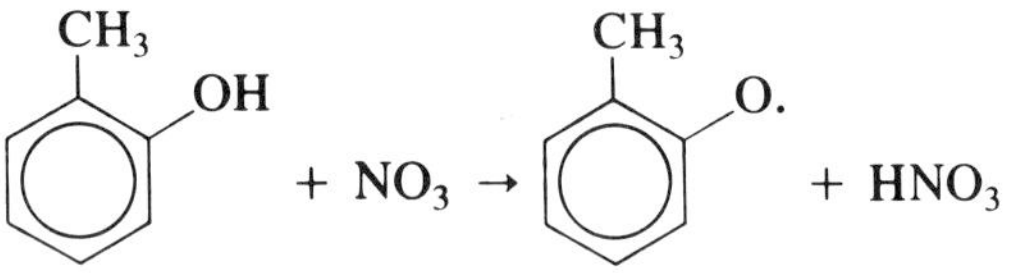
(129)

The room temperature rate constant for the OH–*o*-cresol reaction is 4.0×10^{-11} cm^3 $molecule^{-1}$ s^{-1} (Atkinson, 1985). Thus OH at a daytime concentration of 1×10^6 cm^{-3} will result in an *o*-cresol lifetime of 7 h, whereas NO_3 at a nighttime concentration of 30 ppt (7×10^8 cm^{-3}) will result in a lifetime of only $\sim$1 min. Atmospheric reactions of the cresols (and phenol) must therefore include NO_3 as well as OH, and the former may indeed be the dominant oxidizing species under many conditions. The reactions of OH and NO_3 with cresols in air have been observed in laboratory studies to lead in part to the formation of nitrocresols (e.g. see Grosjean, 1985; Gery et al., 1985). These can be rationalized on the basis of NO_2 addition to the intermediates formed by OH addition to the ring or abstraction of the hydroxyl hydrogen atom (reaction 129), followed by elimination of water in the case of the OH reaction, or rearrangement in the case of the NO_3 reaction. However, the relative importance of the formation of nitrocresols in ambient air will clearly depend on the NO_2 concentration, and may be less than observed in many laboratory studies. For example, Gery et al. (1985) estimate that the rate constant for addition of NO_2 to the OH-toluene adduct formed in reaction (108) is about 3.3×10^4 faster than that for hydrogen abstraction from the adduct, reaction (109). In air at the relatively high NO_2 concentration of 0.1 ppm, less than 2% of the OH-toluene adduct will react with NO_2 rather than with O_2. The same is likely true of the OH-cresol adducts.

7-H-4. Relative Importance of Photolysis and Attack by OH, HO_2, and NO_3 for Aldehydes in the Atmosphere

The four possible fates of aldehydes in the atmosphere are thus reaction with OH, HO_2, and NO_3 and photolysis. Since formaldehyde is the major aldehydic component of auto exhaust and likely the major aldehyde in ambient air (Lloyd, 1979;

TABLE 7.27. Lifetimes of HCHO with Respect to Photolysis and to Reaction with OH, HO_2 and NO_3 in Typical Relatively Clean and Polluted Atmospheres[a]

Loss Process	Estimated Lifetime
Photolysis[b]	
Noon, January 1	8.6 h
Noon, July 1	3.8 h
Near sunrise/sunset, July 1 ($\theta = 86°$)	174 h
Reaction with OH	
Relatively clean conditions	2.6 days
Moderately polluted conditions	6.2 h
Reaction with HO_2	
Relatively clean conditions	1.5 days
Moderately polluted conditions	3.7 h
Reaction with NO_3	
Relatively clean conditions	80 days
Moderately polluted conditions	8 days

[a] Concentrations under relatively clean conditions were taken to be as follows: [OH] = 5×10^5 cm^{-3}, [HO_2] = 1×10^8 cm^{-3}, [NO_3] = 2.5×10^8 cm^{-3}. Under moderately polluted conditions, each concentration was taken to be a factor of 10 higher.
[b] See Table 3.16 in Chapter 3.

Grosjean, 1982), let us compare its loss by reaction with OH, HO_2, and NO_3 to that by photolysis.

In Section 3-B-2, rates of HCHO photolysis at 40°N latitude were calculated for three different dates and times: noon January 1, noon July 1, and sunrise/sunset July 1. The overall photolytic lifetimes for HCHO at these three times were estimated as 8.6, 3.8, and 174 h, respectively.

Table 7.27 compares the lifetimes of HCHO with respect to photolysis and reaction with OH, HO_2, and NO_3 in relatively clean and in moderately polluted atmospheres using the rate constants given earlier. Comparing these lifetimes to those from photolysis, one can see that during daylight hours photolysis will predominate in the loss of HCHO in relatively clean atmospheres; in polluted atmospheres, reaction with HO_2 and OH will be comparable to photolysis. At night when the OH and HO_2 concentrations are low and photolysis does not occur, NO_3 can contribute to the aldehyde loss. However, this is slow compared to the daytime loss by reaction with OH and HO_2.

REFERENCES

Ackerman, R. A., J. N. Pitts, Jr., and R. P. Steer, "Singlet Oxygen in the Environmental Sciences. VIII. Absolute Rates of Deactivation of $O_2(^1\Delta_g)$ by Terminal Olefins, Tetramethylethylene and Methyl Chloride," *J. Chem. Phys.*, **52**, 1603 (1970).

Ackerman, R. A., J. N. Pitts, Jr., and R. P. Steer, "Concerning the Effect of Pressure on the Rate of Reaction of $O_2(^1\Delta_g)$ with Tetramethylethylene," *Chem. Phys. Lett.*, **12,** 526 (1972).

Adachi, H. and N. Basco, "Spectra of Propylperoxy Radicals and Rate Constants for Mutual Interaction," *Int. J. Chem. Kinet.*, **14,** 1125 (1982).

Adachi, H., N. Basco, and D. G. L. James, "The Ethylperoxy Radical Spectrum and Rate Constant for Mutual Interaction Measured by Flash Photolysis and Kinetic Spectroscopy," *Int. J. Chem. Kinet.*, **11,** 1211 (1979).

Addison, M. C., J. P. Burrows, R. A. Cox, and R. Patrick, "Absorption Spectrum and Kinetics of the Acetylperoxy Radical," *Chem. Phys. Lett.*, **73,** 283 (1980).

Akimoto, H., M. Hoshino, G. Inoue, F. Sakamaki, H. Bandow, and M. Okuda, "Formation of Propylene Glycol-1, 2-Dinitrate in the Photooxidation of a Propylene–Nitrogen Oxides–Air System," *J. Environ. Sci. Health*, **A13 (9),** 677 (1978a).

Akimoto, H., M. Hoshino, G. Inoue, M. Okuda, and N. Washida, "Reaction Mechanism of the Photooxidation of the Toluene–NO_2–O_2–N_2 System in the Gas Phase," *Bull. Chem. Soc. Japan*, **51,** 2496 (1978b).

Akimoto, H., H. Bandow, F. Sakamaki, G. Inoue, M. Hoshino, and M. Okuda, "Photooxidation of the Propylene–NO_x–Air System Studied by Long-Path Fourier Transform Infrared Spectrometry," *Environ. Sci. Technol.*, **14,** 172 (1980).

Altshuller, A. P., W. A. Lonneman, F. D. Sutterfield, and S. L. Kopczynski, "Hydrocarbon Composition of the Atmosphere of the Los Angeles Basin—1967," *Environ. Sci. Technol.*, **5,** 1009 (1971).

Anastasi, C., D. J. Waddington, and A. Woolley, "Reactions of Oxygenated Radicals in the Gas Phase. Part 10. Self-Reactions of Ethylperoxy Radicals," *J. Chem. Soc. Faraday Trans. 1*, **79,** 505 (1983).

Atkinson, R., "Kinetics and Mechanisms of the Gas Phase Reactions of the Hydroxyl Radical with Organic Compounds Under Atmospheric Conditions," *Chem. Rev.*, in press, 1985.

Atkinson, R. and S. M. Aschmann, "Rate Constants for the Reactions of O_3 and OH Radicals with a Series of Alkynes," *Int. J. Chem. Kinet.*, **16,** 259 (1984a).

Atkinson, R. and S. M. Aschmann, "Rate Constants for the Reaction of OH Radicals with a Series of Alkenes and Dialkenes at 295 ± 1°K," *Int. J. Chem. Kinet.*, **16,** 1175 (1984b).

Atkinson, R. and W. P. L. Carter, "Kinetics and Mechanisms of the Gas Phase Reactions of Ozone with Organic Compounds Under Atmospheric Conditions," *Chem. Rev.*, **84,** 437 (1984).

Atkinson, R. and A. C. Lloyd, "Evaluation of Kinetic and Mechanistic Data for Modeling of Photochemical Smog," *J. Phys. Chem. Ref. Data*, **13,** 315 (1984).

Atkinson, R. and J. N. Pitts, Jr., "Rate Constants for the Reaction of OH Radicals with Propylene and the Butenes Over the Temperature Range 297–425°K," *J. Chem. Phys.*, **63,** 3591 (1975).

Atkinson, R., B. J. Finlayson, and J. N. Pitts, Jr., "Photoionization Mass Spectrometer Studies of Gas Phase Ozone–Olefin Reactions," *J. Am. Chem. Soc.*, **95,** 7592 (1973).

Atkinson, R., R. A. Perry, and J. N. Pitts, Jr., "Rate Constants for the Reaction of OH Radicals with Ethylene Over the Temperature Range 299–425°K," *J. Chem. Phys.*, **66,** 1197 (1977).

Atkinson, R., D. A. Hansen, and J. N. Pitts, Jr., "Rate Constants for the Reaction of OH Radicals with CHF_2Cl, CF_2Cl_2, and H_2 Over the Temperature Range 297–434°K," *J. Chem. Phys.*, **63,** 1703 (1975).

Atkinson, R., R. A. Perry, and J. N. Pitts, Jr., "Absolute Rate Constants for the Reaction of OH Radicals with Allene, 1,3-Butadiene, and 3-Methyl-1-butene Over the Temperature Range 299–424°K," *J. Chem. Phys.*, **67,** 3170 (1977).

Atkinson, R., K. R. Darnall, and J. N. Pitts, Jr., "Rate Constants for the Reactions of OH Radicals and Ozone with Cresols at 300 ± 1°K," *J. Phys. Chem.*, **82,** 2759 (1978).

Atkinson, R., K. R. Darnall, A. C. Lloyd, A. M. Winer, and J. N. Pitts, Jr., "Kinetics and Mechanisms of the Reaction of the Hydroxyl Radical with Organic Compounds in the Gas Phase," *Adv. Photochem.*, **11,** 375 (1979).

Atkinson, R., W. P. L. Carter, K. R. Darnall, A. M. Winer, and J. N. Pitts, Jr., "A Smog Chamber and Modeling Study of the Gas Phase NO_x–Air Photooxidation of Toluene and the Cresols," *Int. J. Chem. Kinet.*, **12,** 779 (1980).

Atkinson, R., S. M. Aschmann, W. P. L. Carter, A. M. Winer, and J. N. Pitts, Jr., "Kinetics of the Reactions of OH Radicals with *n*-Alkanes at 299 ± 2°K," *Int. J. Chem. Kinet.*, **14,** 781 (1982a).

Atkinson, R., S. M. Aschmann, W. P. L. Carter, A. M. Winer, and J. N. Pitts, Jr., "Alkyl Nitrate Formation from the NO_x–Air Photooxidations of C_2–C_8 *n*-Alkanes," *J. Phys. Chem.*, **86,** 4563 (1982b).

Atkinson, R., S. M. Aschmann, and W. P. L. Carter, "Rate Constants for the Gas Phase Reactions of OH Radicals with a Series of Bi- and Tri-cycloalkanes at 299 ± 2°K: Effects of Ring Strain," *Int. J. Chem. Kinet.*, **15,** 37 (1983a).

Atkinson, R., W. P. L. Carter, and A. M. Winer, "Effects of Pressure on Product Yields in the NO_x Photooxidations of Selected Aromatic Hydrocarbons," *J. Phys. Chem.*, **87,** 1605 (1983b).

Atkinson, R., C. N. Plum, W. P. L. Carter, A. M. Winer, and J. N. Pitts, Jr., "Rate Constants for the Gas Phase Reactions of NO_3 Radicals with a Series of Organics in Air at 298 ± 1°K," *J. Phys. Chem.*, **88,** 1210 (1984a).

Atkinson, R., S. M. Aschmann, A. M. Winer, and J. N. Pitts, Jr., "Kinetics of the Gas Phase Reactions of NO_3 Radicals with a Series of Dialkenes, Cycloalkenes, and Monoterpenes at 295 ± 1°K," *Environ. Sci. Technol.*, **18,** 370 (1984b).

Atkinson, R., W. P. L. Carter, C. N. Plum, A. M. Winer, and J. N. Pitts, Jr., "Kinetics of the Gas Phase Reactions of NO_3 Radicals with a Series of Aromatics at 296 ± 2°K," *Int. J. Chem. Kinet.*, **16,** 887 (1984c).

Atkinson, R., J. N. Pitts, Jr., and S. M. Aschmann, "Tropospheric Reactions of Dimethyl Sulfide with NO_3 and OH Radicals," *J. Phys. Chem.*, **88,** 1584 (1984d).

Atkinson, R., C. N. Plum, W. P. L. Carter, A. M. Winer, and J. N. Pitts, Jr., "Kinetics of the Gas Phase Reactions of NO_3 Radicals with a Series of Alkanes at 298 ± 2°K," *J. Phys. Chem.*, **88,** 2361 (1984e).

Atkinson, R., S. M. Aschmann, W. P. L. Carter, A. M. Winer, and J. N. Pitts, Jr., "Formation of Alkyl Nitrates from the Reaction of Branched and Cyclic Alkyl Peroxy Radicals with NO," *Int. J. Chem. Kinet.*, **16,** 1085 (1984f).

Atkinson, R., A. M. Winer, and J. N. Pitts, Jr., "Estimation of Nighttime N_2O_5 Concentrations from Ambient NO_2 and NO_3 Radical Concentrations and the Role of N_2O_5 in Nighttime Chemistry," *Atmos. Environ.*, **19,** 000 1985a.

Atkinson, R., E. C. Tuazon, and W. P. L. Carter, "Extent of H-Atom Abstraction from the Reaction of the OH Radical with 1-Butene Under Atmospheric Conditions," *Int. J. Chem. Kinet.*, **17,** 725 (1985b).

Baldwin, A. C., J. R. Barker, D. M. Golden, and D. G. Hendry, "Photochemical Smog. Rate Parameter Estimates and Computer Simulations," *J. Phys. Chem.*, **81,** 2483 (1977).

Balla, R. J., H. H. Nelson, and J. R. McDonald, "Kinetics of the Reactions of Isopropoxy Radicals with NO, NO_2, and O_2," *Chem. Phys.*, **99,** 323 (1985).

Bandow, H., M. Okuda, and H. Akimoto, "Mechanism of the Gas-Phase Reactions of C_3H_6 and NO_3 Radicals," *J. Phys. Chem.*, **84,** 3604 (1980).

Barnes, I., K. H. Becker, E. H. Fink, A. Reimer, F. Zabel, and H. Niki, "FTIR Spectroscopic Study of the Gas-Phase Reaction of HO_2 with H_2CO," *Chem. Phys. Lett.*, **115,** 1 (1985).

Bartlett, P. D. and G. Guaraldi, "Di-*t*-butyl Trioxide and Di-*t*-butyl Tetroxide," *J. Am. Chem. Soc.*, **89,** 4799 (1967).

Batt, L., Comments on "Organic Free Radicals," in *Chemical Kinetic Data Needs for Modeling the Lower Troposphere*, National Bureau of Standards Special Publication 557, pp. 62–64 (1979a).

Batt, L., "The Gas-Phase Decomposition of Alkoxy Radicals," *Int. J. Chem. Kinet.*, **11,** 977 (1979b).

Batt, L., Reactions of Alkoxy Radicals Relevant to Atmospheric Chemistry, in *1st European Symposium on Physico-Chemical Behavior of Atmospheric Pollutants*, Ispra, Italy, 1979, B. Verdino and H. Ott, Eds., Commission of the European Communities, 1980.

Baulch, D. L., R. A. Cox, R. F. Hampson, Jr., J. A. Kerr, J. Troe, and R. T. Watson, ''Evaluated Kinetic and Photochemical Data for Atmospheric Chemistry,'' *J. Phys. Chem. Ref. Data*, **9,** 295 (1980).

Baulch, D. L., R. A. Cox, R. F. Hampson, Jr., J. A. Kerr, J. Troe, and R. T. Watson, ''Evaluated Kinetic and Photochemical Data for Atmospheric Chemistry: Supplement II,'' *J. Phys. Chem. Ref. Data*, **13,** 1259 (1984).

Bayes, K. D., ''Absorption of Sunlight by O_2 and Smog Formation,'' Paper presented at the Sixth Informal Photochemistry Conference, University of California, Davis, CA, June 1964.

Benson, S. W., *Thermochemical Kinetics*, 2nd ed., Wiley, New York, 1976.

Besemer, A. C., ''Formation of Chemical Compounds from Irradiated Mixtures of Aromatic Hydrocarbons and Nitrogen Oxides,'' *Atmos. Environ.*, **16,** 1599 (1982).

Biermann, H. W., G. W. Harris, and J. N. Pitts, Jr., ''Photoionization Mass Spectrometer Studies of the Collisionally Stabilized Product Distribution in the Reaction of OH Radicals with Selected Alkenes at 298°K,'' *J. Phys. Chem.*, **86,** 2958 (1982).

Blumenberg, B., K. Hoyermann, and R. Sievert, Primary Products in the Reactions of Oxygen Atoms with Simple and Substituted Hydrocarbons, in *16th Symposium (International) Combustion*, The Combustion Institute, Pittsburgh, PA, 1977, pp. 841–851.

Bogan, D. J., R. S. Sheinson, R. G. Gann, and F. W. Williams, ''Formaldehyde ($A^1A_2 \rightarrow X^1A_1$) Chemiluminescence in the Gas Phase Reaction of $O_2(a^1\Delta_g)$ Plus Ethyl Vinyl Ether,'' *J. Am. Chem. Soc.*, **97,** 2560 (1975).

Bogan, D. J., R. S. Sheinson, and F. W. Williams, ''Gas Phase Dioxetane Chemistry. Formaldehyde ($A \rightarrow X$) Chemiluminescence from the Reaction of $O_2(^1\Delta_g)$ with Ethene,'' *J. Am. Chem. Soc.*, **98,** 1034 (1976).

Borrell, P., C. J. Cobos, A. E. Croce de Cobos, H. Hippler, K. Luther, A. R. Ravishankara, and J. Troe, ''Radical Association Reactions in Gases at High Pressures,'' *Ber. Bunsenges. Phys. Chem.*, **89,** 337 (1985).

Bottenheim, J. W. and O. P. Strausz, ''Gas Phase Chemistry of Clean Air at 55°N Latitude,'' *Environ. Sci. Technol.*, **14,** 709 (1980).

Bradley, J. N., W. Hack, K. Hoyermann, and H. Gg. Wagner, ''Kinetics of the Reaction of Hydroxyl Radicals with Ethylene and with C_3 Hydrocarbons,'' *J. Chem. Soc. Faraday Trans. 1*, **69,** 1889 (1973).

Burrows, J. P., G. S. Tyndall, and G. K. Moortgat, ''A Study of the N_2O_5 Equilibrium Between 275 and 315 K and Determination of the Heat of Formation of NO_3,'' *Chem. Phys. Lett.*, **119,** 193 (1985).

Buss, R. J., R. J. Baseman, G. He, and Y. T. Lee, ''Reaction of Oxygen Atoms with Ethylene and Vinyl Bromide,'' *J. Photochem.*, **17,** 389 (1981).

Calvert, J. G., ''Interactions of Air Pollutants,'' Proceedings of the Conference on Health Effects of Air Pollutants, Assembly of Life Sciences, National Academy of Sciences, National Research Council, October 3–5, 1973.

Calvert, J. G., F. Su, J. W. Bottenheim, and O. P. Strausz, ''Mechanism of the Homogeneous Oxidation of Sulfur Dioxide in the Troposphere,'' *Atmos. Environ.*, **12,** 197 (1978).

Campbell, I. M. and P. E. Parkinson, ''Rate Constants for Reactions of Hydroxyl Radicals with Ester Vapours at 292°K,'' *Chem. Phys. Lett.*, **53,** 385 (1978).

Cantrell, C. A., W. R. Stockwell, L. G. Anderson, K. L. Busarow, D. Perner, A. Schmeltekopf, J. G. Calvert, and H. S. Johnston, ''Kinetic Study of the NO_3–CH_2O Reaction and Its Possible Role in Nighttime Tropospheric Chemistry,'' *J. Phys. Chem.*, **89,** 139, 4160 (1985).

Carter, W. P. L. and R. Atkinson, ''Atmospheric Chemistry of Alkanes,'' *J. Atmos. Chem.*, in press (1985).

Carter, W. P. L., K. R. Darnall, A. C. Lloyd, A. M. Winer, and J. N. Pitts Jr., ''Evidence for Alkoxy Radical Isomerization in Photooxidations of C_4–C_6 Alkanes Under Simulated Atmospheric Conditions,'' *Chem. Phys. Lett.*, **42,** 22 (1976).

Carter, W. P. L., K. R. Darnall, R. A. Graham, A. M. Winer, and J. N. Pitts, Jr., "Reactions of C_2 and C_4 α-Hydroxy Radicals with Oxygen," *J. Phys. Chem.*, **83,** 2305 (1979a).

Carter, W. P. L., A. C. Lloyd, J. L. Sprung, and J. N. Pitts, Jr., "Computer Modeling of Smog Chamber Data: Progress in Validation of a Detailed Mechanism for the Photooxidation of Propene and *n*-Butane in Photochemical Smog," *Int. J. Chem. Kinet.*, **11,** 45 (1979b).

Carter, W. P. L., A. M. Winer, and J. N. Pitts, Jr., "Major Atmospheric Sink for Phenol and the Cresols: Reaction with the Nitrate Radical," *Environ. Sci. Technol.*, **15,** 829 (1981).

Chang, J. S. and F. Kaufman, "Upper Limits of the Rate Constants for the Reactions of $CFCl_3$ (F-11), CF_2Cl_2 (F-12), and N_2O with OH. Estimates of Corresponding Lower Limits to Their Tropospheric Lifetimes," *Geophys. Res. Lett.*, **4,** 192 (1977).

Clemo, A. R., G. L. Duncan, and R. Grice, "Reactive Scattering of a Supersonic Oxygen-Atom Beam: O + C_2H_4, C_2H_2," *J. Chem. Soc. Faraday Tans. 2*, **78,** 1231 (1982).

Cobos, C. J., H. Hippler, K. Luther, A. R. Ravishankara, and J. Troe, "High-Pressure Falloff Curves and Specific Rate Constants for the Reaction $CH_3 + O_2 \rightleftarrows CH_3O_2 \rightleftarrows CH_3O + O$," *J. Phys. Chem.*, **89,** 4332 (1985).

Cohen, N., "The Use of Transition-State Theory to Extrapolate Rate Coefficients for Reactions of OH with Alkanes," *Int. J. Chem. Kinet.*, **14,** 1339 (1982).

Coomber, J. W. and J. N. Pitts, Jr., "Singlet Oxygen in the Environmental Sciences. VIII. Production of $O_2(^1\Delta_g)$ by Energy Transfer from Excited Benzaldehyde under Simulated Atmospheric Conditions," *Environ. Sci. Technol.*, **4,** 506 (1970).

Cox, R. A. and S. A. Penkett, "Oxidation of Atmospheric SO_2 by Products of the Ozone–Olefin Reaction," *Nature*, **230,** 321 (1971).

Cox, R. A. and S. A. Penkett, "Aerosol Formation from Sulfur Dioxide in the Presence of Ozone and Olefinic Hydrocarbons," *J. Chem. Soc. Faraday Trans. 1*, **68,** 1735 (1972).

Cox, R. A. and G. S. Tyndall, "Rate Constants for Reactions of CH_3O_2 in the Gas Phase," *Chem. Phys. Lett.*, **65,** 357 (1979).

Cox, R. A. and G. S. Tyndall, "Rate Constants for the Reactions of CH_3O_2 with HO_2, NO and NO_2 Using Molecular Modulation Spectrometry," *J. Chem. Soc. Faraday Trans. 2*, **76,** 153 (1980).

Cox, R. A., K. F. Patrick, and S. A. Chant, "Mechanism of Atmospheric Photooxidation of Organic Compounds. Reactions of Alkoxy Radicals in Oxidation of *n*-Butane and Simple Ketones," *Environ. Sci. Technol.*, **15,** 587 (1981).

Criegee, R., "Mechanisms of Ozonolysis," *Angew. Chem. Int. Ed. Engl.*, **14,** 745 (1975).

Cvetanovic, R. J., "Addition of Atoms to Olefins in the Gas Phase," *Adv. Photochem.*, **1,** 115 (1963).

Cvetanovic, R. J., "Chemical Kinetic Studies of Atmospheric Interest," 12th International Symposium on Free Radicals, Laguna, Beach, CA, January 4–9, 1976.

Cvetanovic, R. J. and D. L. Singleton, "Reaction of Oxygen Atoms with Olefins," *Rev. Chem. Intermediates*, **5,** 183 (1984).

Darnall, K. R., W. P. L. Carter, A. M. Winer, A. C. Lloyd, and J. N. Pitts, Jr., "Importance of RO_2 + NO in Alkyl Nitrate Formation from C_4–C_6 Alkane Photooxidations under Simulated Atmospheric Conditions," *J. Phys. Chem.*, **80,** 1948 (1976).

Darnall, K. R., R. Atkinson, and J. N. Pitts, Jr., "Rate Constants for the Reaction of the OH Radical with Selected Alkanes at 300°K," *J. Phys. Chem.*, **82,** 1581 (1978).

Darnall, K. R., R. Atkinson, and J. N. Pitts, Jr., "Observation of Biacetyl from the Reaction of OH Radicals with *o*-Xylene: Evidence for Ring Cleavage," *J. Phys. Chem.*, **83,** 1943 (1979).

Davis, D. D., W. Bollinger, and S. Fischer, "A Kinetics Study of the Reaction of the OH Free Radical with Aromatic Compounds. I. Absolute Rate Constants for Reaction with Benzene and Toluene at 300°K," *J. Phys. Chem.*, **79,** 293 (1975).

Dawson, G. A., J. C. Farmer, and J. L. Moyers, "Formic and Acetic Acids in the Atmosphere of the Southwest U.S.A.," *Geophys. Res. Lett.*, **7,** 725 (1980).

Demerjian, K. L., J. A. Kerr, and J. G. Calvert, "The Mechanism of Photochemical Smog Formation," *Adv. Environ. Sci. Technol.*, **4,** 1 (1974).

Demerjian, K. L., K. L. Schere, and J. T. Peterson, "Theoretical Estimates of Actinic (Spherically Integrated) Flux and Photolytic Rate Constants of Atmospheric Species in the Lower Troposphere," *Adv. Environ. Sci. Technol.*, **10,** 369 (1980).

DeMore, W. B., M. J. Molina, R. T. Watson, D. M. Golden, R. F. Hampson, M. J. Kurylo, C. J. Howard, and A. R. Ravishankara, "Chemical Kinetics and Photochemical Data for Use in Stratospheric Modeling, Evaluation No. 6," JPL Publ. 83-62 (1983).

DeMore, W. B., J. J. Margitan, M. J. Molina, R. T. Watson, D. M. Golden, R. F. Hampson, M. J. Kurylo, C. J. Howard, and A. R. Ravishankara, "Chemical Kinetics and Photochemical Data for Use in Stratospheric Modeling, Evaluation No. 7," JPL Publ. 85-37 (1985).

Drew, R. M., J. A. Kerr, and J. Olive, "Relative Rate Constants of the Gas-Phase Decomposition of the *s*-Butoxy Radical," *Int. J. Chem. Kinet.*, **17,** 167 (1985).

Dumdei, B. E. and R. J. O'Brien, "Toluene's Degradation Products In Simulated Atmospheric Conditions," *Nature*, **311,** 248 (1984).

Dupuis, M., J. J. Wendoloski, and W. A. Lester, Jr., "Electronic Structure of Vinoxy Radical CH_2CHO," *J. Chem. Phys.*, **76,** 488 (1982a).

Dupuis, M., J. J. Wendoloski, T. Takada, and W. A. Lester, Jr., "Theoretical Study of Electrophilic Addition: $O(^3P) + C_2H_4$," *J. Chem. Phys.*, **76,** 481 (1982b).

Eisenberg, W., K. Taylor and R. W. Murray, "Production of Singlet Delta Oxygen by Atmospheric Pollutants," *Carcinogenisis*, **5,** 1095 (1984).

Finlayson, B. J., J. N. Pitts, Jr., and H. Akimoto, "Production of Vibrationally Excited OH in Chemiluminescent Ozone–Olefin Reactions," *Chem. Phys. Lett.*, **12,** 495 (1972).

Finlayson, B. J., J. N. Pitts, Jr., and R. Atkinson, "Low-Pressure Gas-Phase Ozone–Olefin Reactions, Chemiluminescence, Kinetics, and Mechanisms," *J. Am. Chem. Soc.*, **96,** 5356 (1974).

Foote, C. S., Photosensitized Oxidation and Singlet Oxygen: Consequences in Biological Systems, in *Free Radicals in Biology*, Vol. II, Academic Press, New York, 1976, pp. 85–133.

Foote, C. S. and S. Wexler, "Olefin Oxidations with Excited Singlet Molecular Oxygen," *J. Am. Chem. Soc.*, **86,** 3879 (1964).

Frankiewicz, T. C. and R. S. Berry, "Production of Metastable Singlet O_2 Photosensitized by NO_2," *J. Chem. Phys.*, **58,** 1787 (1973).

Fritz, B., V. Handwerk, M. Preidel and R. Zellner, "Direct Detection of Hydroxy-Cyclohexadienyl in the Gas Phase by cw-UV-Laser Absorption," *Ber. Bunsenges. Phys. Chem.*, **89,** 343 (1985).

Gaffney, J. S. and S. Z. Levine, "Predicting Gas Phase Organic Molecule Reaction Rates Using Linear Free-Energy Correlations. I. $O(^3P)$ and OH Addition and Abstraction Reactions" *Int. J. Chem. Kinet.*, **11,** 1197 (1979).

Gaydon, A. G., *Dissociation Energies and the Spectra of Diatomic Molecules*, 3rd ed., Chapman and Hall, London, 1968.

Gery, M. W., D. L. Fox, H. E. Jeffries, L. Stockburger and W. Weathers, "A Continuous Stirred Tank Reactor Investigation of the Gas-Phase Reaction of Hydroxyl Radicals and Toluene," *Int. J. Chem. Kinet.*, **17,** 931 (1985).

Gleason, W. S., A. D. Broadbent, E. Whittle, and J. N. Pitts, Jr., "Singlet Oxygen in the Environmental Sciences. IV. Kinetics of the Reactions of Oxygen ($^1\Delta_g$) with Tetramethylethylene and 2, 5-Dimethylfuran in the Gas Phase," *J. Am. Chem. Soc.*, **92,** 2068 (1970).

Gohre, K., and G. C. Miller, "Photochemical Generation of Singlet Oxygen on Non-Transition-Metal Oxide Surfaces," *J. Chem. Soc., Faraday Trans. 1*, **81,** 793 (1985).

Gohre, K., and G. C. Miller, "Singlet Oxygen Generation on Soil Surfaces," *J. Agric. Food Chem.*, **31,** 1104 (1983).

Graham, R. A. and H. S. Johnston, "The Photochemistry of NO_3 and the Kinetics of the N_2O_5–O_3 System," *J. Phys. Chem.*, **82,** 254 (1978).

Graham, R. A., A. M. Winer, R. Atkinson, and J. N. Pitts, Jr., "Rate Constants for the Reaction HO_2 with HO_2, SO_2, CO, N_2O, *trans*-2-Butene, and 2, 3-Dimethyl-2-butene at 300°K," *J. Phys. Chem.*, **83,** 1563 (1979).

Greiner, N. R., "Hydroxyl Radical Kinetics by Kinetic Spectroscopy. VI. Reactions with Alkanes in the Range 300–500°K," *J. Chem. Phys.*, **53,** 1070 (1970).

Grosjean, D., "Formaldehyde and Other Carbonyls in Los Angeles Ambient Air," *Environ. Sci. Technol.*, **16,** 254 (1982).

Grosjean, D. and S. K. Friedlander, "Formation of Organic Aerosols from Cyclic Olefins and Diolefins," *Adv. Environ. Sci. Technol.*, **10,** 435 (1980).

Grotheer, H.-H., G. Riekert, U. Meier and T. Just, "Kinetics of the Reactions of CH_2OH Radicals with O_2 and HO_2," *Ber. Bunsenges. Phys. Chem.*, **89,** 187 (1985).

Gutman, D. and H. H. Nelson, "Gas-Phase Reactions of the Vinoxy Radical with O_2 and NO," *J. Phys. Chem.*, **87,** 3902 (1983).

Gutman, D., N. Sanders, and J. E. Butler, "Kinetics of the Reactions of Methoxy and Ethoxy Radicals with Oxygen," *J. Phys. Chem.*, **86,** 66 (1982).

Ha, T.-K., H. Kühne, S. Vaccani, and Hs. H. Günthard, "A Theoretical Study of the Stability and the Molecular Electronic Structure of Methylene Peroxide (CH_2O_2)," *Chem. Phys. Lett.*, **24,** 172 (1974).

Hägele, J., K. Lorenz, D. Rhäsa, and R. Zellner, "Rate Constants and CH_3O Product Yield of the Reaction OH + CH_3OH → Products," *Ber. Bunsenges. Phys. Chem.*, **87,** 1023 (1983).

Hansen, D. A., R. Atkinson, and J. N. Pitts. Jr., "Rate Constants for the Reaction of OH Radicals with a Series of Aromatic Hydrocarbons," *J. Phys. Chem.*, **79,** 1763 (1975).

Hanst, P. L. and B. W. Gay, Jr., "Atmospheric Oxidation of Hydrocarbons: Formation of Hydroperoxides and Peroxyacids," *Atmos. Environ.*, **17,** 2259 (1983).

Harding, L. B. and W. A. Goddard III, "Mechanisms of Gas-Phase and Liquid-Phase Ozonolysis," *J. Am. Chem. Soc.*, **100,** 7180 (1978).

Hatakeyama, S., H. Bandow, M. Okuda, and H. Akimoto, "Reactions of CH_2OO and $CH_2(^1A_1)$ with H_2O in the Gas Phase," *J. Phys. Chem.*, **85,** 2249 (1981).

Hatakeyama, S., H. Kobayashi, and H. Akimoto, "Gas-Phase Oxidation of SO_2 in the Ozone-Olefin Reactions," *J. Phys. Chem.*, **88,** 4736 (1984).

Hatakeyama, S., T. Tanonaka, J.-h Weng, H. Bandow, H. Takagi, and H. Akimoto, "Ozone-Cyclohexene Reaction in Air: Quantitative Analysis of Particulate Products and the Reaction Mechanism," *Environ. Sci. Technol.*, **19,** 935 (1985).

Heicklen, J., "The Correlation of Rate Coefficients for H-Atom Abstraction by HO Radicals with C—H Bond Dissociation Enthalpies," *Int. J. Chem. Kinet.*, **13,** 651 (1981).

Hendry, D. G. and R. A. Kenley, "Atmospheric Chemistry of Peroxynitrates," in *Nitrogenous Air Pollutants*, D. Grosjean, Ed., Ann Arbor Science Publishers, Ann Arbor, MI, 1979.

Herron, J. T. and R. E. Huie, "Reactions of $O_2(^1\Delta_g)$ with Olefins and Their Significance in Air Pollution," *Environ.ci. Technol.*, **4,** 685 (1970).

Herron, J. T. and R. E. Huie, "Rate Constants for the Reactions of Atomic Oxygen $O(^3P)$ with Organic Compounds in the Gas Phase," *J. Phys. Chem. Ref. Data*, **2,** 467 (1973).

Herron, J. T. and R. E. Huie, "Stopped-Flow Studies of the Mechanisms of Ozoe–Alkene Reactions in the Gas Phase. Ethylene," *J. Am. Chem. Soc.*, **99,** 5430 (1977).

Herron, J. T. and R. E. Huie, "Stopped-Flow Studies of the Mechanisms of Ozone–Alken Reactions in the Gas Phase. Propene and Isobutene," *Int. J. Chem. Kinet.*, **10,** 1019 (1978).

Herron, J. T., R. I. Martinez, and R. E. Huie, "Kinetics and Energetics of the Criegee Intermediate in the Gas Phase. I. The Criegee Intermediate in Ozone–Alkene Reactions. II. The Criegee Intermediate in the Photooxidation of Formaldehyde, in Alkyldioxy Disproportionation and O + Oxoalkane Addition Reactions," *Int. J. Chem. Kinet.*, **14,** 201, 225 (1982).

Hoshino, M., T. Ogata, H. Akimoto, G. Inoue, F. Sakamaki, and M. Okuda, "Gas Phase Reaction of N_2O_5 with Propylene," *Chem. Lett.*, 1367 (1978a).

Hoshino, M., H. Akimoto, and M. Okuda, "Photochemical Oxidation of Benzene, Toluene, and Ethylbenzene Initiated by OH Radicals in the Gas Phase," *Bull. Chem. Soc. Japan*, **51,** 718 (1978b).

Hoyermann, K. and R. Sievert, "Die Reaktion von OH-Radikalen mit Propen: I. Bestimmung der Primärprodukte bei Niedrigen Drücken," *Ber. Bunsenges. Phys. Chem.*, **83,** 933 (1979).

Hoyermann, K. and R. Sievert, "Elementarreaktionen in der Oxidation von Alkenen," *Ber. Bunsenges. Phys. Chem.*, **87,** 1027 (1983).

Huie, R. E. and J. T. Herron, "Kinetics of the Reaction of Singlet Molecular Oxygen $O_2(^1\Delta_g)$ with Organic Compounds in the Gas Phase," *Int. J. Chem. Kinet.*, **5,** 197 (1973).

Huie, R. E. and J. T. Herron, "Reactions of Atomic Oxygen $O(^3P)$ with Organic Compounds," *Prog. React. Kinet.*, **8,** 1 (1975).

Hunziker, H. E., personal communication, 1984.

Hunziker, H. E. and H. R. Wendt, "Electronic Absorption Spectra of Organic Peroxyl Radicals in the Near Infrared," *J. Chem. Phys.*, **64,** 3488 (1976).

Hunziker, H. E., H. Kneppe, and H. R. Wendt, "Photochemical Modulation Spectroscopy of Oxygen Atom Reactions with Olefins," *J. Photochem.*, **17,** 377 (1981).

Inoue, G. and H. Akimoto, "Laser Induced Fluorescence of the C_2H_3O Radical," *J. Chem. Phys.*, **74,** 425 (1981).

Ishikawa, H., K. Watanabe, and W. Ando, "Photochemical Oxidation of Toluene with Nitrogen Dioxide in the Gas Phase. Effects of Oxygen," *Bull. Chem. Soc. Japan*, **51,** 2174 (1978).

Japar, S. M. and H. Niki, "Gas Phase Reactions of the Nitrate Radical with Olefins," *J. Phys. Chem.*, **79,** 1629 (1975).

Japar, S. M., C. H. Wu, and H. Niki, "Effect of Molecular Oxygen on the Gas Phase Kinetics of the Ozonolysis of Olefins," *J. Phys. Chem.*, **80,** 2057 (1976).

Jeong, K.-M., K.-J. Hsu, J. B. Jeffries, and F. Kaufman, "Kinetics of the Reactions of OH with C_2H_6, CH_3CCl_3, $CH_2ClCHCl_2$, $CH_2ClCClF_2$, and CH_2FCF_3," *J. Phys. Chem.*, **88,** 1222 (1984).

Jolly, G. S., G. Paraskevopoulos, and D. L. Singleton, "Rates of OH Radical Reactions. XII. The Reactions of OH with c-C_3H_6, c-C_5H_{10}, c-C_7H_{14}. Correlation of Hydroxyl Radical Rate Constants with Bond Dissociation Energies," *Int. J. Chem. Kinet.*, **17,** 1 (1985).

Jones, I. T. N. and K. D. Bayes, "Energy Transfer from Electronically Excited NO_2," *Chem. Phys. Lett.*, **11,** 163 (1971).

Jones, I. T. N. and K. D. Bayes, "Formation of $O_2(a^1\Delta_g)$ by Electronic Energy Transfer in Mixtures of NO_2 and O_2," *J. Chem. Phys.*, **59,** 3119 (1973).

Kan, C. S., J. G. Calvert, and J. H. Shaw, "Reactive Channels of the CH_3O_2–CH_3O_2 Reaction," *J. Phys. Chem.*, **84,** 3411 (1980).

Kan, C. S., F. Su, J. G. Calvert, and J. H. Shaw, "Mechanism of the Ozone–Ethene Reaction in Dilute N_2/O_2 Mixtures Near 1-Atm Pressure," *J. Phys. Chem.*, **85,** 2359 (1981).

Kanofsky, J. R., D. Lucas, and D. Gutman, Direct Identification of Free-Radical Products of O-Atom Reactions with Olefins Using High Intensity Molecular Beams, in *14th Symposium (International) Combustion*, The Combustion Institute, Pittsburgh, PA, 1973, pp. 285–293.

Kaufman, M. and J. Sherwell, "Kinetics of Gaseous Hydroperoxyl Radical Reactions," *Prog. React. Kinet.*, **12,** 1 (1983).

Kearns, D. R. "Physical and Chemical Properties of Singlet Molecular Oxygen," *Chem. Rev.*, **71,** 395 (1971).

Kenley, R. A., J. E. Davenport, and D. G. Hendry, "Hydroxyl Radical Reactions in the Gas Phase. Products and Pathways for the Reaction of OH with Toluene," *J. Phys. Chem.*, **82,** 1095 (1978).

Kenley, R. A., J. E. Davenport, and D. G. Hendry, "Gas Phase Hydroxyl Radical Reactions. Products and Pathways for the Reaction of OH with Aromatic Hydrocarbons," *J. Phys. Chem.*, **86,** 2740 (1981).

Kerr, J. A. and D. W. Sheppard, "Kinetics of the Reactions of Hydroxyl Radicals with Aldehydes Studied Under Atmospheric Conditions," *Environ. Sci. Technol.*, **15,** 960 (1981).

Khan, A. U., J. N. Pitts Jr., and E. B. Smith, "Singlet Oxygen in the Environmental Sciences. The Role of Singlet Molecular Oxygen in the Production of Photochemical Air Pollution," *Environ. Sci. Technol.*, **1,** 656 (1967).

Killus, J. P. and G. Z. Whitten, "A Mechanism Describing the Photochemical Oxidation of Toluene in Smog," *Atmos. Environ.*, **16,** 1973 (1982).

Kircher, C. C., J. J. Margitan, and S. P. Sander, "Pressure and Temperature Dependence of the Reaction $NO_2 + NO_3 + M \rightarrow N_2O_5 + M$," *J. Phys. Chem.*, **88,** 4370 (1984).

Kirsch, L. J., D. A. Parkes, D. J. Waddington, and A. Woolley, "Self-Reactions of Isopropylperoxy Radicals in the Gas Phase," *J. Chem. Soc. Faraday Trans. 1*, **74,** 2293 (1978).

Kleindienst, T. E., Harris, G. W., and J. N. Pitts, Jr., "Rates and Temperature Dependencies of the Reaction of OH with Isoprene, Its Oxidation Products and Selected Terpenes," *Environ. Sci. Technol.*, **16,** 844 (1982).

Kleinermanns, K. and A. C. Luntz, "Laser-Induced Fluorescence of CH_2CHO Produced in the Crossed Molecular Beam Reactions of $O(^3P)$ with Olefins," *J. Phys. Chem.*, **85,** 1966 (1981).

Kühne, H., S. Vaccani, T.-K. Ha, A. Baunder, and Hs. H. Günthard, "Infrared-Matrix and Microwave Spectroscopy of the Ethylene–Ozone Gas-Phase Reaction," *Chem. Phys. Lett.*, **38,** 449 (1976).

Kühne, H., S. Vaccani, A. Bauder, and Hs. H. Günthard, "Linear Reactor-Infrared Matrix and Microwave Spectroscopy of the Gas Phase Ethylene Ozonolysis," *Chem. Phys.*, **28,** 11 (1978).

Kühne, H., M. Forster, J. Hulligen, H. Ruprecht, A. Bauder, and H.-H. Günthard, "Linear-Reactor-Infrared-Matrix and Microwave Spectroscopy of the *cis*-2-Butene Gas-Phase Ozonolysis," *Helv. Chim. Acta*, **63,** 1971 (1980).

Kummer, W. A., J. N. Pitts, Jr., and R. P. Steer, "Chemiluminescent Reactions of Ozone with Olefins and Sulfides," *Environ. Sci. Technol.*, **5,** 1045 (1971).

Kummler, R. H. and M. H. Bortner, "Production of $O_2(^1\Delta_g)$ by Energy Transfer from Excited Benzaldehyde," *Environ. Sci. Technol.*, **3,** 944 (1969).

Leighton, P. A., *Photochemistry of Air Pollution*, Academic Press, New York, 1961.

Lenhardt, T. M., C. E. McDade, and K. D. Bayes, "Rates of Reaction of Butyl Radicals with Molecular Oxygen," *J. Chem. Phys.*, **72,** 304 (1980).

Leone, J. A. and J. H. Seinfeld, "Updated Chemical Mechanism for Atmospheric Photooxidation of Toluene," *Int. J. Chem. Kinet.*, **16,** 159 (1984).

Leone, J. A., R. C. Flagan, D. Grosjean, and J. H. Seinfeld, "An Outdoor Smog Chamber and Modeling Study of Toluene-NO_x Photooxidation," *Int. J. Chem. Kinet.*, **17,** 177 (1985).

Lloyd, A. C., "Evaluated and Estimated Kinetic Data for Gas Phase Reactions of the Hydroperoxyl Radical," *Int. J. Chem. Kinet.*, **6,** 169 (1974).

Lloyd, A. C., K. R. Darnall, A. M. Winer, and J. N. Pitts, Jr., "Relative Rate Constants for the Reactions of OH Radicals with Isopropyl Alcohol, Diethyl and Di-*n*-Propyl Ether at $305 \pm 2°K$," *Chem. Phys. Lett.*, **42,** 205 (1976).

Lloyd, A. C., Tropospheric Chemistry of Aldehydes, in *Chemical Kinetic Data Needs for Modeling the Lower Troposphere*, J. T. Herron, R. E. Huie, and J. A. Hodgeson, Eds., National Bureau of Standards Spec. Pub. 557, August 1979, pp. 27–48.

Lonneman, W. A., T. A. Bellar, and A. P. Altshuller, "Aromatic Hydrocarbons in the Atmosphere of the Los Angeles Basin," *Environ. Sci. Technol.*, **2,** 1017 (1968).

Lorenz, K. and R. Zellner, "Kinetics of the Reactions of OH Radicals with Benzene-d_6 and Naphthalene," *Ber. Bunenges. Phys. Chem.*, **87,** 629 (1983).

Lorenz, K. and R. Zellner, "Rate Constants and Vinoxy Product Yield in the Reaction OH + Ethylene Oxide," *Ber. Bunsenges. Phys. Chem.*, **88,** 1228 (1984).

Lorenz, K., D. Rhäsa, R. Zellner, and B. Fritz, "Laser Photolysis-LIF Kinetic Studies of the Reactions of CH_3O and CH_2CHO with O_2 between 300 and 500 K," *Ber. Bunsenges. Phys. Chem.*, **89,** 341 (1985).

Lovas, F. J. and R. D. Suenram, "Identification of Dioxirane (H_2COO) in Ozone–Olefin Reactions Via Microwave Spectroscopy," *Chem. Phys. Lett.*, **51,** 453 (1977).

Malko, M. W. and J. Troe, "Analysis of the Unimolecular Reaction $N_2O_5 + M \rightleftarrows NO_2 + NO_3 + M$," *Int. J. Chem. Kinet.*, **14,** 399 (1982).

Manzanares, E. R., M. Suto, and L. C. Lee, "Effect of NO_2 on the Aerosol Formation Kinetics from the $SO_2 + O_3 + C_2H_4$ Mixture," submitted for publication (1985).

Margitan, J. J., F. Kaufman, and J. G. Anderson, "The Reaction of OH with CH_4" *Geophys. Res. Lett.*, **1,** 80 (1974).

Martinez, R. I. and J. T. Herron, "Gas-Phase Reaction of SO_2 with a Criegee Intermediate in the Presence of Water Vapor," *J. Environ. Sci. Health*, **A16,** 623 (1981).

Martinez, R. I., R. E. Huie, and J. T. Herron, "Mass Spectrometric Detection of Dioxirane, H_2COO, and Its Decomposition Products H_2 and CO, from the Reaction of Ozone with Ethylene," *Chem. Phys. Lett.*, **51,** 457 (1977).

Martinez, R. I., R. E. Huie, and J. T. Herron, "Products of the Reaction of Hydroxyl Radicals with *trans*-2-Butene in the Presence of Oxygen and Nitrogen Dioxide," *Chem. Phys. Lett.*, **72,** 443 (1980).

Martinez, R. I., J. T. Herron, and R. E. Huie, "The Mechanism of Ozone–Alkene Reactions in the Gas Phase. A Mass Spectrometric Study of the Reactions of Eight Linear and Branched-Chain Alkanes," *J. Am. Chem. Soc.*, **103,** 3807 (1981).

Mayrsohn, H., J. H. Crabtree, M. Kuramoto, R. D. Sothern, and S. H. Mano, "Source Reconciliation of Atmospheric Hydrocarbons 1974" *Atmos. Environ.*, **11,** 189 (1977).

Mazur, S. and C. S. Foote, "Chemistry of Singlet Oxygen. IX. A Stable Dioxetane from Photooxygenation of Tetramethoxyethylene," *J. Am. Chem. Soc.*, **92,** 3225 (1970).

McCaulley, J. A., S. M. Anderson, J. B. Jeffries, and F. Kaufman, "Kinetics of the Reaction of CH_3O with NO_2," *Chem. Phys. Lett.*, **115,** 180 (1985).

McMillen, D. F. and D. M. Golden, "Hydrocarbon Bond Dissociation Energies," *Ann. Rev. Phys. Chem.*, **33,** 493 (1982).

Meier, U., H. H. Grotheer, and Th. Just, "Temperature Dependence and Branching Ratio of the CH_3OH + OH Reaction," *Chem. Phys. Lett.*, **106,** 97 (1984).

Meier, U., H. H. Grotheer, G. Riekert, and Th. Just, "Study of Hydroxyl Reactions with Methanol and Ethanol by Laser-Induced Fluorescence," *Ber. Bunsenges. Phys. Chem.*, **89,** 325 (1985).

Michael, J. V., D. F. Nava, R. P. Borkowski, W. A. Payne, and L. J. Stief, "Pressure Dependence of the Absolute Rate Constant for the Reaction OH + C_2H_2 from 228 to 413°K," *J. Chem. Phys.*, **73,** 6108 (1980).

Michael, J. V., D. G. Keil, and R. B. Klemm, "Rate Constants for the Reaction of Hydroxyl Radicals with Acetaldehyde from 244-528 K," *J. Chem. Phys.*, **83,** 1630 (1985).

Morabito, P., and J. Heicklen, "Disproportionation to Combination Ratios of Alkoxy Radicals with Nitric Oxide," *J. Phys. Chem.*, **89,** 2914 (1985).

Morris, E. D., Jr. and H. Niki, "Reaction of the Nitrate Radical with Acetaldehyde and Propylene," *J. Phys. Chem.*, **78,** 1337 (1974).

Morris, E. D., Jr., D. H. Stedman, and H. Niki, "Mass Spectrometric Study of the Reactions of the Hydroxyl Radical with Ethylene, Propylene, and Acetaldehyde in a Discharge Flow System," *J. Am. Chem. Soc.*, **93,** 3570 (1971).

Morrison, B. M. Jr., and J. Heicklen, "The Reactions of HO with CH_2O and of HCO with NO_2," *J. Photochem.*, **13,** 189 (1980).

Murray, R. W. and M. L. Kaplan, "Singlet Oxygen Sources in Ozone Chemistry. Chemical Oxygenations Using the Adducts Between Phosphite Esters and Ozone," *J. Am. Chem. Soc.*, **91,** 5358 (1969).

Nangia, P. S. and S. W. Benson, "The Kinetics of the Interaction of Peroxy Radicals. II. Primary and Secondary Alkyl Peroxy," *Int. J. Chem. Kinet.*, **12,** 43 (1980).

Narita, N. and T. Tezuka, "On the Mechanism of Oxidation of Hyroxycyclohexadienyl Radicals with Molecular Oxygen," *J.Am. Chem. Soc.*, **104,** 7316 (1982).

NASA, "Chemical Kinetic and Photochemical Data for Use in Stratospheric Modeling, Evaluation No.5," July 15 (1982), JPL Publ. 82-57.

Nelson, P. F. and S. M. Quigley, "The Hydrocarbon Composition of Exhaust Emitted from Gasoline-Fueled Vehicles," *Atmos. Environ.*, **18** 79 (1984)

Nicovich, J. M., R. L. Thompson and A. R. Ravishankara, "Kinetics of the Reactions of the Hydroxyl Radical with Xylenes," *J. Phys. Chem.*, **85,** 2913 (1981).

Niki, H., P. D. Maker, C. M. Savage, and L. P. Breitenbach, "Fourier Transform IR Spectroscopic Observation of Propylene Ozonide in the Gas Phase Reaction of Ozone–*cis*-2-Butene–Formaldehyde," *Chem. Phys. Lett.*, **46,** 327 (1977).

Niki, H., P. D. Maker, C. M. Savage, and L. P. Breitenbach "Mechanism for Hydroxyl Radical Initiated Oxidation of Olefin–Nitric Oxide Mixtures in Parts Per Million Concentrations," *J. Phys. Chem.*, **82,** 135 (1978a).

Niki, H., P. D. Maker, C. M. Savage, and L. P. Breitenbach, "Relative Rate Constants for the Reaction of Hydroxyl Radical with Aldehydes," *J. Phys. Chem.*, **82,** 132 (1978b).

Niki, H., P. D. Maker, C. M. Savage, and L. P. Breitenbach, "Fourier Transform Infrared Study of the HO Radical Initiated Oxidation of SO_2," *J. Phys. Chem.*, **84,** 14 (1980a).

Niki, H., P. D. Maker, C. M. Savage, and L. P. Breithenbach, "FTIR Studies of the Cl Atom Initiated Oxidation of Formaldehyde: Detection of a New Metastable Species in the Presence of NO_2," *Chem. Phys. Lett.*, **72,** 71 (1980b).

Niki, H., P. D. Maker, C. M. Savage, and L. P. Breitenbach, "Further IR Spectroscopic Evidence for the Formation of $CH_2(OH)OOH$ in the Gas Phase Reaction of HO_2 With CH_2O," *Chem. Phys. Lett.*, **75,** 533 (1980c).

Niki, H., P. D. Maker, C. M. Savage, and L. P. Breitenbach, "Fourier Transform Infrared Studies of the Self-Reaction of CH_3O_2 Radicals," *J. Phys. Chem.*, **85,** 877 (1981a).

Niki, H., P. D. Maker, C. M. Savage, and L. P. Breitenbach, "A FTIR Study of a Transitory Product in the Gas-Phase Ozone–Ethene Reaction," *J. Phys. Chem.*, **85,** 1024 (1981b).

Niki, H., P. D. Maker, C. M. Savage, and L. P. Breitenbach, "An FTIR Study of Mechanism for the HO Radical Initiated Oxidiation of C_2H_4 in the Presence of NO: Detection of Glycoaldehyde," *Chem. Phys. Lett.*, **80,** 499 (1981c).

Niki, H., P. D. Maker, C. M. Savage, and L. P. Breitenbach, "An FTIR Study of the Isomerization and O_2 Reaction of *n*-Butoxy Radicals," *J. Am. Chem. Soc.*, **85,** 2698 (1981d).

Niki, H., P. D. Maker, C. M. Savage, L. P. Breitenbach, R. I. Martinez, and J. T. Herron, "A Fourier Transform Infrared Study of the Gas-Phase Reactions of O_3 with Chloroethylenes. Detection of Peroxyformic Acid," *J. Phys. Chem.*, **86,** 1858 (1982a).

Niki, H., P. D. Maker, C. M. Savage, and L. P. Breitenbach, "Fourier Transform Infrared Studies of the Self-Reaction of $C_2H_5O_2$ Radicals," *J. Phys. Chem.*, **86,** 3825 (1982b).

Niki, H., P. D. Maker, C. M. Savage, and L. P. Breitenbach, "Atmospheric Ozone–Olefin Reactions," *Environ. Sci. Technol.*, **17,** 312A (1983).

Niki, H., P. D. Maker, C. M. Savage, L. P. Breitenbach, and R. I. Martinez, "Fourier Transform Infrared Study of the Gas-Phase Reaction of $^{18}O_3$ with *trans*-CHCl=CHCl in $^{16}O_2$-Rich Mixtures. Branching Ratio for O-Atom Production Via Dissociation of the Primary Criegee Intermediate," *J. Phys. Chem.*, **88,** 766 (1984a).

Niki, H., P.D. Maker, C. M. Savage, and L. P. Breitenbach, "Fourier Transform Infrared Study of the Kinetics and Mechanism for the Reaction of Hydroxyl Radical with Formaldehyde," *J. Phys. Chem.*, **88,** 5342 (1984b).

Niki, H., P. D. Maker, C. M. Savage, and L. P. Breitenbach, "An FTIR Study of the Cl-Atom Initiated Reaction of Glyoxal," *Int. J. Chem. Kinet.*, **17,** 547 (1985).

Nojima, K., K. Fukaya. S. Fukui, and S. Kanno, "The Formation of Glyoxals by the Photochemical Reaction of Aromatic Hydrocarbons in the Presence of Nitrogen Monoxide,"*Chemosphere*, **5,** 247 (1974).

Norton, R. B., J. M. Roberts, and B. J. Huebert, "Tropospheric Oxalate," *Geophys. Res. Lett.*, **10,** 517 (1983).

O'Brien, R. J., P. J. Green, and R. M. Doty, Comments on Reactions of Aromatic Compounds in the Atmosphere, in *Chemical Kinetic Data Needs for Modeling the Lower Troposphere*, J. T. Herron, R. E. Huie, and J. A. Hodgeson, Eds., National Bureau of Standards, Washington, DC, NBS Spec. Publ. 557, pp. 93–95, August 1979a.

O'Brien, R. J., P. J. Green, R. A. Doty, J. W. Vanderzanden, R. R. Easton, and R. P. Irwin, Interaction of Oxides of Nitrogen and Aromatic Hydrocarbons Under Simulated Atmospheric Conditions, in *Nitrogenous Air Pollutants—Chemical and Biological Implications*, D. Grosjean, Ed., Ann Arbor Science Publishers, Ann Arbor, MI, 1979b, pp. 189–210.

O'Brien, R. J., B. E. Dumdei, S. V. Hummel, and R. A. Yost, "Determination of Atmospheric Degradation Products of Toluene by Tandem Mass Spectrometry," *Anal. Chem.*, **56,** 1329 (1984)

Ohta, T., "Rate Constants for the Reactions of Diolefins with OH Radicals in the Gas Phase. Estimate of the Rate Constants from Those for Monoolefins," *J. Phys. Chem.*, **87,** 1209 (1983).

Ohta, T., H. Bandow, and H. Akimoto, "Gas Phase Chlorine-Initiated Photooxidation of Methanol and Isopropanol," *Int. J. Chem. Kinet.*, **14,** 173 (1982).

O'Neal, H. E. and C. Blumstein, "A New Mechanism for Gas Phase Ozone–Olefin Reactions," *Int. J. Chem. Kinet.*, **5,** 397 (1973).

Overend, R. and G. Paraskevopoulos, "Rates of OH Radical Reactions. 4. Reactions with Methanol, Ethanol, 1-Propanol, and 2-Propanol at 296°K," *J. Phys. Chem.*, **82,** 1329 (1978).

Parkes, D. A., The Role of Alkylperoxy and Alkoxy Radicals in Alkyl Radical Oxidation at Room Temperature, in 15th International Symposium Combustion, Tokyo, 1974, The Combustion Institute, Pittsburgh, PA, 1975, pp. 795–804.

Parkes, D. A., D. M. Paul, C. P. Quinn, and R. C. Robson, "The Ultraviolet Absorption by Alkylperoxy Radicals and Their Mutual Reactions," *Chem. Phys. Lett.*, **23,** 425 (1973).

Penzhorn, R.-D., H. Güsten, U. Schurath, and K. H. Becker, "Quenching of Singlet Molecular Oxygen by Some Atmospheric Pollutants," *Environ. Sci. Technol.*, **8,** 907 (1974)

Perner, D., A. Schmeltekopf, R. H. Winkler, H. S. Johnston, J. G. Calvert, C. A. Cantrell, and W. R. Stockwell, "A Laboratory and Field Study of the Equilibrium $N_2O_5 \rightleftarrows NO_3 + NO_2$," *J. Geophys. Res.*, **90,** 3807 (1985).

Perry, R. A. and D. Williamson, "Pressure and Temperature Dependence of the OH Radical Reaction with Acetylene," *Chem. Phys. Lett.*, **93,** 331 (1982).

Perry, R. A., R. Atkinson, and J. N. Pitts, Jr., "Kinetics and Mechanism of the Gas Phase Reaction of OH Radicals with Aromatic Hydrocarbons Over the Temperature Range 296–473°K," *J. Phys. Chem.*, **81,** 296 (1977a).

Perry, R. A., R. Atkinson, and J. N. Pitts, Jr., "Kinetics and Mechanism of the Gas Phase Reaction of OH Radicals with Methoxybenzene and *o*-Cresol Over the Temperature Range 299–435°K," *J. Phys. Chem.*, **81,** 1607 (1977b).

Perry, R. A., R. Atkinson, and J. N. Pitts, Jr., "Kinetics of the Reactions of OH Radicals with C_2H_2 and CO," *J. Chem. Phys.*, **67,** 5577 (1977c).

Perry, R. A., R. Atkinson, and J. N. Pitts, Jr., "Rate Constants for the Reaction of OH Radicals with Dimethyl Ether and Vinyl Methyl Ether Over the Temperature Range 299–427^{0}K," *J. Chem. Phys.*, **67,** 611 (1977d).

Peterson, J. T., "Calculated Actinic Fluxes (290–700 nm) for Air Pollution Photochemistry Application," U.S. Environmental Protection Agency Report No. EPA-600/4-76-025, June 1976.

Pitts, J. N., Jr., The Role of Singlet Molecular Oxygen in the Chemistry of Urban Atmospheres, in *Chemical Reactions in Urban Atmospheres.*, C. S. Tuesday, Ed., Elsevier, New York 1971.

Pitts, J. N. Jr., A. U. Khan, E. B. Smith, and R. P. Wayne, "Singlet Oxygen in the Environmental Sciences. Singlet Molecular Oxygen and Photochemical Air Pollution," *Environ. Sci. Technol.*, **3,** 241 (1969).

Plum, C. N., E. Sanhueza, R. Atkinson, W. P. L. Carter, and J. N. Pitts, Jr., "OH Radical Rate Constants and Photolysis Rates of α-Dicarbonyls," *Environ. Sci. Technol.*, **17,** 479 (1983).

Plumb, I. C. and K. R. Ryan, "Kinetic Studies of the Reaction of C_2H_5 with O_2 at 295°K," *Int. J. Chem. Kinet.*, **13,** 1011 (1981).

Plumb, I. C. and K. R. Ryan, "Kinetics of the Reactions of CH_3 with $O(^3P)$ and O_2 at 295°K," *Int. J. Chem. Kinet.*, **14,** 861 (1982).

Plumb, I. C., K. R. Ryan, J. R. Steven, and M. F. R. Mulcahy, "Kinetics of the Reaction of CH_3O_2 with NO," *Chem. Phys. Lett.*, **63,** 255 (1979).

Radford H. E., "The Fast Reaction of CH_2OH with O_2," *Chem. Phys. Lett.*, **71,** 195 (1980).

Rånby, B. and J. F. Rabek, Eds., *Singlet Oxygen Reactions with Organic Compounds and Polymers*, Wiley, New York, 1978.

Ravishankara, A. R. and R. L. Mauldin III, "Absolute Rate Coefficient for the Reaction of NO_3 with *trans*-2-Butene," *J. Phys. Chem.*, **89,** 3144 (1985).

Ravishankara, A. R., S. Wagner, S. Fischer, G. Smith, R. Schiff, R. T. Watson, G. Tesi, and D. D. Davis, "A Kinetic Study of the reactions of OH with Several Aromatic and Olefinic Compounds," *Int. J. Chem. Kinet.*, **10,** 783 (1978).

Ravishanakara, A. R., F. L. Eisele, N. M. Kreutter, and P. H. Wine, "Kinetics of the Reaction of CH_3O_2 with NO," *J. Chem. Phys.*, **74,** 2267 (1981).

Rinke, M., and C. Zetzsch, "Rate Constants for the Reactions of OH Radicals with Aromatics: Benzene, Phenol, Aniline, and 1,2,4-Trichlorobenzene," *Ber. Bunsenges. Phys. Chem.*, **88,** 55 (1984).

Ruiz, R. P. and K. D. Bayes, "Rates of Reaction of Propyl Radicals with Molecular Oxygen," *J. Chem. Phys.* **88,** 2592 (1984).

Saltzman, B. E., "Kinetic Studies of Formation of Atmospheric Oxidants," *Ind. Eng. Chem.*, **50,** 677 (1958).

Schaap, A. P., Ed., *Singlet Molecular Oxygen,* Benchmark Papers in Organic Chemistry, Vol. 5, Dowden, Hutchinson and Ross Inc., Stroudsburg, PA, 1976.

Schmidt, V., G.-Y. Zhu, K. H. Becker, and E. H. Fink, "Absolute Rate Constant Measurements of OH Reactions Under Atmospheric Conditions by Laser Photolysis/Dye Laser Fluorescence," Presented at the Third European Symposium on the Physico-Chemical Behaviour of Atmospheric Pollutants, Varese, Italy, April 10–12, 1984.

Schmidt, V., G. Y. Zhu, K. H. Becker, and E. H. Fink, "Study of OH Reactions at High Pressures by Excimer Laser Photolysis-Dye Laser Fluorescence," *Ber. Bunsenges. Phys. Chem.*, **89,** 321 (1985).

Schwartz, W., "Chemical Characterization of Model Aerosols," EPA Report No.EPA-650/3-74-011, August 1974.

Selzer, E. A. and K. D. Bayes, "Pressure Dependence of the Rate of Reaction of Methyl Radicals with O_2," *J. Phys. Chem.*, **87,** 392 (1983).

Semmes, D. H., A. R. Ravishankara, C. A. Gump-Perkins, and P. H. Wine, "Kinetics of the Reactions of Hydroxyl Radical with Aliphatic Aldehydes," *Int. J. Chem. Kinet.*, **17,** 303 (1985).

Shepson, P. B., E. O. Edney, and E. W. Corse, "Ring Fragmentation Radical Reactions on the Photooxidations of Toluene and *o*-Xylene," *J. Phys. Chem.*, **88,** 4122 (1984).

Shepson, P. B., E. O. Edney., T. E. Kleindienst, J. H. Pittman, G. R. Namie and L. T. Cupitt, "The Production of Organic Nitrates from Hydroxyl and Nitrate Radical Reactions with Propylene," *Environ. Sci. Technol.*, **19,** 849 (1985).

Singleton, D. L. and R. J. Cvetanovic, "Temperature Dependence of the Reactions of Oxygen Atoms with Olefins," *J. Am. Chem. Soc.*, **98,** 6812 (1976).

Slagle, I. R., Q. Feng, and D. Gutman, "Kinetics of the Reaction of Ethyl Radicals with Molecular Oxygen from 294 to 1002°K," *J. Phys. Chem.*, **88,** 3648 (1984).

Smith, I. W. M. and R. Zellner, "Rate Measurements of Reactions of OH by Resonance Absorption. Part 2. Reactions of OH with CO, C_2H_4, and C_2H_2," *J. Chem. Soc. Faraday Trans. 2*, **69,** 1617 (1973).

Spicer, C. W. and P. W. Jones, "The Fate of Aromatic Hydrocarbons in Photochemical Smog Systems: Toluene," *J. Air Pollut. Control Assoc.*, **27** 1122 (1977).

Sridharan, U. C. and F. Kaufman, "Primary Products of the O + C_2H_4 Reaction," *Chem. Phys. Lett.*, **102,** 45 (1983).

Steer, R. P., J. L. Sprung, and J. N. Pitts, Jr., "Evidence for the Production of $O_2(^1\Delta_g)$ by Energy Transfer in the Gas Phase," *Environ. Sci. Technol.*, **3,** 946 (1969a).

Steer, R. P., K. R. Darnall, and J. N. Pitts, Jr., "The Base-Induced Decomposition of Peroxyacetylnitrate," *Tetrahedron Lett.*, **43** 3765 (1969b).

Stephens, E. R., F. R. Burleson, "Distribution of Light Hydrocarbons in Ambient Air," *J. Air Pollut. Control Assoc.*, **19,** 929 (1969).

Su, F., J. G. Calvert, J. H. Shaw, H. Niki, P. D. Maker, C. M. Savage, L. P. Breitenbach, "Spectroscopic and Kinetic Studies of a New Metastable Species in the Photooxidation of Gaseous Formaldehyde," *Chem. Phys. Lett.*, **65,** 221 (1979a).

Su, F., J. G. Calvert, and J. H. Shaw, "Mechanism of the Photooxidation of Gaseous Formaldehyde," *J. Phys. Chem.*, **83,** 3185 (1979b).

Su, F., J. G. Calvert, and J. H. Shaw, "A FTIR Spectroscopic Study of the Ozone–Ethene Reaction Mechanism in O_2-Rich Mixtures," *J. Phys. Chem.*, **84,** 239 (1980).

Suto, M., E. R. Manzanares, and L. C. Lee, "Detection of Sulfuric Acid Aerosols by Ultraviolet Scattering," *Environ. Sci. Technol.*, **19,** 815 (1985).

Takagi, H., N. Washida, H. Akimoto, K. Nagasawa, Y. Usui, and M. Okuda, "Photooxidation of *o*-Xylene in the NO–H_2O–Air System," *J. Phys. Chem.*, **84,** 478 (1980).

Takagi, H., N. Washida, H. Akimoto, and M. Okuda, "Observation of 3-Hexene-2,5-Dione in the Photooxidation of 1,2,4-Trimethylbenzene in the NO–H_2O–Air System," *Spectrosc. Lett.*, **15,** 145 (1982).

Tuazon, E. C., R. Atkinson, H. MacLeod, H. W. Biermann, A. M. Winer, W. P. L. Carter, and J. N. Pitts, Jr., "Yields of Glyoxal and Methylglyoxal from the NO_x–Air Photooxidations of Toluene and *m*- and *p*-Xylene," *Environ. Sci. Technol.*, **18,** 981 (1984a).

Tuazon, E. C., E. Sanhueza, R. Atkinson, W. P. L. Carter, A. M. Winer, and J. N. Pitts, Jr., "Direct Determination of the Equilibrium Constant at 298°K for the $NO_2 + NO_3 \rightleftarrows N_2O_5$ Reactions," *J. Phys. Chem.*, **88,** 3095 (1984b).

Tuazon, E. C., H. MacLeod, R. Atkinson, and W. P. L. Carter, "α-Dicarbonyl Yields from the NO_x-Air Photooxidations of a Series of Aromatic Hydrocarbons in Air," to be submitted for publication (1986).

Tully, F. P., A. R. Ravishankana, R. L. Thompson, J. M. Nicovich, R. C. Shah, N. M. Kreutter, and P. H. Wine, "Kinetics of the Reactions of Hydroxyl Radical with Benzene and Toluene," *J. Phys. Chem.*, **85,** 2262 (1981).

Veyret, B., J.-C. Rayez, and R. Lesclaux, "Mechanism of the Photooxidation of Formaldehyde Studied by Flash Photolysis of CH_2O–O_2–NO Mixtures," *J. Phys. Chem.*, **86,** 3424 (1982).

Wadt, W. R. and W. A. Goddard III, "The Electronic Structure of the Criegee Intermediate. Ramifications for the Mechanism of Ozonolysis," *J. Am Chem. Soc.*, **97,** 3004 (1975).

Wahner, A., and C. Zetzsch, "The Reaction of OH with C_2H_2 at Atmospheric Conditions Investigated by CW-UV-Laser-Longpath-Absorption of OH," *Ber. Bunsenges. Phys. Chem.*, **89,** 323 (1985).

Walker, R. W., "Temperature Coefficients for Reactions of OH Radicals with Alkanes between 300 and 1000 K," *Int. J. Chem. Kinet.*, **17,** 573 (1985).

Wang, W. C., M. Suto, and L. C. Lee, "$CH_2OH + O_2$ Reaction Rate Constant Measured by Detecting HO_2 from Photofragment Emission," *J. Chem. Phy.*, **81,** 3122 (1984).

Washida, N., "Reaction of Ethanol and $CH_3CH(OH)$ Radicals with Atomic and Molecular Oxygen," *J. Chem. Phys.*, **75,** 2715 (1981).

Wathne, B., "Measurements of Benzene, Toluene, and Xylenes in Urban Air," *Atmos. Environ.*, **17,** 1713 (1983).

Whitbeck, M., "Photooxidation of Methanol," *Atmos Environ.*, **17,** 121 (1983).

Winer, A. M. and K. D. Bayes, "The Decay of $O_2(a^1\Delta_g)$ in Flow Experiments," *J. Phys. Chem.*, 70, 302 (1966).

Winer, A. M., A. C. Lloyd, K. R. Darnall, and J. N. Pitts, Jr., "Relative Rates Constants for the Reaction of the Hydroxyl Radical with Selected Ketones, Chloroethenes and Monoterpene Hydrocarbons," *J. Phys. Chem.*, **80,** 1635 (1976).

Winer, A. M., A. C. Lloyd, K. R. Darnall, R. Atkinson, and J. N. Pitts, Jr., "Rate Constants for the Reaction of OH Radicals with *n*-Propyl Acetate, *Sec*-Butyl Acetate, Tetrahydrofuran, and Peroxyacetyl Nitrate," *Chem. Phys. Lett.*, **51,** 221 (1977).

Winer, A. M., K. R. Darnall, R. Atkinson, and J. N. Pitts, Jr., unpublished results (1978).

Winer, A. M., R. Atkinson, and J. N. Pitts, Jr., "Gaseous Nitrate Radical: Possible Nighttime Atmospheric Sink for Biogenic Organic Compounds," *Science*, **224,** 156 (1984).

World Meterological Organization, "The Stratosphere 1981 Theory and Measurements, WMO Global Ozone Research and Monitoring Project, Report #11, May 1981.

Wu, C. H., S. M. Japar, and H. Niki, "Relative Reactivities of OH–Hydrocarbon Reactions from Smog Reaction Studies," *J. Environ. Sci. Health*, **All,** 191 (1976).

Zabarnick, S., and J. Heicklen, "Reactions of Alkoxy Radicals with O_2. I. C_2H_5O Radicals," *Int. J. Chem. Kinet.*, **17,** 455 (1985a).

Zabarnick, S., and J. Heicklen, "The Reactions of Alkoxy Radicals with O_2. II. *n*-C_3H_7O Radicals," *Int. J. Chem. Kinet.*, **17,** 477 (1985b).

Zabarnick, S., and J. Heicklen, "Reactions of Alkoxy Radicals with O_2. III. *i*-C_4H_9O Radials," *Int. J. Chem. Kinet.*, **17,** 503 (1985c).

Zellner, R. and K. Lorenz, "Laser Photoysis/Resonance Fluorescence Study of the Rate Constants for the Reactions of OH Radicals with C_2H_4 and C_3H_6," **88,** 984 (1984).

Zetzsch, C., "Predicting the Rate of OH-Addition to Aromatics Using σ^+-Electrophilic Substituent Constants for Mono- and Polysubstituted Benzene," Abstract A-11, XVth Infomal Conference on Photochemistry, Stanford, CA, June 27–July 1, 1982.

Zhang, J., S. Hatakeyama, and H. Akimoto, "Rate Constants of the Reaction of Ozone with *trans*-1,2-Dichloroethene and Vinyl Chloride in Air," *Int. J. Chem. Kinet.*, **15,** 655 (1983).

8 Kinetics and Mechanisms of Gas Phase Reactions of Nitrogen–Containing Tropospheric Constituents

Nitrogen-containing compounds, both inorganic and organic, play extremely important roles in the chemistry of both polluted and clean atmospheres. As discussed in Chapter 5, a wide variety of these species have been identified and measured in ambient air or in simulated atmospheres using environmental chambers. Tables 8.1 and 8.2 summarize some of the inorganic nitrogenous compounds with the range of peak concentrations either observed or anticipated in a polluted urban atmosphere such as the Los Angeles area.

In this chapter we review the chemistry of these nitrogenous compounds. Because many of them are closely linked to organic oxidations as well as to acid deposition, we frequently refer the reader to other parts of the book where specific aspects of the chemistry relevant to these issues are discussed in more detail. All rate constants cited are either from Atkinson and Lloyd (1984) or DeMore et al. (1983, 1985). Unless otherwise stated, they are given for 1 atm pressure and 298°K.

8-A. INORGANIC NITROGENOUS COMPOUNDS

8-A-1. Nitric Oxide, NO, and Nitrogen Dioxide, NO_2

8-A-1a. NO

Nitric oxide, NO, is an important primary pollutant emitted by both mobile and stationary sources. Smaller quantities of NO_2 are also emitted along with the NO (see section 1-D-1). As we have seen, NO is converted in ambient air to NO_2. Thus NO_2 is both a primary and a secondary pollutant, that is, it is emitted directly into the air as well as being formed from chemical reactions of NO.

As discussed in Chapter 1, the oxidation of NO by O_2,

$$2\ NO + O_2 \rightarrow 2\ NO_2 \tag{1}$$

$$k_1^{298°K} = 2.0 \times 10^{-38}\ cm^6\ molecule^{-2}\ s^{-1}$$

(Hampson and Garvin, 1978)

is too slow at typical ambient NO concentrations to be significant; it may occur for a brief period of time in plumes where the NO concentrations are higher as

TABLE 8.1. Concentration Ranges for Some Gaseous Nitrogenous Pollutants Observed in Ambient Air in the California South Coast Air Basin

Compound	Range of Peak Concentrations (ppb)	Time of Peak Concentration
Nitric acid (HNO_3)	⩽50	Mid-afternoon
Peroxyacetylnitrate (PAN)	⩽40	Mid-afternoon
Nitrogen dioxide (NO_2)	100–800	Morning/afternoon
Nitrous acid (HONO)	1–8	Before sunrise
Nitrate radical (NO_3)	0.005–0.430[a]	After sunset

Source: Adapted from Pitts et al., 1983.

[a]Note that the measured concentrations may be too large by ~25% if the larger absorption cross section data of Sander (1985) prove to be correct (see Section 3-C-7b).

TABLE 8.2. Some Compounds Observed in Simulated Atmospheres with Ranges of Concentrations Predicted for Polluted Urban Atmospheres

Compound	Range of Peak Concentrations
Peroxynitric acid (HO_2NO_2)	⩽1 ppb
Peroxyalkyl nitrates (RO_2NO_2)	⩽100 ppt
Dinitrogen pentoxide (N_2O_5)	⩽15 ppb[a]

Source: Adapted from Pitts et al., 1983.

[a]Atkinson et al., 1985.

they exit a stack or tailpipe before significant dilution with the surrounding air has occurred.

The conversion of NO to NO_2 in ambient air involves the chain oxidation of organics initiated primarily by the free radical OH. For example, Fig. 8.1 shows a reaction sequence in which OH attacks propane to form a propyl radical and water. The propyl radical then reacts with O_2 to form an alkylperoxy radical and this oxidizes NO to NO_2:

$$RO_2 + NO \xrightarrow[a]{} RO + NO_2$$
$$\xrightarrow[b]{} RONO_2 \qquad (2)$$
$$k_2 = 7.6 \times 10^{-12}\ \text{cm}^3\ \text{molecule}^{-1}\ \text{s}^{-1}$$

As discussed in detail in Section 7-D-2a, reaction (2a) predominates over (2b), with (2b) becoming significant for the larger ($\geq C_4$) alkoxy radicals. The alkoxy radical (RO) formed in (2a) undergoes a hydrogen abstraction by O_2 to form HO_2 and an aldehyde (see Section 7-D-3a); the HO_2 then can oxidize a second NO to NO_2, reforming OH in the process:

$$HO_2 + NO \rightarrow OH + NO_2 \qquad (3)$$
$$k_3^{298°K} = 8.3 \times 10^{-12}\ \text{cm}^3\ \text{molecule}^{-1}\ \text{s}^{-1}$$

Thus, in this cycle, two molecules of NO have been oxidized to NO_2 and OH has been regenerated to carry on further oxidations. The propane has been oxidized to propionaldehyde which can be further oxidized by OH in a cycle similar to that of Fig. 8.1 in which NO is simultaneously converted to NO_2 and acetaldehyde is formed. The original carbon atoms ultimately reside in CO and CO_2, and a number of NO molecules have been oxidized to NO_2.

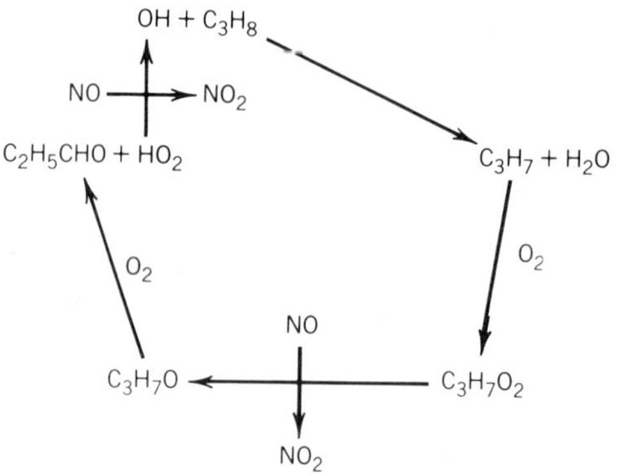

FIGURE 8.1. Oxidation of propane initiated by the hydroxyl radical. Only the reaction path from abstraction of a primary hydrogen atom is shown for simplicity, although the abstraction of primary hydrogens is slower than secondary hydrogen atoms.

NO also reacts rapidly with O_3:

$$NO + O_3 \rightarrow NO_2 + O_2 \tag{4}$$
$$k_4^{298^\circ K} = 1.8 \times 10^{-14} \text{ cm}^3 \text{ molecule}^{-1} \text{ s}^{-1}$$

Thus significant concentrations of NO and O_3 are not found simultaneously in the same air mass; in addition, O_3 does not build up during a photochemical air pollution episode until the NO concentration has fallen to low values (see Chapter 1). This reaction has considerable implications for the development of control strategy options for photochemical oxidants (see Chapter 10).

Other species that react with NO in ambient air in addition to HO_2, RO_2, and O_3, include OH, alkoxy radicals (RO), and nitrate radicals (NO_3).

Reaction of NO with OH and alkoxy radicals produces nitrous acid and, in part, organic nitrites, respectively:

$$NO + OH \xrightarrow{M} HONO \tag{5}$$

$$NO + RO \underset{a}{\rightarrow} RONO$$
$$\underset{b}{\rightarrow} R_1R_2CO + HNO \tag{6}$$

At atmospheric pressure and 298°K, reaction (5) is in the fall-off region between second and third order; Atkinson and Lloyd (1984) recommend under these conditions a rate constant of $k_5 = 6.8 \times 10^{-12}$ cm^3 molecule^{-1} s^{-1}. As discussed in Section 7-D-3d, they also recommend $k_6 \simeq 3 \times 10^{-11}$ cm^3 molecule^{-1} s^{-1} under the same conditions, with ~20% of the reaction proceeding via (6b) for C_2 and larger radicals which have an abstractable hydrogen, and $\leq$ 20% for CH_3O. As discussed in Section 7-D-3d, this value for k_6 is consistent with the one direct measurement available for RO = $(CH_3)_2CHO$. Reactions (5) and (6a) do not permanently remove NO from the atmosphere since HONO and RONO photolyze rapidly (see Section 3-C-5). Recall that as discussed in Sections 3-C-5 and 7-D-3a, recent studies by Heicklen and coworkers suggest that the quantum yields for alkyl nitrite photolysis may be in the range of 0.2–0.4, much less than the unit quantum yields which have generally been assumed. As discussed in Section 3-D, HONO in particular is believed to act as a major OH source at sunrise in urban atmospheres.

NO reacts very rapidly with NO_3:

$$NO + NO_3 \rightarrow 2\, NO_2 \tag{7}$$
$$k_7^{298^\circ K} = 3.0 \times 10^{-11} \text{ cm}^3 \text{ molecule}^{-1} \text{ s}^{-1}$$

Thus significant concentrations of NO_3 can only build up when little NO is present.

In summary, NO can react in ambient air with HO_2 and RO_2, as well as with

O_3, OH, RO, and NO_3. The reactions with HO_2 and RO_2 are a central part of the chain reactions in which organics are oxidized and NO is converted to NO_2 in both clean and polluted atmospheres. The reaction of NO with O_3 controls the development of the O_3 peak in polluted urban areas. The reactions of NO with OH and in part with RO radicals act as temporary nighttime storages for NO, since the major products HONO and RONO rapidly photolyze at dawn. Finally, the reaction with NO_3 controls to a great extent the concentrations of NO_3 in nighttime air masses.

Table 8.3 summarizes these reactions of NO and gives estimated lifetimes for NO with respect to each reaction under conditions characteristic of moderately to heavily polluted urban areas.

8-A-1b. NO_2

There are a number of potential fates of NO_2 in the atmosphere, including photolysis as well as reaction with OH, O_3, NO_3, $O(^3P)$, HO_2, RO_2, and RO:

$$NO_2 + h\nu\ (\lambda \leq 430 \text{ nm}) \rightarrow NO + O(^3P) \tag{8}$$

TABLE 8.3. Loss Processes for NO in Ambient Air and Estimated Lifetimes of NO with Respect to Each Reaction Under Conditions Characteristic of Moderate to Heavily Polluted Urban Areas

Reactant	k^a (cm^3 $molecule^{-1}$ s^{-1})	Peak $[X]^b$ (number cm^{-3})	$\tau_{NO}{}^c$	Comments
HO_2	8.3×10^{-12}	2×10^9	60 s	Central reaction in clean and polluted air
RO_2	7.6×10^{-12}	3×10^9	44 s	Central reaction in clean and polluted air
O_3	1.8×10^{-14}	5×10^{12}	11 s	Central reaction in clean and polluted air
OH	6.8×10^{-12}	1×10^7	4 h	Can act as temporary nighttime NO storage since product photolyzes rapidly
RO	3×10^{-11}	1×10^4	39 days	Can act as temporary nighttime NO storage since product photolyzes rapidly
NO_3	3.0×10^{-11}	9×10^9	4 s	Controls concentration of NO_3 at night

[a] Rate constants at 298°K, 1 atm pressure.

[b] Typical peak concentrations predicted using a propene-n-butane model with an initial reactive hydrocarbon centration of 0.60 ppm and an initial NO_x of 0.060 ppm, except for NO_3 where the experimentally observed peak value of 430 ppt has been used (see Chapter 5).

[c] Lifetime of NO with respect to each reaction; $\tau_{NO} = 1/k_x[X]$.

$$NO_2 + OH \xrightarrow{M} HONO_2 \tag{9}$$

$$\begin{aligned} NO_2 + O_3 &\underset{a}{\rightarrow} NO_3 + O_2 \\ &\underset{b}{\rightarrow} NO + 2O_2 \end{aligned} \tag{10}$$

$$NO_2 + NO_3 \overset{M}{\rightleftarrows} N_2O_5 \tag{11}$$

$$\rightarrow NO + NO_2 + O_2 \tag{12}$$

$$\begin{aligned} NO_2 + O(^3P) &\underset{a}{\rightarrow} NO + O_2 \\ &\underset{b}{\overset{M}{\rightarrow}} NO_3 \end{aligned} \tag{13}$$

$$NO_2 + HO_2 \overset{M}{\rightleftarrows} HO_2NO_2 \tag{14}$$

$$NO_2 + RO_2 \overset{M}{\rightleftarrows} RO_2NO_2 \tag{15}$$

$$\begin{aligned} NO_2 + RO &\underset{a}{\overset{M}{\rightarrow}} RONO_2 \\ &\underset{b}{\rightarrow} HONO + R_1R_2CO \end{aligned} \tag{16}$$

The photolysis of NO_2, reaction (8), is the only known anthropogenic source of O_3 via (8) followed by (17):

$$O(^3P) + O_2 \xrightarrow{M} O_3 \tag{17}$$

A possible exception to this is decomposition of chlorine-containing Criegee biradicals (see Section 7-E-2).

The third-order rate constant recommended for the reaction (17) of $O(^3P)$ with O_2 by the NASA group (DeMore et al., 1983) at a temperature T is:

$$k_{17} = (6.0 \pm 0.5) \times 10^{-34} \left(\frac{T}{300}\right)^{-2.3 \pm 0.5} \text{cm}^6 \text{ molecule}^{-2} \text{ s}^{-1}$$

At 300°K and 1 atm of air, the effective bimolecular rate constant becomes $k_{17} = 1.5 \times 10^{-14}$ cm^3 molecule^{-1} s^{-1}. Using $[O_2] = 0.21$ atm $= 5.2 \times 10^{18}$ molecules cm^{-3}, the lifetime of an oxygen atom formed by NO_2 photolysis at the earth's surface is ~ 13 μs.

The absorption cross sections, quantum yields, and photolysis rates for NO_2 in the atmosphere are discussed in Section 3-C-3.

The photolytic rate constant obviously depends on solar intensity and spectral distribution and hence on such factors as latitude, season, solar zenith angle, and the degree of cloud cover. However, in ambient air, typical photolytic rate constants up to $\sim 9 \times 10^{-3}$ s^{-1} have been measured corresponding to NO_2 lifetimes of ≥ 2 min (Demerjian et al., 1980; Dickerson et al., 1982; Parrish et al., 1983). For example, near Raleigh, North Carolina, the photolytic rate constant for NO_2

was measured to peak at $\sim 8 \times 10^{-3}$ s^{-1} on a day with no clouds in late April 1975, but at $\sim 3 \times 10^{-3}$ s^{-1} on a day in the same period when the skies were overcast and some rain occurred in the morning. Thus a typical lifetime for NO_2 with respect to photolysis is $\simeq (4 \times 10^{-3}\ s^{-1})^{-1} \simeq 4$ min.

Reaction (9) with OH is important in forming nitric acid in the atmosphere and hence in contributing to acid rain and fogs. This occurs primarily during daylight hours since OH is produced in photochemical reactions. At 1 atm and 300°K, Atkinson and Lloyd (1984) recommend an effective bimolecular rate constant of $k_9[M] = 1.1 \times 10^{-11}$ cm^3 molecule^{-1} s^{-1}. At an OH concentration of 5×10^6 radicals cm^{-3} which might be typical of a peak concentration in polluted urban atmospheres, the lifetime of NO_2 with respect to reaction (9) is then ~ 5 h. Thus photolysis is faster than reaction with OH by approximately two orders of magnitude. However, as discussed in more detail in Section 11-B-1a, reaction (9) is still a major source of nitric acid in the troposphere.

Reaction (10) of NO_2 with O_3 is an important reaction in the troposphere when there are significant O_3 and NO_2 concentrations present. Thus it is the source of the nitrate radical which is now believed to play a significant role in the oxidation of both natural and anthropogenic organics, and in the formation of HNO_3 in the troposphere (see Chapters 7, 11, and 14). Because reaction (10) does not require light, it contributes to nighttime NO_x chemistry. The recommended rate constant for reaction (10) at 298°K is $k_{10} = 3.2 \times 10^{-17}$ cm^3 molecule^{-1} s^{-1} (DeMore et al., 1983), so that the lifetime of NO_2 at an O_3 concentration of 0.08 ppm is 4.4 h.

While reaction (10) has been accepted in the past as proceeding essentially entirely to produce $NO_3 + O_2$, that is, reaction (10a), a small contribution from (10b), producing $NO + 2O_2$, cannot be excluded. For example, Cantrell et al. (1985) suggest, based on computer kinetic simulation studies of the complex N_2O_5–NO_3–NO_2–HCHO–air system, that reaction (10b) accounts for $\sim 3\%$ of the O_3–NO_2 reaction.

Reaction (11) of NO_3 with NO_2 can also occur at night under conditions where NO_3 is formed and not removed rapidly by other processes such as reaction with NO. The dinitrogen pentoxide (N_2O_5) decomposes back to $NO_2 + NO_3$, so that (11) is in fact an equilibrium reaction. The seven most recent measurements reported for the equilibrium constant for reactions (11, −11) are given in Table 8.4.

Graham and Johnston measured, in a complex kinetic system, the product of $K_{11,-11}$ and the rate constant for the reaction of O_3 with NO_2. The value of $K_{11,-11}$ they originally reported and the slightly lower value derived from their data using the currently recommended value for the $O_3 + NO_2$ rate constant are both given in Table 8.4.

That of Malko and Troe (1982) is based on an analysis of the kinetic data in the literature for the forward and reverse reactions. The room temperature value of Tuazon et al. (1984a) was the first direct determination based on measuring the concentrations of NO_2, NO_3, and N_2O_5 in an equilibrium mixture and is in good agreement with the estimate of Kircher et al. (1984) based on their measurements of the rate constant for reaction (11) and literature values of k_{-11}. Like Kircher et

TABLE 8.4. Reported Values of the Equilibrium Constant for the Reaction NO_3 + $NO_2 \rightleftarrows N_2O_5$ at 298°K and Its Temperature Dependence

$K_{11,-11}$ at 298°K (cm^3 $molecule^{-1}$)	Temperature Dependence	Reference
2.3×10^{-11} [a]	$1.19 \times 10^{-27}\ e^{11,178/T}$	Graham and Johnston, 1978
2.2×10^{-11} [b]	$1.07 \times 10^{-27}\ e^{11,194/T}$	Graham and Johnston, 1978
1.87×10^{-11}	$1.33 \times 10^{-27}\ (T/300)^{0.32}\ e^{11,080/T}$	Malko and Troe, 1982
3.44×10^{-11}	n.d.[c]	Tuazon et al., 1984a
3.26×10^{-11}	$9.39 \times 10^{-28}\ e^{11,350/T}$	Kircher et al., 1984
$(2.2 - 4.0) \times 10^{-11}$	n.d.	Smith et al., 1985
1.8×10^{-11}	n.d.	Perner et al., 1985
2.5×10^{-11}	$8.13 \times 10^{-29}\ e^{11,960/T}$	Burrows et al., 1985

[a] Reported by Graham and Johnston, 1978.

[b] Adjusted for more recent recommendation for the rate constant for O_3 + NO_2; see discussion in text.

[c] n.d. = Not determined.

al., Smith and coworkers (1985) measured the rate constant for the forward reaction (11) and estimated $K_{11,-11}$ using literature values for k_{-11}; the range for $K_{11,-11}$ given by these researchers (Table 8.4) reflects different literature values chosen for k_{-11}.

Perner and coworkers (1985) carried out laboratory studies of the equilibrium (11, −11) in which they measured the concentrations of NO_2 and NO_3 by DOAS (see Section 5-D-2) and N_2O_5 by FTIR (see Section 5-C-2). They report six values for $K_{11,-11}$ in mixtures with NO_2 and O_3 in nitrogen; in four of the six experiments either formaldehyde or acetaldehyde was present. This group also derived values for the equilibrium constant from atmospheric measurements of the NO_2 and NO_3; due to the greater uncertainty in these values and the lower temperatures at which they were obtained, only the average of the six laboratory measurements is given in Table 8.4. Burrows and coworkers (1985) also measured the NO_2, NO_3, and N_2O_5 concentrations directly in a laboratory system using UV/visible absorption spectrometry.

The wide range in the reported values of $K_{11,-11}$ at room temperature is interesting, particularly in those systems where NO_2, NO_3, and N_2O_5 were all directly measured.

One cautionary note is appropriate regarding the values of the equilibrium constant which were determined by measuring the nitrate radical concentration using its visible absorption; this lies in the larger absorption cross section for NO_3 at 662 nm reported recently by Sander (1985) (see Section 3-C-7b). If these larger cross sections prove to be correct, then in the equilibrium constant studies the absorption cross sections used to convert the experimentally determined absorbances into concentrations are too small, the calculated nitrate radical concentrations will be too large and the equilibrium constant reported will be too small. In that case, the values of the equilibrium constant reported by Tuazon et al. (1984a), Perner et al. (1985), and Burrows et al. (1985) as cited in Table 8.4 will have to be increased by approximately 25%.

Elucidating the correct value for $K_{11,-11}$ is especially important in light of the dependence of most of the rate constants reported to date for nitrate radical reactions with organics on this equilibrium constant (see Section 7-C-2).

The rate constants for the forward reaction (11) of NO_2 with NO_3 and the reverse thermal decomposition of N_2O_5, reaction (−11), have been determined in a number of studies. Although at 1 atm pressure, the forward reaction (11) approaches the high-pressure limit, the falloff curves must be used to estimate k_{11} at atmospheric pressure. A recent study of k_{11} reported the low- and high-pressure limiting rate constants [see Eq. (C), Chapter 4] k_0 and k_∞ in N_2 as $k_0(T) = 4.5 \times 10^{-30}\ (T/300)^{-3.4}$ cm^6 molecule^{-2} s^{-1}, and $k_\infty(T) = 1.65 \times 10^{-12}\ (T/300)^{-0.4}$ cm^3 molecule^{-1} s^{-1} (Kircher et al., 1984), based on F_C obtained from $F_C = 0.565 - 0.000697T$. At 1 atm pressure and 298°K, the rate constant $k_{11} = 1.3 \times 10^{-12}$ cm^3 molecule^{-1} s^{-1}, assuming O_2 and N_2 have the same third-body efficiencies. The work of Kircher et al. (1984) is recommended by Atkinson and Lloyd (1984) for k_{11} as a function of temperature and pressure. Recent work on reaction (11) is in good agreement with this (Borrell et al., 1985; Smith et al., 1985; Burrows et al., 1985). The rate constant for the reverse reaction (−11) can then be obtained from $K_{eq} = k_{11}/k_{-11}$ using the expressions for K_{eq} in Table 8.4.

The reaction of NO_2 with NO_3 can also yield the products $NO + NO_2 + O_2$ [reaction (12)]. At 298°K, $k_{12} = 4.0 \times 10^{-16}$ cm^3 molecule^{-1} s^{-1} (Graham and Johnston, 1978). Thus this channel is approximately four orders of magnitude slower than that giving N_2O_5 and hence is relatively unimportant; it does, however, place an upper limit on the NO_3 lifetime.

NO_3 concentrations up to 430 ppt have been observed at night in heavily polluted urban atmospheres (Atkinson et al., 1985). Under these conditions the lifetime of NO_2 with respect to reaction with NO_3 would be only about 1 min. Thus at night when photolysis does not occur and OH and NO concentrations are low, but O_3 is present, reaction (10) forming NO_3 and (11) forming N_2O_5 can be very important.

Reaction (13) of NO_2 with ground-state oxygen atoms has a two-body route producing $NO + O_2$ and a three-body path producing NO_3. The rate constant for the bimolecular reaction at 300°K is $k_{13a} = 9.3 \times 10^{-12}$ cm^3 molecule^{-1} s^{-1}, while that for the termolecular reaction is $k_0 = 9 \times 10^{-32}$ cm^6 molecule^{-2} s^{-1} and $k_\infty = 2.2 \times 10^{-11}$ cm^3 molecule^{-1} s^{-1} (DeMore et al., 1983); at 1 atm pressure,

$k_{13b} = 2.0 \times 10^{-12}$ cm^3 molecule^{-1} s^{-1} (Atkinson and Lloyd, 1984). Although the concentration of $O(^3P)$ has not been measured in ambient air, it is expected, based on calculations, to peak at $\sim 8 \times 10^4$ cm^{-3} under conditions typical of a moderately to heavily polluted urban atmosphere (see Table 8.5). The lifetime of NO_2 with respect to reaction (13) with $O(^3P)$ at this concentration is ~ 13 days, too slow to be of significance in the troposphere.

Reaction (14) of NO_2 with HO_2 forms peroxynitric acid. However, this compound is thermally unstable and rapidly decomposes back to $HO_2 + NO_2$. The effective bimolecular rate constants for the formation of HO_2NO_2 and for its decomposition recommended by Atkinson and Lloyd (1984) for 1 atm in air and 298°K are based on the work of Graham et al. (1977, 1978) and Sander and Peterson (1984):

$$k_{14} = 1.4 \times 10^{-12} \text{ cm}^3 \text{ molecule}^{-1} \text{ s}^{-1}$$

$$k_{-14} = 0.085 \text{ s}^{-1}$$

Modeling calculations suggest peak HO_2 concentrations under moderate to heavily polluted conditions of $\sim 2 \times 10^9$ cm^{-3} (see Table 8.3). Considering only the forward reaction, this would correspond to an NO_2 lifetime of ~ 6 min. However, the lifetime of the HO_2NO_2 with respect to decomposition back to $HO_2 + NO_2$ under the same conditions is only ~ 12 s. Thus HO_2NO_2 in effect falls apart as rapidly as it is formed at these temperatures and hence is of no consequence in terms of removal of NO_2 from the atmosphere. However, as discussed in more detail below, at lower temperatures, HO_2NO_2 can provide a storage mechanism for NO_2, regenerating it when the temperature rises.

The reaction of alkylperoxy radicals (RO_2) with NO_2 forms alkylperoxy nitrates, RO_2NO_2. However, like HO_2NO_2, these compounds are thermally unstable and decompose back to reactants. Thus Atkinson and Lloyd (1984) recommend, at 1 atm in air at 298°K, an effective second-order rate constant of $k_{15} = 7 \times 10^{-12}$ cm^3 molecule^{-1} s^{-1} for all but methyl peroxy radicals. CH_3O_2 is sufficiently small that the reaction with NO_2 is in the fall off region between second and third order; Atkinson and Lloyd recommend for R = CH_3 that $k_0 = 2.2 \times 10^{-30}$ cm^6 molecule^{-2} s^{-1} and $k_\infty = 7 \times 10^{-12}$ cm^3 molecule^{-1} s^{-1} at 298°K, so that $k_{15} = 4.7 \times 10^{-12}$ cm^3 molecule^{-1} s^{-1}. However, the lifetime of RO_2NO_2 is expected to be ≤ 1 s at 298°K and 1 pressure, so that this reaction is also not a permanent sink for NO_2.

Finally, NO_2 can react with alkoxy radicals to form alkyl nitrate ($RONO_2$) or, if an abstractable hydrogen is available, nitrous acid (HONO) and a carbonyl compound. As discussed in Section 7-D-3e, the abstraction route (16b), appears to be negligible at room temperature, although more work is needed to confirm this. The effective bimolecular rate constant at 1 atm in air recommended by Atkinson and Lloyd (1984) is 1.5×10^{-11} cm^3 molecule^{-1} s^{-1}. At an alkoxy radical concentration of 1×10^4 cm^{-3} (Table 8.3) predicted for a moderate to heavily polluted atmosphere, the lifetime of NO_2 with respect to reaction (16) is ~ 77 days. Thus,

TABLE 8.5. Peak Rates of NO_2 Loss by Various Reactions in the Troposphere Under Moderate to Heavily Polluted Urban Atmosphere Conditions

Reactant or Loss Process	k^a (cm^3 molecule^{-1} s^{-1})	Peak $[X]^b$ (cm^{-3})	$\tau_{NO_2}{}^c$	Comments
Photolysis	$\leqslant 9 \times 10^{-3d}$	—	$\geqslant$2 min	Depends on latitude, season, solar zenith angle, etc.; see Chapter 3
OH	1.1×10^{-11}	5×10^6	5 h	Major source of nitric acid in troposphere; occurs in daytime
O_3	3.2×10^{-17}	2×10^{12}	4 h	Important nighttime reaction of NO_x
NO_3	1.3×10^{-12}	1.1×10^{10}	1 min	Lifetime is lower limit since equilibrium is attained
$O(^3P)$	1.1×10^{-11}	8×10^4	13 days	Unimportant in troposphere
HO_2	1.4×10^{-12}	2×10^9	6 min	Not a permanent sink for NO_2 since product thermally decomposes back to reactants
RO_2	7×10^{-12}	3×10^9	48 s	Not a permanent sink for NO_2 since product thermally decomposes back to reactants
RO	1.5×10^{-11}	1×10^4	77 days	Too slow to be significant sink for NO_2

[a] Effective bimolecular rate constant at 298°K in 1 atm of air.

[b] Peak concentrations of X predicted using a computer model with an initial reactive hydrocarbon concentration of 0.60 ppm and an initial NO_x of 0.060 ppm; see Section 7-B.

[c] Lifetime of NO_2 with respect to reaction with each X; $\tau_{NO_2} = 1/k_x[X]$.

[d] Units of photolytic rate constant are s^{-1}.

this reaction is too slow to contribute significantly to the removal of NO_2 from the troposphere.

Table 8.5 summarizes the relative importance of these loss processes for NO_2. Kinetically, the most important processes which lead to a permanent loss of NO_2 (rather than acting as a temporary storage as HO_2NO_2 and RO_2NO_2 do) are photolysis and reaction with OH during the day, and at night reaction with O_3 and NO_3.

A possible non-photochemical reaction of NO_2, in addition to those in Table 8.3, which may occur in marine environments, is the reaction of NO_2 with NaCl (Finlayson-Pitts, 1983). Stoichiometrically, this reaction may be represented as

$$2\ NO_2 + NaCl \rightarrow ClNO + NaNO_3 \tag{18}$$

although it undoubtedly occurs in several steps involving single collisions of NO_2 with the surface of an NaCl particle. This is discussed in more detail in Section 12-D-1. However, it is not known how fast reaction (18) is nor how the presence of water vapor and co-pollutants affect it.

8-A-1c. Leighton Relationship Between NO, NO_2, and O_3

In an atmosphere containing only NO, NO_2, and air, that is, no organics, the reactions controlling the concentrations of NO and NO_2 are (8), (17), and (4):

$$NO_2 + h\nu\ (\lambda \lesssim 430\ \text{nm}) \rightarrow NO + O(^3P) \tag{8}$$

$$O(^3P) + O_2 \xrightarrow{M} O_3 \tag{17}$$

$$O_3 + NO \rightarrow NO_2 + O_2 \tag{4}$$

Reactions (8), (17), and (4) are the basis of what is commonly known as the *Leighton relationship*, after Philip Leighton who wrote the definitive monograph *Photochemistry of Air Pollution*, published in 1961:

$$\frac{[O_3][NO]}{[NO_2]} = \frac{k_8}{k_4}$$

According to this relationship, the ratio of concentrations of O_3, NO, and NO_2 at any time t in an air mass should be a constant given by ratio of rate constants for photolysis of NO_2 and for the reaction of NO with O_3. Since k_8 changes with the angle of the sun (see Chapter 3), this ratio of concentrations is also expected to change during the day.

The Leighton relationship can be easily derived assuming that reactions (8), (17), and (4) describe the chemistry in an air mass and that a steady state is set up in which O_3 and oxygen atoms are being continuously formed and destroyed. If O_3 is indeed in a steady state, then its concentration is not changing with time and Eq. (A) holds (see Chapter 4):

$$\frac{d[O_3]}{dt} = 0 = k_{17}[O][O_2][M] - k_4[NO][O_3] \quad (A)$$

The steady state concentration of O_3 is thus given by

$$[O_3] = \frac{k_{17}[O][O_2][M]}{k_4[NO]} \quad (B)$$

If the oxygen atom concentration can also be considered to be in a steady state, then

$$\frac{d[O]}{dt} = 0 = k_8[NO_2] - k_{17}[O][O_2][M]$$

Hence

$$[O] = \frac{k_8[NO_2]}{k_{17}[O_2][M]} \quad (C)$$

Substituting Eq. (C) into Eq. (B) one obtains Eq. (D):

$$[O_3] = \frac{k_8[NO_2]}{k_4[NO]} \quad (D)$$

or rearranging,

$$\frac{[O_3][NO]}{[NO_2]} = \frac{k_8}{k_4} \quad (E)$$

The Leighton relationship is more than simply an interesting result of the three basic reactions responsible for O_3 levels in the troposphere. It also has a practical application in that, if it holds, it can be used in computer models of tropospheric chemistry to cut down on computation time. Thus, instead of carrying out numerical integration procedures separately to obtain $[O_3]$, [NO], and $[NO_2]$, if two of the three concentrations are known, the third can be obtained using Eq. (E).

This relationship has been tested in a number of field and theoretical modeling studies [see Calvert and Stockwell (1983) for a summary of these]. It appears to hold under many, although not all, conditions. Deviations occur when loss processes for O_3 other than reaction (4) become significant so that O_3 is no longer in a steady state. The additional loss processes include reactions with NO_2, alkenes, and the radicals HO_2 and OH (see Chapters 7, 8, and 11). In addition, photolysis of O_3 to produce electronically excited oxygen atoms, $O(^1D)$, which react further with water vapor to produce OH can contribute to the deviations from Eq. (E) observed in the field. Finally, at sunset and sunrise, deviations are expected because the rate of photolysis of NO_2 is sufficiently low that the steady state assumption on oxygen atoms and O_3 is not valid (Calvert and Stockwell, 1983).

In general, the Leighton relationship is expected to hold when sufficient NO is present that reaction (4) is the major loss for O_3, for example, at relatively low hydrocarbon to NO ratios.

8-A-2. Nitrate Radical, NO_3, Dinitrogen Pentoxide, N_2O_5, and Dinitrogen Trioxide, N_2O_3

8-A-2a. NO_3

The nitrate radical is formed from the reaction (10) of NO_2 with O_3:

$$NO_2 + O_3 \rightarrow NO_3 + O_2 \tag{10}$$

In the mid-1970's, Niki and coworkers first showed in laboratory experiments that gaseous nitrate radicals reacted rapidly with simple olefins and aldehydes (see Section 7-E-1), and commented on the atmospheric implications of their findings. Thus they estimated that in dry smog, peak levels of NO_3 could reach as high as 0.01% of the O_3 concentration, and noted that "with the large difference in the rate constants for O_3 and NO_3 it becomes apparent that olefin consumption by both species are roughly comparable" (Morris and Niki, 1974; Japar and Niki, 1975).

The direct spectroscopic observation of the nitrate radical in polluted urban air and at concentrations up to 350 ppt (Platt et al., 1980a), when taken with the rate constants of Niki and coworkers, provided the first experimental proof that the nitrate radical did indeed play a key role in the nighttime tropospheric chemistry of simple olefins. Furthermore, since NO_3 radicals were found to react very rapidly with phenols and cresols, this process was clearly a major nighttime sink for these pollutants and one which transformed NO_3 into nitric acid (Carter et al., 1981c).

Finally, on the basis of the known reactions in simulated polluted atmospheres, and the actual ambient concentrations of NO_3 radicals just measured, it was suggested that hydrolysis of N_2O_5 (formed from $NO_2 + NO_3$) to produce nitric acid was a sink for NO_3 (Platt et al., 1980a), and that it could "contribute to the decrease in the pH of precipitation associated with NO_x pollution and represent an important nighttime sink for NO_x" (Pitts et al., 1981). Since then, this has been confirmed in a number of experimental as well as theoretical modeling studies (e.g., see Jones and Seinfeld, 1983; Richards, 1983; Russell et al., 1985).

Once formed, NO_3 can react with NO_2

$$NO_2 + NO_3 \overset{M}{\rightleftarrows} N_2O_5 \tag{11}$$

$$\rightarrow NO + NO_2 + O_2 \tag{12}$$

or with NO

$$NO + NO_3 \rightarrow 2\,NO_2$$

$$k_{19} = 3.0 \times 10^{-11}\ \text{cm}^3\ \text{molecule}^{-1}\ \text{s}^{-1} \tag{19}$$

In addition, as discussed in Chapters 1, 7, 11, and 14, NO_3 has been shown to react rapidly with organics, including alkenes, cresols, and biogenics such as dimethylsulfide. Finally, as discussed in Section 3-C-7b, it photolyzes rapidly:

$$NO_3 + h\nu\ (\lambda \lesssim 670\ \text{nm}) \xrightarrow{a} NO + O_2$$
$$\xrightarrow{b} NO_2 + O(^3P) \qquad (20)$$

As a result, detectable concentrations of NO_3 only build up at night when O_3 and NO_2 are both present and when the concentrations of species with which it can react rapidly, such as NO and organics, are low.

In and near urban areas if there are sufficient sources of NO at the earth's surface at night that O_3 cannot build up due to reaction (4)

$$NO + O_3 \rightarrow NO_2 + O_2 \qquad (4)$$

then any NO_3 formed would be rapidly destroyed via (19). For example, at an NO concentration of 5 ppb, the lifetime of NO_3 with respect to reaction (19) is only 0.3 s! However, as discussed in Section 1-D-2 (see Fig. 1.6) and Richards (1983), O_3 is known to be present at higher elevations above urban areas at night; in these regions NO cannot co-exist due to the rapid reaction (4) and hence NO_3 formation can occur in such regions aloft. In cleaner regions where the NO emissions are small, O_3 and hence NO_3 may also build up close to the earth's surface.

The mechanisms and rate constants for the reactions of NO_3 with alkanes, alkenes, aromatics, and aldehydes are discussed in Chapter 7, and those for biogenics, including dimethyl sulfide, are found in Chapter 14. For all but the alkenes and sulfides, the reactions appear to involve hydrogen abstraction to form HNO_3,

$$RH + NO_3 \rightarrow R + HNO_3 \qquad (21)$$

one way in which the nitrate radical plays a role in acid deposition. The relative contribution of these reactions to the total formation of HNO_3 in the troposphere is discussed in detail in Section 11-B.

The lifetime of NO_3 in ambient air has been observed to depend on the relative humidity (RH), falling rapidly to $\lesssim 10$ min at RH $\gtrsim 50\%$ (Platt et al., 1984; see Section 11-B-1). Platt and co-workers suggest that this could be due to the reaction of NO_3 directly with liquid water present on surfaces and in ambient particles, a reaction suggested earlier by Chameides and Davis (1983). This is supported by modeling studies which suggest that the *sticking coefficient*, that is, the fraction of collisions with the surface which result in sorption of the molecule from the gas phase, need only be $\gtrsim 10^{-3}$ for this to be an important process (Heikes and Thompson, 1983; Seigneur and Saxena, 1984; Russell et al., 1985). Alternatively, the reaction of N_2O_5 with water discussed below may at least partially explain this relationship between NO_3 and relative humidity since NO_3 and N_2O_5 are in equilibrium, reaction (11), which can shift rapidly as one of the reacting species is removed.

Several other reactions of NO_3 have been suggested (Cantrell et al., 1985), including thermal decomposition at ambient temperatures,

$$NO_3 \xrightarrow{M} O_2 + NO$$

and reaction with HO_2

$$NO_3 + HO_2 \rightarrow HNO_3 + O_2$$

Computer simulation to match product concentration–time curves in a complex O_3–NO_2–HCHO–air mixture suggested that the NO_3–HO_2 reaction has a rate constant of $\sim(2.5 \pm 1.5) \times 10^{-12}$ cm^3 molecule^{-3} s^{-1} (Cantrell et al., 1985). In addition, they suggest that the NO_3 thermal decomposition could occur with a rate constant as large as 2×10^{-3} s^{-1} at one atmosphere pressure and 298°K, which corresponds to a half-life of only 8 min. This seems to be a rather large rate constant since N_2O_5 in equilibrium with NO_3 and NO_2 in one atmosphere of air in laboratory systems has been observed to decay with a rate constant of $\sim 1.5 \times 10^{-4}$ s^{-1} (Atkinson et al., 1984b), an order of magnitude slower than they propose NO_3 would thermally decay. Clearly both reactions need further experimental investigation.

8-A-2b. N_2O_5

As discussed in Section 8-A-1b, NO_3, NO_2, and N_2O_5 are in equilibrium:

$$NO_2 + NO_3 \overset{M}{\rightleftarrows} N_2O_5 \qquad (11,\text{-}11)$$

The equilibrium constant for this reaction is uncertain; Table 8.4 shows seven reported values which span a factor of 1.9 at room temperature. The equilibrium is established rapidly ($\lesssim 1$ min) at 298°C and 1 atm, so that any loss process for N_2O_5 is in effect a loss process for NO_3; thus, if N_2O_5 is removed, the equilibrium (11,-11) will rapidly shift to the right, consuming NO_3.

From observed concentrations of NO_2 and NO_3 in ambient air Atkinson et al. (1985) have estimated the maximum nighttime N_2O_5 concentrations that would be in equilibrium with them. Concentrations of N_2O_5 as high as 15 ppb were calculated. Since N_2O_5 can hydrolyze to nitric acid (see below) or react slowly with polycyclic aromatic hydrocarbons (see Chapter 13), the significant predicted concentrations suggest that N_2O_5 may play an important role in tropospheric chemistry.

A major loss process for N_2O_5 in ambient air, in addition to its decomposition (-11) back to $NO_2 + NO_3$, may be reaction with water:

$$N_2O_5 + H_2O \rightarrow 2\ HNO_3 \qquad (22)$$

This reaction was first directly studied in the gas phase by Morris and Niki (1973). From the decay rates of N_2O_5 in the presence of varying water vapor concentrations, an upper limit to the rate constant $k_{22} \leq 1.3 \times 10^{-20}$ cm^3 molecule^{-1} s^{-1} was derived. In a more recent study in which the N_2O_5 decay was monitored and the formation of gas phase HNO_3 also followed, it was found that less than two molecules of HNO_3 were formed in the gas phase for each molecule of N_2O_5 which disappeared (Tuazon et al., 1983a). This was attributed to the contribution of a homogeneous, gas phase contribution to reaction (22), as well as a heterogeneous component which occurred on the surfaces of the reaction chamber. An upper limit for the homogeneous rate constant k_{22} was found to be $(1.3 \pm 0.2) \times 10^{-21}$ cm^3 molecule^{-1} s^{-1}, a factor of 10 lower than the earlier value.

Even with this very low rate constant for the homogeneous hydrolysis of N_2O_5, reaction (22) can still be a significant source of HNO_3 in the troposphere. Relative rates of HNO_3 formation via the reaction of OH with NO_2, NO_3 abstractions from organics, and N_2O_5 hydrolysis are discussed in detail in Section 11-B. However, it appears now that HNO_3 formation via N_2O_5 hydrolysis may be comparable to the OH + NO_2 reaction or, under some conditions, possibly exceed it (Russell et al., 1985).

8-A-2c. N_2O_3

N_2O_3 can be formed under laboratory conditions at high concentrations both in the gas phase, via reaction (23),

$$NO + NO_2 \rightleftarrows N_2O_3 \tag{23}$$

$$K_{eq} = 0.53 \text{ atm}^{-1} \quad \text{(Stull and Prophet, 1971)}$$

and in solution, via reaction (24),

$$2HONO_{(aq)} \rightleftarrows N_2O_3 + H_2O \tag{24}$$

$$K_{eq} = 3.0 \times 10^{-3} \text{ L mole}^{-1} \quad \text{(Markovits et al., 1981)}$$

Under laboratory conditions, N_2O_3 reacts rapidly with organics in solution and perhaps in the gas phase (Williams, 1983), forming, for example, nitroso-nitrite derivatives ($-\underset{NO}{\overset{|}{\underset{|}{C}}}-\underset{ONO}{\overset{|}{\underset{|}{C}}}-$) with alkenes. However, concentrations of NO, NO_2, and HONO under atmospheric conditions are too low for reactions (23) and (24) to be significant sources of N_2O_3 in the gas phase or in atmospheric water droplets. For example, even at the relatively high concentrations of 0.1 ppm NO and 0.1 ppm NO_2, the equilibrium gas phase concentration of N_2O_3 is only $\sim 5 \times 10^{-3}$ ppt. At one of the highest gas phase HONO concentrations observed to date in the atmosphere, 8 ppb (Section 5-G), the solution phase concentration of HONO in equilibrium with it would be only $\sim 4 \times 10^{-7}$ mole L^{-1}, using a Henry's law constant for physical solubility (exclusive of acid–base equilibria) of $H = 49$ moles L^{-1} atm^{-1} (Table 11.3, Chapter 11). At this HONO concentration, the N_2O_3 in equilibrium with it via reaction (24) would only be $\sim 5 \times 10^{-16}$ moles L^{-1}. Even with very large rate constants for the reaction of N_2O_3 with organics in solution, the reactions of other species such as OH will likely predominate (see Section 11-A-2).

In short, unless there are unrecognized sources of N_2O_3 in the atmosphere, this species is unlikely to play a significant role in tropospheric chemistry.

8-A-3. Nitrogen-Containing Acids: Nitrous Acid, HONO, Nitric Acid, HNO_3, and Peroxynitric Acid, HO_2NO_2

8-A-3a. HONO

As discussed in Section 3-C-5, nitrous acid photolyzes very rapidly and is a major source of OH radicals at dawn in polluted urban areas (Section 3-D-1). It is believed at the present time that all other loss processes are negligible compared to photolysis. Although OH reacts rapidly with HONO,

$$OH + HONO \rightarrow H_2O + NO_2$$

$$k_{25} = 6.6 \times 10^{-12}\ \text{cm}^3\ \text{molecule}^{-1}\ \text{s}^{-1} \quad \text{(Cox et al., 1976)} \qquad (25)$$

significant concentrations of OH are only present when photolysis occurs to produce it, but under these same conditions, HONO itself photolyzes. For example, at solar zenith angles of 0° and 86°, respectively, the lifetimes for HONO photolysis are only ~10 min and 5 h, respectively. For OH concentrations of 5×10^5 cm^{-3} and 5×10^6 cm^{-3}, respectively, the lifetime of HONO with respect to reaction (25) is 3.5 days and 8 h, respectively.

Ambient air measurements of gaseous nitrogenous species suggest that NO_3 may react with HONO, in part because the time–concentration profiles for HONO and NO_3 were observed to have an inverse relationship, that is, peak concentrations of the two did not co-exist (Pitts et al., 1984b, 1985b). While abstraction of a hydrogen atom from HONO by NO_3 seems reasonable by analogy to NO_3–organic reactions (Chapter 7), this process has not as yet been confirmed experimentally.

The mechanisms by which HONO is formed in the atmosphere are not well established at the present time, although concentrations up to 8 ppb have been observed in polluted urban atmospheres (Platt et al., 1980b; Harris et al., 1982). The recombination of OH with NO,

$$OH + NO \xrightarrow{M} HONO \qquad (26)$$

is in the falloff region between second and third order at 1 atm pressure and 298°K; Atkinson and Lloyd recommend that under these conditions, $k_{26} = 6.6 \times 10^{-12}$ cm^3 molecule^{-1} s^{-1}. Reaction (26) thus produces a small steady state HONO concentration because of the rapid photolysis of HONO. Two other possible reactions are the surface-catalyzed reactions (27) and (28):

$$NO + NO_2 + H_2O \rightleftarrows 2\ HONO \quad (27)$$

$$2\ NO_2 + H_2O \rightleftarrows HONO + HNO_3 \quad (28)$$

Lee and Schwartz (1981) have shown that at the concentrations of NO and NO_2 present in the atmosphere, if the water is in the form of liquid droplets, then reactions (27) and (28) in the liquid phase are too slow to contribute to NO and NO_2 removal. The gas phase rates are also slow. However, there is evidence that reactions with the overall stoichiometry shown may occur if surfaces (e.g., of the reaction vessel) are present to *catalyze* the reactions. Thus reaction (27) has been observed in small laboratory systems and, indeed, is used as a source of HONO for spectroscopic calibrations and so on. The equilibrium constant is $K_{27} = 5.6 \times 10^{-20}$ cm^3 molecule^{-1} s^{-1} at 298°K (Chan et al., 1976). There is ample evidence that the reaction is heterogeneous in nature, that is, occurs largely on the surface of the reaction vessel, and hence the rate constants k_{27} and k_{-27} will vary from system to system. A study by Kaiser and Wu (1977) reaffirmed the contribution of a heterogeneous component and placed upper limits on the homogeneous part of k_{27}(homogeneous) $\leq 4.4 \times 10^{-40}$ cm^6 molecule^{-2} s^{-1} and k_{-27}(homogeneous) $\leq 1 \times 10^{-20}$ cm^3 molecule^{-1} s^{-1}, respectively.

Reaction (28) has been suggested to explain the observation of the formation of gas phase HONO in environmental chambers in the dark when both NO_2 and H_2O are present (Sakamaki et al., 1983; Pitts et al., 1984a). Thus HONO production was observed, and its rate of formation was approximately first order in the NO_2 concentration and in H_2O. The yields of HONO were $\leq$50% of the NO_2 reacted; while NO was observed during the later stages of the reaction, ~50% of the NO_2 reacted could not be detected in the gas phase products, suggesting that a nitrogen-containing product such as HNO_3 remained adsorbed to the chamber surface. However, this is speculation at present, and there is no definitive evidence for HNO_3 formation. Until such confirmation of HNO_3 formation is obtained, the occurrence of reaction (28) as written must be considered tentative.

NO_2 at concentrations of 5–12 ppm released into a mobile laboratory at a RH of 31–60% has also been shown to lead to the formation of HONO at a rate of 0.25 ppb min^{-1} per ppm of NO_2 (Pitts et al., 1985a). Rather surprisingly, this rate of HONO formation, normalized to the NO_2 concentration, is within the same order of magnitude as that observed in environmental chambers. With some assumptions concerning ventilation rates of HONO decay in typical homes, Pitts and co-workers suggest that steady state concentrations of HONO as high as 15 ppb, double the peak observed to date in ambient air (see Section 5-G), could be formed indoors if the NO_2 concentration is 1 ppm. Since NO_2 concentrations this high can be attained due to emissions from heating and cooking, it is clear that HONO may be formed as a secondary indoor air pollutant. This is of particular concern because HONO is known to form carcinogenic nitrosamines upon reaction with secondary amines.

A fourth mechanism for HONO formation in urban atmospheres has been hypothesized recently by Stockwell and Calvert (1983) on the basis of modeling studies. They propose that the reaction of NO_3 with formaldehyde produces HO_2 through the well-known reaction sequence

$$NO_3 + HCHO \rightarrow HNO_3 + HCO \quad (29)$$

$$HCO + O_2 \rightarrow HO_2 + CO \quad (30)$$

and that HONO is formed in the subsequent reaction of HO_2 with NO_2:

$$HO_2 + NO_2 \rightarrow HONO + O_2 \quad (31)$$

In studies of the $HO_2 + NO_2$ reaction, this path, reaction (31), producing HONO + O_2 was shown to be negligible compared to the addition reaction (14) giving HO_2NO_2. Upper limits to the rate constant k_{31} of 3×10^{-15} cm^3 molecule^{-1} s^{-1} (Howard, 1977) and 1×10^{-15} cm^3 molecule^{-1} s^{-1} (Graham et al., 1977, 1978), respectively, have been assigned.

Stockwell and Calvert used $k_{31} = 3 \times 10^{-15}$ cm^3 molecule^{-1} s^{-1} and suggest that, even with this relatively small rate constant, reaction (31) may explain much of the observed HONO in ambient air. However, based on measured concentrations of HONO and NO_3 in ambient air, Pitts et al. (1984b) have shown that the Stockwell–Calvert mechanism predicts HONO concentrations approximately an order of magnitude below those observed (Pitts et al., 1984b). Further modeling studies by Killus and Whitten (1985) have suggested that the ambient HONO data can be satisfactorily rationalized without reaction (31).

Two other mechanisms of HONO formation have been suggested by Heikes and Thompson (1983): (1) the abstraction of allylic hydrogen atoms from unsaturated hydrocarbons by NO_2 and (2) decomposition of peroxyacetyl nitrate (PAN) to the nitrite ion (NO_2^-), followed by volatilization to the gas phase as HONO. Pryor and Lightsey (1981) showed that unsaturated lipids in solution undergo hydrogen abstraction at low (ppm) NO_2 concentrations,

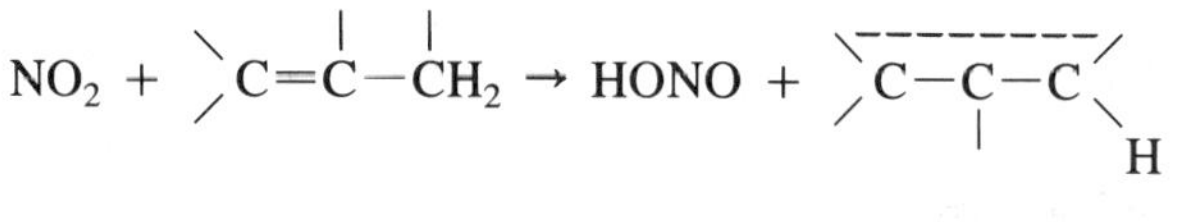
(32)

whereas NO_2 adds to the double bond at high concentrations. However, this appears not to occur significantly in the gas phase with the hydrocarbons typically present in ambient air (Atkinson et al., 1984a).

The alkaline hydrolysis of PAN has been known for years to produce the nitrite ion:

$$CH_3\overset{\overset{\displaystyle O}{\|}}{C}OONO_2 + 2\,OH^- \rightarrow CH_3\overset{\overset{\displaystyle O}{\|}}{C}O^- + O_2 + NO_2^- + H_2O \quad (33)$$

The studies of PAN hydrolysis are reviewed by Stephens (1969). However, there is no clear evidence at the present time that reaction (33) does occur to an appreciable extent in atmospheric droplets; indeed, this seems somewhat unlikely due to the generally acidic nature of atmospheric particles and fogs (see Chapters 11 and 12).

Unpublished studies indicate that PAN is soluble in acidic water samples and in rainwater and that it decays with lifetimes from a few minutes to over 1 h, depending on the temperature; however, the products appear to be nitrate and organics, rather than nitrite (Spicer et al., 1983b). This is consistent with recent data on PAN hydrolysis, (see Section 11-A-2a) in which a lifetime with respect to hydrolysis of ~ 40 min was reported, with NO_3^- as the product (Holdren et al., 1984).

In short, although the abstraction of allylic hydrogens by NO_2 and the hydrolysis of PAN cannot be ruled out as contributors to HONO formation at present, there is no clear quantitative experimental evidence for their occurrence.

At least a portion of the HONO observed in ambient air may be directly emitted from sources rather than being formed by chemical reactions in the air; that is, HONO may be in part a primary pollutant. Thus HONO has been observed in the exhaust of some automobiles which emit relatively high concentrations of NO_x (Pitts et al., 1984c).

In short, while HONO is clearly a key species in the troposphere through its photolysis to produce OH, its sources in ambient air are highly uncertain. This is clearly an area for further work.

8-A-3b. HNO_3

Mechanisms of formation of HNO_3 in the atmosphere include the reaction of OH with NO_2, hydrogen abstraction by NO_3 from organics, and N_2O_5 hydrolysis, which we discussed earlier in this chapter and in Section 11-B. Gaseous nitric acid has a relatively long *chemical* lifetime and, as a result, can act as a terminator for tropospheric chain oxidations. Thus, as we saw in Section 3-C-6, HNO_3 absorbs only weakly in the actinic UV and hence photolysis is slow. For example, using the absorption cross sections and actinic flux data in Chapter 3, the lifetimes of HNO_3 at solar zenith angles of 0° and 86°, respectively, can be calculated to be about 20 days and 11 y, respectively. However, nitric acid undergoes rapid deposition at the earth's surface (see Chapter 11-D-2) as well as absorption into water droplets, which contributes to a shorter overall lifetime in the atmosphere than expected from purely chemical processes.

The OH radical reacts with HNO_3,

$$OH + HNO_3 \rightarrow H_2O + NO_3 \tag{34}$$

but again the reaction is slow. While most recent studies (e.g. Smith et al., 1984; Jolly et al., 1985) have reported a room temperature rate constant of $\sim 1.4 \times 10^{-13}$ cm^3 molecule^{-1} s^{-1}, several others obtain a lower value, $\sim 8.5 \times 10^{-14}$ cm^3 molecule^{-1} s^{-1} (e.g., see Lamb et al., 1984; Connell and Howard, 1985;

Devolder et al., 1984). Even with the larger value, at an OH concentration of 1 $\times 10^6$ cm^{-3}, this corresponds to a lifetime for HNO_3 with respect to reaction (34) of ~136 days. Reaction (34) is interesting mechanistically, however, in that it has a negative activation energy and low preexponential factor; as reviewed by Smith et al. (1984) and Lamb et al., 1984, these can be rationalized in terms of the formation of a bound intermediate which can dissociate into products or back into reactants, or alternatively, in terms of an elementary bimolecular reaction to which transition state theory is applied with certain assumptions concerning the transition state.

Connell and Howard (1985) have shown that the alternate set of products in the OH + HNO_3 reaction, H_2O_2 + NO_2, account for an upper limit of 1% of the overall reaction at 300°K.

As discussed in more detail in Section 11-C, HNO_3 can react with gaseous ammonia to produce particulate ammonium nitrate:

$$HNO_3 + NH_3 \rightleftarrows NH_4NO_3 \tag{35}$$

Since this is an equilibrium reaction, HNO_3 can be reformed in the atmosphere when NH_4NO_3 dissociates back to reactants; this occurs rapidly for both the solid and solutions of NH_4NO_3 (Larson and Taylor, 1983). It is noteworthy that HNO_3 and the base NH_3 have been observed simultaneously in ambient air in a number of studies (e.g., see Doyle et al., 1979; Harrison and Pio, 1983; Hildemann et al., 1984); while the co-existence of an acid and a base may seem odd, the concentrations were in accord with what is expected from the known equilibrium constants for reaction (35).

Ammonium nitrate is deliquescent, so that at 25°C, and below 62% relative humidity, it exists as a solid, but at RH > 62%, it exists as a solution. The HNO_3–NH_3–NH_4NO_3 system has been treated thermodynamically to predict the NH_4NO_3 dissociation constant, K:

$$K = P_{HNO_3}P_{NH_3} \tag{F}$$

where P_{HNO_3} and P_{NH_3} are the equilibrium partial pressures of HNO_3 and NH_3 in the gas phase (e.g., see Stelson et al., 1979; Stelson and Seinfeld, 1982a,b). The magnitude of this dissociation constant and its dependence on temperature, relative humidity, and pH determines the concentrations of gaseous HNO_3 and NH_3 which can co-exist.

For solid NH_4NO_3 in equilibrium with gaseous HNO_3 and NH_3 (i.e., at RH < 62%), the dependence of the dissociation constant on the temperature T(°K) is given by Eq. (G) (Stelson and Seinfeld, 1982a):

$$\ln K = 84.6 - \frac{24{,}220}{T} - 6.1 \ln\left(\frac{T}{298}\right) \tag{G}$$

In Eq. (G), K is in units of (ppb)^2.

Above 62% RH at 25°C, ammonium nitrate exists in solution so that solution thermodynamics must be taken into account in deriving the equilibrium constant K. Figure 8.2 shows calculated values of K as a function of relative humidity for temperatures from 0 to 50°C (Stelson and Seinfeld, 1982a). It is seen that the dissociation constant is smaller for solution NH_4NO_3 than the solid; these researchers point out that this may explain why relatively more NH_4NO_3 evaporates from filters at low relative humidities compared to high ones. The discontinuity shown by the dotted line in Fig. 8.2 is not real, but they suggest likely results from some inconsistencies in the data used for the calculations.

Of course, for a *saturated* solution of NH_4NO_3 surrounding a solid NH_4NO_3 core, the equilibrium constant must be the same as for solid NH_4NO_3, Eq. (G), since the solution must be in equilibrium with both the solid and the gas.

The relationship between K and relative humidity does not change significantly with pH over the pH range 1–7 (Stelson and Seinfeld, 1982b).

Such calculations have been applied to conditions in ambient air to predict the product $K = P_{HNO_3}P_{NH_3}$ (e.g., see Doyle et al., 1979; Harrison and Pio, 1983; Hildemann et al., 1984). The experimentally measured product ($P_{HNO_3}P_{NH_3}$) may,

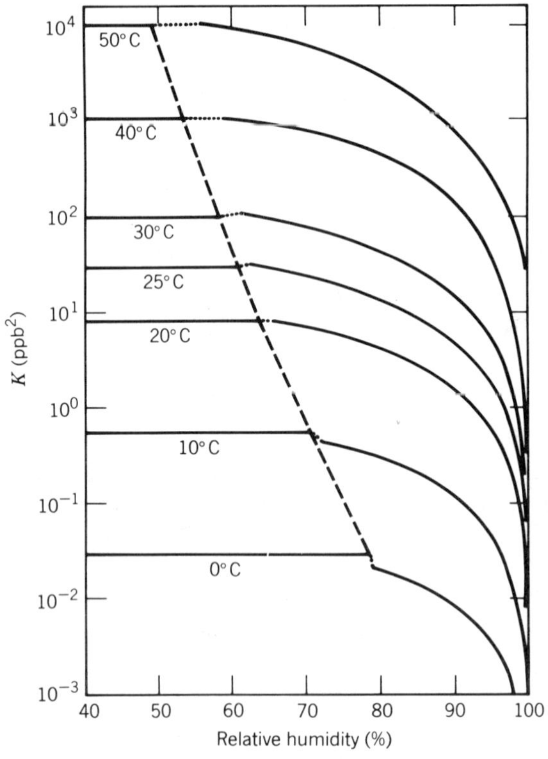

FIGURE 8.2. NH_4NO_3 dissociation constant as a function of relative humidity. Discontinuity at dotted line, which represents the relative humidity of deliquescence at the various temperatures, is not real, but due to artifacts in calculations (see text) (from Stelson and Seinfeld, 1982a).

of course, be smaller than the calculated value of K if insufficient HNO_3 and NH_3 exist to form NH_4NO_3. Figure 8.3, for example, shows the experimentally determined product of the HNO_3 and NH_3 concentrations and the dissociation constant calculated for the ambient conditions, for 2 days at one site in California. Good agreement between the calculated and measured values is seen on the first day, with the calculated value of K exceeding the measured concentration product at the peak on the second day.

Similarly, the ambient concentrations of gas phase HNO_3 and particulate nitrate measured simultaneously in some locations inland from the ocean have been shown to be in accordance with the thermodynamic predictions both below and above the deliquescence point of NH_4NO_3 (Grosjean, 1982). Samples taken near the ocean may not follow this (Hildemann et al., 1984) because of excess nitrate from the reaction of NaCl in sea salt aerosols with HNO_3 or NO_2 to form $NaNO_3$ (see Section 12-D-1).

Of course, in the real atmosphere, particles generally contain other species, particularly sulfate, in addition to NH_4NO_3. The results of calculations for these more complex systems are discussed in Section 12-D-2.

Because most homogeneous gas phase chemical reactions of HNO_3 in the troposphere are slow, it is removed from the atmosphere primarily by absorption into cloud or fog water or rainwater (e.g., see Durham et al., 1981; Jacob and Hoff-

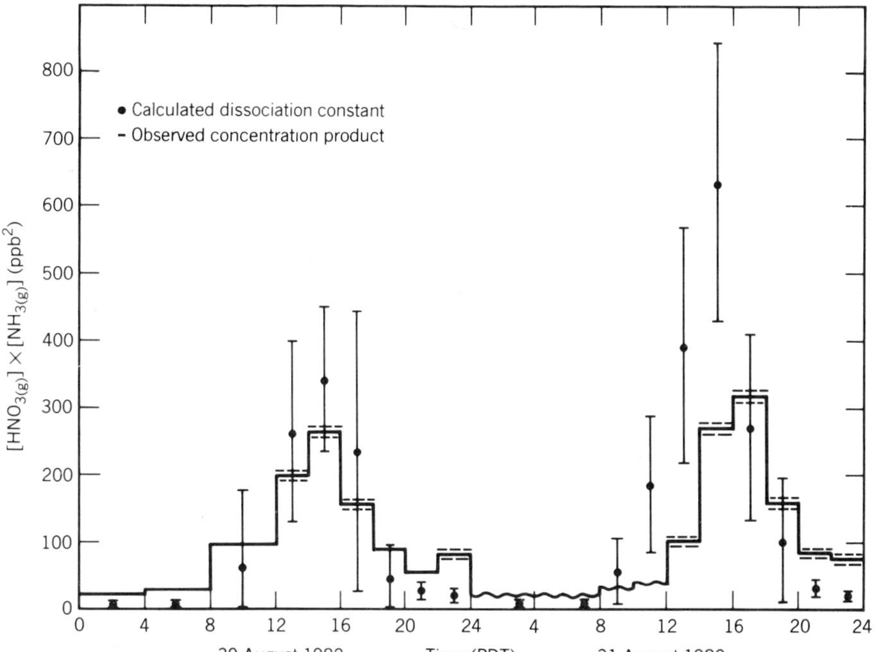

FIGURE 8.3. Experimentally measured concentration product $[HNO_3][NH_3]$ (in units of ppb^2) and calculated value of $K = [HNO_3][NH_3]$ for two days at Rubidoux, California (from Hildemann et al., 1984).

man, 1983) or by deposition at the earth's surface. The high solubility of HNO_3 ensures that the absorption processes are relatively fast, which contributes to the observation of significant concentrations of HNO_3, as well as H_2SO_4, in acid rain and fogs (see Chapter 11).

8-A-3c. HO_2NO_2

The only reaction known to form HO_2NO_2 is the reaction (14) of HO_2 with NO_2. As discussed earlier, the thermal decomposition of HO_2NO back to reactants is fast:

$$HO_2NO_2 \overset{M}{\rightleftarrows} HO_2 + NO_2 \qquad (-14, 14)$$

Atkinson and Lloyd (1984) recommend $k_{-14} = 1.3 \times 10^{14}\ e^{-10,418/T}\ s^{-1}$ at atmospheric pressure, based on the work of Graham et al. (1977, 1978). At 298°K, $k_{-14} = 0.085\ s^{-1}$, which corresponds to an HO_2NO_2 lifetime with respect to thermal decomposition of only ~12 s, much shorter than that for photolysis or reaction with OH.

However, at lower temperatures and pressures, the lifetime of HO_2NO_2 is significantly longer. Table 8.6 gives the rate constants and lifetimes of HO_2NO_2 at several temperatures, calculated using the expression above for k_{-14}. While the lifetime is short at room temperature, at the freezing point or below it is of the order of minutes to hours. As a result, in colder climates in the winter, HO_2NO_2 may act as a temporary storage reservoir for NO_2, releasing it as the temperature rises (e.g. see Dickerson, 1985).

As discussed in Section 3-C-7a, HO_2NO_2 absorbs light at $\lambda \lesssim 330$ nm; the absorption cross sections are sufficiently large that Molina and Molina (1981) es-

TABLE 8.6. Rate Constants[a] and Lifetimes for the Thermal Decomposition of HO_2NO_2 at Various Temperatures

Temperature (°C)	$k\ (s^{-1})$	τ^b
−30	3.1×10^{-5}	9 h
−20	1.7×10^{-4}	1.6 h
−10	8.1×10^{-4}	20 min
0	3.5×10^{-3}	5 min
10	1.3×10^{-2}	75 s
20	4.7×10^{-2}	21 s
30	1.5×10^{-1}	7 s
40	4.6×10^{-1}	2 s

[a] Using $k = 1.3 \times 10^{14}\ e^{-10,418/T}\ s^{-1}$ (see discussion in text).

[b] $\tau = 1/k$.

timate that the lifetime of HO_2NO_2 with respect to photolysis will be of the order of a day.

The OH radical reacts with HO_2NO_2. The products are assumed to be H_2O + O_2 + NO_2, although this has not been confirmed experimentally;

$$OH + HO_2NO_2 \rightarrow H_2O + O_2 + NO_2 \tag{36}$$

The rate constant and its temperature dependence have not been well established, likely due in part to the difficulty in obtaining pure samples of HO_2NO_2. At room temperature, $k_{36} = 5.2 \times 10^{-12}$ cm^3 molecule^{-1} s^{-1} has been reported recently (Smith et al., 1984; Lamb et al., 1984) and is in reasonable agreement with several earlier measurements. At an OH concentration of 1×10^6 cm^{-3}, the lifetime of HO_2NO_2 with respect to reaction (36) is ~2 days, similar to that for photolysis.

8-A-4. Ammonia, NH_3

Ammonia is produced by biological processes and hence exists in both clean and polluted atmospheres. It does not absorb light in the actinic UV and hence cannot photolyze. It can react with OH:

$$NH_3 + OH \rightarrow NH_2 + H_2O$$
$$k_{37}^{298°K} = 1.6 \times 10^{-13} \text{ cm}^3 \text{ molecule}^{-1} \text{ s}^{-1} \tag{37}$$

At an OH concentration of 1×10^6 cm^{-3}, the lifetime of NH_3 with respect to reaction (37) would be ~72 days.

The fate of the NH_2 radical formed is not known with certainty. Somewhat surprisingly, NH_2 does not undergo a three-body addition reaction with O_2 to form NH_2OO; thus an upper limit of 1.5×10^{-36} cm^6 molecule^{-2} s^{-1} has been placed on the rate constant for this reaction (Patrick and Golden, 1984). However, Patrick and Golden suggest, on thermochemical kinetic grounds, that a bimolecular reaction with O_2

$$NH_2 + O_2 \rightarrow OH + HNO \tag{38}$$

is energetically possible and could occur with a rate constant as high as 10^{-18} cm^3 molecule^{-1} s^{-1}. Based on the limited kinetic data available, Patrick and Golden suggest that, if this is the case, reaction (38) could be the major removal process for NH_2 in the troposphere; reactions with NO and NO_2 were suggested as possible competing processes at the high concentrations of polluted urban areas.

NH_3 plays a significant role in atmospheric chemistry because it is the only common base present in the atmosphere which is highly soluble. As discussed above with respect to HNO_3, it can neutralize this acid, forming ammonium nitrate. Similarly, it can be scavenged into atmospheric water droplets and neutralize acids such as H_2SO_4 contained or formed therein (Jacob and Hoffman, 1983); as a result, NH_4^+ is a very common constituent found in atmospheric aerosols and fogs.

8-B. ORGANIC NITROGENOUS COMPOUNDS

8-B-1. Peroxyacetyl Nitrate and Its Higher Homologues

The formation of peroxyacetyl nitrate, commonly known as PAN, was first reported by Stephens and co-workers in laboratory systems in 1956 (Stephens et al., 1956a,b) and subsequently in ambient air (Scott et al., 1957). PAN is known to cause eye irritation and is a plant phytotoxicant. Since that time PAN has been identified in many locations around the world, in urban, rural, and "clean" atmospheres (Temple and Taylor, 1983). For example, at rural locations in the continental United States, PAN was observed *in nearly every sample*, with average concentrations of ~0.5–0.7 ppb (Spicer et al., 1983b). Even in the clean air over the Pacific Ocean, mean concentrations of PAN of 0.032 ppb have been observed (Singh and Salas, 1983).

PAN is formed from the reaction of acetyl radicals with O_2 and NO_2:

$$CH_3\overset{\overset{\displaystyle O}{\|}}{C} + O_2 \rightarrow CH_3\overset{\overset{\displaystyle O}{\|}}{C}OO \tag{39}$$

$$CH_3\overset{\overset{\displaystyle O}{\|}}{C}OO + NO_2 \rightleftarrows CH_3\overset{\overset{\displaystyle O}{\|}}{C}OONO_2 \tag{40, −40}$$

As discussed in Chapter 7, there are a variety of sources of the acetyl radical ($CH_3\overset{\overset{O}{\|}}{C}$); these include acetaldehyde oxidation, for example, by OH or NO_3 (see Chapter 7),

$$CH_3CHO + OH \rightarrow CH_3\overset{\overset{\displaystyle O}{\|}}{C} + H_2O \tag{41}$$

$$CH_3CHO + NO_3 \rightarrow CH_3\overset{\overset{\displaystyle O}{\|}}{C} + HNO_3 \tag{42}$$

as well as the photolysis or oxidation of certain C_2 and larger organics. For example, the photolysis of biacetyl produces acetyl radicals:

$$CH_3\overset{\overset{\displaystyle O}{\|}}{C}\overset{\overset{\displaystyle O}{\|}}{C}CH_3 + h\nu \rightarrow 2\ CH_3\overset{\overset{\displaystyle O}{\|}}{C} \tag{43}$$

Similary, the oxidation of 2-butene, for example, by OH, produces acetaldehyde as a product so that the photooxidation of this alkene in the presence of NO_x gives PAN.

The atmospheric oxidation of ethane has also been suggested as a significant PAN source in the troposphere and lower stratosphere through the reactions (44)–(47)

$$C_2H_6 + OH \rightarrow C_2H_5 + H_2O \quad (44)$$

$$C_2H_5 + O_2 \xrightarrow{M} C_2H_5O_2 \quad (45)$$

$$C_2H_5O_2 + NO \rightarrow C_2H_5O + NO_2 \quad (46)$$

$$C_2H_5O + O_2 \rightarrow CH_3CHO + HO_2 \quad (47)$$

followed by (41) or (42) and (39) and (40) (Aiken et al., 1983).

Finally, ethyl nitrite photolysis in oxygen has been used to generate PAN in laboratory studies (Stephens, 1969) likely due to the following sequence of reactions:

$$C_2H_5ONO + h\nu \rightarrow C_2H_5O + NO \quad (48)$$

$$C_2H_5O + O_2 \rightarrow CH_3CHO + HO_2 \quad (49)$$

$$HO_2 + NO \rightarrow OH + NO_2 \quad (50)$$

followed by reactions (41), (39), and (40).

As discussed in Section 4-A-7c, PAN is thermally unstable and decomposes back into peroxyacetyl radicals and NO_2:

$$CH_3\overset{\overset{\displaystyle O}{\|}}{C}OONO_2 \rightarrow CH_3\overset{\overset{\displaystyle O}{\|}}{C}OO + NO_2 \quad (-40)$$

Thus, like HO_2NO_2, an equilibrium exists between the formation and decomposition of PAN:

$$CH_3\overset{\overset{\displaystyle O}{\|}}{C}OO + NO_2 \rightleftarrows CH_3\overset{\overset{\displaystyle O}{\|}}{C}OONO_2 \quad (40, -40)$$

The rate constant for decomposition of PAN, k_{-40}, has been determined in several studies, which are reviewed by Atkinson and Lloyd (1984), who recommend the following expression:

$$k_{-40} = 1.95 \times 10^{16}\, e^{-13{,}543/T}\ \text{s}^{-1}$$

TABLE 8.7. Rate Constants[a] and Lifetimes for the Thermal Decomposition of PAN at Various Temperatures

Temperature (°C)	k (s^{-1})	τ^b
−30	1.2×10^{-8}	2.6 yr
−20	1.1×10^{-7}	105 days
−10	8.4×10^{-7}	14 days
0	5.6×10^{-6}	50 h
10	3.2×10^{-5}	8.6 h
20	1.6×10^{-4}	1.7 h
30	7.6×10^{-4}	22 min
40	3.2×10^{-3}	5 min

[a] Using $k = 1.95 \times 10^{16}\, e^{-13,543/T}\ s^{-1}$ (see discussion in text).
[b] $\tau = 1/k$.

Table 8.7 shows the rate constants k_{-40} at various temperatures calculated from this expression, as well as the corresponding lifetimes of PAN. The value of k_{-40} calculated from this expression is in excellent agreement with a recent measurement of $3.3 \times 10^{-4}\ s^{-1}$ at 297°K (Niki et al., 1985).

These lifetimes for PAN will only be realized if the peroxyacetyl radical formed on the decomposition of PAN is removed rapidly compared to its reaction with NO_2 to form PAN. This is the case in the presence of NO, since $CH_3\overset{\overset{O}{\|}}{C}OO$ reacts in a manner analogous to that for simple RO_2 radicals (see Section 7-D-2a):

$$CH_3\overset{\overset{O}{\|}}{C}OO + NO \rightarrow CH_3\overset{\overset{O}{\|}}{C}O + NO_2 \qquad (51)$$

Atkinson and Lloyd (1984) recommend use of the rate constants $k_{40} = 4.7 \times 10^{-12}$ and $k_{51} = 7.1 \times 10^{-12}\ cm^3\ molecule^{-1}\ s^{-1}$, respectively, at 298°K and 1 atm pressure. Thus as long as the concentration ratio $[NO]/[NO_2] \gtrsim 7$, then 10% or less of the peroxyacetyl radicals will reform PAN.

PAN acts as a storage reservoir for NO_x (Singh and Hanst, 1981). When the $[NO_2]/[NO]$ ratio is high and/or the temperature low, the decomposition reaction (−40) is inhibited and hence PAN can build up. However, when fresh NO is introduced into the air mass, or the temperature rises significantly, it can decompose and release $CH_3\overset{\overset{O}{\|}}{C}OO + NO_2$. This acts as a non-photolytic free radical source and hence can be important in nighttime chemistry. Thus the $CH_3\overset{\overset{O}{\|}}{C}OO$ released can react in a sequence such as those discussed in Chapter 7, leading to the oxidation of NO to NO_2, the oxidation of organics, and so on:

$$CH_3\overset{\overset{O}{\|}}{C}OO + NO \rightarrow CH_3\overset{\overset{O}{\|}}{C}O + NO_2 \quad (51)$$

$$CH_3\overset{\overset{O}{\|}}{C}O \rightarrow CH_3 + CO_2 \quad (52)$$

$$CH_3 + O_2 \overset{M}{\rightarrow} CH_3O_2 \quad (53)$$

$$CH_3O_2 + NO \rightarrow CH_3O + NO_2 \quad (54)$$

$$CH_3O + O_2 \rightarrow HCHO + HO_2 \quad (55)$$

$$HO_2 + NO \rightarrow OH + NO_2 \quad (56)$$

$$OH + NO \overset{M}{\rightarrow} HONO \quad (57)$$

$$OH + NO_2 \overset{M}{\rightarrow} HNO_3 \quad (58)$$

$$OH + \text{organics} \rightarrow HCHO, CH_3CHO, \text{etc.} \quad (59)$$

This sequence forms photochemically active species such as HCHO and HONO which can photolyze at dawn, accelerating photochemical smog formation.

This suggestion of Hendry and Kenley (1979) that PAN can act as an accelerator for photochemical smog formation has been confirmed experimentally in environmental chamber studies. Thus Fig. 8.4 shows the concentration–time profiles for propene and O_3 in the presence of increasing concentrations of added PAN (Carter et al., 1981a). As the concentration of added PAN increases, the rates of formation of O_3 and the decay of propene both increase. PAN may also contribute to O_3 formation downwind of urban areas by transporting NO_x and then decomposing to release NO_2, which forms O_3.

The thermal decomposition of PAN is believed to represent the only significant chemical sink for PAN in the atmosphere. Its absorption cross sections in the actinic UV are small (Section 3-C-6) and its reaction with OH and O_3 are slow (Wallington et al., 1984; Pate et al., 1976).

It has been suggested that PAN may serve as a useful indicator of photochemical air pollution caused by long-range transport (e.g., Hov, 1984). While elevated O_3 levels are often used for this purpose, the existence of highly variable natural sources of O_3, such as stratospheric injection and photooxidations in the clean troposphere (see Section 14-A), makes assessment of the contribution of anthropogenic sources somewhat uncertain if O_3 is used.

A higher homologue of PAN, peroxypropionyl nitrate (PPN), is often observed along with PAN in ambient air. Figure 8.5, for example, shows a gas chromatogram (see Section 5-F) of an air sample taken in Harwell, England, in 1973 by Penkett, Sandalls, and Lovelock (1975); this was the first reported measurement

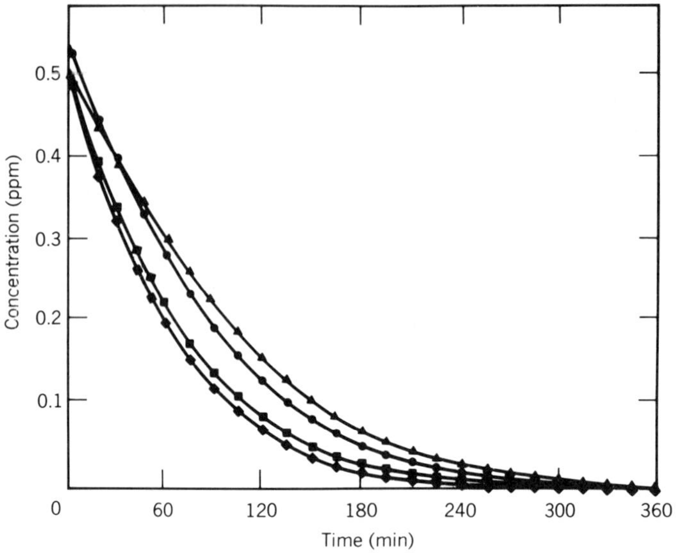

(a)

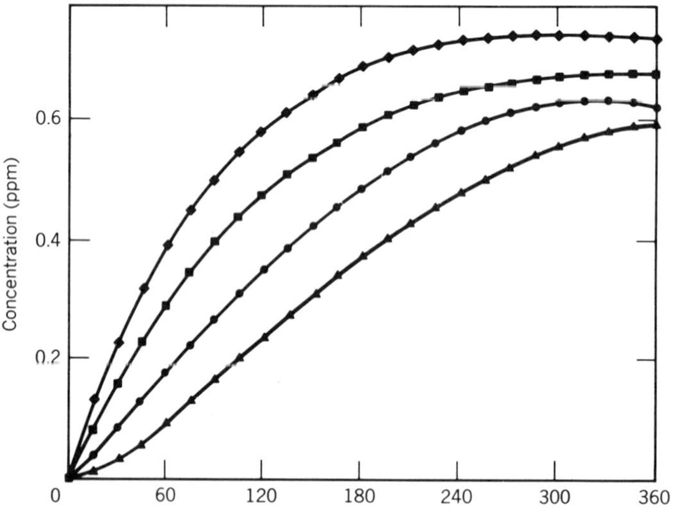

(b)

FIGURE 8.4. Concentration–time profiles in an environmental chamber for (*a*) propene and (*b*) O_3 as a function of increasing initial concentrations of PAN. Temperature ~30°C, relative humidity ~60%, [NO] = 0.26 ppm, $[NO_2]$ = 0.26 ppm, $[C_3H_6]$ = 0.5 ppm. ▲, No added PAN; ●, 0.06 ppm PAN added; ■, 0.13 ppm PAN added; ♦, 0.26 ppm added PAN (from Carter et al., 1981a).

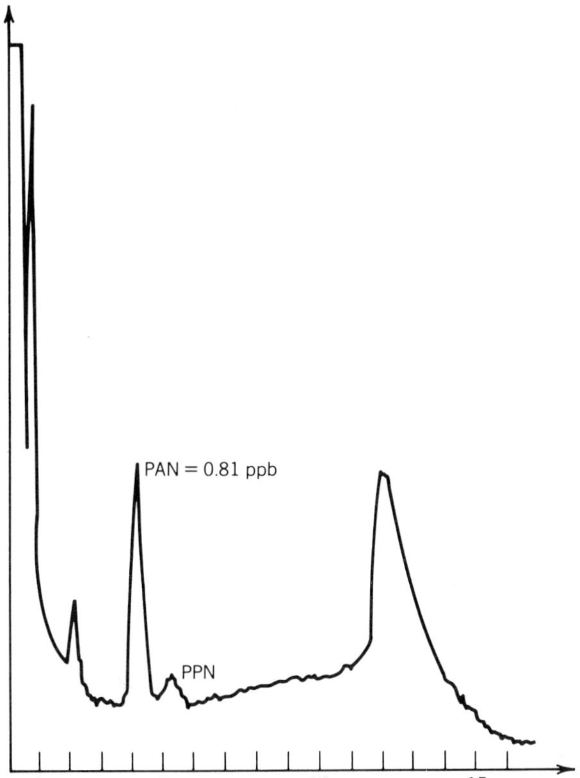

FIGURE 8.5. Gas chromatogram showing presence of PAN and PPN in Harwell, England, in 1973. This is the first reported measurement of PAN in Europe (from Penkett et al., 1975.)

of PAN in Europe. In addition to the strong PAN peak, a smaller one due to PPN is also seen.

PPN is formed in a series of reactions analogous to those forming PAN:

$$C_2H_5\overset{\overset{\displaystyle O}{\|}}{C} + O_2 \rightarrow C_2H_5\overset{\overset{\displaystyle O}{\|}}{C}OO \tag{60}$$

$$C_2H_5\overset{\overset{\displaystyle O}{\|}}{C}OO + NO_2 \rightleftarrows \underset{\text{PPN}}{C_2H_5\overset{\overset{\displaystyle O}{\|}}{C}OONO_2} \tag{61}$$

It is typically present at concentrations approximately an order of magnitude less than those of PAN.

A peroxyacyl nitrate of particular interest is peroxybenzoyl nitrate (PBzN). It is thought to be about two orders of magnitude stronger than PAN in terms of eye irritation (Heuss and Glasson, 1968). Since its first synthesis by Heuss and Glasson in 1968, it has been observed in a number of systems where benzaldehyde or precursor aromatic hydrocarbons such as toluene are the reactant (Appel, 1973; Gay et al., 1976; Niki et al., 1979). It too is formed in an analogous manner to PAN. For example, in the chlorine-atom-initiated oxidation of benzaldehyde in the presence of NO_2, the following reactions occur (Gay et al., 1976; Niki et al., 1979):

$$Cl_2 + h\nu \rightarrow 2\,Cl \tag{62}$$

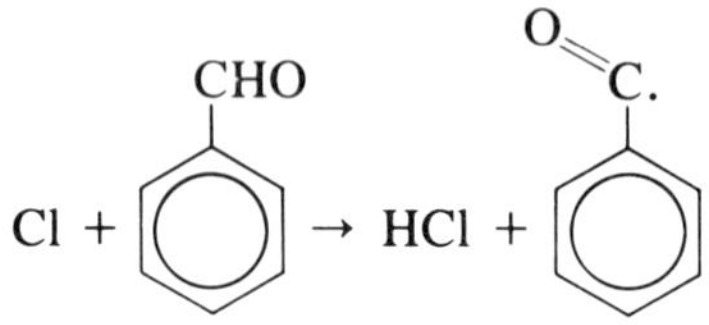

(63)

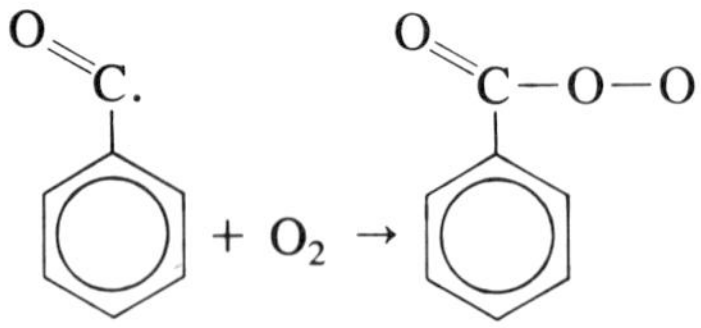

(64)

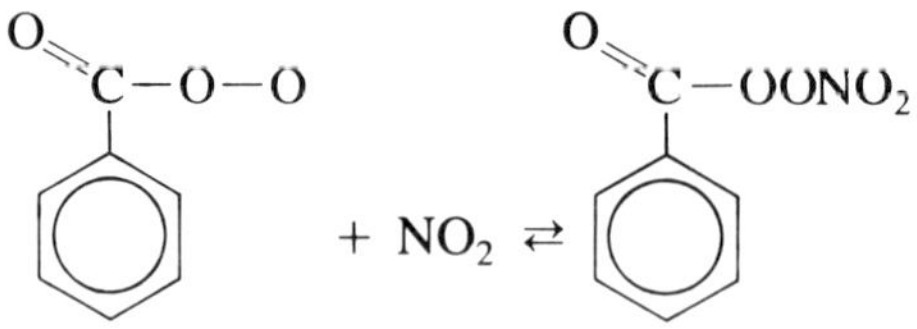

(65)

PBzN

Other peroxy nitrates have also been synthesized in laboratory studies (Edney et al., 1979a,b). However, only PAN and PPN have been observed to date in ambient air. A brief search for PBzN in the San Francisco Bay area did not give positive results, and an upper limit of 0.07 ppb was established for its concentration during that study (Appel, 1973). The lack of identification of peroxy nitrates other than PAN and PPN may be due to the thermal instability and/or low concentrations of the other compounds.

8-B-2. Alkyl Nitrates

Alkyl nitrates are formed in the reactions of alkoxy radicals with NO_2:

$$R_1\!-\!\underset{R_2}{\overset{H}{C}}\!-\!O + NO_2 \xrightarrow{a} R_1\!-\!\underset{R_2}{\overset{H}{C}}ONO_2 \tag{66}$$

$$\xrightarrow{b} HONO + R_1R_2CO$$

As discussed in detail in Section 7-D-3e, it is believed that the abstraction path (66b) is negligible under ambient conditions.

A second route to alkyl nitrate formation is the reaction of larger alkylperoxy radicals, that is, RO_2, with $R \geq C_3$, with NO:

$$\begin{aligned} RO_2 + NO &\underset{a}{\rightarrow} RO + NO_2 \\ &\underset{b}{\rightarrow} RONO_2 \end{aligned} \tag{67}$$

As discussed in Section 7-D-2a, as much as 30% of the reaction proceeds through (67b) for $R = C_8H_{17}$.

These alkyl nitrates serve as an effective sink for free radicals, because they are quite stable thermally and have relatively small absorption cross sections in the actinic UV (see Section 3-C-6). Thus they might be considered as one set of *end products* of NO_x and hydrocarbon reactions in the atmosphere.

8-B-3. Hydrazines

Hydrazines are used as fuels, for example, in the space shuttle and as a source of emergency power in the F-16 fighter plane. As a result of the industrial and fuel uses of hydrazines, with their accompanying transport and storage, there is the possibility of emissions to the atmosphere of these compounds and hence there has been interest over the last few years in their atmospheric reactions.

Hydrazines do not photolyze in the actinic UV and hence reactions with OH and O_3 must be considered. The rate constants for reaction of OH with N_2H_4 and CH_3NHNH_2 are $(6.1 \pm 1.0) \times 10^{-11}$ and $(6.5 \pm 1.3) \times 10^{-11}$ cm^3 molecule^{-1} s^{-1}, respectively, essentially independent of temperature over the range 298–424°K (Harris et al., 1979). At an OH concentration of 1×10^6 cm^{-3}, the lifetimes of both N_2H_4 and CH_3NHNH_2 will be ~4–5 h. Harris and co-workers estimate that the rate constant for the OH reaction with 1,1-dimethylhydrazine reaction is ~$(5 \pm 2) \times 10^{-11}$ cm^3 molecule^{-1} s^{-1}, so that its lifetime with respect to OH will also be ~6 h.

Reaction with O_3 is also relatively fast. Tuazon et al. (1981a) estimate that the rate constant for the N_2H_4–O_3 reaction at 294–297°K is ~1×10^{-16} cm^3 molecule^{-1} s^{-1}. This corresponds to a lifetime of about 1 h at an O_3 concentration of 0.1 ppm. The rate constants for the O_3–CH_3NHNH_2 and O_3–$(CH_3)_2NHNH_2$ reactions were too fast to measure under their experimental conditions; thus the reactions of ~1–3 ppm O_3 with 0.4–4 ppm CH_3NHNH_2 and ~2 ppm O_3 with ~0.2–2 ppm $(CH_3)_2NNH_2$ were complete in less than 2–3 min. From these data, the rate constants must be $>10^{-15}$ cm^3 molecule^{-1} s^{-1}, and thus the lifetime of these two hydrazines must be less than 7 min at 0.1 ppm O_3.

The mechanism of the reaction of hydrazines with O_3 has been investigated using FTIR (Tuazon et al., 1981a; Carter et al., 1981b). In the case of N_2H_4, the major product was H_2O_2, and N_2O appeared as a minor product. For example, the reaction of 2.37 ppm N_2H_4 with 2.4 ppm O_3 gave 0.48 ppm H_2O_2 and 0.062 ppm N_2O. The following tentative mechanism was proposed:

Initiation

$$H_2NNH_2 + O_3 \rightarrow H_2N\text{—}\dot{N}H + OH + O_2 \quad (68)$$

Propagation

$$H_2N\text{—}\dot{N}H + O_2 \rightarrow HN{=}NH + HO_2 \quad (69)$$

$$HN{=}NH + O_3 \rightarrow HN{=}\dot{N} + OH + O_2 \quad (70)$$

$$H_2NNH_2 + OH \rightarrow H_2N\text{—}\dot{N}H + H_2O \quad (71)$$

Product Formation

$$HN{=}\dot{N} \rightarrow H + N_2 \quad (72)$$

$$H + O_2 \xrightarrow{M} HO_2 \quad (73)$$

$$HO_2 + HO_2 \rightarrow H_2O_2 + O_2 \quad (74)$$

According to this mechanism, most of the nitrogen in the hydrazine would form N_2, which would not have been detected in this system. The diazene HN=NH would be expected to react with OH radicals:

$$HN{=}NH + OH \rightarrow HN{=}\dot{N} + H_2O \quad (75)$$

and ultimately form N_2 and H_2O_2 via reactions (72)–(74).

This reaction has been used as a non-photolytic OH source for kinetic studies because of the production of OH in the initial reaction (68) between O_3 and N_2H_4 (Tuazon et al., 1983b). As discussed in Section 4-C, it is difficult to determine rate constants for OH reactions under atmospheric conditions with compounds that photolyze rapidly; thus OH is usually generated using photolysis of species such as CH_3ONO, so that the simultaneous photolysis of the reactant of interest occurs in competition with the OH reaction. However, since the discovery that OH is produced in the dark by O_3–hydrazine reactions, the O_3–N_2H_4 reactions have been successfully used to generate relative rate constants for OH reactions with photochemically labile compounds (Tuazon et al., 1983b).

Figure 8.6 shows an FTIR spectrum taken during the studies by Tuazon et al. (1981a) of the reaction of CH_3NHNH_2 with O_3 (the absorption bands from NH_3 which forms from the slow decay of the hydrazine in the dark have been subtracted out from the spectra). After initial injection of O_3, with the hydrazine present in excess, the observed products were methyl hydroperoxide (CH_3COOH), diazomethane (CH_2N_2), H_2O_2, methyl diazene ($CH_3N{=}NH$), HCHO, CH_3OH, and traces of CH_3ONO_2. After a second injection of O_3 into the system so that O_3 was in excess, $CH_3N{=}NH$ and CH_2N_2 disappear, and higher yields of CH_3OOH, CH_3OH, and HCHO result. Ninety-two percent of the initial carbon atoms could be accounted for in the observed products, but at least 95% of the initial nitrogen in CH_3NHNH_2 could not be found, indicating it likely formed N_2.

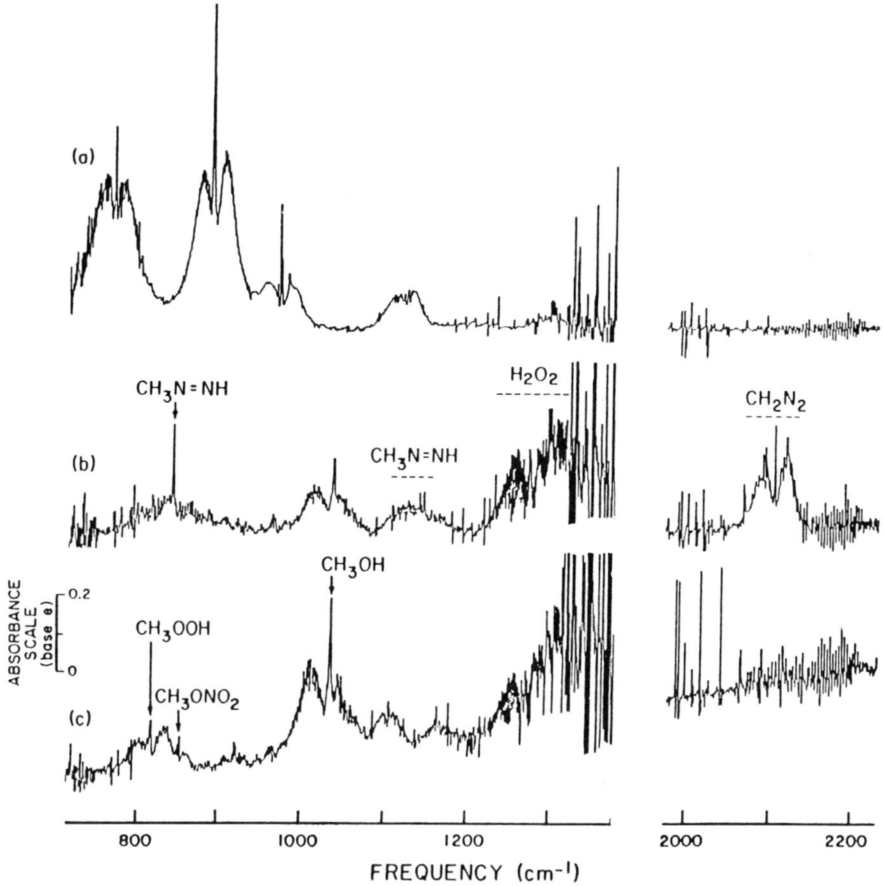

FIGURE 8.6. FTIR spectra taken during reaction of CH_3NHNH_2 with O_3. (*a*) 3.68 ppm CH_3NHNH_2 before reaction, (*b*) 2 min after injection of 2.8 ppm O_3, (*c*) 2.8 ppm O_3 injected 38 min after first injection; spectrum taken 2 min after second injection. Resolution, 1 cm^{-1}; pathlength, 460 m. NH_3 absorptions have been subtracted from (*a*) and (*b*), and both NH_3 and O_3 absorptions from (*c*) (from Tuazon et al., 1981a).

The reaction mechanism proposed by Tuazon et al. (1981a, 1982) to explain these results is analogous to that for N_2H_4, that is, reactions (68)–(75). In this case the intermediate $CH_3N{=}NH$ formed in (69) was identified by FTIR. Reactions (76) and (77) explain the formation of CH_2N_2:

$$CH_3N{=}NH + O_3 \xrightarrow{O_2} CH_2N_2 + OH + O_2 + HO_2 \qquad (76)$$

$$CH_3N{=}NH + O_3 \rightarrow CH_2N_2 + H_2O + O_2 \qquad (77)$$

In excess O_3, CH_2N_2 can decay via (78):

$$CH_2N_2 + O_3 \rightarrow HCHO + O_2 + N_2 \qquad (78)$$

The formation of CH_3OOH, HCHO, CH_3OH, and CH_3ONO_2 is expected from secondary reactions of the methyl radical formed in place of a hydrogen atom in reaction (72):

$$CH_3 + O_2 \xrightarrow{M} CH_3O_2 \tag{79}$$

$$CH_3O_2 + HO_2 \rightarrow CH_3OOH + O_2 \tag{80}$$

$$2\,CH_3O_2 \underset{a}{\rightarrow} HCHO + CH_3OH + O_2$$
$$\underset{b}{\rightarrow} 2\,CH_3O + O_2 \tag{81}$$

$$CH_3O + O_2 \rightarrow HCHO + HO_2 \tag{82}$$

$$CH_3O + NO_2 \xrightarrow{M} CH_3ONO_2 \tag{83}$$

The reaction of 1,1-dimethylhydrazine with O_3 gave dimethylnitrosamine as the major product in ~60% yields within 2–3 min reaction time (Tuazon et al., 1981a). Minor products were HCHO, H_2O_2, HONO, and perhaps NO_x. This different set of products was attributed to the lack of an abstractable hydrogen on one nitrogen of the hydrazine. Thus assuming the initiation step is the same as for N_2H_4 and CH_3NHNH_2, that is,

$$(CH_3)_2NNH_2 + O_3 \rightarrow (CH_3)_2N{-}\dot{N}H + OH + O_2 \tag{84}$$

the propagation step corresponding to reaction (69)—hydrogen abstraction by O_2—cannot occur. The addition of O_2 to a nitrogen centered radical is known in the case of NH_2 to be very slow (see Section 8-A-4) and is thus not anticipated to occur here either. An alternate fate of the radical formed in (84) proposed by Tuazon and co-workers is reaction with O_3, which would be expected to lead to dimethylnitrosamine formation via (85) and (86):

$$(CH_3)_2N{-}\dot{N}H + O_3 \rightarrow (CH_3)_2N{-}\overset{\overset{\displaystyle H}{|}}{N}O + O_2 \tag{85}$$

$$(CH_3)_2N{-}\overset{\overset{\displaystyle H}{|}}{N}{-}O + O_2 \rightarrow (CH_3)_2NNO + HO_2 \tag{86}$$

H_2O_2 would be expected from the $HO_2 + HO_2$ reaction (74) and the other products via secondary reactions of the OH radicals produced in (84).

An alternate mechanism involves the reaction of the dimethylhydrazyl radical with O_2,

$$(CH_3)_2N\dot{N}H + O_2 \rightarrow (CH_3)_2\overset{+}{N}{=}\overset{-}{N} + HO_2 \tag{87}$$

followed by the formation of the nitrosamine by reaction with O_3:

$$(CH_3)_2\overset{+}{N}{=}\overset{-}{N} + O_3 \rightarrow (CH_3)_2NNO + O_2 \tag{88}$$

However, as discussed by Tuazon et al. (1982), neither this nor the first mechanism described above are completely consistent with the experimental data. For example, the H_2O_2 yields were much smaller than expected from reactions (84)–(86) followed by self-reaction of the HO_2 to form H_2O_2. However, the observed reactant stoichiometries are not entirely consistent with reactions (87) and (88).

Clearly, the mechanisms involved are complex and warrant further study.

Studies of the 1,1-dimethylhydrazine reactions in air in the presence of NO have also been carried out and the yields and products have been shown to be somewhat similar to those observed in the reaction with O_3 (Carter et al., 1981b). Figure 8.7 shows the concentration–time profiles for the reactants and selected products during a typical irradiation. The major product in the early stages of the reaction was an unidentified compound with an IR absorption band at $\sim 988\ cm^{-1}$, possibly nitrosohydrazine which could be formed through the reaction sequence (89) and (90):

$$(CH_3)_2N{-}NH_2 + OH \rightarrow (CH_3)_2N{-}\dot{N}H + H_2O \quad (89)$$

$$(CH_3)_2N{-}\dot{N}H + NO \rightarrow (CH_3)_2N{-}N\begin{smallmatrix} \diagup NO \\ \diagdown H \end{smallmatrix} \quad (90)$$

Nitrosohydrazine

Its rapid disappearance after all the reactant hydrazine has been reacted is likely due to photolysis, since *N*-nitrosamines are known to photolyze rapidly in sunlight (see Section 8-B-4b). Nitrosamine formation was attributed to the $O_3{-}(CH_3)_2NNH_2$ reaction discussed above, while the formation of a nitramine $[(CH_3)_2N{-}NO_2]$ and nitrous oxide are due to the further oxidation of the nitrosamine discussed in more detail in Section 8-B-3b.

NO_2 has also been shown to react with 1,1-dimethylhydrazine in air forming HONO and tetramethyltetrazine-2, $(CH_3)_2NN{=}NN(CH_3)_2$ (Tuazon et al., 1983c). The reaction is also hypothesized to involve abstraction of a hydrogen from the weak N—H bond by NO_2, forming HONO. The tetramethyltetrazine-2 is hypothesized to be formed by the addition of NO_2 to the $(CH_3)_2NNH$ radical, followed by decomposition to $(CH_3)_2N_2$ + HONO and the self-recombination of the $(CH_3)_2N_2$ radicals (Tuazon et al., 1982). The apparent overall rate constant for the reaction was $2.3 \times 10^{-17}\ cm^3\ molecule^{-1}\ s^{-1}$ so that the lifetime of 1,1-dimethylhydrazine at an NO_2 concentration of 0.1 ppm would be ~ 5 h. Since the lifetimes with respect to 0.1 ppm O_3, or 1×10^6 OH radicals cm^{-3} are ~ 7 min and 6 h, respectively, the reaction of NO_2 can contribute to the atmospheric reactions of the hydrazine only at low O_3 levels.

Hydrazine, monomethylhydrazine and asymmetrical dimethylhydrazine have also been shown to react rapidly ($k > 10^{-15}\ cm^3\ molecule^{-1}\ s^{-1}$) with HNO_3 in the gas phase to form the corresponding hydrazinium nitrate aerosols (Tuazon et

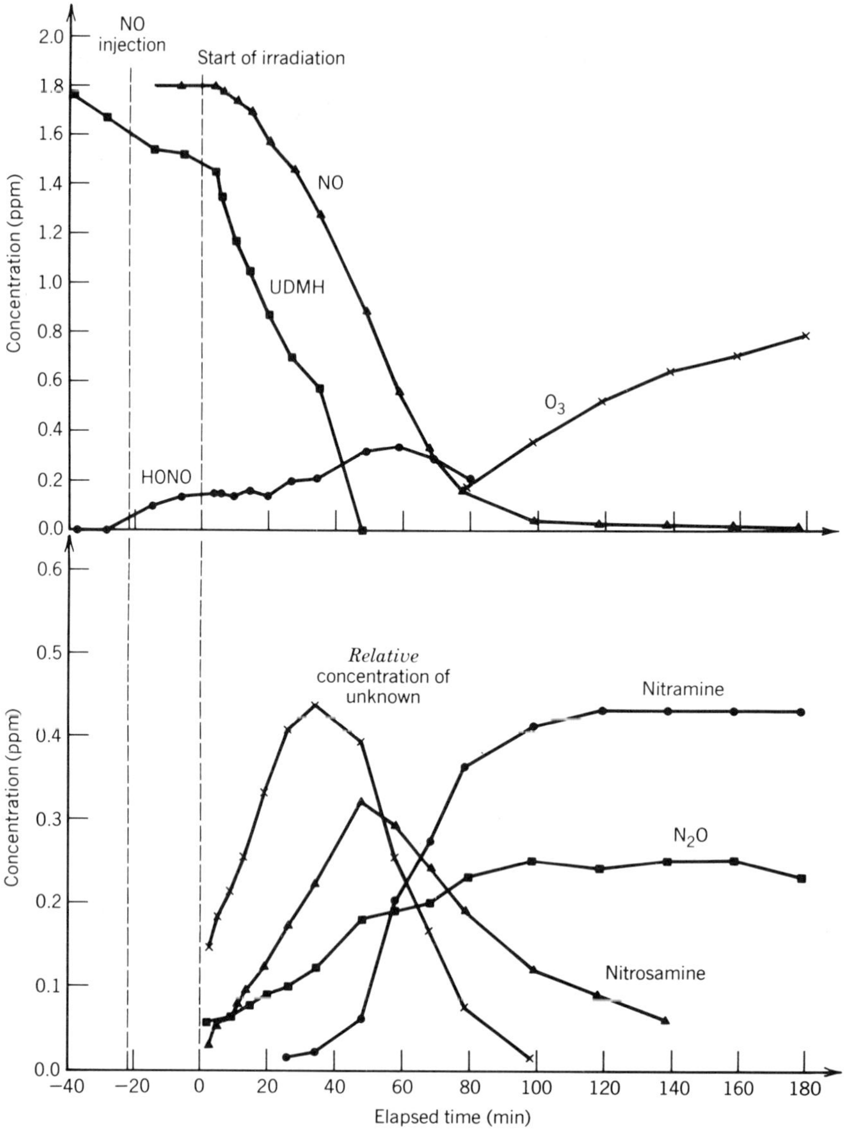

FIGURE 8.7. Concentration–time profiles for the reactants and selected products in the reaction of 1.5 ppm 1,1-dimethylhydrazine (UDMH) with 1.8 ppm NO in air in the presence of light at 38°C and ~4% relative humidity. The curve for the unknown reflects only the qualitative shape of its profile, since only the total absorbance at 988 cm^{-1}, which peaked at ~ 0.3, could be determined (from Carter et al., 1981b).

al., 1982). At an HNO_3 concentration of 10 ppb, the lifetime of the hydrazines with respect to these reactions would be $\lesssim$1.1 h.

While the proposed mechanisms for the hydrazine oxidations account for the observed products and reactant stoichiometry, Tuazon and co-workers caution that further work is needed to confirm these initial findings and proposed mechanisms. This remains an interesting area for future research.

8-B-4. Amines

8-B-4a. Simple Alkyl Amines

Amines are emitted from industrial activities, feedlot operations, waste incineration, sewage treatment, and so on, although relatively few data are available on their atmospheric concentrations in a variety of locations (Mosier et al., 1973).

Aliphatic amines only absorb light below 250 nm and hence do not photolyze in the troposphere (Calvert and Pitts, 1966). However, they do react reasonably rapidly with OH; as seen in Table 8.8, at an OH concentration of 1×10^6 cm^{-3}, the lifetime of the simple amines will be in the range 4–13 h.

Several studies of the photooxidation of simple alkyl amines under simulated atmospheric conditions have been carried out using FTIR (Hanst et al., 1977; Pitts et al., 1978; Tuazon et al., 1978). Figures 8.8 and 8.9 show the concentration–time profiles for 0.5 ppm diethylamine and triethylamine, respectively, when introduced into an outdoor environmental chamber (see Chapter 6) with ~0.08 ppm NO and 0.17 ppm NO_2 at relative humidities from ~24 to ~50%. In these runs the chamber was left in the dark for several hours by covering it with a black cover; after ~2 h, the cover was removed exposing the chamber contents to natural sunlight.

Both amines disappeared rather rapidly in the dark initially, and in both cases diethylnitrosamine $(C_2H_5)_2NNO$, was formed, the yield being ~3% for the diethylamine and ~1% for triethylamine. This was attributed by Hanst and co-workers (1977) to the reaction of the amines with nitrous acid, which is known to be formed upon injection of NO_x with water into environmental chambers (see Section 8-A-3a and Chapter 6):

$$R_2NH + HONO \rightleftarrows R_2NNO + H_2O \tag{91}$$

The amine concentrations tend to plateau after ~1 h in the dark. At least part of this dark reaction may occur on the walls of the reaction chamber (Pitts et al., 1978); whether such reactions can occur in the atmosphere is uncertain because of the possibility that large amounts of surfaces will be needed to catalyze the reaction and the limiting effect of the HONO concentration.

TABLE 8.8. Rate Constants for OH Reactions with Some Simple Aliphatic Amines at Room Temperature and Calculated Lifetimes Under Atmospheric Conditions

Amine	10^{11} k (cm^3 molecule^{-1} s^{-1})	τ^a	Reference
CH_3NH_2	2.20 ± 0.22	13 h	Atkinson et al., 1977
$(CH_3)_2NH$	6.54 ± 0.66	4 h	Atkinson et al., 1978
$(CH_3)_3N$	6.09 ± 0.61	5 h	Atkinson et al., 1978
$C_2H_5NH_2$	2.77 ± 0.28	10 h	Atkinson et al., 1978

[a] $\tau = 1/k[OH]$, where $[OH] = 1 \times 10^6$ cm^{-3}.

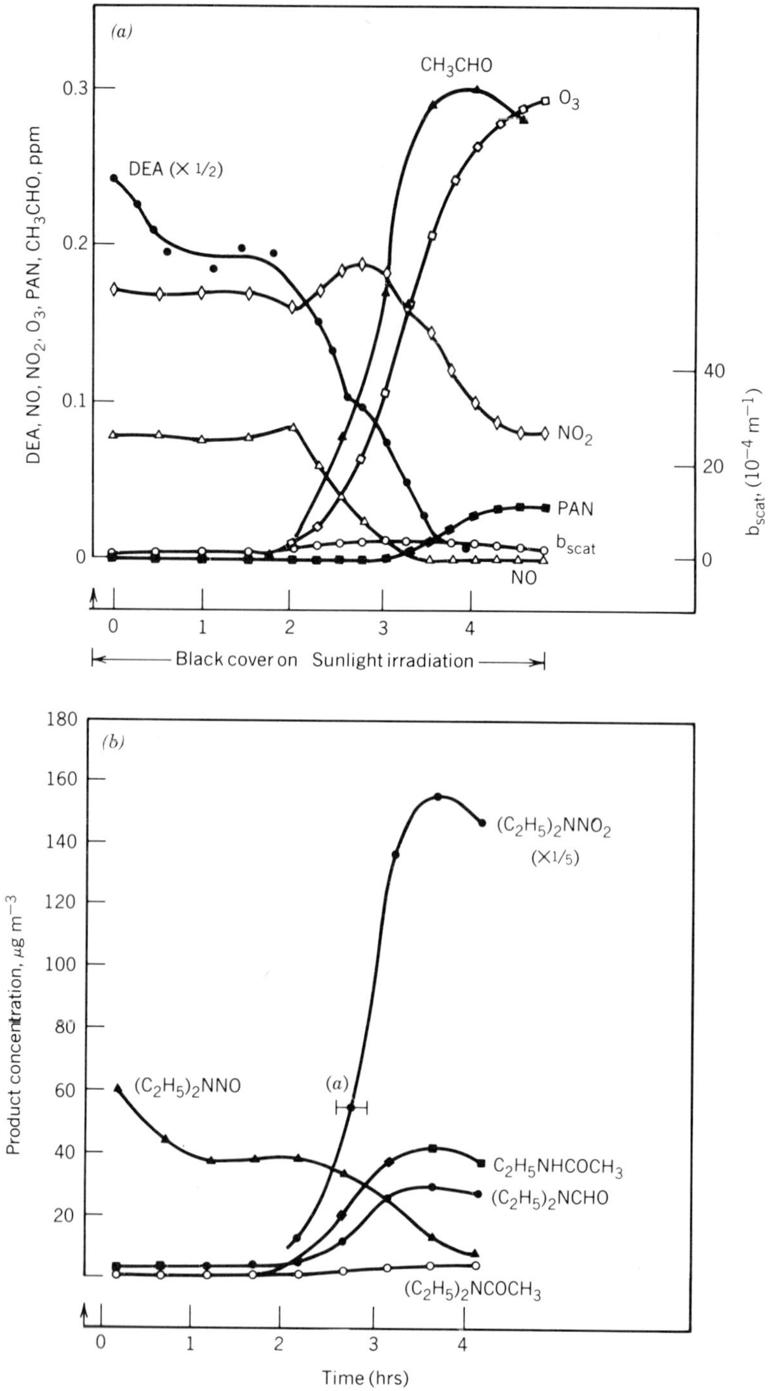

Figure 8.8. Concentration–time profiles in diethylamine (DEA)–NO_x–air photooxidation. (*a*) DEA, NO, NO_2, O_3, acetaldehyde, PAN, and aerosol light-scattering parameter, b_{scat} (see Chapter 12). (*b*) Diethylnitramine, ethylacetamide, diethylformamide, diethylacetamide, and diethylnitrosamine. Horizontal bar (*a*) shows the length of sampling time. NO_2 profiles corrected for known interference from PAN but may include some contribution from diethylnitramine and other nitrogenous products. Reactants irradiated after 2 h in dark (from Pitts et al., 1978).

After about 2 h in the dark, the mixtures were irradiated; the amines rapidly disappeared and a variety of products were formed including O_3 and PAN. In both cases, acetaldehyde and diethylnitramine, $(CH_3)_2NNO_2$, were major products, and significant amounts of ethylacetamide, $C_2H_5NHCOCH_3$, and diethylformamide, $(C_2H_5)_2NCHO$, were formed. In the case of diethylamine, the nitrosamine formed in the dark decayed rapidly in the sunlight, whereas it increased to a second peak in the triethylamine reaction (Fig. 8.10).

These products for $(C_2H_5)_3N$ photooxidation were attributed by Pitts and coworkers to reactions which they point out are typical of OH-initiated atmospheric oxidations:

$$(C_2H_5)_3N + OH \rightarrow (C_2H_5)_2N\dot{C}HCH_3 \qquad (92)$$

As discussed in Section 7-C-1, abstraction of a secondary hydrogen is expected to predominate over that of a primary hydrogen. The subsequent reactions shown in Scheme I are all similar to those outlined in Chapter 7 for hydrocarbons. The much higher yields of the nitramine were postulated to be due to the relatively rapid photolysis anticipated for the nitrosamine compared to the nitramine (see below) and the relative concentrations of NO and NO_2 present. It was also suggested that the nitrosamine may form the nitramine through thermal and/or photochemical processes.

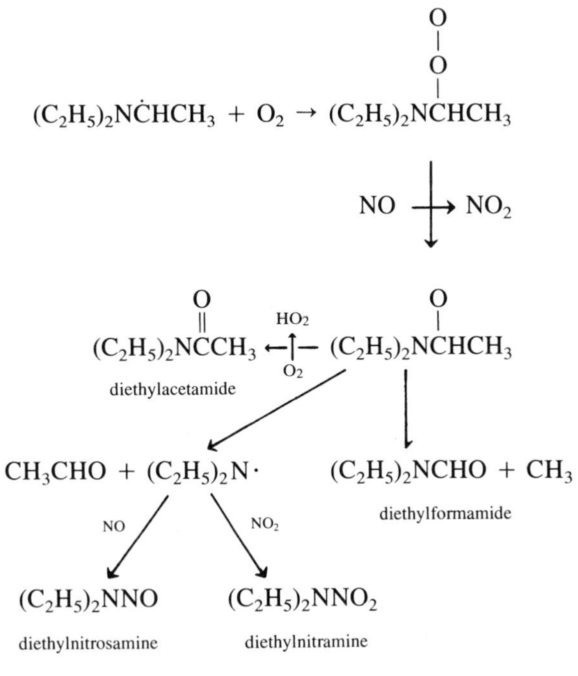

Scheme I

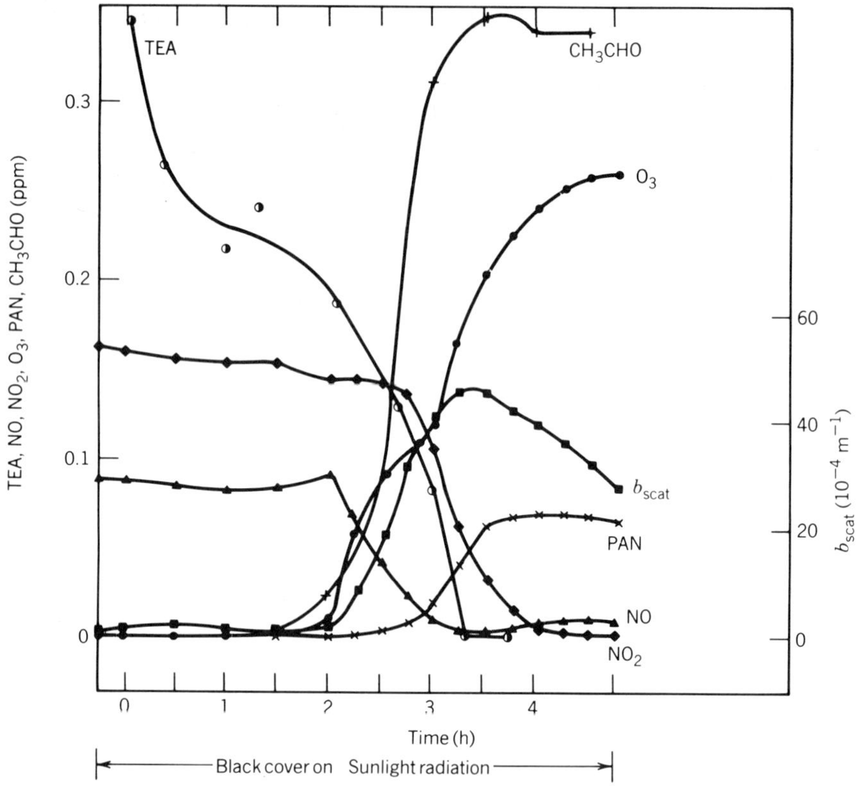

(a)

FIGURE 8.9. Concentration–time profiles in triethylamine (TEA)–NO_x–air photooxidation. (*a*) TEA, NO, NO_2, O_3, acetaldehyde, PAN, and b_{scat}. (*b*) Diethylformamide, diethylnitramine, ethylacetamide, diethylacetamide, and diethylnitrosamine. Concentration of unidentified compound of MW = 87 estimated assuming same mass spectrometer response factor as diethylacetamide (from Pitts et al., 1978).

The very large yields of acetaldehyde compared to the other products could not be explained by Scheme I alone, suggesting that the excess CH_3CHO may be due to OH abstraction of a primary hydrogen from a terminal methyl group in the amine, and/or secondary reactions of the nitrogen-containing stable products and the diethylamine radical formed as shown in Scheme I.

For diethylamine, OH attack can occur not only at a C—H bond but at the N—H bond as well:

$$(C_2H_5)_2NH + OH \rightarrow (C_2H_5)_2N\cdot + H_2O \tag{93}$$

Attack at N—H bonds as well as C—H bonds is supported by the trends in rate constants for OH with a series of simple amines (Atkinson et al., 1978). This attack at N—H bonds was suggested as the primary reason that the molar yields

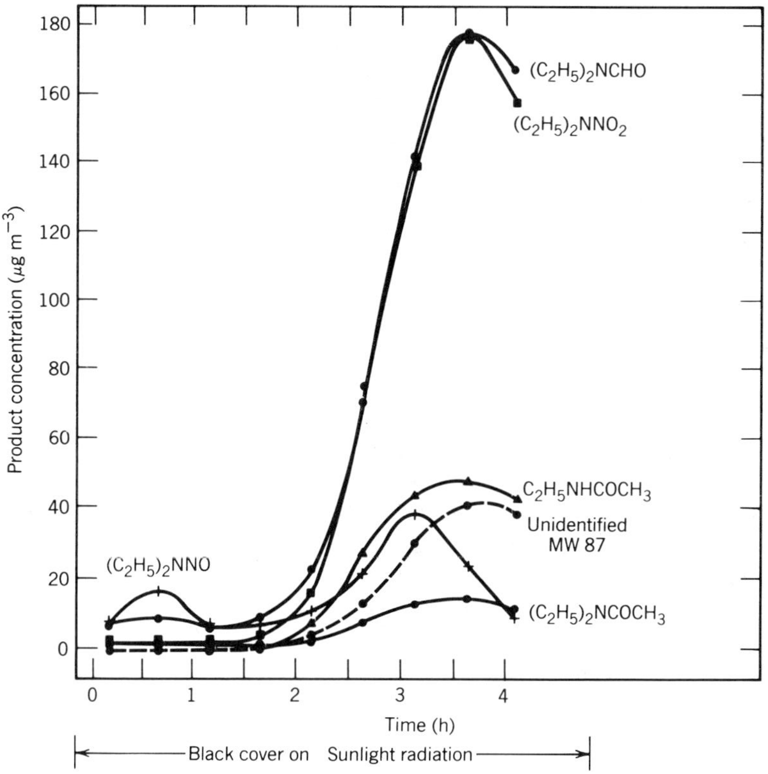

(b)

FIGURE 8.9. (*Continued*)

of $(C_2H_5)_2NNO_2$ from the diethylamine reaction were ~4.5 times those from the trimethylamine photooxidation.

It is noteworthy that the nitrosamines and some nitramines such as $(CH_3)_2NNO_2$ are carcinogenic in experimental animals. Thus the reaction of amines in ambient air may generate volatile toxic organics (see Section 10-D). These amines also are highly photochemically reactive in the traditional sense; thus NO is rapidly converted to NO_2, the organic disappears, and O_3 and PAN are formed. In addition, substantial quantities of light-scattering aerosols are formed.

The photooxidations of the methyl analogs—dimethylamine, $(CH_3)_2NH$, and trimethylamine, $(CH_3)_3N$—have also been studied using FTIR (Hanst et al., 1977; Pitts et al., 1978; Tuazon et al., 1978). Lower yields of dimethylnitrosamine were formed from these compounds in the dark in the presence of air and NO_x. Upon irradiation, the nitrosamine disappeared in the dimethylamine system, and only traces of it were formed on irradiation of the trimethylamine system.

Products identified in the dimethylamine reaction were similar to those for its ethyl analogue and included dimethylnitramine [$(CH_3)_2NNO_2$], nitrous acid

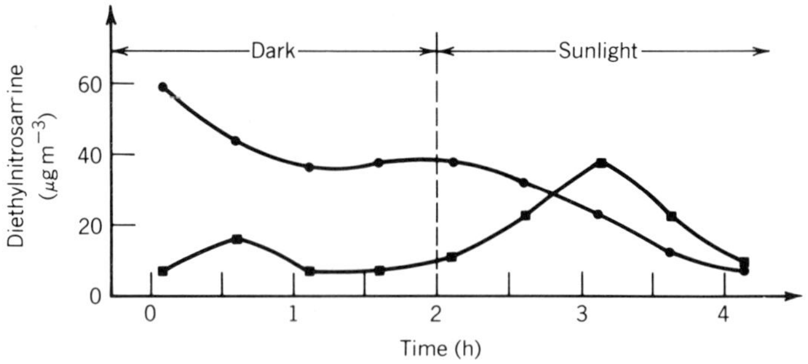

FIGURE 8.10. Concentration–time profiles of diethylnitrosamine from DEA (●) and TEA (■) (from Pitts et al., 1978).

(HONO), tetramethylhydrazine $[(CH_3)_2NN(CH_3)_2]$, CO, HCHO, and small amounts of dimethylformamide and methylformamide. The usual manifestations of photochemical smog formation, including NO to NO_2 conversion and formation of O_3 and aerosols, were also observed. Mechanisms of reaction similar to those discussed above for the ethylamine are likely.

The photooxidation of the primary amine CH_3NH_2 has not been studied extensively under simulated atmospheric conditions. Atkinson et al. (1978) suggest that OH attack at C—H bonds predominate here and that formaldehyde and/or an amide may ultimately be formed from its atmospheric oxidation through the reaction shown in Scheme II.

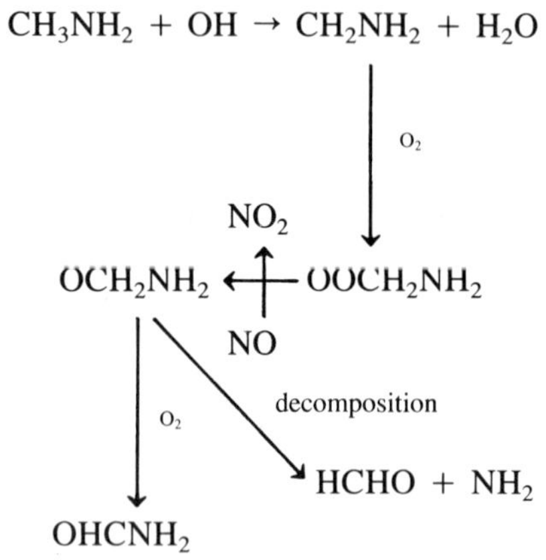

Scheme II

The ambient concentrations of amines are thought to be generally fairly small, so that their overall contribution to photochemical activity in urban atmospheres is likely negligible. However, it is interesting that two nitrogen-containing compounds anticipated from amine photooxidations in ambient air—diethylnitrosamine and dimethylformamide—have been identified in urban and industrial atmospheres (Pellizzari, 1977). However, these may have been in part or wholly from direct primary emissions; certainly the latter was the case for at least one finding of dimethylnitrosamine in ambient air (Fine et al., 1979; Fine, 1980).

8-B-4b. Substituted Amines: N-Nitrosamines, Nitramines, and Diethylhydroxylamine

As we have seen, *N*-nitrosamines can be produced from reactions of some of the simple amines under simulated atmospheric conditions. As discussed by Fine and co-workers (Fine et al., 1979; Fine, 1980), there are also a number of sources that can emit these compounds into the air. These include leather tanneries, rocket fuel, and tire and amine factories. In addition, tobacco smoke and vapors from cooking bacon have been shown to give nitrosamines. These sources, mechanism of formation in the environment, methods of analysis, and human exposure to *N*-nitroso compounds are reviewed in detail by Fine (1980).

Nitrosoamines absorb light in the actinic UV and photolyze readily. For example, Hanst and co-workers (1977) photolyzed 1–3 ppm dimethylnitrosamine, $(CH_3)_2NNO$, in air and found that it rapidly disappeared in less than 1 h to form a compound later identified as dimethylnitramine, $(CH_3)_2NNO_2$ (Pitts et al., 1978). Similarly, Tuazon et al. (1984b) estimate that the half-life of dimethylnitrosamine in air at a latitude of 34°N at the equinox is only ~5 min. On the other hand, nitramines appear to have very low light absorption cross sections and hence do not photolyze readily.

Figure 8.11, for example, shows the absorption spectra of diethylnitrosamine and diethylnitramine in methanol solution; also shown is the actinic cutoff in the troposphere at 290 nm. It is seen that the nitrosamine has a small absorption peak in the actinic UV whereas the nitramine absorption cuts off sharply around 300 nm.

Tuazon and co-workers (1984b) studied the photolysis of *N*-nitrosodimethylamine and of dimethylnitramine in air in an environmental chamber, as well as the reactions of these two compounds with OH and O_3. The photolysis at $\lambda \geq 290$ nm was found to proceed with a primary quantum yield of unity, in agreement with earlier work at the single wavelength 363.5 nm (Geiger et al., 1981).

$$(CH_3)_2NNO + h\nu\ (\lambda \geq 290\ \text{nm}) \rightarrow (CH_3)_2N + NO \tag{94}$$

These runs were carried out in excess O_3 to scavenge the NO formed in (94) so that the nitrosamine did not reform. Under these conditions, the major products were dimethylnitramine (~65% yield), with smaller amounts of CH_3NO_2 and HCHO (~35% yield each). These products are expected from reactions (95)–(99):

$$(CH_3)_2N + NO_2 \rightarrow (CH_3)_2NNO_2 \tag{95}$$

$$(CH_3)_2N + NO_2 \rightarrow HONO + CH_3{-}N{=}CH_2 \tag{96}$$

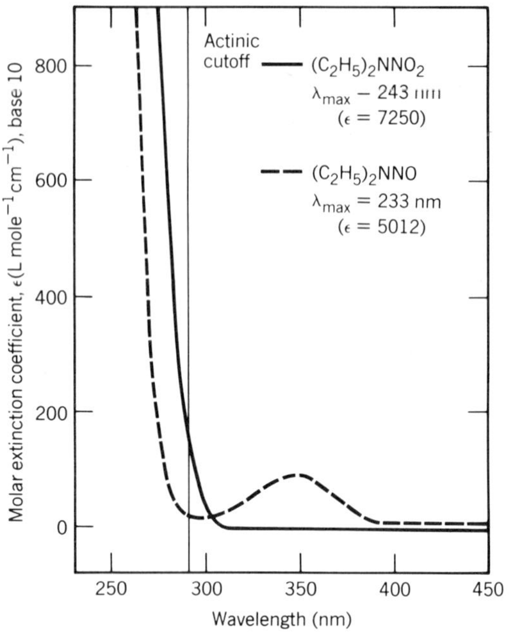

FIGURE 8.11. Absorption spectra of $(C_2H_5)_2NNO_2$ and $(C_2H_5)_2NNO$ in methanol solution (from Pitts et al., 1978).

$$(CH_3)_2N + O_2 \rightarrow HO_2 + CH_3{-}N{=}CH_2 \tag{97}$$

$$CH_3{-}N{=}CH_2 + O_3 \rightarrow CH_3NOO + HCHO \tag{98}$$

$$CH_3NOO \rightarrow CH_3NO_2 \tag{99}$$

Although methylmethlyeneamine, $CH_3{-}N{=}CH_2$, was not observed in this study, it was postulated as an intermediate on the basis of studies by Lindley et al. (1979) who observed it when $(CH_3)_2N$ radicals were formed in air. Lindley and co workers showed that $k_{97}/k_{95} = (3.9 \pm 0.3) \times 10^{-7}$ and $k_{96}/k_{95} = 0.22 \pm 0.06$.

It is interesting that $(CH_3)_2N$ did not appear to react significantly with O_3 in this system; Tuazon et al. (1984b) estimated that the rate constant for reaction with O_3 must be at least two orders of magnitude less than that for reaction (95) with NO_2.

In ambient air where other pollutants are also present, other reactions of the radical $(CH_3)_2N$ must also be considered. For example, it has been suggested that the amino radical can react with aldehydes (Grosjean et al., 1978):

$$\begin{array}{c} R_1R_2N + HCHO \rightarrow R_1R_2N{-}\overset{\displaystyle H}{\overset{|}{C}}HO \\ \downarrow O_2 \\ R_1R_2NCHO + HO_2 \end{array} \tag{100}$$

$$R_1R_2N + CH_3CHO \rightarrow R_1R_2N-\underset{\displaystyle |\atop\displaystyle O}{\overset{\displaystyle H\atop\displaystyle |}{C}}-CH_3$$

$$\downarrow O_2 \qquad (101)$$

$$R_1R_2N\underset{\displaystyle \|\atop\displaystyle O}{C}CH_3 + HO_2$$

Table 8.9 shows the rate constants at 298°K for the reactions of $(CH_3)_2NNO$ and $(CH_3)_2NNO_2$ with OH and O_3, respectively, as well as the estimated lifetimes with respect to these oxidants. It is seen that reaction with O_3 is too slow in both cases to be important in the atmosphere. For the nitrosamine, reaction with OH is also slow relative to photolysis, so that the latter will be essentially its sole atmospheric fate. For the nitramine, however, where photolysis is slow, reaction with OH will predominate; since this reaction is not extremely rapid, the nitramine is expected to persist in the troposphere for several days once it is formed.

The mechanism of reaction of these compounds with OH has not been investigated in detail. Tuazon and co-workers (1984b) suggest that the initial step is abstraction of a hydrogen from a C—H bond,

$$(CH_3)_2N-NO_x + OH \rightarrow CH_2\overset{\displaystyle CH_3\atop\displaystyle |}{N}-NO_x + H_2O \qquad (102)$$

where $x = 1$ or 2, followed by the usual reactions of the free radical formed with O_2, NO, and so on.

One substituted amine that has received significant attention in recent years is diethylhydroxylamine (DEHA), $(C_2H_5)_2NOH$, because of its proposed use as an additive to ambient air to control photochemical smog (Jayanty et al., 1974; Heicklen, 1976, 1981a,b, 1985). Presumably, it functions as a free radical scavenger, thus acting as a terminator in atmospheric chain oxidations. Typical stable end products of a DEHA-NO_x-hydrocarbon photooxidation in the mTorr reactant concentration range are, for example, in the case of C_2H_4 oxidation, CH_3CHO and its

TABLE 8.9. Rate Constants for the Reactions of $(CH_3)_2NNO$ and $(CH_3)_2NNO_2$ with OH and O_3 at 298°K and Estimated Atmospheric Lifetimes

Molecule	k^{OH} (cm^3 $molecule^{-1}$ s^{-1})	$\tau^{OH\,a}$	k^{O_3} (cm^3 $molecule^{-1}$ s^{-1})	$\tau^{O_3\,b}$
$(CH_3)_2NNO$	$(3.0 \pm 0.4) \times 10^{-12}$	4 days	$\leqslant 1 \times 10^{-20}$	$\geqslant 1.3$ yr
$(CH_3)_2NNO_2$	$(4.5 \pm 0.5) \times 10^{-12}$	3 days	$\leqslant 3 \times 10^{-21}$	$\geqslant 4.2$ yr

Source: Tuazon et al., 1984b.

[a] $\tau^{OH} = 1/k^{OH}[OH]$, where $[OH] = 1 \times 10^6\ cm^{-3}$.

[b] $\tau^{O_3} = 1/k^{O_3}[O_3]$, where $[O_3] = 2.5 \times 10^{12}\ cm^{-3}$ (0.1 ppm).

TABLE 8.10. Rate Coefficients for DEHA Reactions in the Gas Phase in the Temperature Range 298–308°K

Reactant	k (cm^3 molecule^{-1} s^{-1})	Reference
HONO	$\sim 2 \times 10^{-18}$	Niki, 1977
NO_2	4.5×10^{-18}	Jayanty et al., 1974
O_3	$>1 \times 10^{-15}$[a]	Olszyna & Heicklen, 1976
SO_2	$<1 \times 10^{-19}$	Landreth et al., 1975
O_2	1.7×10^{-23}	Caceres et al., 1978
HO	1.0×10^{-10}	Gorse et al., 1977
HO_2	3.3×10^{-13}	Gorse et al., 1978
H	1.1×10^{-11}	Gorse et al., 1978
H	7×10^{-11}	Varas et al., 1977
C_2H_5	1.2×10^{-15}	Abuin et al., 1978
$O(^3P)$	6.6×10^{-12}	Filby and Güsten, 1978

Source: Heicklen, 1981a.

[a] Overall rate coefficient, undoubtedly a chain reaction. The initiating rate coefficient may be smaller.

oxidation products (HCHO, CO, and CO_2), N_2O, HONO, C_2H_5OH, $C_2H_5ONO_2$, and $C_2H_5NO_2$.

Table 8.10 summarizes some of the rate constants for $(C_2H_5)_2NOH$ reactions with species of interest in the atmosphere; it reacts quite rapidly with OH and with O_3.

The mechanism of oxidation of this amine proposed by Heicklen and co-workers involves free radical attack at the weak NO—H bond, for example, for OH,

$$OH + (C_2H_5)_2NOH \rightarrow (C_2H_5)_2NO + H_2O \qquad (103)$$

The intermediate nitroxide radical is removed by abstraction of an α hydrogen atom to form a nitrone:

$$R + (C_2H_5)_2NO \rightarrow C_2H_5\overset{\overset{\displaystyle O}{\uparrow}}{N}{=}CHCH_3 + RH \qquad (104)$$

(R is a radical species). The fate of the nitrone is not known, although $C_2H_5NO_2$ and CH_3CHO appear to be the stable oxidation products (Heicklen, 1981a).

Reaction with O_3 is proposed to occur via (105) and (106):

$$O_3 + (C_2H_5)_2NOH \rightarrow OH + O_2 + (C_2H_5)_2NO \qquad (105)$$

$$O_3 + (C_2H_5)_2NO \rightarrow OH + O_2 + C_2H_5\overset{\overset{\displaystyle O}{\uparrow}}{N}{=}CHCH_3 \qquad (106)$$

The use of DEHA to inhibit photochemical air pollution is not generally believed to be a viable control strategy option. For example, addition of 0.05–0.5 ppm DEHA to polluted ambient air, with or without an added hydrocarbon mixture, led to significant *increases* in the rates of formation and peak concentrations of O_3, PAN, and light-scattering particles (Pitts et al., 1977). Only at very high concentrations of added DEHA ($\sim$2 ppm) were the usual manifestations of photochemical smog inhibited for a 6 h irradiation; the effects at larger irradiation times or on alternating light/dark periods characteristic of multi-day exposures are not known. However, since the odor threshold of DEHA is $\sim$0.5 ppm, addition of such high concentrations is impractical. This and other problems with the proposed use of DEHA as a control strategy are discussed by Cupitt (1981).

8-C. DISTRIBUTION OF NITROGEN BETWEEN ORGANIC AND INORGANIC COMPOUNDS

Given the many different nitrogen compounds, both inorganic and organic, which can be formed in the atmosphere, it is worthwhile to ask which forms of oxidized nitrogen actually predominate in ambient air and in studies carried out in environmental chambers. A review of this field to the mid-1970s is given by Spicer (1977).

8-C-1. Ambient Air

Atmospheric studies of the distribution of nitrogen compounds have been hampered to some extent by the lack of sensitive, specific, accurate, and artifact-free methods for measuring some of the end products such as nitric acid and particulate nitrate. However, over the last decade, significant progress has been made in the area of analytical techniques (see Chapters 5 and 12), allowing more extensive studies to be carried out. However, artifacts have undoubtedly been present in many of the earlier studies and may still occur; for example, it is difficult to sample for particulate nitrate without absorbing some HNO_3 from the gas phase, leading to a positive artifact (Section 12-E-4e). The severity of such artifacts will depend, of course, on the particular pollutant concentrations during sampling, as well as the sampling methods used. Thus the reader should be cautioned that some of the results cited below may be best interpreted as approximate estimates of the distributions rather than quantitative, accurate measurements.

Most studies of the distribution of oxidized nitrogen in ambient air have shown that PAN and HNO_3 are the nitrogenous products present in the greatest concentrations. Smaller amounts of particulate nitrate are present. Organic nitrates seem to account for very little of the oxidized nitrogen. For example, Spicer and coworkers have studied the distribution of nitrogen between NO, NO_2, HNO_3, PAN, and particulate nitrate at a number of locations in the continental United States, as well as in urban plumes. Figures 8.12 and 8.13 show the pollutant profiles for some of these pollutants as a function of the time of day in two rather different locations. Figure 8.12 shows the data for Claremont, California, a city in the Los

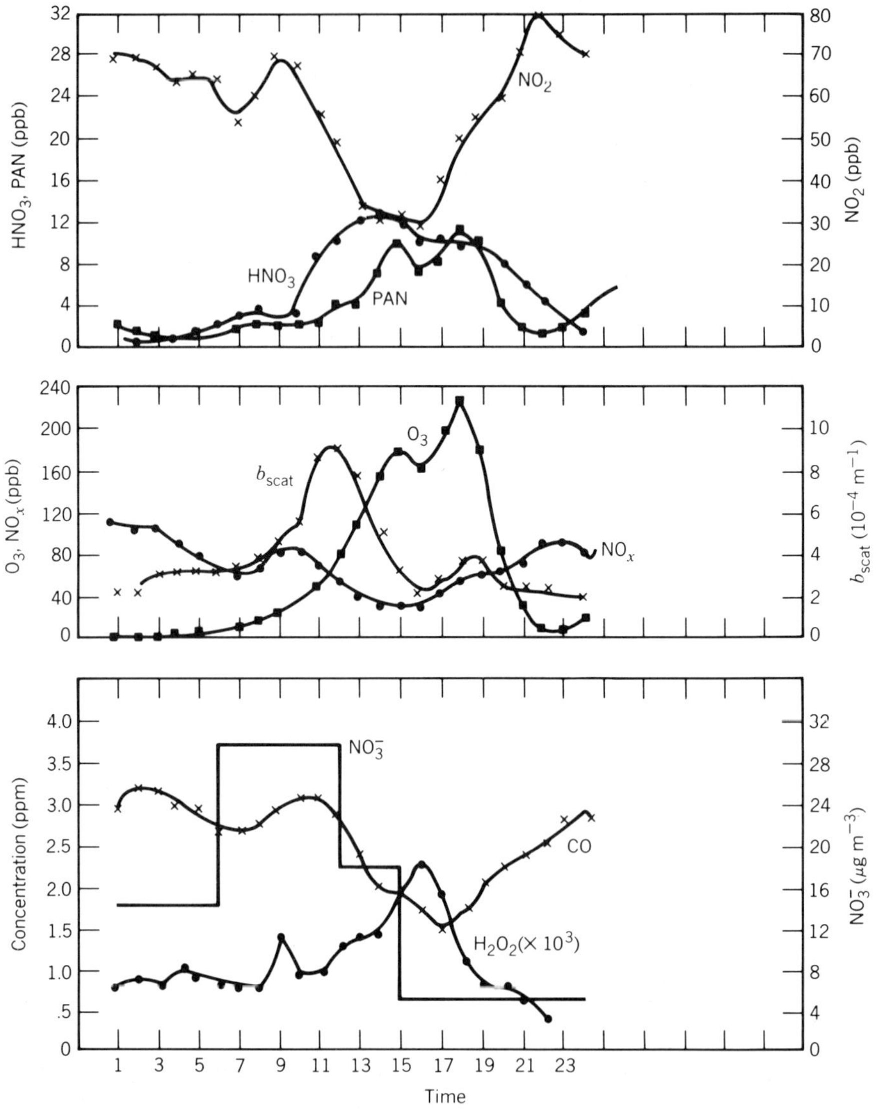

FIGURE 8.12. Pollutant profiles for September 1, 1979 in Claremont, California; NO_3 is particulate nitrate (from Spicer, 1982a).

Angeles air basin about 50 km east of downtown Los Angeles, while Fig. 8.13 shows the data for a rural site in the New Jersey Pine Barrens about 80 km downwind of Philadelphia. Two of the oxidized products, PAN and HNO_3, peak in the mid-afternoon at both sites. However, PAN and HNO_3 formed a smaller fraction of the total oxidized nitrogen at the more urban Claremont site than at the rural New Jersey site. For example, at ~1600 EDT at the New Jersey site when the PAN and HNO_3 concentrations peaked, they accounted for 63% of the gaseous

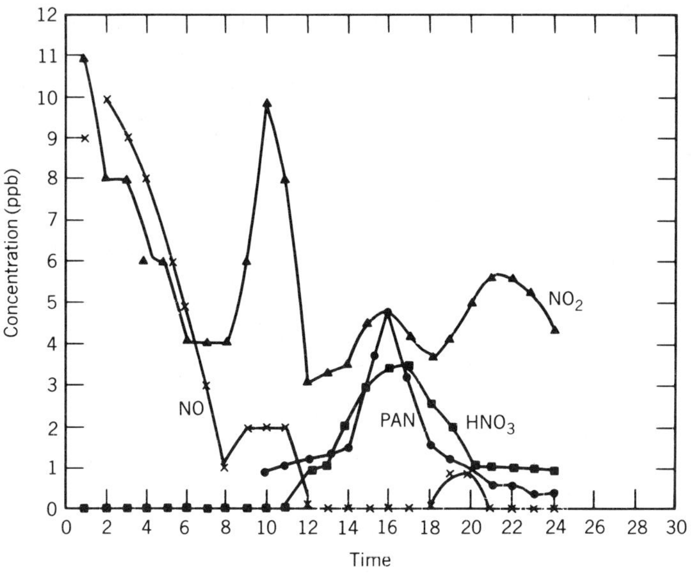

FIGURE 8.13. Oxidized nitrogen distributed downwind of Philadelphia, July 22, 1979 (from Spicer, 1982a).

oxidized nitrogen, that is, $(PAN + HNO_3)/(NO + NO_2 + PAN + HNO_3) = 0.63$. On the other hand, in Claremont, the PAN and HNO_3 accounted for, at most, ~40% of the gaseous oxidized nitrogen. The relative amounts of PAN and HNO_3 also differed; the ratio PAN/HNO_3 was ~1.3 at the New Jersey site at the peak, but only ~0.8 at the first PAN peak in Claremont. These differences were attributed by Spicer to the greater distance and thus longer reaction times between the upwind urban source area and the New Jersey sampling site; these longer reaction times would result not only in greater oxidation but also longer times for the deposition of pollutants to occur. Thus if HNO_3 undergoes both wet and dry deposition faster than PAN, which seems to be the case (see Chapter 11), longer travel times to the sampling sites will result in relatively less HNO_3—or a higher PAN/HNO_3 ratio.

A parameter that expresses the extent of oxidation of NO and NO_2 to oxidized products in ambient air is the fractional conversion F_n defined by

$$F_n = \frac{HNO_3 + PAN + \text{particulate nitrate}}{NO + NO_2 + HNO_3 + PAN + \text{particulate nitrate}}$$

One might expect higher values of F_n in air masses that have had longer reaction times and during periods of high photochemical activity (e.g., summer compared to winter) and lower values in urban areas where there are fresh sources of NO and NO_2. This is indeed what is found; for example, at West Covina, about 40 km downwind of downtown Los Angeles, F_n averaged ~0.11 over a 5 week period, whereas at a rural site in New Jersey about 75 km downwind of Philadephia $F_n \simeq 0.22$. In the urban plume from Boston which was tracked out over the Atlantic Ocean, F_n was as large as 0.7 at 7 h travel time (Spicer, 1982a,b).

Figure 8.14 clearly demonstrates the relationship between photochemical activity and the fractional conversion to PAN and HNO_3. In this case, particulate nitrate

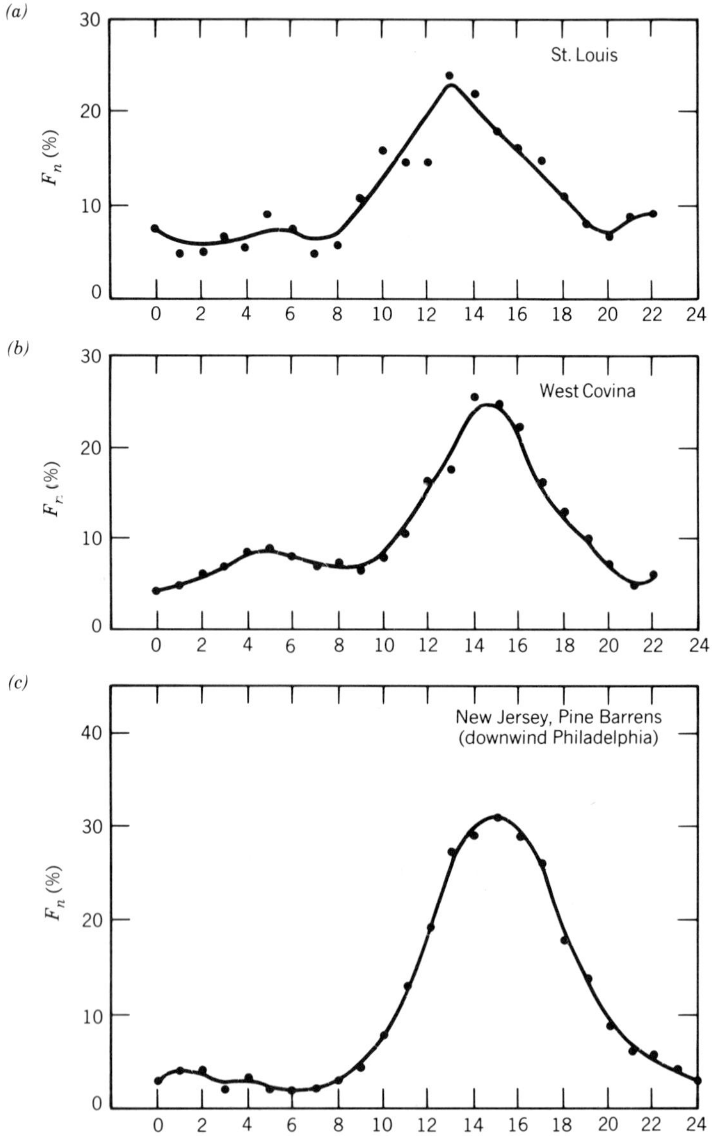

FIGURE 8.14. Composite diurnal profiles of F_n at three locations: (*a*) St. Louis, Missouri, (*b*) West Covina, California, and (*c*) Pine Barrens, New Jersey (from Spicer, 1982a).

data were not available so that F_n represents the ratio (HNO_3 + PAN)/(NO + NO_2 + HNO_3 + PAN). The three sites for which data are shown in Fig. 8.14 are quite different in their nature, yet all three show the maximum conversion to HNO_3 and PAN during the time usually associated with peak photochemical activity in the atmosphere.

These results are consistent with studies showing that F_n is relatively low during "clean" periods in the Los Angeles area but rises during smog episodes. For example, Grosjean (1983) found that F_n was ~2–8% on relatively clean days in Claremont, California, but rose to ~40% during smog episodes when oxidant concentrations were much higher.

Table 8.11 summarizes the percentage of the total measured oxidized nitrogen for which each of the pollutants NO, NO_2, HNO_3, PAN, and particulate nitrate is responsible, averaged (for most data) over several weeks of sampling in the summertime during daylight hours. Spicer (1982a) points out that the data from the 1979 studies are the most reliable due to improvements in analytical techniques over the period encompassing these studies. NO_2 clearly accounts for the largest percentage at all sites, with NO accounting for $\gtrsim$30% only at the two sampling sites located in city centers. PAN was responsible for between 2 and 13% of the oxidized nitrogen, HNO_3 for up to 17% (although it was below the detection limit in two cases), and particulate nitrate $\lesssim$2%, except at Rubidoux and Claremont. The higher values at the latter two locations were attributed to the presence of local sources of NH_3 which would form particulate ammonium nitrate by reaction with HNO_3.

The results are consistent with other studies of the distribution of nitrogen-containing compounds in ambient air. For example, Grosjean (1983) showed that PAN was responsible for 30–80% of the NO_x oxidation products, with PAN generally accounting for more of the nitrogen (~60%) than HNO_3 + particulate nitrate on smoggy days. It is noteworthy that the concentration of one alkyl nitrate, CH_3ONO_2, which was measured simultaneously, was small ($\leq$5 ppb) compared to PAN and HNO_3.

While such studies have established the importance of PAN as a product of NO_x oxidation in ambient air, it is important to remember that PAN does not serve as a chemical sink in the same way as HNO_3. As discussed in Section 8-B-1, PAN thermally decomposes to regenerate NO_2 as well as $CH_3\overset{\overset{O}{\|}}{C}OO$ radicals; thus it can act as an important non-photolytic source of free radicals at night, as well as accelerating the formation of photochemical smog at dawn.

8-C-2. Simulated Atmospheres

The results of smog chamber studies confirm the general conclusions from ambient air analysis regarding the distribution of oxidized nitrogen. Thus PAN and HNO_3 are observed to be the major nitrogen-containing products; however, particulate nitrate concentrations tend to be lower than observed in ambient air, likely because

TABLE 8.11. Oxidized Nitrogen Distribution at Selected Locations[a] (Percent of Total)

Location:	West Covina, CA	Phoenix, AZ	Rubidoux, CA	St. Louis, MO	Quail Farm, NJ	Claremont, CA[d]
Year:	1973	1977	1976	1973	1979	1979
Type:	Urban	Urban	Suburban	Urban	Rural	Urban
Distance from Source:	40 km from Los Angeles	0 km	85 km from Los Angeles	0 km	80 km from Philadelphia	50 km from Los Angeles
NO	23	45	<3	33	11	5
NO_2	57	51	73	45	59	69
HNO_3	6	*b*	*b*	12	17	12
PAN	13	2	10	9	11	6
Particulate NO_3^-	0.5[c]	2	12	1[c]	—	7
PAN/HNO_3	2.2			0.8	0.6	0.2–0.5[c,e,f]

Source: Adapted from Spicer (1982a).

[a]Based on 8 h daylight samples unless otherwise specified.

[b]Generally below instrument detection limit.

[c]24 Hour averages.

[d]Only 8 days of data available.

[e]Tuazon et al., 1981b.

[f]Ratio of peak concentrations was 0.3–0.8.

chamber studies normally do not include NH_3 which, as discussed earlier, forms particulate ammonium nitrate with HNO_3.

Obtaining good nitrogen mass balances in smog chambers is difficult because of the tendency for HNO_3 to adsorb on the walls of the chamber. Thus over a typical 6–10 h chamber run, much of the HNO_3 formed in the gas phase may disappear to the chamber walls, necessitating substantial corrections to the data in order to estimate the total HNO_3 formed.

One recent example of a study in which nitrogenous products were measured and corrected for HNO_3 loss to the walls is that of Spicer (1983a). Figure 8.15 shows the concentration–time profiles for NO, NO_2, O_3, PAN, and gaseous HNO_3 in a typical chamber run. HNO_3 peaks at ~2–3 h and then slowly decreases, even though NO_2 was still present and PAN and O_3 were increasing; this decrease is attributed to surface loss of HNO_3, with some contribution from dilution of the system during the run.

Figure 8.16 shows the concentrations of the major nitrogen-containing products as a function of reaction time during a typical smog chamber experiment. Curve I is the concentration of NO + NO_2 expected if no reaction occurred and the concentrations decreased only from dilution of the chamber contents. Curve III is the sum of the measured concentrations of NO, NO_2, PAN, and HNO_3. The difference between the measured concentrations and those expected from the concentrations of the initial reactions increases significantly during the run; at 6 h reaction time, less than half of the initial nitrogen is present in measured gas phase products.

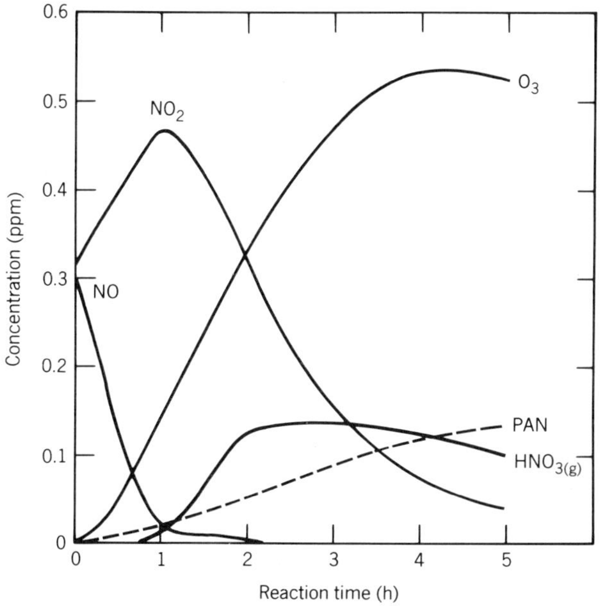

FIGURE 8.15. Profiles of selected reactants and products for a typical smog chamber run. Initial concentrations were 9.30 ppmC NMHC, 0.30 ppm NO, and 0.32 ppm NO_2 (from Spicer, 1983a).

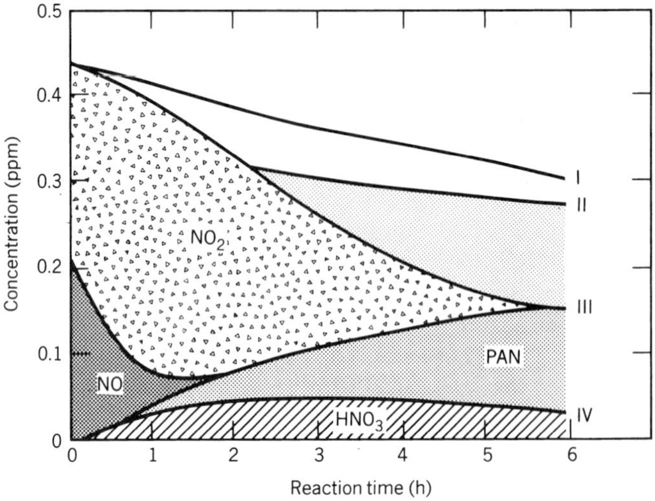

FIGURE 8.16. Cumulative plot of oxidized nitrogen compounds for a typical smog chamber run. Initial concentrations were 4.55 ppmC NMHC, 0.21 ppm NO, and 0.227 ppm NO_2 (from Spicer, 1983a).

In separate studies, the rate of HNO_3 loss to the chamber walls was determined. Using this rate, Spicer calculated the amount of HNO_3 expected to be adsorbed on the chamber walls; this is shown by the shaded area between curves II and III. When this adsorbed HNO_3 is taken into account, only ~10% of the nitrogen is unaccounted for at the end of the run.

It is interesting to compare the fraction of the total NO_x present as (PAN + HNO_3 + particulate nitrate) to the values of this fraction, F_n, observed in ambient air studies. Unlike ambient air work, environmental chamber studies allow one to control the initial reactant concentrations and, in particular, the ratio of hydrocarbon to NO_x concentrations which is a critical factor in determining the concentration–time profiles and products (see Chapters 6 and 10). As might be expected, the value F_n appears to depend on the non-methane hydrocarbon (NMHC) to NO_x ratio. Figure 8.17 shows F_n as a function of the NMHC/NO_x ratio after 3 h of irradiation; F_n increases approximately linearly with this ratio up to NMHC/NO_x $\simeq$ 20. At NMHC/NO_x $\simeq$ 5–10, typical of many urban areas, $F_n \simeq$ 0.25–0.50, that is, 25–50% of the initial NO_x has been converted to products within ~3 h. These values are consistent with those observed in ambient air studies discussed earlier.

The distribution of nitrogen between PAN and HNO_3 also depends on the NMHC/NO_x ratio. Figure 8.18 shows the ratio of the final concentration of PAN to HNO_3 (sum of both gaseous and adsorbed) as a function of initial NMHC/NO_x. The increase in PAN relative to HNO_3 as the NMHC/NO_x ratio increases is understandable because increasing hydrocarbon concentrations will increase the concentration of $CH_3\overset{\overset{\displaystyle O}{\|}}{C}OO$ radicals which form PAN, and hence make the

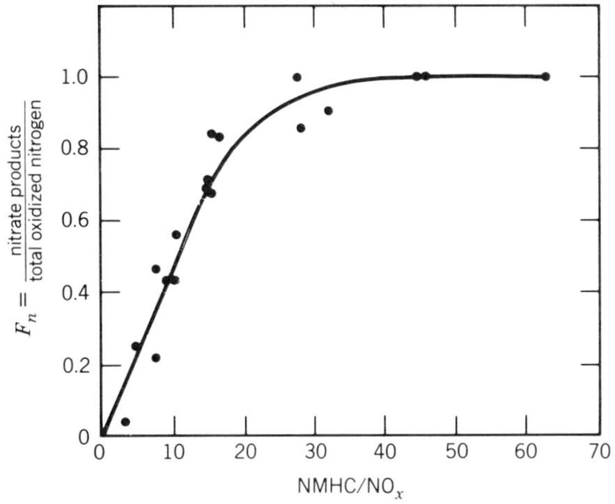

FIGURE 8.17. F_n at 3 h versus NMHC/NO$_x$ in a series of smog chamber experiments (from Spicer, 1983a).

$CH_3\overset{\overset{O}{\|}}{C}OO + NO_2$ reaction more competitive with the $OH + NO_2$ reaction forming HNO_3. At typical urban NMHC/NO$_x$ ratios of ~5–10, PAN/HNO$_3$ ≃ 0.2–0.4, that is, HNO_3 exceeds PAN by factors of ~2–5. The observed PAN/HNO$_3$ ratio in ambient air is often found to be somewhat greater than this (Table 8.11). This may be due to a number of factors such as the influence of fresh emissions and the faster removal of HNO_3 compared to PAN via deposition and particle formation.

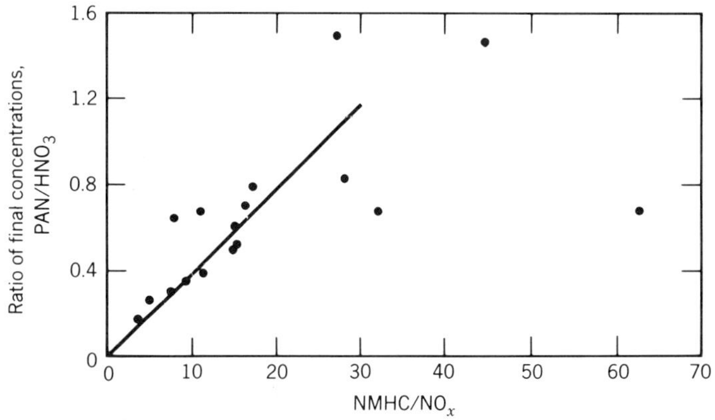

FIGURE 8.18. Ratio of final concentrations of PAN to HNO_3 (PAN/HNO$_3$) versus initial NMHC/NO$_x$ in a series of smog chamber experiments. The HNO_3 includes both that in the gas phase and that estimated to be adsorbed on the chamber walls (from Spicer, 1983).

REFERENCES

Abuin, E., M. V. Encina, S. Diaz, and E. A. Lissi, "On the Reactivity of Diethylhydroxylamine Toward Free Radicals," *Int. J. Chem. Kinet.*, **10,** 677 (1978).

Aikin, A. C., J. R. Herman, E. J. R. Maier, and C. J. McQuillan, "Influence of Peroxyacetyl Nitrate (PAN) on Odd Nitrogen in the Troposphere and Lower Stratosphere," *Planet. Space Sci.*, **31,** 1075 (1983).

Appel, B. R., "A New and More Sensitive Procedure for Analysis of Peroxybenzoyl Nitrate," *J. Air Pollut. Control Assoc.*, **23,** 1042 (1973).

Atkinson, R. and A. C. Lloyd, "Evaluation of Kinetic and Mechanistic Data for Modeling of Photochemical Smog," *J. Phys. Chem. Ref. Data*, **13,** 315 (1984).

Atkinson, R., R. A. Perry, and J. N. Pitts, Jr., "Rate Constants for the Reaction of the OH Radical with CH_3SH and CH_3NH_2 Over the Temperature Range 299–426°K," *J. Chem. Phys.*, **66,** 1578 (1977).

Atkinson, R., R. A. Perry, and J. N. Pitts, Jr., "Rate Constants for the Reactions of the OH Radical with $(CH_3)_2NH$, $(CH_3)_3N$, and $C_2H_5NH_2$ Over the Temperature Range 298–426°K," *J. Chem. Phys.*, **68,** 1850 (1978).

Atkinson, R., S. M. Aschmann, A. M. Winer, and J. N. Pitts, Jr., "Gas Phase Reactions of NO_2 with Alkenes and Dialkenes," *Int. J. Chem. Kinet.*, **16,** 697 (1984a).

Atkinson, R., C. N. Plum, W. P. L. Carter, A. M. Winer, and J. N. Pitts, Jr., "Rate Constants for the Gas-Phase Reactions of Nitrate Radicals with a Series of Organics in Air at 298 ± 1 K," *J. Phys. Chem.*, **88,** 1210 (1984b).

Atkinson, R., A. M. Winer, and J. N. Pitts, Jr., "Estimation of Nighttime N_2O_5 Concentrations from Ambient NO_2 and NO_3 Radical Concentrations and The Role of N_2O_5 in Nighttime Chemistry," *Atmos. Environ.*, **19,** 000 (1985).

Borrell, P., C. J. Cobos, A. E. Croce de Cobos, H. Hippler, K. Luther, A. R. Ravishankara, and J. Troe, "Radical Association Reactions in Gases at High Pressures," *Ber. Bunsenges. Phys. Chem.*, **89,** 337 (1985).

Burrows, J. P., G. S. Tyndall, and G. K. Moortgat, "A Study of the N_2O_5 Equilibrium Between 275 and 315 K and Determination of the Heat of Formation of NO_3," *Chem. Phys. Lett.*, **119,** 193 (1985).

Burrows, J. P., G. S. Tyndall and G. K. Moortgat, "Absorption Spectrum of NO_3 and Kinetics of the Reactions of NO_3 with NO_2, Cl, and Several Stable Atmospheric Species at 298°K," *J. Phys. Chem.*, **89,** 4848 (1985).

Caceres, T., E. A. Lissi, and E. Sanhueza, "Autooxidation of Diethylhydroxylamine," *Int. J. Chem. Kinet.*, **10,** 1167 (1978).

Calvert, J. G. and J. N. Pitts, Jr., *Photochemistry*, Wiley, New York, 1966.

Calvert, J. G. and W. R. Stockwell, "Deviations from the O_3–NO–NO_2 Photostationary State in Tropospheric Chemistry," *Can. J. Chem.*, **61,** 983 (1983).

Cantrell, C. A., W. R. Stockwell, L. G. Anderson, K. L. Busarow, D. Perner, A. Schmeltekopf, J. G. Calvert, and H. S. Johnston, "Kinetic Study of the NO_3–CH_2O Reaction and Its Possible Role in Nighttime Tropospheric Chemistry," *J. Phys. Chem.*, **89,** 139, 4160 (1985).

Carter, W. P. L., A. M. Winer, and J. N. Pitts, Jr., "Effect of Peroxyacetyl Nitrate on the Initiation of Photochemical Smog," *Environ. Sci. Technol.*, **15,** 831 (1981a).

Carter, W. P. L., E. C. Tuazon, A. M. Winer, and J. N. Pitts, Jr., Gas Phase Reactions of *N,N*-Dimethylhydrazine with Ozone and NO_x in Simulated Atmospheres, in *N-Nitroso Compounds*, R. A. Scanlan and S. R. Tannenbaum, Eds., ACS Symposium Series No. 174, American Chemical Society, Washington, DC, 1981b, pp. 117–131.

Carter, W. P. L., A. M. Winer, and J. N. Pitts Jr., "Major Atmospheric Sink for Phenol and the Cresols: Reaction with the Nitrate Radical," *Environ. Sci. Technol.*, **15,** 829 (1981).

Chameides, W. L. and D. D. Davis, The Coupled Gas-Phase/Aqueous-Phase Free Radical Chemistry of a Cloud, in *Precipitation Scavenging, Dry Deposition, and Resuspension*, H. R. Pruppacher, R. G. Semonin, and W. G. N. Slinn, Eds., Elsevier, New York, 1983, pp. 431–443.

Chan, W. H., R. J. Nordstrom, J. G. Calvert, and J. H. Shaw, "Kinetic Study of HONO Formation and Decay Reactions in Gaseous Mixtures of HONO, NO, NO_2, H_2O, and N_2," *Environ. Sci. Technol.*, **10,** 674 (1976).

Connell, P. S. and C. J. Howard, "Kinetics Study of the Reaction of OH + HNO_3," *Int. J. Chem. Kinet.*, **17,** 17 (1985).

Cox, R. A., R. G. Derwent, and P. M. Holt, "Relative Rate Constants for the Reactions of OH Radicals with H_2, CH_4, CO, NO, and HONO at Atmospheric Pressure and 296°K," *J. Chem. Soc. Faraday Trans. I*, **72,** 2031 (1976).

Cupitt, L., "Control of Photochemical Smog by Diethylhydroxylamine," *Atmos. Environ.*, **15,** 416 (1981).

Demerjian, K. L., K. L. Schere, and J. T. Peterson, "Theoretical Estimates of Actinic (Spherically Integrated) Flux and Photolytic Rate Constants of Atmospheric Species in the Lower Troposphere," *Adv. Environ. Sci. Technol.*, **10,** 369 (1980).

DeMore, W. B., M. J. Molina, R. T. Watson, D. M. Golden, R. F. Hampson, M. J. Kurylo, C. J. Howard, and A. R. Ravishankara, "Chemical Kinetics and Photochemical Data for Use in Stratospheric Modeling, Evaluation No. 6," JPL Publication 83-62, September 15, 1983.

DeMore, W. B., J. J. Margitan, M. J. Molina, R. T. Watson, D. M. Golden, R. F. Hampson, M. J. Kurylo, C. J. Howard, and A. R. Ravishankara, "Chemical Kinetics and Photochemical Data for Use in Stratospheric Modeling, Evaluation No. 7," JPL Publication 85-37, 1985.

Devolder, P., M. Carlier, J. F. Pauwels, and L. R. Sochet, "Rate Constant for the Reaction of OH with Nitric Acid: A New Investigation by Discharge Flow Resonance Fluorescence," *Chem. Phys. Lett.*, **111,** 94 (1984).

Dickerson, R. R., "Reactive Nitrogen Compounds in the Artic," *J. Geophys. Res.*, **90,** 10739 (1985).

Dickerson, R. R., D. H. Stedman, and A. C. Delany, "Direct Measurements of Ozone and Nitrogen Dioxide Photolysis Rates in the Troposphere," *J. Geophys. Res.*, **87,** 4933 (1982).

Doyle, G. J., E. C. Tuazon, R. A. Graham, T. M. Mischke, A. M. Winer, and J. N. Pitts, Jr., "Simultaneous Concentrations of Ammonia and Nitric Acid in a Polluted Atmosphere and Their Equilibrium Relationship to Particulate Ammonium Nitrate," *Environ. Sci. Technol.*, **13,** 1416 (1979).

Durham, J. L., J. H. Overton, Jr., and V. P. Aneja, "Influence of Gaseous Nitric Acid on Sulfate Production and Acidity in Rain," *Atmos. Environ.*, **15,** 1059 (1981).

Edney, E. O., J. W. Spence, and P. L. Hanst, Peroxy Nitrate Air Pollutants: Synthesis and Thermal Stability, in *Nitrogenous Air Pollutants*, D. Grosjean, Ed., Ann Arbor Science, Publishers, Ann Arbor, MI, 1979a, pp. 111–135.

Edney, E. O., J. W. Spence, and P. L. Hanst, "Synthesis and Thermal Stability of Peroxy Alkyl Nitrates," *J. Air Pollut. Control Assoc.*, **29,** 741 (1979b).

Filby, W. G. and H. Güsten, "Rate Constants for the Reaction of Oxygen Atoms with Some Potential Photosmog Inhibitors," *Atmos. Environ.*, **12,** 1563 (1978).

Fine, D. H., G. S. Edwards, I. S. Krull, and M. H. Wolf, *N*-Nitroso Compounds in the Air Environment, in *Nitrogenous Air Pollutants,* D. Grosjean, Ed., Ann Arbor Science Publishers, Ann Arbor, MI, 1979, pp. 55–64.

Fine, D. H., "*N*-Nitroso Compounds in the Environment," *Adv. Environ. Sci. Technol.*, **10,** 40 (1980).

Finlayson-Pitts, B. J., "The Reaction of NO_2 with NaCl and Atmospheric Implications of NOCl Formation," *Nature,* **306,** 676 (1983).

Gay, B. W., Jr., R. C. Noonan, J. J. Bufalini, and P. L. Hanst, "Photochemical Synthesis of Peroxyacyl Nitrates in Gas Phase via Chlorine–Aldehyde Reaction," *Environ. Sci. Technol.*, **10,** 82 (1976).

Geiger, G., H. Stafast, U. Bruhlmann, and J. R. Huber, "Photodissociation of Dimethylnitrosamine," *Chem. Phys., Lett.*, **79,** 521 (1981).

Gorse, R. A., Jr., R. R. Lii, and B. B. Saunders, "Hydroxyl Radical Reactivity with Diethylhydroxylamine," *Science,* **197,** 1365 (1977).

Gorse, R. A., Jr., R. R. Lii, and B. B. Saunders, "Reactivity of Diethylhydroxylamine, DEHA," Paper presented at the 71st Air Pollution Control Association Annual Meeting, Paper 78-59.3, Houston, Texas, 1978.

Graham, R. A. and H. S. Johnston, "The Photochemistry of NO_3 and the Kinetics of the N_2O_5–O_3 System," *J. Phys. Chem.*, **82,** 254 (1978).

Graham, R. A., A. M. Winer, and J. N. Pitts, Jr., "Temperature Dependence of the Unimolecular Decomposition of Pernitric Acid and Its Atmospheric Implications," *Chem. Phys. Lett.*, **51,** 215 (1977).

Graham, R. A., A. M. Winer, and J. N. Pitts, Jr., "Pressure and Temperature Dependence of the Unimolecular Decomposition of HO_2NO_2," *J. Chem. Phys.*, **68,** 4505 (1978).

Grosjean, D., "The Stability of Particulate Nitrate in the Los Angeles Atmosphere," *Sci. Total Environ.*, **25,** 263 (1982).

Grosjean, D., "Distribution of Atmospheric Nitrogenous Pollutants at a Los Angeles Area Smog Receptor Site," *Environ. Sci. Technol.*, **17,** 13 (1983).

Grosjean, D., K. Van Cauwenberghe, J. Schmid, and J. N. Pitts, Jr., "Formation of Nitrosamines and Nitramines by Photooxidation of Amines Under Simulated Atmospheric Conditions," *Proceedings of the 4th Joint Conference on Sensing of Environmental Pollutants,* November 6–11, 1977, New Orleans, Louisiana, American Chemical Society, Washington, D.C. 1978, pp. 196–199.

Hampson, R. F., Jr. and D. Garvin, "Reaction Rate and Photochemical Data for Atmospheric Chemistry—1977," National Bureau of Standards Spec. Publ. 513, May 1978.

Hanst, P. L., J. W. Spence, and M. Miller, "Atmospheric Chemistry of *N*-Nitroso Dimethylamine," *Environ. Sci. Technol.*, **11,** 403 (1977).

Harris, G. W., R. Atkinson, and J. N. Pitts, Jr., "Kinetics of the Reactions of the OH Radical with Hydrazine and Methylhydrazine," *J. Phys. Chem.*, **83,** 2557 (1979).

Harris, G. W., W. P. L. Carter, A. M. Winer, J. N. Pitts, Jr., U. Platt, and D. Perner, "Observations of Nitrous Acid in the Los Angeles Atmosphere and Implications for Predictions of Ozone-Precursor Relationships," *Environ. Sci. Technol.*, **16,** 414 (1982).

Harrison, R. M. and C. A. Pio, "An Investigation of the Atmospheric HNO_3–NH_3–NH_4NO_3 Equilibrium Relationship in a Cool, Humid Climate," *Tellus,* **35B,** 155 (1983).

Heicklen, J., *Atmospheric Chemistry,* Academic Press, New York, 1976.

Heicklen, J., "Control of Photochemical Smog by Diethylhydroxylamine," *Atmos. Environ.*, **15,** 229 (1981a).

Heicklen, J., "Authors Reply to 'Control of Photochemical Smog by Diethylhydroxylamine," *Atmos. Environ.*, **15,** 420 (1981b).

Heicklen, J. "The Inhibition of Photochemical Smog-IX. Model Calculations of the Effect of $(C_2H_5)_2$ NOH on C_2H_4–NO_x Atmospheres," *Atmos. Environ.*, **19,** 597 (1985).

Heikes, B. G. and A. M. Thompson, "Effects of Heterogeneous Processes on NO_3, HONO, and HNO_3 Chemistry in the Troposphere," *J. Geophys. Res.*, **88C,** 10, 883 (1983).

Hendry, D. G. and R. A. Kenley, Atmospheric Chemistry of Peroxynitrates, in *Nitrogenous Air Pollutants,* D. Grosjean, Ed., Ann Arbor Science Publishers, Ann Arbor, MI, 1979, pp. 137–148.

Heuss, J. M. and W. A. Glasson, "Hydrocarbon Reactivity and Eye Irritation," *Environ. Sci. Technol.*, **2,** 1109 (1968).

Hildemann, L. M., A. G. Russell, and G. R. Cass, "Ammonia and Nitric Acid Concentrations in Equilibrium with Atmospheric Aerosols: Experiment vs. Theory," *Atmos Environ.*, **18,** 1737 (1984).

Holdren, M. W., C. W. Spicer, and J. M. Hales, "Peroxyacetyl Nitrate Solubility and Decomposition Rate in Acidic Water," *Atmos. Environ.*, **18,** 1171 (1984).

Hov, O., "Modelling of the Long-Range Transport of Peroxyacetyl Nitrate to Scandinavia," *J. Atmos. Chem.*, **1,** 187 (1984).

Howard, C. J., "Kinetics of the Reaction of HO_2 with NO_2," *J. Chem. Phys.*, **67,** 5258 (1977).

Jacob, D. J. and M. R. Hoffman, "A Dynamic Model for the Production of H^+, NO_3^-, and SO_4^{2-} in Urban Fog," *J. Geophys. Res.*, **88,** 6611 (1983).

Japar, S. M. and H. Niki, "Gas-Phase Reactions of the Nitrate Radical with Olefins," *J. Phys. Chem.*, **79,** 1629 (1975).

Jayanty, R. K. M., R. Simonaitis, and J. Heicklen, "The Inhibition of Photochemical Smog. III. Inhibition by Diethylhydroxylamine, *N*-Methylaniline, Triethylamine, Diethylamine, Ethylamine, and Ammonia," *Atmos. Environ.*, **8,** 1283 (1974).

Jolly, G. S., G. Paraskevopoulos, and D. L. Singleton, "The Question of a Pressure Effect in the Reaction OH + HNO_3. A Laser Flash-Photolysis Resonance-Absorption Study," *Chem. Phys. Lett.*, **117,** 132 (1985).

Jones, C. L. and J. H. Seinfeld, "The Oxidation of NO_2 to Nitrate—Day and Night," *Atmos. Environ.*, **17,** 2370 (1983).

Kaiser, E. W. and C. H. Wu, "A Kinetic Study of the Gas Phase Formation and Decomposition Reactions of Nitrous Acid," *J. Phys. Chem.*, **81,** 1701 (1977).

Killus, J. P. and G. Z. Whitten, "Behaviour of Trace NO_x Species in the Nighttime Urban Atmosphere," *J. Geophys. Res.*, **90,** 2430 (1985).

Kircher, C. C., J. J. Margitan, and S. P. Sander, "Pressure and Temperature Dependence of the Reaction $NO_2 + NO_3 + M \rightarrow N_2O_5 + M$," *J. Phys. Chem.*, **88,** 4370 (1984).

Lamb, J. J., M. Mozurkewich, and S. W. Benson, "Negative Activation Energies and Curved Arrhenius Plots. 3. OH + HNO_3 and OH + HNO_4," *J. Phys. Chem.*, **88,** 6441 (1984).

Landreth, R., L. Stockburger III, and J. Heicklen, "The Reaction of SO_2 with *N,N*-Diethylhydroxylamine," Center of Air Environment Studies Report No. 399-75, Pennsylvania State University, 1975.

Larson, T. V. and G. S. Taylor, "On the Evaporation of Ammonium Nitrate Aerosol," *Atmos. Environ.*, **17,** 2489 (1983).

Lee, Y.-N. and S. E. Schwartz, "Evaluation of the Rate of Uptake of Nitrogen Dioxide by Atmospheric and Surface Liquid Water," *J. Geophys. Res.*, **86,** 11,971 (1981).

Lindley, C. R. C., J. G. Calvert, and J. H. Shaw, "Rate Studies of the Reactions of the $(CH_3)_2N$ Radical with O_2, NO, and NO_2," *Chem. Phys. Lett.*, **67,** 57 (1979).

Malko, M. W. and J. Troe, "Analysis of the Unimolecular Reaction $N_2O_5 + M \rightleftarrows NO_2 + NO_2 + M$," *Int. J. Chem. Kinet.*, **14,** 399 (1982).

Markovits, G. Y., S. E. Schwartz, and L. Newman, "Hydrolysis Equilibrium of Dinitrogen Trioxide in Dilute Acid Solution," *Inorg. Chem.*, **20,** 445 (1981).

Molina, L. T. and M. J. Molina, "UV Absorption Cross Sections of HO_2NO_2 Vapor," *J. Photochem.*, **15,** 97 (1981).

Morris, E. D., Jr. and H. Niki, "Reaction of Dinitrogen Pentoxide with Water," *J. Phys. Chem.*, **77,** 1929 (1973).

Morris, E. D., Jr. and H. Niki, "Reaction of the Nitrate Radical with Acetaldehyde and Propylene," *J. Phys. Chem.*, **78,** 1337 (1974).

Moiser, A. R., C. E. Andre, and F. G. Viets, Jr., "Identification of Aliphatic Amines Volatilized from Cattle Feedyard," *Environ. Sci. Technol.*, **7,** 642 (1973).

Niki, H., unpublished data, 1977.

Niki, H., P. D. Maker, C. M. Savage, and L. P. Breitenbach, Fourier Transform Infrared (FTIR) Studies of Gaseous and Particulate Nitrogenous Compounds, in *Nitrogenous Air Pollutants*, D. Grosjean, Ed., Ann Arbor Science Publishers, Ann Arbor, MI, 1979, pp. 1–16.

Niki, H., P. D. Maker, C. M. Savage, and L. P. Breitenbach, "An FTIR Spectroscopic Study of the

Reactions Br + CH_3CHO → HBr + CH_3CO and $CH_3C(O)OO$ + NO_2 ⇄ $CH_3C(O)OONO_2$ (PAN)," *Int. J. Chem. Kinet.*, **17,** 525 (1985).

Olszyna, K. and J. Heicklen, "The Inhibition of Photochemical Smog. VI. The Reaction of O_3 with Diethylhydroxylamine," *Sci. Total Environ.*, **5,** 223 (1976).

Parrish, D. D., P. C. Murphy, D. L. Albritton, and F. C. Fehsenfeld, "The Measurement of the Photodissociation Rate of NO_2 in the Atmosphere," *Atmos. Environ.*, **17,** 1365 (1983).

Pate, C. T., R. Atkinson, and J. N. Pitts, Jr., "Rate Constants for the Gas Phase Reaction of Peroxyacetyl Nitrate with Selected Atmospheric Constituents," *J. Environ. Sci. Health*, **A11,** 19 (1976).

Patrick, R. and D. M. Golden, "Kinetics of the Reactions of NH_2 Radicals with O_3 and O_2," *J. Phys. Chem.*, **88,** 491 (1984).

Pellizzari, E. D., "The Measurement of Carcinogenic Vapors in Ambient Atmospheres," National Technical Information Service Report No. PB 269–582, June 1977.

Penkett, S. A., F. J. Sandalls, and J. E. Lovelock, "Observations of Peroxyacetyl Nitrate (PAN) in Air in Southern England," *Atmos. Environ.*, **9,** 139 (1975).

Perner, D., A. Schmeltekopf, R. H. Winkler, H. S. Johnston, J. G. Calvert, C. A. Cantrell, and W. R. Stockwell, "A Laboratory and Field Study of the Equilibrium N_2O_5 ⇄ NO_3 + NO_2," *J. Geophys. Res.*, **90,** 3807 (1985).

Pitts, J. N., Jr., J. P. Smith, D. R. Fitz, and D. Grosjean, "Enhancement of Photochemical Smog by *N,N′*-Diethylhydroxylamine in Polluted Ambient Air," *Science*, **197,** 255 (1977).

Pitts, J. N., Jr., D. Grosjean, K. Van Cauwenberghe, J. P. Schmid, and D. R. Fitz, "Photooxidation of Aliphatic Amines Under Simulated Atmospheric Conditions: Formation of Nitrosamines, Nitramines, Amides, and Photochemical Oxidant," *Environ. Sci. Technol.*, **12,** 946 (1978).

Pitts, J. N., Jr., G. W. Harris, and A. M. Winer, "Spectroscopic Measurements of the NO_3 Radical and of HONO in the Atmosphere: Implications for Tropospheric Free Radical Chemistry," Abstract D2, 15th International Symposium on Free Radicals, Keltic Lodge, Ingonish Beach, Nova Scotia, Canada, June 2–7, 1981.

Pitts, J. N., Jr., A. M. Winer, G. W. Harris, W. P. L. Carter, and E. C. Tuazon, "Trace Nitrogenous Species in Urban Atmospheres," *Environ. Health Perspect.*, **52,** 153 (1983).

Pitts, J. N., Jr., E. Sanhueza, R. Atkinson, W. P. L. Carter, A. M. Winer, G. W. Harris, and C. N. Plum, "An Investigation of the Dark Formation of Nitrous Acid in Environmental Chambers," *Int. J. Chem. Kinet.*, **16,** 919 (1984a).

Pitts, J. N., Jr., H. W. Biermann, R. Atkinson, and A. M. Winer, "Atmospheric Implications of Simultaneous Nighttime Measurements of NO_3 Radicals and HONO," *Geophys. Res. Lett.*, **11,** 557 (1984b).

Pitts, J. N., Jr., H. W. Biermann, A. M. Winer, and E. C. Tuazon, "Spectroscopic Identification and Measurements of Gaseous Nitrous Acid in Dilute Auto Exhaust," *Atmos. Environ.*, **18,** 847 (1984c).

Pitts, J. N., Jr., T. J. Wallington, H. W. Biermann, and A. M. Winer, "Identification and Measurement of Nitrous Acid in an Indoor Environment," *Atmos. Environ.*, **19,** 763 (1985a).

Pitts, J. N., Jr., H. W. Biermann, T. J. Wallington, A. M. Winer, H. MacLeod, and R. Atkinson, "Simultaneous Measurements of Selected Gaseous Nitrogenous and Oxidant Species in Los Angeles Photochemical Smog," unpublished data (1985b).

Platt, U., D. Perner, A. M. Winer, G. W. Harris, and J. N. Pitts, Jr., "Detection of NO_3 in the Polluted Troposphere by Differential Optical Absorption," *Geophys. Res. Lett.*, **7,** 89 (1980a).

Platt, U., D. Perner, G. W. Harris, A. M. Winer, and J. N. Pitts, Jr., "Observations of Nitrous Acid in an Urban Atmosphere by Differential Optical Absorption," *Nature*, **285,** 312 (1980b).

Platt, U. F., A. M. Winer, H. W. Biermann, R. Atkinson, and J. N. Pitts, Jr., "Measurement of Nitrate Radical Concentrations in Continental Air," *Environ. Sci. Technol.*, **18,** 365 (1984).

Pryor, W. A. and J. W. Lightsey, "Mechanisms of Nitrogen Dioxide Reactions: Initiation of Lipid Peroxidation and the Production of Nitrous Acid," *Science*, **214,** 435 (1981).

Richards, L. W., "Comments on 'the Oxidation of NO_2 to Nitrate—Day and Night,'" *Atmos. Environ.*, **17,** 397 (1983).

Russell, A. G., G. J. McRae, and G. R. Cass, "The Dynamics of Nitric Acid Production and the Fate of Nitrogen Oxides," *Atmos. Environ.*, **19,** 893 (1985).

Sakamaki, F., S. Hatakeyama, and H. Akimoto, "Formation of Nitrous Acid and Nitric Oxide in the Heterogeneous Dark Reaction of Nitrogen Dioxide and Water Vapor in a Smog Chamber," *Int. J. Chem. Kinet.*, **15,** 1013 (1983).

Sander, S. P., *J. Phys. Chem.*, submitted for publication (1985).

Sander, S. P., and M. E. Peterson, "Kinetics of the Reaction $HO_2 + NO_2 + M \rightarrow HO_2NO_2 + M$," *J. Phys. Chem.*, **88,** 1566 (1984).

Scott, W. E., E. R. Stephens, P. L. Hanst, and R. C. Doerr, "Further Developments in the Chemistry of the Atmosphere," Paper presented at the 22nd Midyear Meeting of the American Petroleum Institute, Division of Refining, Philadelphia, Pennsylvania, May 14, 1957.

Seigneur, C. and P. Saxena, "A Study of Atmospheric Acid Formation in Different Environments," *Atmos. Environ.*, **18,** 2109 (1984).

Singh, H. B. and P. L. Hanst, "Peroxyacetyl Nitrate (PAN) in the Unpolluted Atmosphere: An Important Reservoir for Nitrogen Oxides," *Geophys. Res. Lett.*, **8,** 941 (1981).

Singh, H. B. and L. J. Salas, "Peroxyacetyl Nitrate in the Free Troposphere," *Nature*, **302,** 326 (1983).

Smith, C. A., L. T. Molina, J. J. Lamb, and M. J. Molina, "Kinetics of the Reaction of OH with Pernitric and Nitric Acids," *Int. J. Chem. Kinet.*, **16,** 41 (1984).

Smith, C. A., A. R. Ravishankara, and P. H. Wine, "Kinetics of the Reaction $NO_2 + NO_3 + M$ at Low Pressures and 298 K," *J. Phys. Chem.*, **89,** 1423 (1985).

Spicer, C. W., "The Fate of Nitrogen Oxides in the Atmosphere," *Adv. Environ. Sci. Technol.*, **7,** 163 (1977).

Spicer, C. W., "The Distribution of Oxidized Nitrogen in Urban Air," *Sci. Total Environ.*, **24,** 183 (1982a).

Spicer, C. W. "Nitrogen Oxide Reactions in the Urban Plume of Boston," *Science*, **215,** 1095 (1982b).

Spicer, C. W., "Smog Chamber Studies of NO_x Transformation Rate and Nitrate/Precursor Relationships," *Environ. Sci. Technol.*, **17,** 112 (1983a).

Spicer, C. W., M. W. Holdren, and G. W. Keigley, "The Ubiquity of Peroxyacetyl Nitrate in the Continental Boundary Layer," *Atmos. Environ.*, **17,** 1055 (1983b).

Stelson, A. W. and J. H. Seinfeld, "Relative Humidity and Temperature Dependence of the Ammonium Nitrate Dissociation Constant," *Atmos. Environ.*, **16,** 983 (1982a).

Stelson, A. W. and J. H. Seinfeld, "Relative Humidity and pH Dependence of the Vapor Pressure of Ammonium Nitrate–Nitric Acid Solutions at 25°C," *Atmos. Environ.*, **16,** 993 (1982b).

Stelson, A. W., S. K. Friedlander, and J. H. Seinfeld, "A Note on the Equilibrium Relationship Between Ammonia and Nitric Acid and Particulate Ammonium Nitrate," *Atmos. Environ.*, **13,** 369 (1979).

Stephens, E. R., "The Formation, Reactions, and Properties of Peroxyacyl Nitrates (PANs) in Photochemical Air Pollution," *Adv. Environ. Sci.*, **1,** 119 (1969).

Stephens, E. R., P. L. Hanst, R. C. Doerr, and W. E. Scott, "Reactions of Nitrogen Dioxide and Organic Compounds in Air," *Ind. Eng. Chem.*, **48,** 1498 (1956a).

Stephens, E. R., W. E. Scott, P. L. Hanst, and R. C. Doerr, "Recent Developments in the Study of the Organic Chemistry of the Atmosphere," *J. Air Pollut. Control Assoc.*, **6,** 159 (1956b).

Stockwell, W. R. and J. G. Calvert, "The Mechanism of NO_3 and HONO Formation in the Nighttime Chemistry of the Urban Atmosphere," *J. Geophys. Res.*, **88,** 6673 (1983).

Stull, D. R. and H. Prophet, Eds., *JANAF Thermochemical Tables*, 2nd ed., NSRDS-NBS37, June 1971.

Temple, P. J. and O. C. Taylor, "World-Wide Ambient Measurements of Peroxyacetyl Nitrate (PAN) and Implications for Plant Injury," *Atmos. Environ.*, **17,** 1583 (1983).

Tuazon, E. C., A. M. Winer, R. A. Graham, J. P. Schmid, and J. N. Pitts, Jr., "Fourier Transform Infrared Detection of Nitramines in Irradiated Amine–NO_x Systems," *Environ. Sci. Technol.*, **12,** 954 (1978).

Tuazon, E. C., W. P. L. Carter, A. M. Winer, and J. N. Pitts, Jr., "Reactions of Hydrazines with Ozone Under Simulated Atmospheric Conditions," *Environ. Sci. Technol.*, **15,** 823 (1981a).

Tuazon, E. C., A. M. Winer, and J. N. Pitts, Jr. "Trace Pollutant Concentrations in a Multiday Smog Episode in the California South Coast Air Basin by Long Pathlength Fourier Transform Infrared Spectroscopy," *Environ. Sci. Technol.*, **15,** 1232 (1981b).

Tuazon, E. C., W. P. L. Carter, R. V. Brown, R. Atkinson, A. M. Winer, and J. N. Pitts, Jr., "Atmospheric Reaction Mechanisms of Amine Fuels," Report No. ESL-TR-82-17, U.S. Air Force Engineering and Services Center, Tyndall Air Force Base, Florida, March 1982.

Tuazon, E. C., R. Atkinson, C. N. Plum, A. M. Winer, and J. N. Pitts, Jr., "The Reaction of Gas Phase N_2O_5 with Water Vapor," *Geophys. Res. Lett.*, **10,** 953 (1983a).

Tuazon, E. C., W. P. L. Carter, R. Atkinson, and J. N. Pitts, Jr., "The Gas-Phase Reaction of Hydrazine and Ozone: A Nonphotolytic Source of OH Radicals for Measurement of Relative OH Radical Rate Constants," *Int. J. Chem. Kinet.*, **15,** 619 (1983b).

Tuazon, E. C., W. P. L. Carter, R. V. Brown, A. M. Winer, and J. N. Pitts, Jr., "Gas-Phase Reaction of 1,1-Dimethylhydrazine with Nitrogen Dioxide," *J. Phys. Chem.*, **87,** 1600 (1983c).

Tuazon, E. C., E. Sanhueza, R. Atkinson, W. P. L. Carter, A. M. Winer, and J. N. Pitts, Jr., "Direct Determination of the Equilibrium Constant at 298°K for the $NO_2 + NO_3 \rightleftarrows N_2O_5$ Reactions," *J. Phys. Chem.*, **88,** 3095 (1984a).

Tuazon, E. C., W. P. L. Carter, R. Atkinson, A. M. Winer, and J. N. Pitts, Jr., "Atmospheric Reactions of *N*-Nitrosodimethylamine and Dimethylnitramine," *Environ. Sci. Technol.*, **18,** 49 (1984b).

Varas, A., H. Sandoval, E. A. Lissi, S. Diaz, and M. V. Encina, "Gas Phase Reactions of Diethylhydroxylamine," 9th Meeting of the Chilean Chemical Society, Jahuel, Chile, 1977.

Wallington, T. J., R. Atkinson, and A. M. Winer, "Rate Constants for the Gas Phase Reaction of OH Radicals with Peroxyacetyl Nitrate (PAN) at 273 and 297°K," *Geophys. Res. Lett.*, **11,** 861 (1984).

Williams, D. L. H., Nitrosation Mechanisms, in *Advances in Physical and Organic Chemistry*, Vol. 19, V. Gold and D. Bethell, Eds., Academic Press, New York, 1983, pp. 381–428.

PART 5

Photochemical Air Pollution: Mechanisms of Formation and Chemical Basis of Control Strategy Options for Oxidant and Gaseous Airborne Toxic Chemicals

9 Overall Reaction Mechanisms for the Formation of Ozone and Its Co-Pollutants in the Simulated and Real Troposphere: Chemical Kinetic Submodels

9-A. INTRODUCTION

Relating the concentrations of secondary pollutants impacting a given geographical location, under a given set of meteorological conditions, to the emissions of primary pollutant precursors is a major goal of atmospheric scientists. One means of accomplishing this is to construct a mathematical model which is validated by comparison of its predictions to the experimentally observed pollutant concentrations. Typically, the chemical submodel is first validated against data obtained under controlled conditions, independent of meteorology and emissions, using smog chambers (see Chapter 6). Subsequently, model validation for a given air basin or region is carried out by comparison of the model predictions to ambient air data for that region. The validated model may then be used to estimate the effects of various control strategies.

The three major components, or submodels, of such overall airshed models are: (1) the emissions of the primary pollutants, including their chemical nature, locations, and temporal variations; (2) the meteorological and topographical features of the region, including such parameters as temperature, relative humidity, wind speed and direction, atmospheric stability, inversion height, surface elevation, and other terrain features; and (3) the chemistry, including both the kinetics and mechanisms, of the reactions converting the primary pollutants into secondary pollutants. Figure 9.1 summarizes the elements of these major components of an air quality model.

In this chapter we describe several typical chemical submodels used to describe the formation and fates of the major and significant minor pollutants characteristic of photochemical reactions in the real and simulated troposphere.

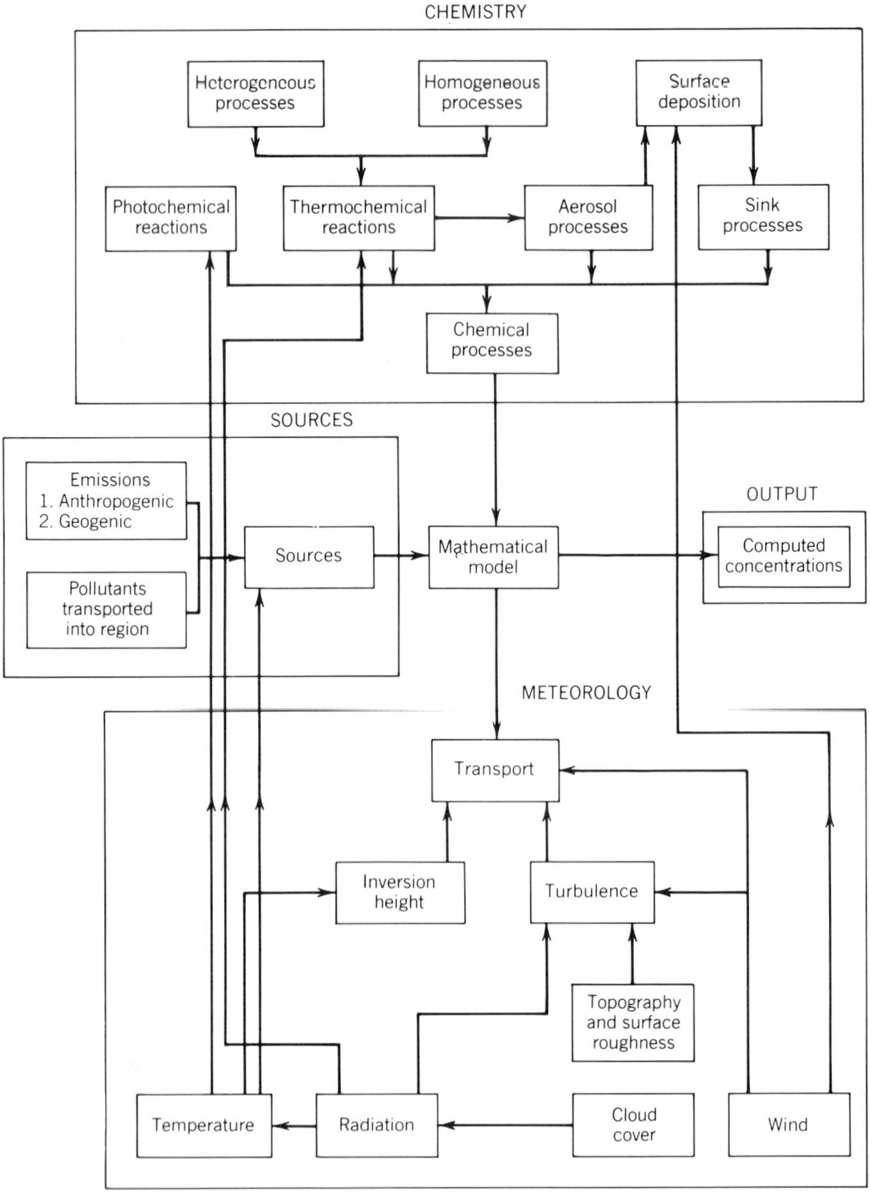

FIGURE 9.1 Elements of a typical airshed model (from McRae et al., 1982a).

We do not treat here models for acid deposition explicitly; however, the chemical submodels discussed generally form a major part of acid deposition chemical submodels, since the species oxidizing SO_2 and NO_x to sulfuric and nitric acids, respectively, are also involved in oxidant formation. The reader is referred to Chapter 11 for a discussion of the chemistry of acid deposition, including a brief treatment of the chemical portion of acid deposition models.

9-B. PHOTOCHEMICAL AIR POLLUTION: EXPLICIT CHEMICAL MECHANISMS

Our fundamental knowledge of tropospheric chemistry increased rapidly in the late 1960's and early 1970's. This was due in part to the confirmation of the importance of the OH radical in tropospheric chemistry as well as to the development of many specific, sensitive, and accurate techniques for studying reaction kinetics and mechanisms. Based on these developments, it became possible to describe in some detail the individual chemical reactions occurring in irradiated mixtures of NO_x with NMOC (non-methane organic compounds). A list of all reactions of the individual primary and secondary pollutants, as well as of the reactive intermediates [e.g., OH, $O(^3P)$], including their kinetics and products, is known as an *explicit chemical mechanism*. Two of the first explicit chemical mechanisms are those described by Niki, Daby and Weinstock (1972), and Demerjian, Kerr and Calvert (1974).

One recent example of such a mechanism is that of Leone and Seinfeld (1984) for toluene–NO_x and toluene–benzaldehyde–NO_x irradiated mixtures in air; a portion of this is shown in Table 9.1. (See also Leone and Seinfeld, 1985, and Leone et al., 1985).

For some reactions where an intermediate was produced which in air rapidly formed other products, the intermediates are not shown. For example, the abstraction of a hydrogen atom from toluene produces an alkyl radical,

$$C_6H_5CH_3 + OH \rightarrow C_6H_5CH_2 + H_2O \tag{1}$$

which rapidly adds O_2 to form an alkyl peroxy radical,

$$C_6H_5CH_2 + O_2 \rightarrow C_6H_5CH_2O_2 \tag{2}$$

These two reactions have been written in the condensed form of reaction (48) in Table 9.1.

Even for this relatively simple system consisting of only one or two reactant organics, a total of 102 reactions were needed to describe the chemistry.

Several problems with introducing such explicit mechanisms into airshed or long-range transport models become apparent immediately. First, as we have seen in earlier chapters, many of the rate constants and, indeed, intermediates and products of the individual reactions are not known. Second, even if all rate constants, intermediates, and products of the reactions were known (clearly *not* the case, as seen in Chapters 7 and 8), the amount of computer time required for numerical integration of the rate equations associated with each of the hundreds, if not thousands, of individual hydrocarbons found in ambient air would be prohibitive.

As a result, these explicit chemical mechanisms are not included in urban airshed models; rather, they are used primarily as a means of investigating the importance of various reaction pathways under conditions typical of the troposphere, by comparing the model predictions to the results of smog chamber experiments. Once the smog chamber data and the predictions of an explicit chemical mechanism

TABLE 9.1. Portion of an Explicit Chemical Mechanism

Reaction	Rate Constant[a]
Inorganic Reactions	
(1) $NO_2 + h\nu \rightarrow NO + O(^3P)$	0.35–0.40 min^{-1}
(2) $O(^3P) + O_2 \rightarrow O_3$	2.6×10^1
(3) $O_3 + NO \rightarrow NO_2 + O_2$	2.7×10^1
(4) $O(^3P) + NO_2 \rightarrow NO + O_2$	1.4×10^4
(5) $O_3 + NO_2 \rightarrow NO_3 + O_2$	4.7×10^{-2}
etc.	
Aldehyde Reactions and PAN Formation	
(30) $CH_3CHO + h\nu \xrightarrow{2O_2} CH_3O_2 + HO_2 + CO$	*b*
(31) $CH_3CHO + OH \xrightarrow{O_2} CH_3CO_3 + H_2O$	2.4×10^4
(32) $CH_3O_2 + NO \rightarrow NO_2 + CH_3O$	1.1×10^4
etc.	
α-Dicarbonyl Chemistry	
(43) $CH_3COCHO + OH \xrightarrow{O_2} CH_3COO_2 + CO + H_2O$	2.5×10^4
(44) $CH_3COCHO + h\nu \xrightarrow{2O_2} CH_3COO_2 + HO_2 + CO$	*c*
(45) $CH_3COCHO + h\nu \xrightarrow{2O_2} CH_3O_2 + HO_2 + 2CO$	*c*
etc.	
Toluene Abstraction Pathway	
(48) $C_6H_5CH_3 + OH \xrightarrow{O_2} C_6H_5CH_2O_2 + H_2O$	7.5×10^2
(49) $C_6H_5CH_2O_2 + NO \xrightarrow{O_2} NO_2 + C_6H_5CHO$	9.0×10^3
(50) $C_6H_5CH_2O_2 + NO \rightarrow C_6H_5CH_2ONO_2$	1.0×10^3
etc.	
Toluene Addition Pathway	
(64) $C_6H_5CH_3 + OH \rightarrow C_6H_5(CH_3)OH$	8.7×10^3
(65) $C_6H_5(CH_3)OH + O_2 \rightarrow C_6H_4(CH_3)OH + HO_2$	1.0×10^1
(66) $C_6H_5(CH_3)OH + NO_2 \rightarrow C_6H_4(CH_3)NO_2 + H_2O$	4.4×10^4
etc.	
Conjugated α-Dicarbonyl Chemistry	
(82) $OHCCH{=}CHCHO + OH \xrightarrow{O_2} OHCCH{=}CHC(O)O_2 + H_2O$	4.4×10^4
(83) $OHCCH{=}CHC(O)O_2 + NO \xrightarrow{O_2} OHCCH{=}CHO_2 + NO_2 + CO_2$	1.0×10^4
etc.	

Source: Leone and Seinfeld, 1984.

[a] In units of ppm^{-1} min^{-1} unless otherwise stated.

[b] $k_{30} = 8.4 \times 10^{-4}\, k_1$.

[c] $(k_{44} + k_{45}) = 0.019\, k_1$.

(including all rate constants) are in good agreement, the mechanism can then be used as a starting point for the development of more condensed mechanisms for use in urban airshed models. These condensed chemical submodels are discussed in the following section.

9-C. PHOTOCHEMICAL AIR POLLUTION: LUMPED CHEMICAL MECHANISMS

In urban airshed models, condensed chemical mechanisms are used to reduce the number of reactions and species to something that can be reasonably handled simultaneously with the complex meteorology and emissions. In these *lumped* mechanisms, the inorganic chemistry is commonly left in the form of explicit reactions because there are a relatively small number of inorganic species and reactions to be considered. The chemistry of organics, however, is treated by grouping or "lumping" together a number of reactions and/or chemical species. The overall rate constant and the products of the lumped reactions are chosen by the modeler to be representative of that group of reactions or reactants.

Table 9.2 shows a portion of one lumped model where the organics are lumped according to classical types of compounds. Specifically, six organics or classes of organics are considered: ALK = alkanes, C_2H_4 = ethene, OLE = olefins (alkenes), ARO = aromatics, HCHO = formaldehyde, and RCHO = aldehydes larger than formaldehyde. The reactions of these hydrocarbons with $O(^3P)$, OH, and, in the case of the alkenes, O_3, produce alkylperoxy radicals (RO_2), alkoxy radicals (RO), and peroxyacetyl radicals (RCO_3).

The lumped reactions in Table 9.2 are developed on the basis of our understanding of the reaction mechanisms in air. For example, reaction (23) is a combination of the following steps:

$$\mathrm{HCHO + OH \rightarrow HCO + H_2O} \tag{3}$$

$$\mathrm{HCO + O_2 \rightarrow HO_2 + CO} \tag{4}$$

$$\text{Net:}\quad \mathrm{HCHO + OH \xrightarrow{O_2} HO_2 + H_2O + CO}$$

For the reactions of the larger organics, where our understanding of the reaction mechanisms involved is far from complete (see Chapter 7), there is, of course, much greater uncertaintly in the lumped reactions. For example, the reaction (30) of alkenes with O_3 is shown in Table 9.2 to form aldehydes and the free radicals HO_2, RO_2, OH, and RO, where (a_1)–(a_6) represent the stoichiometric coefficients. As discussed in Section 7-E-2, however, the fraction of the overall reaction which produces free radicals at 1 atm in air and the detailed mechanisms involved are

TABLE 9.2. Portion of a Lumped Chemical Sub-model Used for Modeling an Urban Airshed

Reaction	Rate Constant[a]
(1) $NO_2 + h\nu \rightarrow NO + O(^3P)$	0.339 min^{-1}[b]
(2) $O(^3P) + O_2 + M \rightarrow O_3 + M$	$3.9 \times 10^{-6}\ e^{510/T}$
⋮	⋮
(21) $HCHO + h\nu \rightarrow 2HO_2 + CO$	0.00163 min^{-1}
(22) $HCHO + h\nu \rightarrow H_2 + CO$	0.00296 min^{-1}
(23) $HCHO + OH \rightarrow HO_2 + H_2O + CO$	1.9×10^4
(24) $RCHO + h\nu \rightarrow RO_2 + HO_2 + CO$	0.00145
(25) $RCHO + OH \rightarrow RCO_3$	2.6×10^4
(26) $C_2H_4 + OH \rightarrow RO_2$	1.2×10^4
(27) $C_2H_4 + O \rightarrow RO_2 + HO_2$	1.2×10^3
(28) $OLE + OH \rightarrow RO_2$	8.9×10^4
(29) $OLE + O \rightarrow RO_2 + RCO_3$	2.2×10^4
(30) $OLE + O_3 \rightarrow (a_1)RCHO + (a_2)HCHO + (a_3)HO_2 + (a_4)RO_2 + (a_5)OH + (a_6)RO$	0.136
(31) $ALK + OH \rightarrow RO_2$	4.7×10^3
(32) $ALK + O \rightarrow RO_2 + OH$	99.8
(33) $ARO + OH \rightarrow RO_2 + RCHO$	1.6×10^4
(16) $RO_2 + NO \rightarrow RO + NO_2$	1.2×10^4
(18) $NO_2 + OH \rightarrow HNO_3$	1.5×10^4
(42) $RCO_3 + NO_2 \rightarrow PAN$	2.07×10^3
(43) $PAN \rightarrow RCO_3 + NO_2$	$4.77 \times 10^{16}\ e^{-12516/T}$
(44) $NO_2 + NO_3 \rightarrow N_2O_5$	$2.19 \times 10^2\ e^{861/T}$
(46) $N_2O_5 + H_2O \rightarrow 2HNO_3$	1.5×10^{-5}
⋮	⋮
(52) $2RO_2 \rightarrow 2RO$	196

Source: McRae et al., 1982a,b.

[a] In units of $ppm^{-1}\ min^{-1}$ unless otherwise stated.

[b] Average of daylight hours.

only now beginning to be elucidated. As a result, the values of the stoichiometric coefficients are highly uncertain.

For a detailed discussion of how such lumped mechanisms are developed from our knowledge of the detailed reaction mechanism, the reader is referred to papers by Seinfeld and co-workers (Falls and Seinfeld, 1978; Falls et al., 1979) on which the mechanism in Table 9.2 is based.

While the reaction scheme in Table 9.2 can thus be rationalized in terms of explicit mechanisms, lumping by its nature involves a great deal of flexibility and hence judgment in choosing products representative of the reactions of a whole class of hydrocarbons.

In addition to the uncertainties in products, there is a great deal of flexibility of choice in the kinetics chosen for the lumped reactions. Alkenes reacting with O_3, for example, have rate constants that increase by two orders in magnitude in going from ethene to *cis*-2-butene (Section 7-E-2). The greater the concentrations of more reactive alkenes in the air mass, the larger the rate constant chosen for reaction (30) should be. Thus, McRae et al. (1982a,b) used rate constants for the reactions in Table 9.2 which are weighted by the relative number of moles of individual compounds within each class of organics.

Of course, as the reactions proceed and hydrocarbons are consumed, the composition of the remaining hydrocarbons shift. If the rate constants for the organic reactions reflect the mix, then they should change with irradiation time. The severity of this problem is reduced by the continuous injection of fresh reactants during the day.

In summary, use of highly explicit chemical submodels in airshed or long-range transport models is not practical. A viable alternative is to lump some of the reactants, intermediates, and individual reactions together to form a much more compact mechanism. In doing so, however, a number of parameters must be somewhat arbitrarily assigned. In validating the model (see below), these adjustable parameters are varied until a good match to the observed data is obtained. Good fits to the observations cannot, however, always be taken as evidence of the validity of the mechanism, given the large number of variables in the models themselves.

While Table 9.2 is one example of a lumped chemical submodel, other forms of lumping are used as well. For example, in the *carbon bond mechanism* (CBM), the reactants are lumped not in terms of the usual classification of organics (as in Table 9.2), but in terms of the bonding of the atoms in the reactants. This CBM formulation (Whitten and Hogo, 1977; Whitten et al., 1980) divides the carbon atoms of the organics into four classes based on their chemical bonding:

1. Single-bonded carbon atoms (e.g., alkanes) represented as PAR.
2. *Fast* doubly bonded atoms (e.g., olefins, except ethylene) represented as OLE.
3. *Slow* doubly bonded atoms (e.g., aromatics and ethylene) represented as ARO.
4. Carbonyl carbon atoms (i.e., aldehydes and ketones) represented as CAR.

Table 9.3 is a portion of the carbon bond mechanism. This approach leads to a grouping of reactant organics which is very similar to that of other mechanisms such as that in Table 9.2, for example. In addition, the same degree of "flexibility" in assigning the products and kinetics of the organic reactions is present here as was discussed above with reference to other lumped mechanisms.

In some models, a combination of an explicit mechanism with a relatively few "lumped" reactions is used; an example is the ERT model, developed by Environmental Research and Technology (Atkinson et al., 1982; Lloyd et al., 1983; Atkinson and Lloyd, 1984).

TABLE 9.3. A Portion of the Carbon Bond Mechanism

Reaction	Rate Constant (ppm^{-1} min^{-1})
$NO_2 + h\nu \rightarrow NO + O$	k_1[a]
$O + O_2 (+ M) \rightarrow O_3 (+ M)$	2.08×10^{-5}
$O_3 + NO \rightarrow NO_2 + O_2$	25.2
$O + NO_2 \rightarrow NO + O_2$	1.34×10^4
$O_3 + NO_2 \rightarrow NO_3 + O_2$	5×10^{-2}
$NO_3 + NO \rightarrow NO_2 + NO_2$	2.5×10^4
$NO_3 + NO_2 + H_2O \rightarrow 2HNO_3$	2.0×10^{-3}
$HNO_2 + h\nu \rightarrow NO + OH$	$0.19\ k_1$
$NO_2 + OH \rightarrow HNO_3$	1.4×10^4
$NO + OH \rightarrow HNO_2$	1.4×10^4
$CO + OH \xrightarrow{O_2} CO_2 + HO_2$	4.5×10^2
$OLE + OH \xrightarrow{O_2} CAR + CH_3O_2$	3.8×10^4
$PAR + OH \xrightarrow{O_2} CH_3O_2 + H_2O$	1.3×10^3
$ARO + OH \xrightarrow{O_2} CAR + CH_3O_2$	8×10^3
$OLE + O \xrightarrow{2O_2} HC(O)O_2 + CH_3O_2$	5.3×10^3
$PAR + O \xrightarrow{O_2} CH_3O_2 + OH$	20
$ARO + O \xrightarrow{2O_2} HC(O)O_2 + CH_3O_2$	37
$ARO + NO_3 \rightarrow$ products (aerosol)	1.0×10^2
$OLE + O_3 \xrightarrow{O_2} \alpha[HC(O)O_2] + HCHO + OH$	1.5×10^{-2}
$CAR + OH \xrightarrow{O_2} HC(O)O_2 + H_2O$	1.0×10^4
$CAR + h\nu \xrightarrow{2O_2} \alpha HC(O)O_2 + \alpha HO_2 + (1 - \alpha)CO$	$6.0 \times 10^{-3}\ k_1$

Source: Whitten and Hogo, 1977; Whitten et al., 1980.
[a]Varies with light intensity.

9-D. VALIDATION OF CHEMICAL SUBMODELS OF PHOTOCHEMICAL AIR POLLUTION

Once a chemical mechanism (i.e., chemical submodel), whether it be explicit or lumped, has been formulated for the photochemical processes in irradiated NMOC–NO_x mixtures, it must be validated. That is, the concentration–time profiles for the primary and secondary pollutants predicted by the model must be compared to experimental observations. These experimental data are usually smog chamber results using known NMOC–NO_x mixtures. Where agreement is not obtained between the predicted and measured concentrations, appropriate adjustments to the model input are made until the agreement is acceptable.

Given the uncertainties in interpretation of smog chamber data (see Chapter 6), validation of even the explicit chemical submodels involves some personal judgment on the part of the modeler. For example, an explicit chemical mechanism for mixtures of propene and/or *n*-butane with NO_x was developed by Carter et al. (1979) and validated using data from the evacuable smog chamber described in Chapter 6. From this comparison between the mechanistic predictions and the smog chamber data, it became evident that a chamber-dependent unknown source of OH radicals was necessary to fit the observed pollutant profiles (see Chapter 6).

Figure 9.2 shows some typical smog chamber data for an *n*-butane–NO_x run. The dashed lines are the model predictions using the known gas phase chemistry and no chamber-dependent radical source. Clearly, there is a large discrepancy between the observations and model predictions even for the major primary and secondary pollutants such as *n*-butane, NO, NO_2, and O_3. The solid lines are the

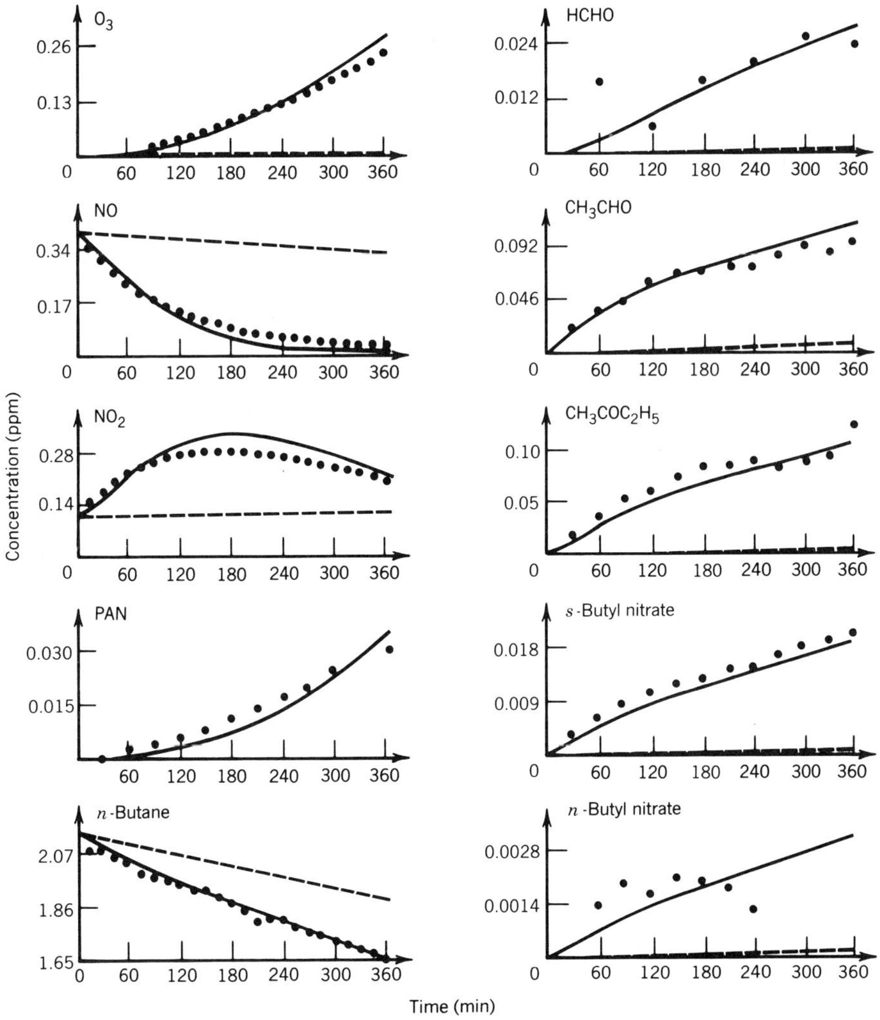

FIGURE 9.2 Comparison of experimental and calculated concentration–time profiles for measured species in an *n*-butane–NO_x–air smog chamber experiment: •, Experimental data; ——, standard model calculation, k_{131} = 0.45 ppb min^{-1}; ---, calculation, k_{131} = 0, where k_{131} is the constant rate of OH production from an unknown source, characteristic of the chamber (from Carter et al. 1979).

model predictions assuming there is a constant source of OH radicals of 0.45 ppb min^{-1}. A good fit to most of the data is obtained with the latter assumption.

Although assuming a constant OH flux gives good fits to the chamber data, the source of the OH is unknown. Because it appears to depend not only on the chamber but on the chamber history (see Chapter 6), it was found necessary with this explicit model to adjust the OH flux from this unknown source for each run; the modelers chose to adjust it to match the rates of consumption of NO and of formation of O_3. Similarly, some arbitrary adjustments of the rate constant for the reaction

$$N_2O_5 + H_2O \rightarrow 2\ HNO_3 \tag{5}$$

were carried out, since this reaction is surface catalyzed (see Section 8-A-2b) and thus dependent on the chamber history.

In addition to these adjustable parameters representing chamber-specific variables, a number of other parameters are treated as adjustable due to the lack of knowledge of the specific mechanisms and kinetics of a number of relevant reactions. For example, the solid line in Fig. 9.3 shows the fit to the O_3 profile of a propene–NO_x run if it is assumed that O_3 reacts with propene either (a) to produce stabilized Criegee biradicals or (b) to produce the free radicals HO_2 and CH_3 with the ratio of the rate constants being $k_a/k_b = 4$. The line represented by long dashes in Fig. 9.3 is the predicted O_3 profile if $k_a/k_b = 2$ (i.e., if it is assumed that more free radicals are produced in the O_3–propene reaction). In both cases, a constant flux of OH of 0.4 ppb min^{-1} was assumed. Interestingly, relatively good fits to the O_3 data could be restored with $k_a/k_b = 2$ if the chamber-specific OH flux was assumed to fall to 0.2 ppb min^{-1} (short dashed line).

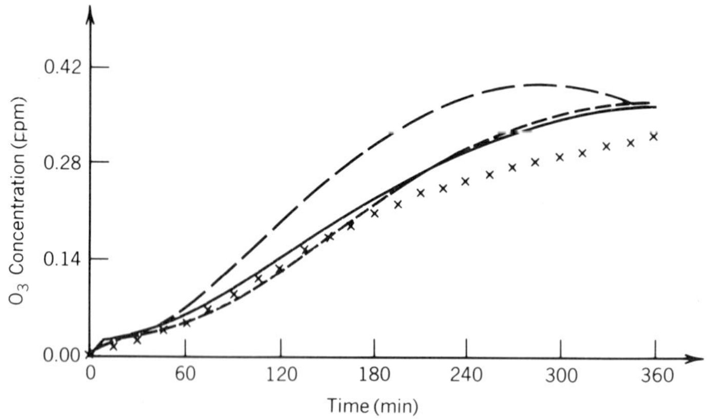

FIGURE 9.3. Effect of varying radical input from the ozone–olefin reaction and the unknown chamber radical source on ozone concentration–time profiles; ×, Experimental data; ——, standard model calculation, $k_{131} = 0.4$ ppb min^{-1}, $k_{42}/k_{43} = 4$; — —, calculated, $k_{131} = 0.4$ ppb min^{-1}, $k_{42}/k_{43} = 2$; - - -, calculated, $k_{131} = 0.2$ ppm/min, $k_{42}/k_{43} = 2$, where k_{131} is as defined for Fig. 9.2 and k_{42} and k_{43} are the rate constants for the ozone–olefin reaction to produce stabilized Criegee biradicals (42) or the free radicals HO_2 and CH_3 (43), respectively (from Carter et al. 1979).

We show this to illustrate that, even with explicit chemical mechanisms, there are a number of adjustable parameters reflecting uncertainties either in our knowledge of the specific reactions or in the chamber-specific processes that enter into model validation. As seen in Fig. 9.3, good fits to the observed data can generally be obtained using more than one combination of adjustable parameters. Thus good fits to experimental data are not necessarily an indication of the correctness of the model. This is especially true of lumped models where the number of adjustable inputs is substantial.

This point is illustrated in a comparison by Dunker et al. (1984) of four chemical mechanisms used in models for ambient air. Dunker and co-workers compared the O_3, NO_2, and PAN concentrations predicted by the four mechanisms under the same set of conditions; they found significant differences between the models for predictions of the O_3 and PAN concentrations (see more detailed discussion in Chapter 10). These differences were attributed in part to assumptions made by the modelers in validating their mechanisms against smog chamber data. Thus chamber-specific effects such as the unknown radical source are omitted from the mechanism found to give the best fits to the chamber data when the model is applied to ambient air. Different assumptions for these chamber-specific effects will alter the models in different ways when these effects are omitted for application to ambient air, and thus lead to different model predictions for O_3 and other secondary pollutants such as PAN. Dunker and co-workers, for example, suggest that the different treatments of the unknown radical source and of aldehyde photolysis rates in validating the mechanisms against smog chamber data are significant factors in the differences in the model predictions when applied to ambient air.

Recently Leone and Seinfeld (1985) have analyzed six reaction mechanisms used to describe the chemistry of photochemical smog, and identified some of the reasons for the discrepancies in the model predictions which occur when these mechanisms are applied to three different sets of initial conditions. In addition to the assumptions built into each mechanism, e.g. regarding the way reactions or species are "lumped," major uncertainties were associated with our current lack of understanding of the mechanism of photooxidation of aromatics (see Section 7-G-1b) and with how to treat the unknown radical source in smog chambers (see Section 6-F).

Despite these problems, explicit mechanisms in the hands of knowledgeable chemists who have an appreciation for the uncertainties in experimental data can be used to elucidate certain aspects of chemical mechanisms and even kinetics. For example, irradiation of mixtures of *n*-alkanes larger than C_4 with NO_x in an evacuable smog chamber produced larger yields of alkyl nitrates than could be explained using the explicit chemical mechanism current at that time (Darnall et al., 1976). It was known that methylperoxy radicals reacted with NO to give methoxy radicals and NO_2:

$$CH_3O_2 + NO \rightarrow CH_3O + NO_2 \tag{6}$$

Alkoxy radicals were known to form alkyl nitrates via

$$RO + NO_2 \rightarrow RONO_2 \tag{7}$$

However, as discussed in Section 7-D-3, the larger alkoxy radicals undergo rapid reaction with O_2 as well as decomposition, isomerization, and reaction with NO in competition with their reaction with NO_2. Applying the best available kinetics and mechanisms for these processes, Darnall and co-workers predicted yields of nitrates which were substantially less than those observed. After an examination of all possible sources of the nitrates, it was concluded than the nitrate data could be matched only if the larger alkylperoxy radicals react with NO not only via reactions analogous to (6), but also via (8) which includes addition, rearrangement, and, finally, stabilization to the nitrate:

$$RO_2 + NO \rightarrow (ROONO)^* \rightarrow (RONO_2)^* \stackrel{M}{\rightarrow} RONO_2 \tag{8}$$

In short, if the modeler exercises sufficient care both in interpretation of the experimental data and in the use of adjustable parameters, useful chemical information on some reactions can be derived from the interplay of explicit mechanisms and smog chamber data.

Validation of the lumped chemical submodels used in airshed models follows the same general procedure as that described for the explicit mechanisms. However, the flexibility in fitting observed data using adjustable parameters is generally greater than for explicit mechanisms because of the uncertainties discussed earlier which are necessarily introduced in the lumping procedure.

Until the late 1970's, our knowledge of nighttime chemistry was sparse. However, with the detection and measurement of the nitrate radical in ambient air (see Chapter 5-D-2) and kinetic studies of its reactions (see Chapter 7), came the recognition that both NO_3 and N_2O_5 may be involved in significant nighttime chemistry. For example, as discussed in Section 8-A-2 and Section 11-B-1c, the hydrolysis of N_2O_5 may be a significant source of nitric acid in ambient air. In addition, formation of species such as HONO can occur through NO_x reactions in air at night (See Section 8-A-3a), and can then act as photoinitiators when the sun rises. Indeed, the formation of such compounds is thought to contribute to the increasing severity of photochemical smog from one day to the next which is often observed in multi-day episodes. The complexity of taking into account nighttime chemistry as well as the diurnally varying emissions and meteorology is such that relatively successful modeling of multi-day episodes has only been carried out over the last few years; an example of this by McRae and Seinfeld (1983) is described in Section 10-B-4c.

REFERENCES

Atkinson, R. and A. C. Lloyd, "Evaluation of Kinetic and Mechanistic Data for Modeling of Photochemical Smog," *J. Phys. Chem. Ref. Data*, **13,** 315 (1984).

Atkinson, R., A. C. Lloyd, and L. Winges, "An Updated Chemical Mechanism for Hydrocarbon/NO_x/SO_2 Photooxidations Suitable for Inclusion in Atmospheric Simulation Models," *Atmos. Environ.*, **16,** 1341 (1982).

Carter, W. P. L., A. C. Lloyd, J. L. Sprung, and J. N. Pitts, Jr., "Computer Modeling of Smog Chamber Data: Progress in Validation of a Detailed Mechanism for the Photooxidation of Propene and *n*-Butane in Photochemical Smog," *Int. J. Chem. Kinet.*, **11,** 45 (1979).

Darnall, K. R., W. P. L. Carter, A. M. Winer, A. C. Lloyd, and J. N. Pitts, Jr., "Importance of RO_2 + NO in Alkyl Nitrate Formation from C_4–C_6 Alkane Photooxidation Under Simulated Atmospheric Conditions," *J. Phys. Chem.*, **80,** 1948 (1976).

Demerjian, K. L., J. A. Kerr, and J. G. Calvert, "The Mechanism of Photochemical Smog Formation," *Adv. Environ. Sci. Technol.*, **4,** 1 (1974).

Dunker, A. M., S. Kumar, and P. H. Berzins, "A Comparison of Chemical Mechanisms Used in Atmospheric Models," *Atmos. Environ.*, **18,** 311 (1984).

Falls, A. H. and J. H. Seinfeld, "Continued Development of a Kinetic Mechanism for Photochemical Smog," *Environ. Sci. Technol.*, **12,** 1398 (1978).

Falls, A. H., G. J. McRae, and J. H. Seinfeld, "Sensitivity and Uncertainty of Reaction Mechanisms for Photochemical Air Pollution," *Int. J. Chem. Kinet.*, **11,** 1137 (1979).

Leone, J. A. and J. H. Seinfeld, "Updated Chemical Mechanism for Atmospheric Photooxidation of Toluene," *Int. J. Chem. Kinet.*, **16,** 159 (1984).

Leone, J. A. and J. H. Seinfeld, "Comparative Analysis of Chemical Reaction Mechanisms for Photochemical Smog," *Atmos. Environ.*, **19,** 437 (1985).

Leone, J. A., R. C. Flagan, D. Grosjean, and J. H. Seinfeld, "An Outdoor Smog Chamber and Modeling Study of Toluene-NO_x Photooxidation," *Int. J. Chem. Kinet.*, **17,** 177 (1985).

Lloyd, A. C., R. Atkinson, F. W. Lurmann, and B. Nitta, "Modeling Potential Impacts from Natural Hydrocarbons-I. Development and Testing of a Chemical Mechanism for the NO_x-Air Photooxidations of Isoprene and α-Pinene Under Ambient Conditions," *Atmos. Environ.*, **17,** 1931 (1983).

McRae, G. J., and J. H. Seinfeld, "Development of a Second Generation Mathematical Model for Urban Air Pollution. II. Evaluation of Model Problems," *Atmos. Environ.*, **17,** 501 (1983).

McRae, G. J., W. R. Goodin, and J. H. Seinfeld, "Mathematical Modeling of Photochemical Air Pollution," Final Report to the California Air Resources Board, Contract No. A5-046-87 and A7-187-30, April 27, 1982a.

McRae, G. J., W. R. Goodin, and J. H. Seinfeld, "Development of a Second-Generation Mathematical Model for Urban Air Pollution—I. Model Formulation," *Atmos. Environ.*, **16,** 679 (1982b).

Niki, H., E. E. Daby, and B. Weinstock, "Mechanisms of Smog Reactions," *Adv. Chem. Series*, **113,** 16 (1972).

Whitten, G. Z. and H. Hogo, "Mathematical Modeling of Simulated Photochemical Smog," EPA Report No. EPA-600/3-77-011 (1977).

Whitten, G. Z., H. Hogo, and J. P. Killus, "The Carbon-Bond Mechanism: A Condensed Kinetic Mechanism for Photochemical Smog," *Environ. Sci. Technol.*, **14,** 690 (1980).

10 Chemical Bases for Strategies for the Control of Photochemical Oxidants and Volatile Toxic Organic Chemicals

> Establishing quantitative relationships between emissions and air quality is perhaps the most fundamental (and frustrating) problem facing air pollution science. (California Air Resources Board, 1981)

One ultimate goal of studying the individual chemical reactions in atmospheric processes involved in oxidant formation, acid deposition, and the atmospheric persistence and fate of volatile toxic air pollutants is well summarized in the above statement. In this chapter we examine the relationship between emissions and air quality and the implications for control of ozone and other secondary pollutants via the control of the primary pollutant precursors non-methane hydrocarbons (NMHC) and NO_x. The concept of organic reactivity is introduced, including a brief discussion of the potential contribution of so-called *natural* emissions of NO_x and hydrocarbons, which are treated in more detail in Chapter 14.

Finally, we apply the principles developed throughout this book to estimate the atmospheric lifetimes and fates of some volatile toxic organics. While the compounds treated by no means comprise a comprehensive list, they serve to illustrate how their lifetimes and fates are controlled by reactive species which are also central to photochemical oxidant control.

For further details on photochemical oxidants and methods of developing control strategies for it, the reader is referred to the U.S. EPA documents "Procedures for Quantifying Relationships Between Photochemical Oxidants and Precursors," published in 1977 and 1978, and the series "International Conference on Oxidants, 1976—Analysis of Evidence and Viewpoints," Parts I–VIII (1977). The most recent assessment of the formation, fates, and impacts of photochemical oxidants is the revised EPA document "Air Quality Criteria for Ozone and Other Photochemical Oxidants", expected to be published in 1986.

The term NMHC, non-methane hydrocarbons, has been used extensively in the past to designate reactive organic compounds. However, not only hydrocarbons but also other organics such as aldehydes are important in atmospheric chemistry. Therefore, we use non-methane organics or NMOC interchangeably with NMHC throughout this chapter.

Finally, as discussed in Section 5-A, the term *photochemical oxidant* includes those species which oxidize I^- in the KI measurement method; while O_3 is the major oxidant, under some conditions others such as NO_2 may also contribute. However, we use oxidant and O_3 interchangeably in this chapter.

The basis for control of O_3 and other secondary pollutants consists essentially of two different approaches: (1) use of observations in field studies and (2) the development and application of various mathematical models that relate the concentrations of the secondary pollutants formed to the initial primary pollutant concentrations. For many of these models, validation of the chemical mechanism against smog chamber data (Chapter 6) is an integral part of their development, and we briefly address the interplay between these models and smog chamber data. We discuss the approach based on field studies in Section 10-A and that based on modeling in Section 10-B.

10-A. FIELD OBSERVATIONS OF AMBIENT LEVELS OF OZONE AND ITS PRECURSORS, NMOC AND NO_x

Field observations of O_3 and its precursors NMOC and NO_x were first used in what is known as the *Appendix J Method*, as a simplified and relatively inexpensive basis for developing oxidant control strategies. The name of this method is derived from the publication in Appendix J of the Federal Register on August 14, 1971 of a strategy for controlling oxidant. This method is based on field data from several urban air sheds for non-methane hydrocarbons from 6:00 to 9:00 a.m., and the maximum 1 h average oxidant concentration observed at those stations later in the day. Figure 10.1 plots these maximum observed oxidant concentrations as a function of the 6:00–9:00 a.m. NMHC (in ppmC); no simple correlation between the two exists. However, one can draw an envelope around these data to obtain a worst case estimate, shown by the solid line in Fig. 10.1

From the upper limit curve in Fig. 10.1, one can predict the percentage control of NMHC required to meet the federal oxidant standard[1] for various daily observed maximum oxidant concentrations. This *modified rollback* approach is shown in Fig. 10.2. The relationship between ambient levels of oxidant and NMHC is not assumed to be linear, but rather to follow a curve.

This approach for controlling oxidant clearly has some major uncertainties associated with it:

- It deals only with NMHC and ignores the role of NO_x, even though $O(^3P)$ atoms formed by NO_2 photolysis are the only known source of ozone from

[1] The standard was 0.08 ppm oxidant for 1 h when Appendix J was published; it is now 0.12 ppm O_3 for 1 h (see Section 1-D-4).

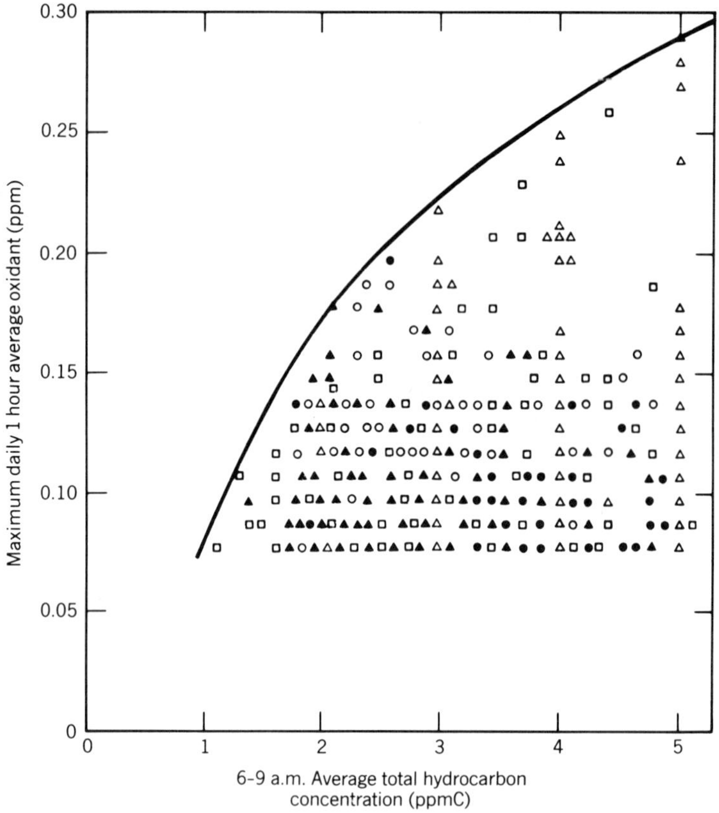

FIGURE 10.1. Maximum daily oxidant as function of early morning total hydrocarbons observed in 1966–1968 at CAMP stations in □ Denver, ● Cincinnati, △ Los Angeles, ○ Philadelphia, ▲ Washington (from Schuck et al., 1970).

anthropogenic precursors, and NO_2 is itself by definition an oxidant (see Section 5-A).

- The 6:00–9:00 a.m. concentration of NMHC at one location is not expected to necessarily bear any relationship to the oxidant concentration hours later at the same location because of transport; the air mass containing the NMHC and NO_x from 6:00–9:00 a.m. will have moved downwind by the time the oxidant concentrations peak at that location. Thus the observed oxidant levels should reflect the 6:00–9:00 a.m. primary pollutant concentrations at some point upwind.
- By using the upper limit curve of Fig. 10.1, oxidant control strategies are based on *worst case* observations for the specific locations where the data of Fig. 10.1 were collected. Some of these worst case measurements may have been highly unusual. In addition, areas having meteorology and emissions which are either more or less conducive to oxidant formation than

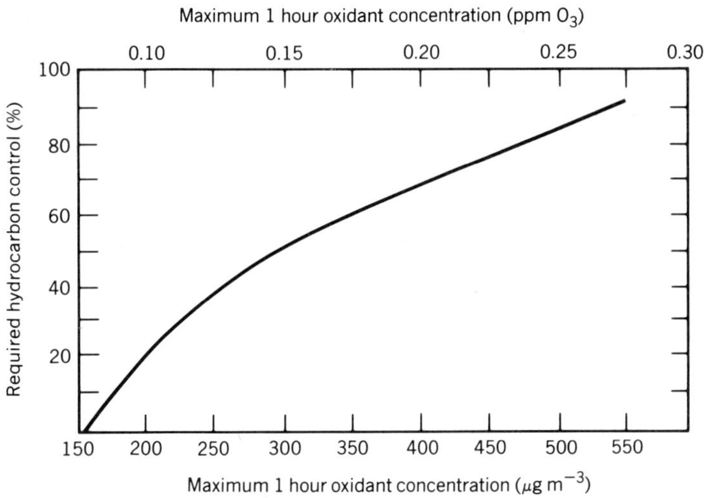

FIGURE 10.2. Required hydrocarbon emission control as a function of maximum 1 h photochemical oxidant concentrations using envelope in Fig. 10.1 (from Dimitriades, 1977).

those used in constructing Fig. 10.1 may not necessarily follow the same curve.

- Even the curve drawn in Fig. 10.1 is subject to substantial uncertainty; the low end is based on measurements of low pollutant concentrations which were difficult to measure accurately, and the high end is based on relatively few data points. In addition, as discussed in Chapter 5, much of the data on which Fig. 10.1 is based may have been rather inaccurate due to measurement problems.
- Since the data in Fig. 10.1 were all taken in central urban areas, application of the curve derived from it (Fig. 10.2) to downwind residential and rural areas is questionable.
- This method treats only photochemical oxidant and not the host of other secondary pollutants which may also be of concern.

Despite these uncertainties, and in the absence of other control strategy methods, this approach has been very useful in developing the initial basis for photochemical oxidant control. Indeed, given the uncertainties associated with the other methods discussed below, this approach has proved to be a helpful and practical first step.

10-B. MODELING STUDIES

There are a variety of mathematical models used to describe the relationship between emissions of primary pollutants and the resulting concentrations of secondary pollutants produced by chemical reactions in the air during transport and dispersion. These range from *linear rollback* and simple empirical models based on observations in ambient air, to complex mathematical airshed models which incorporate emissions, meteorology, chemical reactions, and pollutant sinks (e.g., deposition). In this section we examine the major types of models for predicting the concentration of pollutants in ambient air.

10-B-1. Linear Rollback

As the name implies, linear rollback is based on the assumption that pollutant concentrations will decrease proportionally to a decrease in the precursor emissions. For a pollutant such as CO, for example, the percentage reduction in emissions required to meet air quality standard for CO (A_{CO}), in a region that currently has observed concentrations as high as C_{CO}, is given by the following:

$$\text{percent reduction in emissions} = \frac{(C_{CO} - B_{CO}) - (A_{CO} - B_{CO})}{C_{CO} - B_{CO}} \times 100$$

$$= \frac{C_{CO} - A_{CO}}{C_{CO} - B_{CO}} \times 100$$

Here B_{CO} is the background (i.e. clean air) concentration of CO, which must be taken into account.

For a *non-reactive* pollutant such as CO, linear rollback provides a reasonable control strategy providing the location, temporal distribution, and relative strengths of the sources do not change. However, for a secondary pollutant such as O_3 which is formed through chemical reactions of primary pollutants, the complexity of the chemistry makes this approach a gross oversimplification. In addition, since both NMOC and NO_x are involved in oxidant formation, a decision must be made as to which pollutant—or both—linear rollback should be applied.

10-B-2. Simple Empirical Models

Simple empirical models are based on correlating observed pollutant concentrations in ambient air at a given location to measured relevant variables. The variables used are generally meteorological in nature, since these play such a large role in determining pollutant concentrations.

An example of such an empirical model is that used by the South Coast Air Quality Management District in southern California to predict air quality daily in a number of locations in the Los Angeles area and east of it. The model is based on the use of a number of parameters, mainly meterological, which are typically measured early in the day. These parameters are then entered into the model to predict peak ozone concentrations and visibility at a number of locations later in the day.

The basis for the ozone prediction is described by Zeldin and Thomas in a 1975 paper for a specific location, San Bernardino, California, which is approximately 70 miles downwind of Los Angeles. Points from 0 to 10 (10 being the maximum with respect to oxidant formation) were assigned for each of the following categories:

1. Stability (°C) $= (T_{850\,\text{mb}} - T_{\text{sfc}}) + (T_t - T_b)$

   where T_t = temperature of the inversion top
   T_b = temperature of the inversion base
   $T_{850\,\text{mb}}$ = temperature at an altitude corresponding to a pressure of 850 mb
   T_{sfc} = surface temperature

   This parameter represents the extent of stability of the atmosphere with respect to mixing even after the inversion has broken (first term) plus the strength of the inversion (second term).
2. Temperature (°C) at the height corresponding to a total pressure of 950 mb (~1800 ft). This apparently correlates to the intensity of solar radiation which drives the photochemical reactions.
3. Base height of the inversion layer. This is a measure of the volume of air into which the pollutant emissions are mixed; the lower the base height, the smaller the volume and hence the higher the concentrations expected.
4. Total pressure gradient from the coast to a desert location inland in southern California. This is a measure of the potential for transport of pollutants from the heavily populated Los Angeles area into the less densely populated inland area.
5. Day of the week. The points assigned to this category reflect differences in emissions from day to day. In this area where vehicular emissions accounted at that time for a large portion of the total emissions, traffic is least on Sundays and greatest on Fridays.
6. Month of the year. The category reflects the different solar irradiation intensities in the summer months compared to the winter months.

The points for each category were then added to predict the peak concentration of O_3 in pphm expected inland in the city of San Bernardino on a given day.

This simple empirical model was based on using peak observations of O_3 in San Bernardino and empirically developing a system that would reproduce these historical data. It was then applied in a predictive manner to forecast, early in the morning, the air quality for a particular day. Similar approaches based primarily on local meteorological phenomena have since been developed and applied by others as well (e.g., see Lin, 1982).

Such empirical models are useful for short-term predictions in particular locations, for example, as in the model described above. However, a historical data base must be available to develop the model, which is by its nature, site specific. In addition, such models may not be valid if the distribution of the sources changes (e.g., from widely spaced mobile sources to a few large point sources) or if the mix of pollutants (e.g., nature of the reactive hydrocarbons) changes. Finally, the accuracy of their predictions drops substantially if attempts are made to apply them more than a day or so in advance. While such empirical models are useful for short-term forecasting, they are not capable of long-term predictions of the effects of various control strategy options.

However, such an approach can be used to estimate historical trends in oxidant values corrected for meteorological differences or adjusted to similar meteorological conditions. Chock and co-workers, for example, have used a similar technique to examine the trend in O_3 under fixed meteorological conditions in the South Coast Air Basin of California (Chock et al., 1982; Kumar and Chock, 1984). They find that when meteorological differences are taken into account, the trend in oxidant peaks is to higher values at the eastern, downwind portion of the Los Angeles basin and lower values at the western, upwind end. This is in qualitative agreement with expectations based on isopleths (i.e., contours) of O_3 against initial NMOC and NO_x concentrations in irradiated NMOC–NO_x mixtures (see below).

10-B-3. Simple Mathematical Models

The next step in terms of model sophistication is simple mathematical models which are not directly based on ambient air measurements, but rather are developed *a priori*. One of the most widely used is the Gaussian plume model, originally developed for and primarily applied to simple plumes from point sources. A second approach, which is more complex but still simpler than the complex airshed models, is the box model. One model that has recently received much attention and hence is discussed here, EKMA, is a type of box model.

10-B-3a. Gaussian Plume Models

The basic Gaussian plume model describes the concentration of pollutants in a plume downwind from a point source. Although its most common application remains to plumes from point sources, it is occasionally applied to emissions from line (e.g., highways) and area (e.g., urban area) sources as well.

The name is based on the distribution of pollutants across the plume at any location, which is assumed to be Gaussian in shape (Fig. 10.3). The concentration of the pollutant [X] at various points in space, at coordinates (x, y, z) downwind from the source, can be calculated using this model if the source strength and plume height, wind speed, and atmospheric stability are known. For example, for stable conditions, or unlimited vertical mixing, the concentration of X downwind of the source can be calculated from Eq. (A):

$$[\mathrm{X}] = \frac{10^6 R}{V} \frac{e^{-y^2/2\sigma_y^2}}{2\pi\sigma_y\sigma_z} [e^{-(z-H)^2/2\sigma_z^2} + e^{-(z+H)^2/2\sigma_z^2}] \qquad \text{(A)}$$

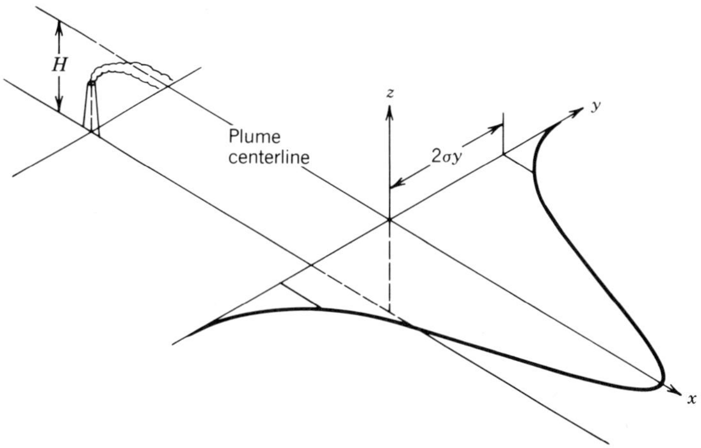

FIGURE 10.3. Schematic diagram of Gaussian plume model. For clarity in this simplified diagram, only the horizontal dispersion in the *y* direction, perpendicular to the wind direction, is shown (from Stern et al., 1984).

[X] is the pollutant concentration in $\mu g\ m^{-3}$ at a point with coordinates (x, y, z), where x is the downwind coordinate, y is the horizontal (crosswind) distance, z is the vertical distance, R is the emission rate of X from the source ($g\ s^{-1}$), V is the wind speed ($m\ s^{-1}$), H is the effective height of emission (in m), which is the stack height plus the height of plume rise (Fig. 10.3), and σ_y and σ_z (in units of m) are the standard deviations of the horizontal and vertical distributions, respectively, downwind of the source. These standard deviations, which describe the spread of the plume, depend on meteorological parameters. The most common approach to determining σ_y and σ_z is based on the work of Pasquill (1961, 1974) and Gifford (1961, 1976). Stability categories (A–F) for the atmosphere were developed based on wind speed, insolation, and extent of cloud cover; plots of σ_y and σ_z as a function of the downwind distance (X) were generated for six stability classes. These plots can then be used to estimate σ_y and σ_z for a given set of meteorological conditions. For further details, the reader is referred to the book by Stern et al. (1984).

The initial Gaussian plume models dealt with non-reactive pollutants such as CO. However, these models have also been modified to include pollutants whose chemical reactions can be described by simple first-order decays. To obtain the pollutant concentration at a given location due to a number of sources, the concentrations predicted from each source individually are simply summed.

A major disadvantage of the Gaussian approach is its inability to take into account easily the chemical reactions involved in the formation of secondary pollutants such as O_3, where simple addition of the contributions from each point source is not valid. There are also other problems; for example, variations in the meteorological parameters over the area considered cannot be easily accounted for in the model. Thus this model has been applied primarily to non-reactive pollutants from well-defined plumes and point sources, rather than to predicting secondary pollutant formation over an air basin.

10-B-3b. Box Models

A second type of simple model which has been applied to predict pollutant concentrations is known as a box model. The air mass over a region is treated as a box into which pollutants are emitted and undergo chemical reactions (Fig. 10.4). Transport into and out of the box by meteorological processess and dilution is taken into account.

The box model is closely related to the more complex airshed models described in Section 10-B-4 in that it is based on the conservation of mass equation and includes chemical submodels which represent the chemistry more accurately than Gaussian models, for example. However, it is less complex and hence requires less computation time. It has the additional advantage that it does not require the detailed emissions, meteorological, and air quality data needed for input and validation of the airshed models. However, the resulting predictions are correspondingly crude, especially in terms of spatial and temporal resolution.

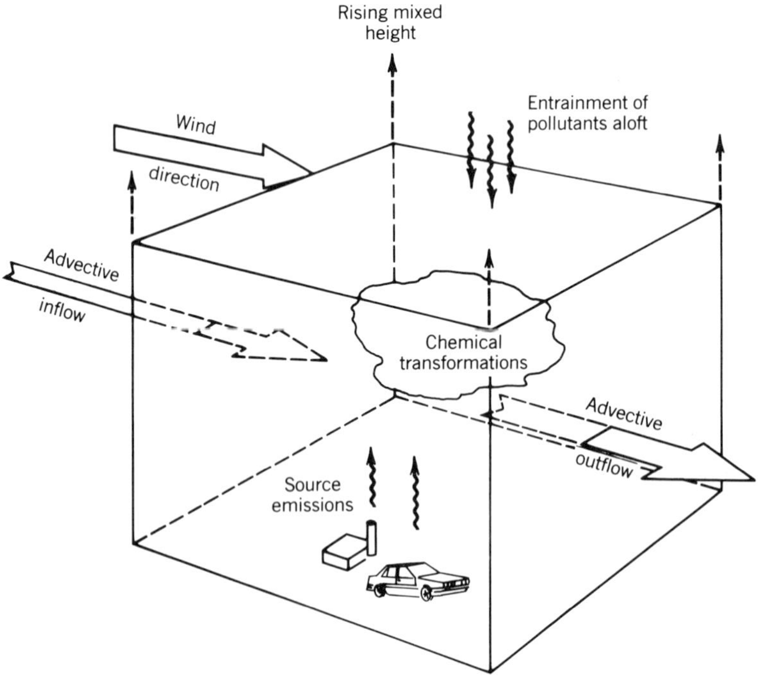

FIGURE 10.4. Schematic diagram showing basic elements of a simple box model (from Schere and Demerjian, 1978).

10-B-3c. EKMA Approach

One variant of the box model which is widely discussed and used at present is the so-called EKMA technique (EKMA = *E*mpirical *K*inetic *M*odeling *A*pproach) (Dodge, 1977a,b; Dimitriades and Dodge, 1983).

The EKMA approach is based on the generation of a series of O_3 isopleths using a model called the ozone isopleth plotting package (OZIPP). These isopleths are three-dimensional plots of daily maximum hourly average O_3 concentrations generated in mixtures with various initial NMOC and NO_x concentrations. Figure 10.5 shows a set of such isopleths. They were calculated using an approach analogous to that used in box models; that is, a box of air containing the initial concentrations of NMHC and NO_x is assumed to be diluted while chemical reactions are simultaneously converting NMHC and NO_x to O_3 and other secondary pollutants. The following assumptions were made in deriving Fig. 10.5:

- The hydrocarbon mix was assumed to consist of propene and butane in a 1:3 ratio (as carbon). Aldehydes were assumed to account for 5% (as carbon) of the initial NMHC.
- Initial NO_2 concentrations were 25% of the initial NO_x.
- The rate constants for photolysis which reflect the sunlight intensity were changed each hour to reflect the diurnal variation in sunlight from 8:00 a.m.

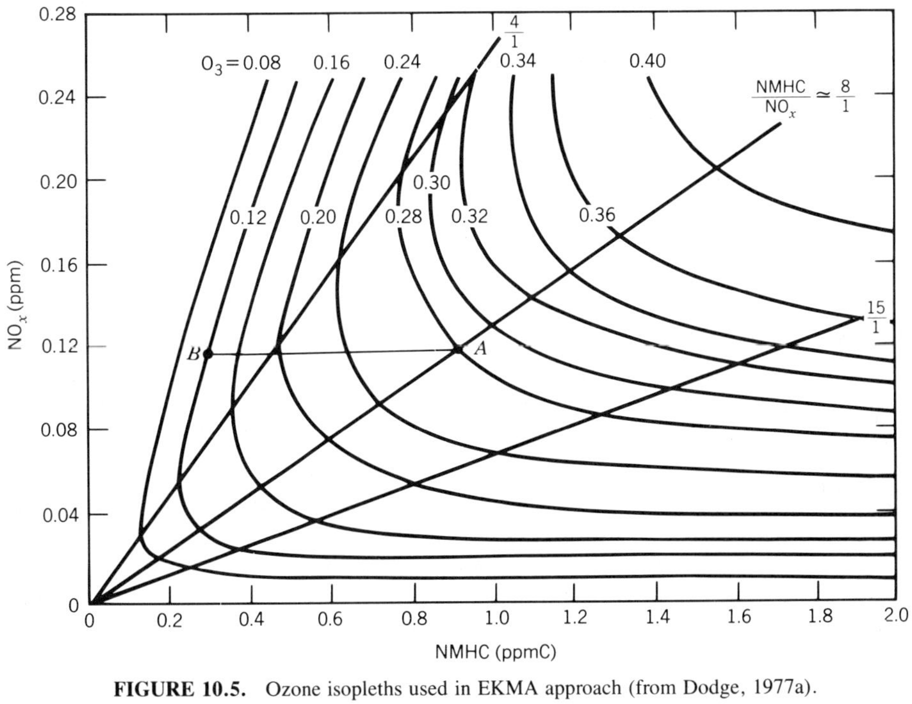

FIGURE 10.5. Ozone isopleths used in EKMA approach (from Dodge, 1977a).

to 5:00 p.m. (local daylight savings time) for the summer solstice at 34°N latitude (representing approximately Los Angeles, Phoenix, Dallas, and Atlanta).

- No addition of fresh pollutants occurred during the simulation.
- Dilution was assumed to occur at a rate of 3% h^{-1} until 3:00 p.m. with no dilution afterward.
- *Background* or transported O_3 was assumed to be negligible.

The chemical mechanism first used with the EKMA model was that of Dodge (1977a,b). The carbon bond mechanism discussed in Section 9-C has been used more recently in EKMA; other chemical mechanisms using a *lumped* approach (Section 9-C) can also be inserted. These mechanisms are first validated against smog chamber data by adjusting the model to obtain best fits to the experimentally determined primary and secondary pollutant concentrations. The chamber-specific effects such as the unknown radical source are then omitted from the model to predict the maximum O_3 concentrations for various initial NMHC and NO_x, that is, to predict the isopleths in Fig. 10.5.

These isopleths illustrate the importance of both NMHC and NO_x as well as their ratio in formulating O_3 control strategies. If constructed in three dimensions, a ridge would be observed along the diagonal running from the lower left to the upper right of the diagram. If one lowers both NMHC and NO_x simultaneously, but keeps their ratio constant at the ridge line value, the predicted O_3 maximum drops.

Different air basins or locations correspond to different sides of the ridge line, depending on the NMHC/NO_x ratio. This ratio and its value relative to the ridge line have important control implications. If one is on the lower right portion of the isopleth diagram (i.e., with NMHC/NO_x greater than the ridge line value), the isopleths run almost parallel to the horizontal axis. Thus the oxidant concentrations are not very sensitive to NMHC control if NO_x is kept constant. However, because ozone formation under these conditions is limited by the available NO_x, reducing NO_x while keeping NMHC levels constant results in reduced O_3. In addition, reducing both NMHC and NO_x simultaneously while keeping their ratio constant also reduces O_3.

A very different type of dependence of O_3 on changes in NMHC and NO_x is seen if the initial NMHC/NO_x ratio corresponds to the region of Fig. 10.5 above and to the left of the ridge line. Here, reducing NMHC with constant NO_x leads to significant decreases in the O_3 maximum. The same is true of decreasing both NMHC and NO_x simultaneously, keeping their ratio constant, except at very low values of the ratio which are not often encountered in the atmosphere.

However, in contrast to the situation at higher NMHC/NO_x ratios, reducing NO_x while keeping NMHC constant leads to *higher* O_3 concentrations until the ridge line is reached. After this, decrease in NO_x are accompanied by decreases in O_3.

The general shape of the O_3 isopleths in Fig. 10.5 can be rationalized on the basis of our knowledge of the chemistry involved in O_3 formation. At high NMHC/

NO_x ratios, that is, to the right of the ridge line, oxidant formation is limited by the amount of the precursor NO_2 which is available to form O_3 via photolysis. Thus oxidant in this region is especially sensitive to decreases in NO_x.

At NMHC/NO_x ratios less than that at the ridge line, the formation of O_3 is limited not by NO_x but by the time available for O_3 formation and the rate at which it is formed. This rate is strongly influenced by the concentrations of free radicals involved in the conversion of NO to NO_2, and these free radical concentrations are sensitive to the NMHC concentrations. That is, at the low free radical concentrations characteristics of small NMHC/NO_x ratios, the conversion of NO to NO_2 is slow and a peak in the O_3 concentration may not occur before sunset when the production of radicals such as OH and HO_2 effectively ceases. The maximum O_3 concentrations can then be determined by the number of hours of irradiation before sunset.

Figure 10.5 can be used to predict the changes in NMHC and NO_x needed to produce a given change in the peak O_3 concentration. For example, suppose a particular city is characterized by a NMHC/NO_x ratio of 8 : 1, and a *design value*[2] of O_3 of 0.28 ppm (i.e., point *A* on Fig. 10.5). If no change in ambient NO_x is anticipated, and one wants to reduce the peak O_3 from 0.28 to the federal air quality standard of 0.12 ppm, one must reduce NMHC to point *B* in Fig. 10.5 or by approximately 67%.

The term *empirical* in the title EKMA comes from the use of *observed* O_3 peaks to examine various control strategy options. The *kinetics modeling* portion of the title reflects the use of a chemical submodel validated against smog chamber data to generate the isopleths of Fig. 10.5.

There are a number of assumptions in this approach which may alter the control predictions and hence render them of limited applicability, at least to certain locations for which the model assumptions inherent in Fig. 10.5 may not be valid. For example, the sunlight intensity may be less or greater than that assumed, the reactivity of the NMHC mix may not be the same as the model input, the dispersion rates may not be 3% h^{-1}, and so on. To overcome some of these problems, a city-specific version of EKMA is available in which isopleth diagrams are generated using data more characteristic of that location. In addition, some procedures are available for taking transport of pollutants into account.

As is the case with the more complex airshed models (see Section 10-B-4), the specific chemical mechanism assumed can have a significant effect on the model predictions. As discussed in Chapter 9, built into chemical mechanisms are assumptions concerning how to treat chamber-specific effects when validating the model.

For example, a comparison of the effects of using different chemical mechanisms was carried out by Dunker and co-workers (1984) who compared the predictions of four different models under the same assumed set of meteorological conditions, initial pollutant mix, and emissions. The initial pollutant mix and emissions corresponded to those in the EKMA procedure, and isopleths of O_3 as well

[2] The design value was defined as the second highest hourly O_3 concentration observed at that location.

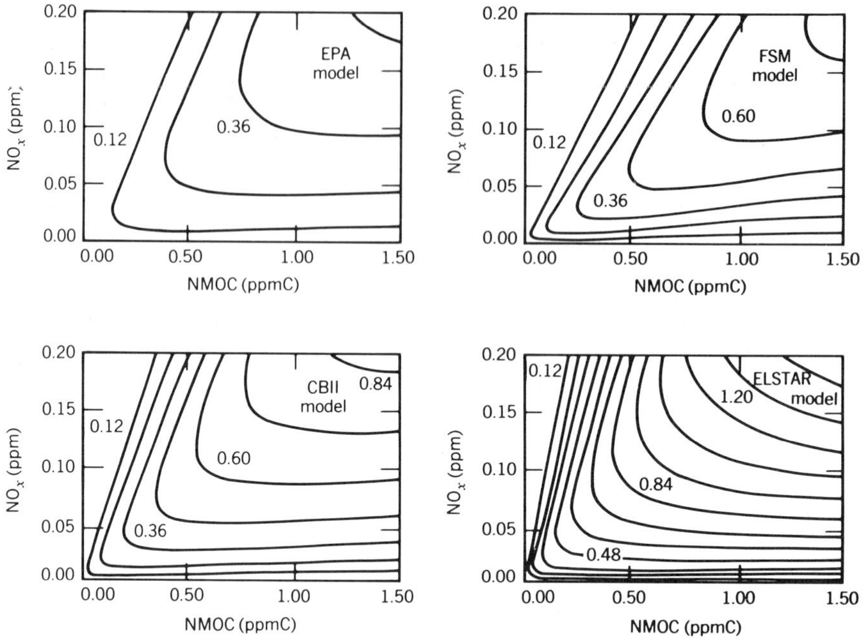

FIGURE 10.6. Isopleths of the maximum hourly average ozone concentrations (ppm) calculated for one set of initial conditions using four different models (from Dunker, 1984).

as NO_2 and PAN were generated for both standard and city-specific EKMA conditions. Figure 10.6 shows the isopleths of maximum hourly O_3 concentrations calculated by the four different models; clearly there are significant differences. The isopleths for the maximum hourly average PAN concentrations predicted by the four models also differed significantly, while those for NO_2 were similar in all four cases.

A second major difference in the model predictions was the time to reach the maximum hourly O_3 concentrations (Fig. 10.7). This and the differences in maximum O_3 and PAN are shown by Dunker and co-workers to be at least partly attributable to different assumptions in the aldehyde photolysis rates (which produce the free radicals to drive the NO to NO_2 conversion, etc.) and the unknown radical source in smog chambers (Chapter 6) which are used to generate the data for the model validation.

These differences in chemical mechanisms lead to substantial differences in the percentage reduction in NMOC needed to meet the federal air quality standard for ozone. Figure 10.8 shows the predicted percentage reductions in NMOC as a function of the accompanying assumed NO_x reduction for three different NMOC/NO_x ratios, 5, 10, and 15, which encompass typical values in urban areas; clearly, there are significant differences in the predicted NMOC reductions needed to reach the federal air quality standards for O_3, depending on the assumed chemical mechanism.

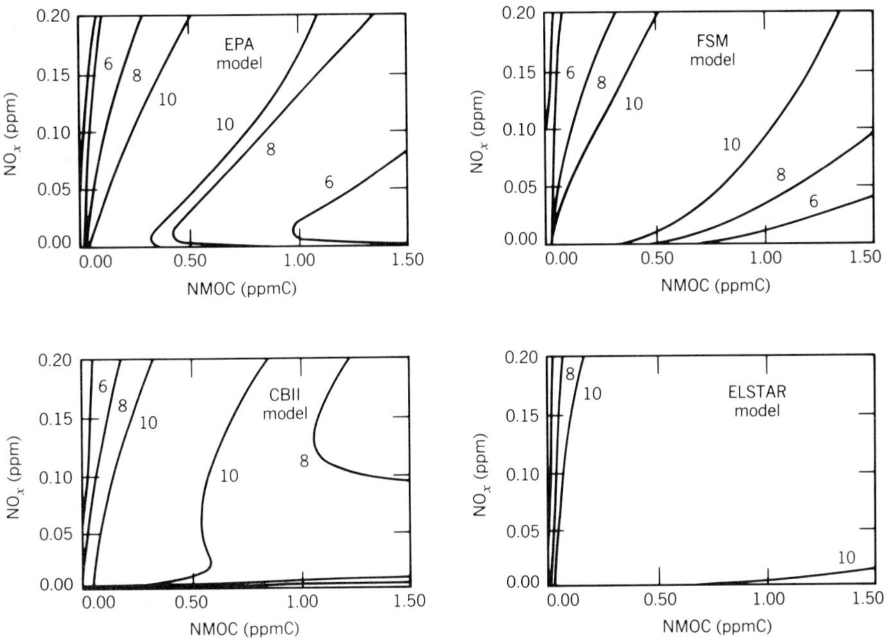

FIGURE 10.7. Isopleths of the time in hours needed to reach the maximum hourly average ozone concentration predicted by four different models for one set of initial conditions (from Dunker et al., 1984).

10-B-3d. Application of O_3 Isopleths to Urban Airsheds

Isopleth diagrams such as those in Fig. 10.5 confirm the ideas advanced in the 1960's, that under some conditions control of a primary pollutant, specifically NO_x, could actually lead to increased concentrations of a secondary pollutant. The major chemical reason for this behavior is the reaction of NO with O_3:

$$NO + O_3 \rightarrow NO_2 + O_2 \tag{1}$$

At high NO_x concentrations, as O_3 forms it will react with the NO present, thus inhibiting the formation of appreciable concentrations of O_3. In addition, high NO_x levels can terminate radical reactions, for example, via reaction (2)

$$NO_2 + OH \xrightarrow{M} HNO_3 \tag{2}$$

This perhaps initially surprising result—that secondary pollutant concentrations may increase in locations when control of primary pollutants is instituted or increased—has been the focus of considerable controversy for years and is the basis for suggestions that, for further reductions in ozone levels, NO_x controls in certain areas, for example, southern California, either not be strengthened further or even be weakened.

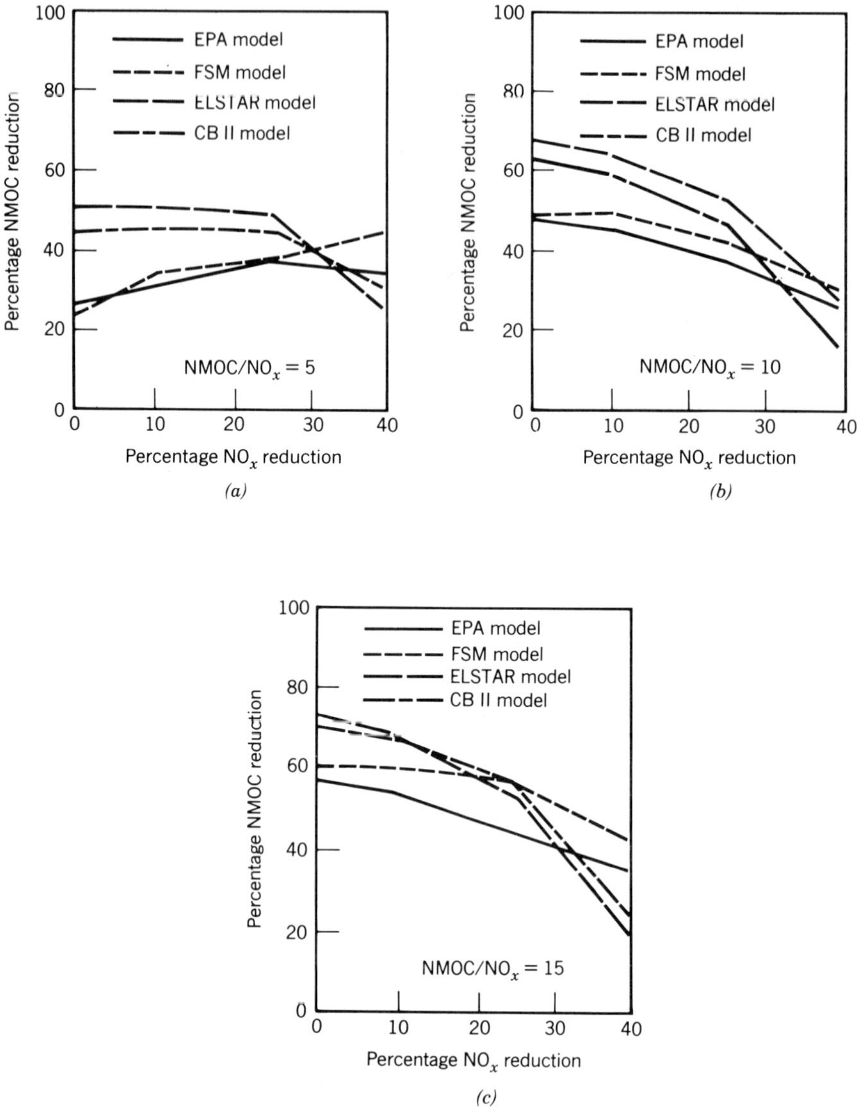

FIGURE 10.8. Predicted reductions in NMOC needed to reach the federal air quality O_3 standard of 0.12 ppm if the current ozone concentration is 0.16 ppm for three different NMOC/NO_x ratios as a function of concurrent % NO_x reduction (from Dunker et al., 1984).

As discussed in detail by Pitts et al. (1983), we see a number of problems with such an approach:

- The effects of such a strategy will clearly depend on the particular NMHC/NO_x ratio (i.e., on which side of the ridge line a particular location is found) and thus will vary from region to region.

- The isopleths are really only applicable to urban areas and ignore the impacts on downwind urban and rural areas. Thus increased NO_x will lead to increased NO_2 and hence O_3 when integrated over an entire air basin. Areas somewhere downwind of the urban centers will therefore be subjected to increased O_3 levels.
- This strategy focuses only on O_3 production. It neglects the fact that increased NO_x is expected to lead to increased concentrations of species such as HONO, which we have seen, can act as photoinitiators in multi-day smog episodes. Additionally, smog chamber and modeling studies show that increased NO_x leads to increased formation of other pollutants such as nitric acid, which is a key component of acid fogs and rain (see Chapter 11). Increased formation of other secondary pollutants such as nitroarenes, which may be mutagens and/or carcinogens (Chapter 13), may also occur. However, while these other secondary pollutants are of concern, their production cannot be quantitatively related to the ozone yields (Altshuller, 1983).
- Increased concentrations of NO_2 are expected in the urban area, and NO_2 itself has significant health and other effects (see Air Quality Criteria Document for NO_2).
- The most severe air pollution episodes generally occur in multi-day episodes. In these episodes, characterized by very stable meteorological conditions, the peak concentrations of O_3 and other secondary pollutants generally increase from one day to the next. This is attributed to the buildup of reactive species such as HONO, HCHO, and so on, which act as initiators when the sun rises and thus increase the rate of the photooxidations. Neither smog chamber studies nor modeling efforts have been very successful at reproducing such multi-day episodes. In addition, few experimental data from ambient air studies on the concentrations of such photoinitiators are available to test model predictions for these important species.

While a number of scientists concur with our concerns regarding the need for strict control of both NO_x and NMOC emissions, agreement on this point is by no means universal. The reader is referred to articles by scientists from General Motors Research Laboratories (Chock et al., 1981, 1983; Klimisch and Heuss, 1983; Kumar and Chock, 1984) for differing viewpoints.

In summary then, while isopleths such as shown in Fig. 10.5 can be used as a *semi-quantitative* guide in formulating control strategies, one must exercise caution in extrapolating them to ambient air in large urban areas. In particular, the results not only in urban areas but also in downwind regions must be considered, and emphasis should be placed not solely on O_3 but on other associated co-pollutants such as HNO_3 as well. Furthermore, the impacts of proposed controls on multiday episodes must be taken into account.

Finally, the isopleths are based on data from conventional NMOC-NO_x photooxidations. In some settings, e.g. in some industrial areas, an unusual pollutant mix may exist, whose reactivity depends not only on the NMOC-NO_x chemistry discussed in detail throughout this book, but also in part on the chemistry associated with the industrial emissions. For example, where there are sources of molecular chlorine, Cl_2, a significant portion of the hydrocarbon loss may be due to attack by Cl atoms formed on photolysis of the Cl_2 (Hov, 1985), in addition to attack by OH. In such situations, the products formed may also be somewhat unusual; for example, in laboratory studies, a variety of chlorinated organics have been identified in irradiated mixtures of ethene in air containing Cl_2 (Laffond et al., 1985). In such cases, the isopleths developed for NMOC-NO_x mixtures may not be applicable.

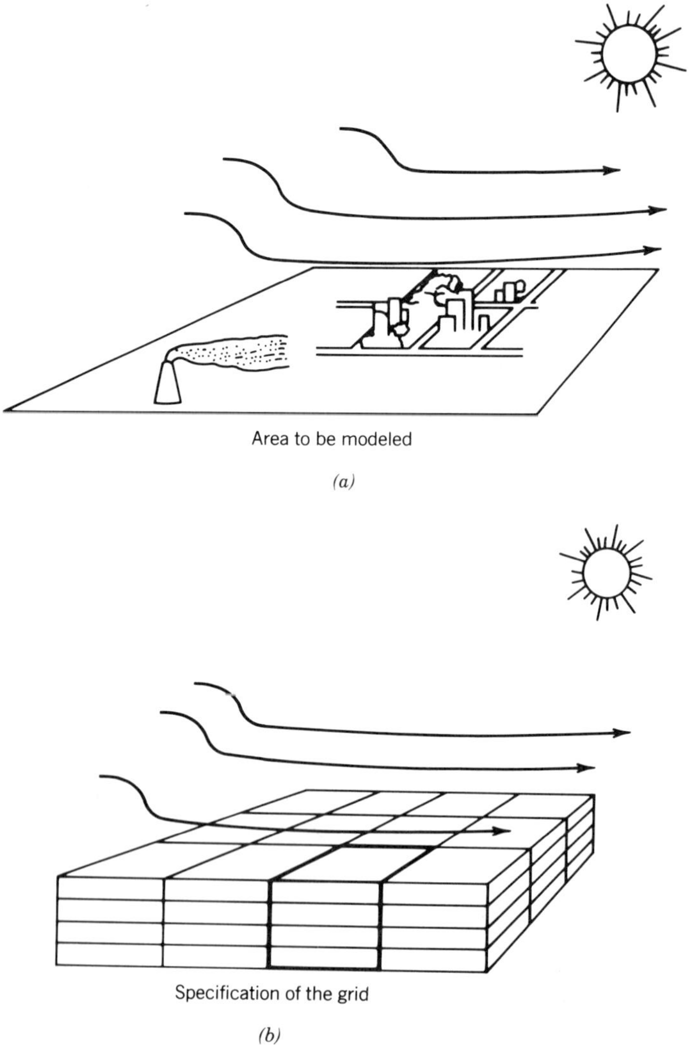

FIGURE 10.9. Schematic illustration of the grid used and treatment of atmospheric processes in one Eulerian airshed model (from Ames et al., 1978).

10-B-4. Complex Airshed Models

Complex airshed models attempt to incorporate into one mathematical model realistic descriptions of the specific emissions, meteorology, chemistry, and pollutant sinks occurring in an air basin. There are two major types of complex airshed models in use today: Eulerian and Lagrangian.

10-B-4a. Eulerian Models

In Eulerian models, pollutant concentrations are calculated at a fixed geographical location at specified times. In practice, the location is an area of 1 km^2 or more. The pollutant concentrations are predicted from their initial concentrations in that area, the new emissions, the transport in and out, dilution, and, finally, the chemical reactions specified by the chemical submodel. By carrying out such calculations for a number of different locations (i.e., small areas) pollutant concentrations can be predicted over an air basin as a function of time. Figure 10.9 is a schematic of the grid and treatment of atmospheric processes used in one Eulerian type of airshed model.

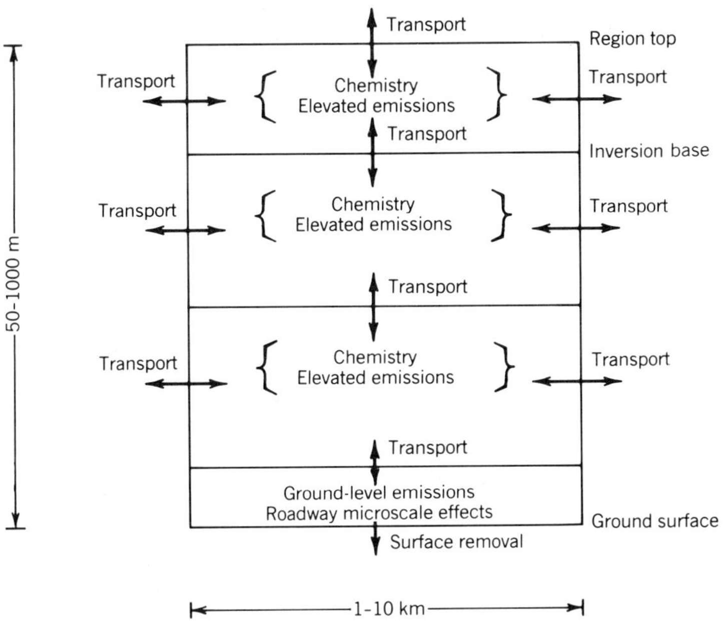

Atmospheric processes treated in a column of grid cells

(c)

FIGURE 10.9. *(Continued)*

One advantage of Eulerian models is the geographical resolution they provide. Thus the different effects of controlling sources of equal strength but located in different places (e.g., point versus mobile sources) can be readily elucidated using Eulerian models. In addition, such questions as the number of people exposed to various pollutants can be examined using population statistics. A major disadvantage is the relatively great expense required to develop the necessary model inputs and to run these models.

10-B-4b. Lagrangian Models

The second major type of model is a Lagrangian model. These models consider one column of air and follow its trajectory as predicted by the prevailing meteorology (Fig. 10.10). As this air parcel containing certain initial pollutant concentrations moves, it is subjected to fresh pollutant emissions, dilution, and chemical reactions. Such models are very useful for examining the impacts of new pollutant sources and are generally less expensive to run than Eulerian models.

10-B-4c. Use and Validation of Airshed Models

Use of airshed models requires substantial input data, including meteorological measurements (e.g., wind speed and temperature profiles, atmospheric stability), over the entire region of interest, and emissions data, which specify both spatially and temporally the emission of pollutants by the classes chosen for the chemical submodel. In addition, suitable initial and boundary conditions must be chosen. It is not difficult to see that major uncertainties can arise in model predictions due to uncertainties in the input data, as well as to approximations and assumptions in the chemical, meteorological, and emissions submodels.

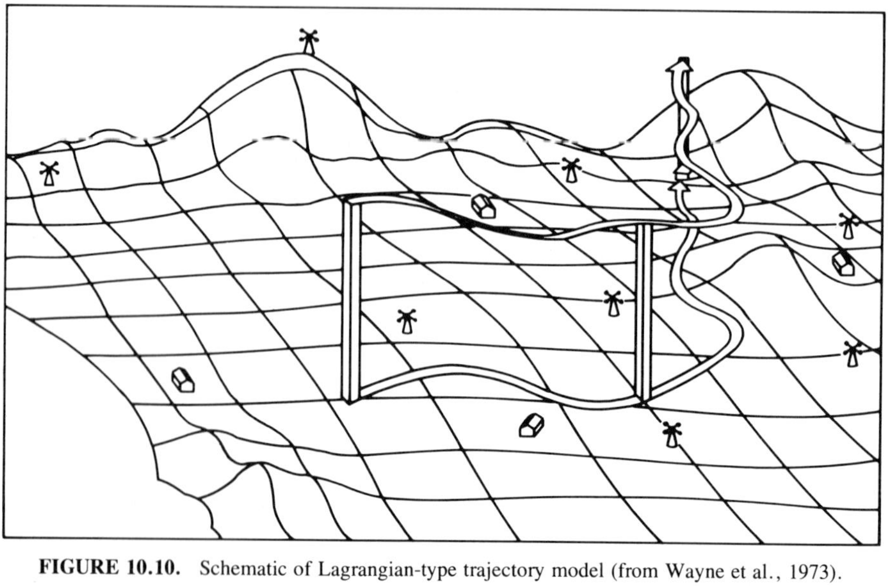

FIGURE 10.10. Schematic of Lagrangian-type trajectory model (from Wayne et al., 1973).

Validation of these complete airshed models requires comparing model predictions to observed pollutant concentrations in the air basin under study. The model uncertainties and approximations are such that good agreement with observations for *all* the major pollutants of concern under *all* conditions is rarely obtained.

Figures 10.11–10.14 illustrate typical comparisons between observed ambient measurements and the predictions of two different models (Chock et al., 1981; McRae and Seinfeld, 1983). Figures 10.11 and 10.12 show the results for NO and CO of one Lagrangian model developed by Chock and co-workers (1981) and

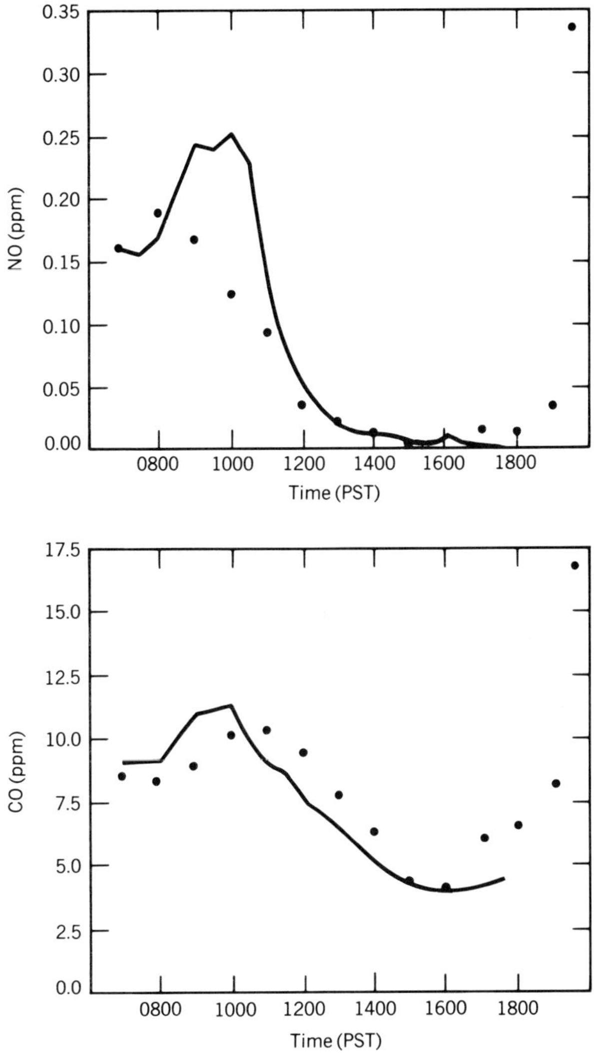

FIGURE 10.11. Diurnal profiles for NO and CO in the Los Angeles basin on October 15, 1973. Observed concentrations are shown by circles and model predictions by solid lines (from Chock et al., 1981).

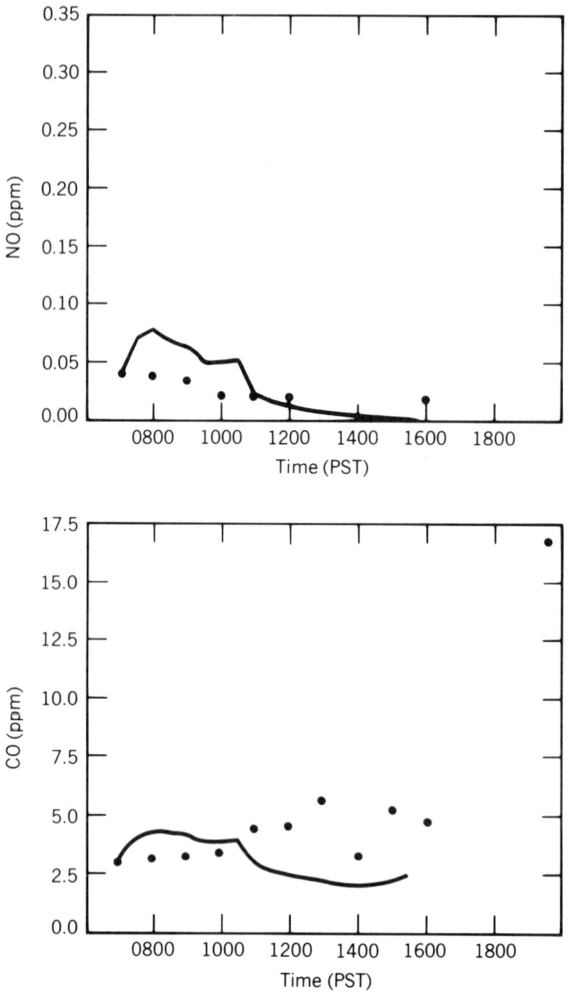

FIGURE 10.12. Diurnal profiles for NO and CO in the Los Angeles basin on July 31, 1973. Observed concentrations are shown by circles and model predictions by solid lines (from Chock et al., 1981).

applied to an air parcel for conditions chosen to be characteristic of the Los Angeles basin on two days in 1973. Note that CO and NO are primary pollutants which presumably are the most straightforward to model. However, the absolute concentrations and the concentration-time profile for CO on July 31, 1973 (Fig. 10.12) are not predicted well by their model.

Figures 10.13 and 10.14 show results of another modeling study by McRae and Seinfeld (1983) for O_3 and NO_2 on June 26 and 27, 1974 in downtown Los Angeles (Fig. 10.13) and simultaneously in a location downwind (Fig. 10.14). While the

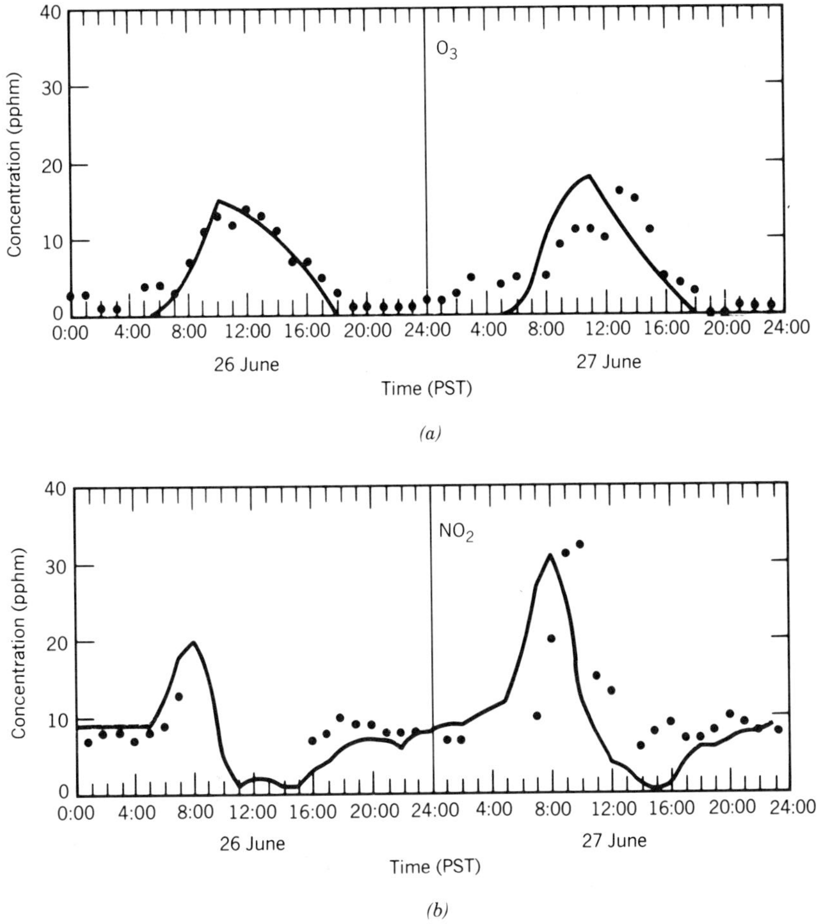

FIGURE 10.13. Observed (●) and model predicted (—) concentrations of (*a*) O_3 and (*b*) NO_2 in downtown Los Angeles, June 26 and 27 1974 (from McRae and Seinfeld, 1983).

absolute values of the predicted O_3 and NO_2 concentrations are not always in good agreement with the ambient air measurements, considering that O_3 and NO_2 are secondary pollutants and that this is a multiday episode, the model generally performed very well indeed, at both the central and downwind sites. In any case, these examples illustrate for the reader typical results of such models at their state of development when published.

Once such models have been validated for a particular region, the inputs can then be changed to assess the effects of various control strategy options. However, there are often significant disagreements between the absolute concentrations predicted by airshed models and those observed. For example, in evaluating three airshed models, Schreffler and Schere (1982) found that the standard deviations of

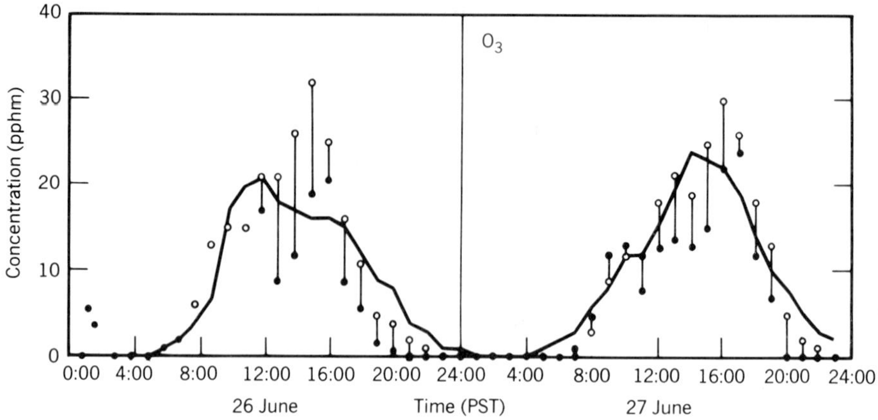

(*a*)

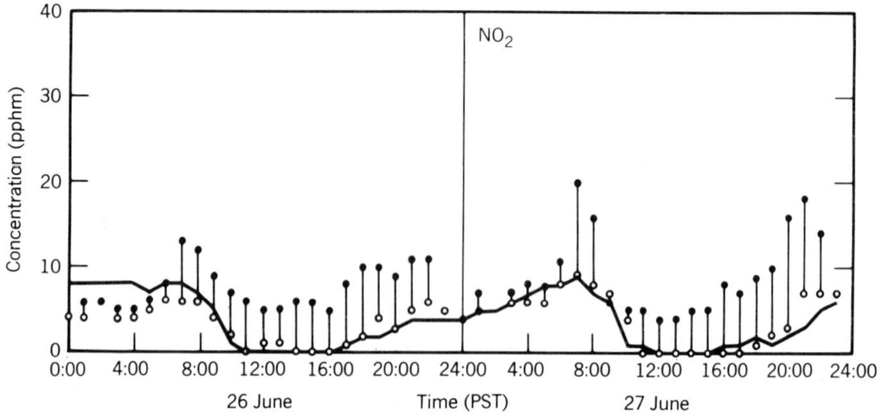

(*b*)

FIGURE 10.14. Observed (○, ●) and model predicted (—) concentrations of (*a*) O_3 and (*b*) NO_2 at a downwind site, Riverside, California, June 26 and 27 1974: ●, Observed at Air Pollution Control District station; ○, observed at California Air Resources Board modeling site 1200 m away (from McRae and Seinfeld, 1983).

the differences between observed O_3 maxima and predicted concentrations were large, ranging from 0.04 to 0.1 ppm for O_3 peaks of 0.19 to 0.26 ppm. Thus such models are best used on a relative basis. That is, the percentage change in pollutant concentrations is probably more reliably predicted by current airshed models than the absolute concentrations.

10-C. REACTIVITY OF ORGANIC COMPOUNDS IN IRRADIATED NMHC–NO_x–AIR SYSTEMS

As discussed above, emission control predictions are sensitive to a number of parameters, including the composition of the initial mixture of hydrocarbons. This is not unexpected since we know that alkenes, for example, react much faster than alkanes with OH radicals. Therefore, one can anticipate that characteristics of photochemical air pollution such as O_3 formation will manifest themselves faster and to a greater extent for alkenes than for alkanes.

An example of the effects of different mixtures of hydrocarbons on the predicted formation of the secondary pollutants O_3, PAN, and sulfate is found in Fig. 10.15. These predictions are based on a simplified box model used to describe London, England. The organics assumed in the *standard* run of the model include 35 different hydrocarbons and oxygenates. Figure 10.15 shows the model predictions for the standard conditions of the model (curves marked s). Also shown (curves marked pb) are the model predictions if the NMHC portion of the 35 organics is reclassified and treated as 25% propylene and 75% butane on a carbon atom basis, as the EKMA model does, for example; clearly a large difference results in the model predictions for O_3, PAN, and sulfate. Thus the reactivity of the individual organics comprising the NMHC is important in secondary pollutant formation, as Hov and Derwent (1981) have shown here.

10-C-1. Typical Reactivity Scales

This sensitivity of secondary pollutant concentrations to the chemical composition of the initial organic mixture was recognized in the earliest studies of photochemical air pollution (e.g., see Haagen-Smit et al., 1953). Out of this recognition grew the concept of *reactivity scales* which could be used to rank organics in terms of their potential for oxidant production.

A number of different parameters have been used to rank organics by their reactivity, including observed rates of reaction, product yields, and effects observed from irradiated NMHC–NO_x mixtures. For example, the rates of O_3 or NO_2 formation or the hydrocarbon loss have been used to develop reactivity scales, as have the yields or dosages of products such as O_3 and PAN in smog chamber studies. Finally, effects such as plant damage, eye irritation, or visibility reduction have been shown to be useful parameters.

With the recognition of the importance of OH radicals in atmospheric chemistry has come an additional suggestion (Darnall et al., 1976; Wu et al., 1976) for a reactivity scale for organics that do not photolyze in actinic radiation. This scale is based on the fact that, for most hydrocarbons, attack by OH is responsible for the vast majority of the hydrocarbon consumption, and this process leads to the free radicals (HO_2, RO_2) that oxidize NO to NO_2, which then forms O_3. Even for alkenes, which react with O_3 at significant rates, consumption by OH still predom-

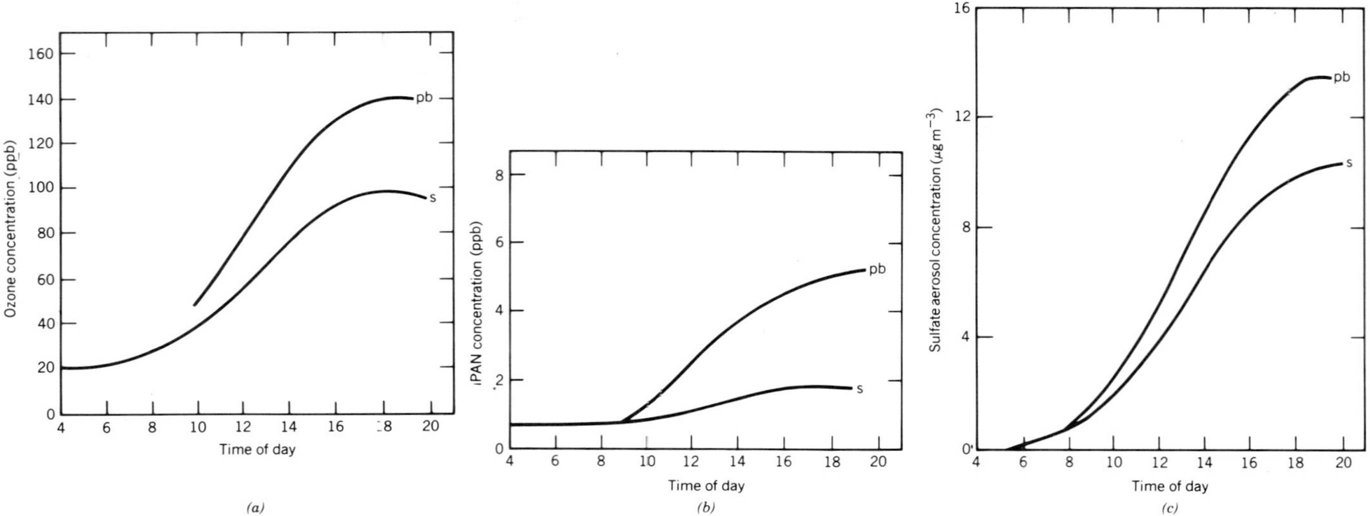

FIGURE 10.15. Diurnal variation in (*a*) ozone, (*b*) PAN, and (*c*) sulfate aerosol calculated with the AERE Harwell fixed point urban model for various model assumptions about the hydrocarbon precursors. "s" is the standard run. In "pb" the London NMHC emissions are grouped as 25% C_3H_6 and 75% C_4H_{10} on a carbon basis (from Hov and Derwent, 1981).

inates in the early portion of the irradiation before O_3 has formed. It has therefore been suggested that the rate constant for reaction between OH and the hydrocarbon should reflect the overall reactivity of the hydrocarbon. The faster the hydrocarbon reacts with OH, the faster HO_2 and RO_2 are produced (and in higher concentrations) and the faster NO is oxidized to NO_2, and O_3 forms. The rate constant for initial OH attack should thus be related to other measures of reactivity of the hydrocarbon (e.g., the rates of NO_2 formation, hydrocarbon loss, and O_3 formation). As discussed below, while this approach is useful for some hydrocarbons, it has significant disadvantages as well.

10-C-2. Typical Classification of Anthropogenic Organics

In general, regardless of the reactivity scale chosen, the reactivity of hydrocarbons tends to be in the order:

[alkenes with internal double bonds] > [di- and trialkyl aromatics, terminal alkenes] > ethylene

> [monoalkyl aromatics] > C_5 and larger alkanes > C_2—C_5 alkanes

Table 10.1 shows the grouping of some hydrocarbons according to their reactivity on the basis of the eye irritation produced when they react in irradiated mix-

TABLE 10.1. Hydrocarbon Reactivity According to Eye Irritation

Hydrocarbon	Eye Irritation Reactivity	Hydrocarbon	Eye Irritation Reactivity
Butane	0	*m*-Xylene	2.9
Hexane	0	1,3,5-Trimethyl	
Isooctane	0.9	benzene	3.1
tert-Butylbenzene	0.9	1-Hexene	3.5
Benzene	1.0	Propylene	3.9
Ethylene	1.0	Ethylbenzene	4.3
1-Butene	1.3	Toluene	5.3
Tetramethylethylene	1.4	*n*-Propylbenzene	5.4
cis-2-Butene	1.6	Isobutylbenzene	5.7
Isopropylbenzene	1.6	*n*-Butylbenzene	6.4
sec-Butylbenzene	1.8	1,3-Butadiene	6.9
2-Methyl-2-butene	1.9	α-Methylstyrene	7.4
trans-2-Butene	2.3	Allylbenzene	8.4
o-Xylene	2.3	β-Methylstyrene	8.9
p-Xylene	2.5	Styrene	8.9

Source: Heuss and Glasson, 1968.

TABLE 10.2. Summary of Organic Reactivities Using Various Reactivity Scales

	Response on 0–10 Scale						
Substance or Subclass	Ozone or Oxidant	Peroxyacyl Nitrate	Formaldehyde	Aerosol	Eye Irritation	Plant Damage	Averaged Response
C_4–C_5 paraffins	0	0	0	0	0	0	0
Acetylene	0	0	0	0	0	0	0
Benzene	0	0	0	0	0	0	0
C_6^+ paraffins[a]	0–4	0[b]	0[b]	0	0[b]	0	1
Toluene (and other monoalkylbenzenes)	4	[c]	2	2	4	0–3	3
Ethylene	6	0	6	1–2	5	+[d]	4
Terminal alkenes[e]	6–10	4–6	7–10	4–8	4–8	6–8	7
Diolefins	6–8	0–2	8–10	10	10	0[b]	6
Dialkyl- and trialkyl benzenes	6–10	5–10	2–4	+[d]	4–8	5–10	6
Internally double-bonded olefins[f]	5–10	8–10	4–6	6–10	4–8	10	8
Aliphatic aldehydes	5–10	+[d]	+[d]	ND[c]	+[d]	+[d]	—

Source: Altshuller, 1966.

[a] Averaged over straight chain and branched chain paraffins.

[b] Very small yields or effects may occur after long irradiations.

[c] Not reported in this study.

[d] Effect noted experimentally, but data insufficient to quantitate.

[e] Includes measurements on propylene through 1-hexene, 2-ethyl-1-butene, and 2,4,4-trimethyl-1-pentene.

[f] Includes measurements on straight chain butenes through heptenes with double bond in 2 and 3 position, 2-methyl-2-butene, 2,3-dimethyl-2-butene, cyclohexene.

TABLE 10.3. Proposed Reactivity of Typical Hydrocarbons Based on OH Rate Constants

Class	k_{OH}[a] (25°C)	$\tau_{1/2}^{OH}$[b]	Typical Hydrocarbons
I	$\leq 8 \times 10^{-14}$	$\geq$ 100 days	CH_4
II	$(8\text{–}80) \times 10^{-14}$	10–100 days	Acetylene, ethane, benzene
III	$(8\text{–}80) \times 10^{-13}$	24 h to 100 days	Ethene, propane, toluene
IV	$(8\text{–}80) \times 10^{-12}$	2–24 h	Propene, *o*; *m*; *p*-xylene, 1,2,4- and 1,2,5-trimethyl benzene
V	$\geq 8 \times 10^{-11}$	0.2–2 h	2-Methyl-2-butene, *d*-limonene

Source: Adapted from Darnall et al., 1976.
[a] Units of cm^3 $molecule^{-1}$ s^{-1}.
[b] Half-lives in atmosphere with respect to reaction with OH assuming $[OH] = 1 \times 10^6$ radicals cm^{-3}.

tures with NO_x in air; products formed in the reaction cause the eye irritation. Table 10.2 shows a summary of the reactivity of various groups of organics using a variety of reactivity scales, including product yields (oxidants, PAN, formaldehyde, aerosols), eye irritation, and plant damage. Table 10.3 shows the results of grouping hydrocarbons on the basis of their OH rate constants.

The reactivities derived using two different scales are generally in agreement, but there are some notable exceptions. For example, non-methane alkanes are predicted to be relatively more important on the basis of OH reactivity than the other scales would imply. This arises because the OH scale ignores important mechanistic aspects of reactions in irradiated NMHC–NO_x mixtures. Thus the initial rate of OH attack does not reflect whether that reaction ultimately leads to the generation of free radicals and hence to continued photooxidation; if the organic does not do so, that is, if it acts as an inhibitor of the chain photooxidations by removing OH, the initial rate of OH attack may assign a mistakenly high reactivity. In addition, the OH reactivity scale also does not take into account the nature of the products formed. For example, organics producing highly photolabile species that photolyze to form free radicals may be more reactive overall than indicated by the initial rate of OH attack. In addition, if products such as plant phytotoxicants or lachrymators are formed, the reactivity on the OH scale may be too low. However, despite these potential deficiencies, the reactivity of organic mixtures in ambient air assessed using the OH reactivity scale, has been found to give results which are generally consistent with the reactivity based on ozone formation (e.g. see Uno et al., 1985).

10-D. PREDICTIONS OF THE ATMOSPHERIC LIFETIMES AND FATES OF VOLATILE TOXIC ORGANIC COMPOUNDS

The potential importance of volatile toxic chemicals in ambient air is being increasingly recognized. Many of the species of potential concern are organics whose persistence and chemical fates in the atmosphere are determined by the chemistry discussed throughout this book. We therefore illustrate here the application of the principles of atmospheric chemistry to the determination of the atmospheric lifetimes and fates of typical toxic organic chemicals.

Included are hydrocarbons, such as benzene, and substituted hydrocarbons containing halogen or oxygen atoms. Polycyclic aromatic hydrocarbons are treated in Chapter 13 and nitrogenous compounds such as the nitrosamines in Chapter 8. Those halogenated organics which are long lived in the troposphere (and hence survive to reach stratosphere) are treated in Chapter 15.

There is no universally agreed upon, clear definition of a toxic air pollutant (TAP). Goldstein (1983) points out that, up to 1983, the major criteria applicable to a TAP seemed to be that it was measurable in air, was produced by anthropogenic activities, and was not a criteria air pollutant such as O_3. Of the thousands of species present in ambient air to which these criteria might apply, by 1983 the U.S. Environmental Protection Agency (EPA) had listed seven as hazardous air pollutants under Section 112 of the Clean Air Act: inorganic arsenic, asbestos, radionuclides, beryllium, mercury, benzene, and vinyl chloride; ethylene oxide and chloroform will likely be added to this list. Others are currently being evaluated and new federal legislation to control airborne toxics is under consideration (Dowd, 1984, 1985).

In 1983, a law (Assembly Bill 1807) was passed by the legislature in California and signed by the governor setting forth a procedure for the identification and control of toxic air pollutants. In the bill, a *toxic air contaminant* (TAC; for consistency we continue to use TAP) is defined as an air pollutant that may cause or contribute to an increase in mortality, or an increase in serious illness, or that may pose a present or potential hazard to human health.

Table 10.4 shows those chemicals (which includes both gaseous and condensed phase species) under consideration in California in 1985; *Level 1 compounds* are those judged to be of concern in terms of risk of harm to public health, amount or potential amount of emissions, manner of usage, atmospheric persistence, and ambient concentrations, *and* for which sufficient data are available to consider listing as a toxic air contaminant. *Level 2 compounds* are those considered to be of *potential* concern. Clearly, a wide variety of metals and organics have the potential of being classified as a TAP.

As part of the risk *assessment* of various TAP, which leads to risk *management* decisions concerning the control of TAP, it is essential to know, in addition to potential health effects, the emissions, manner of usage of the chemical, the atmospheric lifetime and fate, and ambient concentrations. [A review of risk assessment procedures is found in the National Academy of Sciences report (1983).] Thus a knowledge of atmospheric chemistry and its applications to specific TAP is essential to risk assessment and ultimately to risk management.

We will not treat in detail the emissions, manner of usage, or ambient concentrations of the typical TAP whose atmospheric lifetimes and fates we examine in more detail below. However, depending on the specific compound, the sources can be many and varied and can include both stationary point and area sources as

TABLE 10.4. Potential Toxic Air Contaminants Under Consideration in California in 1985

Level 1 Compounds	Level 2 Compounds
Asbestos	Acetaldehyde
Benzene	Acrolein
Cadmium	Acrylonitrile
Carbon tetrachloride	Alkyl chloride
Chloroform	Benzyl chloride
Chromium	Beryllium
Ethylene dibromide	Chlorobenzene
Ethylene dichloride	Chloroprene
Ethylene oxide	Cresol
Formaldehyde	*p*-Dichlorobenzene
Inorganic arsenic	Dialkyl nitrosamines
Nickel	1,4-Dioxane
Polycyclic aromatic hydrocarbons	Epichlorohydrin
Polychlorinated biphenyls	Hexachlorocyclopentadiene
PCD-dioxins	Maleic anhydride
Vinyl chloride	Methyl bromide
Inorganic lead	Mercury
Manganese	Nitrobenzene
Methyl chloroform	Nitrosomorpholine
Methylene chloride	Phenol and chlorinated phenols
Perchloroethylene	Phosgene
Radionuclides	Propylene oxide
Trichloroethylene	Vinylidene chloride
	Xylene

well as mobile sources (Dowd, 1984; Thomson et al., 1985). For example, in California, over 90% of the emissions of benzene are estimated to be associated with the production and use of gasoline, with 83% due to vehicular exhaust emissions (California Air Resources Board and Department of Health Services, 1984).

10-D-1. Review of Potential Atmospheric Fates of Toxic Air Pollutants

Recall from Sections 3-B and 7-B that the possible atmospheric fates of TAP are:

- Photolysis, if the substance absorbs light in the actinic UV ($\lambda \gtrsim 290$ nm).
- Attack during daylight hours by OH radicals.
- Attack by O_3, especially if the molecule contains C=C double bonds.
- Attack at night by NO_3 radicals.

For each of the *selected* TAP examined below, we apply the principles developed earlier to estimate which process (or processes) is most important. Thus the lifetime (τ) of the TAP with respect to removal by that process under typical tropospheric conditions is calculated ($\tau = 1/k[X]$ where X is the second reactant and k is the rate constant in air at 298°K; see Section 4-A-6); additionally, potential reaction mechanisms and products are briefly discussed.

Many of the rate constants needed for risk assessment evaluations (i.e., those with OH, O_3, and NO_3) of all TAP of potential concern have not been measured, especially for NO_3 radical reactions. As discussed briefly in Section 7-C-1, and in detail by Atkinson (1985), a number of methods have been put forward for predicting rate constants for the reactions of OH with compounds that have not yet been studied experimentally. These predicted rate constants are typically accurate to within a factor of 5 and often (e.g., for the alkanes) within a factor of 2.

Although a correlation can be drawn between the O_3 rate constants and those for the corresponding reactions of the organic with $O(^3P)$, and a lesser extent OH, quantitative estimates based on such relationships may not be reliable. Thus Atkinson and Carter (1984) point out that such correlations are only accurate if the reactions being correlated have similar mechanisms. Specifically, since $O(^3P)$ and OH reactions with alkenes occur via addition to the double bond to form a radical species and O_3 adds to form a non-radical molozonide (Section 7-E-2), the reactions are not strictly analogous.

Nevertheless, based on the trend in experimentally measured rate constants for the reactions of O_3 with a series of simple alkenes and those with various substituents, methods have been proposed for estimating unknown rate constants for the reactions of O_3 with more complex alkenes. These are discussed in detail by Atkinson and Carter (1984); they generally give estimates that are accurate to within a factor of ~3.

The sparcity of data on NO_3 reactions at the present time precludes accurate estimation of rate constants for its reactions.

Potential toxic air pollutants treated below include representatives of various classes of organics: benzene, vinyl chloride, ethylene dibromide, trichloroethene, acrolein, cresols, and pesticides and herbicides. The atmospheric lifetimes and fates of formaldehyde are discussed in Section 7-H-4.

10-D-2. Benzene

Benzene does not absorb in the actinic region; however, it reacts fairly rapidly with OH with a room temperature rate constant of $k = 1.28 \times 10^{-12}$ cm^3 molecule^{-1} s^{-1} (Atkinson, 1985). Thus at a 24 h average OH concentration 1×10^6 radicals cm^{-3}, characteristic of a polluted atmosphere, the lifetime of C_6H_6 with respect to OH attack is

$$\tau = \frac{1}{k[\mathrm{OH}]} = \frac{1}{(1.28 \times 10^{-12}\ \mathrm{cm^3\ molecule^{-1}\ s^{-1}}) \times (1 \times 10^6\ \mathrm{radicals\ cm^{-3}})}$$

$$= 7.8 \times 10^5\ \mathrm{s} \quad \text{or} \quad \sim 9\ \text{days}$$

Its reactions with NO_3 and O_3 are sufficiently slow, with rate constants $\lesssim 2 \times 10^{-17}$ and $< 10^{-20}$ cm^3 molecule^{-1} s^{-1}, respectively (Atkinson et al., 1984; Atkinson, 1985), that the lifetime of benzene with respect to each is $\gtrsim 235$ days and ≥ 470 days respectively, for NO_3 at 100 ppt and O_3 at 0.1 ppm. Hence only attack by OH is of concern in assessing the atmospheric fate and persistence of benzene.

As discussed in Section 7-G-1, at ambient temperatures, OH primarily adds to the aromatic ring. The fate of the adduct is not well established, however. Elimination of a hydrogen atom to form phenol occurs part of the time (note that phenol is also on the list in Table 10.4 of potential TAP). Ring cleavage to form oxygenated products such as glyoxal, methylglyoxal, and biacetyl has also been observed (see Atkinson, 1985, and references therein). However, as discussed in Section 7-G, the poor mass balance in many studies of aromatic oxidations suggests that as yet unidentified products are also formed.

10-D-3. Vinyl Chloride

Because vinyl chloride, $CH_2{=}CHCl$, contains a double bond, in the atmosphere it might be expected to react with OH, O_3, and perhaps NO_3. It does not absorb light in the actinic UV, hence photolysis need not be considered. Rate constants for reaction with OH and O_3 have been reported in several studies (Atkinson, 1985; Atkinson and Carter, 1984); while there are insufficient data to establish the rate constants with certainty, they appear to be approximately 6.6×10^{-12} and 2.5×10^{-19} cm^3 molecule^{-1} s^{-1}, respectively. The lifetime with respect to reaction with OH at a concentration of 1×10^6 cm^{-3} is thus ~ 42 h, and with respect to O_3 at 0.1 ppm is ~ 19 days; clearly, attack by OH is expected to dominate over that by O_3 as a sink for removal of vinyl chloride from the atmosphere. No kinetic data for NO_3 radical reactions with vinyl chloride are available, and studies of NO_3 reactions are not yet sufficiently advanced that estimates can be made.

As for other alkenes (Section 7-E-1), the rapid reaction with OH involves an initial addition to the double bond. Although subsequent elimination of the chlorine atom is thermodynamically possible, Atkinson (1985) suggests that the available kinetic data indicate this is a relatively minor pathway. Thus the following reactions seem most likely:

$$\begin{array}{l} \mathrm{OH + CH_2{=}CHCl \rightarrow HOCH_2CHCl \xrightarrow{O_2} HOCH_2CH(O{-}O)Cl} \\ \qquad\qquad\qquad\qquad\qquad\qquad\qquad \mathrm{NO \rightarrow NO_2} \\ \qquad\qquad\qquad\qquad\qquad\qquad\qquad \downarrow \\ \qquad\qquad\qquad\qquad\qquad\qquad\qquad \mathrm{HOCH_2CH(O)Cl} \end{array} \qquad (3)$$

The observation of $\overset{\overset{\displaystyle O}{\|}}{HCCl}$ as a product (Pitts et al., 1984) suggests that decomposition of the alkoxy radical occurs to some extent:

$$HOCH_2\overset{\overset{\displaystyle O}{|}}{C}HCl \rightarrow HOCH_2 + H\overset{\overset{\displaystyle O}{\|}}{C}Cl \tag{4}$$

The CH_2OH radical would then be expected to form $HCHO + HO_2$ as discussed in Section 7-E-1b:

$$CH_2OH + O_2 \rightarrow HCHO + HO_2 \tag{5}$$

However, other possible fates of the alkoxy radical and the intermediates and products should be examined.

10-D-4. Ethylene Dibromide

Ethylene dibromide (EDB), CH_2BrCH_2Br, is currently used as a lead scavenging agent in leaded gasoline and was employed as a pesticide until 1983 when the EPA canceled its use. Because it is a substituted alkane, its reaction with O_3 should be slow ($k < 10^{-23}$ cm^3 molecule^{-1} s^{-1}), as is also expected for attack by NO_3 (Section 7-C-2); thus since EDB does not absorb actinic UV, reaction with OH is again expected to be the major atmospheric fate. Using a rate constant of 2.5×10^{-13} cm^3 molecule^{-1} s^{-1} for the OH–EDB reaction (Howard and Evenson, 1976), an EDB lifetime of ~46 days at 1×10^6 cm^{-3} OH is predicted. This long half-life means that EDB can become widely dispersed.

The reaction of OH with EDB is expected to proceed via hydrogen atom abstraction, with the subsequent formation of an alkoxy radical:

$$\begin{array}{l} CH_2BrCH_2Br + OH \rightarrow CHBrCH_2Br \\ \qquad\qquad\qquad\qquad\quad \downarrow O_2 \\ \qquad\qquad\qquad\qquad\quad OOCHBrCH_2Br \\ \qquad\qquad\qquad\qquad\quad NO \dashv\!\!\rightarrow NO_2 \\ \qquad\qquad\qquad\qquad\quad OCHBrCH_2Br \end{array} \tag{6}$$

As in the case of the alkoxy radical formed from vinyl chloride reactions in air, little is known about the reactions of this radical. However, as discussed in Section 7-D-3, reaction with O_2 and possibly decomposition are potential atmospheric fates:

$$\mathrm{O{-}C(Br)(H)CH_2Br + O_2 \rightarrow O{=}C(Br)CH_2Br + HO_2} \quad (7)$$

$$\mathrm{O{-}C(Br)(H)CH_2Br \xrightarrow[a]{} HC({=}O)Br + CH_2Br}$$

$$\mathrm{\xrightarrow[b]{} HC({=}O)CH_2Br + Br} \quad (8)$$

The C—Br bond is sufficiently weak (~65 kcal $\mathrm{mole^{-1}}$) that the decomposition path to produce bromine atoms may be significant. However, the relative importance of these reactions and the intermediates and products formed remains to be established.

10-D-5. Trichloroethene

Based on its rate constants for reaction with OH, 2.36×10^{-12} $\mathrm{cm^3\ molecule^{-1}\ s^{-1}}$ (Atkinson, 1985) and with O_3, $<3 \times 10^{-20}$ $\mathrm{cm^3\ molecule^{-1}\ s^{-1}}$ (Atkinson and Carter, 1984), lifetimes of TCE with respect to reaction with OH (1×10^6 $\mathrm{cm^{-3}}$) and O_3 (0.1 ppm) are ~5 days and ~157 days, respectively. While the NO_3 reaction rate constant has not been measured, with TCE's electron-withdrawing chlorine substituents it should be less than that for ethene, which is 1.1×10^{-16} $\mathrm{cm^3\ molecule^{-1}\ s^{-1}}$ (Section 7-E-4). At an NO_3 concentration of 100 ppt (2.5×10^9 $\mathrm{cm^{-3}}$), the lifetime of trichloroethene will thus be ≥42 days, too long to compete with OH.

As discussed for vinyl chloride, addition of OH to the double bond, followed by the formation of an alkoxy radical, is expected in ambient air. Addition at the hydrogen-substituted carbon would be as follows:

$$\mathrm{OII + CCl_2{=}CHCl \rightarrow CCl_2CHClOH \xrightarrow{O_2} OOCCl_2CHClOH \xrightarrow{NO \rightarrow NO_2} OCCl_2CHClOH} \quad (9)$$

Few data are available on the fate of the haloalkoxy radical in air, but it would be expected to decompose in part to form phosgene ($COCl_2$), which is also a potential TAP (Table 10.4) (but which would photolyze quite rapidly in sunlight):

$$OCCl_2CHClOH \rightarrow HOCHCl + COCl_2 \qquad (10)$$

Initial addition of OH at the other carbon would lead, through a similar reaction sequence, to the alkoxy radical $HOCCl_2\overset{\overset{\displaystyle O}{|}}{C}HCl$ which, upon decomposition, would form $H\overset{\overset{\displaystyle O}{\|}}{C}Cl$:

$$HOCCl_2\overset{\overset{\displaystyle O}{|}}{C}HCl \rightarrow HOCCl_2 + H\overset{\overset{\displaystyle O}{\|}}{C}Cl \qquad (11)$$

Both $H\overset{\overset{\displaystyle O}{\|}}{C}Cl$ and $COCl_2$ have been observed as products in laboratory studies (Pitts et al., 1984), but the yields are less than unity. Thus other reactions of the chlorinated alkoxy radicals must occur; they warrant investigation.

10-D-6. Acrolein

Acrolein, $CH_2{=}CHCHO$, reacts with OH and O_3 with room temperature rate constants of 2.0×10^{-11} and $(3–7) \times 10^{-19}$ cm^3 $molecule^{-1}$ s^{-1}, respectively (Atkinson and Carter, 1984; Atkinson, 1985). This corresponds to lifetimes of ~14 h and ~9 days for OH at 1×10^6 cm^{-3} and O_3 at 0.1 ppm. No data are available for the NO_3 reaction. Again, OH is the major *chemical* pathway for removal. However, in addition to reaction with OH, acrolein also absorbs light in the actinic region and hence photolysis must also be considered. In a laboratory study of UV irradiated mixtures of gaseous acrolein and O_2 in the Torr pressure range, the rate of photooxidation of this intense lachrymator to give gaseous products was slow. However, its photoreactivity at ppm-ppb levels in simulated atmospheres should be determined experimentally.

Two possible mechanisms of OH attack must be considered—abstraction of the aldehydic hydrogen and addition to the double bond:

$$OH + CH_2{=}CHCHO \rightarrow H_2O + CH_2{=}CHCO \qquad (12)$$

$$\rightarrow HOCH_2CHCHO, \quad \text{or } CH_2CH(OH)CHO \qquad (13)$$

While no definitive results are available, the kinetic data suggest, by comparison to other OH reactions, that hydrogen abstraction will account for most of the reaction (Atkinson, 1985). The fate of the unsaturated species $CH_2{=}CHCO$ is unknown.

10-D-7. Cresols

The cresols (*o*-, *m*-, *p*-) react rapidly both with OH and NO_3 (Section 7-H). For example, the rate constants for reaction of *o*-cresol with OH and NO_3 are 4.0×10^{-11} and 2.2×10^{-11} cm^3 molecule^{-1} s^{-1}, respectively. This leads to predicted lifetimes of ~7 h and ~20 s, respectively, at concentrations of 1×10^6 cm^{-3} OH and 100 ppt (2.5×10^9 cm^{-3}) NO_3. The reaction with O_3 is too slow [$k \simeq (3\text{–}6) \times 10^{-19}$ cm^3 molecule^{-1} s^{-1}, $\tau \simeq 10$ days at 0.1 ppm] to be important. Thus this is one of the cases where a nighttime reaction of NO_3 may determine the atmospheric lifetime and fate of a TAP.

As discussed in Section 7-H-3, the reaction of NO_3 with cresols is believed to give HNO_3 and a phenoxy radical. The phenoxy radical appears to be stable against further oxidation in air, and in the presence of NO_2 has been observed to form nitrophenols (Niki et al., 1979).

10-D-8. Pesticides and Herbicides

The many pesticides and herbicides in use today have a wide variety of chemical structures. A detailed discussion of these and their expected atmospheric lifetime and fates is beyond the scope of this book. However, the reader will find articles by Woodrow, Crosby, and Seiber (1983) and Crosby (1979, 1983), and the references contained therein, helpful in reviewing the photochemistry, products, and lifetimes observed in both laboratory systems and in field studies.

Surprisingly little work has been carried out on the atmospheric reactions of pesticides, other than photolysis. Thus, although some of the products of oxidation have been established for some pesticides under laboratory and, in some cases, field conditions, the mechanisms involved are not. Recently however, a detailed study of the kinetics and products of the reactions of ppm concentrations of *cis* and *trans*-1,3-dichloropropene in air with O_3 and OH radicals, was carried out in Teflon bags and in a 5800 L evacuable chamber (Tuazon et al., 1984). Commercial formulations of these isomers are widely used as an insecticide fumigant (registered trademarks include D-D, Telone etc.) against soil nematodes, and due to their high volatility, on application a significant fraction of 1,3-dichloropropene volatilizes into the atmosphere.

The products from the reaction of O_3 with the two isomers, determined by longpath FTIR spectroscopy, were formyl chloride and chloroacetaldehyde, as well as chloroacetic acid, HCl, CO, CO_2 and formic acid. Attack by OH radicals gave formyl chloride and chloroacetaldehyde with unit yields. Rate constants for reaction with O_3 and OH were for *cis*-$CH_2ClCH{=}CHCl$, 1.5×10^{-19} (O_3 reaction) and 7.7×10^{-12} cm^3 molecules^{-1} s^{-1} (OH reaction), and for *trans*-$CH_2ClCH{=}CHCl$, 6.7×10^{-19} (O_3 reaction) and 1.3×10^{-11} cm^3 molecule^{-1} s^{-1} (OH reaction) respectively. These correspond to lifetimes of 77 days (*cis*) and 17 days (*trans*) for ozone at 1×10^6 molecules cm^{-3} and 1.5 days (*cis*) and 21 h (*trans*) for OH at a concentration of 1×10^6 cm^{-3}.

The only significant loss process for the saturated compound 1,2-dichloropropane which is an important component of the fumigant formulations D-D and Telone, was found to be attack by OH radicals; the lifetime at an OH concentration of 1×10^6 cm^{-3} would be long, $\geq$452 days.

The mechanism of the reaction of the 1,3-dichloropropenes with the OH radical is analogous to that already discussed in this chapter for vinyl chloride and trichloroethene; the mechanism of their reaction with ozone is analogous to those discussed in Section 7-E-2.

This study again illustrates the importance and utility of an integrated approach to the atmospheric chemistry of volatile organics, whether they are of anthropogenic or biogenic origin, and whether benign or toxic.

In a field study, *p*-nitrophenol has been detected over a field treated with parathion (Woodrow et al., 1977); while highly speculative, this may be a result of OH addition to the ring which is known to occur with simple aromatics (Section 7-G-1):

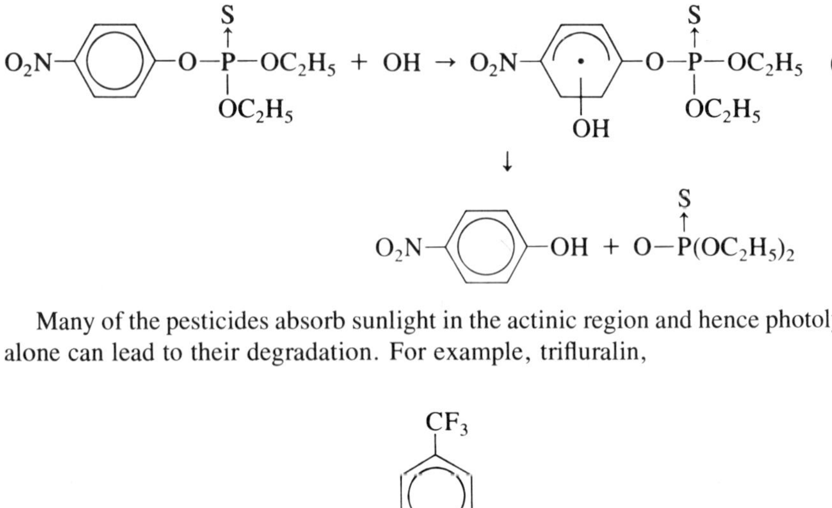

(14)

Many of the pesticides absorb sunlight in the actinic region and hence photolysis alone can lead to their degradation. For example, trifluralin,

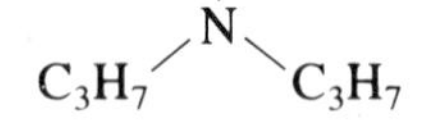

trifluralin

absorbs light throughout the visible region of the spectrum (Zepp, 1982). In the presence of light, its half-life has been observed under laboratory conditions to be approximately 2 h in the absence of O_3, dropping to about three-quarters of an hour when O_3 was added (Woodrow et al., 1983); since the reaction with O_3 in the dark was slow, this suggests that both photolysis and attack by OH (generated by O_3 photolysis, see Section 3-D-1) are important.

While very few studies of the individual reactions of OH, O_3, and NO_3 with pesticides have been carried out, reactions of structurally related compounds may aid in assessing the atmospheric lifetimes and fates. For example, as a first approximation nitrobenzene might serve as a model compound for pesticides such as trifluralin, and herbicides such as butralin:

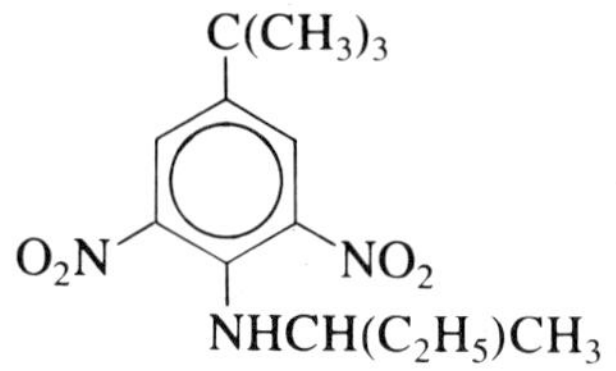

butralin

A potentially important process in the atmospheric fates of pesticides and herbicides is their absoption into atmospheric water droplets and their reactions in the solution phase. For example, the herbicide thiobencarb

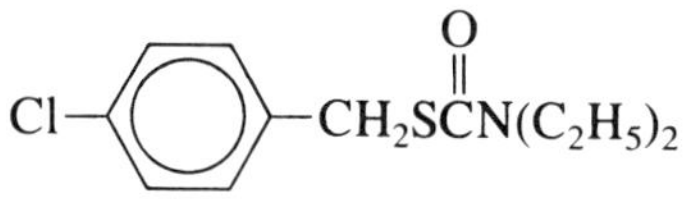

photodecomposes only slowly in aqueous solutions exposed to sunlight, but much more rapidly when H_2O_2 is present (Draper and Crosby, 1981). Several of the products identified were characteristic of solution phase OH reactions; this is expected since H_2O_2 photolyzes to OH in solution (see Section 3-C-12).

However, as discussed in Section 11-A-2a, the chemistry of atmospheric aqueous droplets, which contain oxidants, free radicals, metals, and organics, is complex and not well understood. Clearly, both vapor phase and solution phase reactions of pesticides and herbicides must be considered in assessing their atmospheric fates. Ideally such reactions should be carried out under experimental conditions approximating those in real polluted atmospheres, e.g. in air containing realistic levels of the agricultural chemical, the appropriate co-pollutants, and under irradiation by actinic UV/visible light as well as in the dark.

10-D-9. Summary

In summary, the atmospheric lifetimes and fates of TAP can be estimated by considering their lifetimes for reaction with OH, O_3, NO_3, and with respect to photolysis. Those species having long lifetimes with respect to each process (e.g., ethylene dibromide) will survive long enough to become well dispersed from the original source(s). Those with relatively short lifetimes with respect to at least one of the possible atmospheric fates may react relatively close to the source. However, as we have seen, during these reactions they may form other TAP. Thus an understanding not only of the kinetics of reaction of TAP, but also of the mechanisms and products, is extremely important.

REFERENCES

Altshuller, A. P., "An Evaluation of Techniques for the Determination of the Photochemical Reactivity of Organic Emissions," *J. Air Pollut. Control Assoc.*, **16,** 257 (1966).

Altshuller, A. P., "Measurements of the Products of Atmospheric Photochemical Reactions in Laboratory Studies and in Ambient Air-Relationships Between Ozone and Other Products," *Atmos. Environ.*, **17,** 2383 (1983).

Ames, J., T. C. Myers, L. E. Reid, D. C. Whitney, S. H. Golding, S. R. Hayes, and S. D. Reynolds, "The User's Manual for the SAI Airshed Model," Report to the EPA on Contract No. 68-02-2429 with the U.S. Environmental Protection Agency, August 18, 1978.

Atkinson, R., "Kinetics and Mechanisms of the Gas Phase Reactions of the Hydroxyl Radical with Organic Compounds Under Atmospheric Conditions," *Chem. Rev.*, in press (1985).

Atkinson, R. and W. P. L. Carter, "Kinetics and Mechanisms of the Gas-Phase Reactions of Ozone with Organic Compounds Under Atmospheric Conditions," *Chem. Rev.*, **84,** 437 (1984).

Atkinson, R., W. P. L. Carter, C. N. Plum, A. M. Winer, and J. N. Pitts, Jr., "Kinetics of the Gas-Phase Reactions of NO_3 Radicals with a Series of Aromatics at 296 $\pm$ 2°K," *Int. J. Chem. Kinet.*, **16,** 887 (1984).

California Air Resources Board, Research Division, "Status Report: Ozone Modeling for the South Coast Air Basin," Draft Report presented October 22, 1981 to the Air Resources Board, p. 4.

California Air Resources Board and Department of Health Services, "Report to the Scientific Review Panel on Benzene," November 1984.

Calvert, J. G. and J. N. Pitts, Jr. *Photochemistry*, Wiley, New York, 1966.

Chock, D. P., A. M. Dunker, S. Kumar, and C. S. Sloane, "Effect of NO_x Emission Rates on Smog Formation in the California South Coast Air Basin," *Environ. Sci. Technol.*, **15,** 933 (1981).

Chock, D. P., S. Kumar, and R. W. Herrmann, "An Analysis of Trends in Oxidant Air Quality in the South Coast Air Basin of California," *Atmos. Environ.*, **11,** 2615 (1982).

Chock, D. P., A. M. Dunker, S. Kumar, and C. S. Sloane, "Letter to the Editor," *Environ. Sci. Technol.*, **17,** 58 (1983).

Crosby, D. G., The Significance of Light Induced Pesticide Transformations, in *Advances in Pesticide Science*, Part 3, H. Geissbühler, Ed., Pergamon Press, New York, 1979, pp.568–576.

Crosby, D. G., Atmospheric Reactions of Pesticides, in *Pesticide Chemistry: Human Welfare and the Environment*, Vol. 3, J. Miyamoto and P. C. Kearney, Eds., Pergamon Press, New York, 1983, pp. 327–332.

Darnall, K. R., A. C. Lloyd, A. M. Winer, and J. N. Pitts, Jr., "Reactivity Scale for Atmospheric Hydrocarbons Based on Reaction with Hydroxyl Radical," *Environ. Sci. Technol.*, **10,** 692 (1976).

Dimitriades, B., "Oxidant Control Strategies. Part 1. Urban Oxidant Control Strategy Derived from Existing Smog Chamber Data," *Environ. Sci. Technol.*, **11,** 80 (1977).

Dimitriades, B. and M. Dodge, Eds., *Proceedings of the Empirical Kinetic Modeling Approach (EKMA) Validation Workshop*, EPA Report No. EPA-600/9-83-014, August 1983.

Dodge, M. C., Combined Use of Modeling Techniques and Smog Chamber Data to Derive Ozone-Precursor Relationships, in *Proceedings of the International Conference on Photochemical Oxidant Pollution and Its Control*, Vol. II, Dimitriades, Ed., (1977a), EPA-600/3-77-001b, pp. 881–889.

Dodge, M. C., "Effect of Selected Parameters on Predictions of a Photochemical Model," U.S. Environmental Protection Agency Report No. EPA-60013-77-048, June 1977b.

Dowd, R. M., "EPA's Air Toxics Study," *Environ. Sci. Technol.*, **18,** 373A (1984).

Dowd, R. M., "Proposed Air Toxics Legislation," *Environ. Sci. Technol.*, **19,** 580 (1985).

Draper, W. M. and D. G. Crosby, "Hydrogen Peroxide and Hydroxyl Radical: Intermediates in Indirect Photolysis Reactions in Water," *J. Agric. Food Chem.*, **29,** 699 (1981).

Dunker, A. M., S. Kumar, and P. H. Berzins, "A Comparison of Chemical Mechanisms Used in Atmospheric Models,"*Atmos. Environ.*, **18,** 311 (1984).

Gifford, F. A., Jr., "Use of Routine Meteorological Observations for Estimating Atmospheric Dispersion,"*Nucl. Saf.*, **2,** 47 (1961).

Gifford, F. A., Jr., "Turbulent Diffusion-Typing Schemes: A Review,"*Nucl. Saf.*, **17,** 68 (1976).

Goldstein, B. D., "Toxic Substances in the Atmospheric Environment, A Critical Review," *J. Air Pollut. Control Assoc.*, **33,** 454 (1983).

Haagen-Smit, A. J., C. E. Bradley, and M. M. Fox, "Ozone Formation in Photochemical Oxidation of Organic Substances," *Ind. Eng. Chem.*, **45**, 2086 (1953).

Heuss, J. M. and W. A. Glasson, "Hydrocarbon Reactivity and Eye Irritation," *Environ. Sci. Technol.*, **2**, 1109 (1968).

Howard, C. J. and K. M. Evenson, "Rate Constants for the Reactions of OH with Ethane and Some Halogen Substituted Ethanes at 296°K," *J. Chem. Phys.*, **64**, 4303 (1976).

Hov, O., "The Effect of Chlorine on the Formation of Photochemical Oxidants in Southern Telemark, Norway," *Atmos. Environ.*, **19,** 471 (1985).

Hov, O. and R. G. Derwent, "Sensitivity Studies of the Effects of Model Formulation on the Evaluation of Control Strategies for Photochemical Air Pollution Formation in the United Kingdom," *J. Air Pollut. Control Assoc.*, **31,** 1260 (1981).

Klimisch, R. and J. Heuss, "The Role of NO_x in Smog and Acid Rain: 25 Years of a Pollutant Looking for a Problem?" *Environ. Forum*, August (1983), pp. 28–32.

Kumar, S. and D. P. Chock, "An Update on Oxidant Trends in the South Coast Air Basin of California," *Atmos. Environ.*, **18,** 2131 (1984).

Laffond, M., P. Foster, R. Massot, and R. Perraud, "Étude D'Une Atmosphere de Travail Interactions Chlore-Ethene en Atmosphere Simulé," *Atmos. Environ.*, **19,** 1277 (1985).

Lin, G.-Y., "Oxidant Prediction by Discriminant Analysis in the South Coast Air Basin of California," *Atmos. Environ.*, **16,** 135 (1982).

McRae, G. J. and J. H. Seinfeld, "Development of a Second Generation Mathematical Model for Urban Air Pollution. II. Evaluation of Model Performance," *Atmos. Environ.*, **17**, 501 (1983).

National Academy of Sciences, *Risk Assessment in the Federal Government: Managing the Process*, National Research Council, Washington, DC, 1983.

Niki, H., P. D. Maker, C. M. Savage, and L. P. Breitenbach, Fourier Transform Infrared (FTIR) Studies of Gaseous and Particulate Nitrogenous Compounds, in *Nitrogenous Air Pollutants*, Ann Arbor Science Publishers, Ann Arbor, MI, 1979, pp. 1–16.

Pasquill, F. "The Estimation of the Dispersion of Windborne Material," *Meteorol. Mag.*, **90,** 33 (1961).

Pasquill, F., *Atmospheric Diffusion*, 2nd ed., Halstead Press, New York, 1974.

Pitts, J. N., Jr., A. M. Winer, R. Atkinson, and W. P. L. Carter, "Comment on 'Effect of Nitrogen Oxide Emissions on Ozone Levels in Metropolitan Regions,' 'Effect of NO_x Emission Rates on Smog Formation in the California South Coast Air Basin' and 'Effect of Hydrocarbon and NO_x on Photochemical Smog Formation Under Simulated Transport Conditions' " *Environ. Sci. Technol.*, **17,** 54 (1983).

Pitts, J. N., Jr., R. Atkinson, A. M. Winer, H. W. Biermann, W. P. L. Carter, H. MacLeod and E. C. Tuazon, "Formation and Fate of Toxic Chemicals in California's Atmosphere," Final Report to California Air Resources Board, Contract No. A2-115-32, July 1984.

Schere, K. L. and K. L. Demerjian, A Photochemical Box Model for Urban Air Quality Simulation, in *Proceedings of the 4th Joint Conference on Sensing of Environmental Pollutants*, American Chemical Society, Washington, DC, 1978, pp. 427–433.

Schreffler, J. H. and K. L. Schere, "Evaluation of Four Urban-Scale Photochemical Air Quality Simulation Models,"EPA Report No. EPA-600/S3-82-043, September 1982.

Schuck, E. A., A. P. Altshuller, D. S. Barth, and G. B. Morgan, "Relationship of Hydrocarbons to Oxidants in Ambient Atmospheres," *J. Air Pollut. Control Assoc.*, **20**, 297 (1970).

Stern, A. C., R. W. Boubel, D. B. Turner, and D. L. Fox, *Fundamentals of Air Pollution*, 2nd ed., Academic Press, New York, 1984.

Thomson, V. E., A. Jones, E. Haemisegger, and B. Steigerwald, "The Air Toxics Problem in the United States: An Analysis of Cancer Risks Posed by Selected Air Pollutants," *J. Air. Pollut. Control Assoc.*, **35,** 535 (1985).

Tuazon, E. C., R. Atkinson, A. M. Winer, and J. N. Pitts, Jr., "A Study of the Atmospheric Reactions of 1,3-Dichloropropene and Other Selected Organochlorine Compounds," *Arch. Environ. Contam. Toxicol.*, **13,** 691 (1984).

Uno, I., S. Wakamatsu, R. A. Wadden, S. Konno, and H. Koshio, "Evaluation of Hydrocarbon Reactivity in Urban Air," *Atmos. Environ.*, **19,** 1283 (1985).

U.S. Environmental Protection Agency, "Uses, Limitations, and Technical Basis of Procedures for Quantifying Relationships Between Photochemical Oxidants and Precursors," U.S. EPA Report No. 450/2-77-021a, November 1977.

U.S. Environmental Protection Agency, "Procedures for Quantifying Relationships Between Photochemical Oxidants and Precursors: Support Documentation," U.S. EPA Report No. 450/2-77-0216, February 1978.

U.S. Environmental Protection Agency, "International Conference on Oxidants, 1976—Analysis of Evidence and Viewpoints. Part I. Definition of Key Issues," U.S. EPA Report No. 600/3-77-113, October 1977; "Part II. The Issue of Reactivity," U.S. EPA Report No. 600/3-77-114, October 1977; "Part III. The Issue of Stratospheric Ozone Intrusion," U.S. EPA Report No. 600/3-77-115, December 1977; "Part IV. The Issue of Natural Organic Emissions," U.S. EPA Report No. 600/3-77-116, October 1977; "Part V. The Issue of Oxidant Transport," U.S. EPA Report No. 600/3-77-117, November 1977; "Part VI. The Issue of Air Quality Simulation Model Utility," U.S. EPA Report No. 600/3 77 118, November 1977; "Part VII. The Issue of Oxidant/Ozone Measurement," U.S. EPA Report No. 600/3-77-119, October 1977; "Part VIII. The Issue of Optimum Oxidant Control Strategy," U.S. EPA Report No. 600/3-77-120, November 1977; "Air Quality Criteria for Ozone and Other Photochemical Oxidants," to be published, 1986.

Wayne, L. G., A. Kokin and M. I. Weisburd, "Controlled Evaluation of the Reactive Environmental Simulation Model (REM), "Vol. I, EPA Document No. R4-73-013a, NTIS Publication PB220456/8, February 1973.

Woodrow, J. E., J. N. Seiber, D. G. Crosby, K. W. Moilanen, C. J. Soderquist, and C. Mourer, "Airborne and Surface Residues of Parathion and Its Conversion Products in a Treated Plum Orchard Environment," *Arch. Environ. Contam. Toxicol.*, **6,** 175 (1977).

Woodrow, J. E., D. G. Crosby, and J. N. Seiber, "Vapor Phase Photochemistry of Pesticides," *Res. Rev.*, **85,** 111 (1983).

Wu, C. H., S. M. Japar, and H. Niki, "Relative Reactivities of HO–Hydrocarbon Reactions from Smog Reaction Studies," *J. Environ. Sci. Health, Environ. Sci. Eng.*, **A11,** 191 (1976).

Zeldin, M. D. and D. M. Thomas, "Ozone Trends in the Eastern Los Angeles Basin Corrected for Meteorological Variations," Paper Presented at the International Conference on Environmental Sensing and Assessment, Las Vegas, NV, September 14–19, 1975.

Zepp, R. G. Experimental Approaches to Environmental Photochemistry, in *Handbook of Environmental Chemistry*, Vol. 2/Part B, O. Hutzinger, Ed., Springer-Verlag, New York, 1982.

PART 6

Acid Deposition

11 Formation of Sulfuric and Nitric Acids in Acid Rain and Fogs

Rain with pH in the range 4–5 is common in Europe and the northeast United States and highly acidic precipitation with pH < 4 has been observed in a number of locations around the world. As discussed in Section 11-F, fogs with pH values of 2 or less have also been observed recently. From our perspective as atmospheric chemists, relevant questions about acid deposition include the following: What acids are present? In what concentrations? What are their precursors? How are they formed? Does oxidation occur in the gas phase, in aerosol fog, cloud, or rain droplets, or on the surfaces of solids or in all of these systems? The answers to such questions are essential inputs into the chemical kinetic computer submodels of airshed and long-range transport (LRT) models describing the fates of SO_2 and NO_x and their reaction products in the natural and polluted troposphere.

In this chapter, we examine the wide range of chemical reactions proposed to explain the conversion of gaseous SO_2 to sulfuric acid aerosols and sulfates, and gaseous NO_x to nitric acid and nitrates. In addition, other factors relevant to the chemistry of formation of acids in the atmosphere, such as meterorology and the nature and relative importance of biogenic versus anthropogenic emissions of precursors to these acids, are briefly discussed (Section 11-D).

Despite improvements in our understanding of homogeneous gas phase reactions of SO_2 and NO_x, major gaps still exist even for these relatively straightforward gaseous systems. Unfortunately, even less is known about their tropospheric conversions to acids within rain and cloud droplets, fogs, and mists, as well as on moist surfaces, although major advances are now being made in elucidating reaction products and mechanisms in these complex systems.

For a detailed treatment of various aspects of the acid deposition problem, including observed historical trends in the pH of precipitation and the relationship to trends in emissions, buffering capacities of various geographical areas, as well as the range of biological and ecological effects of acid deposition, the reader is referred to the National Research Council reports of 1981 and 1983, the series on Acid Precipitation edited by Teasley (1984), the book edited by D'Itri (1982), articles by Chamberlain et al. (1981), Ember (1981), Cowling (1982), Johnson and Siccama (1983), the Proceedings of the International Symposium ''Sulfur in the Atmosphere'' (Husar et al., 1978a), the Proceedings of the NATO Conference on Effects of Acid Precipitation on Vegetation and Soils (Hutchinson and Havas, 1980), and the Final Report of the Norwegian SNSF Project (Overrein et al., 1980).

Within the last several years, rapid forest decline has been observed in some areas of Europe, especially in West Germany. As discussed by Schutt and Cowling (1985), this *Waldsterben,* observed in 8% of the total forest area in West Germany in 1982, 34% in 1983, and 50% in 1984, has affected a wide variety of trees, both softwood and hardwood, with such symptoms as a decrease in growth, abnormal growth, or water-stress. The cause is not known; acid deposition *alone* is probably not responsible, yet the effects seem to be related to air pollution (Schutt and Cowling, 1985; Kiester, 1985). As a result, the term *atmospheric deposition* may be a more accurate descriptor of the cause of Waldsterben, and indeed of effects on various parts of the ecosystem observed in other locations around the world. Thus, while we discuss the chemistry of acid deposition in this chapter, the reader should be aware that many of the effects attributed in the past solely to acid deposition may be due to a combination of air pollutants acting in concert, rather than to a single species, e.g. acids plus O_3.

11-A. RATES OF OXIDATION OF SO_2 IN THE TROPOSPHERE

The combustion of sulfur-containing fossil fuels leads to emissions of SO_2 which are proportional to the fuels' sulfur contents. Generally, only small amounts of other sulfur compounds (e.g., SO_3, H_2SO_4, sulfates) are emitted simultaneously. On the East Coast of the United States, sulfur compounds dominate in acid deposition, whereas on the West Coast nitrogeneous species generally predominate except in areas of high local emissions of SO_2.

Once emitted, the gaseous SO_2 is oxidized in the plume and/or ambient atmosphere to sulfuric acid aerosols or sulfates by reactions occurring in the gas phase, in the liquid phase, on the surfaces of solids, or combinations of all three.

Field studies of the rate of oxidation of SO_2 have been carried out both in plumes from point sources and in ambient air in urban and rural areas. In power plant plumes, oxidation rates of $\lesssim 10\%\ h^{-1}$ have been reported in many studies (Gillani et al., 1981; Newman, 1981), although much higher rates are often observed when the plume passes through a cloud or fog bank (e.g., see Eatough et al., 1984). Rates of conversion generally appear to be higher in the summer than in winter, and higher at noon compared to nighttime; for example, Richards and co-workers (1981) reported maximum rates of SO_2 conversion of $0.8\%\ h^{-1}$ in the summer and $0.2\%\ h^{-1}$ in the winter for the plume from a coal-fired power plant. Similarly, Lusis and co-workers (1978) reported SO_2 oxidation rates generally less than $0.5\%\ h^{-1}$ in February and in the early morning in June, but $1\text{–}3\%\ h^{-1}$ at midday in June from a coal-fired power in Alberta, Canada. Noontime conversion rates in a power plant plume were found to be $1\text{–}4\%\ h^{-1}$ compared to nighttime rates of $<0.5\%\ h^{-1}$ (Gillani 1978; Husar et al., 1978b; Forrest et al., 1981).

These effects of season and time of day suggest the importance of photochemistry, and perhaps temperature, in the oxidation. As discussed in more detail below with respect to the individual reactions, this does not necessarily imply that the oxidation reactions themselves are photochemical in nature, but rather that they may involve oxidants such as H_2O_2 which are formed through photochemical processes. This is supported by observations that SO_2 oxidation rates in the plume tend to increase when it mixes with surrounding ambient air containing photochemical oxidants. For example, Gillani and co-workers (Gillani et al., 1981) have shown that the observed rate of conversion of SO_2 in a power plant plume under dry (relative humidity <75%) summertime conditions can be fitted to an equation of the form

$$k = (0.03 \pm 0.01)\, R\, \Delta Z_p [O_3]$$

where R is the solar radiation intensity, ΔZ_p is the verticle spread of the plume (presumably related to the plume dilution), and $[O_3]$ is the background ozone concentration.

In ambient air, rates of oxidation of SO_2 up to ~30% h^{-1} have been observed in such diverse locations as Budapest, Hungary (Mészáros et al., 1977), and St. Louis, Missouri (Breeding et al., 1976; Alkezweeny and Powell, 1977). Again, rates generally seem to be higher during the day than at night, and in the summer compared to winter.

It should be noted, however, that there is some recent evidence suggesting that significant SO_2 to sulfate conversion can occur at night. Thus Cass and Shair (1984) observed SO_2 oxidation in one air parcel during the summer in the Los Angeles basin at rates of 4.5–10.8% h^{-1}; much of the travel time of the trajectory of this air parcel prior to arrival at the measurement site was over the ocean at night, suggesting that under some conditions, at least, nighttime oxidation can be significant. Liquid phase reactions in clouds were suggested by Cass and Shair as a possible contributing factor.

As discussed in connection with plume chemistry, significant rates of oxidation in ambient air at night may arise from the simultaneous presence of oxidizing copollutants such as H_2O_2 and O_3. Thus rates of oxidation of SO_2 both in plumes and in ambient air are generally higher when hydrocarbons and oxides of nitrogen, along with the host of secondary pollutants they produce, are also present. For example, Breeding and co-workers (1976) observed a downward trend in the half-life of SO_2 as the non-methane hydrocarbon (NMHC) concentration in ambient air increased; presumably when NMHC levels are higher, so are oxides of nitrogen and the associated oxidants such as H_2O_2 and O_3.

In both plumes and ambient air, the presence of liquid water in aerosols, clouds, and fog is now accepted to be an important factor in determining the overall rate of conversion of SO_2. These water droplets provide a medium in which aqueous phase reactions can occur (see Section 11-A-2), and it is now believed that such condensed phase reactions contribute significantly to the SO_2 oxidation under some conditions. For example, a "burst of sulfate formation" has been observed in power plant plumes that pass through a cloud layer, and increased rates of SO_2 conversion in such plumes are generally observed at higher relative humidities, >75% (Hegg and Hobbs, 1982, 1983; Gillani and Wilson, 1983; Gillani et al., 1981, 1983; Eatough et al., 1984). Similarly, in ambient air studies in the Ohio River Valley, evidence for significant production of both sulfuric and nitric acids in clouds was obtained (Lazus et al., 1983).

In summary, significant rates of SO_2 oxidation have been observed both in power plant plumes and in ambient air. Sunlight intensity, the presence of oxidants and/or oxidant precursors, relative humidity, and the presence of fogs and clouds all appear to be related to the observed conversion rates. We now discuss possible reactions in the gas phase, liquid phase, and on surfaces which may contribute to these observed rates of conversion.

11-A-1. Homogeneous Gas Phase Reactions

11-A-1a. Photooxidation

Gaseous SO_2 absorbs sunlight in the actinic UV and undergoes two allowed electronic transitions resulting in strong absorption bands in the region 240 nm < λ < 330 nm (see Section 3-C-4):

$$SO_2(^1A_1) + h\nu\ (240 < \lambda < 330\text{ nm}) \rightarrow SO_2(^1A_2) \quad (1a)$$

$$\rightarrow SO_2(^1B_1) \quad (1b)$$

SO_2 can also undergo a "forbidden" transition to the $^3B_1$ state: this results in a much weaker absorption band with a long-wavelength cutoff of ~400 nm.

$$SO_2(^1A_1) + h\nu\ [340 < \lambda < 400\text{ nm}) \rightarrow SO_2(^3B_1) \quad (1c)$$

In contrast to NO_2 which photolyzes to NO + O at $\lambda \lesssim 430$ nm, none of these states are photodissociative. Thus, in terms of the tropospheric photochemistry of gaseous SO_2, the question revolves around the fates of the three electronically excited molecules, ($^1A_2$), ($^1B_1$), and the longer-lived ($^3B_1$) species.

In air the excited singlet states are rapidly deactivated either back to the ground state or to the relatively long-lived $^3B_1$ state. Interestingly, while one can conceive of a variety of possible chemical reactions of the triplet SO_2 molecules with a range of atmospheric constituents, their ultimate fate in air seems to be almost exclusively physical quenching to the ground state by molecular O_2. Furthermore, although singlet oxygen is formed in this energy transfer process, the rate is slow and the yield of $^1O_2$ relatively small compared to that from other sources (see Section 7-E-5).

Thus, while SO_2 absorbs sunlight in the troposphere at a significant rate, by far the predominant fate of the excited singlet and triplet states formed is physical deactivation. Since the direction photooxidation of gaseous SO_2 to H_2SO_4 aerosol in air is negligible, one must consider other atmospheric mechanisms for this important gas to particle conversion.

11-A-1b. Hydroxyl Radical

There are a number of candidate reactions for the oxidation of ground-state SO_2, for example, with O_3 or with such labile species as OH, HO_2, CH_3O_2, and NO_3 radicals, the Criegee biradical R_1R_2COO, and $O(^3P)$ atoms. However, all available evidence suggests the only gas phase process that is fast and efficient enough to account for most of the sulfuric acid aerosol formed by gas phase processes is reaction of SO_2 with the OH radical

$$OH + SO_2 \xrightarrow{M} HOSO_2 \tag{2}$$

Interestingly, the adduct $HOSO_2$, produced by this gas phase reaction of OH with SO_2, has been directly observed in an argon matrix at 11 K by means of FTIR spectroscopy (Hashimoto et al., 1984).

At atmospheric pressure, reaction (2) is in the falloff region between third and second order. The effective bimolecular rate constant recommended for 1 atm and 25°C is $k_2^{bi} = 9 \times 10^{-13}$ cm^3 molecule^{-1} s^{-1} (Atkinson and Lloyd, 1984) with an uncertainty of approximately $\pm 50\%$. For an average OH concentration of 1×10^6 radicals cm^{-3}, the natural lifetime of SO_2 with respect to this one gas phase process will be ~ 13 days.

Thus OH radicals generated during the irradiation of NMOC–NO_x–air mixtures in simulated and real atmospheres not only drive the formation of O_3 in photochemical smog but also efficiently attack SO_2 and NO_2, producing sulfuric acid aerosol and nitric acid, respectively. This chemical linkage between two major atmospheric phenomena which, until recent years, were sometimes considered as being relatively independent, has important control strategy implications for both photochemical oxidant *and* the deposition of sulfuric and nitric acids. Thus, as we have seen in Chapters 3 and 10, the concentration of OH radicals in ambient air depends on the intensity and wavelength distribution of the actinic UV and on the absolute and relative amounts of non-methane hydrocarbons and NO_x present in the polluted air mass (e.g., the NMOC/NO_x ratio). Therefore, strategies for reducing O_3 which focus on strict controls of both organics and NO_x will likely have the additional benefit of reducing the formation of sulfuric acid as well as nitric acid and associated nitrogenous species.

While the kinetics of the elementary reaction (2) are reasonably well defined, and it is known that a significant fraction of the $HOSO_2$ radicals formed ultimately winds up as sulfuric acid aerosol, the rates and products of the reactions of $HOSO_2$ in polluted atmospheres are not well understood.

Part of the problem is the complexity introduced by the number and variety of possible reaction partners for the free radical $HOSO_2$. These include O_2, NO, NO_2, SO_2, peroxy radicals such as HO_2 and RO_2, and water vapor. The importance of understanding these reactions is not simply academic. Thus in some often quoted chemical kinetic–computer models, the fate of $HOSO_2$ radicals is described by the simplified overall reaction sequence

$$\begin{aligned} OH + SO_2 &\xrightarrow{M} HOSO_2 \\ HOSO_2 &\rightarrow\rightarrow\rightarrow H_2SO_4 \end{aligned} \tag{2}$$

However, if significant amounts of other products are formed from $HOSO_2$ radicals, and/or a chain process is involved in ultimately forming H_2SO_4, this widely used simplifying assumption could lead to serious errors in the predictions of such airshed models of H_2SO_4 formation. These could drastically affect the so-called *linearity* of the processes relating reductions in SO_2 emissions to reductions in H_2SO_4 and sulfate depositions at sites downwind (see Section 11-E).

On the basis of theoretical and experimental considerations, Davis et al. (1979) proposed that in the troposphere and lower stratosphere the predominant fate of the $HOSO_2$ adduct is reaction with O_2:

$$HOSO_2 + O_2 \xrightarrow{M} \underset{(\mathbf{I})}{HOSO_2O_2} \tag{3a}$$

They further suggested that, in agreement with Benson's thermodynamic calculations (Benson, 1978), the $HOSO_2O_2$ radical is relatively stable and will react readily with atmospheric water vapor to form the hydrated radical adduct, (**II**):

$$\underset{(\mathbf{I})}{HOSO_2O_2} + H_2O(g) \rightarrow \underset{(\mathbf{II})}{HOSO_2O_2(H_2O)} \tag{4}$$

Reaction (4) is followed by further addition of H_2O molecules to (**II**) until hydrated *clusters* of the form $HOSO_2O_2(H_2O)_x$ are produced. These clusters may react with NO, SO_2, and HO_2 radicals by *quasi-heterogeneous* processes. The products could then ultimately lead to aerosol species (including H_2SO_4) by both condensation and coagulation. In some of the proposed steps, a second SO_2 molecule is oxidized so that one OH radical ultimately oxidizes more than one SO_2.

The adduct $HOSO_2O_2$ has been proposed to react in a manner similar to alkylperoxy radicals, for example, to oxidize NO to NO_2:

$$HOSO_2O_2 + NO \rightarrow HOSO_2O + NO_2 \tag{5}$$

However, if the concentrations of NO are very low, radical–radical reactions (e.g., with a second $HOSO_2O_2$ radical, RO_2, HO_2, etc.) may take place. As discussed above, hydration of the $HOSO_2O_2$ radical may precede such reactions in the atmosphere (Davis et al., 1979).

Formation of H_2SO_4 from the radical $HOSO_2O$ can be envisaged via a number of reactions, including hydrogen abstraction from hydrocarbons,

$$HOSO_2O + -\overset{\displaystyle H}{\underset{|}{\overset{|}{C}}}- \rightarrow HOSO_2OH + -\overset{\displaystyle \cdot}{\underset{|}{C}}- \quad (6)$$

or perhaps reaction with NO or NO_2, followed by hydrolysis:

$$\begin{array}{ccccc} HOSO_2O & \xrightarrow{NO} & HOSO_2ONO & \xrightarrow{H_2O} & HOSO_2OH + HONO \\ \big\downarrow{\scriptstyle NO_2} & & & & \\ HOSO_2ONO_2 & \xrightarrow{H_2O} & HOSO_2OH + HNO_3 & & \end{array} \quad (7)$$

Attempts to observe nitrosylsulfuric acid ($HOSO_2ONO$) and nitrylsulfuric acid ($HOSO_2ONO_2$) under experimental conditions favorable to their formation were unsuccessful (Niki et al., 1980), suggesting either that they are not formed or that they rapidly undergo further reactions (e.g., with H_2O).

Subsequently, Stockwell and Calvert (1983) proposed an alternate pathway for reaction (3),

$$HOSO_2 + O_2 \xrightarrow{M} HO_2 + SO_3 \quad (3b)$$

and presented experimental evidence for its existence. Thus they observed that the concentration of OH radicals in an irradiated HONO–CO–SO_2–NO_x–air system was not sensitive to the SO_2 concentration in the range 0–172 ppm. This led them to conclude that chain propagation of OH radicals must be occurring through the sequence

$$OH + SO_2 \xrightarrow{M} HOSO_2 \quad (2)$$

$$HOSO_2 + O_2 \rightarrow HO_2 + SO_3 \quad (3b)$$

$$HO_2 + NO \rightarrow OH + NO_2 \quad (8)$$

Meagher and co-workers (1984) also concluded, based on smog chamber studies of irradiated SO_2–propene–butane–NO_x–H_2O mixtures that the OH + SO_2 reaction leads 98.5% of the time to HO_2 formation; in order to explain their O_3 yields, they also suggested that one of the intermediates in the OH–SO_2 oxidation such as $HOSO_3$ must react with O_3.

In direct support of the proposal of Stockwell and Calvert (1983), Margitan (1984) recently reported the rate constant for reaction (3b) at 298°K to be $(4 \pm 2) \times 10^{-13}$ cm^3 molecule^{-1} s^{-1}; the lifetime of the $HOSO_2$ adduct is thus expected to be only 0.5 μs at the earth's surface.

The sulfur trioxide formed in (3b) would react rapidly with water vapor to form H_2SO_4:

$$SO_3 + H_2O \rightarrow H_2SO_4 \quad (9)$$

The rate constant for reaction (9) has not been well established. The results of flow tube studies by Castleman et al. (1975) gave a second-order rate constant of $\sim 10^{-12}$ cm^3 molecule^{-1} s^{-1}; however, the possible rapid reaction of SO_3 with water adsorbed on the walls of the flow tube may indicate that this is an upper limit to k_9. A flash photolysis study, where the loss of SO_3 in the presence of water was followed, gave a preliminary value of $k_9 \simeq 1 \times 10^{-13}$ cm^3 molecule^{-1} s^{-1} (Westberg et al., 1985); this is based on an estimated water concentration and thus has a large uncertainty associated with it. However, at 50% RH and 25°C, the lifetime of SO_3 with respect to reaction (9) with H_2O would only be ~ 13 μs, even with this smaller value of the rate constant. Theoretical studies suggest that there is a large energy barrier to formation of sulfuric acid from the SO_3–H_2O adduct which is expected to be an intermediate in reaction (9). Based on their calculations, Chen and Plummer (1985) suggest that the SO_3–H_2O adduct may dissociate back to reactants with about the same probability as it rearranges to sulfuric acid. If this is indeed the case, the kinetics of sulfuric acid formation in reaction (9) may be considerably more complex than if it were a simple bimolecular reaction as written.

In any event, in this reaction sequence (2, 3b, 8) more than one SO_2 would be oxidized per initial OH radical that initiated the chain process.

Calvert and coworkers suggest that one problem with their mechanism is that the key step, reaction (3b), is estimated to be ~ 6 kcal mole^{-1} endothermic. However, this problem may be overcome if the adduct $HOSO_2$ becomes hydrated before it reacts with O_2, as Friend et al. (1980) have suggested:

$$HOSO_2 + H_2O \rightarrow HOSO_2(H_2O) \quad (10)$$

Calvert and Stockwell propose that this is indeed the case, and that the overall reaction, written in their model as

$$HO + SO_2\ (+O_2,\ H_2O) \rightarrow H_2SO_4 + HO_2 \quad (11)$$

is both exothermic and efficient. One notes that the concept of hydration of HSO_x radicals is present in both the mechanisms of Davis and co-workers and Calvert and Stockwell, except that in the former one hydrates the $HOSO_2O_2$ radical and in the latter the $HOSO_2$ radical.

However, as discussed above, Margitan (1984) measured a rate constant for reaction (3b) which is moderately fast; since his studies were carried out in the absence of water, reaction (10) could not have occurred in his system. He suggests that estimates of the OH–SO_2 binding energy on which the 6 kcal endothermicity for reaction (3b) was based, may be slightly high, leading to too high an estimate for the endothermicity of reaction (3b). Thus water does not seem to be necessary for reaction (3b) to occur, although presumably hydration of the intermediates in the reaction could still take place in the atmosphere and change the kinetics from those observed in a water-free system.

Based on these studies showing the production of HO_2 in the reaction of OH with SO_2 in the presence of O_2, the currently accepted concept of SO_2 *termination* of OH chains via reaction (2) must be modified and existing chemical kinetic–computer models revised accordingly. This should result in a greater degree of linearity between proposed reductions in SO_2 emissions and deposition of H_2SO_4 downwind (National Research Council, 1983).

11-A-1c. Criegee Biradical

A second possible oxidant of gaseous SO_2 in polluted air is the stabilized Criegee biradical RCHOO, formed in ozone–olefin reactions. As discussed in detail in Section 7-E-2, these reactions produce excited Criegee biradicals, some of which can be collisionally deactivated in air. The resulting stabilized species, which has two radical centers, is expected to react with a variety of atmospheric constituents.

The classic experiments of Cox and Penkett in 1971 and 1972 first established that SO_2 was rapidly oxidized when ozone and olefins were present simultaneously. Since in the gas phase O_3 itself reacts very slowly with SO_2 [$k < 8 \times 10^{-24}$ cm^3 $molecule^{-1}$ s^{-1} (Hampson and Garvin, 1978)], the oxidation must be attributed to intermediates produced in the ozone–olefin reaction, including the Criegee biradical, for example,

$$CH_3CHOO + SO_2 \rightarrow CH_3CHO + SO_3 \qquad (12)$$

Since the original work of Cox and Penkett, a number of pieces of experimental evidence have been reported that support the hypothesis that a stabilized Criegee biradical is the reactive species in ozone–olefin–SO_2–air systems. For example, Cox and Penkett (1972) showed that the addition of water vapor inhibits the SO_2 oxidation. This is reasonable since, as discussed in Chapter 7, H_2O has been shown to react with the Criegee biradical.

Furthermore, addition of SO_2 to ozone–olefin mixtures leads to a decrease in the stable products which are attributable to secondary reactions of the Criegee biradical. Thus Niki et al. (1977) showed that addition of SO_2 to a mixture of *cis*-2-butene, formaldehyde, and ozone led to a decrease in the amount of the propylene ozonide formed, and this decrease was equal to the amount of SO_2 consumed. This is as expected from the following competitive processes:

$CH_3CHOO + HCHO \rightarrow$ 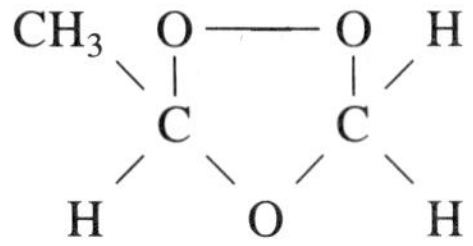 (13)

$$CH_3CHOO + SO_2 \rightarrow CH_3CHO + SO_3 \qquad (12)$$

$$\rightarrow CH_3CHO_2SO_2$$

The yield of sulfuric acid in these ozone–olefin reactions will obviously depend on the fraction of stabilized biradicals produced. For example, at 1 atm in air, the yields of H_2SO_4 from the reaction of ozone with a series of 14 olefins varied from ~3% to 40% of the ozone consumed; this was interpreted as reflecting the yields of the stabilized biradicals in the ozone–olefin reactions (Hatakeyama et al., 1984).

Absolute (and relative) rate constants for these and other reactions of the biradical which may compete with (12) in ambient air (e.g., with NO) are not well established. However, as discussed in Section 7-E-2b, relative rate constants for the reactions of the Criegee biradical CH_2OO with NO_2 and H_2O compared to that with SO_2 have been reported recently by Lee and coworkers. One estimate of the range in which k_{12} falls is 3×10^{-15}–2×10^{-11} (Herron et al., 1982). At all but the higher end of this wide range of values, reaction (12) cannot be a major route for SO_2 oxidation under most atmospheric conditions. However, reaction (12) is distinguished from many other SO_2 oxidations because it can occur at night.

11-A-1d. Ground-State Oxygen Atoms, O(^3P)

The reaction of O(^3P) atoms with SO_2,

$$SO_2 + O(^3P) \xrightarrow{M} SO_3 \tag{14}$$

followed by the hydrolysis of SO_3 to form H_2SO_4 is expected to be unimportant under typical tropospheric conditions. Thus, although the effective bimolecular rate constant at 1 atm in air and 25°C is moderately fast, $k_{14}^{bi} = 3.3 \times 10^{-14}$ cm^3 molecule^{-1} s^{-1} (Atkinson and Lloyd, 1984), the concentration of O(^3P) atoms in ambient air is expected to be only $\lesssim 10^5$ atoms cm^{-3}.

However, reaction (14) may be important in special situations, for example, for a short period of time as stack gases from coal-burning power plants enter the atmosphere as plumes. Thus, if relatively high NO_2 concentrations are also present, in sunlight significant concentrations of O(^3P) may be produced at the stack exit from photodissociation of NO_2; this could lead to a measurable rate of SO_2 oxidation via (14). However, plume dilution in a well-mixed atmosphere will quickly lower the rate of reaction (14) to negligible levels.

11-A-1e. Other Oxidizing Species

As discussed earlier with regard to the Criegee biradical, the rate constant for the gas phase O_3–SO_2 reaction is too slow to make this process significant in the gas phase. [This is not true in the liquid phase, however (*vide infra*).]

Similarly, oxidation by free radicals other than OH, such as HO_2, RO_2, and RO, is also believed to be negligible under most atmospheric conditions. Thus the rate constants for reaction of HO_2 and several different RO_2 radicals are small ($\lesssim 10^{-18}$ cm^3 molecule^{-1} s^{-1}), except possibly for CH_3O_2.

The conflicting data for $CH_3O_2 + SO_2$ are discussed by Calvert and Stockwell (1984). They suggest that if this reaction occurs by a reversible addition to form a $CH_3O_2SO_2$ adduct (rather than by an oxygen atom transfer to form CH_3O +

SO_3), then the *effective* rate constant will depend on the specific conditions. That is, the $CH_3O_2SO_2$ adduct will decompose back to reactants if a species such as NO is not available to react with $CH_3O_2SO_2$, leading to a low overall rate of loss of the reactants. However, if the effective rate constant is as large as 1×10^{-14}, then the reaction would be comparable to the OH–SO_2 reaction in highly polluted atmospheres. This and the reaction(s) of Criegee biradicals with SO_2 are clearly areas that need further detailed study.

While reaction with CH_3O to form CH_3OSO_2 appears to be quite fast ($k \simeq 5 \times 10^{-13}$ cm^3 molecule^{-1} s^{-1}), CH_3O is so rapidly removed by its reaction with the large amounts of molecular O_2 present ($k \simeq 1.3 \times 10^{-15}$ cm^3 molecule^{-1} s^{-1}) that reaction with SO_2 will be negligible. For example, the relative rates of reaction of CH_3O with SO_2 and O_2 even at the relatively high SO_2 concentration of 200 ppb (5×10^{12} molecules cm^{-3}) is given by

$$\frac{k_{CH_3O-SO_2}[SO_2]}{k_{CH_3O-O_2}[O_2]} = \frac{5 \times 10^{-13} \times 5 \times 10^{12}\ \mathrm{s}^{-1}}{1.3 \times 10^{-15} \times 5 \times 10^{18}\ \mathrm{s}^{-1}} = 4 \times 10^{-4}$$

Table 11.1 shows typical estimated peak concentrations of the various possible oxidizing species we have discussed under conditions simulating a moderate to heavily polluted urban atmosphere. Also included are the rate constants for their

TABLE 11.1. Gas Phase Oxidizing Species and Their Calculated Contribution to the Net Loss of SO_2 Through Homogeneous Gas Phase Reactions Under Conditions Typical of Those in a Simulated Moderate to Heavily Polluted Atmosphere

Oxidizing Species	Typical Peak Concentrations[a] (number cm^{-3})	k^b (1 atm, 25°C) (cm^3 molecule^{-1} s^{-1})	Loss of SO_2 (% h^{-1})
OH	1×10^7	9×10^{-13}	3.2
O_3	5×10^{12}	$<8 \times 10^{-24}$	$<1 \times 10^{-5}$
R_1R_2COO	1×10^6	7×10^{-14}	3×10^{-2}
$O(^3P)$	8×10^4	6×10^{-14}	2×10^{-3}
HO_2	2×10^9	$<1 \times 10^{-18}$	$<7 \times 10^{-4}$
RO_2	3×10^9	$<1 \times 10^{-18}$[c]	$<1 \times 10^{-3}$
RO	1×10^4	5×10^{-13}	2×10^{-3}

[a]Concentrations are the peak predicted concentrations for each species using the propylene–*n*-butane model of Atkinson and Lloyd (1984) with an initial reactive hydrocarbon concentration of 0.60 ppm and an initial NO_x of 0.060 ppm.

[b]Rate constants are from Calvert and Stockwell (1984) except for reaction with OH and the Criegee biradical R_1R_2COO where the value recommended by Atkinson and Lloyd (1984) was used.

[c]As discussed in the text, Calvert and Stockwell suggest, based on the literature, that the mechanism of the $CH_3O_2 + SO_2$ reaction involves a reversible addition to form $CH_3O_2SO_2$. In a polluted atmosphere containing NO, NO_2, and so on, the $CH_3O_2SO_2$ adduct may not decompose back to reactants, but react further. The effective CH_3O_2–SO_2 rate constant may then be much greater than 10^{-18}

reactions with SO_2 and, using Eq. (D) below (or the comparable equations for species other than OH), the corresponding loss of SO_2 in % h^{-1}.

Clearly, OH is by far the major reactant with SO_2 in the gas phase. However, all the gas processes taken together cannot account for the experimentally determined rates of oxidation of SO_2 of 10% h^{-1} or more, observed in some field experiments. This strongly suggests that, at least under some atmospheric conditions, oxidation in the aqueous phase and/or on surfaces must also be important, or even dominant.

The rates expected from the gas phase reactions can be computed in the units of % h^{-1} for comparison to rates of oxidation observed in field studies (see above), if the rate constants and the concentrations of oxidizing species are known.

For example, let us assume that the concentration of OH is constant. Then the rate of oxidation of SO_2 by OH,

$$SO_2 + OH \xrightarrow{M} HOSO_2 \tag{2}$$

is given by

$$-\frac{d[SO_2]}{dt} = k_2^{bi}[OH][SO_2] \tag{A}$$

Integrating, one obtains

$$\frac{[SO_2]}{[SO_2]_0} = e^{-k_2^{bi}[OH]t} \tag{B}$$

where $[SO_2]$ and $[SO_2]_0$ are the concentrations at time t and $t = 0$, respectively. The percentage change per hour in the ambient SO_2 concentration is thus given by

$$\begin{aligned} \%\ h^{-1} &= -100\left(\frac{[SO_2]_{t=1\ h} - [SO_2]_0}{[SO_2]_0}\right) \\ &= 100\left(1 - \frac{[SO_2]_{t=1\ h}}{[SO_2]_0}\right) \end{aligned} \tag{C}$$

Substituting into Eq. (C) from (B) one obtains

$$\%\ h^{-1} = 100\ (1 - e^{-k_2^{bi}[OH]t}) \tag{D}$$

where the time t is one hour expressed in the same time units used for the rate constant k_2. Given similar assumptions, the same type of relationship, of course, holds true for other oxidizing species as well.

11-A-2. Homogeneous Aqueous Phase Reactions

The presence of aqueous droplets in the form of aerosols, clouds, fogs, and rain in the troposphere (Pruppacher and Klett, 1978) offers another phase in which oxidation of SO_2 can occur. Thus sulfur dioxide gas dissolves in water to set up equilibria similar to those of CO_2 (*vide infra*):

$$SO_{2(g)} + H_2O \rightleftarrows SO_2 \cdot H_2O \qquad H_{15} = 1.242\ M\ \text{atm}^{-1} \tag{15}$$

$$SO_2 \cdot H_2O \rightleftarrows HSO_3^- + H^+ \qquad K_{16} = 1.32 \times 10^{-2}\ M \tag{16}$$

$$HSO_3^- \rightleftarrows SO_3^{2-} + H^+ \qquad K_{17} = 6.42 \times 10^{-8}\ M \tag{17}$$

[The Henry's law constant H_{15} and the equilibrium constants K_{16} and K_{17} are from Maahs (1982).] As a result, *dissolved SO_2* really includes three chemical species: hydrated SO_2 ($SO_2 \cdot H_2O$), the bisulfite ion (HSO_3^-), and the sulfite ion (SO_3^{2-}). The predominate form depends on the acidity of the solution in which SO_2 dissolves. Figure 11.1 shows the concentrations of the three species as a function of

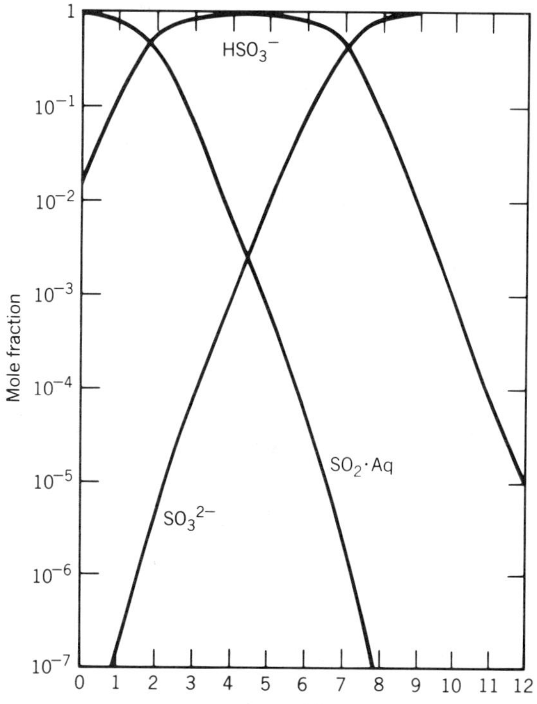

FIGURE 11.1. Mole fraction of sulfur species in solution at different acidities (from Martin and Damschen, 1981).

pH, and we see that over the pH range most typical of atmospheric droplets, 2–6, most of the dissolved SO_2 is in the form of bisulfite ion (HSO_3^-).

Because of the different forms in which dissolved SO_2 exists in solution, the oxidation state (i.e., +4) is often used to denote all these forms of SO_2 taken together, that is,

$$S(IV) = SO_2 \cdot H_2O + HSO_3^- + SO_3^{2-}$$

The oxidized form of sulfur (i.e., sulfuric acid and sulfate) is in the +6 oxidation state and hence is commonly referred to as S(VI). [While reactions (15)–(17) are commonly taken to represent the S(IV) equilibria in aqueous systems, it should be noted that particularly at high S(IV) concentrations, the equilibria may be somewhat more complex, involving for example, the formation of bisulfite ion dimers (Rhee and Dasgupta, 1985)].

The individual reactions in the equilibria represented by (15)–(17) are relatively fast (Martin, 1984). For example, the rate constant for dissociation of hydrated SO_2, k_{16}, is 3.4×10^6 s^{-1} so that the half-life for dissociation of the hydrated SO_2 is only 0.2 μs. Similarly, the second ionization, reaction (17), occurs on time scales of a millisecond or less (Schwartz and Freiberg, 1981). Thus regardless of which of the three species, $SO_2 \cdot H_2O$, HSO_3^- or SO_3^{2-}, is the actual reactant in any particular oxidation, the equilibria will be re-established relatively rapidly under laboratory conditions, and likely under atmospheric conditions as well. The latter is complicated by such factors as the size of the droplet, the efficiency with which gaseous SO_2 striking a droplet surface is absorbed, the chemical nature of the aerosol surface, and so on; for example, the presence of an organic surface film on the droplet could hinder the absorption of SO_2 from the gas phase.

As expected from the equilibria (15)–(17) and Le Chatelier's principle, the more acidic the droplet, the more the equilibria will shift to the left, that is, the less the dissolved SO_2. Figure 11.2 shows the range of dissolved S(IV) concentrations expected in aqueous solutions which are in equilibrium with SO_2 in the gas phase at concentrations of 0.2–200 ppb, and over a pH range of 0–6. It is seen that a wide range of concentrations, from $\sim 10^{-9}$ to 10^{-3} mole L^{-1}, of S(IV) are anticipated, depending on the pH and on the concentration of SO_2 in the gas phase. As expected, the concentrations fall as the pH falls.

This dependence of the S(IV) concentrations on the pH of the droplet plays an important role in determining which oxidant dominates the S(IV) oxidation. As discussed in more detail below, the rates of the various aqueous phase reactions show different dependencies on pH. Some have rate coefficients that increase with increasing pH (e.g., O_3) while others (e.g., H_2O_2) show the opposite trend.

In the first case, both the kinetics and solubility of S(IV) vary with pH in the same direction. As a result, the overall rate of production of S(VI) by such reactions typically shows a strong pH dependence, so that it is often only at higher pH values that they represent important oxidation pathways for S(IV). Such oxidative pathways may initially contribute significantly to acid formation in a relatively neutral droplet; however, as more and more acid is formed, the rate of formation

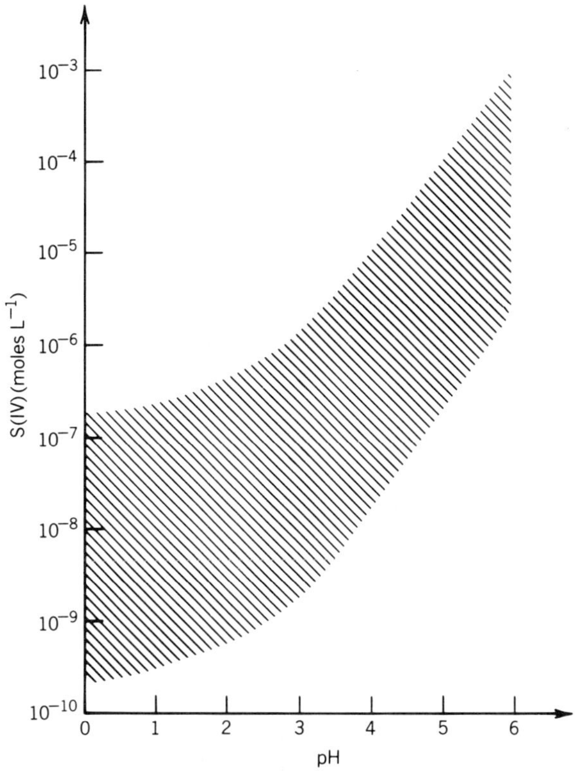

FIGURE 11.2. Range of expected aqueous S(IV) concentrations as a function of acidity for gas phase SO_2 concentrations of 0.2–200 ppb (from Martin, 1984).

of S(VI) by these reactions will decrease. This *self-quenching* effect may lead either to a lowered net oxidation of SO_2 in the atmosphere or to other pathways that do not show this pH dependence becoming relatively more important.

On the other hand, if the rate coefficient decreases with increasing pH, then the reaction rate and solubility of S(IV) work in opposite directions. The net result in terms of S(VI) production may be, as in the case of H_2O_2, a relatively small dependence on pH. Such oxidations may thus contribute in a relatively constant fashion over the entire range of pH values of atmospheric interest.

The equilibria discussed above apply to SO_2 dissolved in pure water, and these have commonly been used for calculations of the concentrations of S(IV) in atmospheric droplets. However, recent measurements of the concentration of S(IV) in fog and cloud water in the Los Angeles areas show that these concentrations are far in excess of what is expected based only on equilibria (15)–(17) (e.g., Munger et al., 1983, 1984; Richards et al., 1983). Water droplets in the atmosphere, especially in or near urban areas, do not consist of pure water; they contain species such as aldehydes (Grosjean and Wright, 1983) and Fe^{3+} which are known to form complexes in solution with the bisulfite or sulfite ions. [For a detailed discussion of S(IV) complexes with transition metal ions and organics, see Huie and Peterson (1983) and Eatough and Hansen (1983).]

Table 11.2 shows some of the equilibria and associated equilibrium constants used by Hoffmann and co-workers (Jacob and Hoffmann, 1983) to calculate the concentration–time profiles of various species in fog water droplets under conditions characteristic of polluted sites. Of particular interest are the hydroxymethanesulfonate ions formed by the reactions of formaldehyde with bisulfite and sulfite ions in solution (Boyce and Hoffmann, 1984):

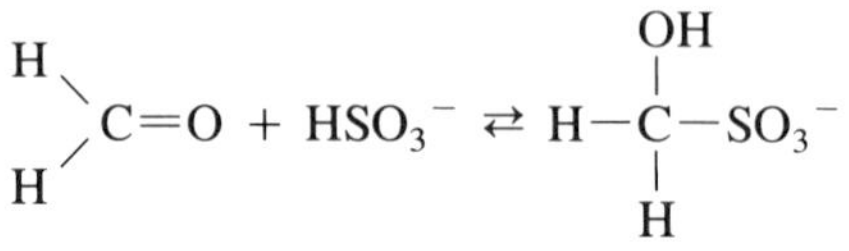

$$\begin{array}{c}H\\ \end{array}\!\!\!\!\!\!\!\!\!\!\!\!\!\!\!\!\!\!C\!=\!O + HSO_3^- \rightleftarrows H-\underset{\underset{H}{|}}{\overset{\overset{OH}{|}}{C}}-SO_3^- \qquad (18a)$$

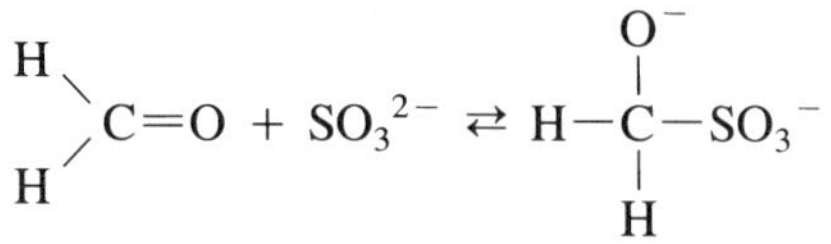

$$\begin{array}{c}H\\ \end{array}\!\!\!\!\!\!\!\!\!\!\!\!\!\!\!\!\!\!C\!=\!O + SO_3^{2-} \rightleftarrows H-\underset{\underset{H}{|}}{\overset{\overset{O^-}{|}}{C}}-SO_3^- \qquad (18b)$$

This complex formation (and the corresponding reaction with larger aldehydes) is at least partially responsible for the high observed S(IV) concentrations in fog and cloud water mentioned above. As seen in Table 11.2, reactions (R12) and (R13), the ions formed by addition of HSO_3^- and SO_3^{2-} and formaldehyde participate in acid-base equilibria. $CH_2(OH)SO_3^-$ is a sufficiently weak acid that, at the pH values typical of clouds and fogs, most of the adducts will exist in this monovalent form (Munger et al., 1984). However, as the adduct forms, SO_2 is drawn from the gas phase into solution, acidifying it via reaction (16).

The kinetics of HCHO–S(IV) adduct formation and dissociation in the pH range 0–3.5 have been shown to be relatively slow (Boyce and Hoffmann, 1984). In fact, under typical conditions, Munger et al. (1984) have shown that adduct formation is sufficiently slow that the solution phase concentrations of S(IV) and HCHO may not be in equilibrium with the gas phase concentrations, which can change more rapidly.

Although the S(IV)–aldehyde adducts are stable toward oxidation, one or more of the oxidation processes for HSO_3^- or SO_3^{2-} described below are likely to be much faster under typical fog and cloud conditions than adduct formation. Thus both processes can occur in parallel as long as the S(IV) is continuously replenished from the gas phase.

With this caveat in mind concerning possible complex formation, we examine potential oxidants for S(IV) in solution. These include O_2, O_3, H_2O_2, free radicals such as OH and HO_2, and oxides of nitrogen (e.g., NO, NO_2, HONO, HNO_3). Metal catalysis may play a role in some of these reactions.

In considering potential oxidants for S(IV), there are two major factors to be considered in assessing the potential contribution of each to the net aqueous phase oxidation. The first is the aqueous phase concentration of the species, and the second is the reaction kinetics, that is, the rate constant and its pH and temperature

TABLE 11.2. S(IV) Equilibria in Aqueous Solutions Containing Aldehydes and Fe^{3+}

Reaction Number	Reaction	pK^{a}	$\Delta H^{\circ}_{298.15^{\circ}K}$ (kcal $mole^{-1}$)	Reference[b]
(R1)	$H_2O_{(l)} \leftrightarrows H^+ + OH^-$	14.00	13.35	SM
(R2)	$SO_{2(g)} + H_2O \leftrightarrows SO_2 \cdot H_2O$	−0.095	−6.25	SM
(R3)	$SO_2 \cdot H_2O \leftrightarrows H^+ + HSO_3^-$	1.89	−4.16	SM
(R4)	$HSO_3^- \leftrightarrows H^+ + SO_3^{2-}$	7.22	−2.23	SM
(R5)	$HNO_{3(g)} \leftrightarrows H^+ + NO_3^-$	−6.51	−17.3	SW
(R6)	$HNO_{2(g)} \leftrightarrows HNO_{2(l)}$	−1.7	−9.5	SW
(R7)	$HNO_{2(l)} \leftrightarrows H^+ + NO_2^-$	3.29	2.5	SW
(R8)	$CO_{2(g)} + H_2O \leftrightarrows CO_2 \cdot H_2O$	1.47	−4.85	SM
(R9)	$CO_2 \cdot H_2O \leftrightarrows H^+ + HCO_3^-$	6.37	1.83	SM
(R10)	$HCO_3^- \leftrightarrows H^+ + CO_3^{2-}$	10.33	3.55	SM
(R11)	$CH_2O_{(g)} + H_2O \leftrightarrows CH_2O \cdot H_2O$	−3.85	−12.85	LB
(R12)	$HOCH_2SO_3H \rightleftarrows H^+ + HOCH_2SO_3^-$	$<0^{c}$	U[d]	R
(R13)	$HOCH_2SO_3^- \rightleftarrows H^+ + {}^-OCH_2SO_3^-$	11.7	U	SA
(R14)	$NH_{3(g)} + H_2O \rightleftarrows NH_3 \cdot H_2O$	−1.77	−8.17	SM
(R15)	$NH_3 \cdot H_2O \rightleftarrows NH_4^+ + OH^-$	4.77	0.9	SM
(R16)	$O_{2(g)} \rightleftarrows O_{2(l)}$	2.90	−3.58	P
(R17)	$H_2O_{2(g)} \rightleftarrows H_2O_{2(l)}$	−4.85	−14.5	MD
(R18)	$O_{3(g)} \rightleftarrows O_{3(l)}$	2.03	−5.04	L-B
(R19)	$CaHCO_3^+ \rightleftarrows Ca^{2+} + HCO_3^-$	11.6	−2.78	SM
(R20)	$CaSO_{4(l)} \rightleftarrows Ca^{2+} + SO_4^{2-}$	2.30	−1.65	SM
(R21)	$NaSO_4^- \rightleftarrows Na^+ + SO_4^{2-}$	0.70	−2.23	SM
(R22)	$FeSO_4^+ \rightleftarrows Fe^{3+} + SO_4^{2-}$	4.20	5.4	SM
(R23)	$Fe(SO_4)_2^- \rightleftarrows Fe^{3+} + 2SO_4^{2-}$	5.60	U	SM
(R24)	$FeCl^{2+} \rightleftarrows Fe^{3+} + Cl^-$	1.40	−7.91	SM
(R25)	$FeOH^{2+} \rightleftarrows Fe^{3+} + OH^-$	12.30	0.04	SM
(R26)	$Fe(OH)_2^+ \rightleftarrows Fe^{3+} + 2OH^-$	23.3	U	SM
(R27)	$Fe(OH)_3 \rightleftarrows Fe^{3+} + 3OH^-$	39.0	20.7	SM
(R28)	$Fe_2(OH)_2^{4+} \rightleftarrows 2Fe^{3+} + 2OH^-$	25.7	16.2	SM
(R29)	$FeSO_3^+ \rightleftarrows Fe^{3+} + SO_3^{2-}$	10.0^{e}	U	H
(R30)	$MnSO_{4(l)} \rightleftarrows Mn^{2+} + SO_4^{2-}$	2.30	−3.39	SM
(R31)	$MnCl^+ \rightleftarrows Mn^{2+} + Cl^-$	1.10	−8.01	SM
(R32)	$HSO_4^- \rightleftarrows H^+ + SO_4^{2-}$	2.20	−4.91	SM

Source: Jacob and Hoffmann, 1983.

[a] K is in M atm^{-1} or M^n. Temperature is 298°K.

[b] Reference code: SM = Sillén and Martell (1964), SW = Schwartz and White (1981), LB = Ledbury and Blair (1925), SA = Sørensen and Andersen (1970), MD = Martin and Damschen (1981), L-B = Landolt-Börnstein (1976), R = Roberts et al. (1971), P = Perry (1963); H = data from Hoffmann Laboratory.

[c] The pK for this reaction is very low.

[d] U = Unknown; $\Delta H = 0$ is assumed in the calculation.

[e] Highly uncertain; values from ≤6.7 to 18 reported in the literature.

dependencies. As a first approximation to the aqueous phase concentrations, Henry's law constants can be applied; values for some potentially important species are given in Table 11.3. It must be noted, however, that as discussed above for S(IV) this approach may lead to low estimates if complex formation occurs in solution. On the other hand, high estimates may result if equilibrium between the gas and liquid phases is not established, for example, if an organic film inhibits the gas-to-liquid transfer (see Section 12-D-2e).

Note that the Henry's law constants given in Table 11.3 are generally for 25°C. At lower temperatures, the values will, of course, be larger due to the increased solubility. While the rate constants for most reactions, especially those in the liquid phase, decrease with decreasing temperature (see Section 4-A-7), the increased reactant concentrations tend to counterbalance this effect. Thus rates of reactant loss and product formation are not as sensitive to temperature as the rate constant alone.

In the discussion below, we focus on the kinetic studies of S(IV) oxidations in aqueous solutions. However, it must be recognized that the oxidation itself is only

TABLE 11.3. Henry's Law Coefficients (*H*) of Some Atmospheric Gases Dissolving in Liquid Water at 25°C

Gas	H (mole L^{-1} atm^{-1})	Reference
O_2	1.3×10^{-3}	Loomis, 1928
NO	1.9×10^{-3}	Loomis, 1928
C_2H_4	4.9×10^{-3}	Loomis, 1928
NO_2[a]	1×10^{-2}	Schwartz and White, 1983
O_3	1.3×10^{-2}	Briner and Perrottet, 1939
N_2O	2.5×10^{-2}	Loomis, 1928
CO_2[b]	3.4×10^{-2}	Loomis, 1928
SO_2[b]	1.24	Maahs, 1982
HONO[b]	49	Schwartz and White, 1981
NH_3[b]	62	Van Krevelen et al., 1949
H_2CO	6.3×10^3	Blair and Ledbury, 1925
H_2O_2	$(0.7–1.0) \times 10^5$	Martin and Damschen, 1981
	1.4×10^5	Yoshizumi et al., 1984[c]
	6.9×10^4	Hwang and Dasgupta, 1985[d]
HNO_3	2.1×10^5	Schwartz and White, 1981
HO_2	$(1–3) \times 10^3$	Schwartz, 1984b
PAN	5	Holdren et al., 1984
CH_3SCH_3	0.56	Dacey et al., 1984

Source: Adapted from Schwartz, 1984a.

[a] Physical solubility; reacts with liquid water.

[b] Physical solubility exclusive of acid–base equilibria.

[c] At 20°C.

[d] Temperature dependence also reported as $H = \exp[7.92 \times 10^3/T(°K) - 15.44]$.

one portion of a sequence of processes which leads from gas phase SO_2 to aqueous phase sulfate. The sequence of steps, depicted in Fig. 11.3, is as follows:

1. Transport of the gas to the surface of the droplet.
2. Transfer of the gas across the air–liquid interface.
3. Formation of aqueous phase equilibria of the dissolved species, for example, (15)–(17) in the case of SO_2.
4. Transport of the dissolved species from the surface to the bulk aqueous phase of the droplet.
5. Reaction in the droplet.

Schwartz and Freiberg (1981) have calculated the rates of these processes for SO_2 and expressed them in terms of *characteristic times* τ, which for step 5, chemical

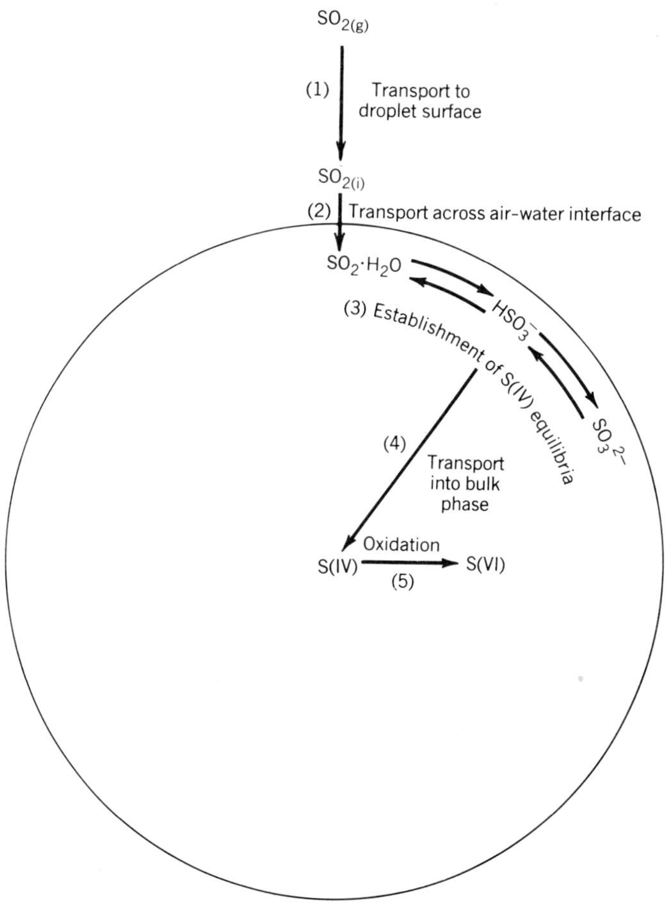

FIGURE 11.3. Schematic of steps involved in the transfer of SO_2 from the gas phase to the aqueous phase of an atmospheric water droplet and its oxidation in the liquid phase. $SO_{2(g)}$ = gas phase SO_2, $SO_{2(i)}$ = SO_2 at water–gas interface.

reaction, is the natural lifetime discussed in Section 4-A-6. For steps 1-4, the characteristic time is the time to establish the appropriate steady-state or equilibrium for the process involved; for example, for step 1, it is the time to establish a steady-state concentration of the gas in the air surrounding the droplet. Seinfeld (1986) discusses in detail calculation procedures for these characteristic times. A brief summary of the results of Schwartz and Freiberg (1981) for steps 1–4 is as follows:

1. *Transport of gas to the surface.* Assuming mixing occurs by molecular diffusion rather than by mechanical or convective processes, the characteristic times for gas phase diffusion to the surface are in the range 10^{-10}–10^{-4} s for droplets with radii from 10^{-5} to 10^{-2} cm, respectively.

2. *Transfer of the gas across the air–liquid interface.* The time for the phase equilibrium to be established across the interface if no reaction is occurring depends on the Henry's law constant for the gas dissolving in the solution; the larger the value of this constant, the larger is the characteristic time for establishing equilibrium across the air-liquid interface because more of the gas must cross the interface in order for the equilibrium to be established. This characteristic time also depends on the sticking coefficient, that is, the fraction of collisions with the surface which result in absorption of the molecule. Assuming this is unity (i.e., no "bounce-off"), the time to establish such equilibria for gases of atmospheric interest is of the order of $<10^{-8}$ to 10^{-1} s over a droplet pH range of 2–6. Of course, if the molecule is absorbed into the droplet on only a small fraction of the collisions, this time will be much longer.

3. *Formation of the S(IV)–H_2O equilibria.* As discussed above, this occurs on a time scale of milliseconds or less.

4. *Transport of the dissolved species within the aqueous phase.* Because diffusion in liquids is much slower than in gases, the characteristic times for diffusion within the droplet itself are much greater (by about four orders of magnitude) than for diffusion of the gas to the droplet surface (again assuming mixing only by molecular diffusion). Thus the times are $\sim 10^{-6}$–1 s for droplets with radii from 10^{-5} to 10^{-2} cm, respectively.

As discussed in detail below, the fastest atmospheric reactions of SO_2 are believed to be with H_2O_2 and perhaps with O_3 at higher pH values. Under extreme conditions of large droplets (>10 μm) and very high oxidant concentrations, the chemical reaction times may approach those of diffusion, particularly in the aqueous phase. In this case, mass transport may become limiting. However, it is believed that under most conditions typical of the troposphere, this will not be the case and the chemical reaction rate will be rate determining in the S(IV) aqueous phase oxidation.

Many experimental studies of the rates of oxidation of S(IV) in solution have used either bulk solutions or droplets that are very large compared to those found in the atmosphere. In addition, reactant concentrations in excess of atmospheric levels have often been used for analytical convenience. The use of large droplets increases the diffusion times, while higher reactant concentrations speed up the aqueous phase chemical reaction rates. The combination of these two factors can lead to a situation where the rates of the diffusion processes, either of the gas to the droplet surface or more likely within the aqueous phase itself, become comparable to, or slower than, the chemical reaction rate. If this is not recognized, the observed rates may be attributed in error to the intrinsic chemical reaction rate. Indeed, as discussed by Freiberg and Schwartz, (1981), this problem appears to have arisen in more than one study of S(IV) aqueous phase oxidation reported in the literature.

In the atmosphere, suspended aqueous solutions are present in the form of aerosols, clouds, fogs, and rain. However, these have different water contents (i.e., g of $H_2O_{(l)}$ per m^3 of air). As discussed in detail in Chapter 12, fine particles ($\lesssim 2\ \mu m$ diameter) emitted directly into the air or formed by chemical reactions can remain suspended for long periods of time. Many of these particles contain water, either in the form of dilute aqueous solutions or as thin films covering an insoluble core; as much as 50% of the mass may be liquid water (Brock and Durham, 1984). Since the total particulate mass in this size range per m^3 of air is typically of the order of 10^{-4} g m^{-3}, the liquid water content due to these small particles is of this order of magnitude.

Clouds, fogs, and rain, however, have much greater liquid water contents and thus have the potential for contributing more to atmospheric aqueous phase oxidations. Clouds typically have liquid water contents of the order of ~ 1 g m^{-3}, with droplet diameters of the order of 5–50 μm; the number concentration and size distribution depend on the type of cloud. Fogs, on the other hand, have smaller liquid water contents (~ 0.1 g m^{-3}) and smaller droplet diameters, generally ~ 0.5–10 μm (Pruppacher and Klett, 1978). Raindrops are, of course, much larger than cloud or fog droplets, with diameters of ~ 0.2–3 mm and correspondingly large liquid water contents. Because of their size, they remain suspended in the atmosphere for only minutes enroute to the earth's surface.

While the volume of liquid water present is much larger in clouds and fogs than that in fine particles, the solute concentrations in the latter may be much higher, and this may serve to increase the rate of SO_2 (and NO_x) aqueous phase oxidations. More importantly, as discussed in Section 11-F, these fine particles are believed to serve as sites for the condensation of water vapor, leading to the formation of fogs and clouds.

The liquid water content of an air mass plays a role in determining the oxidation rate of SO_2 in aqueous atmospheric droplets. This can be seen by developing an expression for the rate of oxidation of SO_2 (in % h^{-1}) in the liquid phase. This conversion rate can be calculated from a knowledge of the gas and aqueous phase reactant concentrations, the solution rate constant, the Henry's law constants (Table 11.3), and the liquid water content of air. In 1 m^3 of air, the rate of formation of S(VI) in the aqueous phase is given by

$$\frac{d[\mathrm{S(VI)}]}{dt} = k[\mathrm{X}][\mathrm{S(IV)}]\ V \qquad \text{(E)}$$

where k is the solution phase rate constant (L $mole^{-1}$ s^{-1}), [X] and [S(IV)] are the aqueous phase concentrations of the oxidant and S(IV), respectively, in units of moles per liter of solution, and V is the volume of liquid water, that is, of aqueous solution available, per m^3 of air. The rate of S(VI) formation is then expressed in moles per m^3 of air per second.

To express this rate of oxidation in % h^{-1} consistent with the units in which the results of field studies are normally reported (see above), one needs to divide this rate by the total number of moles of S(IV) per m^3 of air, convert the unit of time from s^{-1} to h^{-1}, and multiply by 100 to convert the fraction to percent. The gas phase SO_2 in a cubic meter of air is given, according to the ideal gas law, by

$$\left(\frac{n}{V}\right)_{SO_2} = \frac{1000\, P_{SO_2}}{RT} \tag{F}$$

where the factor of 1000 converts from L to m^3. The concentration of S(IV) in solution can be calculated using Henry's law in combination with a knowledge of the total concentration of total dissolved S(IV) relative to dissolved SO_2. Thus the dissolved SO_2 concentration is given by

$$[SO_2]_{aq} = H_{SO_2} P_{SO_2} \tag{G}$$

where H is the Henry's law constant based on physical solubility (Table 11.3) and P is the gas phase pressure of SO_2. The total concentration of S(IV) in solution, taking into account the acid–base equilibria reactions (15)–(17), is then given by

$$[S(IV)]_{aq} = \eta\, H_{SO_2} P_{SO_2} \tag{H}$$

where η is the ratio of the dissolved S(IV) concentration to that of dissolved SO_2. If there are V liters of liquid water per m^3 of air, then the total number of moles of S(IV) contained in the atmospheric water droplets found in 1 m^3 of air becomes

$$\text{moles of aqueous S(IV)} = \eta\, H_{SO_2} P_{SO_2}\, V \tag{I}$$

The total number of moles of S(IV) in a cubic meter of air, including both gas and aqueous phases, is thus given by

$$\text{Total S(IV) per m}^3 \text{ of air} = \frac{1000\, P_{SO_2}}{RT} + \eta\, H_{SO_2} P_{SO_2}\, V \tag{J}$$

Combining (E) and (J), the rate of oxidation of SO_2 in % h^{-1} which occurs in aqueous solution in the atmosphere is given by

$$\%\ h^{-1} = \left[\frac{100\, k[X][S(IV)]\, V}{1000\, P_{SO_2}/RT + \eta\, H_{SO_2} P_{SO_2}\, V}\right] \times 3600\, \frac{s}{h} \tag{K}$$

Equation (K) applies as long as the partial pressure of SO_2 in the gas phase, P_{SO_2}, is measured simultaneously with the solution concentration of S(IV).

With these comments regarding the characteristics of atmospheric aqueous phase oxidations in mind, we now turn to an overview of the chemistry in atmospheric aqueous systems. Subsequently, we summarize the kinetics of S(IV) oxidation in solution by a series of individual potential atmospheric oxidants.

11-A-2a. Overview of Chemistry in Atmospheric Aqueous Systems (Aerosols, Clouds, Fogs, and Rain)

Until approximately 1980, very little was known about the species present or chemistry occurring in tropospheric aqueous droplets which include aerosol particles (Chapter 12), clouds, fogs, and rain. However, with the recognition of the acid deposition problem, and the results of field studies discussed earlier which established that significant SO_2 oxidation takes place in the liquid phase, far greater emphasis has been put on understanding the chemistry occurring in these aqueous systems. We briefly describe in this section some of the chemistry believed to be important, and in the subsequent sections, we apply this to the oxidation of S(IV). As we shall see, much of the data is speculative. However, the reader will find the reviews of Graedel and Weschler (1981), Graedel and Goldberg (1983), Graedel et al. (1985b), Weschler et al. (1985) and Bielski et al. (1985), particularly useful. Unless otherwise stated, the rate constants cited below are from those reviews and the reader should consult them for the original references.

O_3 and H_2O_2 can both be important components of atmospheric aqueous droplets. O_3 is moderately soluble in water, and H_2O_2 is highly soluble (Table 11.3) so that they can be absorbed into solution from the gas phase. As discussed in Chapter 5-F, sensitive and specific spectroscopic monitoring techniques for gas phase H_2O_2 are only now being developed; hence few data are available on the concentrations of gaseous H_2O_2 in ambient air, although some measurements have been carried out in Toronto, Ontario, Canada (Mackay and Schiff, 1985). Using Henry's law constants, Kelly et al. (1985) estimate, based on their measurements of H_2O_2 in the aqueous phase of nonprecipitating clouds of known liquid water content, that gas phase H_2O_2 concentrations in these clouds will be ≤ 1 ppb, with a median value of 0.1 ppb.

However, there are also believed to be sources of these two oxidants in solution. As discussed in Section 3-C-12, some photolysis reactions (e.g., the photolysis of NO_3^- and NO_2) may produce ground-state oxygen atoms, $O(^3P)$, in aerated solutions and lead to O_3 via the same reaction as occurs in the gas phase, that is, addition to O_2.

$$O + O_2 \rightarrow O_3 \qquad k_{19} = 3.0 \times 10^9 \text{ L mole}^{-1} \text{ s}^{-1} \tag{19}$$

H_2O_2 is known to be present in rain, clouds, and natural waters (Zika et al., 1982; Daum et al., 1983; Yoshizumi et al., 1984; Zafiriou et al., 1984; Kelly et al., 1985). A number of different methods have been used for measuring H_2O_2 in atmospheric aqueous solutions, for example, luminol chemiluminescence (Kok et al., 1978), the scopoletin-horseradish peroxidase-phenol method (Zika et al., 1982), and a fluorescence technique (POHPAA method) based on the formation of a fluorescent dimer of *p*-hydroxyphenyl acetic acid (Lazrus et al., 1985). Thus, Kelly et al. (1985) measured the total peroxide content in water from precipitating clouds, nonprecipitating clouds, and rain, using the POHPAA technique, and found concentrations in the range from less than 0.1 to over 100 μM in agreement with earlier studies. To test for the possible contribution of organic hydroperoxides which respond in this assay with the same sensitivity as H_2O_2 (Lazrus et al., 1985), the enzyme catalase was used to selectively destroy H_2O_2 in some cloudwater samples. (Organic hydroperoxides are destroyed much more slowly than H_2O_2 by catalase). In more than half of the 48 cloudwater samples tested, $\leq 5\%$ of the total peroxide signal remained when catalase was used, indicating that the peroxide was essentially all H_2O_2. In the remaining samples, the residual signal after the catalase treatment corresponded to concentrations of 0.3–4.0 μM peroxide. Because there are several possible alternate causes other than the presence of organic peroxides for these signals remaining after catalase treatment, Kelly and coworkers state that while these measurements suggest the existence of organic peroxides in cloudwater, the technique cannot unambiguously confirm this.

In addition to absorption from the gas phase, H_2O_2 can be produced in situ by both dark and photochemical reactions. Thus when organics are present, H_2O_2 is formed photochemically in the presence of actinic radiation (Cooper and Zika, 1983; Draper and Crosby, 1983). This has been suggested to occur via the formation of the superoxide ion, O_2^-, which then reacts to give H_2O_2, via reaction (20):

$$O_2^- + H_3O^+ \rightarrow H_2O + HO_2 \qquad k_{20} \simeq 5 \times 10^{10}\ \text{L mole}^{-1}\ \text{s}^{-1} \tag{20}$$

followed by

$$2HO_2 \rightarrow H_2O_2 + O_2 \qquad k_{21} = 8.3 \times 10^5\ \text{L mole}^{-1}\ \text{s}^{-1} \tag{21}$$

Note that we show the superoxide ion as O_2^- throughout for simplicity, and do not explicitly show it as $O_2^{\dot{-}}$. Superoxide ion can also give H_2O_2 via reaction with the hydroperoxyl free radical in the presence of acids:

$$O_2^- + HO_2 \rightarrow HO_2^- + O_2 \xrightarrow[\text{fast}]{H^+} H_2O_2 \qquad k_{22} = 9.7 \times 10^7\ \text{L mole}^{-1}\ \text{s}^{-1} \tag{22}$$

There has been a great deal of interest in the chemistry and photochemistry of natural waters (oceans, rivers, lakes), and a number of processes have been hypothesized there which could potentially occur in atmospheric droplets as well. For example, data on the photolysis of NO_2^- and NO_3^- (Section 3-C-12) were derived from studies of natural waters. In natural waters, the production of H_2O_2 via reactions such as (23) and (24) has been suggested (Zafiriou, 1983):

$$O_2^- + X^{n+} \rightleftarrows XO_2^{(n-1)+} \xrightarrow{2H+} X^{(n+1)+} + H_2O_2 \tag{23}$$

$$O_2^- + Org_{red} \rightarrow H_2O_2 + Org_{ox} \tag{24}$$

In Eqs. (23) and (24), X^{n+} is a redox active species and Org_{red} and Org_{ox} are the reduced and oxidized forms, respectively, of an organic; whether such reactions might occur to some extent in aqueous atmospheric systems is not clear.

Potential sources of the superoxide ion itself are uncertain. Zafiriou (1983) suggests that in natural waters it could be formed via reactions of electronically excited organics with O_2:

$$\text{organic chromophore } (S_0) + h\nu \rightarrow S_1 \overset{\text{ISC}}{\rightsquigarrow} T_1 \tag{25}$$

$$\text{chromophore } (T_1) + O_2 \rightarrow O_2^- + \text{chromophore}^+ \tag{26}$$

Thus, natural waters have been shown to contain light-absorbing species which can energy transfer from a triplet state to organics and to O_2 (Zepp et al., 1985). [Note, however, that singlet excited oxygen (Section 7-E-5) was assumed to be the product; whether O_2^- can be formed in such an energy transfer is not clear.] Other possible sources suggested include the reaction of unstable reduced metals or autooxidizable organics with O_2 (Zafiriou et al., 1984), for example, reactions (27) and (28):

$$\text{Fe(II)} + O_2 \rightarrow \text{Fe(III)} + O_2^- \tag{27}$$

$$\text{hydroquinones} + O_2 \rightarrow \text{semiquinones} + O_2^- \tag{28}$$

Similarly, metal (M)–liquid (L) complexes are potential sources of O_2^- (Hoffmann and Jacob, 1984):

$$ML^{n+} + O_2 \rightarrow ML^{(n+1)+} + O_2^- \tag{29}$$

Reactions of O_3 with OH^- and HO_2^- also produce O_2^-:

$$O_3 + OH^- \rightarrow HO_2 + O_2^- \tag{30}$$

$$k_{30} = 3.7 \times 10^2 \text{ L mole}^{-1} \text{ s}^{-1}$$

$$O_3 + HO_2^- \rightarrow OH + O_2^- + O_2 \tag{31}$$

$$k_{31} = 2.8 \times 10^6 \text{ L mole}^{-1} \text{ s}^{-1}$$

The rate constant given for reaction (30) may be too high; thus Staehelin and Hoigné (1982) suggest that $k_{30} \simeq (70 \pm 7)$ L mole^{-1} s^{-1}, with the higher values reported in the literature being due to interference from unrecognized radical chain reactions. While reactions (30) and (31) were advanced to explain the enhanced formation of H_2O_2 in the aqueous solution when O_3 is bubbled into it, recent studies by Heikes (1984) rule out reaction (30) as being important; thus both the rate of destruction of O_3 and the rate of H_2O_2 formation in bubblers were observed to increase as the acidity of the aqueous solution increased, which is the opposite of what is expected from reaction (30). The formation of H_2O_2 in the bubbler was attributed to surface reactions of ozone to form reactive species which hydrolyze to give intermediates (e.g. HO_2, HO_2^-, OH, HO_3^+) which then form H_2O_2. (Heikes et al., 1982; Zika and Saltzman, 1982).

Reaction (30) will clearly become less important as the acidity of the solution increases and the OH^- concentration falls. HO_2 in aqueous solution is in effect a weak acid:

$$HO_2 \rightleftarrows H^+ + O_2^- \qquad K_{eq} = 2.1 \times 10^{-5} \text{ moles L}^{-1} \quad \text{(Bielski, 1978)} \tag{32}$$

The equilibrium (32) will shift to the left as the pH falls; thus $[HO_2]/[O_2^-] \approx 10$ at pH 3.7, but ≈ 1 at a pH of 4.7.

If electrons are produced in solution, they will react quickly ($k_{33} = 1.88 \times 10^{10}$ L mole^{-1} s^{-1} at pH 7.0) with O_2 to produce the superoxide ion (Zika, 1981):

$$O_2 + e_{aq}^- \rightarrow O_2^- \tag{33}$$

Aquated electrons have been suggested to be produced via photoionization of aromatics as well as bombardment by cosmic rays (Swallow, 1969); more recently, photochemically induced transition metal (M) charge transfer to solvent reactions of the type (34) have been suggested (Zika, 1981):

$$M^{n+}(H_2O)_n + h\nu \rightarrow M^{(n+1)+}(H_2O)_n + e_{aq}^- \tag{34}$$

H_2O_2 may also be photochemically produced on the surface of a semiconductor metal oxide such as Fe_2O_3 (hematite). The possibility of such photoassisted surface reactions in the atmosphere was first examined by Calvert in 1956. For example, Fe_2O_3 particles suspended in a bisulfite solution in the presence of O_2 rapidly oxidize the S(IV) to S(VI) in the presence of light, whereas no oxidation occurs in the dark (Frank and Bard, 1977). It has been suggested that this is due to the absorption of light by Fe_2O_3 and the migration of excited electrons in the conduction band to the particle surface where they react with O_2 and H^+ to form H_2O_2:

$$O_2 + 2e^{-\cdot}_{surface} + 2H^+ \xrightarrow{\text{surface}} H_2O_2 \tag{35}$$

The H_2O_2 then oxidizes the S(IV) (see Section 11-A-2e). However, since unfiltered radiation from a xenon lamp and quartz cells were used, significant UV radiation was present, and hence it is not certain whether such reactions occur in the troposphere where only light of $\lambda > 290$ nm is present. In addition, it is not clear whether sufficient concentrations of the solids exist in the atmosphere for this to be a significant process.

Finally, potential dark sources of H_2O_2 also exist in atmospheric droplets. For example, the oxidation of organics by O_3 in water produces H_2O_2 (Hoigné and Bader, 1976), although this is thought to play a minor role in the overall H_2O_2 production in atmospheric droplets (Graedel and Goldberg, 1983). The metal-catalyzed autooxidation of organics is another potential dark source of H_2O_2 (Hoffmann and Boyce, 1983):

$$M^{2+} + O_2 \rightleftarrows M^{3+}{-}O_2^- \tag{36}$$

$$\begin{aligned} M^{3+}{-}O_2^- + RH &\xrightarrow[a]{} M^{2+} + R + HO_2 \\ &\xrightarrow[b]{} M^{3+}{-}O_2^-H + R \\ &\xrightarrow[c]{} M^{3+} + R^- + HO_2 \\ &\xrightarrow[d]{} M^{3+} + R + HO_2^- \end{aligned} \tag{37}$$

These reactions can be followed by the formation of H_2O_2 from HO_2 and HO_2^- as discussed above:

$$2HO_2 \rightarrow H_2O_2 + O_2 \tag{21}$$

$$HO_2^- + H^+ \rightarrow H_2O_2 \tag{22}$$

Given that there are a number of potential sources of O_3 and H_2O_2 in aqueous atmospheric droplets, one might anticipate both the chemistry and photochemistry of these highly reactive species to occur. The photochemistry, summarized in reactions (38) and (39), is discussed in Section 3-C-12:

$$O_3 + h\nu\ (\lambda \lesssim 320\ \text{nm}) \xrightarrow{H_2O} H_2O_2 + O_2 \tag{38}$$

$$H_2O_2 + h\nu\ (\lambda \lesssim 380\ \text{nm}) \longrightarrow 2OH \tag{39}$$

The major loss processes for O_3 and H_2O_2 in atmospheric droplets will obviously depend on the presence and concentrations of other species in solution. As discussed in Section 4-D, the total concentration of dissolved species can vary from very low numbers characteristic of remote areas to very high values in urban areas, which occur as water evaporates from fog droplets, leaving a very concentrated solution (see Section 11-F). We therefore briefly summarize here the reactions of O_3 and H_2O_2 of potential importance in atmospheric drops under conditions typical of continental (including coastal) areas where nitrogen, sulfur, and carbon-containing species are present. However, we do not try to assess the importance of each quantitatively since this will depend on the particular conditions.

As discussed earlier, O_3 can react with OH^- and HO_2^-, reactions (30) and (31) to produce superoxide ion:

$$O_3 + OH^- \rightarrow HO_2 + O_2^- \tag{30}$$

$$O_3 + HO_2^- \rightarrow OH + O_2^- + O_2 \tag{31}$$

In addition, reactions with free radical OH and HO_2 in solution can occur:

$$O_3 + OH \rightarrow HO_2 + O_2 \tag{40}$$

$$k_{40} = 1.1 \times 10^8 \text{ L mole}^{-1} \text{ s}^{-1}$$

$$O_3 + HO_2 \rightarrow OH + 2O_2 \tag{41}$$

Reaction (40) has been shown to produce HO_2 in aqueous solutions and thus gives the same products as in the gas phase (Sehested et al., 1984). The rate constant for reaction (41) is subject to controversy. Although $k_{41} = 4.0 \times 10^6$ L mole^{-1} s^{-1} was cited by Graedel and Weschler (1981), based on the literature values, Sehested et al. (1984) have reported that the rate constant is $< 10^4$ L mole^{-1} s^{-1}.

Free radical OH is produced by the photolysis of H_2O_2 and possibly of Fe(III) complexes (Section 3-C-12) as well as by thermal reactions such as (41). In urban areas, nitrite and S(IV) will also be present and O_3 can react with these:

$$O_3 + NO_2^- \longrightarrow NO_3^- + O_2 \tag{42}$$

$$k_{42} = 1.6 \times 10^5 \text{ L mole}^{-1} \text{ s}^{-1}$$

$$O_3 + HSO_3^- \xrightarrow{OH^-} SO_4^{2-} + H_2O + O_2 \tag{43}$$

(See Section 11-A-2d.)

O_3 can react, of course, with organics such as alkenes in solution, and there is an extensive literature on the rates and mechanisms in various solvents, including water. However, these reactions are usually not included in models of reactions in atmospheric droplets, presumably because the low to moderate rate constants and low organic concentrations together yield low overall rates compared to other possible fates of the O_3.

While H_2O_2 photolyzes in solution (Section 3-C-12), other reactions including (44) and (45) have been suggested to be important and, in fact, to predominate over photolysis under many conditions:

$$H_2O_2 + CO_3^{\dot{-}} \rightarrow HCO_3^- + HO_2 \quad (44)$$

$$k_{44} = 8.0 \times 10^5 \text{ L mole}^{-1} \text{ s}^{-1}$$

$$H_2O_2 + OH \rightarrow HO_2 + H_2O \quad (45)$$

$$k_{45} = 2.7 \times 10^7 \text{ L mole}^{-1} \text{ s}^{-1}$$

$$H_2O_2 + HSO_3^- \rightarrow HSO_4^- + H_2O \quad (46)$$

(See Section 11-A-2e)

$$H_2O_2 + HONO \xrightarrow{H_3O^+} NO_3^- + H_3O^+ \quad (47)$$

$$k_{47} = (4.6 \times 10^3[H^+]) \text{ L mole}^{-1} \text{ s}^{-1}$$

(Damschen and Martin, 1983)

The carbonate radical ion $CO_3^{\dot{-}}$ has been proposed to arise from the reaction of superoxide with the bicarbonate ion (Chameides and Davis, 1982; Graedel and Goldberg, 1983):

$$O_2^- + HCO_3^- \rightarrow HO_2^- + CO_3^{\dot{-}} \quad (48)$$

However, Schwartz (1984b) argues that there is a little experimental evidence to support the occurrence of (48), and that reaction (44), which reforms HCO_3^- from the radical ion, compensates for (48).

Whether the superoxide ion reacts with O_3

$$O_2^- + O_3 \rightarrow O_3^- + O_2 \quad (49)$$

is uncertain at present because of conflicting reports in the literature, but Schwartz (1984b) points out that if reaction (49) were rapid, it could play a significant role in cloud water chemistry.

Chameides and Davis (1982) first suggested that free radicals such as OH and HO_2 could be absorbed from the gas phase into aqueous solutions in the atmosphere and that this gas phase scavenging mechanism could be a major source of these free radicals and hence of H_2O_2 in cloud water. A major uncertainty in estimating the importance of gas phase scavenging of free radicals compared to their generation by reactions in the droplet is a lack of knowledge of the *accommodation coefficient* (α) for these species; α is the fraction of collisions of the free radical with the droplet surface which results in absorption of the radical from the gas phase. Schwartz (1984b), using his calculated value for the Henry's law coefficient for HO_2 of (1–3) $\times 10^3$ mole L^{-1} atm^{-1} at 25°C, has confirmed that if $\alpha \geq 10^{-3}$, the scavenging of HO_2 from the gas phase could be a major sink for gas phase HO_2. For value of $\alpha \ll 10^{-2}$, mass transport of HO_2 across the air–water interface becomes limiting, illustrating the importance of determining values of α. Schwartz also established that for $\alpha \geq 10^{-4}$, significantly different results are obtained if one assumes that the uptake of HO_2 is reversible rather than irreversible; with the assumption that HO_2 can go from the aqueous phase into the gas phase as well as the reverse, the gas phase HO_2 concentration is predicted to be substantially greater than if HO_2 is absorbed irreversibly. Because of the coupling of HO_2 with OH in the presence of NO_x in the gas phase, gas phase OH concentrations are also predicted to be greater if the HO_2 absorption is reversible.

While the relatively little work that has been carried out on gas phase scavenging of reactive species has concentrated on OH and HO_2, it is possible that other species, such as RO_2, NO_3, and so on, may also be scavenged efficiently. For example, in a related gas–aqueous system, the transfer of NO_3, CH_3OOH, and CH_3O_2 from the air to the interfacial layer at the surface of the sea has been suggested to be important in the chemistry of this layer (Thompson and Zafiriou, 1983). Clearly, this is an area requiring further study.

Graedel et al. (1985a,b) have examined the potential role of transition metal ions (TMI) in the homogeneous aqueous phase chemistry of atmospheric droplets and propose that these may play a major role in the chemistry of the oxygen and hydrogen species discussed above. They suggest that reactions of manganese, iron, nickel, and copper should be considered in atmospheric aqueous droplets. Thus as discussed in Chapter 3-C-12, they have shown that some of the Fe(III) complexes in aqueous solution absorb light in the actinic region and may produce OH free radicals (Graedel et al., 1985a,b; Weschler et al., 1985); indeed, their model calculations suggest this could be the major source of OH in atmospheric droplets during the day. They suggest that at night, TMI may continue to contribute to OH production via Fenton's reaction:

$$Fe^{2+} + H_2O_2 \rightarrow Fe^{3+} + OH^- + OH$$

This also serves to recycle the Fe(II) back to Fe(III) as do other reactions such as

$$Fe^{2+} + O_2^- \xrightarrow{2H_2O} Fe^{3+} + H_2O_2 + 2OH^-$$

The oxidation of organics in atmospheric droplets may also involve transition metal ions. For example, Weschler et al. (1985) suggest that Fe^{2+} may react with hydroperoxides in a manner analogous to that with H_2O_2, to produce Fe^{3+} + RO + OH^-, and that organics such as aldehydes may form coordination complexes with TMI, facilitating oxidation of the organic.

In short, transition metal ions may contribute significantly to the free radical and ion chemistry of atmospheric aqueous droplets as well as to the oxidation of organics in these systems; the reader should consult the papers by Graedel et al. (1985a,b) and Weschler et al. (1985) for a detailed discussion of the reactions.

Figure 11.4 summarizes some of the major reactions between the hydrogen- and oxygen-containing species believed to play a role in aqueous atmospheric droplets.

In addition to the hydrogen–oxygen chemistry, there is also, of course, nitrogen chemistry occurring in the aqueous phase. Table 11.4 and Fig. 11.5 summarize

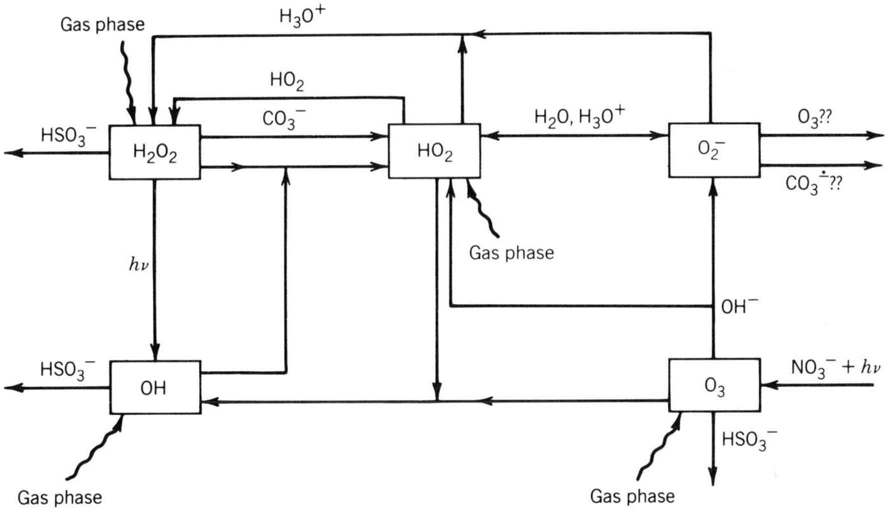

FIGURE 11.4. Schematic of interrelationship between major oxygen- and hydrogen-containing species believed to play a role in the chemistry of aqueous droplets in the atmosphere. Wiggley lines from "gas phase" indicate that the species may be absorbed from the gas phase into solution (adapted from Graedel and Goldberg, 1983).

TABLE 11.4. Chemistry of Nitrogenous Species Suggested to be Important in Aqueous Systems

Reactions	Rate Expression (in moles L^{-1} s^{-1}) or Equilibrium Constant
$NO + OH \rightarrow HONO$	1.0×10^{10} [OH][NO]
$NO + NO_2 \xrightarrow{H_2O} 2HONO$	3.0×10^{7} $[NO][NO_2]$
$2NO_2 \xrightarrow{2H_2O} HONO + H_3O^+ + NO_3^-$	7.0×10^{7} $[NO_2]^2$
$HONO \rightleftarrows H^+ + NO_2^-$	$K_{eq} = 10^{-3.29}$
$HONO + h\nu \rightarrow OH + NO$	3.7×10^{-5} [HONO]
$HONO + OH \rightarrow H_2O + NO_2$	1.0×10^{9} [HONO][OH]
$HONO + H_2O_2 \rightarrow NO_3^- + H_3O^+$	4.6×10^{3} $[H^+][H_2O_2]$[HONO]
$NO_2^- + h\nu \xrightarrow{H_2O} NO + OH + OH^-$	6.3×10^{-6} $[NO_2]$
$NO_2^- + OH \rightarrow NO_2 + OH^-$	1.0×10^{10} $[NO_2]$[OH]
$NO_2^- + O_3 \rightarrow NO_3^- + O_2$	5.0×10^{5} $[O_3][NO_2^-]$
$NO_2^- + CO_3^- \rightarrow NO_2 + CO_3^{2-}$	4.0×10^{5} $[NO_2^-][CO_3^-]$
$NO_3^- + h\nu \rightarrow NO_2^- + O$	1.8×10^{-7} $[NO_3^-]$
$NO_2 + OH \rightarrow NO_3^- + H^+$	1.3×10^{9} $[NO_2]$[OH]
$2NO_2 + HSO_3^- \xrightarrow{4H_2O} 3H_3O^+ + 2NO_2^- + SO_4^{2-}$	2.0×10^{6} $[NO_2]$[S(IV)][a]
$RO_2 + NO \xrightarrow{H_2O} RC(OH)_2 + NO_2$	1.0×10^{8} $[RO_2]$[NO]
$NH_3 + H^+ \rightleftarrows NH_4^+$	$K_{eq} = 10^{9.25}$

Source: Graedel and Goldberg, 1983; Young and Lurmann, 1984.
[a]Preliminary data; see Lee and Schwartz (1983).

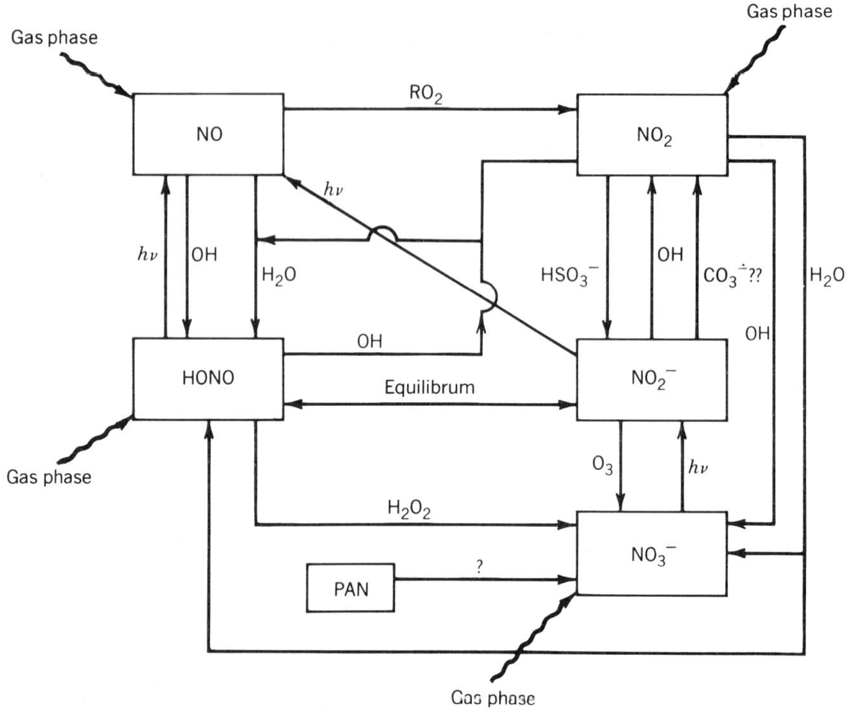

FIGURE 11.5. Schematic of interrelationship between major nitrogenous species believed to play a major role in atmospheric aqueous systems. Wiggley lines from "gas phase" indicate that the species may be absorbed into solution from the gas phase (adapted from Graedel and Goldberg, 1983).

these reactions. The rate expressions and equilibrium constants in Table 11.4 are those recommended by Young and Lurmann (1984) and Graedel and Goldberg (1983); these should be consulted for the original references.

The photochemistry of HONO and the nitrate and nitrite ions is reviewed in Section 3-C-12. There remain uncertainties in the rates and mechanisms of some of the reactions in Table 11.4 and, in some cases, whether they even occur to any significant extent. For example, as discussed above, Schwartz and (1984b) suggests that the carbonate radical ion ($CO_3^{\dot{-}}$) does not play a significant role in atmospheric solution chemistry; if this is the case, the NO_2^- to NO_2 conversion via $CO_3^{\dot{-}}$ would clearly not be important.

It is possible that reactive intermediates not shown in Table 11.4 are formed and may play a role in aqueous phase chemistry. For example, Damschen and Martin (1983) suggest that the relatively long-lived and strongly oxidizing intermediate peroxynitrous acid (HOONO), formed in the HONO–H_2O_2 reaction, could be present in significant concentrations and hence react with other species.

PAN may be absorbed from the gas phase and contribute to nitrate in solution via its hydrolysis (Holdren et al., 1982, 1984; Spicer et al., 1983; Lee et al., 1983):

$$PAN \xrightarrow{H_2O} NO_3^- + RC(O)O^- + H^+ \tag{50}$$

A preliminary rate constant in the range (1–4) $\times$ 10^{-4} s^{-1} has been reported (Lee et al., 1983; Holdren et al., 1984); this is consistent with a measured lifetime of a few minutes to over an hour, depending on the temperature, for PAN in water samples (Spicer et al., 1983). However, due to the limited solubility of PAN in water (Table 11.3), this is not believed to be a significant source of nitrate in atmospheric aqueous droplets (Seigneur and Saxena, 1984).

As discussed in detail by Graedel and Weschler (1981), organics may also contribute to the chemistry of atmospheric aqueous systems. For example, as discussed earlier, the presence of aldehydes in fog and cloud water and the complex formation with S(IV) have been invoked to explain the high concentrations of S(IV) greatly in excess of what is expected for a Henry's law equilibrium with pure water.

Aldehydes and ketones react in aqueous solution to form gem-diols (Buschmann et al., 1980, 1982):

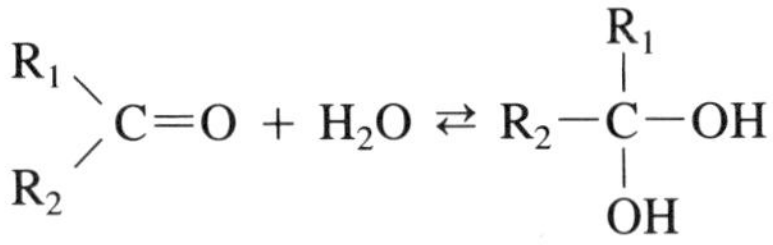

(51)

The hydration reaction (51) is very rapid and, being acid catalyzed, occurs even more rapidly in acidic fogs and clouds than in neutral solutions; for example, at a pH of 3, the hydration of the small aldehydes occurs in about 1 s (Buschmann et al., 1982). The hydration constant for this equilibrium,

$$K_H = \frac{[R_1R_2C(OH)_2]}{[R_1R_2CO]}$$

depends on the nature of the carbonyl compound. For HCHO, the aldehyde typically present in the highest concentrations in the atmosphere (Grosjean, 1982; Grosjean and Wright, 1983), $K_H = 2.5 \times 10^3$ (Schecker and Schulz, 1969), so that, in solution, it exists predominantly in the form of the diol, $CH_2(OH)_2$. However, for the larger aldehydes, K_H is smaller. For CH_3CHO, $K_H = 1.2$, whereas for the straight chain C_3–C_6 aldehydes, $K_H = 0.8$ (Buschmann et al., 1980, 1982); thus the aldehyde and diol exist in comparable concentrations in aqueous solutions for C_2 and larger aldehydes.

The diol form of carbonyl compounds does not absorb actinic UV and hence does not undergo photochemistry in the troposphere (see Section 3-C-12). Chameides and Davis (1983b) suggest that the major fate of the diols may be reaction with OH free radicals to form the corresponding carboxylic acid.

As seen in Chapter 7, the gas phase kinetics of reaction of a variety of organics with species such as OH, O_3, and NO_3 are becoming reasonably well established; however, much less is known about the mechanisms of their reactions. Unfortunately, in aqueous solutions, not only are the mechanisms ill defined but relatively few kinetic data exist. Radical reactions similar to those in the gas phase might be expected, for example, hydrogen abstraction from alkanes by OH and NO_3, followed by formation of alkyl peroxy radicals that can oxidize NO as in the gas phase:

$$\underset{(NO_3)}{RH + OH} \rightarrow R\cdot + \underset{(HNO_3)}{H_2O} \tag{52}$$

$$R + O_2 \rightarrow RO_2 \tag{53}$$

$$RO_2 + NO \rightarrow RO + NO_2\cdot \tag{54}$$

The potential for the formation of organic hydroperoxides thus exists via chain termination reactions such as (55):

$$RO_2 + HO_2 \rightarrow ROOH + O_2 \tag{55}$$

These hydroperoxides would reform free radicals via photolysis:

$$ROOH + h\nu \rightarrow RO + OH \tag{56}$$

The alkoxy radical (RO) could then be oxidized by O_2, as in the gas phase, to the aldehyde which, as we have seen, exists in solution to a large extent in the hydrolyzed glycol form [$RCH(OH)_2$].

Graedel and Goldberg (1983) also suggest that organic hydroperoxides may oxidize S(IV):

$$ROOH + HSO_3^- \rightarrow HSO_4^- + ROH \tag{57}$$

Whether chloride ions play a significant role in atmospheric droplets is not clear. Significant concentrations of chloride ions are found in coastal areas particularly where NaCl aerosols are produced by wave action in the ocean (see Section 12-D-1). In addition, HCl can be scavenged from the gas phase. The levels of HCl in ambient air are not well established but are probably in the low-ppb range (Okita et al., 1974). It is produced from the reaction of sea salt aerosols with acids and perhaps of NO_2 followed by secondary reactions (see Section 12-D-1), by hydrolysis of lead halides (e.g., PbBrCl) emitted from automobiles, and by incineration of waste. As a strong acid, it is completely ionized in solution.

The chloride ion itself does not react quickly with various species, but, as in the gas phase, chlorine atoms are highly reactive (see Chapter 15). Thus if Cl^- can be converted to the Cl atom, then it could play a role in atmospheric aqueous systems.

The major chlorine reactions considered by Graedel and Goldberg (1983) are the following:

$$Cl^- + OH \rightleftarrows ClOH^- \tag{58}$$

$$ClOH^- + H_3O^+ \xrightarrow{H_2O} Cl + 2H_2O \tag{59}$$

$$Cl + H_2O_2 \rightarrow HCl + HO_2 \tag{60}$$

$$Cl + HO_2 \rightarrow HCl + O_2 \tag{61}$$

In addition, the photolysis of ClO^- and its reaction with H_2O_2 were considered. However, using the available rate constants, an efficient pathway for the conversion of Cl^- to Cl was not found, and Graedel and Goldberg concluded that solution phase chlorine chemistry would not be important (see Section 11-A-2c, however).

It is possible that oxides of nitrogen such as NO_3 interact with Cl^- to produce Cl radicals; for example, Thompson and Zafiriou (1983) suggest reaction (62) may occur in seawater and in aerosols, clouds, and so on:

$$NO_3 + Cl^- \rightarrow Cl + NO_3^- \tag{62}$$

However, this remains speculative.

In providing this brief overview of the potential reactions occurring in the aqueous phase in the atmosphere, we have not considered the possibility of metal catalysis. As discussed in Section 12-D, there are a variety of trace metals in atmospheric aerosols and these may play a key role in catalyzing certain of the aqueous phase reactions. For example, as discussed by Hoffmann and Boyce (1983) and Graedel et al. (1985b), trace metals and species such as OH, HO_2, and H_2O_2 may co-exist in many aqueous systems via reaction sequences such as the following:

$$Cu^{2+} + h\nu \xrightarrow{H_2O} Cu^+ + OH + H^+ \tag{63}$$

$$Cu^+ + O_2 \rightleftarrows CuO_2^+ \tag{64}$$

$$CuO_2^+ + H^+ \rightarrow Cu^{2+} + HO_2 \tag{65}$$

$$Cu^+ + HO_2 \rightarrow Cu^{2+} + HO_2^- \tag{66}$$

$$HO_2^- + H^+ \rightarrow H_2O_2 \tag{67}$$

The role of trace metals in catalyzing the oxidation of S(IV) has been reviewed in detail by Hoffmann and co-workers (Hoffmann and Boyce, 1983; Hoffmann and Jacob, 1984; Graedel et al., 1985b); this is discussed briefly in Section 11-A-2c.

11-A-2b. O_2: Uncatalyzed Oxidation

Whether a significant rate of oxidation of S(IV) by O_2 occurs in the liquid phase without metal catalysis is subject to considerable controversy. Figure 11.6 shows the results of some experimental studies of the pseudo-first-order rate constant k_0 as a function of pH for the oxidation of S(IV) by O_2, where k_0 is defined as

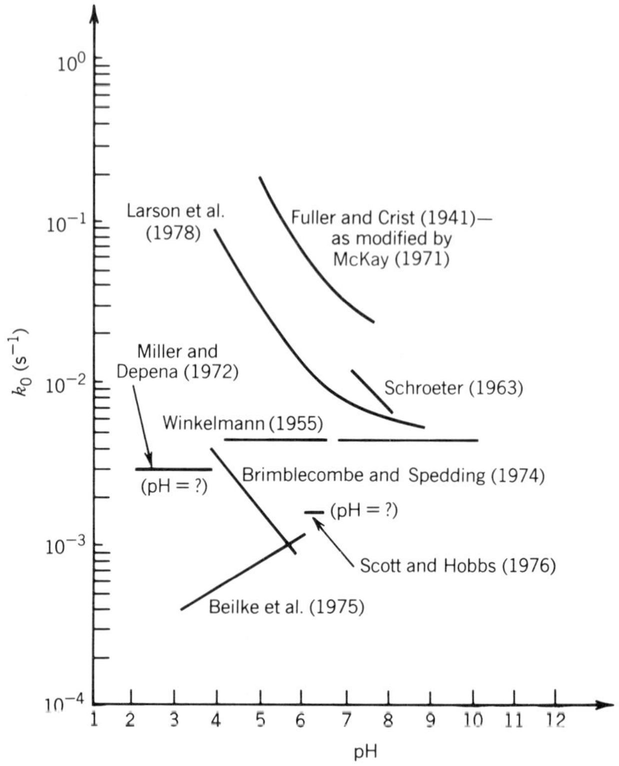

FIGURE 11.6. Results of some experimental studies of the pseudo-first-order rate coefficient (k_0) for the uncatalyzed liquid phase oxidation of SO_2 in the presence of oxygen (from Hegg and Hobbs, 1978).

$$k_0 = \frac{1}{[SO_3^{2-}]} \frac{d[SO_4^{2-}]}{dt} \tag{L}$$

It is seen that not only is there no agreement as to the absolute values of the rate constant, but even the *direction* of the pH dependence is not uniformly agreed upon!

This wide disagreement in the literature reflects to a great extent the experimental difficulties in studying S(IV) oxidations. These difficulties include the following:

1. Some metal ions are known to catalyze this (and other) reactions of S(IV) and these metals may be present as impurities either in the water used or in the reagents themselves.
2. Some organic compounds (e.g., toluene) act as inhibitors for S(IV) oxidation, and these too may be present in small concentrations as impurities.

3. Mass transport problems may have been present in some laboratory studies so that the observed overall rates of oxidation of S(IV) from which rate constants were extracted may have been limited by mass transport and the true solution kinetics were not observed; this may occur particularly at higher reactant concentrations where the chemical reactions are speeded up so they are no longer the rate-determining step. This again illustrates the importance of studying reactions under conditions as close to atmospheric as possible.

Such experimental difficulties are not limited, of course, to the O_2 oxidation but may have been present in other studies of S(IV) oxidation as well. Although the O_2 oxidation is a particularly striking case, there are other examples of substantial discrepancies in the experimental results of S(IV) oxidation which may also be due to such experimental problems.

If the highest rates measured for the uncatalyzed oxidation of S(IV) by O_2 shown in Fig. 11.6 are used, this reaction may be a significant source of acid formation in droplets at pH > 5. It has been suggested that these rates are erroneously high because of catalysis by trace metal impurities. The controversy may be somewhat academic since it is highly unlikely that *laboratory pure* water will exist in urban atmospheres.

11-A-2c. O_2: Catalyzed Oxidation

As reviewed in detail by Martin (1984), Hoffmann and Boyce (1983), and Hoffmann and Jacob (1984), the oxidation of SO_2 by O_2 is known to be catalyzed by a number of metal ions including Fe^{3+} and Mn^{2+}. However, the absolute magnitudes of the increases are not agreed upon. In the case of Fe^{3+}, the rate decreases with pH for pH values from 4 to 0; this, combined with the pH dependence of S(IV) in solution, leads to a strongly declining rate of S(VI) production in solution as the pH falls.

The rate expression changes above a pH of 4 from first to second order in S(IV); there is considerable disagreement in the literature as to the absolute rate coefficients at pH > 4 and even as to the direction of the pH dependence. These discrepancies may be due to the fact that the concentration of Fe^{3+} in solution falls rapidly at pH ≥ 4 because of the formation of hydroxy complexes such as $Fe(OH)^{2+}$, $Fe(OH)_2{}^{+}$, and so on (Hegg and Hobbs, 1978).

Mn^{2+} also catalyzes the O_2–S(IV) reaction. The kinetics are complex, and there is some evidence that the rate law changes at S(IV) concentrations of $\sim 10^{-6}$ moles L^{-1} (Martin, 1984). Since the concentration of S(IV) is determined partly by the droplet pH (Fig. 11.2), the importance of the initial pH of the atmospheric droplets in determining reaction rates and pathways is again seen. The most recent studies show a Mn^{2+} catalyzed rate coefficient that increases with pH in the 0–3 range. Martin (1984) suggests that, in this pH range, the following rate expression applies:

$$\frac{-d[\mathrm{S(IV)}]}{dt} = 25\,\frac{[\mathrm{Mn}^{2+}][\mathrm{S(IV)}]}{[\mathrm{H}^+]}\ \frac{\text{mole}}{\text{L s}}$$

This overlaps well with the rate expression for the pH range 3–7.5 of Ibusuki and Barnes (1984), which, expressed in terms of the HSO_3^- concentration (Martin, 1984), is

$$\frac{-d[\mathrm{S(IV)}]}{dt} = 5000\ [\mathrm{Mn}^{2+}][\mathrm{HSO_3}^-]\quad \text{moles L}^{-1}\ \text{s}^{-1}$$

Since the rate in the pH range 3–7.5 follows the HSO_3^- concentration (Fig. 11.1), it is relatively constant in the pH range 3–6 and then falls at higher pH values.

A synergism between Fe^{3+} and Mn^{2+} (i.e., a rate enhancement in the presence of both ions which is greater than that expected from the sum of the individual catalyzed rates) has been observed; the removal of S(IV) is approximately 3–10 times faster than expected if no synergism were present (Martin, 1984). However, the rate under atmospheric conditions may still not approach that for oxidation by other species; for example, the net rate of oxidation of SO_2 in the presence of a laboratory aerosol containing ammonium sulfate, Fe^{3+} and Mn^{2+} ions was observed to be only 0.02% h^{-1} (Kleinman et al., 1985).

A variety of other metal ions of atmospheric interest have not been studied in detail to date under conditions relevant to the atmosphere. However, metals having at least two accessible oxidation states, so that reversible oxidations and reductions can occur, are most likely to play roles in the S(IV) oxidation; thus, in addition to iron and manganese discussed above, the Cu(II)/Cu(I), Co(II)/Co(III), and V(III)/V(IV) pairs may be important (Hoffmann and Boyce, 1983).

The mechanisms of the metal-catalyzed oxidation of S(IV) are not well established. The suggestion of dark reactions involving free radical chains initiated by a metal (M^{n+}) in which a series of one-electron transfers occurs dates back to at least 1934 (Bäckstrom, 1934):

Initiation

$$\mathrm{M}^{n+} + \mathrm{SO_3}^{2-} \rightarrow \mathrm{M}^{(n-1)+} + \mathrm{SO_3}^{\dot{-}} \tag{68}$$

Propagation

$$\mathrm{SO_3}^{\dot{-}} + \mathrm{O_2} \rightarrow \mathrm{SO_5}^{\dot{-}} \tag{69}$$

$$\mathrm{SO_5}^{\dot{-}} + \mathrm{SO_3}^{2-} \rightarrow \mathrm{SO_5}^{2-} + \mathrm{SO_3}^{\dot{-}} \tag{70}$$

Oxidation

$$\mathrm{SO_5}^{2-} + \mathrm{SO_3}^{2-} \rightarrow 2\mathrm{SO_4}^{2-} \tag{71}$$

Termination

$$2SO_3^{\dot{-}} \rightarrow S_2O_6^{2-} \tag{72}$$

$$SO_3^{\dot{-}} + SO_5^{\dot{-}} \rightarrow S_2O_6^{2-} + O_2 \tag{73}$$

$$SO_5^{\dot{-}} + SO_5^{\dot{-}} \rightarrow S_2O_8^{2-} + O_2 \tag{74}$$

Since then, similar free radical mechanisms have been hypothesized (see Hoffmann and Boyce, 1983, and Hoffmann and Jacob, 1984, for a review of these).

However, these mechanisms have often been found to be inconsistent with some experimental data such as the kinetic order of the reaction in each reactant (see Section 4-A-3). As a result, alternate mechanisms have been hypothesized. These include polar reaction mechanisms and photoassisted mechanisms.

The polar mechanisms involve the initial formation of a metal–sulfite complex and binding of O_2 to this complex, followed by two electron transfers; for example, Bassett and Parker (1951) suggested the following mechanism:

$$Mn^{2+} + SO_3^{2-} \rightleftarrows MnSO_3^{0} \tag{75}$$

$$MnSO_3^{0} + SO_3^{2-} \rightleftarrows Mn(SO_3)_2^{2-} \tag{76}$$

$$Mn(SO_3)_2^{2-} + O_2 \rightleftarrows Mn(SO_3)_2O_2^{2-} \tag{77}$$

$$Mn(SO_3)_2O_2^{2-} \rightarrow Mn^{2+} + 2SO_4^{2-} \tag{78}$$

Similar polar mechanisms have been hypothesized to explain the oxidation of S(IV) by other metals as well (Hoffmann and Boyce, 1983; Hoffmann and Jacob, 1984).

As discussed in Section 11-A-2a, H_2O_2 is produced during the metal-catalyzed autooxidation of organics and this H_2O_2 can oxidize S(IV) (see below). However, for organics that can form a metal–ligand complex, there are other potential mechanisms of oxidation of S(IV) involving the metal–ligand complex itself (Boyce et al., 1983). The mechanism is not clear but may involve complexation of the S(IV) and O_2 followed by a two-electron transfer and hydrolysis to H_2SO_4. Less likely mechanisms including a one-electron transfer chain reaction are discussed by Hoffmann and Boyce (1983). Interestingly, some of these metal–ligand catalytic oxidations may be accelerated by light (Boyce et al., 1983).

Whether suitable ligands are present in atmospheric water droplets in sufficient concentrations that such metal–ligand complexes could contribute significantly to S(IV) oxidations in the atmosphere is not known.

Finally, some of the metal-catalyzed S(IV) oxidations have been observed to occur at least in part via photochemical mechanisms. In homogeneous aqueous systems, for example, the oxidation of S(IV) at wavelengths above ~300 nm was not observed unless Fe^{3+} was present (Luňák and Vepřek-Šiška, 1976); since S(IV) in aqueous solution does not absorb light significantly in the actinic region (see Section 3-C-12), the photooxidation was attributed to absorption of light by an

Fe^{3+}–S(IV) complex. Hoffmann and Boyce (1983) have suggested a mechanism consistent with these findings; it includes a photochemically induced inner-sphere electron transfer in an Fe(III)–S(IV)–O_2 complex and a photoreduction of Fe(III) to Fe(II).

Catalysis by non-metals such as Cl^- and Br^- has also been suggested in the S(IV) → S(VI) oxidation based on studies at fairly high reactant concentrations (Clarke and Williams, 1983; Clarke and Radojevic, 1983). Clarke and co-workers suggest that in sea salt aerosols at high relative humidities (>75%), where the aerosol exists as a liquid droplet, the presence of high Cl^- concentrations could result in much higher oxidation rates of S(IV) than observed in pure water. While the mechanism is unknown, they suggest two possibilities. The first is an oxygen atom transfer from one of the intermediates in the Bäckström mechanism,

$$SO_5^{\dot{-}} + Cl^- \rightarrow SO_4^{\dot{-}} + ClO^- \qquad (79)$$

followed by the reduction of $SO_4^{\dot{-}}$ to SO_4^{2-}. Alternatively, they suggest electron transfer such as

$$SO_4^{\dot{-}} + Cl^- \rightarrow SO_4^{2-} + Cl \qquad (80)$$

Similar interactions between Cl and sulfur oxidation in solution have been proposed by Chameides and Davis (1983a).

11-A-2d. Ozone

Although ozone reacts only very slowly with SO_2 in the gas phase, in the liquid phase the reaction is rapid. Indeed, this rapid reaction may be responsible for a negative interference of SO_2 when measuring O_3 in the atmosphere using the KI technique (see Chapter 5).

The mechanism of the reaction may involve the production of free radicals by O_3, followed by their reaction with S(IV) (see Section 11-A-2f), or more likely an ionic mechanism (Maahs, 1983) which can be expressed as

$$HSO_3^- + OH^- + O_3 \rightarrow SO_4^{2-} + H_2O + O_2 \qquad (81)$$

The concentration of O_3 in the liquid phase in equilibrium with a gas phase concentration of 40 ppb, using a Henry's law constant of 1.3×10^{-2} M atm^{-1}, is 5×10^{-10} moles L^{-1} at 25°C. Although this concentration is about six orders of magnitude less than that for dissolved O_2, oxidation of S(IV) by O_3 is much more important under most conditions because of the much higher rate constant for reaction.

There seems to be reasonable agreement that from pH = 1 to 3, the rate coefficient for this reaction increases rather moderately with pH. For example, from pH = 1 to 3, the rate constant increases by approximately one order of magnitude (Fig. 11.7), whereas over the same pH range, the Fe^{3+} catalyzed oxidation by O_2 increases by approximately two orders of magnitude (Martin, 1984).

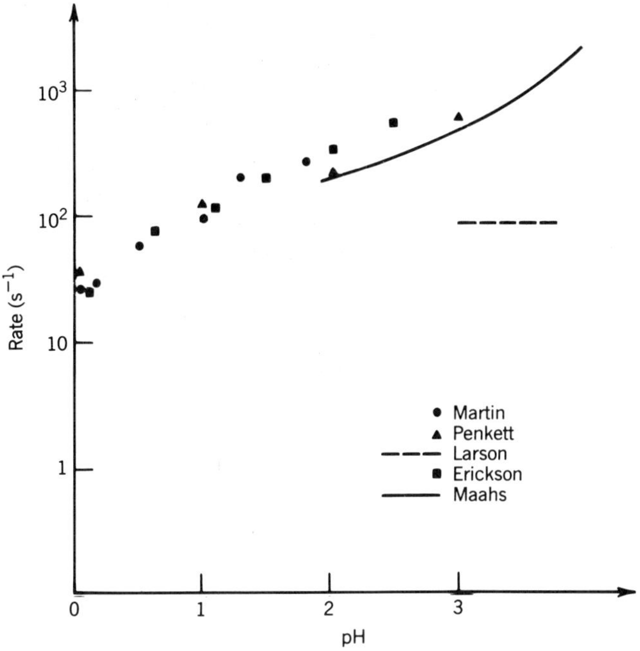

FIGURE 11.7. Pseudo-first-order rates for ozone loss by reaction with S(IV) in solution as a function of pH. $[O_3] = 1 \times 10^{-4}$ M and $[S(IV)] = 5 \times 10^{-4}$ M (from Martin, 1984).

Recent work at higher pH values (3.0–6.2) shown partly by the solid line in Fig. 11.7 suggests that the rate expression over the pH range 1.0–6.2 can be represented as

$$\text{rate (mole L}^{-1}\text{ s}^{-1}) = -\frac{d[O_3]}{dt} = (k_a + k_b[OH^-])[HSO_3^-][O_3]$$

where $k_a = 3.8 \times 10^5$ M^{-1} s^{-1}, and $k_b = 1.05 \times 10^{16}$ M^{-2} s^{-1} at 25°C (Maahs, 1983).

The direction of the pH dependence, combined with the solubility of S(IV), results in a strongly decreasing rate of production of S(VI) as the pH falls; thus this reaction is likely only important at pH $\gtrsim 4.5$. Small amounts of catalysis by Fe^{3+} and Mn^{2+} have been reported, although agreement is not universal on this point, and the catalytic effect, if present, is relatively small.

11-A-2e. Hydrogen Peroxide and Organic Peroxides

Hydrogen peroxide has been shown to oxidize S(IV) relatively rapidly in solution. Furthermore, because it is a highly soluble compound, even gas phase concentrations in the low ppb range lead to significant concentrations in the liquid phase. For example, using the Henry's law constant for H_2O_2 of 1×10^5 M atm^{-1}, a 1 ppb gas phase concentration would produce at equilibrium at 25°C an aqueous phase concentration of 1×10^{-4} moles L^{-1}. This is approximately six orders of magnitude greater than the solution phase concentrations of O_3 expected under ambient conditions! This relatively large concentration anticipated in the solution phase is a significant factor in making H_2O_2 the currently favored candidate for the greatest oxidation of S(IV) in the aqueous phase (see below).

A second factor in favor of the importance of H_2O_2 in the S(IV) oxidation is the pH dependence of the rate coefficient. Figure 11.8 shows the results of some rate studies summarized by Martin (1984) where k_0 is defined as

$$k_0 = \frac{1}{[H_2O_2][S(IV)]} \frac{d[S(VI)]}{dt} \tag{M}$$

Other recent studies (McArdle and Hoffmann, 1983; Kunen et al., 1983) are also in agreement with the data of Fig. 11.8. It is seen that in contrast to most of the other oxidations discussed earlier, k_0 decreases as the pH increases at pH ≥ 1.5. This is in the opposite direction to the S(IV) solubility; the result is that the overall

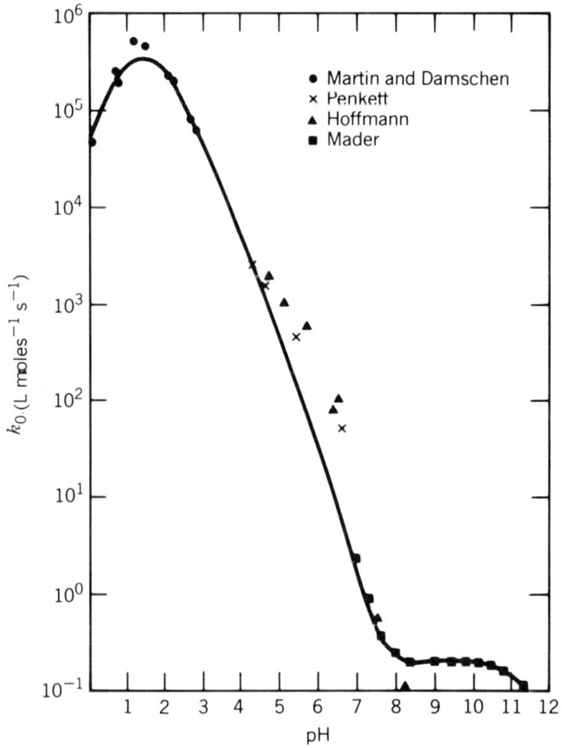

FIGURE 11.8. k_0 in expression $d[S(VI)]/dt = k_0 [H_2O_2] [S(IV)]$; effect of buffer removed and results converted to 25°C (from Martin and Damschen, 1981).

rate of production of S(VI) from this reaction is relatively independent of pH over a wide pH range of interest in the atmosphere (see below).

As discussed in Section 11-A-2a, a number of alternate sources of H_2O_2 in atmospheric droplets have been proposed in addition to dissolution from the gas phase, and H_2O_2 has been measured in atmospheric droplets.

Organic hydroperoxides have also been proposed as potential oxidants of S(IV) in solution (Graedel and Goldberg, 1983):

$$ROOH + HSO_3^- \rightarrow HSO_4^- + ROH \tag{82}$$

$$k_{82} \simeq 7.2 \times 10^4 \text{ L mole}^{-1} \text{ s}^{-1}$$

Both methyl hydroperoxide (CH_3OOH) and peracetic acid ($CH_3C(O)OOH$) have been shown to oxidize S(IV) in the aqueous phase (Lind and Lazrus, 1983). However, it is not clear that there are sufficient concentrations of organic hydroperoxides in aqueous atmospheric droplets for this to contribute significantly compared to oxidation by H_2O_2. (See for example, the discussion in Section 11-A-2a above and Kelly et al., 1985).

While it is generally assumed that oxidants such as H_2O_2 are formed from products of photochemical reactions either in the gas phase (see Section 3-D and Chapter 7) or in aqueous droplets in the atmosphere (see Section 11-A-2a), Benner et al. (1985) have shown that incomplete combustion of fuels such as propane produces an oxidant or oxidants, including H_2O_2, which can oxidize SO_2 to S(VI). Thus both primary and secondary sources of atmospheric H_2O_2 may exist; the relative importance of direct emissions versus production via secondary reactions in the atmosphere is not known.

11-A-2f. Oxidizing Free Radicals

As discussed earlier, free radicals in the liquid phase such as OH and HO_2 are expected to be produced through several processes, including heterogeneous scavenging of OH and HO_2 from the gas phase, and these may play a role in S(IV) oxidation (Graedal and Weschler, 1981; Graedel and Goldberg, 1983; Chameides and Davis, 1982; Schwartz, 1984b). For example, the oxidation of S(IV) by O_3 has been suggested to occur via free radical pathways since, as seen in Section 11-A-2a, O_3 can produce free radicals such as OH in aqueous solutions (Hoigné and Bader, 1975; Penkett et al., 1979). The mechanism is uncertain but may include steps such as the following (Graedel and Goldberg, 1983; Young and Lurmann, 1984):

$$HSO_3^- + OH \rightarrow H_2O + SO_3^- \tag{83}$$

$$k_{83} = 9.5 \times 10^9 \text{ L mole}^{-1} \text{ s}^{-1}$$

$$SO_3^- + O_2 \rightarrow SO_5^- \tag{84}$$

$$k_{84} = 1.0 \times 10^8 \text{ L mole}^{-1} \text{ s}^{-1}$$

$$2SO_5^- \rightarrow SO_4^- + SO_4^{2-} + O_2 \tag{85}$$

$$k_{85} = 1.0 \times 10^{11} \text{ L mole}^{-1} \text{ s}^{-1}$$

$$SO_4^- + HCO_3^- \rightarrow SO_4^{2-} + H_3O^+ + CO_3^{\dot{-}} \tag{86}$$

$$k_{86} = 9.0 \times 10^{6} \text{ L mole}^{-1} \text{ s}^{-1}$$

$$SO_4^- + H_2O_2 \rightarrow HSO_4^- + HO_2 \tag{87}$$

$$k_{87} = 1.2 \times 10^{7} \text{ L mole}^{-1} \text{ s}^{-1}$$

$$SO_4^- + Cl^- \rightarrow SO_4^{2-} + Cl \tag{88}$$

$$k_{88} = 2 \times 10^{8} \text{ L mole}^{-1} \text{ s}^{-1}$$

Modeling studies by Chameides and Davis (1982; 1983a) using an accommodation coefficient for absorption of all soluble species, including OH and HO_2, of 0.01, suggest that as much as 45% of the aqueous phase conversion of S(IV) to S(VI) could be due to free radical oxidation; however, this may be an overestimate since reactive hydrocarbons were not included in the model, and these compete for the OH (Seigneur and Saxena, 1984).

Thus, while our understanding of the aqueous phase chemistry and interactions with gas phase species (e.g., absorption of OH from the gas phase) is meager, these could contribute to the aqueous phase oxidation of S(IV).

11-A-2g. Oxides of Nitrogen

The oxides of nitrogen—NO, NO_2, HONO, and HNO_3—have all been suggested as possible oxidizing agents for dissolved S(IV); however, the reactions of HNO_3 and NO are too slow to be significant (Martin, 1984; Schwartz, 1984a).

In aqueous solutions, nitrous acid reacts with S(IV) at a reasonable rate with the rate constant k, defined by

$$k = \frac{-1}{[\text{N(III)}][\text{S(IV)}]} \frac{d[\text{S(IV)}]}{dt} \tag{N}$$

increasing as the pH decreases (Martin, 1984). However, the levels of gaseous HONO observed in ambient air (~1–8ppb) (see Chapters 5 and 8) taken with the Henry's law constant for HONO (Table 11.3) yield aqueous concentrations too low to contribute substantially to the aqueous phase S(IV) oxidation.

For example, using a Henry's law constant for HONO of 49 M atm^{-1}, a gas phase concentration of 1 ppb would result in a solution phase concentration of only 4.9×10^{-8} moles L^{-1}, compared to an anticipated H_2O_2 solution phase concentration of 10^{-4} moles L^{-1} at the same gas phase concentration. The rate constants also favor the H_2O_2 reaction; at a pH of 3.0, that for oxidation of H_2O_2 is approximately a factor of 10^4 larger than that for reaction with HONO. Thus, the combination of concentrations and rate constants makes HONO unlikely to be a significant S(IV) oxidant in solution unless other oxidants such as O_3 or H_2O_2 are absent.

Whether dissolved NO_2 contributes significantly to the S(IV) oxidation in solution is uncertain at present. As seen from the Henry's law constant in Table 11.3, NO_2 is relatively insoluble; thus at a gas phase concentration of 10 ppb, the equilibrium aqueous phase concentration is only 10^{-10} moles L^{-1}. However, Schwartz and co-workers (Schwartz, 1984a) have inferred from literature data that the rate constants for the NO_2 reaction with HSO_3^- and SO_3^{2-} may be sufficiently large that the NO_2–S(IV) reaction could be significant. In addition, preliminary data indicate that the overall reaction represented by

$$2NO_2 + HSO_3^- \xrightarrow{4H_2O} 3H_3O^+ + 2NO_2^- + SO_4^{2-}$$

may occur sufficiently rapidly that it could be important at higher pH values (Lee and Schwartz, 1983; Lee, 1984). This may be catalyzed in solution by the presence of carbon particles (Schryer et al., 1983). Further support for an aqueous phase oxidation of S(IV) by NO_2 comes from cloud chamber studies where significant sulfate production was observed when NO_2 was present even at concentrations as low as 5 ppb NO_2 (Gertler et al., 1984).

Other reactions, such as the reaction of NO_3 scavenged from the gas phase with S(IV), have also been suggested (Chameides and Davis, 1983a). Clearly, such reactions, including possible catalysis by metals, warrant further experimental studies.

11-A-3. Heterogeneous Reactions on Surfaces

Adsorption of SO_2 on the surfaces of solids, followed by its oxidation on the surface, may provide a third route for the formation of sulfuric acid. The solid surfaces may be suspended in the gas phase or in atmospheric droplets, as discussed in detail by Chang and Novakov (1983).

It is known that SO_2 in air is oxidized on both graphite and soot particles, with water vapor enhancing the sulfate formation on the surfaces (Hulett et al., 1972; Novakov et al., 1974). Whether this surface oxidation is enhanced by the presence of oxidants other than air (e.g., O_3 or NO_2) is subject to controversy. For example, under certain conditions some researchers report an enhancement of the rate of uptake of SO_2 due to the presence of the co-pollutants O_3 and NO_2 (Cofer et al., 1981; Britton and Clarke, 1980); however others have not observed such an effect (Baldwin, 1982), particularly at low gas concentrations more closely approximating those in the atmosphere (Cofer et al., 1984).

Carbon surfaces of various types have been the subject of most studies. However, other types of surfaces, including fly ash, dust, MgO, V_2O_5, Fe_2O_3, and MnO_2, have also been shown to oxidize SO_2 and/or remove it from the gas phase (Hulett et al., 1972; Judeikis et al., 1978; Liberti et al., 1978; Barbaray et al., 1977, 1978). As expected, the rate of removal depends on the nature of the particular surface, the presence of co-pollutants such as NO_2, and, as in the case of carbonaceous surfaces, on the relative humidity. The increase with increasing water vapor suggests that oxidation of the SO_2 may occur in a thin film of water on the surface of the solid.

Carbonaceous particles suspended in aqueous solutions can also act as sites for efficient SO_2 oxidation. Thus Novakov and co-workers have studied the kinetics of such processes with respect to the concentrations of O_2, S(IV), and carbon particles, as well as the pH and temperature dependencies (Brodzinsky et al., 1980; Chang et al., 1981; Benner et al., 1982). Extrapolation of their results to atmospheric conditions led them to suggest that such heterogeneous reactions could be important in the aqueous phase oxidation of SO_2.

Actually, it is difficult today to evaluate quantitatively the importance of such heterogeneous reactions in the overall oxidation of S(IV). Thus the experimental data base for such processes is relatively sparse (Schryer, 1982). Furthermore, their rates will depend on the physical and chemical natures of the surfaces involved, including specific surface areas, yet these are not well understood, especially for highly complex environmental gas–liquid–solid systems. For example, the rates of oxidation of SO_2 at 80% relative humidity on two different samples of fly ash obtained from two coal-fired power plants differed by more than an order of magnitude (Dlugi and Güsten, 1983). Even in laboratory systems the nature of relatively simple surfaces such as carbon depends on the history of the material.

However, the available evidence suggests that such heterogeneous reactions should be considered as potential contributors to the overall oxidation of S(IV), especially close to sources where particle concentrations and hence available surface areas are relatively high (see Chapter 12). Baldwin (1982), for example, estimates that the loss of gaseous SO_2 due to interactions with particle surfaces at a particle density of 100 $\mu g\ m^{-3}$ could be as high as 1% h^{-1}. Such rates are unlikely to be sustained for long periods of time due to saturation of the surface, if the surface is dry. However, Chang and Novakov (1983) point out that when the surface is wet, the active sites are constantly regenerated as the sulfate formed on the surface dissolves in the surrounding aqueous solution.

In addition to the dark oxidation of S(IV) on surfaces, there may be photochemically induced processes as well. For example, irradiation of aqueous supensions of solid α-Fe_2O_3 (hematite) containing S(IV) with light of $\lambda > 295$ nm resulted in the production of Fe(II) in solution (Faust and Hoffmann, 1985). This reductive dissolution of the hematite has been attributed to the absorption of light by surface Fe(III)–S(IV) complexes followed by a ligand-to-metal charge transfer reaction forming Fe(II) and oxidized states of sulfur. Faust and Hoffmann suggest that Mn(IV) oxides may undergo similar photochemical surface reactions since the band gap energy is sufficiently small that absorption of light in the actinic region can occur, yet they are insoluble in water.

These studies were carried out in the absence of O_2; it is not clear whether such reactions will also occur when O_2 is present. However, the potential contribution of such surface photochemical reactions is intriguing and clearly an area for further study.

11-A-4. Relative Importance of Various SO_2 Oxidation Mechanisms

The fractions of SO_2 oxidation which proceed in the gas phase, in the liquid phase, or on surfaces (see below) is not accurately known for a variety of meteorological conditions and, as noted earlier, geographical locations. As discussed earlier, S(IV) → S(VI) oxidation rates well exceeding the maximum expected for homogeneous gas phase reactions have been observed frequently, and these have generally been attributed to the liquid phase reactions described above. The heterogeneous surface reactions are usually assigned a lesser role.

However, as we have seen, there are extensive uncertainties in the rates of various aqueous phase and heterogeneous reactions, as well as in whether they might be catalyzed by various metals present in the environment. Thus, an accurate quantitative estimate of the contributions of various reactions to the overall S(IV) oxidation is not possible at present.

However, considering first only the aqueous phase and heterogeneous oxidations, one can use the rate data discussed above to obtain a crude estimate of the relative importance of the various reactions. Figure 11.9 shows one such estimate for a gas phase SO_2 concentration of 5 ppb by Martin (1984) for 25°C under conditions which the author chose as typical of the atmosphere, using his preferred rate constants and pH dependencies. It was assumed that there were no limitations on the rates of oxidation due to mass transport; as discussed in detail by Freiberg and Schwartz (1981), this assumption is justified except for very large droplets ($>10\ \mu m$) and high pollutant concentrations (e.g., O_3 at 0.5 ppm) where the aqueous phase reactions are very fast. It was also assumed that the aqueous phase present in the atmosphere was a cloud with a liquid water content (V) of 1 g per m^3 of air. As seen in Eq. (K), the latter factor is important in the aqueous phase rates of conversion of S(IV); thus the actual concentrations of iron, manganese, and so on in the liquid phase and hence the kinetics of the reactions depend on the liquid water content. The concentrations used by Martin for each oxidant or catalyst are shown in the right margin of Fig. 11.9.

Only the oxidation by H_2O_2 is relatively independent of pH. This arises because the rate coefficient for the reaction and the solubility of S(IV) show opposite trends with pH. For the other species, the effects of the S(IV) solubility and the pH dependence of the kinetics work in the same direction, leading to a strong overall dependence on pH. The uncatalyzed oxidation of O_2 is not shown, because it is generally believed to be unimportant compared to the other mechanisms in real atmospheric droplets containing ''impurities'' such as metals which will act as catalysts.

The relative importance of gas phase versus liquid phase oxidation of SO_2 is, of course, likely to depend on the particular meteorological conditions, for example, the relative humidity and whether clouds or fog are present. As discussed at the beginning of the chapter with regard to field studies, faster rates of SO_2 oxidation have been observed in both plumes and ambient air at high relative humidities or when clouds or fogs were present and this has generally been attributed

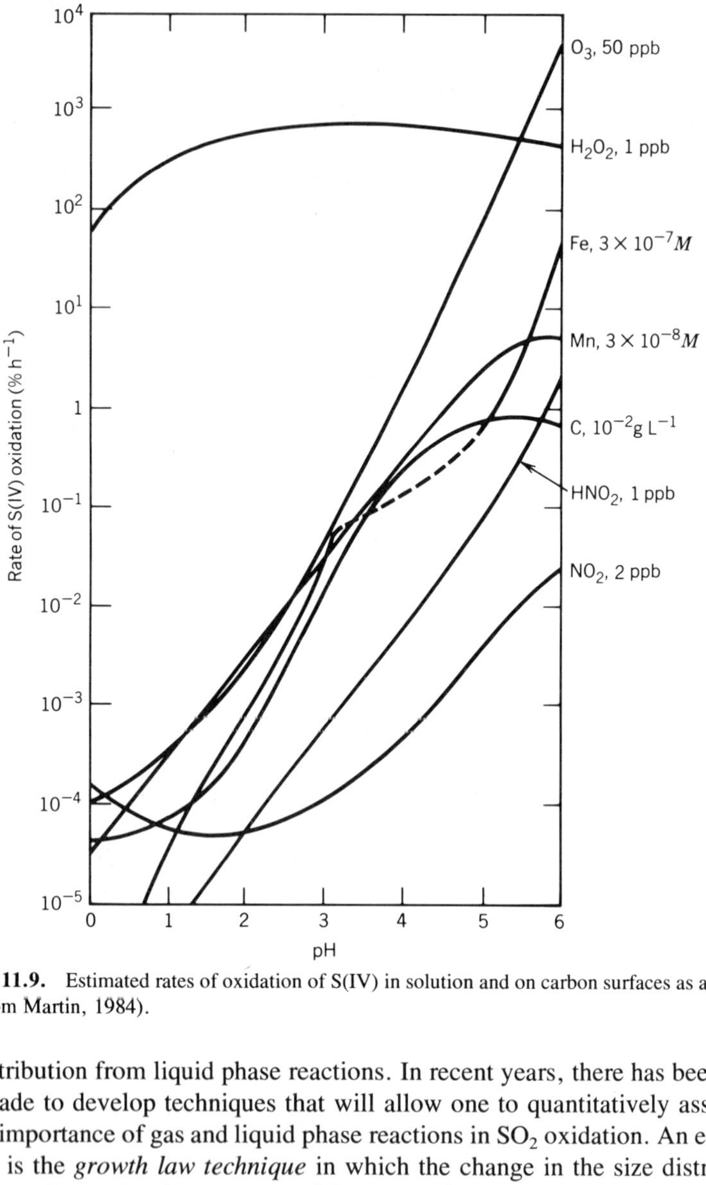

FIGURE 11.9. Estimated rates of oxidation of S(IV) in solution and on carbon surfaces as a function of pH (from Martin, 1984).

to a contribution from liquid phase reactions. In recent years, there has been some effort made to develop techniques that will allow one to quantitatively assess the relative importance of gas and liquid phase reactions in SO_2 oxidation. An example of these is the *growth law technique* in which the change in the size distribution of an aerosol growing by gas-to-particle conversion is followed with time. The particulate diameter growth rates are then compared to those expected theoretically for various mechanisms. This method is discussed in more detail in Section 12-C.

Application of this technique to urban aerosols in the Columbus, Ohio, area indicated that oxidation of SO_2 in the liquid phase at rates up to 12% h^{-1} predominated at relative humidities (RH) $\gtrsim$ 75%, whereas gas phase oxidation up to 5% h^{-1} provided the most important mechanism at lower RH (McMurry and Wilson, 1983). In the Los Angeles area, the aerosol phase oxidation at higher RH occurred in the larger (0.5 μm) droplets, whereas smaller aerosols (0.2 μm) resulted from the homogeneous oxidation of SO_2 in the gas phase (Hering and Freidlander, 1982). Subsequent scavenging of these aerosols by fog, cloud, or rain water may then lead to acid wet deposition.

In plume studies, estimates have also been made of the relative importance of gas and liquid phase reactions. For example, 40% or more of the total oxidation of SO_2 in plumes from coal-fired power plants has been attributed to liquid phase oxidation under conditions where clouds, fogs, rain, or wetted aerosols were present (Gillani and Wilson, 1983).

11-B. OXIDATION OF NO_2 TO NITRIC ACID

Until recently, the oxidation of NO_x in power plant plumes and in ambient air did not receive as much attention as SO_2 with regard to acid deposition because of the emphasis on the problems in Europe and the northeast of the United States where SO_x makes the greatest contribution to acid deposition. However, in power plant plumes, rates of conversion of NO_x from ~0.2 up to 12% h^{-1} have been observed with the rates being much greater in midday than at night (Hegg and Hobbs, 1979; Forrest et al., 1981; Richards et al., 1981). Smog chamber simulations of power plant plume chemistry suggest that the major products are PAN and nitric acid with lesser amounts of particulate nitrate; as expected for photochemical oxidations, the rate depended on the hydrocarbon and NO_x composition of the air and the NMHC/NO_x ratio. In addition, rates of conversion were higher at higher relative humidities (Spicer et al., 1981).

Studies of NO_x conversion rates in so-called *urban plumes*, that is, in ambient air that has passed over urban areas, show that rates of conversion of NO_x of <5% h^{-1} up to 24% h^{-1} occur with most of the NO_x being converted to HNO_3 and PAN (Chang et al., 1979; Spicer, 1980, 1982a,b).

One note of caution is warranted with regard to studies in which particulate nitrate was measured, especially those carried out before 1980; as discussed in Section 12-E-4e, nitrate is especially sensitive to artifact formation via the adsorption of HNO_3 from the gas phase onto filters. Thus the early particulate nitrate data should probably be taken as upper limits to the true particulate nitrate concentrations.

The reactions of oxides of nitrogen are discussed in detail in Chapter 8. We summarize here only those reactions that are believed to result in the formation of nitric acid or nitrates.

11-B-1. Homogeneous Gas Phase Reactions

11-B-1a. Hydroxyl Radical

As discussed in Chapter 1, NO and smaller amounts of NO_2 are formed during high-temperature combustion. The predominant atmospheric fate of the NO emitted is oxidation to NO_2 which can then react with OH during the day:

$$NO_2 + OH \xrightarrow{M} HNO_3 \tag{89}$$

This reaction is relatively rapid; the effective second-order rate constant at 1 atm in air is $k_{89}^{bi} = 1.1 \times 10^{-11}$ cm^3 molecule^{-1} s^{-1} at 25°C (Atkinson and Lloyd, 1984), approximately ten times that for the SO_2–OH reaction.

For a typical tropospheric OH concentration of 1×10^6 radicals cm^{-3}, the lifetime of NO_2 with respect to reaction (89) is only ~1 day, compared to 13 days for SO_2 under the same conditions. This relatively rapid oxidation of NO_2 to HNO_3 implies that nitric acid formation should occur closer to the source than sulfuric acid formation. (However, if significant concentrations of NO_2 are tied up in the form of PAN, for example, which can travel downwind before decomposing thermally, the effective lifetime of NO_2 may be extended; see Chapter 8).

11-B-1b. Nitrate Radical

A second potentially important source of HNO_3 involves reactions of the nitrate radical (NO_3) with some organics. In the presence of NO_2 and O_3, the nitrate radical is formed and is in equilibrium with N_2O_5:

$$O_3 + NO_2 \rightarrow NO_3 + O_2 \tag{90}$$

$$k_{90} = 3.2 \times 10^{-17} \text{ cm}^3 \text{ molecule}^{-1} \text{ s}^{-1}$$

(Atkinson and Lloyd, 1984)

$$NO_3 + NO_2 \overset{M}{\rightleftarrows} N_2O_5 \tag{91}$$

(See Section 8-A-2 for a discussion of the equilibrium constant K_{91}.) As discussed in Chapter 7, NO_3 reacts relatively rapidly with a variety of organics. In the case of alkanes and aldehydes, the reaction is believed to proceed by hydrogen abstraction to form nitric acid.:

$$NO_3 + RH \rightarrow R + HNO_3 \tag{92}$$

$$NO_3 + RCHO \rightarrow RCO + HNO_3 \tag{93}$$

In polluted urban atmospheres, alkane concentrations (excluding CH_4) are typically 100 ppb (2.5×10^{12} molecules cm^{-3}) (Stephens and Burleson, 1969; Lonneman et al., 1974), and formaldehyde and acetaldehyde concentrations are typically in the range around 20 ppb and 10 ppb, respectively (Lloyd, 1979; Grosjean, 1982). Using published rate constants for NO_3 reactions with alkanes, HCHO, and CH_3CHO (Atkinson et al., 1984a,b), and an NO_3 concentration of 100 ppt (2.5×10^9 cm^{-3}), the calculated nighttime rates of HNO_3 formation for these three compounds or classes of compounds are 0.05, 0.06, and 0.18 ppb h^{-1}, respectively, leading to a net rate of HNO_3 formation of ~0.3 ppb h^{-1} from hydrogen abstraction by NO_3.

This can be compared to the daytime rate of formation of HNO_3 by reaction (89) with the hydroxyl radical. At an NO_2 concentration of 50 ppb and OH at an average concentration of 1×10^6 cm^{-3}, characteristic of moderately polluted atmospheres, reaction (89) leads to a daytime acid formation rate of 2 ppb h^{-1}. Thus in polluted atmospheres, as a crude estimate, ~15% of the HNO_3 formed may be due to hydrogen abstraction reactions of the nitrate radical at night.

The rates of HNO_3 formation cited above are calculated using second-order kinetics as discussed in Chapter 4. Thus for the alkane reaction (92), for example, the rate of HNO_3 formation is given by

$$\frac{d[HNO_3]}{dt} = k_{92}[NO_3][RH]$$

A mid-range value of k_{92} for a series of alkanes is $\sim 5 \times 10^{-17}$ cm^3 molecule^{-1} s^{-1} (Atkinson et al., 1984b). Thus

$$\begin{aligned}\frac{d[HNO_3]}{dt} &= (5 \times 10^{-17}\ \text{cm}^3\ \text{molecule}^{-1}\ \text{s}^{-1}) \\ &\quad \times (2.5 \times 10^9\ \text{molecules cm}^{-3}) \\ &\quad \times (2.5 \times 10^{12}\ \text{molecules cm}^{-3}) \\ &= 3.1 \times 10^5\ \text{molecules cm}^{-3}\ \text{s}^{-1}\end{aligned}$$

Using 2.5×10^{10} molecules cm^{-3} = 1 ppb and 3600 s = 1 h, this rate of formation becomes 0.045 ppb h^{-1}. The rates of formation from the other reactions can be calculated in an analogous manner using the appropriate rate constants and concentrations.

While these calculated rates of formation are useful for an initial assessment of the relative importance of various contributing processes, it must be kept in mind that the reactions will proceed only as long as the reactants are available. Thus if all the NO_3 reacts and is not replenished, then the available NO_3 will limit how much HNO_3 is actually formed from these reactions.

11-B-1c. N_2O_5 Hydrolysis

Dinitrogen pentoxide, N_2O_5, is formed in reactions (90) and (91). Its hydrolysis is thus a potential source of nitric acid (Platt et al., 1980):

$$N_2O_5 + H_2O \rightarrow 2\ HNO_3 \qquad (94)$$

$$k_{94} \leq 1.3 \times 10^{-21}\ \text{cm}^3\ \text{molecule}^{-1}\ \text{s}^{-1}$$

(Tuazon et al., 1983)

The rate-determining step in the above sequence is the hydrolysis reaction (94), where only an upper limit can be placed on the rate constant. Because reaction (94) appears to be so slow, this reaction sequence, (90), (91), and (94), has often been ruled out as a significant source of nitric acid.

However, several points must be kept in mind. First, even with k_{94} as low as 1.3×10^{-21} cm^3 $molecule^{-1}$ s^{-1}, this reaction sequence can contribute to HNO_3 formation in the atmosphere. Thus Tuazon and co-workers (1983) calculate that under conditions typical of downwind locations in the Los Angeles area during the night (NO_2 at 3 ppb and NO_3 at 100 ppt), reactions (90), (91), and (94) would result in an HNO_3 formation rate of 0.3 ppb h^{-1} at 50% relative humidity. These rates of HNO_3 formation are comparable to those from the nitrate radical reactions with alkanes and aldehydes.

Second, the upper limit to k_{94} refers to the reaction of N_2O_5 with gaseous water vapor; the dissolution of N_2O_5 into a water droplet followed by reaction with the liquid water or the reaction of N_2O_5 with adsorbed H_2O on surfaces may be significantly faster.

Support for the importance of this reaction sequence comes from studies of the concentration of NO_3 in various locations throughout the Northern Hemisphere (e.g., see Platt et al., 1980, 1984; Atkinson et al., 1985). As discussed in Section 5-D, NO_3 concentrations have been observed under a variety of conditions. From the known kinetics of reaction (90) and the observed ambient O_3, NO_2, and NO_3 concentrations, lifetimes can be calculated for NO_3. These lifetimes show a strong dependence on the relative humidity (RH) as seen in Fig. 11.10, such that detectable concentrations of NO_3 (i.e., $\geq$1 ppt) are rarely observed at RH values above

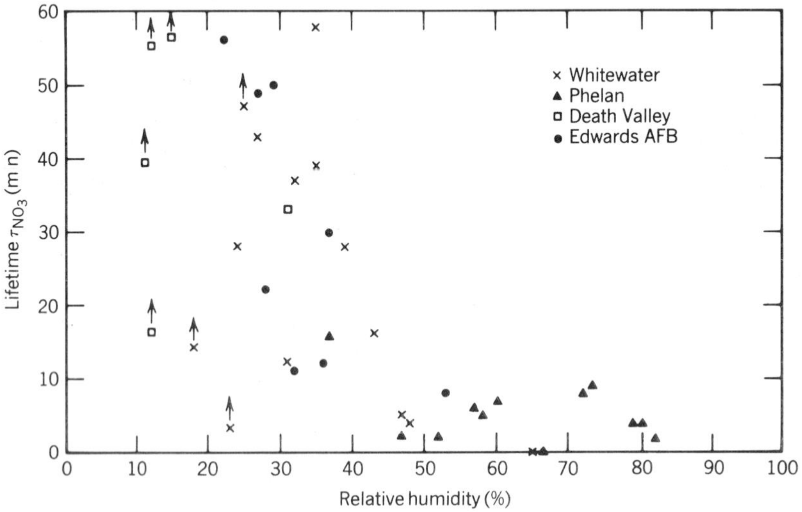

FIGURE 11.10. Lifetime of NO_3 as a function of relative humidity at several locations in California. The points marked with an upward arrow indicate that these represent lower limits to the lifetimes (from Platt et al., 1984).

50%. This has been interpreted as evidence for the reaction of N_2O_5 and/or NO_3 itself with water either in the gas phase or as a liquid, for example, on the surfaces of particles or on the ground.

Modeling studies also support this mechanism. Thus Jones and Seinfeld (1983) showed, using computer modeling, that the measured concentrations and diurnal profile of NO_3 were consistent with the mechanism of Richards (1983) which is based on reactions (90), (91), and (94), as first proposed by Platt et al. (1980).

In a second modeling study, Russell and co-workers (1984, 1985) showed that, despite the slow nature of reaction (94), if the competing removal processes for NO_3 (e.g., by NO) are also slow, reactions (90), (91), and (94) may indeed be a significant source of nitric acid. Figure 11.11 shows the rates of nitric acid production in air parcel predicted by Russell and co-workers by reactions (89), (90), (91), and (94) and by the reactions of NO_3 with aldehydes:

$$NO_3 + HCHO \xrightarrow[a]{O_2} HNO_3 + HO_2 + CO \tag{95}$$

$$NO_3 + RCHO \xrightarrow[b]{O_2} HNO_3 + RCO_3 \tag{93}$$

during one 24 h day for conditions typical of a smog episode in the Los Angeles basin. It is seen that reactions (90), (91), and (94) can be very important. The rate constant used for reaction (94) was the upper limit of Tuazon et al. (1983), 1.3×10^{-21} cm^3 molecule^{-1} s^{-1}); while increasing or decreasing it from this value by approximately an order of magnitude changed the relative contribution of reaction (94), the overall yield of HNO_3 changed by less than 20%. Thus changing the value of k_{94} resulted primarily in a redistribution of the amount of HNO_3 formed by each reaction, rather than a change in the total HNO_3 formed.

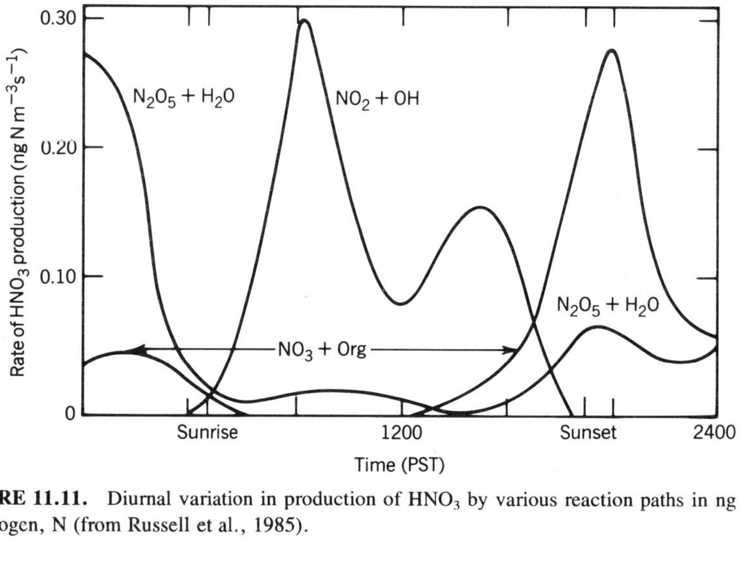

FIGURE 11.11. Diurnal variation in production of HNO_3 by various reaction paths in ng m^{-3} s^{-1} as nitrogen, N (from Russell et al., 1985).

11-B-2. Aqueous Phase Reactions

In addition to the gas phase reactions discussed above, some aqueous phase reactions have been suggested as well. These were summarized in Table 11.4 (Section 11-A-2a). For example, reaction (96) of NO_2 with water has been invoked in the past:

$$2\ NO_{2(g)} + H_2O_{(l)} \leftrightarrows 2\ H^+ + NO_3^- + NO_2^- \tag{96}$$

This reaction is now known to be too slow (Schwartz, 1984a) in aqueous solution at ambient NO_2 concentrations to be a significant source of HNO_3. This slow reaction rate is due to two factors: first, the low solubility of NO_2 (see Table 11.3) and second, the second-order dependence of the rate on the NO_2 concentration. Thus while (96) can be rapid at high NO_2 partial pressures, the rate rapidly drops as the concentration approaches ambient levels.

For a detailed treatment of aqueous phase reactions of the oxides of nitrogen, the reader is referred to reviews by Schwartz (1984a) and Schwartz and White (1981, 1983).

However, the possibility of an increase in the measured rate via catalysis, possibly involving surfaces, cannot be ruled out at the present time. The evidence for a surface-catalyzed version of (96) is discussed in detail in Chapters 6 and 8. Briefly, the formation of significant amounts of nitrous acid has been observed when NO_x is injected into smog chambers in the dark. This HONO then acts as a source of OH when irradiation is initiated. While the source of the HONO is uncertain, some evidence (e.g., the observed stoichiometry of HONO and NO_2) suggests a possible contribution from (96). However, it is important to note that nitrate has never been conclusively identified as the second product in these systems.

Of course, the N_2O_5 hydrolysis discussed above may occur not only in the gas phase but in the liquid phase as well (i.e., N_2O_5 may first be dissolved into a droplet and then hydrolyze to HNO_3). No data are available concerning this reaction. However, as we have seen above, even at the slow rates for the homogeneous gas phase reaction, N_2O_5 is one of the likely precursors to nitric acid.

Finally, the absorption of the nitrate radical into aqueous droplets followed by hydrolysis has been suggested as a potential source of nitrate in solution (Heikes and Thompson, 1983; Chameides and Davis, 1983a; Platt et al., 1984). Indeed, Seigneur and Saxena (1984) suggest this could be a major source of aqueous nitrate under certain conditions if the accommodation coefficient (α) for absorption of NO_3 into the droplet is sufficiently high, $\gtrsim 0.01$. If the nitrate radical is absorbed into solution, an alternate means of forming nitric acid is hydrogen atom abstraction from an organic, similar to the gas phase reactions (see Section 11-B-1b above).

Results of some cloud chamber studies suggest that H_2O_2 may oxidize NO_2 rapidly to nitrate in cloud water droplets; thus at the relatively high H_2O_2 concentrations of 60–70 ppb, NO_2 conversion rates of $\sim 8\%\ h^{-1}$ were observed (Gertler et al., 1984). On the other hand, Lee (1984) has reported that the NO_2–H_2O_2

reaction is too slow to be significant, with a rate constant of ~ 10 L mole^{-1} s^{-1}; the reactions of NO_2 with O_2 and O_3 were also shown to be too slow to be significant.

The addition of 200 ppb O_3 to NO_2 in the cloud chamber study of Gertler et al. (1984) also resulted in nitrate formation, although the authors attribute this primarily to the gas phase reaction (90) of O_3 with NO_2 to form the nitrate radical, followed by secondary reactions to give nitrate as discussed above.

In conclusion, some measurements of the chemical composition of precipitation and the surrounding air mass suggest that significant amounts of nitric acid may be generated in the aqueous phase in the atmosphere (e.g. see Lazrus et al., 1983; Misra et al., 1985) as we saw is the case for sulfuric acid under many conditions.

11-C. SULFURIC AND NITRIC ACIDS: COMPARISON AND CONTRAST

Once formed in the atmosphere, sulfuric and nitric acids show quite different behavior both physically and chemically. Nitric acid is more volatile and thus can exist in significant concentrations in the gas phase, while sulfuric acid has a very low vapor pressure ($< 10^{-7}$ atm) under ambient conditions and hence exists in the form of particles (Roedel, 1979).

Both nitric acid and sulfuric acid can react with bases present in the atmosphere to form salts. One such base frequently present in substantial concentrations is NH_3. When HNO_3 reacts with NH_3, an equilibrium is established:

$$HNO_3 + NH_3 \leftrightarrows NH_4NO_3 \tag{97}$$

The ammonium nitrate exists as a solid if the relative humidity is less than that of deliquescence; if the relative humidity is higher, it exists in the aqueous phase. The dissociation constant $K_{97} = P_{NH_3}P_{HNO_3}$ is a sensitive function of temperature (see Fig. 11.12*a*); at temperatures greater than $\sim 35°C$, little ammonium nitrate is expected under typical ambient conditions. K_{97} is also somewhat dependent on the relative humidity, as shown in Fig. 11.12*b* (Stelson and Seinfeld, 1982a,b).

Because of this dynamic equilibrium between HNO_3, NH_3, and NH_4NO_3, which is discussed in more detail in Section 8-A-3b, nitric acid can relatively easily revolatilize even after forming the ammonium salt; no analogous chemical and physical changes exist for sulfuric acid. [However, dissolved S(IV) compounds in aerosols may be volatile (Richards et al., 1983).].

The physical and chemical properties of nitric acid and sulfuric acids are thus very different. However, their chemistries are intertwined via the OH radical and the other oxidants involved in converting NO_x and SO_2 into the acids. For example, it has been proposed that in ambient atmospheres where significant concentrations of both NO_x and SO_2 are present, increasing NO_x will decrease the gas phase rate of formation of sulfuric acid. This is because increasing NO_x increases the rate of reaction (89), thus lowering the concentration of gas phase OH. This results in a lowered rate of oxidation of SO_2 by OH in the gas phase via reaction (2).

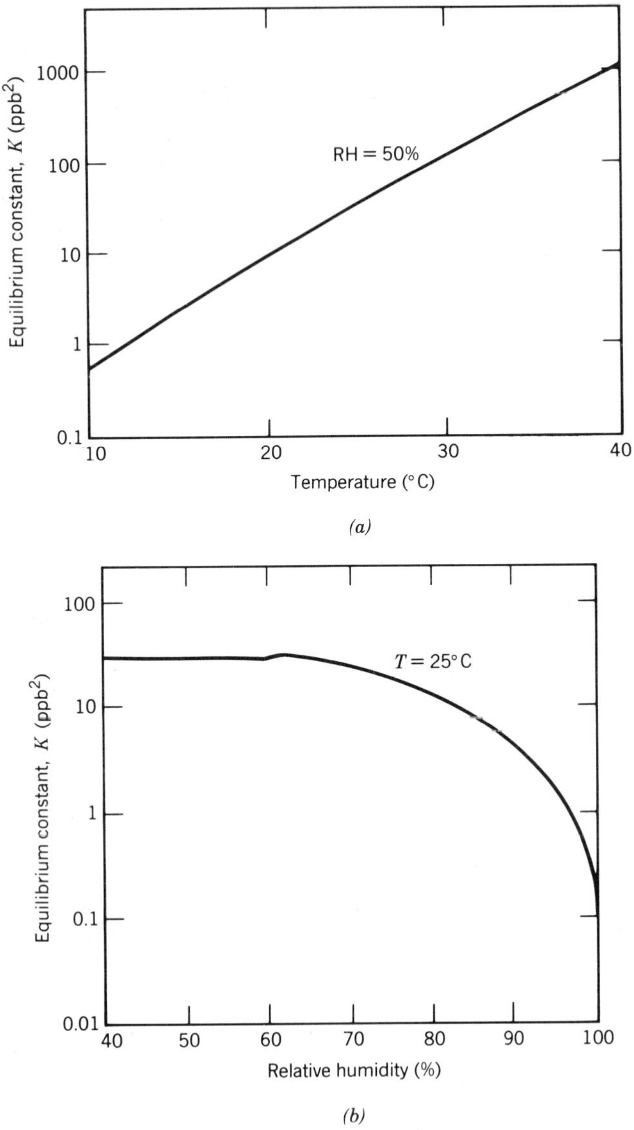

FIGURE 11.12. Dissociation constant for NH_4NO_3 (in ppb^2) as a function of (*a*) temperature (RH = 50%) and (*b*) relative humidity (T = 25°C) (from Russell et al., 1983, 1984).

Increasing NO_x may not only alter the rate of the gas phase oxidation of SO_2, but perhaps the rate of oxidation in aqueous droplets as well (Calvert and Stockwell, 1983). For example, H_2O_2 is formed in the atmosphere by the recombination of HO_2 radicals either as the free radical itself,

$$2HO_2 \xrightarrow{M} H_2O_2 + O_2 \tag{98}$$

or in a hydrated form (see Section 3-D),

$$HO_2 + H_2O \leftrightarrows HO_2 \cdot H_2O \tag{99}$$

$$HO_2 \cdot H_2O + HO_2 \rightarrow H_2O_2 + O_2 + H_2O \tag{100}$$

At low hydrocarbon/NO_x ratios, that is, at high NO_x concentrations, reactions (98) or (99) plus (100), do not occur to a significant extent because of the competition for HO_2 by NO:

$$HO_2 + NO \rightarrow OH + NO_2 \tag{101}$$

At high NO_x then, little gas phase H_2O_2 should be formed. If absorption of gas phase H_2O_2 into aqueous droplets is the major source of H_2O_2 in the liquid phase, and if this reaction contributes as strongly to the SO_2 oxidation as Fig. 11.9 suggests, then the rate of sulfate formation in the liquid phase may also fall at high NO_x concentrations.

If deposition of SO_2 on surfaces (see below) is occurring at a constant rate in competition with its oxidation, more of the SO_2 will end up deposited and less transformed into sulfuric acid in the atmosphere. (Of course, if deposited SO_2 forms acid on the surface, the *net* result in terms of effects may be the same.)

Hydrocarbon–NO_x chemistry may thus control not only the gas phase processes leading to acid formation in the atmosphere, but also the liquid phase processes and possibly the surface oxidation.

11-D. OTHER CONSIDERATIONS

11-D-1. Role of Meteorology

While we have concentrated in this chapter on the chemistry involved in the formation of acids from their gas phase precursors in the atmosphere, the extremely important role played by meteorology must also be recognized. Meteorological processes transport the acid precursors, during which time they react (i.e., over large distances) to form acids. As a result, acid deposition has been thought to be characterized by long-range transport (eg. ~1000 km) from source to receptor. It should be noted, however, that the results of some recent studies in the eastern United States and Canada suggest that a significant portion of the acid deposition in some locations is due to local sources, rather than long-range transport from distant ones (Shaw, 1982).

Meteorology, however, also plays important roles other than mere transport of pollutants as they react. As we have seen, the formation of acids from their precursors involves reactions not only in the gas phase, but also in the liquid phase (e.g., in clouds, fogs, dews) and perhaps on the surfaces of solids as well. Which phase is most important in the oxidation of sulfur and nitrogen compounds depends on the particular meteorological conditions, for example, whether clouds are present. Cloud chemistry may be especially important because clouds can process large volumes of air through vertical circulation, which can be in the range of ~0.3 to several meters per second.

The intensity of sunlight is another major meteorological factor in the formation of acids from their precursors because of the importance of photochemical reactions.

Temperature, of course, is important because of its effects on the kinetics of reactions, and the Henry's law constants for dissolved gases, as well as on the volatility of water droplets in the atmosphere.

11-D-2. Deposition Processes

Once formed, these acids can be deposited at the earth's surface in two ways. Thus deposition can be classed as *dry deposition* or *wet deposition*, depending on the phase in which pollutant strikes the earth's surface and is taken up. Thus pollutants may be dissolved in clouds, fog, rain, or snow; when these water droplets impact the earth's surface (which includes not only soil, but grass, trees, buildings, etc.), it is termed *wet deposition*. However, pollutants in the form of either gases or small particles can also be transported to ground level and absorbed and/or adsorbed by materials there without first being dissolved in atmospheric water droplets; this is called *dry deposition*. It should be noted that the surface itself may be wet or dry; the term dry deposition only refers to the mechanism of transport to the surface, not to the nature of the surface itself.

Because of the highly variable nature of precipitation events, quantitatively estimating wet deposition of pollutants is difficult. In addition to meteorological factors, such parameters as the solubility of the pollutant in ice, snow, and rain and how this varies with temperature and pH, the size of the water droplets, and the number present must also be considered; for example, snow may scavenge HNO_3 more efficiently than rain (Chang, 1984).

As an approximation, the rate of wet deposition of a pollutant is sometimes taken as λC where C is the pollutant concentration and λ is known as a *washout coefficient* which is proportional to the precipitation intensity (Shaw, 1984).

Dry deposition can also be a very important mechanism for removing pollutants from the atmosphere in the absence of precipitation. Indeed, even in such places as eastern England the ratio of dry to wet removal of SO_2 has been estimated to be ~2:1 (Davies and Mitchell, 1983). If this is the case, then in arid and semiarid regions such as much of the western United States dry deposition is clearly important.

Dry deposition is usually characterized by a *deposition velocity*, V_g, which is defined as the flux (F) of the species S to the surface divided by the concentration [S] at some reference height h:

$$V_g = \frac{F}{[S]} \tag{O}$$

The amount of the species deposited per unit area per second in a geographical location, that is, the flux, can be calculated if the deposition velocity and the pollutant concentration are known. The deposition velocity is also frequently related to a resistance r:

$$V_g = \frac{1}{r} \tag{P}$$

By analogy to electrical systems, the resistance r can be thought of as being comprised of several components. For convenience, two such components are often defined, a surface resistance (r_{surf}), which depends on the affinity of the surface for the species, and a gas phase resistance (r_{gas}), which depends on the micrometeorology that transports the gas to the surface:

$$r = r_{\text{gas}}(h) + r_{\text{surf}} \tag{Q}$$

The gas phase resistance depends on the height (h) above the earth's surface, as does the concentration of the pollutant; as a result the deposition velocity is also a function of height. A reference height of 1.0 m over land is usually used.

Figure 11.13 schematically depicts the dry deposition of a pollutant on the leaf of a plant in the form of resistances. In this case, the surface resistance r_{surf} has been broken down even further into a combination of parallel and series resistances (r_b, r_c, and r_s) in the layer of air immediately adjacent to the surface where molecular and Brownian diffusion processes are important. Since leaves may absorb pollutants either through stomata or through the cuticles, the absorption into the leaf is represented by two parallel resistances, r_c for the cuticular resistance and r_s for the stomatal resistance which is in series with a mesophyllic resistance r_m.

The relative importance of gas phase and surface resistances depends on the

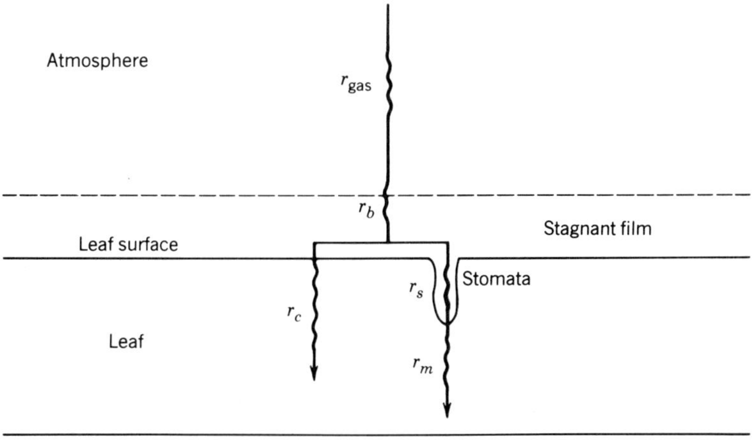

FIGURE 11.13. Schematic diagram of dry deposition of a pollutant onto the surface of a leaf, using the concept of resistances (r) for deposition (from Shaw, 1984).

nature of the pollutant and the surface as well as the meteorology (Shaw, 1984; Unsworth et al., 1984). The gas phase resistance (r_{gas}) is determined by the vertical eddy diffusivity which depends on the evenness of the surface and the meteorology, for example, wind speed, solar surface heating, and so on. The surface resistance (r_{surf}) depends on the detailed characteristics of the surface (e.g., type, whether leaf, building soil, snow, wetness, etc.; see Dasch, 1985; Vandenberg and Knoerr, 1985) as well as on the nature of the pollutant being deposited; thus the resistance r_b in the film of air immediately adjacent to the surface depends on the shape of the surface and the molecular diffusivity of the depositing species, while the resistances r_c and r_s reflect the resistance to adsorption by the surface, which depends on both the nature of the surface itself and on the depositing species. For highly reactive gases, the surface resistance may be sufficiently small so that transport to the surface becomes rate limiting; for example, the surface resistance for deposition of HNO_3 on grass during the day has been shown to be approximately zero (Huebert and Robert, 1985).

Because of the effects of physical, chemical, biological, and meteorological parameters on the "resistances," the deposition velocity (V_g) defined by Eq. (P) also depends on these. As a result, deposition velocities reported for various pollutants show a wide range, from several hundredths to several cm s^{-1}, depending on the conditions during the measurement. For example, the mean value of V_g for SO_2 over grass in one study was 0.56 cm s^{-1} for dry grass, but 0.93 cm s^{-1} for wet grass (Davies and Mitchell, 1983). A diurnal variation in deposition velocities has been reported with daytime values being greater than those at night (Hicks et al., 1983); presumably both r_{gas} and r_{surf} increase under nighttime conditions. For example, peak values of O_3 deposition of up to 1 cm s^{-1} during midday but less than 0.1 cm s^{-1} in the evening have been reported (Droppo, 1985).

For particles, V_g depends on the particle size; thus a minimum in V_g is generally observed at particle diameters around 0.5 μm, ~0.01 cm s^{-1} being a typical minimum value for grasses and grass-like surfaces (McMahon and Denison, 1979; Hicks and Garland, 1983).

Table 11.5 gives some typical ranges of deposition velocities reported for various pollutants and surfaces. The wide ranges given in Table 11.5 reflect a combination of experimental uncertainties as well as real differences due to meteorology, nature of the surface, diurnal variations, and so on. The overall uncertainty in the appropriate value of the deposition velocity to use under a given set of circumstances can thus be quite large; for example, Sievering (1984) has estimated the uncertainty in the dry deposition of small particles ($D < 1$ μm) to water surfaces to be about two orders of magnitude!

For further details of dry deposition, its measurement, and typical values of V_g for various pollutants, the reader is referred to the review article by McMahon and Dension (1979), the volume edited by Pruppacher, Semonin, and Slinn (1983), and the description and results of a field intercomparison study reported in the *Journal of Geophysical Research* (1985).

In short, while we have concentrated in this chapter on the chemistry that converts SO_2 and NO_x to sulfuric and nitric acids, respectively, dry deposition of both

TABLE 11.5. Typical Ranges of Deposition Velocities V_g Reported for Some Pollutants and Surfaces[a]

Species	Surface	Deposition Velocity[a] V_g (cm s^{-1})
O_3	Soil, short grass	0.10–2.10
	Grass, soil, water	0.47–0.55
	Maize, soybean field	0.20–0.84
NO	Soil, cement	0.10–0.20
NO_2	Soil, cement	0.30–0.80
	Alfalfa	1.90
PAN	Grass, soil	0.14–0.30
	Alfalfa	0.63
HNO_3	Grassy field[b]	1.0–4.7
SO_2	Grass[c]	0.1–4.5
	Pine forest[d]	0.1–1.0
Particulate sulfur	Frozen bare soil[e]	g
	Deciduous forest, wintertime[e]	g
	Grassy pasture, drought[e]	0.17–0.24
	Pine forest[e]	0.48–0.90
	Grass[e]	0.02–0.42
	Grassy field[f]	0.1–1.2
COS	Soil[h]	$(3.1–5.7) \times 10^{-4}$

[a]Except where stated elsewhere, from McRae and Russell, 1984, and references therein.
[b]Huebert and Robert, 1985.
[c]Davis and Wright, 1985.
[d]Garland, 1978, and references therein.
[e]Wesely et al., 1985.
[f]Feely et al., 1985.
[g]Indistinguishable from zero.
[h]Kluczowski et al., 1985.

the gaseous precursors and the acids may be significant, exceeding wet deposition under some conditions.

11-D-3. Is All Acid Deposition Necessarily Due to "Pollution"?

Until recently, it was assumed that *natural* or *clean* water would have a pH of about 5.65 due to the presence of ~335 ppm of atmospheric carbon dioxide CO_2 which dissolves to form bicarbonate (HCO_3^-) and carbonate (CO_3^{2-}) ions, as well as H^+:

$$CO_2 + H_2O \leftrightarrows CO_2 \cdot H_2O \qquad H_{102} = 3.4 \times 10^{-2}\ M\ \text{atm}^{-1} \tag{102}$$

$$CO_2 \cdot H_2O \leftrightarrows H^+ + HCO_3^- \qquad K_{103} = 2.9 \times 10^{-7}\ M \tag{103}$$

$$HCO_3^- \leftrightarrows H^+ + CO_3^{2-} \qquad K_{104} = 2.7 \times 10^{-11}\ M \tag{104}$$

However, even clean atmospheres have a variety of species present which can lead to pH values different from that expected from the above equilibria. For example, there are substantial natural emissions of both sulfur and nitrogen compounds (Cullis and Hirschler, 1980; Stedman and Shetter, 1983). Taking sulfur as an example, hydrogen sulfide (H_2S) and dimethyl sulfide (CH_3SCH_3) are produced from biogenic processes. In addition, sulfur compounds are emitted during volcanic eruptions (principally in the form of SO_2). Sea spray also contributes to the natural sulfur cycle in the form of sulfate particles, some of which become deposited on land.

In considering relative impacts on air quality of anthropogenic and natural sources, one must take into account not only the relative source strengths on a global scale, but also their temporal and spatial distributions and chemical form. In the case of sulfur, for example, the biogenic contributions tend to be widely dispersed, whereas the anthropogenic emissions ($\sim$90% of which is due to combustion of coal and petroleum) tend to be concentrated in urban and industrial areas. In addition, the biogenic releases are primarily H_2S and CH_3SCH_3 which must be oxidized in the atmosphere through a number of steps, in part to sulfuric acid (see Section 14-C), whereas combustion of fossil fuels emits SO_2 which is already partially oxidized.

In short, some contribution to sulfuric acid deposition is expected from natural sources, but estimating the relative contribution at a particular location is not straightforward. Complicating the situation is the fact that there are also natural sources of bases such as ammonia, magnesium, and calcium which can neutralize acids produced in the atmosphere. The presence of such neutralizing species can mitigate the effects of acid formation in the atmosphere (e.g., see Munger, 1982). As discussed in detail by Morgan (1982), the term *base neutralizing capacity* is therefore often used to express the net concentration (after neutralization) of acids stronger than CO_2 in atmospheric solutions (e.g., rain). For example, for rain containing sulfuric and nitric acids as well as the neutralizing components $CaCO_3$ and NH_3, the base neutralizing capacity (BNC) referenced to CO_2 is given by

$$[\text{BNC}]_{CO_2} = 2[H_2SO_4] + [HNO_3] - 2[CaCO_3] - [NH_3] \tag{R}$$

Experimentally, this can be determined by titration to the CO_2 reference pH of 5.65.

While sulfuric and nitric acids are at present believed to constitute the major components of acid deposition on populated continents, weaker acids are undoubtedly present as well, their nature and concentrations depending on the particular geographical location. For example, organic acids are thought to be formed in gas phase reactions (see Chapter 7) and some will dissolve in atmospheric droplets; alternatively, oxidation of dissolved organics may occur within the droplet itself (Chameides, 1984). While weak acids may contribute less to the observed pH of precipitation than strong acids, their ultimate capacity for proton production (i.e., their base neutralizing capacity) may be important under some circumstances. For example, in some locations in remote areas of the world, formic and acetic acids have been found to account for a large fraction of the total acidity in the precipitation. Thus at Katherine, Australia, these acids accounted for ~63% of the total acidity measured during part of the 1981–1982 wet season (Keene et al., 1983). In other remote locations, however, insignificant levels of organic acids were observed. Thus the contribution of organics such as acetic and formic acids to acid deposition appears to be highly variable and may be significant compared to sulfuric and nitric acids in remote locations and perhaps in some urban areas.

Acids other than organic acids may also be present. For example, as discussed earlier (Section 11-A-2a), small (ppb) concentrations of hydrochloric acid have been measured in some locations in the gas phase.

Because of the presence of bases in the atmosphere, the potential contributions of natural sources to the formation of acids, and the possible presence of acids other than sulfuric and nitric acids, pH alone is not a good indicator of anthropogenically derived acid deposition. The net effect on the acidity of precipitation in various regions will clearly depend on the particular conditions. For example, Charlson and Rodhe (1982) have suggested that in some cases pH values of 4 or even less could be produced without any contribution from anthropogenic sources.

While an a priori estimate of the pH of rain, clouds, or fogs at particular locations due to natural and/or anthropogenic emissions is thus not possible, we do not wish to leave the reader with the impression that the problem of acid deposition in Europe, Scandinavia, and North America can be attributed primarily to natural processes. Indeed, the weight of the evidence at present strongly supports the overwhelming contribution of human activities to acid deposition in these industrialized areas.

The chemical complexity involved in acid deposition has a direct bearing on the issue of control of acid rain. Thus in the United States legislative bills to control acid rain in the eastern United States through control of upwind SO_2 emissions have been proposed. A major uncertainty associated with the enactment of such legislation has been the quantitative effects on acid deposition of reducing the precursor emissions. Specifically, if SO_2 emissions are reduced by some percentage, will acid deposition at a *particular location* decrease by a corresponding amount? As discussed in more detail by Finlayson-Pitts and Pitts (1982, 1984), and in a National Research Council document (1983), it appears that one can at least state that a reduction in SO_2 will give a proportional decrease in the *overall*, total acid deposition integrated over all locations if everything else remains constant. This and other aspects of linearity are discussed in the following section.

11-E. CHEMICAL SUBMODELS USED IN LONG-RANGE TRANSPORT MODELS

The phenomenon of acid rain has generally been assumed to result from the long range transport (LRT) of pollutants. In the United States for example, LRT from the Midwest and South to the northeastern states has been suggested as a major cause of acid precipitation there. Because of the complex meteorology involved in such LRT, early computer models for acid rain which incorporated the emissions, meteorology, and chemistry only included simple chemical submodels; indeed, until about 1980, the chemistry of SO_2 in these models was often treated in terms of only one lumped reaction,

$$SO_2 \xrightarrow{k_{105}} H_2SO_4 \qquad (105)$$

with the value of k_{105} chosen somewhat arbitraily based on field observations and chemical intuition.

The recognition that the major oxidants for SO_2 are likely photochemical in nature and that oxidation may occur in the gas phase, in the liquid phase, and on surfaces has led to the realization that in order to predict rates of oxidation of SO_2 and thus of formation of H_2SO_4, one must therefore be able to predict the concentrations of the oxidants produced in both the gas and liquid phases in irradiated hydrocarbon–NO_x mixtures. As discussed elsewhere in this book, major uncertainties exist in such predictions.

Despite these uncertainties, it is worthwhile to consider on a very simplified basis the effects of hydrocarbon–NO_x–SO_2 chemistry on acid deposition. The first attempt at this was by Rodhe, Crutzen, and Vanderpol in 1981. They incorporated simplified RHC–NO_x chemistry into an LRT model for sulfuric acid formation and deposition. Table 11.6 shows the chemical submodel used by Rodhe and co-workers. In this model, H_2SO_4 was formed via the gas phase reaction of SO_2 with OH, or via a reaction intended to simulate oxidation of SO_2 by H_2O_2 in a cloud. Because the rate of the latter was unknown due to uncertainties in the concentrations of aqueous phase H_2O_2 and the rate constant, a rate was arbitrarily chosen so that the gas phase and aqueous phase reactions proceeded at roughly equal rates for typical summertime OH concentrations. Formation of HNO_3 was assumed to proceed only by the reaction of OH with NO_2.

Ethylene was taken to represent all reactive hydrocarbons, and several organic reactions were lumped together, resulting in the reactions of Table 11.6. For example, reaction (R5) represents the organic portion of the following reactions:

$$CH_4 + OH \rightarrow CH_3 + H_2O \qquad (106)$$

$$CH_3 + O_2 \xrightarrow{M} CH_3O_2 \qquad (107)$$

$$CH_3O_2 + NO \rightarrow CH_3O + NO_2 \qquad (108)$$

$$CH_3O + O_2 \rightarrow HCHO + HO_2 \qquad (109)$$

TABLE 11.6. Chemical Reactions Considered in the Rodhe Model. Reaction Rates Are Given in Units of cm^3 $molecules^{-1}$ s^{-1} and Dissociation Rates in s^{-1}

(R1)	$O_3 + h\nu \rightarrow O_2 + O(^1D)$	$j_1 = 4.1(-6)^a$
(R2)	$O(^1D) + M \rightarrow O + M$	$k_2 = 5(-11)$
(R3)	$O(^1D) + H_2O \rightarrow 2OH$	$k_3 = 2.3(-10)$
(R4)	$CO + OH \rightarrow \dashrightarrow HO_2 + CO_2$	$k_4 = 3(-13)$
(R5)	$CH_4 + OH \rightarrow \dashrightarrow CH_2O + HO_2$	$k_5 = 5(-15)$
(R6)	$C_2H_4 + OH \rightarrow \dashrightarrow HO_2 + 2CH_2O$	$k_6 = 9(-12)$
(R7)	$CH_2O + OH \rightarrow \dashrightarrow CO + HO_2$	$k_7 = 1.2(-11)$
(R8)	$CH_2O + h\nu \rightarrow \dashrightarrow CO + 2HO_2$	$j_8 = 1.1(-5)$
(R9)	$CH_2O + h\nu \rightarrow \dashrightarrow CO + H_2$	$j_9 = 2.6(-5)$
(R10)	$2HO_2 \rightarrow H_2O_2 + O_2$	$k_{10} = 2.5(-12)$
(R11)	$HO_2 + OH \rightarrow H_2O + O_2$	$k_{11} = 5.1(-11)$
(R12)	$H_2O_2 + OH \rightarrow HO_2 + H_2O$	$k_{12} = 6.9(-13)$
(R13)	$H_2O_2 + h\nu \rightarrow 2OH$	$j_{13} = 3.3(-6)$
(R14)	$NO_2 + h\nu \rightarrow NO + O$	$j_{14} = 4(-3)$
(R15)	$NO + O_3 \rightarrow NO_2 + O_2$	$k_{15} = 1.4(-14)$
(R16)	$NO + HO_2 \rightarrow OH + NO_3$	$k_{16} = 8(-12)$
(R17)	$NO_2 + OH \rightarrow HNO_3$	$k_{17} = 8(-12)$
(R18)	$SO_2 + OH \rightarrow \dashrightarrow H_2SO_4$	$k_{18} = 6(-13)$
(R19)	$SO_2 + H_2O_2 + \text{Cloud} \rightarrow H_2SO_4$	$k_{19} = 6(-16)$

Source: Rodhe et al., 1981.

[a]Numbers in parentheses represent powers of 10.

Inclusion of such detailed hydrocarbon–NO_x chemistry in an LRT model of necessity required treating the meteorology in less detail, and a simplified box model was used.

As pointed out by Rodhe and co-workers, there are a number of assumptions inherent in this chemical package, as well as in other parts of the model. However, keeping these in mind, they calculated the deposition of sulfuric acid at various points downwind under conditions they judged to be representative of those in northern Europe in 1955 and 1975. At travel times of 10 h, for example, the concentration of sulfuric acid was predicted to increase by only 30% over this 20 yr period, despite an increase in SO_2 emissions of ~70%. In fact, most of the calculated 30% increase in H_2SO_4 is due to an assumed increase in the *direct* rate of emission of H_2SO_4; if this direct emission of acid had been assumed to remain constant, the model predictions after 10 h would show a net *decrease* in H_2SO_4 formation despite the large increase in SO_2! This was attributed to a large increase in NO_x (~225%) emissions which occurred simultaneously; this lowers the OH concentrations predicted by the model and hence the rate of formation of H_2SO_4 via reaction (R18) with OH.

Additional predictions of this model can be summarized as follows:

- Nitric acid is predicted to be formed faster than sulfuric acid. This arises because OH is the sole oxidant assumed for NO_2 and a major one for SO_2.

Since the rate constant for OH + NO_2 is an order of magnitude larger than that for OH + SO_2, nitric acid formation is predicted to be a more local phenomenon than sulfuric acid formation.

- An increase in reactive hydrocarbons (RHC) is predicted to increase the rate of production of both sulfuric and nitric acids. This is expected on the basis of our knowledge of RHC–NO_x chemistry since under certain conditions more RHC leads to increased free radical (e.g., OH) concentrations (see Chapter 10).
- Reductions in the deposition of H_2SO_4 is predicted *not* to be directly proportional to reductions in SO_2 emissions *if* RHC and NO_x change simultaneously.

As seen above, these results can be rationalized on the basis of the chemistry input to the model. Of course, as emphasized by Rodhe and co-workers, the chemistry itself was very simplified. Despite these uncertainties and assumptions, this first attempt at incorporating RHC–NO_x–SO_2 chemistry into an LRT model illustrated the importance of including detailed chemistry in these models. Since then, a number of comprehensive programs directed to developing LRT models that incorporate detailed chemistry in both the gas and liquid phases have been initiated in laboratories around the world. For example, models are currently being developed in the United States by the National Center for Atmospheric Research (the RADM model, 1985), by Environmental Research and Technology (Venkatram et al., 1984) and by Systems Application Inc. (Seigneur and Saxena, 1984; Seigneur et al., 1984, 1985).

An example of these more comprehensive models is that of Seigneur and Saxena (1984) who developed a model that incorporates 80 gas phase reactions, 27 liquid phase reactions, and 12 equilibria describing the gas–liquid partitioning of gases; they applied it to various conditions including non-precipitating clouds, raining clouds, and fog to estimate the major oxidation processes for S(IV) and NO_x. One very interesting result was that if the radical scavenging accommodation coefficient for NO_3 is sufficiently high ($\alpha \gtrsim 0.01$), scavenging of NO_3 into droplets followed by conversion to nitrate could contribute significantly to nitrate formation.

The question of *non-linearity* of emissions and acid deposition suggested, for example, by the results of Rodhe and coworkers discussed earlier, is an extremely important one. Thus current modeling efforts are focusing to a large extent on whether a 50% reduction in emissions of SO_2 would lead to a 50% reduction in sulfate deposition at specific points downwind. For example, if the ambient concentration of SO_2 significantly exceeds that of its oxidants such as H_2O_2, and the oxidants are not rapidly regenerated, then the formation of sulfate may be limited by the amount of oxidant available; in the extreme case where all of the oxidant becomes depleted by reaction with SO_2, lowering SO_2 emissions with all else remaining constant would not alter the total sulfate formation and deposition. If either or both NMOC and NO_x change simultaneously, thus changing the oxidant (including free radical) concentrations, the situation is even more complex. Thus it is not surprising that non-linear relationships between SO_2 emissions and sulfate deposition at locations downwind have been predicted by many models (e.g. see Lee and Shannon, 1985; Scire and Venkatram, 1985; Seigneur et al., 1984, 1985).

However, it is important to recognize that uncertainties in the model inputs result in uncertainties in the model outputs. For example, as discussed in Section 11-A-1b, it was only recently recognized that the reaction of SO_2 with OH under atmospheric conditions (i.e. in the presence of O_2) forms HO_2 which regenerates OH, making model predictions more linear. In addition, as we saw in Section 3-C-12 and Section 11-A-2, there are significant gaps in our understanding of solution phase photochemistry and thermal reactions under atmospheric conditions.

Interestingly, in a study of the relationship between SO_2 emissions from nonferrous metal smelters in the western United States and sulfate concentrations in precipitation in the Rocky Mountain states over a four year period, Oppenheimer et al. (1985) indeed found that the two were linearly related. Furthermore, studies had shown a similar direct link between SO_2 emissions and the sulfate concentrations in aerosol particles (see Chapter 12) sampled over a smaller geographical area (Eldred et al., 1983). Oppenheimer and coworkers concluded that "emissions from nonferrous metal smelters are linearly related to concentrations of sulfate in precipitation at stations remote from these sources." If correct, this would be the first such *experimental* evidence for the linearity of emissions and deposition at specific downwind locations. However, agreement is not universal on the conclusions drawn by Oppenheimer et al. (Ember, 1985). For example, even if Oppenheimer et al. (1985) are correct for this specific atmospheric system and geographical region, their conclusions may not be applicable to other locations with differing emission patterns, meteorology and chemistry. The ultimate resolution of this important issue will be most interesting.

11-F. ACID FOGS

The processes that lead to the formation of sulfuric and nitric acids in precipitation also occur in fogs. The prime difference between fogs and clouds which is relevant to the chemistry is the liquid water content; in fogs, the liquid water content is of the order of 0.1 g m^{-3}, whereas in clouds, it is approximately an order of magnitude larger. This smaller liquid water content in fogs, combined with the relatively rapid condensation and evaporation processes which occur as the fog forms and dissipates, can result in urban area fogs that are very acidic.

For example, Hoffmann and co-workers (Waldman et al., 1982; Munger et al., 1983; Jacob et al., 1985) recently reported the results of studies of fogs formed in several California locations, including the Los Angeles basin. pH values of the fog as low as 2.22 were reported, although even lower values (i.e. a pH of 1.69 at an urban site near the Pacific Ocean) have been recorded since then; independent studies by Brewer and co-workers (1983) have confirmed these values. Not only were the fogs much more acidic than precipitation in the same areas and previously reported values for fogs in other locations, but the concentrations of other anions and cations such as SO_4^{2-}, NO_3^-, and NH_4^+ were also much higher, by one to two orders of magnitude. The temporal change in concentrations of the ions as the fogs formed and then dissipated, the nature of the ions present, and the fact that the most concentrated fogs were observed after days of dense haze (i.e., high particle concentrations) suggested that condensation and evaporation of water vapor on preexisting aerosol particles were the major processes controlling the fog water concentrations; this has been supported by fog water measurements in less polluted areas where the observed acidities were lower, pH $\simeq$ 4.3–6.4 (Fuzzi et al., 1984). This model is shown schematically in Fig. 11.14 where the interrelationship of gas phase chemistry, aerosol formation, and acid fogs is seen.

In some cases, both H^+ and NO_3^- increased in concentration simultaneously while other ionic concentrations were decreasing, suggesting either that HNO_3 was being scavenged from the gas phase or that nitric acid production was occurring in the fog droplets, for example, through N_2O_5 hydrolysis. Subsequent modeling studies (Jacob and Hoffmann, 1983) supported the former, that is, scavenging of gas phase HNO_3. These studies also suggested that the major oxidants for S(IV) in the droplet were, in decreasing order of importance, Mn^{2+}- and Fe^{3+}-catalyzed oxidation by O_2, oxidation by H_2O_2, and finally, reaction with O_3.

Such fogs can be highly concentrated and thus be a possible cause of health and/or plant effects. For example, Hoffmann and co-workers (Munger et al., 1983) have estimated the sulfate ion concentrations in the London fog during the 1952 smog episode. While the pollutants that were present and their concentrations are not known, 4000 excess deaths occurred in a 5 day period. Hoffmann and co-workers estimate that the sulfate concentration was approximately 11–46 milliequivalents per liter (meq L^{-1}) during this air pollution episode. This is of the same order of magnitude as those measured recently in acid fogs in the Los Angeles area where sulfate concentrations as high as 5 meq L^{-1} have been observed. While no controlled studies of the health effects of the acid fogs, especially in concert with the co-pollutants that exist during such episodes, have been reported,

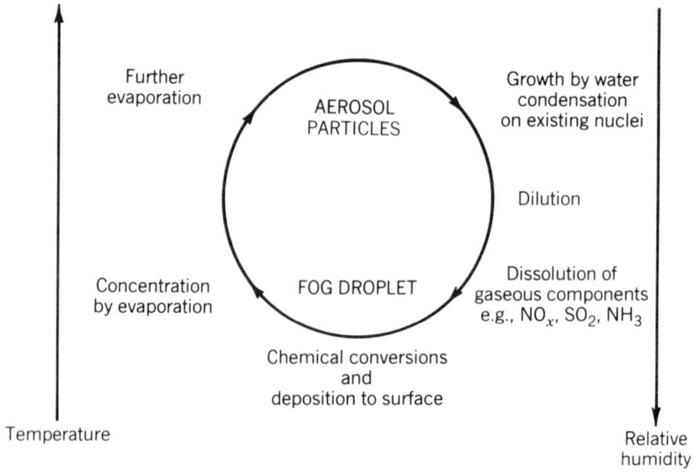

FIGURE 11.14. Schematic of role of atmospheric aerosols in fog formation and evaporation (from Munger et al., 1983).

based on our present knowledge, it is clearly important to investigate this area (Hoffmann, 1984). Additionally, plant damage from acid mists with pH $\lesssim$ 3 has been observed (Granett and Musselman, 1984), and damage to buildings, statues, and metal surfaces is expected (Hoffmann, 1984).

The net contribution of acid fogs to the total deposition of acids at the earth's surface may be relatively small. Thus evaporation of the fog may occur prior to deposition, leaving an aerosol similar to that on which the water vapor originally consensed. In addition, the smaller particle size in fogs compared to clouds and raindrops results in a much smaller total volume of solution in the form of fogs and in slower settling rates. However, given their potential for adverse effects on health, vegetation, and materials, the relative importance of acid fogs may be rather high.

REFERENCES

Alkezweeny, A. J. and D. C. Powell, "Estimation of Transformation Rate of SO_2 to SO_4 from Atmospheric Concentration Data," *Atmos. Environ.*, **11,** 179 (1977).

Atkinson, R. and A. C. Lloyd, "Evaluation of Kinetic and Mechanistic Data for Modeling of Photochemical Smog," *J. Phys. Chem. Ref. Data*, **13**, 315 (1984).

Atkinson, R. and W. P. L. Carter, "Kinetics and Mechanisms of the Gas-Phase Reactions of Ozone with Organic Compounds under Atmospheric Conditions," *Chem. Rev.*, **84,** 437 (1984).

Atkinson, R., C. N. Plum, W. P. L. Carter, A. M. Winer, and J. N. Pitts, Jr., " Rate Constants for the Gas Phase Reactions of NO_3 Radicals with a Series of Organics in Air at 298 $\pm$ 1°K," *J. Phys. Chem.*, **88,** 1210 (1984a).

Atkinson, R., C. N. Plum, W. P. L. Carter, A. M. Winer, and J. N. Pitts, Jr., "Kinetics of the Gas Phase Reactions of NO_3 Radicals with a Series of Alkanes at 298 $\pm$ 1°K," *J. Phys. Chem.*, **88,** 2361 (1984b).

Atkinson, R., A. M. Winer, and J. N. Pitts, Jr., "Estimation of Nighttime N_2O_5 Concentrations from Ambient NO_2 and NO_3 Radical Concentrations and the Role of N_2O_5 in Nighttime Chemistry," *Atmos. Environ.* **19,** in press (1985).

Bäckström, H. L. J., "Der Kettenmechanismus bei der Autoxydation von Natrium-Sulfitlösungen," *Z. Phys. Chem.*, **25B,** 122 (1934).

Baldwin, A. C., "Heterogeneous Reactions of Sulfur Dioxide with Carbonaceous Particles,"*Int. J. Chem. Kinet.*, **14,** 269 (1982).

Barbaray, B., J.-P. Contour, and G. Mouvier, "Sulfur Dioxide Oxidation Over Atmospheric Aerosol–X-Ray Photoelectron Spectra of Sulfur Dixoide Adsorbed on V_2O_5 and Carbon," *Atmos. Environ.*, **11,** 351 (1977).

Barbaray, B., J.-P. Contour, and G. Mouvier, "Effects of Nitrogen Dioxide and Water Vapor on Oxidation of Sulfur Dioxide Over V_2O_5 Particles," *Environ. Sci. Technol.*, **12,** 1294 (1978).

Bassett, H. and W. G. Parker, "The Oxidation of Sulphurous Acid,"*J. Chem. Soc.*, 1540 (1951).

Benner, W. H., R. Brodzinsky, and T. Novakov, "Oxidation of SO_2 in Droplets Which Contain Soot Particles," *Atmos. Environ.*, **16,** 1333 (1982).

Benner, W. H., P. M. McKinney, and T. Novakov, "Oxidation of SO_2 in Fog Droplets by Primary Oxidants," *Atmos. Environ.*, **19,** 1377 (1985).

Benson, S. W., "Thermonchemistry and Kinetics of Sulfur-Containing Molecules and Radicals," *Chem. Rev.*, **78,** 23 (1978).

Bielski, B. H. J., "Reevaluation of the Spectral and Kinetic Properties of HO_2 and O_2^- Free Radicals," *Photochem. Photobiol.*, **28,** 645 (1978).

Bielski, B. H. J., D. E. Cabelli, R. L. Arudi, and A. B. Ross, "Reactivity of HO_2/O_2^- Radicals in Aqueous Solution," *J. Phys. Chem. Ref. Data*, **14,** 1041 (1985).

Blair, E. W. and W. Ledbury, "Partial Formaldehyde Vapor Pressures of Aqueous Solutions of Formaldehyde, I," *J. Chem. Soc.*, **127,** 26 (1925).

Boyce, S. D. and M. R. Hoffmann, "Kinetics and Mechanism of the Formation of Hydroxymethanesulfonic Acid at Low pH," *J. Phys. Chem.*, **88,** 4740 (1984).

Boyce, S. D., M. R. Hoffmann, P.A. Hong, and L. M. Moberly, "Catalysis of the Autooxidation of Aquated Sulfur Dioxide by Homogeneous Metal-Phthalocyanine Complexes," *Environ. Sci. Technol.*, **17,** 602 (1983).

Breeding, R. J., H. B. Klonis, J. P. Lodge, Jr., J. B. Pate, D. C. Sheesley, T. R. Englert, and D. R. Sears, "Measurements of Atmospheric Pollutants in the St. Louis Area," *Atmos. Environ.*, **10,** 181 (1976).

Brewer, R. L., R. J. Gordon, L. S. Shepard, and E. C. Ellis, "Chemistry of Mist and Fog from the Los Angeles Area," *Atmos. Environ.*, **17,** 2267 (1983).

Briner, E. and E. Perrottet, "Détermination des Solubilités de l'Ozone dans l'Eau et dans Une Solution Aqueuse de Chlorure de Sodium: Calcul des Solubilités de l'Ozone Atmosphérique dans les Eaux," *Helv. Chim. Acta*, **22,** 397 (1939).

Britton, L. G. and A. G. Clarke, "Heterogeneous Reactions of Sulfur Dioxide and SO_2/NO_2 Mixtures with a Carbon Soot Aerosol," *Atmos. Environ.*, **14,** 829 (1980).

Brock, J. R. and J. L. Durham, Aqueous Aerosol as a Reactant in the Atmosphere, in *SO_2, NO, and NO_2 Oxidation Mechanisms: Atmospheric Considerations*, J. G. Calvert, Ed., Acid Precipitation Series, J. I. Teasley, Series Ed., Butterworth, Boston, 1984, pp. 209–249.

Brodzinsky, R., S. G. Chang, S. S. Markowitz, and T. Novakov, "Kinetics and Mechanism for the Catalytic Oxidation of Sulfur Dioxide on Carbon in Aqueous Suspensions," *J. Phys. Chem.*, **84,** 3354 (1980).

Buschmann, H.-J., H.-H. Füldner, and W. Knoche, "The Reversible Hydration of Carbonyl Compounds in Aqueous Solution. Part I. The Keto/Gem-Diol Equilibrium," *Ber. Bunsenges. Phys. Chem.*, **85,** 41 (1980).

Buschmann, H.-J., E. Dutkiewicz, and W. Knocke, "The Reversible Hydration of Carbonyl Compounds in Aqueous Solution. Part II. The Kinetics of the Keto/Gem-Diol Transition," *Ber. Bunsenges. Phys. Chem.*, **86,** 129 (1982).

Calvert, J. G., "Photoactivated Surface Reactions," in Air Pollution Foundation Report No. 15, November 1956, pp. 91–112.

Calvert, J. G., Ed., *SO_2, NO and NO_2 Oxidation Mechanisms: Atmospheric Considerations*, Acid Precipitation Series, J. I. Teasley, Series Ed., Butterworth, Boston, 1984, pp. 1–62.

Calvert, J. G. and W. R. Stockwell, "Acid Generation in the Troposphere by Gas-Phase Chemistry," *Environ. Sci. Technol.*, **17,** 428A (1983).

Calvert, J. G. and W. R. Stockwell, The Mechanisms and Rates of the Gas Phase Oxidations of Sulfur Dioxide and the Nitrogen Oxides in the Atmosphere, in *SO_2, NO, and NO_2 Oxidation Mechanisms: Atmospheric Considerations*, Acid Precipitation Series, J. I. Teasley, Series Ed., Butterworth, Boston, 1984, pp. 1–62.

Cass, G. R. and F. H. Shair, "Sulfate Accumulation in a Sea Breeze/Land Breeze Circulation System," *J. Geophys. Res.*, **89**(D1), 1429 (1984).

Castleman, A. W., Jr., R. E. Davis, H. R. Munkelwitz, I. N., Tang, and W. P. Wood, " Kinetics of Association Reactions Pertaining to H_2SO_4 Aerosol Formation," *Int. J. Chem. Kinet. Symp. 1*, 629 (1975).

Chamberlain, J., H. Foley, D. Hammer, G. MacDonald, O. Rothaus, and M. Ruderman, " The Physics and Chemistry of Acid Precipitation," Technical Report NO. JSR-81-25, Prepared for the U.S. Department of Energy, November 1981.

Chameides, W. L., "The Photochemistry of a Remote Marine Stratiform Cloud," *J. Geophys. Res.*, **89,** 4739 (1984).

Chameides, W. and D. D. Davis, "The Free Radical Chemistry of Cloud Droplets and Its Impact Upon the Composition of Rain," *J. Geophys. Res.*, **87,** 4863 (1982).

Chameides, W. L. and D. D. Davis, The Coupled Gas-Phase/Aqueous-Phase Free Radical Chemistry of a Cloud, in *Precipitation Scavenging, Dry Deposition, and Resuspension*, H. R. Pruppacher, R. G. Semonin, and W. G. N. Slinn, Eds,. Elsevier, New York, 1983a, pp. 431–443.

Chameides, W. L. and D. D. Davis, " Aqueous-Phase Source of Formic Acid in Clouds,"*Nature*, **304,** 427, (1983b).

Chang, S.-G. and T. Novakov, "Role of Carbon Particles in Atmospheric Chemistry," *Adv. Environ. Sci. Technol.*, **12,** 191 (1983).

Chang, S. G., R. Toossi, and T. Novakov, "The Importance of Soot Particles and Nitrous Acid in Oxidizing SO_2 in Atmospheric Aqueous Droplets," *Atmos. Environ.*, **15,** 1287 (1981).

Chang, T. Y., "Rain and Snow Scavenging of HNO_3 Vapor in the Atmosphere," *Atmos. Environ.*, **18,** 191 (1984).

Chang, T. Y., J. M. Norbeck, and B. Weinstock, "An Estimate of the NO_x Removal Rate in an Urban Atmosphere," *Environ. Sci. Technol.*, **13,** 1534 (1979).

Charlson, R. J. and H. Rodhe, "Factors Controlling the Acidity of Natural Rainwater," *Nature*, **295,** 683 (1982).

Chen, T. S. and P. L. Moore Plummer, "Ab Initio Investigation of the Gas-Phase Reaction $SO_3 + H_2O \rightarrow H_2SO_4$," *J. Phys. Chem.*, **89,** 3689 (1985).

Clarke, A. G. and P. T. Williams, "The Oxidation of Sulfur Dioxide in Electrolyte Droplets," *Atmos. Environ.*, **17,** 607 (1983).

Clarke, A. G. and M. Radojevic, "Chloride Ion Effects on the Aqueous Oxidation of SO_2," *Atmos. Environ.*, **17,** 617 (1983).

Cofer, W. R., III, D. R. Schryer, and R. S. Rogowski, "The Oxidation of SO_2 on Carbon Particles in the Presence of O_3, NO_2 and N_2O," *Atmos. Environ.*, **15,** 1281 (1981).

Cofer, W. R., III, D. R. Schryer, and R. S. Rogowski, "Oxidation of SO_2 by NO_2 and O_3 on Carbon: Implications to Tropospheric Chemistry," *Atmos. Environ.*, **18,** 243 (1984).

Cooper, W. J. and R. G. Zika, "Photochemical Formation of Hydrogen Peroxide in Surface and Ground Waters Exposed to Sunlight," *Science*, **220,** 711 (1983).

Cowling, E. B., "Acid Precipitation in Historical Perspective," *Environ. Sci. Technol.*, **16,** 110A (1982).

Cox, R. A. and S. A. Penkett, "Oxidation of SO_2 by Oxidants Formed in the Ozone–Olefin Reaction," *Nature*, **230,** 321 (1971).

Cox, R. A. and S. A. Penkett, "Aerosol Formation from Sulphur Dioxide in the Presence of Ozone and Olefinic Hydrocarbons," *J. Chem. Soc. Faraday Trans. I*, **68,** 1735 (1972).

Cullis, C. F. and M. M. Hirschler, "Atmospheric Sulfur: Natural and Man-Made Sources," *Atmos. Environ.*, **14,** 1263 (1980).

Dacey, J. W. H., S. G. Wakeham, and B. L. Howes, "Henry's Law Constants for Dimethylsulfide in Freshwater and Seawater," *Geophys. Res. Lett.*, **11,** 991 (1984).

Damschen, D. E. and L. R. Martin, "Aqueous Aerosol Oxidation of Nitrous Acid by O_2, O_3, and H_2O_2," *Atmos. Environ.*, **17,** 2005 (1983).

Dasch, J. M., "Direct Measurement of Dry Deposition to a Polyethylene Bucket and Various Surrogate Surfaces," *Environ. Sci. Technol.*, **19,** 721 (1985).

Daum, P. H., S. E. Schwartz, and L. Newman, Studies of the Gas- and Aqueous-Phase Composition of Stratiform Clouds, in *Precipitation Scavenging, Dry Deposition, and Resuspension*, Vol. 2, *Precipitation Scavenging*, H. R. Pruppacher, R. G. Semonin, and W. G. N. Slinn, Eds., Elsevier, New York, 1983, pp. 31–44.

Davies, T. D. and J. R. Mitchell, Dry Deposition of Sulphur Dioxide onto Grass in Rural Eastern England with Some Comparisons with Other Forms of Sulphur Deposition, in *Precipitation Scavenging, Dry Deposition, and Resuspension*, Vol. 2, *Dry Deposition and Resuspension*, H. R. Pruppacher, R. G. Semonin, and W. G. Slinn, Eds., Elsevier, New York, 1983, pp. 795–804.

Davis, C. S. and R. G. Wright, "Sulfur Dioxide Deposition Velocity by a Concentration Gradient Measurement System," *J. Geophys. Res.*, **90,** 2091 (1985).

Davis, D. D., A. R. Ravishankara, and S. Fischer, "SO_2 Oxidation Via the Hydroxyl Radical: Atmospheric Fate of HSO_x Radicals," *Geophys. Res. Lett.*, **6,** 113 (1979).

D'Itri, F. M., Ed., *Acid Precipitation: Effects on Ecological Systems*, Ann Arbor Science Publishers, Ann Arbor, MI, 1982.

Dlugi, R. and H. Güsten, "The Catalytic and Photocatalytic Activity of Coal Fly Ashes," *Atmos. Environ.*, **17,** 1765 (1983).

Draper, W. M. and D. G. Crosby, "The Photochemical Generation of Hydrogen Peroxide in Natural Waters," *Arch. Environ. Contam. Toxicol.*, **12,** 121 (1983).

Droppo, J. G., Jr., " Concurrent Measurements of Ozone Dry Deposition Using Eddy Correlation and Profile Flux Methods," *J. Geophys. Res.*, **90,** 2111 (1985).

Durham, J. L., Ed., *Chemistry of Particles, Fogs, and Rain*, Acid Precipitation Series, J. I. Teasley, Series Ed., Butterworth, Boston, 1984.

Eatough, D. J. and L. D. Hansen, "Organic and Inorganic S(IV) Compounds in Airborne Particulate Matter," *Adv. Environ. Sci. Technol.*, **12,** 221 (1983).

Eatough, D. J., R. J. Arthur, N. L. Eatough, M. W. Hill, N. F. Mangelson, B. E. Richter, L. D. Hansen, and J. A. Cooper, "Rapid Conversion of $SO_{2(g)}$ to Sulfate in a Fog Bank," *Environ. Sci. Technol.*, **18,** 855 (1984).

Eldred, R. A., L. L. Ashbaugh, T. A. Cahill, R. G. Flocchini, and M. L. Pitchford, "Sulfate Levels in the Southwest During the 1980 Copper Smelter Strike," *J. Air Pollut. Control Assoc.*, **33,** 110 (1983).

Ember, L. R., "Acid Pollutants: Hitchhikers Ride the Wind," *Chem. Eng. News*, September 14, 1981, pp. 20–31.

Ember, L., "Rain Sulfates in West Linked to Sulfur Emissions," *Chem. Eng. News.*, September 2, 1985, pp. 18–19.

Faust, B. C. and M. R. Hoffmann, "Photo-Induced Reductive Dissolution of Hematite (α-Fe_2O_3) by S(IV) Oxyanions," *Environ. Sci. Technol.*, submitted for publication (1985).

Feely, H. W., D. C. Bogen, S. J. Nagourney, and C. C. Torquato, "Rates of Dry Deposition Determined Using Wet/Dry Collectors," *J. Geophys. Res.*, **90,** 2161 (1985).

Fieser, L. F. and M. Fieser, *Advanced Organic Chemistry*, Reinhold, New York, 1961, p. 697.

Finlayson-Pitts, B. J. and J. N. Pitts, Jr., "An Assessment of the Atmospheric Chemistry of Oxides of Sulfur and Nitrogen: Acid Deposition and Ozone," Report to the Office of Technology Assessment of the U.S. Congress, November 1982.

Finlayson-Pitts, B. J. and J. N. Pitts, Jr., Atmospheric Processes (Appendix C), in *Acid Rain and Transported Air Pollutants, Implications for Public Policy*, U.S. Congress, Office of Technology Assessment, OTA-0-204, Washington, DC, June 1984.

Forrest, J., R. W. Garber, and L. Newman, "Conversion Rates in Power Plant Plumes Based on Filter Pack Data: The Coal-Fired Cumberland Plume," *Atmos. Environ.*, **15,** 2273 (1981).

Frank, S. N. and A. J. Bard, "Heterogeneous Photocatalytic Oxidation of Cyanide and Sulfite in Aqueous Solutions at Semiconductor Powers," *J. Phys. Chem.*, **81,** 1484 (1977).

Freiberg, J. E. and S. E. Schwartz, "Oxidation of SO_2 in Aqueous Droplets: Mass Transport Limitation in Laboratory Studies and the Ambient Atmosphere," *Atmos. Environ.*, **15,** 1145 (1981).

Friend, J. P., R. A. Barnes, and R. M. Vasta, "Nucleation by Free Radicals from the Photooxidation of Sulfur Dioxide in Air," *J. Phys. Chem.*, **84,** 2423 (1980).

Fuzzi, S., R. A. Castillo, J. E. Jiusto, and G. G. Lala, "Chemical Composition of Radiation Fog Water at Albany, New York, and Its Relationship to Fog Microphysics," *J. Geophys. Res.*, **89,** 7159 (1984).

Garland, J. A., "Dry and Wet Removal of Sulphur from the Atmosphere," *Atmos. Environ.*, **12,** 349 (1978).

Gertler, A. W., D. F. Miller, D. Lamb, and U. Katz, Studies of Sulfur Dioxide and Nitrogen Dioxide Reactions in Haze and Cloud, in *Chemistry of Particles, Fogs, and Rain*, J. L. Durham, Ed.,

Acid Precipitation Series, Vol. 2, J. I. Teasley, Series Ed., Butterworth, Boston, 1984, pp. 131–160.

Gillani, N. V., "Project MISTT: Mesoscale Plume Modeling of the Dispersion, Transformation and Ground Removal of SO_2," *Atmos. Environ.*, **12,** 569 (1978).

Gillani, N. V. and W. E. Wilson, "Gas-to-Particle Conversion of Sulfur in Power Plant Plumes. II. Observations of Liquid-Phase Conversions," *Atmos. Environ.*, **17,** 1739 (1983).

Gillani, N. V., S. Kohli, and W. E. Wilson, "Gas-to-Particle Conversion of Sulfur in Power Plant Plumes. I. Parameterization of the Conversion Rate for Dry, Moderately Polluted Ambient Conditions," *Atmos. Environ.*, **15,** 2293 (1981).

Gillani, N. V., J. A. Colby, and W. E. Wilson, "Gas-to Particle Conversion of Sulfur in Power Plant Plumes. III. Parameterization of Plume-Cloud Interactions." *Atmos. Environ.*, **17,** 1753 (1983)

Graedel, T. E. and C. J. Weschler, " Chemistry within Aqueous Atmospheric Aerosols and Raindrops," *Rev. Geophys. Space Phys.*, **19,** 505 (1981).

Graedel, T. E. and K. I. Goldberg, "Kinetic Studies of Raindrop Chemistry. I. Inorganic and Organic Processes," *J. Geophys. Res.*, **88,** 10,865 (1983).

Graedel, T. E., C. J. Weschler, and M. L. Mandlich, "The Influence of Transition Metal Complexes on Atmospheric Droplet Acidity," *Nature*, **317,** 240 (1985a).

Graedel, T. E., M. L. Mandlich, and C. J. Weschler, "Kinetic Model Studies of Atmospheric Droplet Chemistry 2. Homogeneous Transition Metal Chemistry in Raindrops," *J. Geophys. Res.*, **90,** in press (1985b).

Granett, A. L. and R. C. Musselman, "Simulated Acidic Fog Injures Lettuce," *Atmos. Environ.*, **18,** 887 (1984).

Grosjean, D., "Formaldehyde and Other Carbonyls in Los Angeles Ambient Air," *Environ. Sci. Technol.*, **16,** 254 (1982).

Grosjean, D. and B. Wright, "Carbonyls in Urban Fog, Ice Fog, Cloudwater, and Rainwater," *Atmos. Environ.*, **17,** 2093 (1983).

Hampson, R. F., Jr. and D. Garvin, Eds., *Reaction Rate and Photochemical Data for Atmospheric Chemistry — 1977,* NBS Spec. Publ. 513, May 1978.

Hashimoto, S., G. Inoue, and H. Akimoto, "Infrared Spectroscopic Detection of the $HOSO_2$ Radical in Argon Matrix at 11°K," *Chem. Phys. Lett.*, **107,** 198 (1984).

Hatakeyama, S., H. Kobayashi, and H. Akimoto, " Gas-Phase Oxidation of SO_2 in the Ozone–Olefin Reactions," *J. Phys. Chem.*, **88,** 4736 (1984).

Hegg, D. A. and P. V. Hobbs, "Oxidation of Sulfur Dioxide in Aqueous Systems with Particular Reference to the Atmosphere," *Atmos. Environ.*, **12,** 241 (1978).

Hegg, D. A. and P. V. Hobbs, " Some Observations of Particulate Nitrate Concentrations in Coal-Fired Power Plant Plumes," *Atmos. Environ.*, **13,** 1715 (1979).

Hegg, D. A. and P. V. Hobbs, "Measurements of Sulfate Production in Natural Clouds," *Atmos. Environ.*, **16**, 2663 (1982); *ibid*, **17,** 2632 (1983).

Heikes, B. G., "Aqueous H_2O_2 Production from O_3 in Glass Impingers," *Atmos. Environ.*, **18,** 1433 (1984).

Heikes, B. G. and A. M. Thompson, "Effects of Heterogeneous Processes on NO_3, HONO, and HNO_3 Chemistry in the Troposphere," *J. Geophys. Res.*, **88,** 10,883 (1983).

Heikes, B. G., A. L. Lazrus, G. L. Kok, S. M. Kunen, B. W. Gandrud, S. N. Gitlin, and P. D. Sperry, "Evidence for Aqueous Phase Hydrogen Peroxide Synthesis in the Troposphere," *J. Geophys. Res.*, **87**(C4), 3045 (1982).

Hering, S. V. and S. K. Friedlander, " Origins of Aerosol Sulfur Size Distributions in the Los Angeles Basin," *Atmos. Environ.*, **16,** 2647 (1982).

Herron, J. T., R. I. Martinez, and R. E. Huie, "Kinetics and Energetics of the Criegee Intermediate in the Gas Phase. I. The Criegee Intermediate in Ozone–Alkene Reactions," *Int. J. Chem. Kinet.*, **14,** 201 (1982).

Hicks, B. B. and J. A. Garland, Overview and Suggestions for Future Research on Dry Deposition, in *Precipitation Scavenging, Dry Deposition, and Resuspension*, Vol. 2., *Dry Deposition and Resuspension*, H. R. Pruppacher, R. G. Semonin, and W. G. Slinn, Eds., Elsevier, New York, 1983, pp. 1429–1433.

Hicks, B. B., M. L. Wesely, R. L. Coulter, R. L. Hart, J. L. Durham, R. E. Speer, and D. H. Stedman, An Experimental Study of Sulfur Deposition to Grassland, in *Precipitation Scavenging, Dry Deposition, and Resuspension*, Vol. 2, *Dry Deposition and Resuspension*, H. R. Pruppacher, R. G. Semonin and W. G. Slinn, Eds., Elsevier, New York, 1983, pp. 933–942.

Hoffmann, M. R., "Response to Comment on 'Acid Fog,' " *Environ. Sci. Technol.*, **18,** 61 (1984).

Hoffmann, M. R. and S. D. Boyce, "Catalytic Autooxidation of Aqueous Sulfur Dioxide in Relationship to Atmospheric Systems," *Adv. Environ. Sci. Technol.*, **12,** 147 (1983).

Hoffmann, M. R. and D. J. Jacob, Kinetics and Mechanisms of Catalytic Oxidation of Dissolved Sulfur Dioxide in Aqueous Solution: An Application to Nighttime Fog Water Chemistry, in *SO_2, NO, and NO_2 Oxidation Mechanisms: Atmospheric Considerations*, Acid Precipitation Series, Vol. 3., J. I. Teasley, Series Ed., Butterworth, Boston, 1984, pp. 101–172.

Hoigné, J. and H. Bader, "Ozonation of Water: Role of Hydroxyl Radicals as Oxidizing Intermediates," *Science*, **190,** 782 (1975).

Hoigné, J. and H. Bader, "Role of Hydroxyl Radical Reactions in Ozonization Processes in Aqueous Solutions," *Atmos. Environ.*, **10,** 377 (1976).

Holdren, M. W., G. F. Ward, G. W. Keigley, and C. W. Spicer, "Preliminary Investigation of the Effects of Peroxyacetyl Nitrate Precipitation Chemistry," Battelle Pacific Northwest Laboratories, Seattle, 1982.

Holdren, M. W., C. W. Spicer, and J. M. Hales, "Peroxyacetyl Nitrate Solubility and Decomposition Rate in Acidic Water," *Atmos. Environ.*, **18,** 1171 (1984).

Huebert, B. J. and C. H. Robert, "The Dry Deposition of Nitric Acid to Grass" *J. Geophys. Res.*, **90,** 2085 (1985).

Huie, R. E. and N. C. Peterson, "Reaction of Sulfur (IV) with Transition-Metal Ions in Aqueous Solutions," *Adv. Environ. Sci. Technol.*, **12,** 117 (1983).

Hulett, L. D., T. A. Carlson, B. R. Fish, and J. L Durham, Studies of Sulfur Compounds Adsorbed on Smoke Particles and Other Solids by Photoelectron Spectroscopy, in *Determination of Air Quality*, G. Mamantov and W. D. Shults, Eds., Plenum Press, New York, 1972, pp. 179–187.

Husar, R. B., J. P. Lodge, Jr., and D. J. Moore, Eds., "Sulfur in the Atmosphere," Proceedings of the International Symposium, Durbrovnik, Yugoslavia, September 7–14, 1977, *Atmos. Environ.*, **12**(1-3) (1978a).

Husar, R. B., D. E. Patterson, J. D. Husar, N. V. Gillani, and W. E. Wilson, "Sulfur Budget of a Power Plant Plume," *Atmos. Environ.*, **12,** 549 (1978b).

Hutchinson, T. C. and M. Havas, Eds., Effects of Acid Precipitation on Terrestial Ecosystems, in *Proceedings of the NATO Conference on Effects of Acid Precipitation on Vegetation and Soils*, Toronto, Ontario, Canada, May 21–27 1978, Plenum Press, New York, 1980.

Hwang, H. and P. K. Dasgupta, "Thermodynamics of the Hydrogen Peroxide–Water System." *Environ Sci. Technol.*, **19,** 255 (1985).

Ibusuki, T. and H. M. Barnes, "Manganese (II) Catalyzed Sulfur Dioxide Oxidation in Aqueous Solution at Environmental Concentrations," *Atmos. Environ.*, **18,** 145 (1984).

Jacob, D. J. and M. R. Hoffmann, "A Dynamic Model for the Production of H^+, NO_3^-, and SO_4^{2-} in Urban Fog," *J. Geophys. Res.*, **88C,** 6611 (1983).

Jacob, D. J., J. M. Waldman, J. W. Munger, and M. R. Hoffmann, "Chemical Composition of Fogwater Collected along the California Coast," *Environ. Sci. Technol.*, **19,** 730 (1985).

Johnson, A. H. and T. G. Siccama, "Acid Deposition and Forest Decline," *Environ. Sci. Technol.*, **17,** 294A (1983).

Jones, C. L. and J. H. Seinfeld, "The Oxidation of NO_2 to Nitrate—Day and Night," *Atmos. Environ.*, **17,** 2370 (1983).

Journal of Geophysical Research, Vol. 90, No. D1, pp. 2075–2165 (1985).

Judeikis, H. S., T. B. Stewart and A. C. Wren, "Laboratory Studies of Heterogeneous Reactions of SO_2," *Atmos. Environ.*, **12,** 1633 (1978).

Keene, W. C., J. N. Galloway, and J. D. Holden, Jr., "Measurement of Weak Organic Acidity in Precipitation from Remote Areas of the World," *J. Geophys. Res.*, **88C,** 5122 (1983).

Kelly, T. J., P. H. Daum, and S. E. Schwartz, "Measurements of Peroxides in Cloudwater and Rain," *J. Geophys. Res.*, **90,** 7861 (1985).

Kiester, E. Jr., "A Deathly Spell is Hovering Above the Black Forest," *Smithsonian*, **16,** 211 (1985).

Kleinman, M. T., R. F. Phalen, R. Mannix, M. Azizian, and R. Walters, "Influence of Fe and Mn Ions on the Incorporation of Radioactive $^{35}SO_2$ by Sulfate Aerosols," *Atmos. Environ.*, **19,** 607 (1985).

Kluczewski, S. M., K. A. Brown, and J. N. B. Bell, "Deposition of Carbonyl Sulphide to Soils," *Atmos. Environ.*, **19,** 1295 (1985).

Kok, G. L., T. P. Holler, M. B. Lopez, H. A. Nachtrieb, and M. Yuan, "Chemiluminescent Method for Determination of Hydrogen Peroxide in the Ambient Atmosphere," *Environ. Sci. Technol.*, **12,** 1072 (1978).

Kunen, S. M., A. L. Lazrus, G. L. Kok, and B. G. Heikes, "Aqueous Oxidation of SO_2 by Hydrogen Peroxide," *J. Geophys. Res.*, **88,** 3671 (1983).

Landolt-Börnstein, *Zahlenwerte und Funktionen. Gleichgewicht der Absorption von Gasen in Flüssigkeiten von niedrig en Dampfdruck,* 6th ed., Vol. 4, Part 4, Sec. C, Springer-Verlag, Heidelberg, Germany, 1976.

Lazrus, A. L., P. L. Haagenson, G. L. Kok, B. J. Huebert, C. W. Kreitzberg, G. E. Likens, V. A. Mohnen, W. E. Wilson, and J. W. Winchester, "Acidity in Air and Water in a Case of Warm Frontal Precipitation," *Atmos. Environ.*, **17,** 581 (1983).

Lazrus, A. L., G. L. Kok, S. N. Gitlin, J. A. Lind, and S. E. McLaren, "Automated Fluorometric Method for Hydrogen Peroxide in Atmospheric Precipitation," *Anal. Chem.*, **57,** 917 (1985).

Ledbury, W. and E. W. Blair, 'The Partial Formaldehyde Vapor Pressure of Aqueous Solutions of Formaldehyde. Part II," *J. Chem. Soc.*, **127,** 2832 (1925).

Lee, Y.-N., "Atmospheric Aqueous-Phase Reactions of Nitrogen Species," Paper Presented at the Conference on Gas–Liquid Chemistry of Natural Waters, Brookhaven National Laboratory, Upton, New York, April 1984.

Lee, I.-Y. and J. D. Shannon, "Indications of Nonlinearities in Processes of Wet Deposition," *Atmos. Environ.*, **19,** 143 (1985).

Lee,Y.-N. and S. E. Schwartz, Kinetics of Oxidation of Aqueous Sulfur (IV) by Nitrogen Dioxide, In *Precipitation Scavenging, Dry Deposition and Resuspension*, Vol. 1, H. R. Pruppacher, R. G. Semonin, and W. G. N. Slinn, Eds., Elsevier, New York, 1983. pp. 453–470.

Lee, Y.-N., G. S. Senum, and J. S. Gaffney, "Peroxyacetyl Nitrate (PAN) Stability, Solubility, and Reactivity—Implications for Tropospheric Nitrogen Cycles and Precipitation Chemistry," Paper Presented at the Fifth International Conference of the Commission of Atmospheric Chemistry and Global Pollution Symposium on Tropospheric Chemistry, Oxford, England, August 28–September 2, 1983.

Liberti, A., D. Brocco, And M. Possanzini, "Adsorption and Oxidation of Sulfur Dioxide on Particles," *Atmos. Environ.*, **12,** 225 (1978).

Lind, J. A., and A. L. Lazrus, "Aqueous Phase Oxidation of Sulfur IV by Some Organic Peroxides," *EOS Trans.*, Am. Geophys. Union, **64,** 670 (1983).

Lloyd, A. C., Tropospheric Chemistry of Aldehydes in *Chemical Kinetics Data Needs for Modeling the Lower Troposphere*, J. T. Herron, R. E. Huie, and J. A. Hodgeson, Eds., National Bureau of Standards Spec. Publ. 557, August 1979.

Lonneman, W. A., S. L. Kopczynski, P. E. Darley, and F. D. Sutterfield, "Hydrocarbon Composition of Urban Air Pollution," *Environ. Sci. Technol.*, **8,** 229 (1974).

Loomis, A. G., Solubilities of Gases in Water, in *International Critical Tables*, Vol. III, McGraw-Hill, New York, 1928, pp. 255–261.

Luňák, S. and J. Vepřek-Šiška, "Photochemical Autooxidation of Sulfite Catalyzed by Iron (III) Ions," *Coll. Czech. Chem. Commun.*, **41,** 3495 (1976).

Lusis, M. A., K. G. Anlauf, L. A. Barrie, and H. A. Wiebe, "Plume Chemistry Studies at a Northern Alberta Power Plant," *Atmos. Environ.*, **12** 2429 (1978).

Maahs, H. G., Sulfur-Dioxide/Water Equilibria Between 0° and 50°C: An Examination of Data at Low Concentrations, in *Heterogeneous Atmospheric Chemistry*, D. R. Schryer, Ed., Geophysical Monograph 26, American Geophysical Union, Washington DC, 1982, pp. 187–195.

Maahs, H. G., "Kinetics and Mechanism of the Oxidation of S(IV) by Ozone in Aqueous Solution with Particular Reference to SO_2 Coversion in Nonurban Tropospheric Clouds," *J. Geophys. Res.*, **88,** 10,721 (1983).

Mackay, G. I. and H. I. Schiff, personal communication, (1985).

Margitan, J. J., "Mechanism of the Atmospheric Oxidation of Sulfur Dioxide. Catalysis by Hydroxyl Radicals," *J. Phys. Chem.*, **88,** 3314 (1984).

Martin, L. R., Kinetic Studies of Sulfite Oxidation in Aqueous Solutions, in *SO_2, NO, and NO_2 Oxidation Mechanisms: Atmospheric Consideration,* Acid Precipitation Series, J. I. Teasley, Series Ed., Butterworth, Boston, 1984, pp. 63–100 and references therein.

Martin, L. R. and D. E. Damschen, "Aqueous Oxidation of Sulfur Dioxide by Hydrogen Peroxide at Low pH," *Atmos. Environ.*, **15,** 1615 (1981).

McArdle, J. V. and M. R. Hoffmann, "Kinetics and Mechanism of the Oxidation of Aquated Sulfur Dioxide by Hydrogen Peroxide at Low pH," *J. Phys. Chem.*, **87,** 5425 (1983).

McMahon, T. A. and P. J. Denison, "Empirical Atmospheric Deposition Parameters—A Survey," *Atmos. Environ.*, **13,** 571 (1979).

McMurry, P. H. and J. C. Wilson, "Droplet Phase (Heterogeneous) and Gas Phase (Homogeneous) Contributions to Secondary Ambient Aerosol Formation as Functions of Relative Humidity," *J. Geophys. Res.*, **88C,** 5101 (1983).

McRae, G. J. and A. G. Russell, Dry Deposition of Nitrogen-Containing Species, in *Deposition Both Wet and Dry*, B. B. Hicks, Ed., Chapter 9, pp.153–193, Acid Precipitation Series, J. I. Teasley, Series Ed., Butterworth, Boston, 1984.

Meagher, J. F., K. J. Olszyna, and M. Luria, "The Effect of SO_2 Gas Phase Oxidation on Hydroxyl Smog Chemistry," *Atmos. Environ.*, **18,** 2095 (1984).

Mészáros, E., D. J. Moore, and J. P. Lodge, Jr., "Sulfur Dioxide–Sulfate Relationships in Budapest," *Atmos. Environ.*, **11,** 345 (1977).

Misra, P. K., W. H. Chan, D. Chung, and Al. J. S. Tang, "Scavenging Ratios of Acidic Pollutants and Their Use in Long-Range Transport Models," *Atmos. Environ.*, **19,** 1471 (1985).

Morgan, J. J., Factors Governing the pH Availability of H^+ and Oxidation Capacity of Rain, in *Atmospheric Chemistry*, Springer-Verlag, New York, 1982, pp. 17–40.

Munger, J. W., "Chemistry of Atmospheric Precipitation in the North-Central United States: Influence of Sulfate, Nitrate, Ammonia, and Calcareous Soil Particulates," *Atmos. Environ.*, **16,** 1633 (1982).

Munger, J. W., D. J. Jacob, J. M. Waldman, and M. R. Hoffmann, "Fogwater Chemistry in an Urban Atmosphere," *J. Geophys. Res.*, **88C,** 5109 (1983).

Munger, J. W., D. J. Jacob, and M. Hoffmann, "The Occurrence of Bisulfite–Aldehyde Addition Products in Fog- and Cloudwater," *J. Atmos. Chem.*, **1,** 335 (1984).

National Center for Atmospheric Research, "The NCAR Eulerian Regional Acid Deposition Model," NCAR Technical Note NCAR/TN-256+STR, June, 1985.

National Research Council, *Atmosphere–Biosphere Interactions: Toward a Better Understanding of the Ecological Consequences of Fossil Fuel Combustion*, National Academy Press, Washington, DC, 1981.

National Research Council, *Acid Deposition: Atmospheric Processes in Eastern North America*, National Academy Press, Washington, DC, 1983.

Newman, L., ''Atmospheric Oxidation of Sulfur Dioxide: A Review as Viewed from Power Plant and Smelter Plume Studies,'' *Atmos. Environ.*, **15,** 2231 (1981).

Niki, H., P. D. Maker, C. M. Savage, and L. P. Breitenbach, ''Fourier Transform IR Spectroscopic Observation of Propylene Ozonide in the Gas Phase Reaction of Ozone–*cis*-2-Butene–Formaldehyde,'' *Chem. Phys. Lett.*, **46,** 327 (1977).

Niki, H., P. D. Maker, C. M. Savage, and L. P. Breitenbach, ''Fourier Transform Infrared Study of the OH Radical Initiated Oxidation of SO_2,'' *J. Phys. Chem.*, **84,** 14 (1980).

Novakov, T., S. G. Chang and A. B. Harker, ''Sulfates as Pollution Particles: Catalytic Formation on Carbon (Soot) Particles,'' *Science*, **186,** 259 (1974).

Okita, T., K. Kaneda, T. Yanaka, and R. Sugai, ''Determination of Gaseous and Particulate Chloride and Fluoride in the Atmosphere,'' *Atmos. Environ.*, **8,** 927, (1974)

Oppenheimer, M., C. B. Epstein, and R. E. Yuhkne, ''Acid Deposition, Smelter Emissions, and the Linearity Issue in the Western United States,'' *Science*, **229,** 859 (1985).

Overrein, L. N., H. M. Seip, and A. Tollan, ''Acid Precipitation—Effects on Forest and Fish,'' Final Report of the SNSF Project, 1972–1980, RECLAMO, Oslo, Norway, 1980.

Penkett, S. A., B. M. R. Jones, K. A. Brice, and A. E. J. Eggleton, ''The Importance of Atmospheric Ozone Hydrogen Peroxide in Oxidizing Sulfur Dioxide in Cloud and Rainwater,'' *Atmos. Environ.*, **13,** 123 (1979).

Perry, J. M., *Chemical Engineers Handbook*, 4th ed., McGraw-Hill, New York, 1963.

Platt, U., D. Perner, A. M. Winer, G. W. Harris, and J. N. Pitts Jr., ''Detection of NO_3 in the Polluted Troposphere by Differential Optical Absorption,'' *Geophys. Res. Lett.*, **7,** 89 (1980).

Platt, U., A. M. Winer, H. Biermann, R. Atkinson, and J. N. Pitts, Jr., ''Measurement of Nitrate Radical Concentrations in Continental Air,'' *Environ. Sci. Technol.*, **18,** 365 (1984).

Pruppacher, H. R. and J. D. Klett, *Microphysics of Clouds and Precipitation, D. Reidel Publishing.*, Dordrecht, Holland, 1978.

Pruppacher, H. R., R. G. Semonin, and W. G. N. Slinn, *Precipitation Scavenging, Dry Deposition, and Resuspension,* Vol. 2, *Dry Deposition and Resuspension*, Proceedings of the Fourth International Conference, Santa Monica, California, November 29–December 3, 1982, Elsevier, New York, 1983.

Rhee, J.-S. and P. K. Dasgupta, ''The Second Dissociation Constant of $SO_2 \cdot H_2O$,'' *J. Phys. Chem.*, **89,** 1799 (1985).

Richards, L. W., ''Comments on 'The Oxidation of NO_2 to Nitrate—Day and Night,' '' *Atmos. Environ.*, **17,** 397 (1983).

Richards, L. W., J. A. Anderson, D. L. Blumenthal, A. A. Brandt, J. A. McDonald, N. Watus, E. S. Macias, and P. S. Bhardwaja, ''The Chemistry, Aerosol Physics, and Optical Properties of a Western Coal-Fired Power Plant Plume,'' *Atmos. Environ.*, **15,** 2111 (1981).

Richards, L. W., J. A. Anderson, D. L. Blumenthal, J. A. McDonald, G. L. Kok, and A. L. Lazrus, ''Hydrogen Peroxide and Sulfur (IV) in Los Angeles Cloud Water,'' *Atmos. Environ.*, **17,** 911 (1983).

Roberts, J. D., R. Stewart, and M. C. Caserio, *Organic Chemistry*, W. A. Benjamin, Menlo Park, CA 1971.

Rodhe, H., P. Crutzen, and A. Vanderpol ''Formation of Sulfuric and Nitric Acid in the Atmosphere During Long Range Transport,'' *Tellus*, **33,** 132 (1981).

Roedel, W., ''Measurements of Sulfuric Acid Saturation Vapor Pressure: Implications for Aerosol Formation by Heteromolecular Nucleation,'' *J. Aerosol Sci.*, **10,** 375 (1979).

Russell, A. G., G. J. McRae, and G. R. Cass, 'Mathematical Modeling of the Formation and Transport of Ammonium Nitrate Aerosol,'' *Atmos. Environ.*, **17,** 949 (1983).

Russell, A. G., G. J. McRae, and G. R. Cass, Acid Deposition of Photochemical Oxidation Products—

A Study Using a Lagrangian Trajectory Model, in *Air Pollution Modeling and Its Application III*, C. De Wispelaere, Ed., Plenem Press New York, 1984, pp. 539–564.

Russell, A. G., G. J. McRae, and G. R. Cass, "The Dynamics of Nitric Acid Production and the Fate of Nitrogen Oxides," *Atmos. Environ.*, **19,** 893 (1985).

Schecker, H. G. and G. Schulz, 'Untersuchungen zur Hydratationskinetik von Formaldehyd in wä-Briger Lösung," *Z. Phys. Chem.*, **65,** 221 (1969).

Schryer, D. R., Ed., *Heterogeneous Atmospheric Chemistry*, Geophysical Monograph 26, American Geophysical Union, Washington, DC, 1982.

Schryer, D. R., R. S. Rogowski, and W. R. Cofer III, "The Reaction of Nitrogen Oxides with SO_2 in Aqueous Aerosols," *Atmos. Environ.*, **17,** 666 (1983).

Schutt, P. and E. B. Cowling, "Waldsterben, A General Decline of Forests in Central Europe: Symptoms, Development and Possible Causes," *Plant Disease*, **69,** 548 (1985).

Scire, J. S. and A. Venkatram, "The Contribution of In-Cloud Oxidation of SO_2 to Wet Scavenging of Sulfur in Convective Clouds," *Atmos. Environ.*, **19,** 637 (1985).

Schwartz, S. E., Gas–Aqueous Reactions of Sulfur and Nitrogen Oxides in Liquid–Water Clouds, in *SO_2, NO, and NO_2 Oxidation Mechanism: Atmospheric Considerations*, J. G. Calvert, Ed., Acid Precipitation Series, Vol 3, J. I. Teasley, Ed., Butterworth, Boston, 1984a, pp. 173–208, and reference therein.

Schwartz, S. E., "Gas- and Aqueous-Phase Chemistry of HO_2 in Liquid Water Clouds," *J. Geophys Res.*, **89,** 11,589 (1984b).

Schwartz, S. E. and J. E. Freiberg, 'Mass-Transport Limitation to the Rate of Reaction of Gases in Liquid Droplets: Application to Oxidation of SO_2 in Aqueous Solutions," *Atmos. Environ.*, **15,** 1129 (1981).

Schwartz, S. E. and W. H. White, "Solubility Equilibria of the Nitrogen Oxides and Oxyacids in Dilute Aqueous Solution," *Adv. Environ. Sci. Eng.*, **4,** 1 (1981).

Schwartz, S. E. and W. H. White, "Kinetics of Reactive Dissolution of Nitrogen Oxides into Aqueous Solution," *Adv. Environ. Sci. Technol.*, **12,** 1 (1983).

Sehested, K., J. Holcman, E. Bjergbakke, and E. J. Hart, "A Pulse Radiolytic Study of the Reaction OH + O_3 in Aqueous Medium," *J. Phys. Chem.*, **88,** 4144 (1984).

Seigneur, C. and P. Saxena, "A Study of Atmospheric Acid Formation in Different Environments," *Atmos. Environ.*, **18,** 2109 (1984).

Seigneur, C., P. Saxena, and P. M. Roth, "Computer Simulations of the Atmospheric Chemistry of Sulfate and Nitrate Formation," *Science*, **225,** 1028 (1984).

Seigneur, C., P. Saxena, and V. A. Mirabella, "Diffusion and Reaction of Pollutants in Stratus Clouds: Application to Nocturnal Acid Formation in Plumes," *Environ. Sci. Technol.*, **19,** 821 (1985).

Seinfeld, J. H., *Atmospheric Chemistry and Physics of Air Pollution*, Wiley, New York, 1986.

Shaw, R. W., "Deposition of Atmospheric Acid from Local and District Sources at a Rural Site in Nova Scotia," *Atmos. Environ.*, **16,** 337 (1982).

Shaw, R. W., The Atmosphere as Delivery Vehicle and Reaction Chamber for Acid Precipitation, in *Meterological Aspects of Acid Rain*, C. M. Bhumralker, Ed., in Acid Precipitation Series, Vol. 1, J. L. Teasley, Series Ed., Butterworth, Boston, 1984, pp. 33–55.

Sievering, H., "Small-Particle Dry Deposition on Natural Waters: Modeling Uncertainty," *J. Geophys. Res.*, **89,** 9679 (1984).

Sillén, G. H. and A. E. Martell, Stability Constants of Metal–Ion Complexes, Spec. Publ. 17, Chemical Society, London, 1964.

Sørensen, P. E. and V. S. Anderson, "The Formaldehyde–Hydrogen Sulphite System in Alkaline Aqueous Solution: Kinetics, Mechanism, and Equilibria," *Acta Chem. Scand.*, **24,** 1301 (1970).

Spicer, C. W., The Rate of NO_x Reactions in Transported Urban Air, in *Studies in Environmental Science*, Vol. 8, M. M. Benarie, Ed., Elsevier, New York, 1980, pp.181–186.

Spicer, C. W., 'The Distribution of Oxidized Nitrogen In Urban Air, ''*Sci. Total Environ.*, **24,** 183 (1982a).

Spicer, C. W., ''Nitrogen Oxide Reactions in the Urban Plume of Boston,'' *Science*, **215,** 1095 (1982b).

Spicer, C. W., G. M. Sverdrup, and M. R. Kuhlman, 'Smog Chamber Studies of NO_x Chemistry in Power Plant Plumes,'' *Atmos. Environ.*, **15,** 2353 (1981).

Spicer, C. W., M. W. Holdren, and G. W. Keigley, ''The Ubiquity of Peroxyacetyl Nitrate in the Continental Boundary Layer,'' *Atmos. Environ.*, **17,** 1055 (1983).

Staehelin, J. and J. Hoigné, ''Decomposition of Ozone in Water: Rate of Initiation by Hydroxide Ions and Hydrogen Peroxide,'' *Environ. Sci. Technol.*, **16,** 676 (1982).

Stedman, D. H. and R. E. Shetter, ''The Global Budget of Atmospheric Nitrogen Species,'' *Adv. Environ. Sci. Technol.*, **12,** 411 (1983).

Stelson, A. W. and J. H. Seinfeld, ''Relative Humidity and Temperature Dependence of the Ammonium Nitrate Dissociation Constant,'' *Atmos. Environ.*, **16,** 983 (1982a).

Stelson, A. W. and J. H. Seinfeld, ''Relative Humidity and pH Dependence of the Vapor Pressure of Ammonium Nitrate–Nitric Acid Solutions at 25°C,'' *Atmos. Environ.*, **16,** 993, (1982b).

Stephens, E. R. and F. R. Burleson, ''Distribution of Light Hydrocarbons in Ambient Air,'' *J. Air Pollut. Control Assoc.*, **19,** 929 (1969).

Stockwell, W. R. and J. G. Calvert, 'The Mechanism of the HO–SO_2 Reaction,'' *Atmos. Environ.*, **17,** 2231 (1983).

Swallow, A. J., ''Hydrated Electrons in Seawater,'' *Nature*, **222,** 369 (1969).

Teasley, J. I., Ed., *Acid Precipitation Series*, Vols. 1–9, Butterworth, Boston, 1984.

Thompson, A. M. and O. C. Zafiriou, ''Air–Sea Fluxes of Transient Atmospheric Species,'' *J. Geophys. Res.*, **88,** 6696 (1983).

Tuazon, E. C., R. Atkinson, C. N. Plum, A. M. Winer, and J. N. Pitts, Jr., ''The Reaction of Gas Phase N_2O_5 with Water Vapor,'' *Geophys. Res. Lett.*, **10,** 953 (1983).

Unsworth, M. H., A. S. Heagle, and W. W. Heck, ''Gas Exchange in Open-Top Field Chambers. I. Measurement and Analysis of Atmospheric Resistances to Gas Exchange,'' *Atmos. Environ.*, **18,** 373 (1984).

Vandenberg, J. J. and K. R. Knoerr, ''Comparison of Surrogate Surface Techniques for Estimation of Sulfate Dry Deposition,'' *Atmos. Environ.*, **19,** 627 (1985).

Van Krevelen, D. W., P. J. Hoftijzer, and F. J. Huntjens, ''Composition and Vapor Pressures of Aqueous Solutions of Ammonia, Carbon Dioxide, and Hydrogen Sulfide,' *Rec. Trav. Chim. Pays-Bas*, **68,** 191, (1949).

Venkatram, A., J. Scire, and J. Pleim, ''ADOM Model Development Program. Vol. 1: Approach to the Evaluation of the ADOM Model,'' Report No. ERT P-B 866-100, 1984.

Waldman, J. M., J. W. Munger, D. J. Jacob, R. C. Flagan, J. J. Morgan, and M. R. Hoffmann, ''Chemical Composition of Acid Fog.'' *Science*, **218,** 677 (1982).

Weschler, C. J., M. L. Mandlich, and T. E. Graedel, ''Speciation, Photosensitivity, and Reactions of Transition Metal Ions in Atmospheric Droplets,'' *J. Geophys. Res.*, **90,** in press (1985).

Wesely, M. L., D. R. Cook, R. L. Hart, and R. E. Speer, ''Measurements and Parameterization of Particulate Sulfur Dry Deposition Over Grass,'' *J. Geophys. Res.*, **90,** 2131 (1985).

Westberg, K., J. F. Bott, J. Coffer, S. Durso, and J. Holloway, ''A Study of the Formation of Sulfuric Acid Aerosols,'' Aerospace Corp. Report No. ATR-85(7899)-1, February 25, 1985.

Yoshizumi, K., K. Aoki, I. Nouchi, T. Okita, T. Kobayashi, S. Kamakura, and M. Tajima, Measurements of the Concentration in Rainwater and of the Henry's Law Constant of Hydrogen Peroxide,'' *Atmos. Environ.*, **18,** 395 (1984).

Young, J. R. and F. W. Lurmann, ''ADOM/TADAP Model Development Program, Vol. 7, Aqueous Phase Chemistry,'' 1984, Environmental Research & Technology, Inc., 975 Business Center Circle, Newbury, Park, CA 91320.

Zafiriou, O. C., Natural Water Photochemistry , in *Chemical Oceanography*, J. P. Riley and R. Chester Eds., Academic Press, London, 1983, Vol. 8, pp. 339–379.

Zafiriou, O. C., J. Joussot-Dubien, R. G. Zepp, and R. G. Zika, ''Photochemistry of Natural Waters,'' *Environ. Sci. Technol.*, **18,** 358A (1984).

Zepp, R. G., P. F. Schlotzhauer, and R. M. Sink, 'Photosensitized Transformations Involving Electronic Energy Transfer in Natural Waters: Role of Humic Substances,'' *Environ. Sci. Technol.*, **19,** 74 (1985).

Zika, R. G., Marine Organic Photochemistry, in *Marine Organic Chemistry*, E. K. Duursma and R. Dawson, Eds., Elsevier Oceanography Series 31, Elsevier, New York, 1981, pp. 299–325.

Zika, R. G. and E. S. Saltzman, ''Interaction of Ozone and Hydrogen Peroxide in Water: Implications for Analysis of H_2O_2 in Air,'' *Geophys. Res. Lett.*, **9,** 231 (1982).

Zika, R., E. Saltzman, W. L. Chameides, and D. D. Davis, ''H_2O_2 Levels in Rainwater Collected in South Florida and the Bahama Islands,'' *J. Geophys. Res.*, **87,** 5015 (1982).

PART 7

Particulate Matter

12 PARTICULATE MATTER IN THE ATMOSPHERE: PRIMARY AND SECONDARY PARTICLES

12-A. SOME DEFINITIONS

Particles, or particulate matter, may be solid or liquid, with diameters between ~0.002 μm and ~100 μm. The lower end of the size range is not sharply defined because there is no accepted criterion at which a cluster of molecules becomes a particle. However, particles with diameters of ~0.002 μm have been measured and this is the smallest size detectable by condensation nuclei counters (see below). The upper end corresponds to the size of fine drizzle or very fine sand; these particles are so large that they quickly fall out of the atmosphere and hence do not remain suspended for significant periods of time. There are, of course, larger particles produced in the atmosphere (e.g., raindrops, ~1 mm and hail, ~1–20 mm), but their rapid fallout precludes, for all practical purposes, their inclusion in the definition of atmospheric particles. As we shall see, the most important particles with respect to atmospheric chemistry and physics are in the 0.002–10 μm range.

Aerosols are defined as relatively stable suspensions of solid or liquid particles in a gas. Thus aerosols differ from particles in that an aerosol includes both the particles and the gas in which they are suspended.

We discuss in this chapter the physical and chemical characteristics of particles found in the troposphere, as well as methods used to measure their concentrations and size distributions. Particles may be either directly emitted into the atmosphere or formed there by chemical reactions; we refer to these as primary and secondary particles, respectively. The relative importance of primary and secondary particles will clearly depend on the phenomena examined, the geographical location with its particular mix of emissions, and on the atmospheric chemistry. For example, in areas with extensive wood burning, much of the particulate matter is primary in nature. In contrast, at the peak of one severe photochemical smog episode, well over half of the particulate mass was attributed to secondary reactions in the atmosphere (Grosjean and Friedlander, 1975).

There are a number of properties of particles which are important to their role in atmospheric processes. These include, in addition to their number concentration, their mass, size, chemical composition, and aerodynamic and optical properties. Of these, size is the most important; it not only reflects the nature of the source of the particles (see below) but also relates to their health effects (Bates et al., 1966) and to their aesthetic and climatic effects via their light scattering properties (Section 12-B-3).

Atmospheric particles are usually referred to as having a radius or a diameter, implying they are spherical. However, many particles in the atmosphere have quite irregular shapes for which geometrical radii and diameters are not meaningful. Some means of expressing the size of such particles is essential since many important properties of the particle such as volume, mass, and settling velocity depend on the size. In practice, the size of such irregularly shaped particles is expressed in terms of some sort of *equivalent* or *effective* diameter which depends on a physical, rather than a geometrical, property.

There are several different types of effective diameters. The most commonly used is the aerodynamic diameter, D_a, which is defined as the diameter of a sphere of unit density (1 g cm^{-3}) which has the same terminal falling speed in air as the particle under consideration. This effective diameter is particularly useful because it can determine residence time in the air, it reflects the various regions of the respiratory system in which particles of different sizes become deposited, and several instruments directly measure this parameter (Section 12-E-3). D_a is given by Eq. (A):

$$D_a = D_g k \sqrt{\frac{\rho_p}{\rho_0}} \tag{A}$$

D_g is the geometric diameter, ρ_p is the density of the particle neglecting the buoyancy effects of air, ρ_0 is the reference density (1 g cm^{-3}), and k is a shape factor which is 1.0 in the case of a sphere. Because of the effect of particle density on the aerodynamic diameter, a spherical particle of high density will have a larger

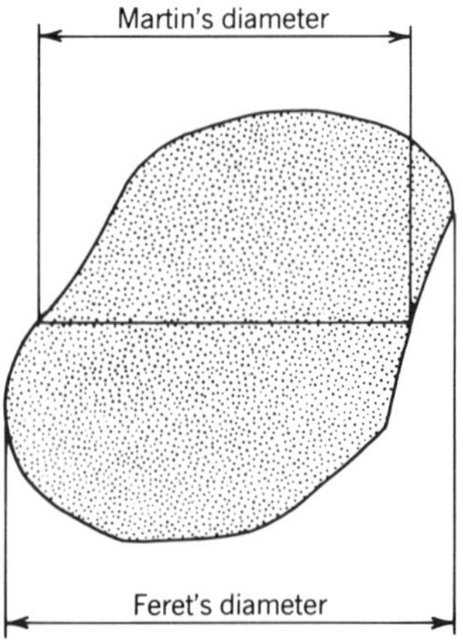

FIGURE 12.1. Illustration of definition of Martin's diameter and Feret's diameter, respectively, for a non-spherical particle (from Herdan, 1960).

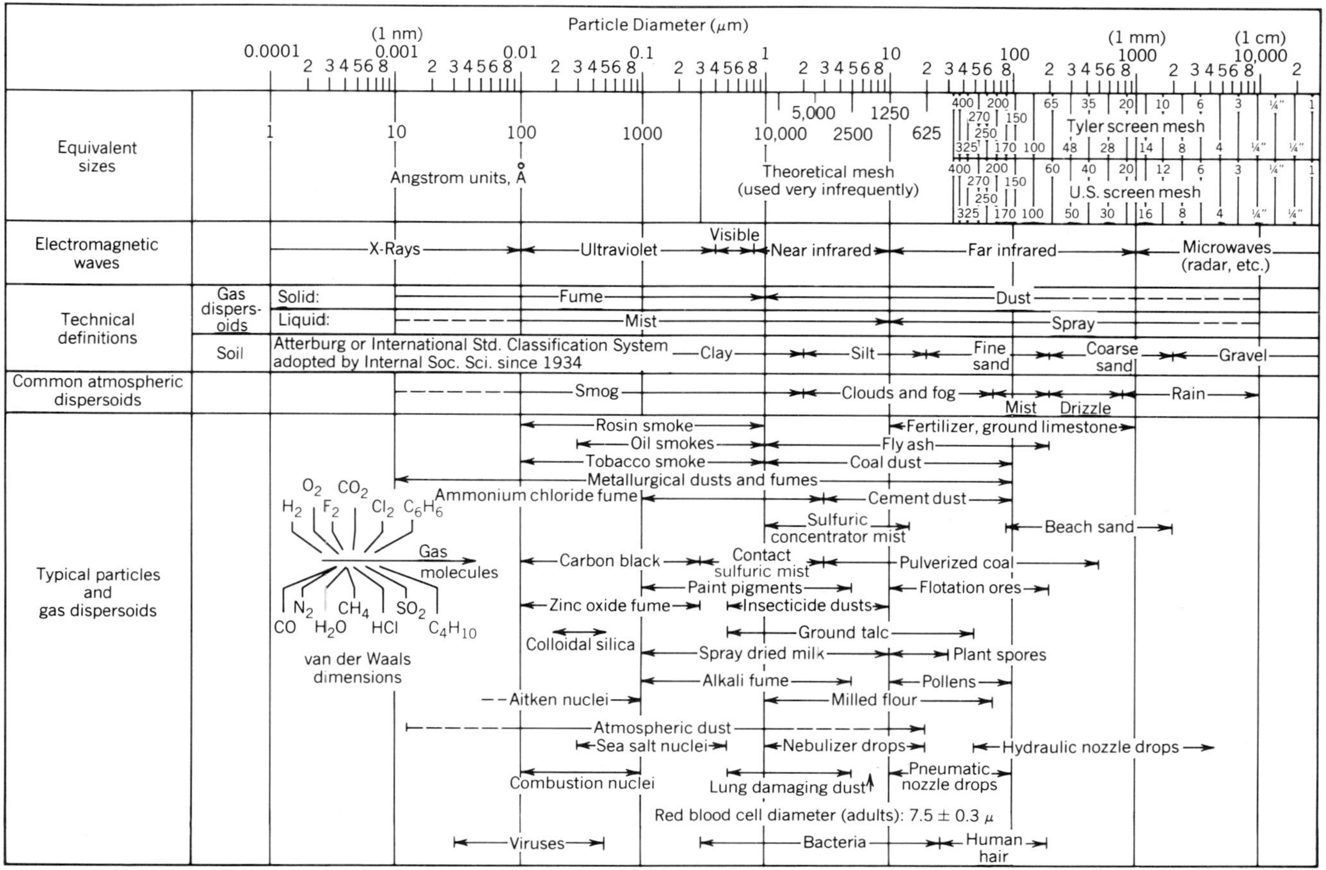

FIGURE 12.2. Some characteristics of particles and aerosols in ambient atmospheres and industrial settings (from Lapple, 1961).

aerodynamic diameter than its geometric diameter. However, for most substances, $\rho_p \lesssim 10$ so that the difference is less than a factor of ~ 3 (Lawrence Berkeley Laboratory, 1979). Particle densities are often lower than bulk densities of pure substances due to voids, pores, and cracks in the particles.

Throughout this chapter, we use the term *diameter* with the understanding that it is the aerodynamic diameter of the particle unless stated otherwise.

Other types of diameters which are are based on the particle dimensions (Hinds, 1982) are not generally as commonly used as the aerodynamic diameter in atmospheric work. For example, the projected area diameter (or projected diameter) is the diameter of a circle with the same area as the profile of a particle viewed normal to its position of greatest stability. Two obsolete (in atmospheric chemistry) diameters based on geometrical–statistical principles are Martin's and Feret's diameters. As shown in Fig. 12.1, Martin's diameter is the length of the line positioned to divide the particle into equal areas, whereas Feret's diameter is the distance between two tangents at the opposite ends of the particle. Martin's and Feret's diameters are known as statistical diameters because, for a collection of particles, the average Martin's or Feret's diameter is assumed to be a statistically adequate measure of the average size of the particles. For a given collection of particles, Martin's diameter is usually slightly less than the projected area diameter, and Feret's is slightly greater.

Figure 12.2 summarizes some of the characteristics of particles and aerosols encountered in both environmental and industrial atmospheres.

This chapter treats those aerosol phenomena which are known or believed to be important in atmospheric chemistry. For treatment of related, but specialized, topics, several good references are available. The classic work on aerosol physics is *The Mechanics of Aerosols* by the late N. A. Fuchs (1964). The chemical engineering aspects of aerosols are emphasized in S. K. Friedlander's book, *Smoke, Dust, and Haze* (1977); several topics including photophoresis, thermophoresis, and coagulation are included in *Aerosol Science* edited by C. N. Davies (1966); health considerations are treated in *Inhalation Studies* by R. F. Phalen (1984) and in *Pulmonary Toxicology of Respirable Particles* (Sanders et al., 1980). Finally, several works deal with laboratory generation and characterization of aerosols, including those by Hinds (1982), Mercer (1973), Willeke (1980), and Liu (1976). For a recent overview of a variety of aspects of particle measurement and generation techniques, sources, composition and size distributions, effects, and basic properties and behavior, the reader is referred to the *Proceedings of the First International Aerosol Conference* held in 1984 (Liu et al., 1984).

12-B. PHYSICAL PROPERTIES

12-B-1. Size Distributions

12-B-1a. Number, Mass, Surface, and Volume Distributions

The atmosphere, whether in remote or urban areas, always contains significant concentrations of particles, up to 10^8 cm^{-3}. These may have diameters anywhere within the entire range from 0.002 to ~100 μm. Because the size of atmospheric particles plays such an important role in both their chemistry and physics in the atmosphere, as well as in their effects, it is important to know the distribution of sizes. We thus consider first how these size distributions are characterized.

An obvious way to express the distribution of particle sizes found in the atmosphere would be to plot, in the form of a histogram, the number (ΔN) of particles found in certain arbitrarily chosen intervals of diameter (ΔD), for example, from 0.002 to 0.01 μm, 1–10 μm, and so on. However, since in the atmosphere there tends to be a much greater number of small particles in relation to large particles, a linear plot of ΔN against D would give what would appear to be a narrow spike at the origin whose details could not be distinguished. This can be seen in Fig. 12.3 where typical impactor data from Wesolowski et al. (1980) for a sample consisting of Arizona road dust have been plotted; the original data in the form of a mass distribution have been converted to a number distribution using certain simplifying assumptions. To circumvent this problem (and for other reasons as well) the horizontal axis is usually scaled in logarithmic intervals—log D—so that several orders of magnitude in D can be clearly seen, as shown in Fig. 12.4.

A second problem in expressing the size distribution of aerosols is that the intervals of diameter over which it is experimentally convenient to measure the number of particles is not constant either in terms of D or log D. For example, as discussed in Section 12-E-3a, a four-stage Lundgren impactor is often used to

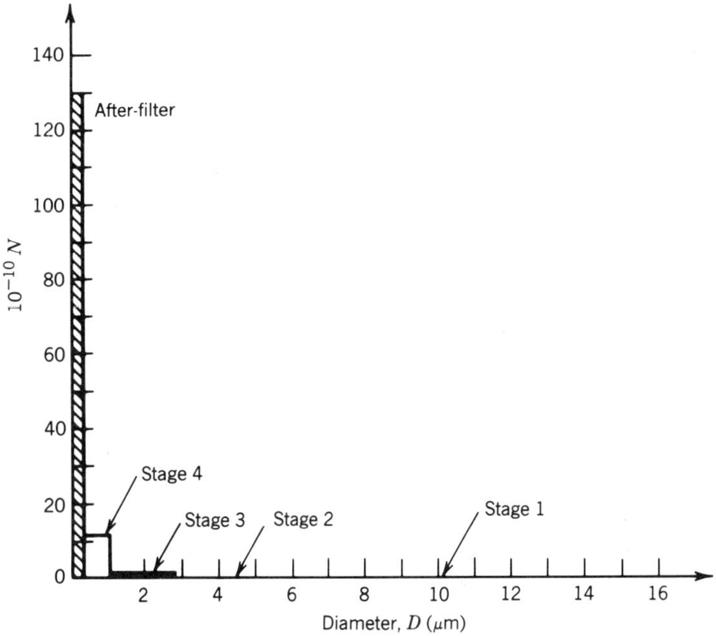

FIGURE 12.3. Plot of number of particles (N) against D (D = aerodynamic diameter) determined using a four-stage impactor with the cutoff points given in Table 12.1. It has been assumed that the particles are spherical with density 2.6 g cm^{-3}.

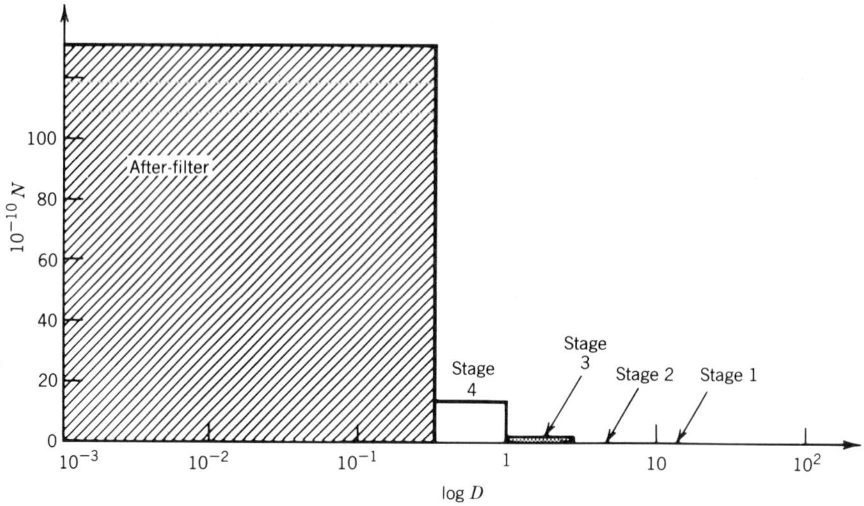

FIGURE 12.4. Plot of number of particles (N) against log D (D = aerodynamic diameter). Data same as Fig. 12.3.

measure the number of particles in certain size ranges. The size intervals are characterized by the 50% cutoff point for each stage, where the 50% cutoff point is defined as the diameter of spheres of unit density, 50% of which are collected by that stage of the impactor. For example, a typical set of 50% cutoff points is 8.0 (stage 1), 4.0 (stage 2), 1.5 (stage 3), and 0.5 μm (stage 4), respectively, with particles smaller than 0.5 μm being collected on an after-filter. One might then use the approximation that each stage captures particles ranging from a diameter corresponding to midway between its cut point and that of the next higher and lower stages. Using these assumptions, the ranges of particle diameters captured by each stage for a typical impactor are given in Table 12.1; also given are the diameter intervals ΔD as well as Δ log D for each stage. It is seen that neither the intervals in terms of diameter nor in terms of the logarithm of the diameter are balanced. As a result, a plot of ΔN versus D or log D will look something like those shown in Figs. 12.3 and 12.4, where the data were generated using an impactor with the cutoff points given in Table 12.1.

These give a somewhat distorted picture of the size distribution, however, because the height of any bar, that is, the number of particles, depends on the width of the interval taken, that is, ΔD or Δ log D. To give a more physically descriptive picture of the size distribution, a modified plot of the number of particles normalized for the width of the diameter interval is used; that is, the number of particles per unit size interval is plotted on the vertical axis. Using log D as the horizontal axis then, a normalized plot is one of $\Delta N/\Delta$ log D against log D, where ΔN is the number of particles in that interval of Δ log D; Fig. 12.5 is an example of this type of plot using the data in Figs. 12.3 and 12.4. The area under each rectangle then gives the number of particles in that size range.

TABLE 12.1. Typical 50% Cutoff Points and Approximate Range of Particle Diameters Captured by Each Stage in a Four-Stage Impactor

Stage	50% Cutoff Point (μm)	Approximate Range of Particle Diameters Captured by that Stage (μm)	ΔD (μm)	Δ log D (μm)
1	8.0	14.0[a]–6.0	8.0	0.37
2	4.0	6.0–2.8	3.2	0.33
3	1.5	2.8–1.0	1.8	0.45
4	0.5	1.0–0.3[b]	0.7	0.52
After-filter	<0.5	0.3–0.001[b]	0.3	2.48

Source: Wesolowski et al., 1980.

[a]Assuming a cutoff of 20 μm for the sampling probe.

[b]Assuming the after-filter collects all particles $\gtrsim$ 0.001 μm diameter.

Rather than showing histograms, a smooth curve is usually drawn through the data. Figure 12.6, for example, shows one such curve of $\Delta N/\Delta \log D$ against log D for a typical urban model aerosol; to emphasize the wide range of numbers involved, a log scale has been used for the vertical axis.

Such data are also frequently plotted in the form of a cumulative distribution curve, which is a plot of the fraction of particles having diameters less than, or greater than, certain values, against D. A typical plot for a group of particles all having diameters $\leq$ 50 μm is shown in Fig. 12.7. As expected, the graph reaches

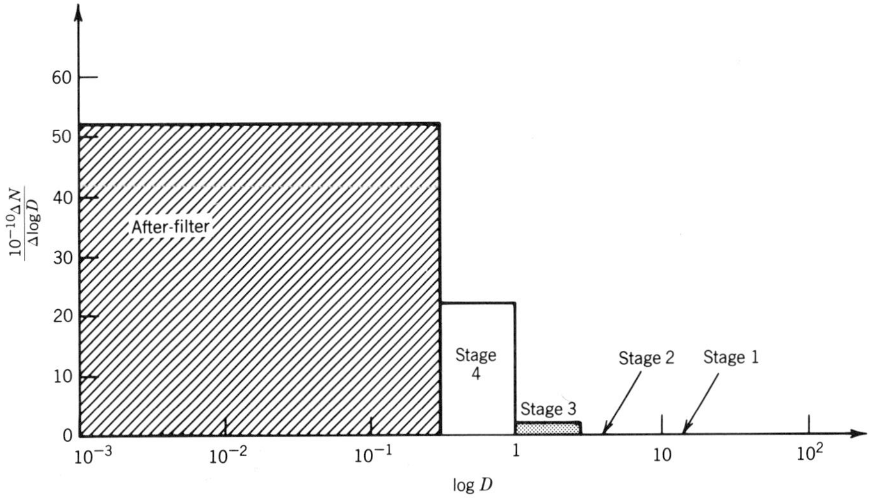

FIGURE 12.5. "Normalized" plot of $\Delta N/\Delta \log D$ versus log D for data shown in Figs. 12.3 and 12.4.

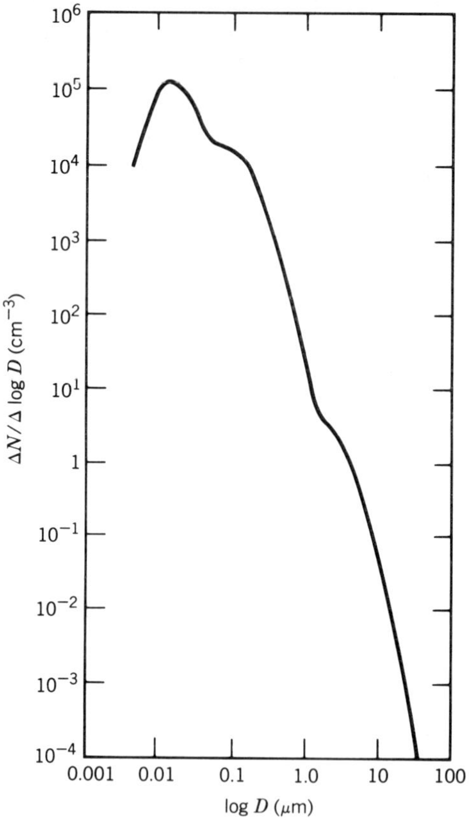

FIGURE 12.6. Plot of number distribution $\triangle N/\triangle \log D$ versus log D for a typical urban model aerosol (from Whitby and Sverdrup, 1980).

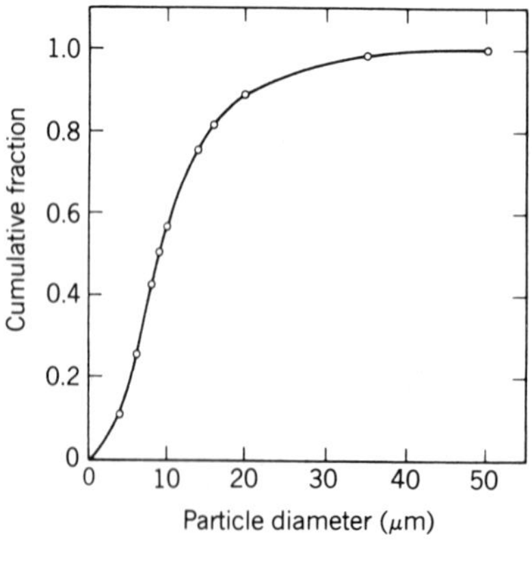

FIGURE 12.7. Typical cumulative distribution curve for a group of particles with diameters $\leq$ 50 μm (from Hinds, 1982).

a plateau at ~50 μm where the fraction of particles having diameters less than this is 1.0.

However, it is not only the number of particles in each size interval which is of interest, but also how other properties such as mass, volume, and surface area are distributed among the various size ranges. For example, the United States Environmental Protection Agency's air quality standards for particulates are currently expressed in terms of mass of particulate matter per unit volume of air; however, restricting the standard to only those particles with diameters $\lesssim$ 10 μm (which are small enough to penetrate beyond the average adult nose when inhaled) is currently under consideration (Federal Register, 1984). It is thus important to know the mass distribution of atmospheric particulate matter. In a given air sample, there may be a large number of small particles which, because of their size, contribute little to the total mass, along with a few large particles which are responsible for most of the measured mass. If the latter are larger than 10 μm, then they would be ignored in terms of meeting the proposed size-based particulate air quality standard.

Similarly, surface and volume distributions are important when considering reactions of gases at the surface of particles or reactions occurring within thc particles themselves, for example, the oxidation of SO_2 to sulfate.

Because of this need to know how the mass, surface, and volume are distributed among the various particle sizes, distribution functions for these parameters (i.e., mass, surface, and volume) are also commonly used for atmospheric aerosols in a manner analogous to the number distribution discussed above. That is, $\Delta m/\Delta \log D$, $\Delta S/\Delta \log D$, or $\Delta V/\Delta \log D$ are plotted against D on a log scale, where Δm, ΔS, and ΔV are the mass, surface area, and volume, respectively, found in a given size interval; again the area under these curves gives the total mass, surface, or volume in the interval considered. Figure 12.8 shows the surface and volume distributions for the number distribution shown in Fig. 12.6; also shown is the same number distribution for comparison, where the vertical axis is now linear (rather than logarithmic as in Fig. 12.6).

When the particulate data are plotted as mass, surface, or volume distributions, an important characteristic of typical urban aerosols emerges clearly. As seen in the number distribution of Fig. 12.8*a*, there is a large peak at ~0.02 μm, and a slight "knee" in the curve around 0.1 μm.

The volume distribution for the same aerosol (Fig. 12.8*c*) shows two strong peaks, one in the 0.1–1.0 μm range and the second in the 1–10 μm region, with a minimum in the 1–2 μm region. The surface distribution (Fig. 12.8*b*) shows a major peak in the vicinity of 0.1 μm with smaller peaks in the region between 0.01 and 0.1 μm and between 1 and 10 μm.

The "knee" observed in the number distribution suggests that the observed curve may be a combination of two different distributions, shown by the dotted lines in Fig. 12.8*a*. While the multimodal nature of atmospheric aerosols had been suggested earlier (e.g., Junge, 1963) Whitby and co-workers, in a classic set of papers (Whitby et al., 1972a,b; Husar et al., 1972), were the first to establish and

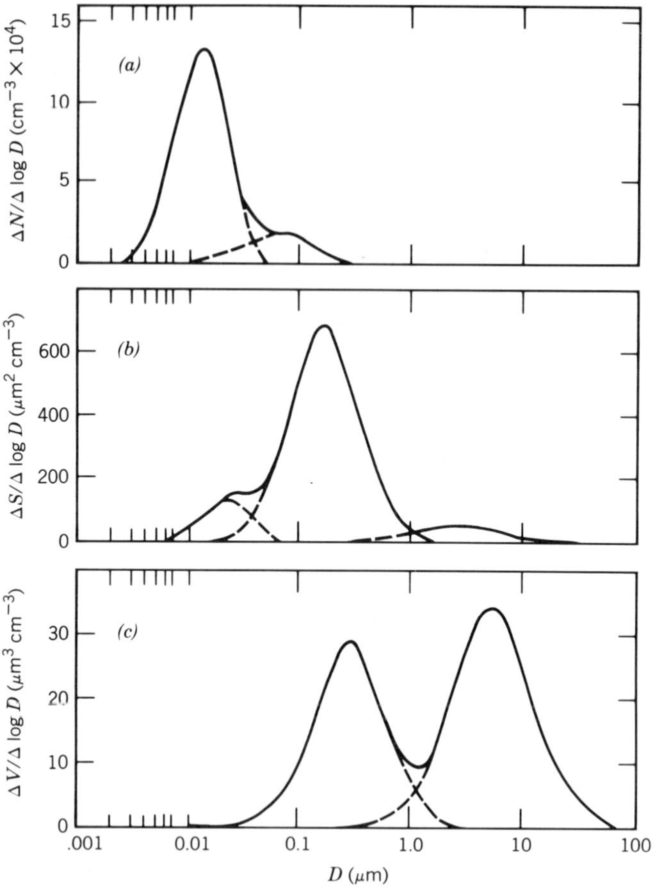

FIGURE 12.8. Number, surface, and volume distributions for a typical urban model aerosol (from Whitby and Sverdrup, 1980).

explore in detail the significance of such a bimodal distribution in terms of the origins, chemical characteristics, and removal processes of the two groups of particles of different sizes. Although instrument artifacts are a confounding factor, it is now widely accepted that atmospheric aerosols usually occur in specific size groupings that are different in their origins and properties.

Indeed, the surface and volume distributions shown in Fig. 12.8*b* and 12.8*c* suggest three distinct groups of particles contributing to this atmospheric aerosol. To environmental scientists, particles with diameters $\gtrsim$ 2.5 μm are usually identified as *coarse particles* and those with diameters $\lesssim$ 2.5 μm are called *fine particles*. The fine particle mode typically includes most of the total number of particles and a large fraction of the mass, for example, about one-third of the mass in non-urban areas and about one-half in urban areas. The fine particle mode can be further broken down into particles with diameters between approximately 0.08

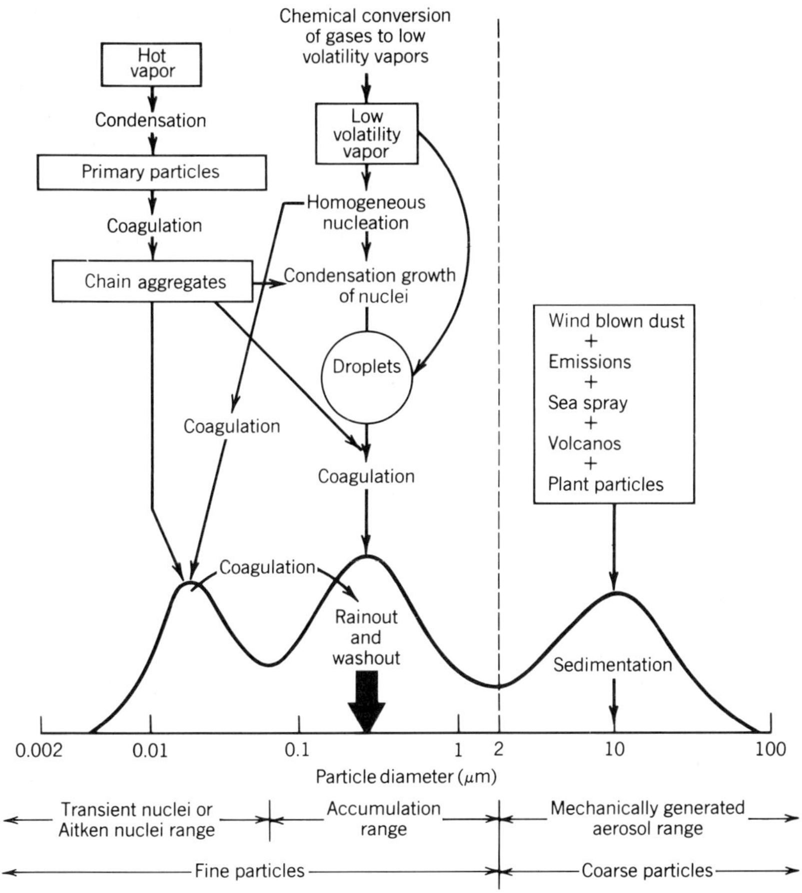

FIGURE 12.9. Schematic of an atmospheric aerosol size distribution showing the three modes, the main source of mass for each mode, and the principal processes involved in inserting mass into and removing mass from each mode (from Whitby and Sverdrup, 1980).

and 1–2 μm, known as the accumulation range, and those with $D \lesssim 0.08$ μm, known as the transient or Aitken nuclei range.

Figure 12.9 summarizes these three ranges as well as the major sources and removal processes for each one. Although the vertical axis is not shown, it could in theory be any of the distributions discussed above, that is, number, mass, surface, or volume; however, in practice, no more than two modes are usually seen in any one type of size distribution.

Particles in the coarse particle range are usually produced by mechanical processes such as grinding, wind, or erosion. As a result, they are relatively large and hence settle out of the atmosphere by sedimentation, except on windy days where fallout is balanced by reentrainment. They can also be removed by washout. In the atmosphere, transport of coarse particles over long distances can occur, however, by convective processes (National Research Council, 1979). Chemically, their composition reflects their source, and hence one finds predominantly inorganics such as sand, sea salt, and so on in this range. Because the sources and sinks are different from the smaller modes, the occurrence of particles in this mode tends to be only weakly associated with the fine particle mode. One notes that the majority of biological particles, spores, pollens, and so on tend to be in the coarse particles.

Particles in the accumulation range with diameters from ~ 0.08 to $\sim 1\text{–}2\ \mu m$ typically arise from condensation of low volatility vapors (as occurs following combustion) and from coagulation of smaller particles in the nuclei range either with themselves or with the larger particles in the accumulation range. The latter is usually the most important mechanism since the coagulation rates for particles in the nuclei range with the larger particles in the accumulation range are often much larger than for self-coagulation of the small particles; this occurs because of the high mobility of the smaller particles combined with the larger target area of the bigger particles. Because of the nature of their source, particles in the accumulation range generally contain far more organics than the coarse particles (other than biologic particles), as well as soluble inorganics such as NH_4^+, NO_3^-, SO_4^{2-}, and so on.

Particles in the accumulation range tend to represent only a small portion of the total particle number (e.g., 5%) but a significant portion (e.g., 50%) of the aerosol mass. Because they are too small to settle out rapidly (see Section 12-B-2), they are removed relatively slowly by incorporation into cloud droplets followed by rainout, or by washout during precipitation. Alternatively, they may be carried to surfaces by eddy diffusion and advection and undergo dry deposition (see Section 11-D-2). As a result, they have much longer lifetimes than coarse particles. This long lifetime, combined with their effects on visibility and penetration and deposition in the respiratory tract, makes them of great interest in atmospheric chemistry.

The smallest particles, $D \lesssim 0.08\ \mu m$, known as Aitken nuclei, arise from ambient temperature gas-to-particle conversion and combustion processes in which hot, supersaturated vapors are formed and subsequently undergo condensation. These particles act as nuclei for the condensation of low-vapor-pressure gaseous species, causing them to grow toward the accumulation range; alternatively, these nuclei may grow larger by coagulation. This range contains most of the total number of particles but relatively little of the total mass because of their small size. The lifetime of these particles is short, sometimes on the order of minutes, due to their rapid coagulation.

Combustion processes (e.g., power plants, diesel engines) may produce particles not only in the Aitken nuclei range, but in the accumulation range as well. The relative numbers of particles produced in the Aitken nuclei range compared to the accumulation range depend on the nature of the combustion process (e.g., fuel, operating conditions), as well as the conditions of dilution. Figure 12.10 shows the surface distribution of particles produced by the combustion of several

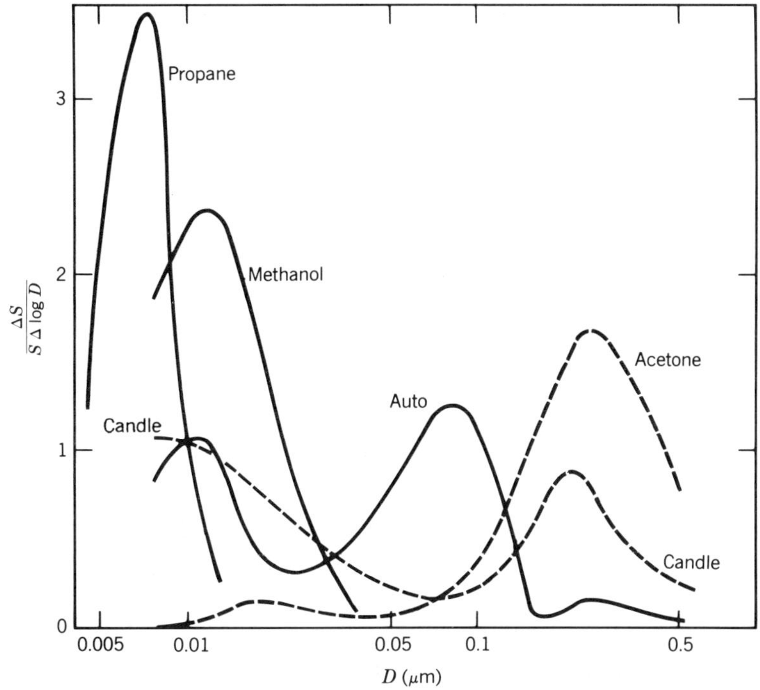

FIGURE 12.10. Surface distribution of particles from the combustion of several organics and from automobiles and a candle (from National Research Council, 1979).

organic compounds, as well as by automobiles and by a burning candle. The "dirtier" flames (e.g., the candle and the acetone flame) produce significant numbers of particles in the accumulation mode, while the cleaner flames produce Aitken nuclei.

The effect of dilution on the particle size can be seen in Fig. 12.11 which shows the calculated number distribution of particles produced by a light duty diesel engine at various aging times assuming that the exhaust is diluted by a factor of 10 upon exiting the tailpipe. This dilution factor is typical of many dilution tunnels used in studies of automobile exhaust. It is clear from Fig. 12.11 that, although most particles are produced initially in the Aitken nuclei range, the particles rapidly grow into the accumulation range by coagulation. (The number concentration, of course, drops dramatically as coagulation occurs, as shown in Fig. 12.11.) However, under roadway conditions, dilution factors of more than 10^3 are typical; when combined with atmospheric dispersion processes which further dilute the original aerosol, such rapid coagulation is not expected under real driving conditions and hence Aitken nuclei would be expected to predominate under typical roadway conditions. The effect of dilution seen in Fig. 12.11 is important to keep in mind in carrying out laboratory studies of combustion-generated particles; as

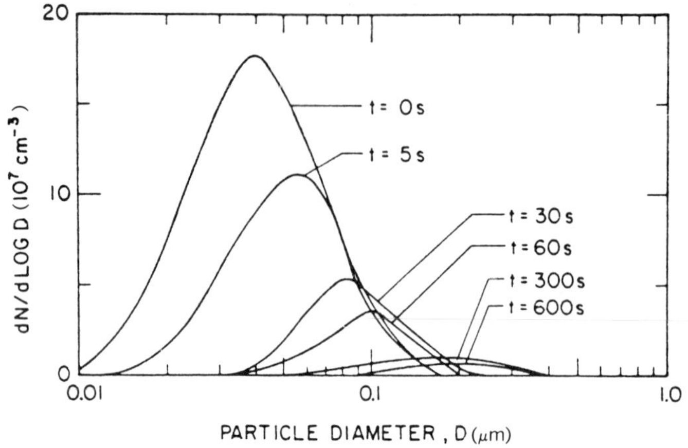

FIGURE 12.11. Calculated number distribution for a diesel exhaust aerosol diluted with air by a factor of 10, at various aging times (from Kittelson and Dolan, 1980).

Kittelson and Dolan (1980) point out, the rapid coagulation expected in typical laboratory tunnel experiments could give a particle size distribution that is not truly characteristic of roadway conditions.

Based on over 10,000 size distributions for atmospheric aerosols from a variety of locations at and above the earth's surface, Whitby and co-workers have postulated that there are seven basic types of aerosols encountered in the atmosphere. These are summarized in Table 12.2.

The categorization shown in Table 12.2, while based on numerous experimental observations, can be rationalized on the basis of the sources of the three particle modes shown in Fig. 12.9. For example, any air mass exposed to the products of combustion should contain enhanced concentrations of Aitken nuclei; thus the Aitken nuclei concentration (ANC) rises in going from clean continental air to average continental air which is influenced by anthropogenic activities. The ANC is even higher when local sources contribute, or in urban areas, since combustion is inevitably present.

Figure 12.12, for example, shows a linear-log plot of the volume distribution for four types of background aerosols, three of which have been influenced by human activities. The background shows few Aitken nuclei since it is uninfluenced by combustion processes (except for naturally occurring fires); however, significant coarse particulate volume is present due to natural mechanical processes such as wind resuspension and erosion. The addition of automobile emissions increases the Aitken nuclei mode to a discernible level in Fig. 12.12 and also increases the two larger modes; the larger increase in the coarse particle mode is due to resuspension of dust by the motion of the cars. The background average curve in Fig. 12.12 reflects the fact that anthropogenic activities impact essentially everywhere over the continents. Especially noteworthy in Fig. 12.12, however, is the aerosol influenced by an urban plume; here large volumes occur in the accumulation range due to the presence of Aitken nuclei which can grow into the accumulation range through condensation and coagulation.

TABLE 12.2. Seven Basic Categories of Atmospheric Aerosols Proposed by Whitby and Co-workers Based on Size Distributions

Type of Aerosol	Typical Approximate Aitken Nuclei Concentration (number^{-3})	Other Characteristics of Size Distribution
Marine surface background	400	Low and reasonably constant nuclei and accumulation mode concentrations
Clean continental background	< 1000	Fine particle volume $\leq$ 2 μm^3 cm^{-3}; course particle volume $\approx$ 5 μm^3 cm^{-3}; anthropogenic contributions negligible
Average continental background	4000–6000	Fine particle volume $\approx$ 5 μm^3 cm^{-3}; coarse particle volume $\approx$ 25 μm^3 cm^{-3}; mixture of background and anthropogenic aerosols
Background plus aged urban plumes	4000–6000	Fine particle volume $\approx$ 45 μm^3 cm^{-3}; coarse particle volume $\approx$ 25 μm^3 cm^{-3}
Background plus local sources	10^4–10^5	Fine particle volume $\approx$ 5 μm^3 cm^{-3}; coarse particle concentrations often higher by ~30% than average continental background
Urban average	10^5	Fine particle volume $\approx$ 30–40 μm^3 cm^{-3}; coarse particle concentrations same as for average continental background
Urban plus freeway	>> 10^5	Nuclei mode volumes $\approx$ 5–10 μm^3 cm^{-3}, which is about 10 times that for average urban aerosol and ~ 100 times that for the average continental background; coarse particle concentrations also elevated compared to average urban aerosol

Source: Whitby and Sverdrup, 1980.

With respect to the above discussion, a note of caution should be sounded; thus scientists in several disciplines, including meteorology, aerobiology, and pure aerosol physics, may not utilize these tri-modal concepts.

12-B-1b. Application of Log-Normal Distributions to Atmospheric Aerosols

Ideally, one would like to describe various size distributions by some relatively simple mathematical function. Because there is no single theoretical basis for a particular function to describe atmospheric aerosols, various empirical matches have been carried out to the experimentally observed size distributions; some of these are discussed in detail elsewhere (e.g., see Hinds, 1982). Out of the various

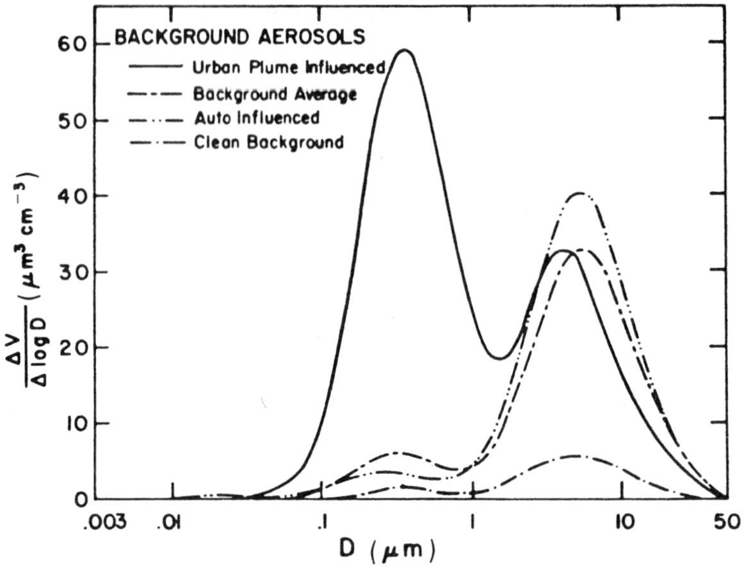

FIGURE 12.12. Linear-log plot of the volume distributions for background aerosols influenced by different types of anthropogenic activities (from Whitby and Sverdrup, 1980).

mathematical distribution functions for fitting aerosol data, the log-normal distribution (Atchinson and Brown, 1957; Patel et al., 1976) has emerged as the mathematical function that most frequently provides a sufficiently good fit, and hence we briefly discuss its application to the size distribution of atmospheric aerosols.

Most readers will be familiar with the bell-shaped normal distribution plotted in Fig. 12.13. When applied to the size distribution of particles, for example, such a distribution is fully characterized by the arithmetic mean $\overline{D}$ and the standard derivation σ, where σ is defined such that 68% of the particles have sizes in the range $\overline{D} \pm \sigma$. In the log-normal distribution the *logarithm* of the diameter D is assumed to have a normal distribution. (Either logarithms to the base 10 or to the

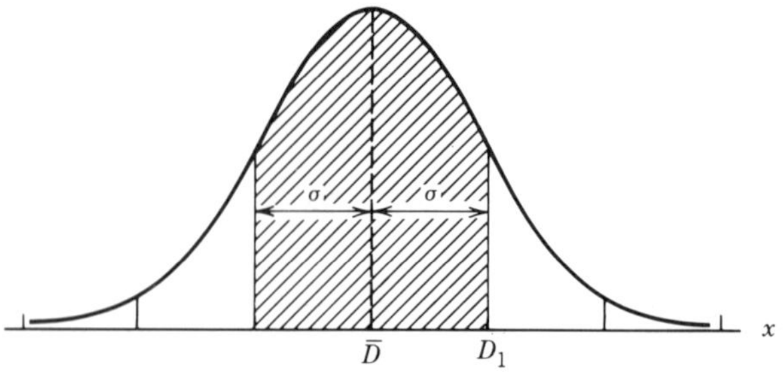

FIGURE 12.13. Meaning of standard deviation for a normal distribution. Hatched area = 68% of total area under curve.

base e can be used, but since the latter is more common, we follow through the discussion using natural logs, ln.) This distribution is expressed as

$$\frac{dN}{d \ln D} = \frac{N_T}{\sqrt{2\pi} \ln \sigma_g} \exp\left[-\frac{(\ln D - \ln \overline{D}_{gN})^2}{2(\ln \sigma_g)^2}\right] \tag{B}$$

where N is the number of particles having diameters whose logarithms are between $\ln D$ and $\ln D + d \ln D$, N_T is the total number of particles, σ_g is known as the geometric standard deviation, and $\overline{D}_{gN}$ is the geometric number mean diameter, which for this distribution is equal to the number median diameter, defined as the diameter for which half the number of particles are smaller and half are larger. A typical plot of Eq. (B) is shown in Fig. 12.14, where $(dN/N_T)/d \ln D$ is plotted against the diameter on a logarithmic scale.

The geometric number mean diameter, $\overline{D}_{gN}$, is related to the arithmetic mean of $\ln D$:

$$\ln \overline{D}_{gN} = \frac{\sum n_j \ln D_j}{N_T} \tag{C}$$

Here n_j is the number of particles in a group whose diameters are centered around D_j. Thus $\ln \overline{D}_{gN}$ is really a weighted value of $\ln D$, where the weighting is by the number of particles in that size interval.

The meaning of the geometric standard deviation σ_g can be seen by referring to the meaning of standard deviation for a typical *normal* distribution. In this case, as shown in Fig. 12.13, the standard deviation is a measure of the spread about the mean ($\overline{D}$) and is defined such that there is a 68% probability of a particle having

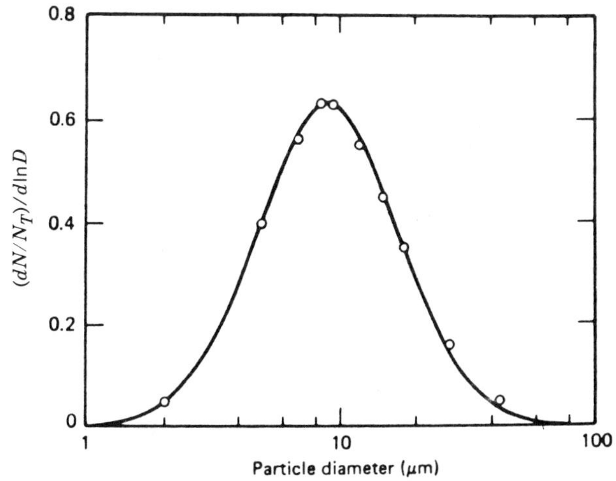

FIGURE 12.14. Frequency distribution curve (logarithmic size scale) (from Hinds, 1982).

a diameter in the range $\overline{D} \pm \sigma$, and a 95% probability of it having a diameter in the range $\overline{D} \pm 2\sigma$. σ can be calculated for the normal distribution using Eq. (D),

$$\sigma = D_1 - \overline{D} \tag{D}$$

where $\overline{D}$ is the diameter for which 50% of the particles have smaller diameters and 50% have larger, and D_1 ($= \overline{D} + \sigma$) is the diameter for which there is an 84% probability that a particle will have a diameter equal to or less than this value.

For a *log-normal* distribution, σ_g is still a measure of spread of the distribution, but it has a slightly different definition because ln D, rather than D, is assumed to have a normal distribution. For a log-normal distribution σ_g is defined by Eq. (E), analogous to (D) above:

$$\ln \sigma_g = \ln D_1 - \ln \overline{D} = \ln \frac{D_1}{\overline{D}} \tag{E}$$

Thus

$$\sigma_g = \frac{D_1}{\overline{D}} \tag{F}$$

and 68% of the distribution is between D_g/σ_g and $D_g\sigma_g$. σ_g is a dimensionless parameter which must be ≥ 1.0 since $D_1 \geq \overline{D}$.

Although the log-normal distribution in Eq. (B) was given in terms of the distribution of the numbers of particles as a function of size, it can also be applied to the size distribution of the other properties of interest, that is, mass, surface, or volume. In these cases, the *average* size used to characterize the distribution is known as the geometric *mass* mean diameter, the geometric *surface* mean diameter, and geometric *volume* mean diameter, respectively. Like the geometric number mean size, $\overline{D}_{gN}$, these are weighted average diameters, where the weighting factors are mass, surface, and volume, respectively.

Let us examine the meaning of these commonly used parameters in more detail. Assume that there are j groups of particles, each group having a representative diameter d_j; in group j, the total mass of the n_j particles is m_j, the surface area is s_j, and the volume is v_j. The geometric mass mean diameter, $\overline{D}_{gM}$ is given by

$$\ln \overline{D}_{gM} = \frac{\Sigma\ m_j \ln d_j}{M} \tag{G}$$

where M is the total mass in the sample, $M = \Sigma\ m_j$. Assuming that the particles are all perfectly smooth and spherical with density ρ, then the volume of each particle in group j is $v_j = \pi d_j^3/6$ and the total volume of the group j is $n_j\ \pi d_j^3/6$. The mass of particles in group j is then given by $m_j = n_j \pi \rho d_j^3/6$. Similarly, the total mass M of all the particles in all groups is $M = \Sigma(n_j \pi \rho d_j^3/6)$. Thus Eq. (G) becomes

$$\ln \overline{D}_{gM} = \frac{\frac{\pi\rho}{6} \sum (n_j d_j^3) \ln d_j}{\frac{\pi\rho}{6} \sum n_j d_j^3} = \frac{\sum n_j d_j^3 \ln d_j}{\sum n_j d_j^3} \qquad \text{(H)}$$

Because mass and volume are directly related with ρ as the constant of proportionality, the expression for the geometric volume mean diameter, $\overline{D}_{gV}$, is the same as that for $\overline{D}_{gM}$ given in Eq. (H).

Similarly, the geometric surface mean diameter, $\overline{D}_{gS}$, is defined as the average diameter weighted according to the surface areas in the various groups:

$$\ln \overline{D}_{gS} = \frac{\sum s_j \ln d_j}{S} = \frac{\sum n_j d_j^2 \ln d_j}{\sum n_j d_j^2} \qquad \text{(I)}$$

using area $= \pi D^2$ and where S is the total surface area of all particles in all groups.

The surface mean diameter (as opposed to the *geometric surface* mean diameter) is also a very useful parameter, since

$$\overline{D}_{gS} = \frac{\sum s_j d_j}{S} = \frac{\sum n_j d_j^3}{\sum n_j d_j^2} = \frac{6}{\rho}\frac{M}{S} \qquad \text{(J)}$$

Thus from the total mass (M) of the particulate sample and its total surface area (S) (which can be determined experimentally; see Section 4-E-3), the surface mean diameter of the particles can be calculated if their density (ρ) is known.

The log-normal distribution can be plotted on a special type of paper to give a linear graph from which the basic parameters characterizing the distribution, $\overline{D}_{gN}$ and σ_g, can be readily determined. Thus a log-normal distribution gives a linear plot on what is known as log-probability paper shown in Fig. 12.15. The vertical axis is a logarithmic scale of particle diameter; the horizontal axis is a special nonlinear cumulative probability (probit) scale, on which is plotted the percentage of particles smaller than the diameter given on the vertical scale. Thus log-probability plots are analogous to *linear* plots of cumulative fraction versus particle diameter as in Fig. 12.7. Note that the horizontal axis of the log-probability plot is expanded at its ends, but compressed in the center around the 50% region. Graph paper in the log-probability format is commercially available and hence relatively convenient to use.

To construct a log-probability graph using the graph paper designed for this purpose, one only needs to establish the number of particles in each size interval, and then convert these numbers to a percentage of the total number of particles.

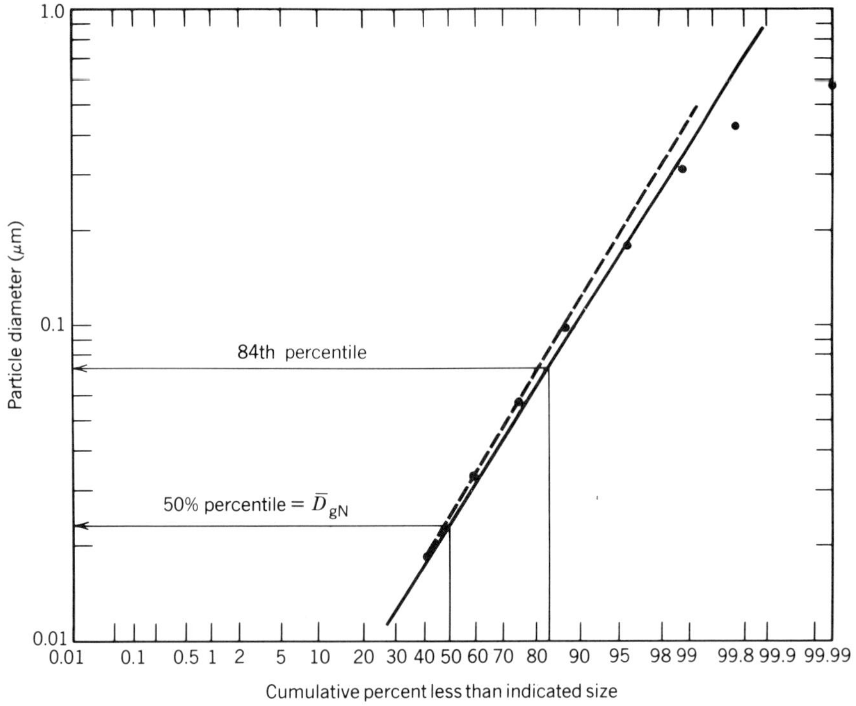

FIGURE 12.15. A typical log-probability graph using ambient air data from Sverdrup et al. (1975) (Table 12.3). The dotted line shows the fit obtained giving more weight to the majority of the particles lying between the 10th and 90th percentiles (see text).

The cumulative percentage, that is, the total percent less than a given size, can then be determined, starting with smallest diameters.

Table 12.3 gives size distribution data for particles with diameters $\leq 0.56\ \mu m$ in the Mohave Desert, California, in 1972 (Sverdrup et al., 1975). The cumulative percentage, that is, the percentage of the particles having diameters less than the upper limit of each size range, is also given. Figure 12.15 is a log-probability graph of these data. The linearity of the plot below $\sim 0.3\ \mu m$ supports the assumption that this size distribution follows a log-normal distribution in this size range; above $\sim 0.3\ \mu m$, this log-normal distribution is overlapped by the one describing the next higher size range and hence a deviation from linearity is observed.

However, a note of caution relevant to all such fitting procedures should be sounded here; the fact that a log-probability plot seems reasonably linear when experimental error is taken into account does not mean that the log-normal distribution is in fact necessarily a good fit to the data. This is illustrated in Fig. 12.16 where a straight line appears to fit the data in the log-probability plot reasonably well in part (*a*); however, as seen in part (*b*), the original data are not closely matched by a log-normal distribution. Thus, although the assumption of a log-

TABLE 12.3. Size Distribution Measured for an Aerosol in the Mohave Desert, California

Size Interval[a] (μm)	Characteristic Diameter (μm)	Number	Cumulative Percentage Less Than Upper Bound of Size Interval
0.01–0.0178	0.0133	527	40.9
–0.0316	0.0237	249	60.2
–0.0562	0.0422	203	75.9
–0.100	0.075	156	88.0
–0.178	0.133	88	94.9
–0.316	0.237	52	98.9
–0.422	0.365	11.4	99.78
–0.562	0.487	2.82	100.00
		Σ 1289.2	100%

Source: Sverdrup et al., 1975.

[a]Top of one range is bottom of the next one.

normal distribution and the use of log-probability paper to test it is common, the experimentalist should keep in mind that it is an approximation that should be applied with some caution. Also when fitting a line, more weight should be given to the data points between the 10th and 90th percentiles, since they represent the main portion of the particles (Fig. 12.15).

If the log-probability plot is found to be reasonably linear and therefore the assumption of a log-normal distribution justified, then the values of $\overline{D}_{gN}$ and σ_g in Eq. (B) can be obtained from the plot in the following manner. Recall that the geometric count mean diameter, $\overline{D}_{gN}$, is equal to the count median diameter for a log-normal distribution, and the median is that value of the diameter for which 50% of the particles have larger diameters and 50% have smaller ones. The median diameter can be determined by reading the diameter corresponding to the 50% point on the horizontal axis, as shown in Fig. 12.15. Since $\overline{D}_{gN}$ is equal to the median, $\overline{D}_{gN}$ is now known.

Obtaining the geometric standard deviation, σ_g, from log-probability plots such as Fig. 12.15 can be accomplished by returning to the definition of σ_g for a log-normal distribution, $\sigma_g = D_1/\overline{D}$ [Eq. (F)]. Recall that D_1 is the diameter lying one standard deviation above the mean diameter for a normal distribution. In terms of cumulative percentages, $\overline{D}$ is the diameter corresponding to 50%, whereas D_1 is the diameter corresponding to a probability of 68/2 = 34% above $\overline{D}$, that is, to a cumulative probability of 84%. Thus in Fig. 12.15, D_1 is obtained by reading the particle diameter on the vertical scale which corresponds to a cumulative percentage of 84% on the horizontal axis. The geometric standard deviation is then given by

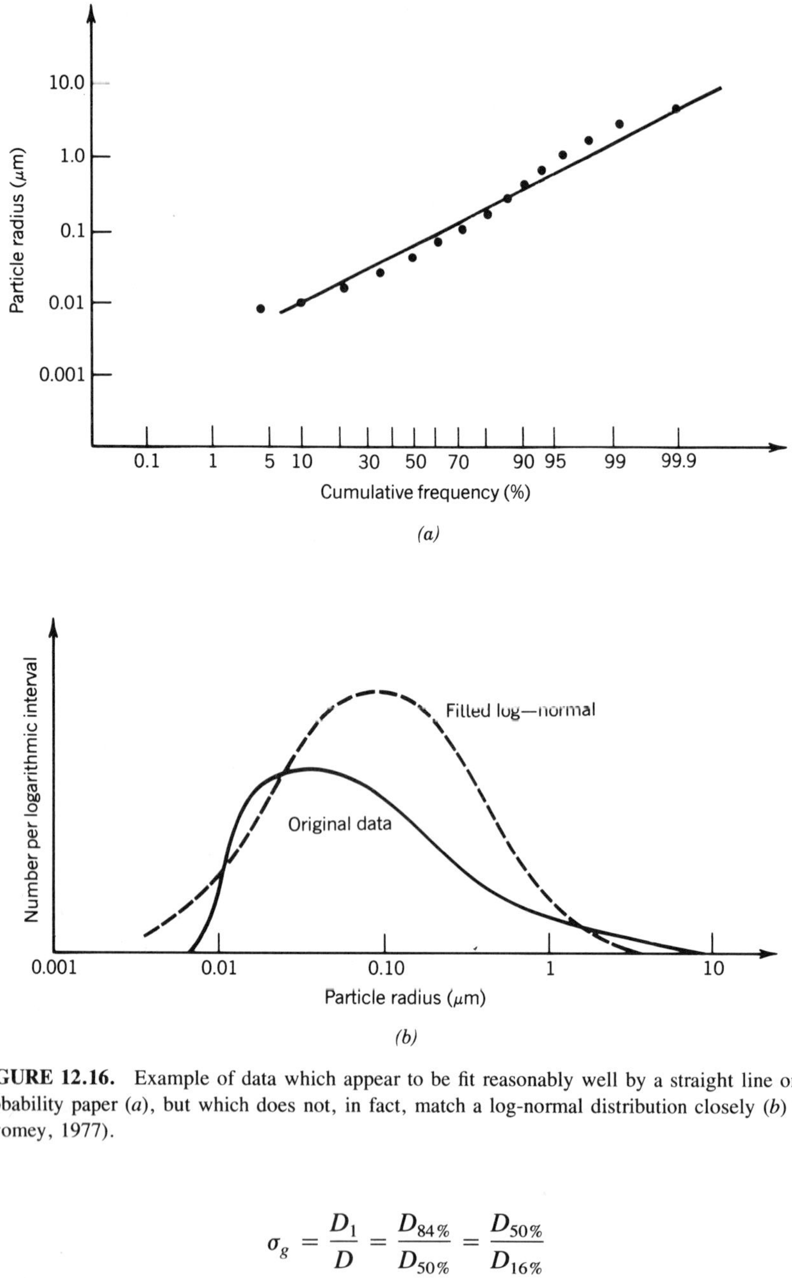

FIGURE 12.16. Example of data which appear to be fit reasonably well by a straight line on log-probability paper (*a*), but which does not, in fact, match a log-normal distribution closely (*b*) (from Twomey, 1977).

$$\sigma_g = \frac{D_1}{D} = \frac{D_{84\%}}{D_{50\%}} = \frac{D_{50\%}}{D_{16\%}} \tag{K}$$

For example, for the data in Fig. 12.15, $D_{84\%} = 0.073$ μm, $D_{50\%} = \overline{D}_{gN} = 0.023$ μm, and $\sigma_g = 3.2$. Once $\overline{D}_{gN}$ and σ_g are known, the log-normal distribution is completely defined.

The spread of a distribution, that is, the size of σ_g, is related to the slope of log-probability plots. This can be seen from Eq. (K); the larger the slope, the larger is $D_{84\%}/D_{50\%}$ and the larger is σ_g, that is, the spread of the distribution. In the limit of very narrow spreads, the slope approaches zero corresponding to a monodisperse aerosol.

An advantage of applying the log-normal distribution to atmospheric aerosols, in addition to the linearity of log-probability plots, is that the value of the geometric standard deviation, σ_g, is the same for a given sample for all types of distributions—count, mass, surface, and volume. It is only the value of the geometric mean diameter which changes depending on the property used for the distribution. This can be seen in Fig. 12.17 which shows a count and mass distribution for a hypothetical log-normal sample; the spread, σ_g, is seen to be the same for each distribution, but the geometric mean diameters, $\overline{D}_{gN}$ and $\overline{D}_{gM}$, are quite different.

On a log-probability plot, since the slope of the line is related to σ_g, the different types of distributions give lines with the same slopes but displaced from each other, as seen in Fig. 12.18 where log-probability plots of the data in Fig. 12.17 are shown.

In practice, when one measures the size distribution of aerosols using the techniques discussed in Section 12-E-3, one normally measures one parameter, for example, number or mass, as a function of size. For example, cascade impactor data usually give the mass of particles by size interval. These can be plotted on log-probability paper to give the geometric mass mean diameter, which applies only to the mass distribution, and σ_g, which as discussed above is the same for all types of log-normal distributions for this one sample. Given the geometric mass mean diameter ($\overline{D}_{gM}$) in this case and σ_g, an important question is whether the other types of mean diameters (i.e., number, surface, volume) can be determined

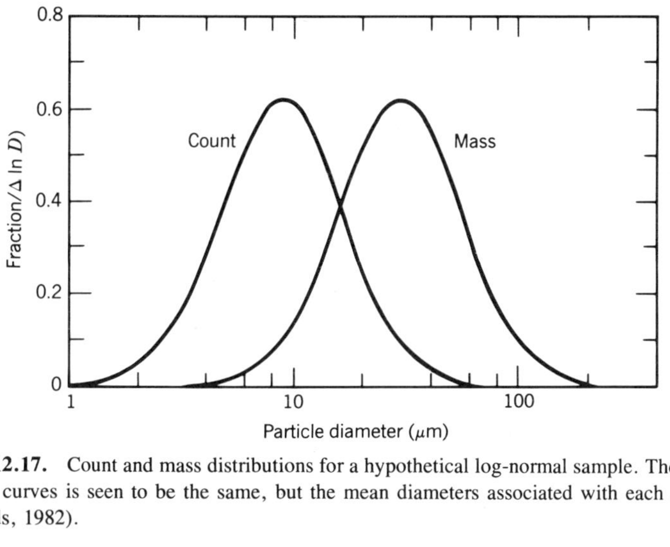

FIGURE 12.17. Count and mass distributions for a hypothetical log-normal sample. The spread, σ_g, of the two curves is seen to be the same, but the mean diameters associated with each are different (from Hinds, 1982).

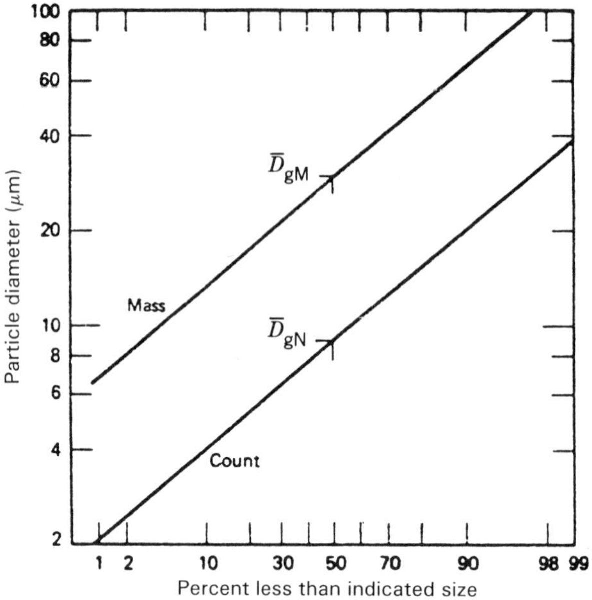

FIGURE 12.18. Log-probability plot of count and mass distributions shown in Fig. 12.17 for a hypothetical sample. $\overline{D}_{gN}$ = geometric count mean diameter, $\overline{D}_{gM}$ = geometric mass mean diameter. Horizontal axis is cumulative percent of total mass for mass distribution, and cumulative percent of the total number for the number distribution (from Hinds, 1982).

from these data, or if separate experimental measurements are required. The answer is that these other types of mean diameters can indeed be calculated for smooth spheres whose density is independent of diameter. The conversions are carried out using equations developed for fine particle technology in 1929 by Hatch and Choate.

These Hatch–Choate equations are of the form

$$\frac{d}{\text{number median diameter}} = \exp\,[b(\ln^2 \sigma_g)] \qquad \text{(L)}$$

where d is the average diameter to be determined using a known value of the number median diameter, σ_g is the geometric standard deviation determined from the log-probability plot of the available data, and b is a constant whose value is determined by the type of average diameter, d, which is to be calculated. Recall that for a log-normal distribution, the number median diameter is equal to the geometric number mean diameter, $\overline{D}_{gN}$. Thus Eq. (L) becomes

$$\frac{d}{\overline{D}_{gN}} = \exp\,[b(\ln^2 \sigma_g)] \qquad \text{(M)}$$

TABLE 12.4. Values of the Constant b in the Hatch–Choate Equations for Converting the Count Geometric Mean Diameter to Mass, Surface, or Volume Mean Diameters

Type of Mean Diameter to Be Calculated	Geometric Mean b[a]	Mean b[a]
Mass	3	3.5
Surface	2	2.5
Volume	3	3.5

[a]See Eq. (L).

Table 12.4 gives the values of b to be used for converting the count geometric mean diameter to the other types of geometric or mean diameters, respectively. For example, if the data in Fig. 12.18 for the number distribution were available, but the mass distribution data shown there were not, then geometric mass mean diameter $\overline{D}_{gM}$ could be calculated as follows:

$$\frac{\overline{D}_{gM}}{\overline{D}_{gN}} = \exp\,(3.0\,\ln^2\sigma_g) = \exp\,[3.0\,(\ln^2 1.89)] = 3.4$$

$$\overline{D}_{gM} = 3.4 \times \overline{D}_{gN} = 3.4 \times 9.0\ \mu\text{m} = 30\ \mu\text{m}$$

σ_g has been calculated from D_1 and $\overline{D}$ [eq. (K)] as described above.

While we have concentrated here on the geometric means weighted by number, mass, surface, or volume, respectively, other types of "average" diameters are also sometimes cited in the literature. One commonly used is the diameter of average mass, which is defined as the diameter such that the mass of this particle multiplied by the total number of particles gives the total mass. The geometric mean diameters discussed here can also be converted to these types of diameters using a form of the Hatch–Choate equations. The reader is referred to the discussion by Hinds (1982) for definition of these other types of diameters and their conversion to the geometric mean diameters discussed here.

Whitby and co-workers have indicated that the many atmospheric aerosol size distributions which they have measured under a variety of conditions and at many locations can be fit reasonably well assuming three additive log-normal distributions corresponding to the Aitken nuclei range, the accumulation range, and the coarse particle range, respectively, which are described in Section 12-B-1a. Each of these log-normal distributions has its own characteristic value of σ_g as well as, of course, average diameters. For example, Fig. 12.8 contains the number, surface, and volume distributions for a typical urban aerosol; these were calculated to be consistent with the sum of two, or in the case of the surface distribution, three, additive components. These components are shown by the dashed lines in

Fig. 12.8. From these separate distributions, geometric mean diameters and σ_g could be found assuming smooth spheres using the methods described above. Table 12.5 summarizes the parameters derived by Whitby and Sverdrup (1980) based on the log-normal distributions in the three size ranges in Fig. 12.8.

If all the data over all size ranges in Fig. 12.8 were plotted on a log-probability plot, the multi-modal nature of the combined data would appear as ''knees'' in the line. For example, Fig. 12.19 shows a log-probability plot for a hypothetical bimodal particle distribution. Since the average diameters associated with the two modes are quite different, the location of the line vertically on the plot changes at the knee. If the spread of the two distributions, that is, σ_g, is also different, then the slope will also change at the knee. Some caution should be exercised, however, because ''knees'' can appear as artifacts in some aerosol measurement methods (see Section 12-E).

Figure 12.20 is a log-probability plot for the volume distribution data of Fig. 12.8*c*. (Note that this is plotted with the axes reversed from the log-probability plots given earlier; however, all else is the same.) The presence of two knees in the plot, one well-defined around 1 μm and one less obvious around 0.1 μm, suggests that the experimental curve is the sum of three different lines in the diameter ranges $\lesssim 0.1$ μm, 0.1–1 μm, and $\gtrsim 1$ μm, respectively. That is, the volume distribution data are consistent with additive contributions from three different log-normal distributions having different average diameters. In addition, the slopes of the lines in these three regions appear slightly different, suggesting that the geometric standard deviation is also different for the three contributing distributions. This is borne out by the data in Table 12.5 which were derived from the plot in Fig. 12.8*c*.

TABLE 12.5. Summary of Parameters $\overline{D}_g$ and σ_g for the Three Additive Log-Normal Distributions Characterizing Data in Fig. 12.8

Type of Distribution Used	Mode	Log-Normal Distribution $\overline{D}^a$ (μm)	σ_g
Number	Aitken nuclei	$\overline{D}_{gN} = 0.013$	1.7
	Accumulation	$\overline{D}_{gN} = 0.069$	2.03
Surface	Aitken nuclei	$\overline{D}_{gS} = 0.023$	*b*
	Accumulation	$\overline{D}_{gS} = 0.19$	*b*
	Coarse particle	$\overline{D}_{gS} = 3.1$	2.15
Volume	Aitken nuclei	$\overline{D}_{gV} = 0.031$	*b*
	Accumulation	$\overline{D}_{gV} = 0.31$	*b*
	Coarse particle	$\overline{D}_{gV} = 5.7$	*c*

Source: Whitby and Sverdrup, 1980.

[a]Note that the diameter—number, surface, or volume–changes depending on the type of distribution used, while σ_g remains constant for each of the three modes. See text,

[b]Same as derived from the number distribution.

[c]Same as derived from the surface distribution.

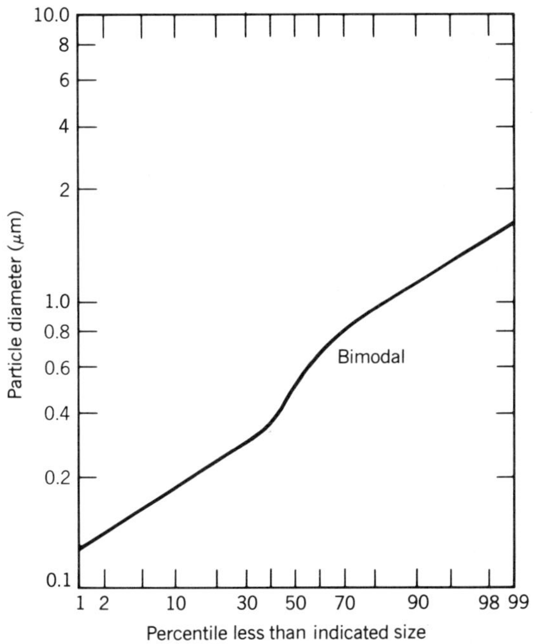

FIGURE 12.19. Typical log-probability plot for a hypothetical bimodal aerosol distribution (from Hinds, 1982).

12-B-2. Particle Motion

One of the important properties of particles which contributes to both the observed size distribution and the number concentration of aerosols in the atmosphere is the motion they undergo when suspended in air. This includes gravitational settling and Brownian diffusion.

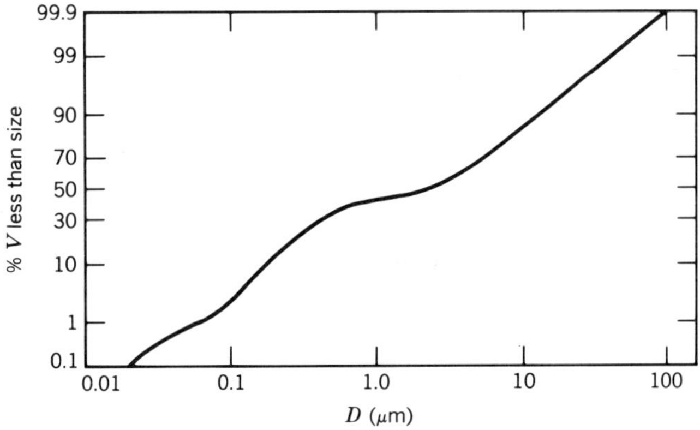

FIGURE 12.20. Log-probability plot for the volume size distribution data given in Fig. 12.8*c*. Note the two "knees" in the line around 0.1 μm and 1 μm, respectively (from Whitby and Sverdrup, 1980).

12-B-2a. Gravitational Settling

In the absence of other forces, relatively large particles, $\gtrsim 0.5$ μm in diameter, tend to move with the flow of air because they are subjected to significant frictional *drag* if they move faster or slower than the surrounding gas. However, in the free troposphere, particles are subjected to gravitational forces. They also can be subjected to electrical forces, for example, in nature as well as in the course of detection and measurement (see Section 12-E). When such forces are applied, the particle moves relative to the gas and hence is subjected to a resistance force. Stokes' law gives the force (F_R) acting on smooth spherical particles due to the laminar flow of air over them:

$$F_R = 3\pi\eta vD \tag{N}$$

η is the gas viscosity, v is the particle velocity relative to the gas, and D is the particle diameter. When a force such as gravity is applied to the particle, it speeds up until the frictional force equals the applied force; it then moves with a constant velocity known as the *terminal velocity*.

One can apply Stokes' law to atmospheric particles to calculate how fast they will settle out of the air when subjected to gravity alone. Thus the terminal settling velocity occurs when the frictional and gravitational forces are balanced, that is,

$$F_R = F_{\text{gravity}} = mg \tag{O}$$

where m is the mass of the particle and g is the acceleration due to gravity (9.8 m s^{-2} at sea level). One can apply the relationship between mass, volume ($\frac{1}{6}\pi D^3$), and density (ρ) of the particle. Equations (N) and (O) combine to give Eq. (P):

$$\frac{\pi D^3 \rho g}{6} = 3\pi\eta vD$$

or

$$v = \frac{D^2 \rho g}{18\,\eta} \quad (\text{for } D \gtrsim 1.5\ \mu\text{m}) \tag{P}$$

The settling velocity thus increases with the density of the particle and with the square of its diameter. In developing Eq. (P), the buoyancy effect of air which tends to lower the effective particle density has been ignored since it is much smaller than the particle density; it can be included if desired by replacing ρ by $\rho_p - \rho_{\text{air}}$, where ρ_p is the particle density and ρ_{air} is the air density (1.2×10^{-3} g cm^{-3} at 20°C and 1 atm pressure).

The expression given in Eq. (P) for the terminal settling velocity only applies to particles with diameters $\gtrsim 1.5$ μm because its derivation is based on the assumption that the relative speed of the air at the surface of the particle is zero.

However, as the particle becomes smaller, the air molecules appear less as a continuous fluid and more as discrete molecules separated by space through which the particles can "slip." The net effect is that the particles can move faster than predicted by Eq. (P) due to this slipping between gas molecules. To correct for this effect, a correction factor must be applied to the resistance force predicted by Stokes' law, Eq. (N). The correction factor is a number greater than 1. Thus Eq. (N) is modified to

$$F_R = \frac{3\pi\eta v D}{C} \tag{Q}$$

and the settling velocity becomes

$$v = \frac{D^2 \rho g C}{18\eta} \tag{R}$$

The correction factor, first derived by Millikan (1923), is given by Eq. (S):

$$C = 1 + \frac{\ell}{D}\left[2.514 + 0.800 \exp\left(-0.55\frac{D}{\ell}\right)\right] \tag{S}$$

This is often called the Cunningham correction factor, although Cunningham's original correction was of the form $1 + A(\ell/D)$ and did not include the exponential term. ℓ is the mean free path between air molecule collisions defined as the average distance traveled between collisions with another molecule; it can be easily calculated using simple kinetic molecular theory, and for air at 1 atm pressure and 20°C, ℓ is 0.066 μm. For large particles, $D \gtrsim 1.5$ μm; the term in brackets in Eq. (S) is ≤ 0.10 and hence Eq. (P) is a good approximation. For particles between ~0.1 and 1.5 μm, the third (exponential) term in Eq. (S) is small relative to 2.5, and hence the Cunningham correction factor can be approximated by

$$C = 1 + \frac{2.514\,\ell}{D} \quad (\text{for } 0.1\ \mu\text{m} \lesssim D \lesssim 1.5\ \mu\text{m}) \tag{T}$$

Figure 12.21 shows the Cunningham slip correction factors for air at 1 atm pressure and 20°C; it is seen that for the smallest particles, there is a significant correction to the speeds calculated using Eq. (P).

The assumptions inherent in the use of Stokes' law (e.g., relatively low speeds), which normally apply to atmospheric particles, are discussed in more detail by Hinds (1982) and Fuchs (1964). For our purposes, we need only recognize that the settling velocities for atmospheric particles calculated using Eq. (P) or (R) are large for particles $\gtrsim 10$ μm diameter. Figure 12.22 shows the settling velocities of spherical particles with $\rho = 1$ g cm^{-3} in still air at 0°C and 1 atm pressure as a function of particle diameter. It is seen that particles with diameters of the order

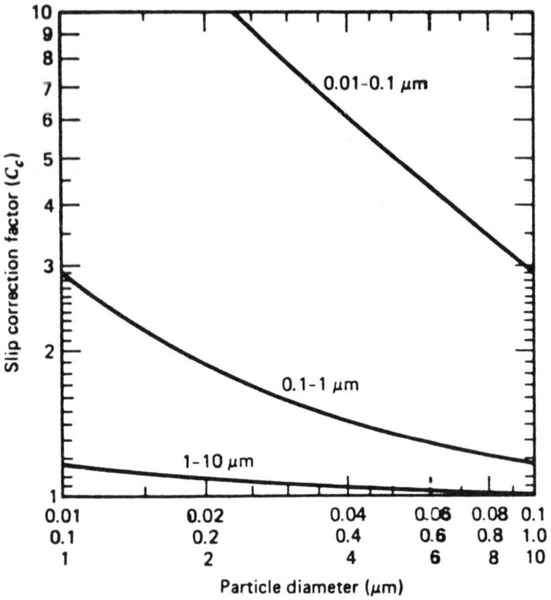

FIGURE 12.21. Cunningham correction factors to be applied in calculating terminal settling velocities of particles (from Hinds, 1982).

of 10 μm or greater have settling velocities $\gtrsim$ 0.1 cm s^{-1} and hence will settle out of the atmosphere relatively rapidly. However, those particles with diameters $\lesssim$ 1 μm will remain suspended for long periods of time and hence can participate in atmospheric transformations.

12-B-2b. Brownian Diffusion

Small particles do not settle via gravity at a significant rate, but they do undergo Brownian diffusion. The classic example of Brownian diffusion is the random zig-zag motion of smoke particles in air which can be observed because of light scattering by the particles. In fact, in the absence of convection, particles $\lesssim$ 0.1 μm diameter are transported primarily by Brownian diffusion; this is primarily responsible for the rapid coagulation of particles in the Aitken nuclei range.

The rates of Brownian diffusion can be quantified by reference to Fig. 12.23. In the left section (*A*) of the box of cross-sectional area 1 cm^2, there are *N* particles cm^{-3}, while the right section (*B*) is initially empty. Particles (or gas molecules) will always tend to diffuse from a region of higher concentrations to one of lower concentrations. The rate at which they diffuse depends on the concentration gradient, *dN*/*dx*; the larger the gradient, the faster the rate of diffusion. This is the basis of the well-known Fick's first law of diffusion:

$$J = -D\frac{dN}{dx} \tag{U}$$

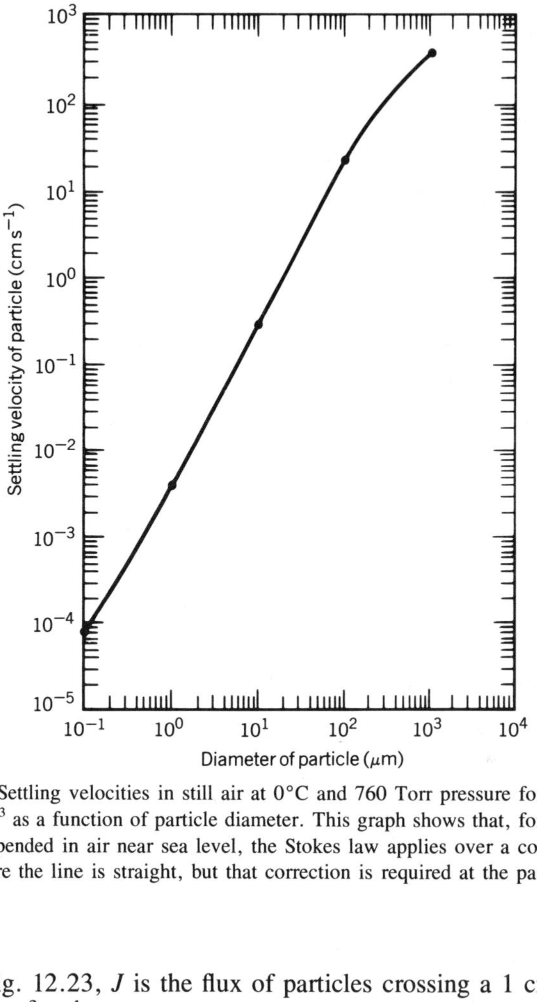

FIGURE 12.22. Settling velocities in still air at 0°C and 760 Torr pressure for particles having a density of 1 g cm^{-3} as a function of particle diameter. This graph shows that, for spherical particles of unit density suspended in air near sea level, the Stokes law applies over a considerable range of particle sizes, where the line is straight, but that correction is required at the particle size extremes (from LBL, 1979).

As shown in Fig. 12.23, J is the flux of particles crossing a 1 cm^2 plane in 1 s (i.e., number cm^{-2} s^{-1}). The constant $\mathbf{D}$ is known as the diffusion coefficient and is simply the proportionality constant relating the flux to the concentration gradient. (Fick's first law applies not only to particles but also to gas and liquid molecules; see Section 4-D-2.)

Intuitively, one might expect that the rate of diffusion would increase with temperature, and decrease with increasing gas viscosity and particle size. Indeed, this is observed to be the case because all these parameters contribute to the diffusion coefficient $\mathbf{D}$, given by

$$\mathbf{D} = \frac{kTC}{3\pi\eta D} \qquad \text{(V)}$$

k is the Boltzmann constant (1.38×10^{-23} J deg^{-1}), T the temperature (in °K), C the Cunningham correction factor given by Eq. (S), η the viscosity of the gas

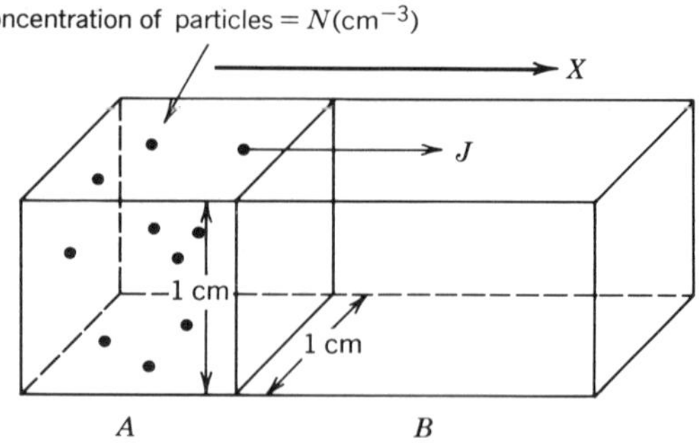

FIGURE 12.23. Schematic diagram showing diffusion of particles from a region of high concentration (side A) to a region of low concentration (side B). J = flux of particles passing through the plane dividing A and B.

in which the particles are suspended, and D is the particle diameter. [The reader is referred to Hinds (1982) for a derivation of Eq. (V).] For large particles where $C \simeq 1$ (i.e., the slip correction is negligible), the rate of diffusion varies inversely with the particle diameter. For very small particles, the rate varies with $1/D^2$, which contributes to making diffusion a major transport mechanism for particles $\lesssim 0.1$ μm. It is this Brownian diffusion which helps to carry small particles through the boundary layer to surfaces where they may stick on impact, a process known as dry deposition and discussed in Section 11-D-2.

The relative importance of Brownian diffusion and gravitational settling in the deposition of particles may be seen by calculating the total deposition of particles onto a horizontal surface by these two processes in a given period of time under certain conditions. Table 12.6 shows the results of such a calculation for particle

TABLE 12.6. Cumulative Deposition of Unit Density Particles onto a Horizontal Surface from Unit Aerosol Concentrations[a] During 100 s by Diffusion and Gravitational Settling

	Cumulative Deposition		
Diameter (μm)	Diffusion (number cm^{-2})	Settling (number cm^{-2})	Ratio Diffusion/Settling
0.001	2.5	6.5×10^{-5}	3.8×10^{4}
0.01	0.26	6.7×10^{-4}	390
0.1	2.9×10^{-2}	8.5×10^{-3}	3.4
1.0	5.9×10^{-3}	0.35	1.7×10^{-2}
10	1.7×10^{-3}	31	5.5×10^{-5}
100	5.5×10^{-4}	2500	2.2×10^{-7}

Source: Hinds, 1982.

[a]This assumes an aerosol concentration of 1 particle cm^{-3} outside the gradient region.

diameters from 0.001 to 100 μm, assuming spherical particles of unit density with a constant concentration of 1 particle cm^{-3} outside the gradient region; also shown is the ratio of the number of particles deposited by diffusion compared to the number deposited by gravitational settling. At a diameter of ~ 0.2 μm, the two mechanisms become equal, with diffusion greatly exceeding gravitational settling for particles in the Aitken nuclei range. (Note that if the *mass* deposited were calculated, the results would be quite different.)

It should be kept in mind that these calculated rates of diffusion and gravitational settling are only applicable to still air. In fact, in the atmosphere the air is rarely still and is usually undergoing some degree of turbulent motion. In this case, the transport of particles becomes more complex and faster due to the velocity gradients and contorted patterns of air flow; however, a discussion of this is outside the scope of this book.

12-B-3. Light Scattering and Absorption and Its Relationship to Visibility Reduction

12-B-3a. Light Scattering and Absorption

As discussed in Section 3-A, solar radiation passing through the atmosphere to the earth's surface is both scattered and absorbed by gases and particles. The intensity of radiation striking the surface can be expressed in the form of a Beer–Lambert type of law (see Section 2-A-2):

$$\frac{I}{I_0} = e^{-b_{\text{ext}}\ell} \tag{W}$$

where I_0 and I are the incident and transmitted light intensities respectively, ℓ is the pathlength of the light beam, and b_{ext} is known as the extinction coefficient and has units of $(\text{length})^{-1}$. This extinction coefficient, representing the total reduction in light intensity due to scattering and absorption of light by gases and particles, is the sum of two terms,

$$b_{\text{ext}} = b_g + b_p \tag{X}$$

where b_g is the extinction due to gases and b_p that due to particles. Each of these terms can be broken down into contributions from light scattering and absorption so that Eq. (X) becomes

$$b_{\text{ext}} = b_{ag} + b_{sg} + b_{ap} + b_{sp} \tag{Y}$$

where b_{ag} and b_{sg} are the light extinctions due to *g*aseous *a*bsorption and *s*cattering, and b_{ap} and b_{sp} are those due to *p*articulate *a*bsorption and *s*cattering.

Scattering and absorption of light by gases have already been discussed in Chapters 2 and 3. In terms of the absorption of visible light in the troposphere by gases, only NO_2 is believed to contribute significantly. However, light absorption by NO_2 is usually much less than the total light scattering and absorption by particles. Thus when NO_2 is present in sufficient concentrations to absorb light, the atmosphere usually also contains relatively high concentrations of other pollutants, including particles, as well. As a result, light scattering and absorption by particles usually exceed light absorption by NO_2 in most situations. For example, in studies of Denver's *brown cloud*, it was shown that of the total extinction excluding Rayleigh scattering by gases, 7% was due to light absorption by NO_2, and 93% was due to light scattering and absorption by particulate matter (Groblicki et al., 1981; Waggoner et al., 1983). [The exception to this may be power plant plumes containing high levels of NO which rapidly oxidize to NO_2 upon dilution with air (Hegg et al., 1985)] Thus we concentrate here on the absorption and scattering of light by particles.

Historically, the symbol b has been used both for the total extinction coefficient and for the individual contributions due to light absorption and scattering by particles and gases. However, the Radiation Commission of the International Association of Meteorology and Atmospheric Physics (IAMAP, 1978) has recommended use of the symbol σ instead. Thus the reader may find in the literature the terms σ_{ag}, σ_{sg}, σ_{ap}, and σ_{sp}, respectively, rather than b_{ag}, b_{sg}, b_{ap}, and b_{sp}. However, in keeping with the common practice in the literature in this area, and to avoid confusion with light absorption cross sections by gases defined in Section 2-A-2, we use the symbol b.

The scattering of light by particles falls into three regions depending on the size of the particles relative to the wavelength (λ) of the light: (1) particle diameter $D \ll \lambda$, known as Rayleigh scattering, (2) $D \simeq \lambda$, known as Mie scattering, and (3) $D \gg \lambda$. Since we are concerned here only with visible and near ultraviolet light in the actinic region, that is, 290 nm $\lesssim \lambda \lesssim$ 750 nm, the first case corresponds to particles with $D \lesssim 0.03$ μm and the third to particles with $D \gtrsim 10$ μm. Particles with sizes between these two extremes fall in the second category where $D \simeq \lambda$; as we have seen, this is the most important size regime for atmospheric particles.

A common convention used when discussing light scattering as a function of particle size is to define a dimensionless size parameter α, which is the ratio of the circumference of the particle to the wavelength of the incident light:

$$\alpha = \frac{\pi D}{\lambda} \tag{Z}$$

Very small particles ($\alpha \ll 3$) behave like gaseous molecules in scattering light and hence produce Rayleigh scattering described in Section 3-A-2. Because the particles or molecules undergoing this type of scattering are small relative to the incident wavelength, the entire species is subjected at any instant of time to what appears to be a uniform electromagnetic field; this creates a dipole that oscillates with the changing electromagnetic field of the light wave and reradiates the energy in all directions. Thus Rayleigh scattering is symmetric in the forward and backward directions relative to the incident light beam and, as we saw in Chapter 3, varies as λ^{-4}.

Very large particles, on the other hand, that is, those with $D >> \lambda$ ($\alpha >> 3$), undergo geometric scattering, where the light beam refracted through the particle can be treated using classical optics. Between these two regimes where $D \simeq \lambda$ ($\alpha \simeq 3$) much more complex light scattering occurs, known as Mie scattering.

Because particles undergoing Mie scattering have dimensions of the same order as the wavelength of the incident light, the electromagnetic field of the light wave is not uniform over the entire particle at one instant of time, and a three-dimensional charge distribution is set up in the scattering particle. In 1908 Mie developed the solutions for scattering of light of wavelength λ by a homogeneous sphere of diameter D. As shown in Fig. 12.24, light is considered to be incident on the sphere and to be scattered at various angles θ to the direction of the unscattered beam. The incident and scattered beams are shown as the combination of two independent polarized beams; one (I_1) has its electric vector perpendicular to the scattering plane defined by the incident and the scattered beams, and the other (I_2) is parallel to it. The intensity of light at a distance R and a scattering angle θ from the particle is given by

$$I(\theta, R) = \frac{I_0 \lambda^2 (i_1 + i_2)}{8\pi^2 R^2} \tag{AA}$$

where I_0 is the intensity of the incident light beam (taken as unpolarized) and i_1 and i_2 are known as the Mie intensity parameters for the perpendicular polarized and parallel polarized components of the scattered light, respectively. The Mie intensity parameters are a complex function of the refractive index (m) of the scatterer, the size parameter (α), and the scattering angle (θ). For further mathematical details and descriptions of Mie scattering, the reader is referred to the books by Van de Hulst (1957), Kerker (1969), and Bohren and Huffman (1983).

The refractive index of a material, m, is defined as the ratio of the speed of

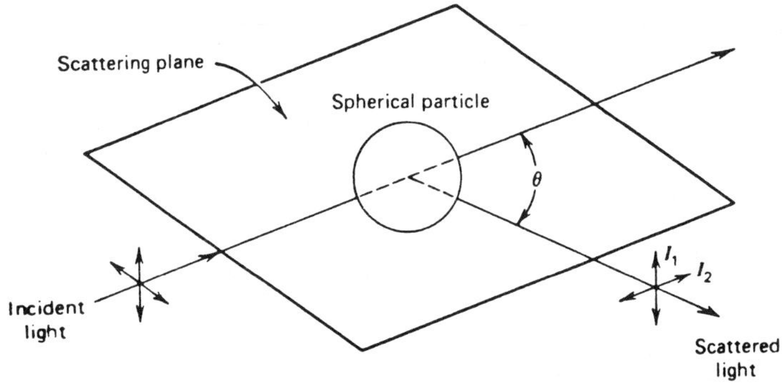

FIGURE 12.24. Diagram showing scattering angle, scattering plane, and the polarized components of scattered light (from Hinds, 1982).

light (c) in a vacuum to that (v) in the material, that is, $m = c/v$. Because light travels more slowly in materials than in air, $m > 1$. The index of refraction, m_a, of materials that absorb light as well as scatter it is expressed in the form of a complex number (See Hinds, 1982)

$$m_a = m_r(1 - ai)$$

where $i = \sqrt{-1}$, m_r is the real refractive index and a is a constant which is directly proportional to the absorption coefficient of the material as well as proportional to the wavelength. Values of the index of refraction for $\lambda = 589$ nm of some materials either found in the atmosphere or used to calibrate instruments which measure particle size using light scattering (see below) are given in Table 12.7. A typical refractive index for a dry aerosol which absorbs light in the atmosphere is $1.5 - 0.02\,i$ (Covert et al., 1980).

Unlike Rayleigh scattering which occurs equally in the forward and backward directions, Mie scattering is predominantly in the forward direction, except for the

TABLE 12.7. Index of Refraction at 589 nm for Some Species Found in the Atmosphere or Used for Instrument Calibration[a]

Species	Index of Refraction
Vacuum	1.0
Water vapor	1.00025
Air	1.00029
Water (liquid)	1.333
Ice	1.309
Rock salt	1.544
Sodium chloride in aqueous solutions	1.342–1.378[b]
Sulfuric acid in aqueous solutions	1.339–1.437[c]
Benzene	1.5011
α-Pinene	1.465
d-Limonene	1.471
Nitrobenzene	1.550
Dioctylphthalate	1.49
Oleic acid	1.46
Polystyrene latex	1.6
Carbon	$1.59 - 0.66\,i$[d,f]
Iron	$1.51 - 1.63\,i$[d]
Magnetite (Fe_3O_4)	$2.58 - 0.58\,i$[e]
Copper	$0.62 - 2.63\,i$[d]

[a]Data from the *Handbook of Chemistry and Physics*, unless otherwise noted (~20–25°C).

[b]For solution densities from 1.035 to 1.189.

[c]For solution densities from 1.028 to 1.811.

[d]From Hinds, 1982.

[e]From Huffman and Stapp, 1973.

[f]At $\lambda = 491$ nm.

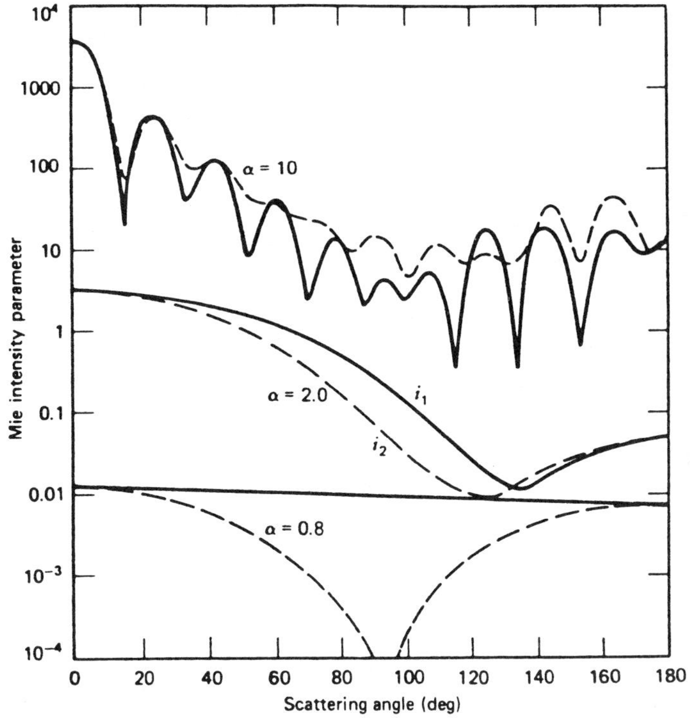

FIGURE 12.25. Mie intensity parameters versus scattering angle for water droplets ($m = 1.333$) having $\alpha = 0.8$, 2.0, and 10.0. Solid lines are i_1, and dashed lines are i_2 (from Hinds, 1982).

smallest particles. This can be seen in Fig. 12.25 which shows the Mie intensity parameters i_1 and i_2 as a function of the scattering angle θ for three different values of the size parameter α defined by Eq. (Z), assuming the droplets are composed entirely of liquid water (i.e., $m = 1.333$). For $\alpha = 2.0$ and 10, (i.e., $D/\lambda = 0.64$ and 3.2) both i_1 and i_2 fall from a maximum value at $\theta = 0$ as θ increases. For smaller particles with $\alpha = 0.8$ (i.e., $D/\lambda = 0.26$), i_2 initially falls as θ increases, but then rises again as θ approaches 180°, corresponding to backscattering; for these particles i_1 decreases only slightly from 0° to 180°. However, in all cases, i_1 and i_2, and hence the scattered light intensity [Eq. (AA)], show their maximum values at $\theta = 0°$, corresponding to forward light scattering.

The variation of scattered light intensity with θ as typified by Fig. 12.25 clearly becomes more complex as the particle size increases, with sharp oscillations seen at $\alpha = 10$. However, recall that this is for a spherical homogeneous particle of a fixed size and for monochromatic light (e.g., a laser); when the particle is irregular in shape, these oscillations are far less prominent. This is also true for a group of particles of various sizes, that is, a polydisperse aerosol, where the overall scattering observed is the sum of many different contributions from particles of various sizes. Finally, non-monochromatic light and fluctuations in polarization also help to smooth out the oscillations.

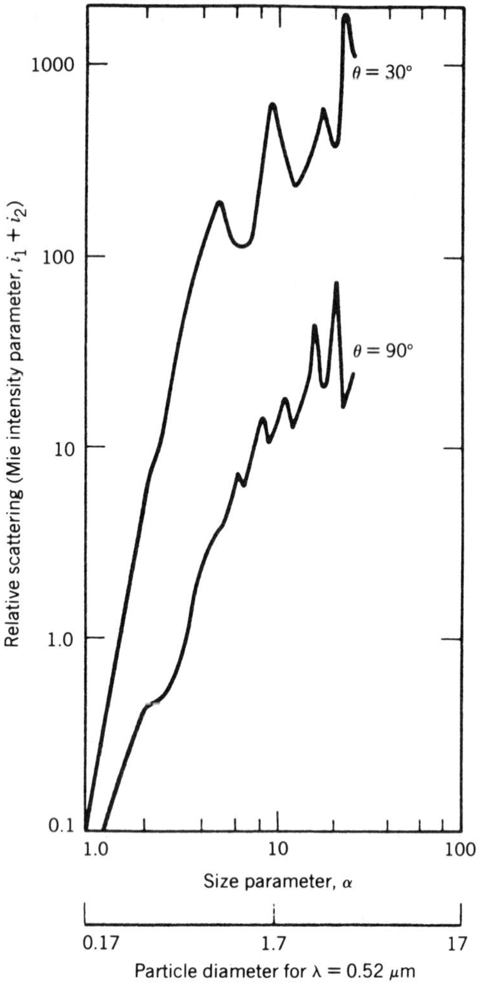

FIGURE 12.26. Relative scattering [Mie intensity parameter $(i_1 + i_2)$] versus size parameter for water droplets ($m = 1.33$) at scattering angles of 30° and 90° (from Hinds, 1982).

The dependence of Mie scattering on particle size can be seen in Fig. 12.26 which shows the sum of the Mie intensity parameters $(i_1 + i_2)$ as a function of the size parameter α for two scattering angles, $\theta = 30°$ and $\theta = 90°$, respectively. It is seen that Mie scattering generally increases with size over this range of values of α and, as seen in Fig. 12.25, scattering is more pronounced in the forward direction (i.e., at smaller values of θ).

It is also noteworthy that in Fig. 12.26 the function becomes smooth and approaches a variation with D^6 as the size parameter decreases toward small values. This is expected, since in the limit of very small particles or molecules, Mie theory reduces to Raleigh scattering, which, as seen in Section 3-A-2, varies with D^6.

Not only scattering, but also absorption of light by particles can also occur in the atmosphere; the radiant energy absorbed is then converted to heat. As discussed in Section 12-B-3b, graphitic carbon is believed to be the species responsible for most of the light absorption occurring in typical urban atmospheres; although iron oxides such as magnetite also absorb light strongly (Table 12.7), their concentrations are not believed to be sufficiently high to contribute significantly to light absorption in urban areas.

Values of b_{sp}, the extinction coefficient for light scattering by particles, measured in ambient air using an integrating nephelometer (see Section 12-E-2e) range from $\sim 10^{-3}\ m^{-1}$ in highly polluted urban areas to $\sim 10^{-7}\ m^{-1}$ in remote locations such as the Mauna Loa Observatory in Hawaii. The extinction coefficient for particulate absorption, b_{ap}, is smaller, from $\sim 10^{-4}$ to $\lesssim 10^{-8}\ m^{-1}$. However, the relative contributions of light scattering and absorption at any particular location will clearly depend on the nature of the sources; this is discussed in more detail with examples in Section 12-B-3b.

12-B-3b. *Relationship of Light Scattering and Absorption to Visibility Reduction*

One of the most evident manifestations of anthropogenic air pollution is the production of a *haze* which causes a reduction in visibility, that is, in *visual range*. Visual range is defined as the distance at which a black object can just be distinguished against the horizon. Two factors enter into visual range: visual acuity and contrast. In the daytime atmosphere, particles reduce the contrast perceived by an observer by scattering light from the object out of the line of sight to the observer's eyes; simultaneously, sunlight is scattered into the line of sight, making dark objects appear lighter. The result is a decrease in the contrast between the object and the horizon. At night, scattering of light out of the visual path decreases the contrast and hence the source intensity becomes a factor in visual range as well.

The Koschmieder equation has been shown to approximate the change in contrast of an object with distance away from an observer (Middleton, 1952); note that it has a form similar to that of the Beer–Lambert law:

$$\frac{C}{C_0} = e^{-b_{\text{ext}}L} \tag{BB}$$

In Eq. (BB), C_0 is the contrast relative to the horizon (or background) of an object seen at the observation point itself, that is, at a distance $L = 0$, and C is the contrast at the distance L. The contrast is defined as the ratio of the brightness of the object (B_0) to that of the horizon or background (B_H) minus one:

$$C = \frac{B_O}{B_H} - 1 \tag{CC}$$

For example, a black object at zero distance has a brightness of zero (i.e., it absorbs all the visible light) and hence has a C_0 of -1. b_{ext} in Eq. (BB) is the total extinction as defined above in Eq. (W). Observers typically can differentiate objects on the horizon if $C/C_0 \simeq 0.02–0.05$. A contrast of 0.02, corresponds, using Eq. (BB), to a visual range L_V of

$$L_V = \frac{\ln C_0/C}{b_{\text{ext}}} = \frac{3.9}{b_{\text{ext}}} \tag{DD}$$

For a contrast of 0.05, $L_V = 3.0/b_{\text{ext}}$.

In a clean particle-free atmosphere, some light scattering occurs due to the Rayleigh scattering by gases. For this scattering, $b_{sg} = 1.5 \times 10^{-5}\ \text{m}^{-1}$ integrated over the solar spectrum at sea level and 25°C (Ouimette et al., 1981). If light absorption by gases is negligible [as it usually is unless NO_2 is present (Charlson et al., 1972)], then $b_{\text{ext}} = b_{sg} = 1.5 \times 10^{-5}\ \text{m}^{-1}$, and the visual range is ~200–260 km for contrasts of 0.05–0.02. While this is an approximation that depends on the nature of the object and on the observer, it does give some idea of the visual range that can be expected in clean air.

Visual ranges can vary from hundreds of kilometers in remote areas to only a few kilometers in heavily polluted urban areas. In the latter case, most of the loss in visibility is due to light scattering and, to some extent, light absorption by suspended particulate matter.

It should be noted that the definition of visual range in Eq. (DD) is not always in accord with visual ranges reported from qualitative sightings of surrounding landmarks, as is done, for example, at airport observation towers. There are a number of factors which might influence this, such as the targets not being black or there being differences between various observers. In general, the airport visual ranges are less than those predicted from Eq. (DD) (Stevens et al., 1983; Waggoner, 1983; Lodge, 1983). Indeed, based on airport visual range observations, Ozkaynak et al. (1985) suggest that the use of the coefficient 3.9 in equation (DD) is optimistic, and that a value less than half that, 1.8, may be more appropriate in urban areas.

Most of the light scattering by particles in the atmosphere is due to particles in the size range 0.1–1 μm as shown by calculations carried out during World War II for screening smoke particle sizes (Sinclair, 1950). This can be seen in Fig. 12.27 which shows the scattering coefficient of a single particle per unit volume as a function of the particle diameter for spheres with a refractive index of 1.50 and light of wavelength 550 nm. The portion of the total extinction coefficient due to particle scattering, b_{sp}, can be obtained by multiplying the curve in Fig. 12.27 by the particle volume size distribution, that is, by the curve of $\Delta V/\Delta \log D$ versus $\log D$.

For example, Fig. 12.28*a* shows the measured volume distribution of one ambient aerosol sample. When this volume distribution is multiplied by the size distribution of the scattering coefficient per unit volume in Fig. 12.27, one obtains the calculated curve for light scattering in Fig. 12.28*b*. It is seen that the particles in the 0.1–1 μm diameter range, that is, in the accumulation mode, are clearly expected to predominate the light scattering.

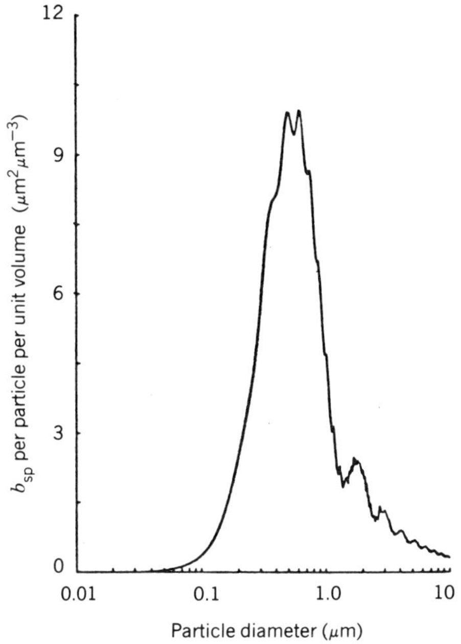

FIGURE 12.27. Scattering coefficient per particle divided by particle volume plotted as a function of diameter. The particles are assumed to be spheres of refractive index 1.50 and the light has λ = 550 nm (from Waggoner and Charlson, 1976).

This is supported by the excellent correlation between the total aerosol volume of particles with diameters in the 0.1–1 μm range and the experimentally determined values of b_{sp} obtained using a nephelometer, as shown in Fig. 12.29. The slopes of lines such as that in Fig. 12.29, however, depend critically on the nature and history of the air mass and can vary by more than a factor of 10 from clean, non-urban air to highly polluted air in the vicinity of sources. For example, Sverdrup and Whitby (1980) have shown that the ratio of submicron aerosol volume to b_{sp}, which corresponds to the slope of the line in Fig. 12.29, varies from 5 to 80, depending on the nature of the air mass.

Consistent with the relationship between the aerosol fine particle volume and the particle scattering coefficient, a number of studies have shown that the fine particle mass and b_{sp} are also related. Figure 12.30 shows the scattering coefficient observed in studies in Denver, Colorado, in 1978 by Groblicki and co-workers (1981) as a function of the observed mass in the fine and coarse particle ranges, respectively. It is seen that a good linear relationship exists between b_{sp} and the fine particle mass (FPM) but not between b_{sp} and the coarse particle mass. This has been observed in a number of areas ranging from pristine to industrial, with the ratio of the scattering coefficient to the fine particle mass concentration (b_{sp}/FPM) being approximately 3 in many areas (Waggoner et al., 1981).

As discussed above, Mie scattering by particles depends on the index of refraction (m) and hence on their chemical composition, which is treated in Section 12-

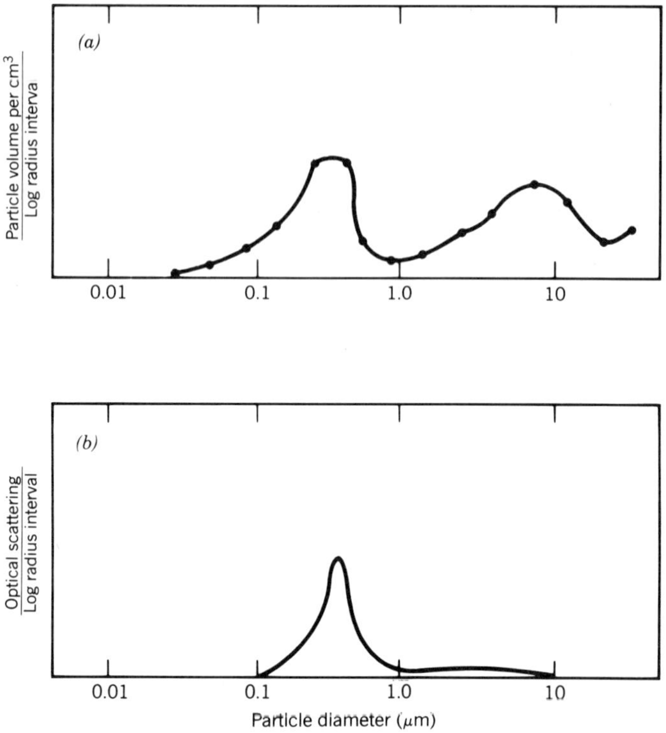

FIGURE 12.28. (*a*) Aerosol particle size distribution measured at Pomona during 1972 State of California Air Resources Board ACHEX program. (*b*) Calculated optical scattering by particles, b_{sp}, for measured size distribution (from Waggoner and Charlson, 1976).

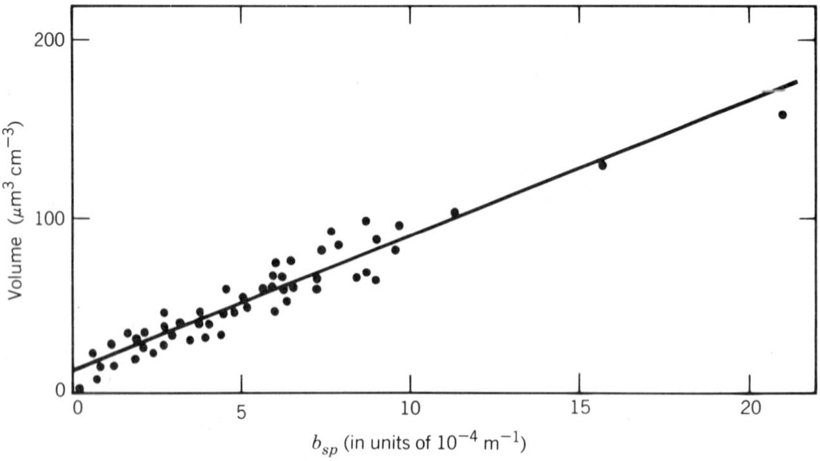

FIGURE 12.29. Plot of measured aerosol fine particle volume (including only those particles of 0.1–1.0 μm diameter) versus measured b_{sp}. Measurements were part of State of California Air Resources Board ACHEX program (from Waggoner and Charlson, 1976; data supplied by Dr. Clark of North American Rockwell).

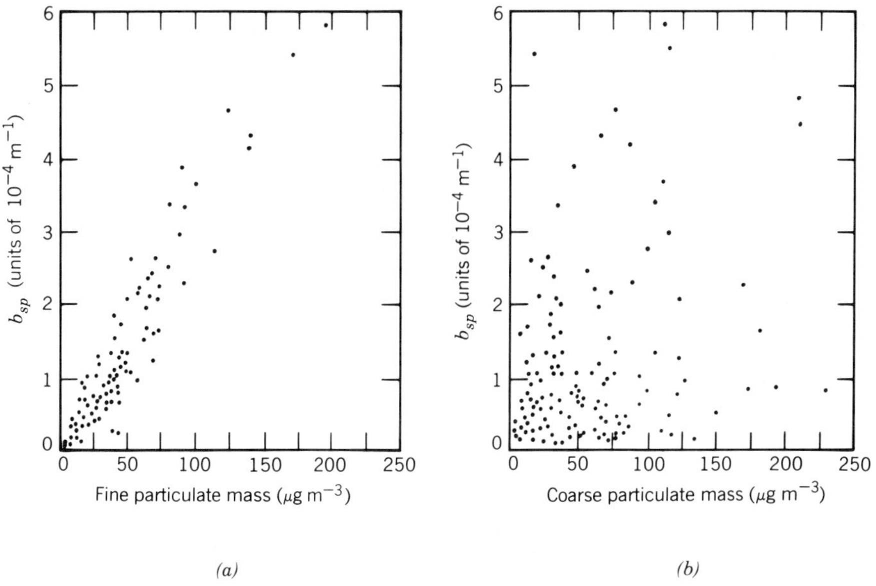

FIGURE 12.30. Correlation of b_{sp} with fine and coarse particulate mass (from Groblicki et al., 1981).

D below. Particles found in the atmosphere generally contain a variety of organics and inorganics with differing values of m; however, it has been found that light scattering is more closely related to the presence and concentration of certain components than to others. Thus in a number of studies, the chemical composition of the fine particle fraction has been determined along with observed scattering coefficient b_{sp}. A best fit to these data is then found in the form of an equation such as (EE),

$$b_{sp} = a_0 + \sum a_i M_i \tag{EE}$$

where M_i is the mass concentration of the ith chemical species, a_i is the mass scattering coefficient for this species, and a_0 is a constant for that data set. In many studies, it has been found that b_{sp} is most highly correlated with the sulfate ion and nitrate ion concentrations, and usually with carbon as well. For example, Table 12.8 shows the mass scattering coefficients (a_i) for sulfate, nitrate, carbonaceous aerosol, and the remainder of the mass, Δmass, (i.e., total mass–sulfate–nitrate–carbonaceous aerosol) reported in some studies. Nitrate has been recognized increasingly as making an important contribution to light scattering; indeed, note that the value reported for the mass scattering coefficient for nitrate seems to have increased steadily as a function of year (Table 12.8). Appel and co-workers (1985) point out that this may be due to the formation, during sampling, of nitrate on filters from gas phase nitric acid (see Section 12-E-4e); in this case, the particulate nitrate concentration would be overestimated and hence the value of the mass scattering coefficient underestimated. Since this artifact formation of particulate

TABLE 12.8. Some Reported Mass Scattering Coefficients (a_i) in Units of $m^2\ g^{-1}$ for Fine Particles Containing Sulfate, Nitrate, and Carbon in Various Locations

Location	Sulfate as $(NH_4)_2SO_4$	Nitrate as NH_4NO_3	Carbonaceous Aerosol[a]	Δ mass	Reference
Los Angeles (LA)	5.4	$2.0 + 4.6\ \mu^{2}$[c]	N.S.[b]	2.0	White and Roberts, 1977
Denver	5.9	2.5	3.2	1.5	Groblicki et al., 1981
Detroit	6.2	N.S.	3.1	1.7	Wolff et al., 1982
China Lake, CA	4.3	*d*	1.5	1.0[e]	Ouimette and Flagan, 1982
Portland, OR	4.9	4.4	5.0	2.1	Shah et al., 1984
Average of San Jose, LA and Riverside	$3.6 + 5.9\ \mu$	$4.9 + 4.8\ \mu$	4.5[a]	0.4[f]	Appel et al., 1985[g]
Western Netherlands	4.8	8.6	n.d.[h]	4.7	Diederen et al., 1985

Source: Adapted from Shah et al., 1984.

[a]Combination of organic and elemental carbon except in studies of Appel et al., (1985) where the coefficient is for elemental carbon only.

[b]N.S. = not significant.

[c]μ = relative humidity/100.

[d]Present only at very low concentrations in these samples.

[e]In addition to sulfate, carbonaceous aerosol, and the remainder, b_{sp} was found to be correlated to the crustal species Fe, Ca, and Si with $a_i = 2.4\ m^2\ g^{-1}$.

[f]Not significantly different from zero at $p = 0.90$.

[g]Data adjusted to reflect $(NH_4)_2SO_4$ and NH_4NO_3 stoichiometry. Correlation of b_{sp} with coarse sulfate was also found with $a_i = 13.4\ m^2\ g^{-1}$; this may be due to a correlation of coarse sulfate with some other efficient light scatterer such as sea salt aerosol.

[h]n.d. = not determined.

nitrate has been recognized and addressed in the later studies, the higher values of the mass scattering coefficient for nitrate are probably more accurate. In this case, nitrate is of equal or greater importance compared to sulfate in light scattering.

In addition to sulfate, nitrate, and carbon, some correlation of b_{sp} to the remainder of the aerosol mass after these components have been subtracted out is generally observed.

While sulfate, nitrate, and carbon compounds thus appear to be most highly correlated to light scattering by atmospheric particles, it must be kept in mind that these components must be in the size range which most efficiently scatters light (Fig. 12.27) if they are to be effective. For example, in some sulfate episodes in St. Louis where sharp increases in aerosol sulfur were observed, there was no corresponding increase in light scattering; this was attributed to the formation of H_2SO_4 by the gas phase oxidation of SO_2, followed by its incorporation into particles too small to scatter light efficiently (Huntzicker et al., 1984). Similarly, in one study in Los Angeles, b_{sp} was not found to be significantly correlated to the carbonaceous aerosol even though it accounted for a significant fraction of the total aerosol mass; this was also rationalized on the basis that most of the carbon was in particles which were too small to scatter light efficiently (White and Roberts, 1977).

A major factor found to affect light scattering by particles is the ambient relative humidity (RH) which changes the size and the refractive index of particles. Thus, as discussed in Section 12-D, atmospheric aerosols are hygroscopic, taking up and releasing water as the RH changes. This is not surprising since many of the chemical components are themselves deliquescent in pure form; for example, sodium chloride, the major component of sea salt, deliquesces at an RH of 75%, whereas ammonium sulfate, $(NH_4)_2SO_4$, and ammonium nitrate, NH_4NO_3, deliquesce at 79% RH and 62%, respectively. The uptake of water with increasing RH causes an increase in both mass and radius, and a decrease in refractive index; the net effect of all these factors is an increase in light scattering. This can be seen in Fig. 12.31 where the light scattering coefficient $b_{\text{scat}} = b_{sp}$, measured using an integrating nephelometer, for some ambient particles was found to increase linearly with the liquid water content of the aerosols.

In some studies, the contribution of water associated with the particles to light scattering has been explicitly investigated by adding a term b_{sw} to Eq. (Y):

$$b_{\text{ext}} = b_{ag} + b_{sg} + b_{ap} + b_{sp} + b_{sw} \tag{FF}$$

One example, Figs. 12.32 and 12.33, shows the variation of b_{sw} with relative humidity in Denver air as a function of the time of day, as well as on different

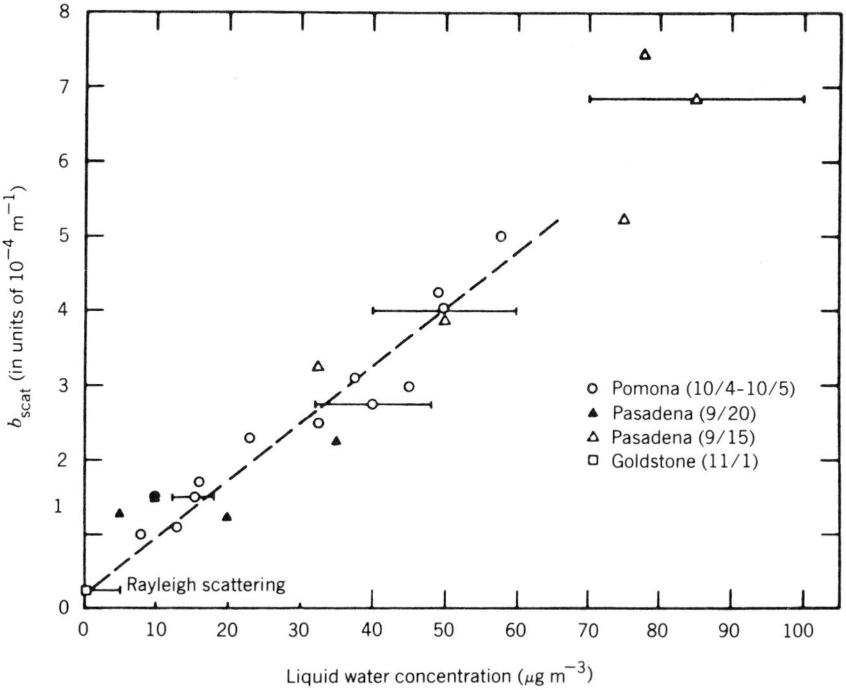

FIGURE 12.31. Comparison of light-scattering coefficient with the liquid water concentration in atmospheric aerosols at some locations in California in 1972 (from Ho et al., 1974).

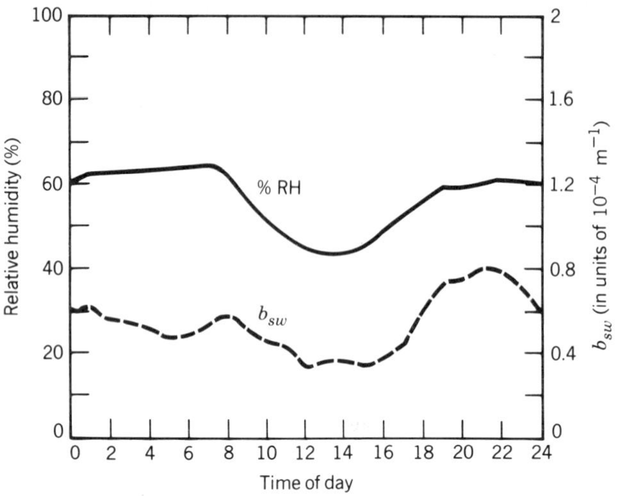

FIGURE 12.32. Average diurnal variation of b_{sw} and relative humidity (from Groblicki et al., 1981).

days. With the exception of three days when it snowed, b_{sw} and the relative humidity follow each other closely (Groblicki et al., 1981).

Both simple and complex expressions for the dependence of b_{sw} on the relative humidity such as those in Eqs. (GG) and (HH) have been used to match atmospheric observations (c.g., see White and Roberts, 1977; Cass, 1979; Groblicki, et al., 1981; Trijonis, 1982; Appel et al., 1985; Sloane and Wolff, 1985):

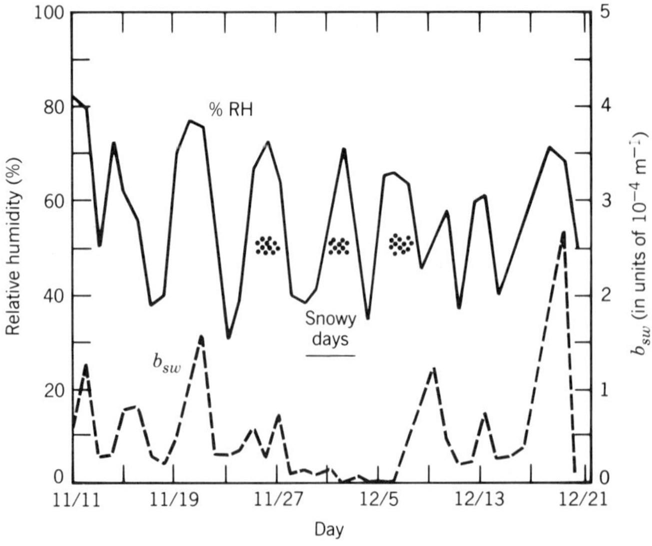

FIGURE 12.33. Temporal comparison of b_{sw} and relative humidity (from Groblicki et al., 1981).

$$b_{sw} = \frac{a[SO_4^{2-}]}{1 - \mu} + \frac{b[NO_3^-]}{1 - \mu} \quad \text{(GG)}$$

$$b_{sw} = a + b[SO_4^{2-}]\,\mu^x + c[NO_3^-]\,\mu^y \quad \text{(HH)}$$

In these expressions, a, b, and c are constants for a given aerosol determined by finding the best fit to the data, and $\mu = 0.01 \times \%RH$. While correlations have not been extensive, it is clear that relative humidity plays a major role in light scattering by particles.

Not only light scattering, but also light absorption by particles can occur. It is generally agreed that the major contributor to light absorption is *black* or *graphitic* carbon, often referred to as elemental carbon. However, as discussed by Chang et al. (1982), carbon particles in the air are made up of a number of crystallites 20–30 Å in diameter, with each crystallite consisting of several carbon layers having the hexagonal structure of graphite, Fig. 12.34. Because of the presence of defects, dislocations, and discontinuities, there are unpaired electrons which constitute active sites in the carbon; during the formation of the carbon particle in combustion processes, these active sites can react with gases to incorporate other elements such as oxygen, nitrogen, and hydrogen into the structure. Thus *elemental*, *black*, or *graphitic* carbon found in atmospheric particles is not chunks of highly structured pure graphite, but rather is a related, but more complex, three-dimensional array of carbon with small amounts of other elements.

Being black, this atmospheric carbon is a strong absorber of visible radiation and is believed to be the major species responsible for light absorption by particles. Its contribution to light extinction varies geographically and temporally as well due to the distribution of the combustion sources which produce graphitic carbon. For example, as seen in Table 12.9, wood-burning fireplaces and diesels are major sources of elemental carbon; thus areas with large numbers of these two sources would be expected to have more graphitic carbon in the atmospheric aerosol, and hence more light absorption. Where wood burning is significant, more particulate graphitic carbon would be expected in winter than in summer.

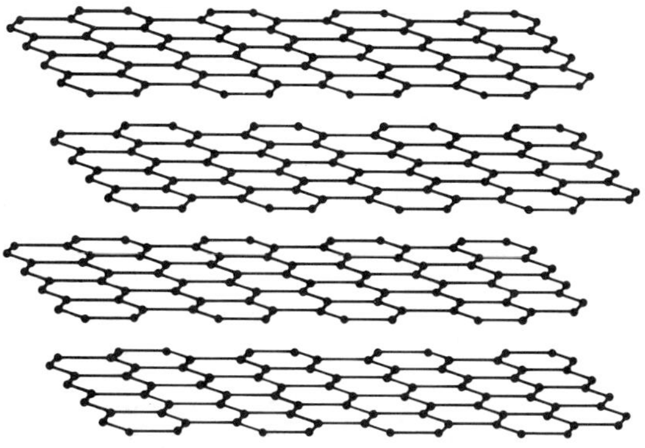

FIGURE 12.34. Structure of elemental carbon.

TABLE 12.9. Fuel-Specific Emission Rates (g carbon per kg fuel)

Source	Organic Carbon	Elemental Carbon
Furnace[a]		
Normal	0.00037	0.00018
Rich	0.0070	0.12
Fireplace		
Hardwood	4.7	0.39
Softwood	2.8	1.3
Automobiles[b]		
Precatalyst		
Sea level FTP	0.040	0.014
High altitude FTP	0.240	0.130
Catalyst		
Sea level FTP	0.014	0.011
High altitude FTP	0.028	0.033
Diesel		
Sea level FTP	0.89	3.4
High altitude FTP	0.96	2.8

Source: Muhlbaier and Williams, 1982.

[a] 5.18×10^7 J per kg natural gas.

[b] Based on the actual mileage of the individual cars tested.

Table 12.10 gives the particulate absorption coefficients b_{ap} for different types of locations from urban-industrial to remote. In urban-industrial areas, b_{ap} varies from approximately 50% to more than 100% of b_{sp}, whereas in remote areas, b_{ap} is much less, $\leq 15\%$ b_{sp}. This is not surprising since the combustion sources producing graphitic carbon tend to be in urban-industrial areas.

Graphitic carbon also scatters light; in one study, its mass scattering coefficient was found to be approximately the same as that for sulfate (Appel et al., 1985). Because of its contribution to both scattering and absorption of light, graphitic carbon tends to play a proportionally much greater role in light extinction than its contribution to the particulate mass would suggest. For example, in Denver, graphitic carbon was found to represent 15% of the fine particle mass, but contribute ~35% to the total light extinction (Groblicki et al., 1981); at Zilna Mesa, Arizona, during one sampling period, graphitic carbon comprised 1.4% of the total aerosol mass and 3.5% of the fine particle mass but was responsible for 15% of the light extinction (Ouimette and Flagan, 1982). Similarly, in one Los Angeles area study, graphitic carbon was found to represent ~8.5–10% of the fine particle mass but to account for ~14–21% of the total light extinction (Pratsinis et al., 1984).

The fraction of the total carbon in atmospheric particulate matter which is in the form of graphitic carbon varies substantially, depending on the location and

TABLE 12.10. Some Mean Light Absorption (b_{ap}) and Scattering (b_{sp}) Coefficients Observed in a Variety of Locations[a]

Type of Location	Site	b_{ap} (10^{-4} m^{-1})	b_{sp} (10^{-4} m^{-1})	$\frac{b_{ap}}{b_{sp}}$
Urban-industrial	Seattle, WA	0.27	0.30	0.90
	Portland, OR	0.86	1.00	0.86
	Denver, CO	0.60	0.56	1.07
	Phoenix, AR	1.18	2.09	0.56
Urban-residential	Seattle, WA	0.10	0.50	0.20
	St. Louis, MO	0.37	1.58	0.23
Remote	Mauna Loa Observatory, Hawaii	0.00045	0.009	0.05
	Abastumani Observatory, USSR	0.057	0.46	0.12

[a]Data from Waggoner et al., 1981.

likely on the season as well. Thus the ratio of graphitic to total carbon has been observed to be 0.15–0.20 in rural and remote areas, increasing to ~0.3–0.6 in suburban and urban areas (Wolff et al., 1982; Gray et al., 1984; Grosjean, 1984; Pratsinis et al., 1984).

12-C. MECHANISMS OF AEROSOL FORMATION

The formation of secondary particulate matter by chemical reactions in the atmosphere may occur by several different mechanisms as is depicted schematically in Fig. 12.35. These include: (1) reaction of gases to form low-vapor-pressure products (e.g., the reaction of cycloalkenes with ozone to give multi-functional oxygenated products), which combine with other condensable molecules or molecular clusters to form new particles, or which condense on preexisting particles, (2) reaction of gases on the surfaces of existing particles to form condensed phase products (e.g., the reaction of NO_2 or HNO_3 with sea salt particles to form $NaNO_3$), and (3) chemical reactions within the aerosol itself (e.g., SO_2 oxidation to sulfate; see Section 11-A-2).

The condensation of low-vapor-pressure species to form a particle is known as homogeneous nucleation or self-nucleation. Recall that the vapor pressure of a substance over the curved surface of a droplet is greater than over a flat surface of

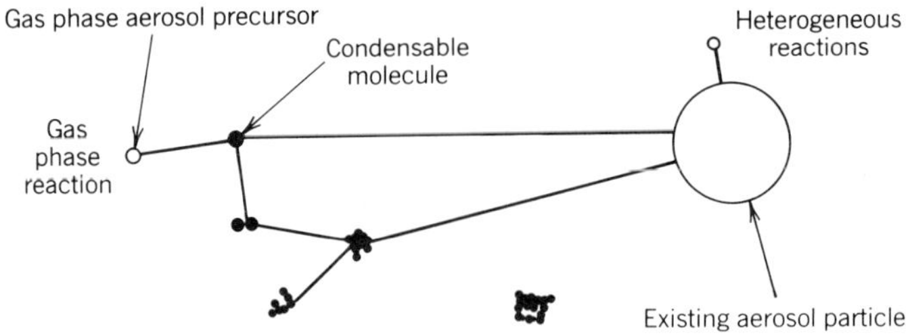

FIGURE 12.35. A schematic diagram showing aerosol formation by different chemical mechanisms. Condensable molecules ($D \simeq 0.5$ nm) formed by chemical reactions involving gas phase precursors can combine with other condensable molecules or molecular clusters to form new particles, or condense on a preexisting aerosol ($D \geq 0.01$ μm) causing them to grow. Alternatively, aerosol precursor gases can react to form a secondary aerosol on the surface of, or within, existing aerosol particles (from McMurry and Wilson, 1982).

the same substance (e.g., see Adamson, 1973). The smaller the radius of the droplet, the higher is the vapor pressure over the droplet surface. For example, for pure water at 25°C, the vapor pressure is only 0.1% greater over a droplet of 1 μm radius compared to that for a flat surface, but is 11% greater if the radius is 0.01 μm. This raises the question as to how homogeneous nucleation can occur at all, since the first very small droplets formed would tend to evaporate rapidly. The explanation lies in the formation of molecular clusters of molecules which occur as molecules collide in the gas phase. When the system becomes supersaturated, the concentration of the condensable species increases, as does that of the clusters. The clusters grow by the sequential attachment of molecules until they reach a critical diameter (D^*) above which the droplets are stable and grow, and below which they evaporate (Friedlander, 1977). The critical diameter is given by

$$D^* = \frac{4\gamma\bar{v}}{kT \ln s} \tag{II}$$

where γ is the surface tension of the chemical forming the particle, $\bar{v}$ is the molecular volume, k is the Boltzmann constant, T is the temperature (°K), and s is the saturation ratio, defined as the ratio of the actual vapor pressure to the equilibrium vapor pressure at that temperature.

In experiments to simulate photochemical smog formation starting with filtered, particle-free air, such homogeneous nucleation forms particles. For example, Fig. 12.36 shows the concentration of nuclei as a function of time when smoggy ambient air, which has been filtered to remove particles, is irradiated with sunlight (Husar and Whitby, 1973). It is seen that the number of particles rises quickly and dramatically (by about four orders of magnitude), peaks, and then falls. The initial sharp rise is due to nucleation of low-vapor-pressure products of gas phase reactions. However, as the number of nuclei increases, two processes set in which tend to limit the formation of new nuclei. These are scavenging of the molecular clusters by the nuclei before they have reached the critical diameter D^* needed to form

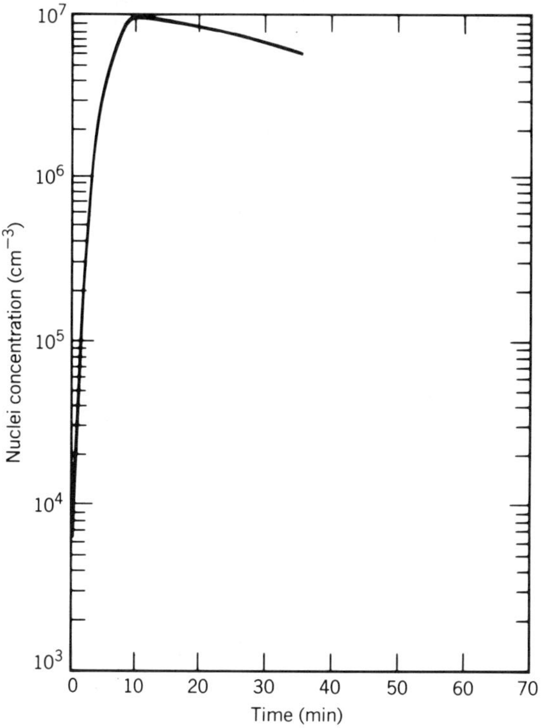

FIGURE 12.36. Change of total particle number concentration in irradiated, initially particle-free air. Initial concentration rise is due to nucleation and the subsequent decay to coagulation (from Husar and Whitby, 1973).

a stable nucleus, and coagulation of the nuclei to form larger particles. When the rates of these heterogeneous processes become equal to the rate of nuclei formation, the number of nuclei remains constant. Subsequently, the total number of particles decreases as coagulation occurs. This time evolution of homogeneous nucleation is described in detail by Friedlander (1983).

Heterogeneous condensation is secondary aerosol formation by the scavenging of the low-vapor-pressure products by preexisting particles. If the concentration of particles is sufficiently high, this can dominate over the formation of new nuclei via homogeneous nucleation. The simultaneous occurrence of homogeneous and heterogeneous nucleation in atmospheric systems has been treated by Friedlander (1978, 1980).

Different mechanisms of aerosol growth give rise to different *growth laws*, which are expressions relating the change in particle volume or diameter with time (i.e., dV/dt or dD/dt) to the particle diameter (Friedlander, 1977; Heisler and Friedlander, 1977). Because the different mechanisms of aerosol formation described above give rise to different growth laws, one can test a set of experimental data to see which of the mechanisms are consistent with the data.

Table 12.11 shows the prediction of some simple theoretical growth laws for

TABLE 12.11. Dependence of Particle Growth Rates on Particle Diameter (*D*) Predicted by Some Simple Theoretical Growth Laws for Three Different Mechanisms of Secondary Aerosol Formation

	Dependence of Growth Rate on D^a	
Mechanism	Volume Growth Rate dV/dt	Diameter Growth Rate dD/dt
Gas phase reaction		
$D \ll$ mean free path (λ) (free molecule regime)	D^2	D°
$D \gg$ mean free path (λ) (continuum regime)	D	D^{-1}
$D \simeq$ mean free path (λ) (transition regime)	b	b
Surface reaction[c]	D^2	D°
Reaction inside particle volume	D^3	D

Source: McMurry and Wilson, 1982.

[a]For example, the growth law for a gas phase reaction in the free molecule regime predicts that $dV/dt \propto D^2$ and $dD/dt \propto D^\circ$, that is, it is independent of D.

[b]Complex expression involving the Knudsen number (= mean free path/radius); see McMurry and Wilson (1982).

[c]Assuming reaction rate on surface is slow compared to the rate of transport of gases to the surface.

the dependence of both the change in volume with time and the change in diameter with time on the particle diameter for the three major mechanisms of aerosol formation. It is seen that, depending on the mechanism of secondary aerosol formation, the diameter growth rate can show a dependence on diameter varying from D^{-1} to D, with the dependence of the volume growth rate varying from D to D^3. In the first case of particle growth by the condensation of the products of gas phase reactions, the growth rates are determined by rates of condensation (see Friedlander, 1977, for a detailed treatment). For reactions on particle surfaces, the particle surface area determines the particle growth rate as long as the reaction rate is slow compared to the rate of transport of the gases to the surface; this is generally the case since the number of collisions of a gas with a surface required for reaction or adsorption of the gas is generally quite large (see Section 4-E). For reactions within the aqueous phase of the particle itself, the volume growth rate is proportional to particle volume (i.e., $\propto D^3$) if the reaction rate is uniform throughout the droplet. (If the reaction is fast compared to the transport of the gas within the droplet, then the reaction occurs in a "shell" on the outer part of the droplet and the growth laws are similar to those for reaction on the surface; the same holds true if reaction occurs in a liquid film surrounding a solid core).

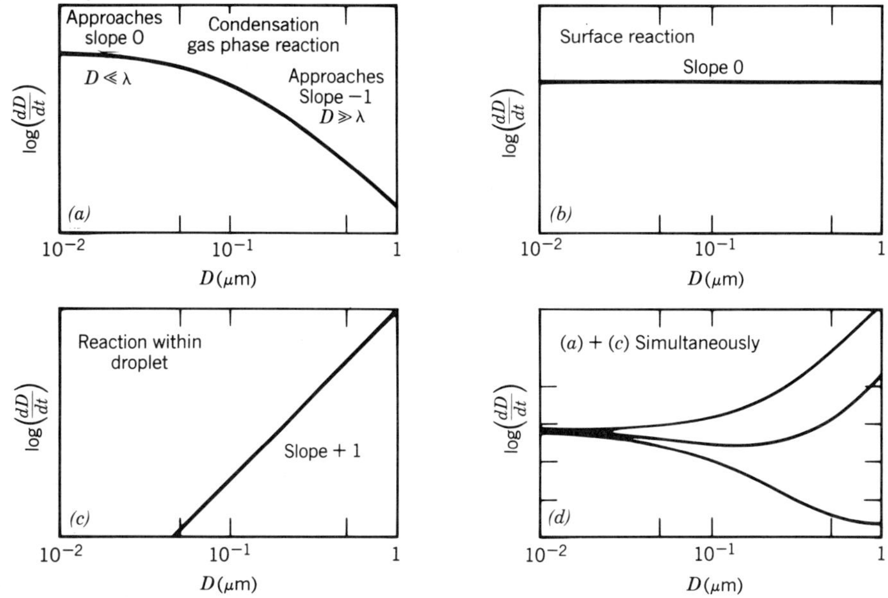

FIGURE 12.37. Log plot showing the diameter dependence of particle diameter growth rates for several chemical conversion mechanisms. The curves in (*d*) represent growth laws which might be observed if aerosol growth by condensation and droplet phase reactions were occurring simultaneously. The three lines represent different relative rates of these two processes (from McMurry and Wilson, 1982).

Figure 12.37 (parts *a*, *b*, and *c*) shows the form of the diameter growth rates predicted for the cases in Table 12.11, as well as for the case where two mechanisms are occurring simultaneously in different proportions (part *d*). It is clear that the dependencies of the growth rates on diameter are sufficiently different that they can be used to differentiate between various mechanisms.

Application of these theoretical growth laws requires that the assumptions used in deriving them are met by the experimental data. Besides the assumptions discussed above, these growth laws are also based on the supposition that the growth of particles is due solely to gas-to-particle conversion and not to coagulation processes or to fresh *primary* emissions of particles from sources.

In several studies of ambient air particles and of laboratory model systems, these conditions have been satisfactorily met and the data have been used to derive the growth laws. For example, Heisler and Friedlander (1977) carried out experiments in which hydrocarbons and NO_x were added to unfiltered ambient air (i.e., containing particles) in a Teflon bag, and these were irradiated with sunlight. The particle growth rates were followed as a function of time. Typical volume growth rates for the case where 2.02 ppm cyclohexene, 0.34 ppm NO, and 0.17 ppm NO_2 were added are shown in Fig. 12.38. The mean free path is ~0.07 μm in air at room temperature and atmospheric pressure, so that for the case of gas phase reactions, the conditions are in the continuum–transition regime (Table 12.11). The

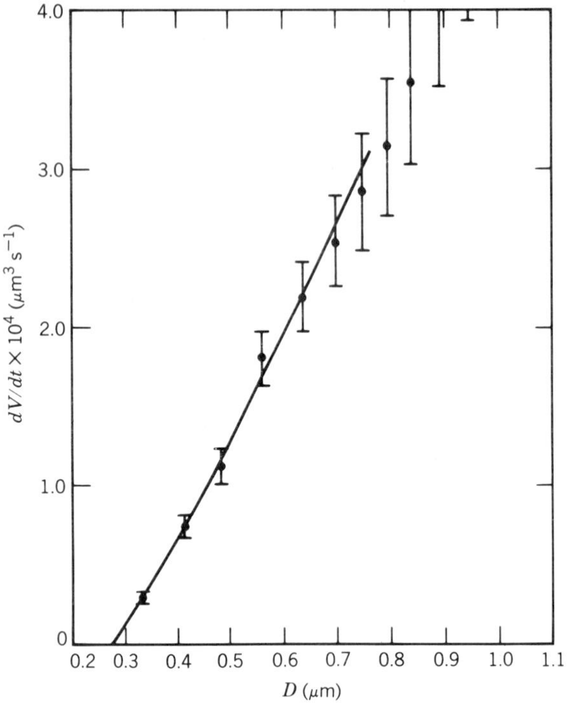

FIGURE 12.38. Particle volume growth rates as a function of diameter with 2.02 ppm cyclohexene, 0.34 ppm NO, and 0.17 ppm NO_2 were added to unfiltered ambient air and irradiated. Solid line is best least squares fit assuming gas phase reaction followed by condensation (from Heisler and Friedlander, 1977).

solid line is the best least squares fit for the case where aerosol growth is due to gas phase reactions followed by condensation of low-vapor-pressure products; consistent with Table 12.11, the growth rate varies directly with the particle diameter when the diameters are sufficiently large to be in the continuum regime. Similar results have been found in studies of aerosol formation from organics such as *o*-cresol, nitrocresol, 1-heptene, and methyl sulfide irradiated in air with NO_x, and in some experiments, ammonia (McMurry and Grosjean, 1985); in these studies, evidence was also found for evaporation of the smaller particles after the production of new particles had stopped, but while aerosol mass concentrations were still increasing due to the gas phase reactions.

This approach to elucidating mechanisms of secondary aerosol formation has also been applied to ambient air. For example, Fig. 12.39 shows typical volume distributions for ambient particulates observed in the Great Smoky Mountains in Tennessee in 1978 at three different times on one day (McMurry and Wilson, 1982). From such data, the volume and diameter growth rates could be obtained as a function of diameter; the data plotted in this form are shown in Fig. 12.40. Also shown are the best fit growth laws for a surface reaction and for a droplet

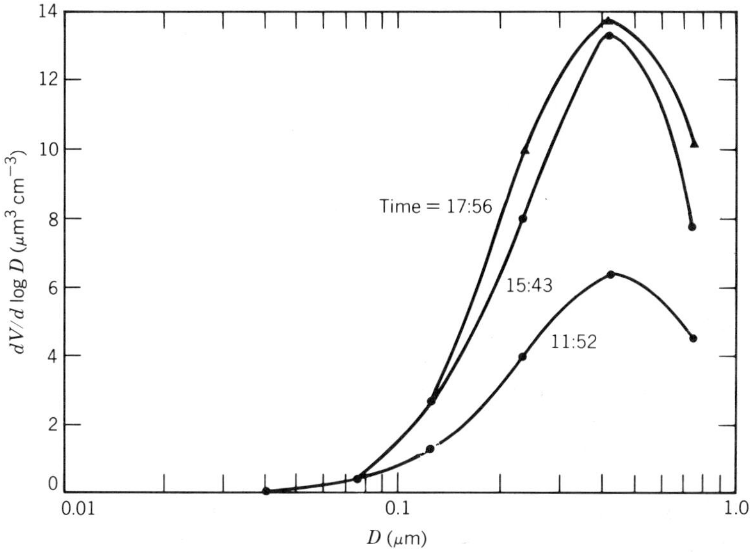

FIGURE 12.39. Aerosol volume distributions measured in the Great Smoky Mountains on August 26, 1978. Note that a considerable amount of secondary aerosol appeared in the 0.5–1.0 μm size range where condensational growth is not expected (from McMurry and Wilson, 1982).

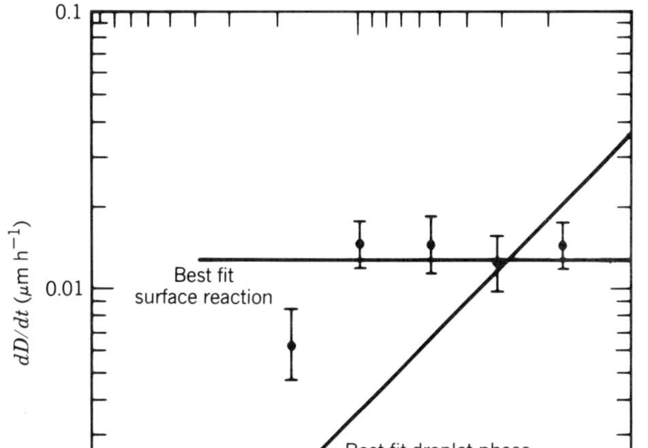

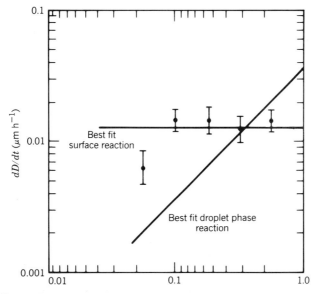

FIGURE 12.40. Particle diameter growth rates as a function of particle diameter for aerosol data measured on August 26, 1978 in the Great Smoky Mountains (see Fig. 12.39). Least squares best fit growth laws for surface and droplet phase reactions are shown. The data are consistent with surface-limited growth (from McMurry and Wilson, 1982).

phase reaction; in this case, the surface mechanism provides the best fit to the data. Recall, however, that this could also mean that reaction is occurring within a thin film on the outside of the droplet.

More complex diameter growth rate behavior than that shown in Figs. 12.38 and 12.40 have been observed in ambient air. For example, Fig. 12.41 shows one set of data for the diameter growth rates for the St. Louis urban plume. The best fit to the data in this case was found by a combination of condensation of gas phase reaction products and reaction inside the particles. The best fit suggested that 75% of the secondary aerosol growth was due to the condensation mechanism and 25% was due to droplet phase reactions.

It is interesting that even in the relatively strongly oxidizing urban atmosphere of Los Angeles, application of these growth laws suggests that not only do both chemical reactions in the aerosol phase and homogeneous reactions in the gas phase contribute to the formation of particulate sulfur, but that the aerosol phase reactions may often predominate; the occurrence of two different oxidation mechanisms is consistent with the observation of two types of fine particle sulfur having different mass median diameters, the smaller one (0.2 ± 0.02 μm) being due to homogeneous gas phase oxidation of SO_2 and the larger (0.54 ± 0.07 μm) being due to reactions in the aerosol phase (Hering and Friedlander, 1982).

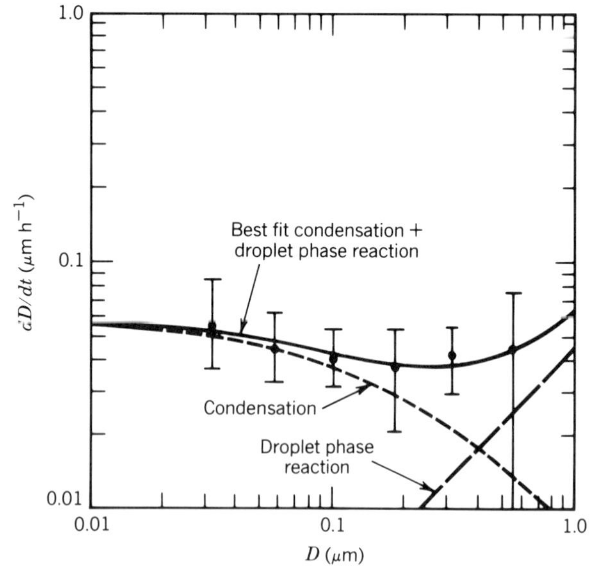

FIGURE 12.41. Particle diameter growth rates from data obtained with an instrumented aircraft in the St. Louis urban plume during the MISTT project in July 1975. In this case, all growth rates are based on data from an electrical aerosol analyzer. The solid line through the data points represents the least squares best fit growth law for simultaneous condensational growth and droplet phase reactions (from McMurry and Wilson, 1982).

12-D. CHEMICAL COMPOSITION OF AMBIENT PARTICULATE MATTER

As described in detail in Section 12-B-1, ambient particulate matter is usually distributed multi-modally. For convenience, we discuss three modes, with the understanding that important exceptions may exist in the environment. The smallest particles, known as the Aitken or transient nuclei range ($D \lesssim 0.08$ μm), are contributed by combustion processes and are primarily observed in significant concentrations near such sources. The middle accumulation range ($0.08 \lesssim D \lesssim 2.5$ μm) has contributions from coagulation of nuclei and the scavenging of low-vapor-pressure products of gas phase reactions, as well as from some combustion processes; these two ranges are said to constitute the fine particle fraction. The coarse particle range ($D \gtrsim 2.5$ μm) has contributions from mechanical processes such as grinding and wind erosion.

Because of these different sources of fine and coarse particles, their chemical compositions also tend to differ.

12-D-1. Coarse Particles

Coarse particles are generally comprised of soil elements over the continents and sea salt elements over the oceans. Thus over land, the major elements found in the coarse particle fraction are Si, Al, Ca, Fe, K, Ti, Mn, and Sr, and these are often in the same ratio as they are found in the earth's crustal material. For example, Table 12.12 shows the mean elemental abundances as ratios to the most abundant element, Si, in coarse particles (2.5–15 μm diameter) from rural sites in the Ohio River Valley; also shown are the ratios for crustal materials. Except for Ca, the agreement between the two is satisfactory, indicating that crustal material is responsible for these elements in the coarse particle mode. If the contribution of crustal limestone is included, the Ca ratio is also consistent with a crustal source.

TABLE 12.12. Mean Elemental Abundances as a Ratio to Si in the Coarse Particle Mode in Rural Ohio River Valley Particles and in Crustal Materials

	Ratio				
Sample	$\frac{Al}{Si}$	$\frac{Ca}{Si}$	$\frac{Ti}{Si}$	$\frac{Fe}{Si}$	$\frac{K}{Si}$
Coarse particles from Ohio River Valley[a]	0.36	0.26	0.02	0.14	0.07
Crustal average[b]	0.29	0.15	0.02	0.20	0.07

Source: Shaw and Paur, 1983.

[a]Average of three sites along a 590 km path.

[b]Based on Taylor, 1964.

In marine areas, aerosols characteristic of sea salt are found. Wave action entrains air and forms bubbles which rise to the surface. As they rise, dissolved organics may become adsorbed on them. The bubbles burst on reaching the surface, producing small droplets that are ejected into the air. Two types of drops have been distinguished—jet drops and film drops. Jet drops are produced from the jet of water which rises from the bottom of the collapsing bubble; film drops are produced from the bursting of the bubble water film. Particles with a wide range of sizes, from less than 0.1 μm to greater than 100 μm, are formed.

It might be assumed that initially the composition of the liquid drops would approximate that of seawater, given in Table 12.13. As evaporation of water in the droplet into the surrounding air mass occurs, the salt becomes more concentrated. If the relative humidity is below the deliquesence point of NaCl (75%), complete evaporation may occur, leaving a solid salt particle. Along with the salt is any organic material originally associated with the bubble. Thus, marine-derived sterols, fatty alcohols, and fatty acid salts have been shown to be enriched in the surface microlayer of the sea (the first 150 μm) compared to their concentrations in bulk seawater, particularly in bubble interfacial microlayer samples which occur at the top 1 μm of the surface (Schneider and Gagosian, 1985). Since this top

TABLE 12.13. Composition of Sea Salt Particles in Clean Atmospheres

Species	Percent by Weight
Al	0.00046–0.0055
Ba	0.00014
Br	0.19
C (non-carbarbonate)	0.0035–0.0087
Ca	1.16
Cl	55.04
Cu	—
Fe	0.00005–0.0005
I	0.00014
K	1.1
Mg	3.69
Mn	0.0000025–0.000025
Na	30.61
NH_4^+	0.0000014–0.000014
NO_3^-	0.000003–0.002
Pb	0.000012–0.000014
Si	0.00014–0.0094
SO_4^{2-}	7.68
V	0.0000009
Zn	0.000014–0.000040

Source: Miller et al., 1972. Based on the assumption that the composition is the same as that of seawater.

microlayer is believed to be involved in the aerosol formation process from the sea, adsorbed organics are expected on the surfaces of such sea salt particles. These organic layers may carry with them enriched concentrations of cations such as Fe^{3+} as well (Thomsen, 1983).

These sea salt particles may play a major role in the global distribution and fluxes of a number of elements. For example, Fogg and Duce (1985) suggest that they are a major source of atmospheric boron.

Salt particles whose chemical composition reflects that of seawater have been observed in many "clean air" locations. However, in a number of other cases, the ratio of the elemental mass of some elements to that of sodium differs from that in seawater. It has been suggested that ion fractionation may occur during particle formation at the surface; as discussed below, chemical reactions may also occur on the particle surface, releasing elements such as chlorine to the gas phase.

As might be expected, the salt concentration and to some extent the size distribution depend on the meteorology, especially wind speed which drives the wave action. For a detailed discussion of sea salt aerosol, the reader is referred to Blanchard and Woodcock (1980).

Most of the elements found in coarse particles over land or sea are involatile and relatively chemically inert; however, for a few elements, such as Cl, this is not true. As early as 1956, Junge noted in marine aerosols on the Florida coast that the particles contained nitrate and that the NO_3^-/Cl^- ratio was highest when the wind direction was from the land and lowest when it was from the ocean; presumably the breeze from the land contained anthropogenic pollutants which could react with the particles. Since then, numerous investigators have observed such a chloride ion deficiency relative to the sodium concentration, and it has generally been ascribed to reactions of acids such as sulfuric and nitric acids (e.g. see Hitchcock et al., 1980) with NaCl to produce gaseous HCl, for example,

$$HNO_3 + NaCl \rightarrow NaNO_3 + HCl \quad (1)$$

$$H_2SO_4 + 2NaCl \rightarrow Na_2SO_4 + 2HCl \quad (2)$$

The conditions under which HCl formed in acidified sodium chloride droplets would be expected to enter the gas phase have been treated by Clegg and Brimblecombe (1985). Cadle and co-workers (Robbins et al., 1959; Cadle and Robbins, 1960) observed that NaCl aerosols in the presence of 0.1–100 ppm NO_2 at relative humidities of 50–100% lost chloride ion from the particulate phase. They ascribed this to the formation of nitric acid by NO_2, followed by reaction (1). Shroeder and Urone (1974) subsequently suggested that NO_2 could react directly with NaCl to produce gaseous nitrosyl chloride, NOCl, which they observed using infrared spectroscopy; stoichiometrically, this is represented as

$$2NO_2 + NaCl \rightarrow NaNO_3 + NOCl \quad (3)$$

However, due to the relatively low analytical sensitivity for NOCl using IR in a 10 cm gas cell, very high NO_2 concentrations (Torr range) were used. Extrapolation to atmospheric conditions could not be carried out reliably because of the significant amount of N_2O_4 present in equilibrium with NO_2 under these conditions:

$$2NO_2 \rightleftarrows N_2O_4 \qquad (4)$$

Subsequently, Finlayson-Pitts (1983) established, using mass spectrometry, that NOCl was produced at low (5 ppm) NO_2 concentrations, so reaction (3) does appear viable in the atmosphere; however, whether it is sufficiently fast to complete with the acid reactions (1) and (2) is not known and is currently under investigation. In any event, the reaction of NO_2 and/or acids with NaCl seems to explain the release of chlorine to the gas phase; whether it is initially released as HCl or NOCl, (or both) is not known.

Similar reactions involving bromine and iodine in sea salt particles have also been suggested based on measurements of their concentrations in sea salt aerosols (Moyers and Duce, 1972a,b; Duce et al., 1973; Moyers and Colovos, 1974).

12-D-2. Fine Particles

12-D-2a. Observed Ionic and Elemental Composition

Because fine particles generally arise from gas-to-particle conversions and combustion (see Section 12-B-1), their chemical composition is expected to be quite different from that of the soil- and ocean-derived coarse particles. Thus the major chemical components of fine particles are sulfate, nitrate, organic and elemental carbon, and ammonium ions. They also contain a variety of trace metals from the combustion processes.

Both the absolute concentrations and the relative proportions of these species in the particles depend on a number of factors such as the nature of the emissions into the air mass, photochemical activity (e.g., solar intensity), and meteorology (e.g., relative humidity). For example, Table 12.14 shows the average aerosol composition for both fine (defined in these studies as $D \leq 2.5\ \mu m$) and coarse particles ($2.5 < D < 15\ \mu m$) at a rural, forested site and at an urban site. It is seen that the products expected from gas-to-particle conversion of SO_2, NO_x, and NH_3, that is, SO_4^{2-}, NO_3^-, and NH_4^+, are all found at significant concentrations in the fine particle phase. The one exception to this in Table 12.14 is nitrate in the Houston aerosol, where the concentrations are higher in the coarse than in the fine particles. The authors suggest that this is due to the alkaline nature of the coarse particles, which efficiently condenses acidic HNO_3 from the gas phase; as discussed below, under the appropriate conditions, nitrate may also revolatilize from fine particles as HNO_3 and subsequently combine on the larger alkaline particles. In addition to the nitrogen and sulfur compounds, particulate carbon, which may be either primary or secondary in nature, was concentrated in the fine particle fraction. As expected, the crustal elements such as Si, Al, Ca, and so on were concentrated in coarse particles.

TABLE 12.14. Average Aerosol Composition (ng m^{-3}) for Fine and Coarse Particles at a Rural, Forested Location (Great Smoky Mountains, Tennessee) and an Urban Location (Houston, Texas)

	Smoky Mountains		Houston[a]	
	Fine[b]	Coarse[b]	Fine[b]	Coarse[b]
Total Mass	24,000 ± 3000	5600 ± 3000	42,500 ± 4250	27,200 ± 2700
SO_4^{2-}	12,000 ± 1300	NA[c]	16,700 ± 1380	1100 ± 200
NO_3^-	300 ± 300	NA	250 ± 260	1800 ± 260
NH_4^+	2280 ± 390	NA	4300 ± 390	<190
H^+	114	NA	67	<1
C (organic)	2220 ± 400	1200 ± 400	NA	NA
C (elemental)	1100 ± 800	<100	NA	NA
C (total)	3300 ± 600	1300 ± 600	7600 ± 500	3300 ± 500
Al	20 ± 18	195 ± 101	95 ± 60	1400 ± 420
Si	38 ± 10	580 ± 262	200 ± 60	3800 ± 1000
S	3744 ± 218	204 ± 187	NA	NA
Cl	<10	7 ± 4	19 ± 6	330 ± 21
K	40 ± 3	108 ± 30	120 ± 7	180 ± 21
Ca	16 ± 1	322 ± 73	150 ± 8	3100 ± 160
Ti	<6	18 ± 5	<8	48 ± 14
V	<4	<5	NA	NA
Mn	NA	NA	13 ± 2	23 ± 3
Fe	28 ± 2	118 ± 9	170 ± 9	730 ± 40
Ni	1 ± 0.5	1 ± 0.5	3 ± 1	5 ± 1
Cu	3 ± 0.7	<5	16 ±2	14 ± 2
Zn	9 ± 1	<4	102 ± 6	68 ± 5
As	2.2 ± 1	<1	NA	NA
Se	1.4 ± 0.3	0.2 ± 0.2	NA	NA
Br	18 ± 1	5 ± 0.4	70 ± 4	39 ± 3
Pb	97 ± 5	14 ± 1	483 ± 23	127 ± 10

Source: Dzubay et al., 1982, and Stevens et al., 1984.

[a]Samples collected during the daytime.

[b]Fine and coarse particles were defined in this study as having aerodynamic diameters in the ranges 0–2.5 and 2.5–15 μm, respectively.

[c]NA = not analyzed.

There are a variety of potential sources of the trace metals found in ambient particulate matter. These include the combustion of coal and oil (including gasoline), wood burning, waste incineration, and metal mining and production. There are also natural sources of metals, such as windblown dusts, sea salt, forest fires, volcanic emissions, and emissions from vegetation (Nriagu, 1979). While the metals in particles produced by mechanical processes (e.g., windblown dust) are mainly in large particles (Section 12-D-1), those produced by the other processes (e.g., combustion) are primarily in fine particles.

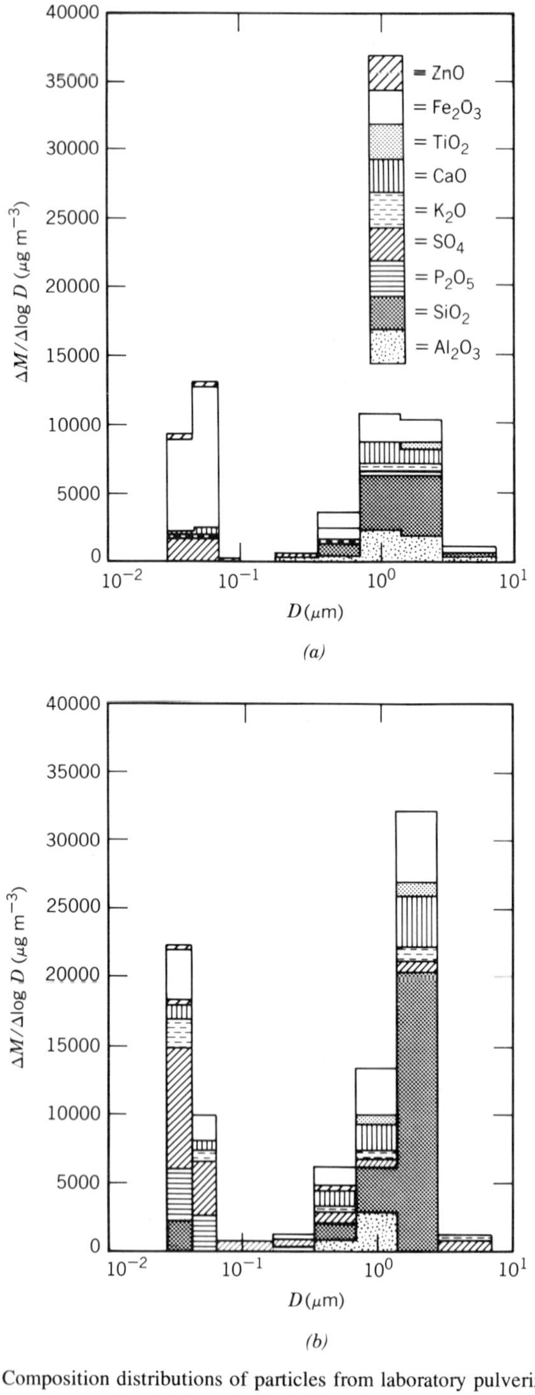

FIGURE 12.42. Composition distributions of particles from laboratory pulverized coal combustion experiments. Temperature was (*a*) 1171°K and (*b*) 1349°K (from Taylor and Flagan, 1982).

For example, Fig. 12.42 shows the size distribution of some of the major species found in fly ash from pulverized coal combustion in laboratory experiments at two combustion temperatures. At the lower temperature, the smallest particles ($D \lesssim 0.1$ μm) consist primarily of iron and sulfur, whereas at the higher temperature, potassium, phosphorous, silicon, and calcium are also present. Thus the composition of particles from combustion can vary significantly not only with the type of fuel, but also with the operating conditions.

In addition to the species shown in Fig. 12.42, there are a whole host of trace elements present in combustion-derived particulate matter. For example, Coles et al. (1979) measured 42 minor and trace elements in fly ash from the stack of a coal-fired power plant. However, elements are not distributed equally in all particle sizes; for example, fine particles are enriched in such elements as As, Cd, Ga, Mo, Pb, Sb, Se, W, and Zn (Natusch et al., 1974; Coles et al., 1979; Fisher, 1980). This is consistent with finding As, Pb, Se, and Zn primarily in fine particles in ambient air (Table 12.14).

The concentrations of both fine and coarse particles were greater in the urban Houston area than in the rural area (Table 12.14). This is expected since there are larger primary emissions compared to the rural area, larger concentrations of oxidant and other species such as OH to carry out gas-to-particle conversions, and greater mechanical activity (e.g., automobile traffic) to suspend coarse particles such as soil dust.

Despite the difference in the nature of the sites, sulfate comprised the largest fraction of the total fine particle mass, ~50% in the Great Smoky Mountains and ~40% in Houston. Total carbon was the next highest component, ~15% of the fine fraction at the rural site and ~18% at the urban site.

However, in many areas at certain times the nitrate concentration in fine particles often exceeds that in the coarse particles. In addition, the aerosol nitrate concentration frequently exceeds that of sulfate. Table 12.15, for example, shows the average fine and coarse particulate sulfate and nitrate concentrations in three locations in California. In two of the locations, fine particle nitrate exceeds coarse particle nitrate, and total particulate nitrate exceeds the sulfate.

For a detailed treatment of the reaction mechanisms oxidizing NO_x to nitrate and SO_2 to sulfate, the reader is referred to Chapters 8 and 11, respectively. The

TABLE 12.15. Average Concentrations (μg m^{-3}) of Sulfate and Nitrate in Fine and Coarse Particles at Three Locations in California[a]

Species	San Jose	Riverside	Los Angeles
Fine sulfate	1.66	6.97	13.0
Fine nitrate	3.38	13.5	7.88
Coarse sulfate	0.61	1.46	1.80
Coarse nitrate	2.11	8.28	7.93

[a]Data from Appel et al., 1985; fine particles have $D \lesssim 2.2$ μm.

equilibria between gas phase NH_3 and HNO_3 and particulate phase NH_4^+ and NO_3^- in the absence of other species are discussed in Chapter 8.

In a number of studies, the elemental composition of aerosol particles has been related to their source, and it has been suggested that certain elements or ratios of elements may serve as tracers for various sources (e.g. see Gordon, 1980; Rahn and Lowenthal, 1984; Shah et al., 1985). Thus V and Ni are indicative of oil combustion, and elevated concentrations of elements such as As and Se are usually associated with coal burning and smelter operations. The ratio of Mn to V has been suggested as an indicator of coal burning (Rahn, 1981). Electron microscopic studies of particles have shown that at least in one episode at Whiteface Mountain in New York, elevated Mn/V ratios were associated with coal fly ash particles, observed simultaneously with high sulfate concentrations (Webber et al., 1985).

The relative concentrations of rare earth elements in fine particles have also been suggested as being indicative of emissions from oil-fired power plants and from refineries; thus the ratio of lanthanum to samarium is much higher in emissions from these sources than found in particles formed from the earth's crust (Olmez and Gordon, 1985).

While there has been a great deal of interest in using the elemental composition of particles as tracers of sulfate and nitrate from sources located large distances upwind, caution must be exercised in using elements or ratios of their concentrations as indicators of such long-range transport. For example, element concentrations and ratios may change from that at the source as the air mass travels over downwind sources. For a discussion of this and some of the other potential problems, see Thurston and Laird (1985) and Rahn and Lowenthal (1985).

12-D-2b. Chemical Forms of Inorganic Species in Particles

A knowledge of the overall ionic and elemental composition of ambient particles is essential to our understanding of their sources and chemistry, yet it leaves unanswered the question of the chemical form in which important species such as sulfate and nitrate exist; for example, does sulfate exist as $(NH_4)_2SO_4$ and/or $(NH_4)HSO_4$ as suggested by the molar concentrations of NH_4^+, H^+, and SO_4^{2-} in some locations? Charlson and co-workers (1974a,b) have addressed this problem using an instrument which they developed known as a humidograph.

This method is based on the different deliquescence properties of various salts found in ambient air (Table 12.16). At the deliquescence point, the transition from a solid particle to a liquid droplet occurs suddenly, marking the onset of growth of the particle size. Figure 12.43, for example, shows the calculated change in particle size for four different sulfate aerosols, consisting of H_2SO_4, $(NH_4)HSO_4$, $(NH_4)_2SO_4$, and $(NH_4)_3H(SO_4)_2$ (letovicite), respectively, as a function of relative humidity (RH). At the deliquescence point in each case the particle suddenly increases in size and then grows in a continuous fashion as the RH increases.

The uptake of water by the particles as the RH is increased can be followed by measuring the light scattering of the aerosol with an integrating nephelometer (see Section 12-E). Figure 12.44 shows the light scattering coefficient b_{sp} (as a ratio to

TABLE 12.16. Deliquescence Points of Some Salts Commonly Found in Ambient Air at 25°

Composition	Deliquescence Humidity (%)
$(NH_4)_2SO_4$	79.5
$(NH_4)HSO_4$	39.0
$(NH_4)_3H(SO_4)_2$	69.0
$2NH_4NO_3 \cdot (NH_4)_2SO_4$	56.4
NaCl	75.7
KCl	84.3
NaCl–KCl	73.8

Source: Tang, 1980a.

that at 30%) as a function of relative humidity for a pure H_2SO_4 aerosol and for one to which NH_3 has been added. b_{sp} increases monotonically for the hygroscopic H_2SO_4 particles. However, when NH_3 is added, converting the aerosol to $(NH_4)_2SO_4$, a sharp inflection point is seen at ~80% RH, as expected for ammonium sulfate. One can thus distinguish between H_2SO_4 and $(NH_4)_2SO_4$ by the

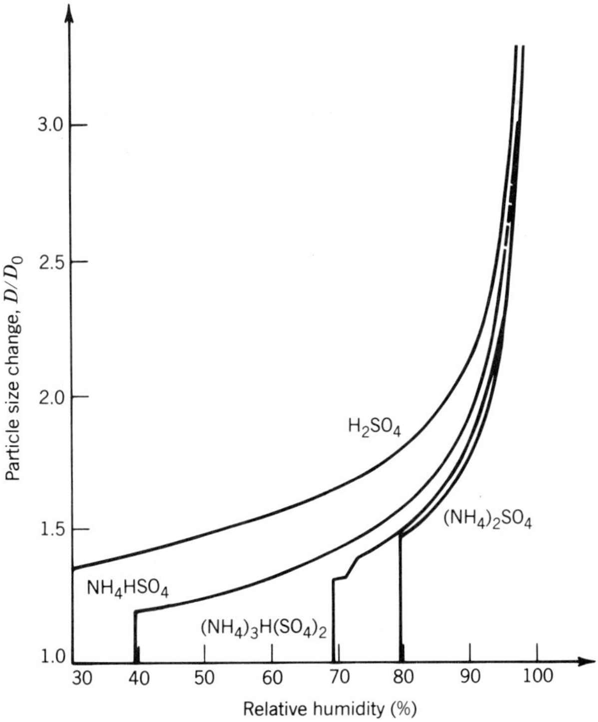

FIGURE 12.43. Calculated changes in particle size as a function of relative humidity at 25°C from particles with four different chemical compositions (from Tang, 1980a).

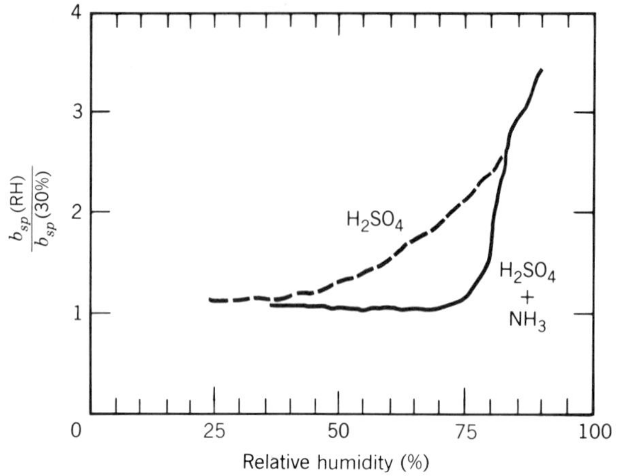

FIGURE 12.44. Humidogram for laboratory H_2SO_4 aerosol (dashed) and for the reaction product of H_2SO_4 and NH_3 (solid). The ordinate is the ratio of light scattering coefficient due to particulate matter at the given relative humidity [b_{sp}(RH)] to the light-scattering coefficient at 30%. The abscissa is relative humidity in percent. The inflection point in the $(NH_4)_2SO_4$ curve corresponds to the humidity at which all the particles have been converted to droplets. The increase starting at ~72% is due either to the presence of submicron particles which have a higher solubility than bulk solubility due to curvature or to mixing characteristics of the flow system (from Charlson et al., 1974a).

dependence of the light scattering properties on RH, with and without added NH_3. While this has the advantage of providing an in situ measurement of the chemical nature of the sulfates, the method responds mainly to those particle sizes that scatter light efficiently (i.e., ~0.1–1 μm); fortunately, this is also the range of greatest importance for health and visibility effects. When the particles are not pure sulfates, but mixtures with other species, the $(NH_4)_2SO_4$ must be at least 30–50 mole % in order to show a clear inflection point in the humidogram.

Using this humidograph technique, the presence of $H_2SO_4/(NH)_4HSO_4$ or $(NH_4)_2SO_4$ as a major component of ambient light scattering aerosols has been established in a number of locations. $(NH_4)_2SO_4$ was frequently observed, suggesting that there is often sufficient NH_3 present in ambient air to completely neutralize the sulfuric acid (Weiss et al., 1977). A modified version of the humidograph, in which a heating and cooling cycle is included to differentiate various salts based on their thermal properties, has also been applied recently to estimate the molar ratio of NH_4^+ to SO_4^{2-} (Cobourn et al., 1978; Weiss et al., 1982; Larson et al., 1982; Rood et al., 1985).

It is clear, from both in situ measurements such as these and other methods such as filter extractions and analyses (e.g., see Appel et al., 1980a, 1982) that, depending on the available ammonia, sulfate may exist as H_2SO_4, $(NH_4)HSO_4$, or $(NH_4)_2SO_4$. However, it is likely that other forms of sulfate also exist. For example, letovicite, $(NH_4)_3H(SO_4)_2$, has been identified in ambient particles by X-ray diffraction (Brosset et al., 1975) and its presence has also been detected using the humidograph technique (Weiss et al., 1977). Other sulfate salts, including

TABLE 12.17. Some Compounds Observed in Aerosols by a Roadway at Argonne National Laboratory Using X-Ray Diffraction

SiO_2	$K_2Sn(SO_4)_2$
$CaCO_3$	$(NH_4)_2Co(SO_4)_2 \cdot 6H_2O$
$CaMg(CO_3)_2$	$(NH_4)_3H(SO_4)_2$ (letovicite)
$CaSO_4 \cdot 2H_2O$	$3(NH_4NO_3) \cdot (NH_4)_2SO_4$
$(NH_4)_2Pb(SO_4)_2$	$2(NH_4NO_3) \cdot (NH_4)_2SO_4$
$(NH_4)_2Ca(SO_4)_2 \cdot H_2O$	$NH_4MgCl_3 \cdot 6H_2O$
$(NH_4)HSO_4$	NaCl
$(NH_4)_2SO_4$	$(NH_4)_2Ni(SO_4)_2 \cdot 6H_2O$

Source: Tani et al., 1983.

mixed ammonium nitrate–ammonium sulfate double salts have been detected using X-ray diffraction (Tani et al., 1983); Table 12.17 shows the compounds observed at a roadside station, while Table 12.18 shows those observed in a forest clearing near State College, PA, relatively distant from any direct emission sources.

We show these data to emphasize that sulfates and nitrates *may* in fact exist in forms other than the simple ammonium salts. However, whether or not they *do* exist in the atmosphere in the complex forms shown in Tables 12.17 and 12.18 is not clear because of potential artifacts in the sampling and analysis procedures. As Tani et al. (1983) point out, both physical changes (e.g., recrystallization) and chemical reactions may occur during or following impaction of the aerosols on the collector, leading to the formation of compounds not initially present in the ambient aerosol.

While the existence of mixed sulfate–nitrate salts in the atmosphere is thus somewhat uncertain, theoretical studies of sulfate–nitrate equilibria do indeed predict the formation of many of these species under atmospheric conditions. A number of treatments of NH_3–HNO_3–H_2SO_4 equilibria under atmospheric conditions have been carried out (e.g., see Tang, 1980b; Tanner, 1982; Stelson and Seinfeld, 1982; Saxena et al., 1983; Bassett and Seinfield, 1983, 1984; Spann and Richardson, 1985). Table 12.19 shows the equilibria included in one such treatment of

TABLE 12.18. Some Compounds Observed in Aerosols in a Forested Area, State College, PA, Using X-Ray Diffraction

$(NH_4)_2SO_4$
$(NH_4)_3H(SO_4)_2$ (letovicite)
$(NH_4)HSO_4$
$2(NH_4NO_3) \cdot (NH_4)_2SO_4$
$(NH_4)_2Pb(SO_4)_2$

Source: Tani et al., 1983.

TABLE 12.19. Chemical Reactions Occurring in the Sulfate[a]–Nitrate–Ammonium System

$H_2SO_{4(g)} \rightleftarrows H^+ + HSO_4^-$
$H_2SO_{4(g)} \rightleftarrows 2H^+ + SO_4^{2-}$
$HNO_{3(g)} \rightleftarrows H^+ + NO_3^-$
$NH_{3(g)} + H_2SO_{4(g)} \rightleftarrows NH_4^+ + HSO_4^-$
$2NH_{3(g)} + H_2SO_{4(g)} \rightleftarrows 2NH_4^+ + SO_4^{2-}$
$NH_{3(g)} + HNO_{3(g)} \rightleftarrows NH_4^+ + NO_3^-$
$NH_{3(g)} + H_2SO_{4(g)} \rightleftarrows (NH_4)HSO_{4(s)}$
$3NH_{3(g)} + 2H_2SO_{4(g)} \rightleftarrows (NH_4)_3H(SO_4)_{2(s)}$
$2NH_{3(g)} + H_2SO_{4(g)} \rightleftarrows (NH_4)_2SO_{4(s)}$
$NH_{3(g)} + HNO_{3(g)} \rightleftarrows NH_4NO_{3(s)}$
$5NH_{3(g)} + 3HNO_{3(g)} + H_2SO_{4(g)} \rightleftarrows (NH_4)_2SO_4 \cdot 3NH_4NO_{3(s)}$
$4NH_{3(g)} + 2HNO_{3(g)} + H_2SO_{4(g)} \rightleftarrows (NH_4)_2SO_4 \cdot 2NH_4NO_{3(s)}$
$H_2O_{(g)} \rightleftarrows H_2O_{(l)}$

Source: Bassett and Seinfeld, 1983; the reader should consult the original reference for the equilibrium constant expressions and values.

[a] $H_2SO_{4(g)}$ is included for the thermodynamic calculations, although its gas phase concentrations are negligible due to its low vapor pressure.

solids and solutions containing pure NH_4NO_3 or mixtures of NH_4^+, NO_3^-, and SO_4^{2-} in equilibrium with gaseous NH_3 and HNO_3 at typical atmospheric concentrations. (H_2SO_4 exists in ambient air in solution with water; in this solution, it has an extremely low vapor pressure so that its gas phase concentrations are negligible.) With the simplifying assumptions that (1) only these species are present (e.g., ignoring species such as NH_4Cl), (2) the system is in thermodynamic equilibrium (i.e., the reactions are sufficiently fast to attain equilibrium immediately), and (3) the sulfate size distribution is fixed to an assumed log-normal distribution, Seinfeld and co-workers treated six cases with different total concentrations of NH_3, HNO_3, and H_2SO_4 at relative humidities of 50 or 90%. The Kelvin effect, that is, the greater vapor pressure of the volatile components NH_3 and HNO_3 over highly curved surfaces compared to flat ones, was included in the calculations. The conditions for these six cases are summarized in Table 12.20, and Fig. 12.45 shows the predicted distribution of the components as a function of particle size.

Several noteworthy results arise from Fig. 12.45. First, note in case 1 that NH_4^+ and NO_3^- tend to be present in the highest concentrations in fairly large particles, ~0.1–1 μm diameter. This results from the Kelvin effect; thus the partial pressures of NH_3 and HNO_3 over a small particle will be greater than over a large particle. When particles of various sizes are present as in the atmosphere, the net effect is volatilization of NH_3 and HNO_3 from the smaller particles and condensation onto the larger ones. Because sulfate is non-volatile, mass transfer of this species does not occur, resulting in different mass median diameters for sulfate and nitrate.

This explains why sulfate and nitrate are often observed to have different size distributions in the same air mass. For example, in the ACHEX study (California Aerosol Characterization Experiment), the mass median diameter in the fine par-

TABLE 12.20. Conditions Used for Thermodynamic Analysis of Aerosol Composition as a Function of Particle Size

Case	Total NH_3 ($\mu g\ m^{-3}$)	Total HNO_3 ($\mu g\ m^{-3}$)	Total H_2SO_4 ($\mu g\ m^{-3}$)	RH (%)
1	20	20	30	90
2	20	20	30	50
3	5	20	30	90
4	5	20	30	50
5	5	15	20	50
6	20	20	5	50

Source: Bassett and Seinfeld, 1984.

ticle fraction for nitrate was often larger than that for sulfate, by as much as a factor of 4 (Appel et al., 1980b). This was not expected at the time since both are formed by gas-to-particle conversions. However, the increased vaporization of NH_3 and HNO_3 over very small particles is now seen to be consistent with the ambient observations. This may also be partially responsible for the observation of higher nitrate concentrations in the larger particles in Houston (Table 12.14).

When the RH in case 1 is lowered from 90% to 50% (case 2 of Table 12.20), the solid particle resulting is calculated to be a mixture of $(NH_4)_2SO_4$ and $(NH_4)_2SO_4 \cdot 2NH_4NO_3$ (Fig. 12.45*b*). Note that, consistent with the Kelvin effect, the nitrate salt is predicted to be only in the larger particles. The presence of the mixed salt $(NH_4)_2SO_4 \cdot 2NH_4NO_3$ in the 0.3–1 μm diameter region is predicted to control the gas phase concentrations of NH_3 and HNO_3 under these conditions through the equilibrium:

$$(NH_4)_2SO_{4(s)} + 2NH_{3(g)} + 2HNO_{3(g)} \leftrightarrows (NH_4)_2SO_4 \cdot 2NH_4NO_{3(s)}$$

As discussed in detail by Seinfeld and co-workers, there are clearly a number of assumptions built into such calculations which may not hold in ambient air. However, such calculations are important because they represent a further refinement of our understanding of the chemical nature of atmospheric aerosols. The major emphasis on ambient measurements to date has been on gross chemical analysis of the ionic and elemental species present. The next step is clearly to examine the molecular form of these species and their size dependence under the appropriate conditions to gain a more detailed understanding of the interplay between gas phase and particle chemistry.

Although metals such as Al, Ca, Fe, Mg, and Pb are known to be present in atmospheric aerosols, their chemical form is also uncertain. Fe_2O_3, Fe_3O_4, Al_2O_3, and $AlPO_4$ have been identified in roadside particulate matter (Biggins and Harrison, 1980). Ca and Mg may exist in the form of oxides (i.e., CaO, MgO), although in the presence of water, Stelson and Seinfeld (1981) suggest that, based

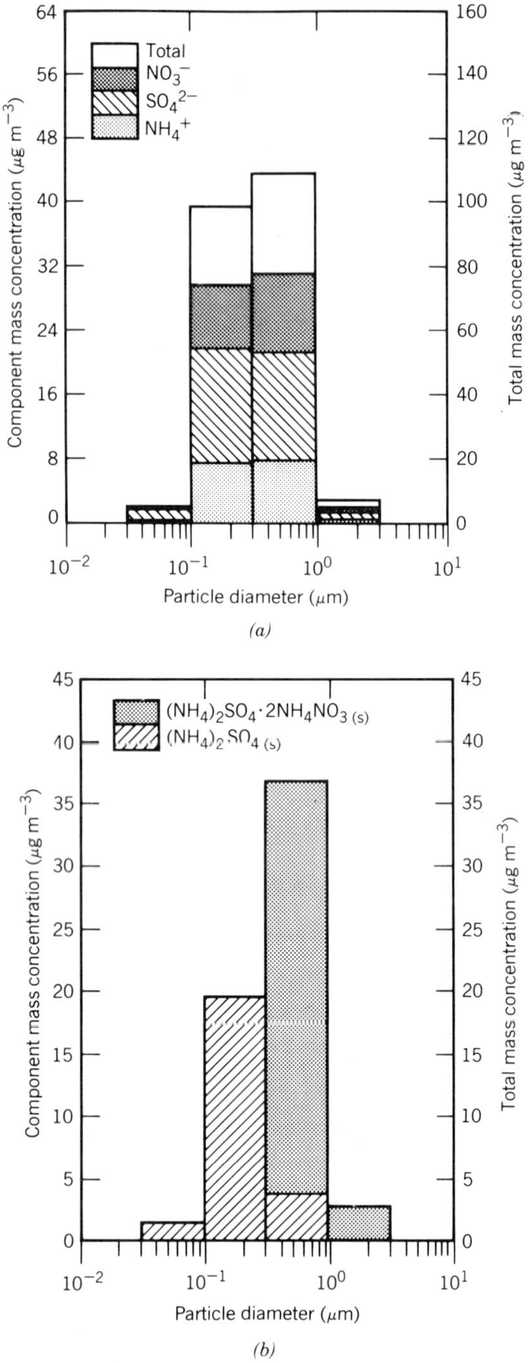

FIGURE 12.45. Predicted distribution of the components as a function of particle size for the cases in Table 12.20: (*a*) case 1, RH = 90%, (*b*) case 2, RH = 50%, (*c*) case 3, RH = 90%, (*d*) case 4, RH = 50%, (*e*) case 5, RH = 50%, (*f*) case 6, RH = 50%. The sulfate size distribution is assumed, but the rest are determined by the chemical equilibria in the system (from Bassett and Seinfeld, 1984).

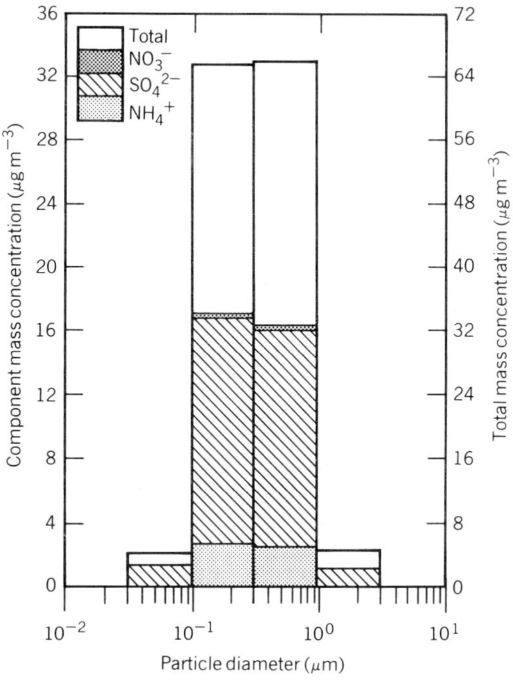

(c)

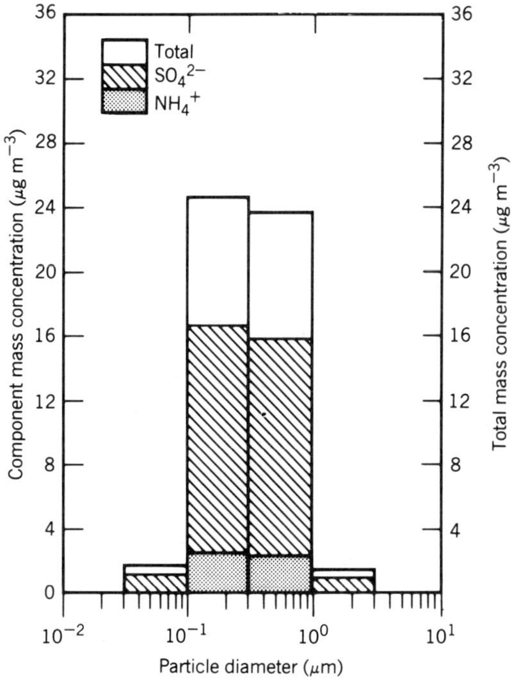

(d)

FIGURE 12.45. *(Continued)*

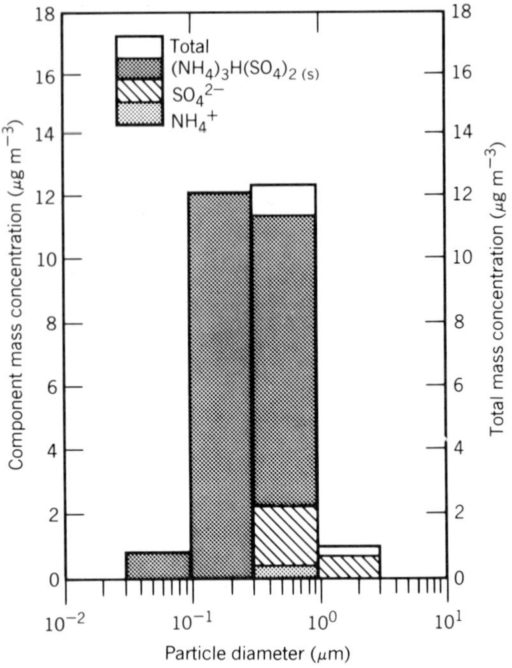

(e)

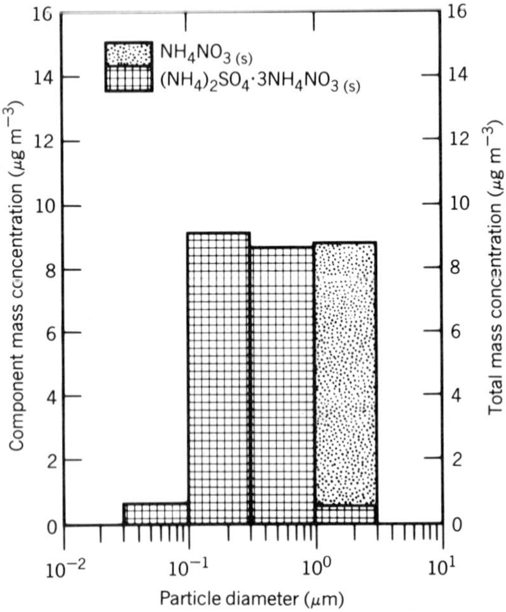

(f)

FIGURE 12.45. *(Continued)*

on equilibrium considerations, CaO and MgO should react to form their hydroxides, $Ca(OH)_2$ and $Mg(OH)_2$, respectively. Similarly, the oxides Na_2O and K_2O should form NaOH and KOH when water is present. Lead has been observed in roadside particulate matter in a wide variety of forms, such as $PbSO_4$, Pb_3O_4, $PbSO_4 \cdot (NH_4)_2SO_4$, $PbO \cdot PbSO_4$, $2PbCO_3 \cdot Pb(OH)_2$, $2PbBrCl \cdot NH_4Cl$, PbBrCl, $(PbO)_2PbBrCl$, $3Pb_3(PO_4)_2 \cdot PbBrCl$, and elemental lead (Biggins and Harrison, 1980; Post and Buseck, 1985).

12-D-2c. Particulate Organics in Non-urban Aerosols

In addition to sulfate, nitrate, and metals, carbon-containing compounds are also found in fine particles in significant concentrations (Table 12.14). These include a variety of organic compounds such as long chain hydrocarbons, polycyclic aromatic hydrocarbons, and organics containing oxygen, nitrogen, and sulfur. Because of the importance of polycyclic aromatic hydrocarbons and the substantial increase in our knowledge of the sources and atmospheric transformations of these compounds in the last decade, they are discussed separately in Chapter 13.

Organics are important components of particulate matter, whether in urban, rural, or remote areas. Indeed, the organic composition of non-urban aerosols is surprisingly complex (Hahn, 1980; Duce et al., 1983). For example, Table 12.21 lists some of the organics found in the ether-extractable portion of aerosol particles collected over the southern portion of the North Atlantic; aliphatics, aromatics, polar compounds, as well as organic acids and bases have been identified. When it is recognized that less than 10% of the total mass of the ether-extractable organics could be identified, and that other compounds not soluble in ether are present, it is clear that this extensive listing is still likely only a minor portion of the total organics in the aerosol. Indeed, in addition to the compounds shown in Table 12.21, fatty alcohols, fatty ketones, fatty acid esters, and fatty acid salts have also been found in non-urban aerosols (e.g., see Gagosian et al., 1981; Simoneit and Mazurek, 1982; Zafiriou et al., 1985).

As is the case in urban areas (see Section 12-D-2d), most of the particulate organic carbon is believed to reside in fine particles.

Compounds within a group with either even or odd numbers of carbon atoms are often found to predominate. For example, Gagosian and co-workers (1981) found that odd carbon *n*-alkanes in aerosol particles of Enewetak Atoll tended to predominate, whereas the fatty alcohols, fatty acid esters, and fatty acid salts tended to have even numbers of carbons. Such preferences for even or odd carbon compounds are believed to reflect the source of the organic. Thus C_{23}–C_{35} straight chain alkanes having a preference for odd carbon numbers are believed to arise from epicuticular waxes of vascular plants. The *n*-fatty alcohols (C_{24}–C_{32}) and *n*-fatty acids (C_{20}–C_{32}) with a preference for even carbon numbering are also believed to be from plant waxes. On the other hand, petroleum-derived hydrocarbons include long chain alkanes from C_1 to C_{35} with no carbon number preference, branched and cyclic isomers of the alkanes, and isoprenoid hydrocarbons (Simoneit and Mazurek, 1982).

TABLE 12.21. Individual Compounds Indentified in the Ether-Extractable Organic Matter from Aerosol Particles Collected in the Southern North Atlantic

	Concentration ($ng\ m^{-3}$ air at STP)
I. Aliphatic Hydrocarbons	
n-Decane	0.55
n-Undecane	1.9
n-Dodecane	0.92
n-Tridecane	0.85
n-Tetradecane	1.2
n-Pentadecane	0.70
n-Hexadecane	0.58
n-Heptadecane	1.0
n-Octadecane	0.69
n-Nonadecane	2.1
n-Eicosane	0.01
n-Heneicosane	0.61
n-Docosane	1.3
n-Tricosane	0.81
n-Tetracosane	0.24
n-Pentacosane	0.10
n-Hexacosane	0.39
n-Heptacosane	0.01
n-Octacosane	<0.01
n-Nonacosane	<0.01
n-Triacontane	<0.01
Total	14.0
II. Aromatic Hydrocarbons	
Naphthalene	2.7
2-Methylnaphthalene	1.8
Acenaphthalene	0.38
Acenaphthene	0.25
Fluorene	0.31
Phenanthrene	0.05
Anthracene	0.49
2-Methylanthracene	2.2
Fluoranthene	1.4
Pyrene	0.24
1,2-Benzofluorene	0.78
Chrysene	0.37
1,2-Benzoanthracene	0.28
3,4-Benzopyrene	<0.01
1,2-Benzopyrene	<0.01
Perylene	<0.01
Total	11.3

TABLE 12.21. (*Continued*)

	Concentration (ng m^{-3} air at STP)
III. Polar Compounds	
Coumarine	0.04
peri-Naphthenone	0.11
Xanthone	0.22
Anthrone	0.04
Flavone	0.07
Benzoanthrone	0.24
Carbazole	0.04
Total	0.76
IV. Organic Acids	
Pelargonic	2.5
Capric	3.5
Hendecanoic	6.1
Lauric	2.0
Tridecanoic	1.6
Myristic	1.3
Pentadecanoic	3.8
Palmitic	5.3
Margaric	2.1
Stearic	2.4
Oleic	1.7
Linoleic	3.1
Linolenic	1.1
Nonadecanoic	1.1
Arachidic	1.1
Heneicosanoic	0.01
Behenic	0.82
Total	39.5
V. Organic Bases	
Quinoline	<0.01
iso-Quinoline	<0.01
Aniline	<0.01
Indole	0.20
2-Methylindole	0.20
7-Methylindole	0.17
2,3-Dimethylindole	0.14
2,5-Dimethylindole	0.15
2,4-Dimethylquinoline	0.41
2,6-Dimethylquinoline	0.02
2,8-Dimethylquinoline	0.78

TABLE 12.21. ***(Continued)***

	Concentration ($ng\ m^{-3}$ air at STP)
α-Naphthylamine	<0.01
Benzo(h)quinoline	<0.01
Acridine	<0.01
Phenanthridine	<0.01
Total	2.1
Total identified	67.7

Source: Hahn, 1980; adapted from Ketseridis et al., 1976.

In addition to these *non-viable* organics, there are, of course, a whole host of viable species, such as fungi, bacteria, pollen, yeasts, and viruses, also present in the atmosphere. For a detailed review of the variety of organics found in aerosols in various locations and their concentrations, the reader is referred to the review by Duce et al., (1983).

12-D-2d. Particulate Organics in Urban Aerosols

In urban atmospheres, the organics characteristic of plant waxes, resin residues, and so on, as well as the long chain hydrocarbons characteristic of petroleum residues are found (Simoneit and Mazurek, 1982). However, a variety of smaller, multi-functional compounds characteristic of gas-to-particle conversions have also been observed. In a photochemically oxidizing atmosphere such as that of Los Angeles, for example, difunctionally substituted alkane derivatives of the type

$$X{-}(CH_2)_n{-}X \text{ and } X{-}(\overset{\overset{CH_3}{|}}{C}H_2)_{n-1}{-}X$$

with $n = 1$–5 have been observed in a number of studies. The substituent X can be —COOH, —CHO, —CH_2OH, —CH_2ONO, —COONO, or —$COONO_2$. For example, Table 12.22 shows some of the difunctional species identified in submicron aerosols using high-resolution mass spectrometry (MS) of ambient particulate samples introduced into the MS by slow heating from 20 to 400°C; these compounds were present primarily in the submicron fraction of the particles, suggesting they were secondary in nature, that is, were formed from chemical reactions in the atmosphere. Subsequently, even larger as well as branched chain dicarboxylic acids (Table 12.23) were observed in Los Angeles aerosol using solvent extraction followed by capillary column gas chromatography and MS with chemical or electron impact ionization (Grosjean et al., 1978).

Monosubstituted alkanes of the type

$$CH_3{-}(CH_2)_n{-}X$$

TABLE 12.22. Some Difunctionally Substituted Alkane Derivatives Found in Submicron Ambient Particles in Urban Air

Compound	n
$HOOC(CH_2)_nCOOH$	1–5
$HOOC(CH_2)_nCHO$	3–5
$HOOC(CH_2)_nCH_2OH$	3–5
$HOOC(CH_2)_nCH_2ONO$ or $CHO(CH_2)_nCH_2ONO_2$	3–5
$CHO(CH_2)_nCH_2OH$	3–5
$CHO(CH_2)_nCHO$	3–5
$HOOC(CH_2)_nCOONO$ or $CHO(CH_2)_nCOONO_2$	3–5
$CHO(CH_2)_nCOONO$	3,4
$HOOC(CH_2)_nCOONO_2$	4,5
$HOOC(CH_2)_nCH_2ONO_2$	3,4

Source: Schuetzle et al., 1975; Cronn et al., 1977.

were not observed, which was attributed to their vapor pressures being sufficiently large that they exist mainly in the gas phase. However, paper chromatography of an aerosol sample from Los Angeles did show a spot where monocarboxylic acids would be expected; the infrared (IR) spectrum (Fig. 12.46) indicated the simultaneous presence of organic nitrate groups, suggesting that at least a portion of these acids were also multi-functional (O'Brien et al., 1975b). From the relative IR peak intensities, approximately one acid molecule in nine also contained a nitrate group. While this would suggest that monocarboxylic acids comprised the

TABLE 12.23. Some Alkyl Dicarboxylic Acids Identified in Ambient Particles in Urban Air

$HOOC(CH_2)_nCOOH$ (n = 1–8)	
CH_3 $HOOCCH{-}COOH$	
CH_3 $HOOCCHCH_2COOH$	
CH_3 $HOOCCHCH_2CH_2COOH$	CH_3 $HOOCCH_2CHCH_2COOH$
$(CH_3)_2$ $HOOC(CH_2)_3COOH$[a]	
CH_3 $HOOCCHCH_2CH_2CH_2COOH$	CH_3 $HOOCCH_2CHCH_2CH_2COOH$

Source: Grosjean et al., 1978.

[a]Isomer of 2-dimethylglutaric acid not identified.

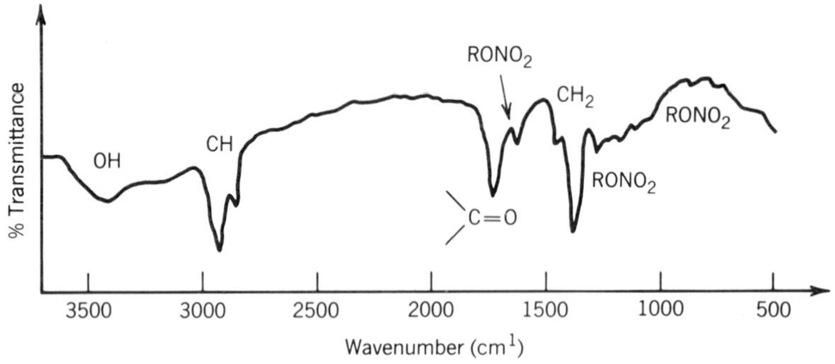

FIGURE 12.46. Infrared spectrum of monocarboxylic acid spot of an aerosol sample taken in downtown Los Angeles. A seventy-two hour sample was taken commencing September 15, 1973 (from O'Brien et al., 1975b).

remainder, it is not clear what these acids are, since stearic acid, which corresponds to the average molecular weight determined for this spot, is not sufficiently acidic to give a positive chromatographic response with the indicator used.

It is possible that the compounds giving this monocarboxylic acid spot contain other substituents which were not differentiated by infrared spectroscopy. For example, the hydroxy-substituted monocarboxylic acids 6-hydroxyhexanoic acid and 5-hydroxyhexanoic acid have since been identified in particles from the Los Angeles area (Cronn et al., 1977).

Also observed are monosubstituted aromatics of the type

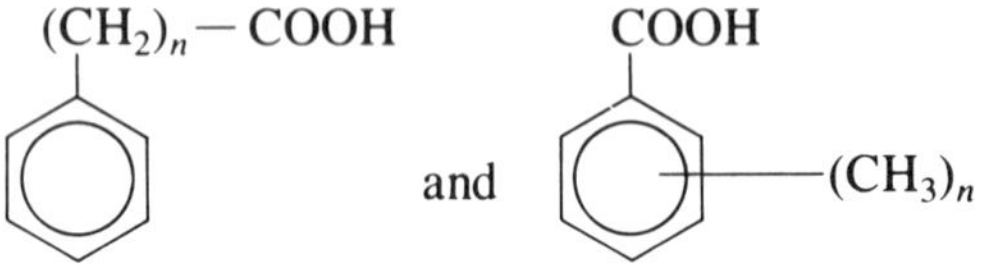

with $n = 0$–3. However, the substituents —CHO and —CH_2OH were not observed as was the case for the alkane derivatives; in addition, no disubstituted aromatics of the type

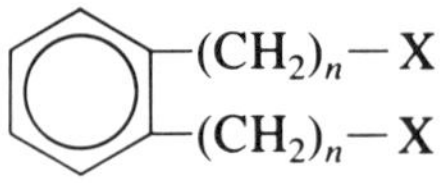

were seen (Schuetzle et al., 1975).

Other oxygenated organic components reported in the photochemical atmosphere of Los Angeles include dihydroxybenzene and phthalates (Appel et al., 1980c). The latter may arise from direct emissions, since phthalates are used as plasticizers.

Of course, in addition to these secondary organics from gas-to-particle conversion, there are primary carbonaceous particulates emitted directly from sources. For example, Rosen et al. (1980) estimate that between 65 and 85% of the carbonaceous particulate mass in two different air basins in California and in the Chicago area was due to primary particulate emissions.

The nature and size of the multi-functional and usually highly oxidized organics shown in Tables 12.22 and 12.23, are consistent with their production from atmospheric reactions of unsaturated hydrocarbon precursors. The most likely precursors appear to be cyclic olefins and perhaps dialkenes which can be oxidized to give stable species containing two substituted groups in the same molecule. For example, the reaction of O_3 with cyclohexene is expected to lead initially to the excited Criegee biradical:

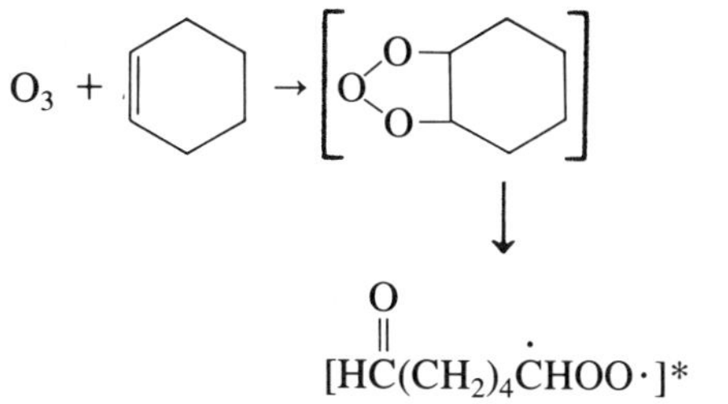

As discussed in detail in Section 7-E-2, the fraction of the excited biradical stabilized under tropospheric conditions is uncertain even for relatively simple alkenes. In addition, the decomposition paths of the biradical are unknown. However, one can envision reactions such as the following leading from the excited biradical to stable multi-functional compounds:

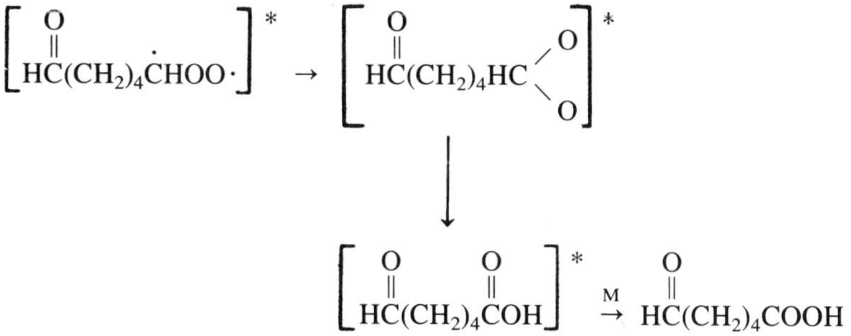

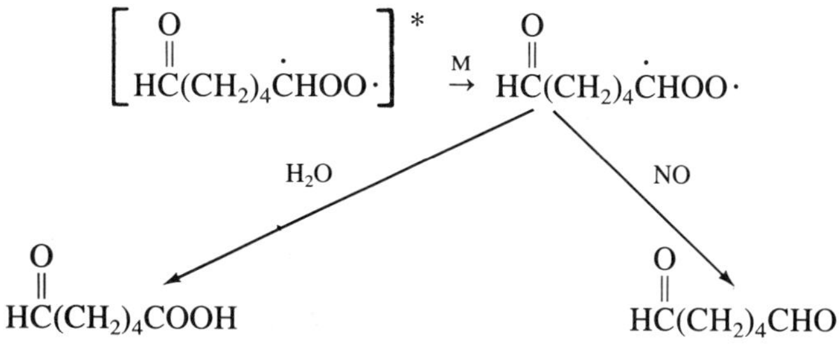

6-oxohexanoic acid

Adipaldehyde

Recently Hatakeyama et al. (1985) studied the reaction of cyclohexene with ozone in air and estimate that at ppm reactant concentrations, 13 $\pm$ 3% of the initial cyclohexene was converted to particulate phase organics. They identified both adipaldehyde and 6-oxohexanoic acid as major particulate phase products, as expected from the reaction scheme above.

Thus while the sources of the multi-functional oxygenated secondary particulate organics are not known, it is reasonable that cycloalkenes and dialkenes serve as precursors. Additional evidence for this is found in Table 12.24 which shows the vapor pressures of the least volatile oxidation products of a series of alkenes; also shown is the mimimum alkene concentration needed to form the least volatile product at a concentration in excess of its saturation concentration, assuming complete conversion of the precursor to the product. The concentrations of even very long chain alkenes required to form condensable acids exceed those commonly observed in ambient air. Only for the cyclic alkenes and the dialkene are the concentrations required to form condensed phase secondary particles in the same range as those typically found. For example, cyclopentene, cyclohexene, and 1-methycyclohexene have been observed in ambient air at concentrations in the ~1–10 ppb range (Neligan, 1962; Stephens and Burleson, 1967, 1969; Calvert, 1976). Few data are available on dialkenes larger than C_6 in ambient air.

Prager and co-workers (1960) were the first to show that cyclic alkenes and dialkenes at ppm concentrations formed light scattering aerosols when irradiated with NO_x in air. Since then, a number of studies have confirmed these conclusions and established the nature of the organic products in the condensed phase. For

TABLE 12.24. Lowest Ambient Olefin Concentration Required to Form the Corresponding Condensable Species in Excess of Its Saturation Concentration

Olefinic Precursor	Least Volatile Photooxidation Product	Product Vapor Pressure (Torr)	Minimum Precursor Concentration[a]
Propylene	Acetic acid	16	21,000 ppm
1-Butene	Propionic acid	4	5,200 ppm
1-Hexene	Pentanoic acid	0.25	327 ppm
1-Heptene	Hexanoic acid	0.02	26 ppm
1-Octene	Heptanoic acid	$\sim 9 \times 10^{-3}$	~12 ppm
1-Decene	Nonanoic acid	$\sim 6 \times 10^{-4}$	~0.8 ppm
1-Tridecene	Dodecanoic acid	10^{-5}	13 ppb
Cyclopentene	Glutaric acid	2×10^{-7}	~0.3 ppb
Cyclohexene	Adipic acid	6×10^{-8}	0.08 ppb
1,7-Octadiene	Adipic acid	6×10^{-8}	0.08 ppb
3-Methylcyclohexene	Methyladipic acid	$\sim 2 \times 10^{-8}$	~0.03 ppb

Source: Grosjean and Friedlander, 1980.

[a]Lower limits calculated assuming complete conversion of the precursor to the least volatile of the possible products.

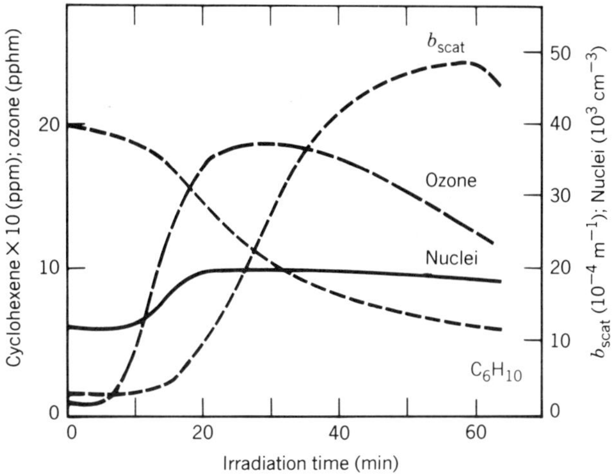

FIGURE 12.47. Ozone, hydrocarbon, condensation nuclei, and aerosol light-scattering coefficient, b_{scat}, as a function of irradiation time for a typical outdoor smog chamber run (from Grosjean and Friedlander, 1980).

example, Fig. 12.47 shows the cyclohexene, ozone, and nuclei concentrations as a function of time when 2.0 ppm cyclohexene, 0.34 ppm NO, and 0.17 ppm NO_2 were irradiated in the presence of ambient air in Pasadena, CA, at 24°C and 40% RH. (The nuclei present initially are those in the ambient air.) Also shown is the light scattering coefficient (b_{scat}) measured using an integrating nephelometer; nuclei as well as aerosol in the light scattering range are generated.

Laboratory studies of the organic condensed phase products of model hydrocarbon–NO_x–air systems have established that multi-functional compounds such as those observed in ambient air are indeed formed. Schwartz (1974) carried out one of the first detailed product analyses of the organic component of particles formed in the oxidation of ~10 ppm of cyclohexene, toluene, or α-pinene in the presence of ~2–5 ppm NO_x in air. The portion of the aerosol which was extractable into methylene chloride but which was insoluble in water was analyzed in order to focus on the aerosol products that had not undergone extensive secondary oxidations, in the hopes that their mechanisms of formation could be more easily related to the parent hydrocarbon. Table 12.25 shows the products tentatively identified in these studies. These are all multi-functional compounds with structures analogous to those observed in ambient aerosols (Tables 12.22 and 12.23).

Since then, a number of studies of model systems have confirmed that dialkenes, cyclic alkenes, and aromatics form substituted monocarboxylic acids, dicarboxylic acids, and organic nitrates in the condensed phase (e.g. see O'Brien et al., 1975a; Grosjean and Friedlander, 1980; Dumdei and O'Brien, 1984). For example, Table 12.26 shows the products identified in the particles formed in the cyclohexene–NO_x–ambient air system. In addition to the products identified by Schwartz (Table 12.25), adipic acid, $COOH(CH_2)_4COOH$, and 6-nitratohexanoic acid, $COOH(CH_2)_4CH_2ONO_2$, have been identified as major products; a variety

TABLE 12.25 Some Organic Aerosol Products Tentatively Identified from Organic-NO_x-Air Irradiations

Reactant Hydrocarbon	Organic Compounds Tentatively Identified in Less Polar Fractions of Aerosol Extractions
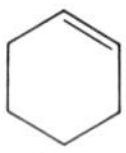 Cyclohexene	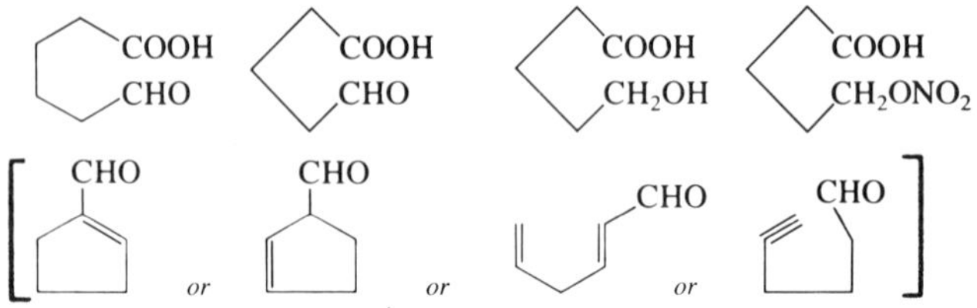
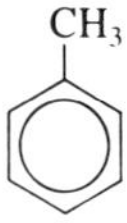 Toluene	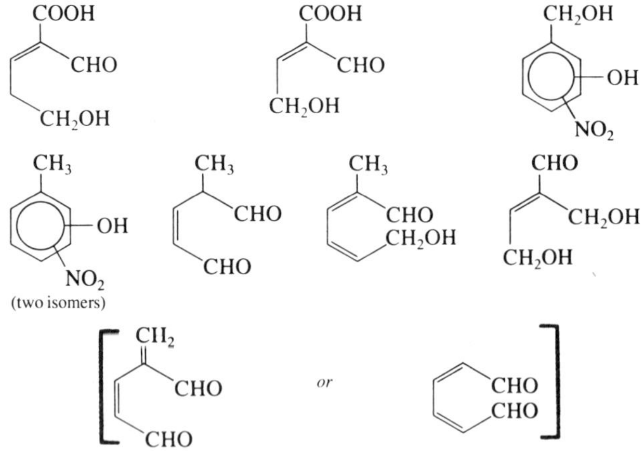
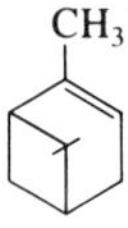 α-Pinene	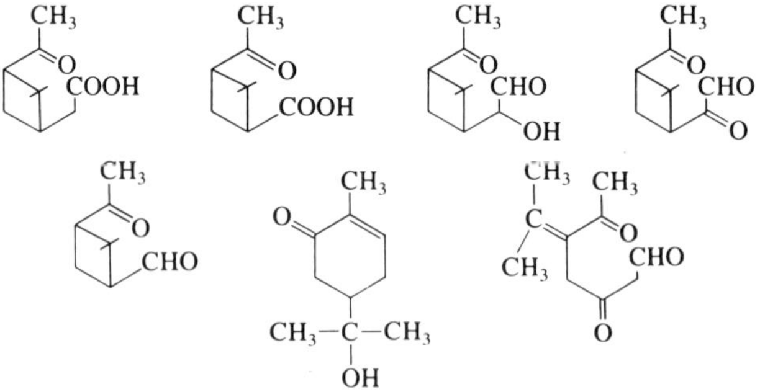

Source: Schwartz, 1974.

of other multi-functional oxygenates were also identified, all of which have been found in ambient air (Tables 12.22 and 12.23).

The products have either the same number of carbon atoms or one less than the parent hydrocarbon. This is not surprising based on our present knowledge of mechanisms of alkene oxidation (see Section 7-E). As discussed earlier, the reaction of cyclohexene with ozone is expected to produce difunctional compounds containing

TABLE 12.26. Compounds Identified in Particles from the Cyclohexene–NO_x—Ambient Air Oxidation

Product	Formula
Adipic acid	$COOH-(CH_2)_4COOH$
6-Nitratohexanoic acid	$COOH-(CH_2)_4CH_2ONO_2$
6-Oxohexanoic acid	$COOH-(CH_2)_4CHO$
6-Hydroxyhexanoic acid	$COOH-(CH_2)_4CH_2OH$
Glutaric acid	$COOH-(CH_2)_3COOH$
5-Nitratopentanoic acid	$COOH-(CH_2)_3CH_2ONO_2$
5-Oxopentanoic acid	$COOH-(CH_2)_3-CHO$
5-Hydroxypentanoic acid	$COOH-(CH_2)_3CH_2OH$
Glutaraldehyde	$CHO-(CH_2)_3CHO$

Source: Grosjean and Friedlander, 1980.

six carbon atoms such as $HCO(CH_2)_4$ CHO and $H\overset{\overset{O}{\|}}{C}(CH_2)_4COOH$. The formation of C_5 products has been observed in the ozone-cyclohexene reaction by Hatakeyama et al. (1985), including adipic acid, glutaraldehyde, 5-oxopentanoic acid, and glutaric acid. These can also be rationalized on the basis of known mechanisms of atmospheric oxidations. For example, one of the possible fates of the excited Criegee biradical formed in the reaction with O_3 is decomposition; in the case of the biradical from cyclohexene, two possible paths are the following:

$$\left[H\overset{\overset{O}{\|}}{C}(CH_2)_4\dot{C}HOO\cdot\right]^* \rightarrow \left[H\overset{\overset{O}{\|}}{C}(CH_2)_4COOH\right]^*$$

$$\rightarrow H\overset{\overset{O}{\|}}{C}(CH_2)_3CH_2 + CO_2 + H$$

$$\rightarrow H\overset{\overset{O}{\|}}{C}(CH_2)_3CH_2 + CO + OH$$

The five-carbon alkyl radical can undergo a series of reactions in air containing NO_x, such as the following to form glutaraldehyde:

$$H\overset{\overset{O}{\|}}{C}(CH_2)_3CH_2 + O_2 \rightarrow H\overset{\overset{O}{\|}}{C}(CH_3)_2CH_2OO \xrightarrow[NO]{NO_2 \uparrow} H\overset{\overset{O}{\|}}{C}(CH_2)_3CH_2O$$

$$H\overset{\overset{O}{\|}}{C}(CH_2)_3CH_2O \xrightarrow{O_2 \;\rightarrow\; HO_2} H\overset{\overset{O}{\|}}{C}(CH_2)_3\overset{\overset{O}{\|}}{C}H$$

The glutaraldehyde, $CHO(CH_2)_3CHO$, presumably acts as the source of the C_5 compounds containing carboxylic acid groups (Table 12.26) through its further oxidation.

Aromatics such as toluene also give significant amounts of condensed phase products. The observed products and the proposed reaction mechanisms leading to them are discussed in Section 7-G.

Only the larger simple alkenes give aerosol formation when irradiated with NO_x in air, likely because the oxidation products of the smaller alkenes have high vapor pressures (Table 12.24). Thus their aerosol forming ability compared to the cyclic alkenes and dialkenes is small and they are not believed to be major contributors to secondary ambient aerosol formation (National Research Council, 1977).

In short, the same types of aerosol organic products have been identified both in model systems and in polluted urban ambient air. While the reaction mechanisms leading to the observed products are to a great extent speculative at present, they can, in general, be rationalized based on the oxidation of analogous compounds.

In addition to organic carbon, elemental carbon can also be an important component of urban aerosols, especially in areas where significant wood burning and use of diesel automobiles occur (Table 12.9). The importance of elemental carbon in light absorption and scattering is discussed in Section 12-B-3.

One particularly interesting aspect of aerosol composition is the presence of formaldehyde, often in concentrations in great excess (e.g., $\sim 10^3$ higher) of that expected on the basis of a simple Henry's law calculation (e.g., see Klippel and Warneck, 1980). This is believed to be due to the formation of complexes, for example, with S(IV) (see Section 11-A-2) and possibly with alcohols, phenols, and amines. Higher carbonyls have also been observed in fog water (Grosjean and Wright, 1983); since atmospheric aerosols may serve as nuclei for fog formation (see Section 11-F), it is possible that some of these larger carbonyls originated in a pre-existing aerosol.

12-D-2e. *Surfactants in Aerosols*

The presence of long chain organics having one or more polar functional groups (e.g., the carboxylic acids and the nitrates) in the condensed phase suggests that these may act as surfactants in aqueous atmospheric aerosols, forming an organic coating over the surface of the aerosol. Molecules that have long chain ($\gtrsim C_5$), non-polar groups attached to polar tails, such as those in Table 12.27, can form a surface film on droplets by lining up with the polar ends in the water and non-polar, hydrophobic ends projecting into air as shown schematically in Fig. 12.48. Note, however, that this representation is simplified in that such an orderly arrangement will not apply to all surface-active compounds; for example, the presence of a double bond as in the case of oleic acid, $CH_3(CH_2)_7{-}CH{=}CH{-}(CH_2)_7COOH$, gives the molecule a "crooked" shape which requires more surface area per molecule.

The possibility of such organic films being formed on aerosol particles in the atmosphere, as well as on fog, cloud, and rain droplets, and snowflakes has been

TABLE 12.27. Structures of Some Classes of Surface-Active Molecules Found in the Atmosphere

Class of Compounds	Structure
Alcohol	$R{-}CH_2OH$
Acid	$R{-}C({=}O){-}OH$
Aldehyde	$R{-}C({=}O){-}H$
Ketone	$R{-}C({=}O){-}R'$
Ester	$R{-}C({=}O){-}OR''$
Amine	$R{-}NH_2$

Source: Gill et al., 1983.

discussed in detail by Gill and co-workers (1983). As seen from our earlier discussions on the types of organics which have been observed in both urban and non-urban aerosols, there is no question that surface-active species which can form organic films on water do exist in the atmosphere. However, Gill and co-workers (1983) have estimated, based on the limited data available, that only in aerosol particles does sufficient surface-active organic clearly exist to form a film around the particle. The presence of surface-active films on cloud droplets, snowflakes and raindrops is unlikely.

There is some evidence for the existence of organic films on the surfaces of particles. For example, electron micrographs of haze aerosol taken by Husar and Shu (1975) show droplets that are "wrinkled" in appearance; they suggested this was due to "haze aerosol" droplets being coated with an organic layer which collapsed when the water in the particle evaporated under vacuum.

The role of organic films on droplets in the atmosphere is speculative at present; Gill et al. (1983) suggest that they may have some or all of the following effects:

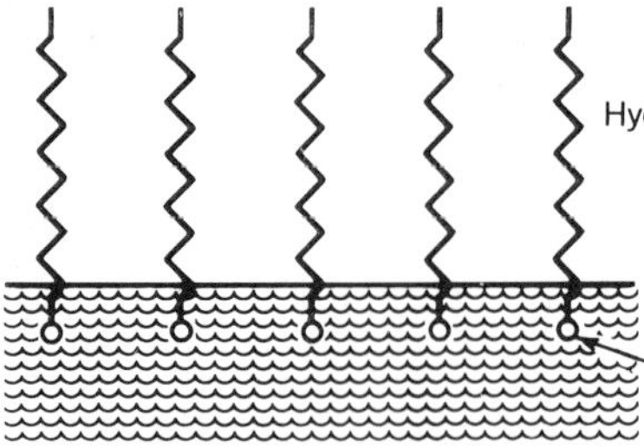

FIGURE 12.48. The orientation of surface-active organic molecules at the water surface (from Gill et al., 1983).

1. Reduction of the rate of evaporation of water from the droplets.
2. Inhibition of the transport of stable molecules and of highly reactive free radicals such as OH and HO_2 from the gas phase into the droplet [possibly reducing their roles in atmospheric aqueous phase oxidations (Chameides and Davis, 1982; Graedel and Weschler, 1981)].
3. Reduction of the efficiency with which the particles are scavenged by larger cloud and rain droplets.

Thus the presence of organic films may increase the lifetime of such particles in the atmosphere compared to those expected if the films were not present (Toossi and Novakov, 1985).

There is some experimental evidence under conditions approaching those in the atmosphere (e.g., small droplets and ppm reactant concentrations) which suggests that the rate of evaporation is reduced by organic films. Thus Chang and Hill (1980) exposed water drops of ~ 19 μm diameter to a stream of air containing the alkene decene (53–220 ppm) and O_3, (20 ppm); the rate of water evaporation was reduced by the presence of decene and O_3, and a major product of the reaction, nonanol, also reduced the evaporation rate when added separately.

12-D-2f. Form of Nitrogen and Sulfur in Particles

In addition to hydrocarbons and their oxygen-containing derivatives, organics containing nitrogen and sulfur have also been identified in ambient particulate matter. For example, polynuclear aza-heterocyclics (aza-arenes) are commonly observed in ambient particulate samples (e.g., see Cautreels and Van Cauwenberghe, 1976; Dong et al., 1977). In addition, as we have seen, some of the oxygen-containing species are multi-functional with nitrate, nitrite, peroxynitrate, and peroxynitrite groups attached (Table 12.22). However, Novakov and co-workers have shown that other chemical forms of nitrogen are present as well. Thus, using X-ray photoelectron spectroscopy (ESCA), they observed reduced nitrogen species tentatively identified as amines, amides, and possibly nitriles (Novakov et al., 1972; Gundel et al., 1979; Chang et al., 1982). These compounds, which they call N_x species, often comprise a major portion of the particulate nitrogen; for example, in particles collected in Berkeley, California, in November 1976, of the total nitrogen present, 50% was identified as the reduced N_x species. Much of this appears to be in the form of organic amides which can be hydrolyzed to the acid and NH_4^+:

$$\mathrm{R{-}\underset{\underset{\displaystyle O}{\|}}{C}{-}NH_2 + H_2O \xrightarrow{H^+} R{-}\underset{\underset{\displaystyle O}{\|}}{C}{-}OH + NH_4^+}$$

This suggests that caution should be exercised in interpreting NH_4^+ concentrations obtained using extraction procedures. Thus in one set of samples, ~85% of the N_x species originally present were removed by water extraction while, simultaneously, the NH_4^+ concentration in the extract increased proportionally. Evidence for the presence of amides in ambient particulates has also been found using high-resolution mass spectroscopy (Cronn et al., 1977).

In addition to the reduced N_x species, a relatively volatile form of ammonium, suggested to be ammonium chloride (NH_4Cl) which has been identified in ambient particles (Schuetzle et al., 1973; Cronn et al., 1977) or, alternatively, adsorbed NH_3, has been observed by ESCA.

Finally, Kneip et al. (1983) have presented evidence that *n*-nitroso compounds are present in ambient particulate matter in New York City.

While sulfur has been known to exist as sulfate in particles for many years, particulate sulfur in other oxidation states has also been observed in ambient aerosols. These have been tentatively identified as adsorbed SO_2, SO_3^{2-}, S^0 (elemental sulfur), and possibly two kinds of S^{2-}. In addition, a species designated as adsorbed SO_3 was observed, although it was likely a sulfate compound whose chemical shift was similar to adsorbed SO_3, since SO_3 is expected to react rapidly with water in air (Section 11-A). Sulfate was the predominant form of sulfur in all samples, and these other forms did not appear consistently in all samples at all times and locations (Craig et al., 1974; Novakov et al., 1977).

Sulfur-containing heterocyclics have also been identified in ambient particles (e.g., see Dong et al., 1977). In addition, dimethyl sulfate and monomethyl sulfuric acid have been detected in particles as well as in the gas phase in power plant plumes and in ambient air (Durham et al., 1984; Eatough et al., 1985); whether these are primary pollutants (i.e., directly emitted from sources) or secondary pollutants (i.e., formed by chemical reactions in air) is unknown. H_2SO_3 and CH_3SO_3H have also been identified in ambient particulate matter (Panter and Penzhorn, 1980).

Eatough and Hansen (1983) review in detail the forms of both inorganic and organic sulfur found in airborne particles.

12-D-3. *Fraction of Total Atmospheric Burden of Carbon, Nitrogen, and Sulfur Found in the Particulate Phase*

Having considered now the chemical composition of atmospheric particulate matter, and in earlier chapters the chemical reactions responsible for gas-to-particle conversion, it is of interest to consider how much of the total mass of sulfur,

TABLE 12.28. Some Gas–Particle Distribution Factors for Carbon (f_C), Nitrogen (f_N), and Sulfur (f_S) Measured in the Los Angeles Area

Location	Carbon[a] f_C	Nitrogen[b] f_N	Sulfur[c] f_S
Riverside, CA[d]	0.021–0.035	0.17–0.41	0.21–0.77
Pasadena, CA[e]	0.009–0.061	0.007–0.17	0.13–0.47
West Los Angeles, CA[f]	0.01	0.02	0.17
Claremont, CA[g]	n.d.[h]	0.01–0.19	n.d.[h]

[a] f_C = Particulate organic carbon/(particulate organic carbon + gas phase non-methane hydrocarbons) in $\mu g\ m^{-3}$ as C.

[b] $f_N = NO_3^-/(NO_3^- + NO_x)$ in $\mu g\ m^{-3}$ as NO_2; should be taken as an upper limit due to possible experimental artifacts as discussed in text.

[c] $f_S = SO_4^{2-}/(SO_4^{2-} + SO_2)$ in $\mu g\ m^{-3}$ as SO_2; should be taken as upper limit due to possible experimental artifacts as discussed in text.

[d] Twenty-four hour samples taken in 6 month period from May to October 1975; from Grosjean et al., 1976a.

[e] One hour samples; from Grosjean and Friedlander, 1975.

[f] From Grosjean et al., 1976b.

[g] From Grosjean, 1983.

[h] n.d. = not determined.

nitrogen, and carbon in the atmosphere is in the gas phase and how much is in particles. This is commonly expressed in the form of gas–particle distribution factors, f, defined as the ratio of mass of the species in the particulate phase divided by the total mass in both the gas and particulate phases. Since the chemical forms and hence masses are different in the two phases (e.g., sulfur is commonly SO_2 in the gas phase and SO_4^{2-} in the particulate phase), the masses are expressed in the form of a common species; for example, sulfate (molecular weight 96) in the condensed phase can be expressed as SO_2 (molecular weight 64) by multiplying the measured mass by 64/96.

Table 12.28 shows some gas–particle distribution factors for carbon (f_C), nitrogen (f_N), and sulfur (f_S) measured in the Los Angeles area. Typically, less than 5% of the total carbon is present in particles; much larger fractions of the total nitrogen and sulfur tend to be in the form of particles, especially in the downwind regions (e.g., Riverside) where the air mass is sufficiently aged that more gas-to-particle conversion has taken place.

One note of caution is in order here, however; as discussed in detail in Section 12-E-4, it has been recognized recently that the quantitative measurement of particulate nitrate and sulfate, and to a lesser extent carbon, can be subject to significant sampling artifacts. Thus the Riverside data for nitrogen and sulfur should be regarded as upper limits since glass fiber filters were used, and these have since been shown to give a positive artifact; however, the Claremont data may represent lower limits because Teflon filters were used for particulate nitrate, and these may have a negative artifact. In any case, it is clear that a substantial portion, at least

20%, of nitrogen and sulfur may exist in particles at certain times and under certain conditions.

12-E. ANALYTICAL TECHNIQUES

Techniques for the sampling and analysis of atmospheric particulate matter have been the subject of considerable research, resulting in the development of methods for measuring the physical and chemical properties of aerosols over a wide range of sizes. Since many books and monographs have been published on this subject, we do not attempt to treat in detail all the methods that have been or are being used. Rather, this section is intended to given an overview of the methods and their capabilities. For further details, the reader is referred to the volumes published by Lawrence Berkeley Laboratory (1979), edited by Liu (1976), Dennis (1976), by Lundgren et al. (1979) and by Liu et al., (1984), and the books by Mercer (1973), Cadle (1975), and Hinds (1982).

12-E-1. Sampling and Collection of Particulate Matter

12-E-1a. Sampling

The first steps in analysis of the physical and chemical properties of atmospheric particulate matter are sampling, that is, obtaining a representative sample over the desired size range, and collection, that is, separating the particles from air. During sampling and collection, such parameters as humidity, temperature, and particle concentration should be controlled in order to maintain the sample integrity. For example, as we have seen in Chapters 8 and 11, ammonium nitrate particles are sensitive to both temperature and relative humidity and if these are different in the instrument than in ambient air, the sample may not be truly representative.

Sampling of particles presents some special problems compared to gases. The larger mass of particles results in a much greater inertia, so that when the gas flow curves sharply the particles tend to go straight ahead. High or low inlet velocities as well as bends in tubing used to sample for particles can thus lead to significant particle size bias and should be avoided. Criteria for unbiased particle sampling are given by Mercer (1973). In addition, the sampling lines should be as short as possible to minimize particle loss by gravitational settling and turbulent deposition. Losses can also occur on the sampling surfaces if an electrostatic charge is allowed to build up. The overall efficiency of any particular inlet is a function of sampler geometry, particle size distribution and charge, wind direction and speed, and sampling flow rate. Additional information on the design and principles of various aerosol sampling inlets are discussed by Mercer (1973), Liu and Pui (1981), Hinds (1982), and in the *Handbook on Aerosols* edited by Dennis (1976).

Because of the EPA proposal to change the primary air quality standard for particulate matter to include only those particles with diameters $\leq 10\ \mu m$ (PM_{10}) rather than total suspended particles (TSP) as is now the case (Federal Register, 1984), there has been a great deal of work carried out to design and evaluate sampling inlets which will exclude particles larger than 10 μm. For example, the results of an intercomparison field study of such PM_{10} inlets is described by Rodes et al. (1985). Preliminary studies indicate that the ratio of PM_{10} to TSP may be reasonably constant (≃0.49), so that if the change is made to this new standard, it may be possible to correlate past TSP data with PM_{10} data taken in the future (Rodes and Evans, 1985).

12-E-1b. Collection

Collection techniques are based on filtration, gravitational and centrifugal sedimentation, inertial impaction and impingement, diffusion, interception or on electrostatic or thermal precipitation. The choice of method depends on a number of parameters such as the composition and size of the particles, the purpose of the sample, and the acceptable sampling rates. Table 12.29 summarizes some of the commonly used methods and the size ranges over which they are effective.

Filters. Filters collect liquid and solid particles by mechanisms including diffusion, impaction, interception, electrostatic attraction, and sedimentation onto the filter while allowing the gas to pass through. The types commonly used in atmospheric particulate collection are fibrous mats, porous sheets, or membranes.

Different filters have unique characteristics, which include the collection efficiency as a function of particle size, the pressure drop at a given flow velocity, and types of reactions that occur on the filter surfaces. Perhaps somewhat surprisingly, sieving action is not the only filtration mechanism. The major filtering action is due to forces that bring the particles into contact with the filter surface where they may stick; these include impaction, interception, diffusion, sedimentation,

TABLE 12.29. Some Commonly Used Methods of Collection of Ambient Particulate Matter

Method	Approximate Range of Diameters[a] (μm)
Filters	>0.003
Sedimentation collectors	
Gravitational	≳10
Centrifugal	0.1–10
Impactors	
Atmospheric pressure	≥0.5
Low pressure	≥0.05
Precipitators	
Electrostatic	0.05–5
Thermal	0.005–5

[a]The upper size ranges are usually related to inlet losses which prevent large particles from reaching the sampling surface.

and electrostatic attraction. At larger particle diameters and high flows, impaction is efficient, while at smaller diameters and flows, collection by diffusion to the surface is important; this increased efficiency at large and small diameters results in a minimum at ~0.3 μm in the curve of collection efficiency against particle diameter at the usual sampling rate per unit surface (Fig. 12.49). The minimum can be shifted by changing conditions. The quoted efficiency of typical HEPA filters is usually significantly better than 90% at the minimum, for example, 99%.

Fibrous mat type filters include the frequently used paper (cellulose) fiber filter, for example, the *Whatman paper filter*, and glass fiber filters. A common fibrous mat filter used for sample collection and air cleaning is known as the HEPA filter (*h*igh *e*fficiency *p*articulate *a*ir filter) and is made of a combination of cellulose and mineral fibers.

Perhaps the most well-known and widely used type of fibrous mat filter is the *high-volume* filter, commonly referred to as *hi-vol*. Figure 12.50 is a schematic of a typical hi-vol filter system with the shelter to protect the filter from rain and so on. The hi-vol method of collecting particles is currently the U.S. Environmental Protection Agency's reference method for measuring total suspended particles in ambient air. However, as mentioned in Chapter 1, a new size-fractionated standard for particulate matter is currently under consideration and, if adopted, will result in the promulgation of a new reference method.

In the hi-vol, air is drawn through an 8 in. by 10 in. glass fiber filter at a flow rate of 1.1–1.7 $m^3\ min^{-1}$, resulting in the collection of particles $\gtrsim$ 0.3 μm with nearly 100% efficiency. While the shelter is intended to allow particles up to 100 μm in size to reach the filter, in practice the cutoff may occur at smaller sizes because of the effect of the shelter; the actual cutoff depends on the particular meteorological conditions (e.g., wind speed). If desired, the upper end of the size range may be restricted by preceding the filter with a cyclone (see below).

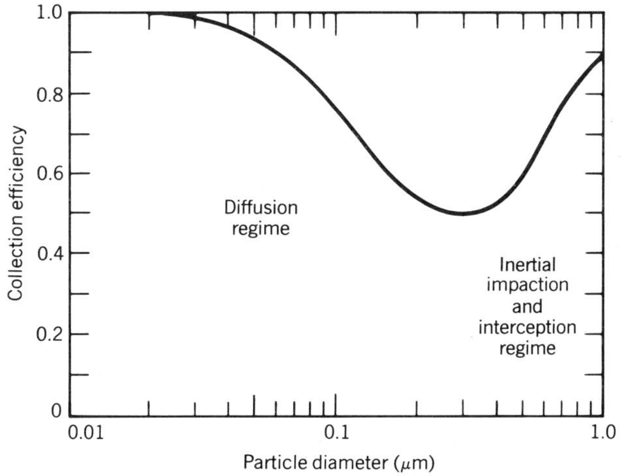

FIGURE 12.49. Particle collection efficiency as a function of particle size for a typical filter (from Liu and Kuhlmey, 1977).

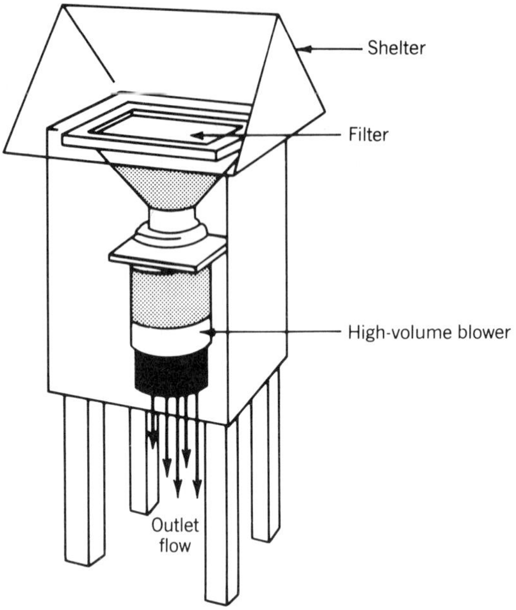

FIGURE 12.50. Schematic of hi-vol particulate sampler (from Lawrence Berkeley Laboratory, 1979).

A modification of the hi-vol filter to increase the total air flow allows the collection of sufficient particulate matter in relatively short time periods (e.g., 2 h) to carry out chemical analysis; this is important for studies of the diurnal variation of various chemical components of the aerosol, as well as for minimizing sampling artifacts (see Section 12-E-4e). Figure 12.51 is a schematic of such a modified hi-vol; it contains four filters connected in series and has a total flow approximately 16 times that of a normal hi-vol, allowing a corresponding reduction in sampling time for a given sample size (Fitz et al., 1983).

Porous materials and membranes used as filters have a number of small, often tortuous, pores. This type of filter includes sintered glass filters, organic membrane filters, and silver membrane filters. Two types of membrane filters are nucleopore and millipore filters, named after their principal manufacturers. Nucleopore filters are thin films with smooth surfaces and straight, uniform cylindrical pores made by irradiating a thin polycarbonate plastic sheet in contact with a uranium sheet with slow neutrons. The neutrons cause fission of ^{235}U and the resulting fragments produce ionization tracks through the plastic; these tracks are then chemically etched to a desired and uniform size using a sodium hydroxide solution.

Millipore filters have twisted, interconnecting pores that are much more complex than those in nucleopore filters. They are available in different materials such as Teflon, polycarbonate, quartz, silver and cellulose acetate.

Membrane filters are particularly useful when surface analytical techniques, such as optical and electron microscopy and X-ray fluorescence analysis, are to be used

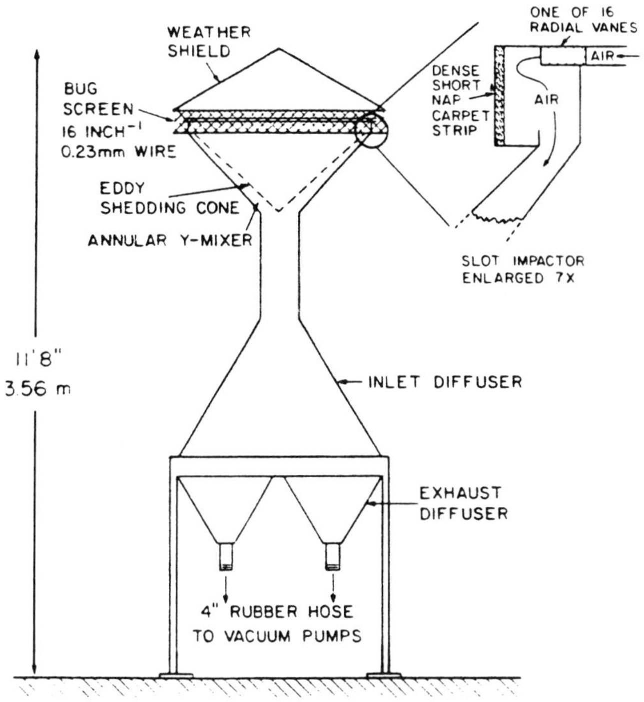

FIGURE 12.51. Schematic of an ultra-high volume sampler (from Fitz et al., 1983).

subsequent to collection, because most of the particles remain on the surface of the filter.

The filter efficiency may also depend on the pressure drop across the filter, as seen for the nucleopore filter in Fig. 12.52. Other types of filters will have different collection characteristics as a function of pressure drop. This can be important if samples are collected over long periods of time because increased particulate loading may clog the filter, leading to an increase pressure drop and hence a change in collection efficiency as a function of collection time.

Filter sampling is also accompanied by potential reactions of pollutant gases with the particles on the filter or with the filter medium (including binders which are used in some filters) during sampling (see Section 12-E-4e for a discussion of filter artifacts) and the absorption of water from humid air. In the first case, conversion of gaseous SO_2 and HNO_3 to particulate sulfate and nitrate, respectively, has been observed on some filters. Not only will this give misleading results with respect to the total mass of particulate matter sampled, but it may make results based on sulfate and nitrate analysis uninterpretable. This problem is important because sulfate and nitrate are major components of ambient aerosols (Section 12-D-2) and are the prime anionic contributors to acid rain and fogs (Chapter 11).

Some filters, especially paper filters, are hygroscopic and thus tend to adsorb

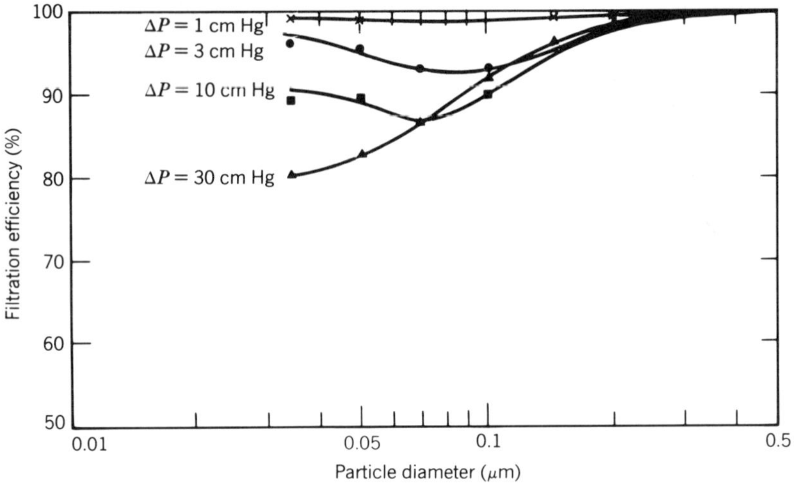

FIGURE 12.52. Particle collection efficiency as a function of particle size for different pressure drops across a typical filter (from Liu and Kuhlmey, 1977).

water vapor from humid air. Essentially all filters will change weight as a function of humidity due to adsorbed water on filter surfaces. This necessitates weighing blank filters at the ambient relative humidity in order to obtain accurate particulate mass measurements. Glass fiber filters are relatively (but not entirely) insensitive to humidity, which is a major reason they have been used in the hi-vol reference method. However, even here the particulate matter collected on the filter may be hygroscopic and adsorb or desorb water. To minimize this problem, hi-vol filters are equilibrated at temperatures between 15 and 35°C and in air with a relative humidity ≤ 50% for 24 h prior to weighing before and after sampling.

Other problems with collection using filters, such as interference of impurities contained in the filter itself with chemical analysis of the collected particles, are discussed in detail in the LBL (1979) report.

Sedimentation Collectors. These collectors are used primarily for large particles (≳2.5 μm), that is, those in the coarse particle range. They include collection by gravitational sedimentation (e.g., dustfall jars) as well as by centrifugal sedimentation which allows collection in the submicrometer range (e.g., centrifuges, cyclone collectors).

Gravitational sedimentation only collects the large particles which settle out of the atmosphere fairly quickly. This *dustfall* generally consists of particles that are relatively large and, as such, are not particularly relevant to the focus of this book. Thus dustfall collectors will not be discussed further.

The principle of centrifugal collection is, of course, well known (Mercer, 1973). Collection of particles using centrifugation involves passing the aerosol at a controlled rate through a rapidly spinning air mass. Collection of particles in ranges as small as ~0.1–1 μm has been reported using this technique. The cyclone collector, a modification of the centrifuge technique, is based on bringing the air samples into a stationary cylindrical vessel at high velocity; a vortex is formed by the entry of the air tangential to the length of the vessel and particles in this vortex are subjected to a centrifugal force that depends on their size (Fig. 12.53). As a result, particles of different sizes are deposited at different locations along the length of the cyclone separator.

Cyclone collectors have been applied to size distribution measurements by using a series of cyclones in parallel, each having a different cut-size, followed by filters; the particles collected on the after-filters and on a filter operated without a cyclone collector can be analyzed and mass and/or chemical information as a function of size range generated knowing the size cutoff characteristics of the individual cyclones (Lippmann and Kydonieus, 1970; Lee et al., 1985).

However, cyclone collectors are most commonly used as precollectors to remove larger particles (~3–30 μm diameter) before the air sample enters a device such as an impactor designed for the measurement of particles in smaller size ranges.

Impactors. Impactors are based on the principle that particles in an airstream will tend to continue in a straight line due to their inertia when the flow of air bends sharply; if a surface to which they can adhere is present, they will impact on it and may stick. In practice, a collection plate is placed in the flow of air,

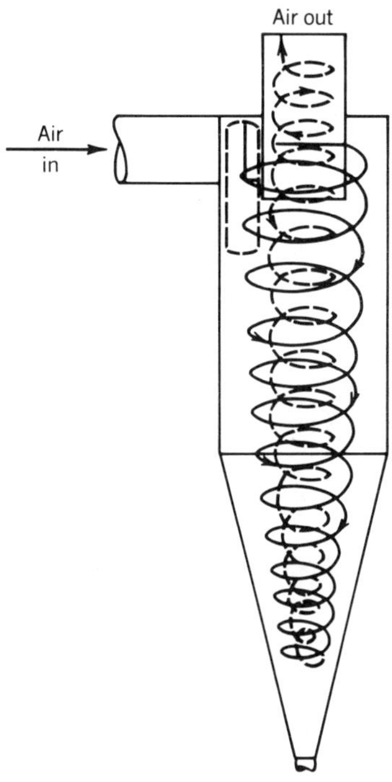

FIGURE 12.53. Schematic diagram of one type of cyclone collector (from Ayer and Hochstrasser, 1979).

causing the gas flow to stream around the obstacle; particles, however, may strike the plate and stick. Obviously, the larger the particle, the greater its inertia and the greater the impaction on the plate.

The impaction efficiency (η) for particles depends directly on the particle diameter (D), the flow velocity of the air (V), and the particle density (ρ); it varies inversely with the gas viscosity (μ) and with a parameter (D_b) that is representative of the impactor's physical dimensions (e.g., the inlet nozzle diameter) and that is related to the curvature of the air stream:

$$\eta = \frac{D^2 V \rho}{18 \mu D_b} \tag{JJ}$$

Thus, the impaction efficiency should be greatest for larger, denser particles and higher flow velocities.

In practice, multi-staged impactors called cascade impactors are commonly used to achieve size-fractionated collection of particles. These and other types of impactors (e.g., virtual impactors) used for size fractionation are discussed in Section 12-E-3.

Two problems with particle collection by impactors are *bounce-off* and *reentrainment*. Reentrainment is the resuspension of a previously collected particle from the surface into the gas flow due either to the motion of the air over the surface or to impact of an incoming particle. When a particle strikes a surface, if it does not stick, it can bounce off back into the gas stream, break into fragments, or cause a previously adsorbed particle to be knocked off into the gas stream; in all three cases the collection efficiency is lowered and the net effect is referred to as *bounce-off*. To minimize these problems, the surface of the impactor is often coated with a soft, energy-absorbing substance such as oil, water, grease, resin, or paraffin, which helps to absorb the kinetic energy of the striking particle; a summary of the types of agents used to minimize bounce-off and reentrainment is given by Marple and Willeke (1979) and Cahill (1979).

While the use of soft surfaces would seem to be mandated by the above discussion of bounce-off problems, there are a number of disadvantages to coating the impactor surfaces with a substance such as grease. For example, it makes accurate mass determinations difficult and can introduce such a large background of certain chemicals that the chemical analysis of these elements in the particles becomes difficult. In addition, with such surfaces one cannot use chemical analytical techniques which only probe the upper surface layer (Section 12-E-4) because the coating surrounds some of the collected particles.

Impingers are a type of impaction device in which impaction of the particle occurs against a liquid rather than a solid surface. Air is sucked in through a tube whose exit is a nozzle facing a splash plate and which is submerged in the liquid. The particles captured in the liquid may then be analyzed by wet chemical methods or by evaporation of the liquid followed by various analytical techniques. This technique is used in industrial hygiene sampling to collect most of the particles ~1 μm and larger. It suffers from the simultaneous collection of soluble gases and water vapor, and the cooling which can occur can cause condensation to form new particles within the impinger itself.

Electrostatic Precipitation. Electrostatic precipitators operate on the principle of the attraction of a charged particle for an oppositely charged collector. They have been used for both collecting particles for further analysis as well as for controlling particulate emissions from sources. In one common design, the particles in air can be charged if introduced into a cylindrical chamber containing a wire down the axis of the cylinder which is at a high negative voltage (e.g., 5–50 kV) relative to the walls of the chamber. A corona discharge is set up around the wire and this produces ions; the negatively charged ions are attracted to the positively charged outer walls. These ions collide with the particles in the air, charging them and causing them to move to the outer walls to be captured there. In place of the corona discharge, ions may also be generated using radioactive bombardment of the particles.

While electrostatic precipitators have relatively high collection efficiencies (99–100%) over a wide range of particle sizes (~0.05–5 μm), there are a number of disadvantages. These include the lack of size information, particle reentrainment due to sparking, and practical problems such as high cost and shock hazards. For a complete discussion of these collectors, see Mercer (1973).

Thermal Precipitation Particles in the size range ~0.005–5 μm can also be collected using a method based on the thermophoresis effect. A particle exposed to a temperature gradient tends to move away from the warmer region and toward the cooler region with a speed proportional to the temperature gradient. Figure 12.54 shows one design of a collection device based on thermal precipitation. The

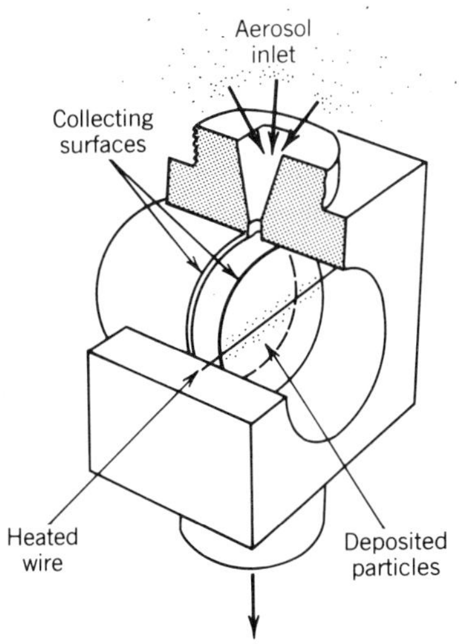

FIGURE 12.54. Schematic diagram of a typical collecting device based on thermal precipitation (from Giever, 1972).

air containing particles passes slowly between two glass slides separated by ~0.15–1.5 mm and between which is a heated wire. This gives rise to a temperature gradient between the wire and slides of ~90–900°C mm^{-1}. The particles move toward the cooler slides and are collected in a band on them.

A major advantage of the thermal precipitator is the high collection efficiency (approaching 100%) for particles over a wide range of sizes. However, its use is restricted to small flow rates (<1 L min^{-1}) and particles of low volatility. In addition, thermal heating can produce artifacts, and in some instruments, the collected particles are not uniformly distributed by mass and size along the collection strip (Mercer, 1973).

12-E-2. Mass Measurement

The total mass of particles per unit volume of air is one measure of their contribution to atmospheric chemical and physical processes. However, as discussed earlier, only the smaller particles are believed to contribute significantly to health and visibility effects, so that separation of particulate matter by size prior to mass determination is important. (See Section 12-E-3 for a discussion of size-fractionation methods.)

Methods of mass measurement include gravimetric methods, beta (β)-ray attenuation, piezoelectric methods, light and electron microscopic methods, and optical methods. An additional technique, the contact charge method, which is not in widespread use, is discussed in the LBL (1979) report.

12-E-2a. Gravimetric Methods

The most straightforward method of determining the particle loading of the atmosphere is to weigh a collection substrate such as a filter before and after sampling. However, care must be taken to be sure that both temperature and relative humidity are carefully controlled when weighing both the loaded and clean substrate. As discussed above, some filters and/or the collected particles are hygroscopic and, unless care is taken to equilibrate them at a fixed temperature and relative humidity, the change in water content may completely mask the change in mass due to the particles. In addition, such problems as forces due to static electricity on the filter which interfere with accurate weight measurements must be controlled. Finally, particulate loading can change the sampling air flow rate and lead to large errors in determining the actual volume of air sampled.

12-E-2b. Beta Ray Attenuation

Beta (β) particle beams (electrons) emitted from a radioactive source are attenuated when they pass through a filter on which particulate matter has been collected. (Beta rather than alpha particle or gamma rays are used because alpha particles do not pentrate typical thicknesses of filter well and gamma rays are too penetrating and hence would require large sample thicknesses.) Figure 12.55 shows a schematic of a β-ray attenuation device, which consists essentially of a β source such

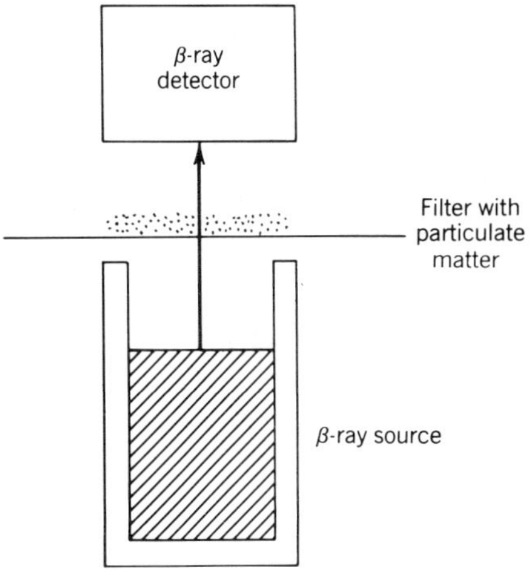

FIGURE 12.55. Schematic diagram of a typical β-ray attenuation device for measuring particulate mass.

as ^{14}C, a β detector, and a means of positioning the filter paper containing the particulate matter between the two. The ratio of the transmission of β-rays through a clean and loaded portion of the filter, respectively, is related to the particle loading via a Beer-Lambert type of relationship:

$$\frac{I}{I_0} = e^{-\mu X} \qquad \text{(KK)}$$

I_0 and I are the intensities of the β-rays that have passed through the clean and loaded portion of the filter, respectively, X is the thickness of the deposit, and μ is an attenuation constant that is approximately proportional to the density (ρ) of the material deposited. The mass per unit area deposited on the filter, given by ρX, is the parameter desired in this measurement. Rearranging Eq. (KK), one obtains

$$\ln \frac{I_0}{I} = \left(\frac{\mu}{\rho}\right)\rho X \qquad \text{(LL)}$$

The parameter μ/ρ is a constant known as the *mass absorption coefficient*; with the assumption that this is independent of the type of absorbing particles (an assumption that generally holds well enough to cause $\lesssim 10\%$ uncertainty), the value of $\ln(I_0/I)$ is directly related to the parameter of interest, ρX = mass per unit area.

Beta gauges, or beta absorption monitors as they are called, are particularly useful for stack monitoring on a short time scale (e.g. 10 min). They have also been successfully applied in a number of ambient air studies (e.g., see Macias and Husar, 1976) and have been shown to give results comparable to gravimetric methods (Jaklevic et al., 1981; Courtney et al., 1982). As one might expect, the presence of radioactive aerosols can produce artifacts with such detectors.

12-E-2c. *Piezoelectric Microbalance*

The piezoelectric microbalance is a type of resonant frequency device. The piezoelectric effect is the development of a charge on some crystals such as quartz when a stress is applied; the stress may be mechanical (e.g., added weight) or electrical. Such crystals may be used as part of a resonant circuit to provide very stable, narrow-band frequencies; the quartz crystal is plated on two sides with a thin conducting layer and leads are connected to the resonant circuit so the crystal replaces an *LC* network. The obtained frequency of vibration (υ_0) depends on a number of parameters of the crystal but is usually ~5–10 MHz. However, if a mass (Δm) becomes attached to one side of the crystal, it changes the resonant frequency by an amount $\Delta\nu_0$ such that

$$\frac{\Delta\nu_0}{\nu_0} = \frac{\Delta m}{m} \qquad \text{(MM)}$$

as long as the increase in mass Δm is much smaller than the mass (m) of the active part of the crystal.

Particulate matter from ambient air can be deposited on the crystal in various ways, for example, by using it as an impaction device (Chuan, 1976). The mass of the collected particles can then be determined by following the change in the frequency. Alternatively, a reference crystal held at the same temperature and pressure as the crystal on which the particles are collected can be used, and the difference in frequencies between the two crystals can be determined.

The piezoelectric microbalance is very sensitive, capable of detecting $\sim 10^{-8}$–10^{-9} g. The particles collected on the crystal surface can be chemically analyzed after collection using surface-sensitive techniques (Section 12-E-4). One limitation is possible overloading of the crystal; thus when the collected mass reaches ~0.5–1% of the mass per unit of the crystal, the surface must be cleaned.

12-E-2d. *Electron and Optical Microscopy*

If the particulate sample is composed of similar particles with simple shapes and known densities, then from a count of the number and size of the particles, the mass can be calculated. Counting the particles and measuring their sizes can be done by optical or electron microscopy, the former for particles with diameters from ~0.4 μm to several hundred microns, and the latter for particles from ~0.001 μm and larger. Clearly these methods do not provide a direct measurement of mass, and their use for such measurements depends critically on having particles of simple shapes and known densities. In addition, the particles must be on the surface of the substrate and form less than a monolayer to minimize overlap of the particles. For electron microscopy, the sample and substrate must also be able to be subjected to high vacuum, heat, and electron bombardment without degradation over a period of time sufficient to make the measurement.

Because of these restrictions, these methods are rarely used to measure particulate mass, but, as discussed below, they are very useful for size measurements, shape analysis, and detection of relatively chemically inert particles such as asbestos.

12-E-2e. *Optical Methods*

As discussed in Section 12-B-3b and seen in Fig. 12.30, the light scattering coefficient for particles, b_{sp}, has been found to be linearly correlated to the mass of fine particles in many cases. Thus an indirect means of measuring fine particle mass is measurement of b_{sp}, if the constant of proportionality between b_{sp} and the fine particle mass is known. However, this method may not be expected to be very accurate for *total* mass determinations, since b_{sp} does not appear to be correlated to the mass of coarse particles (Fig. 12.30) which may contribute the most to the mass (see below).

Light scattering is measured using an integrating nephelometer (Fig. 12.56) which measures light scattering through all angles (i.e., b_{scat}) (Ruby, 1985). This light scattering is due to Rayleigh scattering by gases as well as scattering by particles (Section 12-B-3):

$$b_{scat} = b_{sg} + b_{sp}$$

However, in many areas at sea level, $b_{sp} >> b_{sg}$ so that b_{scat} determined using an integrating nephelometer is approximately equal to b_{sp}, the light scattering by particles.

Air containing the particles flows slowly through the measuring chamber where a flash lamp produces a pulse of white light. A photomultiplier measures the intensity of the scattered light. Rayleigh scattering by gases can be used to calibrate the instrument. For example, the scattering coefficient of helium is calculated to be 3.0×10^{-7} m^{-1}, whereas that of clean air is 2.8×10^{-5} m^{-1} at 460 nm (Charlson et al., 1967).

In some studies of ambient air, a correlation between total particulate mass loading or aerosol volume and b_{scat} has been established. However, the correlation

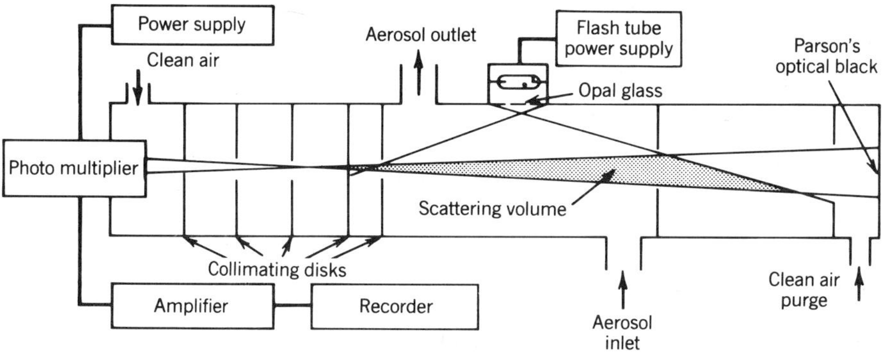

FIGURE 12.56. Schematic diagram of an integrating nephelometer (from Charlson et al., 1967).

has generally such large uncertainties that light scattering has not proved to be a useful technique for accurate mass measurements; for example, Sverdrup and Whitby (1980) have shown that the ratio of submicron aerosol volume to b_{scat} varied by a factor of 16 in samples taken in California in 1972 and 1973, with the variation depending primarily on the size distribution. Not only would the size distribution be expected to affect b_{scat}, but other parameters such as the chemical composition and relative humidity (RH), particularly above ~70% RH (Covert et al., 1972), will also affect b_{scat}.

Of course, as expected from the discussion in Section 12-B-3, b_{scat} does correlate to another parameter which is important for estimating the aesthetic effects of air pollution—visibility.

Light direction and ranging (LIDAR) is a third optical technique whose results bear some relationship to the aerosol mass loading. LIDAR is based on the measurement of backscattering of light as a function of distance away from the observer. It thus operates on the same principle as RADAR and is a remote, rather than point, monitoring technique. Typically, a pulse of light from a laser is sent into the atmosphere and the backscattered light is measured by a photomultiplier. This backscattered light arrives at the detector at various delay times after the original pulse, depending on the distance from the scattering element to the detector; obviously the farther away the scattering occurs, the longer it takes to reach the photomultiplier. Taking into account the decrease in intensity with $1/r^2$, where r is the distance to the point of backscattering, a plot of backscattered intensity against delay time (i.e., against distance from the detector) can be developed. From this, the average backscattering coefficient for the particle size distribution can be determined.

Although the extent of backscattering of light depends on the particulate loading, it is also very sensitive to the size distribution. As a result, like the other optical methods, it is not a direct means of measuring mass loadings and will be most useful for such estimates where the size distribution and composition of the aerosol is well defined and/or relatively constant.

The reader is referred to the monographs edited by Hinkley (1976) and Killinger Mooradian (1983) for examples of the applications of LIDAR to atmospheric measurements. Several other types of optical sizing instruments commonly used are discussed by Mercer (1973) and Hinds (1982).

12-E-3. Size Measurement

Because of the importance of the particle size distribution with respect to air chemistry, meteorology, health effects, visibility, and source identification, most studies of ambient aerosols involve determination of the particulate mass and chemical composition as a function of size. In this section, we discuss the techniques commonly used to determine size distributions. Many of them are variants of the methods used to collect particles (see Section 12-E-1) and the reader should consult that section.

In the following discussion, the reader is cautioned to keep in mind that atmospheric particles are not all spherical nor even necessarily simple in shape. Thus, as discussed in Section 12-A, the term *size* cannot be uniquely defined for atmospheric particles. As a result, a measurement of the distribution of sizes using an impactor which is based on inertial characteristics, for example, may not give the same results as a size measurement based on optical techniques which use light scattering. With this caveat in mind, let us examine the most commonly used methods of determining the size distribution of atmospheric aerosols.

The primary methods used to measure particle size distributions for particles with diameters $\gtrsim 0.2$ μm are inertial methods (e.g., impactors) and optical methods (e.g., optical counters). For very small particles (Aitken nuclei) and up to ~ 0.2 μm, diffusion separators, electrical mobility analyzers, and condensation nuclei counters are used.

12-E-3a. Inertial Methods: Impactors

Cascade and virtual impactors are the major inertial methods used to size-fractionate ambient particles. Cyclones are also used, but, as discussed earlier, these are normally employed as prefiltration devices for other instruments rather than as a means to divide the particulate matter into a number of fractions by particle size; as a result we shall not discuss these further here.

Cascade Impactors. Impactors have been used to obtain different size fractions of ambient particles in the range of diameters ~ 0.5 to 30μ. The range can be extended down to 0.05 μm by operating some of the later stages at reduced pressures (Hering et al., 1978, 1979). The cascade impactor, as its name implies, is a series of impactor plates connected in series or in parallel (Fig. 12.57). The diameters of the nozzles or slits above each impactor plate become increasingly smaller as the air moves through the impactor so that the air moves increasingly faster through these orifices and smaller and smaller particles impact on the plates [see Eq. (JJ) above].

The Lundgren impactor, the Anderson sampler, the Mercer impactor, and the University of Washington Mark III impactor are examples of cascade impactors with various designs and numbers of stages [see Marple and Willeke (1979) for a summary of commercially available impactors]. Some low sampling rate impactors have single nozzles for each stage whereas others have multiple nozzles; there are also several different types of impaction surfaces in use, as shown in Fig. 12.58. For example, Fig. 12.59 shows the design of a Lundgren cascade impactor. Before each of the five stages, which consist of cylinders covered with a removable collection surface, are nozzles or slits of varying sizes. The cylinders can be rotated with time, providing time resolution as well as size resolution. An after-filter is normally used to collect particles not collected on the impactor stages.

Table 12.30 gives the 50% cutoff points for this impactor for a flow rate of 3 cfm and spheres of unit density; the 50% cutoff point, sometimes called the fractionation size, is defined as the diameter of particles in a monodisperse (i.e., single

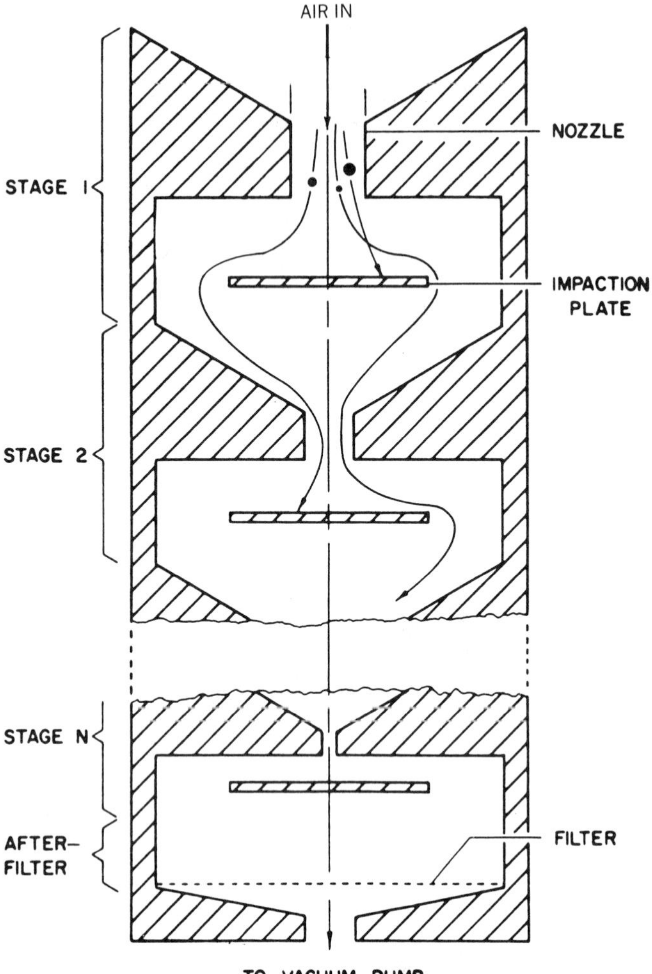

FIGURE 12.57. Principle of operation of cascade impactor (from Marple and Willeke, 1979).

size) aerosol, 50% of which is collected by the device or, in this case, by the impactor stage. Because this cutoff diameter decreases as the flow rate of air increases, it is essential to specify the flow rate under which the measurements were made and, of course, to keep it constant during an experiment.

The first two stages, 1A and 1B, are shown in Table 12.30 as having identical size cutoffs. The reason for the use of two identical stages (i.e., two identical entrance slit widths) in series was to minimize the effects of particle bounce-off which, as mentioned earlier, can be a major problem with the use of impactors. In this particular case, stage 1A had a dry impaction surface whereas 1B was coated with sticky polyethylene to minimize bounce-off.

The importance of using sticky impaction surfaces can be seen from Table 12.31

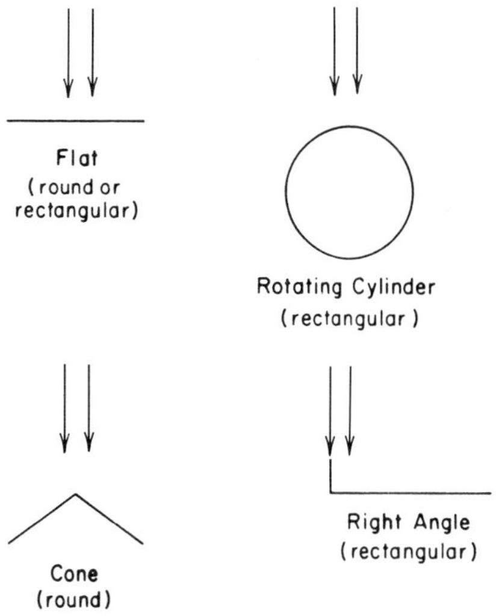

FIGURE 12.58. Schematic of some different types of impaction surfaces used in impactors (from Marple and Willeke, 1976).

which shows the effect of using soft (i.e. sticky) versus hard (i.e. dry) surfaces for the collection of dry particles in two different samples, coal dust and road dust. In this case, all impaction surfaces were either soft or hard, respectively. Clearly, the use of hard surfaces where bounce-off is greatest gives a very different result for the particle size distribution; in the hard case, about 70% of the particulate matter makes it past all impactor stages to the after-filter.

The extent of bounce-off depends on the nature of the particles themselves. The problem may be less severe, for example, for hygroscopic particles which will have a tendency to stick to the impaction surface.

Virtual Impactors. The virtual impactor is a modified type of impactor, an example of which is shown in Fig. 12.60; one commonly used type of virtual impactor is known as the dichotomous sampler. The basis of virtual impactors is that the air stream impacts against a mass of relatively still air rather than against

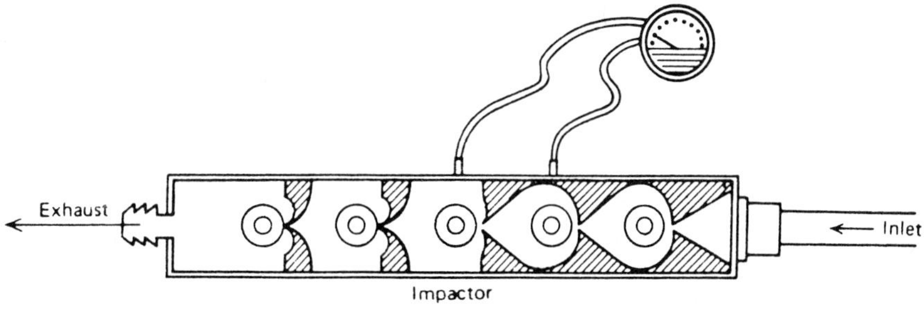

FIGURE 12.59. Schematic diagram of Lundgren cascade impactor (from Wesolowski et al., 1980).

TABLE 12.30. Size Ranges of the Five-Stage Lundgren Cascade Impactor Shown in Fig. 12.59

Stage	50% Cutoff Point[a] (μm)
1A	8.0
1B	8.0
2	4.0
3	1.5
4	0.5
After-filter	<0.5

Source: Wesolowski et al., 1980.

[a]For a flow rate of 3 cfm and unit density spheres.

a plate. The inertia of the particles carries them into the still air mass which is slowly withdrawn through a filter to collect the particles. This type of impactor avoids the problem of particle reentrainment from the impaction surface caused by air motion over the collected particles or by dislodging due to collisions of incoming particles with the impactor surface. It also avoids the problem of bounce-off or of using greases which may interfere with subsequent chemical analysis.

12-E-3b. Optical Methods: Single Particle Optical Counters and Light and Electron Microscopy

A second approach to determining the size distribution of ambient aerosols relies on optical methods. Optical counters, light microscopy, and electron microscopy fall under this heading.

TABLE 12.31. Comparison of Relative Masses on Soft (Sticky) and Hard (Dry) Impaction Surfaces for a Lundgren Impactor

	Coal Dust (1–5 μm)		Arizona Road Dust (0–15 μm)	
Stage[a]	Sticky	Dry	Sticky	Dry
1A[a]	1.0	1.5	3.5	3.2
2	10.9	9.5	13.8	9
3	42.1	10.5	35.7	9.6
4	29.4	6.3	19.5	12.3
After-filter	15.2	72	27.6	66

Source: Wesolowski et al., 1980.

[a]Stage 1B was coated with silicon grease without a supporting film and hence could not be weighed.

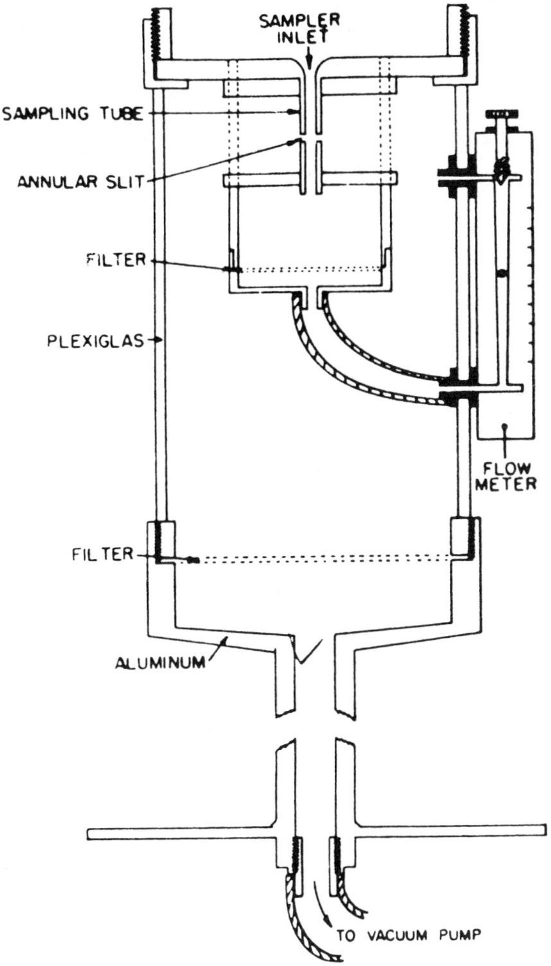

FIGURE 12.60. Schematic diagram of a virtual impactor (from Conner, 1966).

Single Particle Optical Counters. These instruments are used to measure particles in the ~0.1–10 μm range by measuring the amount of light scattered by a single particle (Martens and Keller, 1968). As discussed in Section 12-B-3, the amount of this Mie scattering depends not only on the refractive index but also on the radius of the particle; hence the intensity of scattered light is a measure of the particle size. Assuming that the particles are spherical, smooth, and of known refractive index, one can calculate, using Mie theory, the intensity of scattered light of wavelength λ at various angles (θ) to the incident beam for a particle of a given size. Integrating over all scattering angles and wavelengths (since "white" incandescent sources are normally used in these instruments), one obtains the theoretical response of the single particle counter, that is, the curve of scattered light

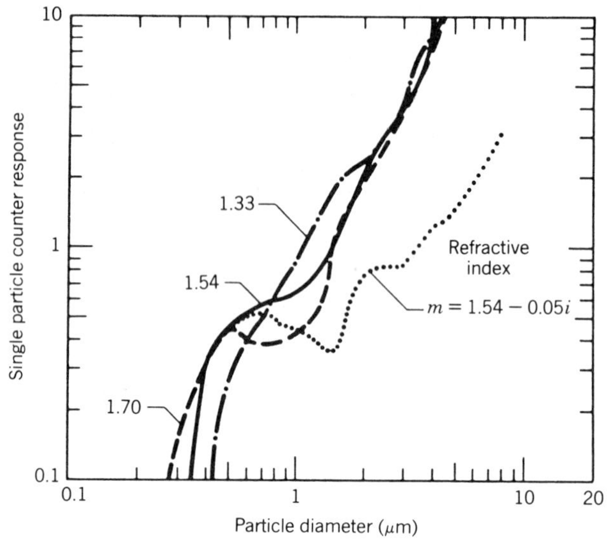

FIGURE 12.61. Theoretical response of a typical single particle counter (from Whitby and Willeke, 1979; data from Cooke and Kerker, 1975).

intensity as a function of the particle diameter. Typical theoretical response curves are shown in Fig. 12.61 (Cooke and Kerker, 1975).

Figure 12.62 shows two types of optical systems used for single particle counters. In (*a*), the light scattered perpendicular to the incident beam is measured, whereas in (*b*), the light scattered into a cone about the illuminating axis is detected; light traps are used to minimize the contribution of unscattered light from the incident beam.

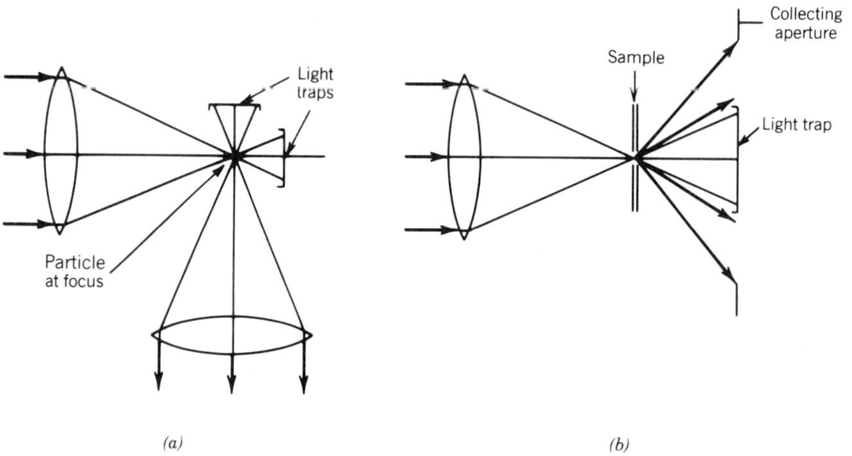

FIGURE 12.62. Schematic diagrams of typical optical geometries used in single particle counters (from Whitby and Willeke, 1979).

Calibration of these single particle counters is usually carried out using monodisperse polystryene latex or polyvinyl latex spheres which are available in sizes from ~0.1 to 3 μm and have a refractive index of 1.6; alternatively, aerosols with lower refractive indices may be generated from liquids such as dioctylphthalate ($m = 1.49$). Whitby and Willeke (1979) discuss the importance of instrument calibration using standardized aerosols with an index of refraction as close as possible to the sample being measured; since the refractive index of atmospheric particles varies from 1.33 for water to 1.7 for minerals, they recommend using a calibration aerosol with $m \simeq 1.5$. Because light scattering is very dependent on the particle shape, when measuring irregularly shaped particles such as coal dust, the instrument should be calibrated with aerosols generated from the same material. Figure 12.63, for example, shows the instrument response as a function of particle diameter for an ideal calibration aerosol of dioctylphthalate and for coal dust particles.

Potential problems with using single particle counters in ambient measurements and ways to minimize these are discussed in detail by Whitby and Willeke (1979).

Optical counters allow relatively rapid measurements of the size distribution and, unlike some of the other methods of size fractionation, include volatile particles in the measurement. However, some care must be taken in interpreting the detailed shape of the size distribution spectrum because of some anomalies which have been observed; for example, around the 1 μm region, interference from light which is reflected or refracted from the front and back of the particle gives a "knee"

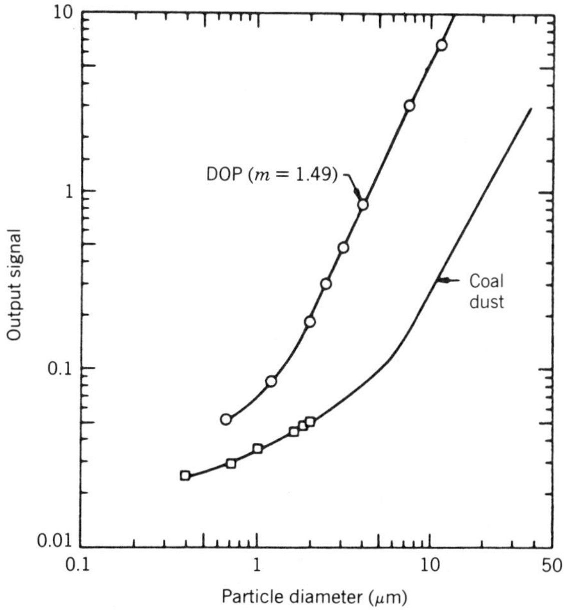

FIGURE 12.63. Experimental calibration curves for a commercial single particle counter and two types of calibration aerosols: dioctylphathalate (DOP) and coal dust (from Whitby and Willeke, 1979).

in many calibration curves of number of particles versus their diameter (LBL, 1979).

Light and Electron Microscopy. As discussed above, light and electron microscopy can be used to measure particles in the ranges ~0.4–100 μm and ~0.001 μm or larger, respectively. Because of the instrumental requirements, these are usually not routine monitoring techniques. However, unlike other methods, they give detailed information on particle shapes. In addition, chemical composition information can be obtained using scanning electron microscopy (SEM) where the beam causes the sample to emit fluorescent X-rays that have energies characteristic of the elements in the sample. Thus a map showing the distribution of elements in the sample can be produced as the electron beam scans the sample.

12-E-3c. Electrical Mobility Analyzer

The techniques described above for obtaining the size distribution of particles apply to particles in the accumulation and coarse particle ranges. However, as we have seen, smaller particles in the Aitken nuclei range are believed to play an important role in the atmosphere. The electrical mobility analyzer, developed by Whitby and co-workers at the University of Minnesota, allows one to measure particles in this important range, ~0.005 to ~1 μm.

Figure 12.64 illustrates the principles of the electrical mobility analyzer. The essential components are the aerosol charger, the mobility analyzer, and the detector (shown in Fig. 12.64 as the current collecting filter). The air containing the

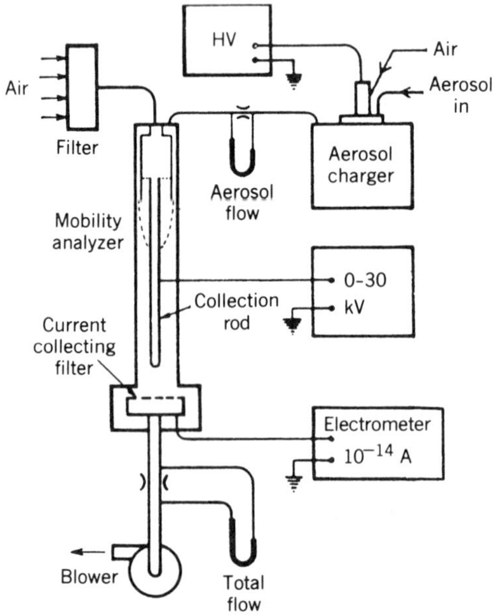

FIGURE 12.64. One type of electrical mobility analyzer (from Whitby and Clark, 1966).

particulate matter is first introduced into the aerosol charger where the particles are given an electrical charge, all of the same (e.g., positive) sign; a corona discharge is usually used to produce ions which become attached to the particles. The particles, now charged, are introduced as a thin layer around the outside of the tubular mobility analyzer. Clean air flows down the central portion of the tube between the layer of ambient aerosol at the walls and the collection rod in the center of the tube. A negative voltage is applied to the collection rod, causing the positively charged particles to move from the outer wall through the clean air to the collection rod.

The particles with the highest mobilities reach the collection rod first and are removed from the gas stream; those that do not reach the rod before the flow passes out of the region of the electric field pass through to a detector and are measured. Increasing the voltage on the collection rod increases the number of charged particles which reach it before passing out of the field and hence decreases the number reaching the detector. The relationship of the particle count at the detector to the voltage in the analyzer is thus dependent on the particle mobility in the analyzer, which depends on particle size. Thus size distributions can be obtained by studying the detector output as a function of collection rod voltage. The detector may be a current sensing device, as in Fig. 12.64, or other type such as a condensation nuclei counter.

Despite size-dependent losses in the commercial versions of this instrument (which are handled by empirical correction factors), the properly calibrated instrument is useful in defining the relative numbers of particles over a wide range of size intervals (Lundgren et al., 1979; Liu et al., 1982).

Details of the calibration, use, performance, and artifactual problems are given in a proceedings entitled *Aerosol Measurement* (Lundgren et. al., 1979); this also shows data for the mobility distributions for monodisperse aerosols.

12-E-3d. Condensation Nuclei Counter

Very small particles in the Aitken range act as condensation nuclei for the formation of larger particles in a supersaturated vapor. If these very small particles are injected into air which is supersaturated with water or another vapor such as an alcohol, the vapor condenses on them to form droplets. In the condensation nuclei counter (CNC), supersaturation of the air containing these particles is achieved by an expansion of air at a relative humidity near saturation, causing cooling of the sample well below the dew point; water then condenses on the nuclei to form larger droplets. These can be counted as is done in absolute nuclei counters, for example, by measuring the pulses of scattered light by a single droplet as it passes through the viewing volume. Alternatively, the particles can be measured using techniques such as total light extinction or scattering. In this case (sometimes called photoelectric nuclei counters), calibration against some other reference is required.

CNCs are applicable in the size range from $\gtrsim 0.001$ to ~ 0.15 μm, although the response of commercially available CNCs appears to depend on the particle size and composition for diameters $\lesssim 0.05$ μm (Liu and Pui, 1979). Further size fractionation within this broad range can be obtained by using the CNC technique as a detector for other sizing methods such as the diffusion separator (see below).

12-E-3e. Diffusion Separator

As discussed in Section 12-B-2, small particles with diameters $\lesssim 0.05\ \mu m$ undergo diffusion via Brownian motion sufficiently rapidly that this can be used to separate particles. The smaller particles diffuse more rapidly (Fig. 12.65). Thus the aerosol can be passed through a tube in which the smaller particles diffuse more rapidly to the walls and are removed there, leaving the larger, more slowly diffusing particles to pass through. Variation of residence time in the tubes by varying the flow rates and tube lengths leads to different size cutoffs (but not high resolution). Hence size fractionation of small particles can be achieved using such diffusion separators. The particles exiting the tube can be measured using techniques such as the CNC. The design and testing of a typical diffusion battery–CNC apparatus is described by Sinclair et al. (1979) and Raes and Reineking (1985).

A major advantage of such diffusion separators is that particles in the range 0.002–0.005 μm can be measured, a range in which the electrical mobility analyzer is quite insensitive and hence requires quite a high number concentration to give reliable data.

12-E-3f. Summary

In summary, no one technique is capable of measuring the size distribution of atmospheric aerosols from the smallest to the largest diameters of interest, a range

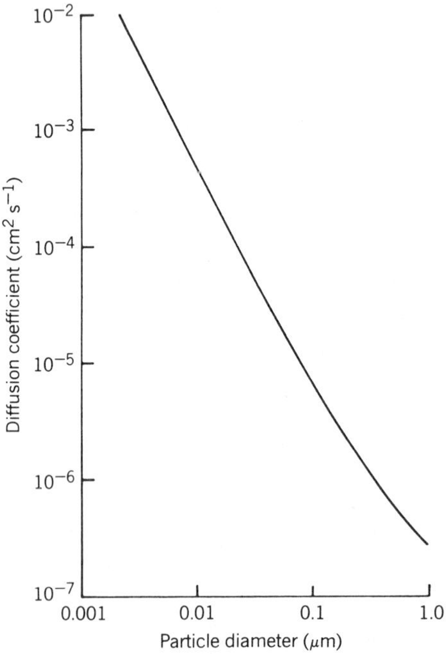

FIGURE 12.65. Dependence of the diffusion coefficient of particles as a function of particle diameter from 0.001 to 1.0 μm (from Sinclair et al., 1979).

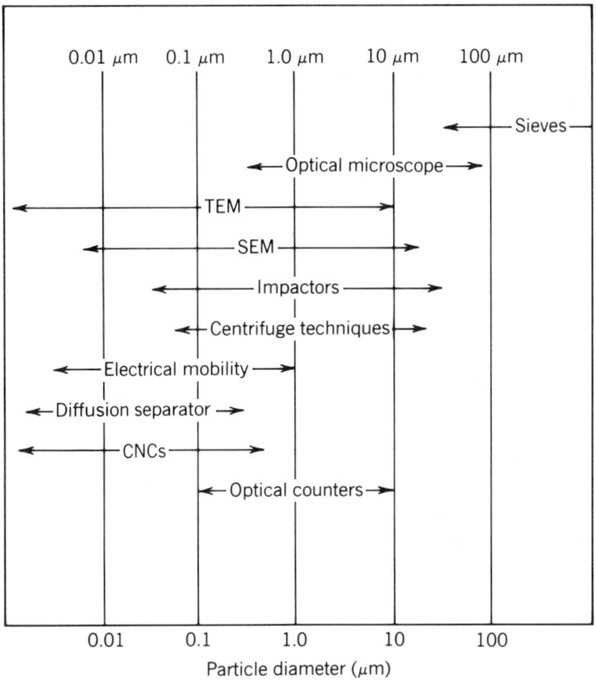

FIGURE 12.66. Summary of size ranges covered by various analytical techniques for atmospheric aerosols. TEM = transmission electron microscopy; SEM = scanning electron microscope (adapted from Hinds, 1982).

covering approximately five orders of magnitude. However, a combination of the methods can be used to provide information over this range. Figure 12.66 summarizes the size ranges covered by the various available techniques. One must be careful not to overestimate the accuracy or precision of any simple device. In-depth discussions of particle sizing are found in books by Mercer (1973), Hinds (1982), Dennis (1976), Phalen (1984), and Liu et al. (1984).

12-E-3g. Generation of Calibration Aerosols

Calibration of the instruments to measure the size distribution requires methods of generating aerosols of well-defined sizes. Generating aerosols of known characterisics, both physical and chemical, is also necessary for carrying out studies of aerosol effects (e.g., on health or visibility) under well-defined experimental conditions.

A monodisperse aerosol is one with a narrow size distribution which, for lognormal distributed particles, usually means a geometric standard deviation of about 1.2 or smaller. Monodisperse particles usually are expected to have simple shapes and uniform composition with respect to size. A polydisperse aerosol, on the other hand, is one containing a wide range of particle sizes, but which may otherwise be homogenous in terms of the basic physical and chemical properties that are not related to size. The term heterodisperse is also used occasionally; this describes aerosols varying widely in physical and chemical characteristics, as well as size.

As discussed in detail by Raabe (1976), an investigator's use of the terms monodisperse and polydisperse aerosols may depend on the particular properties of importance in the study; thus an aerosol may be comprised of particles of the same size, that is, be monodisperse with respect to size, but may vary in settling speed due to variations in density, that is, be polydisperse with respect to settling speed.

Instruments used to measure the size distribution of aerosols can be calibrated directly by injecting aerosols of fixed, known characteristics or by comparison against an instrument that has been previously calibrated. The first method is known as *primary calibration*, and monodisperse aerosols are commonly used in this case. In the case of *secondary calibration*, against another instrument, polydisperse aerosols are often used. Thus methods of generating both monodisperse as well as polydisperse aerosols are needed.

There are a number of techniques for generating aerosols, and these are discussed in detail in the LBL report (1979) and in volumes edited by Willeke (1980), Liu (1976), and Liu et al. (1984). We briefly review here the major methods currently in use; these include nebulizers, vibrating orifices, spinning disks, the electrical mobility analyzer discussed earlier, dry powder dispersion, and condensation of vapors from the gas phase.

Nebulizers. Aerosols may be produced by atomizing liquids or suspensions of solids in liquids. Nebulizers are a type of atomizer in which both large and small particles are initially produced, but in which the large particles are removed by impaction within the nebulizer. As a result, only particles with diameters $\lesssim 10$ μm exit most nebulizers.

There are two basic means of generating particles from liquids in nebulizers: compressed air or ultrasonic vibration. Figure 12.67 shows one relatively simple type of compressed air nebulizer. The compressed air shoots out of a small orifice at high velocity, creating a reduced pressure in the region of the orifice; a feed tube connected to the liquid through a small opening is subjected to this region of lowered pressure (the Venturi effect) and hence liquid is drawn up from the reservoir and exits as a thin stream. The flow of high-velocity air striking the liquid stream breaks it up into small droplets and carries this aerosol toward the exit. The larger droplets are removed by impaction on the curved wall, and the smaller particles exit the device. Detailed descriptions of other types of compressed air nebulizers which differ somewhat in design are found in Raabe (1976), Mercer (1973), and Hinds (1982).

These compressed air nebulizers produce polydisperse aerosols. After the aerosol is produced, the size distribution may change due to evaporation of liquid from the droplets. In addition, the particles may be electrically charged due to an ion imbalance in the droplets as they form; if such charges become further concentrated due to evaporation, the particle may break up into smaller particles. Thus electrical neutralization of the aerosol, for example, by exposure to a radioactive source, is usually necessary to prevent electrostatic effects from dominating the particulate motion, coagulation, and other behavior.

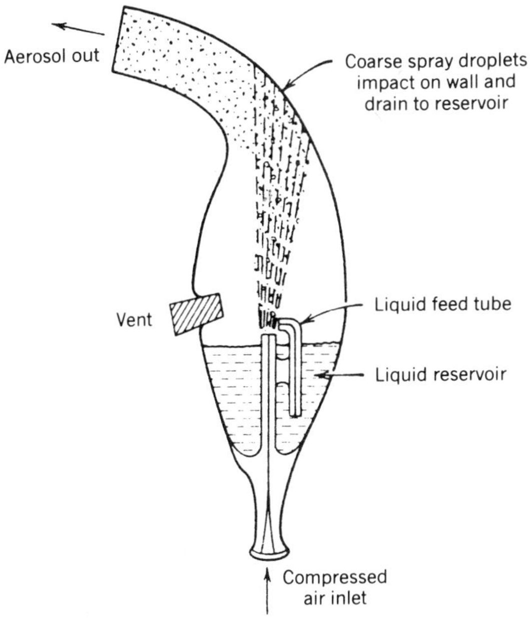

FIGURE 12.67. Schematic diagram of compressed air nebulizer (from Hinds, 1982).

The second type of nebulizer uses ultrasonic vibrations to generate the aerosol. Figure 12.68 is schematic of one such device. In this case, a piezoelectric crystal is excited at a frequency typically in the range of 2 kHz to 3 MHz to produce mechanical oscillations which are transmitted through the coupling fluid to the aerosol generator reservoir that contains the liquid to be aerosolized. These intense compressions and rarefactions produce a geyser effect in the generator reservoir, producing small liquid droplets which are carried off by a flow of air. The size distribution of the aerosol produced depends on the frequency of the acoustical field, on the nature of the coupling fluid, and on the flow rate of the air transporting the aerosol out of the generator; the latter is a factor because of the rapid rate of coagulation in the generator reservoir due to the high concentration of droplets produced initially.

Nebulization may be used to produce suspensions of liquid droplets in air by using the pure liquid as the fluid or by using liquids with low vapor pressures dissolved in volatile solvents which then evaporate off the particle. Suspensions of solid particles in air may also be generated using the nebulization of suspensions of insoluble materials (e.g., insoluble plastic particles suspended in organic solvents, aqueous colloidal suspensions, e.g., of ferric hydroxide, etc.) or of soluble materials dissolved in water (e.g., salts in water). Drying of the aerosol after its generation is an important factor in the final aerosol produced since this may alter both the physical and chemical nature of the particles; for example, rapid drying may produce low-density particles that are basically hollow shells formed by crystallization on the surface of the drying droplet.

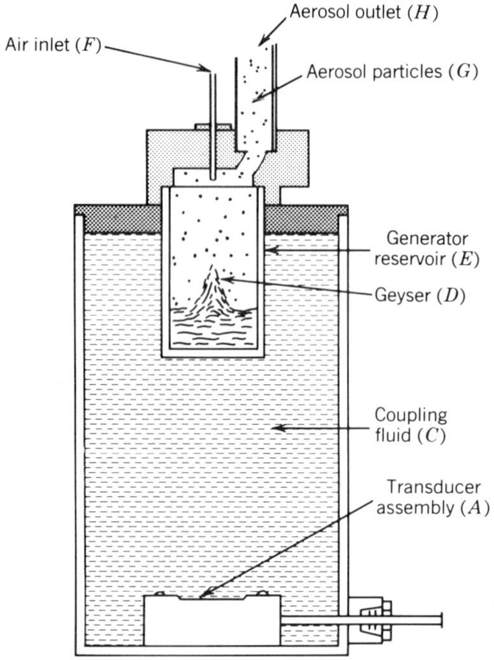

FIGURE 12.68. Sectional schematic view of an operating ultrasonic aerosol generator showing transducer assembly (*A*) generating an acoustic field in the coupling fluid (*C*) creating an ultrasonic geyser (*D*) in the generator reservoir (*E*) and air entering at (*F*) carrying away aerosol (*G*) through the aerosol outlet (*H*) (from Raabe, 1976).

Although these devices produce polydisperse aerosols, monodisperse aerosols can be generated by following the nebulizer with a size fractionating device such as an aerosol centrifuge. Alternatively, nebulizers can be used to produce monodisperse aerosols of solid particles if suspensions of particles of one size are used as the generating fluid; for example, polystyrene and polyvinyl toluene latex beads of uniform size from 0.1 to 3.5 μm are commercially available in water suspensions. However, as discussed in detail by Raabe (1976), care must be taken to ensure that the vast majority of droplets formed contain only one sphere; otherwise, when the liquid evaporates, clusters of spheres will be formed. In addition, care must be taken to be sure they do not carry significant electrical charge (an anionic surfactant is added to the suspension to inhibit coagulation of the spheres). Finally, under some conditions, significant concentrations of other small particles may be simultaneously produced due to drying of empty droplets that contain impurities in the suspending liquid.

Vibrating Orifice Generator. Another common method of generating particles with diameters 0.5–50 μm is the Berglund–Liu (1973) vibrating orifice generator shown schematically in Fig. 12.69. The solution to be aerosolized is pumped through a small orifice 5–20 μm in diameter. The orifice is oscillated by a piezo-

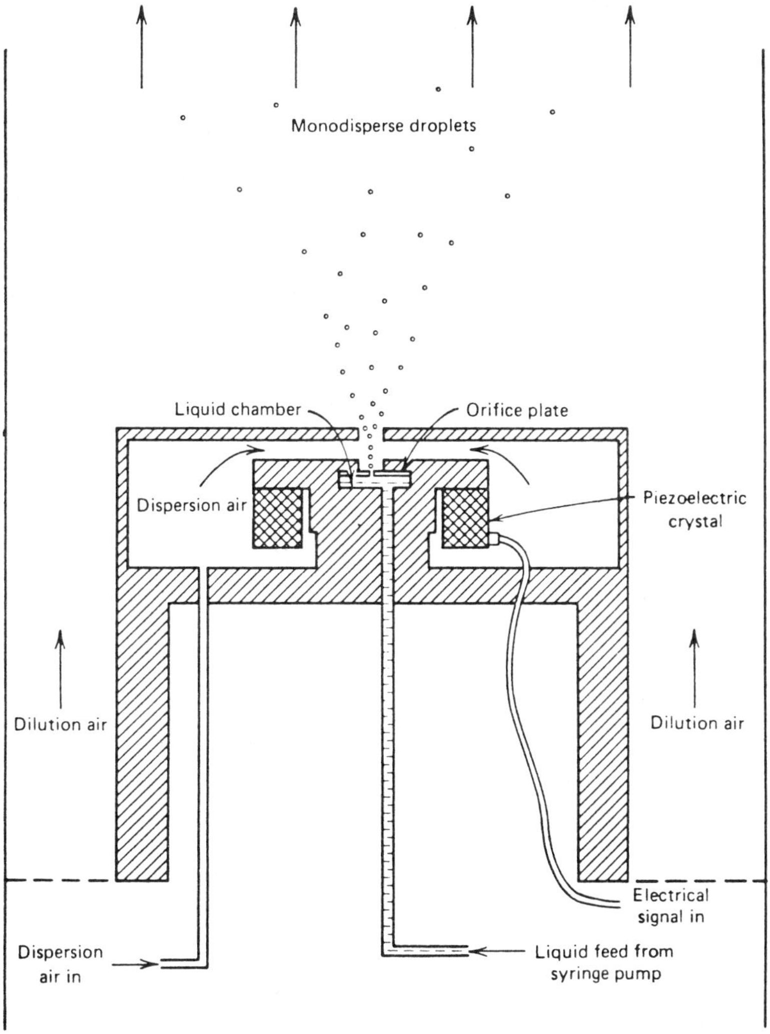

FIGURE 12.69. Schematic diagram of vibrating orifice aerosol generator (from Hinds, 1982).

electric crystal so that the liquid steam is broken on each oscillation forming a small liquid particle which is carried away in a stream of air. These droplets are then mixed with more air to dry the particles. The rate of droplet formation is equal to the oscillation frequency (f) of the piezoelectric crystal. From the volumetric flow rate of the liquid and the frequency f, the volume of the individual drops and particle diameter can be calculated.

Spinning Disk Generator. A third means of producing aerosols is the spinning disk aerosol generator, shown schematically in Fig. 12.70. The liquid is fed to the

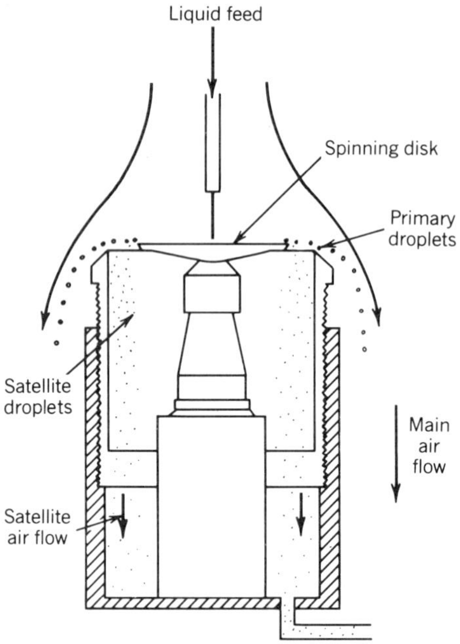

FIGURE 12.70. Schematic drawing of a spinning disk generator used to produce monodisperse aerosols (from Raabe, 1976).

center of a spinning disk and then moves by centrifugal force to the outer edge. It accumulates at the edge until there is sufficient liquid that the centrifugal force exceeds the surface tension forces and a droplet of liquid is thrown off. Some smaller "satellite droplets" are also produced, but these are separated from the primary drops, which are thrown out further by using a flow of air. Monodisperse aerosols with diameters $\gtrsim 0.5$ μm, and more typically in the range $\sim$20–30 μm, are produced, the size being determined by the radius of the disk and the speed of the rotation (Mercer, 1973).

Differential Mobility Analyzer. Monodisperse aerosols with diameters ≤ 0.5 μm can be generated using the principle of the electrical mobility analyzer discussed in Section 12-E-3c (Liu, 1976). Thus a polydisperse aerosol is first produced using a nebulizer and the particles are dried. They are then charged using a radioactive source; the central electrode has a slot to collect particles with a narrow range of electrical mobilities, (i. e., sizes), thus generating monodisperse particles.

Dry Powder Dispersion. When an aerosol consisting of solid particles is required, they are usually generated by the dispersion of a dry powder. One of the most common devices which is particularly effective for dry, hard materials such as silica is known as the *Wright dust feed* (Wright, 1950), shown schematically in Fig. 12.71. The dust is packed into a plug and a sharp blade is used to scrape dust off the surface of the plug; the particles are then swept away by a stream of air along the outer edge of the scraper blade and exit through a jet onto an impactor

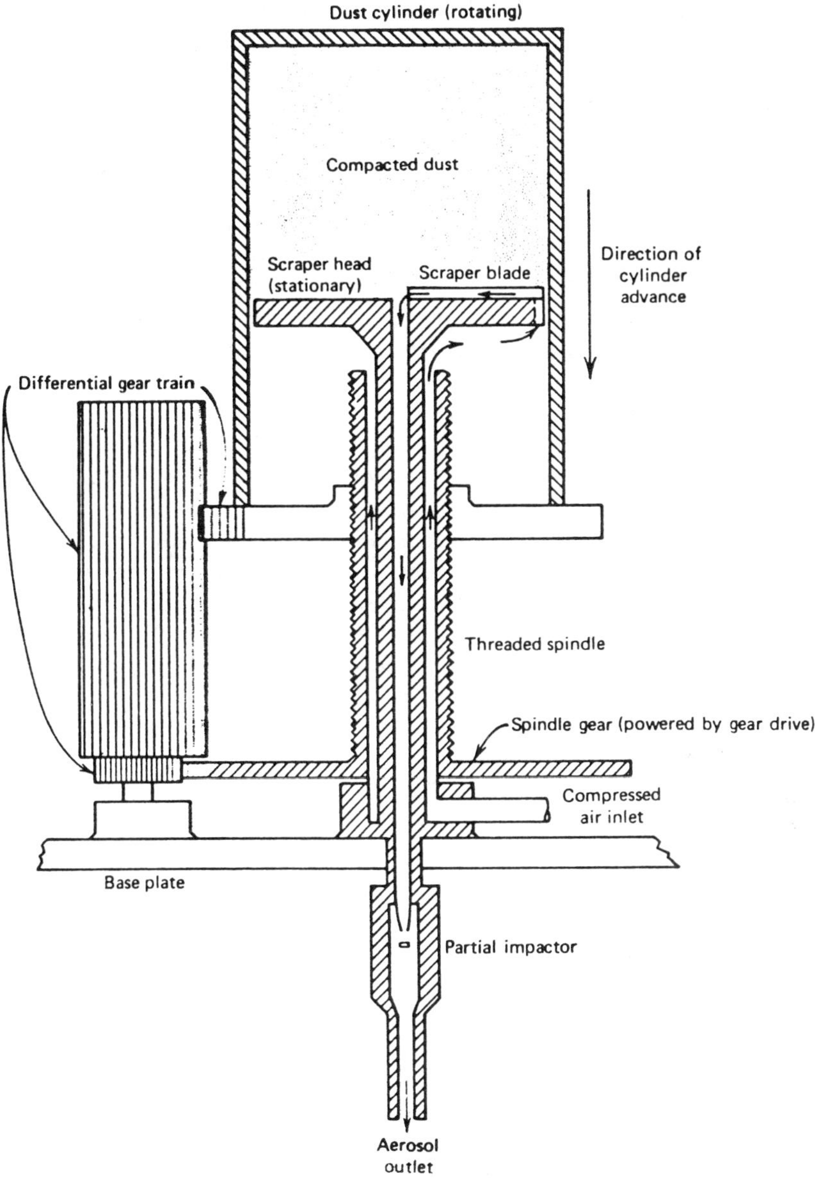

FIGURE 12.71. Diagram of Wright dust feed mechanism (from Hinds, 1982).

to break up particle clusters. The aerosol produced has particle diameters ≤ 10 μm. Variations on this type of aerosol generator are described by Raabe (1976) and in the LBL report (1979).

Fluidized beds have also been used for generating suspensions of solid particles with diameters in the range of approximately 0.5–40 μm. Air flows through the fluidized bed which contains beads kept suspended by the motion of the air; dust injected into the bed is broken up into small particles and carried out with the air flow (Raabe, 1976).

A device such as an impactor or cyclone is frequently used at the exit of these dry powder dispersion devices to eliminate the large particles. A charge neutralizer is usually used to reduce the electrostatic charges on the dispersed particles.

Condensation. When nuclei are present under conditions of supersaturation, condensation occurs on the nuclei, causing them to grow. If the particle radius is much less than the average distance between the particles, the growth rate of the particle with time is uniform, and since the nuclei are so much smaller than the final particles formed, nuclei of different initial sizes all give rise ultimately to particles of the same size, that is, a monodisperse aerosol, with diameters typically in the range 0.003–1.0 μm.

In practice then, a liquid having a low vapor pressure such as oleic acid or lubricating oils is carried by an inert gas to a heater where it vaporizes. As it leaves the heated area, it condenses to form the aerosol. Different generator designs based on condensation of a supersaturated vapor are discussed in the LBL report (1979).

However, care must be taken to ensure that seed nuclei are present on which condensation can occur. If a very clean system is used in which nuclei are not present, spontaneous nucleation may occur; this process is such that nuclei do not appear uniformly either in space or in time, and the initial particle growth rate depends on the degree of supersaturation. As a result, a polydisperse aerosol is produced under these conditions.

While condensation from a supersaturated vapor can be used to produce liquid particles, it is not as easily applied to the generation of solid particles except those which can be liquefied at modest elevated temperatures. However, these may be generated using different techniques. Methods used in the past have included the vaporization of wires or of salts fused onto wires, the use of "exploding" wires, and the use of electric arcs. For example, heating metal wires in an inert gas such as N_2 to a sufficiently high temperature produces small particles in the condensation nuclei and Atken nuclei range. Thus tungsten wire at ~1000–1200°C produces tungsten metal particles, while nichrome wire gives chromium particles.

Solid salt particles may be produced by fusing the salt onto the wire by immersing the wire in a saturated solution of the salt and passing an electrical current through it. When the wire is heated, salt particles are produced. Theoretical and practical considerations for such generators are described by Fuchs and Sutugin (1970).

The exploding wire method involves putting a large amount of energy into a wire suddenly, causing it to "explode." If O_2 is present, a metal-oxide aerosol is produced, whereas particles of pure metal are formed in an inert atmosphere such as helium. Exploding wire generators and their size distribution charateristics have been discussed by Phalen (1972).

Finally, electric arcs have also been used in some cases to produce solid particles of the electrode material, such as graphite, as have argon plasmas at high temperatures which produce a finely divided aerosol on rapid cooling.

All these methods generally give small ($\lesssim 1\ \mu m$), polydisperse aerosols of the solid particles and, unless rapid air dilution is provided, coagulation leads to large agglomerates of the small primary particles.

Gas phase reactions can also be used to produce products of low volatility which condense to give an aerosol. The reaction of gaseous NH_3 with HCl to form particles of solid ammonium chloride or the reactions of alkenes and aromatics discussed earlier (e.g., see Lee et al., 1984) are typical examples; some other reactions used are reviewed in the LBL report (1979). Such methods tend to give submicron particles.

12-E-4 Determination of Chemical Composition

As discussed in Section 12-D, ambient particulate matter contains inorganic elements and ions, including trace metals, as well as graphitic (elemental) carbon and a wide variety of organic compounds and water. Techniques in common use to measure these species are discussed very briefly here. For further details, the reader should consult basic instrumental analysis texts (e. g., Skoog and West, 1980) and the references at the end of this chapter.

Obtaining a representative sample for chemical analysis is probably the most difficult part of determining the chemical composition of atmospheric particulate matter. An ideal sampling method would give enough sample for accurate analysis in the shortest possible time periods without negative or positive artifacts; unfortunately, no such *ideal* method exists at present. For example, the most common sampling method involves the use of filters of various types; however, as discussed in detail in Section 12-E-4e, significant artifacts may occur.

12-E-4a. Inorganic Elements

Table 12.32 summarizes the major methods used to measure the inorganic elements in atmospheric particulate matter.

Colorimetry. Colorimetric methods, that is, wet chemical methods in which reagents are added to generate a light-absorbing species whose absorbance can be quantified using conventional absorption spectrometry, have been used rather extensively in the past. An example is the measurement of Cl^- in aerosols from remote regions by the mercury thiocyanate method (Huebert and Lazrus, 1980). In this technique, chloride ions react with $Hg(SCN)_2$ in a dioxane–ethanol solution to form $HgCl_2$, $HgCl_4{}^{2-}$, and SCN^-. Upon addition of Fe^{3+} in a nitric acid solution, an orange solution due to $FeSCN^{2+}$ results, whose absorbance can be measured at its 460 nm peak (Iwaski et al., 1956).

While colorimetric methods have the advantages of being relatively inexpensive, simple to carry out, and applicable to a large number of elements, they are increasingly being replaced by other physical techniques. The major reason for this is that, as discussed in Chapter 5, wet chemical methods are more likely to suffer from unrecognized interferences, particularly in complex environmental samples. However, when the aerosol composition is sufficiently well known that

TABLE 12.32. Some Common Methods Used to Measure Inorganic Elements in Ambient Particulate Matter

Element	Analytical Methods
Aluminum	XRF[a], PIXE[b]
Antimony	Col,[c] AA,[d] XRF, NA,[e] ASV[f]
Arsenic	Col, AA, XRF, NA, ASV, MS,[g] PIXE
Beryllium	Col, AA, ES[h]
Bismuth	AA, ASV
Bromine	XRF, NA, MS, PIXE
Cadmium	Col, AA, XRF, NA, ASV, MS, ES
Calcium	XRF, PIXE
Chlorine	XRF, NA, MS, PIXE
Chromium	AA, XRF, NA, MS, ES, PIXE
Cobalt	AA, NA, ES
Copper	AA, XRF, ASV, MS, ES, PIXE
Gallium	PIXE
Iodine	MS
Iron	Col, AA, XRF, NA, ASV, MS, ES
Lead	Col, AA, XRF, ESCA,[i] ASV, MS, ES, PIXE
Magnesium	PIXE
Manganese	Col, AA, XRF, NA, MS, ES, PIXE
Mercury	Col, AA, XRF, NA, ES
Molybdenum	Col, AA, ES, MS
Nickel	AA, XRF, NA, MS, ES, PIXE
Nitrogen	ESCA
Potassium	XRF, PIXE
Rubidium	PIXE
Selenium	Col, AA, NA, ASV, ES, XRF, PIXE, MS
Silicon	XRF, PIXE
Sodium	AA
Strontium	PIXE
Sulfur	XRF, ESCA, MS, PIXE
Tin	AA, XRF, MS, ES
Titanium	AA, XRF, NA, ES, PIXE
Vanadium	AA, XRF, NA, MS, ES, PIXE
Zinc	AA, XRF, NA, ASV, MS, ES, PIXE

Source: Adapted from LBL, 1979.

[a]XRF = X-ray fluorescence analysis.

[b]PIXE = particle-induced X-ray emission

[c]Col = colorimetry.

[d]AA = atomic absorption spectrometry.

[e]NA = neutron activation analysis.

[f]ASV = anodic stripping voltammetry.

[g]MS = mass spectrometry.

[h]ES = emission spectrometry.

[i]ESCA = electron spectroscopy for chemical analysis; also known as XPS = X-ray photoelectron spectroscopy.

one can be confident of the absence of interfering species, colorimetric methods are useful.

X-ray Fluorescence (XRF). The sample is irradiated with monochromatic X-rays; the elements in the sample are excited and emit characteristic X-rays whose wavelength is used to identify the element and whose intensity is related to the amount present. XRF is used primarily for elements heavier than magnesium because of the weak fluorescence of lighter elements and absorption of the X-rays within the particles. This method has been used fairly extensively in recent years for elemental analysis of atmospheric particulate matter (e.g., see Stevens et al., 1978).

Particle-Induced X-Ray Emission (PIXE). Elements heavier than sodium can be analyzed using PIXE. In this method, the sample is bombarded with a beam of particles, usually protons, which excites the elements in the sample, causing them to emit X-rays at wavelengths characteristic of the elements (Johansson et al., 1975). Closely related methods of analysis use other ions such as alpha particles to bombard the sample and induce the X-ray emission. As in the case of XRF, the lighter elements (hydrogen through fluorine) cannot be easily measured with this technique; however, backscattering of alpha particles used to bombard the sample can be measured, and the energy lost in the nuclear recoil can be used to identify the scattering element for these lighter species. These ion-excited X-ray analytical techniques (IXA) are reviewed by Cahill (1980, 1981a,b).

Atomic Absorption Spectrometry (AA). This is a standard laboratory analytical tool for metals. The metal is extracted into solution and then vaporized in a flame. A light beam with a wavelength absorbed by the metal of interest passes through the vaporized sample; for example, to measure zinc, a zinc resonance lamp can be used so that the emission and absorbing wavelengths are perfectly matched. The absorption of the light by the sample is measured and Beer's law is applied to quantify the amount present.

Emission Spectrometry (ES). Emisssion spectrometry is based on the excitation of an element to an upper electronically excited state, from which it returns to the ground state by the emission of radiation. As discussed in Chapter 2, the wavelength emitted is characteristic of the emitted species, and, under the appropriate conditions, the emission intensity is proportional to its concentration. Means of excitation include arcs and sparks, plasma jets, and lasers.

Neutron Activation (NA). The sample is bombarded with neutrons and the radioactivity induced in the sample is then measured. Both beta and gamma radiation can be monitored, but gamma radiation is more frequently used because of the discrete wavelengths associated with emission which can be used to identify the emitter.

Anodic Stripping Voltammetry (ASV). This is an electrochemical technique in which the element to be analyzed is first deposited on an electrode and then redissolved, that is, "stripped," from the electrode to form a more concentrated solution. For example, a drop of mercury hanging from a platinum electrode in a solution containing the species to be measured has been used as the deposition electrode. A potential slightly more negative than the half-wave potential for the ion of interest is applied to deposit the element on the electrode. After deposition of the metal for a given time period, stirring of the solution is stopped and the voltage decreased at a constant rate toward the anodic potential while the anodic current is measured. The peak anodic current, corrected for the residual current, is proportional to the elemental concentration under controlled conditions, for example, fixed deposition time.

Mass Spectrometry (MS). Mass spectrometry is a common method for detecting and measuring organics (see below), but it has also been used for certain inorganic elements and ions as well. For example, Schuetzle et al. (1973) volatilized ambient particulate matter into the source region of a high-resolution mass spectrometer by heating the sample continuously from 20 to 400°C. The elements sulfur, cadmium, and iodine were identified and measured using their masses, ion intensities, and vaporization temperatures.

Electron Spectroscopy for Chemical Analysis (ESCA). ESCA is also known as X-ray photoelectron spectroscopy (XPS). The sample is irradiated with X-rays of a fixed frequency, causing ejection of electrons whose kinetic energy is measured. Conservation of energy dictates that the kinetic energy of the electron plus its binding energy must equal the energy of the exciting photon; since the latter is known and the kinetic energy of the electron is measured, the binding energy can be calculated. Since the binding energies are charateristic of each element, this can be used for elemental analysis. In addition, because the binding energies of inner shell electrons are influenced to some extent by the bonding electrons which determine the oxidation state of the element, ESCA also gives information on the chemical form of the element. For example, as discussed above, ESCA has been used by Novakov and co-workers to elucidate the forms of nitrogen and sulfur in atmospheric particulate matter. The application of ESCA to the analysis of atmospheric particles is discussed in detail by Novakov and co-workers (1977).

12-E-4b. Inorganic Ions

Inorganic ions such as NH_4^+, SO_4^{2-}, and NO_3^- are major components of ambient particulate matter and a wide variety of methods has been used to measure their concentrations (e.g., see the volume edited by Sawicki et al.,1978). A few of the methods most commonly used are summarized in Table 12.33 and discussed briefly below.

Colorimetry. A variety of colorimetric techniques have been used to measure ions such as NH_4^+, SO_4^{2-}, and NO_3^- in ambient particles. For example, nitrate can be measured by reduction to nitrite using hydrazine in the presence of a copper catalyst, followed by its conversion to a colored azo dye which can be measured by its absorbance at 524 nm (Mullin and Riley, 1955). Sulfate has been determined using an exchange reaction between sulfate and a barium–nitrosulfonazo III chelate

TABLE 12.33. Some Common Methods of Measuring the Major Inorganic Ions in Particulate Matter

Ion	Analytical Method
NH_4^+	Col,[a] ESCA,[b] IC,[c] SIE,[d] IR[e]
SO_4^{2-}	Col, ESCA, IC, IR
NO_3^-	Col, ESCA, SIE, IC, CC,[f] IR

Source: Adapted from LBL, 1979.
[a]Col = colorimetry.
[b]ESCA = electron spectroscopy for chemical analysis.
[c]IC = ion chromatography.
[d]SIE = selective ion electrodes.
[e]IR = infrared spectroscopy.
[f]Chemical conversion followed by detection of the product of the NO_3^- reaction; see text.

in aqueous acetonitrile; the chelate has an absorbance peak at 642 nm and hence the decrease in this peak can be followed as a measure of the amount of sulfate present which has exchanged with the chelate (Hoffer et al., 1979). Similarly, NH_4^+ is often measured by the indophenol blue method (Weatherburn, 1967).

Ion Chromatography (IC). Ion chromatography has become one of the most widely used methods for the determination of ion concentrations in ambient particles. As the name implies, ions are separated using ion exchange chromatography and are detected using electrical conductivity. For example, sulfate and nitrate can be separated on a column containing a strong basic resin using a carbonate solution as the eluant. To overcome the high conductivity of the eluant which would mask the signal due to the sulfate and nitrate, the solutions then pass into a suppression column that contains a strong acid resin; this converts the carbonate into CO_2 + H_2O which has a low conductivity and the sulfate and nitrate into their acids which have high conductivities and hence can be easily detected against the suppressed eluant background (Mulik et al., 1976). This *eluant suppression* was the key to the development of IC to measure sulfate and nitrate. Since this first application of IC in ambient aerosols, a variety of anions and cations in ambient aerosols have been separated and measured using this technique. Its application to atmospheric samples is discussed in detail in the book edited by Sawicki et al. (1978).

Selective Ion Electrodes (SIE). Selective ion eletrodes are essentially variants of the well-known pH meter. They are membrane indicator types of electrodes, in which a potential is developed across a membrane in the presence of the ion; the size of the potential is related to the concentration and hence can be used to quantitatively detect and measure the species. However, instead of a glass membrane, as in the pH meter, the membranes consist of organics which are immersible in water. For example, anion-sensitive electrodes use a solution of an anion exchange resin in an organic solvent; the liquid can be held in the form of a gel, for example, in polyvinyl chloride. The ion reacts with the organic membrane setting up an equilibrium between the free ion in solution and the ion bound to the membrane, generating a potential difference which is measured

Membrane electrodes used to measure species as NH_4^+ which are in equilibrium with the gaseous form (i.e., NH_3) in solution are known as gas-sensing electrodes. In this case, the solution to be analyzed is separated from the analyzing solution by a gas-permeable membrane. The gas in the solution to be analyzed diffuses through the membrane and changes the pH of the internal solution, which is monitored using a standard glass electrode.

Infrared Spectroscopy (IR). Stephens and Price used infrared spectroscopy to examine both ambient and laboratory-generated aerosols in the early 1970s. They identified sulfate, nitrate, and ammonium ion absorption bands in ambient particles, as well as bands indicating the presence of organics in diesel exhaust (C—H) and oxidized organics in irradiated hydrocarbon–NO_x mixtures (e.g., $-\overset{\overset{\displaystyle O}{\|}}{C}-$ bands from β-pinene aerosol). Other ions have since been identified using IR; for example, Cunningham and co-workers (1974, 1984) have identified CO_3^{2-}, PO_4^{3-}, and SiO_4^{4-}

Chemical Conversion Methods. Some ions in ambient particles have been monitored by reacting them to form products which are then measured. For example, nitrate collected on nylon filters has been analyzed by converting the nitrate to nitrobenzene and measuring the nitrobenzene by gas chromatography (e.g., see Shaw et al., 1982).

12-E-4c. Total Carbon: Organic Versus Graphitic (Elemental)

The separate determination of organic and elemental carbon in atmospheric particles has been addressed in a number of ways by many workers over a period of years; despite this, there is still no accepted accurate and reliable standard method of sampling and analysis for these important aerosol species.

Sampling for particulate carbon has a number of potential problems; for example, gaseous organic compounds may be adsorbed on the filter, leading to a positive error (Cadle et al., 1983). Various techniques such as the use of diffusion denuders coated with adsorbents (e.g., Al_2O_3) are being investigated to minimize this problem (Appel et al., 1983).

Four major methods have been used to separate the organic and elemental carbon: thermal methods, digestion, extraction, and optical techniques. These are discussed in detail in the volume on particulate carbon edited by Wolff and Klimisch (1982) and in the article by Cadle et al. (1983).

In the thermal methods, the sample is heated to increasingly higher temperatures, with most steps being carried out in the presence of O_2. The basis of this method is that volatile organics will vaporize first, and then other organic compounds will be oxidized. Only at the highest temperatures will graphitic carbon oxidize. The carbon thus ejected into the vapor phase at various temperatures is detected in the form of CO_2 or, alternatively, after catalytic reduction, as CH_4.

For example, in one thermal method, shown in Fig. 12.72, the sample is oxidized and volatilized with an O_2–He mixture at 350°C; the volatilized carbon is oxidized to CO_2 in an MnO_2 bed and reduced to CH_4 so it can be measured using the sensitive technique of flame ionization detection (FID) (Huntzicker et al., 1982). The purge gas is then replaced by pure He and the temperature is raised to 600°C; in this step the remaining organic carbon is volatilized, oxidized to CO_2 by the MnO_2 catalyst, and reduced to CH_4 for measurement. Finally, elemental carbon is determined by heating in O_2–He from 400 to 600°C. In this particular apparatus, a light pipe, He–Ne laser, and photocell are used to monitor the reflectance of the filter as an indication of the changes in graphitic carbon on the filter (see below).

Although relatively fast and simple, such thermal methods suffer from the possibility of carbonization of organics during the heating process; thus elemental carbon can be formed from organic carbon during the analysis, leading to significant errors.

A second approach to analyzing organic and elemental carbon has been to digest the sample in a strongly oxidizing solution (e.g., nitric acid) to remove the organics. The remaining carbon on the filter is then measured using standard methods with the assumption that only graphitic carbon remains on the filter after digestion. Organic carbon is then the difference between the total carbon on the filter before and after digestion, respectively. However, it has been shown that during digestion, some elemental carbon is removed, in addition to organic carbon (Cadle et al., 1983). Thus digestion has no clear advantages over thermal methods.

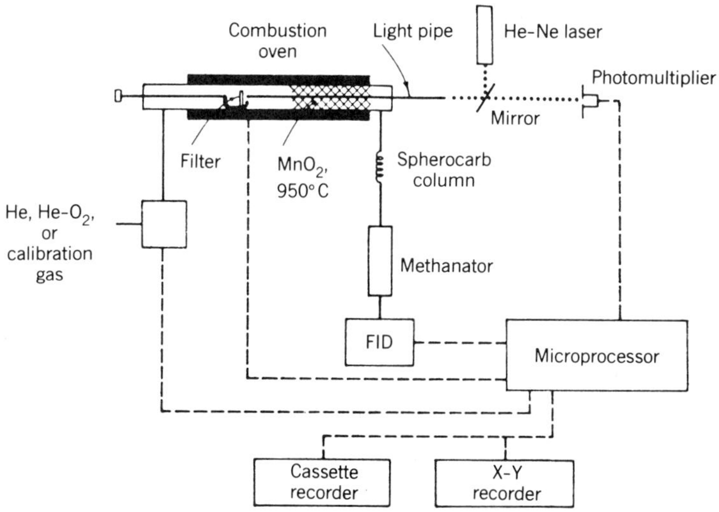

FIGURE 12.72. Schematic diagram of one type of thermal analyzer for organic and graphite carbon (from Huntzicker et al., 1982).

Extraction of the organics from filters using solvents such as a 4:1 benzene–ethanol mixture has also been used. Total carbon analysis of portions of the filter after extraction gives graphitic carbon directly, and organic carbon is obtained by the difference between this and total carbon before extraction. As with thermal and digestion techniques, there are problems in establishing that the organics and elemental carbon are clearly and accurately separated.

There are a variety of optical methods used to measure graphitic carbon alone, the most widely used being visible light absorption or reflectance techniques. Visible light absorption is the basis of what is known as the integrating plate method (IPM) (Lin et al., 1973) shown schematically in Fig. 12.73. Particles are collected on a nucleopore filter and inserted between the light source and the detector; the light transmitted through the filter is compared to that transmitted through a clean filter, that is, one not containing particles. Opal glass is placed between the filter and the detector to transmit an isotropic light flux from scattered and transmitted light through the filter. Scattering of light by the particles which would interfere with an absorption measurement is minimized by using a filter with a refractive index approximately equal to that of the particles.

As might be expected for a measurement based on simple light absorption in a complex sample, there are a number of potential problems. For example, there may be other light-absorbing organics or other species present in the sample; in addition, it is not clear what value should be used for the absorptivity of combustion-derived carbon particles.

Reflectance techniques, like the IPM method, are based on the absorption of visible light by graphitic carbon. However, rather than measuring the decrease in light transmitted through a filter due to absorption, the decrease in light reflected from the carbon-containing surface is measured; the higher the elemental carbon, the more light will be absorbed and the less reflected. Thus log (R_0/R), where R_0 is the reflectance in the absence of carbon and R the reflectance in its presence, has been shown to be linearly related to the elemental carbon concentrations (Delumyea et al., 1980). Because this is a light absorption/reflectance measurement, it suffers from the same types of problems as the IPM. However, it has an advan-

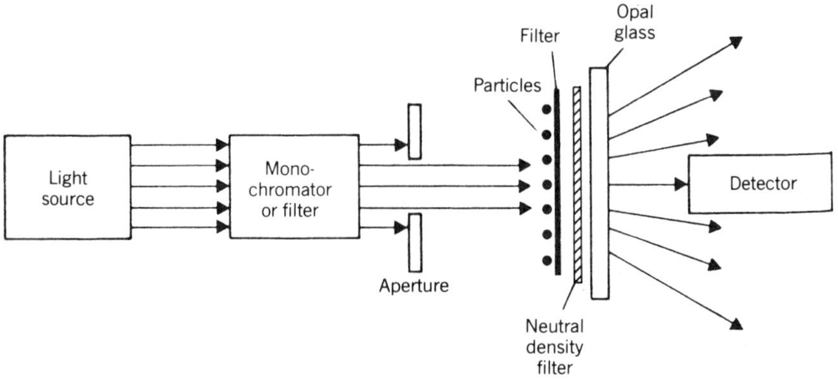

FIGURE 12.73. Schematic of integrating plate method (IPM) for measuring graphite carbon (from Weiss and Waggoner, 1982).

tage in terms of its simplicity. In addition, in some urban areas there are historical records of filter sample reflectances which can be calibrated against more recent methods to examine historical trends in graphitic carbon (e.g., see Cass et al., 1983).

Although carbonate has been observed in some ambient samples (e.g., see Cunningham et al., 1974, 1984), it is generally believed to be present at insignificant concentrations compared to organic and graphitic carbon.

12-E-4d. Speciation of Organics

As seen in Section 12-D, a variety of organics are found in particles in ambient air and in laboratory model systems. The most common means of identification and measurement of these species is mass spectrometry (MS), combined with either thermal separation or solvent extraction and gas chromatographic separation (GC–MS).

Solvent extraction of the sample is most frequently a first step in the analysis of particulate matter. Through the appropriate choice of solvents, the organics can be separated into acid, base, and neutral fractions, polar and non-polar fractions, and so on. This grouping of compounds according to their chemical properties using extraction techniques simplifies the subsequent analysis. Each fraction can then be analyzed by GC–MS, with the GC retention time and the mass spectrum used for identification and measurement.

Two types of ionization sources are in widespread use—electron impact and chemical ionization. The traditional means of ionization by electron impact often causes extensive fragmentation of molecules so that only peaks corresponding to the fragments are seen in the mass spectrum. Particularly in a complex environmental sample, this may preclude positive compound identification. Chemical ionization complements electron impact mass spectra and is particularly useful for establishing the molecular weight of the compound. In chemical ionization sources, an electron beam is used to ionize a reagent gas such as CH_4. The sample is then ionized by collisions with the ionized fragments from CH_4. This often results in relatively strong peaks at masses one greater or one less than the parent peak through reactions such as the following:

$$CH_5{}^+ + MH \rightarrow MH_2{}^+ + CH_4$$

$$C_2H_5{}^+ + MH \rightarrow M^+ + C_2H_6$$

Schuetzle and co-workers (1973, 1975) and Cronn et al. (1977) carried out extensive high-resolution mass spectrometry studies of ambient particles to show the presence of many oxidized multi-functional compounds (Section 12-D-2d). As discussed earlier, the particle sample was heated to sequentially volatilize various compounds into the source region of the MS. Both the temperature at which a compound is vaporized and the mass spectrum are characteristic of a particular compound and hence together could be used for identification.

Other types of mass spectrometry have also been used to examine ambient particulate samples, but these are generally neither well developed nor in widespread use on ambient samples. One such technique is secondary ion mass spectrometry

(SIMS) in which the surface of the sample is bombarded with a beam of ions or neutral atoms which cause ejection of fragments from the surface. The fragments may be neutral atoms or molecules, positively or negatively charged species, electrons or photons. The charged species, that is, the secondary ions, can be analyzed using MS, generating a SIMS spectrum. Elements such as potassium and sodium as well as functional groups such as COOH, sulfates, and nitrates, can be detected by SIMS (e.g., see Klaus, 1983).

12-E-4e Artifact Formation in Particle Sampling

One important aspect of measuring the chemical composition of particles is the possibility of occurrence of artifacts during sampling. Artifacts have been observed for nitrate, sulfate, and organics. Positive errors result from the chemical conversion of gases to particulate phase species (e.g., SO_2 to sulfate) or from the retention of gas phase species such as HNO_3 either on the filter medium itself or on particles previously adsorbed on the filter. Negative errors result from reactions of particulate matter collected on the filter, releasing material into the gas phase; for example, nitrate on the filter can react with gaseous acids such as HCl or acids present in particles (e.g., sulfuric acid) to release HNO_3 to the gas phase. In the case of NH_4NO_3, volatilization can also produce a negative error, because of the easily reversible equilibrium between the solid NH_4NO_3 and gaseous NH_3 and HNO_3 (e.g., see Appel and Tokiwa, 1981):

$$NH_4NO_{3(s)} \leftrightarrows NH_{3(g)} + HNO_{3(g)}$$

Artifacts in sulfate and nitrate formation from gaseous SO_2 and HNO_3 on various types of filter media have been reviewed by Appel and co-workers (1984). In laboratory studies, glass fiber filters were found to retain 8–21% of the SO_2 as sulfate on the filter at 21°C and 80% RH, whereas quartz and Teflon retained 5% or less. This retention seems to be related to the filter alkalinity (Witz, 1985), with the most alkaline filters retaining the greatest amount of SO_2 (Fig. 12.74).

Artifact formation of SO_4^{2-} on nylon filters from SO_2 has also been seen (Japar and Brachaczek, 1984); it was more severe at higher relative humidities (RHs), ranging from <1% at RH less than 20% to 5% for RH greater 55%.

NO_2 conversion to nitrate on glass and quartz filters has also been observed in a number of laboratory studies at ppm NO_2 concentrations (e.g., see Spicer and Schumacher, 1979). At the lower concentrations found in the atmosphere, this appears to represent an insignificant source of filter nitrate, with the possible exception of the case where high concentrations of O_3 are present (Appel et al., 1979).

However, while conversion of NO_2 to nitrate on filters does not appear to be a major source of artifact nitrate under most typical atmospheric conditions, adsorption of gas phase HNO_3 onto the filter medium and/or previously collected particles does give significant artifact nitrate on certain types of filters. Thus in laboratory studies, glass fiber filters retained >94% of the gaseous HNO_3 contained in an airstream passing through the filter, while the quartz filters tested retained

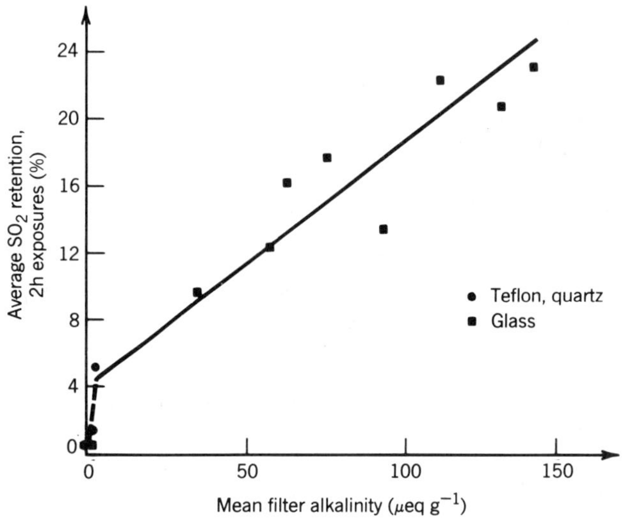

FIGURE 12.74. Scatter diagram of mean SO_2 retention against mean filter alkalinity (2 h trials) (from Appel et al., 1984).

from 33% to ≥99%. Only Teflon filters adsorbed negligible amounts of gaseous HNO_3; however, significant loss of particulate nitrate from Teflon filters has been observed (e.g., see Shaw et al., 1982), which could be due to evaporation from the filter, particularly in response to temperature changes or to reaction of the nitrate with strong acids to release gaseous HNO_3.

There is need for caution in applying such results to atmospheric sampling, however. Thus the results of laboratory studies on HNO_3 in clean air with clean filters does not necessarily imply that the same quantitative results will apply to real atmospheric sampling situations where the filter is holding particulate matter and the air contains a variety of species; in addition, the temperature and relative humidity can vary widely. For example, Appel and co-workers (1984) find that the nitrate concentrations on Teflon and quartz filters from ambient air samples containing HNO_3 were much more similar than would be expected based on the laboratory studies. Because of the possibility of these positive or negative artifacts depending on the filter type, nitrate analyses of samples collected on glass fiber and quartz filters should be considered upper limits to the true particulate nitrate and those collected on Teflon should be considered lower limits.

Artifacts have also been documented in the sampling of organics. For example, the more volatile particulate organics may volatilize during sampling (Van Vaeck et al., 1984), giving a negative artifact. Positive artifacts have also been observed for carbon due to adsorption of gaseous organics on the filter (Stevens et al., 1980; Cadle et al., 1983). In addition, oxidation of organics may occur on the filter itself as air containing such reactive species as O_3 and NO_2 are pulled through the filter; this problem has been investigated particularly with regard to polycyclic aromatic hydrocarbons (e.g., see Pitts et al., 1978) and is discussed in more detail in the following chapter.

REFERENCES

Adamson, A. W., *A Textbook of Physical Chemistry*, Academic Press, New York, 1973.

Appel, B. R. and Y. Tokiwa, "Atomspheric Particulate Nitrate Sampling Errors Due to Reactions with Particulate and Gaseous Strong Acids," *Atoms. Environ.*, **15,** 1087 (1981).

Appel, B. R., S. M. Wall, Y. Tokiwa, and M. Haik, "Interference Effects in Sampling Particulate Nitrate in Ambient Air," *Atmos. Environ.*, **13,** 319 (1979).

Appel, B. R., M., Haik, E. L. Kothny, S. M. Wall, and Y. Tokiwa, "Evaluation of Techniques for Sulfuric Acid and Total Particulate Acidity in Ambient Air," *Atmos. Environ.*, **14,** 559 (1980a).

Appel, B. R., E. L. Kothny, E. M. Hoffer, and J. J. Wesolowski, "Sulfate and Nitrate Data from the California Aerosol Characterization Experiment (ACHEX)," *Adv. Environ. Sci. Technol.*, **10,** 315 (1980b).

Appel, B. R., S. M. Wall, and R. L. Knights, "Characterization of Carbonaceous Materials in Atmospheric Aerosols by High Resolution Mass Spectrometric Thermal Analysis," *Adv. Environ. Sci. Technol.*, **10,** 353 (1980c).

Appel, B. R., E. M. Hoffer, Y. Tokiwa, and E. L. Kothny, "Measurement of Sulfuric Acid and Particulate Strong Acidity in the Los Angeles Basin," *Atmos. Environ.*, **16,** 589 (1982).

Appel, B. R., Y. Tokiwa, and E. L. Kothny, "Sampling of Carbonaceous Particles in the Atmosphere," *Atmos. Environ.*, **17,** 1787 (1983).

Appel, B. R., Y. Tokiwa, M. Haik, and E. L. Kothny, "Artifact Particulate Sulfate, and Nitrate Formation on Filter Media," *Atmos. Environ.*, **18,** 409 (1984).

Appel, B. R., Y. Tokiwa, J. Hsu, E. L. Kothny, and E. Hahn, "Visibility as Related to Atmospheric Aerosol Constituents," *Atmos. Environ.*, **19,** 1525 (1985).

Atchison, J. and J. A. C. Brown, *The Lognormal Distribution*, Cambridge University Press, London, 1957.

Ayer, H.E. and J. M. Hochstrasser, Cyclone Discussion, in *Aerosol Measurement*, D. A. Lundgren, F. S. Harris, Jr., W. H. Marlow, M. Lippmann, W. E. Clark, and M. D. Durham, Eds., University Presses of Florida, Gainesville, FL, 1979, pp. 70–79.

Bassett, M. and J. H. Seinfeld, "Atmospheric Equilibrium Model of Sulfate and Nitrate Aerosols," *Atmos. Environ.*, **17,** 2237 (1983).

Bassett, M. and J. H. Seinfeld, "Atmospheric Equilibrium Model of Sulfate and Nitrate Aerosols—II. Particle Size Analysis," *Atmos. Environ.*, **18,** 1163 (1984).

Bates, D. V., B. R. Fish, T. F. Hatch, T. T. Mercer, and P. E. Morrow, "Deposition and Retention Models for Internal Dosimetry of the Human Respiratory Tract," *Health Physics*, **12,** 173 (1966).

Berglund, R. N. and B. Y. H. Liu, "Generation of Monodisperse Aerosol Standards," *Environ. Sci. Technol.*, **7,** 147 (1973).

Biggins, P. D. E. and R. M. Harrison, "Chemical Speciation of Lead Compounds in Street Dusts," *Environ. Sci. Technol.*, **14,** 336 (1980).

Blanchard, D. C. and A. H. Woodcock, "The Production, Concentration, and Vertical Distribution of the Sea-Salt Aerosol," *Ann. N.Y. Acad. Sci.*, **338,** 330 (1980).

Bohren, C. F. and D. R. Huffman, *Absorption and Scattering by Small Particles*, Wiley, New York, 1983.

Brosset, C., K. Andreasson, and M. Ferm, "The Nature and Possible Origin of Acid Particles Observed at the Swedish West Coast," *Atmos. Environ.*, **9,** 631 (1975).

Cadle, R. D., *The Measurement of Airborne Particles*, Wiley, New York, 1975.

Cadle, R. D. and R. C. Robbins, "Kinetics of Atmospheric Chemical Reactions Involving Aerosols," *Disc. Faraday Soc.*, **30,** 155 (1960).

Cadle, S. H., P. J. Groblicki, and P.A. Mulawa, "Problems in the Sampling and Analysis of Carbon Particulate," *Atmos. Environ.*, **17,** 593 (1983).

Cahill, T. A., Comments on Surface Coatings for Lundgren-Type Impactors, in *Aerosol Measurement*, D. A. Lundgren, F. S. Harris, Jr., W. H. Marlow, M. Lippmann, W. E. Clark, and M. D. Durham, Eds., University Presses of Florida, Gainesville, FL, 1979, pp. 131–134.

Cahill, T. A., "Proton Microprobes and Particle-Induced X-Ray Analytical Systems," *Ann. Rev. Nucl. Part. Sci.*, **30,** 211 (1980).

Cahill, T. A., "Innovative Aerosol Sampling Devices Based Upon PIXE Capabilities," *Nucl. Instrum. Methods*, **181,** 473 (1981a).

Cahill, T. A., "Ion Beam Analysis of Environmental Samples." *Adv. Chem. Ser.*, **197,** 511 (1981b).

Calvert, J. G., "Hydrocarbon Involvement in Photochemical Smog Formation in Los Angeles Atmosphere," *Environ. Sci. Technol.*, **10,** 256 (1976).

Cass, G. R., "On the Relationship Between Sulfate Air Quality and Visibility with Examples in Los Angeles," *Atmos. Environ.*, **13,** 1069 (1979).

Cass, G. R., M. H. Conklin, J. J. Shah, J. J. Huntzicker, and E. S. Macias, "Elemental Carbon Concentrations: Estimation of and Historical Data Base," *Atmos. Environ.*, **18,** 153 (1983).

Cautreels, W. and K. Van Cauwenberghe, "Determination of Organic Compounds in Airborne Particulate Matter by Gas Chromatography-Mass Spectrometry," *Atmos. Environ.*, **10,** 447 (1976).

Chameides, W. L. and D. D. Davis, "The Free Radical Chemistry of Cloud Droplets and Its Impact Upon the Composition of Rain," *J. Geophys. Res.*, **87,** 4863 (1982).

Chang, D. P. Y. and R. C. Hill, "Retardation of Aqueous Droplet Evaportation by Air Pollutants," *Atmos. Environ.*, **14,** 803 (1980).

Chang, S. G., R. Brodzinsky, L. A. Gundel, and T. Novakov, Chemical and Catalytic Properties of Elemental Carbon, in *Particulate Carbon: Atmospheric Life Cycle*, G. T. Wolff and R. L. Klimsch, Eds., Plenum Press, New York, 1982, pp. 159–181.

Charlson, R. J., D. S. Covert, Y. Tokiwa, and P. K. Mueller, Multi-Wavelength Nephelometer Measurements in Los Angeles Smog Aerosol—III, in *Aerosols and Atmospheric Chemistry,* G. M. Hidy, Ed., Academic Press, New York, 1972.

Charlson, R. J., H. Horvath, and R. F. Pueschel, "The Direct Measurement of Atmospheric Light Scattering Coefficient for Studies of Visibility and Pollution," *Atmos. Environ.*, **1,** 469 (1967).

Charlson, R. J., A. H. Vanderpol, D. S. Covert, A. P. Waggoner, and N. C. Ahlquist, "$H_2SO_4/(NH_2)_2SO_4$ Background Aerosol: Optical Detection in the St. Louis Region," *Atmos. Environ.*, **8,** 1257 (1974a).

Charlson, R. J., A. H. Vanderpol, D. S. Covert, A. P. Waggoner, and N. C. Ahlquist, "Sulfuric Acid–Ammonium Sulfate Aerosol: Optical Detection in the St. Louis Region," *Science*, **184,** 156 (1974b).

Chuan, R. L., Rapid Measurement of Particle Size Distribution in the Atmosphere, in *Fine Particles: Aerosol Generation, Measurement, Sampling, and Analysis, B. Y. H. Liu, Ed., Academic Press, New York, 1976, pp. 763–775.*

Clegg, S. L. and P. Brimblecombe, "Potential Degassing of Hydrogen Chloride from Acidified Sodium Chloride Droplets," *Atmos. Environ.*, **19,** 465 (1985).

Cobourn, W. G., R. B. Husar, and J. D. Husar, "Continous *In Situ* Monitoring of Ambient Particulate Sulfur Using Flame Photometry and Thermal Analysis," *Atmos. Environ.*, **12,** 89 (1978).

Coles, D. G., R. C. Ragaini, J. M. Ondov, G. L. Fisher, D. Silberman, and B. A. Prentice, "Chemical Studies of Stack Fly Ash from a Coal-Fired Power Plant," *Environ. Sci. Technol.*, **13,** 455 (1979).

Conner, W. D., "An Inertial-Type Particle Separator for Collecting Large Samples," *J. Air Pollut. Control. Assoc.*, **16,** 35 (1966).

Cooke, D. D. and M. Kerker, "Response Calculations for Light Scattering Aerosol Particle Counters," *Appl. Opt.*, **14,** 734 (1975).

Courtney, W. J., R. W. Shaw, and T. G. Dzubay, "Precision and Accuracy of a Beta Gauge for Aerosol Mass Determinations," *Environ. Sci. Technol.*, **16,** 236 (1982).

Covert, D. S., R. J. Charlson, and N. C. Ahlquist, "A Study of the Relationship of Chemical Composition and Humidity to Light Scattering by Aerosols,' *J. Appl. Meteorol.*, **11,** 968 (1972).

Covert, D. S., A. P. Waggoner, R. E. Weiss, N. C. Ahlquist, and R. J. Charlson, "Atmospheric Aerosols, Humidity, and Visibility," *Adv. Environ. Sci. Technol.*, **10,** 559 (1980).

Craig, N. L., A. B. Harker, and T. Novakov, "Determination of the Chemical States of Sulfur in Ambient Pollution Aerosols by X-Ray Photoelectron Spectroscopy," *Atmos. Environ.*, **8,** 15 (1974).

Cronn, D. R., R. J. Charlson, R. L. Knights, A. L. Crittenden, and B. R. Appel, "A Survey of the Molecular Nature of Primary and Secondary Components of Particles in Urban Air by High Resolution Mass Spectrometry," *Atmos. Environ.*, **11,** 929 (1977).

Cunningham, P. T., S. A. Johnson, and R. T. Yang, "Variations in Chemistry of Airborne Particulate Material with Particle Size and Time," *Environ. Sci. Technol.*, **8,** 131 (1974).

Cunningham, P. T., B. D. Holt, S. A. Johnson, D. L., Drapcho, and R. Kumar, Acidic Aerosols: Oxygen-18 Studies of Formation and Infrared Studies of Occurrence and Neutralization, in *Chemistry of Particles, Fogs, and Rain*, J. L. Durham, Ed., Acid Precipitation Series, J. I. Teasley, Series Ed., Butterworth, Boston, 1984, pp. 53–130.

Davies, C. N., Ed., *Aerosol Science*, Academic Press, London, 1966.

Delumyea, R. G., L.-C.. Chu, and E. S. Macias, "Determination of Elemental Carbon Component of Soot in Ambient Aerosol Samples," *Atmos. Environ.*, **14,** 647 (1980).

Dennis, R., Ed., *Handbook on Aerosols*, Publ. No. TID-26608, National Technical Information Center, Springfield, VA, 1976.

Diederen, H. S. M. A., R. Guicherit, and J. C. T. Hollander, "Visibility Reduction by Air Pollution in the Netherlands," *Atmos. Environ.*, **19,** 377 (1985).

Dong, M. W., D. C. Locke, and D. Hoffmann, "Characterization of Aza-Arenes in Basic Organic Portion of Suspended Particulate Matter," *Environ. Sci. Technol.*, **11,** 612 (1977).

Duce, R. A., W. H. Zoller, and J. L. Moyers, "Particulate and Gasous Halogens in the Antarctic Atmosphere," *J. Geophys. Res.*, **78,** 7802 (1973).

Duce, R. A., V. A. Mohnen, P. R. Zimmerman, D. Grosjean, W. Cautreels, R. Chatfield, R. Jaenicke, J. A. Ogren, E. D. Pellizzari, and G. T. Wallace, "Organic Material in the Gloval Troposphere," *Rev. Geophys. Space Phys.*, **21,** 921 (1983).

Dumdei, B. E. and R. J. O'Brien, "Toluene's Degradation Products Under Simulated Atmospheric Conditions," *Nature*, **311,** 248 (1984).

Durham, J. L., L. J. Spiller, and D. J. Eatough, "Dimethyl and Methyl Hydrogen Sulfate in the Atmosphere," Proceedings of Air Pollution Control Association. Specialty Conference on Environmental Impact of Natural Emissions, Research Triangle Park, North Carolinia, March 8–9, 1984.

Dzubay, T. G., R. K. Stevens, C. W. Lewis, D. H. Hern, W. J. Courtney, J. W. Tesch, and M. A. Mason, "Visibility and Aerosol Composition in Houston, Texas," *Environ. Sci. Technol.*, **16,** 514 (1982).

Eatough, D. J. and L. D. Hansen, "Organic and Inorganic S(IV) Compounds in Airborne Particulate Matter," *Adv. Environ. Sci. Technol.*, **12,** 221 (1983).

Eatough, D. J., V. F. White, L. D. Hansen, N. L. Eatough, and J. L. Cheney, "Identification of Gas Phase Dimethyl Sulfate and Monomethyl Sulfuric Acid in the Los Angeles Atmosphere," submitted for publication (1985).

Federal Register, Vol. 49, No. 55, p. 10408, March 20, 1984.

Finlayson-Pitts, B. J., "The Reaction of NO_2 with NaCl and Atmospheric Implications of NOCl Formation," *Nature*, **306,** 676 (1983).

Fisher, G. L., Size-Related Chemical and Physical Properties of Power Plant Fly Ash, in *Generation of Aerosols and Facilities for Exposure Experiments*, Ann Arbor Science Publishers, Ann Arbor, MI, 1980, pp. 203–214.

Fitz, D. R., G. J. Doyle, and J. N. Pitts, Jr., "An Ultrahigh Volume Sampler for the Multiple Filter Collection of Respirable Particulate Matter," *J. Air Pollut. Control. Assoc.*, **33,** 877 (1983).

Fogg, T. R. and R. A. Duce, "Boron in the Troposphere: Distribution and Fluxes," *J. Geophys. Res.*, **90,** 3781 (1985).

Friedlander, S. K., *Smoke, Dust, and Haze: Fundmentals of Aerosol Behavior*, Wiley-Interscience, New York, 1977.

Friedlander, S. K., "A Note on New Particle Formation in the Presence of an Aerosol," *J. Collid Interface Sci.*, **67,** 387 (1978).

Friedlander, S. K., "Future Aerosols of the Southwest: Implications for Fundamental Aerosol Research," *Ann. N.Y. Acad. Sci.*, **338,** 588 (1980).

Friedlander, S. K., "Dynamics of Aerosol Formation by Chemical Reaction," *Ann. N.Y. Acad. Sci.*, **404,** 354 (1983).

Fuchs, N. A., *The Mechanics of Aerosols*, Pergamon Press, Oxford, 1964.

Fuchs, N. A. and A. G. Sutugin, *Highly Dispersed Aerosols*, Ann Arbor Science Publishers, Ann Arbor, MI, 1970.

Gagosian, R. B., E. T. Pletzer, and O. C. Zafiriou, "Atmospheric Transport of Continentally Derived Lipids to the Tropical North Pacific," *Nature*, **291,** 312 (1981).

Giever, P. M., Ed., *Air Pollution Manual Part I—Evaluation, Vol. I., 2nd, American Industrial Hygiene Association, 1972.*

Gill, P. S., T. E. Graedel, and C. J. Weschler, "Organic Films on Atmospheric Aerosol Particles, Fog Droplets, Cloud Droplets, Raindrops, and Snowflakes," *Rev. Geophys. Space Phys.*, **21,** 903 (1983).

Gordon, G. E., "Receptor Models," *Environ. Sci. Technol.*, **14,** 792 (1980).

Graedel, T. E. and C. J. Weschler, "Chemistry Within Aqueous Atmospheric Aerosols and Raindrops," *Rev. Geophys. Space Phys.*, **19,** 505 (1981).

Gray, H. A., G. R. Cass, J. J. Huntzicker, E. K. Heyerdahl, and J. A. Rau, "Elemental and Organic Carbon Particle Concentrations: A Long Term Perspective,"*Sci. Total Environ.*, **36,** 17 (1984).

Groblicki, P. J., G. T. Wolff, and R. J. Countess, "Visibility Reducing Species in the Denver Brown Cloud. I. *Atmos. Environ.*, **15,** 2473 (1981).

Grosjean, D., "Quantitative Collection of Total Inorganic Atmospheric Nitrate on Nylon Filters," *Anal. Lett.*, **15,** 785 (1982).

Grosjean, D., "Distribution of Atmospheric Nitrogenous Pollutants at a Los Angeles Area Smog Receptor Site," *Environ. Sci. Technol.*, **17,** 13 (1983).

Grosjean, D., "Particulate Carbon in Los Angeles Air," *Sci. Total Environ.*, **32,** 133 (1984).

Grosjean, D. and S. K. Friedlander, "Gas–Particle Distribution Factors for Organic and Other Pollutants in the Los Angeles Atmosphere," *J. Air Pollut. Control Assoc.*, **25,** 1038 (1975).

Grosjean, D. and S. K. Friedlander, "Formation of Organic Aerosols from Cyclic Olefins and Diolefins," *Adv. Environ. Sci. Technol.*, **10,** 435 (1980).

Grosjean, D. and B. Wright, "Carbonyls in Urban Fog, Ice Fog, Cloudwater, and Rainwater," *Atmos. Environ.*, **17,** 2093 (1983).

Grosjean, D., G. J. Doyle, T. M. Mischke, M. P. Poe, D. R. Fitz, J. P. Smith, and J. N. Pitts, Jr., "The Concentration, Size Distribution, and Modes of Formation of Particulate Nitrate, Sulfate, and Ammonium Compounds in the Eastern Part of the Los Angeles Basin," Paper No. 76-20.3, Presented at the 69th Annual Meeting of the Air Pollution Control Association Meeting, Portland, OR, June 27–July 1, 1976a.

Grosjean, D., J. P. Smith, T. M. Mischke, and J. N. Pitts, Jr., Chemical and Physical Transformatons in Urban-Suburban Transport of Air Pollutants, in *Atmospheric Pollution*, M. M. Benarie, Ed., Elsevier, Amsterdam, 1976b, pp. 549–563.

Grosjean, D., K. Van Cauwenberghe, J. P. Schmid, P. E. Kelley, and J. N. Pitts, Jr., "Identification of C_3–C_10 Aliphatic Dicarboxylic Acids in Airborne Particulate Matter," *Environ. Sci. Technol.*, **12,** 313 (1978).

Gundel, L. A., S. G. Chang, M. S. Clemenson, S. S. Markowitz, and T. Novakov, Characterization of Particulate Amines, in *Nitrogenous Air Pollutants. Chemical and Biological Implications*, D. Grosjean, Ed., Ann Arbor Science Publishers, Ann Arbor, MI, 1979, pp. 211–220.

Hahn, J., "Organic Constituents of Natural Aerosols," *Ann. N.Y. Acad. Sci.*, **338,** 359 (1980).

Hatakeyama, S., T. Tanonaka, J.-h Weng, H. Bandow, H. Takagi, and H. Akimoto, "Ozone-Cyclohexene Reaction in Air: Quantitative Analysis of Particulate Products and the Reaction Mechanism," *Environ. Sci. Technol.*, **19,** 935 (1985).

Hatch, T. and S. P. Choate, "Statisical Description of the Size Properties of Non-Uniform Particulate Substances," *J. Franklin Inst.*, **207,** 369 (1929).

Hegg, D. A., P. V. Hobbs, and J. H. Lyons, "Field Studies of a Power Plant Plume in the Arid Southwestern United States," *Atmos. Environ.*, **19,** 1147 (1985).

Heisler, S. L. and S. K. Friedlander, "Gas-to-Particle Conversion in Photochemical Smog: Aerosol Growth Laws and Mechanisms for Organics," *Atmos. Environ.*, **11,** 157 (1977).

Herdan, G., *Small Particle Statistics*, 2nd ed., Academic Press, New York, 1960.

Hering, S. V., R. C. Flagan, and S. K. Friedlander, "Design and Evaluation of New Low-Pressure Impactor. I., *Environ. Sci. Technol.*, **12,** 667 (1978).

Hering, S. V., S. K. Friedlander, J. J. Collins, and L. W. Richards, "Design and Evaluation of a New Low-Pressure Impactor. 2.," *Environ. Sci. Technol.*, **13,** 184 (1979).

Hering, S. V., and S. K. Friedlander, "Origins of Aerosol Sulfur Size Distributions in the Los Angeles Basin," *Atmos. Environ.*, **16,** 2647 (1982).

Hinds, W. C., *Aerosol Technology*, Wiley, New York, 1982.

Hinkley, E. D., Ed., *Laser Monitoring of the Atmosphere*, Topics in Applied Physics, Vol. 14, Springer-Verlag, New York, 1976.

Hitchcock, D. R., L. L. Spiller, and W. E. Wilson, "Sulfuric Acid Aerosols and HCl Release in Coastal Atmospheres: Evidence of Rapid Formation of Sulfuric Acid Particulates," *Atmos. Environ.*, **14,** 165 (1980).

Ho, W. W., G. M. Hidy, and R. M. Govan, "Microwave Measurements of the Liquid Water Content of Atmospheric Aerosols," *J. Appl. Meteorol.*, **13,** 871 (1974).

Hodkinson, J. R., The Optical Measurement of Aerosols, in *Aerosol Science*, C. N. Davies, Ed., Academic Press, New York, 1966.

Hoffer, E. M., E. L. Kothny, and B. R. Appel, "Simple Method for Microgram Amounts of Sulfate in Atmospheric Particulates," *Atmos. Environ.*, **13,** 303 (1979).

Hollowell, C. D., R. D. McLaughlin, and J. A. Stokes, "Current Methods in Air Quality Measurements and Monitoring," Report No. LBL-3223, Lawrence Berkeley Laboratory, Berkeley, CA, November 1974.

Huebert, B. J. and A. L. Lazrus, "Tropospheric Gas-Phase and Particulate Nitrate Measurements," *J. Geophys. Res.*, **85,** 7322 (1980).

Huffman, D. R. and J. L. Stapp, Optical Measurements on Solids of Possible Interstellar Importance, in *Interstellar Dust and Related Topics*, J. M. Greenberg and H. C. Van de Hulst, Eds., Reidel, Boston, 1973.

Huntzicker, J. J., R. L. Johnson, J. J. Shah, and R. A. Cary, Analysis of Organic and Elemental Carbon in Ambient Aerosols by a Thermal–Optical Method, in *Particulate Carbon: Atmospheric Life Cycles*, G. T. Wolff and R. L. Klimisch, Eds., Plenum Press, New York, 1982, pp. 79–88.

Huntzicker, J. J., R. S. Hoffman, and R. A. Cary, "Aerosol Sulfur Episodes in St. Louis, Missouri," *Environ. Sci. Technol.*, **18,** 962 (1984).

Husar, R. B. and K. T. Whitby, "Growth Mechanisms and Size Spectra of Photochemical Aerosols," *Environ. Sci. Technol.*, **7,** 241 (1973).

Husar, R. B. and W. R. Shu, "Thermal Analysis of the Los Angeles Smog Aerosol," *J. Appl. Meteorol.*, **14,** 1558 (1975).

Husar, R. B., K. T. Whitby, and B. Y. H. Liu, "Physical Mechanisms Governing the Dynamics of Los Angeles Smog Aerosol," *J. Colloid Interface Sci.*, **39,** 211 (1972).

IAMAP, "Terminology and Units of Radiation Quantities and Measurements," Radiation Commission of the International Association of Meteorology and Atmospheric Physics, Boulder, CO, 1978.

Iwaski, I., S. Utsumi, K. Hagino, and T. Ozawa, "A New Spectrophotometric Method for the Determination of Small Amounts of Chloride Using the Mercury Thiocyanate Method," *Bull. Chem. Soc. Japan*, **29,** 860 (1956).

Jaklevic, J. M., R. C. Gatti, F. S. Goulding, and B. W. Loo, "A β-Gauge Method Applied to Aerosol Samples," *Environ. Sci. Technol.*, **15,** 680 (1981).

Japar, S. M. and W. W. Brachaczek, "Artifact Sulfate Formation from SO_2 on Nylon Filters," *Atmos. Environ.*, **18,** 2479 (1984)

Johansson, T. B., R. E. Van Grieken, J. W. Nelson, and J. W. Winchester, "Elemental Trace Analysis of Small Samples by Proton Induced X-Ray Emission," *Anal. Chem.*, **47,** 855 (1975).

Junge, C. E., "Recent Investigations in Air Chemistry," *Tellus*, **18,** 127 (1956).

Junge, C. E., *Air Chemistry and Radioactivity*, Academic Press, New York, 1963.

Kerker, M., *The Scattering of Light and Other Electromagnetic Radiation*, Academic Press, New York, 1969.

Ketseridis, G., J. Hahn, R. Jaenicke, and C. Junge, "The Organic Constituents of Atmospheric Particulate Matter," *Atmos. Environ.*, **10,** 603 (1976).

Killinger, D. K. and A. Mooradian, *Optical and Laser Remote Sensing*, Springer Series in Optical Sciences, Vol. 39, Springer-Verlag, New York, 1983.

Kittelson, D. B. and D. F. Dolan, Diesel Exhaust Aerosols, in *Generation of Aerosols and Facilities for Exposure Experiments*, K. Willeke, Ed., Ann Arbor Science Publishers, Ann Arbor, MI, 1980, pp. 337–359.

Klaus, N., "Urban Aerosols Analyzed by Secondary Ion Mass Spectroscopy," *Sci. Total Environ.*, **31,** 263 (1983).

Klippel, W. and P. Warneck, "The Formaldehyde Content of the Atmospheric Aerosol," *Atmos. Environ.*, **14,** 809 (1980).

Kneip, T. J., J. M. Daisey, J. J. Solomon, and R. J. Hershman, "*N*-Nitroso Compounds: Evidence for Their Presence in Airborne Particles," *Science*, **221,** 1045 (1983).

Lapple, C. E., "The Little Things in Life," *SRI Journal, Stanford Research Institute*, **5,** 95 (1961).

Larson, T. V., N. C. Ahlquist, R. E. Weiss, D. S. Covert, and A. P. Waggoner, "Chemical Speciation of H_2SO_4–$(NH_4)_2SO_4$ Particles Using Temperature and Humidity Controlled Nephelometry," *Atmos. Environ*, **16,** 1587 (1982).

Lawrence Berkeley Laboratory (LBL), *Instrumentation for Environmental Monitoring Air*, Vol. 1, Part 2, September 1979.

LBL Report, 1979, see preceding reference.

Lee, K. W., J. A. Gieseke, and W. H. Piispanen, "Evaluation of Cyclone Performance in Different Gases," *Atmos. Environ.*, **19,** 847 (1985).

Lee, L. C., R. L. Day, and M. Suto, "Optical Characteristics and Formation Processes of Aerosols Produced by an Ozone and 1-Butene Reaction," *J. Photochem.*, **26,** 95 (1984).

Lin, C. -I,. M. Baker, and R. J. Charlson, "Absorption Coefficient of Atmospheric Aerosol: A Method for Measurement," *Appl. Opt.*, **12,** 1356 (1973).

Lippmann, M. and A. Kydonieus, "A Multi-Stage Aerosol Sampler for Extended Sampling Intervals," *Am. Ind. Hyg. Assoc. J.*, **31,** 730 (1970).

Liu, B. Y. H., Ed., *Fine Particles: Aerosol Generation Measurement, Sampling, and Analysis*, Academic Press, New York, 1976.

Liu, B. Y. H. and G. A. Kuhlmey, Efficiency of Air Sampling Filter Media, in *X-Ray Fluorescence Analysis of Environmental Samples*, T. Dzubay, Ed., Ann Arbor Science Publishers, Ann Arbor, MI, 1977.

Lui, B. Y. H. and D. Y. H. Pui, "Particle Size Dependence of a Condensation Nuclei Counter," *Atmos. Environ.*, **13,** 563 (1979).

Liu, B. Y. H. and D. Y. H. Pui, " Aerosol Sampling Inlets and Inhalable Particles," *Atmos. Environ.*, **15,** 589 (1981).

Liu, B. Y. H., D. Y. H. Pui, and A. Kapadia, Electrical Aerosol Analyzer: History, Principle, and Data Reduction, in *Aerosol Measurement*, D. A. Lundgren, F. S. Harris, Jr., W. H. Marlow, M. Lippmann, W. E. Clark, and M. D. Durham, Eds., University Presses of Florida, Gainesville, FL, 1979, pp. 341–383.

Liu, B. Y. H., D. Y. H. Pui, R. L. McKenzie, J. K. Agarwal, R. Jaenicke, F. G. Pohl, O. Preining, G. Reischl, W. Szymanski, and P. E. Wagner, "Intercomparison of Different 'Absolute' Instruments for Measurement of Aerosol Number Concentration," *J. Aerosol Sci.*, **13,** 429 (1982).

Liu, B. Y. H., D. Y. H. Pui, and H. J. Fissan, *Aerosols, Science, Technology, and Industrial Applications of Airborne Particles*, Elsevier, New York, 1984.

Lodge, J. P., Jr., "Final Comment to Discussions of Paper Entitled Non-Health Effects of Airborne Particulate Matter,"' *Atmos. Environ.*, **17,** 899 (1983).

Lundgren, D. A., F. S. Harris, Jr., W. H. Marlow, M. Lippmann, W. E. Clark, and M. D. Durham, Eds., *Aerosol Measurement*, University Presses of Florida, Gainesville, FL, 1979.

Macias, E. S. and R. B. Husar, A Review of Atmospheric Particulate Mass Measurement Via the Beta Attenuation Technique, in *Fine Particles: Aerosol Generation, Measurement, Sampling, and Analysis*, B. Y. H. Liu, Ed., Academic Press, New York, 1976, pp. 535–564.

Marple, V. A. and K. Willeke, Inertial Properties: Theory Design, and Use, in *Fine Particles: Aerosol Generation, Measurement, Sampling, and Analysis*, B. Y. H. Liu, Ed., Academic Press, New York, 1976, pp. 411–446.

Marple, V. A. and K. Willeke, Inertial Impactors, in *Aerosol Measurement*, D. A. Lundgren, F. S. Harris, Jr., W. H. Marlow, M. Lippmann, W. E. Clark, and M. D. Durham, Eds., University Presses of Florida, Gainesville, FL, 1979, pp. 90–107.

Martens, A. E. and J. D. Keller, "An Instrument for Sizing and Counting Airborne Particles," *Am. Ind. Hyg. Assoc. J.*, **29,** 257 (1968).

McMurry, P. H. and D. Grosjean, "Photochemical Formation of Organic Aerosols: Growth Laws and Mechanisms," *Atmos. Environ.*, **19,** 1445 (1985).

McMurry, P. H. and J. C. Wilson, "Growth Laws for the Formation of Secondary Ambient Aerosols: Implications for Chemical Conversion Mechanisms," *Atmos. Environ.*, **16,** 121 (1982).

Mercer, T. T., *Aerosol Technology in Hazard Evaluation*, Academic Press, New York, 1973.

Middleton, W. E. K., *Vision Through the Atmosphere*, University of Toronto Press, Toronto, Ontario, Canada, 1952.

Miller, M. S., S. K. Friedlander, and G. M. Hidy, "A Chemical Element Balance for the Pasadena Aerosol," *J. Colloid Interface Sci.*, **39,** 165 (1972).

Millikan, R. A., "The General Law of a Small Spherical Body Through a Gas, and Its Bearing Upon the Nature of Molecular Reflection from Surfaces," *Phys. Rev.*, **22,** 1 (1923).

Moyers, J. L. and G. Colovos, "Discussions—Lead and Bromine Particle Size Distribution in the San Francisco Bay Area," *Atmos. Environ.*, **8,** 1339 (1974).

Moyers, J. L. and R. A. Duce, "Gaseous and Particulate Iodine in the Marine Atmosphere," *J. Geophys. Res.*, **77,** 5229 (1972a).

Moyers, J. L. and R. A. Duce, "Gaseous and Particulate Bromine in the Marine Atmosphere," *J. Geophys. Res.*, **77,** 5330 (1972b).

Muhlbaier, J. L. and R. L. Williams, Fireplaces, Furnaces, and Vehicles as Emission Sources of Paticulate Carbon, in *Particulate Carbon: Atmospheric Life Cycle*, G. T. Wolff and R. L. Klimisch, Eds., Plenum Press, New York, 1982, pp. 185–205.

Mulik, J., R. Puckett, D. Williams, and E. Sawicki, "Ion Chromatographic Analysis of Sulfate and Nitrate in Ambient Aerosols," *Anal. Lett.*, **9,** 653 (1976).

Mullin, J. B. and J. P. Riley, "The Spectrophotometric Determination of Nitrate in Natural Waters with Particular Reference to Sea Water," *Anal. Chim. Acta*, **12,** 464 (1955).

National Research Council, *Ozone and Other Photochemical Oxidants*, National Academy of Sciences Press, Washington, DC, 1977.

National Research Council, *Airborne Particles*, University Park Press, Baltimore, MD, 1979.

Natusch, D. F. S., J. R. Wallace, and C. A. Evans, Jr., "Toxic Trace Elements: Preferential Concentration in Respirable Particles," *Science*, **183,** 202 (1974).

Neligan, R. E. "Hydrocarbons in the Los Angeles Atmosphere," *Arch. Environ, Health*, **5,** 581 (1962).

Novakov, T., P. K. Mueller, A. E. Alcocer, and J. W. Otvos, "Chemical Composition of Pasadena Aerosol by Particle Size and Time of Day. III. Chemical States of Nitrogen and Sulfur by Photoelectron Spectroscopy," *J. Colloid Interface Sci.*, **39,** 225 (1972).

Novakov, T., S. G. Chang, and R. L. Dod, Application of ESCA to the Analysis of Atmospheric Particulates, in *Contemporary Topics in Analytical and Clinical Chemistry*, Vol. 1, D. M. Hercules, G. M. Hieftje, L. R. Snyder, and M. A. Evenson, Eds., Plenum Press, New York, 1977, pp. 249–286.

Nriagu, J. R., "Global Inventory of Natural and Anthropogenic Emissions of Trace Metals to the Atmosphere," *Nature*, **279,** 409 (1979).

O'Brien, R. J., J. R. Holmes, and A. H. Bockian, "Formation of Photochemical Aerosol from Hydrocarbons. Chemical Reactivity and Products," *Environ. Sci. Technol.*, **9,** 568 (1975a).

O'Brien, R. J., J. H. Crabtree, J. R. Holmes, M. C. Hoggan, and A. H. Bockian, "Formation of Photochemical Aerosol from Hydrocarbons. Atmospheric Analysis," *Environ. Sci. Technol.*, **9,** 577 (1975b).

Olmez, I. and G. E. Gordon, "Rare Earths: Atmospheric Signatures for Oil-Fired Power Plants and Refineries," *Science*, **229,** 966 (1985).

Ouimette, J. R. and R. C. Flagan, "The Extinction Coefficient of Multicomponent Aerosols," *Atmos. Environ.*, **16,** 2405 (1982).

Ouimette, J. R., R. C. Flagan, and A. R. Kelso, Chemical Species Contributions to Light Scattering by Aerosols at a Remote Arid Site, in *Atmospheric Aerosol: Source/Air Quality Relationships*, ACS Symposium Series, No. 167, E. S. Macias and P. K. Hopke, Eds., American Chemical Society, Washington, DC, 1981, pp. 125–156.

Ozkaynak, H., A. D. Schatz, G. D. Thurston, R. G. Isaacs, and R. B. Husar, "Relationships between Aerosol Extinction Coefficients Derived from Airport Visual Range Observations and Alternative Measures of Airborne Particle Mass,"*J. Air Pollut. Control Assoc.*, **35,** 1176 (1985).

Panter, R. and R. -D. Penzhorn, "Alkyl Sulfonic Acids in the Atmosphere," *Atmos. Environ.*, **14,** 149 (1980).

Patel, J. K., C. H. Kapedia, and D. B. Owen, *Handbook of Statistical Distributions*, Dekker, New York, 1976.

Phalen, R. F., "Evaluation of an Exploded-Wire Aerosol Generator for Use in Inhalation Studies," *Aerosol Sci.*, **3,** 395 (1972).

Phalen, R. F., *Inhalation Studies: Foundations and Techniques*, CRS Press, Boca Raton, FL, 1984.

Pitts, J. N., Jr., K. A. Van Cauwenberghe, D. Grosjean, J. P. Schmid, D. R. Fitz, W. L. Belser, Jr., G. B. Knudson, and P. M. Hynds, "Atmospheric Reactions of Polycyclic Aromatic Hydrocarbons: Facile Formation of Mutagenic Nitro Derivatives," *Science*, **202,** 515 (1978).

Post, J. E. and P. R. Buseck, "Quantitative Energy-Dispersive Analysis of Lead Halide Particles from the Phoenix Urban Aerosol," *Environ. Sci. Technol.*, **19,** 682 (1985).

Prager, M. J., E. R. Stephens, and W. E. Scott, "Aerosol Formation from Gaseous Air Pollutants," *Ind. Eng. Chem.*, **52,** 521 (1960).

Pratsinis, S., T. Novakov, E. C. Ellis, and S. K. Friedlander, "The Carbon Containing Component of the Los Angeles Aerosol: Source, Apportionment, and Contributions to the Visibility Budget," *J. Air Pollut. Control Assoc.*, **34,** 643 (1984).

Raabe, O. G., The Generation of Aerosols of Fine Particles, in *Fine Particles: Aerosol Generation, Measurement, Sampling, and Analysis*, B. Y. L. Liu, Ed., Academic Press, New York, 1976, pp. 57–110.

Raabe, O. G., Generation and Characterization of Power Plant Fly Ash, in *Generation of Aerosols and Facilities for Exposure Experiments*, K. Willeke, Ed., Ann Arbor Science Publishers, Ann Arbor, MI, 1980, pp. 215–234.

Raes, F. and A. Reineking, "A New Diffusion Battery Design for the Measurement of Sub-20 nm Aerosol Particles: The Diffusion Carrousel," *Atmos. Environ.*, **19,** 385 (1985).

Rahn, K. A., "The Mn/V Ratio as a Tracer of Large-Scale Sources of Pollution Aerosol for the Artic," *Atmos. Environ.*, **15,** 1457 (1981).

Rahn, K. A. and D. H. Lowenthal, "Elemental Tracers of Distant Regional Pollution Aerosols," *Science*, **223,** 132 (1984).

Rahn, K. A. and D. H. Lowenthal, "Reply to Comments of Thurston and Laird, 'Tracing Aerosol Pollution'," *Science*, **227,** 1407 (1985).

Robbins, R. C., R. D. Cadle, and D. L. Eckhardt, "The Conversion of Sodium Chloride to Hydrogen Chloride in the Atmosphere," *J. Meteorol.*, **16,** 53 (1959).

Rodes, C. E. and E. G. Evans, "Preliminary Assessment of 10 μm Particulate Sampling at Eight Locations in the United States," *Atmos. Environ.*, **19,** 293 (1985).

Rodes, C. E., D. M. Holland, L. J. Purdue, and K. A. Rehme, "A Field Comparison of PM_{10} Inlets at Four Locations," *J. Air Pollut. Control Assoc.*, **35,** 345 (1985).

Rood, M. J., T.V. Larson, D. S. Covert, and N. C. Ahlquist, "Measurement of Laboratory and Ambient Aerosols with Temperature and Humidity Controlled Nephelometry," *Atmos. Environ.*, **19,** 1181 (1985).

Rosen, H., A. D. A. Hansen, R. L. Dod, and T. Novakov, "Soot in Urban Atmospheres: Determination by an Optical Absorption Technique," *Science*, **208,** 741 (1980).

Ruby, M. G., "Visibility Measurement Methods: I Integrating Nephelometer," *J. Air Pollut. Control Assoc.*, **35,** 244 (1985).

Sanders, C. C., F. T. Cross, G. E. Dagle, and J. A. Mahaffey, Eds., *Pulmonary Toxicology of Respirable Particles*, CONF-791002, National Technical Information Service, Springfield, VA, 1980.

Sawicki, E., J. D. Mulik, and E. Wittgenstein, Eds., *Ion Chromatographic Analysis of Environmental Pollutants*, Ann Arbor Science Publishers, Ann Arbor, MI, 1978.

Saxena, P., C. Seigneur, and T. W. Peterson, "Modeling of Multiphase Atmospheric Aerosols," *Atmos. Environ.*, **17,** 1315 (1983).

Schneider, J. K. and R. B. Gagosian, "Particle Size Distribution of Lipids in Aerosols off the Coast of Peru," *J. Geophys. Res.*, **90,** 7889 (1985).

Schuetzle, D., A. L. Crittenden, and R. J. Charlson, "Application of Computer Controlled High Resolution Mass Spectrometry to the Analysis of Air Pollutants," *J. Air Pollut. Control Assoc.*, **23,** 704 (1973).

Schuetzle, D., D. Cronn, A. L. Crittenden, and R. J. Charlson, "Molecular Composition of Secondary Aerosol and Its Possible Origin," *Environ. Sci. Technol.*, **9,** 838 (1975).

Schwartz, W., *Chemical Characterization of Model Aerosols*, U.S. Environmental Protection Agency, Report No. EPA-650/3-74-011, August 1974.

Shah, J. J., J. G. Watson, Jr., J. A. Cooper, and J. J. Huntzicker, "Aerosol Chemical Composition and Light Scattering in Portland, Oregon: The Role of Carbon," *Atmos. Environ.*, **18,** 235 (1984).

Shah, J. J., T. J. Kneip, and J. M. Daisey, "Source Apportionment of Carbonaceous Aerosol in New York City by Multiple Linear Regression," *J. Air Pollut. Control Assoc.*, **35,** 541 (1985).

Shaw, R. W., Jr. and R. J. Paur, "Composition of Aerosol Particles Collected at Rural Sites in the Ohio River Valley," *Atmos. Environ.*, **17,** 2031 (1983).

Shaw, R. W., Jr., R. K. Stevens, J. Bowermaster, J. W. Tesch, and E. Tew, "Measurements of Atmospheric Nitrate and Nitric Acid: The Denuder Difference Experiment," *Atmos. Environ.* **16,** 845 (1982).

Shroeder, W. H. and P. Urone, "Formation of Nitrosyl Chloride from Salt Particles in Air," *Environ. Sci. Technol.*, **8,** 756 (1974).

Simoneit, B. R. T. and M. A. Mazurek, "Organic Matter of the Troposphere. II. Natural Background of Biogenic Lipid Matter in Aerosols Over the Rural Western United States," *Atmos. Environ.*, **16,** 2139 (1982).

Sinclair, D., in *Handbook on Aerosols*, U.S. Atomic Energy Commission, Washingtion, DC, 1950.

Sinclair, D., R. J. Countess, B. Y. H. Liu, and D. Y. H. Pui, Automatic Analysis of Submicron Aerosols, in *Aerosol Measurement*, D. A. Lundgren, F. S. Harris, Jr., W. H. Marlow, M. Lippmann, W. E. Clark, and M. D. Durham, Eds., University Presses of Florida, Gainesville, FL, 1979, pp. 544—563.

Skoog, D. A. and D. M. West, *Principles of Instrumental Analysis*, Saunders, Philadephia, 1980.

Sloane, C. S. and G. T. Wolff, "Prediction of Ambient Light Scattering Using a Physical Model Responsive to Relative Humidity: Validation with Measurements from Detroit," *Atmos. Environ.*, **19,** 669 (1985).

Spann, J. F. and C. B. Richardson, "Measurement of the Water Cycle in Mixed Ammonium Acid Sulfate Particles," *Atmos. Environ.*, **19,** 819 (1985).

Spicer, C. W. and P. M. Schumacher, "Particulate Nitrate: Laboratory and Field Studies of Major Sampling Interferences," *Atmos. Environ.*, **13,** 543 (1979).

Stelson, A. W. and J. H. Seinfeld, "Chemical Mass Accounting of Urban Aerosol," *Environ. Sci. Technol.*, **15,** 671 (1981).

Stelson, A. W. and J. H. Seinfeld, "Thermodynamic Prediction of the Water Activity, NH_4NO_3 Dissociation Constant, Density, and Refractive Index for the NH_4NO_3–$(NH_4)_2SO_4$–H_2O System at 25°C," *Atmos. Environ.*, **16,** 2507 (1982).

Stephens, E. R. and F. R. Burleson, "Analysis of the Atmosphere for Light Hydrocarbons," *J. Air Pollut. Control Assoc.*, **17,** 147 (1967).

Stephens, E. R. and F. R. Burleson, "Distribution of Light Hydrocarbons in Ambient Air," *J. Air Pollut. Control Assoc.*, **19,** 929 (1969).

Stephens, E. R. and M. A. Price, "Smog Aerosol: Infrared Spectra," *Science*, **168,** 1584 (1970).

Stephens, E. R. and M. A. Price, "Comparison of Synthetic and Smog Aerosols," *J. Colloid Interface Sci.*, **39,** 272 (1972).

Stevens, R. K., T. G. Dzubay, G. Russwurm, and D. Rickel, "Sampling and Analysis of Atmospheric Sulfates and Related Species," *Atmos. Environ.*, **12,** 55 (1978).

Stevens, R. K., T. G. Dzubay, R. W. Shaw, Jr., W. A. McClenny, C. W. Lewis, and W. E. Silson, "Characterization of the Aerosol in the Great Smoky Mountains," *Environ. Sci. Technol.*, **14,** 1491 (1980).

Stevens, R. K., T. G. Dzubay, C. W. Lewis, and A. P. Altshuller, "Discussions of Paper Entitled 'Non-Health Effects of Airborne Particulate Matter,'" *Atmos. Environ.*, **17,** 899, 903 (1983).

Stevens, R. K., T. G. Dzubay, C. W. Lewis, and R. W. Shaw, Jr., "Source Apportionment Methods Applied to the Determination of the Origin of Ambient Aerosols that Affect Visibility in Forested Areas," *Atmos. Environ.*, **18,** 261 (1984).

Sverdrup, G. M. and K. T. Whitby, "The Variation of the Aerosol Volume to Light-Scattering Coefficient." *Adv. Environ. Sci. Technol.*, **10,** 539 (1980).

Sverdrup, G. M., K. T. Whitby, and W. E. Clark, "Characterization of California Aerosols—III. Aerosol Size Distribution Measurements in the Mohave Desert," *Atmos. Environ.*, **9,** 483 (1975).

Tang, I. N., Deliquescence Properties and Particle Size Change of Hygroscopic Aerosols, in *Generation of Aerosols and Facilities for Exposure Experiments*, K. Willeke, Ed., Ann Arbor Science Publishers, Ann Arbor, MI, 1980a, Chap. 7, pp. 153–167.

Tang, I. N., "On the Equilibrium Partial Pressures of Nitric Acid and Ammonia in the Atmosphere," *Atmos. Environ.*, **14,** 819 (1980b).

Tani, B., S. Siegel, S. A. Johnson, and R. Kumar, "X-Ray Diffraction Investigation of Atmospheric Aerosols in the 0.3–1.0 μm Aerodynamic Size Range," *Atmos. Environ.*, **17,** 2277 (1983).

Tanner, R. L., "An Ambient Experimental Study of Phase Equilibrium in the Atmospheric System: Aerosol H^+, NH_4^+, SO_4^{2-}, NO_3^-–$NH_{3(g)}$, $HNO_{3(g)}$," *Atmos. Environ.*, **16,** 2935 (1982).

Taylor, D. D. and R. C. Flagan, "The Influence of Combustion Operation on Fine Particles from Coal Combustion," *Aerosol Sci. Technol.*, **1,** 103 (1982).

Taylor, S. R., "Abundance of Chemical Elements in the Continental Crust: A New Table," *Geochim. Cosmochim. Acta*, **28,** 1273 (1964).

Thomsen, J., "Enrichment of Na^+, Mn^{2+}, Fe^{3+}, and Cu^{2+} at Water Surfaces Covered by Artificial Multilayer Films," *J. Phys. Chem.*, **87,** 4974 (1983).

Thurston, G. D. and N. M. Laird, "Tracing Aerosol Pollution," *Science*, **227,** 1406 (1985).

Toossi, R. and T. Novakov, "The Lifetime of Aerosols in Ambient Air: Consideration of the Effects of Surfactants and Chemical Reactions," *Atmos. Environ.*, **19,** 127 (1985).

Trijonis, J., "Visibility in California," *J. Air Pollut. Control Assoc.*, **32,** 165 (1982).

Twomey, S., *Atmospheric Aerosols*, Elsevier Scientific Publishing, New York, 1977.

Van de Hulst, H. C., *Light Scattering by Small Particles*, Wiley, New York, 1957.

Van Vaeck, L., K. Van Cauwenberghe, and J. Janssens, "The Gas–Particle Distribution of Organic Aerosol Constituents: Measurement of the Volatilization Artifact in Hi-Vol Cascade Impactor Sampling," *Atmos. Environ.*, **18,** 417 (1984).

Waggoner, A. P., "Reply to Discussions of Paper Entitled 'Non-Health Effects of Airborne Particulate Matter,'" *Atmos. Environ.*, **17,** 900, 903 (1983).

Waggoner, A. P. and R. J. Charlson, Measurements of Aerosol Optical Parameters, in *Fine Particles: Aerosol Generation, Measurement, Sampling, and Analysis*, B. Y. H. Liu, Ed., Academic Press, New York, 1976, pp. 511–533.

Waggoner, A. P., R. E. Weiss, N. C. Ahlquist, D. S. Covert, S. Will, and R. J. Charlson, "Optical Characteristics of Atmospheric Aerosols," *Atmos. Environ.*, **15,** 1891 (1981).

Waggoner, A. P., R. E. Weiss, and N. C. Ahlquist, "The Color of Denver Haze," *Atmos. Environ.*, **17,** 2081 (1983).

Weatherburn, M. W. "Phenol–Hypochlorite Reaction for Determination of Ammonia," *Anal. Chem.*, **39,** 971 (1967).

Webber, J. S., V. A. Dutkiewicz, and L. Husain, "Identification of Submicrometer Coal Fly Ash in a High-Sulfate Episode at Whiteface Mountain, New York," *Atmos. Environ.*, **19,** 285 (1985).

Weiss, R. E. and A. P. Waggoner, Optical Measurements of Airborne Soot in Urban, Rural, and Remote Locations, in *Particulate Carbon: Atmospheric Life Cycle*, G. T. Wolff and R. L. Klimisch, Eds., Plenum Press, New York, 1982, pp. 317–325.

Weiss, R. E., A. P. Waggoner, R. J. Charlson, and N. C. Ahlquist, "Sulfate Aerosol: Its Geographical Extent in the Midwestern and Southern United States," *Science*, **195.** 979 (1977).

Weiss, R. E., T. V. Larson, and A. P. Waggoner, "In Situ Rapid-Response Measurement of H_2SO_4/$(NH_4)_2SO_4$ Aerosols in Rural Virginia," *Environ. Sci. Technol.*, **16,** 525 (1982).

Wesolowski, J. J., A. E. Alcocer, and B. R. Appel, "The Validation of the Lundgren Impactor," *Adv. Environ. Sci. Technol.*, **10,** 125 (1980).

Whitby, K. T. and W. E. Clark, "Electric Aerosol Particle Counting and Size Distribution for the 0.015 to 1 μm Size Range," *Tellus*, **XVIII,** 573 (1966).

Whitby, K. T. and G. M. Sverdrup, "California Aerosols: Their Physical and Chemical Characteristics," *Adv. Environ. Sci. Technol.*, **10,** 477 (1980).

Whitby, K. T. and K. Willeke, Single Particle Optical Counters: Principles and Field Use, in *Aerosol Measurement*, D. A. Lundgren, F. S. Harris, Jr., W. H. Marlow, M. Lippman, W. E. Clark, and M. D. Durham, Eds., University Presses of Florida, Gainesville, FL, 1979, pp. 145–182.

Whitby, K. T., B. Y. H. Liu, R. B. Husar, and N. J. Barsic, "The Minnesota Aerosol Analyzing System Used in the Los Angeles Smog Project," *J. Colloid Interface Sci.*, **39,** 136 (1972a).

Whitby, K. T., R. B. Husar, and B. Y. H. Liu, "The Aerosol Size Distribution of Los Angeles Smog," *J. Colloid Interface Sci.*, **39,** 177 (1972b).

White, W. H. and P. T. Roberts, "On the Nature and Origins of Visibility-Reducing Aerosols in the Los Angeles Air Basin," *Atmos. Environ.*, **11,** 803 (1977).

Willeke, K., Ed., *Generation of Aerosols*, Ann Arbor Science Publishers, Ann Arbor, MI, 1980.

Witz, S., "Effect of Environmental Factors on Filter Alkalinity and Artifact Formation," *Environ. Sci. Technol.*, **19,** 831 (1985).

Wolff, G. T., M. A. Ferman, N. A. Kelley, D. P. Stroup, and M. S. Ruthkosky, "The Relationships Between the Chemical Composition of Fine Particles and Visibility in the Detroit Metropolitan Area," *J. Air Pollut. Control Assoc.*, **32,** 1216 (1982).

Wolff, G. T., P. J. Groblicki, S. H. Cadle, and R. J. Countess, Particulate Carbon at Various Locations in the United States, in *Particulate Carbon: Atmospheric Life Cycles*, G. T. Wolff and R. L. Klimisch, Eds., Plenum Press, New York, 1982, pp. 297–315.

Wolff, G. T. and R. L. Klimisch, Eds., *Particulate Carbon: Atmospheric Life Cycle*, Plenum Press, New York, 1982.

Wright, B. M., "A New Dust-Feed Mechanism," *J. Sci. Instrum.*, **27,** 12 (1950).

Zafiriou, O. C., R. B. Gagosian, E. T. Peltzer, J. B. Alford, and T. Loder, "Air-to-Sea Fluxes of Lipids at Enewetak Atoll," *J. Geophys. Res.*, **90,** 2409 (1985).

13 Chemistry and Mutagenic Activity of Airborne Polycyclic Aromatic Hydrocarbons and Their Derivatives

In Chapter 12 we considered the physical and chemical properties of airborne particles and their measurement, sources, ambient levels, and sinks. However, because of the carcinogenic nature of several representative molecules in their class and their ubiquitous presence in respirable ambient particles (as well as our water and soil environments), we have reserved for this chapter a discussion of the role of polycyclic aromatic hydrocarbons (PAH) in tropospheric chemistry. Although a variety of polycyclic aromatic compounds containing nitrogen, oxygen, and sulfur heteroatoms have also been identified in respirable particles, from both primary sources and ambient air, we focus on PAH and their derivatives, because much more is known about their atmospheric sources, reactions, and sinks.

We cannot attempt here a *comprehensive* treatment of this important class of environmental carcinogens. This would include: formation of PAH in combustion processes, analytical techniques used for identification and quantification of PAH, emission sources, distribution in the air environment, chemical and physical transformations in the atmosphere, deposition on the earth, and finally their transport (and associated transformations) through air, water, and soil interfaces. Accompanying these processes of interest to engineers and physical scientists would be discussions of the biological properties of "model" carcinogenic PAH [e.g., benzo(a)pyrene (BaP)] and their derivatives. Topics would include the biochemical mechanisms by which they are metabolized *in vitro* and *in vivo* into reactive intermediates in bacterial and mammalian systems and a consideration of risk assessment evaluations of the potential impacts of biologically active PAH and associated nitroarenes on certain occupational groups and the general public.

These topics have been treated in monographs, reviews, and assessments, including the "classics," *Polycyclic Hydrocarbons* by Clar (1964) and the NAS report on *Particulate Polycyclic Organic Matter* (1972). The latter gives a comprehensive review and critique of the status of the subject up to about 1971. Subsequent assessments include *Health and Ecological Assessment of Polynuclear Aromatic Hydrocarbons* (Santodonato et al., 1981) and the NAS document *Polycyclic Aromatic Hydrocarbons: Evaluation of Sources and Effects* (1983); the latter has a literature cutoff date of June 1982.

Recent volumes with contributed chapters include *Handbook of Polycyclic Aromatic Hydrocarbons* (Vols. 1 and 2, Bjørseth, 1983; Bjorseth and Ramdahl, 1985), which deals with the more physical–chemical aspects and sources of PAH, and *Environmental Carcinogens: Polycyclic Aromatic Hydrocarbons* (Grimmer, 1983), which focuses more on the biological aspects of PAH. Volumes on nitroarenes include *Nitrated Polycyclic Aromatic Hydrocarbons* (White, 1985) and *Toxicity of Nitroaromatic Compounds* (Rickert, 1985). Also important is the definitive chapter by Hoffman and Wynder (1977), "Organic Particulate Pollutants—Chemical Analysis and Bioassays for Carcinogenicity". The analysis of PAH and their derivatives is considered in detail in *Analytical Chemistry of Polycyclic Aromatic Compounds* by Lee, Novotny, and Bartle (1981) while the characterization of microbial mutagens in complex samples has been addressed by Alfheim, Bjørseth, and Möller (1984b).

Timely and important are the International Agency for Research on Cancer (IARC) Monographs on the *Evaluation of the Carcinogenic Risk of Chemicals to Humans*, especially Volumes 32, 33, and 34 which deal specifically with PAH and some nitroarenes (1983, 1984, and 1985).

A number of symposia have been devoted to the analytical, physical, chemical, and biological aspects of the mutagenicity and carcinogenicity of air pollutants, including the Symposium on *Biological Tests in the Evaluation of Mutagenicity and Carcinogenicity of Air Pollutants with Special Reference to Motor Exhausts and Coal Combustion Products* (Environmental Health Perspectives, **47,** 1983); *Short-Term Bioassays in the Analysis of Complex Environmental Mixtures*, Volumes I, II, and III (Waters et al., 1979, 1981, and 1983); Proceedings of the 1984 Workshop on Genotoxic Air Pollutants (*Environment International Journal*, 1985); and the Battelle series of International Symposia on *Polynuclear Aromatic Hydrocarbons*.

13-A. HISTORICAL

In the late 19th century, unusually high rates of skin cancer were reported for workers in the paraffin refinery, shale oil, and coal tar industries. Subsequently, in the period 1915–1918 Japanese scientists showed that repeatedly painting the ears of rabbits with coal tar induced tumors, and similar treatments with mice produced malignant skin cancers.

The fascinating search for the chemical carcinogen(s) responsible for this activity of coal tar and pitch is documented by Phillips (1983) in "Fifty Years of Benzo(a)pyrene," a commentary in which he describes the pioneering research of Kennaway and his associates at the Institute of Cancer Research in London. This led to the isolation in 1931 of the *coal tar carcinogen*, its subsequent synthesis, and, in 1933, its identification as a new compound, benzo(a)pyrene (BaP), (**I**), that proved to be a very strong animal carcinogen (Cook et al., 1933). Along the way they showed, in 1930, that the related PAH, dibenz(a,h)anthracene (DBA), (**II**), and its 3-methyl derivative produced tumors in animals. This was the first example of *pure* chemical compounds demonstrating carcinogenic activity (Kennaway and Hieger, 1930).

The structure and accepted numbering systems for BaP and DBA are shown below; earlier systems differed, a possible source of confusion in the literature:

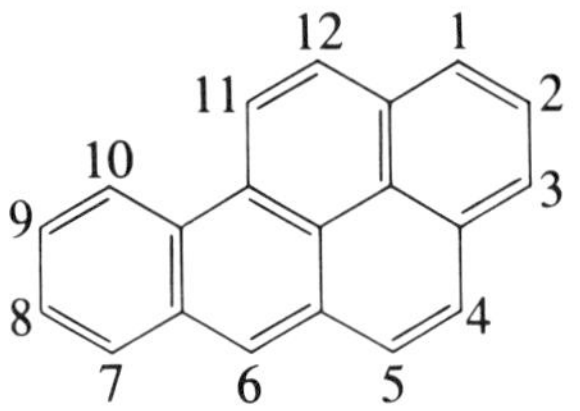

Benzo(a)pyrene
(BaP)
(I)

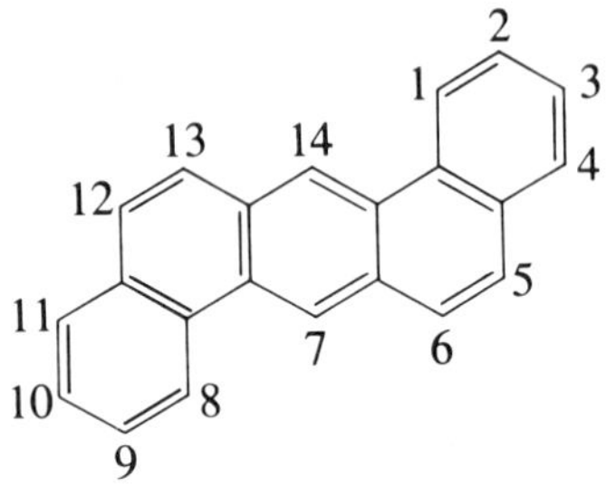

Dibenz(a,h)anthracene
(DBA)
(II)

In addition to the occupational hazard of coal tar, potential environmental hazards were suggested by the discoveries that organic extracts of soot are carcinogenic in experimental animals (Passey, 1922), as are extracts of ambient particulate matter (Leiter et al., 1942). Subsequently, this biological activity was observed with extracts of respirable ambient particulates collected from Los Angeles photochemical smog (Kotin et al., 1954), and then from major centers throughout the world.

These observations are related to the earlier studies on the carcinogenicity of coal tar extracts because many carcinogenic PAH are present in both industrial and ambient air environments. Thus in 1949 BaP was identified in domestic soot (Goulden and Tipler, 1949), while in 1952 it was found in ambient particles collected at ten stations throughout Great Britain (Waller, 1952).

By 1970, BaP and related carcinogenic PAH were recognized as being distributed throughout the world in respirable ambient urban aerosols. Furthermore, they were found in combustion-generated respirable particles collected from such primary sources as motor vehicle exhaust, smoke from residential wood combustion, and fly ash from coal-fired electric generating plants. Associated with these PAH were other particulate polycyclic carcinogens, including their nitrogen analogues the aza-arenes (e.g., carbazoles and indoles) and the benzacridines and dibenzacridines (Sawicki et al., 1965; Sawicki, 1967).

Importantly, the organic extracts of ambient and primary combustion-generated respirable particles exhibited a carcinogenicity in animals significantly greater than could be accounted for by the amounts of *known* carcinogenic polycyclic aromatic compounds determined analytically to be present in the samples; that is, there was an *excess carcinogenicity* (Kotin et al., 1954; Hueper et al., 1962; Epstein et al., 1966). This phenomenon was also observed in studies of the effects of auto exhaust condensate on the lungs of Syrian golden hamsters (Mohr et al., 1976).

Finally, in the early 1970s, benzene extracts of airborne particles collected in the Los Angeles area were reported to have 100–1000 times the cell transformation activity of that which could be attributed to their BaP contents. Indeed, the *polar* (methanol) fraction of these extracts, which contained only about 3% of the total BaP in the particulate sample, displayed an activity equal to the *neutral* benzene extract which contained the remaining 97% of the BaP (Gordon et al., 1973).

Thus by the early 1970s, it was clear that, along with such well-recognized carcinogens as BaP, other as yet unknown chemicals with strong biological activities must also be present in organic extracts of ambient particles and primary particulate organic matter (POM). A key question became: What are their structures, sources, and sinks?

On the basis of these types of studies and a growing concern that community air pollution might contribute to the rising rate of lung cancer, the U.S. National Academy of Sciences established in 1970 a broadly based Panel on Polycyclic Organic Matter. In 1972 this Panel issued their report (NAS, 1972) which attempted

> To interpret, evaluate, and reconcile the immense amount of information available, especially that concerning the carcinogenic effects of POM. (NAS 1972)

While no definite conclusions were drawn by the Panel as to the contributions of ambient POM to an urban–rural difference in lung cancer—and this remains a highly controversial subject—the NAS document is a useful, critical evaluation of the status of the chemical and biological aspects of POM up to about 1971 as well as a review of the pertinent earlier primary literature.

Despite the pronounced scientific and societal interest in the subject [e.g., Section 122a of the Clean Air Act of 1977 (see Section 1-E-2)], progress in isolating and identifying the chemical species in airborne POM responsible for the *excess* biological activity, and their sources, reactions, and sinks, remained relatively slow. However, a dramatic change occurred in the mid-1970s with the introduction of a very sensitive, short-term bioassay for chemicals that are bacterial mutagens (Ames et al., 1973, 1975). The Ames *Salmonella* mutagenicity assay was almost immediately used to test the activities of pure compounds (McCann et al., 1975) and of extracts of complex environmental mixtures.

For example, it was found that organic extracts of fine particles collected from ambient air contained not only the expected promutagens such as BaP (Tokiwa et al., 1976, 1977) but also *direct-acting* mutagens (Pitts et al., 1977; Talcott and Wei, 1977). Subsequently, extracts of respirable particles collected from primary combustion sources such as diesel engines (Huisingh et al., 1979), coal-fired power plants (Chrisp et al., 1978), and wood-burning stoves (Löfroth, 1978; Rudling et al., 1982) were also found to be directly active. Concurrently, certain PAH were shown to react with near ambient levels of NO_2 + HNO_3 and with O_3 in synthetic atmospheres, to form directly mutagenic nitro-PAH and OXY-PAH. It was suggested that such reactions might also occur in ambient air and/or on filters during the collection of airborne POM (Pitts et al., 1978).

These were interesting observations because, in contrast to promutagens which require metabolic activation (symbolized as +S9) to form reactive intermediates that then attack the DNA of the bacteria, the activity of a direct mutagen in the Ames test is expressed without the need for addition of mammalian enzymes (symbolized as −S9). Thus, although the chemical structures of these direct mutagens in ambient POM were unknown in 1977–1978, their importance in contributing to the bacterial mutagenicity, and possible animal carcinogenicity, of POM samples was clear. Subsequently, in several laboratories the Ames test was coordinated with conventional chemical analytical procedures. Such *activity-directed* chemical analyses provided a relatively rapid and inexpensive (compared to animal testing) means of following biologically active PAH and their derivatives, from their presence in complex mixtures through their extraction, fractionation, and ultimate isolation and identification.

In the following sections we summarize some of the physical properties of airborne PAH and discuss their sources, ambient levels, atmospheric transformations, and sinks. We also consider the Ames *Salmonella typhimurium* assay from a chemist's viewpoint and describe its utility in activity-directed chemical analyses to identify the mutagenic species present in primary combustion-generated POM and respirable ambient particles.

13-B. SELECTED PROPERTIES OF SOME ATMOSPHERICALLY RELEVANT POLYCYCLIC AROMATIC HYDROCARBONS

13-B-1. Formation, Structures, and Nomenclature

Polycyclic aromatic hydrocarbons and their heteroatom analogues are formed during the incomplete combustion of organic matter, for example, coal, oil, wood, gasoline, and diesel fuel. Major sources of PAH in developed countries include residential heating by coal and wood, open burning (including forest and agricultural), coke and aluminum production, and motor vehicle exhaust. For perspective, in the United States, PAH emissions from residential wood and coal combustion are estimated at ~700 tons yr^{-1} versus ~1 ton yr^{-1} from coal-fired power plants (Bjørseth and Ramdahl, 1985). The occurrence and surveillance of PAH are discussed by Baum (1978).

The mechanism of formation of PAH (Badger, 1962) involves the production of reactive free radicals by pyrolysis (at ~500–800°C) of fuel hydrocarbons in the chemically reducing zone of a flame burning with an insufficient supply of oxygen. The C_2 fragments, as well as C_1 and higher radicals, combine rapidly in the reducing atmosphere to form partially condensed aromatic molecules. On cooling the reaction mixture these PAH condense from the vapor phase onto co-existing particulate substrates, with a product distribution that generally reflects their thermodynamic stabilities in the oxygen-deficient flame.

Their structures can be built up by fusing several benzene ring systems with each other or with cyclopentadiene. A number of Kekulé's resonance bond structures are possible for a given PAH. The *Fries Rule* states that "the most stable form of a polynuclear hydrocarbon is that in which the maximum number of rings have the benzenoid arrangement of three double bonds" (Fieser and Fieser, 1956). This is illustrated for naphthalene, in which the symmetrical structure (**I**) with two benzenoid rings is more stable than the two asymmetrical, equivalent structures (**II**) and (**IIa**). Thus the latter each has one benzenoid and one quinonoid ring; since the quinonoid system is more reactive than the benzenoid system, structure (**I**) would be expected to be the more stable.

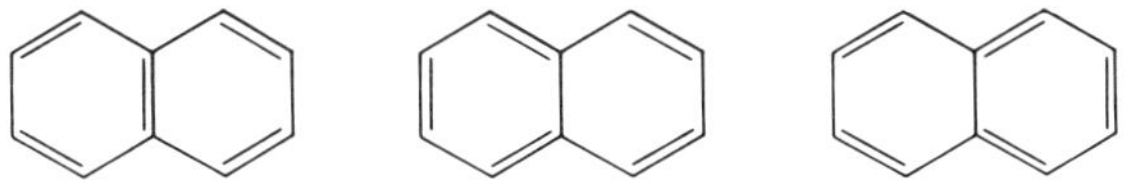

(**I**) (**II**) (**IIa**)

Trivial names are used for some of the simple PAH, for example, anthracene, phenanthrene, pyrene, fluoranthene, and perylene. More complicated structures are defined by their substitution on the basic skeleton, for example, by benzo-, dibenzo-, or naptho- groups. Note that not all PAH formulated this way are fully aromatic, conjugated structures. Some may contain a second hydrogen at a particular carbon position; it is called the *indicated hydrogen* (e.g., 9*H*-fluorene).

The nomenclature we use is that employed in the NAS documents (1972, 1983) and described in detail by Patterson et al. (1960) in *The Ring Index*. In general, to find the correct name of a given PAH one takes the following steps:

1. Define the largest molecular structure for which a trivial name exists in the IUPAC list (International Union of Pure and Applied Chemistry, 1958).
2. Orient this ring system in such a way that:
   (a) The maximum number of rings lie in a horizontal row.
   (b) Draw horizontal and vertical axes through the center of the horizontal row and orient the molecule so that the larger number of rings is located in the upper right quadrant.
3. Number the carbons in a clockwise fashion, starting with the carbon atom that is not part of another ring and is in the most counterclockwise position of the uppermost ring farthest to the right; carbon atoms common to two or more rings are not numbered.
4. Letter the faces of the rings in alphabetical order beginning with "a" for the side between carbon atoms 1 and 2, and continue clockwise around the molecule; ring faces common to two rings are *not* lettered.

An example is fluoranthene:

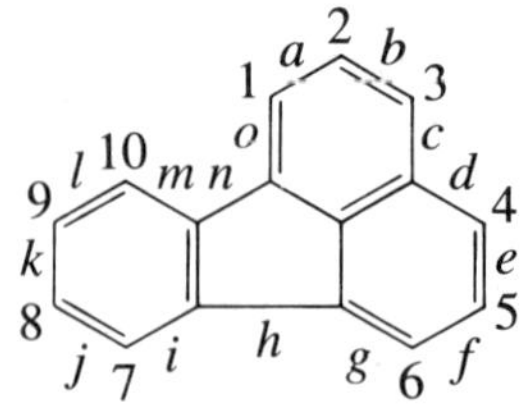

Note that substitution is defined by the bond(s) involved in the fusion. For example, benzo(k)fluoranthene is

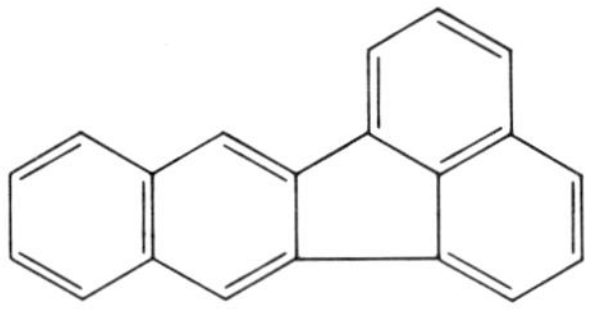

The numbering in the final molecule can be different from the one used in the original compound, since the addition of more rings may change its orientation in the frame of reference. For example, consider anthracene (**I**), 9,10-dimethylanthracene (**II**), and 7,12-dimethylbenz(a)anthracene (**III**):

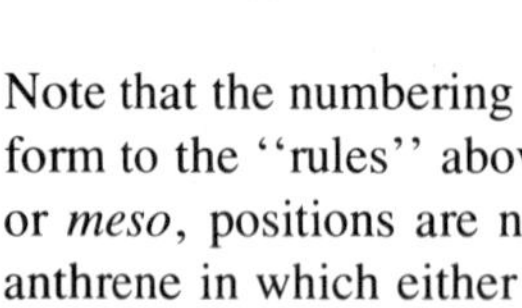

(I)

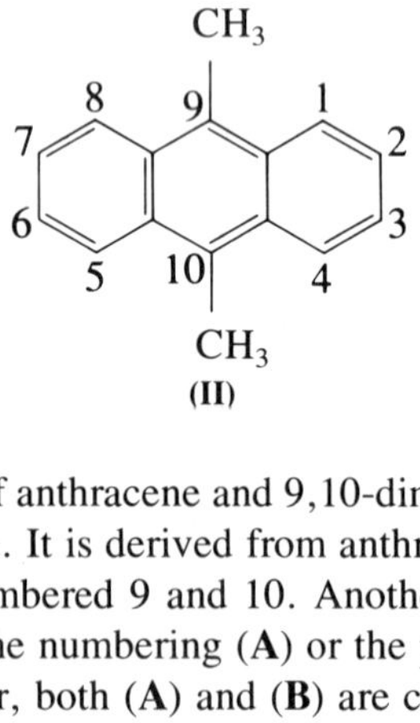

(II)

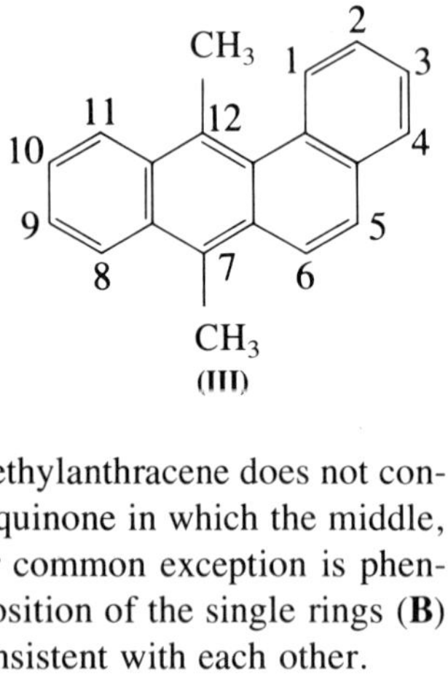

(III)

Note that the numbering of anthracene and 9,10-dimethylanthracene does not conform to the ''rules'' above. It is derived from anthraquinone in which the middle, or *meso*, positions are numbered 9 and 10. Another common exception is phenanthrene in which either the numbering (**A**) or the position of the single rings (**B**) violates the rules; however, both (**A**) and (**B**) are consistent with each other.

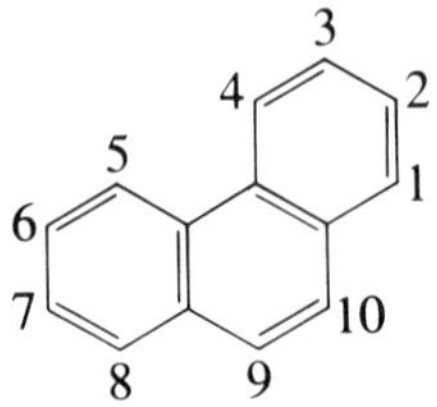

(A)

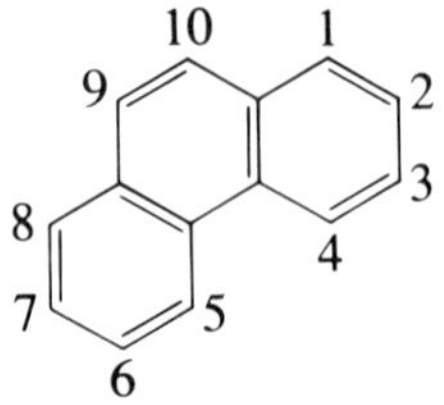

(B)

An indicated hydrogen should be mentioned by carbon number, even when it is further oxidized to a carbonyl group; consider, for example, 7*H*-benz(de)-anthracene (**I**) and 7*H*-benz(de)anthracen-7-one (**II**) (benzanthrone).

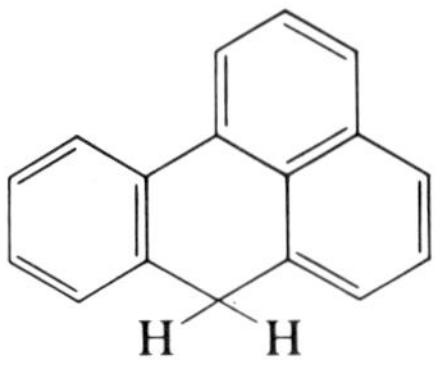

(I)

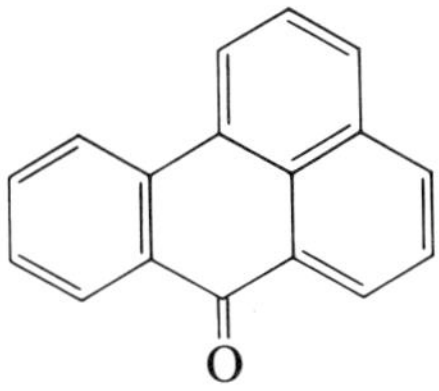

(II)

The nomenclature of the aza-heterocyclic polyaromatic hydrocarbons (containing a ring nitrogen) is similar; for example, the structure of the carcinogen dibenz(a,h)acridine is

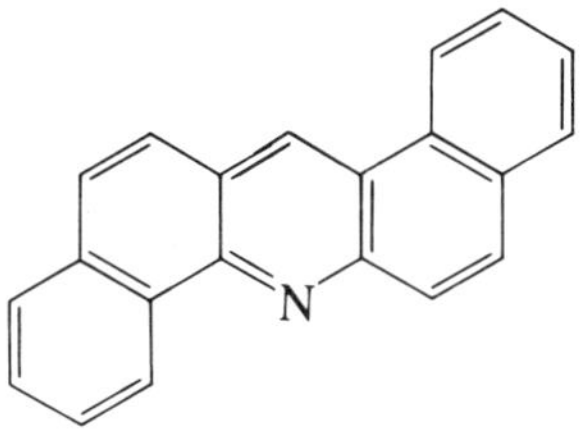

The PAH observed in the atmosphere range from bicyclic species such as naphthalene, present largely in the gas phase, to PAH containing seven or more fused rings, such as coronene, which are present solely adsorbed on particles. Table 13.1 shows the structures, numbering systems, acronyms, molecular weights, and carcinogenic activities (NAS, 1983) of some representative airborne PAH. More extensive compilations of these properties of PAH and PAH derivatives or analogs with nitrogen, oxygen, and sulfur atoms are given in Appendixes 1–5 of *Analytical Chemistry of Polycyclic Aromatic Compounds* (Lee, et al., 1981) and in the NAS reports (1972, 1983).

13-B-2. Vapor Pressure and Distribution Between Gaseous and Solid States

An obviously importanl physical property determining thc distribution of PAH in the air–water–soil environment is the vapor pressure. Table 13.2 gives the vapor pressures at 25°C of some representative PAH ranging from naphthalene to the carcinogen benzo(a)pyrene (BaP); biphenyl is also included as a reference. Most of the values are from Sonnefeld et al. (1983), who also give equations describing their temperature dependencies between 283 and 323°K.

The range in vapor pressure of a factor of $\sim 10^7$ in Table 13.2 is reflected in the fact that, at equilibrium in ambient air at 25°C, naphthalene exists virtually

TABLE 13.1. Structures, Numbering, Acronyms, Molecular Weights, and Carcinogenic Activities of Some Environmentally Relevant PAH

Structural Formula	Name	Molecular Weight	Carcinogenic Activity
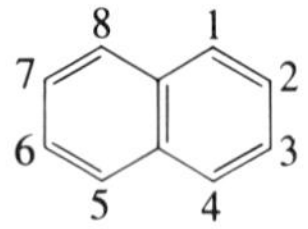	Naphthalene NAP	128.0626	0
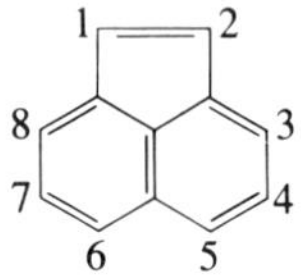	Acenaphthylene ACE	152.0626	0
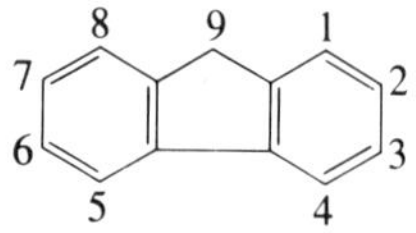	9*H*-Fluorene FLN	166.0783	0
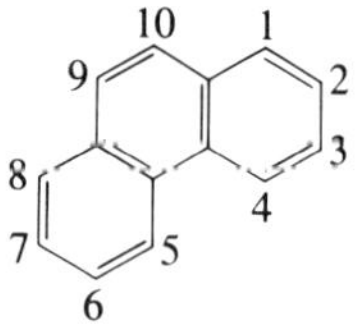	Phenanthrene PHE	178.0783	0
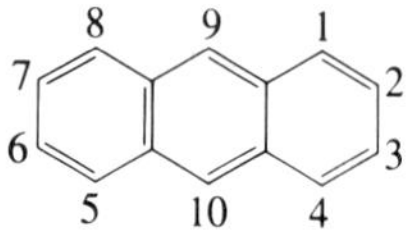	Anthracene ANT	178.0783	0
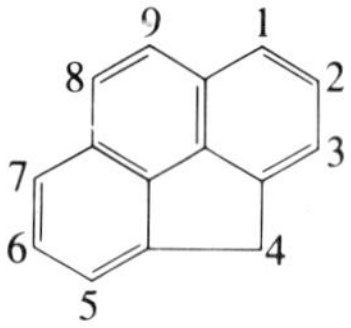	4*H*-Cyclopenta(def)-phenanthrene CPP	190.0783	NA[a]
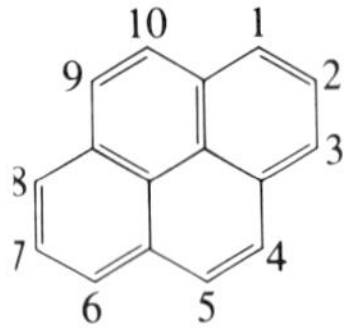	Pyrene PY	202.0783	0

TABLE 13.1. (*Continued*)

Structural Formula	Name	Molecular Weight	Carcinogenic Activity
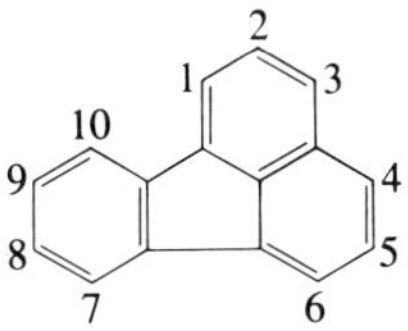	Fluoranthene FL	202.0783	+
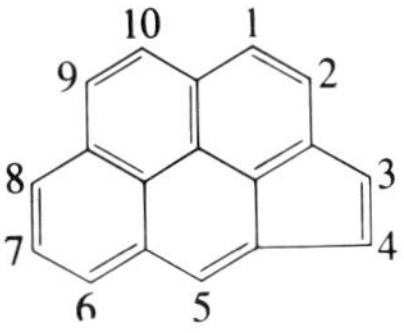	Cyclopenta(cd)pyrene CPY	226.0783	+
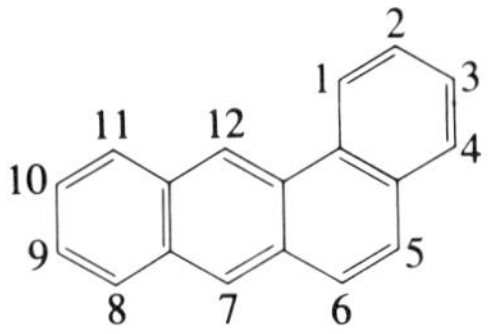	Benz(a)anthracene BaA	228.0939	+
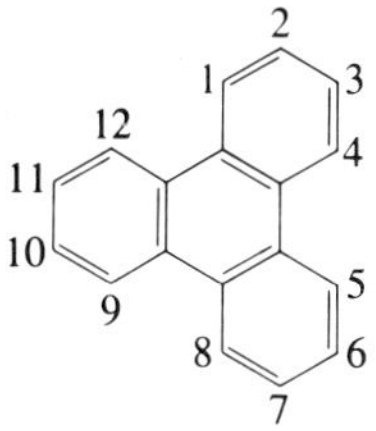	Triphenylene TRI	228.0939	0
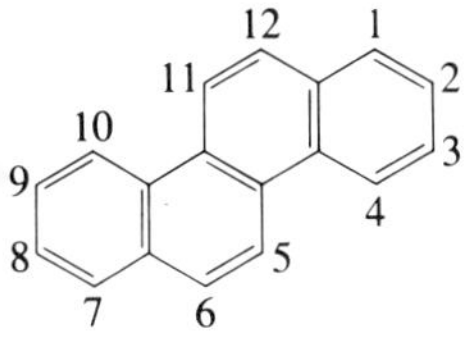	Chrysene CHR	228.0939	0/+
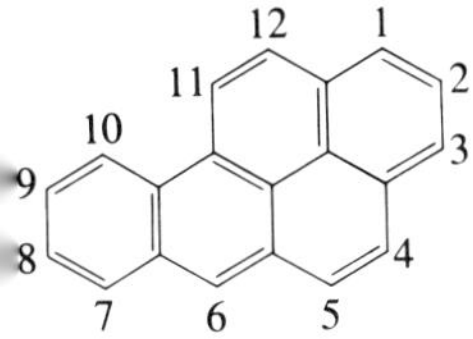	Benzo(a)pyrene BaP	252.0939	++

TABLE 13.1. (*Continued*)

Structural Formula	Name	Molecular Weight	Carcinogenic Activity
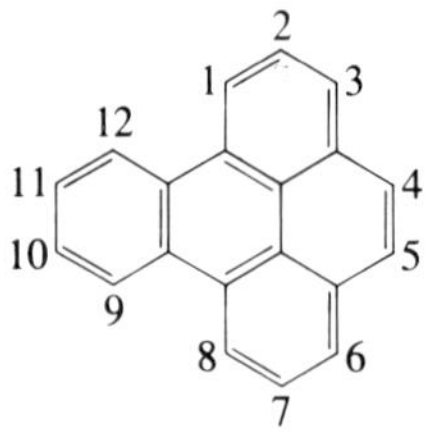	Benzo(e)pyrene BeP	252.0939	0/+
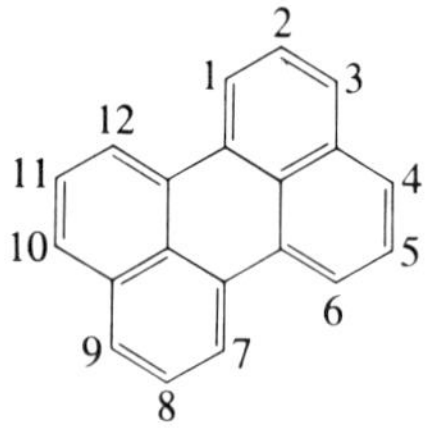	Perylene PER	252.0939	0
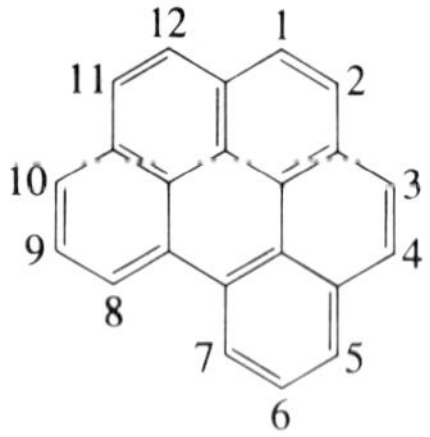	Benzo(ghi)perylene BghiP	276.0939	+
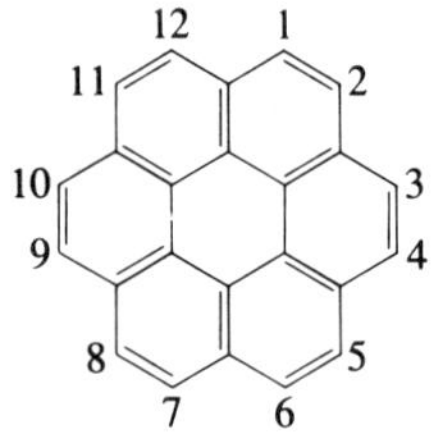	Coronene COR	300.0939	0/+
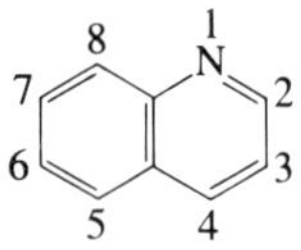	Quinoline	129.0578	+
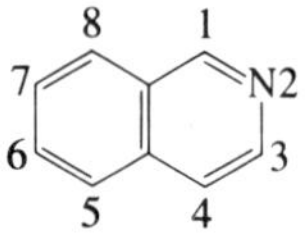	Isoquinoline	129.0578	0

TABLE 13.1. (*Continued*)

Structural Formula	Name	Molecular Weight	Carcinogenic Activity
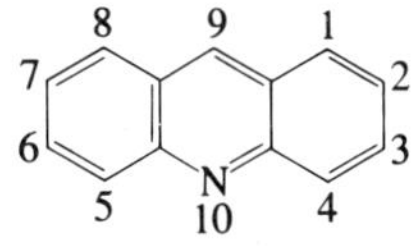	Acridine	179.0735	0
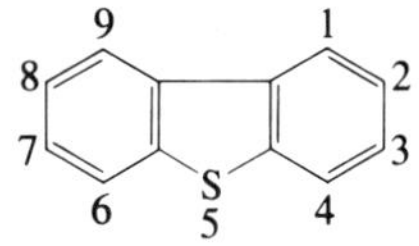	Dibenzothiophene	184.0347	0

Source: Adapted from NAS, 1983.
[a]NA = not available.

TABLE 13.2. Vapor Pressures at 25°C of Some Representative PAH and Biphenyl

Compound	Vapor Prcssurc	
	Pascals	mm Hg
Naphthalene	10.4 ± 0.2	7.8×10^{-2}
Naphthalene-d_8	10.4 ± 0.1	7.8×10^{-2}
Acenaphthylene	$(8.9 \pm 0.2) \times 10^{-1}$	6.7×10^{-3}
Acenaphthene	$(2.87 \pm 0.09) \times 10^{-1}$	2.15×10^{-3}
Fluorene	$(8.0 \pm 0.2) \times 10^{-2}$	6.0×10^{-4}
Phenanthrene	$(1.61 \pm 0.04) \times 10^{-2}$	1.20×10^{-4}
Phenanthrene-d_{10}	$(1.92 \pm 0.5) \times 10^{-2}$	1.44×10^{-4}
Anthracene	$(8.0 \pm 0.2) \times 10^{-4}$	6.0×10^{-6}
Fluoranthene	$(1.23 \pm 0.07) \times 10^{-3}$	9.2×10^{-6}
Pyrene	$(6.0 \pm 0.2) \times 10^{-4}$	4.5×10^{-6}
Benz(a)anthracene	$(2.8 \pm 0.1) \times 10^{-5}$	2.1×10^{-7}
Chrysene[a]	8.5×10^{-7}	6.4×10^{-9}
Benzo(a)pyrene[a]	7.5×10^{-7}	5.6×10^{-9}
Biphenyl[b]	1.19 ± 0.03	8.9×10^{-3}

Source: Sonnefeld et al., 1983, except as indicated.
[a]From Yamasaki et al. (1984).
[b]From Burkhard et al. (1984).

100% in the gas phase, while BaP and other PAH with five and six rings are predominately adsorbed on particulate matter. The intermediate PAH are distributed in both phases. Note, however, that their vapor pressures can be significantly reduced by their adsorption on various types of substrates. This phenomenon, as well as the important effects of temperature, and the initial amounts of PAH present on the surface of the sample being studied are described by Lao and Thomas (1980, and references therein); they are treated theoretically for coal fly ash by Natusch and Tomkins (1978).

The total concentrations of three- or four-ring PAH in ambient air or in vehicle exhaust sampled in dilution tunnels reflect this gas–solid distribution; they cannot be determined accurately by collecting them solely on hi-vol filters. For example, the partitioning between the gas and particulate phase of several environmentally relevant PAH emitted in the exhaust from a gasoline-fueled, stratified-charge engine is shown in Table 13.3 (Schuetzle, 1983). Distribution factors for PAH collected from ambient air are discussed in Section 13-D-2.

Additionally, there are *physical* artifacts. Thus the pressure drop behind a hi-vol filter or cascade impactor contributes to volatilization into the gas phase of the three- and four-ring compounds, to a degree reflecting their vapor pressures. The magnitude of this *blow off* artifact depends on a number of factors including sampling temperature and the volume of air samples (Commins, 1962; Rondia, 1965; Pupp et al., 1974; Lao and Thomas, 1980; Van Vaeck et al., 1984, and references therein).

13-B-3 Solubilities

As seen in Table 13.4, PAH generally have very low solubilities in pure water (Mackay and Shiu, 1977). However, their oxidation to more polar species (e.g., acids, phenols, ketones, etc.) will greatly enhance their solubilities in aqueous

TABLE 13.3. Partitioning Between the Particulate and Gas Phase of Some PAH in the Diluted Exhaust from a Gasoline-Fueled Stratified-Charge Engine[a]

PAH	Partition Coefficient[b]
Anthracene + phenanthrene	0.14–0.27
Fluoranthene	0.81–1.13
Pyrene	0.60–0.91
Benz(a)anthracene + isomers	1.05–1.40
Benzo(a)pyrene	15–27
Coronene	20–32

Source: Schuetzle, 1983.

[a]Collection conditions: 37°C, 20% RH, and exhaust diluted 15:1.

[b]The partition coefficient is the ratio of the masses in the particulate/gas phase.

TABLE 13.4. Solubilities of PAH in Water at 25°C

Compound	Solubility ($\mu g\ L^{-1}$)
Naphthalene	31,700
1-Methylnaphthalene	28,500
2-Methylnaphthalene	25,400
1,3-Dimethylnaphthalene	8,000
1,4-Dimethylnaphthalene	11,400
1,5-Dimethylnaphthalene	3,380
2,3-Dimethylnaphthalene	3,000
2,6-Dimethylnaphthalene	2,000
1-Ethylnaphthalene	10,700
1,4,5-Trimethylnaphthalene	2,100
Biphenyl	7,000
Acenaphthene	3,930
Fluorene	1,980
Phenanthrene	1,290
Anthracene	73
2-Methylanthracene	39
9-Methylanthracene	261
9,10-Dimethylanthracene	56
Pyrene	135
Fluoranthene	260
Benzo(a)fluorene	45
Benzo(b)fluorene	2.0
Chrysene	2.0
Triphenylene	43
Naphthacene	0.6
Benz(a)anthracene	14
7,12-Dimethylbenz(a)anthracene	61
Perylene	0.4
Benzo(a)pyrene	0.05[a]
Benzo(e)pyrene	3.8
Benzo(ghi)perylene	0.3
Coronene	0.1

Source: Adapted from Lee et al., 1981 (with the exception of BaP, data are from Mackay and Shiu, 1977).
[a]Value at 20°C (from Locke, 1974).

solutions. This has major implications when one considers the distribution of PAH and their derivatives through air ↔ water ↔ soil environments.

13-B-4 UV/Visible Absorption and Emission Spectra

All carcinogenic PAH and aza-arenes identified in ambient air as of 1972 have highly structured absorption bands in the ultraviolet (NAS, 1972). In fact, all PAH,

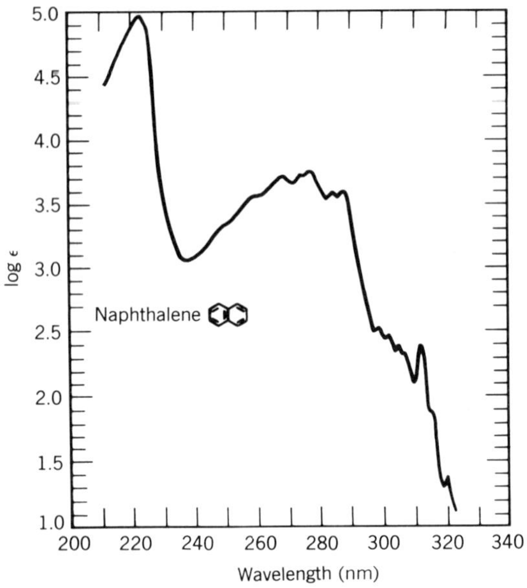

(a)

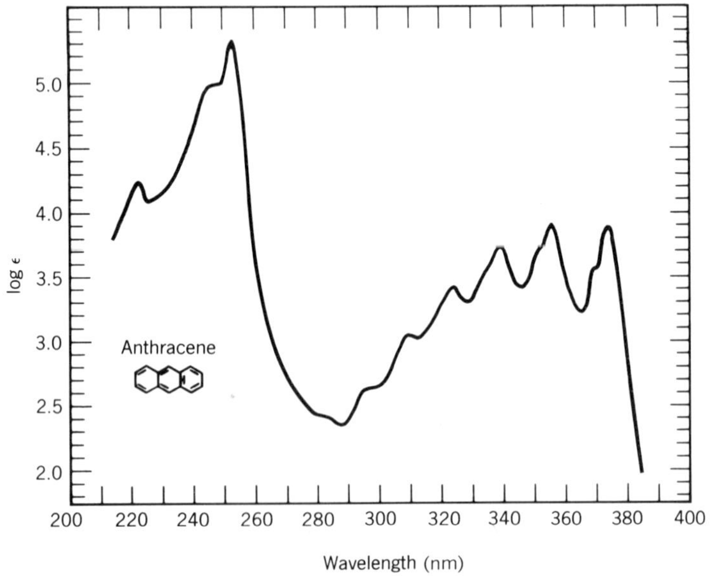

(b)

FIGURE 13.1. UV/visible absorption spectra of some atmospherically relevant PAH: (*a*) naphthalene, (95% ethanol), (b) anthracene (cyclohexane), (*c*) fluoranthene (95% ethanol), and (*d*) pyrene (95% ethanol) ε are to the base 10 (adapted from Friedel and Orchin, 1951).

including naphthalene which has its first pronounced absorption maximum at ~310 nm (ϵ = ~240 L mole^{-1} cm^{-1} in methanol), absorb *actinic* UV radiation; as a consequence of its extended system of conjugated double bonds, perylene absorbs well into the visible (to ~500 nm) and consequently is colored yellow.

The UV/visible absorption spectra of some representative PAH are shown in Fig. 13.1. They have distinct banded structures and, with the exception of benzene

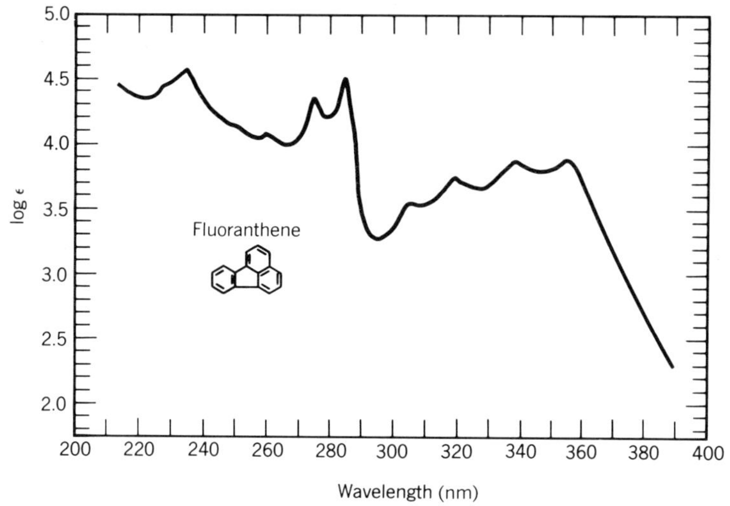

(c)

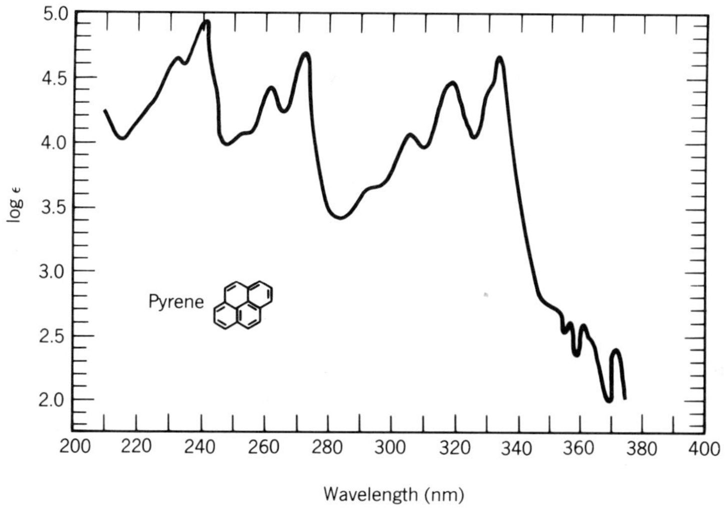

(d)

FIGURE 13.1. (*Continued*)

and naphthalene where the transition is "forbidden" because of symmetry considerations, their first $\Pi \rightarrow \Pi^*$ absorption bands are much more intense than corresponding $n \rightarrow \Pi^*$ transitions in aromatic carbonyl compounds (e.g., benzophenone; see Calvert and Pitts, 1966, and Turro, 1978). Furthermore, for PAH the splitting between the lowest excited singlet state, $^1(\Pi, \Pi^*)$, and the lowest triplet state, $^3(\Pi, \Pi^*)$, is generally much larger than for the corresponding lowest excited (n, Π^*) states in molecules with carbonyl chromophores.

This property and the generally much shorter lifetimes of the excited (Π, Π^*) singlet states of PAH versus the (n, Π^*) states of most aromatic carbonyl compounds have important ramifications vis-à-vis the relative efficiencies of the photophysical and photochemical processes of these two classes of compounds. Thus many PAH important in airborne particles have high quantum yields of fluorescence, a property very useful in detecting their presence in trace amounts in environmental samples. On the other hand, when suspended in rigid matrices at low temperatures, many aromatic carbonyl compounds have high quantum yields of phosphorescence (see Calvert and Pitts, 1966, and Turro, 1978).

As an example, in degassed *n*-heptane at room temperature ϕ_f are: FL (0.35), BaA (0.23), CHR (0.18), BaP (0.60), BeP (0.11), and BghiP (0.29). The carcinogen cyclopenta(cd)pyrene, found in ambient POM and throughout the environment, does not fluoresce. These and other aspects of the fluorescence spectroscopy of environmentally important PAH are well discussed by Heinrich and Güsten (1980).

Going from a non-polar solvent (e.g., cyclohexane) to a polar solvent (e.g., methanol) will shift the (Π, Π^*) bands of PAH to longer wavelengths (red shift). For example, the long-wavelength band of anthracene shifts from 375 to 381 nm in *n*-hexane versus acetonitrile solvents (Wehry, 1983).

With carbonyl compounds, the corresponding shift in the (n, Π^*) band is stronger and toward shorter wavelengths (blue shift). Thus for acetone, λ_{max} for the $n \rightarrow \Pi^*$ transition decreases as follows: hexane (279.0 nm), chloroform (277.0 nm), ethanol (272.0 nm), methanol (270.0 nm), and water (264.5 nm) (see Jaffe and Orchin, 1962). This phenomenon may have environmental implications for the aqueous phase photochemistry of carbonyl compounds absorbing actinic radiation (e.g., their photochemistry in droplets, see Section 3-C-12).

The absorption and emission spectra of organic compounds are altered significantly in going from the gaseous $\rightarrow$ solution $\rightarrow$ solid states. This is illustrated for anthracene in Fig. 13.2, which shows the pronounced red shifts of the 0–0 bands for both absorption and fluorescence and the increased separation of the corresponding 0–0 bands, accompanying the phase changes $A(g) \rightarrow A(soln) \rightarrow A(s)$ (Bowen, 1946).

Similar considerations apply to molecules adsorbed on solid surfaces. Thus Leermakers and co-workers (Leermakers and Thomas, 1965; Nicholls and Leermakers, 1971) showed that in going from naphthalene and anthracene dissolved in pure cyclohexane solvent to their being adsorbed on silica gel–cyclohexane matrices, slight red shifts of ~100–200 cm^{-1} for their (Π, Π^*) bands are observed. For comparison, blue shifts resulting from adsorption on silica gel of molecules with $n \rightarrow \Pi^*$ transitions (e.g., ketones) are much larger, ~1700–2000 cm^{-1}.

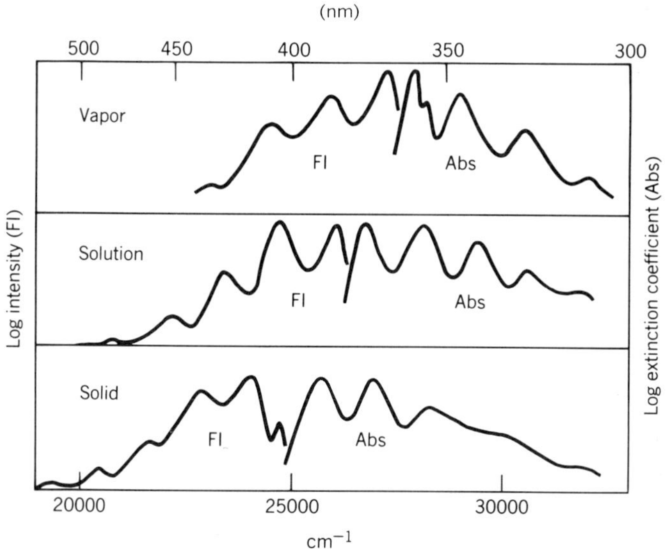

FIGURE 13.2. Absorption and fluorescence spectra of anthracene showing the pronounced red shifts in going from the vapor state to solution in dioxane to solid phases; the separation of the 0–0 bands also increases (from Bowen, 1946).

Such phenomena as the red shifts in the absorption spectra of adsorbed PAH may have important implications to their atmospheric photochemistry; they warrant further investigation.

13-C. ANALYSIS OF AIRBORNE PARTICULATE PAH: A BRIEF OVERVIEW

Ambient particles are chemically very complex; they contain hundreds of organic species, among which are relatively small amounts of PAH (e.g. see Cautreels and Van Cauwenberghe, 1976). As noted earlier, in the 1940s organic extracts of these respirable particles were shown to contain carcinogens, some of which were subsequently identified as BaP and related PAH. These observations led to intense efforts to identify and quantify the carcinogenic PAH, with BaP generally acting as a surrogate compound.

In the 1950s and 1960s, the highly structured spectra and strong transition probabilities of the higher PAH resulted in UV/visible absorption spectrophotometry becoming a widely used technique for their analysis in extracts of ambient POM (e.g., see Sawicki et al., 1960). Assisting in the development of this technique was the classification by Clar (1964) of the UV/visible absorption bands of many PAH. In his terminology, α bands, usually at the long-wavelength end of the spectra, are weak ($\epsilon \sim 10^2$–10^3) and may be partially overlapped by the ρ bands found at somewhat shorter wavelengths. These are more intense ($\epsilon \sim 10^4$) and may show pronounced vibronic structure. Finally, the β bands occur at still shorter wavelength and are very intense ($\epsilon \sim 10^5$). These are illustrated for BaP in cyclohexane in Fig. 13.3.

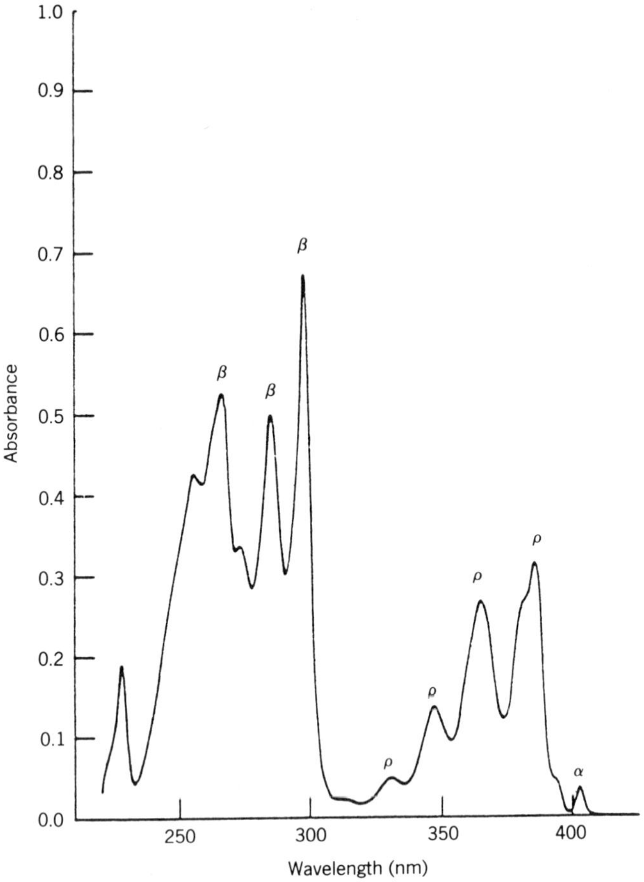

FIGURE 13.3. UV absorption spectrum of BaP (taken at a concentration of 1.0×10^{-5} *M* in cyclohexane with a pathlength of 1 cm) showing the α, ρ, and β bands (after Clar, 1964; from NAS, 1972).

At that time, the technique of fluorescence spectrometry was also being developed and used for analysis of PAH samples. Its major advantage over UV/visible absorption methods was a much greater sensitivity (by a factor of $\sim 10^2$–10^3); a major disadvantage was interference by fluorescing trace impurities. Both optical methods suffered from an inability to analyze complex *mixtures* containing many isomeric PAH (see NAS, 1972, for a review and critique).

Since about 1972, a number of important advances have been made in optical spectroscopy; an excellent review of the current status of optical spectrometric techniques for determination of PAH is given by Wehry (1983). He discusses and critiques UV/visible absorption spectrophotometry, fluorescence and phosphorescence spectroscopy, and IR, FTIR, and Raman vibrational spectroscopy. The use of FTIR to analyze PAH in rigid matrices at low temperatures has interesting possibilities.

In the 1970s, a number of techniques were developed for more efficiently separating mixtures of PAH into their individual components. They include GC, capillary column GC, high-performance liquid chromatography (HPLC), and thin-layer chromatography (TLC). These have been interfaced to (or utilized with) a variety of detection systems, including mass spectrometers and optical detectors based on the UV/visible, absorption, and fluorescence spectra of PAH. Today such combinations as GC–MS, capillary GC–MS, LC–MS, HPLC–UV, and so on are widely used in PAH analysis, and new developments in more efficient separation techniques and more sensitive detectors are occurring at a rapid pace.

In the following sections, we comment only briefly on the current status of the sampling, extraction, fractionation, and identification of PAH in organic extracts of respirable ambient particles and primary combustion-generated POM. Details are to be found in the excellent book *Analytical Chemistry of Polycyclic Aromatic Compounds* (Lee et al., 1981), relevant chapters in *Environmental Carcinogens: Polycyclic Aromatic Hydrocarbons* (Grimmer, 1983) and *Handbook of Polycyclic Aromatic Hydrocarbons* Vols. 1 and 2 (Bjørseth, 1983; Bjørseth and Ramdahl, 1985), reviews such as *Modern Analytical Methods for Environmental Polycyclic Aromatic Compounds* (Bartle et al., 1981), *The Characterization of Microbial Mutagens in Complex Samples-Methodology and Application* (Alfheim et al., 1984b), and the original literature.

13-C-1 Sampling, Extraction, and Fractionation

The collection of ambient particulate matter by a variety of techniques has been discussed in Chapter 12. Until the late 1970s, the classic hi-vol sampler using standard glass fiber (GF) filters was the instrument of choice for collecting sufficient material for PAH analysis. Today they remain in widespread use, preferably with cutoffs at 10–20 μm to reject much of the coarse particulate matter.

However, as noted earlier, they have certain inherent problems. These include physical artifacts due to blow off and chemical artifacts due to reactions of the PAH collected on the filter with gaseous co-pollutants (Pitts et al., 1978; Lee et al., 1980). The latter depend significantly on the type of filter employed, for example, glass fiber (GF) versus Teflon impregnated glass fiber (TIGF).

Finally, use of a filter *alone* overlooks the major contributions of volatile PAH, such as naphthalene, and relatively volatile three- and four-ring compounds to the total burden of airborne PAH; a backup device is required to collect the gaseous species (see Section 13-D-2).

Similar problems can occur with the collection of exhaust particles from motor vehicles, but they can be overcome in part by use of a well-designed exhaust dilution tube and associated sampling instrumentation for gases and particles (see Schuetzle, 1983, and Lee and Schuetzle, 1983). For example, recent Ames assay chromatograms of diesel particles collected in ambient air from heavy duty trucks and from passenger cars on a laboratory dynamometer support this view, in that they had the same general shape, and presumably the same general distribution, of direct mutagens (Salmeen et al., 1985).

Vehicle exhaust dilution tubes are sometimes employed to rapidly collect sufficient amounts of combustion-generated fine particles for the development of new analytical procedures, for example, for polar PAH derivatives.

Figure 13.4 from Lee and Schuetzle (1983) shows their scheme for concurrent collection of gaseous and particulate PAH in vehicle exhaust and the subsequent analytical steps of extraction, solvent evaporation, gravimetric analysis, fractionation, chemical analysis, and biological assay. Except for some differences in sampling techniques, similar procedures are often followed for samples of POM from ambient air.

Precautions should be taken to minimize the time between sample collection and extraction because of evaporation losses of the lighter PAH and reactions of PAH on the filter during storage. For example, 57% of the BaP in a sample of diesel particulate matter was lost after 150 days in storage (Swarin and Williams, 1980). Such loss processes can depend on the nature of the collection filter, for example, GF, TIGF, fluoropore, and Tissuquartz. Of these, for quantitative sampling of vehicle emissions, the Teflon membrane seems superior (Lee and Schuetzle, 1983), while for ambient particles both GF and TIGF filters are commonly used for routine work—however, not without problems of artifacts (Section 13-H-2).

To collect gas phase PAH present in air and in primary emission sources, a variety of polymeric adsorbents have been used; they include XAD-2 (Lee and Schuetzle, 1983), Tenax GC adsorbents (Van Vaeck et al., 1984), and plugs of polyurethane foam (Thrane and Mikalsen, 1981).

Generally, the organic fraction of ambient or primary POM is separated from the inorganic or carbonaceous portion of the particles by extraction into an organic solvent using a Soxhlet apparatus or ultrasonic agitation. A variety of solvents have been employed, singly, in mixtures (e.g., methylene chloride + toluene + methanol), or in succession. The specific solvent employed depends on a number of factors, including the nature of the sample, for example, ambient air or diesel exhaust. Extraction methods have been reviewed by Lee et al., (1981), Lee and Schuetzle (1983), Griest and Caton (1983), and Grimmer (1983); they will not be considered further.

A variety of techniques have been employed to separate effectively POM extracts into a series of subfractions. These include HPLC, liquid–liquid partition, and column liquid chromatography. Use of HPLC instrumentation and procedures for the determination of PAH in a variety of samples (e.g., water, air, sediment, petroleum, and alternate fuels) are reviewed by Wise (1983).

13-C-2. Identification and Measurement of PAH and PAH Derivatives

Instrumental techniques for the characterization, identification, and measurement of PAH in extracts of environmentally relevant samples include: optical spectroscopy, HPLC, GC, capillary column GC, MS, GC–MS, high-resolution mass spectrometry (HRMS), mass spectrometry–mass spectrometry (MS–MS), and various combinations thereof.

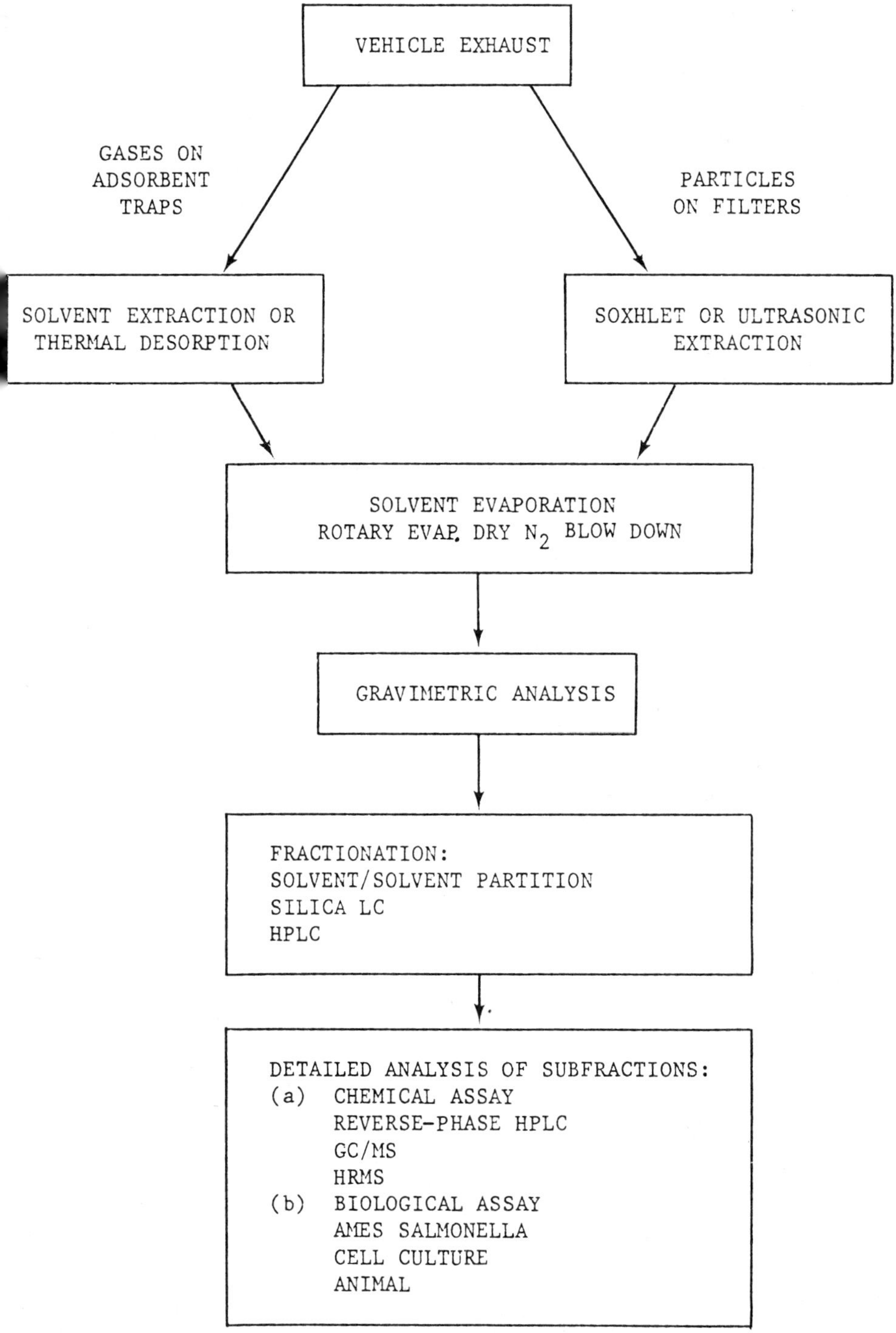

FIGURE 13.4. Scheme for analyzing gaseous and particulate PAH in vehicle exhaust; similar procedures are used for ambient samples (from Lee and Schuetzle, 1983).

Among the recent techniques for the analysis of isomeric PAH is charge exchange, chemical ionization mass spectrometry (CE/CI)MS using a reagent gas mixture of 15% methane in argon (Simonsick and Hites, 1984). This gives abundant $(M + H)^+$ and M^+ ions, and PAH isomers (e.g., the carcinogen BaP versus the non-carcinogen BeP) are identified on the basis of their $(M + H)^+/M^+$ ratios. Analyses of a *reference* sample of ambient particulate matter by (CE/CI)MS and by liquid chromatography with fluorescence detection are compared in Table 13.5; agreement between the two widely different techniques is excellent.

In addition to analyzing PAH in ambient POM, Simonsick and Hites combined capillary column GC with their (CE/CI)MS techniques to analyze an extract of a sample of 0.26 μm diameter carbon black. The chromatogram is shown in Fig. 13.5; their results, using (CE/CI)MS and conventional electron impact mass spectrometry (EI–MS), are in Table 13.6; numbers in the chromatogram refer to specific PAH in the table. The results by the two methods are generally in good agreement, but the authors note that "The response factors of PAH obtained under (CE/CI)MS conditions vary from 0.5 to 2.0, a much larger range than the response factors obtained under EI conditions which are in the vicinity of 1 when compared to their deuterated analogues." Of environmental interest are the number, nature, and levels of the PAH in this sample of respirable, carbon black particles.

Currently of special interest are instruments that will analyze the labile, relatively *non-volatile* polar derivatives of PAH in POM extracts; some of these are direct mutagens (e.g., hydroxynitropyrene). Schuetzle et al. (1985) have evaluated

TABLE 13.5. Comparison of Analyses of a Reference Sample of Ambient Particulate Matter[a] by Reverse Phase Liquid Chromatography with Fluorescence Detection (LC/FD) and by (CE/CI)MS

PAH	M^+	(CE/CI)MS[b] μg/g	LC/FD[c] μg/g
Phenanthrene	178	4.8	4.6
Anthracene	178	0.50	—
Pyrene	202	5.8	6.4
Fluoranthene	202	7.0	7.0
Benz(a)anthracene	228	3.3	2.6
Benzo(e)pyrene	252	3.6	3.6
Benzo(a)pyrene	252	2.8	2.6
Perylene	252	0.57	0.76
Indeno(1,2,3-cd)pyrene	276	3.4	3.3
Benzo(ghi)perylene	276	4.4	3.9

Source: Adapted from Simonsick and Hites, 1984.

[a]SRM 1649; from National Bureau of Standards (NBS), Office of Standard Reference Materials, Washington, DC.

[b]Simonsick and Hites, 1984.

[c]May and Wise, 1984.

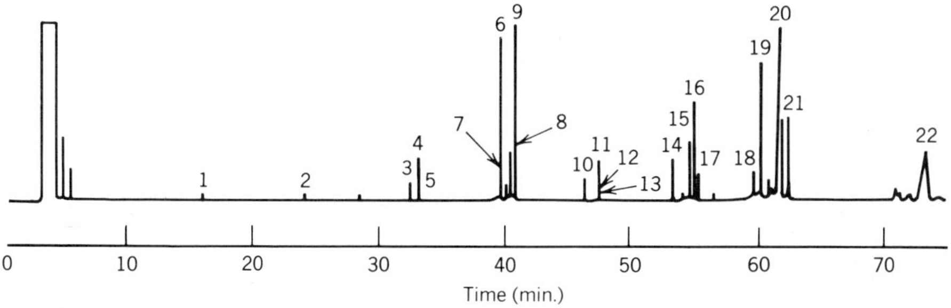

FIGURE 13.5. Temperature programmed (~50–250°C) capillary column gas chromatogram of an extract of carbon black. Conditions: 30 m × 0.25 mm i.d., DB-5 column, hydrogen carrier gas with FID. Numbers on peaks correspond to structures in Table 13.6 (from Simonsick and Hites, 1984).

TABLE 13.6. PAH Content of an Extract of a Sample of 0.26 μm Diameter Carbon Black as Determined by Capillary Column GC Coupled to a (CE/CI)MS System, and to a EI–MS Instrument

Peak	Compound	Concentration (EI) $\mu g\ g^{-1}$	Concentration (CE/CI) $\mu g\ g^{-1}$
1	Naphthalene	4.8	5.2
2	Acenaphthylene	3.5	2.6
3	Dibenzothiophene	17	14
4	Phenanthrene	39	28
5	Anthracene	1.9	0.82
6	Fluoranthene	84	110
7	Acephenanthrylene	(11)[a]	(6.9)
8	Phenanthro(5,6-bcd)thiophene	(27)	(32)
9	Pyrene	170	150
10	Benzo(ghi)fluoranthene	(30)	(34)
11	Cyclopenta(cd)pyrene	(45)	(27)
12	Benz(a)anthracene	4.8	4.2
13	Chrysene	7.2	4.8
14	Benzofluoranthenes	(76)	(71)
15	Benzo(e)pyrene	56	40
16	Benzo(a)pyrene	78	171
17	Perylene	30	20
18	$C_{22}H_{12}$isomer		(26)
19	Indeno(1,2,3-cd)pyrene	280	360
20	Benzo(ghi)perylene	490	500
21	Anthanthrene	160	180
22	Coronene	210	140

Source: Adapted from Simonsick and Hites, 1984.

[a]No response factor available for calculation of concentrations given in parentheses.

a triple stage MS instrument and concluded that the polar derivatives of benz(a)anthracene, for example, carboxylic acids, alcohols, amines, and quinones, can be determined with good selectivity.

Other useful techniques for screening polar PAH derivatives of relatively low vapor pressure include: MS, or MS–MS, with low-energy ionization techniques such as fast atom bombardment (FAB), fast ion bombardment (FIB), field desorption, and static secondary ion mass spectrometry (SSIMS); the latter is highly sensitive (Schuetzle et al., 1985).

To date, use of moving belt HPLC–MS systems to separate and characterize complex mixtures containing low-volatility polar PAH in polar solvents has often proved disappointing. However, a new type of interface incorporating a *thermospray* technique appears promising (Blakely and Vestal, 1983; Covey and Henion, 1983); indeed a number of MS manufacturers are offering commercial units based on this approach. In contrast to conventional HPLC–MS, this technique is most effective for a polar mobile phase (methanol or water) and a polar molecule and has a detection limit of 1–10 pg, $\sim 10^3$ times more sensitive than moving belt HPLC–MS (Schuetzle et al., 1985).

While these are sensitive and useful screening techniques, very efficient chromatographic separations are generally required for unequivocal identification of specific isomers of polar PAH and for large (e.g., five- to nine-ring) PAH in complex extracts. Promising approaches include supercritical fluid chromatography (SFC) (Peaden et al., 1982; Fjeldsted and Lee, 1984) and SFC coupled to a mass spectrometer (Smith et al., 1984).

Another development, employed, for example, in the separation and identification of very large PAH in the neutral fractions of extracts of coal-derived materials, is the coupling of microcolumn liquid chromatography with MS (Novotny et al., 1984; Novotny and Ishii, 1985). An advantage of this system is that separation efficiencies are above 200,000 theoretical plates, permitting the resolution of 170 peaks; this was not possible by conventional HPLC. Furthermore, microcolumn LC has a much higher solute capacity than capillary GC.

Novotny and coworkers have characterized a variety of PAH with five to nine rings ranging from $\sim$250 to 450 in molecular weight. However, they state "given the current state of structure identification methodology, the term 'characterization' more adequately describes our efforts than identification." Thus they emphasize the need for pure standards of PAH in this higher molecular weight range if one is to progress from *characterization* to *identification*. This differentiation in terms is important; it should be kept in mind by all concerned with the analysis of PAH and their derivatives in extracts of complex environmental samples.

13-D. POLYCYCLIC AROMATIC HYDROCARBONS IN AMBIENT AIR

13-D-1. PAH in Airborne POM: Size Distribution and Levels of BaP

A wide variety of PAH and associated aza-arenes are present in ambient particles; for example, over 70 major PAH from two to seven rings were separated and identified by Lao et al. (1973), while Lee et al. (1976) characterized 122 GC peaks (and identified 29 of them) in a different sample. Some of the more common PAH found in ambient POM have been shown in Tables 13.1 and 13.5. A comprehensive listing of airborne PAH is given in Table 13.7 (Lee et al., 1981).

The kinds and levels of particle bound PAH in ambient air have been measured in urban, suburban, rural, and pristine areas throughout the world, with special attention being given to BaP because of its carcinogenic activity (NAS, 1972, 1983; Hoffman and Wynder, 1977; Grimmer, 1983 and references therein). For example, BaP concentrations in ambient air in major cities in 24 countries and seasonal BaP levels from 1965 to 1978 at a variety of sites in Germany have been summarized by Grimmer and Pott (1983); similar information for the United States is contained in the 1983 NAS document.

Because of the wide variety of analytical techniques employed over the last three decades to measure BaP in ambient particles, one should not try to compare quantitatively results by different investigators. However, several general, qualitative observations are valid. Reported concentrations range from a fraction of a ng m^{-3} in unpolluted air parcels to $\sim$100 ng m^{-3} or more in heavily polluted urban areas. BaP levels are generally much higher in the winter than in the summer and in urban communities versus rural areas (for example, see the recent study of Greenberg et al., 1985). Finally, annual averages have dropped dramatically following enactment of clean air legislation. For example, in central London they fell from 46 ng m^{-3} in 1949–1951 to 4 ng m^{-3} in 1972–1973, and in the United States from 7 ng m^{-3} in 1958 to 2 ng m^{-3} in 1970!

From a health perspective, it is important to recognize that in ambient air the highest concentrations of BaP generally are found in respirable fine particles. This is seen in Table 13.8 which shows the size distribution of BaP and coronene in ambient particles collected in Pasadena, California (east of downtown Los Angeles), for sampling periods in October and in December 1976 (Miguel and Friedlander, 1978). Note that the maximum concentration of BaP is only 0.52 ng m^{-3} in December 1976. This reflects a relatively warm climate, residential heating primarily by natural gas rather than coal, and motor vehicle exhaust as the major source.

13-D-2. Gas/Particle Distribution of PAH and Their Long-Range Transport

While the data reported for concentrations of BaP in ambient particles collected on hi-vol filters may be expected to reflect true ambient concentrations, barring major filter artifacts, the reported levels of three- and four-ring PAH in these particles do not represent their true concentrations in the atmosphere. As discussed previously, a major fraction of these species exists in the gas phase (Pupp et al., 1974; Cautreels and Van Cauwenberghe, 1978; Lao and Thomas, 1980; Van Vaeck et al., 1980). The absolute gas phase concentrations of PAH are of interest because in simulated polluted atmospheres gas phase reactions giving products of chemical and biological interest have been observed (see Section 13-I-1).

The system shown in Fig. 13.6 was employed by Thrane and Mikalsen (1981) to determine the absolute concentrations of PAH in ambient air, and their gas/

TABLE 13.7. **PAH Identified in Airborne Particles**

Compound	Molecular Weight
Biphenyl	154
Fluorene	166
Benzindenes	166
Methylbiphenyls	168
Dihydrofluorenes	168
Phenanthrene	178
Anthracene	178
Dihydrophenanthrene	180
Dihydroanthracene	180
1-Methylfluorene	180
2-Methylfluorene	180
9-Methylfluorene	180
Octahydrophenanthrene and/or octahydroanthracene	186
4*H*-Cyclopenta(def)phenanthrene	190
Methylphenanthrene	192
1-Methylphenanthrene	192
2-Methylphenanthrene	192
3-Methylphenanthrene	192
9-Methylphenanthrene	192
Methylanthracene	192
1-Methylanthracene	192
2-Methylanthracene	192
Fluoranthene	202
Pyrene	202
Benzacenaphthylene	202
Methyl-4*H*-cyclopenta(def)phenanthrene	204
Dihydrofluoranthene	204
Dihydropyrene	204
Dimethylphenanthrenes[a] and/or dimethylanthracenes[a]	206
Octahydrofluoranthene	210
Octahydropyrene	210
Methylfluoranthene	216
1-Methylfluoranthene	216
2-Methylfluoranthene	216
3-Methylfluoranthene	216
7-Methylfluoranthene	216
8-Methylfluoranthene	216
Methylpyrene	216
1-Methylpyrene	216
2-Methylpyrene	216
4-Methylpyrene	216
Benzo(a)fluorene	216
Benzo(b)fluorene	216
Benzo(c)fluorene	216
Dimethyl-4*H*-cyclopenta(def)phenanthrene[a]	218

TABLE 13.7. (*Continued*)

Compound	Molecular Weight
Dihydrobenzo(a)fluorene and/or dihydrobenzo(b)fluorene	218
Dihydrobenzo(c)fluorene	218
Trimethylphenanthrenes[b] and/or trimethylanthracenes[b]	220
Benzo(ghi)fluoranthene	226
Benzo(c)phenanthrene	228
Benz(a)anthracene	228
Chrysene	228
Triphenylene	228
Dimethylfluoranthenes[a] and/or dimethylpyrenes[a]	230
Dihydrobenzo(c)phenanthrene	230
Dihydrobenz(a)anthracene and/or dihydrochrysene and/or dihydrotriphenylene	230
Trimethyl-4*H*-cyclopenta(def)phenanthrenes[b]	232
Hexahydrochrysene	234
Methylbenzo(ghi)fluoranthene	240
Dihydromethylbenzo(ghi)fluoranthene	242
Methylbenzo(c)phenanthrenes	242
Methylchrysene	242
1-Methylchrysene	242
2-Methylchrysene	242
3-Methylchrysene	242
6-Methylchrysene	242
Methylbenz(a)anthracenes	242
Methyltriphenylene	242
Trimethylfluoranthenes[b] and/or trimethylpyrenes[b]	244
Tetrahydromethyltriphenylene and/or tetrahydromethylbenz(a)anthracene and/or tetrahydromethylchrysene	246
Benzo(j)fluoranthene	252
Benzo(k)fluoranthene	252
Benzo(b)fluoranthene	252
Benzo(e)pyrene	252
Benzo(a)pyrene	252
Perylene	252
Binaphthyls	254
Dimethylchrysenes[a] and/or dimethylbenz(a)anthracenes[a] and/or	256
Dimethyltriphenylene[a]	256
Methylbenzopyrenes and/or methylbenzofluoranthenes	266
Methylbinaphthyls	268
Methyl-2,2′-binaphthyl	268
3-Methylcholanthrene(3-methylbenz(j)aceanthrylene)	268
Benzo(ghi)perylene	276
Anthanthrene	276
Indeno(1,2,3-cd)fluoranthene	276

TABLE 13.7. (*Continued*)

Compound	Molecular Weight
Indeno(1,2,3-cd)pyrene	276
Dibenzanthracenes	278
Dibenz(a,c)anthracene	278
Pentacene	278
Benzo(b)chrysene	278
Picene	278
Dimethylbenzofluoranthenes[a]	280
Dimethylbenzopyrenes[a]	280
Methylindeno(1,2,3-cd)fluoranthene	290
Methylindeno(1,2-cd)pyrene	290
Methylbenzo(ghi)perylene and/or methylanthanthrene	290
Methyldibenzanthracenes	292
Methylbenzo(b)chrysene and/or methylbenzo(c)tetraphene	292
Methylpicene	292
Coronene	300
Dibenzopyrenes	302
Diphenylacenaphthylene	304
Quaterphenyls	306

Source: Adapted from Lee et al., 1981; see this for references to the original literature.

[a]Could be ethyl.

[b]Could be ethylmethyl or propyl.

TABLE 13.8. Size Distribution of Benzo(a)pyrene and Coronene in Ambient Particles Collected in Pasadena, California

Aerodynamic Diameter of Particle (μm)	Concentration (pg m^{-3})			
	October 25–28, 1976		December 14–17, 1976	
	BaP	Cor	BaP	Cor
0.05–0.075	30	431	83	1200
0.075–0.12	196	1390	523	4220
0.12–0.26	79	460	208	907
0.26–0.50	28	205	67	113
0.5–1.0	36	166	81	218
1.0–2.0	16	107	44	<45
2.0–4.0	16	118	30	<45
>4	21	136	23	<45

Source: Miguel and Friedlander, 1978.

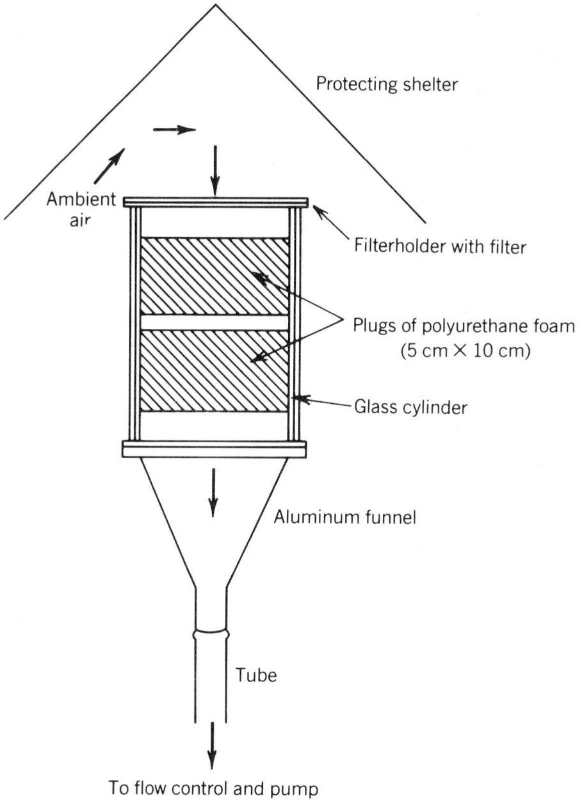

FIGURE 13.6. Schematic diagram of a sampling unit for determining gas/particle distribution of PAH in ambient air (from Thrane and Mikalsen, 1981).

particle phase distribution factors, at several sites in Norway. Their samples were from a background station in south Norway (Birkenes) and from an urban site on a busy street (Radhusgaten) in downtown Oslo in January and February 1979. With their unit, ambient air is pulled at a rate of 25 m^3 h^{-1} through a precleaned GF filter and then through two successive 5 cm thick plugs of polyurethane foam.

As seen in Table 13-9, most of the PAH at the urban site were found at the background station. The averages of the total PAH concentrations (gaseous + solid) were 20 ng m^{-3} versus 1680 ng m^{-3}, respectively; the ranges were 7–40 ng m^{-3} versus 1000–3500 ng m^{-3}.

The influence of long-range transport (LRT) of air parcels on PAH levels of aerosols were evident at Birkenes. Thus clean air coming off the Atlantic from the W/SW and NW had an average total PAH concentration of ~10 ng m^{-3}. However, an air parcel passing over England and Scotland, and one passing over continental Europe, after arriving at Birkenes each averaged ~25 ng m^{-3} of PAH.

TABLE 13.9. Gaseous and Particulate PAH Identified in Ambient Air at a Background Station in Southern Norway (Birkenes) and on a Street in Downtown Oslo

Number[a]	PAH	Background Site	Urban Site
1	Naphthalene	x	x
2	Biphenyl	x	x
3	Fluorene	x	x
4	Dibenzothiophene	x	x
5	Phenanthrene	x	x
6	Anthracene	x	x
7	Carbazole	x	x
8	2-Methylanthracene		x
9	1-Methylphenanthrene		x
10	Fluoranthene	x	x
11	Pyrene	x	x
12	Benzo(a)fluorene	x	x
13	Benzo(b)fluorene	x	x
14	Benz(a)anthracene	x	x
15 + 16	Chrysene/triphenylene	x	x
17	Benzo(b)fluoranthene		x
18	Benzo(e)pyrene	x	x
19	Benzo(a)pyrene	x	x
20	Perylene	x	x
21	Indeno(1,2,3-cd)pyrene	x	x
22 + 23	Dibenz(a,c)/(a,h)anthracene		x
24	Benzo(ghi)perylene	x	x
25	Coronene		x

Source: Thrane and Mikalsen, 1981.

[a]Numbers refer to order of elution from the GC.

These results are generally consistent with those of Bjørseth and co-workers who earlier had provided conclusive evidence for LRT of airborne PAH in similar studies at Birkenes (Lunde and Bjørseth, 1977; Bjørseth et al., 1979). They found PAH concentrations up to 20 times higher in air transported from the European continent and Great Britain compared to air from northern Norway and concluded that industrial or urban plumes may be transported over long distances without undergoing significant degradation or dilution [see Bjørseth and Olufsen (1983) for a review of the LRT of PAH].

The PAH shown in Table 13.9 cover a wide range of volatilities, from predominantly gaseous naphthalene to adsorbed coronene; their average ambient distributions in the wintertime at the urban site in Oslo are shown in Fig. 13.7. As the

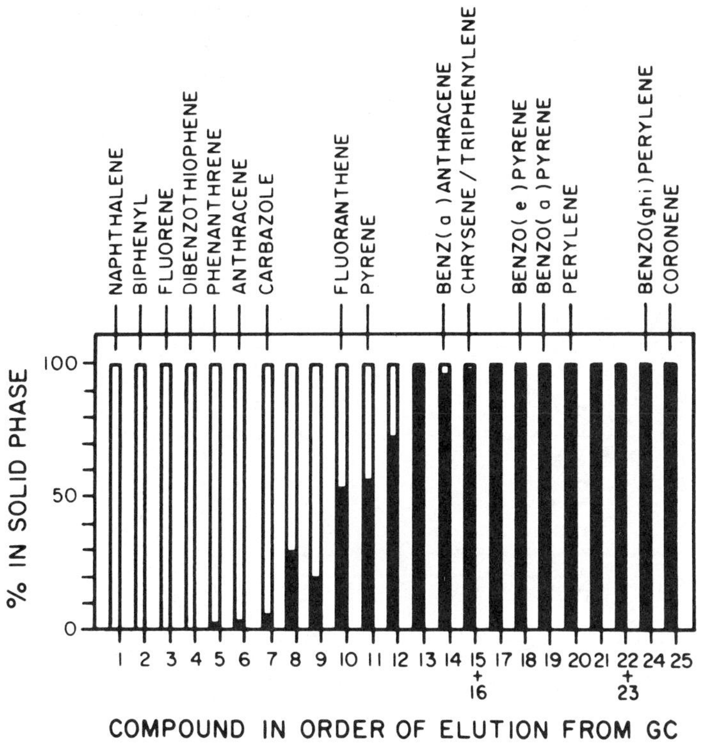

FIGURE 13.7. Average distribution of PAH and biphenyl collected on a GF filter and on the backup polyurethane foam plug, in samples collected on a busy one-way street in downtown Oslo; numbers refer to structures in Table 13.9. Solid bars represent percent of PAH found in the solid phase, open rectangles that in the gas phase (from Thrane and Mikalsen, 1981).

authors note, sampling artifacts such as sublimation and blow off from the filter (Pupp et al., 1974; Van Vaeck et al., 1984) preclude overly quantitative interpretation of the results in Fig. 13.7. However, clearly not only naphthalene but also fluorene and phenanthrene can exist to a significant extent in the gas phase, along with fluoranthene and pyrene. At higher ambient temperatures, their volatility will increase. Similar overall conclusions on PAH distributions in Belgium were reached by Cautreels and Van Cauwenberghe (1978).

The compositions of the total samples of PAH (+ biphenyl) collected at Birkenes and Oslo are given in Fig. 13.8. The profiles are obtained by plotting on the vertical axis (log scale) the concentration of a given PAH as a percentage of the sum of the concentrations of all PAH (gaseous + particulate) versus their order of elution from the GC.

Interestingly, the percent compositions of PY, BaA, BaP, and BghiP are somewhat lower in Birkenes than in downtown Oslo, possibly because of chemical degradation of these more reactive PAH during LRT.

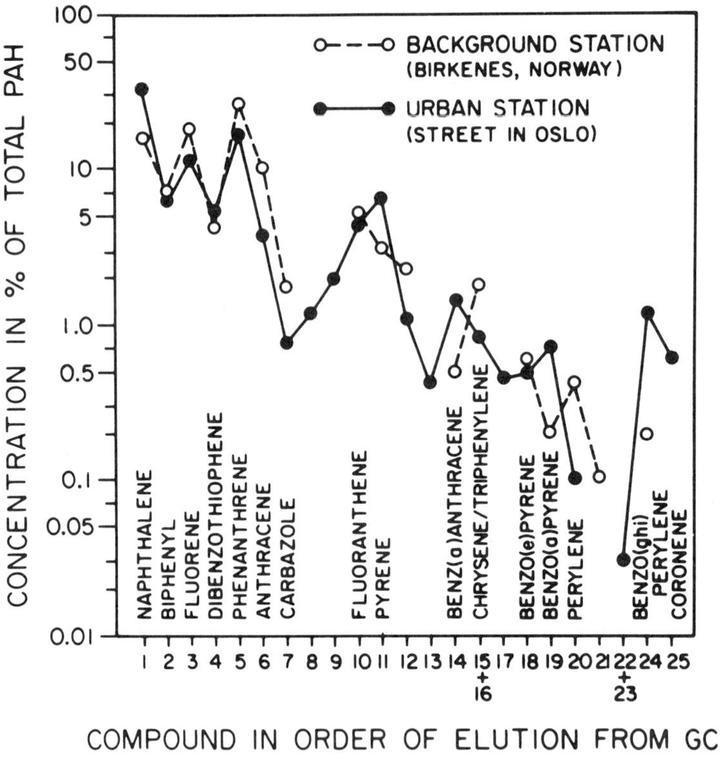

FIGURE 13.8. Profiles of PAH and biphenyl collected at a background station in south Norway and on a one-way street in Oslo, in the fall of 1978. Vertical axis is the log of the concentration of a specific PAH expressed as a percentage of the sum of the concentrations of the individual PAH. The numbers on the horizontal axis specify the PAH in the order they are eluted from the GC; see Table 13.9 for a listing of all structures (from Thrane and Mikalsen, 1981).

13-D-3. Aza-Arenes and S-Heterocycles in Ambient POM

Aza-arenes (N-heterocycles) are products of incomplete combustion of nitrogen-containing substances and organics in the presence of NO_x and are found along with the PAH in respirable particulate matter from ambient air, diesel- and spark-ignition engines, wood smoke, and so on (see Hoffman and Wynder, 1977; Lee et al., 1981, and references therein). (Oxygen-containing PAH in primary emissions and ambient particles are discussed in Sections 13-G-1 and 13-G-2b, respectively.)

For example, in a detailed study of the basic fraction of ambient POM collected in early 1975 in New York City, Dong, Locke, and Hoffman (1977) unequivocally identified over 20 aza-arenes and other N-bases; these are listed in Table 13.10 along with their concentrations in ng m^{-3} of air. The authors of this careful study note the abundance and complexity of the quinoline and isoquinoline fractions relative to the other aza-arenes and suggest the need for more analytical work on the basic fraction, in part because of the carcinogenic properties of quinoline (Hirao et al., 1976).

Much less is known about the atmospheric levels and reactions of aza-arenes than for their parent PAH; these areas warrant further research. Even less is known about particulate S-heterocycles, as seen from the short list in Table 13.11 (Lee et al., 1976).

As seen from the PAH distribution for Oslo in Fig. 13.7 and the profiles in Fig. 13.8, Thrane and Mikalsen identified substantial quantities of carbazole (**I**) and dibenzothiophene (**II**) in samples collected at the background station and downtown Oslo; at the latter site these heterocycles were distributed almost entirely in the gas phase (neglecting possible blow off effects).

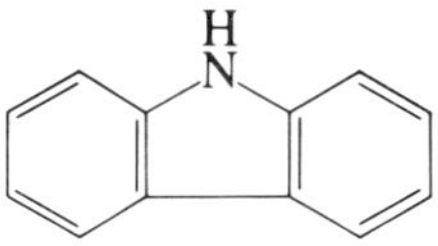

(**I**)

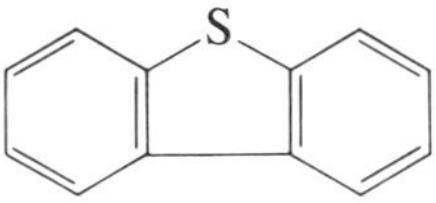

(**II**)

(**III**)

(**IV**)

While the gas phase atmospheric chemistry of these specific N- and S-atom PAH has not been investigated, Atkinson et al. (1984) reported that gaseous pyrrole (**III**) in air reacts very rapidly with OH radicals (1.2×10^{-10} cm^3 $molecule^{-1}$ s^{-1}) giving a lifetime of 2 h at an average OH concentration of 1×10^6 molecules cm^{-3}. Presumably, gaseous carbazole will also react rapidly with OH radicals and have a short lifetime in ambient air. Gaseous thiophene (**IV**) in air reacts considerably more slowly with OH radicals than does pyrrole (Atkinson et al., 1983), and by analogy dibenzothiophene (**II**) would be expected to have a correspondingly longer atmospheric lifetime than carbazole by approximately a factor of 10 if thiophene and carbazole are appropriate model compounds. Corresponding lifetimes for the oxygen analog, furan, lie between pyrrole and thiophene, being ~7 h for OH attack during the day and ~1 h for the nighttime reaction with NO_3 at concentrations of 1×10^6 cm^{-3} for OH and 10 ppt (2.5×10^8 cm^{-3}) for NO_3.

Similar considerations apply to the nighttime reactions of pyrrole and thiophene with NO_3 and to predictions for the analogous processes with carbazole and dibenzothiophene. Rate constants in air at 298°K for pyrrole and thiophene are 4.9×10^{-11} and 3.2×10^{-14} cm^3 $molecule^{-1}$ s^{-1}, giving lifetimes of 1.4 min and 35 h, for an average NO_3 level of 10 ppt (Atkinson et al., 1985a).

In solution, furan, pyrrole, and thiophene are reactive toward electrophilic attack; however, as Nielsen et al. (1983) point out, in heteroatomic compounds with two or more ring systems, the heteroatom effect is diminished. It will be interesting to see if this is also the case for gas phase processes. Clearly, to define accurately the atmospheric fates of gaseous carbazole and dibenzothiophene, one must measure experimentally their gas phase rate constants for attack by OH and NO_3 radicals (see Section 13-I).

TABLE 13.10. **Concentrations of Some Aza-Arenes Identified in the Basic Fraction of Ambient POM Collected in New York City**

Structure	Name	Molecular Weight	Concentration (ng m^{-3})
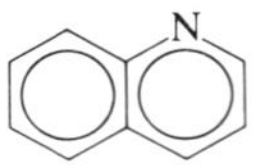	Quinoline	129	0.069
	Methylquinolines	143	0.035
	Dimethylquinolines	157	0.048
	Ethylquinolines	157	0.014
	3 *C*-quinolines	171	0.010
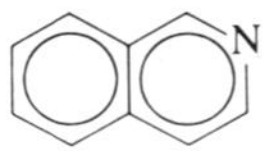	Isoquinoline	129	0.180
	5 or 8-Methylisoquinoline	143	0.310
	Other methylisoquinolines	143	0.076
	Dimethylisoquinolines	157	0.062
	Ethylisoquinolines	157	0.160
	3 *C*-isoquinolines	171	0.028
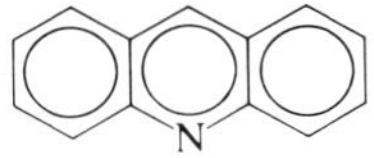	Acridine	179	0.041
	Methylacridines	183	0.007

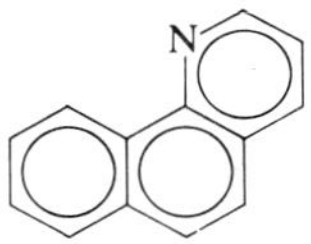	Benzo(h)quinoline	179	0.010
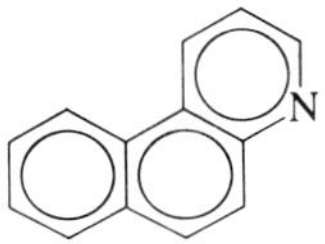	Benzo(f)quinoline	179	0.011
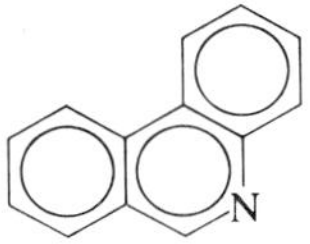	Phenanthridine	179	0.022
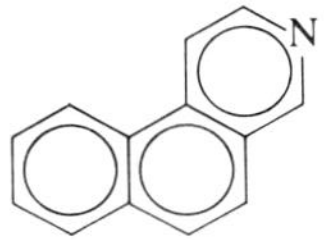	Benzo(f)isoquinoline	179	0.110
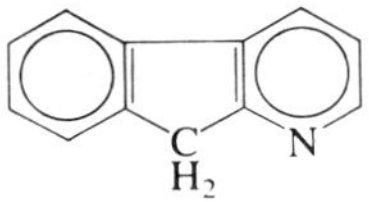	4-Azafluorene (5*H*-indeno-(1,2-b)-pyridine)[a]	167	0.005
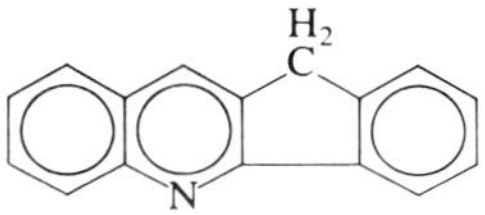	11*H*-indeno(1,2-b)-quinoline	217	Trace

TABLE 13.10. (*Continued*)

Structure	Name	Molecular Weight	Concentration ($ng\ m^{-3}$)
	4-Azapyrene (benzo(lmn)-phenanthridine)	203	0.021[a]
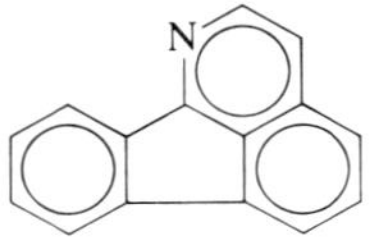	1-Azafluoranthene (indeno-(1,2,3-ij)isoquinoline)	203	0.005[a]
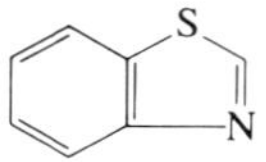	Benzothiazole	135	0.014

Source: Dong et al., 1977.
[a]Includes other isomers.

TABLE 13.11. S-Heterocycles Characterized in Airborne Particulate Matter

Compound	Molecular Weight
Dibenzothiophene	184[a]
Methyldibenzothiophenes	198
Phenanthro(4,5-bcd)thiophene	208
Naphthobenzothiophenes	234
Methylnaphthobenzothiophenes	248

Source: Lee et al., 1976.

[a]Identified predominately in the gas phase at background and urban sites in Norway (Thrane and Mikalsen, 1981).

13-E. THE AMES/*SALMONELLA* TEST: BIOASSAY-DIRECTED CHEMICAL ANALYSES FOR AIRBORNE MUTAGENS

Over 100 in situ, short-term bioassays, using a variety of cell types from bacteria and phage to human cells, now exist for detecting potential chemical mutagens and carcinogens (see the detailed review by Hollstein et al., 1979). Some of these have been employed to evaluate the mutagenicity of ambient fine particles and of primary POM from motor vehicle exhaust and coal and wood combustion (Waters et al., 1979; Waters et al., 1981, 1983; Holmberg and Ahlborg, 1983; Alfheim et al., 1984b).

The most widely employed test to date is that of Bruce Ames and colleagues. This rapid ($\simeq$3 days) and relatively inexpensive assay employs special strains of *Salmonella typhimurium* to detect frameshift or base-pair substitution *reverse* mutations (Ames et al., 1973, 1975; McCann et al., 1975; Ames and McCann, 1976). Thousands of chemicals have been tested by this method and the results—their mutagenic activities—are listed in the Environmental Mutagen Information Center (EMIC) bibliography. These data are available free of charge if one contacts the Center (EMIC, 1985).

It is beyond the scope of this book to discuss the Ames/*Salmonella* assay in any detail, or to consider its effectiveness in predicting the potential carcinogenic activities of chemicals in the air environment (McCann, 1983). However, because of its widespread use as a screening test for the mutagenicity of extracts of POM samples, and because of its success in bioassay-directed chemical analyses for the isolation and identification of chemical mutagens in these complex environmental samples, we briefly outline the test and illustrate some of its analytical applications. Revised methods for the *Salmonella* test are given by Maron and Ames (1983) and Alfheim et al. (1984b).

13-E-1. Chemist's View of the Ames/*Salmonella* Test

13-E-1a. Principle and Procedure

Normal *Salmonella* bacteria (*his*$^+$) do not require added histidine for growth; they can make their own. Ames and colleagues constructed mutant strains (*his*$^-$) that lacked the ability to produce their own histidine and could grow only in histidine-enriched media. However, attack of certain chemicals (i.e., mutagens) at appropriate sites on the bacterial DNA produce *reverse* mutations in which the bacteria are converted back to their wild forms (*his*$^+$); they can now produce the histidine necessary for cell growth, and colonies develop on the plates.

Figure 13.9 is a simplified diagram of the test procedure; it takes about 3 days to complete. Essentially, one adds serial dilutions of the sample, dissolved in DMSO, to a series of tubes each containing agar and one of the several *his*$^-$ Ames test strains of bacteria. The resulting solution is then poured onto petri plates filled with minimal salts agar. A trace of histidine is added, not enough to permit the colonies to form but enough to allow sufficient growth for expression of mutations. The plates are then incubated for ~63 h at 37°C. At the end of that time they are removed from the incubator and any colonies present are counted.

If a significant increase in colonies above background is observed on Test Plate A, one concludes that the environmental sample contained a chemical(s) that is a *direct* mutagen for the particular Ames strain employed. (There is a background caused by *spontaneous his*$^-$ to *his*$^+$ reversions).

If no direct activity is observed on Test Plate A, it does not necessarily mean

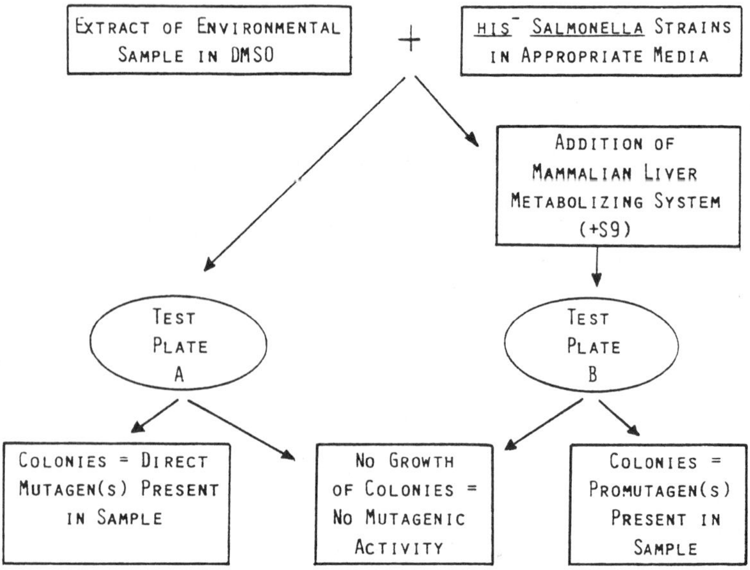

FIGURE 13.9. Diagram of the procedure involved in the Ames test for chemical mutagens.

there are *no* bacterial mutagens in the sample. Thus carcinogenic PAH such as BaP and benz(a)anthracene are *promutagens*; they are inactive per se and must first be metabolized to more reactive intermediates, which then attack the bacterial DNA. This *microsomal activation* is achieved by adding to another portion of the test sample a small amount of an enzyme system (+S9, the right-hand side of Fig. 13.9) which is derived from the liver of rats injected intraperitoneally with the inducer Arochlor (a mixture of polychlorinated biphenyls). If a significant number of colonies (above background) are seen on Test Plate B, then the environmental sample contains promutagens (i.e., activatable mutagens). In the case of primary and ambient airborne POM, both direct and promutagens are present.

One obtains the specific activity of the chemical or mixture of chemicals (i.e., the mutagenic potency in bacteria) from the slope of the quantitative dose–response curve. Activities are usually expressed as revertants (colonies counted) per microgram or per nanomole; depending on the chemical, they cover an enormous range, from <1 to $>5 \times 10^5$ rev μg^{-1} for certain dinitro-PAH.

13-E-1b. "Chemical Clues" from the Salmonella Assay

In addition to differentiating between a *direct*-acting nitroarene and its parent *promutagenic* PAH (e.g., 1-NO_2-BaP versus BaP) by the addition of S9 mix, other valuable clues as to the chemical structures of environmental mutagens can be obtained through use of the various tester strains developed by Ames and his colleagues. Currently recommended strains include TA98 and TA100, both of which contain a plasmid that increases their sensitivities over their parent strains, TA1538 and TA1535, respectively (Maron and Ames, 1983).

It is also chemically useful to recognize that the *his*$^-$ *Salmonella* strain TA1535 is reverted to the *his*$^+$ form, with accompanying growth of colonies on the test plate, by chemicals that are *base-pair substitution* mutagens, for example, β-propiolactone. On the other hand, strain TA98 responds to chemicals that produce *frameshift* mutations; these include, for example, direct-acting nitro-PAH and promutagenic amines and PAH such as 2-aminofluorene and BaP, respectively. Strain TA100 is very sensitive, with a relatively high background of spontaneous *his*$^-$ $\rightarrow$ *his*$^+$ reversions; it detects both base-pair substitution and frameshift mutagens.

The *Salmonella* reversion assay is also highly sensitive to changes in chemical structure, often showing dramatic specificity between various isomers of a substituted PAH. Thus, of the 12 monohydroxy-isomers of BaP, the 6-OH and 12-OH isomers are moderately active and the 1-, 3-, and 7-OH-BaP relatively weak direct-acting frameshift mutagens on strain TA98. The remaining seven isomers are inactive in the absence of metabolic activation (Jerina et al., 1976a, b; Wislocki et al., 1976). Interestingly, in a different short-term assay system using Chinese hamster V79 cells, 12-OH-BaP was inactive (Wislocki et al., 1976), while 6-OH-BaP tested negative for carcinogenicity on mouse skin (Kapitulnik et al., 1976).

The three nitro-isomers of BaP, formed when BaP is nitrated under laboratory conditions or in simulated polluted atmospheres, offer another example. Thus 6-NO_2-BaP is a strong promutagen, but it is not active (<1 rev μg^{-1}) in the absence of S9 mix whereas the 1- and 3-NO_2-isomers have direct activities of 2500 and 5300 rev μg^{-1} (Pitts et al., 1984). For comparison, the activities of BaP and 6-NO_2-BaP were 390 and 1500 rev μg^{-1}, respectively, when S9 mix was added. Examples of recently reported mutagenic activities of some nitro-derivatives of PAH relevant to simulated and real atmospheres are shown in Table 13.12. Comprehensive lists of mutagenic activities for nitroarenes, determined in several laboratories, include those of Rosenkranz and Mermelstein (1983, 1985); assay data for 300 chemicals of a variety of types are given in an early paper by McCann et al. (1975).

Another microbiological clue of chemical importance is derived through use of strains developed by H. Rosenkranz and his associates (Mermelstein et al., 1981). Thus strain TA98NR is an isolate of TA98, deficient in the "classical" bacterial nitroreductase which catalyzes the bioactivation of most mononitro-PAH to their ultimate mutagenic forms. These are generally agreed to be hydroxylamines which can then form adducts with the cellular DNA (Rosenkranz and Mermelstein, 1983). A lower response to TA98NR relative to TA98 indicates the probable presence of mononitro-PAH in the sample (for an excellent review and critique see Rosenkranz and Mermelstein, 1983). For example, the activities on strains TA98 and TA98NR, respectively, of 2-NO_2-FL, 1-NO_2-PY, 2-NO_2-PY, and 2-NO_2-FLN are: 3,900 vs. 930; 2,300 vs. 310; 16,000 vs. 1,600; and 450 vs. 60 rev μg^{-1} (Ramdahl et al., 1985b). Interestingly, it has recently been shown that in studies with mammalian systems, activation of nitro-PAH can occur through another mechanism, that of ring oxidation (El-Bayoumy and Hecht, 1982, 1983; Rosenkranz, 1984, and references therein).

Rosenkranz and his associates also developed another strain of TA98, TA98/1,8-DNP_6. While possessing a functioning nitroreductase, this strain is deficient in a second enzyme which activates the very potent bacterial mutagens, 1,8- and 1,6-dinitropyrene. Thus the fact that a POM extract shows a significant decrease in response on this strain, relative to TA98 and TA98NR, suggests the presence of dinitropyrenes in the sample. For example, extracts of ambient particulate matter show diminished responses in both strains, TA98NR and TA98/1,8-DNP_6. This indicates the probable presence of both mononitro- and dinitro-PAH in the respirable particles. These microbiological clues have since been confirmed chemically in several laboratories (see Section 13-G-2). As an example, the activities of 1,8-dinitropyrene on strains TA98, TA98NR, and TA98/1,8-DNP_6 are 900,000, 990,000, and 21,000 rev μg^{-1}, respectively (Ramdahl et al., 1985b).

Our understanding of the mechanism of action of nitro-PAH on bacterial DNA has progressed to the point where predictive chemical structure–mutagenic activity relationships have been developed (Rosenkranz et al., 1984). Thus Klopman and Rosenkranz (1984) report that certain structural features are essential for mutagenic activity. If these are absent or certain deactivating moieties are also present in the molecule, the nitroarene will be inactive. Their mechanism correctly predicted the activity of 47 out of 53 nitro-PAH, and the authors state that their theoretical, computerized approach should prove valuable in assessing the activities of the over 200 nitro-compounds which have been identified to date in the environment, but

Mutagenicity of Selected Nitro-PAH for *Salmonella typhimurium* TA98

Chemical	Revertants per Microgram (Revertants per Nanomole)							
	Literature Values				Values from One Laboratory[d]			
	−S9		+S9		−S9		+S9	
2-Nitrofluorene	66[a]	(14)			450[f]	(95)		
1-Nitrofluoranthene	2,200[b]	(540)						
2-Nitrofluoranthene					3,900[f]	(960)		
3-Nitrofluoranthene	22,000[b]	(5,400)			80,000[d]	(20,000)		
7-Nitrofluoranthene	300[b]	(70)						
8-Nitrofluoranthene	45,000[b]	(11,000)			100,000[d]	(25,000)		
1,2-Dinitrofluoranthene					6,200[d]	(1,810)		
1-Nitropyrene	1,900[b]	(470)			2,300[f]	(570)		
2-Nitropyrene	9,000[b]	(2,200)			16,000[f]	(3,900)		
1,3-Dinitropyrene	500,000[c]	(140,000)						
1,6-Dinitropyrene	630,000[c]	(180,000)						
1,8-Dinitropyrene	870,000[c]	(250,000)			900,000[f]	(260,000)		
1,3,6-Trinitropyrene	120,000[c]	(41,000)						
1,3,6,8-Tetranitropyrene	41,000[c]	(16,000)						
5-Nitrochrysene	< 2[b]	(< 0.6)	10[b]	(2.7)				
6-Nitrochrysene	< 50[b]	(< 14)	400[b]	(110)				
7-Nitrobenz(a)anthracene	< 1[b]	(< 0.3)	5[b]	(1.4)				
6-Nitrobenzo(a)pyrene	< 5[b]	(< 1.5)	700[b]	(210)	0[e]	0	1,500	(450)
1-Nitrobenzo(a)pyrene					2,500[e]	(740)	3,100	(920)
3-Nitrobenzo(a)pyrene					5,300[e]	(1,600)	2,900	(860)
1-Nitrobenzo(e)pyrene	130[b]	(39)						
3-Nitrobenzo(e)pyrene	3,000[b]	(890)						
3-Nitroperylene	< 100[b]	(< 30)	6,000[b]	(1,800)				

[a]McCoy et al., 1981.
[b]Greibrokk et al., 1985.
[c]Mermelstein et al., 1981.
[d]Statewide Air Pollution Research Center, University of California, Riverside, CA.
[e]Pitts et al., 1984.
[f]Ramdahl et al., 1985b; for reference, quercetin was 12 rev μg^{-1}(TA98,-S9); BaP was 390 rev μg^{-1} (TA98, +59).

which have not yet been tested. Another predictive approach utilizes four structural features of a nitroarene, including the physical dimensions of the aromatic rings, to predict its activity in *Salmonella* assays (Vance and Levin, 1984).

Finally, among the newer tester strains are TA102 and TA104. These are sensitive to a number of mutagens which are not detected, or detected poorly, by TA98 or TA100. Thus TA102 is reverted by hydroperoxides and DNA cross-linking agents such as mitomycin C (Levin et al., 1982; Maron and Ames, 1983).

Strain TA104 has recently been used to demonstrate the mutagenicity of various quinones, PAH derivatives that are widely distributed in nature (Chesis et al., 1984). Additionally, simple carbonyl compounds, α,β-unsaturated aldehydes, and dicarbonyl compounds are reported to be direct-acting mutagens in strain TA104 (Marnett et al., 1985); atmospherically relevant examples include formaldehyde, acrolein, methyl glyoxal, and glyoxal, respectively. Indeed, methyl glyoxal was the most potent of the carbonyl compounds tested in the study. Interestingly, Löfroth and co-workers (1985) have found that extracts of airborne POM are indeed mutagenic on strain TA104.

Use of the Ames test to screen for chemical mutagens in gas phase systems is more complicated than for condensed phases and much less is known about this potentially important area. In one recent study, the mutagenic activity of irradiated toluene-NO_x-H_2O-air mixtures was studied with strains TA98 and TA100, and a positive response was observed (Shepson et al., 1985). No unequivocal evidence was obtained as to the chemical nature(s) of the active species. Another such study with propylene as the hydrocarbon gave different products, of course, but similar overall results (Kleindienst et al., 1985). Methods for determining the bacterial mutagenicities of volatile and semivolatile compounds have recently been assessed by Claxton (1985).

13-E-1c. Examples of Other Short-Term Assay Systems

Several points can now be made from the perspective of atmospheric chemists using the Ames/*Salmonella* mutagenicity assay. It is a highly sensitive means of rapidly detecting the presence of chemical mutagens in complex environmental mixtures. Furthermore, in conjunction with chemical/physical analytical techniques, the test has proven an invaluable aid for the separation, isolation, and proof of structure of biologically active chemicals from complex combustion-related materials, for example, synthetic fuels and diesel soot. For a recent review with a discussion of modifications to the conventional Ames assay, see Alfheim et al. (1984b).

However, we caution that results from a *single*, short-term assay system should not be used to infer possible carcinogenic potencies of individual chemical mutagens or of organic extracts of primary and ambient airborne particulate polycyclic organic matter. Clearly, for this purpose, a battery of different types of short-term tests should be employed (Holmberg and Ahlborg, 1983; Lewtas, 1983; McCann, 1983; Brooks et al., 1984).

One in vitro, short-term test also employs *Salmonella typhimurium* bacteria, but in a *forward* mutation assay (Skopek et al., 1978; Skopek and Thilly, 1983). Using resistance to 8-azaguanine as a genetic marker, it has been employed to determine the mutagenic activities of a variety of combustion products of fossil fuels. For example, phenalen-1-one is a potent mutagen in this forward assay, six times more active than benzo(c)cinnoline (Leary et al. 1983).

Other test systems for gaseous atmospheric mutagens include the *Tradescantia* plant assay which utilizes somatic mutations in stamen hairs. The assay is more complex than the Ames test but has been used to demonstrate the presence of gas phase mutagens in ambient urban air environments (Schairer et al., 1979, 1983) and in environmental chamber exposures to simulated photochemical smog. Both PAN and O_3 in air produced mutagenic responses in this plant system.

13-E-1d. Variability of the Ames Test

Mutagenic activities for a given chemical reported in the literature have varied widely from laboratory to laboratory, and from a single laboratory as experimental procedures have improved over time. This is primarily due to differences in the manner in which the Ames test is carried out and in the nature and purity of the samples being tested.

For example, first reports of the direct mutagenicity of 1-nitropyrene gave activities of over 10,000 rev μg^{-1}, but today they fall in the vicinity of ~1000–2000 rev μg^{-1}. This drop in activity is the result of testing more highly purified 1-NO_2-PY samples with reduced levels of dinitropyrene contaminants. The latter are very powerful mutagens, some with activites over 500,000 rev μg^{-1}. Obviously, given the sensitivity difference of a factor of ~500, a trace of such a dinitropyrene would lead to a spuriously high value for the specific activity of 1-NO_2-PY.

Similar considerations apply to 6-NO_2-BaP which was first reported as a strong or moderate direct mutagen, but after careful purification procedures had essentially no activity in the absence of S9 mix (Pitts et al., 1982a, 1984). The trace impurities responsible for the activity were 1- and 3-NO_2-BaP, both of which are strong mutagens (see Table 13.12).

This question of sample purity is vexing, not only because of the large number of isomeric nitroarenes present in the environment and their huge range of activities, but also because some strongly mutagenic PAH derivatives, for example, certain hydroxynitro-PAH, tend to decompose during sample preparation and GC or MS analysis (Schuetzle, 1985).

With sufficient attention to certain procedural details in the Ames/*Salmonella* test, for example, the uniformity of the soft agar thickness, a high degree of *intralaboratory* precision can be obtained for samples assayed on a given day (Belser et al., 1981; Maron and Ames, 1983). This is evident from Table 13.13 which gives the specific activities (rev μg^{-1}) and mutagen densities (rev m^{-3}) for four identical samples of ambient POM collected concurrently on a *megasampler* developed to collect multiple samples of respirable particles at much higher flow rates than conventional hi-vol samplers (see Section 12-E-1). Standard deviations for the entire procedure—collection of particles, extraction, evaporation to dryness, and subsequent biological testing—were approximately ±10%. Both the specific

TABLE 13.13. Mutagenic Activities on Strain TA98, With and Without Metabolic Activation, of Organic Extracts of Four Identical Samples of Ambient Particles Collected Concurrently Using a Megasampler[a]

Filter Number	Total Extract Mass (mg)	Specific Activity (rev μg^{-1})		Mutagen Density (rev m^{-3})	
		−S9	+S9	−S9	+S9
1	22.47	2.4	2.6	73	79
2	23.26	2.0	2.6	63	81
3	24.46	2.0	2.4	66	79
4	23.25	2.1	2.5	66	78
Average (1–4)	23.26	2.1	2.5	67	79
Standard Deviation (%)	3.5	8.9	3.4	6.3	1.6

Source: Fitz et al., 1983.

[a]October 2–3, 1979, El Monte, California; Teflon coated glass fiber filters.

activities and mutagen densities are typical of average levels in the Los Angeles air basin. For a point of reference recall that an adult human at rest and at light work breathes ~11 m^3 and ~17 m^3 of air, respectively, per 24 h.

The excellent intralaboratory precision obtained for concurrently collected, replicate samples of ambient POM shows that one can obtain good precision if pre cautions are taken to control all important biological, chemical, and physical variables. This fact is also evident from the much better precision obtained in a four-laboratory intercomparison where rigorous control of such variables was implemented (Dunkel, 1979) versus the wider range found in an international comparison study where this apparently was not the case (DeSerres and Ashby, 1981). Further perspective on this issue is given by the intralaboratory study of the reproducibility of the activities of two *standard* mutagens in the Ames test (2-nitrofluorene and 2-aminoanthracene) carried out by Mermelstein et al. (1985) and by the results of a European collaborative study (Grafe et al., 1981).

13-E-1e. Bioassay-Directed Analyses for Chemical Mutagens in Complex Mixtures

Activity-directed analytical procedures are now routinely used to isolate and identify trace quantities of chemical mutagens in mixtures of products of combustion of fossil fuels and in ambient particles (Epler et al., 1979; Guerin et al., 1979; Huisingh et al., 1979; Pellizzari et al., 1979; Epler, 1980; Waters et al., 1979, 1981, 1983; Bjørseth, 1983; Holmberg and Ahlborg, 1983; Battelle Symposia on PAH; Alfheim et al., 1984; Schuetzle et al., 1985, and references therein).

They were also used in early studies of PAH reactions in simulated atmospheres. For example, ozonolysis of BaP produces a complex mixture of products active in the Ames assay without S9 mix. Figure 13.10 shows the overall procedure used to isolate the strong, direct mutagen BaP-4,5-oxide (1750 rev μg^{-1},

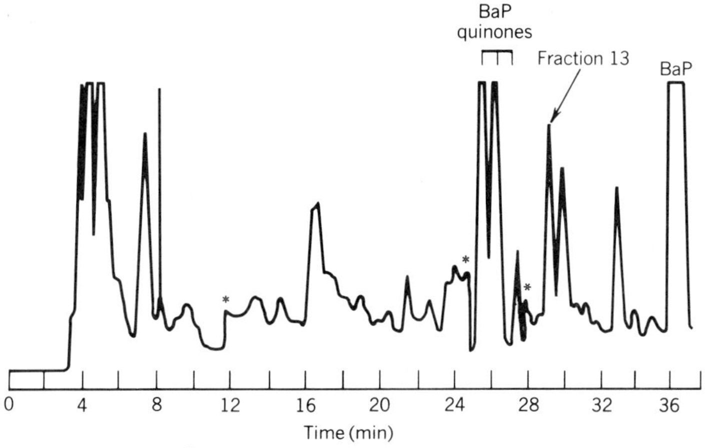

(a)

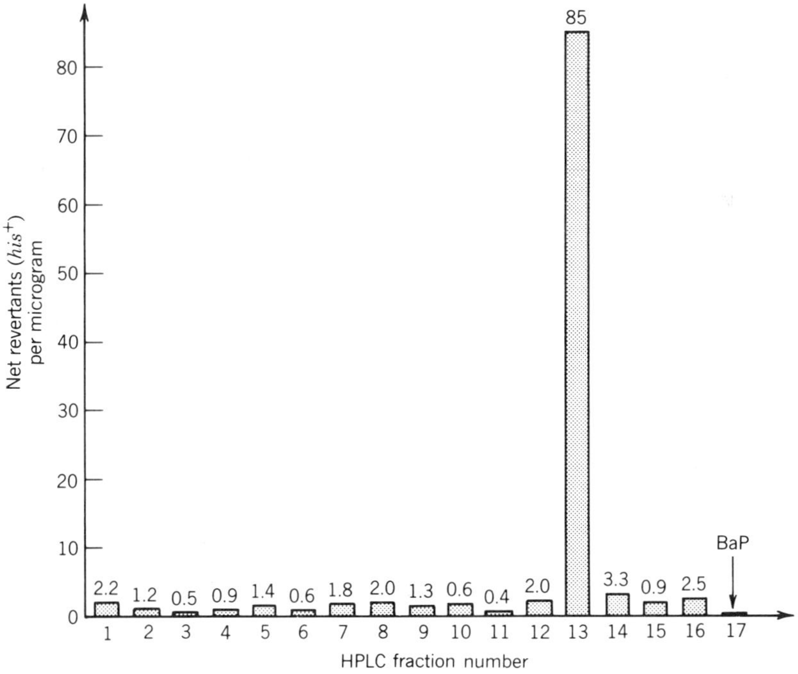

(b)

FIGURE 13.10. (*a*) Preparative HPLC separation of BaP–O_3 products. Asterisk marks scale change. (*b*) Specific mutagenic activity of HPLC fractions of BaP–O_3 products toward TA98, without activation. (*c*) Preparative HPLC separation of fraction 13. N.A., no activity. (*d*) Dose–response curve for fraction 13-4 (TA98) without activation (from Pitts et al., 1980).

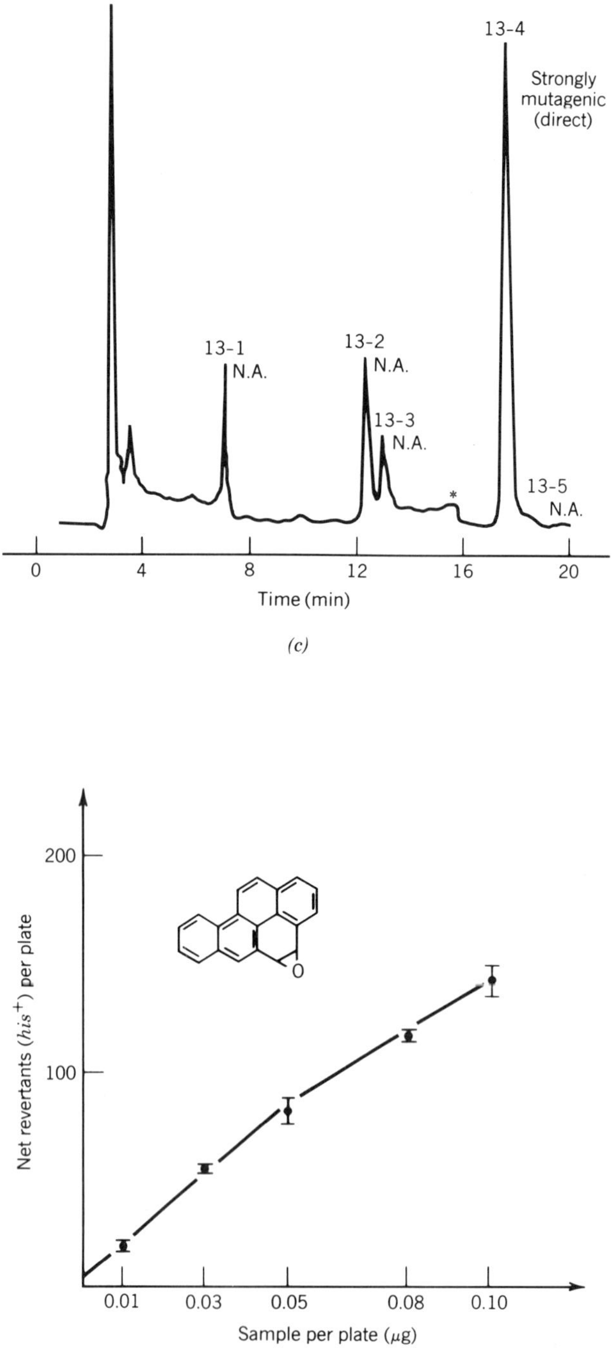

(c)

(d)

FIGURE 13.10. (*Continued*)

TA98) present in HPLC fraction 13-4 and formed in ~1–2% yield. This epoxide was identified by comparisons of its mass spectrum, UV absorbance, and fluorescence with a synthesized, authentic sample (Pitts et al., 1980). Interestingly, it previously had been identified as an animal metabolite of BaP by Wislocki and coworkers (1976), who reported an activity of 1370 rev μg^{-1}(TA98, -S9).

In summary, the Ames/*Salmonella* reversion assay:

- Is sensitive enough to detect trace amounts of a variety of chemical mutagens in chemically complex environmental samples.
- Is capable of being carried out by an "assembly line" procedure so that, with automatic counting and computerized data handling and display, a large number of plates can be handled efficiently (e.g., 600–700 plates per experiment).
- Is capable of a good degree of intralaboratory precision for replicate samples from a given experiment assayed on the same day, and reasonably good reproducibility between intralaboratory assays run on different days.
- Through the use of a variety of strains and S9 mix provides useful biological clues to the chemical identities of environmental mutagens.

The generosity of Professors Ames and Rosenkranz in providing their *Salmonella* strains, on request and without cost, to scientists throughout the world studying chemical mutagens is worthy of appreciative acknowledgment from the scientific community.

13-F. MUTAGENICITY OF RESPIRABLE AMBIENT PARTICLES

In 1976, Tokiwa and his colleagues reported that organic extracts of airborne particles collected in Ohmuta and Fukuoka, Japan, were active in the Ames test when S9 mix was present (Tokiwa et al., 1976, 1977). This was reasonable since it was known then that ambient POM contained promutagens such as BaP and BaA.

Soon thereafter, surprisingly strong *direct* mutagenic activity (i.e., -S9 mix) was found for extracts of POM samples collected in Berkeley, California, and Buffalo, New York (Talcott and Wei, 1977); the Los Angeles air basin (Pitts et al., 1977); Chicago, Illinois (Commoner et al., 1978); Kobe, Japan (Teranishi et al., 1978); New York City (Daisey et al., 1979); Ohmuta and Fukuoka, Japan (Tokiwa et al., 1980); Stockholm, Sweden (Löfroth, 1981); Contra Costa County, California (Wesolowski et al., 1981) as well as for extracts from aerosols transported long ranges (Alfheim and Møller, 1979). For reviews, see Chrisp and Fisher (1980) and Alfheim et al. (1984b).

Furthermore, the direct activity was primarily associated with organic species in respirable particles less than ~2 μm in diameter (Commoner et al., 1979; Pitts et al., 1979; Talcott and Harger, 1980; Tokiwa et al., 1980; Löfroth, 1981). Similar size distributions of direct and activatable mutagenicity were observed for coal fly ash (Fisher et al., 1979).

Reports of the discovery of the presence of hitherto unknown, direct-acting mutagenic species in extracts of respirable ambient POM, vehicle emissions (Wang et al., 1978; Huisingh et al., 1979), coal fly ash (Chrisp et al., 1978), and wood combustion (Löfroth, 1978; Rudling et al., 1982) resulted in a worldwide explosion of research. Prime goals were to isolate and characterize the airborne particulate chemicals responsible for this direct activity, to identify their sources, atmospheric reactions, and sinks, to determine their carcinogenic potencies, and to develop a chemical–biological data base that could be used to develop reliable risk assessment evaluations of their health effects.

Concurrently, interest in mutagenic respirable particles was fueled by the report of the strong, direct mutagenicity displayed by extracts of toners used for office copiers (Löfroth et al., 1980). This was soon shown to be due to the presence of strongly mutagenic nitro-PAH in the carbon black (Rosenkranz et al., 1980) and the problem quickly resolved by changing the conditions for activating the carbonaceous material.

Today the dramatic expansion of research on PAH, related to airborne mutagens and carcinogens, is reflected not only in the literature but also in a rapidly expanding number of symposia. The Proceedings of Battelle's Eighth International Symposium held in 1983, *Polynuclear Aromatic Hydrocarbons: Mechanisms, Methods, and Metabolism* (Cooke and Dennis, 1984), not only has 1464 pages, but there are also seven excellent predecessor volumes and two in preparation!

13-F-1. Ambient Air

The 1977 observations of the direct mutagenicity of extracts of ambient POM have been confirmed in major cities throughout the world. Although there are interlaboratory variations in mutagenicity determinations, some general conclusions have been reached. They include the following:

Most of the activity is in submicron particles, as with BaP, and is almost exclusively due to frameshift-type chemical mutagens. The activities without (−S9) and with (+S9) metabolic activation are generally about the same, but there are some significant differences. For example, ambient POM samples taken near major, heavily traveled streets or freeways during rush-hour traffic show enhanced activity when S9 is added. Presumably, this reflects the influence of unreacted exhaust emissions with high loadings of promutagenic PAH (e.g., BaP). However, after transport of the emissions from street levels downwind and to higher altitudes, this distinction disappears and the direct activities are equal or greater, possibly because of atmospheric reactions creating direct mutagens and/or destroying promutagens (Pitts et al., 1977, 1978; Alfheim et al., 1983).

Major variations in the mutagen densities (rev m^{-3}) and specific activities (rev μg^{-1} of extracted material, TA98, −S9 unless otherwise noted) are seen on samples averaged over a year, a season, a day, and 3 h intervals. Thus typical seasonally averaged levels in Stockholm, New York City, and Contra Costa County, California, are ~20 rev m^{-3} or less (Alfheim et al., 1983; Daisey et al., 1980; and Flessel et al., 1981) respectively. However, levels are generally significantly higher in the winter than summer, ranging in Stockholm, for example, from a peak of ~50 rev m^{-3} in mid-winter to 10 or less in summertime. Wintertime samples in Santiago, Chile, are reported to have ranged from ~40 to 300 rev m^{-3}, in two cases reaching ~1000 rev m^{-3} when S9 mix was added (Tokiwa et al., 1983).

The impact of day-to-day variations in meteorological conditions (e.g., atmospheric stability and wind speed) is illustrated for Oslo, where the activities of particulate samples collected on a day-to-day basis on a heavily traveled street in Oslo ranged from ~10 to 50 rev m^{-3} during January–February 1979 (Alfheim et al., 1983).

The diurnal variation in mutagen densities of samples of ambient POM collected for consecutive, 3 h time intervals on TIGF filters at a downwind site immediately adjacent to a heavily traveled freeway in West Los Angeles, California, is shown in Fig. 13.11. The mutagen densities reflect the freeway traffic and the meteorology, with maxima at mid-morning and early evening rush hours (~100–110 rev m^{-3}) and minima in the early afternoon (~40 rev m^{-3}) and early morning hours (~40 rev m^{-3}).

A comparable sampling station was also operating on the other side of the freeway. Concurrent sampling under appropriate wind conditions showed that the *incremental* burden of mutagenic respirable particles due to light and heavy duty vehicles could reach as high as 50 rev m^{-3} (Pitts et al., 1985c). The levels of mutagenic activity and their diurnal profiles observed in this study in West Los Angeles were similar to those observed earlier at a site in East LA where a maximum of ~120 rev m^{-3} was reached (Pitts et al., 1982b). Clearly, short-term peak mutagen densities can be much higher than 24 h average values commonly reported. Furthermore, as can be seen from the diminished response on strain TA98NR versus TA98 (Fig. 13.11), substantial quantities of nitroarenes appear to be present in POM collected near the freeway. These results are consistent with the original observations of Wang et al. (1980) who first showed the diminished activity on strain TA98NR of extracts of ambient POM collected in Wayne County, Michigan. This mutagenic effect has also been reported for POM collected in Stockholm, Sweden (Löfroth, 1981; Alfheim et al., 1983), for example, and is a general urban phenomenon. Interestingly, a strong correlation between mutagenicity and lead containing fine particles from gasoline-fueled LDMV without catalysts operating in Contra Costa County, California, was recently reported by Flessel and co-workers (1985). They also noted possible contributions to ambient levels of PAH by residential wood combustion.

Additional information on certain chemical constituents in ambient POM is derived from studies using TA98/1,8-DNP_6; this strain is insensitive to dinitropyrenes. Thus, for example, in samples of ambient POM collected at rooftop levels in both Gothenburg and Stockholm, the mutagenic activities on strain TA98 were significantly diminished when strains TA98NR and TA98/1,8-DNP_6 were employed; this suggests that *both* mononitro-PAH and dinitropyrenes were present (Toftgard et al., 1983; Alfheim et al., 1983). Similarly, Tokiwa et al. (1983) found that extracts of POM from Santiago, Chile, were not only strongly mutagenic, but the activity was greatly reduced on strain TA98NR/1,8-DNP_6 versus TA98, sug-

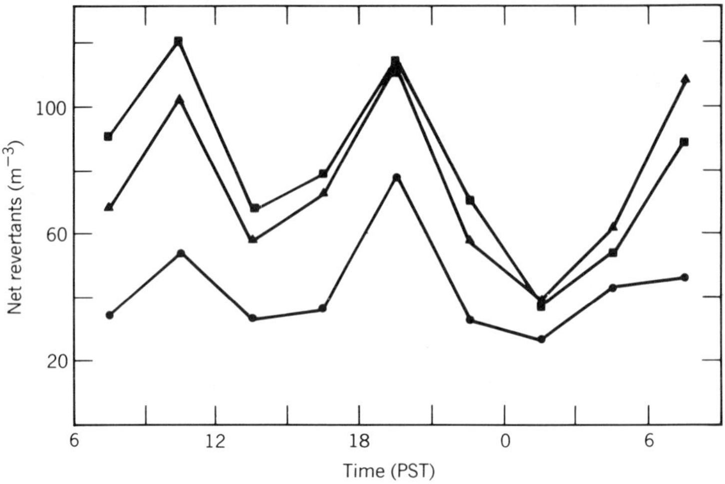

FIGURE 13.11. Diurnal variations in mutagen densities (rev m^{-3}) of 3 h averaged samples of ambient POM collected on TIGF filters with hi-vol samplers at a site downwind of and immediately adjacent to a heavily traveled freeway in West Los Angeles, CA, March 9–10, 1983. Strain TA98; ■, +S9; ▲, −S9; ●, TA98NR (−S9) (from Pitts et al., 1985c).

gesting the presence of dinitropyrenes. This was confirmed by chemical analyses which showed about 0.2 μg of dinitropyrene per gram of particulate matter.

Smoke from residential wood combustion (RWC) can also be a major source of airborne PAH and direct-acting mutagens (Cooper. 1980; Wolff et al., 1981; Rudling et al., 1982). For example, in the winter of 1981, in the town of Elverum, Norway (which has 10,000 inhabitants and is surrounded by forests), RWC contributed about two-thirds of the total PAH emissions, and mutagen densities (TA98, −S9) averaged ~20 rev m^{-3} (Ramdahl et al., 1984a). In this work, as in studies in Denver, Colorado (Wolff et al., 1981), and Oregon (Cooper et al., 1981), carbon-14 was employed as a tracer of contemporary carbon (vs. fossil carbon), while retene, a pyrolysis product of abietic acid which is present in wood resin, was used as a specific marker for RWC (Ramdahl, 1983a).

Today, as a result of the escalating energy costs of the mid-1970's to early 1980s, residential combusion of wood has reached major proportions in many larger cities in the world where it is available for domestic use, and the smoke contributes significantly to the burden of airborne particulate mutagens. Emissions from the combustion of wood and peat in commercial hot water boilers also contain strong direct mutagens, as Löfroth reported in 1978 in one of the early papers discussing use of the *Salmonella*/microsome test to determine the mutagenicity of combustion-generated emissions of POM.

In contrast to emissions from motor vehicles, wood stoves, or small hot water boilers, large power plants fired with pulverized coal and run under efficient conditions of fuel combustion generally emit only small amounts of organic compounds (Fisher, 1983). Thus health concerns focus on the presence of possible metal human carcinogens (such as arsenic, cadmium, and their oxides) which are present in coal and condensed on the surfaces of respirable fly ash particles emitted to the atmosphere (McElroy et al., 1982; Fisher, 1983; Vouk and Piver, 1983; Ahlberg et al., 1983; see the report edited by Holmberg and Ahlborg, 1983).

Residential combustion of wood in an open fireplace has been shown to contribute to indoor air pollution, producing PAH and relatively high concentrations of moderately polar PAH derivatives, and mutagenicity levels (rev m^{-3}) which approximated those found in urban air (Alfheim and Ramdahl, 1984). Interestingly, in this study, the indoor mutagenic burden of smoke particles from the open fireplace proved to be *moderate* compared to that resulting from tobacco smoking in the same room which, in this experiment, was heated by electricity. While the PAH profiles were about the same in both cases, the mutagenicity from smoking was much higher than for wood burning and the authors note that PAH levels cannot be used as an indicator for mutagenicity, a point that also has been made for outdoor air pollution.

Another wintertime study was conducted in Kyoto, Japan, in the living room of the home of a non-smoker. It was heated by an unvented kerosene-fired radiant stove for 5–8 h day^{-1}. Extracts of POM collected *indoors* proved to be much more active on strains TA98 and TA100 than those from outside ambient air (Yamanaka and Maruoka, 1984).

In concluding this section, it is important to note that collecting airborne particulate matter on different types of hi-vol filters (e.g., GF vs. TIGF) or with different types of samplers can lead to significantly different results. For example, in a comparison study of high-volume filtration (HVF) and electrostatic precipitation (ESP) in which samples of urban air were collected simultaneously, Alfheim and Lindskog (1984) found that the concentrations of PAH in the samples of particulate matter were in good agreement, but wide variations in the mutagenic activities of the extracts were observed. Higher activities were seen for samples from TIGF filters than from GF, possibly because of differences in filter artifacts and/or sample degradation. This entire area remains a complex and perplexing issue, important to risk assessment evaluations of combustion-generated POM and worthy of further fundamental and applied research.

13-F-2. Primary Emissions of Direct Mutagens from Motor Vehicle Exhaust

The fact that exhaust emissions are a major source of particulate mutagens was first shown in studies of the respirable-size, exhaust particles emitted by light duty motor vehicles (LDMV) and heavy duty motor vehicles (HDMV) with gasoline and diesel engines (Wang et al., 1978; Huisingh et al., 1979; Ohnishi et al., 1980). Diesels received special emphasis because of their much greater emissions of soot (up to 50–100 times greater than for a gasoline-fueled car with an oxidation catalyst) and because of the powerful direct activity in the *Salmonella* assay displayed by extracts of diesel soot. Additionally, it was predicted that by 1985 the sale of diesel-equipped LDMV would comprise a significant fraction of the new car market, especially in the United States. Although this has not occurred, intense research activity on the chemical composition and mutagenic and carcinogenic activity of exhaust particles continues.

These studies cannot be reviewed in detail here; see the NAS diesel assessment documents (1981, 1982a, 1982b) and the edited books and symposia published on this and related subjects, for example, Lewtas (1982), Bjørseth (1983), Holmberg and Ahlborg (1983), NAS (1983), Rondia et al. (1983), White (1985), and Cooke and Dennis (1984) (and preceeding volumes in this series).

As noted above, exhaust particles from gasoline- or diesel-fueled LDMV and HDMV fall in the submicron size range. Their emission factors, mutagenic activities, and chemical compositions are subject to wide variations, depending on manufacturer and engine types, engine temperatures and mode of operation, emission control systems, fuel, and so on (e.g., see Claxton and Kohan, 1981, Claxton, 1982, and Jensen and Hites, 1983).

Estimates of the specific mutagenic activities of the organic extracts of particles from diesel and gasoline-fueled LDMV vary widely but generally fall in the range from ~1 to 5 rev μg^{-1} with extremes as high as ~20. While there was some disagreement in the earlier literature as to which type of engine produced the most mutagenically potent POM extracts, a recent study in the laboratory and in a highway tunnel indicates that the order of potency of the POM extracts from the lowest to the highest (rev μg^{-1}, TA98, −S9) is: ambient air POM sample < tunnel < heavy duty diesel < light duty diesel < spark ignition (Brooks et al., 1984). However, on a per km (or mile) driven basis, the diesel LDMV have far higher mutagenic emission rates (TA98, −S9) than gasoline-fueled vehicles with catalysts. Thus, in studies of a number of certification vehicles, Claxton and Kohan (1981) estimate the diesel cars emitted ~45 to 800 times as much mutagenic activity per mile as gasoline cars *with catalysts*; spark-ignition cars without controls of course emit significantly higher levels of mutagenic POM per mile than cars with catalysts (Alsberg et al., 1985).

13-G. CHEMICAL COMPOSITION OF AIRBORNE PARTICULATE POLYCYCLIC ORGANIC MATTER: DIRECT-ACTING MUTAGENIC DERIVATIVES OF PAH

The task of determining the identities of the chemical species in ambient and primary combustion-generated POM responsible for its bacterial mutagenicity has been, and continues to be, formidable. Thousands of organic compounds are present in the extracts, their activities range from 0 to 10^6 rev μg^{-1}, and many of them are thermally unstable. However, through bioassay-directed chemical analyses incorporating a range of sophisticated analytical techniques, major advances are being made.

13-G-1. Bioassay-Directed Chemical Analyses of Extracts of Diesel POM

As noted, in the late 1970s when the *Salmonella* assay was first being applied to POM extracts, there were understandably wide inter- and intralaboratory variations between reported activities for standard reference mutagens (e.g., 2-nitrofluorene) and total POM extracts. This led to some misconceptions as to the relative contributions of specific direct mutagens such as 1-nitropyrene to the observed activity of a total POM extract or its fractions. Such problems are being resolved and current *mutagen balances* are much improved; that is, there is better agreement between the sum of the products of the concentration of each chemical mutagen (c_i) times its specific activity (a_i) [$\Sigma c_i a_i$] and the observed total activity of an entire extract or of a given subfraction.

An example of the current state of the art in bioassay-directed chemical analyses is the recent study of diesel POM by Salmeen, Schuetzle, and co-workers at the Ford Motor Company. Their procedures for the analysis of moderately polar and polar fractions of a diesel extract can also be applied to the biological and chemical characterization of ambient POM (Salmeen et al., 1984; Schuetzle et al., 1985).

In the study by Salmeen and co-workers, dichloromethane extracts of POM from passenger diesel cars were separated into 65 fractions by normal phase HPLC. Of these, only six fractions showed pronounced mutagenicity profiles (TA98, −S9); these were attributed to 3- and 8-nitrofluoranthene, 1-nitropyrene, and 1,3-, 1,6-, and 1,8-dinitropyrene. These six nitroarenes accounted for ~30–40% of the total mutagenicity recovered from the HPLC columns. An additional 15–20% of the total recovered activity was found in the polar fraction, accounting for ~20–30% of the extract mass; this fraction was not characterized in this study (*vide infra*). The rest of the total mutagenicity was distributed more or less equally among the remaining fractions; these were discarded.

Table 13.14 shows the concentrations of the six nitroarenes identified in the CH_2Cl_2 extract of one sample and their percent contribution to the total mutagenicity recovered from the columns (about 70% of the activity of the unfractionated sample).

The difficult problem of characterizing the polar mutagens has been addressed by Schuetzle and co-workers (1985). They extracted the particulate matter collected from several heavy duty diesel engines (National Bureau of Standards—Standard Reference Material 1650) with CH_2Cl_2 followed by methanol. The CH_2Cl_2 extract was fractionated on silicic acid and normal phase HPLC (NP-HPLC) was carried out on the fraction eluted from the silicic acid with methanol. Three moderately polar fractions (defined as eluting between 1-nitronaphthalene and 1,6-pyrene quinone in NP-HPLC) contained mono- and dinitro-PAH, including the six cited earlier (Table 13.14). Three polar fractions were further fractionated on a reverse phase C_{18} column.

These polar fractions accounted for a large percentage of the total extract mutagenicity and contained hydroxy-PAH, hydroxynitro-PAH, PAH-quinones, and nitrated heterocyclics.

TABLE 13.14. Concentrations of the Six Nitroarenes Identified in a Dichloromethane Extract of a Sample of Diesel POM and Their Percentages of the Total Recovered Direct Mutagenicity (TA98, −S9) from All Fractions of an HPLC Separation of the Extract

Nitroarene	Concentration (ppm)	Percentage Recovered Mutagenicity[a]
1-Nitropyrene	75 ± 10	
3-Nitrofluoranthene	3 ± 2	6 (1-nitropyrene, 3- and 8-nitrofluoranthene together)
8-Nitrofluoranthene	2 ± 1	
1,3-Dinitropyrene	0.3 ± 0.2	9 (1,3- and 1,6-dinitropyrene together)
1,6-Dinitropyrene	0.4 ± 0.2	
1,8-Dinitropyrene	0.5 ± 0.3	19
Broadly distributed (not assignable)	—	50
Polar fraction	—	16
		100%

Source: Salmeen et al., 1984.

[a]See original paper for specific activities of these nitroarenes used in this calculation. Table 13.12 gives other published values for these specific activities.

Certain of the multi-functional nitrated PAH (e.g., hydroxynitro-acenaphthene, hydroxynitro-9*H*-fluorenone, and hydroxynitro-fluorene) and the nitrated heterocyclic isomers (e.g., nitrobenzocinnoline and nitroquinoline), which were tentatively identified, are particularly interesting because of their relatively high mutagenic activities (TA98, −S9). However, the authors note that specific isomers of the hydroxynitro-PAH, including the hydroxynitropyrenes, are difficult to identify and to assay quantitatively because they decompose readily during sample preparation and GC or GC—MS analysis. New analytical techniques that would reduce this problem are discussed by Schuetzle et al. (1985).

The relative importance of the polar (P) versus the moderately polar (MP) fractions of extracts of ambient POM and particulate matter from light duty and heavy duty diesel engines and wood smoke is shown in Fig. 13.12 (Schuetzle et al., 1985). With the exception of POM from LD diesels, the majority of the total direct-acting mutagenic activity is due to unknown polar mutagens. As we have seen, these have proven difficult to identify but their importance, for example, in ambient air and wood smoke (Alfheim et al., 1984a) is evident. Their isolation and identification has a high priority.

The six nitroarenes identified by Salmeen et al. (1984) as contributing significantly to the HPLC-mutagenicity profile of the moderately polar fraction of a diesel extract are, as they note, only a small fraction of the total of those that have been characterized (e.g., tentatively identified) or identified. Thus Xu and coworkers (1982), based on the HRMS of directly mutagenic fractions from both

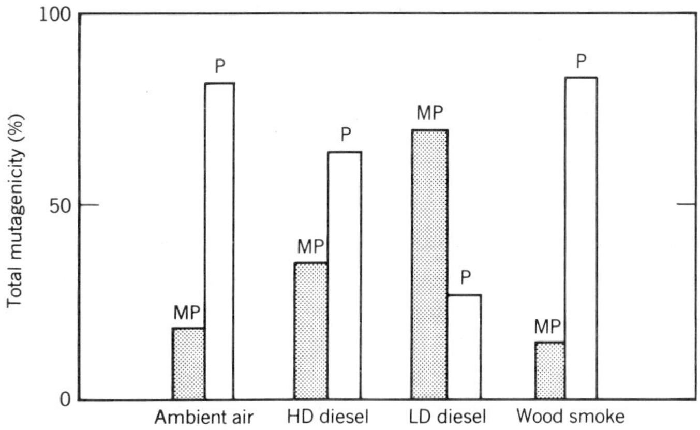

FIGURE 13.12. Distribution of direct-acting mutagenicity (TA98, −S9) between moderately polar (MP) and polar (P) fractions of extracts of particulate matter in ambient air, wood smoke, and in the exhaust from heavy duty (HD) and light duty (LD) motor vehicles. Adapted from Schuetzle et al. (1985); see this for original references.

low- and high-resolution HPLC, tentatively identified more than 50 nitro-PAH in an extract of diesel POM.

Subsequently, Paputa-Peck and co-workers (1983) used capillary column GC with a nitrogen–phosphorus detector (GC–NPD) to characterize and tentatively identify 45 nitro-PAH (and positively identify 17 specific isomers) in HPLC fractions from a CH_2Cl_2 (DCM) extract of POM from an LD diesel. Compounds included alkylated and oxygenated nitro-PAH derivatives as well as nitroheterocycles. Quantitative GC–MS analysis of 1-NO_2-PY, 1,3-DNP, 1,6-DNP, and 1,8-DNP was achieved through use of perdeuterated analogues as internal standards.

The bacterial mutagenicities of many of these nitroarenes in diesel POM have not been determined and there is considerable debate as to whether or not they constitute a significant threat to public health. Serious concerns have been expressed by Rosenkranz (1982, 1984) and Wei and Shu (1983). However, Gibson (1983) has addressed the sources of direct-acting mutagens in airborne POM and concluded that "the health risks from nitroarenes, if any, are not at all certain." To one interested in risk assessments and public health, it is worth reading these widely devergent perspectives of the potential impact of the mutagenic nitroarenes in diesel POM; see also the 1981, 1982, and 1983 NAS documents and the recent article "Health Effects from Light-Duty Diesel Vehicles" by Cuddihy et al. (1984). It is also relevant that several mono- and dinitro-PAH have been shown to be animal carcinogens (Takemura et al., 1974; El-Bayoumy et al., 1982; Sugimura and Takayama, 1983; Ohgaki et al., 1985; King et al., 1985). Additionally, 2-nitronaphthalene is metabolically reduced to the corresponding amine (β-naphthylamine) which is a proven human carcinogen (IARC, 1983, 1984).

One interesting aspect of the problem emerged from the study by Brooks et al. (1984) who tested the activity of extracts of diesel POM in various short-term bioassays. These included cell killing, sister chromatid exchange, the Ames/*Sal-*

monella assay, and several others. They detected genotoxic activity in every particle extract; furthermore, the "potency ranking was similar using different genetic endpoints, and the magnitude of the genotoxic potency was similar."

While the focus of much of the research to date on diesel POM has been on nitro-PAH, a host of oxygenated PAH (OXY-PAH) have been identified, as well as N-heterocycles and their derivatives. Thus, for example, Schuetzle et al. (1981) using high-resolution GC–HRMS, HPLC, and direct-probe HRMS have found hydroxy, ketone, quinone, carboxaldehyde, acid anhydride, and dihydroxy derivatives of PAH in diesel extracts. Additionally, Ramdahl (1983b) has characterized some 60 polycyclic aromatic ketones with MW 180–330 in diesel exhaust.

While most of the attention to date has been on diesel POM, a recent chemical/biological study of the extracts of exhaust particles from a noncatalyst, gasoline-fueled LDMV led to the following conclusions (Alsberg et al., 1985). The crude extracts, and fractions III and IV derived therefrom, gave strong, positive responses to the three biological test systems employed, including direct mutagenicity in the Ames test. In these two fractions, polynuclear aromatic ketones are the most abundant species; however, none of those isolated and tested displayed significant mutagenicity. The most polar fraction V, which held the nitrogen-containing compounds, showed lower effects in all test systems than fractions III and IV. With the exception of the promutagenic PAH found in fraction II, the chemical species responsible for most of the biological activity observed in the three test systems were not identified; however, derivatives of fluoren-9-one and phenalen-1-one or other oxy-PAH in fractions III and IV were suggested as possible candidates.

Finally, Behymer and Hites (1984) have compared the abundance and distribution of certain PAH and OXY-PAH (and their alkyl homologues) emitted by LDMV with diesel and spark-ignition engines. They found that the different engine types produced mixtures of these compounds that were qualitatively and quantitatively similar and they suggest it would be difficult to distinguish diesel from spark-engine emissions once they enter the environment. Among the compounds characterized were carboxaldehydes of naphthalene (and their C_1, C_2, and C_3 alkyl homologues) biphenyl and phenanthrene, and 9*H*-fluoren-9-one and thioxanthene.

13-G-2. Chemical Composition and Related Direct-Acting Mutagens in Ambient POM

The chemical complexity of organic extracts of ambient POM has already been discussed in several sections. We focus here on the PAH derivatives that have been characterized or identified and that may be direct-acting mutagens.

13-G-2a. Nitro-PAH and Derivatives

To date the most intensively studied, for reasons we have discussed, are the mono- and dinitroarenes and, more recently, hydroxynitro-PAH. Initial interest in nitroarenes was triggered by concurrent reports in 1978 from three laboratories that: (1) 3-nitrofluoranthene and 6-nitro-BaP were present in DCM extracts of ambient

POM collected in Prague (Jägar, 1978); (2) automobile exhaust contained direct mutagens including nitro-PAH (Wang et al., 1978); and (3) exposures of BaP, perylene, and pyrene to near ambient levels of gaseous NO_2 + HNO_3 in simulated atmospheres yielded nitro-derivatives, some of which were potent, direct-acting mutagens (Pitts et al., 1978).

Subsequently in Japan, Tokiwa et al. (1981) detected 1-NO_2-PY, 3-NO_2-FL, and 5-nitroacenaphthalene in extracts of both ambient particles and diesel exhaust, as well as dinitropyrenes in POM samples from Santiago, Chile (Tokiwa et al., 1983). In Europe, Nielsen (1983) identified and measured 1-NO_2-PY, 9-nitroanthracene, and 10-nitrobenz(a)anthracene; subsequently, in addition to these nitro-PAH, he and his colleagues identified 3-NO_2-FL, 6-NO_2-BaP, 2-NO_2-PY, and *x*-nitro-4,5-methylene-phenanthrene (*x*-nitro-4*H*-cyclopenta(def)phenanthrene) in POM collected in a rural area of Denmark (Nielsen et al., 1984). The latter compound recently was identifed as 3-nitro-4*H*-cyclopenta(def)phenanthrene (Ramdahl, 1985).

The first mass spectrometric evidence for the presence of nitroarenes in ambient POM was that of Ramdahl and co-workers (1982). They identified nitronaphthalene, 9-nitroanthracene, 3-nitrofluoranthene, 1-nitropyrene, as well as arenecarbonitriles and OXY-PAH, in a sample of airborne particulate matter from St. Louis, Missouri. Concurrently, Gibson (1982) also identified 1-nitropyrene and 6-NO_2-BaP in POM collected in Detroit, Michigan, and Sweetman and co-workers (1982) identified 1-nitropyrene in extracts of POM collected in southern Ontario, Canada.

In a recent chemical/bioassay study at an urban and a rural site in Michigan, 1-nitropyrene, 1,6-dinitropyrene, and 1,8-dinitropyrene were detected by HPLC in moderately polar fractions. However, their combined activities on Ames strain TA98 only accounted for $\sim 3\%$ of the total airborne mutagenicity, most of which was present in the four most polar fractions; the compounds responsible were not identified but are under investigation (Siak et al., 1985).

While most of the nitroarenes (and OXY-PAH) found in ambient POM are also present in primary combustion-generated emissions, for example, from motor vehicles or wood smoke, several are not. Thus the discovery by Scandinavian researchers of 2-nitropyrene (Nielsen et al., 1984) in rural POM (cited above) and the subsequent identification of 2-nitrofluoranthene (and 2-nitropyrene) in urban–suburban POM in southern California (Pitts et al., 1985g) are interesting. The authors of these two studies point out that the presence of these isomers is evidence for atmospheric nitration reactions, especially since the 2-isomers of pyrene and fluoranthene are not artifacts generated during the collection of ambient POM on filters. They also note the need to conduct studies of the environmental fates and biological activities of these 2-nitro isomers; most of the attention to date has been on 1-nitropyrene and 3-nitrofluoranthene.

Due to very close gas chromatographic elution times for 2- and 3-nitrofluoranthene (Pitts et al., 1985g), some mis-identifications have been made in two of the studies mentioned above. Thus, the samples studied by Ramdahl et al. (1982) and Nielsen et al. (1984) have been re-analyzed, and the 3-nitrofluoranthene reported in these studies has been shown, in fact, to be 2-nitrofluoranthene (Ramdahl et al., 1985a). 2-Nitrofluoranthene has also recently been shown to be present in

POM collected in Washington, DC (NBS SRM 1649) by Sweetman et al. (1985a). In all these studies, including reports by Pitts et al. (1985g) and Ramdahl et al. (1985b), 2-nitrofluoranthene is shown usually to be the most abundant nitro-PAH in ambient air, regardless of sampling location.

The contribution to the mutagenicity of ambient POM from 2-nitrofluoranthene has been shown to be as high as 5% of the total mutagenicity (Ramdahl et al., 1985b). This suggests that the mutagenicity caused by this single nitro-PAH could be even higher than the contribution from the highly mutagenic dinitropyrene reported by Siak et al. (1985).

13-G-2b. Oxygenated PAH

Because most of the direct-acting mutagenicity in the moderately polar and polar fractions of primary and ambient POM is still unaccounted for, there is strong interest in the possibility that certain OXY-PAH may be important contributors.

An example of the complexity of the problems is furnished by the detailed studies of König and co-workers (1983) and Ramdahl (1983b). König used a two-step column chromatographic separation of extracts of POM collected in Duisburg, West Germany (Ruhr area), followed by analysis using capillary GC and capillary GC–MS to characterize 38 OXY-PAH; they belonged to five different classes of the compounds: PAH-ketones, quinones, anhydrides, coumarins, and aldehydes. Of this total, the compounds unequivocally identified were: the ketones 9*H*-fluorene-9-one, 11*H*-benzo(a)fluorene-11-one, and 7*H*-benz(de)anthracene-7-one; the quinones anthracene-9,10-dione, benz(a)anthracene-7,12-dione, and naphthacene-5,12-dione, and naphthalene-1,8-dicarboxylic acid anhydride. None of the three coumarins or the five PAH-aldehydes were specifically identified, although the latter were stated to be the first aldehydes to be detected in airborne POM. A major reason is the lack of pure reference substances. As we have already seen, this is a worldwide problem inherent in today's analytical chemistry of PAH derivatives.

13-H. CHEMICAL TRANSFORMATIONS OF PARTICULATE PAH IN SIMULATED AND AMBIENT ATMOSPHERES

A wide variety of PAH are released to the atmosphere adsorbed on the surfaces of combustion-generated respirable particles. The physical and spectroscopic properties of these particulate-bound PAH depend on the nature of the substrate, for example, diesel soot versus wood smoke versus coal fly ash; their chemical reactivities are also affected significantly by their molecular environment. These combustion-generated particles are released into urban atmospheres containing gaseous co-pollutants such as O_3, NO_2, SO_2, HNO_3, PAN, and so on and are concurrently exposed to sunlight and molecular oxygen.

Given this "real world" reaction system, it is clear why definitive, quantitative answers to basic questions on the rates, products, and mechanisms of the atmospheric reactions of particle-bound PAH suspended in urban air have been, and continue to be, elusive. Significant progress has been made in laboratory studies of the rates and products of the reactions of PAH deposited on various specific substrates and exposed in the dark and in the light to simulated polluted atmospheres; however, there remains the major problem of extrapolating the results to conditions that prevail in ambient atmospheres.

With this major caveat in mind, we summarize the results of studies of PAH transformations in simulated and real atmospheres. Details, historical and scientific, as well as references to the literature are found in review/critiques by Nielsen et al. (1983), Pitts (1983), Van Cauwenberghe and Van Vaeck (1983), Nikolaou et al. (1984), and Valerio et al. (1984).

13-H-1. Historical

Based on experiments in the 1950s and 1960s (Kotin et al., 1954; Falk et al., 1956; Tebbens et al., 1966, 1971; Thomas et al., 1968), a Committe of the National Academy of Sciences suggested that gas–particle interactions in exhaust systems, plumes, and ambient air might lead to significant degradation of particle-bound PAH and to the formation of products more polar than the parent PAH (see NAS, 1972). In laboratory simulations and in filter artifact studies, this has proven to be the case. Additionally, it has been demonstrated that the products of such interactions include powerful, direct-acting mutagens.

13-H-2. Nitration of PAH in Simulated Atmospheres

13-H-2a. $NO_2 + HNO_3$

It is now well established that certain PAH, deposited on a variety of substrates and exposed to a range of concentrations of gaseous $NO_2 + HNO_3$ from ppm to near ambient levels, react to form mono- and dinitro-PAH (Pitts et al., 1978, Hughes et al., 1980; Jäger and Hanus, 1980; Tokiwa et al., 1981).

For example, in the first product study in a simulated atmosphere, BaP deposited on a GF filter was exposed for 8 h to air containing 0.25 ppm of NO_2 and traces of gaseous HNO_3. About 20% of the BaP was converted, predominately to 6-NO_2-BaP with smaller amounts of the 1- and 3-isomers (Pitts et al., 1978; Pitts, 1979). Figure 13.13*a* shows the UV/visible absorption spectra of BaP, 6-NO_2-BaP, and a mixture of the 1- and 3-NO_2 isomers.

In this reaction, a promutagen was converted to a stronger promutagen, 6-NO_2-BaP, and two powerful direct mutagens, the 1- and 3-NO_2-BaP isomers (TA98; see Table 13.12).

The generality of the reaction was demonstrated when, under similar laboratory conditions, perylene (a weak promutagen) and pyrene (a non-mutagen) were converted to 3-NO_2-perylene (Fig. 13.13*b*) and, at a slower rate, to 1-nitro-pyrene, respectively; these both proved to be direct-acting mutagens. In a similar experiment, chrysene was not nitrated; this was consistent with their relative rates of nitration in solution, chrysene = 1, pyrene = 4.9, perylene = 22, and BaP (6-position) = 31 (Dewar et al., 1956).

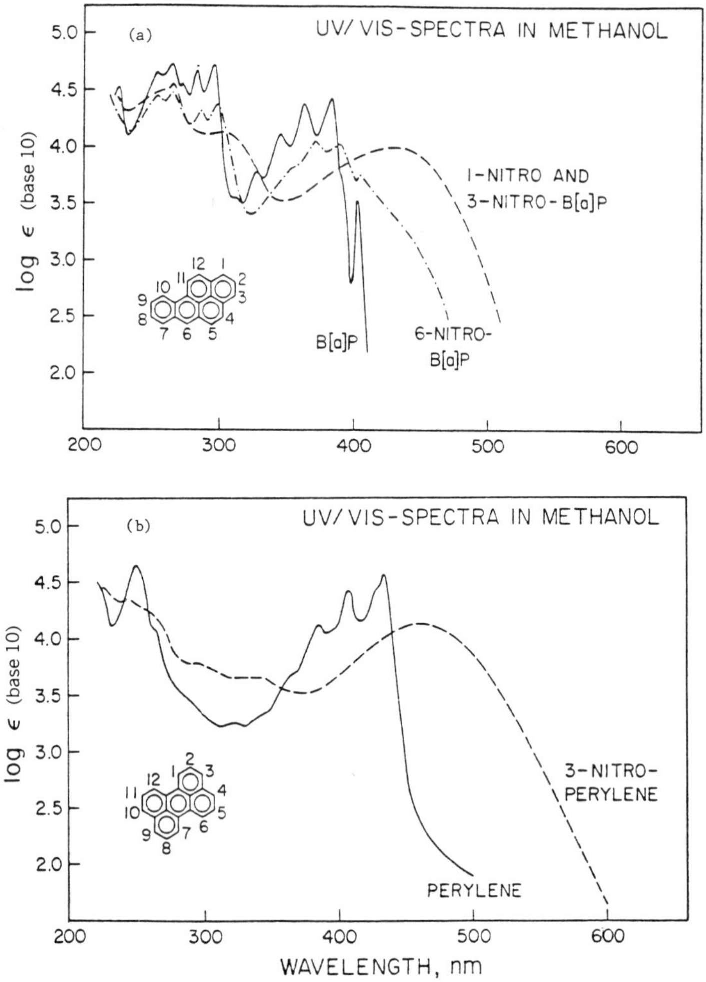

FIGURE 13.13. Absorption spectra in methanol of the mononitro derivatives of BaP (*a*) and perylene (*b*) formed when they are deposited on GF filters and exposed to air containing 1 ppm of NO_2 + 10 ppb HNO_3 in air for 8 h (from Pitts et al., 1978).

The presence of small amounts of gaseous HNO_3 along with the NO_2 was shown to be essential for efficient nitration; apparently it acts as a catalyst and/or reactant. This is an important point when considering apparent anomalies on nitration rates of adsorbed PAH by NO_2. If gaseous HNO_3 is absent, as in the experiments of Grosjean et al. (1983), or if the PAH substrate is basic, as it can be in some combustion systems (Lindskog et al., 1985), nitration rates can be negligible. It should also be kept in mind that FTIR studies have established that many tanks of dilute "pure NO_2" in nitrogen also contain $\geq 1\%$ of gaseous HNO_3 as an impurity.

Similar conclusions were reached by Tokiwa et al. (1981) who reported that direct-acting mutagenic derivatives were formed when pyrene, phenanthrene, fluorene, fluoranthene, and chrysene (and carbazole) were deposited on filter paper and exposed to 10 ppm NO_2. Furthermore, the 24 h nitration yields for pyrene rose from 0.02% with 1 ppm NO_2 in air to 2.85% when traces of HNO_3 were present. An enhancement was also observed when SO_2 was added to the NO_2–air mixture, but it was substantially less than for HNO_3.

In a study of factors influencing the rate of nitration of several adsorbed PAH exposed to 1.33 ppm of NO_2 in air, Jäger and Hanus (1980) found an order of reactivity for PAH adsorbed on substrates: silica gel > fly ash > deactivated aluminum oxide > carbon; the qualitative composition of nitro-PAH products, however, was independent of the carrier.

Recently, six PAH were adsorbed on several substrates and exposed to 0.5 ppm of NO_2 in air containing water vapor and traces of HNO_3. The highest reactivity was observed for PAH adsorbed on silica, with the yields of three nitro-PAH detected on alumina being only 14–24% of those on silica (Ramdahl et al., 1984b). The relative reactivity order was perylene > benzo(a)pyrene > pyrene > chrysene > fluoranthene = phenanthrene. This order is similar to that found in solution phase studies with HNO_3 in acetic anhydride (Dewar et al., 1956), HNO_3 in a mixed solvent of MeOH, H_2O, and dioxane (Nielsen, 1984), and NO_2/N_2O_4 in dichloromethane (Pryor et al., 1984). A similar reaction order was observed for the loss of PAH deposited on soot and exposed to NO_2 in air (Butler and Crossley, 1981).

As noted at the beginning of this section, in ambient air and in exhaust effluent from mobile and stationary sources, the PAH adsorbed on the surface of suspended particulate matter are present in a physically and chemically complex system. Thus while the order of reactivity may be similar to that in solution, as discussed above, the rates and mechanisms of reaction may not necessarily be readily correlated with homogeneous solution phase laboratory studies of aromatic nitrations. The latter are discussed in the book *Aromatic Nitration* by K. Schofield (1980) and will not be considered here.

A comprehensive discussion of the implications of fundamental information on rates, mechanisms, and products of laboratory aromatic nitrations to real world systems is that of Nielsen, Ramdahl, and Bjørseth (1983). Because ambient particles can absorb large quantities of water, they suggest that a plausible reaction system for nitration reactions of PAH in ambient POM involves liquid–solid processes. Thus fundamental laboratory studies in polar solvents have at least a qualitative relevance to urban atmospheres. Nielsen and co-workers explored a number of possible mechanisms of nitroarene formation including those involving radical cations. Recently, Pryor and his colleagues (1984) have provided experimental evidence for such a mechanism being applicable to PAH–NO_2/N_2O_4 reactions in CH_2Cl_2 solutions; they correlated their results with simple molecular orbital theory.

In a later treatment of the subject, Nielsen (1984) suggests a classification of the reactivities of PAH in electrophilic reactions, based on several chemical and spectroscopic parameters (Table 13.15). Note that none of the compounds in Group

TABLE 13.15. Reactivity Scale Developed by Nielsen for the Electrophilic Reactions of PAH[a]

I:	Benzo(a)tetracene, dibenzo(a,h)pyrene, pentacene, tetracene
II:	Anthanthrene, anthracene, benzo(a)pyrene, cyclopenta(cd) pyrene, dibenzo(a,l)pyrene, dibenzo(a,i)pyrene, dibenzo(a,c)tetracene, perylene
III:	Benz(a)anthracene, benzo(g)chrysene, benzo(ghi)perylene, dibenzo(a,e)pyrene, picene, pyrene
IV:	Benzo(c)chrysene, benzo(c)phenanthrene, benzo(e)pyrene, chrysene, coronene, dibenzanthracene, dibenzo(e,l)pyrene
V:	Acenaphthylene, benzofluoranthenes, fluoranthene, indeno(1,2,3-cd)fluoranthene, indeno(1,2,3-cd)pyrene, naphthalene, phenanthrene, triphenylene
VI:	Benzene, biphenyl

Source: Nielsen, 1984.

[a]Reactivity decreases in the order I to VI.

I [with the exception of dibenzo(a,h)pyrene] have as yet been identified in ambient samples, suggesting they may undergo rapid atmospheric transformations. The reactivity generally follows that established for PAH adsorbed on substrates and exposed to near ambient levels of NO_2 + HNO_3 in air.

The nitration of PAH by NO_2 + HNO_3 under simulated ambient conditions also occurs under laboratory conditions approximating *plume* gases, that is, higher concentrations of gases and deposition on coal fly ash as a substrate. Thus Hughes and co-workers (1980) found that BaP and pyrene reacted with 100 ppm of NO_2; the presence of nitric acid (possibly on the surface of the fly ash) enhanced the rate of reaction. The reactions proceeded more rapidly on silica gel than fly ash substrates. For pyrene, both mono- and dinitro isomers were found. Interestingly, on standing for 3 weeks, the mononitropyrene found on silica gel disappeared, apparently having been converted to one of the dinitropyrene isomers. With respect to other plume gases at the 100 ppm level, neither NO nor SO_2 reacted with BaP or pyrene on the several substrates studied; however, they both reacted with SO_3, but the products were not characterized.

In a recent study simulating stack gas sampling, a substantial degradation of the reactive PAH present in soot generated from a propane flame was observed with added NO_2. Additionally, a strong enhancing effect was observed when gaseous HCl was added to the laboratory stack gases, for example, 90% loss of BaP in 1 h with added HCl versus 20% without it (Brorström-Lundén and Lindskog, 1985). Addition of SO_3 also affected the degradation rate. Clearly, in addition to other processes which might occur in the hot effluent stream, under acidic conditions, sampling artifacts may be a major problem in sampling stack gases, often in times as short as 15–30 min or less.

In summary, although much has been learned through laboratory studies on nitrations by $NO_2 + HNO_3$ in simulated ambient atmospheres, the complexity of particle-bound PAH and related substrate effects and the variable and complex nature of the gaseous constituents of urban air pollution preclude a definite answer as to the degree to which such nitrations actually take place in ambient POM. On balance, and in view of the apparent lack of significant reactions of PAH during the long-range transport of aerosols from central Europe to Scandinavia observed by Bjørseth and colleagues (1979), they are probably relatively slow in most air environments. However, this may not always be the case in airsheds with high NO_2 and HNO_3 levels and strongly acidic particles.

That 2-nitropyrene (Nielsen et al., 1984) and then both 2-nitrofluoranthene and 2-nitropyrene (Pitts et al., 1985g; see also Section 13-G-2a) have been identified in POM from rural and urban air but not in primary exhaust emissions or laboratory nitrations with $NO_2 + HNO_3$ suggests that other, more rapid atmospheric nitration mechanisms may be operative. For example, the gas-phase formation of 2-NO_2-fluoranthene can be rationalized by the following reaction scheme.

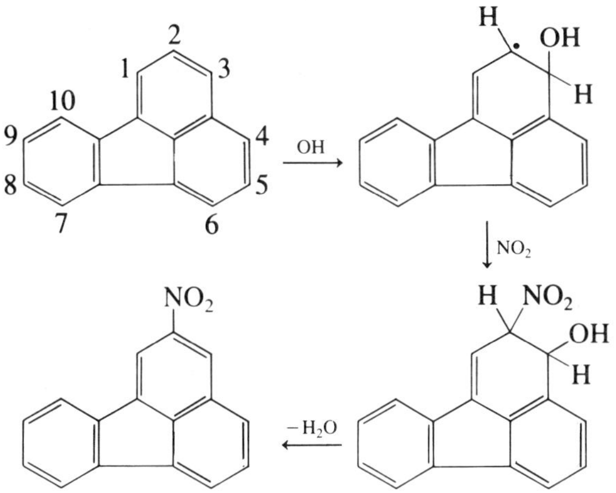

It is derived by analogy from the original gas phase studies of monocyclic aromatics by Kenley et al. (1978) and Atkinson et al. (1980) (see the discussion in Section 7-G-1) and is in accord with the suggestion of Nielsen and co-workers (1983, 1984). Of course the importance of the addition of NO_2 to the hydroxy-adduct of FL will depend on the rates of competing reactions, e.g., possibly with O_2, and on the ambient NO_2 concentration.

In ambient air a significant fraction of the fluoranthene, and pyrene, is in the gas phase. Thus the reaction sequence to form 2-NO_2-pyrene and 2-NO_2-fluoranthene could occur in the gas phase, where OH attack on the gaseous PAH would be expected to be fast (Section 13-I-1), followed by condensation of the 2-nitro-PAH derivative on the POM surfaces. Additionally, a gas phase reaction with N_2O_5 might also account for part of the 2-NO_2-fluoranthene observed (see below).

Finally, there have been experimental confirmations of the original proposal of artifactual formation of nitroarenes during collection of ambient or diesel POM in atmospheres containing NO_2 and HNO_3 (Pitts et al., 1978). Such chemical collection artifacts have recently been reviewed by Schuetzle (1983) and Lee and Schuetzle (1983), and one should consult these sources for original references.

While there are conflicting quantitative results from various laboratories, it is clear that such artifacts can be significant in PAH degradations during sampling of ambient POM with the inlet air "doped" with NO_2 (Brorström et al., 1983).

Similar considerations apply to PAH degradation during the sampling of diesel POM. However, Schuetzle (1983) concludes that artifactual formation of nitro-PAH "is a minor problem" at short sampling times [e.g., 23 min which is one federal test procedure (FTP) driving cycle], at low sampling temperatures (42°C), and in diluted exhaust containing 3 ppm NO_2. Similar considerations apply to sampling ambient POM, that is, the times should be kept as short as possible. Clearly, there is an experimental "conflict of interest" between the desire to minimize PAH degradation during POM collection and the need to acquire adequate amounts of particulate material for bioassay-directed chemical analysis of the sample.

13-H-2b. Nitration of Gaseous and Particulate PAH by Gaseous N_2O_5

As discussed in Section 5-D-2b, polluted ambient air may contain significant levels (up to ~430 ppt) of the NO_3 radical which, together with NO_2, is in equilibrium with gaseous N_2O_5 (Atkinson et al., 1985b). While NO_3 had a low reactivity, recent studies demonstrated that the associated N_2O_5 is a potent nitrating agent for certain PAH.

In the first such investigation, ppm levels of gaseous naphthalene and N_2O_5 in air (in a 5800 L environmental chamber) reacted with a rate constant of ~(2–3) $\times 10^{-17}$ cm^3 $molecule^{-1}$ s^{-1} to form 1- and 2-nitronaphthalene with yields of ~18% and ~7.5%, respectively (Pitts et al., 1985a). This is interesting because of the relatively high levels of gaseous naphthalene in urban air (e.g., see Figs. 13.7 and 13.8) and the fact that the 2-isomer is metabolically reduced in vivo to 2-aminonaphthalene (Johnson and Cornish, 1978), a human carcinogen.

Subsequent chamber studies of the reaction of gaseous N_2O_5 with pyrene and perylene revealed a rapid rate of nitration to form 1-nitropyrene (60–70% yield in 50 min exposure), but surprisingly, no nitroperylene was observed (Pitts et al., 1985b). This is opposite from the reactivity for the nitration of these PAH in solution or adsorbed on solid substrates. However, a recent, more extensive set of chamber experiments on six PAH exposed in a similar manner to N_2O_5 confirmed these initial results; the reaction order was pyrene > fluoranthene > BaP > benz(a)anthracene > perylene > chrysene (Pitts et al., 1985d).

In CCl_4 solutions and ambient temperature, these PAH reacted rapidly with N_2O_5 to form mononitro and dinitro isomers, and the reactivity order was perylene > BaP > benz(a)anthracene ≥ pyrene > fluoranthene ≥ chrysene; interestingly, only the 2-NO_2-isomer of fluoranthene was found (Zielinska et al., 1985a). This is in sharp contrast to the usual distribution of 1-, 3-, 7-, and 8-NO_2 isomers formed in the reaction of N_2O_5 with FL in polar solvents or deposited on solid surfaces; this suggests a different mechanism is operational in CCl_4 at 25°C. Interestingly, it was recently shown that 2-NO_2-FL is also, in fact, the sole mononitroisomer produced in the reaction of *gaseous* FL with N_2O_5 (Sweetman et al., 1985a,b).

These laboratory reactions of PAH with N_2O_5 clearly may have environmental significance. For example, a nighttime gas phase reaction of FL with N_2O_5 is a possible formation pathway for the 2-NO_2-FL observed in ambient POM (*vide supra*). Furthermore, in recent studies of the reactions of wood smoke with added NO_2, O_3, and $NO_2 + O_3$ mixtures (Kamens et al., 1984, 1985), the greatest enhancement in mutagenicity resulted from the $O_3 + NO_2$ mixture with most of the extracted mass and mutagenicity in the most polar fractions. Since the reactions were carried out in the dark, at least part of the increased direct-acting mutagenicity may be due to nitration of pyrene, fluoranthene, and certain other PAH present by gaseous N_2O_5 formed from the added $NO_2 + O_3$ (Pitts et al., 1985b).

Consistent with this is the observation that when reactive PAH on smoke gas-generated particles are exposed in the dark to air containing both 1 ppm NO_2 and 0.5 ppm O_3, they are degraded to a significantly greater extent than when exposed to either NO_2 or O_3 alone (Lindskog et al., 1985).

13-H-2c. Environmental Fate of Nitro-PAH: Photodecomposition

One probable environmental fate of certain nitroarenes is their photodecomposition, ultimately into quinones, and possibly phenolic derivatives. Thus on irradiation, 9-NO_2-anthrancene forms 9, 10-anthraquinone, both in solution and on silica gel.

Furthermore, 6-NO_2-BaP deposited on silica gel also photolyzes to the expected BaP quinones (1,6-, 3,6-, and 6,12-isomers). By analogy, one might expect similar photooxidations on ambient POM. However, such processes are no doubt dependent on the structure of the nitro-PAH and the nature of the substrate surface.

A guide for studying such structure–reactivity relationships is the mechanism Chapman and co-workers (1966) developed for 9-NO_2-anthracene. With this nitroarene, the major primary photochemical act is rearrangement into a nitrite, followed by dissociation into NO and a phenoxy-type radical; quinones are among the final products.

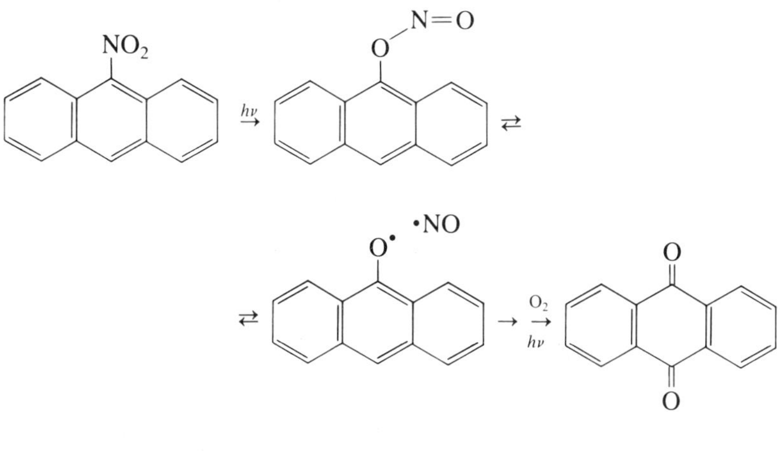

If one applies Chapman's elegant mechanistic theory to a more general case, the isomers of nitro-BaP, some predictions can be made: the 6-NO_2-isomer with 2-*peri*-hydrogens should be photochemically less stable than the 1- and 3-isomers with only 1-*peri*-hydrogcn (Fig. 13.14) (Pitts, 1983). This appears to be the case. Thus 6-NO_2-BaP photodecomposes very readily when irradiated in the solution phase. In contrast, the 1- and 3-isomers with 1-*peri*-hydrogen are far more stable under comparable conditions of irradiation (Zielinska, 1985).

From limited initial studies on the mononitro isomers of BaP, one can infer that during daylight hours any 6-NO_2-BaP on the exposed surface of POM should have a much shorter lifetime than the 1- and 3-isomers and ultimately degrade, in part to quinones. The latter are either very weak mutagens on strain TA1537 (1,6- and 3,6-isomers) or non-mutagens (6,12-quinone). Conversely, the 1- and 3-nitro isomers of BaP should have relatively longer atmospheric lifetimes, at least with respect to photodegradation.

Recently, Benson et al. (1985) found that 1-NO_2-PY deposited on glass photodecomposed in sunlight. There was an initial rapid reaction with a half-life of 14 h, followed by a slower reaction with a half-life of ~500 h. This was accompanied by loss of the nitro-group, the formation of a phenolic derivative and possibly quinones, and a significant reduction in mutagenicity. These results seem consistent with Chapman's mechanism and the results on nitro-BaP isomers, and further emphasize the urgent need for additional studies of the environmental fates of ambient mutagens and carcinogens.

These results also illustrate the need for proper storage and handling of the nitro-PAH used for long-term animal studies, especially those which are very photolabile, such as 6-NO_2-BaP.

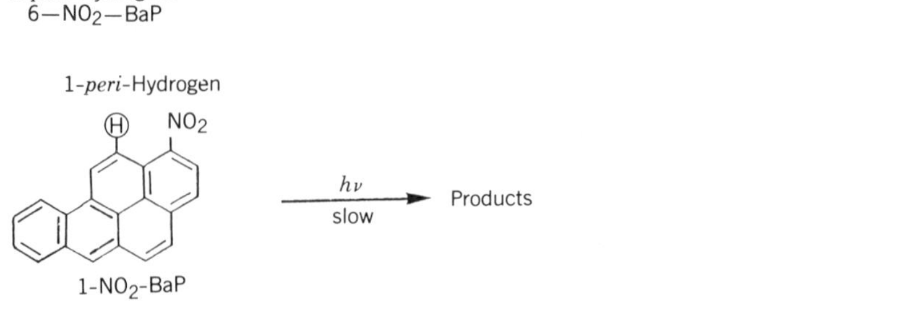

FIGURE 13.14. Possible effects of the molecular environment of the NO_2 group on the photostability of two mononitro isomers of BaP.

13-H-3. Oxidation of PAH in Simulated Atmospheres: Ozonolysis

In addition to nitroarenes, oxidized PAH are possible contributors to the direct-acting frameshift-type mutagenicity of polar fractions of ambient POM (see Fig. 13.12). Historically, interest dates back to the early studies of Kotin et al. on the carcinogenic activity of products of the oxidation of aliphatic hydrocarbons (1956) and of ozonized gasolines (1958). Subsequent studies demonstrated the carcinogenic and cell transformation activities of polar fractions of particulate organics.

From the outset, interest centered on the photooxidations of PAH (Falk et al., 1956), with some of the most relevant early results being those of Tebbens et al. (1966) who studied the chemical modifications in smoke irradiated while passing through a flow chamber. They observed a loss of 35–65% of the BaP and perylene in the original sample. Subsequently, Thomas et al. (1968) found a 60% decrease in the BaP content of soot from the entrance to the exit of a 22 ft long irradiation chamber. However, neither the PAH reaction products nor their biological activities were established.

In 1976, several phenols and epoxides known to be metabolites of BaP in mammalian cells were shown to be direct-acting frameshift mutagens. In part by analogy with these metabolites, it seemed reasonable that, in addition to nitroarenes, certain of the direct mutagens in ambient POM could be formed in atmospheric reactions of BaP (and other PAH) with other pollutants present in photochemical smog such as O_3 and PAN. Additional degradation paths would be PAH photooxidations involving singlet molecular oxygen ($O_2{}^1\Delta$), and free radicals such as OH.

Combined chemical–microbiological studies of BaP exposed to the gaseous portion of filtered ambient photochemical smog, as well as to low levels of O_3 and PAN in pure air, confirmed that direct-acting frameshift mutagens were indeed formed (Pitts et al., 1978). Some had TLC R_f and mass spectra resembling certain of the BaP metabolites formed when BaP was treated with the same S9 mix used in the Ames assay.

The present state of knowledge with regard to the degradation of PAH by ozonolysis is summarized briefly below.

In their initial studies of the reactions of O_3 with PAH deposited on glass surfaces, Lane and Katz (1977) found half-lives of $\approx$ 40 min for BaP exposed in the dark to 0.19 ppm of O_3. In a more extensive study, Katz et al. (1979) deposited nine PAH on TLC plates of cellulose and exposed them in the dark to 0.2 ppm O_3; to simulated sunlight (irradiation by quartz lamp); and to both conditions. As seen from Table 13.16, there are pronounced differences between the PAH in their rates of ozonolysis; the half-lives of BaP and the anthracenes are relatively short, those BeP and pyrene are intermediate, and those of the benzofluoranthenes are long.

The half-life for photooxidation of anthracene was only ~12 min compared to ~70 min for ozonolysis; the benzanthracenes, however, were more stable to light than to ozone. The combination of light and O_3 seemed especially effective with pyrene and the benzofluoranthenes. Products of the BaP–O_3 exposures were the 1,3-, 3,6-, and 6,12-diones, all of which are found in Toronto air (Pierce and Katz,

TABLE 13.16. Half-Lives (in h) of PAH Deposited on Cellulose TLC Plates and Exposed to 0.2 ppm O_3 in the Dark, in Simulated Sunlight, and Both

	Ozonolysis in Dark	Photooxidation	Photooxidation and Ozonolysis
Anthracene	1.2	0.2	0.15
Benz(a)anthracene	2.9	4.2	1.4
Dibenz(a,h)anthracene	2.7	9.6	4.8
Dibenz(a,c)anthracene	3.8	9.2	4.6
Pyrene	15.7	4.2	2.8
Benzo(a)pyrene	0.6	5.3	0.6
Benzo(e)pyrene	7.6	21.1	5.4
Benzo(b)fluoranthene	53	8.7	4.2
Benzo(k)fluoranthene	35	14.1	3.9

Source: Katz et al., 1979.

1976); quinones were also present in urban POM collected in West Germany (König et al., 1983) and Scandinavia (Ramdahl, 1983b).

In another study of the ozonolysis of BaP, Peters and Seifert (1980) found the lifetimes of BaP in POM collected under hi-vol sampling conditions was inversely proportional to the ambient O_3 concentration. This is another example of a chemical filter artifact. Similar results were obtained by Van Vaeck and Van Cauwenberghe (1984) who exposed diesel POM to ppm levels of O_3 under hi-vol sampling conditions and found conversion yields of BaP and benz(a)anthracene of ~60 and 50%, respectively.

A recent study of the factors influencing the reactivity of five PAH (PY, FL, BaA, BaP, and BeP) adsorbed on GF and TIGF filters (and in ambient POM collected near a major freeway) produced the following results: The most reactive under all conditions were pyrene, benz(a)anthracene, and BaP, with conversions of 50–80% in 3 h exposures to 200 ppb of O_3 in air at 1% RH. With the exception of BaP, the reactivities of the PAH on both model substrates were much lower at 50% RH (200 ppb O_3). However, as seen in Table 13.17, the effects of relative humidity on the reactivity of PAH present in samples of *ambient* POM passively exposed in a chamber to 200 ppb O_3 in air were not, in general, pronounced (Pitts et al. 1985e).

In addition to the 1,6-, 3,6-, and 6,12-diones formed as major products of the ozonolysis of BaP, a host of other OXY-PAH are formed as a result of substitution and ring-opening reactions. Some of these which have been tentatively identified are seen in Fig. 13.15 (Van Cauwenberghe et al., 1979). As we noted in Section 13-E-1e, this complex mixture of products contains direct-acting mutagens (TA98, -S9).

Relatively little is known to date about the full range of products, their mechanisms of formation, and their mutagenicities, resulting from the ozonolysis of other PAH; this is a worthy and challenging project. In this vein, among the most fundamental studies of the ozonolysis of solid PAH is that of Wu, Salmeen, and Niki (1984), who studied the reactions of O_3 with perylene and BaP adsorbed on

TABLE 13.17. Effect of Relative Humidity on the Percent PAH Reacted When Ambient Particles Collected on GF and TIGF Filters Were Passively Exposed in a Chamber to 200 ppb of O_3 in Air for Three Hours

	Percent PAH Reacted			
	GF		TIGF	
PAH	1% RH	50% RH	1% RH	50% RH
Pyrene	50	40	40	50
Fluoranthene	20	20	0	20
Benz(a)anthracene	50	40	40	60
Benzo(e)pyrene	0	0	10	20
Benzo(a)pyrene	60	40	40	70

Source: Pitts et al., 1985e.

fused silica plates using fluorescence spectroscopy. Interestingly, they found PAH in dispersed forms reacted much faster than when they were in aggregated states, as inferred from their excimer fluorescence.

13-H-4. Photooxidation of PAH

Polycyclic aromatic hydrocarbons in their pure forms, in synthetic mixtures or in ambient POM, can be photooxidized by actinic UV from the sun or mercury or xenon lamps. Because of the complexity of the various heterogeneous reaction systems investigated to date, interlaboratory comparisons of rates, mechanisms, and products sometimes appear contradictory; this is understandable.

Possible reaction paths in solution phase laboratory systems are outlined in the NAS document (1972). They include photooxidations of PAH by free radical and singlet oxygen mechanisms. For details, the reader is referred to Gollnick and Schenck (1967); Schaap (1976), and Rånby and Rabek (1978).

Substrate effects on the photodegradation of PAH are well documented. Thus Korfmacher and co-workers (1979, 1980a,b) found that adsorption on a specific type of fly ash stabilized certain PAH towards photooxidation. Interestingly, they also found that, in the dark in clean air, fluorene adsorbed on coal fly ash was decomposed by thermal processes (Korfmacher et al., 1981).

In a subsequent study, Dlugi and Güsten (1983) deposited anthracene and phenanthrene on three different substrates, silica gel and coal fly ash from two different power plants. One of the fly ash samples had an acidic surface (pH = 5.65) and the other a basic surface (pH = 9.3). As expected, on irradiation in a smog chamber, the anthracene deposited on the silica gel substrate degraded more rapidly than phenanthrene. However, there was a dramatic increase in the photooxidation rates of both PAH adsorbed on the *acidic* fly ash versus silica gel; furthermore, they were both highly resistant to photodegradation when adsorbed on the basic fly ash. Surface and surface pH effects must be considered when estimating life-

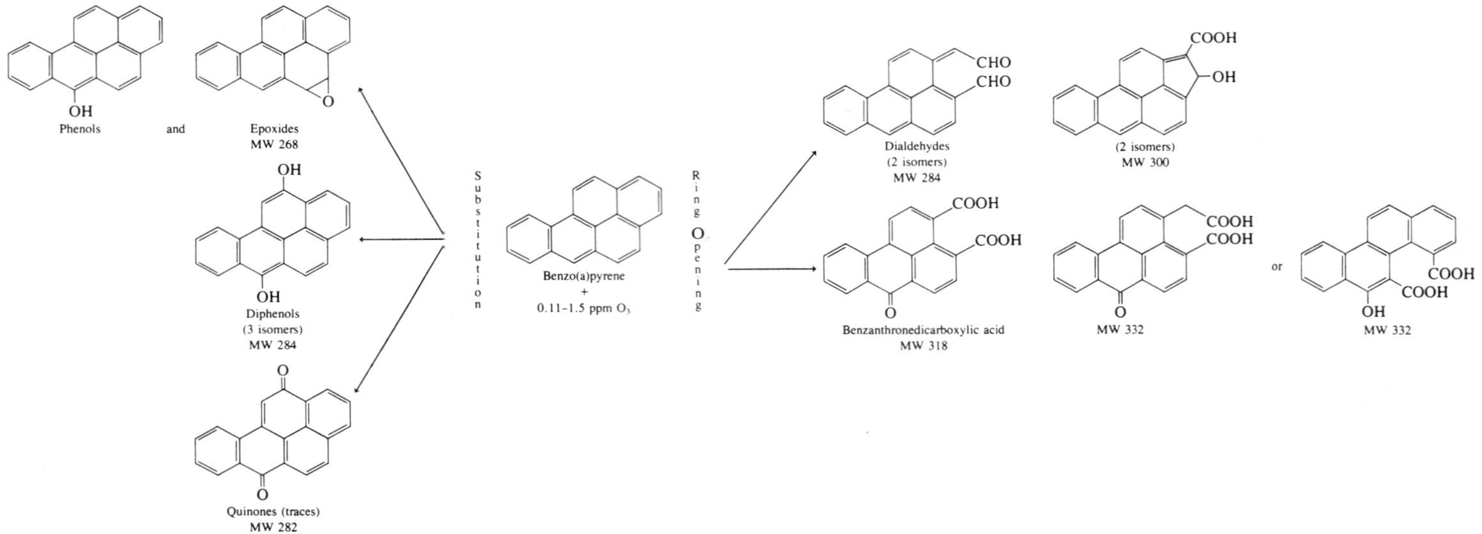

FIGURE 13.15. Major reaction pathways and structures of OXY-PAH tentatively identified by low- and high-resolution MS from the reaction of BaP deposited on a GF filter and exposed to 0.11–1.5 ppm of O_3 in air (from Van Cauwenberghe et al., 1979).

times of irradiated PAH in simulated or real atmospheres (for example, see Behymer and Hites, 1985).

Measurements of the quantum yields of the heterogeneous photodegradation of selected PAH deposited on silica gel and exposed to simulated actinic UV light are of fundamental interest. Thus Blau and Güsten (1981) found absolute quantum yields ranged from 8.5×10^{-2} for anthracene to 5.3×10^{-5} for BaP. However, when the differences in absorption cross sections are factored into the estimation of mean tropospheric lifetimes for these PAH, the lifetimes are 0.02 versus 10 h, respectively, for this particular mode of decomposition. Of course, other competing processes (e.g., ozonolysis) could change these lifetimes.

In another type of fundamental laboratory experiment, Barofsky and Baum (1976) studied in situ the primary photooxidation products of a series of irradiated PAH adsorbed on carbon microneedle emitters and set in a field desorption mass spectrometer. Anthracene, BaA, pyrene, and BaP all gave photoproducts with distinctive mass spectra.

Finally, studies of the photooxidation of anthracene dispersed into ambient POM suggest the involvement of $O_2(^1\Delta)$ as a major environmental process with this PAH (Fox and Olive, 1979). The possibility of ambient PAH acting as photosensitizers to form $O_2(^1\Delta)$ was suggested some years ago (Khan et al., 1967; Pitts et al., 1969) and now seems confirmed in a model system. Thus Eisenberg et al. (1984) have generated $O_2(^1\Delta)$ at atmospheric pressure in air using several model PAH deposited on surfaces as sensitizers. Certain of the sensitizer PAH are then proposed to react with the singlet oxygen they have generated, to form products which are direct-acting mutagens.

In this regard, the experiments of McCoy and Rosenkranz (1980) in which photodynamically generated $O_2(^1\Delta)$ was reacted with chrysene and 3-methylcholanthrene to form products that were direct-acting mutagens are relevant to the entire question of possible mechanisms of atmospheric oxidations of POM to produce biologically active products. A fascinating example is the phototoxicity (particularly of FL and PY) of several PAH dissolved in water, and irradiated, toward several common aquatic organisms, recently reported by Kagan et al. (1985). The possible involvement of singlet oxygen was suggested.

We have focused our discussion on direct-acting mutagenic derivatives of PAH. However, from a biological viewpoint their promutagenic properties may be equally, if not more, important (recall BaP is a promutagen). For example, consider the report of Rappaport et al. (1980) on the anhydride derivative of pyrene isolated from diesel POM. Although this compound is only a very weak direct mutagen in the Ames bacterial assay, it, or other members of this class of compounds, may prove to be active in other in vitro or in vivo test systems. Similar considerations apply to other oxygenated (e.g., lactones) and nitrated (e.g., 3-NO_2-perylene and 6-NO_2-BaP) PAH derivatives. For example, as we have seen, some of the latter show only weak or no direct activity with the Ames *Salmonella* reversion assay system, but they are powerful promutagens and possible carcinogens.

13-I. CHEMICAL TRANSFORMATIONS OF GASEOUS PAH IN SIMULATED ATMOSPHERES

This research area, which until recently had been virtually unexplored, is now becoming active. Largely, this is due to the recognition that substantial fractions of the three- and four-ring PAH exist in the environment as gases (Section 13-D-2). Additionally, they are among the most abundant PAH emitted from motor vehicles and wood-burning stoves.

Theoretical studies have suggested that PAH should react rapidly with OH radicals, and, as discussed above, recent chamber experiments show certain PAH react relatively rapidly with gaseous N_2O_5 to form nitroarenes.

Confirming their predicted high reactivity are recent smog chamber studies. In these, GC and differential optical absorption spectroscopy (DOAS, Section 5-D) were used to determine relative rate constants (and derive absolute values) for the gas phase reactions of OH radicals in 1 atm of air with naphthalene, phenanthrene, and anthracene (Biermann et al., 1985). At 298°K, the values of k_{OH} are (2.35 $\pm$ 0.06), (3.4 $\pm$ 1.2), and for anthracene at 325°K, (11.0 $\pm$ 0.9) $\times$ 10^{-11} cm^3 $molecule^{-1}$ s^{-1}, respectively. They lead to atmospheric lifetimes ([OH] assumed $= 1.0 \times 10^6$ cm^{-3}) of ~12, ~8, and ~3 h for gaseous naphthalene, phenanthrene, and anthracene, respectively.

The reactions almost certainly are initiated by OH addition, favoring the most electron-rich sites on the molecule. By analogy with benzene and its derivatives, if the PAH ring structure remains intact, products such as hydroxy-PAH, nitro-PAH, and hydroxynitro-PAH may be formed in air containing NO_2. Thus, as we have seen above (Section 13-H-2), the observation of 2-NO_2-fluoranthene as a major mononitro-PAH in ambient POM can be rationalized as being formed by the reaction of gaseous FL during the daytime with OH radicals, followed by addition of NO_2, and at night by a somewhat slower, direct reaction with gaseous N_2O_5. Similar considerations could apply to gaseous anthracene, phenanthrene, and anthracene, and indeed with other volatile PAH, including their N-atom and O-atom (and more slowly, S-atom) heterocyclic analogs. Ring cleavage could also occur; clearly, products studies to elucidate the reaction pathways are in order.

In conclusion, we trust that this overview of the sources, ambient levels, mutagenic activities, and atmospheric transformations of combustion-generated PAH and their heteroatom analogs has illustrated not only the complexity of their roles in our environment but also the important and exciting research opportunities that lie ahead. For example, we have seen that the chemical species responsible for *most* of the direct mutagenic activity observed for combustion-generated POM in exhaust plumes as in ambient air, are unknown; studies of their isolation, characterization, reactions, and biological activities are both challenging and relevant. Additionally, elucidation of the rates and mechanisms of their transport through, and interactions in, the air-water-soil interfaces of our ecosystem should have a high priority. Thus there is not only the continuing issue of whether or not such PAH and their transformation products have a significant impact on public health but also such new, intriguing, and highly speculative questions as to whether or not they might play some role in rapid forest decline and other phenomena associated with acidic/atmospheric deposition (see Chapter 11).

REFERENCES

Ahlberg, M., L. Berghem, G. Nordberg, S.-A. Persson, L. Rudling, and B. Steen, "Chemical and Biological Characterization of Emissions from Coal- and Oil-Fired Power Plants," *Environ. Health Perspect.*, **47,** 85 (1983).

Alfheim, I. and A. Lindskog, "A Comparison Between Different High Volume Sampling Systems for Collecting Ambient Airborne Particles for Mutagenicity Testing and for Analysis of Organic Compounds," *Sci. Total Environ.*, **34,** 203 (1984).

Alfheim, I. and M. Møller, "Mutagenicity of Long-Range Transported Atmospheric Aerosols," *Sci. Total Environ.*, **13,** 275 (1979).

Alfheim, I. and T. Ramdahl, "Contribution of Wood Combustion to Indoor Air Pollution as Measured by Mutagenicity in *Salmonella* and Polycyclic Aromatic Hydrocarbon Concentration," *Environ. Mutagen.*, **6,** 121 (1984).

Alfheim, I., G. Löfroth, and M. Møller, "Bioassay of Extracts of Ambient Particulate Matter," *Environ. Health Perspect.*, **47,** 227 (1983).

Alfheim, I., G. Becher, J. K. Hongslo, and T. Ramdahl, "Mutagenicity Testing of High Performance Liquid Chromatography Fractions from Wood Stove Emission Samples Using a Modified *Salmonella* Assay Requiring Smaller Sample Volumes," *Environ. Mutagen.*, **6,** 91 (1984a).

Alfheim, I., A. Bjørseth, and M. Møller, "Characterization of Microbial Mutagens in Complex Samples—Methodology and Application," *Crit. Rev. Environ. Control.*, **14,** 91 (1984b).

Alsberg, T., U. Stenberg, R. Westerholm, M. Strandell, U. Rannug, A. Sundvall, L. Romert, V. Bernson, B. Pettersson, R. Toftgärd, B. Franzen, M. Jansson, J. A. Gustafsson, K. E. Egebäck, and G. Tejie, "Chemical and Biological Characterization of Organic Material from Gasoline Exhaust Particles," *Environ. Sci. Technol.*, **19,** 43 (1985).

Ames, B. N. and J. McCann, Carcinogens Are Mutagens: A Simple Test System, in *Screening Tests in Chemical Carcinogens*, Vol. 12, R. Montesano et al., Eds., International Agency of Research on Cancer, Lyon, France, 1976, p. 493.

Ames, B. N., W. E. Durston, E. Yamasaki, and F. D. Lee, "Carcinogens Are Mutagens: A Simple Test System Combining Liver Homogenates for Activation and Bacteria for Detection," *Proc. Natl. Acad. Sci.*, **70,** 2281 (1973).

Ames, B. N., J. McCann, and E. Yamasaki, "Methods for Detecting Carcinogens and Mutagens with the *Salmonella*/Mammalian-Microsome Mutagenicity Test," *Mutat. Res.*, **31**, 347 (1975).

Atkinson, R., W. P. L. Carter, K. R. Darnall, A. M. Winer, and J. N. Pitts, Jr., "A Smog Chamber and Modeling Study of the Gas Phase NO_x–Air Photooxidation of Toluene and the Cresols," *Int. J. Chem. Kinet.*, **12,** 779 (1980).

Atkinson, R., S. M. Aschmann, and W. P. L. Carter, "Kinetics of the Reactions of O_3 and OH Radicals with Furan and Thiophene at 298 $\pm$ 2°K," *Int. J. Chem. Kinet.*, **15,** 51 (1983).

Atkinson, R., S. M. Aschmann, A. M. Winer, and W. P. L. Carter, "Rate Constants for the Gas Phase Reactions of OH Radicals and O_3 with Pyrrole at 295 $\pm$ 1°K and Atmospheric Pressure," *Atmos. Environ.*, **18,** 2105 (1984).

Atkinson, R., S. M. Aschmann, A. M. Winer, and W. P. L. Carter, "Rate Constants for the Gas Phase Reactions of NO_3 Radicals with Furan, Thiophene, and Pyrrole at 295 $\pm$ 1°K and Atmospheric Pressure," *Environ. Sci. Technol.*, **19,** 90 (1985a).

Atkinson, R., A. M. Winer, and J. N. Pitts, Jr., "Estimation of Nighttime N_2O_5 Concentrations from Ambient NO_2 and NO_3 Radical Concentrations and the Role of N_2O_5 in Nighttime Chemistry," *Atmos. Environ.*, in press (1985b).

Badger, G. M., "Mode of Formation of Carcinogens in Human Environment," *Natl. Cancer Inst. Monogr.*, **9,** 1 (1962).

Barofsky, D. F. and E. J. Baum, "Exploratory Field Desorption Mass Analysis of the Photoconversion of Adsorbed Polycyclic Aromatic Hydrocarbons," *J. Am. Chem. Soc.*, **98,** 8286 (1976).

Bartle, K. D., M. L. Lee, and S. A. Wise, "Modern Analytical Methods for Environmental Polycyclic Aromatic Compounds," *Chem. Soc. Rev.*, **10,** 113 (1981).

Baum, E. J., Occurrence and Surveillance of Polycyclic Aromatic Hydrocarbons, in *Hydrocarbons and Cancer*, Vol. 1, *Environment, Chemistry, and Metabolism*, H. V. Gelboin and P. O. P. Ts'o, Eds., Academic Press, New York, 1978, pp. 45–69.

Behymer, T. D. and R. A Hites, "Similarity of Some Organic Compounds in Spark-Ignition and Diesel Engine Particulate Extracts," *Environ. Sci. Technol.*, **18,** 203 (1984).

Behymer, T. D. and R. A. Hites, "Photolysis of Polycyclic Aromatic Hydrocarbons Adsorbed on Simulated Atmospheric Particulates," *Environ. Sci. Technol.*, **19,** 1004 (1985).

Belser, W. L., Jr., S. D. Shaffer, R. D. Bliss, P. M. Hynds, L. Yamamoto, J. N. Pitts, Jr., and J. A. Winer, "A Standardized Procedure for Quantification of the Ames *Salmonella*/Mammalian Microsome Mutagenicity Test," *Environ. Mutagen.*, **3,** 123 (1981).

Benson, J. M., A. L. Brooks, Y. S. Cheng, T. R. Henderson, and J. E. White, "Environmental Transformation of 1-Nitropyrene on Glass Surfaces," *Atmos. Environ.*, **19,** 1169 (1985).

Biermann, H., H. Mac Leod, R. Atkinson, A. M. Winer, and J. N. Pitts, Jr., "Kinetics of the Gas-Phase Reactions of the Hydroxyl Radical with Naphthalene, Phenanthrene, and Anthracene," *Environ. Sci. Technol.*, **19,** 244 (1985).

Bjørseth, A., Ed., *Handbook of Polycyclic Aromatic Hydrocarbons*, Dekker, New York, 1983.

Bjørseth, A. and B. S. Olufsen, Long-Range Transport of Polycyclic Aromatic Hydrocarbons, in *Handbook of Polycyclic Aromatic Hydrocarbons*, A. Bjørseth, Ed., Dekker, New York, 1983, pp. 507–524.

Bjørseth, A. and T. Ramdahl, Eds., *Handbook of Polycyclic Aromatic Hydrocarbons*, Vol. 2, *Emission Sources and Recent Advances in Analytical Chemistry*, Dekker, New York, 1985.

Bjørseth, A., G. Lunde, and A. Lindskog, "Long-Range Transport of Polycyclic Aromatic Hydrocarbons," *Atmos. Environ.*, **13,** 45 (1979).

Blakely, C. R. and M. L. Vestal, "Thermospray Interface for Liquid Chromatography/Mass Spectrometry," *Anal. Chem.*, **55,** 750 (1983).

Blau, L. and H. Güsten, Quantum Yields of the Photodecomposition of Polynuclear Aromatic Hydrocarbons Adsorbed on Silica Gel, in *Polynuclear Aromatic Hydrocarbons*, M. Cooke, A. J. Dennis, and G. L. Fisher, Eds., Battelle Press, Columbus, OH, 1981, pp. 133–144.

Bowen, E. J., *Chemical Aspects of Light*, 2nd ed., Oxford University Press, Oxford, 1946, p. 162.

Brooks, A. L., A. P. Li, J. S. Dutcher, C. R. Clark, S. J. Rothenberg, R. Kiyoura, W. E. Bechtold, and R. O. McClellan, "A Comparison of Genotoxicity of Automotive Exhaust Particles from Laboratory and Environmental Sources," *Environ. Mutagen.*, **6,** 651 (1984).

Brorström-Lunden, E. and A. Lindskog, "Degradation of Polycyclic Aromatic Hydrocarbons During Simulated Stack Gas Sampling," *Environ. Sci. Technol.*, **19,** 313 (1985).

Brorström, E., P. Greenfelt, and A. Lindskog, "The Effect of Nitrogen Dioxide and Ozone on the Decomposition of Particle-Associated Polycyclic Aromatic Hydrocarbons During Sampling from the Atmosphere," *Atmos. Environ.*, **17,** 601 (1983).

Burkhard, L. P., D. E. Armstrong, and A. W. Andren, "Vapor Pressures for Biphenyl, 4-Chlorobiphenyl, 2,2′,3,3′,5,5′,6,6′-Octachlorobiphenyl, and Decachlorobiphenyl," *J. Chem. Eng. Data*, **29,** 248 (1984).

Butler, J. D. and P. Crossley, "Reactivity of Polycyclic Aromatic Hydrocarbons Adsorbed on Soot Particles," *Atmos. Environ.*, **15,** 91 (1981).

Calvert, J. G. and J. N. Pitts, Jr., *Photochemistry*, Wiley, New York, 1966.

Cautreels, W. and K. Van Cauwenbergh, "Determination of Organic Compounds in Airborne Particulate Matter by Gas Chromatography–Mass Spectrometry," *Atmos. Environ.*, **10,** 447 (1976).

Cautreels, W. and K. Van Cauwenberghe, "Experiments on the Distribution of Organic Pollutants Between Airborne Particulate Matter and the Corresponding Gas Phase," *Atmos. Environ.*, **12,** 1133 (1978).

Chapman, O. L., D. C. Heckert, J. W. Reasoner, and S. P. Thackaberry, "Photochemical Studies on 9-Nitroanthracene," *J. Am. Chem. Soc.*, **88,** 5550 (1966).

Chesis, P. L., D. E. Levin, M. T. Smith, L. Ernster, and B. N. Ames, "Mutagenicity of Quinones: Pathways of Metabolic Activation and Detoxification," *Proc. Natl. Acad. Sci. USA*, **81,** 1696 (1984).

Chrisp, C. and G. L. Fisher, "Mutagenicity of Airborne Particles," *Mutat. Res.*, **76,** 143 (1980).

Chrisp, C. E., G. L. Fisher, and J. E. Lammert, "Mutagenicity of Filtrates from Respirable Coal Fly Ash," *Science*, **199,** 73 (1978).

Clar, E., *Polycyclic Hydrocarbons*, 2 Vols., Academic Press, New York, 1964, 974 pp.

Claxton, L., The Utility of Bacterial Mutagenesis Testing in the Characterization of Mobile Source Emissions: A Review, in *Toxicological Effects of Emissions from Diesel Engines*, J. Lewtas, Ed., Elsevier, New York, 1982 pp. 69-82.

Claxon, L. D., "Assessment of Bacterial Mutagenicity Methods for Volatile and Semivolatile Compounds and Mixtures," *Environ. Int. J.,* **11,** 375 (1985).

Claxton, L. and M. Kohan, Bacterial Mutagenesis and the Evaluation of Mobile-Source Emission, in *Short-Term Bioassays in the Analysis of Complex Environmental Mixtures, II,* M. D. Waters, S. Sandu, J. Lewtas-Huisingh, L. Claxton, and S. Nesnow, Eds., Plenum, New York, 1981 pp. 299-317.

Commins, B. T., "Interim Report on the Study of Techniques for Determination of Polycyclic Aromatic Hydrocarbons in Air," *Natl. Cancer Inst. Monogr.*, **9,** 225 (1962).

Commoner, B., P. Madyastha, A. Bronsdon, and A. J. Vithayathil, "Environmental Mutagens in Urban Air Particulates," *J. Toxicol. Environ. Health*, **4,** 59 (1978).

Commoner, B., A. J. Vithayathil, and P. Dolara, Mutagenic Analysis of Complex Samples of Aqueous Effluents, Air Particulates, and Foods, in *Application of Short-Term Bioassays in the Fractionation and Analysis of Complex Environmental Mixtures*, M. D. Waters, S. Nesnow, J. L. Huisingh, S. Sandhu, and L. Claxton, Eds., Plenum, New York, 1979, pp. 529–570.

Cook, J. W., C. L. Hewett, and I. Hieger, "The Isolation of a Cancer-Producing Hydrocarbon from Coal Tar. Parts I, II, and III," *J. Chem. Soc.*, 395 (1933).

Cooke, M. and A. J. Dennis, Eds., *Polynuclear Aromatic Hydrocarbons: Mechanisms, Methods, and Metabolism*, Eighth International Symposium, Battelle Press, Columbus, OH, 1984.

Cooper, J. A., "Environmental Impact of Residential Wood Combustion Emissions and Its Implications," *J. Air. Pollut. Control Assoc.*, **30,** 855 (1980).

Cooper, J. A., L. A. Currie, and G. A. Klouda, "Assessment of Contemporary Carbon Combustion Source Contributions to Urban Air Particulate Levels Using Carbon-14 Measurements," *Environ. Sci. Technol.*, **15,** 1045 (1981).

Covey, T. and J. Henion, "Direct Liquid Introduction/Thermospray Interface for Liquid Chromatography/Mass Spectrometry," *Anal. Chem.*, **55,** 2275 (1983).

Cuddihy, R. G., W. C. Griffith, and R. O. McClellan, "Health Risks from Light-Duty Diesel Vehicles," *Environ. Sci. Technol.*, **18,** 14A (1984).

Daisey, J. M., L. Hawryluk, T. J. Kneip, and F. Mukai, Mutagenic Activity in Organic Fractions of Airborne Particulate Matter, in *Proceedings-Carbonaceous Particles in the Atmosphere*, T. Novakov, Ed., National Technical Information Service, Springfield, VA, 1979, pp. 187–192.

Daisey, J. M., T. J. Kneip, I. Hawryluk, and F. Mukai, "Seasonal Variations in the Bacterial Mutagenicity of Airborne Particulate Organic Matter in New York City," *Environ. Sci. Technol.*, **14,** 1487 (1980).

DeSerres, F. and J. Ashby, *Evaluation of Short-Term Tests for Carcinogens*, Elsevier/North Holland, Amsterdam, 1981.

Dewar, M. J. S., T. Mole, and E. W. T. Warford, ''Electrophilic Substitution. Part VI. The Nitration of Aromatic Hydrocarbons; Partial Rate Factors and Their Interpretation,'' *J. Chem. Soc.*, 3581 (1956).

Dlugi, R. and H. Güsten, ''The Catalytic and Photocatalytic Activity of Coal Fly Ashes,'' *Atmos. Environ.*, **17,** 1765 (1983).

Dong, M. W., D. C. Locke, and D. Hoffman, ''Characterization of Aza-Arenes in Basic Organic Portion of Suspended Particulate Matter,'' *Environ. Sci. Technol.*, **11,** 612 (1977).

Dunkel, V. C., ''Collaborative Studies on the *Salmonella*/Microsome Mutagenicity Assay,'' *J. Assoc. Off. Anal. Chem.*, **62,** 874 (1979).

Eisenberg, W. C., K. Taylor, and R. W. Murray, ''Production of Singlet Delta Oxygen by Atmospheric Pollutants,'' *Carcinogenesis*, **5,** 1095 (1984).

El-Bayoumy, K. and S. S. Hecht, ''Identification of Mutagenic Metabolites Formed by C-Hydroxylation and Nitroreduction of 5-Nitroacenaphthene in Rat Liver,'' *Cancer Res.*, **42,** 1243 (1982).

El-Bayoumy, K. and S. S. Hecht, ''Identification and Mutagenicity of Metabolites of 1-Nitropyrene Formed by Rat Liver,'' *Cancer Res.*, **43,** 3132 (1983).

El-Bayoumy, K., S. S. Hecht, and D. Hoffmann, ''Comparative Tumor Initiating Activity on Mouse Skin of 6-Nitrobenzo(a)pyrene, 6-Nitrochrysene, 3-Nitroperylene, 1-Nitropyrene, and Their Parent Hydrocarbons,'' *Cancer Lett.*, **16,** 333 (1982).

Environment International Journal, ''Proceedings of the Workshop on Genotoxic Air Pollutants,'' Rougemont, North Carolina, April 24–27, 1984; **11,** 103–418 (1985).

Environmental Mutagen Information Center, 1985. Write c/o Dr. John S. Wasson, Director, P.O. Box Y, Bldg. 9224Y-12, Oak Ridge National Laboratories, Oak Ridge, Tennessee 37831; Telephone (615) 574-7871.

Epler, J. L., The Use of Short-Term Tests in the Isolation and Idenification of Chemical Mutagens in Complex Mixtures, *Chemical Mutagens: Principles and Methods for Their Detection*, Vol. 6, F. J. DeSerres and A. Hollander, Eds., Plenum Press, New York, 1980, pp. 239–270.

Epler, J. L., B. R. Clark, C. Ho, M. R. Guerin, and T. K. Rao, Short-Term Bioassay of Complex Organic Mixtures: Part II, Mutagenicity Testing, in *Application of Short-Term Bioassays in the Fractionation and Analysis of Complex Environmental Mixtures*, M. D. Waters, S. Nesnow, J. Huisingh, S. Sandhu, and L. Claxton, Eds., Plenum Press, New York, 1979, pp. 269–289.

Epstein, S. S., S. Joshi, J. Andrea, N. Mantel, E. Sawicki, T. Stanley, and E. C. Tabor, ''Carcinogenicity of Organic Particulate Pollutants in Urban Air After Administration of Trace Quantities to Neonatal Mice,'' *Nature (London)*, **212,** 1305 (1966).

Falk, H. L., I. Markal, and P. Kotin, ''Aromatic Hydrocarbons. IV. Their Fate Following Emission into the Atmosphere and Experimental Exposure to Washed Air and Synthetic Smog,'' *Arch. Ind. Health*, **13,** 13 (1956).

Fieser, L. F. and M. Fieser, *Organic Chemistry*, 3rd ed., Reinhold, New York, 1956.

Fisher, G. L., ''Biomedically Relevant Chemical and Physical Properties of Coal Combustion Products,'' *Environ. Health Perspect.*, **47,** 189 (1983).

Fisher, G. L., C. E. Chrisp, and O. G. Raabe, ''Physical Factors Affecting the Mutagenicity of Fly Ash from a Coal-Fired Power Plant,'' *Science*, **204,** 879 (1979).

Fitz, D. R., G. J. Doyle, and J. N. Pitts, Jr., ''An Ultrahigh Volume Sampler for the Multiple Filter Collection of Respirable Particulate Matter,'' *J. Air Pollut. Control Assoc.*, **33,** 877 (1983).

Fjeldsted, J. C. and M. L. Lee, ''Capillary Supercritical Fluid Chromatography,'' *Anal. Chem.*, **56,** 619a (1984).

Flessel, C. P., J. J. Wesolowski, S. Twiss, J. Cheng, J. Ondo, N. Monto, and R. Chan, ''Integration of the Ames Bioassay and Chemical Analyses in an Epidemiological Cancer Incidence Study,'' in *Short-Term Bioassays in the Analysis of Complex Environmental Mixtures, II,* M. D. Waters,

S. S. Sandhu, J. L. Huisingh, L. Claxton, and S. Nesnow, Eds., Plenum, New York, 1981, pp. 61–84.

Flessel, C. P., G. N. Quirquis, J. C. Cheng, Kuo-In Chang, E. S. Hahn, S. Twiss, and J. J. Wesolowski, "Sources of Mutagens in Contra Costa County Community Aerosols During Pollution Episodes: Diurnal Variations and Relations to Source Emissions Tracers," *Environ. Int. J.*, **11,** 293 (1985).

Fox, M. A. and S. Olive, "Photooxidation of Anthracene on Atmospheric Particulate Matter," *Science*, **205,** 582 (1979).

Friedel, R. A. and M. Orchin, *Ultraviolet Spectra of Aromatic Compounds*, Wiley, New York, 1951.

Gibson, T. L., "Nitro Derivatives of Polynuclear Aromatic Hydrocarbons in Airborne and Source Particulate Matter," *Atmos. Environ.*, **16,** 2037 (1982).

Gibson, T. L., "Sources of Direct-Acting Nitroarene Mutagens in Airborne Particulate Matter," *Mutat. Res.*, **122,** 115 (1983).

Gollnick, K. and G. O. Schenck, Oxygen as a Dienophile, in *1,4-Cycloaddition Reactions*, J. Hamer, Ed., Academic Press, New York, 1967, pp. 255–344.

Gordon, R. J., R. J. Bryan, J. S. Rhim, C. Demoise, R. G. Wolford, A. E. Freeman, and R. J. Huebner, "Transformation of Rat and Mouse Embryo Cells by a New Class of Carcinogenic Compounds Isolated from Particles in City Air," *Int. J. Cancer*, **12,** 223 (1973).

Goulden, F. and M. M. Tipler, "Experiments on the Identification of 3,4-Benzpyrene in Domestic Soot by Means of the Fluorescence Spectrum," *Br. J. Cancer*, **3,** 157 (1949).

Grafe, A., I. E. Mattern, and M. Green, "A European Collaborative Study of the Ames Assay. I. Results and General Interpretation," *Mutat. Res.*, **85,** 391 (1981).

Greenberg, A., F. Darack, R. Harkov, P. Lioy, and J. Daisey, "Polycyclic Aromatic Hydrocarbons in New Jersey: A Comparison of Winter and Summer Concentrations Over a Two-Year Period," *Atmos. Environ.*, **19,** 1325 (1985).

Greibrokk, T., G. Löfroth, L. Nilsson, R. Toftgard, J. Carlstedt-Duke, and J. Gustafsson, Nitroarenes: Mutagenicity in the Ames *Salmonella*/Microsome Assay and Affinity to the TCDD-Receptor Protein, in *Toxicity of Nitroaromatic Compounds*, D. E. Rickert, Ed., Hemisphere Publishing, Washington DC, 1985, pp. 166–183.

Griest, W. H. and J. E. Caton, Extraction of Polycyclic Aromatic Hydrocarbons for Quantitative Analyses, in *Handbook of Polycyclic Aromatic Hydrocarbons*, A. Bjφrseth, Ed., Dekker, New York, 1983, pp. 95–148.

Grimmer, G., Chemistry, in *Environmental Carcinogens: Polycyclic Aromatic Hydrocarbons*, G. Grimmer, Ed., CRC Press, Boca Raton, FL, 1983, pp. 27–60.

Grimmer, G. and F. Pott, Occurrence of PAH, In *Environmental Carcinogens: Polycyclic Aromatic Hydrocarbons*, G. Grimmer, Ed., CRC Press, Boca Raton, FL, 1983, pp. 61–129.

Grosjean, D., K. Fung, and J. Harrison, "Interactions of Polycyclic Aromatic Hydrocarbons with Atmospheric Pollutants," *Environ. Sci. Technol.*, **17,** 673 (1983).

Guerin, M. R., B. R. Clark, C. Ho, J. L. Epler, and T. K. Rao, Short-Term Bioassays of Complex Organic Mixtures: Part 1, Chemistry, in *Application of Short-Term Bioassays in the Fractionation and Analysis of Complex Environmental Mixtures,* M. D. Waters, S. Nesnow, J. L. Huisingh, S. Sandhu, and L. Claxton, Eds., Plenum Press, New York, 1979, pp. 249–268.

Heinrich, G. and H. Güsten, Fluorescence Spectroscopic Properties of Carcinogenic and Airborne Polynuclear Aromatic Hydrocarbons, in *Polynuclear Aromatic Hydrocarbons: Chemistry and Biological Effects*, A. Bjφrseth and A. J. Dennis, Eds., Battelle Press, Columbus, OH, 1980, pp. 983–1003.

Hirao, K., Y. Shinohara, H. Tsuka, S. Fakushima, M. Takahashi, and N. Ito, "Carcinogenic Activity of Quinoline on Rat Liver," *Cancer Res.*, **36,** 329 (1976).

Hoffman, D. and E. L. Wynder, Organic Particulate Pollutants—Chemical Analysis and Bioassays for Carcinogenicity, in *Air Pollution*, Vol. II, 3rd ed., A. C. Stern, Ed., Academic Press, New York, 1977, pp. 361–455.

Hollstein, M., J. McCann, F. A. Angelosanto, and W. W. Nichols, "Short-Term Tests for Carcinogens and Mutagens," *Mutat. Res.*, **65,** 133 (1979).

Holmberg, B. and U. Ahlborg, Eds., Symposium on Biological Tests in the Evaluation of Mutagenicity and Carcinogenicity of Air Pollutants with Special Reference to Motor Exhausts and Coal Combustion Products, Stockholm, Sweden, February 8-12, 1982; in *Environ. Health Perspect.*, **47,** 1-345 (1983).

Hueper, W. C., P. Kotin, E. C. Tabor, W. W. Payne, H. L. Falk, and E. Sawicki, "Carcinogenic Bioassays on Air Pollutants," *Arch. Pathol. and Lab. Med.*, **74,** 89 (1962).

Hughes, M. M., D. F. S. Natusch, D. R. Taylor, and M. V. Zeller, Chemical Transformations of Particulate Polycyclic Organic Matter, in *Polynuclear Aromatic Hydrocarbons; Chemistry and Biological Effects*, A. Bjϕrseth and A. J. Dennis, Eds., Battelle Press, Columbus, OH, 1980, pp. 1–8.

Huisingh, J., R. Bradow, R. Jungers, L. Claxton, R. Zweidinger, S. Tejada, J. Bumgarner, F. Duffield, M. Waters, V. F. Simmon, C. Hare, C. Rodgriguez, and L. Snow, Application of Bioassay to Characterization of Diesel Particle Emissions, in *Application of Short-Term Bioassay in the Fractionation and Analysis of Complex Environmental Mixtures*, M. D. Waters, S. Nesnow, J. L. Huisingh, S. Sandhu, and L. Claxton, Eds., Plenum Press, New York, 1979, pp. 383–418.

International Agency for Research on Cancer (IARC), Monographs on the *Evaluation of the Carcinogenic Risk of Chemicals to Humans*: Vol. 32, *Polynuclear Aromatic Compounds, Part 1*, 1983; Vol. 33, *Polynuclear Aromatic Compounds, Part 2, Carbon Blacks, Mineral Oils, and Some Nitroarenes,* 1984; Vol. 34, *Polynuclear Aromatic Compounds, Part 3, Industrial Exposures in Aluminum Production, Coal Gasification, Coke Production, and Iron and Steel Foundry,*, 1984.

International Union of Pure and Applied Chemistry, *Nomenclature of Organic Chemistry*, Butterworths, London, 1958.

Jaffee, H. H. and M. Orchin, *Theory and Applications of Ultraviolet Spectroscopy*, Wiley, New York, 1962.

Jäger, J., "Detection and Characterization of Nitro Derivatives of Some Polycyclic Aromatic Hydrocarbons by Fluorescence Quenching After Thin-Layer Chromatography: Application to Air Pollution Analysis," *J. Chromatogr.*, **152,** 575 (1978).

Jäger, J. and V. Hanus, "Reaction of Solid Carrier-Adsorbed Polycyclic Aromatic Hydrocarbons with Gaseous Low-Concentrated Nitrogen Dioxide," *J. Hyg. Epidemiol. Microbiol. Immunol.*, **24,** 1 (1980).

Jensen, T. E. and R. A. Hites, "Aromatic Diesel Emissions as a Function of Engine Conditions," *Anal. Chem.*, **55,** 594 (1983).

Jerina, D. M., R. E. Lehr, H. Yagi, O. Hernandez, P. M. Dansette, P. G. Wislocki, A. W. Wood, R. L. Chang, W. Levin, and A. H. Conney, Mutagenicity of Benzo(a)pyrene Derivatives and the Description of a Quantum Mechanical Model Which Predicts the Ease of Carbonium Ion Formation from Diol Epoxides, in *In Vitro Metabolic Activation in Mutagenesis Testing*, F. J. DeSerres, J. R. Fouts, J. R. Bend, and R. M. Philpot, Eds., Elsevier, Amsterdam, 1976a, pp. 159–177.

Jerina, D. M., H. Yagi, O. Hernandez, P. M. Dansette, A. W. Wood, W. Levin, R. L. Chang, P. G. Wislocki, and A. H. Conney, Synthesis and Biological Activity of Potential Benzo(a)pyrene Metabolites, in *Polynuclear Aromatic Hydrocarbons: Chemistry, Metabolism, and Carcinogenesis*, R. I. Frundenthal and P. W. Jones, Eds., Raven Press, New York, 1976b, pp. 91–113.

Johnson, D. E. and H. H. Cornish, "Metabolic Conversion of 1- and 2-Nitronaphthalene to 1- 2-Naphthylamine in the Rat," *Toxicol. Appl. Pharmacol.*, **46,** 549 (1978).

Kagan, J., E. D. Kagan, I. A. Kagan, and P. A. Kagan, "Do Polycyclic Aromatic Hydrocarbons, Acting as Photosensitizers, Participate in the Toxic Effects of Acid Rain?" Proceedings of Symposium on Aquatic Photochemistry, American Chemical Society National Meeting, Miami Beach, Florida (1985).

Kamens, R. M., G. D. Rives, J. M. Perry, D. A. Bell, R. F. Paylor, Jr., R. G. Goodman, and L. D.

Claxton, "Mutagenic Changes in Dilute Wood Smoke as It Ages and Reacts with Ozone and Nitrogen Dioxide: An Outdoor Chamber Study," *Environ. Sci. Technol.*, **18,** 523 (1984).

Kamens, R., D. Bell, A. Dietrich, J. Perry, R. Goodman, L. Claxton, and S. Tejada, "Mutagenic Transformations of Dilute Wood Smoke Systems in the Presence of Ozone and Nitrogen Dioxide. Analysis of Selected High-Pressure Liquid Chromatography Fractions from Wood Smoke Particle Extracts," *Environ. Sci. Technol.*, **19,** 63 (1985).

Kapitulnik, J., W. Levin, H. Yagi, D. Jerina, and A. Conney, "Lack of Carcinogenicity of 4-, 5-, 6-, 7-, 8-, 9-, and 10-Hydroxybenzo(a)pyrene in Mouse Skin," *Cancer Res.*, **36,** 3625 (1976).

Katz, M., C. Chan, H. Tosine, and T. Sakuma, Relative Rates of Photochemical and Biological Oxidation (*in vitro*) of Polycyclic Aromatic Hydrocarbons, in *Polynuclear Aromatic Hydrocarbons*, P. W. Jones and P. Leber, Eds., Ann Arbor Science Publishers, Ann Arbor, MI, 1979, pp. 171–189.

Kenley, R. A., J. E. Davenport, and D. G. Hendry, "Hydroxyl Radical Reactions in the Gas Phase. Products and Pathways for the Reaction of OH with Toluene," *J. Phys. Chem.*, **82,** 1095 (1978).

Kennaway, E. L. and I. Hieger, "Carcinogenic Substances and Their Fluorescence Spectra," *Br. Med. J.*, 1044 (1930).

Khan, A. U., J. N. Pitts, Jr., and E. B. Smith, "The Role of Singlet Molecular Oxygen in the Production of Photochemical Air Pollution," *Environ. Sci. Technol.*, **1,** 656 (1967).

King, C. M., L. K. Tay, M.-S. Lee, K. Irnaida, and C. Y. Wang, "Tumorigenicity and Metabolism of Dinitropyrenes in the Female CD Rat," presented at the Tenth International Symposium on Polynuclear Aromatic Hydrocarbons, Columbus, OH, October 21–23, 1985.

Kleindienst, T. E., P. B. Shepson, E. O. Edney, L. Cupitt, and L. D. Claxton, "The Mutagenic Activity of the Products of Propylene Photooxidation," *Environ. Sci. Technol.*, **19,** 620 (1985).

Klopman, G. and H. S. Rosenkranz, "Structural Requirements for the Mutagenicity of Environmental Nitroarenes," *Mutat. Res.*, **126,** 227 (1984).

König, J., E. Balfanz, W. Funcke, and T. Romanowski, "Determination of Oxygenated Polycyclic Aromatic Hydrocarbons in Airborne Particulate Matter by Capillary Gas Chromatography and Gas Chromatography/Mass Spectrometry," *Anal. Chem.*, **55,** 599 (1983).

Korfmacher, W. A., D. F. S. Natusch, D. R. Taylor, E. L. Wehry, and G. Mamantov, Thermal and Photochemical Decomposition of Particulate PAH, in *Polynuclear Aromatic Hydrocarbons*, P. W. Jones and P. Leber, Eds., Ann Arbor Science Publishers, Ann Arbor, MI, 1979, pp. 165–170.

Korfmacher, W. A., D. F. S. Natusch, D. R. Taylor, G. Mamantov, and E. L. Wehry, "Oxidative Transformations of Polycyclic Aromatic Hydrocarbons Adsorbed on Coal Fly Ash," *Science*, **207,** 763 (1980a).

Korfmacher, W. A., E. L. Wehry, G. Mamantov, and D. F. S. Natusch, "Resistance to Photochemical Decomposition of Polycyclic Aromatic Hydrocarbons Vapor-Adsorbed on Coal Fly Ash," *Environ. Sci. Technol.*, **14,** 1094 (1980b).

Korfmacher, W. A., G. Mamantov, E. L. Wehry, D. F. S. Natusch, and T. Maurey, "Nonphotochemical Decomposition of Fluorene Vapor-Adsorbed on Coal Fly Ash," *Environ. Sci. Technol.*, **15,** 1370 (1981).

Kotin, P., H. L. Falk, P. Mader, and M. Thomas, "Aromatic Hydrocarbons. I. Presence in the Los Angeles Atmosphere and the Carcinogenicity of Atmospheric Extracts," *Arch. Ind. Hyg.*, **9,** 153 (1954).

Kotin, P., H. L. Falk, and M. Thomas, "Production of Skin Tumors in Mice with Oxidation Products of Aliphatic Hydrocarbons," *Cancer*, **9,** 905 (1956).

Kotin, P., H. L. Falk, and G. J. McCammon, "The Experimental Induction of Pulmonary Tumors and Changes in the Respiratory Epithelium in C57B1 Mice Following Their Exposure to an Atmosphere of Ozonized Gasoline," *Cancer*, **11,** 473 (1958).

Lane, D. A. and M. Katz, The Photomodification of Benzo(a)pyrene, Benzo(b)fluoranthene, and Benzo(k)fluoranthene Under Simulated Atmospheric Conditions, in *Fate of Pollutants in the Air and Water Environments, Part 2*, I. A. Suffet, Ed., Wiley-Interscience, New York, 1977, pp. 137–154.

Lao, R. C. and R. S. Thomas, The Volatility of PAH and Possible Losses in Ambient Sampling, in *Polycyclic Aromatic Hydrocarbons*, A. Bjφrseth and A. J. Dennis, Eds., Battelle Press, Columbus, OH, 1980,pp. 829–839.

Lao, R. C., R. S. Thomas, H. Oja, and L. Dubois, "Application of a Gas Chromatograph–Mass Spectrometer–Data Processor Combination to the Analysis of the Polycyclic Aromatic Hydrocarbon Content of Airborne Pollutants," *Anal. Chem.*, **45,** 908 (1973).

Leary, J. A., A. L. Lafleur, H. L. Liber, and K. Blemann, "Chemical and Toxicologic Characterization of Fossil Fuel Combustion Product Phenalen-1-one," *Anal. Chem.*, **55,** 758 (1983).

Lee, F. and D. Schuetzle, Sampling, Extraction, and Analysis of Polycyclic Aromatic Hydrocarbons from Internal Combustion Engines, in *Handbook of Polycyclic Aromatic Hydrocarbons*, A. Bjφrseth, Ed., Dekker, New York, 1983, pp. 27–94.

Lee F. S. C., W. R. Pierson, and J. Ezike, The Problem of PAH Degradation During Filter Collection of Airborne Particulates—An Evaluation of Several Commonly Used Filter Media, In *Polynuclear Aromatic Hydrocarbons: Chemical and Biological Effects*. A Bjφrseth and A. J. Dennis, Eds., Battelle Press, Columbus, OH, 1980, pp. 543–563.

Lee, M. L., M. V. Novotny, and K. D. Bartle, *Analytical Chemistry of Polycyclic Aromatic Compounds*, Academic, New York, 1981.

Leermakers, P. A. and H. T. Thomas, "Electronic Spectra and Photochemistry of Adsorbed Organic Molecules. I. Spectra of Ketones on Silica Gel," *J. Am. Chem. Soc.*, **87,** 1620 (1965).

Leiter, J. and M. J. Shear, "Quantitative Experiments on the Production of Subcutaneous Tumors in Strain A Mice with Marginal Doses of 3,4-Benzpyrene," *J. Natl. Cancer Inst.*, **3,** 455 (1943).

Leiter, J., M. B. Shimkin, and M. J. Shear, "Production of Subcutaneous Sarcomas in Mice with Tars Extracted from Atmospheric Dusts," *J. Natl. Cancer Inst.*, **3,** 155 (1942).

Levin, D. E., M. Hollstein, M. F. Christman, E. A. Schwiers, and B. N. Ames, "A New *Salmonella* Tester Strain (TA102) with A:T Base Pairs at the Site of Mutation Detects Oxidative Mutagens," *Proc. Natl. Acad. Sci. USA*, **79,** 7445 (1982).

Lewtas, J., Ed., *Toxicological Effects of Emissions from Diesel Engines*, Elsevier, New York, 1982.

Lewtas, J., " Evaluation of the Mutagenicity and Carcinogenicity of Motor Vehicle Emissions in Short-Term Bioassays," *Environ. Health Perspect.*, **47,** 141 (1983).

Lindskog, A., E. Brorström-Lundén, and A. Sjödin, "Transformation of Reactive PAH on Particles by Exposure to Oxidized Nitrogen Compounds and Ozone," *Environ. Int. J.*, **11,** 125 (1985).

Locke, D. C., "Selectivity in Reversed-Phase Liquid Chromatography Using Chemically Bonded Stationary Phases," *J. Chromatogr. Sci.*, **12,** 433 (1974).

Löfroth, G., "Mutagenicity Assay of Combustion Emissions," *Chemosphere*, **7,** 791 (1978).

Löfroth G., Comparison of the Mutagenic Activity in Carbon Particulate Matter and in Diesel and Gasoline Engine Exhaust, in *Short-Term Bioassays in the Analysis of Complex Environmental Mixtures, II*, M. D. Waters, S. Sandhu, J. Lewtas-Huisingh, L. Claxton, and S. Nesnow, Eds., Plenum Press, New York, 1981, pp. 319–336.

Löfroth, G., E. Hefner, I. Alfheim, and M. Møller, "Mutagenic Activity in Photocopies," *Science*, **209,** 1037 (1980)

Löfroth, G., G. Lazaridis, and E. Agurell, "The Use of the *Salmonella*/Microsome Mutagenicity Test for the Characterization of Organic Extracts from Ambient Particulate Matter," *Environ. Int. J.*, **11,** 161 (1985).

Lunde, G. and A. Bjørseth, "Polycyclic Aromatic Hydrocarbons in Long-Range Transported Aerosols," *Nature*, **268,** 518 (1977).

Mackay, D. and W. Y. Shiu, "Aqueous Solubility of Polynuclear Aromatic Hydrocarbons," *J. Chem. Eng. Data.*, **22,** 399 (1977).

Marnett, L. J., H. K. Hurd, M. C. Hollstein, D. E. Levin, H. Esterbauer, and B. N. Ames, "Naturally

Occurring Carbonyl Compounds Are Mutagens in *Salmonella* Tester Strain TA104,'' *Mutat. Res.*, **148,** 25 (1985).

Maron, D. M. and B. N. Ames, ''Revised Methods for the *Salmonella* Mutagenicity Test,'' *Mutat. Res.*, **113,** 173 (1983).

Marty, J. C., M. J. Tissier, and A. Saliot, ''Gaseous and Particulate Polycyclic Aromatic Hydrocarbons from the Marine Atmosphere,'' *Atmos. Environ.*, **18,** 2183 (1984).

May, W. E. and S. A. Wise, ''Liquid Chromatographic Determination of Polynuclear Aromatic Hydrocarbons in Air Particulate Exracts,'' *Anal. Chem.*, **56,** 225 (1984).

McCann, J., ''In Vitro Testing for Cancer-Causing Chemicals,'' *Hosp. Prac.*, 73–85, September (1983).

McCann, J., E. Choi, E. Yamasaki, and B. N. Ames, ''Detection of Carcinogens as Mutagens in the *Salmonella*/Microsome Test: Assay of 300 Chemicals,'' *Proc. Natl. Acad. Sci. USA.*, **72,** 5135 (1975).

McCoy, E. C. and H. S. Rosenkranz, ''Activation of Polycyclic Aromatic Hydrocarbons to Mutagens by Singlet Oxygen: An Enhancing Effect of Atmospheric Pollutants.'' *Cancer Lett.*, **9,** 35 (1980).

McCoy, E. C., E. J. Rosenkranz and R. Mermelstein, ''Nitrated Fluorene Derivatives are Potent Frameshift Mutagens,'' *Mutat. Res.*, **90,** 11 (1981).

McElroy, M. W., R. C. Carr, D. S. Enson, and G. R. Markowski ''Size Distribution of Fine Particles from Coal Combustion,'' *Science,* **215,** 13 (1982).

Mermelstein, R., D. K. Demosthenes, N. Butler, E. C. McCoy, and H. S. Rosenkranz, ''The Extraordinary Mutagenicity of Nitropyrenes in Bacteria,'' *Mutat. Res.*, **89,** 187 (1981).

Mermelstein, R., E. C. McCoy and H. S. Rosenkranz, The Mutagenic Properties of Nitroarenes: Structure–Activity Relationships, in *Toxicity of Nitroarenes*, D. E. Rickert, Ed., Hemisphere Publishing, Washington DC, 1985, Ch. 14.

Miguel, A. H. and S. K. Friedlander, ''Distribution of Benzo(a)pyrene and Coronene with Respect to Particle Size in Pasadena Aerosols in the Submicron Range,'' *Atmos. Environ.*, **12,** 2407 (1978).

Mohr, U., H. Reznik-Schuller, G. Reznik, G. Grimmer, and J. Misfeld, ''Investigations on the Carcinogenic Burden by Air Pollution in Man. XIV. Effects of Automobile Exhaust Condensate on the Syrian Golden Hamster Lung,'' *Zentralbl. Bakteriol. Parasitenkd. Infektionshr. Hyg. Abt. Orig. Reihe. B*, **163,** 425 (1976).

National Academy of Sciences, *Particulate Polycylic Organic Matter*, Committee on Biologic Effects of Atmospheric Pollutants, National Academy Press, Washington, DC, 1972.

National Academy of Sciences, *Health Effects of Exposures to Diesel Exhaust*, National Academy Press, Washington DC, 1981.

National Academy of Sciences, *Diesel Cars: Benefits, Risks, and Public Policy*, National Academy Press, Washington, DC, 1982a.

National Academy of Sciences, *Diesel Technology*, National Academy Press, Washington DC, 1982b.

National Academy of Sciences, *Polycyclic Aromatic Hydrocarbons: Evaluation of Sources and Effects*, Committee on Pyrene and Selected Analogues, National Academy Press, Washington, DC 1983.

Natusch, D. F. S. and B. A. Tomkins, Theoretical Considerations of the Adsorption of Polynuclear Aromatic Hydrocarbon Vapor onto Fly Ash in a Coal-Fired Power Plant, in *Carcinogenesis*, Vol. 3, *Polynuclear Aromatic Hydrocarbons*, P. W. Jones and R. I. Freudenthal, Eds., Raven Press, New York, 1978.

Nicholls, C. H. and P. A. Leermakers, Photochemical and Spectroscopic Properties of Organic Molecules in Adsorbed or Other Perturbing Polar Environments, in *Advances in Photochemistry*, Vol. 8, J. N. Pitts, Jr., G. S. Hammond, and W. A. Noyes, Jr., Eds., Wiley-Interscience, New York, 1971, pp. 315–336.

Nielsen, T., ''Isolation of Polycyclic Aromatic Hydrocarbons and Nitro-Derivatives in Complex Mixtures by Liquid Chromatography,'' *Anal. Chem.*, **55,** 286 (1983).

Nielsen, T., "Reactivity of Polycyclic Aromatic Hydrocarbons Towards Nitrating Species," *Environ. Sci. Technol.*, **18,** 157 (1984).

Nielsen, T., T. Ramdahl, and A. Bjørseth, "The Fate of Airborne Polycyclic Organic Matter," *Environ. Health Perspect.*, **47,** 103 (1983).

Nielsen, T., B. Seitz, and T. Ramdahl, "Occurrence of Nitro-PAH in the Atmosphere in a Rural Area," *Atmos. Environ.*, **18,** 2159 (1984).

Nikolaou, K., P. Masclet, and G. Mouvier, "Sources and Chemical Reactivity of Polynuclear Aromatic Hydrocarbons in the Atmosphere—A Critical Review," *Sci. Total Environ.*, **32,** 103 (1984).

Novotny, M. and D. Ishii, Eds., *Microcolumn Separation Methods*, Elsevier, New York, 1985.

Novotny, M., A. Hirose, and D. Wiesler, "Separation and Characterization of Very Large Neutral Polycyclic Molecules in Fossil Fuels by Microcolumn Liquid Chromatography and Mass Spectrometry," *Anal. Chem.*, **56,** 1243 (1984).

Ohgaki, H., H. Hasegawa, T. Kato, C. Negishi, S. Sato, and T. Sugimura, "Absence of Carcinogenicity of 1-Nitropyrene, Correction of Previous Results, and New Demonstration of Carcinogenicity of 1,6-Dinitropyrene in Rats," *Cancer Lett.*, **25,** 239 (1985).

Ohnishi, Y., K. Kachk, K. Sato, I. Tahara, H. Takeyoshi, and H. Tokiwa, "Detection of Mutagenic Activity in Automobile Exhaust," *Mutat. Res.*, **77,** 229 (1980).

Paputa-Peck, M. C., R. S. Marano, D. Schuetzle, T. L. Riley, C. V. Hampton, T. J. Prater, L. M. Skewes, T. E. Jensen, P. H. Ruehle, L. C. Bosch, and W. P. Duncan, "Determination of Nitrated Polynuclear Aromatic Hydrocarbons in Particulate Extracts by Capillary Column Gas Chromatography with Nitrogen Selective Detection," *Anal Chem.*, **55,** 1946 (1983).

Passey, R. D., "Experimental Soot Cancer," *Br. Med. J.*, **2,** 1112 (1922).

Patterson, A. M., L. T. Capell, and D. F. Walker, *The Ring Index. A List of Ring Systems Used in Organic Chemistry*, 2nd ed., American Chemical Society, Washington, DC, 1960.

Peaden, P. A., J. C. Fjeldsted, M. L. Lee, S. R. Springston, and M. Novotny, "Instrumental Aspects of Capillary Supercritical Fluid Chromatography," *Anal. Chem.*, **54,** 1090 (1982).

Pellizzari, E. D., L. W. Little, C. Sparacino, and T. J. Hughes, Integrating Microbiological and Chemical Testing into the Screening of Air Samples for Potential Mutagenicity, in *Application of Short-Term Bioassays in the Fractionation and Analysis of Complex Environmental Mixtures*, M. D. Waters, S. Nesnow, J. L. Huisingh, S. Sandhu, and L. Claxton, Eds., Plenum Press, New York, 1979, pp. 331–351.

Peters, J. and B. Seifert, "Losses of Benzo(a)pyrene Under the Conditions of High-Volume Sampling," *Atmos. Environ.*, **14,** 117 (1980).

Phillips, D. H., "Fifty Years of Benzo(a)pyrene," *Nature*, **303,** 468 (1983).

Pierce, R. C. and M. Kratz, "Chromatographic Isolation and Spectral Analysis of Polycyclic Quinones. Application to Air Pollution Analysis," *Environ. Sci. Technol.*, **10,** 45 (1976).

Pitts, J. N., Jr., "Photochemical and Biological Implications of the Atmospheric Reactions of Amines and Benzo(a)pyrene," *Phil. Trans. R. Soc. Lond.*, **A290,** 551 (1979).

Pitts, J. N., Jr., "Formation and Fate of Gaseous and Particulate Mutagens and Carcinogens in Real and Simulated Atmospheres," *Environ. Health Perspect.*, **47,** 115 (1983).

Pitts, J. N. Jr., A. U. Khan, E. B. Smith, and R. P. Wayne, "Singlet Oxygen in the Environmental Sciences. Singlet Molecular Oxygen and Photochemical Air Pollution," *Environ. Sci. Technol.*, **3,** 241 (1969).

Pitts, J. N., Jr., D. Grosjean, T. M. Mischke, V. F. Simmon, and D. Poole, "Mutagenic Activity of Airborne Particulate Organic Pollutants," *Toxicol. Lett.*, **1,** 65 (1977).

Pitts, J. N., Jr., K. A. Van Cauwenberghe, D. Grosjean, J. P. Schmid, D. R. Fitz, W. L. Belser, Jr., G. B. Knudson, and P. M. Hynds, "Atmospheric Reactions of Polycyclic Aromatic Hydrocarbons: Facile Formation of Mutagenic Nitro Derivatives," *Science*, **202,** 515 (1978).

Pitts, J. N., Jr., D. Grosjean, T. M. Mischke, V. F. Simmon, and D. Poole, Mutagenic Activity of

Airborne Particulate Organic Pollutants, in *Biological Effects of Environmental Pollutants*, S. D. Lee and J. B. Mudd, Eds., Ann Arbor Science Publishers, Ann Arbor, MI, 1979, pp. 219–235.

Pitts, J. N., Jr., D. M. Lokensgard, P. S. Ripley, K. A. Van Cauwenberghe, L. van Vaeck, S. D. Shaffer, A. J. Thill, and W. L. Belser, Jr., "'Atmospheric' Epoxidation of Benzo(a)pyrene by Ozone: Formation of the Metabolite Benzo(a)pyrene-4,5 oxide," *Science*, **210**, 1347 (1980).

Pitts, J. N., Jr., D. M. Lokensgard, W. Harger, T. S. Fisher, V. Mejia, J. Schuler, G. M. Scorziell, and Y. A., Katzenstein "Mutagens in Diesel Exhaust: Identification and Direct Activities of 6-Nitrobenzo(a)pyrene, 9-Nitroanthracene, 1-Nitropyrene, and 5*H*-Phenanthro(4,5-bcd)pyran-5-one," *Mutat. Res.*, **103**, 241 (1982a).

Pitts, J. N., Jr., W. Harger, D. M. Lokensgard, D. R. Fitz, G. M. Scorziell, and V. Mejia, "Diurnal Variations in the Mutagenicity of Airborne Particulate Organic Matter in California's South Coast Air Basin," *Mutat. Res.*, **104**, 35 (1982b).

Pitts, J. N., Jr., B. Zielinska, and W. P. Harger, "Isomeric Mononitrobenzo(a)pyrenes: Synthesis, Identification, and Mutagenic Activities," *Mutat. Res.*, **140**, 81 (1984).

Pitts, J. N., Jr., R. Atkinson, J. A. Sweetman, and B. Zielinska, "The Gas-Phase Reaction of Naphthalene with N_2O_5 to Form Nitronaphthalenes," *Atmos. Environ.*, **19**, 701 (1985a).

Pitts, J. N., Jr., B. Zielinska, J. A. Sweetman, R. Atkinson, and A. M. Winer, "Reactions of Adsorbed Pyrene and Perylene with Gaseous N_2O_5 Under Simulated Atmospheric Conditions." *Atmos Environ.*, **19**, 911 (1985b).

Pitts, J. N., Jr., J. A. Sweetman, W. Harger, D. Fitz, H. R. Paur, and A. M. Winer, "Diurnal Mutagenicity of Airborne Particulate Organic Matter Adjacent to a Heavily Travelled West Los Angeles Freeway," *J. Air Pollut. Control Assoc.*, **35**, 638 (1985c).

Pitts, J. N., Jr., J. A. Sweetman, B. Zielinska, A. M. Winer, R. Atkinson, and W. P. Harger, "Formation of Nitroarenes from the Reaction of Polycyclic Aromatic Hydrocarbons with Dinitrogen Pentoxide," *Environ. Sci. Technol.*, in press (1985d).

Pitts, J. N., Jr., H. R. Paur, B. Zielinska, J. A. Sweetman, A. M. Winer, T. Ramdahl, and V. Mejia, "Factors Influencing the Reactivity of Polycyclic Aromatic Hydrocarbons Adsorbed on Model Substrates and in Ambient POM with Ambient Levels of Ozone," *Chemosphere*, submitted for publication (1985e).

Pitts, J. N., Jr., J. A. Sweetman, B. Zielinska, A. M. Winer, and R. Atkinson, "Determination of 2-Nitrofluoranthene and 2-Nitropyrene in Ambient Particulate Organic Matter: Evidence for Atmospheric Reactions," *Atmos. Environ.*, **19**, 1601 (1985g).

Pryor, W. A., G. J. Geicker, J. P. Cosgrove, and D. F. Church, "Reaction of Polycyclic Aromatic Hydrocarbons (PAH) with Nitrogen Dioxide in Solution. Support for an Electron-Transfer Mechanism of Aromatic Nitration Based on Correlations Using Simple Molecular Orbital Theory," *J. Org. Chem.*. **49**, 5189 (1984).

Pupp, C., R. C. Lao, J. J. Murray, and R. F. Pottie, "Equilibrium Vapor Concentrations of Some Polycyclic Aromatic Hydrocarbons, As_4O_6 and SeO_2, and the Collection Efficiencies of These Air Pollutants," *Atmos. Environ.*, **8**, 915 (1974).

Ramdahl, T., "Retene—A Molecular Marker of Wood Combustion in Ambient Air," *Nature*, **306**, 580 (1983a).

Ramdahl, T., "Polycyclic Aromatic Ketones in Environmental Samples," *Environ. Sci. Technol.*, **17**, 666 (1983b).

Ramdahl, T., "Analysis of Nitro-4*H*-Cyclopenta(def)phenanthrenes in Ambient Air by Fused Silica Capillary Gas Chromatography/Mass Spectrometry, *J. High Resol. Chromatogr. Chromatogr. Commun.*, **8**, 82 (1985).

Ramdahl, T., G. Becher, and A. Bjørseth, "Nitrated Polycyclic Aromatic Hydrocarbons in Urban Air Particles," *Environ. Sci. Technol.*, **16**, 861 (1982).

Ramdahl, T., J. Schjoldager, L. A. Currie, J. E. Hanssen, M. Møller, G. A. Klouda, and I. Alfheim,

"Ambient Impact of Residential Wood Combustion in Elverum, Norway," *Sci. Total Environ.*, **36,** 81 (1984a).

Ramdahl, T., A. Bjørseth, D. Lokensgard, and J. N. Pitts, Jr., "Nitration of Polycyclic Aromatic Hydrocarbons Adsorbed to Different Carriers in a Fluidized Bed Reactor," *Chemosphere*, **13,** 527 (1984b).

Ramdahl, T., J. A. Sweetman, B. Zielinska, R. Atkinson, A. M. Winer, and J. N. Pitts, Jr., "Analysis of Mononitro Isomers of Fluoranthene and Pyrene by High Resolution Capillary Gas Chromatography/Mass Spectrometry," *J. High Resolution Chromat. Chromat. Commun.*, in press, December (1985a).

Ramdahl, T., J. A. Sweetman, B. Zielinska, W. P. Harger, A. M. Winer, and R. Atkinson, "Determination of Nitrofluoranthenes and Nitropyrenes in Ambient Air and Their Contribution to Direct Mutagenicity," presented at the Tenth Anniversary of the International Symposium on Polynuclear Aromatic Hydrocarbons, Columbus, OH, October 21–23, 1985b.

Rånby, B. and J. F. Rabek, Eds., *Singlet Oxygen Reactions with Organic Compounds and Polymers*, Wiley, New York, 1978.

Rappaport, S. M., Y. Y. Wang, E. T. Wei, R. Saywer, B. E. Watkins, and H. Rappaport, "Isolation and Identification of a Direct-Acting Mutagen in Diesel-Exhaust Particulates," *Environ. Sci. Technol.*, **14,** 1505 (1980).

Rickert, D. E., Ed., *Toxicity of Nitroaromatic Compounds*, Hemisphere Publishing, Washington, DC, 1985.

Rondia, D., "Sur la Volatilité des Hydrocarbures Polycycliques,"*Int. J. Water Air Pollut.*, **9,** 113 (1965).

Rondia, D., M. Cooke, and R. K. Haroz, Eds., *Mobile Source Emissions Including Polycyclic Organic Species*, D. Reidel Publishing, Dordrecht, 1983.

Rosenkranz, H. S., "Direct Acting Mutagens in Diesel Exhausts: Magnitude of the Problem," *Mutat. Res.*, **101,** 1 (1982).

Rosenkranz, H. S., "Mutagenic and Carcinogenic Nitroarenes in Diesel Emissions: Risk Identification," *Mutat. Res.*, **140,** 1(1984).

Rosenkranz, H. S. and R. Mermelstein, "Mutagenicity and Genotoxicity of Nitroarenes: All Nitro-containing Chemicals were Not Created Equal," *Mutat. Res.*, **114,** 217 (1983).

Rosenkranz, H. S. and R. Mermelstein, The Mutagenic and Carcinogenic Properties of Nitrated Polycyclic Aromatic Hydrocarbons, in *Nitrated Polycyclic Aromatic Hydrocarbons*, C. White, Ed., Hüethig Publishing, Heidelberg, West Germany, 1985, pp. 267–297.

Rosenkranz, H. S., E. C. McCoy, D. R. Sanders, M. Butler, D. K. Kiriazides, and R. Mermelstein, "Nitropyrenes: Isolation, Identification, and Reduction of Mutagenic Impurities in Carbon Black and Toners," *Science*, **209,** 1039 (1980).

Rosenkranz, H. S., G. Klopman, V. Changkong, J. Pet-Edwards, Y. Y. Haimes, "Predication of Environmental Carcinogens: A Strategy for the Mid-1980's," *Environ. Mutagen*, **6**, 231 (1984).

Rudling, L., B. Ahling, and G. Löfroth, Chemical and Biological Characterization of Emissions from Combustion of Wood and Wood-Chips in Small Furnaces and Stoves, in *Residential Solid Fuels*, J. A. Cooper and D. Malik, Eds., Oregon Graduate Center, Beaverton, 1982, pp.34–53.

Salmeen, I. T., A. M. Pero, R. Zator, D. Schuetzle, and T. L. Riley, "Ames Assay Chromatograms and the Identification of Mutagens in Diesel Particle Extracts," *Environ. Sci. Technol.*, **18,** 375 (1984).

Salmeen, I. T., R. A. Gorse, Jr., and W. R. Pierson, "Ames Assay Chromatograms of Extracts of Diesel Exhaust Particles from Heavy-Duty Trucks on the Road and from Passenger Cars on a Dynamometer," *Environ. Sci. Technol.*, **19,** 270 (1985).

Santodonato, J., P. Howard, and D. Basu, Eds., "Health and Ecological Assessment of Polynuclear Aromatic Hydrocarbons," *J. Environ. Pathol. Toxicol.* (Special Issue), **5**(1), 1–372 (1981).

Sawicki, E. "Airborne Carcinogens and Allied Compounds," *Arch. Environ. Health*, **14,** 46 (1967).

Sawicki, E., S. P. McPherson, T. W. Stanley, J. Meeker, and W. C. Elbert, ''Quantitative Composition of the Urban Atmosphere in Terms of Polynuclear Aza-Heterocyclic Compounds and Aliphatic and Polynuclear Aromatic Hydrocarbons,'' *Int. J. Air Water Pollut.*, **9,** 515 (1965).

Sawicki, E., T. R. Hauser, and T. W. Stanley, ''Ultraviolet, Visible, and Fluorescence Spectral Analyses of Polynuclear Hydrocarbons,'' *Int. J. Air Pollut.*, **2,** 253 (1960).

Schaap, A. P., Ed., *Singlet Molecular Oxygen*, Benchmark Papers in Organic Chemistry, Vol. 5, Dowden, Hutchinson and Ross Inc., Stroudsburg, PA, 1976.

Schairer, L. A., J. Van't Hof, C. G. Hayes, and R. M. Burton, Measurement of Biological Activity of Ambient Air Mixtures Using a Mobile Laboratory for *in situ* Exposures: Preliminary Results from the *Tradescantia* Plant Test System, in *Application of Short-Term Bioassays in the Fractionation and Analysis of Complex Environmental Mixtures*, M. D. Waters, S. Nesnow, J. Huisingh, S. Sandhu, and L. Claxton, Eds., Plenum, New York, 1979, pp. 419–440.

Schairer, L. A., R. C. Sautkulis, and N. R. Tempel, A Search for the Identity of Genotoxic Agents in the Ambient Air Using the *Tradescantia* Bioassay, in *Short-Term Bioassays in the Analyses of Complex Environmental Mixtures, III*, M. D. Waters, S. S. Sandhu, J. Lewtas, L. Claxton, N. Chernoff, and S. Nesnow, Eds., Plenum Press, New York, 1983.

Schofield, K., *Aromatic Nitration*, Cambridge University Press, Cambridge, MA, 1980.

Schuetzle, D., ''Sampling of Vehicle Emissions for Chemical Analysis and Biological Testing'' *Environ. Health Perspect.*, **47,** 65 (1983).

Schuetzle, D., F. S-C Lee, T. J. Prater and S. B. Tejada, ''The Identification of Polynuclear Aromatic Hydrocarbon (PAH) Derivatives in Mutagenic Fractions of Diesel Particulate Extracts'' *Int. J. Environ. Anal. Chem.* **9,** 93 (1981).

Schuetzle, D., T. E. Jensen, and J. C. Ball, ''Polar Polycyclic Aromatic Hydrocarbon (PAH) Derivatives in Extracts of Particulates: Biological Characterization and Techniques for Chemical Analysis,'' *Environ. Int. J.*, **11,** 169 (1985).

Shepson, P. B., T. E. Kleindienst, E. O. Edney, G. Namie, J. Pittman, L. Cupitt, and L. Claxton, ''The Mutagenic Activity of Irradiated Toluene/NO_x/H_2O/Air Mixtures,'' *Environ. Sci. Technol.*, **19,** 249 (1985).

Siak, J., T. L. Chan, T. L. Gibson, and G. T. Wolff, ''Contribution to Bacterial Mutagenicity from Nitro-PAH Compounds in Ambient Aerosols,'' *Atmos. Environ.*, **19,** 369 (1985).

Simonsick, W. J., Jr. and R. A. Hites, ''Analysis of Isomeric Polycyclic Aromatic Hydrocarbons by Charge-Exchange Chemical Ionization Mass Spectrometry,'' *Anal. Chem.*, **56,** 2749 (1984).

Skopek, T. R. and W. G. Thilly, ''Rate of Induced Forward Mutation at 3 Genetic Loci in *Salmonella typhimurium*,'' *Mutat. Res.*, **108,** 45 (1983).

Skopek, T. R., H. L. Liber, D. A. Kaden, and W. G. Thilly, ''Relative Sensitivities of Forward and Reverse Mutation Assays in *Salmonella typhimurium*,'' *Proc. Natl. Acad. Sci. USA*, **75,** 4465 (1978).

Smith, R. D., J. C. Fjeldsted, and M. L. Lee, ''Supercritical Fluid Chromatography–Mass Spectrometry,'' *Int. J. Mass Spectrom. Ion Phys.*, **46,** 217 (1984).

Sonnefeld, W. J., W. H. Zoller, and W. E. May, ''Dynamic Coupled Column Liquid Chromatographic Determination of Ambient Temperature Vapor Pressures of Polynuclear Aromatic Hydrocarbons,'' *Anal. Chem.*, **55,** 275 (1983).

Sugimura, T. and S. Takayama, ''Biological Actions of Nitroarenes in Short-Term Tests on *Salmonella*, Cultured Mammalian Cells and Cultured Human Tracheal Tissues: Possible Basis for Regulatory Control,'' *Environ. Health Perspec.*, **47,** 171 (1983).

Swarin, S. J. and R. L. Williams, Liquid Chromatographic Determination of Benzo(a)pyrene in Diesel Exhaust Particulates: Verification of the Collection and Analytical Methods, in *Polynuclear Aromatic Hydrocarbons: Chemistry and Biological Effects*, A. Bjørseth and A. J. Dennis, Eds., Battelle Press, Columbus, OH, 1980, pp. 771–790.

Sweetman, J. A., F. W. Karasek, and D. Schuetzle, ''Decomposition of Nitropyrene During Gas

Chromatographic–Mass Spectrometric Analysis of Air Particulate and Fly-Ash Samples, *J. Chromatogr.*, **247,** 245 (1982).

Sweetman, J. A., B. Zielinska, R. Atkinson, A. M. Winer, and J. N. Pitts, Jr., "Nitration Products from the Reaction of Fluoranthene and Pyrene with N_2O_5 and Other Nitrogenous Species in the Gaseous, Adsorbed and Solution Phases: Implications for Atmospheric Transformations of PAH," presented at the Tenth Anniversary of the International Symposium on Polynuclear Aromatic Hydrocarbons, Columbus, OH, October 21–23, 1985a.

Sweetman, J. A., B. Zielinska, R. Atkinson, T. Ramdahl, A. M. Winer, and J. N. Pitts, Jr., "A Possible Formation Pathway for the 2-Nitrofluoranthene Observed in Ambient Particulate Organic Matter," *Atmos. Environ.*, in press (1985b).

Takemura, N., C. Hashida, and M. Terasawa, "Carcinogenic Action of 5-Nitroacenaphthalene," *Br. J. Cancer*, **30,** 481 (1974).

Talcott, R. and W. Harger, "Airborne Mutagens Extracted from Particles of Respirable Size," *Mutat. Res.*, **79,** 177 (1980).

Talcott, R. and E. Wei, "Airborne Mutagens Bioassayed in *Salmonella typhimurium*," *J. Natl. Cancer Inst.*, **58,** 449 (1977).

Tebbens, B. D., J. F. Thomas, and M. Mukai, "Fate of Arenes Incorporated with Airborne Soot," *Am. Ind. Hyg. Assoc. J.*, **27,** 415 (1966).

Tebbens, B. D., M. Mukai, and J. F. Thomas, "Fate of Arenes Incorporated with Airborne Soot: Effects of Irradiation," *Am. Ind. Hyg. Assoc. J.*, **32,** 365 (1971).

Teranishi, K., K. Hamada, and H. Watanabe, "Mutagenicity in *Salmonella typhimurium* Mutants of the Benzene-Soluble Organic Matter Derived from Airborne Particulate Matter and Its Five Fractions," *Mutat. Res.*, **56,** 273 (1978).

Thomas, J. F., M. Mukai, and B. D. Tebbens, "Fate of Airborne Benzo(a)pyrene," *Environ. Sci. Technol.*, **2,** 33 (1968).

Thrane, K. and A. Mikalsen, "High-Volume Sampling of Airborne Polycyclic Aromatic Hydrocarbons Using Glass Fiber Filters and Polyurethane Foam," *Atmos. Environ.*, **15,** 909 (1981).

Toftgard, R., G. Löfroth, J. Carlstedt-Duke, R. Kurl, and J. Gustafsson, "Compounds in Urban Air Compete with 2,3,7,8-Tetrachlorodibenzo-*p*-dioxin for Binding to the Receptor Protein," *Chem. Biol. Interactions*, **46,** 355 (1983).

Tokiwa, H., H. Takeyoshi, K. Morita, K. Takahashi, N. Saruta, and Y. Ohnishi, "Detection of Mutagenic Activity in Urban Air Pollutants," *Mutat. Res.*, **38,** 351 (1976).

Tokiwa, H., Morita, H. Takeyoshi, K. Takahashi, and Y. Ohnishi, "Detection of Mutagenic Activity in Particulate Air Pollutants," *Mutat. Res.*, **48,** 237 (1977).

Tokiwa, H., S. Kitamori, K. Takahashi, and Y. Ohnishi, "Mutagenic and Chemical Assay of Extracts of Airborne Particulates," *Mutat. Res.*, **77,** 99 (1980).

Tokiwa, H., R. Nakagawa, K. Morita, and Y. Ohnishi, "Mutagenicity of Nitro Derivatives Induced by Exposure of Aromatic Compounds to Nitrogen Dioxide," *Mutat. Res.*, **85,** 195 (1981).

Tokiwa, H., S. Kitamori, R. Nakagawa, K. Horkiawa, and L. Matamala, "Demonstration of a Powerful Mutagenic Dinitropyrene in Airborne Particulate Matter," *Mutat. Res.*, **121,** 107 (1983).

Turro, N. J., *Modern Molecular Photochemistry*, Benjamin/Cummings, Menlo Park, CA, 1978.

Valerio, F., P. Bottino, D. Ugolini, M. R. Cimberle, G. Tozzi, A. Frigerio, "Chemical and Photochemical Degradation of Polycyclic Aromatic Hydrocarbons in the Atmosphere," *Sci. Total Environ.*, **40,** 169 (1984).

Van Cauwenberghe, K. and L. Van Vaeck, "Toxicological Implications of the Organic Fraction of Aerosols: A Chemist's View," *Mutat. Res.*, **116,** 1 (1983).

Van Cauwenberghe, K., L. Van Vaeck, and J. N. Pitts, Jr., "Chemical Transformations of Organic Pollutants During Aerosol Sampling," *Adv. Mass Spectrom.*, **8,** 1499 (1979).

Vance, W. A. and D. E. Levin, "Structural Features of Nitroaromatics that Determine Mutagenic Activity in *Salmonella typhimurium*," *Environ. Mutagen.*, **6,** 797 (1984).

Van Vaeck, L., G. Broddin, and K. Van Cauwenberghe, "On the Relevance of Air Pollution Measurements of Aliphatic and Polyaromatic Hydrocarbons in Ambient Particulate Matter," *Biomed. Mass Spectrom.*, **7,** 473 (1980).

Van Vaeck, L. and K. Van Cauwenberghe, "Conversion of Polycyclic Aromatic Hydrocarbons on Diesel Particulate Matter Upon Exposure to PPM Levels of Ozone," *Atmos. Environ.*, **18,** 323 (1984).

Van Vaeck, L., K. Van Cauwenberghe, and J. Janssens, "The Gas–Particle Distribution of Organic Aerosol Constituents: Measurement of the Volatilization Artifact in Hi-Vol Cascade Impactor Sampling," *Atmos. Environ.*, **18,** 417 (1984).

Vouk, V. B. and W. T. Piver, "Metallic Elements in Fossil Fuel Combustion Products. Amounts and Form of Emissions and Evaluation of Carcinogenicity and Mutagenicity," *Environ. Health Perspect.*, **47,** 201 (1983).

Waller, R. E., "The Benzpyrene Content of Town Air," *Br. J. Cancer*, **6,** 8 (1952).

Wang, Y. Y., S. M. Rappaport, R. F. Sawyer, R. E. Talcott, and E. T. Wei, "Direct-Acting Mutagens in Automobile Exhaust," *Cancer Lett.*, **5,** 39 (1978).

Wang, C. Y., Mei-Sie Lee, C. M. King, and P. O. Warner, "Evidence for Nitroaromatics as Direct-Acting Mutagens of Airborne Particles," *Chemosphere*, **9,** 83 (1980).

Waters, M. D., S. Nesnow, J. L. Huisingh, S. Sandhu, and L. Claxton, Eds., *Application of Short-Term Bioassays in the Fractionation and Analysis of Complex Environmental Mixtures*, Plenum Press, New York, 1979.

Waters, M., S. Sandhu, J. Lewtas-Huisingh, L. Claxton, and S. Nesnow, Eds., *Short-Term Bioassays in the Analysis of Complex Environmental Mixtures, II*, Plenum Press, New York, 1981.

Waters, M. D., S. S. Sandhu, J. Lewtas, L. Claxton, N. Chernoff, and S. Nesnow, Eds. *Short-Term Bioassays in the Analysis of Complex Environmental Mixtures, III*, Plenum Press, New York, 1983.

Wehry, E. L., Optical Spectrometric Techniques for Determination of Polycyclic Aromatic Hydrocarbons, in *Handbook of Polycyclic Aromatic Hydrocarbons*, A. Bjørseth, Ed., Dekker, New York, 1983, pp. 323–396.

Wei, E. T. and H. P. Shu, "Nitroaromatic Cacinogens in Diesel Soot: A Review of Laboratory Findings," *Am. J. Public Health*, **73,** 1085 (1983).

Wesolowski, J., P. Flessel, S. Twiss, J. Cheng, R. Chan, L. Garcia, J. Ondo, A. Fong, and S. Lum, "The Chemical and Biological Characterization of Particulate Matter as Part of an Epidemiological Cancer Study," *J. Aerosol Sci.*, **12,** 208 (1981).

White, C. M., Ed., *Nitrated Polycyclic Aromatic Hydrocarbons*, Hüethig Publishing, Heidelberg, West Germany, 1985.

Wise, S. A., High-Performance Liquid Chromatography for the Determination of Polycyclic Aromatic Hydrocarbons, in *Handbook of Polycyclic Aromatic Hydrocarbons*, A. Bjørseth, Ed., Dekker, New York, 1983, pp. 183–256.

Wislocki, P., A. Wood, R. Chang, W. Levin, H. Yagi, O. Hernandez, P. Dansette, D. Jerina, and A. Conney, "Mutagenicity and Cytotoxicity of Benzo(a)pyrene Arene Oxides, Phenols, Quinones, and Dihydrodiols in Bacterial and Mammalian Cells," *Cancer Res.*, **36,** 3350 (1976).

Wolff, G. T., R. J. Countess, P. J. Grolicki, M. A. Ferman, S. H. Cadle, and J. L. Muhlbaier, "Visibility-Reducing Species in the Denver 'Brown Cloud'—II. Sources and Temporal Patterns," *Atmos. Environ.*, **15,** 2485 (1981).

Wu, C. H., I. Salmeen, and H. Niki, "Fluorescence Spectroscopic Study of Reactions Between Gaseous Ozone and Surface-Adsorbed Polycyclic Aromatic Hydrocarbons," *Environ, Sci. Technol.*, **18,** 603 (1984).

Xu, X. B., J. P. Nachtman, Z. L. Jin, E. T. Wei, S. M. Rappaport, and A. L. Burlingame, "Isolation and Identification of Mutagenic Nitro-PAH (Polycyclic Aromatic Hydrocarbon) in Diesel Exhaust Particulates," *Anal. Chim. Acta*, **136,** 163 (1982).

Yamanaka, S. and S. Maruoka, ''Mutagenicity of the Extract Recovered from Airborne Particles Outside and Inside a Home with an Unvented Kerosene Heater,'' *Atmos. Environ.*, **18,** 1485 (1984).

Yamasaki, H., Kuwata, and Y. Kuge, ''Determination of Vapor Pressure of Polycyclic Aromatic Hydrocarbons in the Supercooled Liquid Phase and Their Adsorption on Airborne Particulate Matter,'' *J. Chem. Soc. Japan*, **8,** 1329 (1984).

Zielinska, B., University of California, Riverside, Statewide Air Pollution Research Center, unpublished results (1985).

Zielinska, B., J. A. Sweetman, R. Atkinson, T. Ramdahl, A. M. Winer, and J. N. Pitts, Jr., ''The Reaction of Dinitrogen Pentoxide with Fluoranthene,'' to be submitted for publication (1985a).

Zielinska, B., J. A. Sweetman, and W. Harger, University of California, Riverside, Statewide Air Pollution Research Center, unpublished results (1985b).

PART 8

CHEMISTRY OF THE NATURAL TROPOSPHERE

14 Sources, Atmospheric Lifetimes, and Chemical Fates of Species in the Natural Troposphere

14-A. OZONE AND OXIDES OF NITROGEN

14-A-1. O_3 Concentrations and Sources

Ozone and oxides of nitrogen are important species present in the atmosphere in areas remote from anthropogenic activities and air pollutant emissions. While there is some question as to whether any location on earth is truly untouched by human activities in some way, we refer to the atmosphere in such remote areas as the *natural* troposphere.

A particularly important question in the chemistry of the natural troposphere relevant to rural and urban areas is the concentration of O_3 in the natural troposphere and the processes contributing to it. Background concentrations of O_3 observed in a number of locations around the world typically show average daily 1 h maxima of ~20–60 ppb (Singh et al., 1978). Long-term O_3 data at such sites typically show a yearly cycle with a maximum in the late winter or early spring. For example, Fig. 14.1 shows the long-term O_3 data for six sites considered by Singh and co-workers to be remote. An area being classified as remote does not rule out the possibility of long-range transport of pollutants to these sites; for example, the O_3 peak in August–October 1974 observed at Quillayute, Washington, on the western edge of the Olympic Peninsula (Fig. 14.1) could be due to transport of precursors from Portland, Oregon, or the Seattle area. Indeed, many of the other locations in Fig. 14.1 also seem likely to be influenced by long-range transport. However, many of the data in Fig. 14.1 show a trend to maxima around February–April in all four measures of O_3 plotted while some (e.g. Rio Blanco, CO) show a broad summer maximum. It is interesting to note in Fig. 14.1 that the maximum 1 h O_3 concentration for the month (curve A) occasionally exceeds 80 ppb, which until 1978 was the U.S. federal air quality standard for oxidant. No clear diurnal variation in O_3 was observed at these sites.

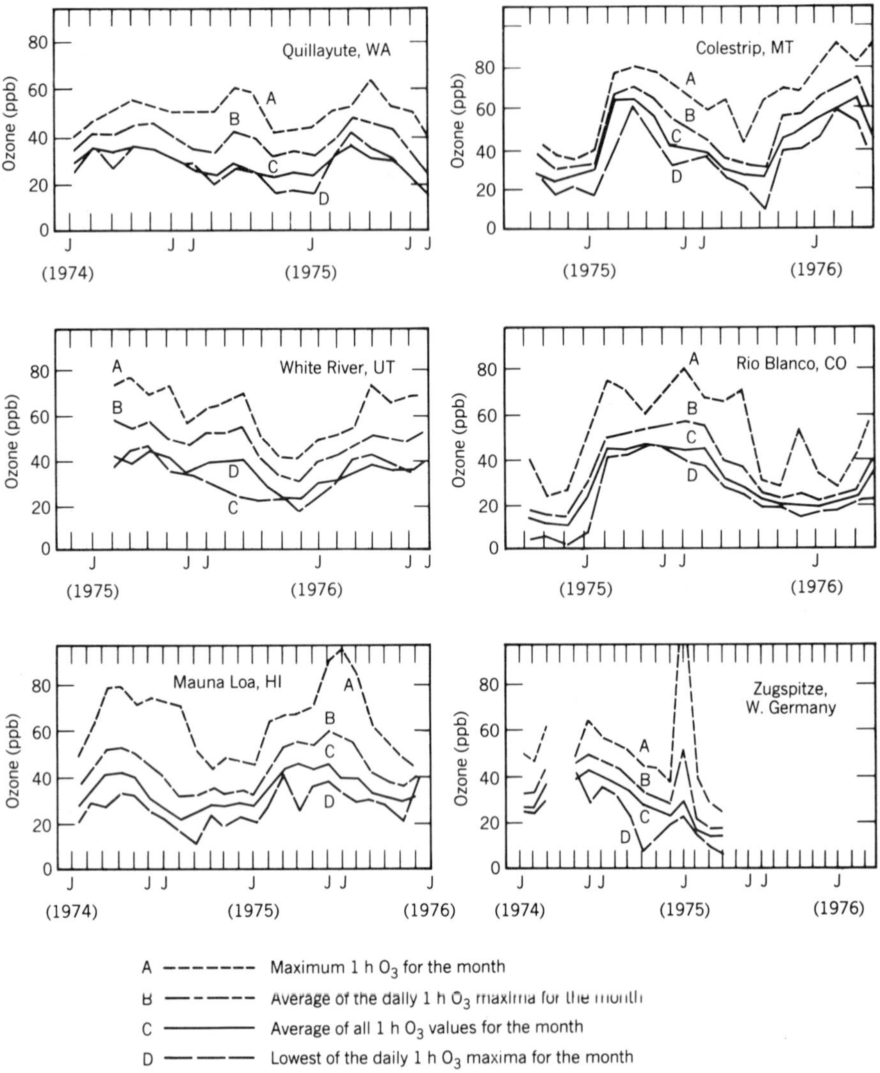

FIGURE 14.1. Long-term variations in atmospheric O_3 concentrations at six "remote" sites. JJ = June and July, J = January. Off-scale measurements at Zugspitze, West Germany, peaked at 196 ppb (from Singh et al., 1978).

Two sources of O_3 in the natural troposphere have been suggested: (1) injection from the stratosphere and (2) photochemical production via NO_x–NMOC (nonmethane organic compounds) reactions in sunlight involving naturally occurring oxides of nitrogen and hydrocarbons, or CO. There has been considerable controversy regarding the relative importance of these two sources over the last few years (Vukovich et al., 1985; Levy et al., 1985; Logan, 1985). Let us briefly review the evidence for each of these processes.

The occasional occurrence of stratospheric injection of O_3 at particular times and in certain locations is now reasonably well accepted. The stratosphere contains relatively high concentrations of O_3 compared to the troposphere; indeed, the troposphere contains only ~10% of the total O_3 in the atmosphere (Crutzen and Gidel, 1983). Although mixing of stratospheric air into the troposphere is inhibited by the temperature increase at the tropopause (see Section 1-B), meteorological phenomena can lead to periodic short-term "breakdowns" of this temperature discontinuity in a particular location, leading to a temporary mixing of stratospheric air containing O_3 into the troposphere (Singh et al., 1980). Such episodes are often characterized by relatively rapid changes in the O_3 concentration with peaks occurring at times not expected from photochemical processes. For example, the data for Zugspitze, West Germany, in Fig. 14.1 showed a maximum 1 h average O_3 concentration of 196 ppb which occurred at midnight on January 8, 1975. Such high O_3 concentrations were not observed in the region surrounding Zugspitze on the previous day, suggesting that photochemical processes combined with transport could not account for this peak. At the same time as the O_3 increased, the concentration of the isotope ^7Be increased; ^7Be has been suggested as a tracer of stratospheric air, although its applicability as a tracer is controversial at present (e.g., see Singh et al., 1980; Dutkiewicz and Husain, 1985). In any event, such rapid changes producing high O_3 peaks at times inconsistent with a photochemical source have been observed at a number of locations (e.g. see Chung and Dann, 1985), and have been taken as evidence for the injection of air from the stratosphere into the troposphere.

The seasonal variation in O_3, with a peak in the late winter and early spring, and the lack of a diurnal variation in the O_3 are also cited as evidence for a non-photochemical source of O_3. Thus exchange of stratospheric and tropospheric air is most effective during late winter and early spring (Danielsen and Mohnen, 1977); in addition, a photochemical source would be expected to peak with the solar intensity—in the summer and in the afternoon.

Tropospheric O_3 concentrations tend to be larger in the northern hemisphere than in the southern hemisphere (Fishman and Crutzen, 1978), which is consistent with model predictions that the downward flux of O_3 from the stratosphere should be larger in the northern hemisphere (Mahlman et al., 1980; Gidel and Shapiro, 1980).

Finally, the O_3 mixing ratio (defined as the ratio of the concentration of O_3 in molecules cm^{-3} to that of air) increases with height in the troposphere at all latitudes and seasons in the northern hemisphere (Chatfield and Harrison, 1977). Since atmospheric mixing tends to carry O_3 down toward the earth's surface, the increase with altitude implies a stratospheric source.

However, while these arguments have been made for the contribution of stratospheric injection of O_3 to the earth's surface, other data strongly suggest that ground-level impacts are infrequent and do not generally lead to O_3 concentrations exceeding 100 ppb. Thus long-range transport of photochemical oxidant precursors from urban areas is well known; for example, long-range transport from western Europe has been postulated to be partially responsible for some of the elevated ozone concentrations observed in Norway (Schjoldager et al., 1978; Schjoldager, 1981).

In addition, aircraft measurements of four stratospheric O_3 intrusion events suggest that more than half of the O_3 mass injected into the upper troposphere by these events is likely mixed and diluted into the troposphere above the 3 km elevation level (Viezee et al., 1983); these researchers suggest that ground-level elevated concentrations of O_3 due to stratospheric intrusions occur less than 1% of the time.

There are a number of indications that O_3 produced in the troposphere by photochemical processes is the most significant factor for some locations and times. First, as discussed throughout this book, NO_x–NMOC reactions are well known to produce O_3 in rural and urban areas; there is no a priori reason to think the same types of reactions might not occur with naturally emitted NO_x and hydrocarbons. Second, the larger O_3 concentrations in the northern hemisphere, which are often cited as evidence for stratospheric injection, may in fact be due to photochemical reactions which involve in some part contributions from anthropogenically emitted pollutants. Third, the broad maxima occurring in the summer at many locations, especially those within a few hundred kilometers of urban areas in Europe and the United States, has been suggested to support the production of ozone via photochemical processes involving anthropogenic emissions (Logan, 1985).

As discussed in Section 1-D-1, anthropogenic emissions in the highly industrialized northern hemisphere exceed those in the southern hemisphere, so that higher O_3 concentrations might be expected in the north if there is a contribution from an anthropogenic source. The lack of a diurnal variation in O_3 also does not rule out photochemical processes; Liu et al. (1980) point out that the photochemical lifetime expected for O_3 is sufficiently long (~10 days) that a large diurnal variation in its concentration would not be expected.

Finally, the major loss processes assumed for O_3 in the natural troposphere are loss at the surface (i.e. dry deposition; see Section 11-D-2) and, to a lesser extent, photochemical reactions (1) and (2a):

$$O_3 + h\nu(\lambda \leq 320 \text{ nm}) \rightarrow O(^1D) + O_2 \qquad (1)$$

$$\begin{aligned} O(^1D) + H_2O &\underset{a}{\rightarrow} 2\,OH \\ &\underset{b}{\rightarrow} O(^3P) + H_2O \end{aligned} \qquad (2)$$

As discussed in Section 3-C-2b, at 50% relative humidity in 1 atm of air at 298°K, approximately 10% of the $O(^1D)$ produced in reaction (1) reacts with water vapor; the rest is deactivated to the ground state, $O(^3P)$, by air. Of the 10% which reacts with H_2O, ≥95% proceeds via reaction (2a) to produce 2 OH. The remaining 5%, reaction (2b), does not represent a net loss since the $O(^3P)$ recombines with O_2 to give O_3:

$$O(^3P) + O_2 \overset{M}{\rightarrow} O_3 \qquad (3)$$

The calculated loss of O_3 at the surface and by photolysis is approximately a factor of 4 larger than the estimated stratospheric flux, implying that another major O_3 source must be present to maintain relatively constant O_3 levels in the troposphere (Liu et al., 1980).

It has been shown recently that O_3 and CO concentration profiles as a function of altitude above the earth's surface frequently show either strong positive or negative correlations (Fishman and Seiler, 1983). For example, Fig. 14.2 shows one set of O_3 and CO profiles with altitude over the Pacific Ocean off the California coast; the major fluctuations observed for each pollutant are similar. Figure 14.3, on the other hand, shows a case where the fluctuations of O_3 and CO are in opposite directions.

Fishman, Seiler, and co-workers attribute the positive CO–O_3 correlations to photochemical production of O_3 in the troposphere and the negative correlations to stratospheric injection. Thus the stratosphere has very low CO mixing ratios (40–50 ppb), so that mixing of stratospheric air into the troposphere should be characterized by lower CO and higher O_3 concentrations. On the other hand, since

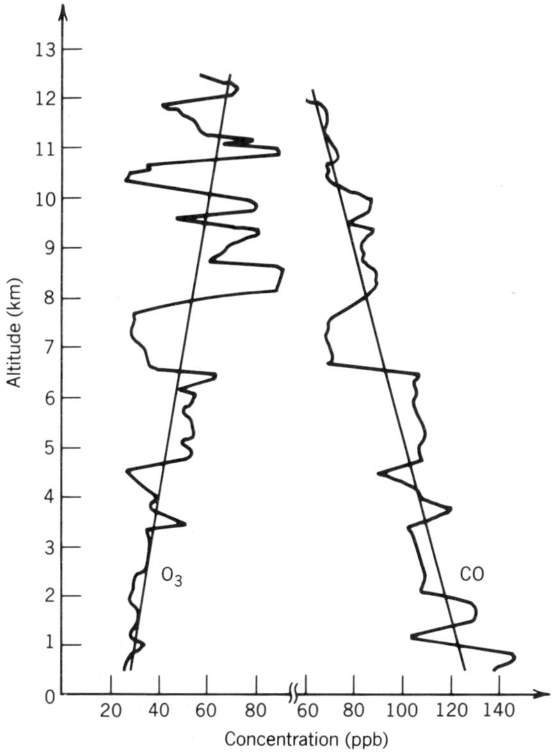

FIGURE 14.2. O_3 and CO profiles measured over the Pacific Ocean (41°N, 126°W) off the California coast on July 27, 1974. The lines through each of the profiles indicate the best fit straight line through each profile (from Fishman and Seiler, 1983).

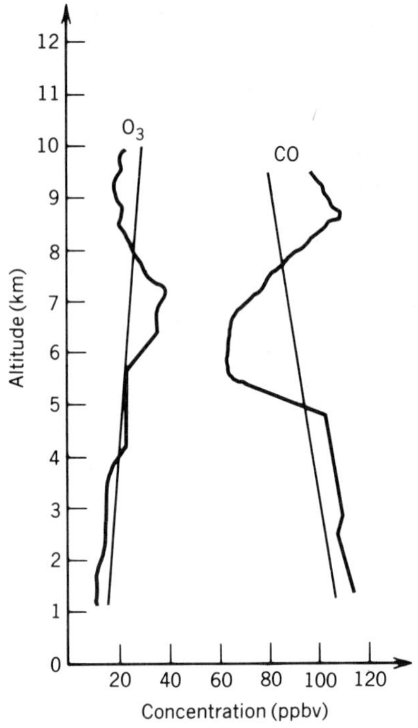

FIGURE 14.3. O_3 and CO profiles measured during descent into Guayaquil, Ecuador, on August 2, 1974 (from Fishman and Seiler, 1983).

the oxidation of CO can lead to the formation of O_3 if sufficient NO is present (see below), photochemical production of O_3 should be associated with higher CO levels. While the correlations between O_3 and CO can thus be suggestive of the source of O_3, Fishman and Seiler (1983) point out that they do not provide conclusive proof.

Figure 14.4 summarizes the correlations found by Fishman and Seiler between CO and O_3 as a function of altitude and latitude. Negative correlations are found for high altitudes at high latitudes in both hemispheres, suggesting that recent stratospheric injection may have occurred. The region of negative correlation close to the earth's surface is attributed to production of CO at the surface where O_3, however, is destroyed.

The strongest positive correlations were found in the middle troposphere between ~20°N and 45°N and are consistent with the larger sources of precursors of O_3, that is, NO_x, NMOC, and CO, in the northern hemisphere compared to the southern. Model calculations by Fishman and Seiler (1983) suggested that ~15–25 ppb O_3 is generated throughout the troposphere north of ~30°N by photochemical processes. This modeling also showed that the increasing O_3 concentrations with altitude were consistent with a photochemical source of O_3 if the surface loss of O_3 was sufficiently large compared to its rate of production.

In summary, it appears likely that O_3 in the natural troposphere has two sources, stratospheric injection and NO_x–NMOC–CO chemistry. The contribution of each

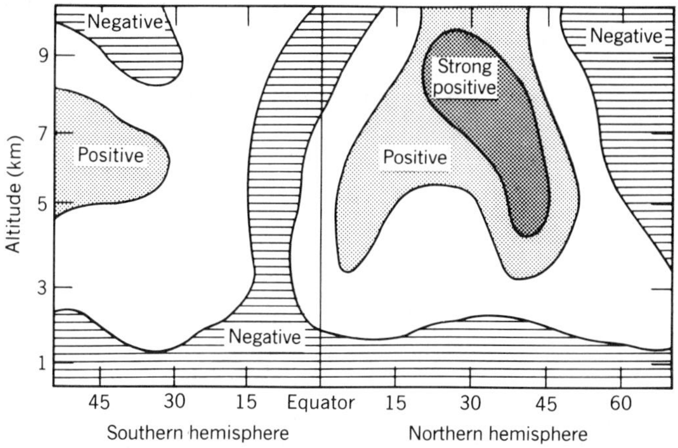

FIGURE 14.4. The latitude–altitude depiction of the regions where O_3 and CO variability was positively or negatively correlated (from Fishman and Seiler, 1983).

varies from location to location, as well as from time to time. However, the larger anthropogenic emissions in the northern hemisphere are expected to lead to a relatively greater importance of the photochemical source in this portion of the globe.

14-A-2. Role of NO in O_3 Production and CO and CH_4 Oxidations

In the natural troposphere, NO_x and CO are produced by natural processes, but their concentrations are much less than those in polluted urban atmospheres. In addition, ~1.4–1.6 ppm CH_4 is present from anaerobic biological processes; smaller amounts of larger hydrocarbons are produced as well (see Section 14-B). The concentration of NO is a critical factor in determining the chemistry of the natural troposphere. In particular, O_3 may be produced *or* destroyed during the oxidation of CO, *or* no change may occur, depending on the NO concentration. Similarly, the oxidation of CH_4 may or may not lead to O_3 production, depending on the NO level. Let us examine the chemistry responsible for this important role of NO.

The sole known oxidant for CO in the atmosphere is the hydroxyl radical. In the natural troposphere a major source of OH is the photolysis of O_3, reaction (1), followed by the reaction (2a) of electronically excited $O(^1D)$ atoms with water. The OH–CO reaction sets off a sequence which produces HO_2:

$$CO + OH \rightarrow H + CO_2 \quad (4)$$

$$H + O_2 \xrightarrow{M} HO_2 \quad (5)$$

[Note that as discussed in Section 4-A-5, the OH–CO reaction proceeds via intermediate complex formation, so that HO_2 likely results from reaction of the complex with O_2, rather than as shown in reactions (4) and (5)]. The fate of the HO_2, however, depends on the ambient concentration of NO. If sufficient NO is present, it converts NO to NO_2, which then photolyzes to produce O_3:

$$HO_2 + NO \rightarrow OH + NO_2 \tag{6}$$

$$NO_2 + h\nu\ (\lambda \leq 430\ \text{nm}) \rightarrow NO + O \tag{7}$$

$$O + O_2 \xrightarrow{M} O_3 \tag{3}$$

The sequence of reactions (4)–(7) and (3) adds up to the following overall stoichiometry:

$$CO + 2\ O_2 \rightarrow CO_2 + O_3$$

That is, one O_3 molecule is produced for each CO oxidized.

On the other hand, if the NO concentration is very low, reaction (6) will not be very fast, and hence either the self-reaction (8) of two radicals or the reaction (9) of HO_2 with O_3 can begin to compete with (6):

$$HO_2 + HO_2 \rightarrow H_2O_2 + O_2 \tag{8}$$

$$HO_2 + O_3 \rightarrow OH + 2\ O_2 \tag{9}$$

Reactions (4), (5), (8), and (10),

$$H_2O_2 + h\nu \rightarrow 2\ OH \tag{10}$$

taken together add up to the following net reaction:

$$2\ CO + O_2 \rightarrow 2\ CO_2$$

Thus O_3 is neither destroyed nor formed in the oxidation of CO.

Finally, if HO_2 reacts with O_3 instead, then the sum of reactions (4), (5), and (9) is

$$CO + O_3 \rightarrow CO_2 + O_2$$

that is, one O_3 molecule is destroyed in the CO oxidation.

The NO concentration at which these various paths become equal can be calculated from a knowledge of the rate constants k_6, k_8, and k_9 for these HO_2 reactions, and typical concentrations of HO_2 and O_3. As discussed in earlier chapters, recommended values of the rate constants at 298°K are as follows:

$$k_6 = 8.3 \times 10^{-12}\ \text{cm}^3\ \text{molecule}^{-1}\ \text{s}^{-1}$$

$$k_8 = 5.6 \times 10^{-12}\ \text{cm}^3\ \text{molecule}^{-1}\ \text{s}^{-1} \quad \text{(at 50\% relative humidity)}$$

$$k_9 = 2.0 \times 10^{-15}\ \text{cm}^3\ \text{molecule}^{-1}\ \text{s}^{-1}$$

In order for the self-reaction (8) of HO_2 to be equal to the reaction (6) of HO_2 with NO,

$$k_6[\mathrm{NO}] = k_8[\mathrm{HO_2}]$$

Taking an HO_2 concentration of 1×10^8 cm^{-3} as being typical of the natural troposphere during the daylight hours (Bottenheim and Strausz, 1980), then

$$[\mathrm{NO}] = \frac{k_8}{k_6}[\mathrm{HO_2}] = \left(\frac{5.6 \times 10^{-12}}{8.3 \times 10^{-12}}\right) \times 1 \times 10^8 \text{ cm}^{-3}$$

$$= 7 \times 10^7 \text{ molecules cm}^{-3} \simeq 3 \text{ ppt}$$

Similarly, for reaction (9) of HO_2 with O_3 to compete with reaction (6) of HO_2 with NO,

$$k_6[\mathrm{NO}] = k_9[\mathrm{O_3}]$$

Taking a typical O_3 concentration of 40 ppb ($\sim 1 \times 10^{12}$ cm^{-3}), then

$$[\mathrm{NO}] = \frac{k_9}{k_6}[\mathrm{O_3}] = \left(\frac{2.0 \times 10^{-15}}{8.3 \times 10^{-12}}\right) \times 1 \times 10^{12} \text{ cm}^{-3}$$

$$= 2.4 \times 10^8 \text{ molecules cm}^{-3} \simeq 10 \text{ ppt}$$

In short, the NO concentration must be $\sim$3–10 ppt or less in order for reactions of HO_2, other than that with NO, to be significant. While such small concentrations of NO have been measured in some locations (Section 5-G), it is not clear in how much of the global troposphere this is the case. Crutzen (1979) suggests that the NO levels in the industrialized northern hemisphere are likely larger than 5 ppt due to anthropogenic emissions, whereas in the southern hemisphere, there may be much lower NO levels; this is consistent with the hypothesis of Fishman and Seiler (1983) that photochemical production of O_3 is more important in the northern than in the southern hemisphere (Fig. 14.4).

Table 14.1 summarizes the possible fates of HO_2 and the resulting impact on O_3 production or destruction during CO oxidation.

Methane oxidation in the natural troposphere occurs by OH attack:

$$CH_4 + OH \rightarrow CH_3 + H_2O \quad (11)$$

$$CH_3 + O_2 \xrightarrow{M} CH_3O_2 \quad (12)$$

Again the fate of the peroxy radical depends on the ambient concentrations of NO. In the presence of sufficient NO, the CH_3O_2 reacts with NO as discussed in earlier chapters:

TABLE 14.1. Possible Fates of HO_2 in the Natural Troposphere and Resulting Impacts on O_3 During CO Oxidation

Fate of HO_2	Impact on O_3 During Oxidation of One of CO Molecule	Approximate Concentration of NO at Which Rate of HO_2 Removal in (6) Equals that in (8) or (9)[a]
$HO_2 + NO \rightarrow OH + NO_2$ (6)	One O_3 formed	—
$HO_2 + HO_2 \rightarrow H_2O_2 + O_2$ (8)	No O_3 formed or destroyed	3ppt
$HO_2 + O_3 \rightarrow OH + 2O_2$ (9)	One O_3 destroyed	10 ppt

[a]Assuming $[HO_2] = 1 \times 10^8$ cm^{-3} and 50% relative humidity, and $[O_3] = 40$ ppb ($\sim 1 \times 10^{12}$ cm^{-3}).

$$CH_3O_2 + NO \rightarrow CH_3O + NO_2 \tag{13}$$

$$CH_3O + O_2 \rightarrow HCHO + HO_2 \tag{14}$$

The HO_2 produced in (14) then oxidizes another NO to NO_2:

$$HO_2 + NO \rightarrow HO + NO_2 \tag{6}$$

Thus, ignoring the further oxidation of HCHO, two molecules of NO have been oxidized to NO_2 and hence two molecules of O_3 can result from photolysis of the NO_2.

However, as in the case of CO oxidation, if the NO level is low, reaction (13) may be sufficiently slow that reaction (15) with HO_2 may compete:

$$CH_3O_2 + HO_2 \rightarrow CH_3OOH + O_2 \tag{15}$$

In this case, O_3 is not formed in the CH_4 oxidation. Taking $k_{13} = 7.6 \times 10^{-12}$ cm^3 molecule^{-1} s^{-1} and $k_{15} = 3 \times 10^{-12}$ cm^3 molecule^{-1} s^{-1} (Atkinson and Lloyd, 1984), removal of CH_3O_2 by reaction (13) is equal to that by (15) at an HO_2 concentration of 1×10^8 cm^{-3} and an NO concentration of ~ 2 ppt.

It is thus expected that hydroperoxides should be formed in the clean troposphere if NO levels are very low. While hydroperoxides have been identified by FTIR in NO_x-free oxidations of hydrocarbons in the laboratory (Hanst and Gay, 1983), they have not been seen to date in the gas phase in the atmosphere. This is likely due to the lack of sufficiently sensitive and specific methods of detection. In addition, they are highly soluble and may be rapidly removed in cloud and rainwater; thus, as discussed in Section 11-A-2a, Kelly et al. (1985) have tentative evidence for the existence of organic hydroperoxides in some cloudwater samples.

In addition to removal by dissolving in atmospheric water droplets, CH_3OOH reacts relatively rapidly with OH (Niki et al., 1983c):

$$OH + CH_3OOH \xrightarrow[a]{} CH_2OOH + H_2O$$

$$\xrightarrow[b]{} CH_3OO + H_2O \qquad (16)$$

$$k_{16a}/k_{16b} = 0.77 \quad \text{(Niki et al., 1983c)}$$

With a daytime OH concentration of $\sim 5 \times 10^5$ cm^{-3} in the natural troposphere and using $k_{16} = 1 \times 10^{-11}$ cm^3 $molecule^{-1}$ s^{-1} (Niki et al., 1983c), the lifetime of CH_3OOH with respect to reaction with OH is ~ 2 days. Thus, depending on the availability of atmospheric water droplets, reaction with OH could contribute to removal of CH_3OOH from the atmosphere.

Figures 14.5 and 14.6 summarize the major chemical reactions interconverting the oxygen, nitrogen, and hydrogen species which play important roles in the chemistry of the natural troposphere (Logan et al., 1981)

14-A-3. NO_x Reactions in Remote Atmospheres

Sources of NO_x in the remote troposphere include lightening, microbial activity in soils, injection from the stratosphere, and the burning of fossil fuels and biomass. As discussed in Section 1-D-1, accurately estimating the contribution from each source is difficult, but it appears that in North America, fossil fuel combustion contributes more than the natural sources by a factor between 3 and 13 (Logan, 1983). Some global source estimates for both natural and anthropogenic emissions are given by Stedman and Shetter (1983), Logan (1983), Crutzen and Gidel (1983), and Liu et al. (1983).

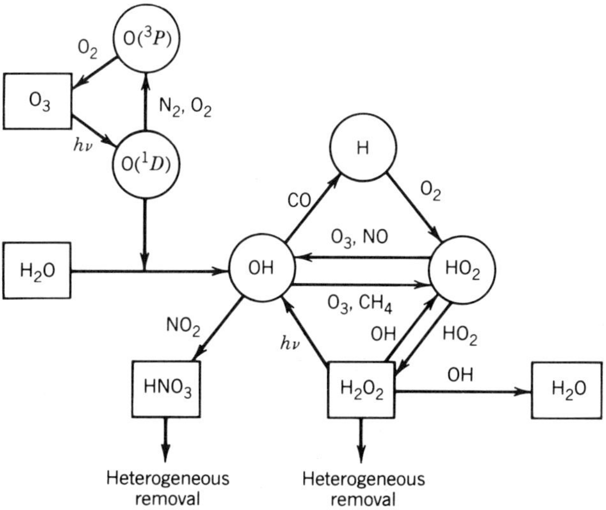

FIGURE 14.5. Major chemical reactions affecting odd hydrogen (OH, H, HO_2, H_2O_2) in the troposphere (from Logan et al., 1981).

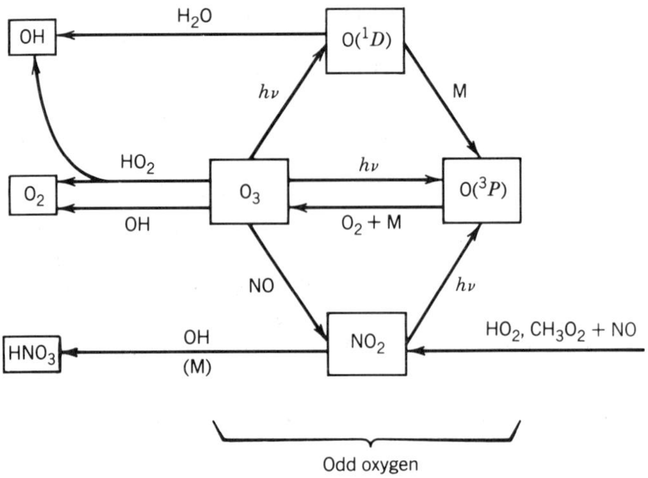

FIGURE 14.6. Major chemical reactions affecting odd oxygen [O_3, $O(^3P)$, $O(^1D)$, NO_2] in the troposphere. Air molecules acting as third bodies are denoted by "M" (N_2, O_2, Ar, H_2O) (from Logan, 1981).

The chemistry of the oxides of nitrogen in the remote troposphere is similar to that in more polluted atmospheres, except that the concentrations are obviously smaller. Thus reactions (17)–(26) and Fig. 14.7 summarize the major chemistry of NO_x in the natural troposphere:

$$NO + O_3 \rightarrow NO_2 + O_2 \tag{17}$$

$$NO_2 + h\nu \rightarrow NO + O(^3P) \tag{18}$$

$$O(^3P) + O_2 \xrightarrow{M} O_3 \tag{19}$$

$$\underset{(HO_2)}{NO + RO_2} \rightarrow \underset{(OH)}{NO_2 + RO} \tag{20}$$

$$OH + NO_2 \xrightarrow{M} HNO_3 \tag{21}$$

$$OH + NO \underset{h\nu}{\overset{M}{\rightleftarrows}} HONO \tag{22}$$

$$NO_2 + O_3 \rightarrow NO_3 + O_2 \tag{23}$$

$$NO_3 + NO_2 \leftrightarrows N_2O_5 \tag{24}$$

$$N_2O_5 + H_2O \rightarrow 2\ HNO_3 \tag{25}$$

$$NO_2 + RO_2 \leftrightarrows RO_2NO_2 \tag{26}$$

The rate constants and role of these reactions in the chemistry of the troposphere are discussed in detail in Chapter 8. Suffice it to say that measurements of trace

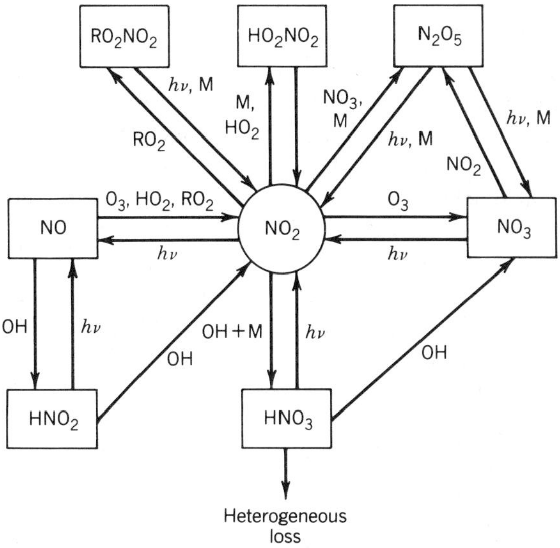

FIGURE 14.7. Summary of the gas phase chemistry of the oxides of nitrogen in the clean troposphere (from Logan, 1983).

species in remote locations support at least qualitatively the importance of reactions (17)–(26). Thus, as discussed in detail above (Section 14-A-1), there is good evidence that at least in some locations at some times a photochemical source [i.e., reactions (18) and (19)] exists for O_3. Similarly, PAN, NO_3, and HNO_3 have been observed in remote locations (see Section 5-G), suggesting the occurrence of reactions (20)–(26). For example, the mean concentration of PAN was observed over the Pacific Ocean to be 32 $\pm$ 24 ppt (Singh and Salas, 1983), while that of HNO_3 was ~50–100 ppt (Huebert and Lazrus, 1980). The nitrate radical concentration has been measured to be 0.3 ppt at 3 km altitude near Mauna Loa, Hawaii (Noxon, 1983) and up to 45 ppt at the earth's surface in Death Valley, California (Platt et al., 1984). Thus, even in relatively remote locations, the oxides of nitrogen expected from reactions (17)–(26) have been observed, consistent with the expectation that the chemistry is similar to that in more polluted environments.

14-B. NATURAL HYDROCARBONS

The characteristic pleasant smell associated with pine forests, flowers, and various shrubs indicates that volatile organic compounds must be emitted naturally by living plants. (Of course, the decay of plants via microbial action and combustion during forest fires also produces organics.) It was suggested by Went in 1960 that the reaction of alkenes emitted by pine trees with background O_3 in the atmosphere was responsible for the production of a blue haze over the Smoky Mountains in Tennessee, from which the mountain range derived its name.

14-B-1. Methane

The single naturally produced organic present in the largest concentration in ambient air throughout the world is methane. CH_4 is produced by anaerobic bacterial fermentation processes in water which contains substantial organic matter, such as swamps, marshes, rice fields, and lakes. In addition, CH_4 is produced by enteric fermentation in mammals as well as by other species. Smaller amounts are emitted into the air via seepage of natural gas from the earth. Forest fires, some of which are due to natural causes, also produce some methane.

Until the late 1970s, it was accepted that the *background* concentration of CH_4 was in the range of 1.4–1.6 ppm in the northern hemisphere and slightly less in the southern hemisphere (Ehhalt and Schmidt, 1978; Heidt and Ehhalt, 1980). However, the concentration appears to have been increasing in recent years at a rate of ~1–2% per year (Graedel and McRae, 1980; Rasmussen and Khalil, 1981, 1983; Blake et al., 1982; Stephens, 1985). Figure 14.8 shows the CH_4 concentrations at locations from the South Pole to the Arctic; all show increasing methane concentrations over this time period. The observation of lower methane concentrations in air bubbles trapped in polar ice cores, where the bubbles are several hundred to several thousand years old (Robbins et al., 1973; Rasmussen et al., 1982a; Craig and Chou, 1982; Rasmussen and Khalil, 1984), suggests that this increase reflects a trend occurring over a long period of time. For example, Rasmussen and Khalil (1984) and Stauffer et al. (1985), based on their ice core data, find that the methane concentration in the atmosphere was relatively constant at ~0.7 ppm until about 150 years ago when it started to rise.

However, direct measurements of atmospheric CH_4 only extend back to 1948. As a result, there is some uncertainty in the trend in concentration before that time. Thus Ehhalt and co-workers (1983) suggest that infrared measurements of the atmospheric CH_4 column density show little or no increase between 1948 and 1965, and that between 1965 and 1975 the increase was only ~0.5% yr^{-1}, much less than the current rate of increase. While the increase observed over the past decade thus seems to be real, there may be some uncertainty in the historical trend before that.

Based on measurements of the ^{14}C content of atmospheric CH_4, the main sources of CH_4 in the atmosphere appear to be aerobic fermentation processes, rather than fossil fuel sources. Thus the ^{14}C content of CH_4 from fermentation is expected to match that of recent wood, whereas ^{14}C in fossil fuels is very small due to the long decay time available as the fossil fuels form. The ^{14}C content of CH_4 samples collected before contamination of the atmosphere with ^{14}C from nuclear explosions suggests that 80% or more of the CH_4 is from the decay of recent organic matter (Ehhalt and Schmidt, 1978).

Table 14.2 gives one estimate of global CH_4 production. The major sources appear to be enteric fermentation in cattle and an aerobic process in paddy fields and wetlands. The "other" category reflects some of the uncertainties in the estimates and includes such newly suggested sources as termites (Zimmerman et al., 1982; Rasmussen and Khalil, 1983). (The strength of the latter source is, however, controversial).

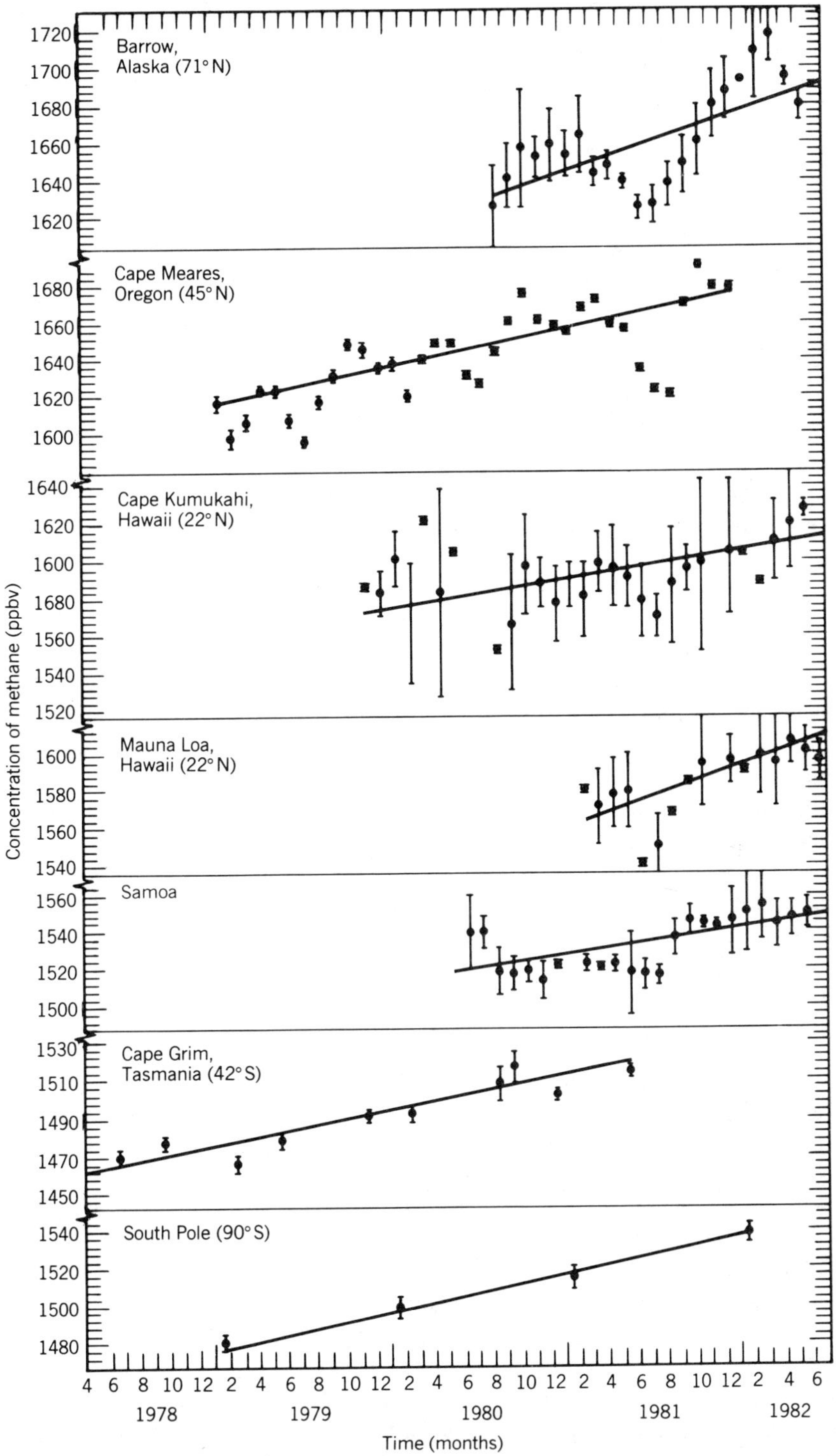

FIGURE 14.8. Monthly average concentrations of atmospheric methane (CH_4) at seven locations ranging in latitude from the Arctic to the South Pole. The vertical bars represent 90% confidence limits of the mean. The solid line through the data points is obtained from a linear least squares technique. The increase in the concentration of CH_4 seems apparent at each site and may be statistically significant (from Khalil and Rasmussen, 1983).

TABLE 14.2. Estimated Global Sources of Methane in 1978

Source[a]	Emissions ($Tg\ yr^{-1}$)
Oceans	13
Wetlands	150
Freshwater lakes	10
Tundra	12
Paddy fields	95
Cattle	120
Biomass burn	25
Direct anthropogenic	40
Other	88
TOTAL ($Tg\ yr^{-1}$)	553

Source: Khalil and Rasmussen, 1983.
[a]The sources are estimated for 1978 and are estimated to increase about 10 Tg every year.

The reason for the recent increase in atmospheric CH_4 concentrations is not clear. One obvious potential explanation is an increase in emissions from one or more of the sources. It has been hypothesized that as the human population increases, the number of cattle and the amount of land devoted to rice fields are also increasing, and hence more CH_4 is produced from these sources. Thus the rise would be caused indirectly by human activities.

Alternatively, an increase in atmospheric CH_4 could be due to a decrease in its *sinks*, rather than an increase in its sources. The major atmospheric sink for CH_4, as discussed above, is reaction with OH:

$$CH_4 + OH \rightarrow CH_3 + H_2O \tag{11}$$

Figure 14.9 summarizes the atmospheric reactions following reaction (11). A decrease in the rate of removal of CH_4 by reaction (11), and hence an increase in the atmospheric CH_4 concentrations, would result if the OH concentration were decreasing. One of the major species controlling the OH concentration in remote areas is CO, since it reacts with OH as well (see above). Chameides et al. (1977) estimate that an annual increase in CO of $\sim 4.5\%$ would decrease the OH concentration to the extent that a CH_4 increase of $\sim 1.4\%\ yr^{-1}$ would result. One recent report that CO is indeed increasing awaits further confirmation and data gathering in locations around the globe (Khalil and Rasmussen, 1984).

In conclusion, Khalil and Rasmussen (1985) estimate that approximately 70% of the increase in ambient methane levels over the last 200 years is probably due to an increase in primary emissions, with the remaining 30% due to a decrease in global OH concentrations.

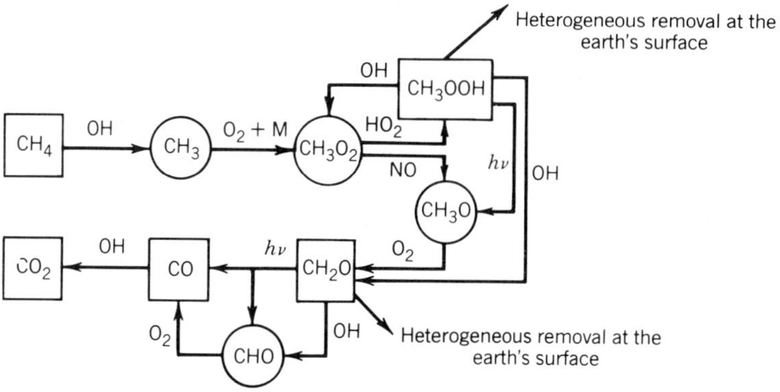

FIGURE 14.9. The atmospheric oxidation of methane (from Logan et al., 1981).

Increased CH_4 could have effects both on atmospheric chemistry and on the earth's radiation balance. Clearly CH_4 plays a major role in global tropospheric chemistry, as discussed in Section 14-A-2; additionally it affects the radiation balance of the earth. Thus it absorbs infrared radiation at 7.7 μm and converts the radiation to heat. Increased CH_4 should thus lead to increased surface temperatures (Wang et al., 1976; Marland and Rotty, 1985; Ramanathan et al., 1985).

A small amount ($\lesssim$ 15%) of the CH_4 diffuses upward to the stratosphere where it can react with OH (Ehhalt and Schmidt, 1978; Crutzen and Schmailzl, 1983). As discussed in Chapter 15, an additional role of CH_4 in stratospheric chemistry is as a terminator for chlorine atom reactions via reaction (27):

$$CH_4 + Cl \rightarrow CH_3 + HCl \tag{27}$$

While HCl remaining in the stratosphere eventually reforms Cl, some of the HCl reaches the troposphere and is rained out, thus permanently removing Cl from the stratosphere.

14-B-2. Sources and Emissions of Larger Hydrocarbons and Other Organics ($\geq C_2$)

Natural sources are responsible for the emissions of larger organics in addition to methane. One review cites 367 different compounds known to be emitted by vegetation (Graedel, 1979a)! These include simple alkanes (ethane, propane, butanes, pentanes, etc.) from natural gas seepage and bacterial fermentation, and from plants such as trees, crops, and grasses. Some of these may play significant but as yet incompletely understood roles in tropospheric chemistry; for example, Singh and Hanst (1981) suggest that ethane and propane can act as major reservoirs for NO_x in the unpolluted troposphere via PAN formation.

Alkenes such as isoprene and the monoterpenes are also emitted by various types of plants and trees. Small quantities of ethene are produced by fruits, flowers, leaves, roots, and tubers, and an aromatic, *p*-cymene, is also produced by some sage and eucalyptus.

A comprehensive review of organics observed in the troposphere, including both gas phase and particulate materials, is found in the article by Duce et al. (1983); this includes the chemical structures, atmospheric concentrations, sources, sinks, and atmospheric transformations of non-methane hydrocarbons observed in both continental and marine areas. The number and chemical structures of organics observed even in remote regions are surprisingly large, including oxygen-containing species such as alcohols, acids, acid salts, and esters.

However, the greatest interest has been in the naturally emitted isoprene and the monoterpenes because of their potential role in oxidant and aerosol formation in both urban and remote regions. As a result, our knowledge of the sources, sinks, and atmospheric reactions of these compounds has increased substantially, whereas relatively little is known about most of the other large organics. We thus concentrate in this section on isoprene and the monoterpenes.

The sources, emission rates, atmospheric reactions, and contribution to O_3 formation of these natural, or *biogenic*, hydrocarbons are reviewed in detail by Dimitriades (1981), Altshuller (1983), and Duce et al. (1983), and in the series edited by Bufalini and Arnts (1981); the reader is urged to consult these articles for the original references and further details.

Table 14.3 lists some of the common natural organics, including isoprene and monoterpenes, found in the troposphere and some of their known sources. Terpenes are a class of organic compounds whose structures are based on isoprene:

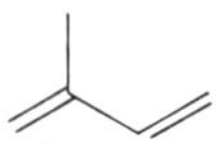

isoprene

Monoterpenes such as those in Table 14.3 have the chemical formula $C_{10}H_{10}$ and structures based on two isoprene units; for example, *d*-limonene can be thought of as consisting of two isoprene molecules:

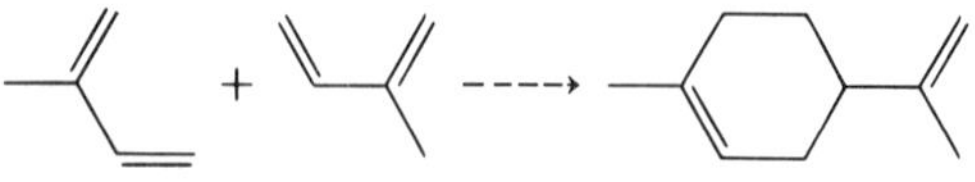

d-limonene

Monoterpenes may be acyclic (e.g., myrcene) or have one (*d*-limonene) or two ring structures (α- and β-pinene, Δ^3-carene).

The emissions from deciduous trees (hardwoods), such as oaks, aspens, willows, and poplars, are comprised mainly of isoprene, whereas coniferous forests (i.e., softwoods), such as pine trees, firs, juniper, spruce, cedars, and redwoods, are mainly monoterpene emitters. Some trees, for example, spruce, sweet gum, and eucalyptus, produce both isoprene and the terpenes (Rasmussen, 1981). A

TABLE 14.3 Some Natural Organics Observed in the Biosphere

Compound	Structure	Occurrence
Isoprene		Measured in ambient air in rural areas; emitted by deciduous trees and shrubs
α-Pinene	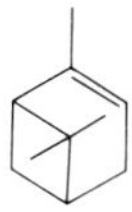	Measured in ambient air in rural areas; emitted by numerous conifers
β-Pinene	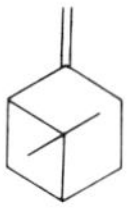	Measured in ambient air in rural areas; emitted by California black sage and general conifers
d-Limonene	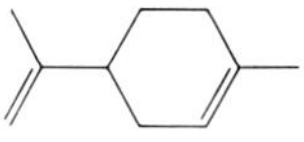	Measured in ambient rural air; emitted by loblolly pine, California black sage, and "disturbed" eucalyptus
p-Cymene	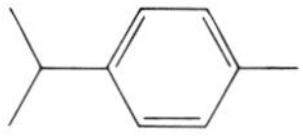	Emitted by California black sage and from "disturbed" eucalyptus foliage
Myrcene	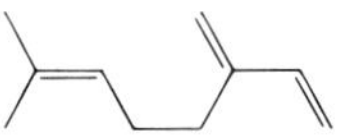	Measured in air above pine needle litter; emitted by loblolly pine, California black sage, and redwood
Δ^3-Carene	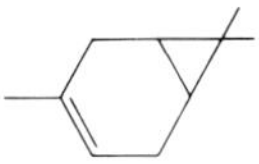	Measured in ambient air; found in the gum turpentines of some pines

Source: Adapted from Arnts and Gay, 1979.

number of organics emitted by plants have not been identified, but their presence has been recognized by the appearance of unknown peaks on gas chromatograms used to measure isoprene and the monoterpenes (Altshuller, 1983). Some of these may contain oxygen; for example, the ether methyl chavicol was identified in the emissions from ponderosa pine (Westberg, 1981) and a variety of oxygen-containing species have been identified in the atmosphere of Russian forests (Isidorov et al., 1985). Figures 14.10, 14.11, and 14.12 show the structures of some monoterpenes and their derivatives which could be formed in their atmospheric oxidations (Graedel, 1979a); clearly a wide variety of compounds, some multi-functional, may be present in the atmosphere due to natural processes (Zimmerman, 1979).

Classification of the compounds in Table 14.3 on the basis of their reactivity with OH (see Section 10-C) puts them in the moderate to high reactivity classes, suggesting that significant emissions of these compounds in the presence of NO_x could lead to a *natural smog*.

These observations have led to increased attention to the role of natural organic emissions in the formation of atmospheric O_3 (and, of course, other manifestations of photochemical air pollution). Specifically, the following issues are of interest:

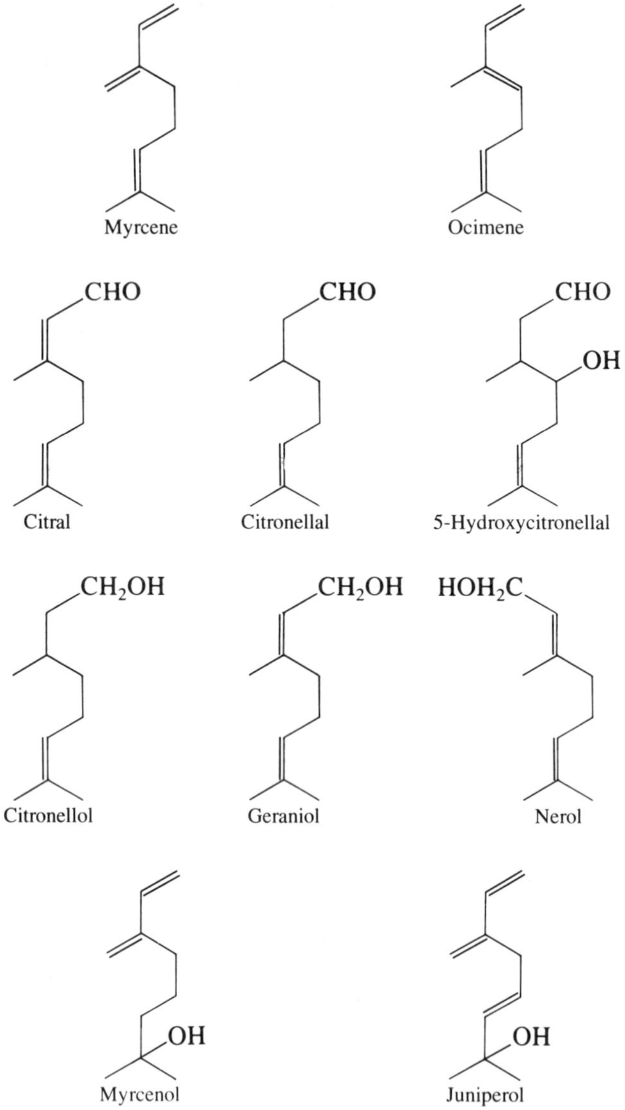

FIGURE 14.10. Structures of some acyclic terpenoids thought to be involved in natural atmospheric processes (from Graedel, 1979a).

- What is the chemical composition of NMHC emitted by various trees, shrubs, grass, and so on, and how much is emitted?
- Do these organics lead to the formation of O_3 and other secondary pollutants in ambient air, and if so, to what extent?

14-B-2a. Emissions of Natural Hydrocarbons

The emission rates of natural hydrocarbons and their chemical composition can be measured by several methods: (1) putting a Teflon bag over the plant and measur-

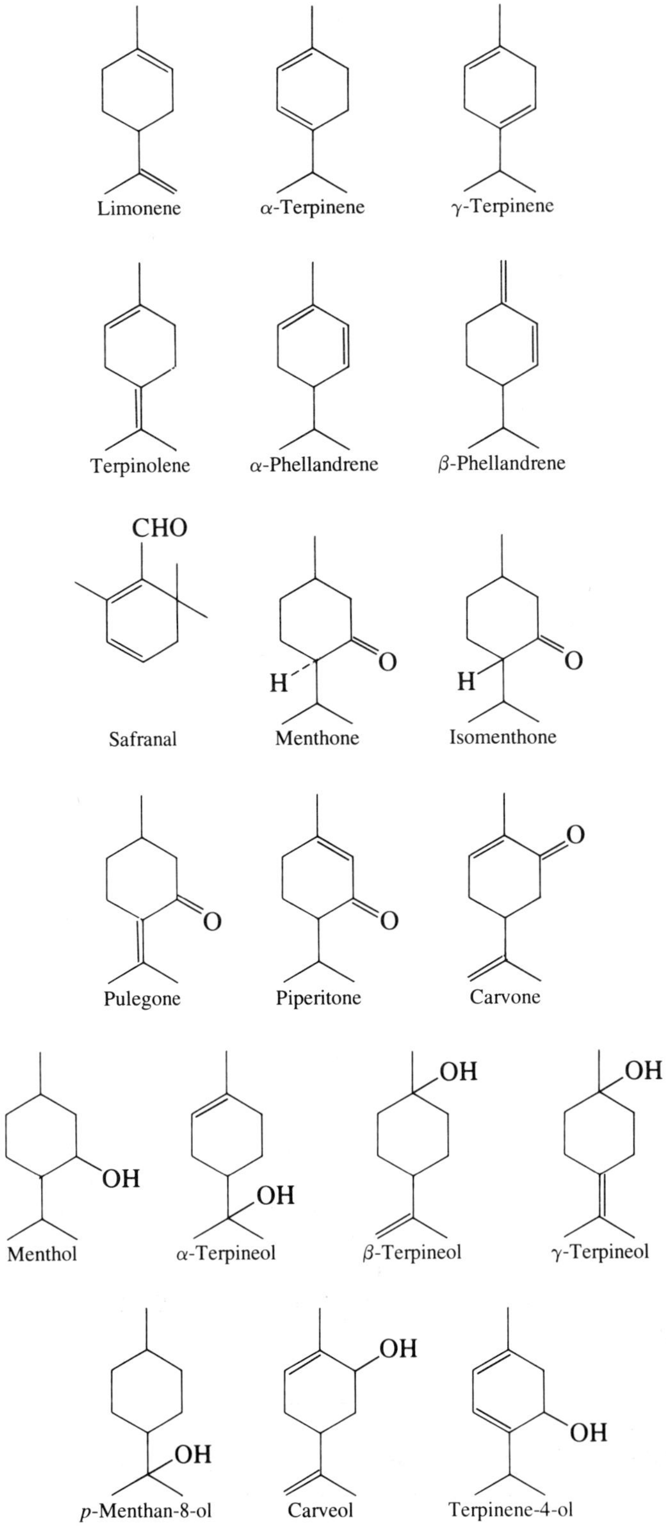

FIGURE 14.11. Structures of some monocyclic hydrocarbons and their derivatives thought to be involved in natural atmospheric processes (from Graedel, 1979a).

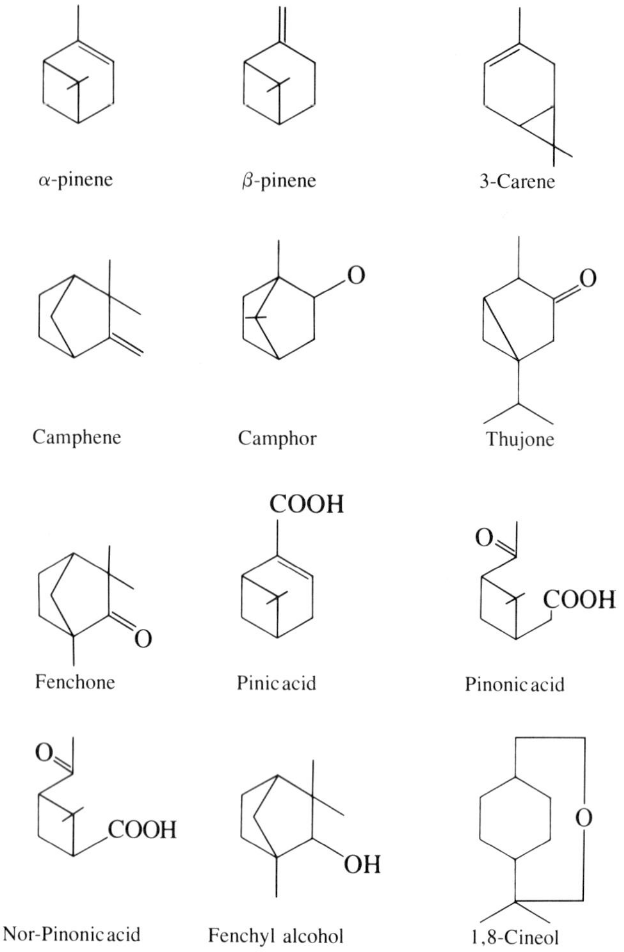

FIGURE 14.12 Structures of some bicyclic hydrocarbons and their derivatives thought to be involved in natural atmospheric processes (from Graedel, 1979a).

ing the hydrocarbons in the bag as a function of time, or (2) measuring the hydrocarbons, as well as meteorological variables, as a function of the vertical distance above a vegetation canopy and relating these to a flux from the canopy.

Both methods intrinsically have substantial uncertainties. For example, enclosing a plant in a bag may perturb its emissions through changes in such parameters as temperature, relative humidity, and light intensity. On the other hand, accurately relating measured hydrocarbon levels to fluxes using meteorological variables is very difficult. However, reasonable agreement on the hydrocarbon fluxes has been obtained when results from the bag technique and the flux method have been compared (Lamb et al., 1985).

In addition to these specific measurement problems, the emission rates may vary with temperature, radiation intensity, relative humidity, plant type, and so on.

For application to air basins, the results from measurements of emission rates from specific plants under a given set of conditions must be extrapolated to the field situation where a variety of vegetation is found. Techniques for doing this are described by Miller and Winer (1984) and Brown and Winer (1985).

Some types of hydrocarbon emissions from vegetation have been shown to depend on both light and temperature (Sanadze and Kalandadze, 1966a,b; Tingey et al., 1979, 1980; Lamb et al., 1985). For example, emissions of isoprene increase with temperature and also increase with light intensity (Tingey et al., 1979). On the other hand, while the emission rates for the monoterpenes—α-pinene, β-pinene, myrcene, *d*-limonene, and β-phellandrene—increase with temperature, they are not altered significantly by variations in light intensity (Tingey et al., 1980; Isidorov et al., 1985). Relative humidity (RH) has also been shown to affect rates of α-pinene emission, with the rate increasing as the RH increases (Lamb et al., 1985).

As expected from the effects of temperature and light intensity, emission rates show seasonal variations as well as latitudinal differences. For example, Altshuller estimated emission rates for isoprene alone and for isoprene plus the monoterpenes as a function of month in the geographic zones in the contiguous United States shown in Fig. 14.13. These estimated emission rates are shown in Fig. 14.14. The largest emission rates are in the south in mid-summer.

In urban areas, the estimated emissions of natural organics are thought to form a relatively small fraction of the total organic emissions. These emission data are supported by measurements of the organics in ambient air in urban areas where those of anthropogenic origin greatly exceed those attributable to natural sources.

For rural areas, however, there is a considerable controversy concerning the relative contributions of natural and anthropogenic emissions. This arises from a fundamental inconsistency between conclusions drawn using estimated emission rates and those based on ambient air data. Application of current emissions data over the continental United States indicates that natural organic emissions are in the range of two to ten times those from anthropogenic sources. However, sampling and analysis of the concentrations of volatile organics in ambient air in rural areas is not consistent with the emissions estimates. Concentrations of isoprene and the monoterpenes in ambient air are typically low, $\sim$1–10 ppbC outside forest canopies and $\sim$10–100 ppbC within them. They usually comprise $\leq 10\%$ of the total hydrocarbons at rural sites.

Altshuller (1983) suggests that a comparison can be made between the ratio of biogenic and anthropogenic emissions and the ratio of the ambient air concentrations of compounds from each source, if a compound or group of compounds characteristic of each type of source can be identified. Clearly, isoprene and the monoterpenes are characteristic of biogenic emissions, and Altshuller suggests using propene as a marker for anthropogenic sources since it reacts with a rate constant within the same order of magnitude as isoprene in the atmosphere, and measurements of its concentrations at rural sites have been carried out. (Choosing an indicator of anthropogenic activity which has about the same reactivity as the biogenics is important; if one reacts significantly faster than the other, their ratio in ambient air will change with reaction time.)

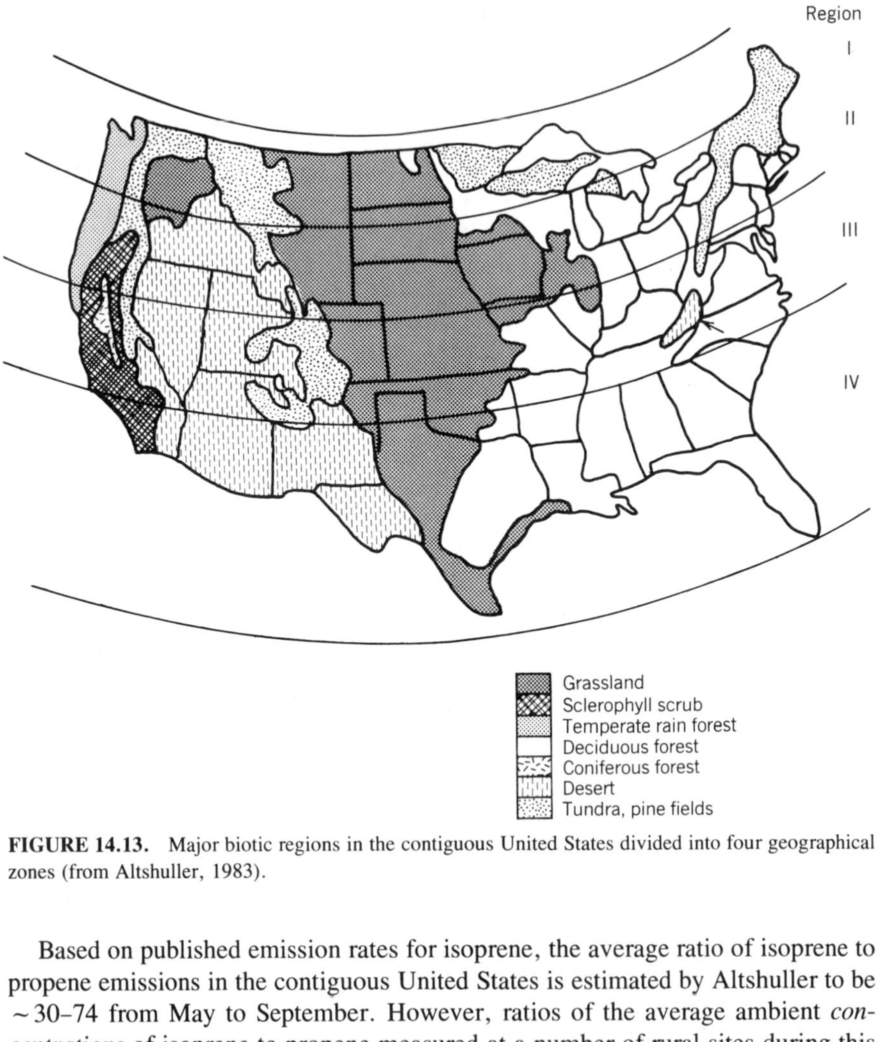

FIGURE 14.13. Major biotic regions in the contiguous United States divided into four geographical zones (from Altshuller, 1983).

Based on published emission rates for isoprene, the average ratio of isoprene to propene emissions in the contiguous United States is estimated by Altshuller to be ~30–74 from May to September. However, ratios of the average ambient *concentrations* of isoprene to propene measured at a number of rural sites during this period of the year vary from ~2 to 9. Similarly, the ratio of emissions of α-pinene to propene was estimated to be in the range of ~3–92 from January to September, whereas the ratio of the average concentrations of these compounds at rural sites was observed to vary from 0.6 to 3.

In summary, the *calculated* ratios of biogenic emissions to anthropogenic emissions are much higher than their *measured* concentration ratios in ambient air. Possible reasons for the large discrepancies are discussed in detail by Dimitriades (1981) and Altshuller (1983). They include errors in the emission inventory estimates, unrecognized rapid atmospheric reactions of the biogenics, and, finally, reaction in and adsorption on the sampling container prior to analysis. Of these, the most likely chemical explanation of the very large discrepancies is the possibility of unrecognized rapid reactions of the biogenics.

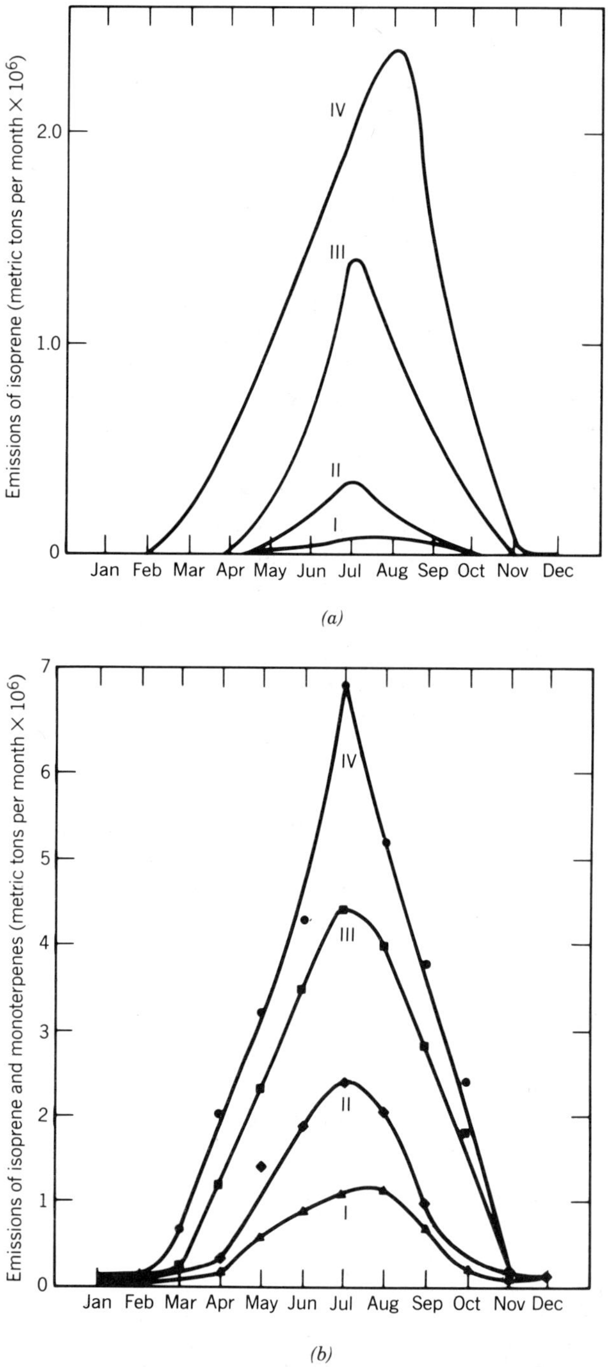

FIGURE 14.14. Emission rates of (*a*) isoprene and (*b*) isoprene plus monoterpenes as a function of month in the four geographic zones in the contiguous United States shown in Fig. 14.13 (from Altshuller, 1983).

Thus it has been shown recently that the nitrate radical, NO_3, formed at night from the $NO_2 + O_3$ reaction, reacts rapidly with isoprene and the monoterpenes. Table 14.4 shows these rate constants for the reactions with NO_3 at $295 \pm 1°K$, as well as the predicted lifetimes of the organics for an NO_3 concentration of 10 ppt, characteristic of a clean atmosphere. Particularly for the monoterpenes, short lifetimes are expected due to reaction with NO_3, and Winer et al. (1984) suggest that this may be at least partially responsible for the observed relatively low ambient concentrations of the biogenics. This explanation seems reasonable in light of the fact that NO_3 reacts with isoprene and α-pinene approximately two and three orders of magnitude faster, respectively, than with propene.

For example, Fig. 14.15 shows the calculated concentration–time profiles for NO_3 and monoterpenes under conditions typical of Death Valley, California, where NO_3 radical concentrations have been measured (see Section 5-D-2b). Inclusion of the monoterpene–NO_3 reaction leads to a predicted rapid decrease in the monoterpene concentration, by ~1–2 orders of magnitude, while the NO_3 concentration profile is not altered to a large extent. The latter is due to the fact that, under these conditions, the formation rate for NO_3 exceeds the monoterpene emission rate. The model predictions for the opposite case, that is, monoterpene emission rates exceeding the NO_3 formation rate, are discussed by Winer et al. (1984).

While we have concentrated here on the chemistry involved in determining the monoterpene concentrations in ambient air, meteorological factors will, of course, also play a major, and often predominant role (e.g. see Roberts et al., 1985).

14-B-2b. Contribution to O_3 Formation

Determining whether natural organic emissions contribute significantly to the formation of secondary pollutants, especially O_3, requires an understanding of their chemistry in irradiated hydrocarbon–NO_x mixtures. Using the reactivity classification for organics based on the rate constants for their reaction with OH (see Section 10-C), isoprene is in the most reactive class and α-pinene falls in the next

TABLE 14.4. Rate Constants for the Reaction of NO_3 with Isoprene and Some Monoterpenes at $295 \pm 1°K$[a]

Compound	10^{12} k (cm^3 $molecule^{-1}$ s^{-1})	τ[b]
Isoprene	0.58	1.9 hr
α-Pinene	6.1	11 min
β-Pinene	2.5	27 min
Δ^3-Carene	10.6	6 min
d-Limonene	13.9	5 min

Source: Atkinson et al., 1984a, 1985.

[a]Adjusted to a value of $K_{eq} = 3.35 \times 10^{-11}$ cm^3 $molecule^{-1}$ for the equilibrium $NO_2 + NO_3 \rightleftarrows N_2O_5$ (see Section 7-C-2).

[b]$\tau = 1/k[NO_3]$, where $NO_3 = 2.5 \times 10^8$ cm^{-3} (10 ppt).

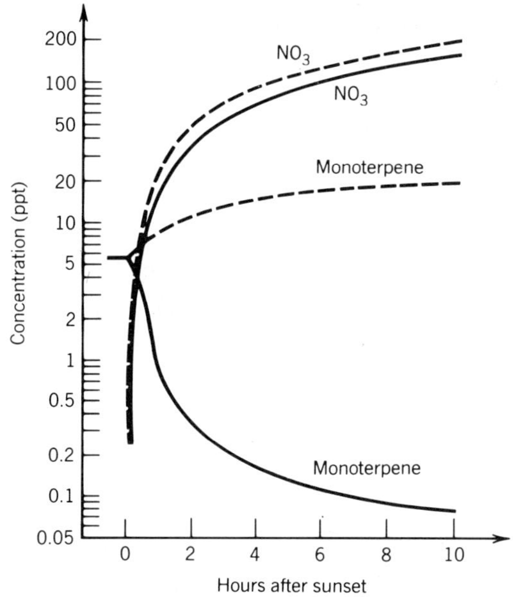

FIGURE 14.15. Calculated concentration–time profiles for NO_3 radicals and monoterpenes for conditions typical of Death Valley. Solid lines are predictions for reaction between NO_3 radicals and monoterpenes; dashed lines are predictions for the assumption that there is no reaction between these species (from Winer et al., 1984).

most reactive class. Both types of compounds are thus expected to react rapidly in ambient air, which is consistent with atmospheric observations. For example, based on measurements of ambient terpene concentrations in a Norwegian forest, Hov et al. (1983) estimated that the lifetime of the terpenes was ~4–8 h.

Examination of chemical structures of some of the typical natural organics suggests that some might react relatively rapidly with O_3 and NO_3, as well as with OH. Table 14.5 shows the rate constants for reaction of isoprene (emitted by deciduous trees) and some monoterpenes (emitted by pine trees) with OH, O_3, and NO_3, as well as their atmospheric lifetimes with respect to typical tropospheric concentrations of these oxidants over the continents under clean air conditions. NO_3 at a concentration of 10 ppt would provide by far the major reaction path compared to O_3 at 0.03 ppm and OH at 5×10^5 radicals cm^{-3} for α-pinene, β-pinene, Δ^3-carene, and *d*-limonene. For isoprene, removal by NO_3 is slightly faster than that by OH. For *d*-limonene, reaction with O_3 also contributes, although significantly less than NO_3. Of course, since NO_3 photolyzes rapidly in actinic radiation (see Sections 3-C-7b and 8-A-2a), significant concentrations of NO_3 occur only at night and hence removal of hydrocarbons by this reaction is a nighttime phenomenon. On the other hand, since significant sources of OH exist only during the daylight hours when photolysis of its precursors occurs, OH reactions are only important in the daytime. O_3, however, can be present both during the day and at night.

TABLE 14.5. Typical Estimated Lifetimes for Some Natural Hydrocarbons in Clean Continental Air with Respect to Oxidation by OH, O_3, and NO_3 at Room Temperature

Molecule	τ_{OH}[a] (k_{OH})[b] Day	τ_{O_3}[a] (k_{O_3})[c] Day + Night	τ_{NO_3}[a] (k_{NO_3})[d] Night
Isoprene	6 h (1.0×10^{-10})	26 h (1.4×10^{-17})	1.9 h (5.8×10^{-13})
α-Pinene	10 h (5.3×10^{-11})	4 h (8.4×10^{-17})	11 min (5.3×10^{-12})
β-Pinene	7 h (7.8×10^{-11})	18 h (2.1×10^{-17})	27 min (2.5×10^{-12})
Δ^3-Carene	6 h ($\sim 9 \times 10^{-11}$)	3 h (1.2×10^{-16})	6 min (1.06×10^{-11})
d-Limonene	4 h (1.5×10^{-10})	35 min (6.4×10^{-16})	5 min (1.39×10^{-11})

[a]For comparison, $\tau_{OH} = 21$ h, $\tau_{O_3} = 33$ h, and $\tau_{NO_3} = 6$ days for propene.
[b]From Atkinson, 1985. The room temperature rate constant is given in parentheses underneath the lifetimes; τ is defined by $\tau = 1/k[X]$, where X = OH, O_3, or NO_3, respectively. The concentration of OH is taken as $\sim 5 \times 10^5$ radicals cm^{-3} for the lifetime calculation.
[c]Rate constants from Atkinson and Carter, 1984; lifetime assumes $[O_3] = 30$ ppb $= 7.5 \times 10^{11}$ cm^{-3}.
[d]Rate constants from Atkinson et al., 1984a; 1985; lifetime assumes $[NO_3] = 10$ ppt $= 2.5 \times 10^8$ cm^{-3}.

Another factor which must be taken into account in comparing various paths for removal of the organic is the time of day during which the organic is emitted. For example, as discussed above, there is some evidence that isoprene is emitted during the day so that OH removal may predominate more than the calculated lifetimes in Table 14.5 would suggest.

Whereas only OH is expected to participate significantly in the daytime oxidation of isoprene, for α-pinene both OH and O_3 are expected to contribute significantly. Thus some naturally emitted organics, at least, may not only act as a source of O_3 via hydrocarbon–NO_x chemistry, but also as a sink via their direct reaction with O_3.

In urban atmospheres, natural emissions of organics are sufficiently low that they are expected to contribute to only a minor extent to the formation of O_3. Furthermore, significant transport of natural hydrocarbons from rural areas into urban centers is unlikely because of their relatively short lifetime with respect to reaction with background OH, O_3, and NO_3 (Table 14.5).

In rural areas, however, this may not be the case, especially given the controversy concerning the relative importance of emissions from natural and anthropogenic sources. In a comprehensive smog chamber study, Arnts and Gay (1979) studied the O_3 forming potential of some natural hydrocarbons. Figure 14.16 shows

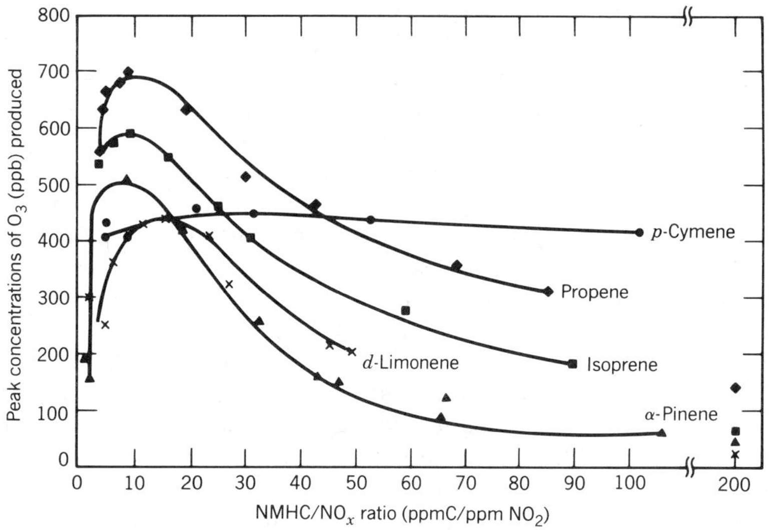

FIGURE 14.16. Effect of hydrocarbon/NO_x ratio on maximum ozone concentration in irradiations of propene and natural organics with NO_x in air at 25°C (from Arnts and Gay, 1979).

the peak O_3 concentration formed as a function of the NMHC/NO_x ratio for four natural organics and propene. At NMHC/NO_x $\lesssim$ 15, which is characteristic of most urban areas, all these compounds result in substantial O_3 formation. At higher ratios of NMHC/NO_x, the peak O_3 concentrations fall because the organic itself reacts with the O_3 formed, and the formation of O_3 becomes NO_x limited.

Most rural areas have NMHC/NO_x ratios greater than 15 : 1; in remote areas, for example, the ratios typically vary from ~400 to 6500. This suggests that even if there are significant natural sources of organics in these areas, they may act as sinks for O_3, rather than as sources.

A note of caution is in order in extrapolating these smog chamber studies to ambient air. They were carried out at concentrations much larger (by a factor of 10–1000) than are observed in rural areas (i.e., ppmC levels, rather than ppbC). The concentration regime chosen for study may have a marked effect on the results and, in particular, on the O_3 levels formed. As stressed by Arnts and Gay, these studies should be repeated under conditions more representative of true atmospheric conditions before firm conclusions are drawn. However, the results of these chamber studies are consistent with modeling studies of the impacts of biogenic emissions. Using a chemical mechanism developed for isoprene and α-pinene by Lloyd et al. (1983) and validated against smog chamber data (see Chapter 9), Lurmann et al. (1983) investigated the impact of natural hydrocarbon emissions on ambient O_3 concentrations using a photochemical box model. They predict that at high NMHC/NO_x ratios typical of rural atmospheres, biogenic hydrocarbons are inefficient producers of O_3 due to the scarcity of NO_x. They also predict, in agreement with Fig. 14.16, that there is an optimum ratio of NMHC/NO_x for O_3 production for various natural hydrocarbons; at lower NMHC/NO_x ratios, O_3 production is limited by NO_x scavenging of O_3, and at higher values, insufficient NO_x is available for O_3 formation.

Although the emphasis in these studies has been on continental areas, modeling studies of the marine atmosphere also suggest that natural emissions do not result in significant O_3 production over the oceans (Graedel, 1979b).

In addition to the O_3-forming potential of natural hydrocarbons, there is a second major area of interest, their aerosol-forming potential. For example, in the Arnts and Gay studies, only 4–44% of the total carbon reacted could be accounted for in the gas phase products, suggesting that 56–96% may have been in the particulate phase or deposited on the surface of the chamber. This is not unexpected since oxygenated monoterpene derivatives typically have vapor pressures about one order of magnitude less than the precursor hydrocarbon (Graedel, 1979a). Such substantial aerosol formation would support the suggestion of Went (1960) that natural organics are responsible for such phenomena as the haze over the Smoky Mountains. However, in other studies, relatively little of the carbon was found in aerosols. For example, Kaiserman and Corse (1981) found that only ~2–3% of the carbon initially present in α- and β-pinene–NO_x mixtures was converted to aerosols; caution is again needed in extrapolating to the atmosphere since high (ppm) concentrations were used in these studies and loss to the walls could not be quantified. However, based on these and other studies, Altshuller (1983) suggests that the contribution of monoterpenes to the fine particle concentration in both rural and urban areas is small.

14-B-2c. Mechanisms of Atmospheric Oxidations

As discussed in Section 7-E-4, the mechanisms of reaction of NO_3 with alkenes have not been fully elucidated, although addition to the double bond seems likely as a first step. Virtually nothing is known about the mechanisms and products of NO_3 reactions with natural hydrocarbons such as isoprene and α-pinene.

The mechanisms of reaction of OH with isoprene and of OH and O_3 with monoterpenes also have not been fully elucidated as yet. For isoprene, the paths for initial addition of OH can be postulated on the basis of the mechanisms of reaction of OH with other alkenes:

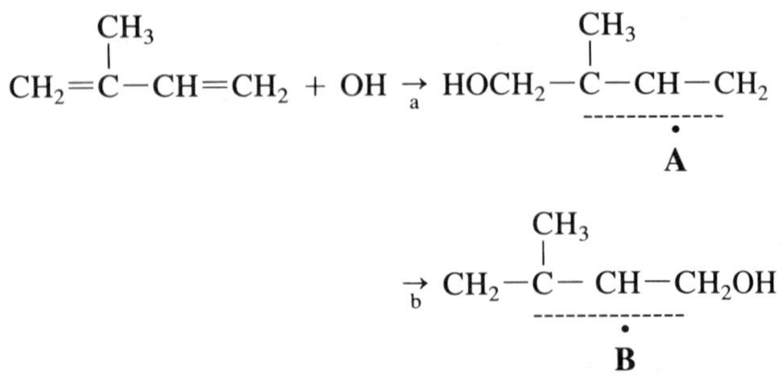

(28)

The free electron is expected to be delocalized as shown over two bonds, because of the presence of the second double bond in the reactant molecule.

Addition of O_2 followed by reaction with NO is expected under atmospheric conditions. For example, in the case of **A**, methyl vinyl ketone production would result from addition of O_2 to the central carbon atom:

$$\underset{\mathbf{A}}{HOCH_2-\underset{\underset{\cdot}{CH_3}}{\overset{CH_3}{C}}-CH-CH_2} + O_2 \rightarrow HOCH_2-\underset{\underset{\underset{\cdot}{O}}{O}}{\overset{CH_3}{C}}-CH=CH_2$$

$$NO_2 \leftarrow | \leftarrow NO \quad (29)$$

$$\underset{\underset{\downarrow}{O_2}}{HOCH_2} + \underset{\text{methyl vinyl ketone}}{CH_3\overset{O}{\overset{\|}{C}}CH=CH_2} \leftarrow HOCH_2-\underset{\underset{\cdot}{O}}{\overset{CH_3}{C}}-CH=CH_2$$

$$HO_2 + HCHO \qquad (90\%)$$
$$+$$
$$HOCH_2-\underset{ONO_2}{\overset{CH_3}{C}}-CH=CH_2$$
$$(10\%)$$

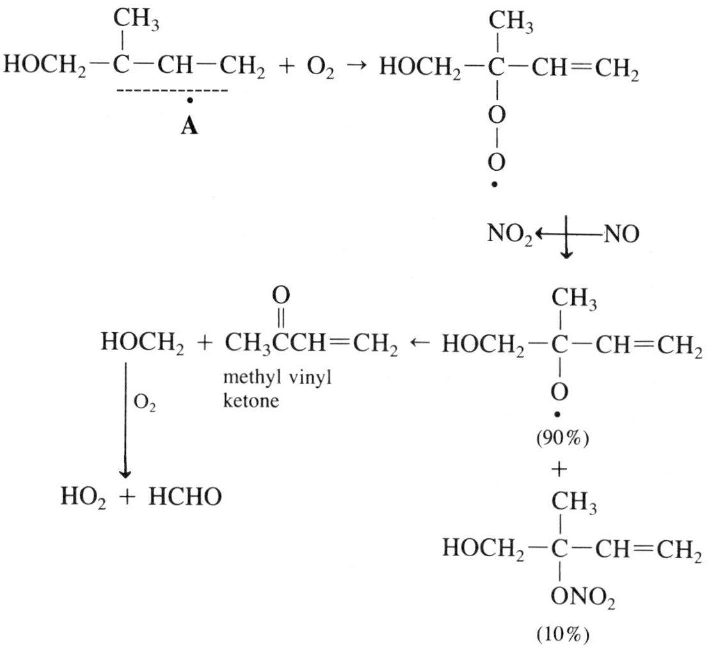

These products, and proposed mechanism, are in agreement with the recent findings of Gu et al. (1985), who find the major products of the OH-isoprene reaction to be methyl vinyl ketone, methacrolein, and 3-methylfuran. Methyl vinyl ketone is also expected from the reaction of ozone with isoprene across the 1,2 double bond:

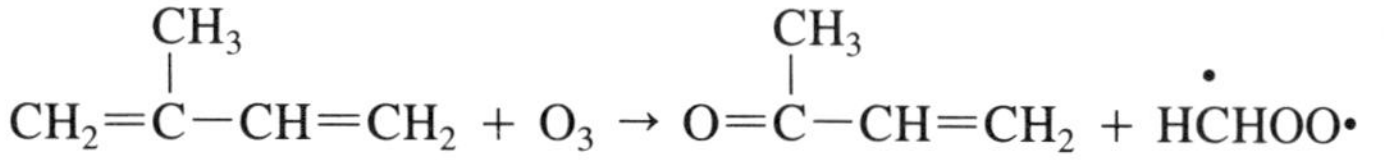

$$CH_2=\overset{CH_3}{\underset{|}{C}}-CH=CH_2 + O_3 \rightarrow O=\overset{CH_3}{\underset{|}{C}}-CH=CH_2 + HCHOO\cdot \quad (30)$$

Methyl vinyl ketone was in fact observed by Arnts and Gay (1979) in their chamber irradiations of isoprene-NO_x mixtures. Methacrolein was also shown to be a product. This could arise from a second path, ozonolysis across the 3,4 double bond:

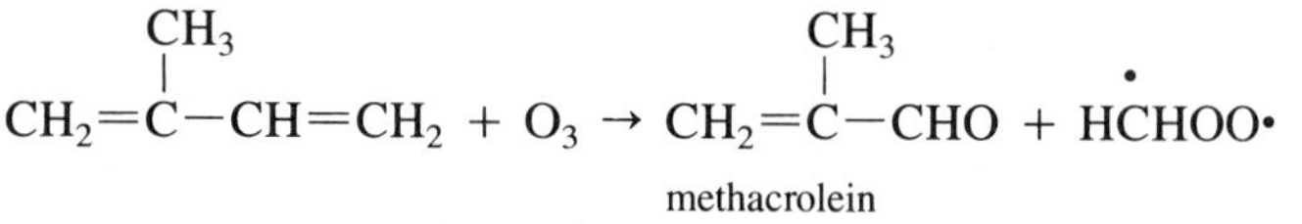

$$CH_2=\overset{CH_3}{\underset{|}{C}}-CH=CH_2 + O_3 \rightarrow \underset{\text{methacrolein}}{CH_2=\overset{CH_3}{\underset{|}{C}}-CHO} + HCHOO\cdot \quad (31)$$

Using FTIR spectroscopy, Niki and co-workers (1983d) estimate that reactions (30) and (31) account for at least 13% and 33%, respectively, of the O_3-isoprene reaction. The remaining two ozonolysis routes produce formaldehyde:

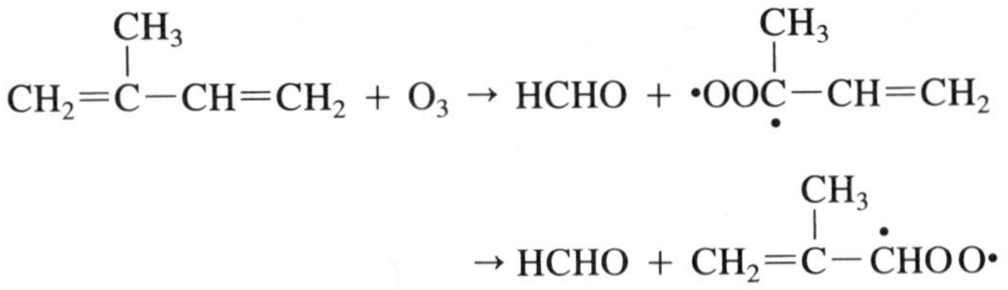

$$CH_2=\underset{\underset{CH_3}{|}}{C}-CH=CH_2 + O_3 \rightarrow HCHO + \cdot OO\underset{\underset{CH_3}{|}}{\overset{\cdot}{C}}-CH=CH_2 \quad (32)$$

$$\rightarrow HCHO + CH_2=\underset{\underset{CH_3}{|}}{C}-\overset{\cdot}{C}HOO\cdot$$

HCHO has been observed as a product but in sufficiently large yields that some of it must be due to secondary reactions (Niki et al., 1983d). The possible fates of the Criegee biradicals in ambient air are discussed in Section 7-E-2.

As noted above, methacrolein is also a product of the reaction of isoprene with OH (Gu et al., 1985). It may be produced by OH addition to form the radical C, followed by reaction with O_2 and NO:

$$CH_2=\underset{\underset{CH_3}{|}}{C}-CH=CH_2 + OH \rightarrow CH_2=\underset{\underset{CH_3}{|}}{C}-\underset{\cdot}{C}H-CH_2OH$$
$$C$$
$$\downarrow O_2$$

$$CH_2=\underset{\underset{CH_3}{|}}{C}-\underset{\underset{ONO_2}{|}}{C}H-CH_2OH + CH_2=\underset{\underset{CH_3}{|}}{C}-\underset{\underset{\overset{\cdot}{O}}{|}}{C}H-CH_2OH \underset{NO}{\overset{NO_2}{\longleftarrow}} CH_2=\underset{\underset{CH_3}{|}}{C}-\underset{\underset{\overset{\cdot}{O}-O}{|}}{C}H-CH_2OH$$

(10%) (90%)

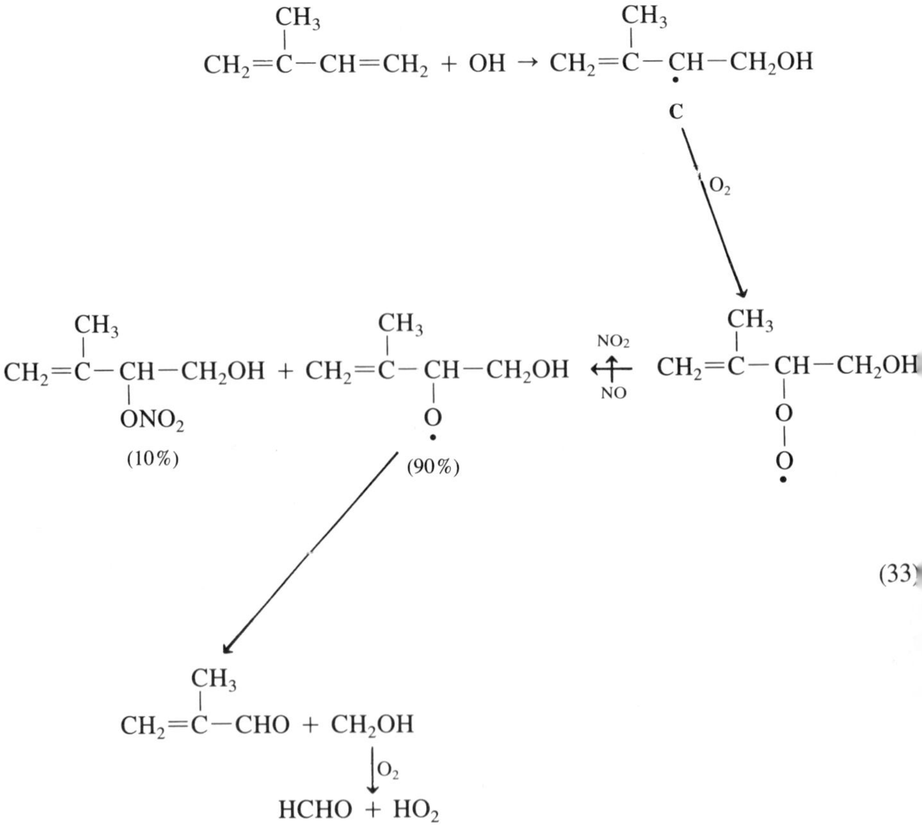

$$(33)$$

$$CH_2=\underset{\underset{CH_3}{|}}{C}-CHO + CH_2OH$$
$$\downarrow O_2$$
$$HCHO + HO_2$$

Clearly, parallel reactions can be written for all OH–olefin radical adducts, leading to a variety of predicted products. While confirmation of the mechanism awaits further experimental work, Lloyd and co-workers (1983) have used the initial reactions cited above along with their best estimates for the fates of the Criegee biradicals and for further reactions of the products methyl vinyl ketone and methacrolein in order to model the atmospheric reactions of isoprene. Using this mechanism they were able to match reasonably well the experimental data from irradiations of α-pinene–NO_x mixtures in an outdoor smog chamber.

Several other modeling studies have also been carried out for isoprene reactions under tropospheric conditions. Killus and Whitten (1984) suggest, based on their work, that isoprene may contribute as much as 50% of the overall reactivity of NMOC in rural air even at relatively low concentrations, that is, ~6% of the total ambient hydrocarbons. Brewer et al. (1984) suggest that at low NO_x concentrations, the fate of the majority of the carbon in isoprene may be the formation of longer chain carbon species such as alkylhydroperoxides and alkylperoxy acids; however, this awaits experimental confirmation.

The mechanisms of reaction of the monoterpenes are subject to even greater uncertainty than those for isoprene. In the study of Arnts and Gay, less than 20% of the carbon atoms reacted could be accounted for by the gas phase products observed by IR spectroscopy. The gaseous products included H_2CO, HCOOH, CO, CO_2, CH_3CHO, PAN, and $(CH_3)_2CO$. Few products were detected by GC. The remaining reacted material was suggested to exist in the aerosol phase (which might become deposited on the wall), or to be so polar that it did not pass through the GC columns. This is not surprising since the products anticipated from reaction with OH and O_3 are expected to be multi-functional oxygenates. The atmospheric chemistry expected for α-pinene, based on what is known for other alkenes (see Chapter 7), is discussed in detail by Lloyd et al. (1983).

The reactions of isoprene and the monoterpenes illustrate one of the deficiencies with the OH reactivity scale (Section 10-C). As discussed earlier, isoprene and the monoterpenes fall into the top two classes of reactivity with respect to the rate constant for reaction with OH. However, as seen in Fig. 14.16, the actual O_3 formed may not reflect this reactivity with OH since many of these organics also react rapidly with O_3, thus reducing its peak observed concentration.

Similarly, with respect to the rate of hydrocarbon loss, *p*-cymene was shown by Arnts and Gay to have low reactivity compared to α-pinene, isoprene, and *d*-limonene. However, as seen in Fig. 14.16, *p*-cymene formed more O_3 than these three compounds at NMHC/NO_x ratios $\gtrsim$ 25. Such discrepancies may be due at least partially to aerosol formation; Bufalini et al. (1976) suggest that aerosol formation removes hydrocarbons from the gas phase, so that further reactions, photolysis, and so on, leading to O_3 formation, do not occur. In this regard, it is noteworthy that pinonaldehyde, a product of the ozone-α-pinene reaction, has been detected in particles from pine forests where α-pinene is the most abundant monoterpene (e.g. see Yokouchi and Ambe, 1985).

14-C. REDUCED SULFUR COMPOUNDS

14-C-1. Chemical Nature and Sources of Emissions

As discussed in Section 1-D-1b, there are substantial natural emissions of sulfur compounds into the troposphere from biological activity in vegetation, soils, and water ecosystems. However, in contrast to anthropogenic emissions, which are almost entirely in the form of SO_2, these natural emissions are predominantly in the form of reduced sulfur compounds.

For example, Adams and co-workers (1981) carried out field measurements of sulfur emissions from soils, water, and vegetation in the eastern and southeastern United States. Varying numbers of different sulfur compounds were identified in the samples, some having only one compound and others having as many as ten. Six compounds positively identified were hydrogen sulfide (H_2S), dimethyl sulfide (CH_3SCH_3), carbon disulfide (CS_2), carbonyl sulfide (COS), methyl mercaptan (CH_3SH), and dimethyl disulfide (CH_3SSCH_3). Four unidentified compounds, when present in the samples, comprised less than 1–2% of the total sulfur emissions. The relative emission strength of these compounds varied from sample to sample, as well as with the different types of biological sources. However, H_2S, CH_3SCH_3, COS, and CS_2 were generally present in the greatest concentrations, with CH_3SH and CH_3SSCH_3, emissions typically being somewhat smaller.

The oceans have also been shown to be a significant source of reduced sulfur compounds. For example, Andreae and co-workers (Andreae and Raemdonck, 1983; Ferek and Andreae, 1983) have shown that the ocean emits significant quantities of CH_3SCH_3 and smaller amounts of COS ($\sim 1\%$ as S compared to CH_3SCH_3). They estimate that the flux of CH_3SCH_3 from the ocean is $\sim 38.5 \times 10^{12}$ g of sulfur per year, compared to the Adams et al. estimate of 13×10^{12} g yr^{-1} of this compound from land sources.

14-C-2. Atmospheric Reactions

Relatively little is known about the mechanisms of oxidation of reduced sulfur compounds in the atmosphere. Cox and Sheppard (1980) reported that SO_2 was a major product in the reactions of OH with COS, CS_2, H_2S, CH_3SH, CH_3SCH_3, and CH_3SSCH_3. Although they did not report yields of SO_2 quantitatively (except for H_2S), it has often been assumed that these compounds are converted first to SO_2 and subsequently to sulfuric acid and sulfates via mechanisms discussed in Chapter 11. However, the results of several recent laboratory studies of these oxidations have suggested that SO_2 is not the sole, nor even predominant, sulfur-containing product in the oxidation of the organic sulfur compounds CH_3SH, CH_3SCH_3, and CH_3SSCH_3.

As for other species in the troposphere, the possible reactions that must be considered are those with O_3, OH, and NO_3, in addition to photolysis. The absorption cross sections of all sulfides in the actinic region $\lambda \geq 290$ nm are small and hence photolysis is negligible (Calvert and Pitts, 1966). The reactions with O_3 are also too slow (Atkinson and Carter, 1984) to be significant under tropospheric conditions, leaving attack by OH and by NO_3 as the major potential oxidation routes.

Table 14.6 gives the rate constants for OH and NO_3 reactions with some of the naturally emitted reduced sulfur compounds. Also given are the lifetimes of the organics with respect to reaction with OH and NO_3 under conditions characteristic

TABLE 14.6 Rate Constants and Lifetimes at Room Temperature for the Reactions of OH and NO_3 with Some Reduced Sulfur Compounds Emitted Biogenically

	OH (Day)		NO_3 (Night)	
Compound	k (cm^3 $molecule^{-1}$ s^{-1})	τ[b]	k (cm^3 $molecule^{-1}$ s^{-1})	τ[d]
H_2S	4.7×10^{-12}[a]	5 days	[f]	—
COS	1.0×10^{-15}[a]	63 yr	[f]	—
CS_2	2.8×10^{-12}[a]	8 days	[f]	—
CH_3SCH_3	6.3×10^{-12}[e]	4 days	9.7×10^{-13}[c]	1.1 h
CH_3SH	3.3×10^{-11}[e]	17 h	1.0×10^{-12}[g]	1.1 h
CH_3SSCH_3	2.0×10^{-10}[e]	2.8 h	0.42×10^{-13}[g]	26 h

[a]DeMore et al., 1985; rate constant for OH + CS_2 calculated for 760 Torr in air.

[b]$\tau = 1/k[OH]$, where $[OH] = 5 \times 10^5$ radicals cm^{-3}.

[c]Atkinson et al., 1984b, adjusted to a value of $K_{eq} = 3.35 \times 10^{-11}$ cm^3 $molecule^{-1}$ for the equilibrium $NO_2 + NO_3 \rightleftarrows N_2O_5$ (see Section 7-C-2).

[d]$\tau = 1/k[NO_3]$, where $[NO_3]$ = 10 ppt = 2.5×10^8 cm^{-3}.

[e]See Atkinson (1985) review.

[f]Not reported.

[g]Mac Leod et al., 1985.

of a relatively clean atmosphere. For CH_3SCH_3 and CH_3SH, attack by NO_3 at night is likely to be a major loss process.

It should be noted that the results of kinetic and mechanistic studies of sulfur compounds are frequently in wide disagreement. This is due at least partly to experimental difficulties in handling sulfur compounds. For example, they tend to adsorb readily on the walls of reaction vessels, and hence heterogeneous wall reactions can contribute to the overall oxidation in laboratory systems.

Mechanisms of reaction of reduced sulfur compounds with OH have not been well established. The OH–H_2S reaction is believed to proceed via hydrogen atom abstraction,

$$OH + H_2S \rightarrow H_2O + SH \tag{34}$$

with the SH radical reacting further to form SO_2 under atmospheric conditions (Cox and Sheppard, 1980).

The OH–CS_2 reaction is similar to the OH–CO reaction (see Section 4-A-5) in that an intermediate complex, suggested by Kurylo (1978), is formed:

$$OH + CS_2 \overset{M}{\rightleftarrows} HOCS_2 \tag{35}$$

The results of kinetic studies (e.g., see Jones et al., 1982; Barnes et al., 1983) suggest that the adduct can react with O_2:

$$HOCS_2 + O_2 \rightarrow \text{products} \tag{36}$$

Thus the rate constant has been shown in a number of studies to increase in the presence of O_2, to have a negative temperature dependence, and to depend on total pressure (DeMore et al., 1985). The subsequent reactions of the adduct are subject to speculation; Cox and Sheppard (1980) suggest that it decomposes to form COS and HS, based on their observation of a yield of COS of ~100%:

$$OH + CS_2 \xrightarrow{M} S{-}\underset{\substack{|\\ O{-}H}}{C}{-}S \xrightarrow{O_2} COS + SH$$

The SH radical would then react to form SO_2, as in the case of the OH–H_2S reaction. This has since been confirmed by Jones et al. (1982) and Barnes et al. (1983).

The mechanisms of the OH reactions with organic sulfides have been studied in detail by several groups (Cox and Sandalls, 1974; Hatakeyama et al., 1982; Grosjean and Lewis, 1982; Niki et al., 1983a; Hatakeyama and Akimoto, 1983; Grosjean, 1984; Mac Leod et al., 1984; Hatakeyama et al., 1985). In the dimethyl sulfide oxidation, SO_2 was observed as a product, but in yields much less than 100%; in most studies, the yield of SO_2 was found to be approximately 20%. Grosjean and Lewis (1982) and Grosjean (1984) report much higher SO_2 yields and suggested the lower yields observed in other studies may have been due to the use of high reactant concentrations. However, recent studies by Hatakeyama et al. (1985) at much lower concentrations also gave SO_2 yields of ~20%, in agreement with their work at higher concentrations. Methanesulfonic acid is a major product of the dimethyl sulfide oxidation, with reported yields ≥50% (Hatakeyama et al., 1982). Minor amounts of $(CH_3)_2SO_2$ and H_2SO_4 are also formed. Unidentified products which are thought to contain both sulfur and nitrogen were observed. Methanesulfonic acid and sulfate are also found in the particles formed in during the dimethyl sulfide oxidation (Grosjean and Lewis, 1982; Grosjean, 1984; Hatakayema et al., 1985).

An initial abstraction of hydrogen is likely the first step:

$$CH_3SCH_3 + OH \rightarrow CH_3SCH_2 + H_2O \tag{37}$$

This appears to be followed by secondary reactions of the CH_3SCH_2 radical such as the following:

$$CH_3SCH_2 + O_2 \rightarrow CH_3SCH_2OO \tag{38}$$

$$CH_3SCH_2OO + NO \rightarrow CH_3SCH_2O + NO_2 \tag{39}$$

$$CH_3SCH_2O \rightarrow CH_3S + HCHO \tag{40}$$

Methyl thionitrite (CH_3SNO) has been observed (Niki et al., 1983a) and may be formed from the reaction of CH_3S with NO:

$$CH_3S + NO \rightarrow CH_3SNO$$

The fates of the CH_3S radical in air are uncertain: it reacts with O_2, but there are two possible fates for the adduct formed:

$$CH_3S + O_2 \rightarrow (CH_3SO_2)^* \underset{a}{\rightarrow} CH_3 + SO_2 \qquad \underset{b}{\rightarrow} CH_3SO_2 \tag{41}$$

The CH_3SO_2 would then form one of the observed products, methanesulfonic acid. While the data of Niki et al. (1983b) and Grosjean (1984) suggest both paths are important, Hatakeyama and Akimoto (1983) find that their data are compatible only with reaction (41b) to form the stabilized adduct.

Hatakeyama and Akimoto (1983) suggest that a second mode of attack of OH on CH_3SCH_3 is important, the addition of OH to the sulfur atom:

$$CH_3SCH_3 + OH \xrightarrow{M} CH_3\overset{\overset{OH}{|}}{S}CH_3 \tag{42}$$

They propose that the adduct decomposes and that secondary reactions then form CH_3SO_3H:

$$CH_3\overset{\overset{OH}{|}}{S}CH_3 \rightarrow CH_3SOH + CH_3 \tag{43}$$

$$CH_3SOH + O_2 \xrightarrow{M} CH_3SO_3H \tag{44}$$

Thus there is agreement that SO_2 is not, in fact, the major sulfur-containing product as previously assumed, and that methanesulfonic acid is a major product of the CH_3SCH_3 oxidation under atmospheric conditions. However, the relative importance of hydrogen atom abstraction versus addition to the sulfur atom is not certain. Atkinson suggests, based on kinetic data for the reactions of OH with a series of alkyl sulfides, that at 1 atm of air, the hydrogen abstraction process accounts for ~ 70% of the overall reaction, and OH addition accounts for the remainder. The mechanism of oxidation of CH_3S radicals under atmospheric conditions is also not well understood.

Hatakeyama and Akimoto (1983) also studied the products of the reactions of OH with CH_3SH and CH_3SSCH_3. In the former case, they postulate the formation of an OH–CH_3SH adduct or complex, which can decompose to give the CH_3S radical:

$$OH + CH_3SH \xrightarrow{M} \left(CH_3\underset{\underset{OH}{|}}{S}H \text{ or } CH_3SH \text{ ---- } OH \right) \rightarrow CH_3S + H_2O \tag{45}$$

Grosjean (1984), based on his study of the photooxidation of CH_3SH in NO_x–air mixtures, also proposed that the OH reaction with CH_3SH produces $CH_3S + H_2O$, although via a direct hydrogen atom abstraction from the weak S—H bond, rather than by adduct formation. Similarly, OH is proposed to add to CH_3SSCH_3, with the adduct decomposing to $CH_3S + CH_3SOH$:

$$CH_3SSCH_3 + OH \rightarrow \underset{\displaystyle \overset{|}{OH}}{CH_3SSCH_3} \rightarrow CH_3SOH + CH_3S \qquad (46)$$

The CH_3S and CH_3SOH then react to give SO_2 and CH_3SO_3H.

Thus, in contrast to earlier assumptions, it now appears that a variety of organic sulfur compounds react to give CH_3SO_3H as a major product, in addition to SO_2. It is noteworthy that methanesulfonic acid has been detected in ambient aerosols (Panter and Penzhorn, 1980; Saltzman et al., 1985).

As discussed above, it is only recently that the importance of NO_3 radicals in both remote and moderately polluted atmospheres has been recognized. While NO_3 reactions with all the sulfur compounds listed in Table 14.6 have not been studied, those that have indicate that NO_3 plays a major role in the nighttime chemistry of some of these as well. For example, as seen in Table 14.6, the lifetime of dimethyl sulfide with respect to reaction with 10 ppt NO_3 is only 1 h, compared to 4 days for the daytime reaction with 5×10^5 cm^{-3} OH.

Little is known about the mechanisms of reaction of the nitrate radical with reduced sulfur compounds. Mac Leod et al. (1985) have studied the reactions of NO_3 with CH_3SH, C_2H_5SH, and CH_3SSCH_3 at 297°K in 1 atm of an 80% N_2 + 20% O_2 mixture, and in 1 atm of pure N_2. In the case of CH_3SH, major products were found to be CH_3ONO_2, SO_2 (~15% molar yield), and CH_3SNO_2, and when O_2 was present, CH_3SSCH_3 and HCHO. In the case of C_2H_5SH, SO_2 (~40% molar yield) was produced, as was a compound tentatively identified as $C_2H_5SNO_2$. The dimethyl disulfide reaction gave HCHO, SO_2 (~60-70% molar yield), CH_3ONO_2, and CH_3SNO_2. In all cases, unidentified products were also observed by FTIR.

The mechanisms involved in these reactions is not clear. MacLeod et al. suggest that NO_3 may initially add to the sulfur atom to form an adduct, and this adduct then decomposes, e.g. in the case of CH_3SH to form ($CH_3S + HNO_3$) or (SO_2 + other products). Secondary reactions of the CH_3S radical such as those discussed above may then give rise to some of the observed products. This is clearly an area requiring further study.

In their studies of dimethyl sulfide concentrations in ocean water and in the air over the oceans, Andreae and Raemdonck (1983) found that the sulfide concentrations in air were much lower over the Atlantic (6.1 ng m^{-3}) and the Gulf of Mexico (9.6 ng m^{-3}) than over the remote Pacific (167 ng m^{-3}). In addition, a diurnal variation in the dimethyl sulfide concentration was observed over the Pacific, with minimum concentrations occurring in midday. This is consistent with loss of the sulfide by reaction with OH, whose concentration peaks when photolysis of its precursors (in this case predominantly O_3) is at a maximum.

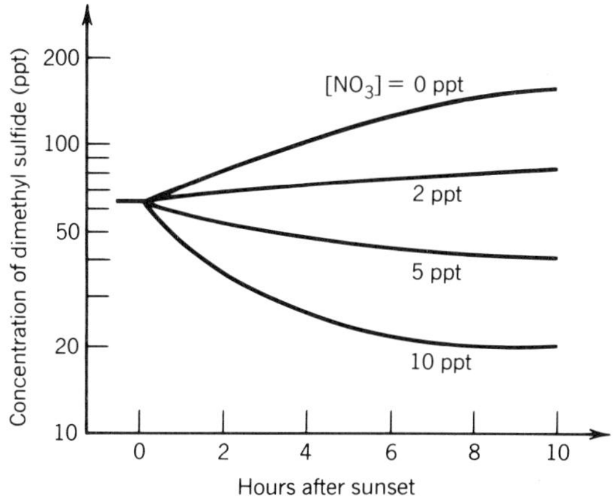

FIGURE 14.17. Calculated concentration–time profiles for dimethyl sulfide for constant NO_3 radical concentrations of 0, 2, 5, and 10 ppt (from Winer et al., 1984).

However, somewhat surprisingly, no diurnal variation in the dimethyl sulfide concentration was observed over the Atlantic Ocean or the Gulf of Mexico. In these air masses, aerosols which indicated some contribution of pollutants from the continent were always present. The lack of a diurnal variation in the CH_3SCH_3 concentration, as well as lower average levels, can then be explained by a combination of daytime removal by OH and nighttime removal by NO_3 where the NO_x precursor to NO_3 arises from the transport of continental anthropogenic emissions.

Winer and co-workers (1984) have carried out modeling studies which show that only relatively small NO_3 concentrations are required to explain the lack of diurnal variation of dimethyl sulfide and the lower concentrations observed. Figure 14.17 shows the calculated concentration–time profiles for CH_3SCH_3 after sunset as a function of NO_3 radical concentrations from 0 to 10 ppt. Clearly, even 2 ppt of NO_3 can significantly lower the dimethyl sulfide concentration. Under conditions characteristic of the remote Pacific, little NO_3 is expected to be present, and hence it should play a very minor role in the CH_3SCH_3 removal.

14-D. HALOGEN-CONTAINING COMPOUNDS

A variety of halogens in both inorganic and organic forms is found in the troposphere, with some being produced in natural processes. The halogenated species, their sources, and reactions in the atmosphere, both natural and man-made, are reviewed by Cicerone (1981).

Methyl chloride, methyl bromide, and methyl iodide are all produced by natural processes, predominantly in the oceans (Lovelock et al., 1973; Lovelock 1975; Singh et al., 1983; Rasmussen, 1982b). In the troposphere, CH_3Cl and CH_3Br are attacked by OH with rate constants at 298°K of 4.36×10^{-14} and 3.85×10^{-14} cm^3 molecule^{-1} s^{-1}, respectively (Atkinson, 1985). The corresponding lifetimes of CH_3Cl and CH_3Br at an OH concentration of 5×10^5 cm^3 are ~ 1.5 yr and 1.6 yr, respectively. The lifetimes are sufficiently long that these molecules can diffuse upward into the stratosphere and play a role in the chemistry of that region (see Chapter 15). Based on the chemistry discussed in Chapter 7, the reactions subsequent to hydrogen abstraction by OH might be expected to be as follows, using CH_3Cl as an example:

$$CH_3Cl + OH \rightarrow CH_2Cl + H_2O \tag{47}$$

$$CH_2Cl + O_2 \rightarrow OOCH_2Cl \tag{48}$$

$$OOCH_2Cl + NO \rightarrow OCH_2Cl + NO_2 \tag{49}$$

$$OCH_2Cl + O_2 \rightarrow HCOCl + HO_2 \tag{50}$$

Methyl iodide, unlike the chlorine and bromine analogues, absorbs light in the actinic region and hence photolyzes to produce iodine atoms:

$$CH_3I + h\nu \rightarrow CH_3 + I \tag{51}$$

The lifetime of CH_3I with respect to photolysis has been estimated to be about 8 days (Chameides and Davis, 1980); while a rate constant for reaction with OH does not appear to be available, if it is similar to that for CH_3Cl and CH_3Br, reaction with OH will be too slow to compete with photolysis.

Several researchers have suggested that the iodine atom produced in reaction (51) and its oxide may play a significant role in tropospheric chemistry (Moyers and Duce, 1972; Zafiriou, 1974; Chameides and Davis, 1980). For example, the following reactions may occur:

$$I + O_3 \rightarrow IO + O_2 \tag{52}$$

$$IO + CO \rightarrow I + CO_2 \tag{53}$$

In examining iodine atmospheric photochemistry in greater detail, Chameides and Davis (1980) suggested that other cycles may also be important, for example, (51) and (52) followed by (54):

$$IO + IO \rightarrow 2I + O_2 \tag{54}$$

Reaction (54) has been shown, at total pressures between 10 and 400 Torr, to proceed by two paths, one pressure independent and the other pressure dependent (Cox and Coker, 1983; Jenkin and Cox, 1985). The pressure-independent path was suggested to occur via (54) and (55),

$$IO + IO \rightarrow I_2 + O_2 \tag{55}$$

with the pressure-dependent path forming I_2O_2:

$$IO + IO \xrightarrow{M} I_2O_2 \tag{56}$$

The IO radical has also been shown to react with NO_2 in a manner analogous to ClO and BrO (see Chapter 15):

$$IO + NO_2 \xrightarrow{M} IONO_2 \tag{57}$$

Reaction (57) is in the falloff region between second- and third-order kinetics in the pressure range 35–400 Torr (Jenkin and Cox, 1985).

Interaction of the iodine oxides with other tropospheric species has been proposed to lead to iodinated compounds such as HOI and $IONO_2$ in the troposphere and to possibly change the NO_2/NO ratio and alter the conversion of HO_2 to OH. However, subsequent measurements of the concentrations of CH_3I in background air have found low levels of CH_3I, typically 1–3 ppt, suggesting that such cycles do not play a major role in O_3 loss in the troposphere or in determining other important chemical characteristics such as the NO_2/NO ratio of the natural troposphere.

Naturally produced particulate halogens present in sea salt aerosols are discussed in Section 12-D-1.

REFERENCES

Adams, D. F., S. O. Farwell, E. Robinson, M. R. Pack, and W. L. Bamesberger, "Biogenic Sulfur Source Strengths," *Environ. Sci. Technol.*, **15,** 1493 (1981).

Altshuller, A. P., "Review: Natural Volatile Organic Substances and Their Effect on Air Quality in the United States," *Atmos. Environ.*, **17,** 2131 (1983).

Andreae, M. O. and Raemdonck, H., "Dimethyl Sulfide in the Surface Ocean and Marine Atmosphere: A Global View," *Science*, **221,** 744 (1983).

Arnts, R. R. and B. W. Gay, Jr., "Photochemistry of Some Naturally Emitted Hydrocarbons," U.S. Environmental Protection Agency Report No. EPA-600/3-79-081, September 1979

Atkinson, R., "Kinetics and Mechanisms of the Gas Phase Reactions of the Hydroxyl Radical with Organic Compounds Under Atmospheric Conditions," *Chem. Rev.*, in press (1985).

Atkinson, R. and W. P. L. Carter, "Kinetics and Mechanisms of the Gas Phase Reactions of Ozone with Organic Compounds Under Atmospheric Conditions," *Chem. Rev.*, **84,** 437 (1984).

Atkinson, R. and A. C. Lloyd, "Evaluation of Kinetic and Mechanistic Data for Modeling of Photochemical Smog," *J. Phys. Chem. Ref. Data*, **13,** 315 (1984).

Atkinson, R., A. M.Winer, and J. N. Pitts, Jr., "Rate Constants for the Gas Phase Reactions of O_3 with the Natural Hydrocarbons Isoprene and α- and β-Pinene," *Atmos. Environ.*, **16,** 1017 (1982).

Atkinson, R., S. M. Aschmann, A. M. Winer, and J. N. Pitts, Jr., "Kinetics of the Gas Phase Reactions of NO_3 Radicals with a Series of Dialkenes, Cycloalkenes, and Monoterpenes at 295 $\pm$ 1 K," *Environ. Sci. Technol.*, **18,** 370 (1984a).

Atkinson, R., J. N. Pitts, Jr., and S. M. Aschmann, "Tropospheric Reactions of Dimethylsulfide with NO_3 and OH Radicals," *J. Phys. Chem.*, **88,** 1584 (1984b).

Atkinson, R., S. M. Aschmann, A. M. Winer, and J. N. Pitts, Jr., "Kinetics and Atmospheric Implications of the Gas-Phase Reactions of NO_3 Radicals with a Series of Monoterpenes and Related Organics at 294 $\pm$ 2K," *Environ. Sci. Technol.*, **19,** 159 (1985).

Barnes, I., K. H. Becker, E. H. Fink, A. Reimer, F. Zabel, and H. Niki, "Rate Constant and Products of the Reaction CS_2 + OH in the Presence of O_2," *Int. J. Chem. Kinet.*, **15,** 631 (1983).

Blake, D. R., E. W. Mayer, S. C. Tyler, Y. Makide, D. C. Montague, and F. S. Rowland, "Global Increase in Atmospheric Methane Concentrations Between 1978 and 1980," *Geophys. Res. Lett.*, **9,** 477 (1982).

Bottenheim, J. W. and O. P. Strausz, "Gas-Phase Chemistry of Clean Air at 55°N Latitude," *Environ. Sci. Technol.*, **14,** 709 (1980).

Brewer, D. A., M. A. Ogliaruso, T. R. Augustsson, and J. S. Levine, "The Oxidation of Isoprene in the Troposphere: Mechanism and Model Calculations," *Atmos. Environ.*, **18,** 2723 (1984).

Brown, D. E. and A. M. Winer, "Estimating Urban Vegetation Cover and Leaf Mass in the Los Angeles Basin with High Altitude and Low Altitude Aerial Photography," *Photogramm. Eng. Remote Sens.*, in press (1985).

Bufalini, J. J. and R. R. Arnts, Eds., *Atmospheric Biogenic Hydrocarbons*, Vols. 1 and 2, Ann Arbor Science Publishers, Ann Arbor, MI, 1981.

Bufalini, J. J., T. A. Walter, and M. M. Bufalini, "Ozone Formation Potential of Organic Compounds," *Environ. Sci. Technol.*, **10,** 908 (1976).

Calvert, J. G. and J. N. Pitts, Jr., *Photochemistry*, Wiley, New York, 1966.

Chameides, W. L. and D. D. Davis, "Iodine: Its Possible Role in Tropospheric Photochemistry," *J. Geophys. Res.*, **85,** 7383 (1980).

Chameides, W. L., S. C. Liu, and R. J. Cicerone, "Possible Variations in Atmospheric Methane," *J. Geophys. Res.*, **82,** 1795 (1977).

Chatfield, R. and H. Harrison, "Tropospheric Ozone. 2. Variations Along a Meridional Band," *J. Geophys. Res.*, **82,** 5969 (1977).

Chung, Y. S. and T. Dann, "Observations of Stratospheric Ozone at the Ground Level in Regina, Canada," *Atmos. Environ.*, **19,** 157 (1985).

Cicerone, R. J., "Halogens in the Atmosphere," *Rev. Geophys. Space Phys.*, **19,** 123 (1981).

Cox, R. A. and G. B. Coker, "Absorption Cross Section and Kinetics of IO in the Photolysis of CH_3I in the Presence of Ozone," *J. Phys. Chem.*, **87,** 4478 (1983).

Cox, R. A. and F. J. Sandalls, "The Photooxidation of Hydrogen Sulphide and Dimethyl Sulfide in Air," *Atmos. Environ.*, **8,** 1269 (1974).

Cox, R. A. and D. Sheppard, "Reactions of OH Radicals with Gaseous Sulphur Compounds," *Nature*, **284,** 330 (1980).

Craig, H. and C. C. Chou, "Methane: The Record in Polar Ice Cores," *Geophys. Res. Lett.*, **9,** 1221 (1982).

Crutzen, P. J., "The Role of NO and NO_2 in the Chemistry of the Troposphere and Stratosphere," *Ann. Rev. Earth Planet. Sci.*, **7,** 443 (1979).

Crutzen, P. J. and L. T. Gidel, "A Two-Dimensional Photochemical Model of the Atmosphere. 2: The Tropospheric Budgets of the Anthropogenic Chlorocarbons, CO, CH_4, CH_3Cl, and the Effect of Various NO_x Sources on Tropospheric Ozone," *J. Geophys. Res.*, **88,** 6641 (1983).

Crutzen, P. J. and U. Schmailzl, "Chemical Budgets of the Stratosphere," *Planet. Space Sci.*, **31,** 1009 (1983).

Danielsen, E. F. and V. A. Mohnen, "Project Dustorm Report: Ozone Transport, in Situ Measurements, and Meteorological Analyses of Tropopause Folding," *J. Geophys. Res.*, **82,** 5867 (1977).

DeMore, W. B., J. J. Margitan, M. J. Molina, R. T. Watson, D. M. Golden, R. F. Hampson, M. J. Kurylo, C. J. Howard, and A. R. Ravishankara, "Chemical Kinetics and Photochemical Data for Use in Stratospheric Modeling, Evaluation No. 7," JPL Publ. No. 85-37 (1985).

Dimitriades, B., "The Role of Natural Organics in Photochemical Air Pollution: Issues and Research Needs," *J. Air Pollut. Control Assoc.*, **31,** 229 (1981).

Duce, R. A., V. A. Mohnen, P. R. Zimmerman, D. Grosjean, W. Cautreels, R. Chatfield, R. Jaenicke, J. A. Ogren, E. D. Pellizzari, and G. T. Wallace, "Organic Material in the Global Troposphere," *Rev. Geophys. Space Phys.*, **21,** 921 (1983).

Dutkiewicz, V. A. and L. Husain, "Stratospheric and Tropospheric Components of ^{7}Be in Surface Air," *J. Geophys. Res.*, **90,** 5783 (1985).

Ehhalt, D. H. and U. Schmidt, "Sources and Sinks of Atmospheric Methane," *Pure Appl. Geophys.*, **116,** 452 (1978).

Ehhalt, D. H., R. J. Zander, and R. A. Lamontagne, "On the Temporal Increase of Tropospheric CH_4," *J. Geophys. Res.*, **88,** 8442 (1983).

Ferek, R. J. and M. O. Andreae, "The Supersaturation of Carbonyl Sulfide in Surface Waters of the Pacific Ocean Off Peru," *Geophys. Res. Lett.*, **10,** 393 (1983).

Fishman, J. and P. J. Crutzen, "The Origin of Ozone in the Troposphere," *Nature*, **274,** 855 (1978).

Fishman, J. and W. Seiler, "Correlative Nature of Ozone and Carbon Monoxide in the Troposphere: Implications for the Tropospheric Ozone Budget," *J. Geophys. Res.*, **88**(C6), 3662 (1983).

Gidel, L. T. and M. Shapiro, "General Circulation Model Estimates of the Net Vertical Flux of Ozone in the Lower Stratosphere and the Implications for the Tropospheric Ozone Budget," *J. Geophys. Res.*, **85,** 4049 (1980).

Graedel, T. E., "Terpenoids in the Atmosphere," *Rev. Geophys. Space Phys.*, **17,** 937 (1979a).

Graedel, T. E., "The Kinetic Photochemistry of the Marine Atmosphere," *J. Geophys. Res.*, **84,** 273 (1979b).

Graedel, T. E. and J. E. McRae, "On the Possible Increase of the Atmospheric Methane and Carbon Monoxide Concentrations During the Last Decade," *Geophys. Res. Lett.*, **7,** 977 (1980).

Grosjean, D., "Photooxidation of Methyl Sulfide, Ethyl Sulfide and Methanethiol," *Environ. Sci. Technol.*, **18,** 460 (1984).

Grosjean, D. and R. Lewis, "Atmospheric Photooxidation of Methyl Sulfide," *Geophys. Res. Lett.*, **9,** 1203 (1982).

Gu, C.-l., C. M. Rynard, D. G. Hendry, and T. Mill, "Hydroxyl Radical Oxidation of Isoprene," *Environ. Sci. Technol.*, **19,** 151 (1985).

Hanst, P. L. and B. W. Gay, Jr., "Atmospheric Oxidation of Hydrocarbons: Formation of Hydroperoxides and Peroxyacids," *Atmos. Environ.*, **17,** 2259 (1983).

Hatakeyama, S. and H. Akimoto, "Reactions of OH Radicals with Methanethiol, Dimethyl Sulfide, and Dimethyl Disulfide in Air," *J. Phys. Chem.*, **87,** 2387 (1983).

Hatakeyama, S., M. Okuda, and H. Akimoto, "Formation of Sulfur Dioxide and Methanesulfonic Acid in the Photooxidation of Dimethyl Sulfide in the Air," *Geophys. Res. Lett.*, **9,** 583 (1982).

Hatakeyama, S., K. Izumi, and H. Akimoto, "Yield of SO_2 and Formation of Aerosol in the Photooxidation of DMS Under Atmospheric Conditions," *Atmos. Environ.*, **19,** 135 (1985).

Heidt, L. E. and D. H. Ehhalt, "Corrections of CH_4 Concentrations Measured Prior to 1974," *Geophys. Res. Lett.*, **7,** 1023 (1980).

Hov, O., J. Schjoldager, and B. M. Wathne, "Measurement and Modeling of the Concentrations of Terpenes in Coniferous Forest Air," *J. Geophys. Res.*, **88,** 10,679 (1983).

Huebert, B. J. and A. L. Lazrus, "Tropospheric Gas-Phase and Particulate Nitrate Measurements," *J. Geophys. Res.*, **85,** 7322 (1980).

Isidorov, V. A., I. G. Zenkevich, and B. V. Ioffe, "Volatile Organic Compounds in the Atmosphere of Forests," *Atmos. Environ.*, **19,** 1 (1985).

Jenkin, M. E. and R. A. Cox, "Kinetics Study of the Reactions $IO + NO_2 + M \rightarrow IONO_2 + M$, $IO + IO \rightarrow$ Products and $I + O_3 \rightarrow IO + O_2$," *J. Phys. Chem.*, **89,** 192 (1985).

Jones, B. M. R., J. P. Burrows, R. A. Cox, and S. A. Penkett, "OCS Formation in the Reaction of OH with CS_2," *Chem. Phys. Lett.*, **88,** 372 (1982).

Kaiserman, M. J. and E. W. Corse, Aerosols and Carbon Balance in the Pinene–NO_x Photochemical System, in *Atmospheric Biogenic Hydrocarbons,* Vol. 2, *Ambient Concentrations and Atmospheric Chemistry*, J. J. Bufalini and P. R. Arnts, Eds., Ann Arbor Science Publishers, Ann Arbor, MI, 1981, pp. 139–159.

Kelly, T. J., P. H. Daum, and S. E. Schwartz, "Measurements of Peroxides in Cloudwater and Rain," *J. Geophys. Res.*, **90,** 7861 (1985).

Khalil, M. A. K. and R. A. Rasmussen, "Sources, Sinks, and Seasonal Cycles of Atmospheric Methane," *J. Geophys. Res.*, **88,** 5131 (1983).

Khalil, M. A. K. and R. A. Rasmussen, "Carbon Monoxide in the Earth's Atmosphere: Increasing Trend," *Science*, **224,** 54 (1984).

Khalil, M. A. K. and R. A. Rasmussen, "Causes of Increasing Atmospheric Methane: Depletion of Hydroxyl Radicals and the Rise of Emissions," *Atmos. Environ.*, **19,** 397 (1985).

Killus, J. P. and G. Z. Whitten, "Isoprene: A Photochemical Kinetic Mechanism," *Environ. Sci. Technol.*, **18,** 142 (1984).

Kurylo, M. J., "Flash Photolysis Resonance Fluorescence Investigation of the Reactions of OH Radicals with OCS and CS_2," *Chem. Phys. Lett.*, **58,** 238 (1978).

Lamb, B., H. Westberg, G. Allwine, and T. Quarles, "Biogenic Hydrocarbon Emissions from Deciduous and Coniferous Trees in the United States," *J. Geophys. Res.*, **90,** 2380 (1985).

Levy, H. II, J. D. Mahlman, W. J. Moxim, and S. C. Liu, "Tropospheric Ozone: The Role of Transport," *J. Geophys. Res.*, **90,** 3753 (1985).

Liu, S. C., D. Kley, M. McFarland, J. D. Mahlman, and H. Levy II, "On the Origin of Tropospheric Ozone," *J. Geophys. Res.*, **85**(C12), 7546 (1980).

Liu, S. C., M. McFarland, D. Kley, O. Zafiriou, and B. Huebert, "Tropospheric NO_x and O_3 Budgets in the Equatorial Pacific," *J. Geophys. Res.*, **88,** 1360 (1983).

Lloyd, A. C., R. Atkinson, F. W. Lurmann, and B. Nitta, "Modeling Potential Ozone Impacts from Natural Hydrocarbons—I. Development and Testing of a Chemical Mechanism for the NO_x–Air Photooxidations of Isoprene and α-Pinene Under Ambient Conditions," *Atmos. Environ.*, **17,** 1931 (1983).

Logan, J. A., "Nitrogen Oxides in the Troposphere: Global and Regional Budgets," *J. Geophys. Res.*, **88,** 10,785 (1983).

Logan, J. A., "Tropospheric Ozone: Seasonal Behavior, Trends, and Anthropogenic Influence," *J. Geophys. Res.*, **90,** 10,463 (1985).

Logan, J. A., M. J. Prather, S. C. Wofsy, and M. B. McElroy, "Tropospheric Chemistry: A Global Perspective," *J. Geophys. Res.*, **86,** 7210 (1981).

Lovelock, J. E., "Natural Hydrocarbons in the Air and in the Sea," *Nature*, **256,** 193 (1975).

Lovelock, J. E., R. J. Maggs, and R. J. Wade, "Halogenated Hydrocarbons in and Over the Atlantic," *Nature*, **241,** 194 (1973).

Lurmann, F. W., A. C. Lloyd, and B. Nitta, "Modeling Potential Ozone Impacts from Natural Hydrocarbons. II. Hypothetical Biogenic HC Emission Scenario Modeling," *Atmos. Environ.*, **17,** 1951 (1983).

MacLeod, H., J. L. Jourdain, G. Poulet, and G. LeBras, "Kinetic Study of Reactions of Some Organic Sulfur Compounds with OH Radicals," *Atmos. Environ.*, **18,** 2621 (1984).

MacLeod, H., S. M. Aschmann, R. Atkinson, E. C. Tuazon, J. A. Sweetman, A. M. Winer, and J.

N. Pitts, Jr., "Kinetics and Mechanisms of the Gas Phase Reactions of the NO_3 Radical with a Series of Reduced Sulfur Compounds," *J. Geophys. Res.*, in press (1985).

Mahlman, J. D., H. Levy II, and W. J. Moxim, "Three-Dimensional Tracer Structure and Behaviour as Simulated in Two Ozone Precursor Experiments," *J. Atmos. Sci.*, **37,** 655 (1980).

Marland, G. and R. M. Rotty, "Greenhouse Gases in the Atmosphere: What Do We Know?," *J. Air Pollut. Control Assoc.*, **35,** 1033 (1985).

Miller, P. R. and A. M. Winer, "Composition and Dominance in Los Angeles Basin Urban Vegetation," *Urban Ecol.*, **8,** 29 (1984).

Moyers, J. L. and R. A. Duce, "Gaseous and Particulate Iodine in the Marine Atmosphere," *J. Geophys. Res.*, **77,** 5229 (1972).

Niki, H., P. D. Maker, C. M. Savage, and L. P. Breitenbach, "An FTIR Study of the Mechanism for the Reaction HO + CH_3SCH_3," *Int. J. Chem. Kinet.*, **15,** 647 (1983a).

Niki, H., P. D. Maker, C. M. Savage, and L. P. Breitenbach, "Spectroscopic and Photochemical Properties of CH_3SNO," *J. Phys. Chem.*, **87,** 7 (1983b).

Niki, H., P. D. Maker, C. M. Savage, and L. P. Breitenbach, "A Fourier Transform Infrared Study of the Kinetics and Mechanism for the Reaction HO + CH_3OOH," *J. Phys. Chem.*, **87,** 2190 (1983c).

Niki, H., P. D. Maker, C. M. Savage, and L. P. Breitenbach, "Atmospheric Ozone–Olefin Reactions," *Environ Sci. Technol.*, **17,** 312A (1983d).

Noxon, J. F., "NO_3 and NO_2 in the Mid-Pacific Troposphere," *J. Geophys. Res.*, **88,** 11,017 (1983).

Panter, R. and R.-D. Penzhorn, "Alkyl Sulfonic Acids in the Atmosphere," *Atmos. Environ.*, **14,** 149 (1980).

Platt, U., A. M. Winer, H. W. Biermann, R. Atkinson, and J. N. Pitts, Jr., "Measurement of Nitrate Radical Concentrations in Continental Air," *Environ. Sci. Technol.*, **18,** 365 (1984).

Ramanathan, V., R. J. Cicerone, H. B. Singh, and J. T. Kiehl, "Trace Gas Trends and Their Potential Role in Climate Change," *J. Geophys. Res.*, **90,** 5547 (1985).

Rasmussen, R. A., A Review of the Natural Hydrocarbon Issue, in *Atmospheric Biogenic Hydrocarbons,* Vol. 1, *Emissions*, J. J. Bufalini and R. R. Arnts, Eds., Ann Arbor Science Publishers, Ann Arbor, MI, 1981, pp. 3–14.

Rasmussen, R. A. and M. A. K. Khalil, "Atmospheric Methane (CH_4): Trends and Seasonal Cycles," *J. Geophys. Res.*, **86,** 9826 (1981).

Rasmussen, R. A. and M. A. K. Khalil, "Global Production of Methane by Termites," *Nature*, **301,** 700 (1983).

Rasmussen, R. A. and M. A. K. Khalil, "Atmospheric Methane in the Recent and Ancient Atmospheres: Concentrations, Trends, and Interhemispheric Gradient," *J. Geophys. Res.*, **89,** 11,599 (1984).

Rasmussen, R. A., M. A. K. Khalil, and S. D. Hoyt, "Methane and Carbon Monoxide in Snow," *J. Air Pollut. Control Assoc.*, **32,** 176 (1982a).

Rasmussen, R. A., M. A. K. Khalil, R. Gunawardena, and S. D. Hoyt, "Atmospheric Methyl Iodide (CH_3I)," *J. Geophys. Res.*, **87,** 3086 (1982b).

Robbins, R. C., L. A. Cavanagh, and L. J. Salas, "Analysis of Ancient Atmospheres," *J. Geophys. Res.*, **78,** 5341 (1973).

Roberts, J. M., C. J. Hahn, F. C. Fehsenfeld, J. M. Warnock, D. L. Albritton, and R. E. Sievers, "Monoterpene Hydrocarbons in the Nighttime Troposphere," *Environ. Sci. Technol.*, **19,** 364 (1985).

Saltzman, E. S., D. L. Savoie, J. M. Prospero, and R. G. Zika, "Atmospheric Methanesulfonic Acid and Non-Sea-Salt Sulfate at Fanning and American Samoa," *Geophys. Res. Lett.*, **12,** 437 (1985).

Sanadze, G. A. and A. M. Kalandadze, "Light and Temperature Curves of the Evolution of C_5H_8," *Fiziol. Rast. Moscow*, **13,** 411 (1966a).

Sanadze, G. A. and A. M. Kalandadze, "Elimination of C_5H_8 Diene from Leaves of Poplar Under Various Conditions of Illumination," *Dokl. Akad. Nauk SSSR*, **168,** 227 (1966b).

Schjoldager, J., "Ambient Ozone Measurements in Norway, 1975–1979," *J. Air Pollut. Control Assoc.*, **31,** 1187 (1981).

Schjoldager, J., B. Sivertsen, and J. E. Hanssen, "On the Occurrence of Photochemical Oxidants at High Latitudes," *Atmos. Environ.*, **12,** 2461 (1978).

Singh, H. B. and P. L. Hanst, "Peroxyacetyl Nitrate (PAN) in the Unpolluted Atmosphere: An Important Reservoir for Nitrogen Oxide," *Geophys. Res,. Lett.*, **8,** 941 (1981).

Singh, H. B. and L. J. Salas, "Peroxyacetyl Nitrate in the Free Troposphere," *Nature*, **302,** 326 (1983)

Singh, H. B., F. L. Ludwig, and W. B. Johnson, "Tropospheric Ozone: Concentrations and Variabilities in Clean Remote Atmospheres," *Atmos. Environ.*, **12,** 2185 (1978).

Singh, H. B., W. Viezee, W. B. Johnson, and F. L. Ludwig, "The Impact of Stratospheric Ozone on Tropospheric Air Quality," *J. Air Pollut. Control Assoc.*, **30,** 1009 (1980).

Singh, H. B., L. J. Salas, and R. E. Stiles, "Methyl Halides in and Over the Eastern Pacific (40°N–32°S)," *J. Geophys. Res.*, **88,** 3684 (1983).

Stauffer, B., G. Fischer, A. Neftel, and H. Oeschger, "Increase of Atmospheric Methane Recorded in Antarctic Ice Core," *Science*, **229,** 1386 (1985).

Stedman, D. H. and R. E. Shetter, "The Global Budget of Atmospheric Nitrogen Species," *Adv. Environ. Sci. Technol.*, **12,** 411 (1983).

Stephens, E. R., "Tropospheric Methane: Concentrations Between 1963 and 1970," *J. Geophys. Res.*, in press (1985).

Tingey, D. T., M. Manning, L. C. Grothhaus, and W. F. Burns, "The Influence of Light and Temperature on Isoprene Emission Rates from Live Oak," *Physiol Plant.*, **47**, 112 (1979).

Tingey, D. T., M. Manning, L. C. Grothaus, and W. F. Burns, "The Influence of Light and Temperature on Monoterpene Emission Rates from Slash Pine," *Plant Physiol.*, **65,** 797 (1980).

Turco, R. P., Stratospheric Ozone Perturbation, in *Stratospheric Ozone*, R. C. Whitten and S. S. Prasad, Eds., Van Nostrand-Reinhold, New York, 1984, Chap. 6.

Viezee, W., W. B. Johnson, and H. B. Singh, "Stratospheric Ozone in the Lower Troposphere—II. Assessment of Downwind Flux and Ground-Level Impact," *Atmos. Environ.*, **17,** 1979 (1983).

Vukovich, F. M., J. Fishman, and E. V. Browell, "The Reservoir of Ozone in the Boundary Layer of the Eastern United States and Its Potential Impact on the Global Tropospheric Ozone Budget," *J. Geophys. Res.*, **90,** 5687 (1985).

Wang, W. C., Y. L. Yung, A. A. Lacis, T. Mo, and J. E. Hansen, "Gereenhouse Effects Due to Man-made Perturbations of Trace Gases," *Science*, **194,** 685 (1976).

Went, F. W., "Blue Hazes in the Atmosphere," *Nature*, **187,** 641 (1960).

Westberg, H. H., Biogenic Hydrocarbon Measurements, in *Atmospheric Biogenic Hydrocarbons,* Vol. 2, *Ambient Concentrations and Atmospheric Chemistry*, J. J. Bufalini and R. R. Arnts, Eds., Ann Arbor Science Publishers, Ann Arbor, MI, 1981, pp. 25–49.

Wine, P. H., N. M. Kneutter, C. A. Gump, and A. R. Ravishankara, "Kinetics of OH Reactions with the Atmospheric Sulfur Compounds H_2S, CH_3SH, CH_3SCH_3, and CH_3SSCH_3," *J. Phys. Chem.*, **85,** 2660 (1982).

Winer, A. M., M. C. Dodd, D. R. Fitz, P. R. Miller, E. R. Stephens, K. Neisess, M. Meyers, D. E. Brown, and C. W. Johnson, "Assembling a Vegetative Hydrocarbon Emission Inventory for the California South Coast Air Basin: Direct Measurement of Emission Rates, Leaf Biomass, and Vegetative Distribution," Paper No. 82-51.6, Presented at the 75th Annual Meeting of the Air Pollution Control Association, New Orleans, LA, June 20–25, 1982.

Winer, A. M., R. Atkinson, and J. N. Pitts, Jr., "Gaseous Nitrate Radical: Possible Nighttime Atmospheric Sink for Biogenic Organic Compounds," *Science*, **224,** 156 (1984).

Yokouchi, Y. and Y. Ambe, "Aerosols Formed from the Chemical Reaction of Monoterpenes and Ozone," *Atmos. Environ.*, **19,** 1271 (1985).

Zafiriou, O. C., "Photochemistry of Halogens in the Marine Atmosphere," *J. Geophys. Res.*, **79,** 2730 (1974).

Zimmerman, P. R., "Testing of Hydrocarbon Emissions from Vegetation Leaf Litter and Aquatic Surfaces, and Development of a Methodology for Compiling Biogenic Emission Inventories," EPA Report No. 450/4-79-004, 1979.

Zimmerman, P. R., J. P. Greenberg, S. O. Wandiga, and P. J. Crutzen, "Termites: A Potentially Large Source of Atmospheric Methane, Carbon Dioxide, and Molecular Hydrogen," *Science*, **218,** 563 (1982).

PART 9

Impact of Tropospheric Chemical Processes on the Stratosphere

15 Interactions Between Tropospheric and Stratospheric Chemistry

The chemistry of oxygen species ($O_x = O_2$, O, O_3) in the upper atmosphere was first quantitatively treated in 1930 by Chapman who suggested that reactions (1)–(4), now known as the *Chapman cycle*, were important:

$$O_2 + h\nu\ (\lambda \lesssim 220\ \text{nm}) \rightarrow 2\ O \tag{1}$$

$$O + O_2 \xrightarrow{M} O_3 \tag{2}$$

$$O + O_3 \rightarrow 2\ O_2 \tag{3}$$

$$O_3 + h\nu \rightarrow O + O_2 \tag{4}$$

The importance of interactions between O_x species and hydrogen- and nitrogen-containing species was later recognized (Bates and Nicolet, 1950; Crutzen, 1970). However, it was only with the suggestion by Johnston (1971) that NO_x emissions injected into the stratosphere from the proposed fleet of SST's could cause a reduction in ozone that research into the chemistry of the stratosphere began to flourish. This interest in the upper atmosphere was spurred on by the hypothesis of Molina and Rowland in 1974 that chlorofluoromethanes released at the earth's surface could diffuse into the stratosphere and also cause a reduction in the ozone concentration.

Since then, research efforts to define the chemistry of the natural stratosphere and the effects of anthropogenic perturbations on it have multiplied. There are, of course, significant differences between conditions in the troposphere, on which this book has focused, and those in the stratosphere. Thus, as seen in Section 3-A-1 (Fig. 3.2 and Table 3.2, Chapter 3), ultraviolet light down to ~180 nm penetrates into the stratosphere and is absorbed by O_2 and O_3; light absorption by O_3 is primarily responsible for the increase in temperature with altitude which characterizes the stratosphere. Additionally, the total pressure in the stratosphere is lower, ~1–100 Torr, and the temperature range is ~210–275°K (see Fig. 1.2, Chapter 1).

Despite these differences, much of our knowledge of tropospheric chemistry can be applied to the stratosphere. Many of the reactions occurring in the stratosphere are either analogous or, in some cases, identical to those occurring in the troposphere; thus the same basic kinetic and mechanistic principles discussed throughout this book for tropospheric reactions apply.

In this chapter, we do not attempt a detailed review of stratospheric chemistry. Indeed, given the explosion of research in this area over the last decade, an entire book could in itself be devoted to this subject. Rather, we concentrate here on the relationship between the chemistry of species in the troposphere and in the stratosphere. As we shall see, compounds that are long-lived in the troposphere can diffuse into the stratosphere and contribute to the chemistry there; conversely, those species having short tropospheric lifetimes do not survive long enough to reach the stratosphere and hence do not play a role in the chemistry of the upper atmosphere.

For detailed reviews of stratospheric chemistry, the reader should consult the National Research Council (1984), and the National Aeronautics and Space Administration documents (1984), the kinetics summaries (DeMore et al., 1983, 1985), and the reviews by Rowland and Molina (1975), Chang and Duewer (1979), Crutzen (1979), Turco (1985a,b), Whitten and Prasad (1985), Wayne (1985), and Bower and Ward (1982). The anticipated effects of stratospheric ozone perturbations on earth are discussed in the earlier National Research Council reviews (1973, 1975, 1976, 1977, 1979, 1982).

15-A. CHLOROFLUOROCARBONS

Chlorofluorocarbons, that is, compounds containing chlorine, fluorine, carbon, and possibly hydrogen, have been used extensively in the industrialized nations in the past decades primarily as propellants in aerosol spray cans, as refrigerants, and as blowing agents, for example, for producing polyurethane foam. Their chemical characteristics have made them ideally suited for such uses in that they are generally non-toxic and chemically inert. Thus they can be used around open flames, and leaks in refrigeration units do not present a health hazard as older units operated on coolants such as SO_2 once did.

The principal chlorofluorocarbons in use are CCl_3F, CCl_2F_2, and $CHClF_2$. These are often referred to as F-11, F-12, and F-22, respectively, after the Dupont trade name Freon; alternatively, the abbreviation CFC-11, -12, or -22 is used for chlorofluorocarbon. The numbers following the "F" reflect the number of hydrogen and fluorine atoms, respectively. Thus the first number is the number of hydrogen atoms, plus one, and the second gives the number of fluorine atoms. For chlorofluorocarbons containing two or more carbon atoms, a three digit numbering system is used; the first digit gives the number of carbons minus one, the second the number of hydrogens plus one, and the third the number of fluorine atoms. Thus CCl_2FCClF_2 is F-113 and CCF_2CClF_2 is F-114.

Table 15.1 gives the distribution of 1977 worldwide and U.S. sales of F-11 and F-12, which are used most extensively, broken down into aerosol propellants, refrigerants, and other uses (e.g., blowing agents). The use for aerosol propellants has been decreasing in recent years due to concern over their potential effects on the stratospheric ozone layer (see below). On the other hand, their use in refrigeration units and as blowing agents has been increasing. It is seen from Table 15.1 that, on a global basis, the use as aerosol propellants accounted for almost half of the sales of F-11 and F-12 in 1977. Essentially all of this is released into the atmosphere, along with some fraction of the compounds put to other uses.

Figure 15.1 shows the estimated annual release rates of F-11 and F-12 from 1952 to 1980. Over this period of about three decades, the release rate into the atmosphere increased dramatically until about 1974, as these compounds have found increasing use in our industrialized society. The measured atmospheric concentrations have also been increasing; Table 15.2 shows the rate of increase in the tropospheric concentrations of F-11, F-12, and F-22, as well as three other potentially important halocarbons, methyl chloroform (CH_3CCl_3), carbon tetrachloride (CCl_4), and methyl chloride (CH_3Cl) (see below). Although the release rates of F-11, F-12, and F-22 temporarily leveled off in the mid-1970s, the concentrations in ambient air continued to rise because steady state had not been reached in the atmosphere (see below).

The chlorofluorocarbons (CFCs) have very long lifetimes in the troposphere. This is a consequence of the fact that they do not absorb light of wavelengths above 290 nm and do not react at significant rates with O_3 and OH. While rate constants for their reactions with NO_3 have not been determined, it is unlikely that these are significant either. In addition to the lack of chemical sinks, there do not appear to be substantial physical sinks; thus they are not very soluble in water and hence are

TABLE 15.1. Distribution of 1977 Sales of Chlorofluorocarbons (F-11 and F-12) in 10^3 Metric Tons and as a Percentage of the Total (Given in Parentheses)

	Use			
Country	Aerosol Propellants	Refrigerants	Other	Total
United States	84.2 (12.0%)	99.9 (14.3%)	64.5 (9.2%)	248.6 (35.5%)
Other Countries	255.7 (36.5%)	79.5 (11.4%)	116.1 (16.6%)	451.0 (64.5%)
World	339.9 (48.6%)	179.4 (25.6%)	180.6 (25.8%)	699.6 (100%)

Source: National Research Council, 1979.

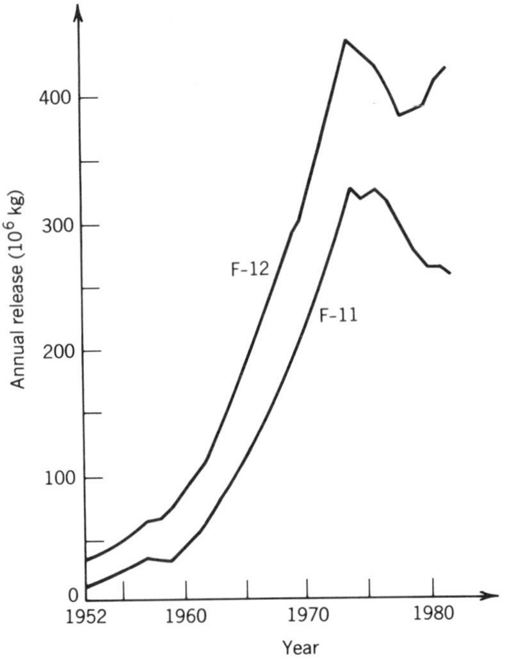

FIGURE 15.1. Estimated annual worldwide releases of F-11 and F-12 from 1952 to 1980. Data from Chemical Manufacturers Association (from National Research Council, 1984).

not removed rapidly by rainout. While laboratory studies have shown that some of the CFCs decompose on exposure to visible and near UV present in the troposphere when the compounds are adsorbed on siliceous materials such as sand (Ausloos et al., 1977; Gäb et al., 1977, 1978), the lifetimes for F-11 and F-12 with respect to these processes have been estimated to be ~540 yr and 1800 yr, re-

TABLE 15.2. Rate of Increase of Some Chlorofluorocarbons and Halocarbons in the Troposphere

Compound	Rate of Increase in Concentration (% yr^{-1})
$CFCl_3$ (F-11)	6
CF_2Cl_2 (F-12)	6
CHF_2Cl (F-22)	>6
CH_3CCl_3	9
CCl_4	2
CH_3Cl	0

Source: National Research Council, 1984.

spectively (National Research Council, 1979). Similarly, an observed thermal decomposition when adsorbed on sand appears to be an insignificant loss process under atmospheric conditions.

As a result, CFCs reside in the troposphere for years, slowly diffusing up across the tropopause into the stratosphere. The lifetime of CFCs in the atmosphere can be estimated using a mass balance approach. Knowing the atmospheric concentrations of CFCs, one can calculate the total amount in the atmosphere. This amount must be the result of a balance between emissions into and loss from the atmosphere. If the emission rates are known, the loss rate required to give the observed atmospheric concentrations can be calculated, and from this, a lifetime obtained (see Section 4-A-6).

Such calculations may be based on either the absolute atmospheric concentrations of CFCs or, alternatively, on the observed relative rates of change in the concentrations. From such calculations, the lifetimes of F-11 and F-12 in the global atmosphere, that is, the time to diffuse to the stratosphere and undergo photolysis, appear to be in the range of 50–150 yr (National Research Council, 1984).

Since the wavelength distribution of solar radiation shifts to shorter wavelengths with increasing altitude (see Section 3-A-2e), the CFCs eventually become exposed to wavelengths of light which they can absorb.

Figure 15.2 shows the absorption cross sections of some halogenated methanes from 160 to 280 nm, and Tables 15.3 and 15.4 give the absorption cross sections

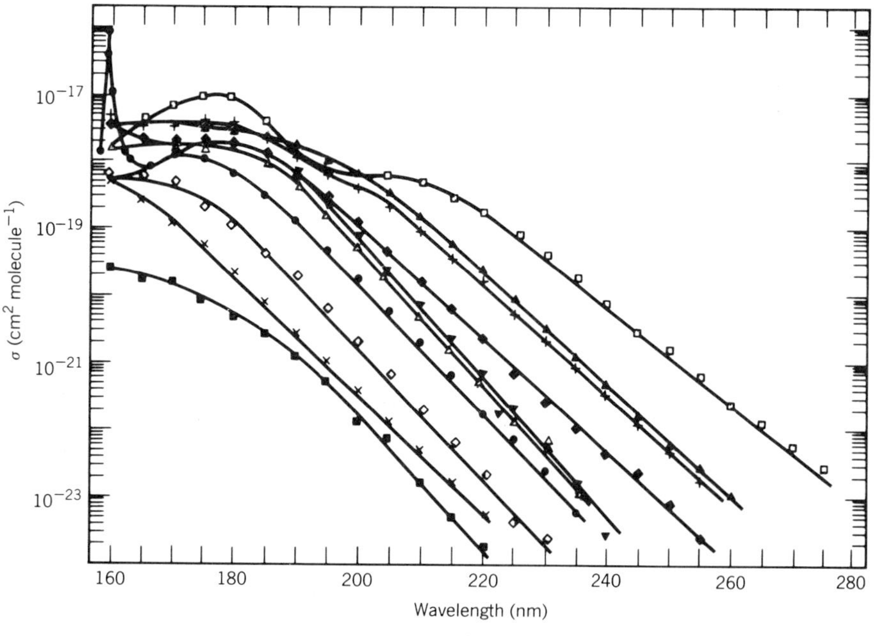

FIGURE 15.2. Semi-logarithmic plot of the absorption cross sections of the halogenated methanes at 298°K: +, $CHCl_3$; △, $CHCl_2F$; ×, $CHClF_2$; ♦, CH_2Cl_2; ◇, CH_2ClF; ●, CH_3Cl; □ CCl_4; ▲, CCl_3F (F-11); ▼, CCl_2F_2 (F-12); ■, $CClF_3$ (from Hubrich and Stuhl, 1980).

TABLE 15.3. Absorption Cross Sections for CCl_3F (F-11)

λ(nm)	$10^{20}\sigma$, base *e* (cm^2 $molecule^{-1}$)	λ(nm)	$10^{20}\sigma$, base *e* (cm^2 $molecule^{-1}$)
170	316	208	21.2
172	319	210	15.4
174	315	212	10.9
176	311	214	7.52
178	304	216	5.28
180	308	218	3.56
182	285	220	2.42
184	260	222	1.60
186	233	224	1.10
188	208	226	0.80
190	178	228	0.55
192	149	230	0.35
194	123	235	0.126
196	99	240	0.0464
198	80.1	245	0.0173
200	64.7	250	0.00661
202	50.8	255	0.00337
204	38.8	260	0.00147
206	29.3		

Source: DeMore et al., 1985.

TABLE 15.4. Absorption Cross Sections for CCl_2F_2 (F-12)

λ(nm)	$10^{20}\sigma$, base *e* (cm^2 $molecule^{-1}$)	λ(nm)	$10^{20}\sigma$, base *e* (cm^2 $molecule^{-1}$)
170	124	200	8.84
172	151	202	5.60
174	171	204	3.47
176	183	206	2.16
178	189	208	1.32
180	173	210	0.80
182	157	212	0.48
184	137	214	0.29
186	104	216	0.18
188	84.1	218	0.12
190	62.8	220	0.068
192	44.5	225	0.022
194	30.6	230	0.0055
196	20.8	235	0.0016
198	13.2	240	0.00029

Source: DeMore et al., 1985.

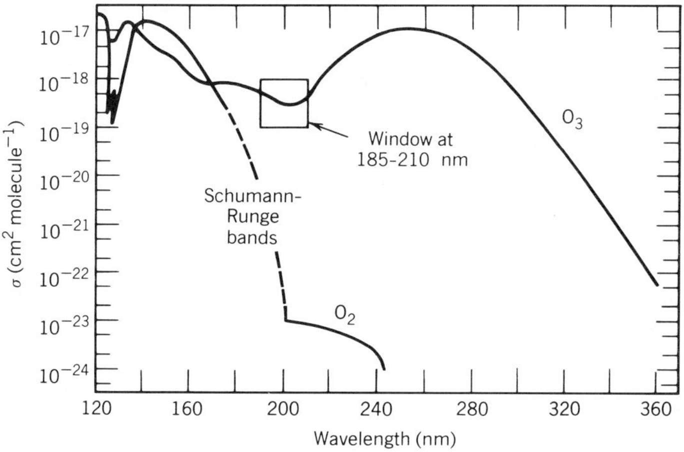

FIGURE 15.3. Absorption cross sections for O_2 and O_3 from 120 to 360 nm, showing the window from ~185 to 210 nm (from Rowland and Molina, 1975).

for F-11 and F-12 from 170 to 260 nm. The absorptions become very weak beyond ~240 nm in the case of F-11, and 220 nm in the case of F-12. Recall from Chapter 3 that both O_3 and O_2 absorb radiation in the ultraviolet. Figure 15.3 shows these absorption cross sections for O_2 and O_3 from 120 to 360 nm; there is a *window* in the overlapping O_2 and O_3 absorptions from ~185 to 210 nm, that is, a region where the total light absorption is in a shallow minimum. By coincidence, this is the region in which the CFCs absorb light (Fig. 15.2).

The C—Cl bond dissociation energy in CF_2Cl_2 is 76 kcal mole^{-1}, whereas that for the strong C—F bond is 110 kcal mole^{-1}. As a result upon light absorption, the weaker C—Cl bond breaks:

$$CF_2Cl_2 + h\nu\ (\lambda \lesssim 240\ \text{nm}) \rightarrow CF_2Cl + Cl \qquad (5)$$

The Cl atom released then reacts in a catalytic chain reaction which leads to the destruction of O_3:

$$Cl + O_3 \rightarrow ClO + O_2 \qquad (6)$$

$$ClO + O \rightarrow Cl + O_2 \qquad (7)$$

$$\text{Net:}\quad O_3 + O \rightarrow 2O_2$$

15-B. COUPLING OF ClO_x CHEMISTRY WITH CH_4, NO_x, and HO_x

The chemistry of the stratosphere involves closely coupled oxygen/hydrogen/chlorine/nitrogen chemistry. There are three major approaches to investigating the chemistry of the stratosphere and the coupling between the various cycles:

1. Laboratory measurements of the rate constants and mechanisms for the fundamental reactions.
2. Computer modeling studies using the data from approach 1 to predict the concentrations of various intermediates and products and the changes in them resulting from anthropogenic emissions such as CFCs.
3. Most importantly, measurements of the concentrations of various species involved in the chemistry, including atoms such as $O(^3P)$ and free radicals such as ClO. Most of the key species postulated in the stratosphere have now been detected and measured (e.g. see National Aeronautics and Space Administration, 1984; National Research Council, 1984; Anderson et al. 1985; Brune et al., 1985; Heaps and McGee, 1985; Larsen et al., 1985).

Before examining some of the model predictions, let us briefly review some of the trace substances believed to be coupled via chemical cycles to CFCs. These include CH_4, NO_x, and N_2O.

As discussed in Section 14-B-1, a small amount ($\lesssim 15\%$) of the CH_4 released at the earth's surface diffuses into the stratosphere where it can act as a terminator for the Cl atom catalyzed destruction of O_3, by forming HCl:

$$CH_4 + Cl \rightarrow CH_3 + HCl \tag{8}$$

Although the HCl can recycle to Cl via reactions with OH,

$$HCl + OH \rightarrow H_2O + Cl \tag{9}$$

some of it diffuses down through the tropopause and is rained out, thereby acting as a permanent Cl sink. The observed increase in CH_4 in the troposphere may therefore result in lowered stratospheric Cl atom concentrations and hence decreased Cl atom destruction of O_3. CH_4 may also contribute to O_3 production in the stratosphere independent of the CFC cycle via the CH_4–OH–NO_x chemistry discussed throughout this book (e.g., see Herman and McQuillan, 1985).

As discussed at the beginning of this chapter, Johnston (1971) drew attention to the potential effects on stratospheric O_3 concentrations of injecting NO_x from supersonic transports (SSTs). NO emitted from SSTs, as well as from subsonic commercial aircraft, can react in a catalytic cycle similar to that subsequently proposed for Cl atoms:

$$NO + O_3 \rightarrow NO_2 + O_2 \tag{10}$$

$$NO_2 + O \rightarrow NO + O_2 \tag{11}$$

$$\text{Net:} \quad O_3 + O \rightarrow 2O_2$$

However, the effect of NO_x emissions on the O_3 concentration depends critically on the altitude of injection. Thus injection at high altitudes, that is, in the mid-

and upper stratosphere, will lead to net O_3 destruction via reactions (10) and (11). However, the opposite effect, that is, a net *increase* in O_3, is predicted if NO is injected into the lower stratosphere and upper troposphere. This seemingly paradoxical result is due to competing processes for the NO_2 formed in (10). In the mid- and upper stratosphere, reaction (11) dominates the NO_2 loss, so that two "odd oxygen" species—O_3 and O—are destroyed in the cycle consisting of (10) and (11). However, in the lower stratosphere and upper troposphere, the O atom concentrations are smaller and the photolysis of NO_2 to form O, which then produces O_3, becomes competitive:

$$NO_2 + h\nu \rightarrow NO + O \tag{12}$$

$$O + O_2 \xrightarrow{M} O_3 \tag{2}$$

Nitrous oxide (N_2O) is emitted in biological processes at the earth's surface, and its emission rate may be modified by anthropogenic activities, for example, the increasing use of fertilizers (National Research Council, 1984). It has no known significant tropospheric reactions, except possibly decomposition when adsorbed on surfaces such as sand; however, this appears to be too slow to be a significant loss process (Rebbert and Ausloos, 1978). Hence, like the CFCs, N_2O slowly diffuses into the stratosphere. Here it either photolyzes to $N_2 + O$ or is converted into NO via reaction (13):

$$O(^1D) + N_2O \underset{a}{\rightarrow} N_2 + O_2$$
$$\underset{b}{\rightarrow} 2NO \tag{13}$$

The NO produced in (13b) can then react via (10) and (11) above to destroy odd oxygen. Since the production of NO, reaction (13b), occurs primarily between 20 and 40 km (National Research Council, 1984), the net effect is a decrease in O_3.

NO_x is closely coupled to the ClO_x reactions. Thus the ClO formed in reaction (6) reacts with NO_2 to produce chlorine nitrate in essentially 100% yield (Burrows et al., 1985),

$$ClO + NO_2 \xrightarrow{M} ClONO_2 \tag{14}$$

which acts as a temporary reservoir for chlorine at night. At dawn, $ClONO_2$ photolyzes, reforming chlorine atoms:

$$ClONO_2 + h\nu \rightarrow Cl + NO_3 \tag{15}$$

Reactions (14) and (15) have been shown in modeling studies (Ko and Sze, 1984) to predict a diurnal variation of ClO consistent with Cl stratospheric observations (Solomon et al., 1984).

In addition to these interactions between ClO_x and NO_x chemistry, there are also important ClO_x–HO_x interactions. Thus Cl can be converted to HCl by reaction (8) with CH_4 and also by reaction with HO_2:

$$Cl + HO_2 \rightarrow HCl + O_2 \tag{16}$$

The reaction of ClO with HO_2 forms HOCl which, along with HCl and $ClONO_2$, is another temporary reservoir for chlorine atoms:

$$ClO + HO_2 \rightarrow HOCl + O_2 \tag{17}$$

However, like $ClONO_2$, HOCl photolyzes to form chlorine atoms.

Figure 15.4 summarizes these interactions between the ClO_x, NO_x, and HO_x chemistry in the stratosphere, and Fig. 15.5 schematically shows the interactions between these species in the troposphere and stratosphere.

Because of this close coupling of the varous chemical cycles, as new data from laboratory experiments have become available, the changing inputs to the models have resulted in substantial changes in the predicted change in column O_3. For example, Fig. 15.6 shows the estimates of the percentage reductions in the steady state total column ozone (i.e., the total O_3 in a column from the earth's surface to the top of the atmosphere) for continuous releases of CFCs at the 1975 emission rates, as a function of year from Molina and Rowland's first paper in 1974 up to 1983. As new inputs have been used with the models, the estimated changes in the total columnar O_3 abundance have been significant. Furthermore, it is evident that the absolute values of the predicted change in column ozone vary from model to model; for example, Wuebbles (1985) has shown that the Lawrence Livermore National Laboratory one-dimensional model predicts *increases* in total column ozone for several years around 1980 as the chemistry input to the model changes.

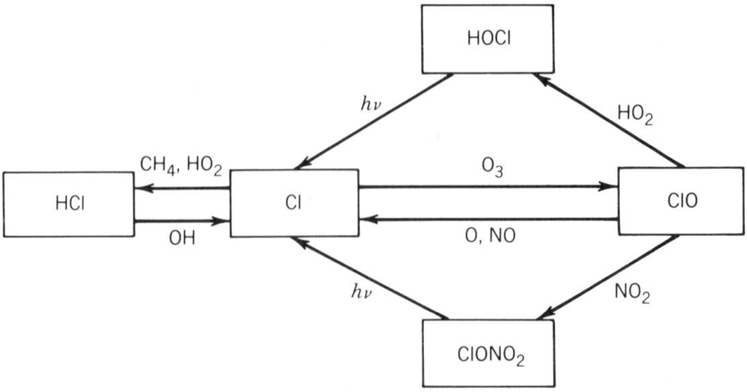

FIGURE 15.4. Major reactions of ClO_x in the stratosphere, showing the interaction between ClO_x, NO_x, and HO_x.

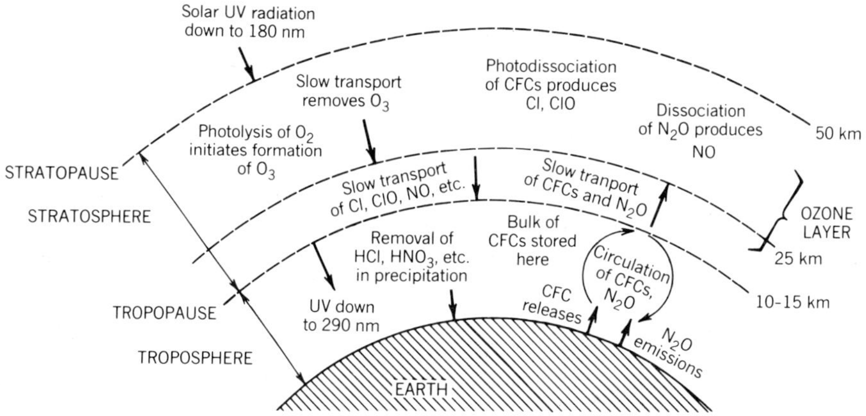

FIGURE 15.5. Schematic diagram of the processes determining the concentration of ozone in the stratosphere (from National Research Council, 1982).

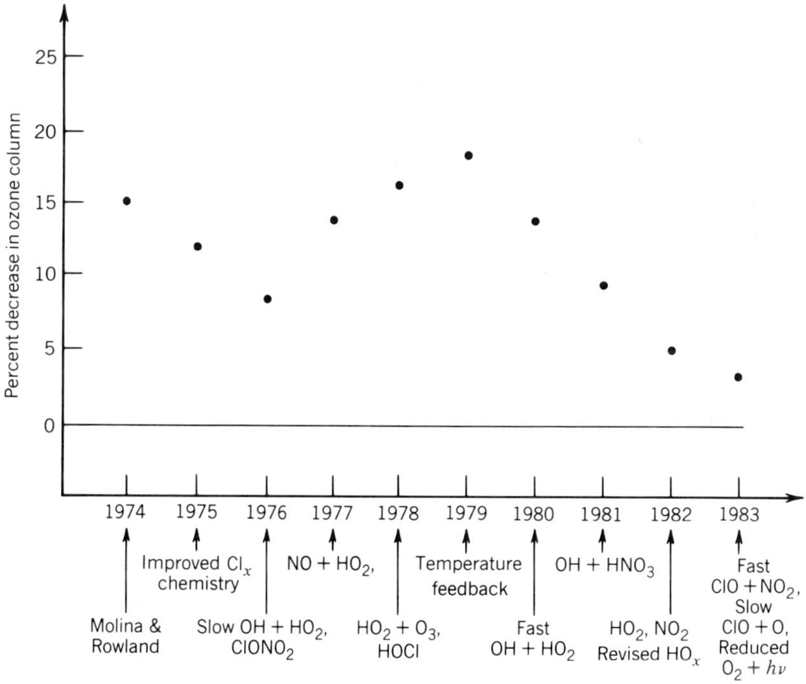

FIGURE 15.6. Estimates of steady state reductions in total column ozone for continuous releases of CFCs at approximately 1975 rates as calculated from chemical models with different input chemistry. Changes in the models and simulation techniques are indicated in chronological order (from National Research Council, 1984; adapted from Turco, 1985a).

To illustrate the coupling between the chemical cycles and how this alters the model predictions, let us take one of the more significant changes—the change in the HO_2 + NO rate constant—and examine why this changed the predicted column abundance of O_3.

Prior to 1973, the room temperature rate constant reported for the NO + HO_2 reaction was $\leq 2 \times 10^{-15}$ cm^3 $molecule^{-1}$ s^{-1}. However, measurements carried out from 1973 to 1977 gave increasing values, with the largest being 8.1×10^{-12} cm^3 $molecule^{-1}$ s^{-1} reported by Howard and Evenson in 1977; this value has since been confirmed by a number of other research groups. When this revised, larger rate constant (including its temperature dependence to correct it to the lower temperatures in the stratosphere) was used as a model input, an *increase* in the predicted loss of total column O_3 due to CFCs, but a *decrease* due to NO_x, resulted (Crutzen and Howard, 1978). This occurs because the rapid removal of NO by reaction with HO_2 shifts the fate of NO away from its reaction (10) with O_3 which destroys O_3. It also changes the $[NO_2]/[NO]$ ratio in favor of NO_2:

$$HO_2 + NO \rightarrow OH + NO_2 \tag{18}$$

Part of the NO_2 then photolyzes, ultimately producing O_3:

$$NO_2 + h\nu \rightarrow NO + O \tag{12}$$

$$O + O_2 \xrightarrow{M} O_3 \tag{2}$$

On the other hand, the increased OH produced in reaction (18) increases the rate of recycling of the HCl back to the catalytically active Cl atom:

$$OH + HCl \rightarrow H_2O + Cl \tag{9}$$

As seen in Fig. 15.6, the change in this one rate constant in the NO_x–HO_x cycle had a dramatic effect on the predicted changes in the total column O_3 concentration, illustrating the importance of the coupled chemistry.

Table 15.5 shows the results of one-dimensional model calculations of the effect of a steady state release of the individual gases discussed above on the total O_3 column, assuming that the gases are acting individually. Also given in Table 15.5 are the estimated atmospheric lifetimes of the pollutants and their principal mode of removal. It is noteworthy that halocarbons other than F-11 and F-12 are thought to contribute significantly to ozone depletion. For example, even though F-22 ($CHClF_2$) has an abstractable hydrogen atom and can thus be removed in the troposphere by reaction with OH, much of what is released survives to reach the stratosphere. Indeed, F-22 is thought to be exceeded only by F-12 and CH_3Cl (see below) in its contribution to the rate of formation of chlorine atoms at altitudes above 40 km, and may in the future become the dominating source if the current global rate of increase of 16% per year in its emissions continues (Fabian et al., 1985).

TABLE 15.5. Sensitivity of Total Column Ozone to Perturbing Influences Assuming Only That Perturbation is Occurring

Trace Gas and Magnitude of Perturbation	Typical Estimate of Ozone Column Change[a] (%)	Pollutant Atmospheric Lifetime (yr)	Principal Mode of Pollutant Removal
CFC-11 and 12 (1980 release rates)	−2 to −4	50–150	Photolysis of CFCs in middle stratosphere
Other halocarbons (2 ppbv Cl_x increase)	~ −1	1–15	Decomposition of tropospheric chlorocarbon reservoir by reaction with OH
Subsonic aircraft (2×10^9 kg NO_2 per yr at 12 km)	~ +1	<1	Conversion to nitric acid and removal at the surface
N_2O (20% increase by 2050)	~ −4	100	Photolysis in stratosphere
CH_4 (doubling)	~ +3	10	Reaction with tropospheric OH
CO_2 (doubling)	+3 to +6	?	Uptake by oceans, sediment, and biosphere

Source: National Research Council, 1984; estimated using one-dimensional photochemical models.
[a]Steady state change.

The only species included in Table 15.5 which has not yet been treated is carbon dioxide. CO_2 can interact with the HO_x–NO_x–ClO_x cycles through its effect on the radiation balance. Although CO_2 acts to warm the troposphere by absorbing infrared (IR) radiation (the greenhouse effect) and converting it to heat, due to a complicated balance between emission and absorption, it has the opposite effect in the stratosphere (National Research Council, 1979, 1984).

Increases in CO_2 due to fossil fuel combustion are thus expected to warm the troposphere, but cool the stratosphere. This cooling effect changes the kinetics of the chemical reactions in the stratosphere (see Section 4-A-7). The net effect of an increase in CO_2 on the kinetics is expected to result in an increase in the O_3 concentration. However, the absorption of light by O_3 in the stratosphere produces heat (see Section 1-B); this increased heating from increased O_3 may partially offset the decreased temperature due to CO_2.

It is important to recognize that the time scale for altering the O_3 column is quite different for the different species in Table 15.5. For example, the long time required for CFCs to diffuse from the earth's surface into the stratosphere means that CFCs released today will have their effect on the O_3 many years later. On the other hand, NO_x released from aircraft directly into the upper atmosphere causes effects that are essentially immediate.

This difference presents difficult control strategy choices in the case of CFCs. Thus if controls on CFC releases are delayed until a real change is detected in the O_3 column, much more severe losses can be anticipated to occur many decades later from CFCs already released into the troposphere but which have not yet reached the stratosphere. On the other hand, the economic cost of acting before confirming that significant O_3 losses will occur can be substantial.

Complicating the control issue is the close coupling of the chemistry of O_x–HO_x–NO_x–ClO_x cycles in the stratosphere. As seen from the predicted individual effects of various gases in Table 15.5 and the earlier discussion, some are expected to increase the total column O_3, and some to decrease it. Thus there may be counterbalancing effects of various species. Indeed, some model results in which the chemistry and increased release rates of some or all of the gases in Table 15.5 have been taken into account suggest that the net effect could even be an *increase* in the total O_3 (National Research Council, 1984)!

One significant result of the modeling effects, however, is that even if a net decrease in total O_3 does not occur, due to counterbalancing effects, the vertical distribution of O_3 may change. Thus, at the present time, it is expected that large decreases (of the order of 40–50%) in O_3 will occur at ~40 km with small increases ($\lesssim 5\%$) occurring in the lower stratosphere, ~20 km. However, these increases are relative to existing O_3 concentrations at those altitudes; since the total pressure, that is, concentration of gas molecules, decreases with altitude, the total *absolute* decrease in the number of O_3 molecules in the upper stratosphere is comparable to the increased number of O_3 molecules in the lower regions. The *overall* effect on the total amount of O_3 might thus be small ($\lesssim 5\%$) although the distribution of O_3 may shift. Such changes in distribution may have the potential for climate modification (National Aeronautics and Space Administration, 1984).

It should be noted that these model predictions typically assume constant release rates for CFCs. If significant increases in the release rates occur so that the concentration of chlorine exceeds that of NO_x, much larger decreases in total column O_3 are predicted by all models. Thus the effect is anticipated to be non-linear in emissions. At large chlorine concentrations the NO_x is predicted to be titrated out; at this point the effects of ClO_x will be relatively much greater (e.g. see Prather et al., 1984).

Finally, it should be noted that decreases in the total column ozone due to decreases in stratospheric ozone may be partially compensated by increases in ozone in the troposphere; for example, Logan (1985) estimates that approximately 20–30% of the decrease in stratospheric ozone over middle and high latitudes of the northern hemisphere could be compensated for by what appears to be a trend to increasing ozone in the troposphere in these areas.

15-C. OTHER HALOGEN-CONTAINING COMPOUNDS

In addition to the CFCs, a number of other halogen-containing organics are emitted into the troposphere by natural as well as anthropogenic processes. These include such compounds as methyl chloroform (CH_3CCl_3), carbon tetrachloride (CCl_4), dichloromethane (CH_2Cl_2), perchloroethene (C_2Cl_4), trichloroethene (C_2HCl_3), methyl chloride (CH_3Cl), methyl iodide (CH_3I), and methyl bromide (CH_3Br). All but the methyl halides are emitted from anthropogenic sources. For example, methyl chloroform, perchloroethene, and trichloroethene are used as solvents and degreasing agents. Carbon tetrachloride is used as a solvent, as a reagent in the manufacture of F-11 and F-12, and as a grain fumigant in marine shipments. The estimated annual emission rates in 1979 for these compounds are shown in Table 15.6, as well as for F-11 and F-12 for comparison; clearly substantial quantities of these other halocarbons are released.

The dominant source of methyl chloride in the atmosphere is believed to be natural in origin, with much of it, along with methyl iodide and methyl bromide, being produced by the oceans. Whether CH_3Cl is a product of biological processes or perhaps of the reaction of chloride ions with methyl iodide is not known (Singh et al., 1983).

An important difference between the CFCs such as F-11 and F-12 and the other compounds in Table 15.6 is the presence of abstractable hydrogen atoms and/or reactive C=C double bonds. Thus OH attack on such species is much faster than for F-11 and F-12, and, as a result, their tropospheric lifetimes are much shorter. Table 15.7 gives the room temperature rate constants for OH reactions with these molecules, as well as the corresponding lifetimes for an OH concentration of 1×10^6 radicals cm^{-3} during the day. In contrast to F-11 and F-12, whose lifetimes in the troposphere are determined by the rate of transport into the stratosphere, these compounds have finite lifetimes with respect to reaction with OH. As a result, some are removed in the troposphere, with the balance being transported into the stratosphere, where they too may produce chlorine atoms on photolysis.

TABLE 15.6. Annual Emission Rates of Some Halocarbons, as Well as F-11 and F-12 in 1979

Halocarbon	1979 Emission Rate (10^6 kg)
F-11 ($CFCl_3$)	283.6
F-12 (CF_2Cl_2)	340.9
Carbon tetrachloride (CCl_4)	118.0
Methyl chloroform (CH_3CCl_3)	522.3
Trichloroethene (C_2HCl_3)	459.1
Perchloroethene (C_2Cl_4)	667.0
Dichloromethane (CH_2Cl_2)	356.0

Source: Gidel et al., 1983.

TABLE 15.7. Rate Constants for OH Reactions at 25°C with Some Halocarbons and Corresponding Atmospheric Lifetimes Under Relatively Clean Conditions

Halocarbon	k^a (cm^3 molecule^{-1} s^{-1})	τ^b
CH_3CCl_3	1.2×10^{-14}	2.6 yr
C_2HCl_3	2.2×10^{-12}	5 days
C_2Cl_4	1.7×10^{-13}	68 days
CH_2Cl_2	1.4×10^{-13}	83 days
CH_3Cl	4.3×10^{-14}	270 days

[a]From DeMore et al., 1983, 1985.

[b]$\tau = 1/k[OH]$, where $[OH] = 1 \times 10^6$ cm^{-3}.

Modeling calculations indicate that some of these halocarbons have already contributed significantly to the depletion of O_3 in the stratosphere, and others may play important roles in the future. For example, Gidel and co-workers (1983) suggest that the naturally produced methyl chloride accounts for ~25% of the total chlorine in the stratosphere. In addition, they predict that the total O_3 depletion between 1960 and 1983 was ~3%, and, of this 3%, 2.5% was due to F-11, F-12, and CCl_4, with the remainder being due to methyl chloroform. They suggest that prior to 1970, almost all of the decrease in O_3 was due to CCl_4, and that by 1995, CCl_4 and methyl chloroform will account for about half of the total O_3 depletion due to halocarbons. (Of course, as discussed earlier, any such decrease may be counterbalanced by increases due to CH_4, NO_x, etc.).

In short, halocarbons other than the CFCs are anticipated to contribute to stratospheric O_3 depletion even though they have significant tropospheric removal rates compared to the CFCs.

Although the emphasis in the last decade has been on chlorine-containing compounds, it has been suggested that other halogens may also play a role in the interaction between tropospheric and stratospheric chemistry (see Cicerone, 1981, for a review of halogens in the atmosphere). In particular, brominated organics are emitted into the troposphere through both natural and anthropogenic processes. The major natural source is believed to be marine biological activity (Lovelock, 1975) which produces methyl bromide, the major bromine-containing organic in the remote troposphere (Singh et al., 1977). Anthropogenic sources of organic bromine compounds include their use as soil fumigants (CH_3Br), gasoline additives ($C_2H_4Br_2$), and flame retardants (e.g., $CBrF_3$ and $CBrF_2CBrF_2$).

The tropospheric chemistry of CH_3Br and other organic bromine compounds is similar to that of the chlorine-containing organics. Thus CH_3Br with its abstractable hydrogen atoms is attacked by OH in the troposphere with a mean lifetime of ~1 yr for an OH concentration of 1×10^6 cm^{-3} (Yung et al., 1980). Alternatively, it can diffuse into the stratosphere where it not only continues to react with OH but also photolyzes to produce Br atoms which participate in catalytic O_3 destruction via reactions analogous to reactions (6) and (7).

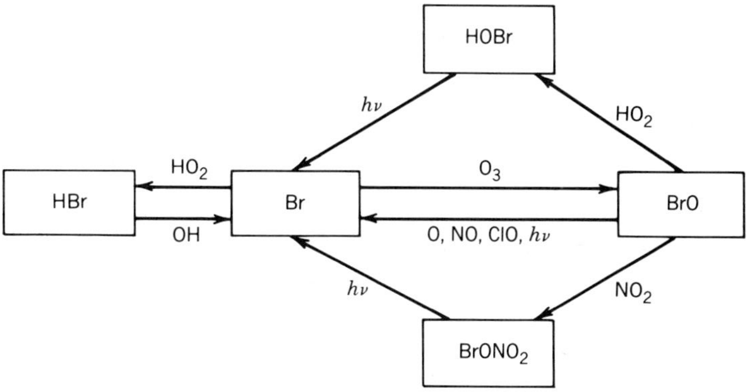

FIGURE 15.7. Schematic diagram of major bromine reactions in the stratosphere.

Figure 15.7 summarizes the major bromine atom reactions in the stratosphere. They are similar to those of chlorine (Fig. 15.4) with a few exceptions. For example, Br does not abstract hydrogen from CH_4, as the reaction is endothermic; thus the formation of HBr from atomic bromine involves primarily reaction with HO_2 which is slower than the corresponding HO_2 + Cl reaction. In addition, BrO photolyzes much faster than ClO and hence photolysis can recycle BrO into Br atoms. As a result of these differences compared to ClO, Br is relatively more efficient than Cl at O_3 destruction because it spends more time in the catalytically active atomic Br form and less in the form of HBr.

A particularly interesting aspect of stratospheric bromine chemistry is the possibility of synergistic interactions between the bromine and chlorine cycles via reaction (19):

$$BrO + ClO \rightarrow Br + Cl + O_2 \tag{19}$$

Yung et al. (1980) suggest that this ClO_x–BrO_x interaction could lead to enhanced O_3 destruction in the lower stratosphere.

REFERENCES

Anderson, J. G., N. L. Hazen, B. E. McLaren, S. P. Rowe, C. M. Schiller, M. J. Schwab, L. Solomon, E. E. Thompson, and E. M. Weinstock, "Free Radicals in the Stratosphere: A New Observational Technique," *Science*, **228,** 1309 (1985).

Ausloos, P., R. E. Rebbert, and L. Glasgow, "Photodecomposition of Chloromethanes Adsorbed on Silica Surfaces," *J. Res. Natl. Bur. Stand. U.S.*, **82,** 1 (1977).

Bates, D. R. and M. Nicolet, "The Photochemistry of Atmospheric Water Vapor," *J. Geophys. Res.*, **55,** 301 (1950).

Bower, F. A. and R. B. Ward, Eds., *Stratospheric Ozone and Man*, Vols. 1 and 2, CRC Press, Boca Raton, FL, 1982.

Brune, W. H., E. M. Weinstock, M. J. Schwab, R. M. Stimpfle, and J. G. Anderson, "Stratospheric ClO: In-Situ Detection with a New Approach," *Geophys. Res. Lett.*, **12,** 441 (1985).

Burrows, J. P., D. W. T. Griffith, G. K. Moortgat, and G. S. Tyndall, "Matrix Isolation Fourier Transform Infrared Study of the Products of the Reaction between ClO and NO_2," *J. Phys. Chem.*, **89,** 266 (1985).

Chang, J. S. and W. H. Duewer, "Modeling Chemical Processes in the Stratosphere," *Ann. Rev. Phys. Chem.*, **30,** 443 (1979).

Chapman, S., "A Theory of Upper-Atmospheric Ozone," *Memoirs Roy. Meteorol. Soc.*, **III,** 103 (1930).

Cicerone, R. J., "Halogens in the Atmosphere," *Rev. Geophys. Space Phys.*, **19,** 123 (1981).

Crutzen, P. J., "The Influence of Nitrogen Oxides on the Atmospheric Ozone Content," *Q. J. R. Meteorol. Soc.*, **96,** 320 (1970).

Crutzen, P. J., "Chlorofluoromethanes: Threats to the Ozone Layer," *Rev. Geophys. Space Phys.*, **17,** 1824 (1979).

Crutzen, P. J. and C. J. Howard, "The Effect of the HO_2 + NO Reaction Rate Constant on One Dimensional Model Calculations of Stratospheric Ozone Perturbations," *Pure Appl. Geophys.*, **116,** 498 (1978).

DeMore, W. B., M. J. Molina, R. T. Watson, D. M. Golden, R. F. Hampson, M. J. Kurylo, C. J. Howard, and A. R. Ravishankara, "Chemical Kinetics and Photochemical Data for Use in Stratospheric Modeling, Evaluation No. 6," JPL Publ. No. 83-62, September 15, 1983.

DeMore, W. B., J. J. Margitan, M. J. Molina, R. T. Watson, D. M. Golden, R. F. Hampson, M. J. Kurylo, C. J. Howard, and A. R. Ravishankara, "Chemical Kinetics and Photochemical Data for Use in Stratospheric Modeling, Evaluation No. 7," JPL Publ. No. 85-37, 1985.

Fabian, P., R. Borchers, B. C. Kruger, S. Lal, and S. A. Penkett, "The Vertical Distribution of $CHClF_2$ (CFC-22) in the Stratosphere," *Geophys. Res. Lett.*, **12,** 1 (1985).

Gäb, S., J. Schmitzer, H. W. Thamm, H. Parlar, and F. Korte, "Photomineralization Rate of Organic Compounds Adsorbed on Particulate Matter," *Nature*, **270,** 331 (1977).

Gäb, S., J. Schmitzer, H. W. Thamm, and F. Korte, "Mineralization of Chlorofluorocarbons in the Sunlight of the Troposphere," *Ang. Chem. Int. Ed. (Engl.)*, **17,** 366 (1978).

Gidel, L. T., P. J. Crutzen, and J. Fishman, "A Two-Dimensional Photochemical Model of the Atmosphere. 1: Chlorocarbon Emissions and Their Effect on Stratospheric Ozone," *J. Geophys. Res.*, **88,** 6622 (1983).

Heaps, W. S. and T. J. McGee, "Progress in Stratospheric Hydroxyl Measurement by Balloon-Borne LIDAR," *J. Geophys. Res.*, **90,** 7913 (1985).

Herman, J. R. and C. J. McQuillan, "Atmospheric Chlorine and Stratospheric Ozone. Nonlinearities and Trends," *J. Geophys. Res.*, **90,** 5721 (1985).

Howard, C. J. and K. M. Evenson, "Kinetics of the Reaction with HO_2 and NO," *Geophys. Res. Lett.*, **4,** 437 (1977).

Hubrich, C. and F. Stuhl, "The Ultraviolet Absorption of Some Halogenated Methanes and Ethanes of Atmospheric Interest," *J. Photochem.*, **12,** 93 (1980).

Johnston, H. S., "Reduction of Stratospheric Ozone by Nitrogen Oxide Catalysts from Supersonic Transport Exhaust," *Science*, **173,** 517 (1971).

Ko, M. K. W. and N. D. Sze, "Diurnal Variation of ClO: Implications for the Stratospheric Chemistries of $ClONO_2$, HOCl, and HCl," *J. Geophys. Res.*, **89,** 11,619 (1984).

Larsen, J. C., C. P. Rinsland, A. Goldman, D. G. Murcray, and F. J. Murcray, "Upper Limits for Stratospheric H_2O_2 and HOCl from High Resolution Balloon-Borne Infrared Solar Absorption Spectra," *Geophys. Res. Lett.*, **12,** 663 (1985).

Logan, J. A., "Tropospheric Ozone: Seasonal Behavior, Trends, and Anthropogenic Influence," *J. Geophys. Res.*, **90,** 10, 463 (1985).

Lovelock, J. E., "Natural Halocarbons in the Air and in the Sea," *Nature*, **256,** 193 (1975).

Molina, M. J. and F. S. Rowland, "Stratospheric Sink for Chlorofluoromethanes: Chlorine Atom Catalyzed Destruction of Ozone," *Nature*, **249,** 810 (1974).

National Aeronautics and Space Administration, *Present State of Knowledge of the Upper Atmosphere*, January 1984.

National Research Council, *Biological Impacts of Increased Intensities of Solar Ultraviolet Radiation*, Report PB-215, 524, DOT. ESB Ad Hoc Panel to Examine the Biological Impacts of Increased Intensities of Solar Ultraviolet Radiation, National Academy of Sciences, Washington, DC, 1973.

National Research Council, *Environmental Impact of Stratospheric Flight: Biological and Climatic Effects of Aircraft Emissions in the Stratosphere.* Climatic Impact Committee, National Academy of Sciences, Washington, DC, 1975.

National Research Council, *Halocarbons: Environmental Effects of Chlorofluoromethane Release.* Committee on Impacts of Stratospheric Change, Assembly of Mathematical and Physical Sciences, National Academy of Sciences, Washington, DC, 1976.

National Research Council, *Medical and Biologic Effects of Environmental Pollutants: Ozone and Other Photochemical Oxidants.* Committee on Medical and Biologic Effects of Environmental Pollutants, Assembly of Life Sciences, National Academy Press, Washington, DC, 1977.

National Research Council, *Stratospheric Ozone Depletion by Halocarbons: Chemistry and Transport*, Panel on Chemistry and Transport. Committee on Impacts of Stratospheric Change, Assembly of Mathematical and Physical Sciences, National Academy of Sciences, Washington, DC, 1979.

National Research Council, *Causes and Effects of Stratospheric Ozone Reduction: An Update.* Committee on Chemistry and Physics of Ozone Depletion and the Committee on Biological Effects of Increased Solar Ultraviolet Radiation, Environmental Studies Board, Commission on Natural Resources, National Academy Press, Washington, DC, 1982.

National Research Council, *Causes and Effects of Changes in Stratospheric Ozone: Update, 1983*, Committee on Causes and Effects of Changes in Stratospheric Ozone, Update, 1983, Environmental Studies Board, Commission on Physical Sciences, Mathematics, and Resources, National Academy Press, Washington, DC, 1984.

Prather, M. J., M. B. McElroy, and S. C. Wofsy, "Reductions in Ozone at High Concentrations of Stratospheric Halogens," *Nature*, **312,** 227 (1984).

Rebbert, R. E. and P. Ausloos, "Decomposition of N_2O Over Particulate Matter," *Geophys. Res. Lett.*, **3,** 761 (1978).

Rowland, F. S. and M. J. Molina, "Chlorofluoromethanes in the Environment," *Rev. Geophys. Space Phys.*, **13,** 1 (1975).

Singh, H. B., L. Salas, H. Shigeishi, and A. Crawford, "Urban-Nonurban Relationships of Halocarbons. SF_6, N_2O, and Other Atmospheric Trace Constitutents," *Atmos. Environ.*, **11,** 819 (1977).

Singh, H. B., L. J. Salas, and R. E. Stiles, "Methyl Halides in and Over the Eastern Pacific (40°N–32°S)," *J. Geophys. Res.*, **88,** 3684 (1983).

Solomon, P. M., R. deZafra, A. Parrish, and J. W. Barrett, "Diurnal Variation of Stratospheric Chlorine Monoxide: A Critical Test of Chlorine Chemistry in the Ozone Layer," *Science*, **224,** 1210 (1984).

Turco, R. P., Stratospheric Ozone Perturbations, in *Ozone in the Free Atmosphere*, R. C. Whitten and S. S. Prasad, Eds., Van Nostrand Reinhold, New York, 1985a, Chap. 5.

Turco, R. P., The Photochemistry of the Stratosphere, in *The Photochemistry of Atmospheres, Earth, the Other Planets, and Comets*, J. S. Levine, Ed. Academic Press, New York, 1985b.

Wayne, R. P., *Chemistry of Atmospheres,* Clarendon Press, Oxford, 1985.

Whitten, R. C. and S. S. Prasad, Eds., *Ozone in the Free Atmosphere*, Van Nostrand Reinhold, New York, 1985.

Wuebbles, D., "1-D Model Predictions at Cruise Altitudes," *Upper Atmospheric Programs Bulletin*, Issue No. 85-1 (1985).

Yung, Y. L., J. P. Pinto, R. T. Watson, and S. P. Sander, "Atmospheric Bromine and Ozone Perturbations in the Lower Stratosphere," *J. Atmos. Sci.*, **37,** 339 (1980).

APPENDIX I
Enthalpies of Formation of Some Gaseous Molecules, Atoms, and Free Radicals at 298°K[a]

Species	ΔH_f° (298) (kcal mole^{-1})	Species	ΔH_f° (298) (kcal mole^{-1})
H	52.1	C_3H_8	−24.8
$O(^3P)$	59.6	CH_3	34.8
$O(^1D)$	104.9	C_2H_5	25.7
$O_2(^1\Delta)$	22.5	n-C_3H_7	22.6
$O_2(^1\Sigma)$	37.5	*i*-C_3H_7	18.2
O_3	34.1	C_2H_4	12.4
HO	9.3	C_2H_3	72[d]
HO_2	3.5[b]	C_3H_6	4.8
H_2O	−57.8	HCO	9.0
H_2O_2	−32.6	HCHO	−26.0
		HCOOH	−90.5
NH_3	−11.0	CH_3O	3.5
NH_2	44.2	CH_3O_2	5.5[e]
NH	82.0	CH_2OH	−6.2
NO	21.6	CH_3OH	−48.0
NO_2	7.9	CH_3OOH	−31
NO_3	17.4[c]	CH_3ONO	−15.6
N_2O	19.6	CH_3ONO_2	−28.6
N_2O_4	2.2	CH_3CO	−5.8
N_2O_5	2.7	CH_3CHO	−39.7
HNO_2	−19.0	C_2H_5O	−4.1
HNO	23.8	C_2H_5OH	−56.2
HNO_3	−32.3	CH_2CH_2OH	−13.2
HO_2NO_2	−13	$C_2H_5O_2$	−1.8
		CH_3CO_2	−49.6
CH_4	−17.9		
C_2H_6	−20.0	CF_2Cl_2	−117.9

Species	ΔH_f° (298) (kcal mole^{-1})	Species	ΔH_f° (298) (kcal mole^{-1})
CCl_3F	−68.1	CS_2	28.0
$CHClF_2$	−115.6	CS	65.0
CO	−26.4	CH_3SCH_3	−8.9
CO_2	−94.1	CH_3SSCH_3	−5.8
H_2S	−4.9	OCS	−33.9
SH	34.9	HCl	−22.1
SO_2	−70.9	ClO	24
SO_3	−94.6	HOCl	−19
HSO_3	−115.0	$ClONO_2$	6.3

[a]Unless otherwise noted, from D. L. Baulch, R. A. Cox, R. F. Hampson, Jr., J. A. Kerr, J. Troe, and R. T. Watson, *J. Phys. Chem. Ref. Data*, **13,** 1259 (1984).

[b]From L. G. S. Shum and S. W. Benson, *J. Phys. Chem.*, **87,** 3479 (1983).

[c]From C. C. Kircher, J. J. Margitan, and S. P. Sander, *J. Phys. Chem.*, **88,** 4370 (1984); J. P. Burrows, G. S. Tyndall, and G. K. Moortgat, *Chem. Phys. Lett.*, **119,** 193 (1985).

[d]From R. B. Sharma, N. N. Semo, and W. S. Koski, *Int. J. Chem. Kinet.*, **17,** 831 (1985).

[e]From O. Kondo and S. W. Benson, *J. Phys. Chem.*, **88,** 6675 (1984).

APPENDIX II
Summary of Room Temperature Rate Constants[a] (298°K) for Reactions of Some Organic Compounds with OH, O_3, and NO_3 at One Atmosphere in Air

Compound	OH	O_3	NO_3[d]
Alkanes			
CH_4	8.4(−15)[b]	*c*	n.d.[i]
C_2H_6	2.7(−13)	*c*	n.d.
C_3H_8	1.2(−12)	*c*	n.d.
n-C_4H_{10}	2.5(−12)	*c*	3.6(−17)
i-C_4H_{10}	2.4(−12)	*c*	5.2(−17)
n-C_5H_{12}	4.1(−12)	*c*	4.3(−17)
Cyclohexane	7.4(−12)	*c*	7.2(−17)
Alkenes			
C_2H_4	8.5(−12)	1.8(−18)	*e*
C_3H_6	2.6(−11)	1.1(−17)	7.5(−15)
1-C_4H_8	3.1(−11)	1.1(−17)	9.7(−15)
cis-2-C_4H_8	5.6(−11)	1.3(−16)	3.4(−13)
trans-2-C_4H_8	6.4(−11)	2.0(−16)	3.8(−13)
2,3-Dimethyl-2-butene	1.1(−10)	1.2(−15)	6.1(−11)
Cyclohexene	6.7(−11)	*e*	5.1(−13)
Isoprene	1.0(−10)	1.4(−17)	5.8(−13)
α-Pinene	5.3(−11)	*e*	6.1(−12)
β-Pinene	7.8(−11)	2(−17)	2.5(−12)
Alkynes			
C_2H_2	7.8(−13)	≲1(−19)	n.d.

Compound	OH	O_3	NO_3[d]
Aromatics			
Benzene	1.3(−12)	*f*	≤2.0(−17)
Toluene	6.2(−12)	*f*	3.2(−17)
o-Xylene	1.5(−11)	*f*	2.0(−16)
m-Xylene	2.5(−11)	*f*	1.4(−16)
p-Xylene	1.5(−11)	*f*	2.5(−16)
Ethylbenzene	7.5(−12)	*f*	n.d.
n-Propylbenzene	5.7(−12)	*f*	n.d.
Phenol	2.8(−11)	n.d.	3.8(−12)
o-Cresol	4.0(−11)	*g*	2.2(−11)
m-Cresol	5.7(−11)	*g*	1.7(−11)
p-Cresol	4.4(−11)	*g*	2.3(−11)
Oxygen- Sulfur- and Nitrogen-Containing Compounds			
HCHO	9.0(−12)	*h*	5.8(−16)
CH_3CHO	1.6(−11)	*h*	2.4(−15)
C_2H_5CHO	2.0(−11)	*h*	n.d.
C_6H_5CHO	1.3(−11)	*h*	2.0(−15)
CH_3COCH_3	2.3(−13)	*h*	n.d.
$CH_3COC_2H_5$	1.0(−12)	*h*	n.d.
CH_3SH	3.3(−11)	n.d.	1.0 (−12)
C_2H_5SH	4.7(−11)	n.d.	n.d.
CH_3SCH_3	6.3(−12)	<8(−19)	9.7(−13)
CH_3SSCH_3	2.0(−10)	n.d.	4.2(−14)
H_2S	4.7(−12)	n.d.	n.d.
Thiophene	9.5(−12)	<6(−20)	n.d.
CH_3CN	2.1(−14)	≤1.5(−19)	n.d.

[a]Units of cm^3 $molecule^{-1}$ s^{-1}; see text for references.

[b]$8.4(-15) = 8.4 \times 10^{-15}$.

[c]All O_3–alkane rate constants $\lesssim 10^{-23}$ cm^3 $molecule^{-1}$ s^{-1}.

[d]All except the value for *trans*-2-butene based on a value of $K_{eq} = 3.35 \times 10^{-11}$ cm^3 $molecule^{-1}$ for the equilibrium $NO_2 + NO_3 \overset{M}{\rightleftarrows} N_2O_5$. The rate constants are linearly proportional to the value of K_{eq}, which is uncertain to a factor of ~2. To adjust these rate constants to other values of K_{eq}, the reader should multiply by the appropriate ratio; for example, to adjust to $K_{eq} = 1.87 \times 10^{-11}$ cm^3 $molecule^{-1}$, multiply the rate constants in this table by 1.87/3.35 = 0.56.

[e]Not cited due to large uncertainties; see text.

[f]Monocyclic aromatics (except styrene) all react slowly ($k \lesssim 10^{-20}$ cm^3 $molecule^{-1}$ s^{-1}).

[g]Not cited due to large uncertainties; reported values are of the order of 10^{-18} to 10^{-19} cm^3 $molecule^{-1}$ s^{-1}.

[h]Reactions with oxygen-containing compounds not containing double bonds are slow; $k \lesssim 10^{-20}$ cm^3 $molecule^{-1}$ s^{-1}.

[i]n.d. = not determined.

APPENDIX III
Approximate Bond Dissociation Energies[a]

Bond Broken	Bond Dissociation Energy (kcal mole^{-1})
CH_3—H	105
C_2H_5—H	98
n-C_3H_7—H	98
i-C_3H_7—H	95
s-C_4H_9—H	96
t-C_4H_9—H	93
cyclopropyl—H	106
cyclobutyl—H	97
cyclopentyl—H	95
cyclohexyl—H	96
cycloheptyl—H	93
$HOCH_2$—H	94
$C_6H_5CH_2$—H	88
C=C—H	110
C_6H_5—H	111
C │ C=CC—H	86
C=C—C(C)—H	83
C=C—C$(C)_2$—H	77
C≡C—H	132
CH_3—CH_3	90
CH_3—C_2H_5	86
C_2H_5—C_2H_5	82
n-C_3H_7—CH_3	87
i-C_3H_7—CH_3	86
s-C_4H_9—CH_3	85
t-C_4H_9—CH_3	84
n-C_3H_7—C_2H_5	82
i-C_3H_7—C_2H_5	81

Bond Broken	Bond Dissociation Energy (kcal mole^{-1})
s-$C_4H_9{-}C_2H_5$	80
$C_6H_5{-}CH_3$	102
$C_6H_5{-}C_2H_5$	97
$C_6H_5{-}n$-C_3H_7	98
$C{=}C{-}CH_3$	101
$C{=}CC{-}CH_3$	74
$\overset{\overset{C}{\vert}}{C}{=}CC{-}CH_3$	73
$C{\equiv}C{-}CH_3$	126
$C{\equiv}CC{-}CH_3$	76
$CH_3{-}OH$	92
$C_2H_5{-}OH$	92
$C_6H_5CH_2{-}OH$	81
$CH_3O{-}H$	104
$C_2H_5O{-}H$	104
n-$C_3H_7O{-}H$	103
$C_6H_5O{-}H$	87
$CH_3O{-}CH_3$	83
$CH_3O{-}C_2H_5$	82
$C_6H_5O{-}CH_3$	64
$CH_3O{-}OCH_3$	38
$C_2H_5O{-}OC_2H_5$	38
$CH_3\overset{\overset{O}{\Vert}}{C}O{-}H$	106
$C_2H_5\overset{\overset{O}{\Vert}}{C}O{-}H$	106
n-$C_3H_7\overset{\overset{O}{\Vert}}{C}O{-}H$	106
$H\overset{\overset{O}{\Vert}}{C}{-}H$	87
$CH_3\overset{\overset{O}{\Vert}}{C}{-}H$	86
$C_6H_5\overset{\overset{O}{\Vert}}{C}{-}H$	87
$OHC{-}CH_3$	83
$OHC{-}C_6H_5$	96
$CH_3\overset{\overset{O}{\Vert}}{C}{-}CH_3$	81
$CH_3\overset{\overset{O}{\Vert}}{C}{-}C_6H_5$	94

Bond Broken	Bond Dissociation Energy (kcal $mole^{-1}$)
$CH_3C(=O)-C(=O)CH_3$	67
HS—H	91
CH_3S-H	91
RS—H	91
CH_3-SH	74
C_2H_5-SH	71
$t\text{-}C_4H_9-SH$	68
C_6H_5-SH	87
CH_3S-CH_3	77
$CH_3S-C_2H_5$	73
$CH_3S-n\text{-}C_3H_7$	74
$CH_3SO_2-CH_3$	67
RS_2-H	70
RS_2-CH_3	57
RS—SR	72
CH_3-NH_2	85
$C_2H_5-NH_2$	82
CH_3NH-H	100
CH_3NH-CH_3	82
$CH_3NH-C_2H_5$	80
$(CH_3)_2N-H$	92
$(CH_3)_2N-CH_3$	76
$(CH_3)_2N-C_2H_5$	72
C_6H_5NH-H	88
$C_6H_5NH-CH_3$	71
$C_6H_5NH-C_2H_5$	69
H_2NCH_2-H	93
CH_3-CN	122
C_6H_5-CN	131
CH_3-F	110
CH_3-Cl	85
CH_3-Br	71
CH_3-I	57
CF_2Cl-F	123
CF_2Cl-Cl	76
CF_2Cl-Br	65
$CFCl_2-F$	110
$CFCl_2-Cl$	73
HO—H	119
HO—OH	51
HOO—H	87
HO_2-NO_2	23

Bond Broken	Bond Dissociation Energy (kcal $mole^{-1}$)
$H—NO_2$	78
HO—NO	49
$H—ONO_2$	101
$HO—NO_2$	49
$HO—ONO_2$	39
CS—S	103
OS—O	130
HS—SH	66
$NH_2—H$	107
CN—H	124
$CH_3—NO$	40
$CH_3—NO_2$	61
i-$C_3H_7—NO$	37
i-$C_3H_7—NO_2$	59
RO—NO	41
$RO—NO_2$	41

[a]From D. F. McMillen and D. M. Golden, *Ann. Rev. Phys. Chem.*, **33,** 493 (1982).

Author Index

(Numbers in parentheses give the number of citations on the page)

Subject Index